The Ribosome

Structure, Function, Antibiotics, and Cellular Interactions

The Ribosome

Structure, Function, Antibiotics, and Cellular Interactions

Edited by

Roger A. Garrett
RNA Regulation Centre, Institute of Molecular Biology,
Copenhagen University,
Copenhagen, Denmark

Stephen R. Douthwaite
RNA Regulation Centre, Institute of Molecular Biology,
Odense University,
Odense, Denmark

Anders Liljas
Molecular Biophysics, Centre for Chemistry and Chemical Engineering,
Lund University, Lund, Sweden

Alistair T. Matheson
Division of Continuing Studies, University of Victoria,
Victoria, British Columbia, Canada

Peter B. Moore
Department of Chemistry and Department of Molecular Biophysics and Biochemistry,
Yale University, New Haven, Connecticut

Harry F. Noller
Center for Molecular Biology of RNA, Sinsheimer Laboratories,
University of California, Santa Cruz, California

Washington, DC

American Society for Microbiology
1752 N Street NW
Washington, DC 20036

Library of Congress Cataloging-in-Publication Data

The Ribosome: structure, function, antibiotics, and cellular interactions/edited by Roger A. Garrett . . . [et al.].
p. cm.
Includes bibliographical references and index.
ISBN 1-55581-184-1 (hc)
1. Ribosomes. I. Garrett, Roger A.

QH603.R5 R48 2000
571.6′58—dc21 99-049979

Cover figure: See Figure 5 on page 157.

CONTENTS

IV. Ribosomal Modeling

V. rRNA Modification and Protein Signals

VI. Probing of Functional Sites

CONTRIBUTORS

Ilana Agmon ■ Department of Structural Biology, Weizmann Institute, 76100 Rehovot, Israel (3)

Rajendra K. Agrawal ■ Health Research Inc., at Wadsworth Center, and Department of Biomedical Sciences, State University of New York at Albany, Empire State Plaza, Albany, NY 12201-0509 (5, 6, 15)

Irina Alimov ■ Department of Biochemistry and Molecular Biology and Program in Molecular and Cellular Biology, University of Massachusetts, Amherst, MA 01003 (10)

Salam Al-Karadaghi ■ Molecular Biophysics, Center of Chemistry and Chemical Engineering, Lund University, Box 124, SE-221 00 Lund, Sweden (7, 29)

Gregers R. Andersen ■ Institute of Molecular and Structural Biology, University of Aarhus, Gustav Wieds Vej 10C, DK-8000 Aarhus C, Denmark (27)

Alexey L. Arkov ■ Department of Molecular Genetics (Box 11), The University of Texas M. D. Anderson Cancer Center, 1515 Holcombe Blvd., Houston, TX 77030 (41)

John F. Atkins ■ Department of Human Genetics, University of Utah, 15N 2030E Room 7410, Salt Lake City, UT 84112-5330 (30)

Tamar Auerbach ■ Department of Structural Biology, Weizmann Institute, 76100 Rehovot, Israel, and Department of Biochemistry and Pharmacology, FU-Berlin, Takustr. 3, 14195 Berlin, Germany (3)

Horacio Avila ■ Max Planck Institute for Molecular Genetics, Ihnestr. 73, 14195 Berlin, Germany (3)

Jean-Pierre Bachellerie ■ Laboratoire de Biologie Moléculaire Eucaryote du C.N.R.S., Université Paul-Sabatier, 118 route de Narbonne, 31062 Toulouse Cédex, France (17)

Juan P. G. Ballesta ■ Centro de Biología Molecular "Severo Ochoa," CSIC and UAM, Canto Blanco, 28049 Madrid, Spain (12)

Nenad Ban ■ Department of Molecular Biophysics and Biochemistry and Howard Hughes Medical Institute, Yale University, New Haven, CT 06520 (2)

Heike Bartels ■ Department of Structural Biology, Weizmann Institute, 76100 Rehovot, Israel (3)

Anat Bashan ■ Department of Structural Biology, Weizmann Institute, 76100 Rehovot, Israel (3)

Mark Bayfield ■ J. W. Wilson Laboratory, Department of Molecular and Cellular Biology and Biochemistry, Brown University, Providence, RI 02912 (19)

William S. Bennett ■ Max Planck Research Unit for Ribosomal Structure, Notkestr. 22603 Hamburg, Germany (3)

Gregor Blaha ■ Max-Planck-Institut für Molekulare Genetik, AG Ribosomen, Ihnestraße 73, D-14195 Berlin, and GKSS Research Center, Max-Planck-Straße, 21502 Geesthacht, Germany (26)

Scott C. Blanchard ■ Department of Structural Biology, Stanford University School of Medicine, Stanford, CA 94305-5421 (34)

Alexey A. Bogdanov ■ Department of Chemistry and Belozersky Institute of Physico-Chemical Biology, Moscow State University, Moscow 119899, Russia (21)

Letizia Brandi ■ Laboratory of Genetics, Department of Biology MCA, University of Camerino, 62032 Camerino (MC), Italy (39)

Richard Brimacombe ■ Max-Planck-Institut für Molekulare Genetik, AG-Ribosomen, Ihnestraße 73, 14195 Berlin, Germany **(14)**

Delbert Brod ■ Department of Chemistry and Biochemistry, The University of Texas at Austin, Austin, TX 78712-1167 **(24)**

Richard H. Buckingham ■ UPR 9073 du CNRS, Institut de Biologie Physico-Chimique, 13 Rue Pierre et Marie Curie, Paris 75005, France **(44)**

Douglas J. Bucklin ■ Division of Biological Sciences, The University of Montana, Missoula, MT 59812 **(22)**

Yuri Bukhtiyarov ■ DuPont Pharmaceuticals Co., Experimental Station E400/3410, Wilmington, DE 19880 **(23)**

James M. Bullard ■ Division of Biological Sciences, The University of Montana, Missoula, MT 59812 **(22)**

Nils Burkhardt ■ Max-Planck-Institut für Molekulare Genetik, AG Ribosomen, Ihnestraße 73, D-14195 Berlin, and BAYER AG-Wuppertal, Gebäude 405, Abteilung MST, D-42096 Wuppertal, Germany **(26)**

Malcolm S. Capel ■ Biology Department, Brookhaven National Laboratory, Upton, NY 11973 **(1)**

David A. Case ■ Department of Molecular Biology, The Scripps Institute, 10550 N. Torrey Pines Rd., La Jolla, CA 92037 **(15)**

Enrico Caserta ■ Laboratory of Genetics, Department of Biology MCA, University of Camerino, 62032 Camerino (MC), Italy **(39)**

Jamie Cate ■ Center for Molecular Biology of RNA, Sinsheimer Laboratories, University of California–Santa Cruz, Santa Cruz, CA 95064 **(13)**

Jérôme Cavaillé ■ Laboratoire de Biologie Moléculaire Eucaryote du C.N.R.S., Université Paul-Sabatier, 118 route de Narbonne, 31062 Toulouse Cédex, France **(17)**

Yuen-Ling Chan ■ Department of Biochemistry and Molecular Biology, The University of Chicago, Chicago, IL 60637 **(38)**

Natalya S. Chernyaeva ■ Department of Molecular Genetics (Box 11), The University of Texas M. D. Anderson Cancer Center, 1515 Holcombe Blvd., Houston, TX 77030 **(41)**

William M. Clemons, Jr. ■ Department of Biochemistry, University of Utah School of Medicine, Salt Lake City, UT 84132 **(1, 8)**

Graeme L. Conn ■ Department of Chemistry, Johns Hopkins University, Baltimore, MD 21218 **(11)**

Barry S. Cooperman ■ Department of Chemistry, University of Pennsylvania, Philadelphia, PA 19104-6323 **(23)**

Carl C. Correll ■ Department of Biochemistry and Molecular Biology, The University of Chicago, Chicago, IL 60637 **(38)**

Gloria Culver ■ Center for Molecular Biology of RNA, Sinsheimer Laboratories, University of California–Santa Cruz, Santa Cruz, CA 95064 **(13)**

Eric Cundliffe ■ Department of Biochemistry, University of Leicester, Leicester LE1 7RH, United Kingdom **(33)**

Marylena Dabrowski ■ Max-Planck-Institut für Molekulare Genetik, AG Ribosomen, Ihnestraße 73, D-14195 Berlin, Germany **(26)**

Albert E. Dahlberg ■ J. W. Wilson Laboratory, Department of Molecular and Cellular Biology and Biochemistry, Brown University, Providence, RI 02912 **(19)**

Kam D. Dahlquist ■ Department of Structural Biology, Stanford University School of Medicine, Stanford, CA 94305-5421 **(34)**

Anne Dallas ■ Center for Molecular Biology of RNA, Sinsheimer Laboratories, University of California–Santa Cruz, Santa Cruz, CA 95064 **(13)**

Mark E. Dalphin ■ Amgen Inc., Amgen Center, Thousand Oaks, CA 91320-1799 **(40)**

Christopher Davies ■ Department of Structural Biology, St. Jude Children's Research Hospital, 332 N. Lauderdale, Memphis, TN 38105, and School of Biological Sciences, University of Sussex, Falmer, Brighton BN1 9QG, United Kingdom **(8)**

Julian Davies ■ Department of Microbiology and Immunology, The University of British Columbia, 6174 University Blvd., Vancouver, British Columbia V6T 1Z3, Canada **(37)**

Natalia Davydova ■ Molecular Biophysics, Center of Chemistry and Chemical Engineering, Lund University, Box 124, SE-221 00 Lund, Sweden (7)

Gundo Diedrich ■ Max-Planck-Institut für Molekulare Genetik, AG Ribosomen, Ihnestraße 73, D-14195 Berlin, and BAYER AG-Wuppertal, Gebäude 405, Abteilung MST, D-42096 Wuppertal, Germany (26)

Vildan Dincbas ■ Department of Cell and Molecular Biology, BMC, Box 596, S-751 24 Uppsala, Sweden (44)

Peining Dong ■ Department of Biochemistry and Molecular Biology and Program in Molecular and Cellular Biology, University of Massachusetts, Amherst, MA 01003 (10)

Olga A. Dontsova ■ Department of Chemistry and Belozersky Institute of Physico-Chemical Biology, Moscow State University, Moscow 119899, Russia (21)

Stephen Douthwaite ■ Department of Molecular Biology, Odense University, DK-5230 Odense M, Denmark (35)

David E. Draper ■ Department of Chemistry, Johns Hopkins University, Baltimore, MD 21218 (11)

Zhanna Druzina ■ Department of Chemistry, University of Pennsylvania, Philadelphia, PA 19104-6323 (23)

Denis Drygin ■ Department of Biochemistry and Molecular Biology and Program in Molecular and Cellular Biology, University of Massachusetts, Amherst, MA 01003 (10)

Thomas N. Earnest ■ Macromolecular Crystallography Facility, Advanced Light Source, Lawrence Berkeley National Laboratory, Berkeley, CA 94720 (13)

Robert G. Eason ■ Department of Structural Biology, Stanford University School of Medicine, Stanford, CA 94305-5421 (34)

Thomas R. Easterwood ■ Department of Biochemistry and Molecular Genetics, University of Alabama at Birmingham, Birmingham, AL 35294-0005 (15)

Måns Ehrenberg ■ Department of Cell and Molecular Biology, BMC, Box 596, S-751 24 Uppsala, Sweden (44)

Edda Einfeldt ■ Max-Planck-Institut für Molekulare Genetik, AG Ribosomen, Ihnestraße 73, D-14195 Berlin, Germany (26)

Irina Eliseikina ■ Molecular Biophysics, Center of Chemistry and Chemical Engineering, Lund University, Box 124, SE-221 00 Lund, Sweden (7)

Dominique Fourmy ■ Department of Structural Biology, Stanford University School of Medicine, Stanford, CA 94305-5421 (34)

François Franceschi ■ Max Planck Institute for Molecular Genetics, Ihnestr. 73, 14195 Berlin, Germany (3)

Joachim Frank ■ Health Research, Inc., at the Wadsworth Center, Howard Hughes Medical Institute, and Department of Biomedical Sciences, State University of New York at Albany, Empire State Plaza, Albany, NY 12201-0509 (5, 6, 15)

David Freistroffer ■ Department of Cell and Molecular Biology, BMC, Box 596, S-751 24 Uppsala, Sweden (44)

Jonathan Gallant ■ Genetics Department, University of Washington, Box 357360, Warehouse of the Muses, 1959 NE Pacific St.–J-205, Seattle, WA 98195-7360 (31)

Maria Garber ■ Institute of Protein Research, Russian Academy of Sciences, 142292 Pushchino, Moscow Region, Russia (7)

Roger A. Garrett ■ RNA Regulation Centre, Institute of Molecular Biology, Copenhagen University, DK-1307 Copenhagen K, Denmark (36)

Raymond F. Gesteland ■ Department of Human Genetics, University of Utah, 15N 2030E Room 7410, Salt Lake City, UT 84112-5330 (30)

Apostolos G. Gittis ■ Department of Biophysics, Johns Hopkins University, Baltimore, MD 21218 (11)

Marco Gluehmann ■ Max Planck Research Unit for Ribosomal Structure, Notkestr. 85, 22603 Hamburg, Germany (3)

Wendy T. Grace ■ Division of Biological Sciences, The University of Montana, Missoula, MT 59812 (22)

Robert A. Grassucci ■ Health Research, Inc., at the Wadsworth Center and Howard Hughes Medical Institute, Empire State Plaza, Albany, NY 12201-0509 (5)

Rachel Green ■ Center for Molecular Biology of RNA, Sinsheimer Laboratories, University of California–Santa Cruz, Santa Cruz, CA 95064 (13)

Steven T. Gregory ■ J. W. Wilson Laboratory, Department of Molecular and Cellular Biology and Biochemistry, Brown University, Providence, RI 02912 (19)

Barbara Greuer ■ Max-Planck-Institut für Molekulare Genetik, AG-Ribosomen, Ihnestrasse 73, 14195 Berlin, Germany (14)

Claudio O. Gualerzi ■ Laboratory of Genetics, Department of Biology MCA, University of Camerino, 62032 Camerino (MC), Italy (39)

Esther Guarinos ■ Centro de Biología Molecular "Severo Ochoa," CSIC and UAM, Canto Blanco, 28049 Madrid, Spain (12)

Anatoly Gudkov ■ Institute of Protein Research, Russian Academy of Sciences, Pushchino, Moscow Region, Russia (29)

Debraj GuhaThakurta ■ Department of Chemistry, Johns Hopkins University, Baltimore, MD 21218 (11)

Harly A. S. Hansen ■ Max Planck Research Unit for Ribosomal Structure, Notkestr. 85, 22603 Hamburg, Germany (3)

Boyd Hardesty ■ Department of Chemistry and Biochemistry, The University of Texas at Austin, Austin, TX 78712-1167 (24)

Joerg Harms ■ Max Planck Research Unit for Ribosomal Structure, Notkestr. 85, 22603 Hamburg, Germany (3)

Stephen C. Harvey ■ Department of Biochemistry and Molecular Genetics, University of Alabama at Birmingham, Birmingham, AL 35294-0005 (15)

Amy B. Heagle ■ Health Research, Inc., at Wadsworth Center, and Howard Hughes Medical Institute, Empire State Plaza, Albany, NY 12201-0509 (5, 6)

Klas O. F. Hedenstierna ■ Department of Molecular Genetics (Box 11), The University of Texas M. D. Anderson Cancer Center, 1515 Holcombe Blvd., Houston, TX 77030 (41)

Scott P. Hennelly ■ Division of Biological Sciences, The University of Montana, Missoula, MT 59812 (22)

Alan J. Herr ■ Department of Human Genetics, University of Utah, 15N 2030E Room 7410, Salt Lake City, UT 84112-5330 (30)

Valerie Heurgué-Hamard ■ Department of Cell and Molecular Biology, BMC, Box 596, S-751 24 Uppsala, Sweden (44)

Rolf Hilgenfeld ■ Institute of Molecular Biotechnology, Beutenbergstr. 11, D-07745 Jena, Germany (28)

Walter E. Hill ■ Division of Biological Sciences, The University of Montana, Missoula, MT 59812 (22)

Go Hirokawa ■ Department of Microbiology, School of Medicine, University of Pennsylvania, 319B Johnson Pavilion, 3610 Hamilton Walk, Philadelphia, PA 19104 (43)

Tanis Hogg ■ Institute of Molecular Biotechnology, Beutenbergstr. 11, D-07745 Jena, Germany (28)

Lovisa Holmberg ■ Center for Molecular Biology of RNA, Sinsheimer Laboratories, University of California–Santa Cruz, Santa Cruz, CA 95064 (13)

Harumi Hosaka ■ Division of Biological Sciences, Graduate School of Science, Hokkaido University, Sapporo 060-0810, Japan (9)

Diarmaid Hughes ■ Department of Cell and Molecular Biology, BMC, Uppsala University, Box 596, SE-75124 Uppsala, Sweden (29)

Koichi Ito ■ Department of Tumor Biology, The Institute of Medical Science, The University of Tokyo, Minato-ku, Tokyo 108-8639, Japan (42)

Ivaylo Ivanov ■ Department of Human Genetics, University of Utah, 15N 2030E Room 7410, Salt Lake City, UT 84112-5330 (30)

Daniela Janell ■ Max Planck Research Unit for Ribosomal Structure, Notkestr. 85, 22603 Hamburg, Germany (3)

Lihong Jiang ■ Department of Biochemistry and Molecular Biology and Program in Molecular and Cellular Biology, University of Massachusetts, Amherst, MA 01003 (10)
Simpson Joseph ■ Center for Molecular Biology of RNA, Sinsheimer Laboratories, University of California–Santa Cruz, Santa Cruz, CA 95064 (13)
Akira Kaji ■ Department of Microbiology, School of Medicine, University of Pennsylvania, 319B Johnson Pavilion, 3610 Hamilton Walk, Philadelphia, PA 19104 (43)
Detlev Kamp ■ Max-Planck-Institut für Molekulare Genetik, AG Ribosomen, Ihnestraße 73, D-14195 Berlin, Germany (26)
Richard Kao ■ Department of Microbiology and Immunology, The University of British Columbia, 6174 University Blvd., Vancouver, British Columbia V6T 1Z3, Canada (37)
Reza Karimi ■ Department of Cell and Molecular Biology, BMC, Box 596, S-751 24 Uppsala, Sweden (44)
Yoichi Kawazu ■ Department of Tumor Biology, The Institute of Medical Science, The University of Tokyo, Minato-ku, Tokyo 108-8639, Japan (42)
Maggie Kessler ■ Department of Structural Biology, Weizmann Institute, 76100 Rehovot, Israel (3)
Philipp Khaitovich ■ Center for Pharmaceutical Biotechnology-m/c 870, University of Illinois, 900 S. Ashland Ave., Chicago, IL 60607 (20)
Makoto Kimura ■ Laboratory of Biochemistry, Faculty of Agriculture, Kyushu University, Fukuoka 812-8512, Japan (9)
Stanislav V. Kirillov ■ RNA Regulation Centre, Institute of Molecular Biology, Copenhagen University, DK-1307 Copenhagen K, Denmark (36)
Morten Kjeldgaard ■ Institute of Molecular and Structural Biology, University of Aarhus, Gustav Wieds Vej 10C, DK-8000 Aarhus C, Denmark (27)
Gisela Kramer ■ Department of Chemistry and Biochemistry, The University of Texas at Austin, Austin, TX 78712-1167 (24)
Ole Kristensen ■ Molecular Biophysics, Lund University, Box 124, SE-22100 Lund, Sweden (29)
Vassiliki S. Lalioti ■ Centro de Biología Molecular "Severo Ochoa," CSIC and UAM, Canto Blanco, 28049 Madrid, Spain (12)
Laura Lancaster ■ Center for Molecular Biology of RNA, Sinsheimer Laboratories, University of California–Santa Cruz, Santa Cruz, CA 95064 (13)
Anna La Teana ■ Laboratory of Genetics, Department of Biology MCA, University of Camerino, 62032 Camerino (MC), Italy (39)
Eaton E. Lattman ■ Department of Biophysics, Johns Hopkins University, Baltimore, MD 21218 (11)
Martin Laurberg ■ Molecular Biophysics, Lund University, Box 124, SE-22100 Lund, Sweden (29)
Inna N. Lavrik ■ Department of Chemistry and Belozersky Institute of Physico-Chemical Biology, Moscow State University, Moscow 119899, Russia (21)
Wyan-Ching Mimi Lee ■ J. W. Wilson Laboratory, Department of Molecular and Cellular Biology and Biochemistry, Brown University, Providence, RI 02912 (19)
Andrey A. Leonov ■ Department of Chemistry and Belozersky Institute of Physico-Chemical Biology, Moscow State University, Moscow 119899, Russia (21)
Inna Levin ■ Department of Structural Biology, Weizmann Institute, 76100 Rehovot, Israel (3)
Kate Lieberman ■ Center for Molecular Biology of RNA, Sinsheimer Laboratories, University of California–Santa Cruz, Santa Cruz, CA 95064 (13)
Anders Liljas ■ Molecular Biophysics, Center of Chemistry and Chemical Engineering, Lund University, Box 124, SE-221 00 Lund, Sweden (7, 29)
Dale Lindsley ■ Genetics Department, University of Washington, Box 357360, Warehouse of the Muses, 1959 NE Pacific St.–J-205, Seattle, WA 98195-7360 (31)
J. Stephen Lodmell ■ J. W. Wilson Laboratory, Department of Molecular and Cellular Biology and Biochemistry, Brown University, Providence, RI 02912 (19)

Stephen R. Lynch ■ Department of Structural Biology, Stanford University School of Medicine, Stanford, CA 94305-5421 **(34)**

Louise L. Major ■ Department of Biochemistry and Centre for Gene Research, University of Otago, PO Box 56, Dunedin, New Zealand **(40)**

Arun Malhotra ■ Wadsworth Center, P.O. Box 509, Albany, NY 12201-0509 **(15)**

Anuj Mankad ■ J. W. Wilson Laboratory, Department of Molecular and Cellular Biology and Biochemistry, Brown University, Providence, RI 02912 **(19)**

Alexander S. Mankin ■ Center for Pharmaceutical Biotechnology-m/c 870, University of Illinois, 900 S. Ashland Ave., Chicago, IL 60607 **(20)**

John B. Mansell ■ Department of Biochemistry and Centre for Gene Research, University of Otago, PO Box 56, Dunedin, New Zealand **(40)**

Viter Marquez ■ Max-Planck-Institut für Molekulare Genetik, AG Ribosomen, Ihnestraße 73, D-14195 Berlin, Germany **(26)**

Kirill Martemyanov ■ Institute of Protein Research, Russian Academy of Sciences, Pushchino, Moscow Region, Russia **(29)**

Christian Massire ■ Institut de Biologie Moléculaire et Cellulaire du CNRS UPR 9002, 15 rue Descartes, 67084 Strasbourg, France **(30)**

Judy Masucci ■ Genetics Department, University of Washington, Box 357360, Warehouse of the Muses, 1959 NE Pacific St.–J-205, Seattle, WA 98195-7360 **(31)**

Rishi Matadeen ■ Department of Biochemistry, Imperial College of Science, Technology and Medicine, London SW7 2AY, United Kingdom **(4)**

Natalia B. Matassova ■ Institute of Molecular Biology, University of Witten/Herdecke, 58448 Witten, Germany **(25)**

Joanna L. C. May ■ Department of Biochemistry, University of Utah School of Medicine, Salt Lake City, UT 84132 **(1)**

Bryan McIntosh ■ Department of Chemistry and Biochemistry, The University of Texas at Austin, Austin, TX 78712-1167 **(24)**

Chuck Merryman ■ Center for Molecular Biology of RNA, Sinsheimer Laboratories, University of California–Santa Cruz, Santa Cruz, CA 95064 **(13)**

Jeroen Mesters ■ Institute of Molecular Biotechnology, Beutenbergstr. 11, D-07745 Jena, Germany **(28)**

Dagmar Mohr ■ Institute of Molecular Biology, University of Witten/Herdecke, 58448 Witten, Germany **(25)**

Peter B. Moore ■ Department of Chemistry and Department of Molecular Biophysics and Biochemistry, Yale University, New Haven, CT 06520 **(2, 45)**

Florian Mueller ■ Max-Planck-Institut für Molekulare Genetik, AG-Ribosomen, Ihnestrasse 73, 14195 Berlin, Germany **(14)**

Emanuel J. Murgola ■ Department of Molecular Genetics (Box 11), The University of Texas M. D. Anderson Cancer Center, 1515 Holcombe Blvd., Houston, TX 77030 **(41)**

Gregory W. Muth ■ Department of Chemistry, The University of Montana, Missoula, MT 59812 **(22)**

Ivan Nagaev ■ Department of Cell and Molecular Biology, BMC, Uppsala University, Box 596, SE-75124 Uppsala, Sweden **(29)**

Atsushi Nakagawa ■ Division of Biological Sciences, Graduate School of Science, Hokkaido University, Sapporo 060-0810, Japan **(9)**

Yoshikazu Nakamura ■ Department of Tumor Biology, The Institute of Medical Science, The University of Tokyo, Minato-ku, Tokyo 108-8639, Japan **(42)**

Takashi Nakashima ■ Division of Biological Sciences, Graduate School of Science, Hokkaido University, Sapporo 060-0810, Japan **(9)**

Natalia Nevskaya ■ Institute of Protein Research, Russian Academy of Sciences, 142292 Pushchino, Moscow Region, Russia **(7)**

Lisa Newcomb ■ Center for Molecular Biology of RNA, Sinsheimer Laboratories, University of California–Santa Cruz, Santa Cruz, CA 95064 **(13)**

Knud H. Nierhaus ■ Max-Planck-Institut für Molekulare Genetik, AG Ribosomen, Ihnestraße 73, D-14195 Berlin, Germany **(26)**

Stanislav Nikonov ■ Institute of Protein Research, Russian Academy of Sciences, 142292 Pushchino, Moscow Region, Russia (7)
Edward P. Nikonowicz ■ Department of Biochemistry and Cell Biology, Rice University, Houston, TX 77251 (10)
Poul Nissen ■ Department of Molecular Biophysics and Biochemistry, Yale University, New Haven, CT 06520 (2)
Harry F. Noller ■ Center for Molecular Biology of RNA, Sinsheimer Laboratories, University of California–Santa Cruz, Santa Cruz, CA 95064 (13)
Gretel Nusspaumer ■ Centro de Biología Molecular "Severo Ochoa," CSIC and UAM, Canto Blanco, 28049 Madrid, Spain (12)
Jens Nyborg ■ Institute of Molecular and Structural Biology, University of Aarhus, Gustav Wieds Vej 10C, DK-8000 Aarhus C, Denmark (27)
Michael O'Connor ■ J. W. Wilson Laboratory, Department of Molecular and Cellular Biology and Biochemistry, Brown University, Providence, RI 02912 (19, 30)
James Ofengand ■ Department of Biochemistry and Molecular Biology, University of Miami School of Medicine, Miami, FL 33101 (16)
Elena V. Orlova ■ Department of Biochemistry, Imperial College of Science, Technology and Medicine, London SW7 2AY, United Kingdom (4)
Monika Osswald ■ Max-Planck-Institut für Molekulare Genetik, AG-Ribosomen, Ihnestrasse 73, 14195 Berlin, Germany (14)
Frances T. Pagel ■ Department of Molecular Genetics (Box 11), The University of Texas M. D. Anderson Cancer Center, 1515 Holcombe Blvd., Houston, TX 77030 (41)
Tillmann Pape ■ Department of Biochemistry, Imperial College of Science, Technology and Medicine, London SW7 2AY, United Kingdom (4, 25)
Pilar Parada ■ Centro de Biología Molecular "Severo Ochoa," CSIC and UAM, Canto Blanco, 28049 Madrid, Spain (12)
Sebastian Patzke ■ Max-Planck-Institut für Molekulare Genetik, AG Ribosomen, Ihnestraße 73, D-14195 Berlin, Germany (26)
Michael Pavlov ■ Department of Cell and Molecular Biology, BMC, Box 596, S-751 24 Uppsala, Sweden (44)
Herman J. Pel ■ DSM Food Specialties, DSM Gist 426-0295, P.O. Box 1, 2600 MA Delft, The Netherlands (40)
Pawel Penczek ■ Wadsworth Center, Empire State Plaza, Albany, NY 12201-0509 (5, 15)
Moshe Peretz ■ Department of Structural Biology, Weizmann Institute, 76100 Rehovot, Israel (3)
Jorge Perez-Fernandez ■ Centro de Biología Molecular "Severo Ochoa," CSIC and UAM, Canto Blanco, 28049 Madrid, Spain (12)
Marta Pioletti ■ Max Planck Institute for Molecular Genetics, Ihnestr. 73, 14195 Berlin, Germany (3)
Cynthia L. Pon ■ Laboratory of Genetics, Department of Biology MCA, University of Camerino, 62032 Camerino (MC), Italy (39)
Bo T. Porse ■ RNA Regulation Centre, Institute of Molecular Biology, Copenhagen University, DK-1307 Copenhagen K, Denmark (36)
Joseph D. Puglisi ■ Department of Structural Biology, Stanford University School of Medicine, Stanford, CA 94305-5421 (34)
Laing-Hu Qu ■ Biotechnology Research Center, Zhongshan University, Guangzhou 510 275, China (17)
Vasanthi Ramachandiran ■ Department of Chemistry and Biochemistry, The University of Texas at Austin, Austin, TX 78712-1167 (24)
V. Ramakrishnan ■ Department of Biochemistry, University of Utah School of Medicine, Salt Lake City, UT 84132, and Structural Studies Division, MRC Laboratory of Molecular Biology, Hills Rd., Cambridge CB2 2QH, United Kingdom (1, 8)
Hendrik A. Raué ■ Faculty of Science, Division of Chemistry, Department of Biochemistry and Molecular Biology, Institute for Molecular Biological Sciences, BioCentrum Amsterdam, Vrije Universiteit, de Boelelaan 1083, 1081 HV Amsterdam, The Netherlands (18)

Michael I. Recht ■ Department of Structural Biology, Stanford University School of Medicine, Stanford, CA 94305-5421 (34)
Miguel Remacha ■ Centro de Biología Molecular "Severo Ochoa," CSIC and UAM, Canto Blanco, 28049 Madrid, Spain (12)
Luis Reynaldo ■ Third Wave Technologies, Inc., 502 South Rosa Rd., Madison, WI 53719 (11)
Jutta Rinke-Appel ■ Max-Planck-Institut für Molekulare Genetik, AG-Ribosomen, Ihnestrasse 73, 14195 Berlin, Germany (14)
Marina V. Rodnina ■ Institute of Molecular Biology, University of Witten/Herdecke, 58448 Witten, Germany (25)
Kenneth E. Rudd ■ Department of Biochemistry and Molecular Biology, University of Miami School of Medicine, Miami, FL 33101 (16)
Raymond Samaha ■ Center for Molecular Biology of RNA, Sinsheimer Laboratories, University of California–Santa Cruz, Santa Cruz, CA 95064 (13)
Andreas Savelsbergh ■ Institute of Molecular Biology, University of Witten/Herdecke, 58448 Witten, Germany (25)
Markus A. Schäfer ■ Max-Planck-Institut für Molekulare Genetik, AG Ribosomen, Ihnestraße 73, D-14195 Berlin, Germany (26)
Frank Schluenzen ■ Max Planck Research Unit for Ribosomal Structure, Notkestr. 85, 22603 Hamburg, Germany (3)
Hyuk-Soo Seo ■ Department of Chemistry, University of Pennsylvania, Philadelphia, PA 19104-6323 (23)
Petr V. Sergiev ■ Department of Chemistry and Belozersky Institute of Physico-Chemical Biology, Moscow State University, Moscow 119899, Russia (21)
Maria Simitsopoulou ■ Max Planck Institute for Molecular Genetics, Ihnestr. 73, 14195 Berlin, Germany (3)
Ingolf Sommer ■ Max-Planck-Institut für Molekulare Genetik, AG-Ribosomen, Ihnestrasse 73, 14195 Berlin, Germany (14)
Christian M. T. Spahn ■ Max-Planck-Institut für Molekulare Genetik, AG Ribosomen, Ihnestraße 73, D-14195 Berlin, Germany, and Health Research, Inc., at the Wadsworth Center, and Howard Hughes Medical Institute, Empire State Plaza, Albany, NY 12201-0509 (5, 26)
Olga V. Spanchenko ■ Department of Chemistry and Belozersky Institute of Physico-Chemical Biology, Moscow State University, Moscow 119899, Russia (21)
Roberto Spurio ■ Laboratory of Genetics, Department of Biology MCA, University of Camerino, 62032 Camerino (MC), Italy (39)
Catherine L. Squires ■ Department of Molecular Biology and Microbiology, Tufts University School of Medicine, Boston, MA 02111 (19)
Holger Stark ■ Institut für Molekularbiologie und Tumorforschung, Philipps-Universität Marburg, Emil-Mannkopff-Straße 2, 35037 Marburg, Germany (4)
Thomas A. Steitz ■ Department of Molecular Biophysics and Biochemistry, Howard Hughes Medical Institute, and Department of Chemistry, Yale University, New Haven, CT 06520 (2)
Ulrich Stelzl ■ Max-Planck-Institut für Molekulare Genetik, AG Ribosomen, Ihnestraße 73, D-14195 Berlin, Germany (26)
Victor G. Stepanov ■ Institute of Molecular and Structural Biology, University of Aarhus, Gustav Wieds Vej 10C, DK-8000 Aarhus C, Denmark (27)
Rogier Stuger ■ Department of Molecular Cell Physiology, Vrije Universiteit, De Boelelaan 1087, 1081 HV Amsterdam, The Netherlands (18)
Heinrich B. Stuhrmann ■ GKSS Research Center, Max-Planck-Straße, 21502 Geesthacht, Germany, and Institut de Biologie Structurale, 41 Avenue des Martyrs, F-38027 Grenoble, CEDEX 1, France (26)
Isao Tanaka ■ Division of Biological Sciences, Graduate School of Science, Hokkaido University, Sapporo 060-0810, Japan (9)
Masae Taniguchi ■ Division of Biological Sciences, Graduate School of Science, Hokkaido University, Sapporo 060-0810, Japan (9)

Warren P. Tate ■ Department of Biochemistry and Centre for Gene Research, University of Otago, PO Box 56, Dunedin, New Zealand (40)
Søren S. Thirup ■ Institute of Molecular and Structural Biology, University of Aarhus, Gustav Wieds Vej 10C, DK-8000 Aarhus C, Denmark (27)
Charles M. Thompson ■ Department of Chemistry, The University of Montana, Missoula, MT 59812 (22)
Jill R. Thompson ■ J. W. Wilson Laboratory, Department of Molecular and Cellular Biology and Biochemistry, Brown University, Providence, RI 02912 (19)
Antonius C. J. Timmers ■ Laboratoire de Biologie Moléculaire des Relations Plantes-Microorganismes, CNRS-INRA, BP 27, 31326 Castanet-Tolosan Cedex, France (18)
Svetlana Tishchenko ■ Institute of Protein Research, Russian Academy of Sciences, 142292 Pushchino, Moscow Region, Russia (7)
Ante Tocilji ■ Max Planck Research Unit for Ribosomal Structure, Notkestr. 85, 22603 Hamburg, Germany (3)
Jerneja Tomšic ■ Laboratory of Genetics, Department of Biology MCA, University of Camerino, 62032 Camerino (MC), Italy (39)
Tamara Tsalkova ■ Department of Chemistry and Biochemistry, The University of Texas at Austin, Austin, TX 78712-1167 (24)
K. Uma ■ Department of Biochemistry and Molecular Biology and Program in Molecular and Cellular Biology, University of Massachusetts, Amherst, MA 01003 (10)
Makiko Uno ■ Department of Tumor Biology, The Institute of Medical Science, The University of Tokyo, Minato-ku, Tokyo 108-8639, Japan (42)
Marin van Heel ■ Department of Biochemistry, Imperial College of Science, Technology and Medicine, London SW7 2AY, United Kingdom (4)
Margaret S. VanLoock ■ Department of Biochemistry and Molecular Genetics, University of Alabama at Birmingham, Birmingham, AL 35294-0005 (15)
Jan van 't Riet ■ Faculty of Science, Division of Chemistry, Department of Biochemistry and Molecular Biology, Institute for Molecular Biological Sciences, BioCentrum Amsterdam, Vrije Universiteit, de Boelelaan 1083, 1081 HV Amsterdam, The Netherlands (18)
Michael A. Van Waes ■ Division of Biological Sciences, The University of Montana, Missoula, MT 59812 (22)
Birte Vester ■ RNA Regulation Centre, Department of Molecular Biology, University of Copenhagen, DK-1307 Copenhagen K, Denmark (35)
Anton Vila-Sanjurjo ■ J. W. Wilson Laboratory, Department of Molecular and Cellular Biology and Biochemistry, Brown University, Providence, RI 02912 (19)
Serguei N. Vladimirov ■ Department of Chemistry, University of Pennsylvania, Philadelphia, PA 19104-6323 (23)
Uwe von Ahsen ■ Center for Molecular Biology of RNA, Sinsheimer Laboratories, University of California–Santa Cruz, Santa Cruz, CA 95064 (13)
Ruo Wang ■ Schering-Plough Research Institute, 2015 Galloping Hill Rd., K-15-3-3545, Kenilworth, NJ 07033-0539 (23)
Shulamith Weinstein ■ Department of Structural Biology, Weizmann Institute, 76100 Rehovot, Israel (3)
Stephen W. White ■ Department of Structural Biology, St. Jude Children's Research Hospital, 332 N. Lauderdale, Memphis, TN 38105, and Department of Biochemistry, University of Tennessee, 858 Madison Ave., Memphis, TN 38163 (8)
Regine Willumeit ■ GKSS Research Center, Max-Planck-Straße, 21502 Geesthacht, Germany (26)
Daniel N. Wilson ■ Department of Biochemistry and Centre for Gene Research, University of Otago, PO Box 56, Dunedin, New Zealand (40)
Kevin Wilson ■ Center for Molecular Biology of RNA, Sinsheimer Laboratories, University of California–Santa Cruz, Santa Cruz, CA 95064 (13)
Brian T. Wimberly ■ Department of Biochemistry, University of Utah School of Medicine, Salt Lake City, UT 84132, and MRC Laboratory of Molecular Biology, Hills Rd., Cambridge CB2 2QH, United Kingdom (1, 8)

Wolfgang Wintermeyer ■ Institute of Molecular Biology, University of Witten/Herdecke, 58448 Witten, Germany (25)

Ira G. Wool ■ Department of Biochemistry and Molecular Biology, The University of Chicago, Chicago, IL 60637 (38)

Iwona Wower ■ Department of Biochemistry and Molecular Biology and Program in Molecular and Cellular Biology, University of Massachusetts, Amherst, MA 01003 (10)

Iwona K. Wower ■ Department of Animal and Dairy Sciences, Program in Cell and Molecular Biosciences, Auburn University, Auburn, AL 36849-5415 (32)

Jacek Wower ■ Department of Animal and Dairy Sciences, Program in Cell and Molecular Biosciences, Auburn University, Auburn, AL 36849-5415 (32)

Herren Wu ■ Department of Biochemistry and Molecular Biology and Program in Molecular and Cellular Biology, University of Massachusetts, Amherst, MA 01003 (10)

Ada Yonath ■ Department of Structural Biology, Weizmann Institute, 76100 Rehovot, Israel, and Max Planck Research Unit for Ribosomal Structure, Notkestr. 85, 22603 Hamburg, Germany (3)

Kuniyasu Yoshimura ■ Department of Tumor Biology, The Institute of Medical Science, The University of Tokyo, Minato-ku, Tokyo 108-8639, Japan (42)

Satoko Yoshizawa ■ Department of Structural Biology, Stanford University School of Medicine, Stanford, CA 94305-5421 (34)

Jing Yuan ■ Division of Biological Sciences, The University of Montana, Missoula, MT 59812 (22)

Marat Yusupov ■ Center for Molecular Biology of RNA, Sinsheimer Laboratories, University of California–Santa Cruz, Santa Cruz, CA 95064 (13)

Gulnara Yusupova ■ Center for Molecular Biology of RNA, Sinsheimer Laboratories, University of California–Santa Cruz, Santa Cruz, CA 95064 (13)

Robert A. Zimmermann ■ Department of Biochemistry and Molecular Biology and Program in Molecular and Cellular Biology, University of Massachusetts, Amherst, MA 01003 (10)

Jesus Zurdo ■ Centro de Biología Molecular "Severo Ochoa," CSIC and UAM, Canto Blanco, 28049 Madrid, Spain (12)

Christian Zwieb ■ Department of Molecular Biology, The University of Texas Health Science Center at Tyler, 11937 U.S. Highway 271, Tyler, TX 75708-3154 (32)

PREFACE

This book is based primarily on presentations made at a ribosome conference held from 13 to 17 June 1999 in Helsingør, Denmark, which was attended by around 270 researchers. It was a landmark meeting because many recent and important advances were presented for the first time, in particular some of the X-ray crystallographic models of ribosomes and ribosomal subunits. We are presently at a stage where there is real promise of gaining insight into the detailed molecular workings of the ribosomal machine. In the book, the individual authors were asked both to cover their recent work and to summarize the contributions of others in the field.

Major international ribosome meetings are held every few years. They date back to a meeting held at Cold Spring Harbor, N.Y., in 1973, which was followed by meetings in Madison, Wis. (1979), Port Aransas, Tex. (1985), East Glacier Park, Mont. (1989), Berlin, Germany (1992), and Victoria, British Columbia, Canada (1995), and they have covered the main developments within the protein biosynthesis field. Given the high complexity of ribosomal structure and function, this time interval generally guarantees that each meeting will cover many important new developments. Each of these meetings has produced a book summarizing these developments. The books have served as valuable reference works over the years. They are affectionately known by the color of their covers, but since we have now run out of appropriate (single) colors, the present book will be "red and white" (also Danish national colors).

The present meeting was a little more focused than usual, concentrating primarily on the structure and function of bacterial and archaeal ribosomes, with some emphasis on antibiotic action and a little (less than anticipated) on cellular interactions. This bias reflects to a large degree the nature of recent progress. There have been major developments in analyzing and modeling the structure of ribosomes primarily on the basis of cryo-electron microscopy and X-ray diffraction results. Moreover, the crystal structures of the ternary complex of aminoacyl-tRNA–elongation factor Tu–GTP and elongation factor G have been determined recently. Furthermore, we have gained considerable insight into how antibiotics act, especially within the mRNA-decoding site, the GTPase-associated site on the 50S subunit, and the peptidyltransferase center.

My co-organizers (and coeditors) have been most creative and supportive in producing a very successful meeting, and Steve Douthwaite, in particular, was deeply involved in the meeting organization. The Meeting Advisory Committee are thanked for their valuable help, particularly with refereeing the ribosome book chapters, and ASM and Susan Birch's help with the book production is much appreciated. Walt Hill and Al Matheson are also acknowledged for approaching Anders Liljas and me at the last meeting in Victoria and suggesting that we organize this meeting in Scandinavia. Walt Hill has been a constant source of encouragement and inspiration during the intervening 4 years. We are also grateful for the financial support, crucial for organizing such meetings, from the Danish Plasmid Foundation, The Carlsberg Foundation, Løvens Kemisk Fabrik, the International Union of Biochemistry and Molecular Biology, Amersham-Pharmacia Biotech, and the Dupont Pharmaceutical Company.

Roger A. Garrett

I. X-RAY CRYSTALLOGRAPHIC STUDIES OF RIBOSOMAL SUBUNITS

I. X-RAY CRYSTALLOGRAPHIC STUDIES OF RIBOSOMAL SUBUNITS

The dream of analyzing the structure of the ribosome by X-ray diffraction goes back to the 1960s, when the first crystalline arrays of ribosomes were detected in embryos. Progress was slow, however, and it was in the 1980s that high-quality crystals were first produced in vitro. Since then, many of the techniques for analyzing crystals of this large macromolecular complex have been developed, and for the first time, partial models of the bacterial 30S subunit and archaeal 50S subunit were presented at the meeting at resolutions of 5.5 and 5Å, respectively.

The Ribosome: Structure, Function, Antibiotics, and Cellular Interactions
Edited by R. A. Garrett, S. R. Douthwaite, A. Liljas, A. T. Matheson, P. B. Moore, and H. F. Noller

Chapter 1

Progress toward the Crystal Structure of a Bacterial 30S Ribosomal Subunit

V. RAMAKRISHNAN, MALCOLM S. CAPEL, WILLIAM M. CLEMONS, JR., JOANNA L. C. MAY, and BRIAN T. WIMBERLY

In 1991, Yonath and coworkers, after nearly a decade of pioneering work on the crystallization of ribosomes, showed that it was possible to obtain diffraction to beyond 3 Å from crystals of the 50S subunit of *Haloarcula marismortui* (von Böhlen et al., 1991). This work marked a milestone in ribosome crystallography, because it established that in principle an atomic-resolution structure of a ribosomal subunit could be obtained.

In the meantime, work on whole-ribosome crystallography was also carried out with *Thermus thermophilus*. The work on *Thermus* led to crystals of both the small (Trakhanov et al., 1987; Yonath et al., 1988; Yusupov et al., 1988) and large (Yonath and Franceschi, 1998) ribosomal subunits as well as whole 70S ribosomes (Trakhanov et al., 1987). Crystals of the small subunit initially diffracted to 9- to 12-Å resolution, and their quality improved only slightly over the years. However, using a modification of previously reported crystallization conditions, we have obtained crystals of the 30S subunit from *T. thermophilus* that diffract to 3.6-Å resolution (Clemons et al., 1999). These crystals appear to be similar to ones reported independently by Yonath and coworkers recently (Yonath et al., 1998) (see chapter 3). They establish the fact that it may be possible to obtain a nucleotide resolution structure of the 30S subunit.

While these reports showed that it was possible to obtain high-resolution structures of both subunits in principle, the problem of obtaining phase information from such large asymmetric units seemed insurmountable until recently. However, over the last decade, relatively large asymmetric units, such as F1-ATPase (Abrahams et al., 1994) or cytochrome oxidase (Tsukihara et al., 1996), were solved by conventional multiple isomorphous replacement, suggesting that existing crystallographic methods could probably be used to tackle even larger problems, such as ribosomal subunits and whole ribosomes. Moreover, the last decade has seen the advent of better sources in the form of highly stable synchrotron beams, better and more sensitive detectors, and faster computing. A particularly useful advance that allows optimal use of powerful synchrotron sources is the use of cryocooled crystals for data collection (of which Yonath and her colleagues were early advocates).

PHASING THE 30S SUBUNIT

The signal from a heavy-atom derivative depends not only on the size and number of sites, but also on the size of the macromolecule (Crick and Magdoff, 1956). Even with a very large macromolecule, it is possible to get sufficient phasing power with conventional (i.e., single-atom) derivatives provided that there are a sufficient number of sites. However, it was felt in the past that determining the locations of these sites, a necessary condition for phasing, would prove problematic ab initio. Thus, for very large structures, the use of heavy-atom clusters of tantalum or tungsten to obtain initial phases was suggested over 2 decades ago (Blundell and Johnson, 1976). In fact, such clusters were used to obtain low-

V. Ramakrishnan ■ Department of Biochemistry, University of Utah School of Medicine, Salt Lake City, UT 84132, and Structural Studies Division, MRC Laboratory of Molecular Biology, Hills Road, Cambridge CB2 2QH, England. **Malcolm S. Capel** ■ Biology Department, Brookhaven National Laboratory, Upton, NY 11973. **William M. Clemons, Jr., Joanna L. C. May, and Brian T. Wimberly** ■ Department of Biochemistry, University of Utah School of Medicine, Salt Lake City, UT 84132.

resolution phases for the proteasome structure (Knablein et al., 1997), and their use for ribosome crystallography has been advocated by Yonath and coworkers (Thygesen et al., 1996). But, as was clearly shown by the work on the proteasome, the form factors of such heavy-atom clusters become zero at relatively low resolution (6 to 20 Å, depending on the cluster) and have negative values beyond that, so that even at modest resolution the clusters cannot be treated as point "superheavy" atoms, and they have rapidly diminishing phasing power. Of course, if the clusters are ordered in the structure, they can contribute phase information even at high resolution by being treated as a collection of single atoms (Knablein et al., 1997).

Another approach has been to use molecular replacement starting with an electron-microscopic reconstruction model to obtain low-resolution phases (Ban et al., 1998), following an approach first used for virus crystallography. This approach allows the detection of heavy-atom cluster sites by using difference Fouriers from low-resolution phases, so that the problem can be bootstrapped to increasingly higher resolutions. However, work on the 50S subunit (Ban et al., 1998), as well as our own work on the 30S subunit (Fig. 1), shows that it is clearly possible to see peaks from heavy-atom clusters in difference Patterson maps and thereby to calculate phases directly from X-ray data.

Moreover, even large numbers of individual (i.e., nonclustered) heavy-atom sites in a derivative can now be detected from crystallographic data by direct methods (Hauptman, 1997) or new automated search algorithms (Terwilliger and Berendzen, 1999), so the use of clusters is no longer necessary as long as derivatives can be found with a sufficient number of sites to provide good phasing power. Finally, the problem of nonisomorphism can be alleviated by the use of multiwavelength anomalous dispersion (Hendrickson, 1991), in which phase information is obtained by data collection on the same crystal at different wavelengths. However, this presupposes that all of the data will be collected from one crystal or a group of isomorphous crystals. If crystal-to-crystal nonisomorphism is intrinsic rather than the result of soaks in heavy atoms, and if data collection requires the use of many crystals, the nonisomorphism problem will not necessarily be circumvented by multiwavelength anomalous dispersion.

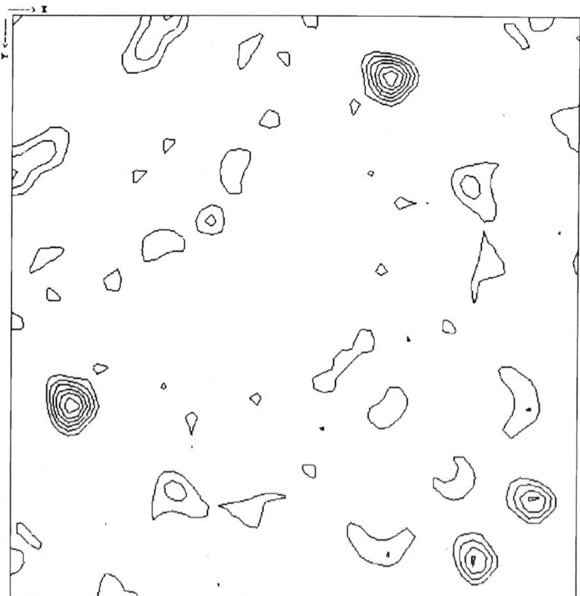

Figure 1. Harker section ($w = 0.5$) of an isomorphous difference Patterson map from a W17 derivative of *T. thermophilus* 30S crystals, showing both a major and a minor site.

INITIAL RESULTS WITH THE 30S SUBUNIT FROM *T. THERMOPHILUS*

Using isomorphous replacement methods and a combination of both heavy-atom clusters and single-atom, multiple-site derivatives, we have obtained a map of the 30S subunit at 5.5-Å resolution. Details of the crystallography and the results of an initial analysis of this map have been published elsewhere (Clemons et al., 1999). Here, we summarize the main findings.

Features Visible at 5.5-Å Resolution

At 5.5-Å resolution, it is clearly possible to see individual strands of RNA in double-stranded RNA, as well as phosphate bumps in the density (Fig. 2a). Alpha helices of proteins are also visible in the structure, so that it is possible to find and to determine the orientations of proteins of known crystal structure in the map (Fig. 2b). Finally, it is possible to see protein-RNA complexes in many cases. These are seen in sufficient detail to determine which part of the protein is interacting with which part of the RNA but not at the level of individual residue-base contacts.

Locations of Proteins of Known Structure

To locate proteins in the electron density of the 30S subunit, the neutron map that shows the centers of mass of the 30S proteins (Capel et al., 1987) was very helpful. Once two or three proteins of known structure had been found, it was possible to quickly identify the others in the map, since their approxi-

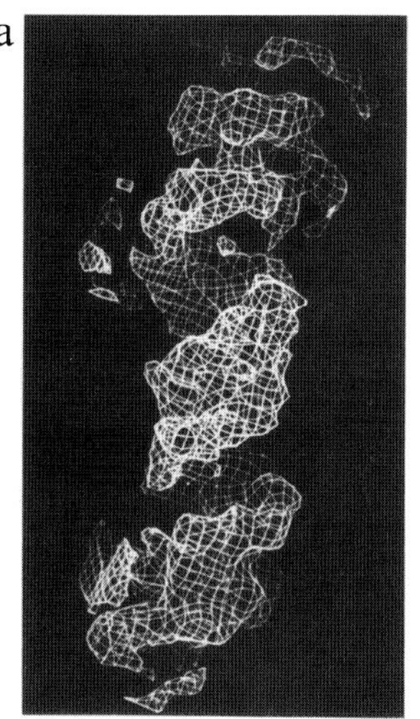

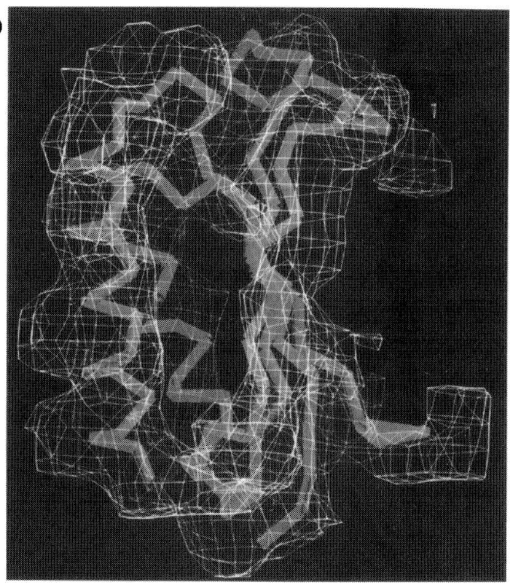

Figure 2. Electron density map at 5.5-Å resolution. (a) A stretch of double-stranded RNA that shows individual strands, major and minor grooves, and bumps that correspond to phosphate groups. (b) Fit of the crystal structure of S6 (Lindahl et al., 1994) to the electron density, showing that individual proteins of known structure can be both located and oriented in the map.

mate locations in the map could be predicted by the neutron data. As a result, we have now been able to place all seven proteins of known structure into the 30S map. In general, the agreement with the neutron map is very good, with a root-mean-squared-deviation of 11.7 Å when S15 is excluded from the comparison. With this superposition, S15 is about 51 Å from its predicted neutron map position. This discrepancy could be due to a single distance, S11 to S15 (Ramakrishnan et al., 1981), which had poor statistics and may have had the effect of placing S15 on the wrong side of an approximately planar group of proteins.

Near the bottom of the subunit, we see a three-helix bundle that we have identified as S20. This identification is based on biochemical data (Powers and Noller, 1995; Ungewickell et al., 1975) and is consistent with a model derived from cryo-electron-microscopic and biochemical data (Mueller and Brimacombe, 1997) but not with the neutron map, which places S20 in the head of the subunit. Our assignment of this density represents a new ribosomal protein structure, albeit at low resolution. The structure is consistent with secondary-structure prediction based on the sequence of S20.

In addition, we see protein density at or near predicted locations for several other proteins.

The Fold of the Central Domain

The central domain of 16S RNA spans nucleotides 565 to 912 (*Escherichia coli* sequence). This part of 16S RNA binds several proteins, including the primary binding proteins S8 and S15. It encompasses the platform as well as part of the body of the subunit and is of considerable functional interest. Moreover, the interactions with rRNA of proteins S8 (Moine et al., 1997; Wu et al., 1994) and S15 (Batey and Williamson, 1996; Serganov et al., 1996) are particularly well characterized. Because the central domain contains a high density of proteins of previously known crystal structure and well-characterized biochemistry, it was possible to determine a fold for it even though our resolution is well below the traditional "atomic" resolution of 3.5 Å required for a new trace.

This was done by first independently fitting double-stranded RNA and proteins of known structure into the electron density. At this point, a number of protein-RNA complexes were apparent, such as the adjacent S6-RNA and S15-RNA complexes. We then used biochemical data on protein-RNA interactions in the ribosome from cross-linking (reviewed in Mueller and Brimacombe, 1997) and footprinting (Powers and Noller, 1995) experiments to assign specific segments of 16S RNA to double-helical density. As a result, it was possible to determine the fold of the central domain, including helix 27 (H27) at the 3′ end, which has been implicated in a functionally important switch (Lodmell and Dahlberg, 1997). The fold of the central-domain RNA is shown in Fig. 3a, and details of the interactions that help to organize it have been described elsewhere (Clemons et al., 1999). Both protein-RNA contacts and RNA-RNA contacts help to stabilize the structure. In particular, the three-way junction formed by the RNA helices 20, 21, and 22, and the interaction of this junction

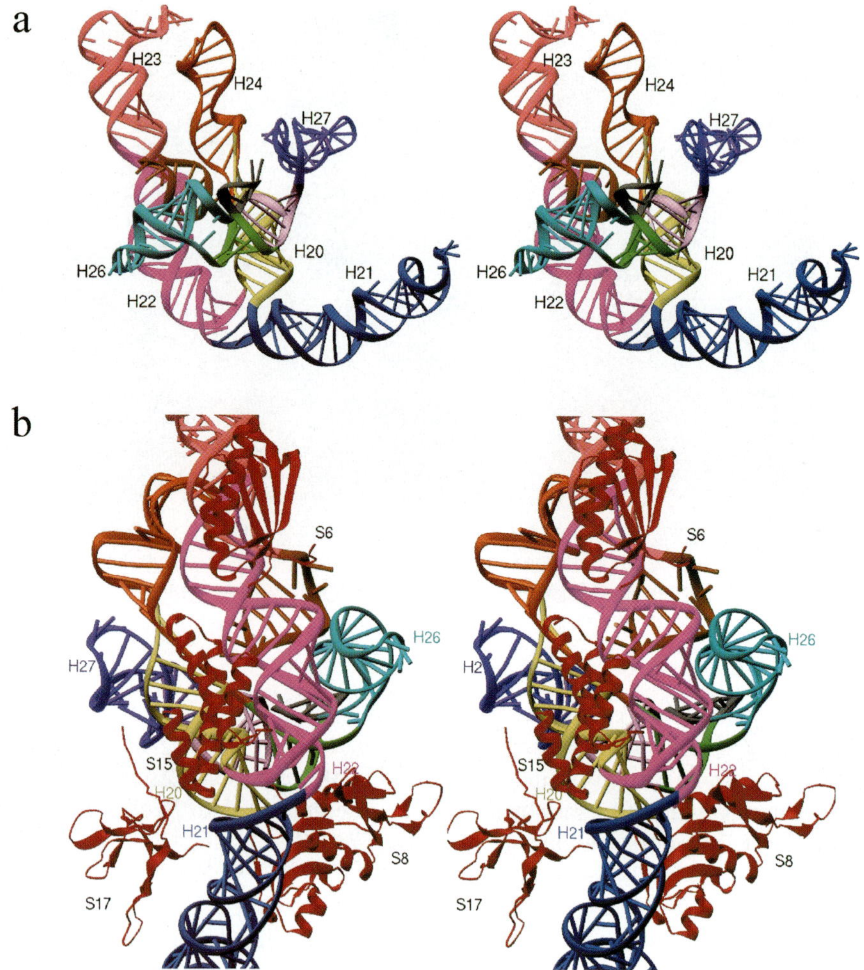

Figure 3. (a) Stereo view of the fold of the central domain of 16S RNA. (b) Stereo view of the three-helix junction formed by helices 20, 21, and 22 in the central domain, with associated proteins. The green helix shows the interaction of the N-terminal helix of S8 with a minor groove of H25.

with proteins S8, S15, and S17, shown in Fig. 3b, are all in agreement with a large body of biochemical data. At the same time, the resolution of the map allows us to indicate which parts of the protein interact with which parts of the RNA, such as the interaction of the N-terminal helix of S8 with a minor groove of H25 (Fig. 3b).

The Platform

The platform of the 30S subunit is a functionally important region that ends in H23 and H24, which are terminated by the 690 and 790 loops, respectively. It is known to be conformationally variable. For example, binding of both 50S (Agrawal et al., 1998) and IF3 (McCutcheon et al., 1999) results in a movement of the platform. An interesting observation from the map is that the platform of the 30S subunit consists of two halves separated by a plane of lesser density, suggesting that the two halves of the platform may move independently during translation.

The H27 Accuracy Switch

H27 is known to be involved in a conformational switch that affects translational accuracy (Lodmell and Dahlberg, 1997). The switch is known to involve two alternative base-pairing schemes of a triplet in H27. In our model, the hairpin loop of H27 packs against the minor groove of the proximal end of H24 in the platform (Fig. 4). If the front half of the platform does move independently, then H27 may well move with it. The other end of H27 makes a very extensive minor-groove-packing interaction with H44 (Fig. 4), just below the decoding site where the A-site tRNA binds (Gornicki et al., 1984). This end of H27 is also close to ribosomal protein S5 and to RNA residues footprinted by streptomycin; both

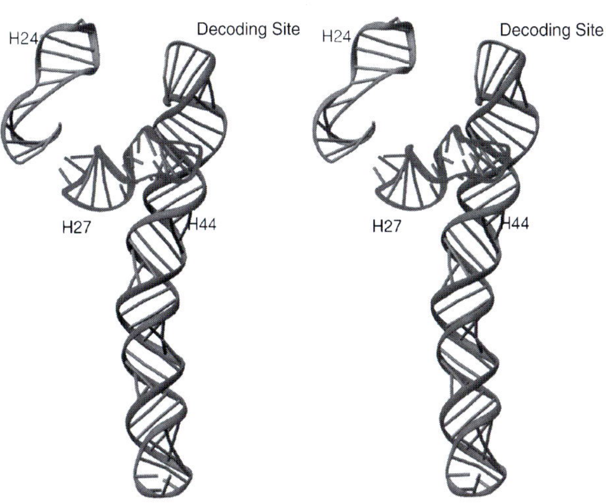

Figure 4. Relative positions of RNA helices 24, 27, and 44 (the 1400/1500 stem-loop) in the structure.

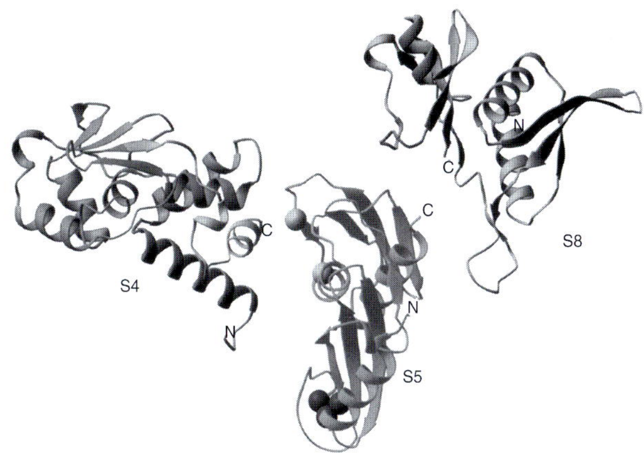

Figure 5. Importance of protein-protein interactions in the 30S subunit. The figure shows relative positions of proteins S4, S5, and S8 in the 30S structure. The spheres on S5 at the S4 interface are the sites of *ram* mutations. Similar mutations on S4 involve deletions of the C-terminal helix, which is also at the S4-S5 interface. The other spheres on the S5 structure are the sites of resistance to spectinomycin.

streptomycin and mutations in S5 influence the effect of the switch in accuracy (Lodmell and Dahlberg, 1997). Previously, it was unknown how the H27 switch affected translational accuracy; now it appears probable that the H27 switch determines translational accuracy by modulating the conformation of the decoding site by a direct interaction, including a possible movement of H44.

Protein-Protein Interactions

Although most proteins seen interact mainly with rRNA, even from the subset of proteins we have identified, we can see an interesting example of protein-protein interactions in the 30S subunit. As shown in Fig. 5, S5 forms an interface with both S4 and S8. The S5-S8 interaction is one of the oldest known (Allen et al., 1979; Lutter et al., 1972), and the positions are consistent with a cross-link from the C-terminal domain of S5 to S8. A more interesting situation is the S4-S5 interface. This interface contains the sites of the *ram* mutations on both proteins (van Acken, 1975; Wittmann-Liebold and Greuer, 1978), suggesting that these mutations disrupt the interface between these two proteins rather than a protein-RNA contact as had previously been suggested (Ramakrishnan and White, 1992).

Overview of the 30S Subunit

We have been able to identify double-stranded RNA segments throughout the subunit. Figure 6 shows these segments, along with the proteins of known structure that we have placed, our model for the central domain, and our putative location and structure for S20.

A striking conclusion from our work so far is the nearly complete absence of proteins from the subunit interface. It is not clear if this conclusion will still hold when all the proteins of both subunits have been identified, but so far, it is consistent with a tritium bombardment experiment, suggesting that there are no proteins at the interface (Agafonov et al., 1997). Since the subunit interface is likely to be the functionally most important part of the ribosome, this supports speculations that a primordial ribosome may have contained no proteins at all.

However, as noted above for the *ram* mutations involving S4 and S5, protein-protein interactions are also important. Therefore, not all of the important

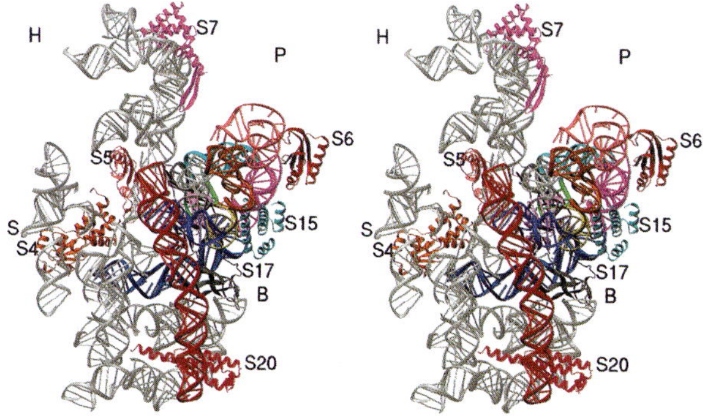

Figure 6. Stereo view of our current interpretation of the 30S subunit at 5.5-Å resolution. The proteins of known structure are shown, along with S20, the central domain of 16S RNA, and H44. Regions that appear to be clearly double-stranded RNA in the map are shown in gray.

surfaces and moving parts in the ribosome today are likely to consist exclusively of RNA.

Conclusions

It is tempting to model much of the ribosome at an intermediate resolution, where one can see double-stranded RNA and recognize known protein structures, or at even-lower resolution, as has been done by others in conjunction with biochemical and electron microscopic data. Such models provide a useful framework for thinking about ribosome function. They certainly provide a global view of the structural changes involved and the approximate locations of important ligands. However, many of the critical functions of the 30S subunit, such as decoding and proofreading and movements such as translocation, can only be understood biochemically when the model has nucleotide and amino acid residue resolution. Many of these functions are likely to involve conformational switches, such as the H27 switch that was identified on the basis of genetic and biochemical data (Lodmell and Dahlberg, 1997). These switches are likely to involve alternative but highly specific contacts, and their biochemical and structural nature can only be revealed at high resolution.

What resolution is enough? A cautionary reminder is the case of tRNA. Although it is well over an order of magnitude simpler than the 30S subunit, the first polynucleotide chain trace of tRNA, at 4-Å resolution, was incorrect near the four-way junction and missed important base triples in the structure. This was subsequently corrected when the resolution improved to 3 Å. We know much more about RNA structure today, but we are also likely to encounter new motifs in extremely complicated contexts, thus raising the probability of mistakes. Therefore, all models at low resolution must be treated with some scepticism at this stage.

However, given the work described here and in other chapters in this volume, there appear to be no fundamental impediments to obtaining a high-resolution structure of each subunit and perhaps of the whole 70S ribosome.

This work was supported by NIH grant GM 44973, the United Kingdom Medical Research Council, and the University of Utah.

REFERENCES

Abrahams, J. P., A. G. Leslie, R. Lutter, and J. E. Walker. 1994. Structure at 2.8 A resolution of F1-ATPase from bovine heart mitochondria. *Nature* 370:621–628.

Agafonov, D. E., V. A. Kolb, and A. S. Spirin. 1997. Proteins on ribosome surface: measurements of protein exposure by hot tritium bombardment technique. *Proc. Natl. Acad. Sci. USA* **94:** 12892–12897.

Agrawal, R. K., P. Penczek, R. A. Grassucci, and J. Frank. 1998. Visualization of elongation factor G on the Escherichia coli 70S ribosome: the mechanism of translocation. *Proc. Natl. Acad. Sci. USA* **95:**6134–6138.

Allen, G., R. Capasso, and C. Gualerzi. 1979. Identification of the amino acid residues of proteins S5 and S8 adjacent to each other in the 30 S ribosomal subunit of Escherichia coli. *J. Biol. Chem.* **254:**9800–9806.

Ban, N., B. Freeborn, P. Nissen, P. Penczek, R. A. Grassucci, R. Sweet, J. Frank, P. B. Moore, and T. A. Steitz. 1998. A 9 Å resolution X-ray crystallographic map of the large ribosomal subunit. *Cell* **93:**1105–1115.

Batey, R., and J. Williamson. 1996. Interaction of the *Bacillus stearothermophilus* ribosomal protein S15 with 16 S rRNA. I. Defining the minimal RNA site. *J. Mol. Biol.* **261:**536–549.

Blundell, T. L., and L. N. Johnson. 1976. *Protein Crystallography.* Academic Press, New York, N.Y.

Capel, M. S., D. M. Engelman, B. R. Freeborn, M. Kjeldgaard, J. A. Langer, V. Ramakrishnan, D. G. Schindler, D. K. Schneider, B. P. Schoenborn, I.-Y. Sillers, S. Yabuki, and P. B. Moore. 1987. A complete mapping of the proteins in the small ribosomal subunit of *Escherichia coli. Science* **238:**1403–1406.

Clemons, W. M., Jr., J. L. C. May, B. T. Wimberly, J. P. McCutcheon, M. Capel, and V. Ramakrishnan. 1999. Structure of a bacterial 30S ribosomal subunit at 5.5 Å resolution. *Nature* **400:**833–840.

Crick, F. H. C., and B. S. Magdoff. 1956. The theory of the method of isomorphous replacement for protein crystals. I. *Acta Crystallogr.* **9:**901–908.

Gornicki, P., K. Nurse, W. Hellmann, M. Boublik, and J. Ofengand. 1984. High resolution localization of the tRNA anticodon interaction site on the Escherichia coli 30 S ribosomal subunit. *J. Biol. Chem.* **259:**10493–10498.

Hauptman, H. 1997. Shake-and-bake: an algorithm for automated solution *ab initio* of crystal structures. *Methods Enzymol.* **277:** 3–13.

Hendrickson, W. A. 1991. Determination of macromolecular structures from anomalous diffraction of synchrotron radiation. *Science* **254:**51–58.

Knablein, J., T. Neuefeind, F. Schneider, A. Bergner, A. Messerschmidt, J. Lowe, B. Steipe, and R. Huber. 1997. Ta6Br(2+)12, a tool for phase determination of large biological assemblies by X-ray crystallography. *J. Mol. Biol.* **270:**1–7.

Lindahl, M., L. A. Svensson, A. Liljas, S. E. Sedelnikova, I. A. Eliseikina, N. P. Fomenkova, N. Nevskaya, S. V. Nikonov, M. B. Garber, T. A. Muranova, A. I. Rykonova, and R. Amons. 1994. Crystal structure of the ribosomal protein S6 from *Thermus thermophilus. EMBO J.* **13:**1249–1254.

Lodmell, J. S., and A. E. Dahlberg. 1997. A conformational switch in Escherichia coli 16S ribosomal RNA during decoding of messenger RNA. *Science* **277:**1262–1267.

Lutter, L. C., H. Zeichhardt, and C. G. Kurland. 1972. Ribosomal protein neighborhoods. I. S18 and S21 as well as S5 and S8 are neighbors. *Mol. Gen. Genet.* **119:**357–366.

McCutcheon, J. P., R. K. Agrawal, S. M. Philips, R. A. Grassucci, S. E. Gerchman, W. M. Clemons, Jr., V. Ramakrishnan, and J. Frank. 1999. Location of translational initiation factor IF3 on the small ribosomal subunit. *Proc. Natl. Acad. Sci. USA* **96:**4301–4306.

Moine, H., C. Cachia, E. Westhof, B. Ehresmann, and C. Ehresmann. 1997. The RNA binding site of S8 ribosomal protein of Escherichia coli: selex and hydroxyl radical probing studies. *RNA* **3:**255–268.

Mueller, F., and R. Brimacombe. 1997. A new model for the three-dimensional folding of Escherichia coli 16 S ribosomal RNA. II. The RNA-protein interaction data. *J. Mol. Biol.* **271**:545–565.

Powers, T., and H. F. Noller. 1995. Hydroxyl radical footprinting of ribosomal proteins on 16S rRNA. *RNA* **1**:194–209.

Ramakrishnan, V., and S. W. White. 1992. Structure of ribosomal protein S5 reveals sites of interaction with 16S RNA. *Nature* **358**:768–771.

Ramakrishnan, V., S. Yabuki, I.-Y. Sillers, D. G. Schindler, D. M. Engelman, and P. B. Moore. 1981. Position of proteins S6, S11 and S15 in the 30 S ribosomal subunit of *Escherichia coli*. *J. Mol. Biol.* **153**:739–760.

Serganov, A. A., B. Masquida, E. Westhof, C. Cachia, C. Portier, M. Garber, B. Ehresmann, and C. Ehresmann. 1996. The 16S rRNA binding site of Thermus thermophilus ribosomal protein S15: comparison with Escherichia coli S15, minimum site and structure. *RNA* **2**:1124–1138.

Terwilliger, T., and J. Berendzen. 1999. Automated MAD and MIR structure determination. *Acta Crystallogr.* **D55**:849–861.

Thygesen, J., S. Weinstein, F. Franceschi, and A. Yonath. 1996. The suitability of multi-metal clusters for phasing in crystallography of large macromolecular assemblies. *Structure* **4**:513–518.

Trakhanov, S. D., M. M. Yusupov, S. C. Agalarov, M. B. Garber, S. N. Ryazantsev, S. V. Tischenko, and V. A. Shirokov. 1987. Crystallization of 70 S ribosomes and 30 S ribosomal subunits from *Thermus thermophilus*. *FEBS Lett.* **220**:319–322.

Tsukihara, T., H. Aoyama, E. Yamashita, T. Tomizaki, H. Yamaguchi, K. Shinzawa-Itoh, R. Nakashima, R. Yaono, and S. Yoshikawa. 1996. The whole structure of the 13-subunit oxidized cytochrome c oxidase at 2.8 A. *Science* **272**:1136–1144.

Ungewickell, E., R. Garrett, C. Ehresmann, P. Stiegler, and P. Fellner. 1975. An investigation of the 16-S RNA binding sites of ribosomal proteins S4, S8, S15, and S20 from Escherichia coli. *Eur. J. Biochem.* **51**:165–180.

van Acken, U. 1975. Proteinchemical studies on ribosomal proteins S4 and S12 from ram (ribosomal ambiguity) mutants of Escherichia coli. *Mol. Gen. Genet.* **140**:61–68.

von Böhlen, K., I. Makowski, H. A. S. Hansen, H. Bartels, Z. Berkovitch-Yellin, A. Zaytzev-Bashan, S. Meyer, C. Paulke, F. Franceschi, and A. Yonath. 1991. Characterization and preliminary attempts for derivatization of crystals of large ribosomal subunits from *Haloarcula marismortui* diffracting to 3 Å resolution. *J. Mol. Biol.* **222**:11–15.

Wittmann-Liebold, B., and B. Greuer. 1978. The primary structure of protein S5 from the small subunit of the Escherichia coli ribosome. *FEBS Lett.* **95**:91–98.

Wu, H., L. Jiang, and R. A. Zimmermann. 1994. The binding site for ribosomal protein S8 in 16S rRNA and spc mRNA from Escherichia coli: minimum structural requirements and the effects of single bulged bases on S8-RNA interaction. *Nucleic Acids Res.* **22**:1687–1695.

Yonath, A., and F. Franceschi. 1998. Functional universality and evolutionary diversity: insights from the structure of the ribosome. *Structure* **6**:679–684.

Yonath, A., C. Glotz, H. S. Gewitz, K. S. Bartels, K. von Böhlen, L. Makowski, and H. G. Wittmann. 1988. Characterization of crystals of small ribosomal subunits. *J. Mol. Biol.* **203**:831–834.

Yonath, A., J. Harms, H. A. Hansen, A. Bashan, F. Schlunzen, I. Levin, I. Koelln, A. Tocilj, I. Agmon, M. Peretz, H. Bartels, W. S. Bennett, S. Krumbholz, D. Janell, S. Weinstein, T. Auerbach, H. Avila, M. Piolleti, S. Morlang, and F. Franceschi. 1998. Crystallographic studies on the ribosome, a large macromolecular assembly exhibiting severe nonisomorphism, extreme beam sensitivity and no internal symmetry. *Acta Crystallogr.* **A54**:945–955.

Yusupov, M. M., S. V. Tischenko, S. D. Trakhanov, S. N. Ryazantsev, and M. B. Garber. 1988. A new crystalline form of 30 S ribosomal subunits from *Thermus thermophilus*. *FEBS Lett.* **238**:113–115.

The Ribosome: Structure, Function, Antibiotics, and Cellular Interactions
Edited by R. A. Garrett, S. R. Douthwaite, A. Liljas, A. T. Matheson, P. B. Moore, and H. F. Noller

Chapter 2

Crystal Structure of the Large Ribosomal Subunit at 5-Angstrom Resolution

NENAD BAN, POUL NISSEN, PETER B. MOORE, and THOMAS A. STEITZ

The ribonucleoproteins called ribosomes were discovered by cytologists in the mid-1950s, and by 1960 it was apparent that they catalyze protein synthesis (Tissieres, 1974). Ribosomes consume aminoacyl transfer RNAs, and the sequences of the proteins they produce are determined by those of the mRNAs with which they interact. Biochemists interested in protein synthesis have long taken it as axiomatic that the mechanism of the ribosomal phase of gene expression will not be fully deciphered until the three-dimensional structure of the ribosome is understood in atomic detail. For that reason, much of the work done on the ribosome since 1960 has been directed at its structure.

Only two techniques could conceivably provide the high-resolution structural information the ribosome field ultimately requires: electron microscopy (EM) and X-ray crystallography. EM played an important role in the discovery of the ribosome and, until very recently, was the source of most of what was known about ribosome morphology, the locations of its active sites, and the positions of its components (Oakes et al., 1990; Stoeffler-Meilicke and Stoeffler, 1990). As chapters 4 and 5 in this volume show, EM continues to make exciting contributions, but it remains to be demonstrated that ribosomes can be visualized at atomic resolution by EM. That X-ray crystallography can provide high-resolution structures for macromolecules has been clear for decades, but its relevance to the ribosome was uncertain for a long time. Molecules must be crystallized before their structures can be solved by X-ray crystallography, and for many years there were no ribosome crystals. In addition, even if ribosome crystals had been available, it was not obvious how rapidly structures would have followed because the asymmetric structure to be determined is very large and the noncrystallographic symmetry that has proven so important in determining the structures of comparably large assemblies, like viruses, does not exist in these particles.

The first indication that ribosomes crystallize was provided in 1966 when two-dimensional ribosome crystals were discovered in the tissues of hypothermic chick embryos (Byers, 1966). About a decade later, Lake and coworkers reported the formation of microscopic ordered arrays of *Escherichia coli* ribosomes in vitro (Clark et al., 1979, 1982), and almost simultaneously, Ada Yonath and her colleagues at the Max-Planck Institute in Berlin prepared the first three-dimensional ribosome crystals large enough for crystallographic study (Yonath et al., 1980).

The first ribosome crystals did not diffract to atomic resolution, but even if they had, it is uncertain what would have come of it in the short term; their analysis would have severely tested the crystallographic technology of the day. The Berlin group continued working on ribosome crystallization, and in 1991, they reported that 3-Å-resolution diffraction can be obtained from frozen crystals of the large ribosomal subunit of *Haloarcula marismortui* (Hope et al., 1989; von Bohlen et al., 1991), though some crystal pathologies still existed (Yonath et al., 1998). Thus, ribosome crystals can be prepared that diffract to a resolution sufficiently high that if their diffraction patterns were phased, the resulting electron den-

Nenad Ban ■ Department of Molecular Biophysics and Biochemistry and Howard Hughes Medical Institute, Yale University, New Haven, CT 06520. **Poul Nissen** ■ Department of Molecular Biophysics and Biochemistry, Yale University, New Haven, CT 06520. **Peter B. Moore** ■ Department of Chemistry and Department of Molecular Biophysics and Biochemistry, Yale University, New Haven, CT 06520. **Thomas A. Steitz** ■ Department of Molecular Biophysics and Biochemistry, Howard Hughes Medical Institute, and Department of Chemistry, Yale University, New Haven, CT 06520.

sity maps could be interpreted in terms of an atomic structure. Thus, the most formidable of the many challenges that remained to be overcome was that of phasing ribosomal diffraction patterns.

THE STATE OF RIBOSOME CRYSTALLOGRAPHY IN 1995

In 1995 a ribosome meeting was held in Victoria, Canada, that influenced the course of the work described below. There, the Berlin group presented a 7-Å-resolution electron density map of the *H. marismortui* large ribosomal subunit (Schlunzen et al., 1995) that was not convincing because it lacked the connected density features expected in low-resolution maps of macromolecules, which therefore suggested that there were phasing problems. Also presented were some remarkable, low-resolution electron density maps of the ribosome obtained by reconstructing EM images of unstained ribosomes embedded in vitreous ice (Frank et al., 1995). The manifest quality of these images suggested that the approach used by Harrison in the early 1970s to initiate the phasing of tomato bushy stunt virus crystals (Jack et al., 1975) might work even better with ribosome crystals. EM images could be used to obtain low-resolution phases for ribosome diffraction patterns by molecular replacement. These low-resolution phases could then be used to locate heavy-atom binding sites in derivatized crystals by difference Fourier analysis, either to check results obtained with the much less powerful difference Patterson map method, which is the only alternative, or to avoid that method altogether. Once heavy atoms were located, phases could be extended to higher resolution by more routine heavy-atom, multiple isomorphous replacement, and anomalous-scattering methods.

LARGE RIBOSOMAL SUBUNIT CRYSTALS FROM *H. MARISMORTUI*

We succeeded in crystallizing large ribosomal subunits from *H. marismortui* under Yonath's conditions (von Bohlen et al., 1991), but the overwhelming majority of the crystals obtained were useless. They were either too small or polycrystalline. Our rate of success in preparing crystals suitable for crystallography has improved with experience, but we still find ourselves working to control this system better. Thus, qualitatively, our experience with this material has been similar to that reported by Yonath and her colleagues (Yonath et al., 1998).

The struggle to obtain good crystals has been further complicated by the sensitivity of these particles to minor changes in crystallization conditions. The first crystals we obtained were orthorhombic crystals, like those reported by Yonath (Shoham et al., 1987; von Bohlen et al., 1991); their space group is $C222_1$, and their unit cell dimensions are as follows: $a = 210$ Å, $b = 300$ Å, $c = 570$ Å. However, monoclinic crystals belonging to the space group $P2_1$ were obtained with many subsequent 50S preparations ($a = 184$ Å; $b = 570$ Å; $c = 184$ Å; $\beta = 108.5°$). We have also obtained crystals having the symmetries C2 and $P222_1$.

PHASING AT 9-Å RESOLUTION

Motivated by the idea of using EM images to start the phasing process, we initiated a collaboration with Joachim Frank and his colleagues in 1996, shortly after obtaining our first data, and the packing of subunits in the *H. marismortui* orthorhombic unit cell was determined by molecular replacement by using their EM image of the *E. coli* large ribosomal subunit (Ban et al., 1998). As it turned out, however, EM-derived phases were not required to locate the heavy atoms in our first derivative. The first compounds used to make heavy-atom derivatives of these crystals were all cluster compounds, which others had shown to be useful for studying crystals with large unit cells (Ladenstein et al., 1987; O'Halloran et al., 1987). One of the first screened was a W_{18} compound obtained from Michael Pope and Gerta Sazani at Georgetown University. It binds preferentially to a single site in the *H. marismortui* large ribosomal subunit that was located by difference Patterson methods. Difference Fourier maps computed by using EM-derived phases confirmed its location, which was reassuring (Ban et al., 1998).

Our elation was short-lived. Crystals grown in the winter of 1996–1997 from new batches of subunits, which appeared to have the same unit cell dimensions and the same symmetry as their predecessors, gave diffraction data that did not agree well with previous data (R_{cross}, ~40%), suggesting that the new crystals were not isomorphous with the old. Worse, when the W_{18} experiment was repeated with these new crystals, the difference Patterson map obtained did not show all the expected Harker section peaks, and we could not even work out how subunits pack these crystals by molecular replacement.

It took six discouraging months to come to the realization that the space group of the new crystals was the monoclinic space group $P2_1$, not the orthorhombic space group $C222_1$. Normally a difference in crystal class and space group like this is instantly diagnosed. There were two reasons it took so long in this case. First, by accident, the reciprocal lattice of the $P2_1$ crystals virtually superimposes on that of the $C222_1$ crystals obtained earlier. Second, the diffraction patterns of these two crystal forms both have the symmetry that is the hallmark of the orthorhombic space group $C222_1$, i.e., three mutually perpendicular mirror planes. Normally, a monoclinic crystal belonging to space group $P2_1$ should have a diffraction pattern with a single mirror plane, but these $P2_1$ crystals are merohedrally twinned, and in this instance, twinning adds two mirror planes to their diffraction patterns. The structure of a twinned crystal cannot be solved unless the data obtained from it can be detwinned. This problem has doubtless plagued Yonath and her colleagues, as it has us, and may account for the difficulties they have encountered in trying to solve the structure of these crystals.

A 20-Å-resolution reconstruction of the *H. marismortui* large subunit produced by Frank and his colleagues in the summer of 1997 helped confirm that the $P2_1$ crystals are twinned. Even at low resolution, the structure of the *H. marismortui* large subunit differs significantly from that of the *E. coli* large subunit. The EM map of the *H. marismortui* large subunit explains the low-resolution *H. marismortui* diffraction data better; a lower R-factor solution is obtained when it is used for molecular replacement (Ban et al., 1998). The EM image of the *H. marismortui* large subunit is so good, in fact, that a molecular replacement solution could be found for the two subunits in the asymmetric unit of the $P2_1$ crystals, even though only twinned data were available. The subunits packing in the $P2_1$ crystals and the $C222_1$ crystals differ only slightly, and once the $P2_1$ packing had been worked out, the nature of the twinning could be determined and the monoclinic W_{18} difference Patterson map could be explained.

An exceptional batch of orthorhombic crystals was produced in the winter of 1997–1998. They were used to search for new heavy-atom derivatives, of which two more were found. By the spring of 1998, a 9-Å-resolution electron density map of the *H. marismortui* large ribosomal subunit could be calculated based on multiple isomorphous replacement–anomalous-scattering (MIRAS) phases (Ban et al., 1998). That both this map and that produced by cryo-EM are valid was indicated by the great similarity that exists between the features of the all-X-ray-phased map and those of the EM reconstruction Frank and his coworkers had produced independently (Ban et al., 1998). Furthermore, the 9-Å-resolution map showed many long rods of density crisscrossing the subunit that have the right-handed twist and the diameter characteristic of the RNA double helix. It should be noted in passing that this 9-Å-resolution electron density map contained no sections resembling the section of the 12-Å-resolution electron density map of these same crystals published by the Berlin group later in 1998 and that the subunit packing in that Berlin map is not the same as that we have determined (Yonath et al., 1998).

PHASING AT 5-Å RESOLUTION

As our experience confirms, heavy-atom cluster compounds are useful for phasing macromolecular diffraction patterns of large macromolecules at low resolution. At low resolution, the contribution they make to the diffraction of the crystals in which they are embedded approximates that of a single atom having an atomic number equal to the sum of the atomic numbers of all the atoms in the cluster. The large signal that results is easy to pick up against the "background" produced by the low-atomic-number material in the rest of the unit cell. Beyond about 9-Å resolution, however, this advantage rapidly diminishes, and the contribution of each cluster approaches that expected for the same number of heavy atoms bound at independent sites. In fact, from a practical point of view, it would be preferable to have an equivalent number of independent single heavy atoms bound because it is easier to describe the disorder associated with bound single heavy atoms than it is to do the same for groups of covalently bonded heavy atoms. Thus, additional effort was clearly going to be required to take the problem from 9-Å to 3-Å resolution.

The methods used to extend the phasing from 9 Å to 5 Å, which are described in detail elsewhere (Ban et al., 1999), can be summarized as follows. First, many of the data sets that contributed to the 9-Å-resolution map had to be re-collected both to improve their accuracy at low resolution and to extend them to higher resolution. Second, an osmium hexamine derivative was found, and by combining data obtained from it with the data for the original three heavy-atom derivatives, MIRAS phases could be obtained that had a satisfactory figure of merit to 7 Å and contained significant phase informative beyond that. Phases were extended beyond 7 Å by combining single-atom anomalous dispersion phases obtained from each derivative with four-derivative MIRAS phases and by intercrystal averaging (Cowtan, 1994; Jones, 1992) with two nonisomorphous orthorhombic data sets and two monoclinic data sets detwinned by different methods. Twinned data sets were detwinned at the intensity level when the twin fraction was below 35% and by back-transforming maps when the twinning was complete. Solvent flipping was also useful (Abrahams and Leslie, 1996).

RESOLUTION

The structure of an ordinary macromolecular crystal is solved when the experimental phases available are accurate to a resolution high enough so that an all-atom model of the molecule's sequence can be fitted into the resulting electron density map. Once this is achieved, phases can be extended to the diffraction limit by refinement of atomic positions and B-factors. Thus, however good or bad the phasing of the initial electron density map, the resolution attributed to the final structure determination is established by the crystal diffraction limit. At this stage in the development of the ribosome problem, refinement is not possible, and resolution is determined by the quality of the experimental phases available, not the crystal diffraction limit, which is about 3 Å.

Operationally, "5-Å resolution" means that the phases available beyond 5 Å are so inaccurate that inclusion of higher-resolution data did not improve the map appearance. This does not mean that the phases available at lower resolutions are perfect, however. It is always the case that the accuracy of experimental phases falls gradually as resolution increases. Thus, experimental maps improve as the resolution of significant phasing increases for two reasons: (i) higher-resolution terms start contributing significantly, which adds structural detail, and (ii) the accuracy of lower-resolution phases often also improves, which reduces error.

Both effects are illustrated in Fig. 1, which shows four different versions of the electron density in a single region of 50S structure. In all of them it is clear that the object being visualized is an RNA segment that is approximately a regular double-helical stem capped by a GNRA tetraloop. Figure 1A shows the way the region looked in the 1998 9-Å-resolution map, and Fig. 1B shows the same region visualized at 9-Å resolution with the phases available now. The improvement is obvious. The region's appearance in today's 5-Å-resolution electron density map is displayed in Fig. 1C, and Fig. 1D shows what that structure would look like if the molecular model superimposed on the electron density was the truth and perfectly accurate phases were available to 5-Å resolution. Some of the differences between the images in Fig. 1C and D reflect the fact that the atomic model shown does not fit the experimental electron density perfectly, but those differences aside, comparison of Fig. 1C and D confirms that the resolution of the current map is indeed in the neighborhood of 5 Å. That said, it is important to realize that the quality of the current map is not uniform over the entire volume of the subunit because of the effects of local disorder (Agrawal et al., 1999).

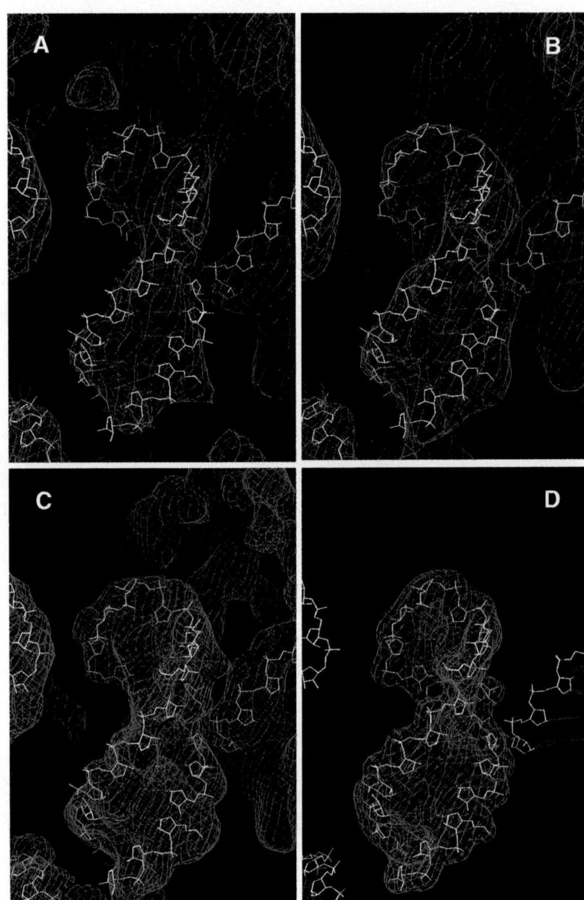

Figure 1. Four versions of the electron density in a specific region of the large ribosomal subunit from *H. marismortui*. (A) The region as it appeared in the 9-Å-resolution map published in 1998 (Ban et al., 1998) with an approximate molecular interpretation supplied. (B) The same region in a map computed with the phases available today, also to 9-Å resolution. (C) The region at the full resolution available today (5 Å). (D) The electron density expected if the atomic structure shown were the correct structure and if the resolution limit was in fact 5 Å. Panel D was computed by assuming a temperature factor of 80 $Å^2$, which is appropriate for the *H. marismortui* crystals under investigation. Molecular models were prepared with the O program, and the illustrations were prepared with both O and RIBBONS (Carson, 1991; Jones et al., 1991).

GENERAL APPEARANCE OF THE 5-Å MAP

In the current map, the RNA backbone appears as a continuous, curved strand that is punctuated in many places by regions of high density separated by spacings, suggestive of successive phosphate groups. In helical regions, base pairs appear as a wall connecting strands; the separation between adjacent base pairs is not yet apparent. The polypeptide backbone cannot be followed at this resolution, but protein is easily distinguished from RNA. Alpha helices are straight rods of appropriate diameter, and the texture of protein electron density is distinctly finer than that of RNA.

A property of the map that is immediately apparent is its overwhelming size (Fig. 2). Even if it were perfectly phased out to the diffraction limit, interpretation would be a daunting task. An interpretation tool we have found useful is a program called

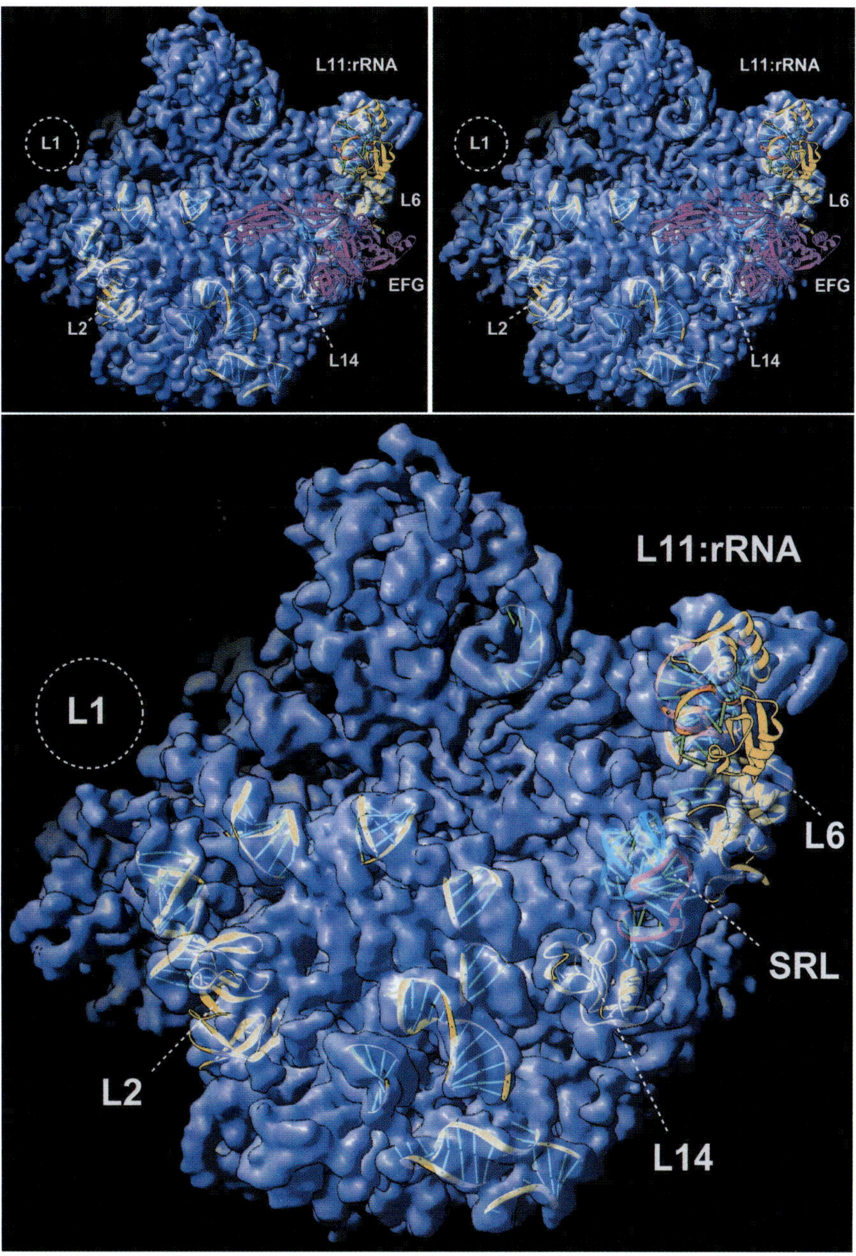

Figure 2. A surface rendering of the large ribosomal subunit in the crown view. (Top) Stereo view of the subunit with EF-G docked to the factor binding center. EF-G is the protein whose ribbon diagram is colored purple. (Bottom) The same view of the subunit with EF-G removed so that the factor binding center can be seen. In both panels, segments of helical RNA (white backbone) are inserted into the density in several places to guide the reader. Also inserted are ribbon diagrams for proteins L2, L6, L11, and L14 (yellow), the SRL (red backbone), two other segments of domain IV (blue backbone), and the thiostrepton binding segment of 23S rRNA (orange backbone). The position where L1 is seen in lower-resolution maps (see the text) is also indicated.

ESSENS (Kleywegt and Jones, 1997), which searches electron density maps for features that correspond to known structures. With a 5-bp RNA double helix as the search template, for example, ESSENS has identified helical segments in this map corresponding to about 500 bp. (Secondary structure models of *H. marismortui* 23S rRNA suggest that the total amount of double helix expected is about 780 bp [Gutell et al., 1993].) These helical segments are distributed throughout the particle, and none are much longer than a single turn of the helix. The atomic structures of a few of the helices identified by ESSENS, which happen to be visible on the subunit interface surface of the subunit, are included in Fig. 2 to help calibrate the scale of the subunit for the reader. Those familiar with the shape of the ribosome as a whole will note the almost-total absence from the image in Fig. 2 of anything that can be identified with the L1 stalk. This is a manifestation of the local-disorder phenomenon mentioned earlier; the L1 stalk is so mobile that its image virtually disappears from high-resolution views of this subunit.

As Fig. 3 demonstrates, it is very easy to distinguish protein from RNA in these maps when the protein density corresponds to a molecule whose three-dimensional structure is already known. In Fig. 3, we see the RNA binding domain of L2 (Nakagawa et al., 1999) adjacent to, and probably interacting with, the RNA stem-loop that happens to form the base of the long helical structure that forms the L1 stalk. Figure 4 provides another example of a region of the map where electron density that corresponds to both a protein of known structure and RNA can be seen (see below).

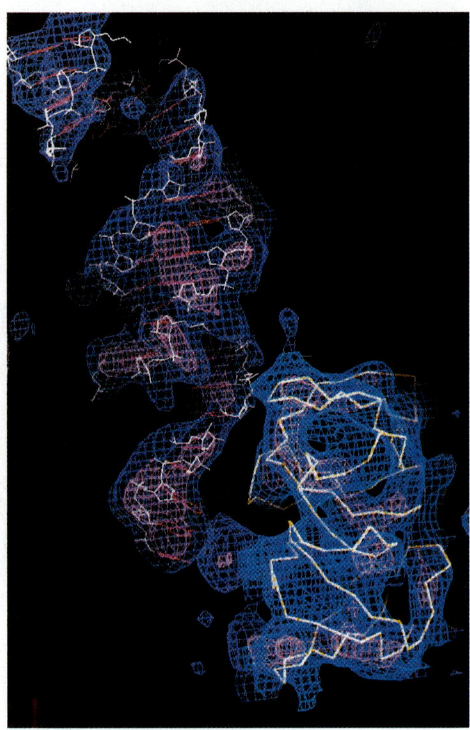

Figure 3. Electron density in the L2 region. Electron density corresponding to the RNA binding domain of L2 is shown adjacent to the stem-loop that forms the bottom of the L1 stalk. It is clear that RNA can easily be distinguished from protein in these maps. (It should be noted that the atomic structures shown in this and all other figures have been fitted into the electron density by hand. No effort has been made to optimize these fits computationally. Furthermore, in every case, the protein structures used are those of homologues of the *H. marismortui* proteins in question, not the structures of the *H. marismortui* proteins themselves, none of which are known. Since the sequences of these homologues are not the same as those of the proteins they are being used to represent, some divergence between the structures used and the electron density into which they are fitted is to be expected.)

STRUCTURE OF THE FACTOR BINDING CENTER

ESSENS can also be used to find less generic structures, such as the sarcin-ricin loop (SRL). The SRL is a critical part of the factor binding center, or GTPase center, of the large ribosomal subunit (Macbeth and Wool, 1999; Wool et al., 1992). It consists of a GNRA tetraloop attached to a cross-strand A stack that adjoins a bulged-G motif, a combination that introduces a distinctive S shape into the backbone of the 5′ side of the SRL sequence (Correll et al., 1998; Szewczak and Moore, 1995). ESSENS identified two regions of electron density that match this motif. The more obvious is at the surface of the subunit on the side away from the 30S interface, behind and below the L7-L12 arm, and the less obvious is on the subunit interface side, immediately below the L7-L12 arm. What the other SRL-like motif is we do not yet know, but a huge body of functional data indicates that the motif on the subunit interface side must be the SRL (Agrawal et al., 1998; Munishkin and Wool, 1997; Wilson and Noller, 1998), and strong support for this view can be found in the electron density map itself.

The SRL is the terminus of stem-loop 95 in 23S rRNA, which is part of domain VI, which includes helices 94 to 97 (Leffers et al., 1987). Ribosomal protein L6, the structure of which is known (Golden et al., 1993), binds to domain VI (Leffers et al., 1988; Uchiumi et al., 1999), and we have found the electron density in the *H. marismortui* map that represents that protein (Fig. 4, bottom). The end of this rather elongated protein that had been identified previously as a likely RNA binding site interacts directly with the SRL. We have also found electron density that probably corresponds to helix 96, the terminal

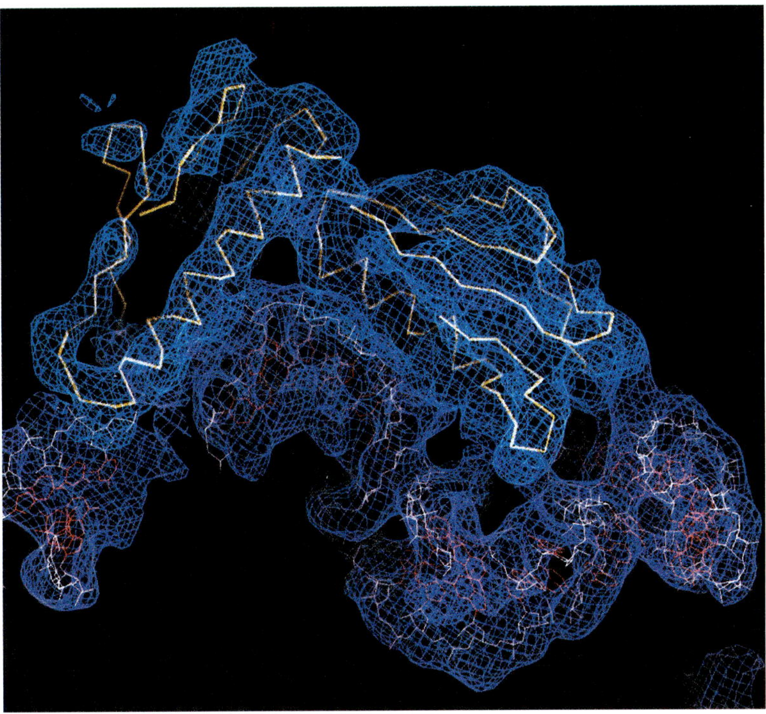

Figure 4. L6 and the rRNA segments with which it interacts. The backbone structure of L6 is shown superimposed on the electron density assigned to it. Four rRNA segments interact with L6, and the SRL is on the far right.

loop of which makes a tetraloop-tetraloop receptor-like interaction with the SRL, and another feature that may be helix 97, which also interacts with L6. (The SRL is the RNA on the upper-left side of Fig. 4, the putative helix 96 is the segment in the middle, and a helix that may correspond to helix 97 is on the right.) Furthermore, on the side of the SRL opposite L6, there is electron density into which the structure of ribosomal protein L14 fits (Davies et al., 1996 and data not shown). Since elongation factor G, (EF-G), which is believed to interact directly with the SRL (Munishkin and Wool, 1997), can be cross-linked to both L6 and L14 when it is ribosome bound (Traut et al., 1986), it must be that the components whose placement we describe here form a large part of the site to which EF-G binds.

Two groups have recently published crystal structures of complexes between L11, or its C-terminal domain, and the segment of 23S rRNA to which it binds (Conn et al., 1999; Wimberly et al., 1999). This rRNA fragment includes base 1067, which is part of the site that binds the antibiotic thiostrepton, and it and its associated proteins are the only ribosomal components other than the SRL known to be intimately associated with factor activity. The thiostrepton RNA-L11 assembly has also been found in the 5-Å-resolution map (Fig. 2), but it is interesting to note that while there is electron density in the map corresponding to the C-terminal domain of L11, which is the one that binds to rRNA, there is none for its N-terminal domain, the part that interacts with thiostrepton. That domain must be highly disordered when the ribosome is in this state.

DOCKING EF-G WITH THE 50S SUBUNIT

Having progressed this far, the temptation to dock EF-G and the EF-Tu ternary complex with the factor binding center is irresistible, and there is a lot of information available to guide such an exercise. For example, there are site-specific hydroxy radical footprinting data that associate specific residues in EF-G with specific residues in 23S rRNA (Wilson and Noller, 1998) and three-dimensional EM images of both EF-G and the EF-Tu ternary complex bound to the ribosome (Agrawal et al., 1998; Frank, 1999; Stark et al., 1997). These data indicate that both factors bind to the 50S subunit at the base of the L7-L12 stalk with their nucleotide binding sites facing the 50S subunit and domain 4 of EF-G or, correspondingly, the anticodon arm of the tRNA in the ternary complex (Nissen et al., 1995) pointed towards the center of the 50S subunit's interface surface, more or less parallel with the lip of the active-site cleft.

The model obtained by using these constraints (Fig. 5) suggests that the L14-SRL-L6 assembly is the binding site for both factors. L6 contacts the G domain of both factors, the GAGA tetraloop of the SRL interacts with their G domains close to the nucleotide binding site, and L14 contacts their second domains. This arrangement puts those sites on EF-G's fifth domain that produce a hydroxy radical footprint on the SRL in appropriate positions (Wilson and Noller, 1998) and accounts for the SRL protections reported by Noller and his colleagues (Moazed et al., 1988). Furthermore, the orientation of the SRL with respect to both factors is such that the damage to factor binding activity known to ensue when its terminal tetraloop is covalently altered by ricin or alpha sarcin is easy to rationalize (Wool et al., 1992), as are the observations that both factors protect the SRL from the action of alpha sarcin (Leffers et al., 1988) and that the SRL binds to EF-G in isolation (Munishkin and Wool, 1997).

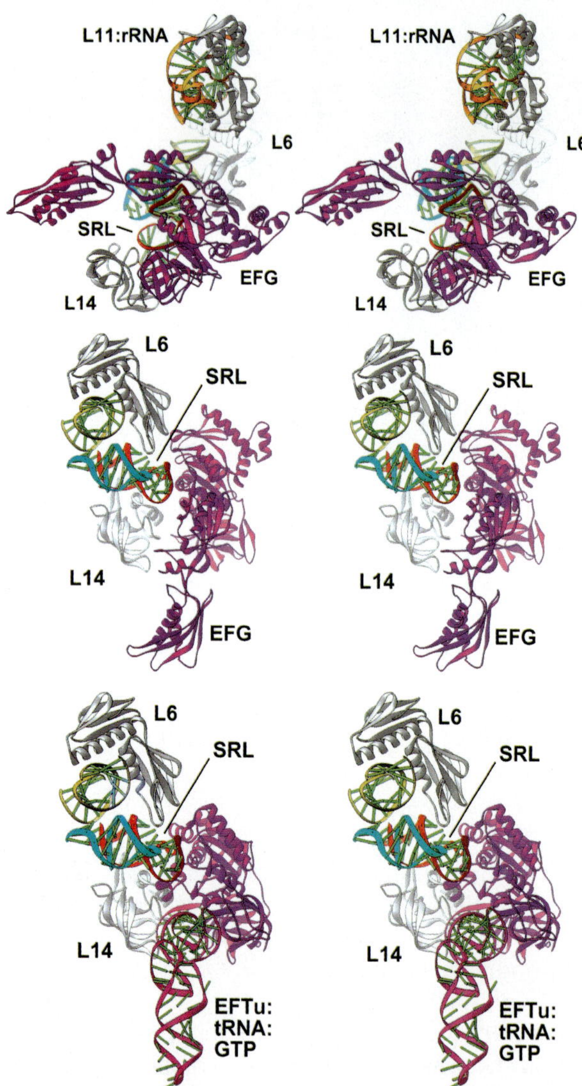

Figure 5. Stereo views of a model for the interaction of EF-G and the EF-Tu·tRNA complex with the factor binding center of the large ribosomal subunit. (Top) A view of the model in the same orientation as in Fig. 2, top. EF-G is shown in purple, the SRL is red, a helix is blue, helix 97 is yellow, and the thiostrepton RNA is orange. All ribosomal proteins are gray. (Middle) The same model as in the top panel, viewed from the top with the L11-rRNA assembly removed so that the interaction proposed to occur between the SRL and EF-G can be visualized. (Bottom) The same view as the middle panel with EF-G replaced by the EF-Tu·tRNA complex. The tRNA backbone is magenta.

The surprise in this model is the −25 Å that separates the 1067 region of 23S rRNA and the closest part of EF-G. If the site where factors bind is that far from L11 and the rRNA sequences to which it binds, why are they so important for factor activity (Porse et al., 1999; Thompson and Cundliffe, 1991), why does EF-G protect residues in the thiostrepton region but not EF-Tu (Moazed et al., 1988), and why is the RNA footprinted from ribosome-bound EF-G (Moazed et al., 1988)? One possibility is that EF-G binds to the large subunit at a position closer to L11 than the one shown here, but we find this option unattractive because it makes the cross-link between EF-G and L14 hard to explain and because the excellent steric complementarity that exists between EF-G and the ribosome in the site we have identified would no longer be taken advantage of. A conformational change that moves L11 and its associated rRNA towards the putative factor binding site seems at least equally likely, and since there is no ribosomal material in the way to prevent it, that kind of motion is possible. In this connection, it is interesting to note that there is a substantial literature indicating that this part of the ribosome does indeed move during protein synthesis (Agrawal et al., 1998; Liljas et al., 1986; Moller et al., 1983; Stark et al., 1997).

We thank Betty Freeborn for the biochemical support she has provided throughout this project and Jeff Hanson for his assistance with data collection. We are also indebted to Robert Sweet and Malcolm Capel of Brookhaven National Laboratory for making data collection possible at the National Synchrotron Light Source, where all the data related to the results discussed above were obtained, to Joachim Frank, Pawel Penczek, and Robert Grassucci of the Wadsworth Center in Albany for the EM they did in support of the initial stages of this work, to I. Tanaka for providing coordinates of L2 prior to publication, and to David Draper and Venkatraman Ramakrishnan for making their coordinates of the L11-thiostrepton RNA complex available.

This investigation is supported by two NIH grants: GM-22778 (to T.A.S.) and GM-54216 (to P.B.M.). N.B. was a Damon Runyon-Walter Winchell Postdoctoral Fellow during part of this study, and P.N. is supported by a grant from the Danish Research Council.

REFERENCES

Abrahams, J. P., and A. G. W. Leslie. 1996. Methods used in the structure determination of bovine mitochondrial F1 ATPase. *Acta Crystallogr.* **D52:**30–42.

Agrawal, R. K., P. Penczek, R. A. Grassucci, and J. Frank. 1998. Visualization of elongation factor G on the *Escherichia coli* 70S ribosome: the mechanism of translocation. *Proc. Natl. Acad. Sci. USA* **95:**6134–6138.

Agrawal, R. K., R. K. Lata, and J. Frank. 1999. Conformational variability in *Escherichia coli* 70S ribosome as revealed by 3D cryo-electron microscopy. *Int. J. Biochem. Cell Biol.* **31:**243–254.

Ban, N., B. Freeborn, P. Nissen, P. Penczek, R. A. Grassucci, R. Sweet, J. Frank, P. B. Moore, and T. A. Steitz. 1998. A 9 A resolution X-ray crystallographic map of the large ribosomal subunit. *Cell* **93:**1105–1115.

Ban, N., P. Nissen, J. Hansen, M. Capel, P. B. Moore, and T. A. Steitz. 1999. Placement of protein and RNA structures into a 5 Å-resolution map of the 50S ribosomal subunit. *Nature* **400:** 841–847.

Byers, B. 1966. Ribosome crystallization induced in chick embryo tissue by hypothermia. *J. Cell Biol.* **30:**C1.

Carson, M. 1991. Ribbons 2.0. *J. Appl. Crystallogr.* **24:**103–106.

Clark, M. W., M. Hammons, J. A. Langer, and J. A. Lake. 1979. Helical arrays of *Escherichia coli* small ribosomal subunits produced *in vitro*. *J. Mol. Biol.* **135:**507–512.

Clark, M. W., K. Leonard, and J. A. Lake. 1982. Ribosomal crystalline arrays of large subunits from *E. coli*. *Science* **216:**999.

Conn, G. L., D. E. Draper, E. E. Lattman, and A. G. Gittis. 1999. Crystal structure of a conserved ribosomal protein RNA complex. *Science* **284:**1171–1174.

Correll, C. C., A. Munishkin, Y. Chan, Z. Ren, I. G. Wool, and T. A. Steitz. 1998. Crystal structure of the ribosomal RNA loop essential for binding both elongation factors. *Proc. Natl. Acad. Sci. USA* **95:**13436–13441.

Cowtan, K. D. 1994. An automated procedure for phase improvement by density modification. *Jt. CCP4 ESF-EACBM Newsl. Protein Crystallogr.* **31:**34–38.

Davies, C., S. W. White, and V. Ramakrishnan. 1996. The crystal structure of ribosomal protein L14 reveals an important organizational component of the translational apparatus. *Structure* **4:** 55–66.

Frank, J. 1999. The ribosome—structure and functional ligand-binding experiments using cryo-electron microscopy. *J. Struct. Biol.* **124:**142–150.

Frank, J., A. Verschoor, Y. Li, J. Zhu, R. K. Lata, M. Rademacher, P. Penczek, R. Grassucci, R. K. Agrawal, and S. Srivastava. 1995. A model of the translational apparatus based on a three-dimensional reconstruction of the *Escherichia coli* ribosome. *Biochem. Cell Biol.* **73:**757–765.

Golden, B. L., V. Ramakrishnan, and S. W. White. 1993. Ribosomal protein L6. Structural evidence of gene duplication from a primitive RNA binding protein. *EMBO J.* **12:**4901–4908.

Gutell, R. R., M. W. Gray, and M. N. Schnare. 1993. A compilation of large subunit (23S and 23S-like) ribosomal RNA structures: 1993. *Nucleic Acids Res.* **21:**3055–3074.

Hope, H., F. Frolow, K. von Bohlen, I. Makowski, C. Kratky, Y. Halfori, H. Danz, P. Webster, K. S. Bartles, H. G. Wittmann, and A. Yonath. 1989. Cryocrystallography of ribosomal particles. *Acta Crystallogr.* **B45:**190–199.

Jack, A., S. C. Harrison, and R. A. Crowther. 1975. Structure of tomato bushy stunt virus. II. Comparison of results obtained by electron microscopy and X-ray diffraction. *J. Mol. Biol.* **97:**163–172.

Jones, T. A. 1992. A set of averaging programs, p. 91–105. *In* E. J. Dodson, S. Glover, and W. Wolf (ed.), *CCP4 Study Weekend, Molecular Replacement*. Daresbury Laboratory, Warrington, United Kingdom.

Jones, T. A., S. Cowan, J.-Y. Zou, and M. Kjeldgaard. 1991. Improved methods for building protein models in electron density maps and the location of errors in these models. *Acta Crystallogr.* **A46:**110–119.

Kleywegt, G. J., and T. A. Jones. 1997. Template convolution to enhance or detect structural features in macromolecular electron-density maps. *Acta Crystallogr.* **D53:**179–185.

Ladenstein, R., A. Bacher, and R. Huber. 1987. Some observations of a correlation between the symmetry of large heavy-atom compounds and their binding sites on proteins. *J. Mol. Biol.* **195:** 751–753.

Leffers, H., J. Kjems, L. Ostergaard, N. Larsen, and R. A. Garrett. 1987. Evolutionary relationships amongst archaebacteria: a comparative study of 23S ribosomal RNAs of a sulphur-dependent extreme thermophile, an extreme halophile, and a thermophilic methanogen. *J. Mol. Biol.* **195:**43–61.

Leffers, H., J. Egeberg, A. Andersen, T. Christensen, and R. A. Garrett. 1988. Domain VI of *Escherichia coli* 23 S ribosomal RNA. Structure, assembly and function. *J. Mol. Biol.* **204:**507–522.

Liljas, A., L. A. Kirsebom, and M. Leijonmarck. 1986. Structural studies of the factor binding domain, p. 379–390. *In* B. Hardesty and G. Kramer (ed.), *Structure, Function and Genetics of Ribosomes*. Springer-Verlag, New York, N.Y.

Macbeth, M. R., and I. G. Wool. 1999. The phenotype of mutations of G2655 in the sarcin/ricin domain of 23S ribosomal RNA. *J. Mol. Biol.* **285:**965–975.

Moazed, D., J. M. Roberston, and H. F. Noller. 1988. Interaction of elongation factors EF-G and EF-Tu with a conserved loop in 23S RNA. *Nature* **334:**362–364.

Moller, W., P. I. Schrier, J. A. Maassen, A. Zanteva, E. Schop, and H. Reinalda. 1983. Ribosomal proteins L7/L12 of *E. coli*. Localization and possible molecular mechanism in translation. *J. Mol. Biol.* **163:**553–573.

Munishkin, A., and I. G. Wool. 1997. The ribosome-in-pieces: binding of elongation factor EF-G to oligoribonucleotides that mimic the sarcin/ricin and thiostrepton domains of 23S ribosomal RNA. *Proc. Natl. Acad. Sci. USA* **94:**12280–12284.

Nakagawa, A., T. Nakashima, M. Taniguchi, H. Hosaka, M. Kimura, and I. Tanaka. 1999. The three-dimensional structure of the RNA-binding domain of ribosomal protein L2; a protein at the peptidyl transferase center of the ribosome. *EMBO J.* **18:** 1459–1467.

Nissen, P., M. Kjeldgaard, S. Thirup, G. Polekhina, L. Reshetnikova, B. F. Clark, and J. Nyborg. 1995. Crystal structure of the ternary complex of Phe-tRNAPhe, EF-Tu, and a GTP analog. *Science* **270:**1464–1472.

Oakes, M. I., A. Scheinman, T. Atha, G. Shankweiler, and J. A. Lake. 1990. Ribosome structure: three-dimensional locations of rRNA and proteins, p. 180–193. *In* W. E. Hill, A. E. Dahlberg, R. A. Garrett, P. B. Moore, D. Schlessinger, and J. R. Warner (ed.), *The Ribosome. Structure, Function, and Evolution*. American Society for Microbiology, Washington, D.C.

O'Halloran, T. V., S. J. Lippard, T. J. Richmond, and A. Klug. 1987. Multiple heavy-atom reagents for macromolecular X-ray structure determination. Application to the nucleosome core particle. *J. Mol. Biol.* **194:**705–712.

Porse, B. T., E. Cundliffe, and R. A. Garrett. 1999. The antibiotic micrococcin acts on protein L11 at the ribosomal GTPase centre. *J. Mol. Biol.* **287:**33–45.

Schlunzen, F., H. A. S. Hansen, J. Thygesen, W. S. Bennett, N. Volkmann, I. Levin, J. Harms, H. Bartles, A. Zaytzev-Bashan, Z. Berkovitch-Yellin, I. Sagi, F. Fransceschi, S. Krumbholtz, M. Geva, S. Weinstein, I. Agmon, N. Boddeker, S. Morlang, R. Sharon, A. Dribin, E. Maltz, M. Peretz, V. Weinrich, and A. Yonath. 1995. A milestone in ribosomal crystallography: the construction of preliminary electron density maps at intermediate resolution. *Biochem. Cell Biol.* **73:**739–749.

Shoham, M., H. G. Wittmann, and A. Yonath. 1987. Single crystals of large ribosomal particles from *Halobacterium marismortui* diffract to 6 A. *J. Mol. Biol.* **193:**819–822.

Stark, H., M. V. Rodnin, J. Rinke-Appel, R. Brimacombe, W. Wintermeyer, and M. van Heel. 1997. Visualization of elongation factor Tu on the *Escherichia coli* ribosome. *Nature* **389:**403–406.

Stoeffler-Meilicke, M., and G. Stoeffler. 1990. Topography of the ribosomal proteins from *Escherichia coli* within the intact subunits as determined by immunoelectron microscopy and protein-protein cross-linking, p. 123–133. *In* W. E. Hill, A. Dahlberg, R. A. Garrett, P. B. Moore, D. Schlessinger, and J. R. Warner (ed.), *The Ribosome. Structure, Function, and Evolution*. American Society for Microbiology, Washington, D.C.

Szewczak, A. A., and P. B. Moore. 1995. The sarcin/ricin loop, a modular RNA. *J. Mol. Biol.* **247:**81–98.

Thompson, J., and E. Cundliffe. 1991. The binding of thiostrepton to 23S ribosomal RNA. *Biochimie* **73:**1131–1135.

Tissieres, A. 1974. Ribosome research: historical background, p. 3–12. *In* M. Nomura, A. Tissieres, and P. Lengyel (ed.), *Ribosomes*. Cold Spring Harbor Laboratory, Cold Spring Harbor, N.Y.

Traut, R. R., D. S. Tewari, A. Sommer, G. R. Gavino, H. M. Olson, and D. G. Glitz. 1986. Protein topography of ribosomal functional domains: effects of monoclonal antibodies to different epitopes in *Escherichia coli* protein L7/L12 on ribosome function and structure, p. 286–308. *In* B. Hardesty and G. Kramer, (ed.), *Structure, Function, and Genetics of Ribosomes*. Springer-Verlag, New York, N.Y.

Uchiumi, T., N. Sato, A. Wada, and A. Hachimori. 1999. Interaction of the sarcin/ricin domain of 23S ribosomal RNA with proteins L3 and L6. *J. Biol. Chem.* **274:**681–686.

von Bohlen, K., I. Makowski, H. A. S. Hansen, H. Bartels, Z. Berkovitch-Yellin, A. Zaytzev-Bushan, S. Meyer, C. Paulke, F. Franceschi, and A. Yonath. 1991. Characterization and preliminary attempts for derivatization of crystals of large ribosomal subunits from *Haloarcula marismortui* diffracting to 3 A resolution. *J. Mol. Biol.* **222:**11–15.

Wilson, K. S., and H. F. Noller. 1998. Mapping the position of translational elongation factor EF-G in the ribosome by directed hydroxyl radical probing. *Cell* **92:**131–139.

Wimberly, B. T., R. Guymon, J. P. McCutcheon, S. W. White, and V. R. Ramakrishnan. 1999. A detailed view of a ribosomal active site: the structure of the L11-RNA complex. *Cell* **97:**491–502.

Wool, I. G., A. Gluck, and Y. Endo. 1992. Ribotoxin recognition of ribosomal RNA and a proposal for the mechanism of translocation. *Trends Biochem. Sci.* **17:**266–269.

Yonath, A., J. Mussig, B. Tesche, S. Lorenz, V. A. Erdmann, and H. G. Wittmann. 1980. Crystallization of the large ribosomal subunits from *Bacillus stearothermophilus*. *Biochem. Int.* **1:**428–435.

Yonath, A., J. Harms, H. A. S. Hansen, A. Bashan, F. Schlunzen, I. Levin, I. Koelln, A. Tocili, I. Agmon, M. Peretz, H. Bartles, W. S. Bennett, S. Krumbholz, D. Janell, S. Weinstein, T. Auerbach, H. Avila, M. Pioletti, S. Morlang, and F. Franceschi. 1998. Crystallographic studies on the ribosome, a large macromolecular assembly exhibiting severe non-isomorphism, extreme beam sensitivity and no internal symmetry. *Acta Crystallogr.* **A54:**945–955.

The Ribosome: Structure, Function, Antibiotics, and Cellular Interactions
Edited by R. A. Garrett, S. R. Douthwaite, A. Liljas, A. T. Matheson, P. B. Moore, and H. F. Noller

Chapter 3

Identification of Selected Ribosomal Components in Crystallographic Maps of Prokaryotic Ribosomal Subunits at Medium Resolution

ANAT BASHAN, MARTA PIOLETTI, HEIKE BARTELS, DANIELA JANELL, FRANK SCHLUENZEN, MARCO GLUEHMANN, INNA LEVIN, JOERG HARMS, HARLY A. S. HANSEN, ANTE TOCILJI, TAMAR AUERBACH, HORACIO AVILA, MARIA SIMITSOPOULOU, MOSHE PERETZ, WILLIAM S. BENNETT, ILANA AGMON, MAGGIE KESSLER, SHULAMITH WEINSTEIN, FRANÇOIS FRANCESCHI, and ADA YONATH

Several crystal types were obtained from intact or complexed ribosomal particles by systematic increase of the homogeneity of the crystalline materials prior to and after crystallization (Table 1) (Yonath et al., 1998). The crystals of the large ribosomal subunits from *Haloarcula marismortui*, H50S, diffract to the highest resolution, 2.7 Å (von Böhlen et al., 1991), but display several undesired properties (Yonath et al., 1998). In contrast, the crystals of the small (T30S) and large (T50S) subunits from *Thermus thermophilus*, which currently diffract to around 3 Å, yielded quality data even when their resolution extended only to low-resolution, 8- to 12-Å (Yonath et al., 1988; Volkmann et al., 1990).

Synchrotron radiation (SR) is required for all steps of the crystallographic analysis, due to the extremely weak diffraction power of the ribosomal crystals. It was found that a moderate SR beam causes minor crystal damage but yields only medium-resolution data. For collecting data at the higher-resolution shells, beyond 5 Å, brighter SR is required. Under these conditions the ribosomal crystals show severe radiation sensitivity even at helium stream temperatures, 15 to 25 K (Krumbholz et al., 1998; Yonath et al., 1998). Nevertheless, procedures overcoming part of the decay problem have been developed for the crystals of the thermophilic ribosomes (Schluenzen et al., 1999). For H50S, however, the situation is more complicated due to the inherent low level of isomorphism and the high heterogeneity of their crystals.

The electron density maps obtained for T30S and H50S reveal recognizable features, resembling those seen in the electron-microscopic (EM) reconstructions of the corresponding particles, obtained either by using averaging techniques of single particles or diffraction from ordered arrays. In this chapter we focus on the electron density map of the small ribosomal subunit. We show features interpreted as ribosomal proteins and rRNA and pinpoint secondary-structure elements. The use of heavy-atom markers for unbiased targeting of surface rRNA (e.g., the 3′ end of the 16S RNA) and for the localization of proteins TS11 and TS13 is highlighted. Efforts to induce controlled conformational changes within the crystals are also discussed.

Anat Bashan, Heike Bartels, Inna Levin, Moshe Peretz, Ilana Agmon, Maggie Kessler, and Shulamith Weinstein ■ Department of Structural Biology, Weizmann Institute, 76100 Rehovot, Israel. **Marta Pioletti, Horacio Avila, Maria Simitsopoulou, and François Franceschi** ■ Max Planck Institute for Molecular Genetics, Ihnestr. 73, 14195 Berlin, Germany. **Daniela Janell, Frank Schluenzen, Marco Gluehmann, Joerg Harms, Harly A. S. Hansen, Ante Tocilji, and William S. Bennett** ■ Max Planck Research Unit for Ribosomal Structure, Notkestr. 85, 22603 Hamburg, Germany. **Tamar Auerbach** ■ Department of Structural Biology, Weizmann Institute, 76100 Rehovot, Israel, and Department of Biochemistry and Pharmacology, FU-Berlin, Takustr. 3, 14195 Berlin, Germany. **Ada Yonath** ■ Department of Structural Biology, Weizmann Institute, 76100 Rehovot, Israel, and Max Planck Research Unit for Ribosomal Structure, Notkestr. 85, 22603 Hamburg, Germany.

Table 1. Crystals of ribosomal particles suitable for crystallographic studies

Source	Grown form[a]	Cell dimensions (Å)	Symmetry	Resolution (Å)
T70S	MPD	524 × 524 × 306	$P4_12_12$	17–22
T70S[b]	MPD	524 × 524 × 306	$P4_12_12$	10–14
T30S-LR	MPD	407 × 407 × 170	$P4_12_12$	9–12
T30S-HR	MPD	407 × 407 × 170	$P4_12_12$	3.0
T50S	AS	495 × 495 × 196	$P4_12_12$	3.4
H50S	PEG	211 × 300 × 567	$C222_1$	2.7

[a] Crystals were grown by vapor diffusion in hanging drops from solutions containing methyl-pentane-diol (MPD), ammonium sulfate (AS), or polyethylene glycol (PEG).
[b] A complex of T70S, 2 molecules of Phe-tRNA^{Phe}, and an oligomer of 35 uridines (as mRNA).

THE CONFORMATION OF THE CRYSTALLINE 30S SUBUNITS

The small ribosomal subunits exhibit the lowest level of stability and the highest level of flexibility among the ribosomal particles (Berkovitch-Yellin et al., 1992; Lata et al., 1996; Harms et al., 1999; Gabashvili et al., 1999). Multiple conformational states were suggested to account for the inconsistencies in locations of selected components revealed by surface probing (Alexander et al., 1994) or by monitoring the ribosomal activity (Weller and Hill, 1992). Indeed, the early T30S crystals (Yonath et al., 1988; Trakhanov et al., 1989) yielded satisfactory data only to 10- to 12-Å resolution (Schluenzen et al., 1995).

Marked improvement was obtained by employing milder conditions for ribosome purification and by inducing conformational adaptations prior to the crystallization or within the crystals, exploiting selected additives or controlled heating. Careful interplay between multimetal clusters, organometallic compounds, and heavy-atom salts yielded four derivatives located in over 50 sites. These, after extensive cross-verifications, led to a 7.2-Å electron density map (Harms et al., 1999) that was later extended to 5.5-Å resolution by the addition of a heavy-atom derivative that diffracts to higher resolution as well as by stepwise addition of phase improvement obtained by density modification procedures (Schluenzen et al., 1999).

Both the 7.2- and 5-Å maps contain morphologies remarkably similar to most of the EM reconstructions of the small subunits with their recognizable features (Fig. 1), showing the traditional division of the 30S subunit into three main parts: a head, a rather narrow neck, and a bulky lower body (Stark et al., 1995; Frank et al., 1995; Lata et al., 1996; Gabashvili et al., 1999). Some regions of the map show tertiary organization resembling the arrangements of RNA duplexes postulated by incorporating biochemical, footprinting, and cross-linking data within envelopes obtained by cryo-EM (Mueller and Brimacombe, 1997).

At the current resolution the accurate definition of the borders of the particle is prone to misinterpretation. Therefore, the relationships between the available EM reconstructions of the different conformers of the 30S subunit (Stark et al., 1995; Gabashvili et al., 1999; Harms et al., 1999) and the crystallographic map were examined. This turned out to be a rather complicated task, since neither of the available models seemed to fully fit the shape obtained by X-ray crystallography (Fig. 1). Conformational changes induced by the crystallization conditions cannot be excluded. However, since the crystalline particles exhibit functional activity within the crystals, it is reasonable to assume that these conformational changes are neither extensive nor destructive.

Reasonable agreement of the packing arrangements and moderate scores were obtained in molecular-replacement searches with cryo-EM reconstructions of the thermophilic particles, namely, that obtained from the T30S particles from the same batch that was used for crystallization and that computed by removing the density assigned to the T50S subunit within the three-dimensional reconstruction of the T70S ribosome (Harms et al., 1999). Slightly higher scores resulted when the reconstruction of *E. coli* 30S (E30S) at a conformation close to that assigned to the small subunit within the E70S ribosomes to which fMet-tRNA was bound (Malhotra et al., 1998) was used. This solution (correlation coefficients calculated with structure factors, 71%; correlation coefficients calculated with agreement factors, 39.2%; and ratios between the correlation coefficient of the best solution and that of the following one, 1.04) was treated as a mask for solvent flattening of the multiple isomorphous replacement (MIR) maps. This synthesis resulted in a map with higher connectivity and more-distinct features, but did not remove the discrepancies between the crystalline and the EM-reconstructed views.

Despite the discrepancies discussed above, all molecular-replacement searches gave rise to two common features: a large internal solvent region and pairing of the 30S particles around the twofold axis.

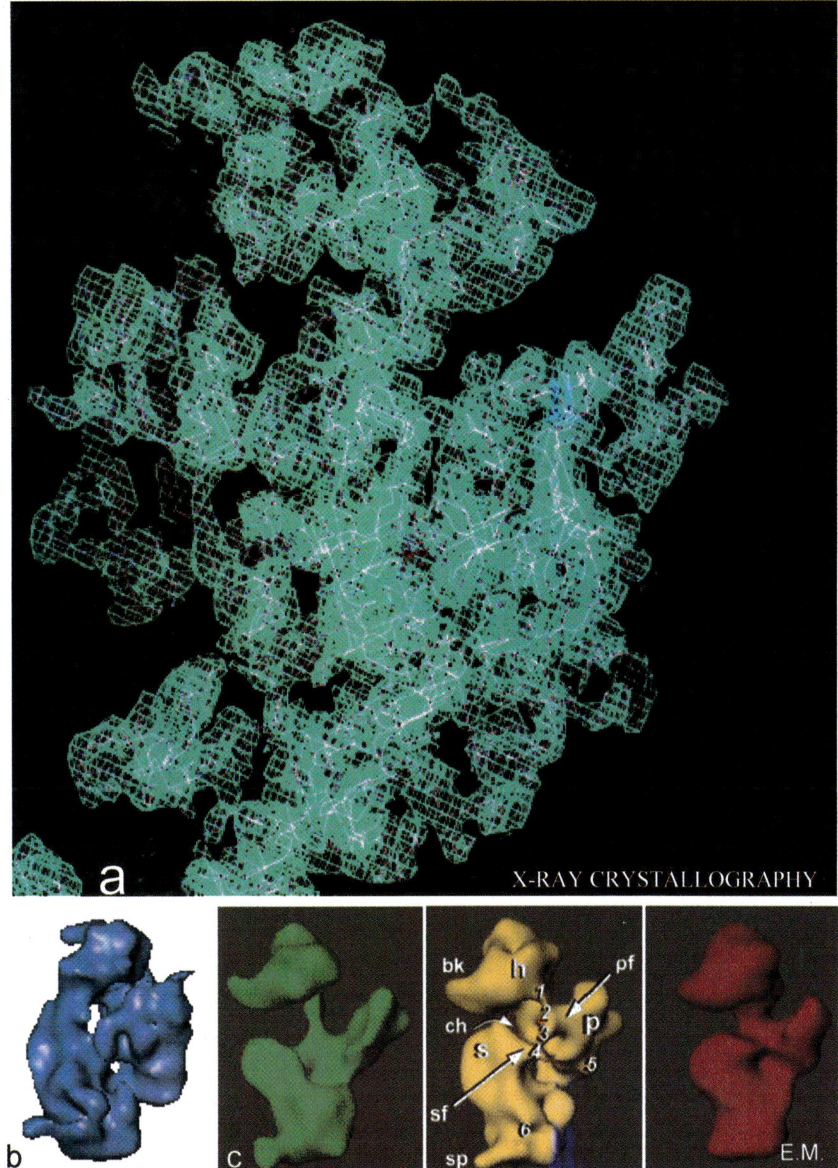

Figure 1. (a) The 7.2-Å map of T30S, contoured at 1.3 σ and sectioned in a direction allowing comparison to the available EM reconstructions of the T70S-bound T30S particle (van Heel and Stark, unpublished) (b) and the E30S conformer resembling that seen in 70S ribosomes to which fMet-tRNA is bound (Gabashvili et al., 1999) (c). A total of 21,000 unique reflections at the range 7.2 to 12.5 Å were included. Four derivatives were used: mercury acetate, platinum tetrachloride, Ta_6Br_{14}, and $C_2Hg_6N_2O_8$. The final statistics were as follows: FOM = 0.671; phasing power = 1.4, R_{cullis} = 0.8 to 0.91. For the 5-Å map, phase information was added from an additional heavy-atom derivative, methylmercury acetate, that was bound to the particle prior to crystallization. A total of 62,037 reflections were used. The final FOM was 0.77, and the phasing power was 1.25. bk, beak; ch, channel; sp, spur; sf, shoulder finger; pf, platform finger; s, shoulder; h, head; p, platform.

The pairing contacts are fairly extensive, so that they are maintained even after the crystals are carefully dissolved. Thus, the majority of the particles seen in samples of dissolved crystals are pairs with a typical butterfly-like shape. So far, it has not been possible to assign functional relevance to the crystallographic pairing or to the previously observed pairing (Zamir et al., 1971; Guerin and Hayes, 1987). It is conceivable that these extensive particle interactions emulate some of the contacts in which the small subunit is involved at the intra-ribosomal subunit interface and/or the contacts needed for the formation of intermediates along the path of the translation cycle.

SELECTED COMPONENTS WITHIN THE 30S SUBUNIT

The 7.2-Å map of T30S reveals a wealth of internal features. These include dense elongated chains that span the 30S particle in various directions and show features similar to those detected in nucleosome core particles at comparable resolution (Richmond et al., 1984). Some of these are traceable either as RNA duplexes or as single strands. The globular regions of lower density could be assigned to folds observed in isolated ribosomal proteins as determined by nuclear magnetic resonance (NMR) and crystallography at atomic resolution.

Uncertainties are associated with the placement of structures determined for isolated ribosomal components into medium- or low-resolution maps of entire ribosomal particles. These stem from the ambiguities which are likely to accompany the fitting of frequently occurring structural motifs (Liljas and Al-Karadaghi, 1997; Ramakrishnan and White, 1998) into medium-resolution maps. An additional source of uncertainties is connected to the long-lasting question of whether the structures determined for individual ribosomal components can represent the conformation within the ribosome. Certainly the conformational variability of the isolated components is not negligible, and the in situ conformations of the individual ribosomal components may be influenced by their proximity to other ribosomal proteins or rRNA. Nevertheless, structural studies showed in several cases that conformation variability is limited to the flexible part of the molecule (Clemons et al., 1998; Draper and Reynaldo, 1999). In addition, in one case the crystal structures of a ribosomal protein from two different sources are almost identical (Hosaka et al., 1997; Wimberly et al., 1997).

The elongated continuous dense regions which span the particle in various directions were traced as RNA chains (Fig. 2). The fitting of the RNA was performed mostly manually, using segments of 6 to 12 bases at a time. In several cases, regions of higher density, interpretable as the backbone phosphates, were detected within the elongated chains. About a third of the regions assigned to the 16S RNA were fitted, with varying levels of confidence. Most of these regions display the known duplex conformation, but in several cases we encountered duplexes with conformations that deviated from that of canonical double helices. Automated searches with the program ESSENS (Jones et al., 1991) were also employed, with UUAGCU as a 6-base template for single-stranded RNA of a duplex conformation and the same segment together with its counterpart, AAUCGA, to represent the duplexes. The segment CGCUACAA in the conformation of bases 69 to 76 of *Saccharomyces cerevisiae* initiator tRNA (Basavappa and Sigler, 1991) was used as a representative of relatively long stretches, and the anticodon region from the same molecule was the template for curved conformations. Although these searches led to some impressive assignments, the risk involved in their exclusive employment cannot be overlooked. Since the ribosomes contain repeating motifs of RNA, such searches are bound to suggest multiple solutions, with no provision for making the best choice. Therefore, manual assignment was preferred.

The globular regions seen in the maps, most of which are of lower average density (e.g., 0.8 σ vs. 1.1 σ for RNA), were found appropriate to accommodate ribosomal proteins. Keeping in mind that such assignments are only partially justified, the known structures of the ribosomal proteins were positioned in the map. This was performed according to the locations suggested in earlier studies, based on immuno-EM, neutron scattering, cross-linking, and modeling (Stöffler and Stöffler-Meilicke, 1986; Capel et al., 1988; Mueller and Brimacombe, 1997). The main chain coordinates, as determined by X-ray crystallography or NMR for the isolated proteins at high resolution, were used as templates.

A large fraction of protein S5 (Ramakrishnan and White, 1992) is made of beta sheets, features that are not readily detectable at medium resolution. Nevertheless, it could be fitted in the 7.2-Å map (Fig. 3). Of interest are the contacts that this protein makes with its neighborhood, which is clearly separated from its own region. This protein is of special interest for our studies, as it is being used as a vehicle for the introduction of a relatively large number of selenium atoms into the small subunit for multiwavelength anomalous dispersion phasing at high resolution (Auerbach et al., in press). Proteins S6 (Lindhal, 1994), S8 (Davies et al., 1996; Nevskaya et al., 1998), and S17 (Golden et al., 1993) were also fitted manually. However, the searches for locations suita-

Figure 2. RNA chains fitted into the 7.2-Å map of T30S (contour levels, 1.25 σ). Except for those shown in the lower right two panels, all of the chains were fitted manually, using building blocks of the canonical A-form RNA duplex (mainly 5 by 2 bases). The two views shown at the lower right were located semi-automatically, using as building blocks the CCA end of the tRNA molecule (Basavappa and Sigler, 1991). The large sphere shown in the center right panel between two helical regions represents a heavy-atom cluster.

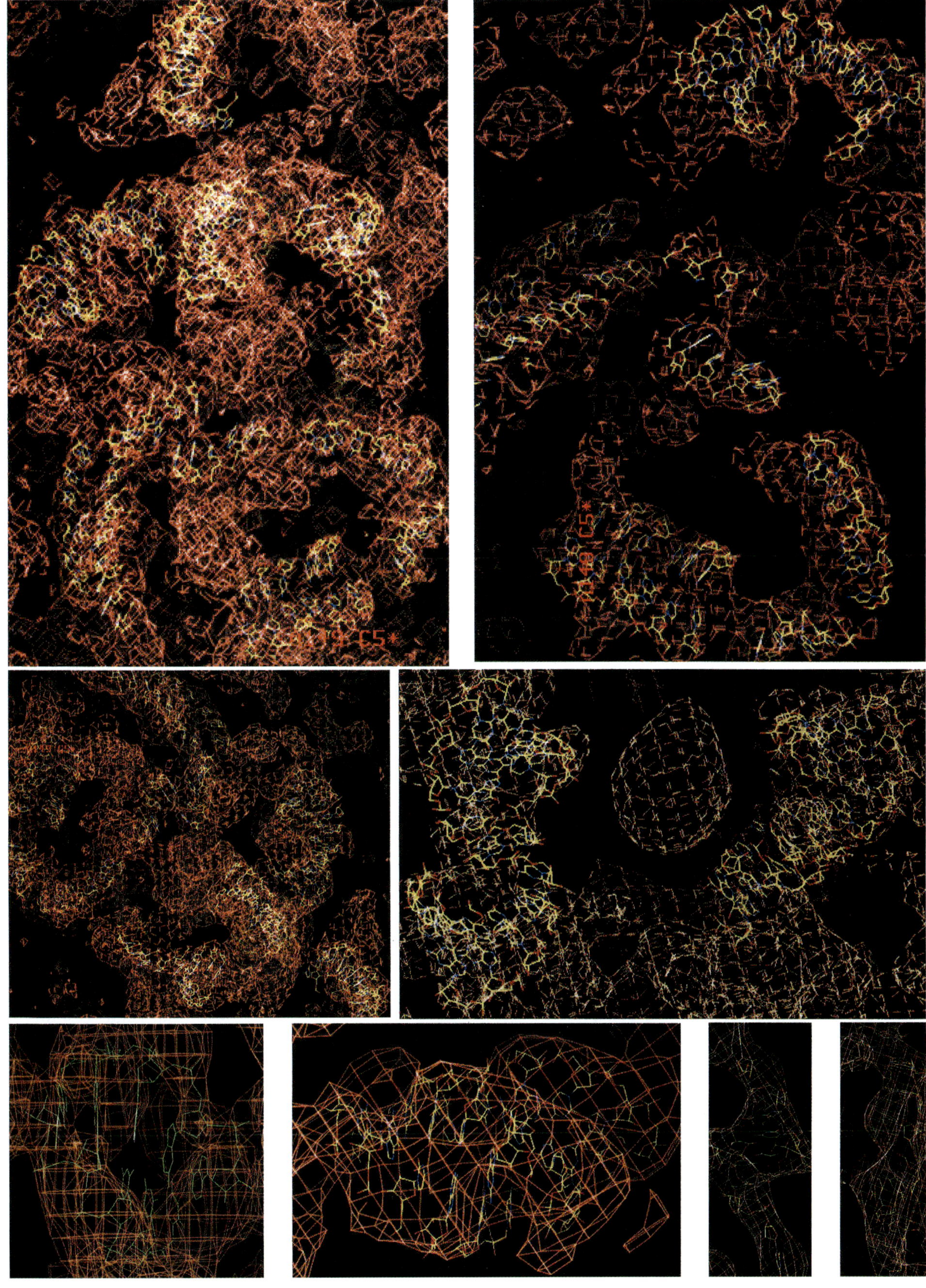

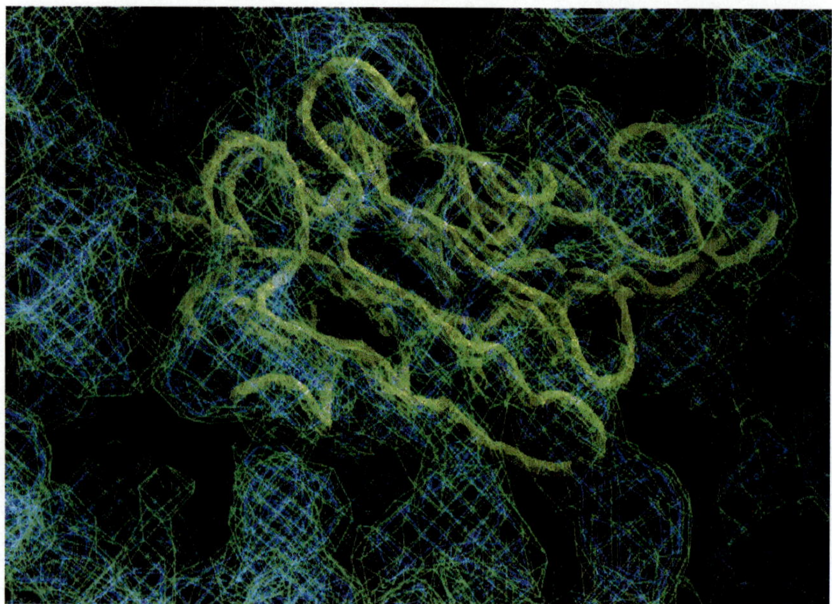

Figure 3. The main chain of the RNA binding domain of the X-ray-determined structure of protein TS5 (shown in yellow-green) overlaid on the 7.2-Å map (contoured at a signal-to-noise ratio of 0.8 [dark green] and at 1.3 σ [blue]). Note that a fair portion of the protein surface is in contact with other prominent features.

ble to host the RNA segments which interact with them were conducted in a semiautomatic fashion. Since no severe collisions were detected, these combined fitting experiments seem to be quite reliable.

The positioning of protein S7, the structure of which was determined by using two bacterial sources, *Bacillus stearothermophilus* and *T. thermophilus* (Hosaka et al., 1997; Wimberly et al., 1997) was performed manually, based on its postulated position (Tanaka et al., 1998). As mentioned above, the RNA duplexes were positioned automatically. These searches revealed a short RNA duplex contacting protein S7 (Fig. 4) that separates into its two strands once it leaves the vicinity of the protein. Interestingly, although the fitting of the protein was performed with only the main chain, the automatic search positioned the RNA so that there is sufficient room for the protein side chains.

S15 is a protein whose structure was determined by NMR and by X-ray crystallography and was shown to possess a fair amount of flexibility at a specific hinge (Clemons et al., 1998). This protein is built of three well-packed alpha-helical chains and one helix, which exhibits structural flexibility. The tentative location suggested for this protein accommodates the main helices well, but less density was found at the location where the flexible arm should be positioned if the crystallographic structure determined in isolation was maintained within the ribosomes (Schluenzen et al., 1999). Therefore, it seems that the structure determined by NMR is closer to the in situ conformation of this protein.

Inspection of the interactions among the various ribosomal components revealed several architectural elements with significant diversity in their modes of recognition, as found for smaller protein-RNA complexes (summarized by Cusack, 1999, and Draper and Reynaldo, 1999). For example, hints of the provision of support by ribosomal proteins for the stabilization of the three-dimensional fold of the rRNA chains were detected in several locations on the map.

Comparisons between the 7.2- and the 5-Å maps showed that the extension of the resolution added some detail (Schluenzen et al., 1999), but its quality is still not sufficient for reliable interpretation at the molecular level. For that, higher resolution and improved phases are essential. As we recently identified several heavy-atom derivatives that diffract to 3.5-Å and seem to possess suitable phasing power, further interpretation has been deferred to later stages.

STRUCTURAL MARKERS TARGETED TO PREDETERMINED SITES

To facilitate unbiased map interpretation, markers inserted in predetermined sites are being exploited. These are composed of heavy-atom compounds, attached either directly to the T30S particle or through carriers that bind to the ribosomal

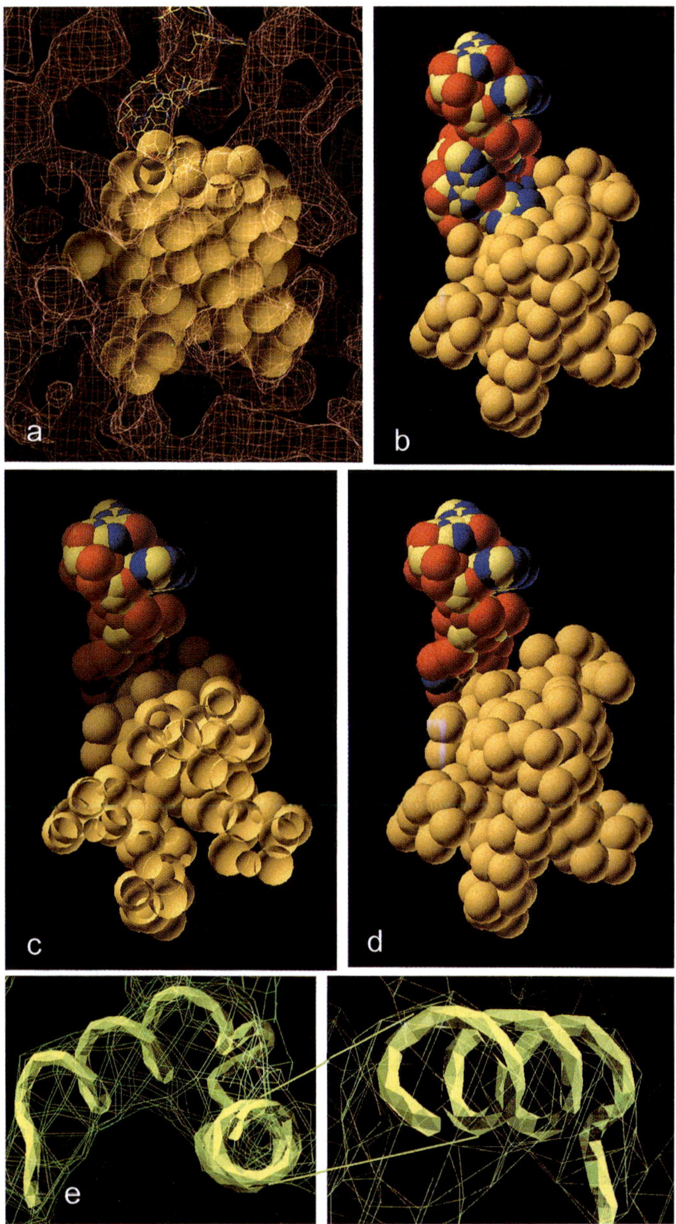

Figure 4. (a to d) About two-thirds of the backbone of protein TS7 fitted into the 7.2-Å map. The remaining third has been removed for clarity. Each dark-yellow sphere represents an amino acid. All atoms of the RNA chains are shown. (a and b) Space-filling presentation of TS7 surrounded by several features, among which one was fitted as an RNA duplex of 2 by 6 bases (shown as sticks in panel a and as a space-filling model in panel b. (c and d) Representations similar to that in panel b, but half a duplex has been removed, and a slice of it is shown, highlighting the contacts between the proteins and the RNA and the space available for the protein side chains (not shown). (e) The quality of the fitting is shown by a ribbon representing the main chain.

particles with high affinity. Candidates for carriers may be antibiotics, cDNA oligomers, charged tRNA molecules, or factors participating in the translation process, and for heavy atoms they are medium-size materials, such as tetrakis(acetoxymercuri)-methane (TAMM), a rather compact compound of four Hg atoms, and a tetrairidium cluster. The latter is composed of an internal core of four iridium atoms, with a diameter of 4.8 Å (Jahn, 1989), surrounded by a shell of organic moieties of chemical composition similar to that of proteins. To reduce the potential flexibility of this marker, its bridging arm was designed with a minimal length and maximal stability. It is slightly longer than the longest amino acid side chain and contains an amide bond. Hence, the expectations for local, rather rigid conformations are legitimate.

Surface Ribosomal Proteins

Labeling studies showed one fully and one partially exposed SH group (Sagi et al., 1995) belonging to proteins TS11 (residue 119) and TS13 (residue 84) (Wada et al., 1999). These were used for cluster binding prior to crystallization. The crystals obtained from the modified T30S particles diffract to 4.5-Å resolution and are isomorphous with the native ones, indicating that in the crystals the motion of the cluster bridging arm is limited, as frequently happens to long side chains in proteins. As expected, the attachment of 1 to 2 equivalents of the tetrairidium cluster yielded a weak derivative, albeit a suitable marker. Thus, two prominent peaks were revealed in difference Fourier maps, constructed with the MIR phases of T30S (Weinstein et al., 1999). These were found to be in accord with the locations suggested by noncrystallographic studies. The minor site is located on the particle's upper part (the head), in a position similar to that of protein S13, as revealed by immuno-EM (Stöffler and Stöffler-Meilicke, 1986), neutron scattering (Capel et al., 1988), and modeling (Mueller and Brimacombe, 1997). The major site, assigned as the cysteine of TS11, is located at the central part of the particle, in a position roughly compatible with that suggested by immuno-EM for protein S11 in E30S as well as by modeling the ribosomal components within cryo-EM reconstructions. However, it deviates by approximately 35 Å from the position assigned to the center of mass of this protein according to neutron scattering and contrast variation. Since the tetrairidium cluster and immuno-EM target the surfaces of the ribosomal particles, whereas the triangulation method approximates the positions of the centers of mass of the ribosomal proteins, such deviation is tolerable. In fact, it is smaller than the inconsistencies of 65 Å between neutron-scattering triangulation studies and those exploiting cDNA for the localization of ribosomal components (Alexander et al., 1994).

cDNA

cDNA oligomers are being exploited for the derivatization of ribosomal crystals and for flagging the locations of the rRNA regions targeted by them, benefiting from earlier attempts at mapping the surface *E. coli* RNA in solution (Tapprich and Hill, 1986; Oakes et al., 1986; Camp and Hill, 1987; Ricker and Kaji, 1991; Hill et al., 1988; Weller and Hill, 1992; Alexander et al., 1994). In these studies synthetic oligodeoxynucleotide probes, complementary to specific rRNA, were hybridized with ribosomal particles, and alterations in activity, binding, and recognition were monitored. We chose the sequences according to their affinities and specificities, based on previous measurements carried out on T30S particles in solution. The lengths of the oligomers (10 to 22 nucleotides) were designed to increase the stability of the expected hybrid double helix.

Cocrystallization and soaking were employed, although such long heavy-atom carriers (they may reach 70 Å) are not commonly used in protein crystallography because of space limitations. In fact, soaking is preferable, since the hybridization, which is an equilibrium process, can be terminated by the shock freezing needed for data collection once it is assumed that all hybrids have been formed. A procedure leading to efficient binding has been developed. First, unmodified oligomers are diffused into native T30S crystals. Their influence on the internal order of the crystals is used to indicate whether the region with which they were supposed to interact is involved in the crystal network or exposed to the solvent. The oligomers that do not cause a substantial resolution drop are modified by heavy atoms, either in their cytidines or at their terminal phosphates. Alternatively, the heavy atoms are attached to thiolated nucleotides. The degree of hybridization of the heavy-atom-modified DNA oligomers is tested in solution, and those that display high affinity are used either for soaking experiments or for hybridization in solution and subsequent cocrystallization (Auerbach et al., in press).

Over a dozen cDNA oligomers were tested, targeting seven regions in the 16S RNA. Among them, those matching the 3′ end of the 16S RNA, as well as two regions near helix 41 (nomenclature adopted from Brimacomb [1995]), did not cause crystal damage, allowing data collection to 3.5 to 3.9 Å. Those directed to the 5′ end and to helix 21 caused a drop in resolution to 6 to 7 Å but yielded data with reasonable quality. The oligomer complementary to the 3′ end (AGAAAGGAGGTGATC) was used for derivatization after the addition of three 6-thio-G, forming a probe with three potential heavy-metal binders. Three TAMM molecules were bound quantitatively via one of their SH groups to each of the three thiolated guanidines. The unbound Hg atoms of the TAMM molecules were masked by small compounds, such as cysteine, prior to binding to the DNA oligomer. Crystallographic analysis led to the location of the TAMM compound in a difference Fourier map, using the 7.2-Å MIR phases. Thus, the proximity of the location of the 3′ end of the 16S RNA was determined (Fig. 5).

FUNCTIONAL ACTIVATION IN PRE- AND POSTCRYSTALLIZATION STATES

The solvent content of the ribosomal crystals, 50 to 75%, falls within the range observed for other macromolecules. However, the distribution of the solvent regions is rather unusual. Sizable continuous solvent regions, with dimensions that may reach over 200 Å, were detected in the maps of T30S, T50S, and H50S (Yonath et al., 1998; Harms et al., 1999). In the crystals of H50S the unusually large continuous solvent region is rather loosely held by one interparticle contact area, leading to the undesired properties of this crystal form. In contrast, the bulky solvent within the crystals of T30S and T50S, over 200 Å in their longer dimension, could be exploited for their improvement. Thus, although in general conformational changes are not induced within crystals of biological macromolecules, because of the limited possibilities of motion imposed by the crystal network, the ribosomal particles could be reactivated within the crystals. In this way the frequency of obtaining high-quality crystals was increased significantly.

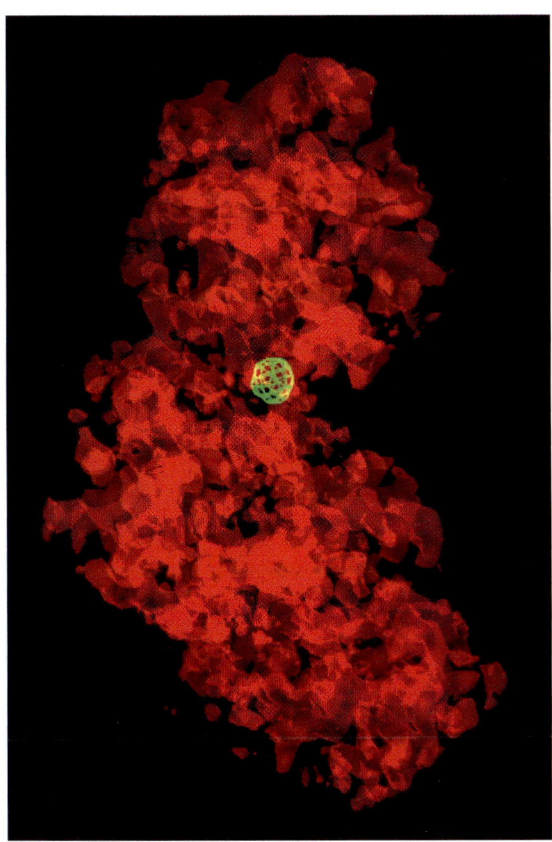

Figure 5. The suggested locations of the 3′ end of the 16S RNA (yellow-green) as detected by the Fourier method overlaid on the part of the 7.2-Å map assigned to T30S.

The Small Ribosomal Subunit

For T30S, high-quality crystals were obtained by activation of the ribosomal particles while in the crystals. With the goal of increasing the proportion of the particles that do not deviate from the preferred crystal conformation (shown to be close to that of active conformers), we are increasing the homogeneity by controlled heating (Zamir et al., 1971). Consequently, increasing proportions of crystals diffracting to around 3 Å were obtained under conditions originally found to yield crystals of moderate resolution, i.e., 4 to 4.5 Å. Along these lines, improvement of crystal quality was also achieved when crystallizing T30S particles trapped chemically at their activation state prior to crystallization or complexed with antibiotics, such as edeine (Szer and Kurylo-Borowska, 1972), known to "freeze" the 30S subunits at a particular conformational state (Moazed and Noller, 1987).

The Large Ribosomal Subunit

Among the numerous crystal forms that were obtained from large ribosomal subunits (Berkovitch-Yellin et al., 1992), two seem to be suitable for crystallographic studies. These are crystals of T50S, which yield high-quality diffraction data but until recently diffracted only to low resolution, 9 to 10 Å (Volkmann et al., 1990), and those of H50S, which give rise to rather problematic data that extend to 2.7 Å (von Böhlen et al., 1991). Owing to the undesirable properties of the latter crystals, i.e., their severe sensitivity to X-ray irradiation and their extremely low level of isomorphism, despite intensive studies carried out by us (Yonath and Franceschi, 1998; Yonath et al., 1998) as well as by others (Ban et al., 1998), so far no heavy-atom derivative that can lead to experimental phasing beyond 5-Å resolution has been found, and no marker or flag could be inserted into the electron density maps. Furthermore, it was shown that the undesirable properties of the H50S crystals become less tolerable with the increase in resolution. Hence, despite the fact that the current MIR with anomalous scattering map shows the overall structure of this particle and its main features, such as the exit tunnel (summarized by Yonath and Franceschi [1998]), as well as finer detail, traceable as RNA chains or ribosomal proteins, the prospects for solving the structure of this particle at molecular resolution are rather slim and depend on the development of new, innovative methods.

The packing diagram of the MIR with anomalous scattering map of H50S provides possible reasons for the odd combination of the properties of these crystals: high resolution accompanied by problematic diffraction. The high resolution may result from the extensive particle interactions that are concentrated in parts of the unit cell. In contrast, there is only a small interparticle contact area along the *c* axis (564 Å), which is surrounded by an extremely large solvent region. The small number of particle interactions may cause the poor isomorphism, the unfavorable crystal habit (plates, made of sliding layers, typically reaching up to 0.5 mm^2 with an average thickness of a few microns in the direction of the *c*-axis), and the variations in the *c*-axis length (567 to 570 Å) as a function of irradiation (Harms et al., 1999). It can also explain how very large clusters, such as those containing 30 tungsten atoms, 110 oxygens atoms, and 5 phosphate atoms [$K_{14}(NaP_5W_{30}O_{110})31H_2O$], diffuse readily into the H50S crystals and why they may introduce subtle nonisomorphism and lead to limited and/or less reliable phase information. In order to facilitate structure determination, efforts aimed at increasing the level of isomorphism of the H50S crystals were made along the lines that were found suitable for T30S. These included inducing controlled rearrangements within the crystals by heat activation and treatment with selected additives, but neither led to significant improvement.

Crystals of T50S Diffract to 3.4 Å

In contrast to the H50S case, attempts at increasing the homogeneity of the crystallized T50S particle led most recently to a marked extension in resolution, from the previous limit of 9 to 10 Å to 3.4 Å. It should be mentioned that even when only low-resolution data could be obtained, the T50S crystals have been the target for crystallographic studies because of the high quality of their diffraction. Thus, Ta_6Br_{14}-soaked crystals led to two sites, readily extracted from isomorphous and anomalous-difference Patterson maps. These were used for initial single isomorphous replacement with anomalous phasing with the following final statistics: R_{cullis} = 0.59 (total) and 0.87 (anomalous) (R_{cullis} is defined as ⟨phase integrated lack of closure⟩/⟨$|F_{PH} - F_P|$⟩, where F_{PH} is the structure factor of the heavy-atom-derivatized crystals and F_P is that of the native protein); phasing power = 1.95; figure of merit (FOM)=0.58. In parallel, molecular replacement studies, performed with cryo-EM reconstructions of these particles, led to a unique solution that subsequently confirmed the main Ta_6Br_{14} sites (Fig. 6) (Harms et al., 1999). As these findings raise expectations for T50S structure determination, studies aimed at following the path of the nascent protein chains have been initiated. To shed light on the dynamic driving forces participating in the movement of newly born proteins, we cocrystallized complexes of T50S with short nascent polypeptides, along the same lines as for H50S (Berkovitch-Yellin et al., 1992), including the attachment of heavy atoms to the incorporated amino acids.

FUTURE PROSPECTS

We have shown that medium-resolution maps can be constructed for ribosomal particles, using experimental phases. These maps exhibit the external shapes and the internal features of the corresponding ribosomal particles, and for significant parts of them feasible interpretations at a level close to molecular resolution can be provided. Furthermore, selected locations on the ribosomal particles were revealed by covalently bound heavy-atom compounds. Thus, despite severe crystallographic problems, the way to structure determination has been paved and electron density maps at close to molecular resolution are emerging.

Intriguing questions associated with the functional flexibility of the small ribosomal subunit may be approached by the crystallization of particles at various conformational states, using tailor-made ligands, such as antibiotics or cDNA, to exposed single-stranded rRNA regions, or by letting these materials diffuse into the already-formed crystals. The observation that the conformation of the crystallized T30S particles is close to that of the active state in T30S has already led to the design of experiments aimed at obtaining higher-quality diffraction by further reactivation of the crystallized particles. Proceeding along these lines, and the examination of the changes in crystal behavior upon binding of selected ligands, should indicate the level of their involvement in functional recognition. Binding of metal compounds to these ligands should enable their unbiased positioning and provide indispensable information for map interpretation. Suitable examples, such as the freezing of the ribosomal structure by antibiotics, were mentioned above. Others will be based on the available biochemical information.

The extension of the resolution of T50S crystals, interesting and encouraging in their own rights, should make a noticeable impact on the future of ribosomal crystallography, since it opens the possibility

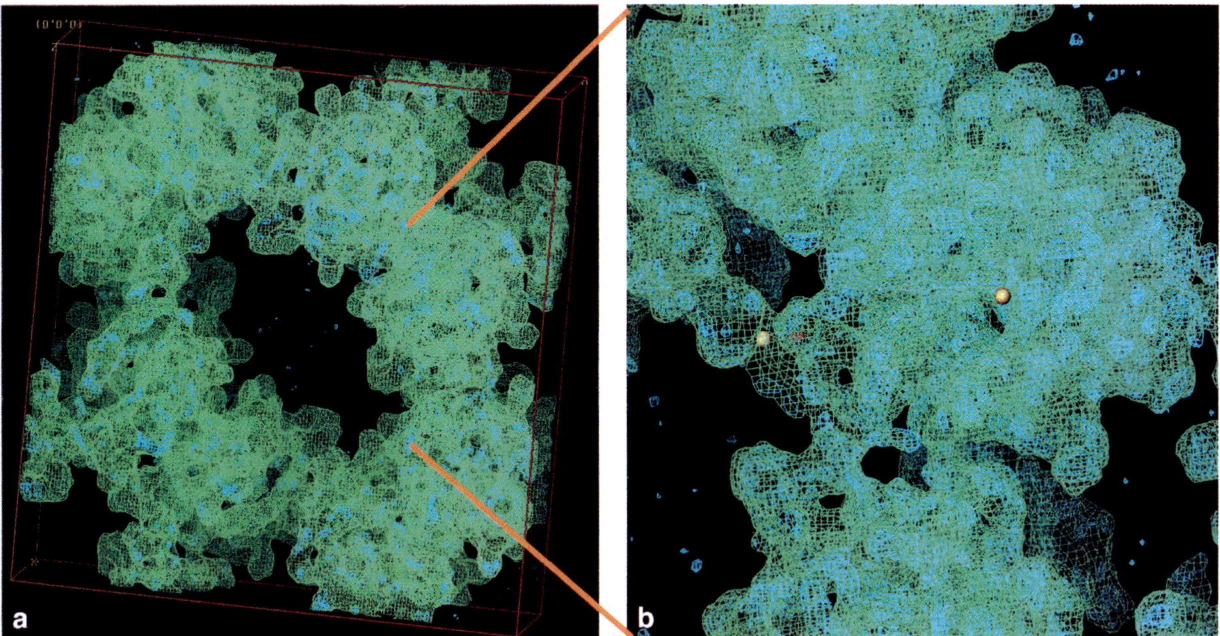

Figure 6. (a) Packing diagram of T50S, assembled by positioning the EM-reconstructed image (Yonath et al., 1998; Yonath and Franceschi, 1998; Harms et al., 1999) in the crystallographic unit cell according to the unique prominent result of the molecular replacement search. A total of 11,000 unique reflections (to 9-Å resolution) were collected from native and Ta_6Br_{14}-derivatized crystals. Six heavy-atom sites were extracted from anomalous-difference Patterson maps (FOM = 0.7188; R_{cullis} = 0.75; phasing power = 1.56). (b) The most prominent Ta_6Br_{14} sites are shown as golden balls (group scatterers with artificial diameters of about 6 Å).

of studying the three ribosomal particles from the same source, thus providing the tools for functional-structural comparisons. In the past we have grown crystals of T70S ribosomes diffracting to rather low resolution (20 to 24 Å) that was extended to 12 to 14 Å once the crystallized material was complexed with a 35-base RNA oligomer (as mRNA) and two charged tRNA molecules (Table 1) (Hansen et al., 1990). Since the packing arrangement of these crystals has been determined by molecular replacement (Harms et al., 1999) and since procedures that may lead to improvement in crystal quality have been developed, there is reason to expect improvement of T70S resolution too.

The growing popularity of ribosomal crystallography is indeed gratifying. This, together with the fruitful interactions with the exciting advances in cryo-EM, is bound to lead to major breakthroughs.

Thanks are given to M. Safro for active participation in phasing; M. Wilchek for indispensable advice; W. Jahn, W. Preetz, and M. Pope for their generous gifts of heavy-atom compounds; M. Van Heel, H. Stark, J. Frank, and I. Gabashvili for providing us with their cryo-EM reconstructions; W. Traub and A. Podjarny for fruitful discussions; and C. Radzwill, H. Burmeister, R. Albrecht, C. Glotz, J. Müssig, C. Paulke, M. Laschever, S. Meier, Y. Halfon, and K. Knaack for their excellent contributions in the different stages of these studies. Data were collected at the European Molecular Biology Laboratory and Max-Planck Society beam lines at The German Electron Synchrotron Facility; Station F1 at Cornell High Energy Synchrotron Source, Cornell University; Stations ID2 and ID13 at the European Synchrotron Radiation Facility, Grenoble, Spain; and Station ID19 at the Advanced Photon Source, Argonne National Laboratory.

Support was provided by the Max-Planck Society, the National Institutes of Health (NIH GM 34360), the German Ministry for Science and Technology (BMBF 05-641EA), and the Kimmelman Center for Macromolecular Assembly at the Weizmann Institute. A.Y. holds the Martin S. Kimmel Professorial Chair.

REFERENCES

Alexander, R. W., P. Muralikrishna, and B. S. Cooperman. 1994. Ribosomal components neighboring the conserved 518–533 loop of 16S rRNA in 30S subunits. *Biochemistry* **33:**12109–12118.

Auerbach, T., T. Pioletti, H. Avila, K. Anagnostopoulos, S. Weinstein, F. Franceschi, and A. Yonath. Genetic and biochemical manipulation of the small ribosomal subunit from *T. thermophilus* HB8. *J. Biomol. Struct. Dyn.*, in press.

Ban, N., B. Freeborn, P. Nissen, P. Penczek, R. A. Graussucci, R. Sweet, F. Frank, P. Moore, and T. Steitz. 1998. The 9 Å resolution X-ray crystallography map of the large ribosomal subunits. *Cell* **93:**1105–1115.

Basavappa, R., and P. B. Sigler. 1991. The 3-dimensional structure of yeast initiator tRNA: functional implication in initiator/elongator discrimination. *EMBO J.* **10:**3105–3110.

Berkovitch-Yellin, Z., W. S. Bennett, and A. Yonath. 1992. Aspects in structural studies on ribosomes. *Crit. Rev. Biochem. Mol. Biol.* **27:**403–444.

Brimacombe, R. 1995. The structure of ribosomal RNA; a three-dimensional jigsaw puzzle. *Eur. J. Biochem.* **230**:365–383.

Camp, D., and W. E. Hill. 1987. Probing E. coli 16S ribosomal-RNA with DNA oligomers to determine functional and structural characteristics of the highly conserved G(530) loop. *FASEB J.* **46**:2216–2221.

Capel, M. S., M. Kjeldgaard, D. M. Engelman, and P. B. Moore. 1988. Positions of S2, S13, S16, S17, S19 and S21 in the 30S ribosomal subunit of *E. coli*. *J. Mol. Biol.* **200**:65–87.

Clemons, W. M., C. Davies, S. White, and V. Ramakrishnan. 1998. Conformational variability of the N-terminal helix in the structure of ribosomal protein S15. *Structure* **6**:429–438.

Cusack, S. 1999. RNA protein complexes. *Curr. Opin. Struct. Biol.* **9**:66–73.

Davies, C., V. Ramakrishnan, and S. W. White. 1996. Structural evidence for specific S8-RNA and S8-protein interactions within the 30S ribosomal subunit: ribosomal protein S8 from *B. stearothermophilus* at 1.9 Å resolution. *Structure* **4**:1093–1104.

Draper, D. E., and L. P. Reynoldo. 1999. RNA binding strategies of ribosomal proteins. *Nucleic Acids Res.* **27**:381–388.

Frank, F., J. Zhu, P. Penczek, Y. Li, S. Srivastava, A. Verschoor, M. Radermacher, R. Grassucci, A. R. Lata, and R. K. Agrawal. 1995. A model of protein synthesis based on cryo electron microscopy of the *E. coli* ribosome. *Nature* **376**:441–444.

Gabashvili, I. S., R. K. Agrawal, R. Grassucci, and J. Frank. 1999. Structure and structural variations of the E. coli 30S ribosomal subunit as revealed by three-dimensional cryo-electron microscopy. *J. Mol. Biol.* **286**:1285–1291.

Golden, B. L., D. W. Hoffman, V. Ramakrishnan, and S. W. White. 1993. Ribosomal protein S17: characterization of the three-dimensional structure by H and N NMR. *Biochemistry* **32**: 12812–12820.

Guerin, M. F., and D. H. Hayes. 1987. Comparison of active and inactive forms of the E. coli 30S ribosomal subunits. *Biochimie* **69**:965–974.

Hansen, H. A. S., N. Volkmann, J. Piefke, C. Glotz, S. Weinstein, I. Makowski, S. Meyer, H. G. Wittmann, and A. Yonath. 1990. Crystals of complexes mimicking protein biosynthesis are suitable for crystallographic studies. *Biochim. Biophys. Acta* **1050**: 1–7.

Harms, J., A. Tocilj, I. Levin, I. Agmon, I. Kölln, H. Stark, M. van Heel, M. Cuff, F. Schlünzen, A. Bashan, F. Franceschi, and A. Yonath. 1999. Elucidating the medium resolution structure of ribosomal particles: an interplay between electron-cryo-microscopy and X-ray crystallography. *Structure* **7**:931–941.

Hill, W. E., D. G. Camp, W. E. Tapprich, and A. Tassanakajon. 1988. Probing ribosomal structure and function using short oligo-deoxy-ribonucleotides. *Methods Enzymol.* **164**:401–419.

Hosaka, H., A. Nakagawa, I. Tanaka, N. Harada, K. Sano, M. Kimura, M. Yao, and S. Wakatsuki. 1997. Ribosomal protein S7: a new RNA-binding motif with structural similarities to a DNA architectural factor. *Structure* **5**:1199–1208.

Jahn, W. 1989. Synthesis of water soluble tetrairidium cluster for specific labelling of proteins. *Z. Naturforsch.* **44b**:79–82.

Jones, T. A., J.-Y. Zou, S. W. Cowan, and M. Kjeldgaard. 1991. Improved methods for building protein models in electron density maps and the location of errors in these models. *Acta Crystallogr.* **A47**:110–119.

Krumbholz, S., F. Schlünzen, J. Harms, H. Bartels, I. Kölln, K. Knaack, W. S. Bennett, P. Bhanumoorthy, H. A. S. Hansen, N. Volkmann, A. Bashan, I. Levin, A. Tocilj, and A. Yonath. 1998. Ribosomal crystallography: cryo protectants and cooling agents. *Periodicum Biologorum* **100**:119–125.

Lata, A. R., R. K. Agrawal, P. Penczek, R. Grassucci, J. Zhu, and J. Frank. 1996. Three-dimensional reconstruction of the E. coli 30S ribosomal subunit in ice. *J. Mol. Biol.* **262**:43–52.

Liljas, A., and S. Al-Karadaghi. 1997. Structural aspects of protein synthesis. *Nat. Struct. Biol.* **4**:767–771.

Lindhal, M., L. A. Svensson, A. Liljas, S. E. Sedelnikova, I. Eliseukina, N. Fomenkova, N. Nevskaya, S. Nikonov, N. Garber, and T. A. Muranova. 1994. Crystal structure of the ribosomal protein S6 from *T. thermophilus*. *EMBO J.* **13**:1249–1254.

Malhotra, A., P. Penczek, R. K. Agrawal, I. S. Gabashvili, R. A. Grassucci, R. Junemann, N. Burkhardt, K. H. Nierhaus, and J. Frank. 1998. E. coli 70S ribosome at 15 A resolution by cryo-electron microscopy: localization of fMet-tRNA$_f^{Met}$ and fitting of L1 protein. *J. Mol. Biol.* **280**:103–115.

Moazed, D., and H. F. Noller. 1987. Interaction of antibiotics with functional sites in 16S ribosomal RNA. *Nature* **327**:389–394.

Mueller, F., and R. Brimacombe. 1997. A new model of the three-dimensional folding of *E. coli* 16S ribosomal RNA. II. The RNA-protein interaction data. *J. Mol. Biol.* **271**:524–544.

Nevskaya, N., S. Tishchenko, A. Nikulin, S. Al-Karadaghi, A. Liljas, B. Ehresmann, C. Ehresmann, M. Garber, and S. Nikonov. 1998. Crystal structure of ribosomal protein S8 from Thermus thermophilus reveals a high degree of structural conservation of a specific RNA binding site. *J. Mol. Biol.* **279**:233–244.

Oakes, M. I., M. W. Clark, E. Henderson, and J. A. Lake. 1986. DNA hybridization electron microscopy: ribosomal RNA nucleotides 1392–1407 are exposed in the cleft of the small subunit. *Proc. Natl. Acad. Sci. USA* **83**:275–279.

Ramakrishnan, V., and S. W. White. 1992. The structure of ribosomal protein S5 reveals sites of interaction with 16S RNA. *Nature* **358**:768–771.

Ramakrishnan, V., and S. W. White. 1998. Ribosomal protein structures: insights into the architecture, machinery and evolution of the ribosome. *Trends Biochem. Sci.* **3**:208–212.

Richmond, T. J., J. T. Finch, B. Rushton, D. Rhodes, and A. Klug. 1984. Structure of the nucleosome core particle at 7 Å resolution. *Nature* **311**:532–537.

Ricker, R. D., and A. Kaji. 1991. Use of single strand DNA oligonucleotide in programming ribosomes for translation. *Nucleic Acids Res.* **19**:6573–6578.

Sagi, I., V. Weinrich, I. Levin, C. Glotz, M. Laschever, M. Melamud, F. Franceschi, S. Weinstein, and A. Yonath. 1995. Crystallography of ribosomes: attempts at decorating the ribosomal surface. *Biophys. J.* **55**:31–41.

Schluenzen F., H. A. S. Hansen, J. Thygesen, W. S. Bennett, N. Volkmann, I. Levin, J. Harms, H. Bartels, A. Bashan, Z. Berkovitch-Yellin, I. Sagi, F. Franceschi, S. Krumbholz, M. Geva, S. Weinstein, I. Agmon, N. Boeddeker, S. Morlang, R. Sharon, A. Dribin, M. Peretz, V. Weinrich, and A. Yonath. 1995. A milestone in ribosomal crystallography: the construction of preliminary electron density maps at intermediate resolution. *J. Biochem. Cell Biol.* **73**:739–749.

Schluenzen, F., M. Gluehmann, D. Janell, I. Levin, A. Bashan, J. Harms, H. Bartels, T. Auerbach, T. Pioletti, H. Avila, K. Anagnostopoulos, H. A. S. Hansen, W. S. Bennett, I. Agmon, M. Kessler, A. Tocilj, M. Peretz, S. Weinstein, F. Franceschi, and A. Yonath. 1999. The identification of selected components in electron density maps of prokaryotic ribosomes at 7 Å resolution. *J. Synth. Radiat.* **6**:928–941.

Stark, H., F. Mueller, E. V. Orlova, M. Schatz, P. Dube, T. Erdemir, F. Zemlin, R. Brimacombe, and M. van Heel. 1995. The 70S *E. coli* ribosome at 23 Å resolution: fitting the ribosomal RNA. *Structure* **3**:815–821.

Stöffler, G., and M. Stöffler-Meilicke. 1986. Immuno electron microscopy on E. coli ribosomes, p. 28–46. *In* B. Hardesty and G. Kramer (ed.) *Structure, Function and Genetics of Ribosomes.* Springer Verlag, Heidelberg, Germany.

Szer, W., and Z. Kurylo-Borowska. 1972. Interactions of edeine with bacterial ribosomal subunits. Selective inhibition of aminoacyl-tRNA binding sites. *Biochem. Biophys. Acta* 259:357–368.

Tanaka, I., A. Nakagawa, H. Hosaka, S. Wakatsuki, F. Mueller, and R. Brimacombe. 1998. Matching the crystallographic structure of ribosomal protein S7 to the 3D model of 16S RNA. *RNA* 4:542–550.

Tapprich, W. E., and W. E. Hill. 1986. Involvement of bases 787–795 of E. coli 16S RNA in ribosomal subunit association. *Proc. Natl. Acad. Sci. USA* 83:556–560.

Trakhanov, S. D., M. M. Yusupove, V. A. Shirokov, M. B. Garber, A. Mitscher, M. Ruff, J.-C. Tierry, and D. Moras. 1989. Preliminary X-ray investigation on 70S ribosome crystals. *J. Mol. Biol.* 209:327–334.

Van Heel, M., and H. Stark. Unpublished data.

Volkmann, N., S. Hottentrager, H. A. S. Hansen, A. Zaytzev-Bashan, R. Sharon, Z. Berkovitch-Yellin, A. Yonath, and H. G. Wittmann. 1990. Characterization and preliminary crystallographic studies on large ribosomal subunits from Thermus thermophilus. *J. Mol. Biol.* 216:239–243.

von Böhlen, K., I. Makowski, H. A. S. Hansen, H. Bartels, Z. Berkovitch-Yellin, A. Zaytzev-Bashan, S. Meyer, C. Paulke, F. Franceschi, and A. Yonath. 1991. Characterization and preliminary attempts for derivatization of crystals of large ribosomal subunits from Haloarcula marismortui, diffracting to 3 Å resolution. *J. Mol. Biol.* 222:11–15.

Wada, T., T. Yamazaki, S. Kuramitsu, and Y. Kyogoku. 1999. Cloning of the RNA polymerase alfa subunit gene from Thermus thermophilus HB8 and characterization of the protein. *J. Biochem.* 125:143–150.

Weinstein, S., W. Jahn, C. Glotz, F. Schlünzen, I. Levin, D. Janell, J. Harms, I. Kölln, H. A. S. Hansen, M. Glühmann, W. S. Bennett, H. Bartels, A. Bashan, I. Agmon, M. Kessler, M. Pioletti, H. Avila, K. Anagnostopoulos, M. Peretz, T. Auerbach, F. Franceschi, and A. Yonath. 1999. Metal compounds as tools for the construction and the interpretation of medium resolution maps of ribosomal particles. *J. Struct. Biol.* 127:141–151.

Weller, J. W., and W. E. Hill. 1992. Probing dynamic changes in rRNA conformation in the 30S subunit of the E. coli ribosome. *Biochemistry* 31:2748–2757.

Wimberly, B.T., S. W. White, and V. Ramakrishnan. 1997. The structure of ribosomal protein S7 at 1.9 Å resolution reveals a beta-hairpin motif that binds double-stranded nucleic acids. *Structure* 5:1187–1198.

Yonath, A., and F. Franceschi. 1998. Functional universality and evolutionary diversity: insights from the structure of the ribosome. *Structure* 6:678–684.

Yonath, A. C. Glotz, H. S. Gewitz, K. Bartels, K. von Boehlen, I. Makowski, and H. G. Wittmann. 1988. Characterization of crystals of small ribosomal subunits, *J. Mol. Biol.* 203:831–833.

Yonath, A., J. Harms, H. A. S. Hansen, A. Bashan, M. Peretz, H. Bartels, F. Schluenzen, I. Koelln, W. S. Bennett, I. Levin, S. Krumbholz, A. Tocilj, S. Weinstein, I. Agmon, M. Piolleti, T. Auerbach, and F. Franceschi. 1998. The quest for the molecular structure of a large macromolecular assembly exhibiting severe non-isomorphism, extreme beam sensitivity and no internal symmetry. *Acta Crystallogr.* 54A:945–955.

Zamir, A., R. Miskin, and D. Elson. 1971. Inactivation and reactivation of ribosomal subunits: amino acyl-transfer RNA binding activity of the 30S subunits of E. coli. *J. Mol. Biol.* 60:347–364.

II. FUNCTIONAL STATES MAPPED BY CRYO-ELECTRON MICROSCOPY

II. FUNCTIONAL STATES MAPPED BY CRYO-ELECTRON MICROSCOPY

Electron microscopy has played an important role in ribosomal research since the early 1960s, but recent developments in cryo-electron microscopy techniques have greatly improved structural resolution and provided detailed views of the structures of the ribosomal subunits. This, in turn, has provided a framework for interpreting the extensive biochemical and genetic data that pertain to ribosomal structure and function and has led to the visualization of mRNAs, tRNAs, and factors on the ribosome.

The Ribosome: Structure, Function, Antibiotics, and Cellular Interactions
Edited by R. A. Garrett, S. R. Douthwaite, A. Liljas, A. T. Matheson, P. B. Moore, and H. F. Noller

Chapter 4

Visualization of the Translational Elongation Cycle by Cryo-Electron Microscopy

TILLMANN PAPE, HOLGER STARK, RISHI MATADEEN, ELENA V. ORLOVA, and MARIN VAN HEEL

A detailed structural knowledge of the complex and dynamic interplay of the ribosome with mRNA, tRNA, and other protein factors is necessary to understand translation in molecular terms. The core of protein synthesis is the elongation cycle. This consists of the decoding of mRNA in the ribosomal A site by the aminoacyl-tRNA·EF-Tu·GTP ternary complex, peptidyl transfer of the nascent peptide onto the accepted aminoacyl-tRNA (aa-tRNA), and translocation of the resulting peptidyl-tRNA from the A site into the P site catalyzed by EF-G·GTP. Simultaneously, the deacylated tRNA moves to the E site, where it dissociates, restoring the initial state of the ribosome, which then enters the next cycle. The two G proteins (EF-Tu and EF-G) ensure the accuracy and speed of elongation at the expense of chemical energy derived from GTP hydrolysis.

Cryo-electron microscopy together with single-particle analysis has enabled direct visualization of the *Escherichia coli* ribosome in different functional states. The complexes are either biochemically stable or trapped by antibiotics which arrest the elongation factors Tu and G in their respective binding sites on the ribosome. Thus, short-lived intermediate states of the ribosome can be resolved during A-site binding of aa-tRNA and during translocation. At present these structures are determined to 10- to 20-Å resolution. This provides us with new insights into the interaction and associated conformational changes of individual molecular components when they are assembled in a fully functional complex. In this chapter we review the functional implications of these findings and propose a comprehensive structural model for the elongation cycle.

CRYO-ELECTRON MICROSCOPY AND IMAGE PROCESSING

Cryo-electron microscopy of individual noncrystallized molecules (single particles) together with angular reconstitution is a standard approach for probing the three-dimensional (3-D) structure of biological macromolecules. The angular reconstitution technique exploits the random orientations of particles within a vitreous ice matrix, allowing the direct extraction of 3-D information from electron-microscopic images (van Heel, 1987). A sample is embedded in amorphous ice by rapidly freezing it in its natural hydrated state (Dubochet et al., 1988). Images are taken, each at a certain defocus, at liquid nitrogen or liquid helium temperatures in an electron microscope with a low-dose protocol (10 $e^-/Å^2$). Good micrographs are selected by optical diffraction and subsequently digitized with a high-resolution densitometer.

The molecular images are then processed within the IMAGIC-5 software system (van Heel et al., 1996). Contrast transfer function correction is applied to micrographs. Images of single particles are selected and aligned. Multivariate statistical analysis (van Heel and Frank, 1981) and automatic classification (van Heel, 1989) are used to sort the data set into groups of similar images, which can be averaged to obtain relatively noise-free characteristic views of the ribosome. The angular reconstitution technique is then used to obtain Euler angle orientations of selected characteristic views. 3-D reconstructions are calculated with the exact filter back-projection algorithm (Harauz and van Heel, 1986; Radermacher,

Tillmann Pape, Rishi Matadeen, Elena V. Orlova, and Marin van Heel ■ Department of Biochemistry, Imperial College of Science, Technology and Medicine, London SW7 2AY, United Kingdom. **Holger Stark** ■ Institut für Molekularbiologie und Tumorforschung, Philipps-Universität Marburg, Emil-Mannkopff-Straβe 2, 35037 Marburg, Germany.

1988). The procedure of alignment, multivariate statistical analysis data compression, classification, and angular reconstitution is applied iteratively to refine the 3-D structure. The resolution of the final 3-D reconstruction is determined by the 3σ Fourier shell correlation criterion (Harauz and van Heel, 1986) by using two independent reconstructions, each based on half of the available good class averages. A summary of the procedure is given in Fig. 1.

ELONGATION FACTORS Tu AND G ON THE TRANSLATING RIBOSOME

After the localization of tRNA molecules in the pre- and posttranslocation state of the ribosome (Stark et al., 1997a), the binding sites of EF-Tu (Stark et al., 1997b, unpublished data) and EF-G in two different states (Stark et al., submitted) on the ribosome were determined (Fig. 2).

EF-Tu, GTP, and aa-tRNA form a ternary complex that binds to the posttranslocation state of the ribosome (Fig. 2A). Following codon recognition, GTP hydrolysis, and a large-scale rearrangement of EF-Tu (Abel et al., 1996; Polekhina et al., 1996), the aa-tRNA is released to the ribosomal A site (Rodnina et al., 1996; Pape et al., 1998, and references therein). Kirromycin stalls the multistep process of A-site binding immediately after GTP hydrolysis and prevents the conformational change of EF-Tu from the GTP- to the GDP-bound form (Rodnina et al., 1995) by binding to the interface of domains 1 and 3 of EF-Tu. Thus, the aa-tRNA remains in complex with EF-Tu and is not released to accommodate in the A site. Consequently, the ternary complex is frozen in the "GTPase state" after codon recognition (Fig. 2B). The structure was first determined to a resolution of 18 Å (Stark et al., 1997b) and has recently been refined to 13 Å (Stark et al., unpublished data). This allowed the clear fitting of the X-ray structure of the ternary complex (Nissen et al., 1995) into the 3-D maps. In these structures the ternary complex spans the gap between the two subunits, with the anticodon domain of the tRNA reaching into the decoding center. Domain 1 (G domain) of EF-Tu is bound both to the C-terminal domain(s) of the L7-L12 stalk and to the 50S body beneath the stalk, whereas domain 2 is oriented so that it almost touches the S12 region on the body of the 30S subunit, which contacts the 530 stem-loop of 16S rRNA. The region underneath the stalk probably contains the α-sarcin stem-loop around position 2660 of 23S RNA, with which EF-Tu appears to interact as indicated by both functional (Tapprich and Dahlberg, 1990) and footprinting (Moazed et al., 1988) data. At higher resolution the density of the ternary com-

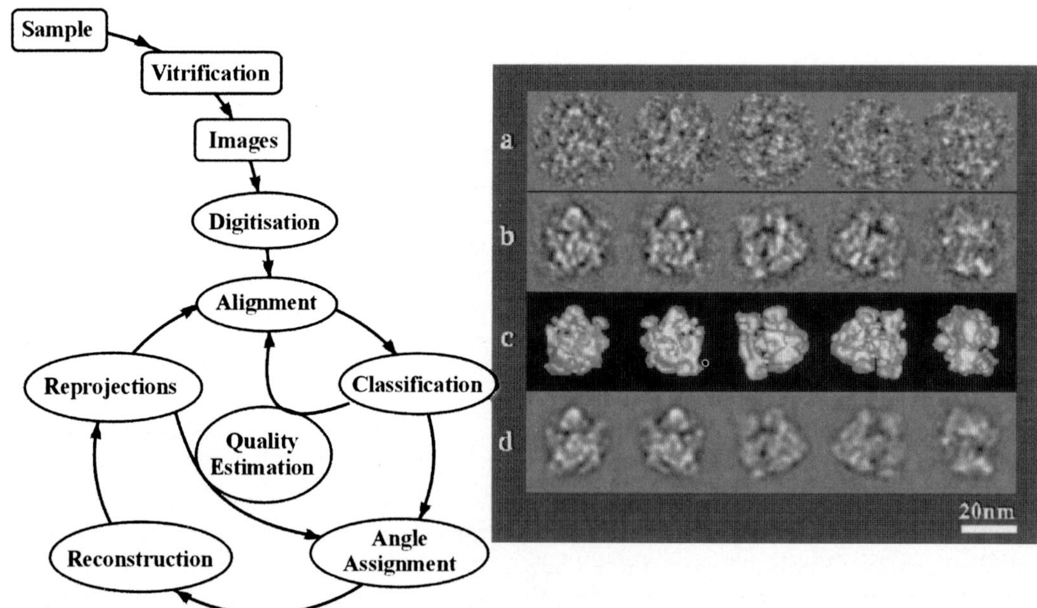

Figure 1. Synopsis of the 3-D reconstruction of the 70S ribosome. Electron-microscopic images of individual ribosomes are aligned by multireference alignment techniques (a), and similar images are grouped into classes by multivariate statistical classification procedures. All images belonging to the same class are averaged to produce noise-reduced class averages or characteristic views (b). An Euler angle orientation is assigned to each class average by the angular reconstitution technique. A 3-D reconstruction is computed (c) and used to produce reprojected images (d), which in turn are used as reference images for realignment of the original raw image data set and for improving the accuracy of the Euler angle determinations.

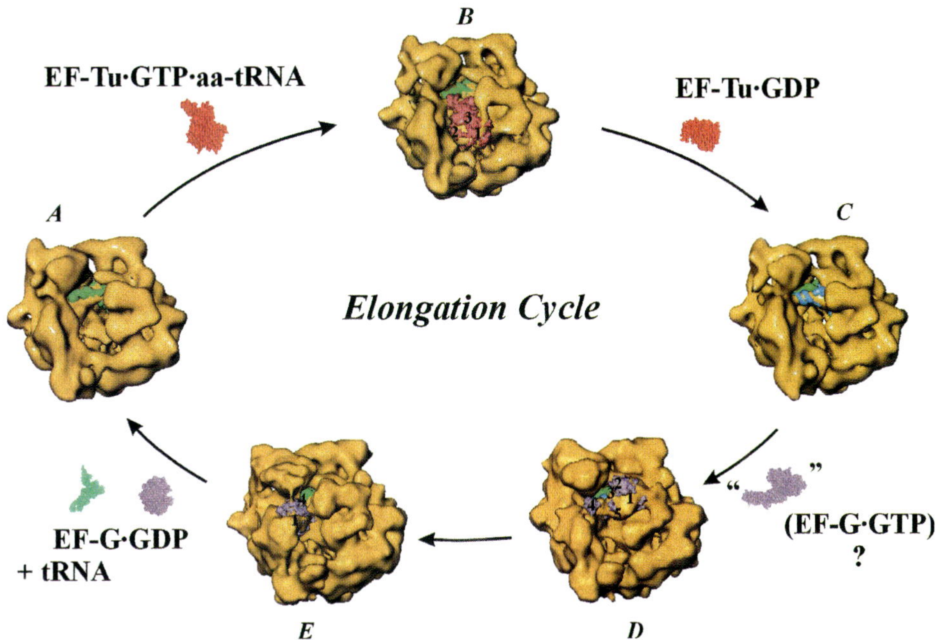

Figure 2. Visualization of the elongation cycle by 3-D reconstructions of different ribosome complexes at ~20-Å resolution. The ternary complex (red) binds to the posttranslocation state (A) with the peptidyl-tRNA in the P site (green) and forms a "GTPase state" on the ribosome (B). After dissociation of EF-Tu·GDP and accommodation of the aa-tRNA (blue) in the A site, peptidyl transfer takes place, leading to the pretranslocation state (C). EF-G·GTP (purple) binding and instantaneous GTP hydrolysis leads to a pre*-translocation state (D), which rearranges after translocation into a post*-translocation state (E). Finally, EF-G·GDP and deacylated tRNA dissociate to give the posttranslocation state again. The numbers indicate the domains of the elongation factors as revealed by fitting the X-ray structures into the respective densities. The views are from the L7-L12 side along the intersubunit space, with the 30S subunit on the left and the 50S subunit on the right. For morphological details, refer to Fig. 3–5. As the EF-G·GTP X-ray structure is unknown, the EF-G·GDP coordinates were used and are therefore shown in quotation marks.

plex clearly suggests a more extended conformation on the ribosome. Domains 2 and 3 of EF-Tu and the aa-tRNA have to be shifted relative to domain 1, indicating the occurrence of local contortions in the transition state ("GTPase state").

After successful delivery of the aa-tRNA to the ribosome, the pretranslocation state (Fig. 2C) has tRNAs bound in the A and P sites. These tRNAs lie close together in the intersubunit space, above the major bridge connecting the subunits. The anticodon domains are deeply buried in the 30S density of the decoding center (neck region), while the flexible CCA ends lie adjacent to the peptide tunnel entrance close to the peptidyltransferase center. The enclosed angle between the plane of the A- and P-site tRNAs is ~50°, and the distance between the elbow regions is 38 Å ("S-type" configuration). The positioning of these tRNAs is in agreement with footprinting (Moazed and Noller, 1989), cross-linking (Döring et al., 1994), and fluorescence energy transfer data (Paulsen et al., 1983).

Upon binding of EF-G·GTP to the pretranslocation state of the ribosome, GTP is instantaneously hydrolyzed (Rodnina et al., 1997). This causes a conformational change in EF-G that induces the formation of a transition state of the ribosome. In the transition state, translocation of the peptidyl-tRNA from the A to the P site takes place. As a result, the ribosome returns to the ground state, and EF-G assumes the GDP-bound conformation. EF-G·GDP and the deacylated tRNA then dissociate from the ribosome, resulting in the posttranslocation state—with peptidyl-tRNA in the P site and an empty A site ready for the next cycle. The conformational change of EF-G upon GTP hydrolysis remains hypothetical, since the EF-G·GTP structure is as yet unknown. The addition of thiostrepton (which binds to the L11-23S RNA region around position 1070 [Porse et al., 1998; Wimberly et al., 1999]) to the pretranslocation state traps the EF-G–ribosome complex in a pre*-translocation state (before the movement of the tRNAs has proceeded to any significant extent) and in a post*-translocation state (with peptidyl-tRNA in the P site and EF-G·GDP still bound to the ribosome). The antibiotic does not affect the ribosome binding of EF-G·GTP or GTP hydrolysis, whereas it strongly inhibits translocation and blocks EF-G turnover after GTP hydrolysis and translocation (Rodnina et al.,

1999). The extra densities in the 3-D reconstructions could be attributed to EF-G by difference maps of the respective states with a control (pretranslocation state of the ribosome in the presence of thiostrepton). The crystal structure of EF-G (Ævarsson et al., 1994; Czworkowski et al., 1994) fits well both in size and in general shape into these densities. In the pre*-translocation state at 18-Å resolution (Fig. 2D), EF-G bridges the cleft between the two subunits. The body of EF-G interacts with the L7-L12 stalk and the base of the stalk (L10, L11). Domain 4 reaches across the intersubunit space and contacts the head-body junction (shoulder) of the 30S subunit, where protein S4 is located (Mueller and Brimacombe, 1997), giving access to the decoding center. There are no other contacts of EF-G with the ribosome in the pre* state. The arrangement of EF-G in the post*-translocation state at 19-Å resolution (Fig. 2E) appears to be entirely different. EF-G has completely rearranged its position on the ribosome; the body of EF-G is closely attached to the 30S subunit, and domain 4 now points into the decoding center. The strong connection of EF-G to L7-L12 prevailing in the pre* state has been largely lost in the post* state, except for density emerging from below the globular head of the L7-L12 stalk that still forms a link with EF-G. Instead, there are several contacts between the body of EF-G and structural elements of the 30S subunit. The detailed arrangement of EF-G around the long axis of the molecule is not defined precisely. Thus, the locations of individual domains in the body of EF-G relative to the ribosome cannot be assigned unambiguously. In the post* state, the overall arrangement of EF-G is similar to, but not identical with, the positioning of EF-G on the ribosome in the presence of fusidic acid with or without bound tRNAs (Agrawal et al., 1998; Wilson and Noller, 1998). Fusidic acid is known to block the elongation cycle by preventing the dissociation of EF-G·GDP from the ribosome after translocation (Willie et al., 1975).

If we compare the position of the ternary complex and EF-G on the ribosome, we observe two different arrangements: one in which both factors are bound to the ribosome and stalled with the help of the respective antibiotic immediately after GTP hydrolysis and the other where EF-G is ready to leave the ribosome after translocation, freeing the A site to accept the next ternary complex. Both arrangements are shown in Fig. 3 as pre* and post* complexes. The density of the ribosome is displayed in a semitransparent manner, and the ternary complex is taken from the kirromycin-stalled complex (Stark et al., 1997b) and placed into the reconstructions of the pre*- and post*-translocation states.

The ternary complex in the Pre* comparison points into the decoding center, while EF-G lies on the outside of the intersubunit space. The two factors are positioned almost orthogonal to each another. In particular, their position relative to the L7-L12 stalk appears different, although there is an interaction between the G domain and the stalk in both complexes. This is in line with previous data implicating L7-L12 both in binding and in GTPase activation of the elongation factors (Traut et al., 1995). Nevertheless, the body of EF-G binds on the ribosome at a higher position than does EF-Tu. This may indicate that other distinctive regions of the 50S subunit besides the L7-L12 stalk play a role in GTPase activation. In the case of EF-Tu, it is possible that the α-sarcin stem-loop contributes additional contacts, whereas for EF-G, interactions with the L11 and 1070 loops of 23S rRNA appear more likely. The molecular mechanism of GTPase activation could therefore be quite different, as reflected in the different timing of the GTP hydrolysis. EF-G seems to undergo a large rearrangement on the ribosome during translocation, which leads to a similar overall arrangement of both factors on the ribosome. In the post* comparison, the position of EF-G is roughly parallel to the ternary complex orientation, with domain 4 of EF-G also pointing into the decoding center. The body of EF-G is in a position slightly different from that of EF-Tu. This may result from the fact that EF-Tu has not yet undergone its conformational change from the GTP- to the GDP-bound form, which leads to its dissociation from the ribosome.

The "molecular mimicry" hypothesis based on the structural similarity of the ternary complex and EF-G·GDP (Nissen et al., 1995) appears to be restricted to the posttranslocation state, when EF-G·GDP leaves the ribosome in a conformation ready to accept the next ternary complex. The overall similarity in the lengths of both molecules may be conserved to guarantee the communication between functional centers on both subunits.

STRUCTURAL BASIS OF ELONGATION

In the context of recent 3-D models of the 16S and 23S rRNAs (Fig. 4) (Mueller and Brimacombe, 1997; Mueller et al., unpublished data) and the kinetic analysis of both elongation factors (Rodnina et al., 1997; Pape et al., 1998, 1999), we suggest a new model for the elongation cycle of translation.

The small subunit is the major actor in the decoding of the mRNA, whereas the large subunit is involved in the catalysis of the chemistry steps, such as GTP hydrolysis and peptidyl transfer. The L7-L12 stalk assumes different conformations, associated

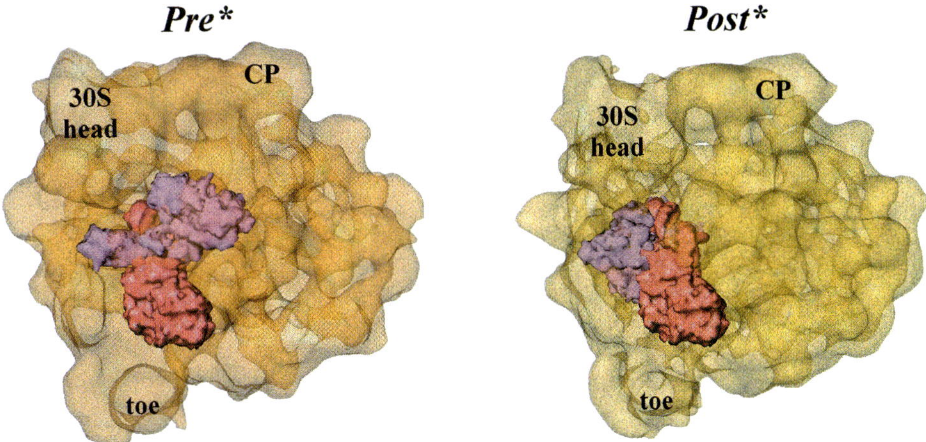

Figure 3. Comparison of the positions of EF-Tu in the ternary complex and EF-G on the ribosome. In the pre* complex the position of the ternary complex (red) is almost orthogonal to the position of EF-G (purple). In the post* complex both factors are pointing into the decoding site, showing similar overall arrangements. The views are the same as in Fig. 2.

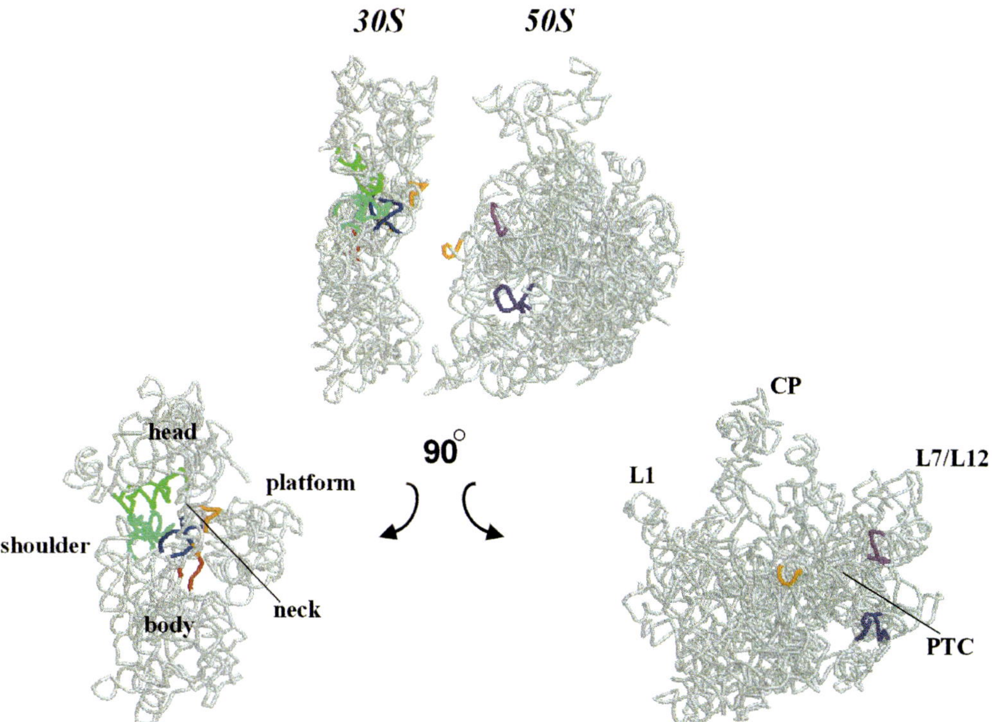

Figure 4. Models of the *E. coli* 16S and 23S rRNAs. rRNA elements possibly involved in elongation are highlighted. In the 30S subunit the decoding site (A site) is depicted in blue, the 530 stem-loop is in cyan, helix 34 is in green, and the 912 region (helix 27) is in red. In the 50S subunit the 1070 loop region (the GTPase-associated center) is shown in magenta and the 2660 stem-loop (α-sarcin stem-loop) is in purple. Shown in gold are the 790 loop of 16S rRNA and the 1920 loop of 23S rRNA, which probably form the major intersubunit bridge. The morphological details are indicated CP, central protuberance; PTC, peptidyltransferase center. (Coordinates of the models courtesy of R. Brimacombe.)

with bending in the flexible hinge region of the L7-L12 dimer, in various functional states throughout translation and might represent the initial contact of both elongation factors with the ribosome. Therefore, in the first encounter of the ternary complex with the ribosome the anticodon domain of aa-tRNA may be preoriented for codon reading. The ribosome is in a conformation in which the affinity of aa-tRNA to the A site is determined mainly by codon-anticodon interaction (hyperaccurate phenotype). This conformation may be associated with the 912-888 conformation of 16S rRNA (Lodmell and Dahlberg, 1997) and will be referred to as "binding" conformation. As a result of successful codon recognition, the ternary complex is stabilized by the formation of nonspecific interactions with the decoding region and leads to the stimulation of GTP hydrolysis in EF-Tu, suggesting a rearrangement of the whole complex into a "productive" conformation. The molecular mechanism by which the signal of codon recognition is transmitted to the G domain of EF-Tu remains unknown. A possible scenario may involve a concerted interplay of tRNA, 16S rRNA, and 23S rRNA (Fig. 4). The conformational change induced by the formation of the codon-anticodon duplex in the decoding center may affect the tRNA and several neighboring regions of the 16S rRNA, involved in decoding. The centrally located 912 region of 16S rRNA, which is positioned at the junction of its three major domains close to three pseudoknot structures (Stern et al., 1989; Allen and Noller, 1989), might switch to the alternative 912-885 conformation. This would lead to the observed stabilization of aa-tRNA in the A site (error-prone phenotype) (Lodmell and Dahlberg, 1997) and could create a change in the main intersubunit bridge between the 790 loop of 16S rRNA and the 1920 loop of 23S rRNA. The contacts of the 50S subunit with the G domain of EF-Tu would alter as a result. Another conformational change could take place in the 530 stem-loop (helix 18), which is conformationally coupled to the decoding center and is oriented towards helix 34 (1046-1067 and 1189-1211). Proteins S12, S5, and S4 bind to these regions and are known to affect the fidelity of decoding (Powers and Noller, 1994; Noller et al., 1996; Zimmermann, 1996). Domain 2 of EF-Tu in the GTPase state is located close to this region and could interact directly with it. Distortions of the tRNA resulting from the conformational change in the decoding center also seem to play a part in conducting EF-Tu into the GTPase state. This is supported by fluorescence accessibility measurements (Rodnina et al., 1994) and cryo-electron-microscopic reconstructions of the ternary complex on the ribosome (Stark et al., 1997b, unpublished data). After GTP hydrolysis, EF-Tu undergoes a dramatic structural change from the GTP to the GDP form, which leads to the release of the 3′ acceptor end of aa-tRNA. This rearrangement may bring the ribosome into the binding conformation for an instant through the contact of domain 2 with the 530 stem-loop. If the codon-anticodon interaction is not complementary, the aa-tRNA is rejected (proofreading) in this state and the ribosome stays in the binding conformation ready to accept a new ternary complex. In the case of correct codon-anticodon interaction, the ribosome returns to the productive conformation as aa-tRNA is accommodated in the A site, resulting in peptide bond formation (Pape, 1998).

EF-G binds to the ribosome in the productive conformation and hydrolyzes GTP immediately. EF-G then lies on the outside of the intersubunit space with domain 4 oriented towards the S4 binding region (530 stem-loop) on the 30S subunit, possibly affecting the decoding center. A conformational change of EF-G may then bring the ribosome back into the binding conformation during translocation. Domain 4 of EF-G probably follows the tRNA movement and prevents back-reaction of the mRNA-tRNA complex. Finally, EF-G·GDP and deacylated tRNA dissociate from the ribosome, thus completing one round of the elongation cycle.

This model implies structural changes of the ribosome mainly in the 30S subunit. Indeed, differences are seen when comparing the pre*- and post*-translocation cryo-electron-microscopic reconstructions (Fig. 5). Several characteristic structural landmarks exhibit significant differences, in particular in the neck, shoulder (connection), and head (second bill and beak). These changes are also seen in the side views of Fig. 2. The cleft between the shoulder and the head is widened in the pre*- compared to the post*-translocation state. It is possible that this is the cleft through which domain 4 of EF-G moves to enter its posttranslocation position. The structures of these regions appear to be flexible, as a recent cryo-electron-microscopic study of isolated 30S subunits has revealed (Gabashvili et al., 1999).

Throughout the translation elongation cycle the ribosome is likely to alternate between two main conformations, a binding and a productive conformation. These two states might be regulated by EF-Tu and EF-G through interaction with similar rRNA regions. The structural dynamics of the ribosome, therefore, seem to be crucial in protein synthesis. The fine molecular details of these changes cannot be assessed at the resolutions achieved. However, with recent improvements in single-particle analysis it is now

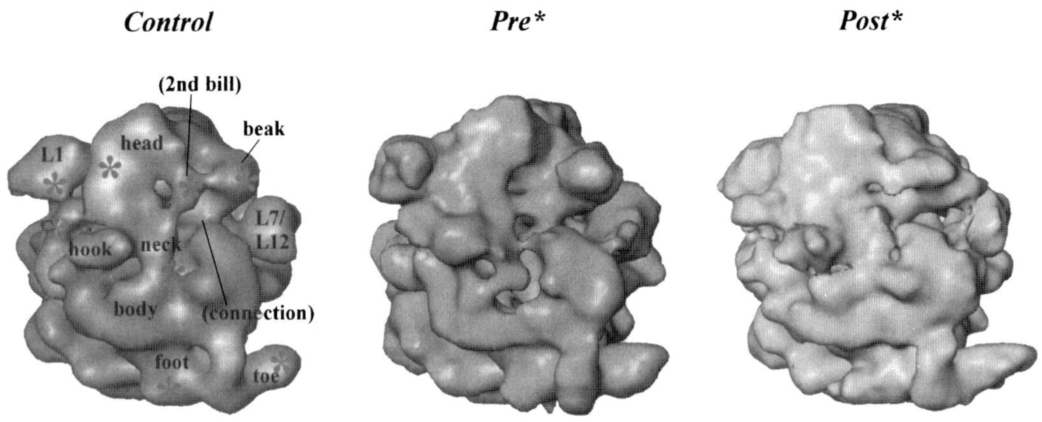

Figure 5. Structural changes of the *E. coli* 30S subunit during translocation viewed from the solvent side. Characteristic morphological features are indicated; those which are not always visible are in parentheses.

possible to obtain reconstructions below 10-Å resolution. This may lead to greater insights into the mechanism of protein synthesis.

REFERENCES

Abel, K., M. D. Yoder, R. Hilgenfeld, and F. Jurnak. 1996. An α to β conformational switch in EF-Tu. *Structure* **4:**1153–1159.

Ævarsson, A., E. Brazhnikov, M. Garber, J. Zheltonosova, Y. Chirgadze, S. al-Karadaghi, L. A. Svensson, and A. Liljas. 1994. Three-dimensional structure of the ribosomal translocase: elongation factor G from *Thermus thermophilus*. *EMBO J.* **13:**3669–3677.

Agrawal, R. K., P. Penczek, R.A. Grassucci, and J. Frank. 1998. Visualisation of elongation factor G on the *Escherichia coli* 70S ribosome: the mechanism of translocation. *Proc. Natl. Acad. Sci. USA* **95:**6134–6138.

Allen, P. N., and H. F. Noller. 1989. Mutations in ribosomal proteins S4 and S12 influence the higher order structure of 16 S ribosomal RNA. *J. Mol. Biol.* **208:**457–468.

Czworkowski, J., J. Wang, T. A. Steitz, and P. B. Moore. 1994. The crystal structure of elongation factor G complexed with GDP, at 2.7 Å resolution. *EMBO J.* **13:**3661–3668.

Döring, T., P. Mitchell, M. Osswald, D. Bochkariov, and R. Brimacombe. 1994. The decoding region of 16S RNA: a cross-linking study of the ribosomal A, P and E sites using tRNA derivatized at position 32 in the anticodon loop. *EMBO J.* **13:** 2677–2685.

Dubochet, J., M. Adrian, J. J. Chang, J. C. Homo, J. Lepault, A. W. McDowall, and P. Schultz. 1988. Cryo-electron microscopy of vitrified specimens. *Q. Rev. Biophys.* **21:**129–228.

Gabashvili, I. S., R. K. Agrawal, R. Grassucci, and J. Frank. 1999. Structure and structural variations of the Escherichia coli 30 S ribosomal subunit as revealed by three-dimensional cryo-electron microscopy. *J. Mol. Biol.* **286:**1285–1291.

Harauz, G., and M. van Heel. 1986. Exact filters for general geometry three-dimensional reconstruction. *Optik* **73:**146–156.

Lodmell, J. S., and A. E. Dahlberg. 1997. A conformational switch in *Escherichia coli* 16S ribosomal RNA during decoding of messenger RNA. *Science* **277:**1262–1267.

Moazed, D., and H. F. Noller. 1989. Intermediate states in the movement of transfer RNA in the ribosome. *Nature* **342:**142–148.

Moazed, D., J. M. Robertson, and H. F. Noller. 1988. Interaction of elongation factors EF-G and EF-Tu with a conserved loop in 23S RNA. *Nature* **334:**362–364.

Mueller, F., and R. Brimacombe. 1997. A new model for the three-dimensional folding of *Escherichia coli* 16S ribosomal RNA. I. Fitting the RNA to a 3D electron microscopic map at 20 Å. *J. Mol. Biol.* **271:**524–544.

Mueller, F., I. Sommer, P. Baranov, R. Matadeen, M. van Heel, and R. Brimacombe. 1999. The 3D arrangement of the RNA in the *E. coli* 50S ribosomal subunit. I. Fitting the 23S and 5S rRNA to a cryo-electron microscopic map of the 70S ribosome at 13 Å resolution. Submitted for publication.

Nissen, P., M. Kjeldgaard, S. Thirup, G. Polekhina, L. Reshetnikova, B. F. Clark, and J. Nyborg. 1995. Crystal structure of the ternary complex of Phe-tRNA^{Phe}, EF-Tu, and a GTP analog. *Science* **270:**1464–1472.

Noller, H. F., T. Powers, P. N. Allen, D. Moazed, and S. Stern. 1996. rRNA and translation: tRNA selection and movement in the ribosome, p. 239–258. *In* R. A. Zimmermann and A. E. Dahlberg, (ed.), *Ribosomal RNA Structure, Evolution, Processing, and Function in Protein Biosynthesis*. CRC Press, Boca Raton, Fla.

Pape, T. 1998. Induced fit in aminoacyl-tRNA selection on the ribosome. Ph.D. thesis. Witten/Herdecke University, Witten, Germany.

Pape, T., W. Wintermeyer, and M. V. Rodnina. 1998. Complete kinetic mechanism of elongation factor Tu-dependent binding of aminoacyl-tRNA to the A site of the *E. coli* ribosome. *EMBO J.* **17:**7490–7497.

Pape, T., W. Wintermeyer, and M. V. Rodnina. 1999. Induced fit in initial selection and proofreading of aminoacyl-tRNA on the ribosome. *EMBO J.* **18:**3800–3807.

Paulsen, H., J. M. Robertson, and W. Wintermeyer. 1983. Topological arrangement of two transfer RNAs on the ribosome. Fluorescence energy transfer measurements between A and P site-bound tRNA^{Phe}. *Nucleic Acids Res.* **10:**2651–2663.

Polekhina, G., S. Thirup, M. Kjeldgaard, P. Nissen, C. Lippmann, and J. Nyborg. 1996. Helix unwinding in the effector region of elongation factor EF-Tu-GDP. *Structure* **4:**1141–1151.

Porse, B. T., I. Leviev, A. S. Mankin, and R. A. Garrett. 1998. The antibiotic thiostrepton inhibits a functional transition within protein L11 at the ribosomal GTPase center. *J. Mol. Biol.* **276:** 391–404.

Powers, T., and H. F. Noller. 1994. The 530 loop of 16S rRNA: a signal to EF-Tu? *Trends Genet.* **10:**27–31.

Radermacher, M. 1988. Three-dimensional reconstruction of single particles from random and non-random tilt series. *J. Electron Microsc. Tech.* **9:**359–394.

Rodnina, M. V., R. Fricke, and W. Wintermeyer. 1994. Transient conformational states of aminoacyl-tRNA during ribosome binding catalysed by elongation factor Tu. *Biochemistry* **33:**12267–12275.

Rodnina, M. V., R. Fricke, L. Kuhn, and W. Wintermeyer. 1995. Codon-dependent conformational change of elongation factor Tu preceding GTP hydrolysis on the ribosome. *EMBO J.* **14:** 2613–2619.

Rodnina, M. V., T. Pape, R. Fricke, L. Kuhn, and W. Wintermeyer. 1996. Initial binding of the elongation factor Tu·GTP·aminoacyl-tRNA complex preceding codon recognition on the ribosome. *J. Biol. Chem.* **271:**646–652.

Rodnina, M. V., A. Savelsbergh, V. I. Katunin, and W. Wintermeyer. 1997. Hydrolysis of GTP by elongation factor G drives tRNA movement on the ribosome. *Nature* **385:**37–41.

Rodnina, M. V., A. Savelsbergh, N. B. Matassova, V. I. Katunin, Y. P. Semenkov, and W. Wintermeyer. 1999. Thiostrepton inhibits the turnover but not the GTPase of elongation factor G on the ribosome. *Proc. Natl. Acad. Sci. USA* **96:**9586–9590.

Stark, H., E. V. Orlova, J. Rinke-Appel, N. Junke, F. Mueller, M. Rodnina, W. Wintermeyer, R. Brimacombe, and M. van Heel. 1997a. Arrangement of tRNAs in pre- and posttranslocational ribosomes revealed by electron cryomicroscopy. *Cell* **88:**19–28.

Stark, H., M. V. Rodnina, J. Rinke-Appel, R. Brimacombe, W. Wintermeyer, and M. van Heel. 1997b. Visualisation of elongation factor Tu on the *Escherichia coli* ribosome. *Nature* **389:** 403–406.

Stark, H., M. V. Rodnina, F. Zemlin, W. Wintermeyer, and M. van Heel. The 13Å structure of the *E. coli* ribosome: conformational changes of elongation factor Tu upon ribosome binding. Unpublished data.

Stark, H., M. V. Rodnina, M. van Heel, and W. Wintermeyer. Large-scale movement of elongation factor G and extensive conformational changes of the ribosome during translocation as visualised by electron cryomicroscopy. Submitted for publication.

Stern, S., T. Powers, L. M. Changchien, and H. F. Noller. 1989. RNA-protein interactions in 30S ribosomal subunits: folding and function of 16S rRNA. *Science* **244:**783–790.

Tapprich, W. E., and A. E. Dahlberg. 1990. A single base mutation at position 2661 in *E. coli* 23S ribosomal RNA affects the binding of ternary complex to the ribosome. *EMBO J.* **9:**2649–2655.

Traut, R. R., D. Dey, D. E. Bochkariov, A. V. Oleinikov, G. G. Jokhadze, B. Hamman, and D. Jameson. 1995. Location and domain structure of *Escherichia coli* ribosomal protein L7/L12: site specific cysteine cross-linking and attachment of fluorescent probes. *Biochem. Cell Biol.* **73:**949–958.

van Heel, M. 1987. Angular reconstitution: a posteriori assignment of projection directions for 3D reconstruction. *Ultramicroscopy* **21:**111–124.

van Heel, M. 1989. Classification of very large electron microscopical image data sets. *Optik* **82:**114–126.

van Heel, M., and J. Frank. 1981. Use of multivariate statistics in analysing the images of biological macromolecules. *Ultramicroscopy* **6:**187–194.

van Heel, M., G. Harauz, E. Orlova, R. Schmidt, and M. Schatz. 1996. A new generation of the IMAGIC image processing system. *J. Struct. Biol.* **116:**17–24.

Willie, G. R., N. Richman, W. P. Godtfredsen, and J. W. Bodley. 1975. Some characteristics of and structural requirements for the interaction of 24,25-dihydrofusidic acid with ribosome-elongation factor G complexes. *Biochemistry* **14:**1713–1718.

Wilson, K. S., and H. F. Noller. 1998. Mapping the position of translational elongation factor EF-G in the ribosome by directed hydroxyl radical probing. *Cell* **92:**131–139.

Wimberly, B. T., R. Guymon, J. P. McCutcheon, S. W. White, and V. Ramakrishnan. 1999. A detailed view of a ribosomal active site: the structure of the L11-RNA complex. *Cell* **97:**491–502.

Zimmermann, R. A. 1996. The decoding domain, p. 277–309. *In* R. A. Zimmermann and A. E. Dahlberg (ed.), *Ribosomal RNA Structure, Evolution, Processing, and Function in Protein Biosynthesis.* CRC Press, Boca Raton, Fla.

The Ribosome: Structure, Function, Antibiotics, and Cellular Interactions
Edited by R. A. Garrett, S. R. Douthwaite, A. Liljas, A. T. Matheson, P. B. Moore, and H. F. Noller

Chapter 5

Cryo-Electron Microscopy of the Translational Apparatus: Experimental Evidence for the Paths of mRNA, tRNA, and the Polypeptide Chain

JOACHIM FRANK, PAWEL PENCZEK, ROBERT A. GRASSUCCI, AMY HEAGLE, CHRISTIAN M. T. SPAHN, and RAJENDRA K. AGRAWAL

As a process that brings together, in close proximity, two linear structures of considerable length (mRNA and the nascent polypeptide chain) and large protein factors (EF-G and aminoacyl-tRNA·EF-Tu·GTP ternary complex), protein synthesis poses a logistic problem of traffic control: how to guarantee uninterrupted, high-precision performance without steric interference and entanglement of the various ligands. It appears from the evidence obtained by cryo-electron microscopy (cryo-EM) that this is achieved in the ribosome by a layout of conduits and openings that allows for a maximum separation of the different traffic flows.

Since cryo-EM visualization provided the first detailed three-dimensional (3-D) images of the ribosome (Fig. 1, top) (Frank et al., 1995a, 1995b; Stark et al., 1995), much work has gone into the mapping of tRNA (Agrawal et al., 1996, 1998b, 1999a; Stark et al., 1997a) and elongation factors (Stark et al., 1997b; Agrawal et al., 1998a, 1999b) bound to the ribosome at various stages of the elongation cycle. From these studies, an approximate pathway of tRNA, from its delivery by EF-Tu to its removal from the ribosome, can be constructed, although we are still far from the point where we can understand the atomic details in the interactions of these ligands with the ribosome. In contrast to this, there has been a great deal of uncertainty about the trajectories of mRNA and the nascent polypeptide chain, which could not be directly visualized, mainly because of resolution limitations. Consequently, their paths through the ribosome could only be conjectured from circumstantial evidence.

A few words are in order about the general feasibility of ligand mapping by cryo-EM. Mapping of any ligand bound to the ribosome requires a control cryomap of the unliganded ribosome prepared under identical buffer conditions. This is a crucial prerequisite, since the buffer conditions have been proven to affect ribosomal conformation as well as the binding affinities and positions of tRNAs bound to the ribosome (Agrawal et al., 1999a; Frank et al., in press [a]). Quite often, the two maps to be compared, those of the ligand-ribosome complex and the control, have different resolutions, and in that case, to avoid spurious results, it is necessary to filter the map with the better resolution to the resolution of the inferior map. Another complication arises from limited occupancy of a particular binding state. By means of image processing, it is difficult to sort out raw images (which have very low signal-to-noise ratios) according to ligand occupancy. Hence, if the fraction of the ligand-ribosome complex cannot be enriched by biochemical means, the ligand will leave a very weak "diluted" trace in the difference map, which tends to be buried in noise. For these reasons, the ligand localizations have varying degrees of accuracy.

Largely based on refinements of existing cryo-EM reconstructions, using increased data sets, we have now been able to obtain concrete evidence for the trajectories through the ribosome, in all cases confirming the earlier conjectures but also yielding potential additional information about conformation.

Joachim Frank, Robert A. Grassucci, Amy Heagle, and Christian M. T. Spahn ■ Health Research, Inc., at the Wadsworth Center and Howard Hughes Medical Institute, Empire State Plaza, Albany, NY 12201-0509. **Pawel Penczek and Rajendra K. Agrawal** ■ Wadsworth Center, Empire State Plaza, Albany, NY 12201-0509.

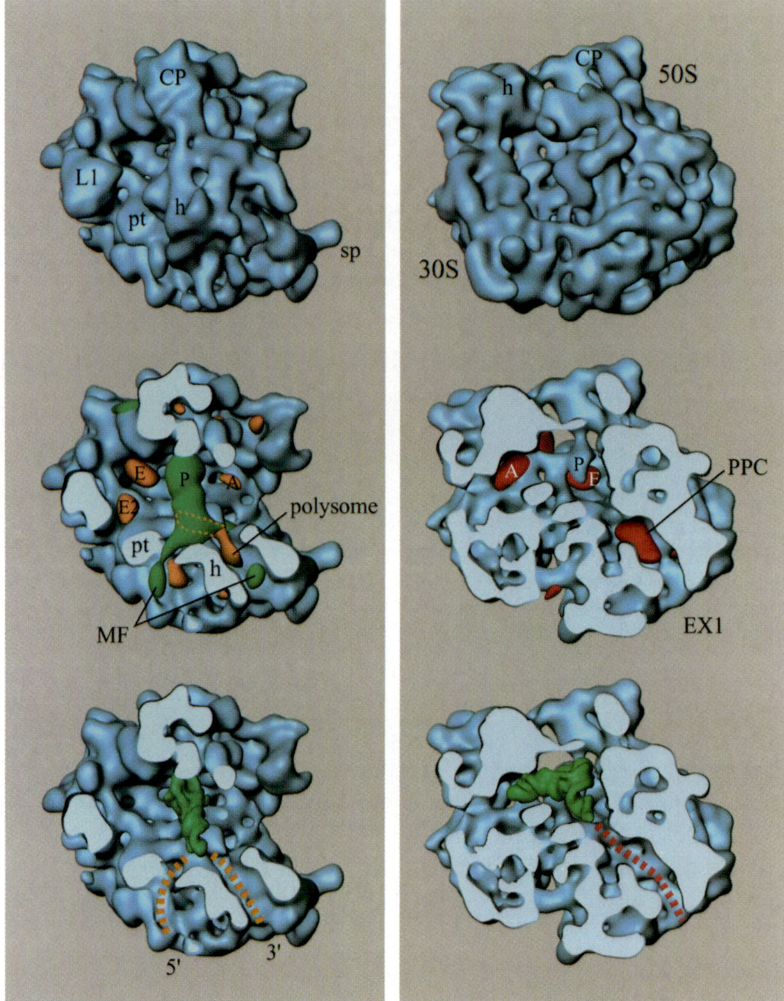

Figure 1. (Top left) Uncut 15-Å map of the fMet-tRNA$_f^{Met}$·ribosome complex (Malhotra et al., 1998) in the view selected for the middle left and bottom left panels. Landmarks, 50S subunit: CP, central protuberance; L1, L1 protein 30S subunit; pt, platform; sp, spur. (Middle left) Experimental evidence for the path of mRNA, obtained from two difference maps, superimposed on the 15-Å map of the ribosome that was cut open to reveal the intersubunit space and the relevant plane of mRNA movement. The color code is as follows: (i) Green indicates the result of subtraction of the mRNA-free control from the fMet-tRNA·ribosome complex map, displayed with reduced threshold. As a result, the difference mass corresponding to the P-site tRNA is enlarged compared to its appearance in the study of Malhotra et al. (1998), but it also extends along a curved path (MF) toward the left, into the space between the platform and the head. Another globular mass probably also related to mRNA occurs in the channel. Its position is slightly above the cutting plane, toward the viewer. The other peaks of this difference map are probably related to conformational changes between the naked ribosome and the tRNA-bound ribosome. (ii) Brown indicates the result of subtraction of the vacant-ribosome map from the map of the polysome. The difference mass follows a curved arc (polysome) through the 30S channel, partially overlapping (dashed outline) and complementing the path of the green MF-mRNA-related difference mass. The other difference peaks are related to low-occupancy A-site tRNA (A), E-site tRNA (E), and E2-site tRNA (E2) and to conformational changes. (Bottom left) Path of mRNA (dashed line) inferred from the experimental difference maps, along with a model of P-site tRNA that was inserted in the position found in the study of Malhotra et al. (1998). (Top right) Uncut 15-Å map of the fMet-tRNA$_f^{Met}$·ribosome complex (Malhotra et al., 1998) in the view selected for the middle right and bottom right panels. h, head of the 30S subunit. (Middle right) Difference mass observed in the tunnel of the ribosome when the map of the polysome is compared with that of the empty ribosome. The difference mass has been overlaid on the 15-Å map cut open along the plane of the tunnel. The mass attributed to partially folded polypeptide chain (PPC) occurs in the widened midsection of the tunnel. The other difference peaks relate to the presence of tRNA at the A site (A), P-site tRNA (P), E-site tRNA (E), and the exit site of the tunnel (EX1). (Note that the side branch of the tunnel, ending in the second tunnel exit side EX2, is not visible in this orientation; cf. Frank et al., 1995.) (Bottom right) Same view of 15-Å map, with X-ray structure of P-site tRNA in fitted position and probable path of polypeptide chain indicated (dashed line).

THE PATH OF mRNA

Based on the topography of the 3-D cryomap and previous experimental findings, an approximately semicircular path was postulated for the mRNA (Frank et al., 1995a; 1995b) leading from the 3′ end into the 30S subunit channel, curving around the neck in the cleft region (see also Lata et al., 1996), and exiting the ribosome through the space between the small-subunit head and platform. Since this proposal was put forth, various lines of indirect support have surfaced, and apparent conflicts have been resolved: localization of both A-site tRNA (Agrawal et al., 1996, 1999a; Stark et al., 1997a) and domain IV of EF-G (Agrawal et al., 1998b) point to a location of the decoding site in the intersubunit region in the immediate vicinity of the channel entrance. Furthermore, it has been found that one of the walls of the channel is *not* yet formed in the unbound 30S subunit, allowing free access for mRNA as it binds to the 30S subunit during initiation (Lata et al., 1996), a result which resolves the conflict posed by the possibility of translation of circular messages (Bretscher et al., 1968). The mRNA entry through the channel has also been incorporated in Brimacombe's recent 16S rRNA model (Müller et al., 1997). Interestingly, site-directed mutagenesis of 16S RNA, combined with chemical probing (Lodmell and Dahlberg, 1997), provides evidence that this region of the 30S subunit may also undergo dynamic changes in the course of the elongation cycle, which has been confirmed by most recent cryo-EM visualizations (Gabashvili et al., 1999).

As for direct visualization, single-stranded RNA is apparently at the threshold of visualization, as indicated by the absence of the single-stranded section of tRNA in the fMet-tRNA$_f^{Met}$·ribosome complex in the density map even at 15-Å resolution (Malhotra et al., 1998). Whether single-stranded mRNA will show up in the map—assuming the basic prerequisite of high occupancy to be met—will depend on the degree of stabilization in a particular conformation. In this regard, a polysome preparation which is close to in vivo conditions probably provides more favorable conditions, through the presence of stabilizing factors.

Two recent 3-D cryomaps of mRNA-programmed ribosomes, compared with maps of the empty ribosome obtained by using corresponding buffer conditions, show evidence of a linear mass distribution along the postulated semicircular pathway of mRNA. For reasons that are unclear but that may have to do with the different stabilizing conditions (see below), neither of the two cryomaps shows the path in its entirety, but when taken together, the two linear mass distributions indeed complement each other.

The first map is the aforementioned reconstruction of the fMet-tRNA$_f^{Met}$·ribosome complex (Malhotra et al., 1998), which was programmed with a 46-nucleotide mRNA fragment (MF-mRNA, containing the codons for methionine and phenylalanine in its middle; see Malhotra et al., 1998) and prepared in a buffer containing polyamines. The comparison with the then-available naked ribosome made in that study was limited by the resolution (25.4 Å) of the naked-ribosome map. Still, there was an indication of a linear mass projecting sideways, off the tRNA anticodon tip, in the direction of the 30S subunit channel. A recent study of a 70S ribosome carrying a genetically inserted tRNA-like RNA fragment (Spahn et al., in press) furnished a higher-resolution (17-Å) map of the vacant ribosome, and the use of this new map in the subtraction produced a linear mass distribution covering the platform side segment of the mRNA path, as well as a mass hovering just at the entrance of the 30S subunit channel (Fig. 1, middle left). The reason why the most prominent visualization occurs on the side of the platform may be related to the composition of MF-mRNA. MF-mRNA has codons AUGUUC in the middle. The flanking regions are repeats of the sequence AAAAG, which have similarity with the purine-rich Shine-Dalgarno sequence and are thus likely to interact with the anti-Shine-Dalgarno region of 16S rRNA located on the platform (Oakes et al., 1990).

The second map is a 22-Å reconstruction of the polysome (Agrawal et al., unpublished [b]), which was isolated from an *Escherichia coli* strain overexpressing β-galactosidase (see below). Consequently, the mass distribution in the 3-D cryomap along the mRNA path should reflect the average of the corresponding mRNA scattering density. Subtraction of the vacant-ribosome map (Spahn et al., in press) in this case reveals a linear mass (Fig. 1, middle left) extending from the channel entrance up to the platform side, where the anticodon of the E site tRNA has been located (Agrawal et al., 1998b, unpublished [b]).

By putting the two segments together and tracing along the central line of the masses, we were able to construct the complete path (Fig. 1, bottom left). Because of the need for robust translocation, there are unlikely to be any sequence-specific differences in the way mRNA is guided through the small subunit, and the path constructed from the two pieces may be characteristic at least for prokaryotes. The length of the whole path, with reference to an entry point and exit point defined in the cut map (Fig. 1, bottom left), is 130 Å, which would account for 38 nucleotides in a natural A-form configuration of mRNA. This is in remarkable agreement with the results obtained by radiolabeling, 39 ± 3 nucleotides (Beyer et al., 1994). It should also be mentioned that the results from recent photoaffinity crosslinking studies of IF3 binding to the 30S subunit are consistent with the semicircular path of mRNA (P. Wollenzien, personal communication).

THE PATH OF THE POLYPEPTIDE CHAIN

For the path of the polypeptide chain, two points on the large ribosomal subunit are fixed by experiment: the starting point, on the CCA end of the P-site tRNA, and the point of exit from the ribosome (Fig. 1, bottom right). The former is fixed by the cryo-EM localization of the P-site tRNA (Malhotra et al., 1998), and the latter is fixed by immuno-EM localization of the nascent polypeptide chain (Bernabeu and Lake, 1982). The data from the immuno-EM experiment are somewhat inaccurate, since they are based on direct interpretation of micrographs, not on a 3-D reconstruction. Nevertheless, a tentative 3-D interpretation of these data can be achieved by orienting the 3-D density map of the ribosome so that its projection matches the profiles of Barnabeu and Lake's immuno-EM ribosome images (not shown). Another related point of reference is given by recent work of Beckmann et al. (1997), where the site of attachment of the Sec61 channel on the ribosome from *Saccharomyces cerevisiae* was located in 3-D. The putative exit site inferred from this experiment agrees with that from the immuno-EM experiment.

The two points of reference, relating to the start and exit sites of the nascent chain, are linked by a prominent topological feature of the ribosome: the so-called tunnel (Yonath et al., 1987; for a summary of earlier work, see Eisenstein et al., 1994). The tunnel is a feature found in all cryo-EM reconstructions (Frank et al., 1995a; Malhotra et al., 1998; Dube et al., 1998). It has also been identified in low-resolution X-ray maps of the 50S subunit from *Haloarcula marismortui* (Ban et al., 1998; Yonath and Franceschi, 1998).

The locations of tunnel entry and exit points closely match those of the experimental reference points described above, suggesting that the tunnel indeed represents the conduit connecting the two points. However, proof of this conjecture requires direct experimental evidence, which has thus far been missing. In the absence of such proof, another hypothesis, advanced by Spirin's group (Ryabova et al., 1988) on the basis of immuno-EM studies, could not be ruled out; according to their hypothesis, the polypeptide chain follows a groove along the back of the 50S subunit. A more recent proposal suggests that such a groove may be covered with nascent chain-associated complex, forming a protective environment (Wang et al., 1995).

We have now found positive evidence in favor of the tunnel hypothesis by analyzing a reconstruction of the ribosome from cryoimages of polysomes (Agrawal et al., unpublished [b]), although the possibility that in some instances, for some polypeptides or for particular modes of the translation process, alternative paths may be used still cannot be ruled out.

In the following discussion, reference will be made to two density maps that show the tunnel as a continuous linear trough of density (Frank et al., 1995a; Malhotra et al., 1998). Both maps represent the ribosome in a state where no polypeptide chain is present. In the first density map, the tunnel is discontinuous—only by an increase in the display threshold can an upper segment be induced to open up. This behavior is typical of features in a size range comparable to, or smaller than, the resolution.

Polysomes were isolated from *E. coli* HB101 containing plasmid pGH101, which overexpresses β-galactosidase. As the active polysome fractions containing high β-galactosidase activity were used for the 3-D cryo-EM visualization, the chances of locating nascent polypeptide chains were near 100%. Each projection depicts the ribosome engaged with a different segment of the polypeptide chain, so the resulting 3-D map must be interpreted as an ensemble average over the whole duration of the translation. Since the width of the unfolded polypeptide chain is not greater than ~5 Å, it is not expected to leave a trace in the 3-D map at the current resolution, even if it occurred in a consistent position throughout the ribosome population. Rather, visualization would require the presence of substantial secondary-structure elements at the same site for a large fraction of the ribosome population.

The difference map, obtained by subtracting the map of the fMet-$tRNA_f^{Met}$·ribosome complex (i.e., the 17-Å map obtained by Agrawal and coworkers [1999a], filtered to 22-Å resolution) from the 22-Å polysome map, shows a mass of density in a 40-Å-long segment of the tunnel (Fig. 1, middle right). Significantly, the observed obstruction of the tunnel occurs *not* at the site of narrowest constriction—where it could be explained as an effect of limited resolution—but at an adjacent site where the tunnel is relatively wide. We interpret the mass as the site where structural elements of secondary and even higher order might be most consistently formed, as the polypeptide chain passes through the ribosome.

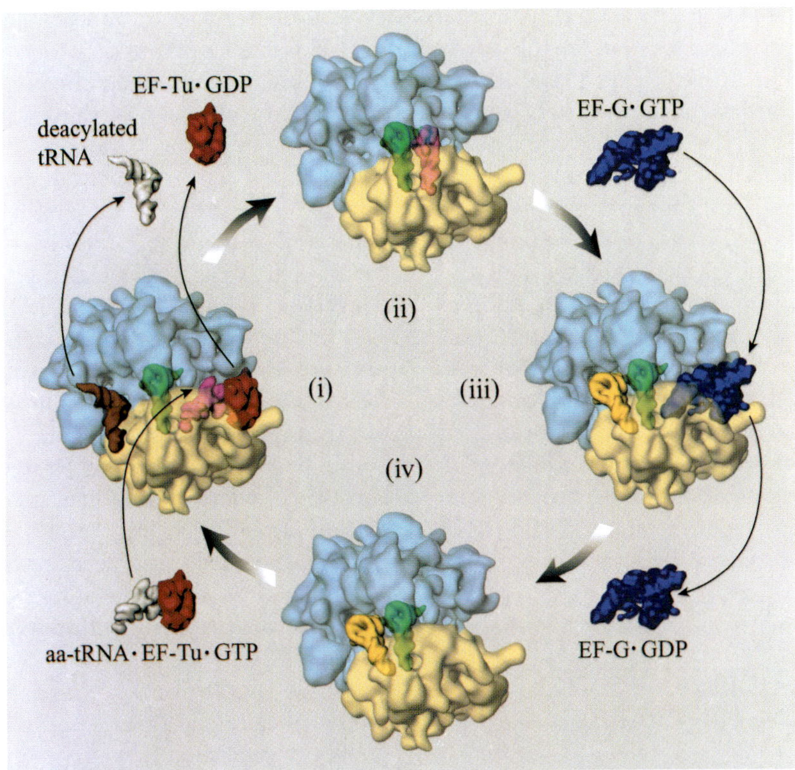

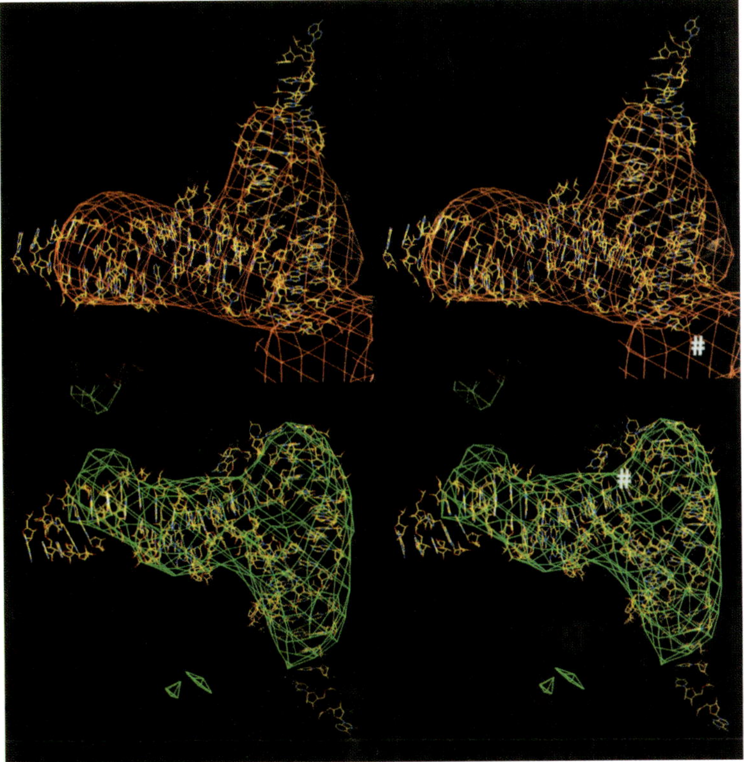

Figure 2. (Top) Schematic diagram of the movements of tRNA and elongation factors during the elongation cycle of protein synthesis. The ribosome map and the 3-D positions of the ligand molecules are based on several cryo-EM visualizations (for an explanation of panels (i) through (iv), see the text). (Bottom) Examples of the fitting of the X-ray structure of tRNA to experimental difference masses: E2 site (above) and P site (below). Density belonging to the neighboring tRNA is marked by "#" in both maps.

The reason why the mass does not extend farther toward the tunnel exit could be that in this lower segment, chaperone-like activity takes place, as suggested by studies of Hardesty and coworkers (see the review in Hardesty et al., 1999), which could result in a less consistent pattern of chain localization.

The occurrence of significant protein folding directly on the ribosome (co-translational folding) has been proposed by Hardesty et al. (1993), and this view has received support from the interpretation of cross-linking studies by Brimacombe's group (Stade et al., 1995). Lim and Spirin (1986) argued on theoretical grounds that the polypeptide chain might be immediately formed in alpha-helical configuration at the peptidyltransferase center. The approximate path that is supported by the finding is indicated in Fig. 1, bottom right. The total length, from the P-site tRNA CCA end to the point marked EX1, is 105 Å.

THE PATH OF tRNA AND THE ELONGATION CYCLE

tRNA bound to the ribosome has been directly visualized by 3-D cryo-EM in various tRNA-ribosome complexes (Agrawal et al., 1996, 1998b, 1999a; Stark et al., 1997a). Visualization of the ribosome-bound tRNA is still a challenging task because the smallest dimension of the molecule is on the order of the resolution of cryo-EM and the occupancy of some tRNA binding states is intrinsically low. Because of these problems, the tRNA is often not fully visible—a truncation occurs in the 3′ end of the acceptor arm, which is single-stranded and probably subject to flexing movements—and the masses in the density map stemming from adjacent tRNAs may be partially fused (Agrawal et al., 1999a).

Largely by putting these direct tRNA visualization results into the context of existing knowledge, the pathway of tRNA through the ribosome could be inferred. Parallel to these studies, efforts have been made to map the binding positions of EF-G (Agrawal et al., 1998a, 1999b) (see chapter 6) and the ternary complex aminoacyl-tRNA·EF-Tu·GTP (Stark et al., 1997b; Agrawal et al., unpublished [a]) on the ribosome.

On the basis of these ligand binding studies, it is now possible to sketch out the complete elongation cycle (Fig. 2, top) in terms of the 3-D binding positions of tRNA and the elongation factors. In each case, the tRNA positions were obtained by fitting the X-ray structure of tRNA into difference masses from several experiments (see the examples in Fig. 2, bottom). In Fig. 2 (top), four states of the elongation cycle are depicted in panels numbered (i) through (iv). These are explained as follows.

(i) Binding of ternary complex and proofreading. The ternary complex is shown in the position obtained in unpublished studies (Agrawal et al., unpublished [a]). From a visual comparison of published maps, our position is identical to that obtained by Stark and coworkers (Stark et al., 1997b). In both studies, kirromycin was used for stabilization of the irreversible binding of the ligand.

In this state, the first deacylated tRNA is ready to leave the ribosome. The position in which the tRNA is depicted, in direct contact with protein L1, is a transitory position ("E2 site") inferred from several cryo-EM visualization experiments (Agrawal et al., 1996 [where it was identified as the "E site"], 1998b, unpublished [b]).

(ii) Pretranslocational state (after incorporation of cognate tRNA into the A site). The A site position of tRNA was based on three experiments: (i) ribosome $(tRNA^{Phe})_3$ complex (Agrawal et al., 1996); (ii) polysome (Agrawal et al., unpublished [b]); and (iii) pretranslocational state (Agrawal et al., 1998b, 1999b, unpublished [b]).

(iii) Binding of EF-G, inducing the posttranslocational state. EF-G is shown in the position obtained by Agrawal and coworkers (1998a), where it was stabilized in the posttranslocational GDP state with the help of fusidic acid. The tRNA formerly at the A site has moved to the P site, while the tRNA formerly in the P site has moved to the E site.

(iv) Posttranslocational state. The P site tRNA is shown in the position found by Malhotra and coworkers (1998), who fitted the X-ray structure of initiator tRNA to the fMet-$tRNA_f^{Met}$·ribosome complex map. The high accuracy of this fitting makes this position the most reliable. Visual comparison with the map published by Stark and coworkers (1997a) suggests that it is very similar to the P-site position found by these authors. The E site position, known from several independent experiments (Agrawal et al., 1998b, 1999a, 1999 [b], unpublished [b]), would allow contact with mRNA.

By interpolating between the different ligand positions shown in the scheme depicted in Fig. 2 (top), we have also composed a brief animation of the molecule movements during the elongation cycle (Frank et al., in press [b]).

This work was supported by NIH grants 1 R01 GM55440 and 1 R37 GM29169.

REFERENCES

Agrawal, R. K., P. Penczek, R. A. Grassucci, Y. Li, A. Leith, K. H. Nierhaus, and J. Frank. 1996. Direct visualization of A-,

P-, and E-site transfer RNAs in the *Escherichia coli* ribosome. *Science* 271:1000–1002.

Agrawal, R. K., P. Penczek, R. A. Grassucci, and J. Frank. 1998a. Visualization of elongation factor G on the *Escherichia coli* 70S ribosome: the mechanism of translocation. *Proc. Natl. Acad. Sci. USA* 95:6134–6138.

Agrawal, R. K., P. Penczek, A. Malhotra, R. A. Grassucci, I. S. Gabashvili, A. B. Heagle, S. Srivastava, N. Burkhardt, R. Jünemann, K. H. Nierhaus, and J. Frank. 1998b. Binding positions of tRNAs in translating *Escherichia coli* ribosomes, p. 717–718. *In* H. A. Calderón Benavides and M. J. Yacamán (ed.), *Proceedings of the 14th International Congress on Electron Microscopy*. Institute of Physics Publishing, Bristol, United Kingdom.

Agrawal, R. K., P. Penczek, R. A. Grassucci, N. Burkhardt, K. H. Nierhaus, and J. Frank. 1999a. Effect of buffer conditions on the position of tRNAs on the 70S ribosome as visualized by cryoelectron microscopy. *J. Biol. Chem.* 274:8723–8729.

Agrawal, R. K., A. B. Heagle, P. Penczek, R. A. Grassucci, and J. Frank. 1999b. EF-G-dependent GTP hydrolysis induces translocation accompanied by large conformational changes in the 70S ribosome. *Nat. Struct. Biol.* 6:643–647.

Agrawal, R. K., N. Burkhardt, R. Grassucci, and K. Nierhaus. Unpublished data [a].

Agrawal, R. K., C. M. T. Spahn, P. Penczek, S. Srivastava, R. A. Grassucci, K. H. Nierhaus, and J. Frank. Unpublished data [b].

Ban, N., B. Freeborn, P. Nissen, P. Penczek, R. A. Grassucci, R. Sweet, J. Frank, P. B. Moore, and T. A. Steitz. 1998. A 9 Å resolution X-ray crystallographic map of the large ribosomal subunit. *Cell* 93:1105–1115.

Beckmann, R., D. Bubeck, R. Grassucci, P. Penczek, A. Verschoor, G. Blobel, and J. Frank. 1997. Alignment of conduits for the nascent polypeptide chain in the ribosome-Sec61 complex. *Science* 278:2123–2126.

Bernabeu, C., and J. A. Lake. 1982. Nascent polypeptide chains emerge from the exit domain of the large ribosomal subunit: immune mapping of the nascent chain. *Proc. Natl. Acad. Sci. USA* 79:3111–3115.

Beyer, A., M. Sikes, and Y. Osheim. 1994. EM methods for visualization of genetic activity from disrupted nuclei. *Methods Cell Biol.* 44:613–630.

Bretscher, M. S. 1968. Direct translation of a circular messenger DNA. *Nature* 220:1088–1091.

Dube, P., M. Wieske, H. Stark, M. Schatz, J. Stahl, F. Zemlin, G. Lutsch, and M. van Heel. 1998. The 80S rat liver ribosome at 25 A resolution by electron cryomicroscopy and angular reconstitution. *Structure* 6:389–399.

Eisenstein, M., B. Hardesty, O. W. Odom, W. Kudlicki, G. Kramer, T. Arad, F. Franceschi, and A. Yonath. 1994. Modeling and experimental study of the progression of nascent proteins in ribosomes, p. 213–246. *In* G. Pifat (ed.), *Biophysical Methods in Molecular Biology*. Balaban Press, Rehovot, Israel.

Frank, J., J. Zhu, P. Penczek, Y. Li, S. Srivastava, A. Verschoor, M. Radermacher, R. Grassucci, R. K. Lata, and R. K. Agrawal. 1995a. A model of protein synthesis based on cryo-electron microscopy of the *E. coli* ribosome. *Nature* 376:441–444.

Frank, J., A. Verschoor, Y. Li, J. Zhu, R. K. Lata, M. Radermacher, P. Penczek, R. Grassucci, R. K. Agrawal, and S. Srivastava. 1995b. A model of the translational apparatus based on a three-dimensional reconstruction of the *Escherichia coli* ribosome. *Biochem. Cell Biol.* 73:757–765.

Frank, J., A. B. Heagle, and R. K. Agrawal. Animation of the dynamic events of the elongation cycle based on cryoelectron microscopy of functional complexes of the ribosome. *J. Struct. Biol.*, in press [b].

Frank, J., P. Penczek, R. K. Agrawal, R. A. Grassucci, and A. B. Heagle. Three-dimensional cryo-electron microscopy of ribosomes. *Methods. Enzymol.*, in press [a].

Gabashvili, I., K. Agrawal, R. Grassucci, C. Squires, A. Dahlberg, and J. Frank. Unpublished data.

Gabashvili, I. S., R. K. Agrawal, R. Grassucci, C. L. Squires, A. E. Dahlberg, and J. Frank. 1999. Major rearrangements in the 70S ribosomal 3D structure caused by a conformational switch in 16S ribosomal RNA. *EMBO J.* 18:6501–6507.

Hardesty, B., O. Odom, W. Kudlicki, and G. Kramer 1993. Extension and folding of nascent peptides on ribosomes, p. 347–358. *In* K. H. Nierhaus, F. Franceschi, A. R. Subramanian, V. A. Erdmann, and B. Wittmann-Liebold (ed.), *The Translational Apparatus*. Plenum Press, New York, N.Y.

Hardesty, B., T. Tsalkova, and G. Kramer. 1999. Co-translational folding. *Curr. Opin. Struct. Biol.* 9:111–114.

Lata, K. R., R. K. Agrawal, P. Penczek, R. Grassucci, J. Zhu, and J. Frank. 1996. Three-dimensional reconstruction of the *Escherichia coli* 30 S ribosomal subunit in ice. *J. Mol. Biol.* 262:43–52.

Lim, V. I., and A. S. Spirin. 1986. Stereochemical analysis of ribosomal transpeptidation. Conformation of nascent peptide. *J. Mol. Biol.* 188:565–574.

Lodmell, J. S., and A. E . Dahlberg. 1997. A conformational switch in *Escherichia coli* 16S ribosomal RNA during decoding of messenger RNA. *Science* 277:1262–1267.

Malhotra, A., P. Penczek, R. K. Agrawal, I. S. Gabashvili, R. A. Grassucci, R. Junemann, N. Burkhardt, K. H. Nierhaus, and J. Frank. 1998. *Escherichia coli* 70 S ribosome at 15 Å resolution by cryo-electron microscopy: localization of fMet-$tRNA_f^{Met}$ and fitting of L1 protein. *J. Mol. Biol.* 280:103–116.

Müller, F., H. Stark, M. van Heel, J. Rinke-Appel, and R. Brimacombe. 1997. A new model for the three-dimensional folding of Escherichia coli 16 S ribosomal RNA. III. The topography of the functional centre. *J. Mol. Biol.* 271:566–587.

Oakes, M. I., L. Kahan, and J. A. Lake. 1990. DNA-hybridization electron microscopy tertiary structure of 16 S rRNA. *J. Mol. Biol.* 211:907–918.

Ryabova, L. A., O. M. Selivanova, V. I. Baranov, V. D. Vasiliev, and A. S. Spirin. 1988. Does the channel for nascent peptide exist inside the ribosome? *FEBS Lett.* 226:255–260.

Spahn, C. M. T., R. A. Grassucci, P. Penczek, and J. Frank. Direct three-dimensional localization and positive identification of RNA helices within the ribosome by means of genetic tagging and cryo-electron microscopy. *Structure*, in press.

Stade, K., N. Junke, and R. Brimacombe. 1995. Mapping the path of the nascent peptide chain through the 23S RNA in the 50S ribosomal subunit. *Nucleic Acids Res.* 23:2371–2380.

Stark, H., F. Müeller, E. V. Orlova, M. Schatz, P. Dube, T. Erdemir, F. Zemlin, R. Brimacombe, and M. van Heel. 1995. The 70S *Escherichia coli* ribosome at 23 Å resolution: fitting the ribosomal RNA. *Structure* 3:815–821.

Stark, H., E. V. Orlova, J. Rinke-Appel, N. Junke, F. Müeller, M. Rodnina, W. Wintermeyer, R. Brimacombe, and M. van Heel. 1997a. Arrangement of tRNAs in pre- and posttranslocational ribosomes revealed by electron cryomicroscopy. *Cell* 88:19–28.

Stark, H., M. V. Rodnina, J. Rinke-Appel, R. Brimacombe, W. Wintermeyer, and M. van Heel. 1997b. Visualization of elongation factor Tu on the *Escherichia coli* ribosome. *Nature* 389:403–406.

Wang, S., H. Saki, and M. Wiedmann. 1995. NAC covers ribosome-associated nascent chains thereby forming a protective environment for regions of nascent chains just emerging from the peptidyl transferase center. *J. Cell Biol.* 130:519–528.

Yonath, A., and F. Franceschi. 1998. Functional universality and evolutionary diversity: insights from the structure of the ribosome. *Structure* 6:679–684.

Yonath, A., K. R. Leonard, and H. G. Wittmann. 1987. A tunnel in the large ribosomal subunit revealed by three-dimensional image reconstruction. *Science* 236:813–816.

The Ribosome: Structure, Function, Antibiotics, and Cellular Interactions
Edited by R. A. Garrett, S. R. Douthwaite, A. Liljas, A. T. Matheson, P. B. Moore, and H. F. Noller

Chapter 6

Studies of Elongation Factor G-Dependent tRNA Translocation by Three-Dimensional Cryo-Electron Microscopy

RAJENDRA K. AGRAWAL, AMY B. HEAGLE, and JOACHIM FRANK

The elongation cycle of protein biosynthesis on the ribosome entails three successive steps: (i) binding of aminoacyl-tRNA to the A site, (ii) peptide bond formation, and (iii) translocation of the A- and P-site tRNAs to the P and E sites, respectively, with the concomitant advance of the mRNA by one codon. Although all three steps can be carried out by the ribosome itself (Pestka, 1969; Gavrilova et al., 1976), two elongation factors (EFs), EF-Tu and EF-G, are employed to facilitate and greatly accelerate steps 1 and 3. EF-Tu is known to deliver aminoacyl-tRNA in the form of the ternary complex (aminoacyl-tRNA·EF-Tu·GTP) to the A site of the elongating ribosome, whereas EF-G promotes the translocation step. Both EF-related events are accompanied by GTP hydrolysis, and the intrinsic GTPase activity of EF-Tu·GTP and EF-G·GTP is stimulated both by the empty ribosome (Nishizuka and Lipmann, 1966; Kawakita et al., 1974) and by the tRNA-bound ribosome (Chinali and Parmeggiani, 1982; Voigt and Nagel, 1993). Whether the energy liberated by GTP hydrolysis is used directly for tRNA binding and translocation or rather for the release of the factors has been a matter of debate for a long time (Spirin, 1985; Nierhaus, 1993); however, a recent kinetic study (Rodnina et al., 1997) has shown that it is the EF-G-dependent GTP hydrolysis that drives the translocation step.

Both EFs are members of the GTPase superfamily (Bourne et al., 1991), and they have partial sequence homology. In overall shape, the crystal structures of EF-G (Ævarsson et al., 1994), EF-G·GDP (Czworkowski et al., 1994; Al-karadaghi et al., 1996), and the ternary complex, EF-Tu·Phe-tRNA·GMPPNP (Nissen et al., 1995), are strikingly similar. In particular, the first two domains of EF-G and EF-Tu are homologous in sequence and similar in tertiary structure, while the other three domains of EF-G (III, IV, and V) appear to mimic the tRNA (the CCA arm, the anticodon arm, and the elbow, respectively) in the ternary-complex structure (Nissen et al., 1995). As the X-ray structures and the surface charge patterns of EF-G and the ternary complex have strong similarity (Nissen et al., 1995), and both EFs have overlapping binding sites on the 50S ribosome (Moazed et al., 1988), the idea of "molecular mimicry" has gathered momentum in the last few years (e.g., Nissen et al., 1995; Nyborg et al., 1996; Abel and Jurnak, 1996).

Recently, using cryo-electron microscopy (cryo-EM) and single-particle three-dimensional (3-D) reconstruction techniques, the binding positions of EF-Tu ternary complex in a kirromycin-stalled ribosome (Stark et al., 1997a; Agrawal et al., unpublished) and EF-G on a naked ribosome (Agrawal et al., 1998a), as well as in pre- and posttranslocational complexes (Agrawal et al., 1999a), have been directly visualized. The results of these studies clearly showed that (i) the G domains of both factors interact with the same region on the 50S subunit, supporting the earlier observations that activation of GTP hydrolysis depends on binding of the factors to a common site on the 50S subunit (Gordon, 1969; Bodley and Lin, 1970) and that both EFs protect some common nucleotides in the α-sarcin loop of the 23S RNA (Moazed et al., 1988); (ii) domains II of both EFs interact with the same region on the 30S subunit; and (iii) the anticodon end of the tRNA of the EF-Tu ternary

Rajendra K. Agrawal ■ Health Research, Inc., at Wadsworth Center, and Department of Biomedical Sciences, State University of New York at Albany, Empire State Plaza, Albany, NY 12201-0509. **Amy B. Heagle** ■ Health Research Inc., at Wadsworth Center, and Howard Hughes Medical Institute, Empire State Plaza, Albany, NY 12201-0509. **Joachim Frank** ■ Health Research, Inc., at Wadsworth Center, Howard Hughes Medical Institute, and Department of Biomedical Sciences, State University of New York at Albany, Empire State Plaza, Albany, NY 12201-0509.

complex and the tip of domain IV of EF-G interact with a common site on the 30S subunit in the A-site region. The binding position of EF-G on the 70S ribosome has also been derived by hydroxyl radical probing (Wilson and Noller, 1998) and found to be in striking agreement with the binding position obtained from the cryo-EM study (Agrawal et al., 1998a, 1999a). The cryo-EM study, however, revealed conformational changes of EF-G upon binding to the ribosome. Information of this type could not be gathered from hydroxyl radical probing.

In the first cryo-EM study (Agrawal et al., 1998a), EF-G was bound to the naked 70S ribosome, whereas in the subsequent study (Agrawal et al., 1999a), complexes of EF-G with the pretranslocational ribosome complex were obtained by using GTP and a nonhydrolyzable GTP analogue, and 3-D cryo-EM maps were determined. Along with binding positions of EF-G in GDP and GTP states, the later study also yielded the positions of A-, P-, and E-site tRNAs in pre- and posttranslocational complexes. In this chapter we present a detailed analysis of cryo-EM results, which reveal conformational changes of both EF-G and the ribosome.

The picture that emerges is one of translocation as the result of a complex cooperation between the ribosome and EF-G. In the following discussion, we will first describe the binding positions of EF-G on the ribosome. The availability of X-ray structures for some of the ribosome domains in immediate contact with EF-G allows the areas of contact to be located with precision. Following this is an account of the conformational changes observed in the ribosome that are triggered by EF-G binding and GTP hydrolysis.

BINDING POSITIONS OF EF-G IN GDP AND GTP STATES

Fusidic acid was used to stabilize the binding of EF-G in the GDP state; its presence, following GTP hydrolysis, prevents the dissociation of EF-G from the ribosome (Willie et al., 1975). In order to study the binding of EF-G in the GTP state, a nonhydrolyzable GTP analogue, GMPP(CH_2)P, was used. Five different complexes were prepared (Table 1): two by using naked ribosomes, one in each state, and the other three by using a pretranslocational complex containing $tRNA^{Phe}$ at the P site and Phe-$tRNA^{Phe}$ at the A site of poly(U)-programmed ribosomes. On the naked ribosomes, EF-G binds efficiently in the presence of both GTP and a nonhydrolyzable GTP analogue, and is found in the same binding position (Agrawal et al., 1999a).

When a pretranslocational complex was used, EF-G bound efficiently only in the presence of GTP, whereas very poor binding was observed in the presence of a nonhydrolyzable GTP analogue. Due to the low binding (~20%), EF-G was not visible in the cryo-EM map and was incapable of inducing translocation (Fig. 1, top). Binding could be enhanced (to ~40%) and partial translocation could be induced by raising the concentration of EF-G (to 1.6 μM) in the latter case. However, the cryo-EM map showed only a fragmented mass of EF-G, whose fragments correspond to the two main anchoring domains, I and II, and also domain III of EF-G (Fig. 1, top).

In the 70S·$(tRNA)_2$·EF-G·GDP·fusidic acid complex (Fig. 1, bottom), in addition to densities corresponding to P- and E-site tRNAs, all five domains of EF-G could be distinctly recognized. The mass of density present on the L1 side of the P-site tRNA, which appears to be fused with the L1 protein (Fig. 1, bottom), may be partly due to a conformational change in the L1 protein region of the ribosome or due to overlapping binding positions of the tRNA at more than one exit site (Agrawal et al., 1998b; Lill and Wintermeyer, 1987).

Subtraction of the 3-D map of the naked 70S control (Frank et al., 1995) from the 3-D map of the 70S·$(tRNA)_2$·EF-G·GMPP(CH_2)P complex, showing only ~20% of EF-G binding, gives a partly fused mass of density (not shown) corresponding to A- and P-site tRNAs. The use of the 70S·fMet-$tRNA_f^{Met}$ complex as a control results in a distinct mass of density on the L12 side of the P-site tRNA (Fig. 1, top). This mass can readily be assigned to the A-site tRNA, as it matches the A-site position obtained from other studies (Agrawal et al., 1996, 1998b, 1999c; Stark et al., 1997b).

The binding position of the anticodon stem of the A-site tRNA and the tip portion of domain IV of EF-G show substantial overlap. However, we note for later discussion that the tip of domain IV is separated by approximately 10 Å (toward the 50S subunit side) from the anticodon end of the tRNA. Subtraction from the 3-D map of the complex that was prepared with a higher concentration of EF-G gives additional masses of density attributable to domains I, II, and III of EF-G (Fig. 1, top) and also a weak mass on the L1 side (not shown), which is visible only at low threshold values.

LOCALIZATION OF RIBOSOMAL DOMAINS ASSOCIATED WITH EF-G BINDING

Several binding sites of various domains of EF-G have been located at the base of the L12 stalk (Agrawal et al., 1998a). Of those, the position of do-

Table 1. 70S ribosome·EF-G complexes used in this study

No.	Complex	No. of projections in 3-D reconstruction	Final resolution (Å)[a]	Observed tRNA positions
1	70S·EF-G·GDP·FA[b]	11,249	18.4	
2	70S·EF-G·GMPP(CH_2)P	8,909	17.9	
3	70S·$(tRNA)_2$·EF-G·GDP·FA	11,253	17.5	P and E
4	70S·$(tRNA)_2$·EF-G[c]·GMPP(CH_2)P	13,057	18.4	A and P
5	70S·$(tRNA)_2$·EF-G[d]·GMPP(CH_2)P	9,912	16.3	A, P, and weak E

[a] For the criterion of resolution used, see Malhotra et al., 1998.
[b] FA, fusidic acid.
[c] EF-G concentration same (0.8 μM) as in 1 to 3.
[d] EF-G concentration higher (1.6 μM).

main V was slightly refined (Fig. 2, top) in the subsequent study (Agrawal et al., 1999a). A detailed description of the proximities of the various domains of EF-G to ribosomal regions was presented in the first study (Agrawal et al., 1998a). In this chapter, we will make use of information from hydroxyl radical probing of EF-G (Wilson and Noller, 1998) on the ribosome to locate positions of various ribosomal RNA fragments on the ribosome (Fig. 2, middle).

Very recently, the complete crystal structure of the 58-nucleotide 23S RNA fragment (nucleotides 1051 to 1108) in complex with the L11 protein has

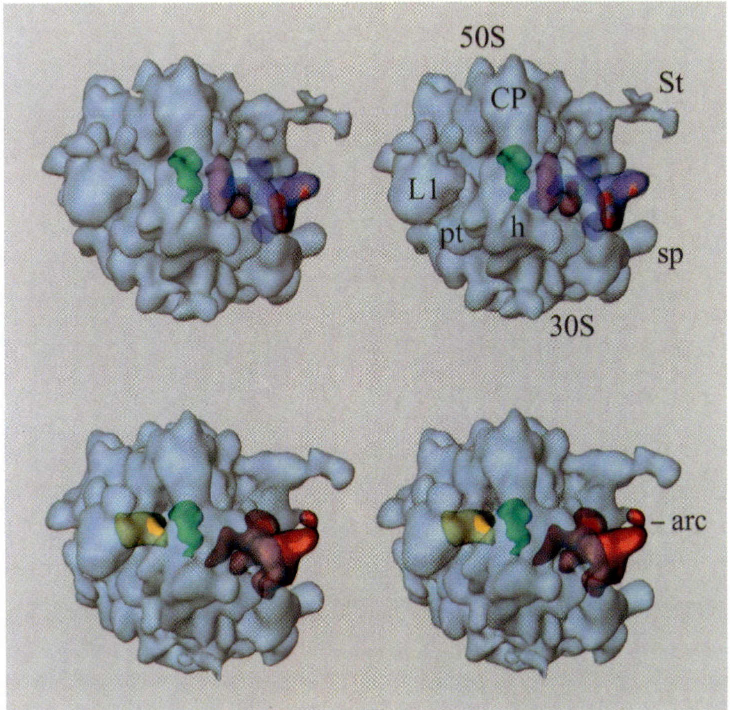

Figure 1. Stereo-view presentation of 3-D cryo-EM maps (transparent blue) of the 70S·EF-G·GMPP(CH_2)P complex (Table 1, complex 2) (top) and the 70S·$(tRNA)_2$·EF-G·GDP·fusidic acid complex (Table 1, complex 3) (bottom). The 30S subunit is below the 50S subunit. In the top panel, the stalk is bifurcated and no connection is formed between the stalk base and EF-G (red and transparent dark blue). In this panel, the difference masses corresponding to A-site (pink) and P-site (green) tRNAs were obtained from the 70S·$(tRNA)_2$·EF-G·GMPP(CH_2)P complex (Table 1, complex 4), prepared with a lower concentration of EF-G (0.8 μM), and the fragmented mass corresponding to domains I, II, and III of EF-G (red) was obtained from the 70S·$(tRNA)_2$·EF-G·GMPP(CH_2)P complex Table 1, complex 5), prepared with a higher concentration of EF-G (1.6 μM) (Agrawal et al., 1999a). The mass shown in transparent dark blue corresponds to EF-G in the 70S·EF-G·GMPP(CH_2)P complex (Table 1, complex 2). Landmarks of the 30S subunit are as follows: h, head; pt, platform; sp, spur. Landmarks of the 50S subunit are as follows: CP, central protuberance; L1, L1 protein; St, stalk. In the bottom panel, the L12 stalk is in an extended conformation and an arc-like connection (arc) is present between the stalk base and EF-G. In addition to EF-G (red), a distinct mass corresponding to the P-site tRNA (green) and a smeared mass corresponding to the E-site tRNAs (yellow) are seen in the intersubunit space.

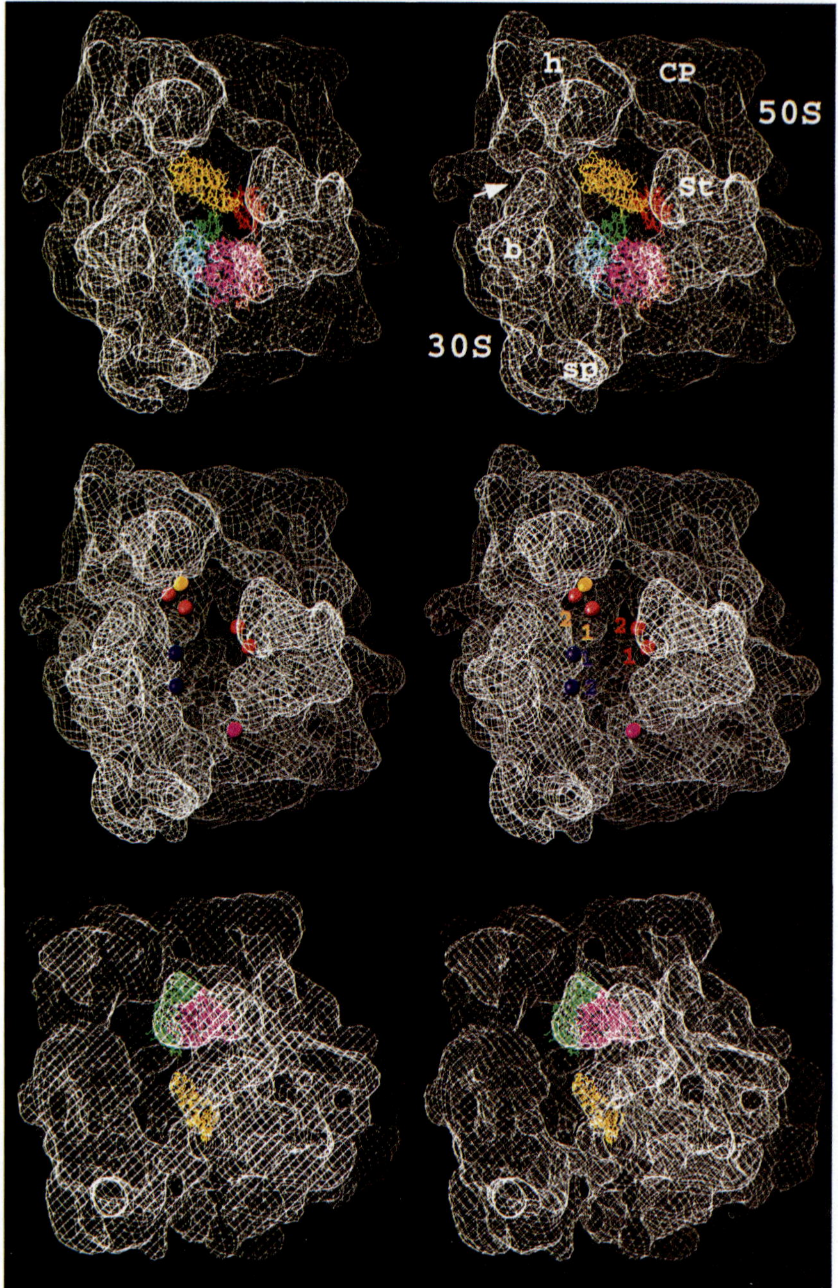

Figure 2. (Top) Stereo-view presentation showing the positions of various domains of EF-G on the ribosome (white wire mesh) obtained by fitting the X-ray crystal structure of EF-G into the mass corresponding to EF-G in the 70S·$(tRNA)_2$·EF-G·GDP·fusidic acid complex. A detailed description of the fitting procedure is provided elsewhere (Agraal et al., 1998a, 1999a). Various domains of the crystal structure of EF-G are shown in different colors: magenta and pink, domain I (magenta, G domain, and pink, G′domain); blue, domain II; green, domain III; yellow, domain IV; and red, domain V. Landmarks of the 30S subunit are as follows: b, body; h, head; sp, spur. Landmarks of the 50S subunit are as follows: CP, central protuberance; St, stalk. The arrow points to the mRNA channel passing through the neck of the 30S subunit (see chapter 5). (Middle) Stereo-view presentation showing the proximity of various amino acids of EF-G that are expected to be close to the specific nucleotide residues of rRNAs (Wilson and Noller, 1998). Amino acid residues are shown as beads (5-Å radius) of different colors. Magenta represents amino acid 196 of domain I (G domain) of EF-G, which is proximal to nucleotides 2650 to 2653 and 2668 to 2669 of 23S RNA. Blue 1 and 2 represent amino acids 301 and 314, respectively, of domain II of EF-G, which are proximal to nucleotides 37 to 39, 441 to 445, 496 to 497, and 537 to 539 and to nucleotides 368 to 370, 384 to 385, and 493 to 497 of 16S RNA, respectively. Yellow represents amino acid 541 of domain IV of EF-G, which is proximal to nucleotides 1213 to 1214 of 16S RNA. Orange 1 and 2 represent amino acids 506 and 585, respectively, of domain IV of EF-G. These two amino acids are proximal to specific residues of both 16S and 23S RNAs, indicating that 23S RNA reaches the decoding region of the 30S subunit, thus strongly supporting Brimacombe's cross-linking data (Mitchell et al., 1992). Amino acid residue 506 is proximal to nucleotides 1228 to 1230 of the 16S RNA and nucleotides 1920 to

been obtained (Wimberly et al., 1999). This complex is the minimum requirement for the EF-G-dependent GTPase activity (Ryan and Draper, 1991). The crystal structure of another fragment of 23S-28S RNA (α-sarcin- and ricin-binding 2660 stem-loop region by *Escherichia coli* 23S rRNA numbering and 4325 stem-loop region by rat 28S rRNA numbering), which is universally conserved and is important for all EF-dependent activities of both prokaryotic (Hausner et al., 1987) and eukaryotic (Wool et al., 1992) ribosomes, is also known (Correll et al., 1998). Hydroxyl radical probing (Wilson and Noller, 1998) indicated that nucleotide 1100 of 23S RNA should be up to 22 Å away from the 650th amino acid within domain V of EF-G, whereas the 1067 nucleotide region, which is also responsible for the binding of the antibiotic thiostrepton (Cundliffe, 1986), can be within a distance of 12 to 36 Å from the 655th amino acid of the same EF-G domain. Using this information, we have docked the X-ray structure of the L11-23S RNA fragment (nucleotides 1051 to 1108) complex (Wimberly et al., 1999) into the 15-Å-resolution map (Malhotra et al., 1998) of the ribosome (Fig. 2, bottom). Similarly, the 196th amino acid, within the G domain of EF-G, can be up to 22 Å away from the stem portion (particularly from nucleotides 2650 to 2653 and 2668 to 2669) of the α-sarcin-ricin stem-loop region, while the 650th amino acid within domain V could be at a 22-Å distance from the loop portion involving nucleotide 2659 of the 23S RNA. This information has been used to dock the X-ray structure of the α-sarcin-ricin stem-loop (Correll et al., 1998) into the ribosome map (Fig. 2, bottom).

Figure 2 (bottom) shows the fitting of these X-ray structures near the base of the L12 stalk. It is interesting to note that the groove-like feature present at the base of the stalk matches a groove observed in the X-ray structure of the L11-23S RNA fragment (nucleotides 1051 to 1108) complex (Wimberly et al., 1999). This feature is the binding site of the antibiotic thiostrepton, which is known to inhibit the translocation step of protein synthesis. Thus, the corresponding structure seen at the base of the stalk in the cryo-EM map must be the interaction site for the antibiotic. Our cryo-EM localization studies (Agrawal et al., 1998a, 1999a) (Fig. 2, top) show that domain V, which we proposed to play the role of communicator between the GTP binding domain and domain IV of EF-G through the ribosomal stalk base (Agrawal et al., 1998a), indeed makes close contact with the groove-like structure.

EF-G BINDING-ASSOCIATED CONFORMATIONAL CHANGES IN THE RIBOSOME

It has been generally accepted that EF-G induces conformational changes in the ribosome which are ultimately responsible for the translocation of tRNA. Burma and coworkers (1985) and Mesters and coworkers (1994) showed that EF-dependent GTPase activity changes the conformation of the naked ribosome. However, in the absence of a 3-D map, the nature of these conformational changes was unknown. Our results conclusively show that a number of ribosomal regions undergo a change of conformation upon EF-G binding. These are primarily located in the vicinity of the binding sites of EF-G, but there is also evidence of dramatic changes at remote locations in both the 30S and the 50S subunits (Agrawal et al., 1998a, 1999a).

Conformational Changes of the 50S Subunit

The L12 protein, which is present in four copies as two dimers, is known to be highly mobile. One of its dimers is folded on the 50S subunit, while the other is present in its extended conformation, visible as a stalk (Traut et al., 1995). Removal of this protein from the ribosome reduces the rate of protein synthesis by an order of magnitude and adversely affects its accuracy (Hamel et al., 1972; Pettersson and Kurland, 1980).

Our results show that in the GDP state, the stalk assumes an extended conformation (Fig. 1, bottom), whereas in the GTP state (Fig. 1, top), the extended stalk splits into two subdomains. This probably represents a configuration in which the C-terminal domains of the L12 dimer that forms the stalk have

1925 of the 23S RNA, whereas amino acid residue 585 is proximal to nucleotides 790, 1229 to 1230, 1339 to 1340, and 1397 to 1400 (decoding region) of the 16S RNA and nucleotides 1921 to 1923 of the 23S RNA. Red 1 and 2 represent amino acids 650 and 655, respectively, of domain V of EF-G, which are proximal to nucleotides 1100 and 2659 and 1065 to 1068 of 23S RNA, respectively. (Bottom) Fitting of the crystal structures of the L11-23S RNA fragment (nucleotides 1051 to 1108) complex (Wimberly et al., 1999) and the α-sarcin-ricin stem-loop (Correll et al., 1998) into the 15-Å-resolution map (Malhotra et al., 1998) of the 70S ribosome. the placement of crystal structures is based on proximities of 23S RNA nucleotides to amino acids of EF-G (see the middle panel), and is strongly supported by the similarity between the structural features of the crystal structure and the cryo-EM map. The L11-23S RNA fragment was refitted in line with the placement of this structure into the 5Å-resolution X-ray map of Ban and coworkers (1999). Green, L11 protein portion of the complex; magenta, 58-nucleotide portion of the same complex; yellow, α-sarcin-ricin stem-loop structure.

separated. This finding corroborates a recent study indicating the independent, mobile nature of the C-terminal domains of the L12 dimer (Hamman et al., 1996). Furthermore, these results point to a long-suspected direct role of the stalk in translocation (Möller et al., 1983; Traut et al., 1986, 1995; Burma et al., 1986; Liljas and Gudkov, 1987). One could argue that the observation of a split stalk (Fig. 1, top) might indicate the existence of two different positions of the stalk, which the ribosome might alternately assume in the GTP state. In such a situation, however, the two positions would show up with reduced density, ruling out this explanation.

Another major difference between the conformations of ribosomes in the GDP and GTP states is that in the GDP state, an arc-like connection between the base of the L7/L12 stalk and the G′ domain (part of domain I) of EF-G (Fig. 1, bottom), which is absent in the GTP state (Fig. 1, top), is formed. This arc-like connection might involve one of the folded C-terminal domains of the L12 stalk. A similar connection is formed upon binding of the EF-Tu ternary complex (Stark et al., 1997a). Thus, this conformation of the stalk is clearly assumed following GTP hydrolysis. In contrast, when the nonhydrolyzable GTP analogue is used, the arc-like connection is not observed, a finding that was also reported with the binding of the EF-Tu ternary complex, using a nonhydrolyzable GTP analogue, to the ribosome (Stark et al., 1997a).

It thus appears that the formation of the arc-like connection is characteristic of the GDP state of the ribosome, while the split form of the stalk is a feature of the GTP state. These changes in conformation are clearly related to EF-G binding and to EF-G-dependent GTP hydrolysis, which appears to be regulated through the base of the stalk, where the GTP binding domain of EF-G interacts (Fig. 2, top). It is worth mentioning in this context that an earlier model of translocation (Burma et al., 1986) proposed the existence of two conformational states of the ribosome, in which the stalk assumes two different positions that are regulated by a conformational switch of the 23S RNA region located at the base of the stalk. The existence of a 23S RNA-based conformational switch in the universally conserved 2660 loop, which is protected from chemical modification by both EFs (Moazed et al., 1988) and is expected to be located at the base of the stalk, was also suggested by Wool and coworkers (1992) for the eukaryotic ribosomes.

In order to identify other conformational changes of the ribosome related to the binding of EF-G and the translocation of tRNAs, the 3-D maps of various ribosome·EF-G complexes were compared with that of the 70S·fMet-$tRNA_f^{Met}$ complex (Malhotra et al., 1998). In addition to the L12 stalk region of the 50S subunit, another conformational change points to the ribosomal region that is proximal to the junction of domains I, III, and V, i.e., on the subunit interface side of the 50S subunit (Agrawal et al., 1999a). Two small lobes on the remotely located, globular L1 protein region were also observed in EF-G-bound ribosome complexes (Fig. 1, top). However, the functional significance of these lobes, which fuse (Fig. 1, bottom) in the presence of tRNA, is as yet unclear.

Conformational Changes of the 30S Subunit

We observe substantial conformational changes in the 30S subunit of all 70S·EF-G complexes relative to the 70S·fMet-$tRNA_f^{Met}$ complex. These changes, which are dependent on the binding of EF-G, are most prominent in the GTP state and are evident as movements in the "spur" and head regions. The angle between the spur and the main body is narrowed, and the head of the 30S subunit moves toward the L1 side (Fig. 3). The capacity of the head region for independent movement is not surprising, as it has minimal connection with the rest of the 30S subunit, through a single-stranded region of 16S RNA (Müller and Brimacombe, 1997). Moreover, the head in the isolated 30S subunit has been found to assume different positions (Lata et al., 1996; Gabashvili et al., 1999a). The region of the 30S subunit that interacts with domain II of EF-G shows a large conformational change. The nature of this induced conformational change, however, is different in GDP- than in GTP-state complexes (Agrawal et al., 1999a). In the GTP

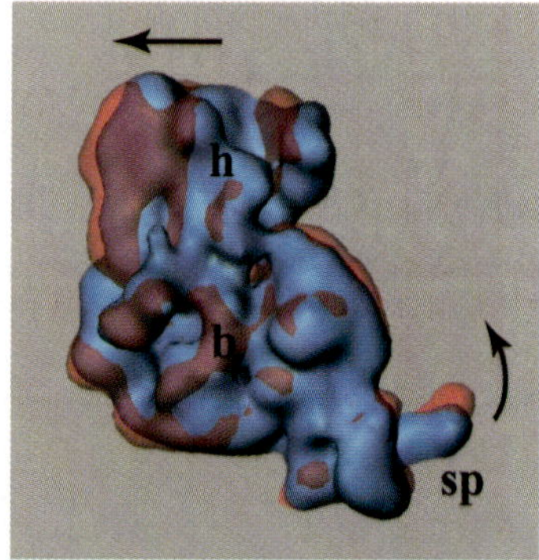

Figure 3. Relative movement of the 30S subunit with respect to the 50S subunit upon EF-G·GMPP(CH_2)P binding to the 70S ribosome. Two volumes, the 70S·fMet-$tRNA_f^{Met}$ complex (Malhotra et al., 1998) (shown as solid blue) and the 70S·EF-G·GMPP(CH_2)P complex (transparent pink), were superimposed so that the 50S subunits of the two volumes were perfectly aligned. The 50S subunits of both volumes have been removed to avoid visual confusion. (Reproduced from Agrawal et al., 1999a.) The landmarks are the same as in Fig. 2, top panel.

state, domain II of EF-G makes a smooth contact along its surface facing the 30S subunit, whereas in the GDP state, two protrusions emerging from the 30S subunit appear to be mainly involved in the contact with domain II of EF-G.

Traditionally, EF-G has been assigned to the large subunit, as far as its role in EF-G-dependent GTP hydrolysis is concerned. However, 3-D mapping of EF-G (Agrawal et al., 1998a, 1999a; Wilson and Noller, 1998) implies that domain IV of EF-G acts on the small subunit (Fig. 2, top) to effect the movement of tRNA from the A site. The 30S subunit regions that interact with the anticodon region of the A-site tRNA, or alternatively, with the tip of domain IV of EF-G, and that make contact with domain II of EF-G appear to be the most flexible domains of the 30S subunit (Agrawal et al., 1999b; Gabashvili et al., 1999a). Furthermore, a recent mutation study of the 16S rRNA (Lodmell and Dahlberg, 1997) supports the notion that the 30S subunit is actively involved in the translocation process. This study shows that there are two alternative base pairings for the triplet sequence of nucleotides 910 to 912 (CUC) either with nucleotides 888 to 890 (GAG) or with the immediately adjacent nucleotides 885 to 887 (GGG) within the 16S rRNA, both of which are required for ribosome function. This 16S rRNA region has been placed (Müller and Brimacombe, 1997) near the junction point of the body and the platform. The movement of the triplet sequence by precisely 3 nucleotides between adjacent complementary triplets and the observed EF-G-dependent conformational change in the same region present an interesting corollary to the process of translocation. Moreover, in a recent 3-D cryo-EM analysis of these mutants (Gabashvili et al., 1999b), a shift in the structural domains of the A-site region has been observed. These findings relate to the earlier proposals (Moazed and Noller, 1989; Agrawal and Burma, 1996) that the relative movement of the two subunits could mediate the translocation event.

CONFORMATIONAL CHANGE OF EF-G AND IMPLICATION FOR THE TRANSLOCATION EVENTS

Fitting of the X-ray structure of EF-G (Czworkowski et al., 1994) into the cryo-EM density (Agrawal et al., 1998a, 1999a) revealed a large conformational change in the X-ray structure upon binding to the ribosome. Both of these studies show that the substructure, comprising domains III, IV, and V, which together mimic the tRNA portion of the EF-Tu ternary complex, moves by ~10 Å and rotates by ~10° with respect to domains I and II, the main anchoring domains of EF-G on the 50S and 30S subunits, respectively. In addition, domain V moves by ~10 Å with respect to domains III and IV. It appears that proper binding of EF-G on the ribosome (Fig. 2, top) is accompanied by two conformational changes, which can be described as pivoting movements (i) at the junction of domains II and III (Agrawal et al., 1998a) and (ii) at the junction of domains IV and V (Agrawal et al., 1999a).

Earlier studies (e.g., Modolell et al., 1975) have shown that EF-G can bind to the pretranslocational complexes in the presence of a nonhydrolyzable GTP analogue and can induce a single round of translocation. However, we observe only a low level of EF-G binding and only partial translocation even at a *higher* concentration of EF-G (Table 1). Furthermore, the lower level of binding yields only a fragmented mass of density which could be assigned to domains I, II, and III, in which domains I and II are at exactly the same place as in other complexes while domain III has a slightly different but overlapping position (Agrawal et al., 1999a) with respect to its position in those complexes. This observation, and the weakened densities in the regions of domains IV and V (Fig. 1, top), are consistent with our earlier suggestion (Agrawal et al., 1998a) that there is a movement at the junction of domains II and III and with the assumption that the substructure formed by domains III, IV, and V assumes different positions in the two subpopulations created by the limited occupancy of the A site (approximately 60%), namely, one (with the A-site occupied) corresponding to the X-ray structure and another (with the A-site unoccupied) corresponding to the EF-G structure observed for the naked ribosomes. Based on all these observations, an advanced, two-step model of translocation was presented (Agrawal et al., 1999a).

The difference in conformation between ribosome-bound EF-G observed by cryo-EM in both GDP and GTP states and the X-ray structure in its GDP state could have two explanations: (i) the conformations of EF-G in the GDP and GTP states are different, and the fact that both appear to be the same in the ribosome-bound state is due to the binding of fusidic acid in the GTP conformation of EF-G, immediately after GTP hydrolysis but before EF-G could come to its GDP conformation, and (ii) the conformational change of the molecule is not directly effected by GTP hydrolysis but rather is an effect brought about by its binding to the ribosome, a passive movement of the mobile substructure (domains III, IV, and V) caused by a conformational change of the ribosome (see below) that is in turn triggered by GTP hydrolysis. The latter explanation would imply that the structure observed by X-ray crystallography reflects one of two conformations that might be sep-

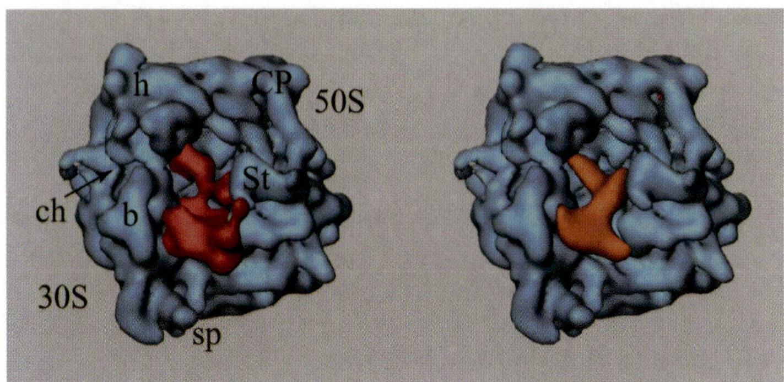

Figure 4. Side by side comparison of binding positions of EF-G (red; adapted from Agrawal et al., 1999a) (left) and ternary complex (orange) of Phe-tRNA^{Phe}, EF-Tu, and GDP (Agrawal et al., unpublished) on the 15-Å-resolution map (Malhotra et al., 1998) of the 70S ribosome. The binding of the ternary complex was arrested by using a kirromycin-stalled ribosome. The landmarks are the same as those introduced in Fig. 1 (top panel) and 2 (top panel). ch, mRNA channel.

arated by a small energy barrier. The reason only one of the conformations is seen could be the result of the strains caused by crystal-packing forces. This would explain Czworkowski and Moore's (1997) inability to find significant differences between the X-ray structure of EF-G·GDP and the solution structure of EF-G·GTP. This picture, however, stands in contrast to the view (Rodnina et al., 1997) that GTP hydrolysis directly causes a conformational change within EF-G, akin to a power stroke.

COMPARISON OF BINDING POSITIONS OF EF-Tu AND EF-G

Since the determination of the crystal structures of EF-G, in its free (Ævarsson et al., 1994) and GDP complex (Czworkowski et al., 1994) forms, and the ternary complex of aminoacyl-tRNA, EF-Tu, and GMPPNP (Nissen et al., 1995), a dynamic picture of the steps involving interaction of EFs with the ribosome has begun to emerge. Close similarity in these structures has prompted a proposal of the existence of molecular mimicry (Nissen et al., 1995; Nyborg et al., 1996; Abel and Jurnak, 1996). Both factors bind to the ribosome in their GTP conformations and are released after the GTP hydrolysis. Cryo-EM results show that the binding positions on the ribosome are very similar (Fig. 4). Strict molecular mimicry would imply that the GTP state conformations of the two molecules should be similar. However, it is the GTP conformation of EF-Tu ternary complex that matches closely with the GDP conformation of EF-G, while the crystal structure of EF-G in its GTP conformation is unknown. Moreover, our study (Agrawal et al., 1999a) clearly shows that the overall conformation of EF-G bound to the naked ribosome is the same in both the GTP and GDP states, corroborating and strengthening the finding of Czworkowski and Moore (1997). In discussing the interactions with the ribosome, one must keep in mind that the two factors have to perform completely different roles in the elongation cycle. One has to deliver a codon-specific aminoacyl-tRNA to the decoding site, so that the anticodon tip of tRNA reaches the mRNA codon. The other has to assure the completion of the translocation step after the peptide bond formation. For this, the tip of domain IV does not have to reach the

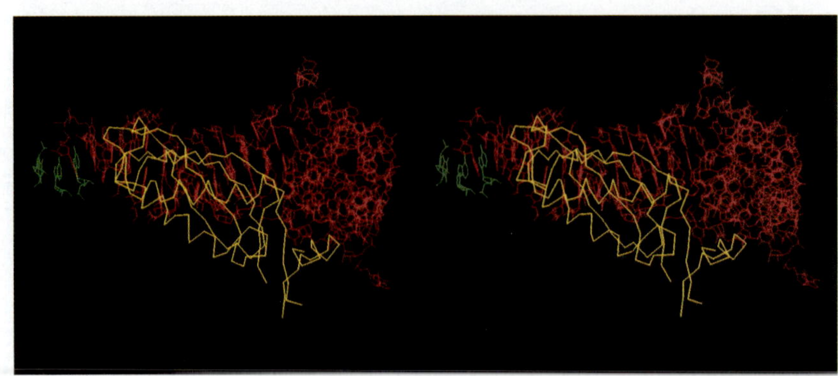

Figure 5. Stereo-view presentation to show the proximity of the tip of domain IV of EF-G (yellow) to the anticodon (green) end of the A-site tRNA (red).

mRNA codon. Indeed, our results show (Fig. 5) that the tip proper of domain IV is separated from the position of the anticodon end of the A-site tRNA by ~10 Å.

The similarity in the observed binding positions of EF-Tu ternary complex and EF-G invites a further comparison with binding positions of release factors, which are known to have high sequence similarity (Nakamura et al., 1996) and thus are expected to interact with the same sites on the ribosome while fulfilling a completely different function of polypeptide release when the ribosome encounters a termination codon. Furthermore, another group of G proteins, e.g., Tet(O), involved in conferring tetracyclin resistance on pathogenic bacteria, are also known to have significant sequence similarity with the EFs, and thus were expected to have similar structures and binding positions on the ribosome. A very recent cryo-EM localization study of Tet(O) protein (Spahn et al., unpublished) reveals that Tet(O) indeed has a structure similar to that of EF-G and binds in the same position on the 70S ribosome. From the comparison of the binding of all these factors to the ribosome, it appears that they use the same anchoring points on the ribosome and make contact with the same GTPase-associated center while differing substantially in their functions.

We thank Venki Ramakrishnan for providing the coordinates of the X-ray structure of the L11-23S RNA fragment complex.

This work was supported by grants from the National Institutes of Health (R37-GM29169 and R01-GM55440) and the National Science Foundation (BIR9219043).

REFERENCES

Abel, K., and F. Jurnak. 1996. A complex profile of protein elongation: translating chemical energy into molecular movement. *Structure* **4**:229–238.

Ævarsson, A., E. Brazhnikov, M. Garber, J. Zheltonosova, Y. Chirgadze, S. Al-Karadaghi, L. A. Svensson, and A. Liljas. 1994. Three-dimensional structure of the ribosomal translocase: elongation factor G from *Thermus thermophilus*. *EMBO J.* **13**:3669–3677.

Agrawal, R. K., and D. P. Burma. 1996. Sites of ribosomal RNAs involved in the subunit association of tight and loose couple ribosomes. *J. Biol. Chem.* **271**:21285–21291.

Agrawal, R. K., P. Penczek, R. A. Grassucci, Y. Li, A. Leith, K. H. Nierhaus, and J. Frank. 1996. Direct visualization of A-, P-, and E-site transfer RNAs in the *Escherichia coli* ribosome. *Science* **271**:1000–1002.

Agrawal, R. K., P. Penczek, R. A. Grassucci, and J. Frank. 1998a. Visualization of elongation factor G on the *Escherichia coli* 70S ribosome: the mechanism of translocation. *Proc. Natl. Acad. Sci. USA* **95**:6134–6138.

Agrawal, R. K., P. Penczek, A. Malhotra, R. A. Grassucci, I. S. Gabashvili, A. B. Heagle, S. Srivastava, N. Burkhardt, R. Jünemann, K. H. Nierhaus, and J. Frank. 1998b. Binding positions of tRNAs in translating *Escherichia coli* ribosomes, p. 717–718. *In* H. A. Calderón Benavides and M. J. Yacamán (ed.), *Proceedings of the 14th International Congress on Electron Microscopy*. Institute of Physics Publishing, Bristol, United Kingdom.

Agrawal, R. K., A. B. Heagle, P. Penczek, R. A. Grassucci, and J. Frank. 1999a. Elongation factor-G-dependent GTP hydrolysis induces translocation accompanied by large conformational changes in the 70S ribosome. *Nat. Struct. Biol.* **6**:643–647.

Agrawal, R. K., K. R. Lata, and J. Frank. 1999b. Conformational variability in *E. coli* 70S ribosome as revealed by 3D cryo-electron microscopy. *Int. J. Biochem. Cell Biol.* **31**:243–254.

Agrawal, R. K., P. Penczek, R. A. Grassucci, N. Burkhardt, K. H. Nierhaus, and J. Frank. 1999c. Effect of buffer conditions on the position of tRNA on the 70S ribosome as visualized by cryo-electron microscopy. *J. Biol. Chem.* **274**:8723–8729.

Agrawal, R. K., N. Burkhardt, R. Grassucci, K. H. Nierhaus, and J. Frank. Unpublished data.

Al-karadaghi, S., A. Ævarsson, M. Garber, J. Zheltonosova, and A. Liljas. 1996. The structure of elongation factor G in complex with GDP: conformational flexibility and nucleotide exchange. *Structure* **4**:555–565.

Ban, N., P. Nissen, J. C. Hansen, M. Capel, P. B. Moore, and T. A. Steitz. 1999. Placement of protein and RNA structures into a 5-Å resolution map of the 50S ribosomal subunit. *Nature* **400**:841–847.

Bodley, J. W., and L. Lin. 1970. Interaction of *E. coli* G factor with the 50S ribosomal subunit. *Nature* **227**:60–61.

Bourne, H. R., D. A. Sanders, and F. McCormick. 1991. The GTPase superfamily: conserved structure and molecular mechanism. *Nature* **349**:117–127.

Burma, D. P., A. K. Srivastava, S. Srivastava, and D. Dash. 1985. Interconversion of tight and loose couple 50S ribosome and translocation in protein synthesis. *J. Biol. Chem.* **260**:10517–10525.

Burma, D. P., S. Srivastava, A. K. Srivastava, S. Mahanti, and D. Dash. 1986. Conformational change of 50S ribosomes during protein synthesis, p. 438–453. *In* B. Hardesty and G. Kramer (ed.), *Structure, Function and Genetics of Ribosomes*. Springer-Verlag, New York, N.Y.

Chinali, G., and A. Parmeggiani. 1982. Differential modulation of the elongation factor-G GTPase activity by tRNA bound to the ribosomal A-site or P-site. *Eur. J. Biochem.* **125**:415–421.

Correll, C. C., A. Munishkin, Y.-L. Chan, Z. Ren, I. G. Wool, and T. A. Steitz. 1998. Crystal structure of the ribosomal RNA domain essential for binding elongation factors. *Proc. Natl. Acad. Sci. USA* **95**:13436–13441.

Cundliffe, E. 1986. Involvement of specific proteins of ribosomal RNA in defined ribosomal functions: a study utilizing antibiotics, p. 587–604. *In* B. Hardesty and G. Kramer (ed.), *Structure, Function and Genetics of Ribosomes*. Springer-Verlag, New York, N.Y.

Czworkowski, J., and P. B. Moore. 1997. The conformational properties of Elongation Factor G and the mechanism of translocation. *Biochemistry* **36**:10327–10334.

Czworkowski, J., J. Wang, T. A. Steitz, and P. B. Moore. 1994. The crystal structure of elongation factor G complexed with GDP, at 2.7D. *EMBO J.* **13**:3661–3668.

Frank, J., J. Zhu, P. Penczek, Y. Li, S. Srivastava, A. Verschoor, M. Radermacher, R. Grassucci, R. K. Lata, and R. K. Agrawal. 1995. A model of protein synthesis based on cryo-electron microscopy of the *E. coli* ribosome. *Nature* **376**:441–444.

Gabashvili, I. S., R. K. Agrawal, R. A. Grassucci, and J. Frank. 1999a. Structure and structural variations of the *E. coli* 30S ribosomal subunit as revealed by three-dimensional cryo-electron microscopy. *J. Mol. Biol.* **286**:1285–1291.

Gabashvili, I. S., R. K. Agrawal, R. A. Grassucci, S. L. Squires, A. E. Dahlberg, and J. Frank. 1999b. Major rearrangements in the 70S ribosomal 3D structure caused by a conformational switch in 16S ribosomal RNA. *EMBO J.* **18**:6501–6507.

Gavrilova, L. P., O. E. Kostiashkina, V. E. Koteliansky, N. M. Rutkevitch, and A. S. Spirin. 1976. Factor-free ("non-enzymic") and factor-dependent systems of translation of polyuridylic acid by *Escherichia coli* ribosomes. *J. Mol. Biol.* **101**:537–552.

Gordon, J. 1969. Hydrolysis of guanosine 5′-triphosphate associated with binding of aminoacyl transfer ribonucleic acid to ribosomes. *J. Biol. Chem.* **244**:5680–5686.

Hamel, E., M. Koka, and T. Nakamoto. 1972. Requirement of an *Escherichia coli* 50S ribosomal protein component for effective interaction of the ribosome with T and G factors and with guanosine triphosphate. *J. Biol. Chem.* **10**:805–814.

Hamman, B. D., A. V. Oleinikov, G. G. Jokhadze, R. R. Traut, and D. M. Jameson. 1996. Rotational and conformational dynamics of *Escherichia coli* ribosomal protein L7/L12. *Biochemistry* **35**:16672–16679.

Hausner, T.-P., J. Atmadja, and K. H. Nierhaus. 1987. Evidence that the G2661 region of 23S rRNA is located at the ribosomal binding sites of both elongation factors. *Biochimie* **169**:911–923.

Kawakita, M., K. Arai, and Y. Kaziro. 1974. Interactions between elongation factor tu-guanosine triphosphate and ribosomes and the role of ribosome-bound transfer RNA in guanosine triphosphatase reaction. *J. Biochem.* **76**:801–809.

Lata, K. R., R. K. Agrawal, P. Penczek, R. Grassucci, J. Zhu, and J. Frank. 1996. Three-dimensional reconstruction of the *Escherichia coli* 30S ribosomal subunit in ice. *J. Mol. Biol.* **262**:43–52.

Liljas, A., and T. Gudkov. 1987. The structure and dynamics of ribosomal protein L12. *Biochimie* **69**:1043–1047.

Lill, R., and W. Wintermeyer. 1987. Destabilization of codon-anticodon interaction in the ribosomal exit site. *J. Mol. Biol.* **196**:137–148.

Lodmell, J. S., and A. E. Dahlberg. 1997. A conformational switch in *Escherichia coli* 16S ribosomal RNA during decoding of messenger RNA. *Science* **277**:1262–1267.

Malhotra, A., P. Penczek, R. K. Agrawal, I. Gabashvili, R. A. Grassucci, N. Burkhardt, R. Jünemann, K. H. Nierhaus, and J. Frank. 1998. *E. coli* 70S ribosome at 15 Å resolution by cryo-electron microscopy: localization of fMet-tRNA$_f^{Met}$ and fitting of L1 protein. *J. Mol. Biol.* **280**:103–116.

Mesters, J. R., A. A. Potapov, J. M. de Graft, and B. Kraal. 1994. Synergism between the GTPase activities of EF-TuAGTP and EF-GAGTP on empty ribosomes: elongation factors as stimulators of the ribosomal oscillation between two conformations. *J. Mol. Biol.* **242**:644–654.

Mitchell, P., M. Oßwald, and R. Brimacombe. 1992. Identification of intermolecular RNA cross-links at the subunit interface of the *Escherichia coli* ribosome. *Biochemistry* **31**:3004–3011.

Moazed, D., and H. F. Noller. 1989. Intermediate states in the movement of transfer RNA in the ribosome. *Nature* **342**:142–148.

Moazed, D., J. M. Robertson, and H. F. Noller. 1988. Interaction of elongation factor EF-G and EF-Tu with a conserved loop in 23S RNA. *Nature* **334**:362–364.

Modolell, J., T. Girbes, and D. Vazquez. 1975. Ribosomal translocation promoted by guanylylimido diphosphate and guanylyl-methylene diphosphonate. *FEBS Lett.* **60**:109–113.

Möller, W., P. I. Schrier, J. A. Maassen, A. Zantema, E. Schop, H. Reinalda, A. F. M. Cremers, and J. E. Mellema. 1983. Ribosomal proteins L7/L12 of *Escherichia coli.* Localization and possible molecular mechanism in translation. *J. Mol. Biol.* **163**: 553–573.

Müller, F., and R. Brimacombe. 1997. A new model for the three-dimensional folding of *Escherichia coli* 16 S ribosomal RNA. 1. Fitting the RNA to a 3D electron microscopic map at 20 angstrom. *J. Mol. Biol.* **271**:524–544.

Nakamura, Y., K. Ito, and L. A. Isaksson. 1996. Emerging understanding of translation termination. *Cell* **87**:147–150.

Nierhaus, K. H. 1993. Solution of the ribosome riddle: how the ribosome selects the correct aminoacyl-tRNA out of 41 similar contestants. *Mol. Microbiol.* **9**:661–669.

Nishizuka, Y., and F. Lipmann. 1966. Comparison of guanosine triphosphate split and polypeptide synthesis with purified *E. coli* system. *Proc. Natl. Acad. Sci. USA* **55**:212–219.

Nissen, P., M. Kjeldgaard, S. Thirup, G. Polekhina, L. Reshetnikova, B. F. C. Clark, and J. Nyborg. 1995. Crystal structure of the ternary complex of Phe-tRNA^{Phe}, EF-Tu, and a GTP analog. *Science* **270**:1464–1472.

Nyborg, J., P. Nissen, M. Kjeldgaard, S. Thirup, G. Polekhina, and B. F. Clark. 1996. Structure of the ternary complex of EF-Tu: macromolecular mimicry in translation. *Trends Biochem. Sci.* **21**:81–82.

Pestka, S. 1969. Studies on the formation of transfer ribonucleic acid-ribosome complexes. VI. Oligopeptide synthesis and translocation on ribosome in the presence and absence of transfer factors. *J. Biol. Chem.* **244**:1533–1539.

Pettersson, I., and C. G. Kurland. 1980. Ribosomal protein L7/L12 is required for optimal translation. *Proc. Natl. Acad. Sci. USA* **77**:4007–4010.

Rodnina, M. V., A. Savelsbergh, V. I. Katunin, and W. Wintermeyer. 1997. Hydrolysis of GTP by elongation factor G drives tRNA movement on the ribosome. *Nature* **385**:37–41.

Ryan, P. C., and D. E. Draper. 1991. Detection of a key tertiary interaction in the highly conserved GTPase center of large subunit ribosomal RNA. *Proc. Natl. Acad. Sci. USA* **88**:6308–6312.

Spahn, C. M. T., G. Blaha, R. K. Agrawal, R. A. Grassucci, S. Connell, D. E. Taylor, K. H. Nierhaus, and J. Frank. Unpublished data.

Spirin, A. S. 1985. Ribosomal translocation: facts and models. *Prog. Nucleic Acid Res. Mol. Biol.* **32**:75–114.

Stark, H., M. Rodnina, J. Rinke-Appel, R. Brimacombe, W. Wintermeyer, and M. van Heel. 1997a. Visualization of elongation factor Tu on the *Escherichia coli* ribosome. *Nature* **389**:403–406.

Stark, H., E. V. Orlova, J. Rinke-Appel, N. Jünke, F. Müller, M. Rodnina, W. Wintermeyer, R. Brimacombe, and M. van Heel. 1997b. Arrangement of tRNAs in pre- and posttranslocational ribosomes revealed by electron cryomicroscopy. *Cell* **88**:19–28.

Traut, R. R., D. S. Tewari, A. Sommer, G. R. Gavino, H. M. Olson, and D. G. Glitz. 1986. Protein topography of ribosomal functional domains: effects of monoclonal antibodies to different epitopes in Escherichia coli protein L7/L12 on ribosome function and structure, p. 286–308. *In* B. Hardesty and G. Kramer (ed.), *Structure, Function and Genetics of Ribosomes.* Springer-Verlag, New York, N.Y.

Traut, R. R., D. Dey, D. E. Bochkarlov, A. V. Oleinikov, G. G. Jokhadze, B. Hamman, and D. Jameson. 1995. Location and domain structure of *Escherichia coli* ribosomal protein L7/L12: site-specific cysteine crosslinking and attachment of fluorescent probes. *Biochem. Cell Biol.* **73**:949–958.

Voigt, J., and K. Nagel. 1993. Regulation of elongation factor G GTPase activity by the ribosomal state. The effects of initiation factors and differentially bound tRNA, aminoacyl-tRNA, and peptidyl-tRNA. *J. Biol. Chem.* **268**:100–106.

Willie, G. R., N. Richman, W. O. Godtfredson, and J. W. Bodley. 1975. Some characteristics of and structural requirements for the interaction of 24, 25-dihydrofusidic acid with ribosome-elongation factor G complexes. *Biochemistry* **14**:1713–1718.

Wilson, K. S., and H. F. Noller. 1998. Mapping the position of translational elongation factor EF-G in the ribosome by direct hydroxyl radical probing. *Cell* **92**:131–139.

Wimberly, B. T., R. Guymon, J. P. McCutcheon, S. W. White, and V. A. Ramakrishnan. 1999. A detailed view of a ribosomal active site: the structure of the L11-RNA complex. *Cell* **97**:491–502.

Wool, I. G., A. Gluck, and Y. Endo. 1992. Ribotoxin recognition of ribosomal RNA and proposal for the mechanism of translocation. *Trends Biochem. Sci.* **17**:266–269.

III. STRUCTURES OF ISOLATED COMPONENTS

III. STRUCTURES OF ISOLATED COMPONENTS

Although an enormous effort has gone into isolating and characterizing ribosomal components over the past 30 years, progress in analyzing their three-dimensional structures was greatly hindered by the tendency of many components to denature and/or aggregate. However, many of these problems have been circumvented over the past decade by using components from thermophilic bacteria. This has led to the determination of the structures of several proteins and a protein-rRNA complex, using nuclear magnetic resonance and X-ray diffraction approaches.

The Ribosome: Structure, Function, Antibiotics, and Cellular Interactions
Edited by R. A. Garrett, S. R. Douthwaite, A. Liljas, A. T. Matheson, P. B. Moore, and H. F. Noller

Chapter 7

Ribosomal Proteins and Their Structural Transitions on and off the Ribosome

SALAM AL-KARADAGHI, NATALIA DAVYDOVA, IRINA ELISEIKINA, MARIA GARBER, ANDERS LILJAS, NATALIA NEVSKAYA, STANISLAV NIKONOV, and SVETLANA TISHCHENKO

It is well known that rRNAs undergo conformational changes when ribosomal proteins bind. These conformational changes are necessary for the proper binding of additional ribosomal proteins (Held et al., 1974; Nierhaus, 1990) and possibly also for ribosomal function. It is reasonable to think that the ribosomal proteins can also change their conformations when they bind to the rRNA even though these transitions are less well characterized. Thus, the structure of a protein observed in isolation may or may not be completely relevant to the situation in the ribosome. We will address the question of what changes in the conformation of ribosomal proteins can be expected between the free form and the one in the ribosome. Some flexibility in these proteins may be necessary for binding to the rRNA and formation of functional ribosomes. In addition, some proteins may need their flexibility for the proper functioning of the ribosome.

The number of structures of ribosomal proteins as well as of fragments of rRNA is rapidly increasing (for reviews see Liljas and Garber, 1995; Liljas and Al-Karadaghi, 1997; Moore, 1998; Nikonov et al., 1998; Ramakrishnan and White, 1998). Some doubt has been expressed about whether the structures observed in isolation are valid for the ribosome (Yonath and Franceschi, 1997). Some ribosomal proteins have been investigated more than once by different methods or from different species, with interesting observations. In addition, the first high-resolution studies of complexes between fragments of rRNA and ribosomal proteins are emerging (Conn et al., 1999; Wimberly et al., 1999). We are now also able to recognize individual proteins in the crystallographic structures of the ribosomal subunits (Ban et al., 1998) (see chapters 1 and 2).

Some differences in the conformations of the proteins have been observed. They may partly depend on the crystal packing or other conditions of the structural studies, they may be due to a flexibility that depends on the lack of interaction with other components in the ribosome, or they may illustrate characteristics of the protein important for ribosome assembly and function. We will briefly review the structural data available, identify similarities and differences, and illustrate some difficulties in using the structures of isolated components for insertion into the structures of whole ribosomes or subunits determined at lower resolution. An awareness of the possible differences in structure is necessary for an appreciation of the usefulness of structural studies of isolated components from a larger system such as the ribosome.

AVAILABLE STRUCTURES

The fraction of ribosomal proteins that has been structurally characterized is now more than one-third of all ribosomal proteins from bacteria (Table 1). In addition, the first structure of a ribosomal protein from an archaeon (L1) has been determined (Nevskaya et al., unpublished). Structural studies of eucaryal translation systems have just begun (Marcotrigiano et al., 1997; Matsuo et al., 1997), and the structural studies of bacterial rRNA are limited to a few percent of the total. The first structure

Salam Al-Karadaghi, Natalia Davydova, Irina Eliseikina, and Anders Liljas ■ Molecular Biophysics, Center of Chemistry and Chemical Engineering, Lund University, Box 124, SE-221 00 Lund, Sweden. **Maria Garber, Natalia Nevskaya, Stanislav Nikonov, and Svetlana Tishchenko** ■ Institute of Protein Research, Russian Academy of Sciences, 142292 Pushchino, Moscow Region, Russia.

Table 1. Relationships of eubacterial proteins and available structures[a]

Protein[b]	Taxonomic range[c]
L1	**AP**ECc
L2	A**P**EC M
L3	APEC
L4	P C
L5	APEC M
L6	A**P**EC M
L9	**P** Cc
L10	APE
L11	A**P**ECc
L12	A**P**ECc
L13	APECc
L14	A**P**EC M
L15	APE c
L16	P C M
L17	P m
L18	APEC
L19	PEC
L20	P C
L21	P Cc
L22	A**P**ECc
L23	APEC
L24	APECc
L25	**P**
L27	PEC
L28	P Cc m
L29	APEC
L30	A**P**E
L31	P C
L32	P
L33	P C
L34	P C
L35	P Cc
L36	P C
S1	**P**ECcM
S2	APEC M
S3	APEC M
S4	A**P**EC M
S5	A**P**EC
S6	**P** C
S7	A**P**EC M
S8	A**P**EC M
S9	APEC
S10	APEC M
S11	APEC M
S12	APEC M
S13	APECcM
S14	APEC M
S15	A**P**EC M
S16	P C m
S17	A**P**ECc m
S18	P C
S19	APEC M
S20	P C
S21	P

[a] Adapted from Amos Bairoch, Geneva (http://www.expasy.ch/cgi-bin/lists?ribosomp.txt).
[b] L7 = L12; L8 = $(L12)_4$:L10; L26 = S20.
[c] A, archea; P, eubacteria; E, eucarya; C, chloroplast encoded; c, chloroplast, nuclear encoded; M, mitochondrion encoded; m, mitochondrion, nuclear encoded. Available structures are shown in boldface.

of a complex of a ribosomal protein with its rRNA binding site was recently determined (Conn et al., 1999; Wimberly et al., 1999).

THE DOMAIN ARRANGEMENT OF RIBOSOMAL PROTEINS

Most proteins are divided into two or more structural domains, each with approximately 100 amino acid residues (Rossmann and Liljas, 1974). These domains are loosely or firmly associated with each other. Sometimes the domains are like beads on a string, but in some cases there are several connections between neighboring domains. The ribosomal proteins follow the same pattern (Table 2). They frequently have only one or two domains, but the largest proteins, S1 and L2, have more. The domains of ribosomal proteins are often significantly smaller than 100 amino acid residues, and those that have several connections between neighboring domains are rather infrequent (S4 and L1).

The ribosomal proteins with two domains sometimes have significantly extended conformations (Table 2). The most prominent examples are L1 from *Methanococcus jannaschii* (Nevskaya et al., unpublished), L6 (Golden et al., 1993b), L9 (Hoffman et al., 1996), and L12 (Liljas and Gudkov, 1987; Gudkov, 1997). In addition, several ribosomal proteins are quite elongated due to extended loops of different forms (Liljas and Al-Karadaghi, 1997). Evidently such elongated structures are able to connect and stabilize elements of the rRNA that are relatively far apart.

STRUCTURAL MOTIFS

The available structures have been related to each other and to other proteins from an early date (Leijonmarck et al., 1988). In discussing the fold of various proteins, the domain fold is the unit of primary interest. Ramakrishnan and White (1998) have done an extensive characterization of the folds of ribosomal proteins. Earlier review have also discussed this question (Liljas and Garber, 1995; Liljas and Al-Karadaghi, 1997; Moore, 1998). A few additional protein structures have emerged since the review by Ramakrishnan and White (1998). They are included in our short review.

The structures of ribosomal proteins display a number of motifs, many of which have been identified previously (Table 3). One of the motifs, RRM, has been observed very frequently, not only in ribosomal proteins but also in translation factors, such as

Table 2. Domain divisions of ribosome proteins as observed from structural analysis

Protein	Species	Domain	Residues (amino acids)	Comments	Total dimensions (Å)	Reference(s)
S1	Homologous protein	Repeated domain	≈70	Repeated 6 times	30 × 25	Bycroft et al., 1997
S4	*Bacillus stearothermophilus*	NTD	1–41	Flexible		
		1	42–91, 180–200	Helical	40 × 50 × 65	Davies et al., 1998
		2	92–179			Marcus et al., 1998
S5	*B. stearothermophilus*	NTD	1–74		45 × 40 × 25	Ramakrishnan and White, 1992
		CTD	75–147			
S6	*T. thermophilus*		1–101		33 × 15 × 15	Lindahl et al., 1994
S7	*B. stearothermophilus*	Main	1–69, 92–155		50 × 30 × 25	Hosaka et al., 1997
	T. thermophilus	β-ribbon	70–91		30 × 35 × 55	Wimberly et al., 1997
S8	*B. stearothermophilus*	NTD	1–63		45 × 40 × 35	Davies et al., 1996a
	T. thermophilus	Connecting loop	64–73	Extended		Nevskaya et al., 1998
		CTD	74–130			
S15	*T. thermophilus*		1–88			Berglund et al., 1997
	B. stearothermophilus				44 × 22 × 18	Clemons et al., 1998
S17	*B. stearothermophilus*		1–85		35 × 35 × 21	Golden et al., 1993a; Jaishree et al., 1996
L1	*T. thermophilus*	I	1–66, 159–228		52 × 39 × 35	Nikonov et al., 1996
	M. jannaschii	II	71–158			Nevskaya et al., unpublished
L2	*B. stearothermophilus*	NTD	1–61			
		II	62–130			Nakagawa et al., 1999
		III	131–194		35 × 27 × 18	
		CTD	195–275			
L6	*B. stearothermophilus*	NTD	1–79		25 × 25 × 60	Golden et al., 1993b
		CTD	80–177			
L9	*B. stearothermophilus*	NTD	1–40		25 × 30 × 82	Hoffman et al., 1996
		Connecting helix	41–73			
		CTD	75–149			
L11	*Thermotoga maritima*	NTD	1–72			Wimberly et al., 1999
		CTD	75–141		40 × 14 × 15	
L12	*E. coli*	NTD	1–37			Bocharov et al., 1996
		Flexible hinge	37–52			Bushuev et al., 1989
		CTD	53–120		29 × 16 × 17	Leijonmarck et al., 1980
L14	*B. stearothermophilus*	Main	1–88		45 × 33 × 24	Davies et al., 1996b
		CTD	103–122			
L22	*T. thermophilus*		1–113	Long β-ribbon	67 × 29 × 18	Unge et al., 1998
L25	*E. coli*		1–93	Twice-repeated motif	37 × 36 × 22	Stoldt et al., 1998
L30	*B. stearothermophilus*					Wilson et al., 1986
	T. thermophilus				33 × 21 × 16	Nikonov et al., 1998

Table 3. Structural classes of ribosomal proteins

Class of structure	Protein
OB fold or cold shock	S1 repeated domain
	S17
	L2 II
SH3-like	L2 III
β-Barrel	L14
	L25
Left hand β-α-β	S5C
RRM	S6
	S8N
	S8C
	L1 I
	L6N
	L6C
	L9N
	L12C
	L22
	L30
Rossmann	L1 II
Four-helix bundle	L12N
Helical	S15
	S4 I
	S7
Homeodomain	L11-C
ETS	S4 II
Unique?	L9C
	L11N

EF-G (domains III and V [Ævarsson et al., 1994; Laurberg et al., unpublished]) and eIF-4E (Marcotrigiano et al., 1997; Matsuo et al., 1997; Liljas and Al-Karadaghi, 1997) and other proteins (Burd and Dreyfuss, 1994). The possible evolutionary connection between these observations has not been clarified. With about two-thirds of the ribosomal proteins remaining to be studied, the possibility of predicting their folding from the general knowledge of protein structure as well as the frequent occurrence of some motifs should be considered.

CONFORMATION DIFFERENCES OF RIBOSOMAL PROTEINS

The extended conformations of some ribosomal proteins could be compared to proteins like calmodulin (Babu et al., 1985; Meador et al., 1992), which has a very elongated structure in one state while the α-helix that separates the two domains becomes bent in another state, with the effect that the protein adopts a more globular structure. One should not be surprised if some ribosomal proteins can undergo similar conformational changes between their states in isolation and when bound to the ribosome. Indeed, conformational differences of ribosomal proteins have been observed by different methods. Furthermore the extended loops observed in several ribosomal proteins (Liljas and Al-Karadaghi, 1997) and the difficulty of observing parts of ribosomal proteins or including regions of the molecules in crystallization experiments also indicate that the molecules are flexible.

S15

For protein S15, the orientation of one helix differs in the nuclear magnetic resonance (Berglund et al., 1997) and X-ray structures (Clemons et al., 1998). Since dimers are formed in the X-ray structure in which the N-terminal helix of one monomer occupies the position this helix has in the monomeric nuclear magnetic resonance structure, it seems likely that the conformation observed by crystallography may not be fully relevant for the ribosome. However, the flexibility of this helix may be an important feature for the situation in the ribosome.

L1

L1 is a two-domain protein. The structure of L1 from *Thermus thermophilus* shows the two domains in close contact (Nikonov et al., 1996). Domain II can be described as an insert in domain I. Thus, there are two connections between the domains. The conserved residues that could be involved in binding to the 23S RNA were found in the interface between the domains and were inaccessible for interaction with rRNA. It was also observed that the contact surface between the domains was very small, only about 250 Å^2. In an investigation of a mutant of L1, it was observed that conserved glycyl residues in the domain connections allowed the domains to open somewhat (Unge et al., 1997). Recently, the structure of L1 from *M. jannaschii* was determined (Nevskaya et al., unpublished). It was found that the domain structures are essentially the same, but the mutual orientation of the domains was dramatically different (Fig. 1). This form of L1 is an extended structure in which the surface likely to bind rRNA is totally exposed. The regions previously observed to be flexible showed the largest conformation differences. It remains to be established whether L1 when bound to the rRNA has this conformation or another.

L11

L11 is particularly interesting, since the structure of its C-terminal domain (CTD) has been investigated both in isolation (Marcus et al., 1997; Xing et al., 1997) and in complex with a fragment of the 23S RNA (Marcus et al., 1997; Hinck et al., 1997; Conn

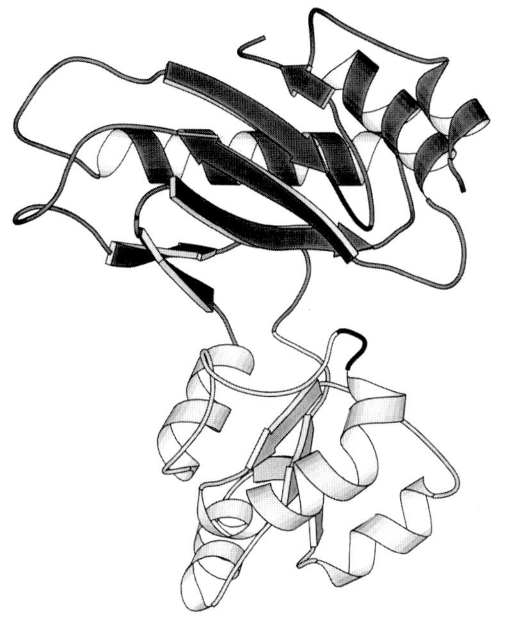

Figure 1. Two conformations of ribosomal protein L1 from *T. thermophilus* (left) and *M. jannaschii* (right). The orientation of the lower domain (domain II) is identical in both models. The residues indicated in black are the totally conserved residues 132 to 135 (GPRG) in the lower domain and 217 to 219 (TMG) in the upper domain. These residues are thought to interact with the 23S rRNA. The rotation of the upper domain opens the space between the domains to allow for rRNA binding in the case of the *M. jannaschii* structure.

et al., 1999; Wimberly et al., 1999). In one case, only the CTD was included in the crystals (Conn et al., 1999), but in the other case both domains were studied in complex with the RNA fragment (Wimberly et al., 1999). A long loop (loop 6) in the CTD was highly flexible when the protein was studied in isolation but firmly bound to the RNA in the complex. It is evident that the CTD is firmly bound to the RNA fragment, whereas the N-terminal domain (NTD) contacts this part of the rRNA more loosely. This is illustrated by the fact that the CTD has a contact area of 1,700 Å^2 with the rRNA whereas the NTD contacts the rRNA by less than 100 Å^2. Furthermore, the temperature factors indicate that the NTD has a significantly greater flexibility than the CTD.

Several observations indicate that antibiotics, particularly thiostrepton and micrococcin, may bind between the NTD and the regions of the 23S rRNA around nucleotides A1067 and A1095. These antibiotics inhibit the function of EF-G. There is a possibility that the binding of thiostrepton affects the structural transition of the NTD between two conformational states and thereby affects the function of EF-G (Porse et al., 1998; Wimberly et al., 1999).

L12

L12 is a classical case in which the highly extended conformation of the protein was observed early (Österberg et al., 1976; Bocharov et al., 1996). The identification of a hinge region, residues 37 to 51, has explained the extreme flexibility of the protein (Cowgill et al., 1984; Bushuev et al., 1984, 1989; Hamman et al., 1996). The hinge connects the NTD, which binds to protein L10, to the CTD (Leijonmarck and Liljas, 1987), which is involved in factor interactions (Kischa et al., 1971). Recent investigations have evidenced the necessity of this long hinge for the proper function of the protein in its interplay with the translational GTPases (Gudkov et al., 1991; Oleinikov et al., 1993). In addition, the flexibility and extension of L12 are probably needed for the different widely separated interactions in the ribosome that the protein is capable of (Traut et al., 1986, 1995). A full understanding of its functional cycle remains to be established.

L25

L25 is one of the proteins that bind to the 5S RNA (Erdmann et al., 1971). As can be seen from Table 1, it is one of the few ribosomal proteins for which no corresponding protein has been found in other groups of organisms. In addition, it is only rarely found in bacteria (http://www.expasy.ch/cgi-bin/lists?ribosomp.txt). In a few bacteria, a different form of the protein has been found. One was TL5 from *T. thermophilus*, which is almost twice as large as L25 from *Escherichia coli* (Gongadze et al., 1993). The N-terminal half of TL5 corresponds to L25, while the whole protein corresponds to a general shock factor, CTC (Gryaznova et al., 1996), first observed in *Bacillus subtilis* (Völker et al., 1994). Whether L25 and TL5 are genuine ribosomal proteins or shock factors remains to be explored. These studies have led to the protein structure as well as to studies of the protein in complex with fragments of the 5S RNA (Stoldt et al., 1998; Gongadze et al., 1999) (see chapter 2).

The core of L25 is a stable β-barrel domain with a motif that is repeated twice (Stoldt et al., 1998). This type of fold has previously been observed in a domain of Gln tRNA synthetase (Rould et al., 1991). L25 has a long loop (residues 9 to 25) that is disordered in the free protein but becomes ordered upon RNA binding (Stoldt et al., 1998).

SUMMARY AND CONCLUSIONS

The structural investigations have clearly established that the ribosomal proteins are formed by stable domains with significant hydrophobic cores that would hardly alter their structures upon binding to the ribosome. Several ribosomal proteins are built of two or more domains, sometimes with significant flexibility between them. Long, more or less flexible loops also frequently occur in ribosomal proteins. It is thus evident that the structures of these proteins can adopt different conformations in isolation and in the ribosome. This flexibility may be required for proper interaction with the appropriate parts of the rRNA in the assembly of the ribosomal subunits. In other cases the flexibility may be needed for the proper functioning of the ribosome.

This work was supported by grants from the Swedish Natural Science Research Council and EC Biotechnology (BIO4-CT97-2188) and from the Russian Academy of Sciences and the Russian Foundation for Basic Research. In addition, these studies were supported in part by the International Research Scholar's award of the Howard Hughes Medical Institute to M.G. and A.L.

REFERENCES

Ævarsson, A., E. Brazhnikov, M. Garber, J. Zheltonosova, Y. Chirgadze, S. Al-Karadaghi, L. A. Svensson, and A. Liljas. 1994. Three-dimensional structure of the ribosomal translocase: elongation factor G from *Thermus thermophilus. EMBO J.* **13:**3669–3677.

Babu, A., J. S. Sack, T. G. Greenhough, C. E. Bugg, A. R. Means, and W. J. Cook. 1985. Three-dimensional structure of calmodulin. *Nature* **315:**37–40.

Ban, N., B. Freeborn, P. Nissen, P. Penczek, R. A. Grassucci, R. Sweet, J. Frank, P. B. Moore, and T. A. Steitz. 1998. A 9Å resolution X-ray crystallographic map of the large ribosomal subunit. *Cell* **93:**1105–1115.

Berglund, H., A. Rak, A. Serganov, M. Garber, and T. Härd. 1997. Solution structure of the ribosomal RNA binding protein S15 from *Thermus thermophilus. Nat. Struct. Biol.* **4:**20–23.

Bocharov, E. V., A. T. Gudkov, and A. S. Arseniev. 1996. Topology of the secondary structure of ribosomal protein L7/L12 from *E. coli* in solution. *FEBS Lett.* **379:**291–294.

Burd, C. G., and G. Dreyfuss. 1994. Conserved structures and diversity of functions of RNA-binding proteins. *Science* **265:** 615–621.

Bushuev, V. N., N. F. Sepetov, and A. T. Gudkov. 1984. Symmetrical structure of the L7 protein dimer. *FEBS Lett.* **178:**101–104.

Bushuev, V. N., A. T. Gudkov, A. Liljas, and N. F. Sepetov. 1989. The flexible region of protein L12 from bacterial ribosomes studied by proton nuclear magnetic resonance. *J. Biol. Chem.* **264:**4498–4505.

Bycroft, M., T. J. P. Hubbard, M. Proctor, S. M. V. Freund, and A. G. Murzin. 1997. The solution structure of the S1 RNA binding domain: a member of an ancient nucleic acid-binding fold. *Cell* **88:**235–242.

Clemons, W. M., C. Davies, S. W. White, and V. Ramakrishnan. 1998. Conformational variability of the N-terminal helix in the structure of ribosomal protein S15. *Structure* **6:**429–438.

Conn, G. L., D. E. Draper, E. E. Lattmann, and A. G. Gittis. 1999. Crystal structure of a conserved ribosomal protein-RNA complex. *Science* **284:**1171–1174.

Cowgill, C., B. Nichols, J. Kenny, P. Butler, E. Bradbury, and R. R. Traut. 1984. Mobile domains in ribosomes revealed by proton nuclear magnetic resonance. *J. Biol. Chem.* **259:**15257–15263.

Davies, C., V. Ramakrishnan, and S. W. White. 1996a. Structural evidence for specific S8-RNA and S8-protein interactions within the 30S ribosomal subunit; ribosomal protein S8 from *Bacillus stearothermophilus* at 1.9Å resolution. *Structure* **4:**1093–1104.

Davies, C., S. W. White, and V. Ramakrishnan. 1996b. The crystal structure of ribosomal protein L14 reveals an important organizational component of the translational apparatus. *Structure* **4:** 55–66.

Davies, C., R. B. Gerstner, D. E. Draper, V. Ramakrishnan, and S. W. White. 1998. The crystal structure of ribosomal protein S4 reveals a two domain molecule with an extensive RNA-binding surface: one domain shows structural homology to the ETS DNA-binding motif. *EMBO J.* **17:**4545–4558.

Erdmann, V. A, S. Fahnestock, K. Higo, and M. Nomura. 1971. Role of 5S RNA in the functions of 50S ribosomal subunits. *Proc. Natl. Acad. Sci. USA* **68:**2932–2936.

Golden, B. L., D. W. Hoffman, V. Ramakrishnan, and S. W. White. 1993a. Ribosomal protein S17: characterization of the three-dimensional structure by ^{1}H and ^{15}N NMR. *Biochemistry* **32:**12812–12820.

Golden, B. L., V. Ramakrishnan, and S. W. White. 1993b. Ribosomal protein L6: structural evidence of gene duplication from a primitive RNA binding protein. *EMBO J.* **12:**4901–4908.

Gongadze, G. M., S. V. Tishchenko, S. E. Sedelnikova, and M. B. Garber. 1993. Ribosomal proteins, TL4 and TL5, from *Thermus thermophilus* form hybrid complexes with 5S ribosomal RNA from different microorganisms. *FEBS Lett.* **386:**46–48.

Gongadze, G. M., V. A. Meshcheryakov, A. A. Serganov, N. P. Fomenkova, E. S. Mudrik, B. H. Jonsson, A. Liljas, S. V. Nikonov, and M. B. Garber. 1999. N-terminal domain, residues 1-91, of ribosomal protein TL5 from *Thermus thermophilus* binds specifically and strongly to the region of 5S rRNA containing loop E. *FEBS Lett.* 451:51–55.

Gryaznova, O. I., N. L. Davydova, G. M. Gongadze, B.-H. Jonsson, M. B. Garber, and A. Liljas. 1996. A ribosomal protein from *Thermus thermophilus* is homologous to a general shock protein. *Biochimie* 78:915–919.

Gudkov, A. T. 1997. The L7/L12 domain of the ribosome: structural and functional studies. *FEBS Lett.* 407:253–256.

Gudkov, A. T., M. Bubunenko, and O. Gryaznova. 1991. Overexpression of L7/L12 protein with mutations in its flexible region. *Biochimie* 73:1387–1389.

Hamman, B. D., A. V. Oleinikov, G. G. Jokhadze, R. R. Traut, and D. M. Jameson. 1996. Rotational and conformation dynamics of *Escherichia coli* ribosomal protein L7/L12. *Biochemistry* 35:16672–16679.

Held, W. A., B. Ballow, S. Mizushima, and M. Nomura. 1974. Assembly mapping of 30S ribosomal proteins from *E. coli.* Further studies. *J. Biol. Chem.* 249:3103–3111.

Hinck, A. P., M. A. Marcus, S. Huang, S. Gizesiek, I. Kurtonovich, D. E. Draper, and D. A. Torchia. 1997. The RNA binding domain of ribosomal protein L11: three-dimensional structure of the RNA-bound form of the protein and its interaction with 23S rRNA. *J. Mol. Biol.* 274:101–113.

Hoffman, D. W., C. S. Cameron, C. Davies, S. W. White, and V. Ramakrishnan. 1996. Ribosomal protein L9: a structure determination by the combined use of X-ray crystallography and NMR spectroscopy. *J. Mol. Biol.* 264:1058–1071.

Hosaka, H., A. Nakagawa, I. Tanaka, N. Harada, K. Sano, M. Kimura, M. Yao, and S. Wakatsuki. 1997. Ribosomal protein S7: a new RNA-binding motif with structural similarities to a DNA architectural factor. *Structure* 5:1199–1208.

Jaishree, T. N., V. Ramakrishnan, and S. W. White. 1996. Solution structure of prokaryotic ribosomal protein S17 by high-resolution NMR spectroscopy. *Biochemistry* 35:2845–2853.

Kischa, K., W. Möller, and G. Stöffler. 1971. Reconstitution of a GTPase activity by a 50S ribosomal protein from *E. coli. Nat. New Biol.* 233:62–63.

Laurberg, M., O. Kristensen, S. Al-Karadaghi, K. Martemyanov, A. T. Gudkov, and A. Liljas. Unpublished data.

Leijonmarck, M., and A. Liljas. 1987. Structure of the C-terminal domain of the ribosomal protein L7/L12 from *Escherichia coli* at 1.7Å. *J. Mol. Biol.* 195:555–579.

Leijonmarck, M., S. Erikson, and A. Liljas. 1980. Crystal structure of a ribosomal component at 2.6Å. *Nature* 286:824–827.

Leijonmarck, M., K. Appelt, J. Badger, A. Liljas, K. S. Wilson, and S. W. White. 1988. Structural comparison of the procaryotic ribosomal protein L7/L12 and L30. *Proteins* 3:243–248.

Liljas, A., and M. Garber. 1995. Ribosomal proteins and elongation factors. *Curr. Opin. Struct. Biol.* 5:721–727.

Liljas, A., and A. T. Gudkov. 1987. The structure and dynamics of ribosomal protein L12. *Biochimie* 69:1043–1047.

Liljas, A., and S. Al-Karadaghi. 1997. Structural aspects of protein synthesis. *Nat. Struct. Biol.* 4:767–771.

Lindahl, M., S. A. Svensson, A. Liljas, S. E. Sedelnikova, I. A. Eliseikina, N. P. Fomenkova, N. Nevskaya, S. V. Nikonov, M. B. Garber, T. A. Muranova, A. I. Rykunova, and R. Amons. 1994. Crystal structure of ribosomal protein S6 from *Thermus thermophilus. EMBO J.* 13:1249–1254.

Marcotrigiano, J., A. C. Gingras, N. Sonenberg, and S. K. Burley. 1997. Cocrystal structure of the messenger RNA 5′ cap-binding protein (eIF4E) bound to 7-methyl-GDP. *Cell* 89:951–961.

Marcus, M. A., A. P. Hinck, S. Huang, D. E. Draper, and D. A. Torchia. 1997. High resolution structure of ribosomal protein L11-C76, a helical protein with a flexible loop that becomes structured upon binding to RNA. *Nat. Struct. Biol.* 4:70–77.

Marcus, M. A., R. B. Gerstner, D. E. Draper, and D. A. Torchia. 1998. The solution structure of ribosomal protein S4Δ41 reveals two subdomains and a positively charged surface that may interact with RNA. *EMBO J.* 17:4559–4571.

Matsuo, H., H. Li, A. M. McGuire, C. M. Fletcher, A. C. Gingras, N. Soneneberg, and G. Wagner. 1997. Structure of translation factor eIF4E bound to m7GDP in interaction with 4E-binding protein. *Nat. Struct. Biol.* 4:717–724.

Meador, W. E., A. R. Means, and F. A. Quiocho. 1992. Target enzyme recognition by calmodulin: 2.4Å structure of a calmodulin-peptide complex. *Science* 257:1251–1255.

Moore, P. B. 1998. The three-dimensional structure of the ribosome and its components. *Annu. Rev. Biophys. Biomol. Struct.* 27:35–58.

Nakagawa, A., T. Nakashima, M. Taniguchi, H. Harumi, M. Kimura, and I. Tanaka. 1999. The three-dimensional structure of the RNA-binding domain of ribosomal protein L2; a protein at the peptidyl transferase center of the ribosome. *EMBO J.* 18: 1459–1467.

Nevskaya, N., S. Tishchenko, A. Nikulin, S. Al-Karadaghi, A. Liljas, B. Ehresmann, C. Ehresmann, M. Garber, and S. Nikonov. 1998. Crystal structure of ribosomal protein S8 from *Thermus thermophilus* reveals a high degree of structural conservation of a specific RNA binding site. *J. Mol. Biol.* 279:233–244.

Nevskaya, N., S. Tishchenko, R. Fedorov, S. Al-Karadaghi, A. Liljas, A. Kraft, W. Piendl, M. Garber, and S. Nikonov. Unpublished data.

Nierhaus, K. 1990. Reconstitution of ribosomes, 161–189. *In* G. Spedding (ed.), *Ribosomes and Protein Synthesis.* IRL Press, Oxford, United Kingdom.

Nikonov, S., N. Nevskaya, I. Eliseikina, N. Fomenkova, A. Nikulin, N. Ossina, M. Garber, B. H. Jonsson, C. Briand, S. Al-Karadaghi, A. Svensson, A. Ævarsson, and A. Liljas. 1996. Crystal structure of the RNA-binding ribosomal protein L1 from *Thermus thermophilus. EMBO J.* 15:1350–1359.

Nikonov, S., N. Nevskaya, R. Fedorov, A. Khairullina, S. Tishchenko, A. Nikulin, and M. Garber. 1998. Structural studies of ribosomal proteins. *Biol. Chem.* 379:795–806.

Oleinikov, A. V., B. Perroud, B. Wang, and R. R. Traut. 1993. Structural and functional domains of *Escherichia coli* ribosomal protein L7/L12 the hinge region is required for activity. *J. Biol. Chem.* 268:917–922.

Österberg, R., B. Sjöberg, A. Liljas, and I. Petterson. 1976. Small-angle X-ray scattering and crosslinking study of the protein L7/L12 from *E. coli* ribosomes. *FEBS Lett.* 66:48–51.

Porse, B. T., I. Leviev, A. S. Mankin, and R. A. Garrett. 1998. The antibiotic thiostrepton inhibits a functional transition within protein L11 at the ribosomal GTPase center. *J. Mol. Biol.* 276: 391–404.

Ramakrishnan, V., and S. W. White. 1992. The structure of ribosomal protein S5 reveals sites of interaction with 16S rRNA. *Nature* 358:768–771.

Ramakrishnan, V., and S. W. White. 1998. Ribosomal protein structures: insights into the architecture, machinery and evolution of the ribosome. *Trends Biochem. Sci.* 23:208–212.

Rossmann, M. G., and A. Liljas. 1974. Recognition of structural domains in globular proteins. *J. Mol. Biol.* 85:177–181.

Rould, M. A., J. J. Perona, and T. A. Steitz. 1991. Structural basis of anticodon loop recognition by glutaminyl-tRNA synthetase. *Nature* 352:213–218.

Stoldt, M., J. Wöhnert, M. Görlach, and L. R. Brown. 1998. The NMR structure of *Escherichia coli* ribosomal protein L25 shows homology to general stress protein and glutaminyl-tRNA synthetase. *EMBO J.* 17:6377–6384.

Traut, R. R., D. Tewari, A. Sommer, G. Gavino, D. Glitz, and B. Wang. 1986. Protein topography of ribosomal functional domains: effect of monoclonal antibodies to different epitopes in *E. coli* protein L7/L12 on ribosome function and structure, p. 286–308. *In* B. Hardesty and G. Kramer (ed.), *Structure, Function and Genetics of Ribosomes.* Springer-Verlag Press, New York, N.Y.

Traut, R. R., D. Dey, D. Bochkariov, A. V. Oleinikov, G. G. Jokhadse, B. D. Hamman, and D. M. Jameson. 1995. Location and domain structure of *Escherichia coli* ribosomal protein L7/L12: site specific cysteine crosslinking and attachment of fluorescent probes. *Biochem. Cell Biol.* 73:949–958.

Unge, J., S. Al-Karadaghi, A. Liljas, B.-H. Jonsson, I. Eliseikina, N. Ossina, N. Nevskaya, N. Fomenkova, M. Garber, and S. Nikonov. 1997. A mutant form of ribosomal protein L1 reveals conformational flexibility. *FEBS Lett.* **411**:53–59.

Unge, J., A. Åberg, S. Al-Kharadaghi, A. Nikulin, S. Nikonov, N. L. Davydova, N. Nevskaya, M. Garber, and A. Liljas. 1998. The crystal structure of ribosomal protein L22 from *Thermus thermophilus.* An RNA binding protein at the polypeptide exit channel. *Structure* **6**:1577–1586.

Völker, U., S. Engelmann, B. Maul, S. Riethdorf, R. Schmid, H. Mach, and M. Hecker. 1994. Analysis of the induction of general stress proteins of *Bacillus subtilis. Microbiology* **140**:741–752.

Wilson, K. S., K. Appelt, J. Badger, I. Tanaka, and S. W. White. 1986. Crystal structure of prokaryotic ribosomal protein. *Proc. Natl. Acad. Sci. USA* **83**:7251–7255.

Wimberly, B. T., S. W. White, and V. Ramakrishnan. 1997. The structure of ribosomal protein S7 at 1.9Å resolution reveals a beta-hairpin motif that binds double stranded nucleic acids. *Structure* **5**:1187–1198.

Wimberly, B. T., R. Guymon, J. P. McCutcheon, S. W. White, and V. Ramakrishnan. 1999. A detailed view of a ribosomal active site: the structure of the L11-RNA complex. *Cell* **97**:491–502.

Xing, Y., D. Guha Takurta, and D. E. Draper. 1997. The RNA binding domain of ribosomal protein L11 is structurally similar to homeodomains. *Nat. Struct. Biol.* **4**:24–27.

Yonath, A., and F. Franceschi. 1997. New RNA recognition features revealed in ancient ribosomal proteins. *Nat. Struct. Biol.* **4**:3–5.

The Ribosome: Structure, Function, Antibiotics, and Cellular Interactions
Edited by R. A. Garrett, S. R. Douthwaite, A. Liljas, A. T. Matheson, P. B. Moore, and H. F. Noller

Chapter 8

Structures of Bacterial Ribosomal Proteins: High-Resolution Probes of the Architecture and Mechanism of the Ribosome

STEPHEN W. WHITE, WILLIAM M. CLEMONS, JR., CHRISTOPHER DAVIES, V. RAMAKRISHNAN, and BRIAN T. WIMBERLY

It has long been recognized that a complete understanding of the mechanism of translation will ultimately depend on determining the molecular structure of the ribosome. The past 4 years have seen a number of exciting advances in ribosome research, and this daunting task is now considered a real possibility. These advances have come from many groups approaching the problem in a holistic manner from several different but coordinated biochemical and biophysical directions. Structural studies have been particularly important to this effort. Most notable has been the dramatic increase in resolution of the three-dimensional images of the ribosome from cryo-electron microscopy (Stark et al., 1997a, 1997b; Malhotra et al., 1998; Frank, 1998) and X-ray crystallography (Ban et al., 1998). These images provide a reliable structural basis for building models of the complex. Another advance has been the determination of the high-resolution structures of ribosomal components and factors. These structures and their active sites represent high-resolution probes with which to investigate and validate the models. Also, the improving images of the whole ribosome are starting to reveal these components at lower resolution, and their high-resolution structures are starting to be incorporated in a modular fashion.

For the past 10 years, our contribution to this effort has been the determination of the structures of individual ribosomal proteins. To date, our work has resulted in 11 of the 18 currently known structures, including some duplication with other groups (Ramakrishnan and White, 1998). Not all ribosomal proteins are of equal importance, either to the function of the ribosome or to determining its structure, and we have been careful to prioritize the proteins for analysis. This strategy has been possible due to our early development of methods for cloning and overexpressing ribosomal protein genes (Ramakrishnan and Gerchman, 1991). Most of the proteins studied contain important sites, such as RNA binding and functional point mutations, that provide the maximal number of structural and mechanistic constraints for developing and interpreting models of the complete subunits. For the past 3 years, we have concentrated on the so-called primary rRNA binding proteins of the 30S subunit, since they initiate the folding of the 16S rRNA and are integral to its three-dimensional conformation. These primary rRNA binding proteins are also the best candidates for future structural studies of protein-RNA complexes, since they bind to their RNA target sites in the absence of other ribosomal components.

An important aspect of this work has always been the insights it has given into the evolution and possible origins of protein structures and motifs. Ribosomal proteins are extremely ancient and may have been among the first proteins that emerged from a possible RNA world. Our results and those of colleagues in the field support these ideas, and most of the ribosomal protein structures are homologous to

Stephen W. White ■ Department of Structural Biology, St. Jude Children's Research Hospital, 332 N. Lauderdale, Memphis, TN 38105, and Department of Biochemistry, University of Tennessee, 858 Madison Ave., Memphis, TN 38163. **William M. Clemons, Jr.** ■ Department of Biochemistry, University of Utah School of Medicine, Salt Lake City, UT 84132. **Christopher Davies** ■ Department of Structural Biology, St. Jude Children's Research Hospital, 332 N. Lauderdale, Memphis, TN 38105, and School of Biological Sciences, University of Sussex, Falmer, Brighton BN1 9QG, United Kingdom. **V. Ramakrishnan and Brian T. Wimberly** ■ Department of Biochemistry, University of Utah School of Medicine, Salt Lake City, UT 84132, and MRC Laboratory of Molecular Biology, Hills Rd., Cambridge CB2 2QH, United Kingdom.

and may be antecedents of modern proteins, in particular RNA and DNA binding proteins.

Here, we summarize our findings from the past 4 years. Emphasis will be placed on describing the structures and putative functional sites of four primary RNA binding proteins that we have determined from the 30S subunit. However, to complement several reports elsewhere in this volume, we will also discuss how our structures have provided information for structural studies of the complete subunits.

STRUCTURAL STUDIES OF 30S RIBOSOMAL PROTEINS S4, S7, S8, AND S15

The four proteins, S4, S7, S8, and S15, together with S17, control the initial stages of the folding of the 16S rRNA molecule and are crucial determinants of the 30S architecture. The high-resolution structures of these proteins represent important landmarks within the 30S subunit and will be invaluable for developing models of the complex. Particularly useful for modeling purposes are the potential sites on each protein for specific interactions with 16S rRNA and other ribosomal proteins. These are tentatively identified by considering the conserved surface regions of each protein. Also useful for identifying the putative RNA binding regions are the sites of mutations on the proteins that produce phenotypes such as antibiotic resistance and defective ribosome assembly. It is now generally accepted that the principal role of the ribosomal proteins is to help fold and maintain the three-dimensional structure of the catalytic rRNA component (Noller et al., 1992), and these mutations within the proteins appear to disrupt the local RNA structure. We had already determined the structure of S17 by using nuclear magnetic resonance (NMR) techniques, and we have now completed the crystal structures of the remaining four proteins.

Ribosomal Protein S4

With a molecular mass of 23 kDa, S4 is one of the largest and most important of the ribosomal proteins. S4 mediates the early folding of the 16S rRNA and controls the assembly of the body of the 30S subunit (Nowotny and Nierhaus, 1988). Its extensive RNA footprint at the junction of five helical segments (helices 3, 4, 16, 17, and 18) within a 460-nucleotide region at the 5′ end of the RNA (Powers and Noller, 1995; Heilek et al., 1995) reflects its role as a focal point for the assembly process. It is located in the body of the 30S subunit close to S3, S5, and S12 (Capel et al., 1987; Stöffler and Stöffler-Meilicke, 1984; Lambert et al., 1983) and is part of the so-called accuracy domain that mediates translational fidelity (Kurland et al., 1990; Powers and Noller, 1991). Specifically, S4 has been identified as a major locus for *ram* mutations that compromise fidelity. S4 also specifically binds to a 110-nucleotide pseudoknot structure within the α operon mRNA and thereby controls the expression of other ribosomal proteins (Spedding and Draper, 1993).

We cloned the S4 gene from the thermophile *Bacillus stearothermophilus* by our usual techniques (Ramakrishnan and Gerchman, 1991) but failed to crystallize the purified intact protein. Based on earlier protease digestion studies of the S4-rRNA complex (Changchien and Craven, 1976), Draper and coworkers have shown that the first 40 amino acids of S4 are apparently flexible and can be deleted without seriously affecting the specific RNA binding properties of the protein. The region of the gene that codes for these residues was removed from the gene, and the truncated protein (S4Δ41) produced excellent crystals that diffract to 1.7-Å resolution. The structure was determined by standard multiple isomorphous replacement (Davies et al., 1998a).

S4Δ41 is slightly elongated, with dimensions of 40 by 50 by 65 $Å^3$, and comprises two domains (Fig. 1a). Domain 1 is entirely α-helical and incorporates the noncontiguous residues 42 to 91 and 180 to 200. It contains four α-helices ($\alpha1$-$\alpha2$-$\alpha3$-$\alpha7$) and appears to be particularly stable due to a cluster of five aromatic residues in the hydrophobic core and several strategically placed salt bridges. Residues 92 to 179 form the α/β domain 2 that contains three α-helices and a five-stranded antiparallel β-sheet with a Greek key topology ($\alpha4$-$\alpha5$-$\beta1$-$\beta2$-$\beta3$-$\alpha6$-$\beta4$-$\beta5$). The three α-helices pack onto one side of the β-sheet, leaving the opposite side of the β-sheet exposed. Domain 2 represents an insertion into domain 1, and in common with most of the larger ribosomal proteins, S4 appears to be a fusion of two small protein motifs. The two domains are packed closely together, but a canyon that encircles two-thirds of the molecule is a characteristic feature of the boundary. Parallel with our crystallographic studies, Torchia and coworkers also determined the S4 structure by NMR techniques. Apart from a possible difference in the relative orientation of the two domains, the structures are essentially identical (Markus et al., 1998).

In previous studies, we have attempted to identify the RNA binding sites on ribosomal proteins by considering sequence conservation of appropriate residues, surface charge, and the sites of functional mutations. Each of these criteria points emphatically to the surface shown in Fig. 1b. This surface, which contains the canyon, is highly basic (Fig. 1c) and extremely conserved in contrast to the opposite surface,

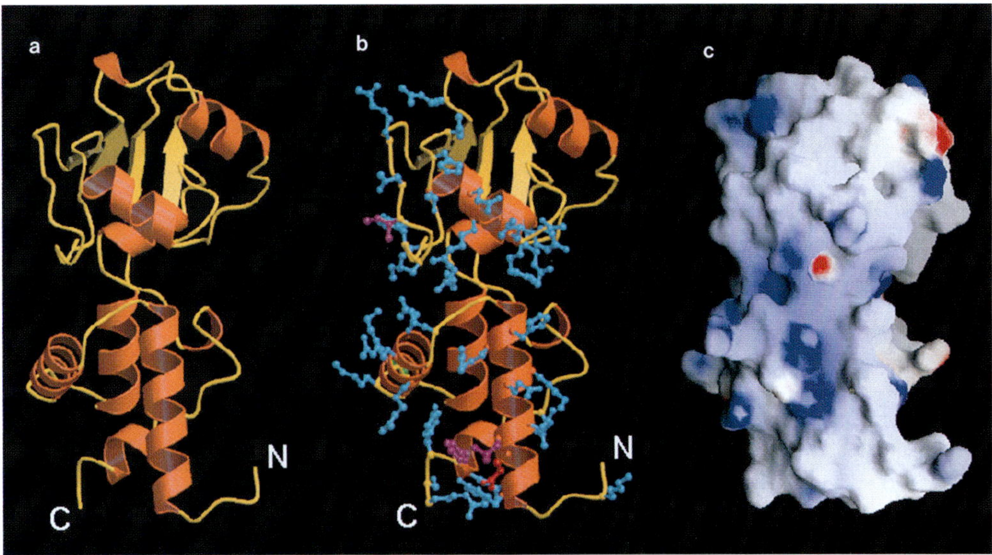

Figure 1. Structure of ribosomal protein S4 from *B. stearothermophilus*. (a) Cartoon showing locations of the α-helices (orange), β-strands (yellow), and loops (yellow). The N and C termini are indicated. (b) Locations of functionally and biochemically important amino acids. Blue, putative RNA binding residues; magenta, conserved surface hydrophobic residues; red, site of point mutation conferring a *ram* phenotype. (c) Surface charge potential as calculated by the program GRASP (Nicholls et al., 1991). Blue, positive; red, negative; white, hydrophobic.

which has none of these features. RNA appears to interact extensively with this surface, a conclusion which is supported by protein-rRNA cross-linking studies (Urlaub et al., 1995, 1997). However, three possible RNA binding patches are evident, including a large one associated with the canyon. It has been shown that the S4 *ram* mutations seriously affect the RNA binding properties of the molecule (Green and Kurland, 1971) and perturb the 16S rRNA structure in vivo (Allen and Noller, 1989). The majority of these mutations are C-terminal truncations of the protein that would compromise the folding of domain 1 rather than domain 2 (Daya-Grosjean et al., 1972). Therefore, domain 1 is likely to be interacting with the 16S rRNA elements associated with the accuracy domain, which serves to orient S4 on the 30S subunit.

Ribosomal Protein S7

Just as S4 controls the assembly of the 30S body, S7 is equally important for the assembly of the 30S head region (Nowotny and Nierhaus, 1988). It has a molecular mass of 17.5 kDa and is located in the head close to proteins S9, S13, S14, and S19 (Capel et al., 1987; Stöffler and Stöffler-Meilicke, 1984; Lambert et al., 1983). Like S4, S7 footprints a large region of 16S rRNA, consistent with its role in early folding (Powers and Noller, 1995). This region is within the 3′ region which is known to constitute the head. S7 appears to bind a substructure within the 16S rRNA comprising two multistem junctions (Dragon and Brakier-Gingras, 1993; Dragon et al., 1994). The functional location of S7 has been determined by a number of cross-linking studies. It is located close to the decoding site (Sylvers et al., 1992; Doring et al., 1994; Muralikrishna and Cooperman, 1994) on its 5′, or exiting, side (Dontsova et al., 1992), and it also appears to be adjacent to the 530 loop of 16S rRNA, which forms part of the accuracy domain (Alexander et al., 1994). The protein also controls the synthesis of other ribosomal proteins in the *str* operon by binding to a defined region of the cistronic mRNA (Saito and Nomura, 1994).

The gene was cloned from *Thermus thermophilus*, and crystals of the full-length protein were grown that diffract to 1.9-Å resolution. The structure was solved using multiwavelength anomalous methods on a mercury derivative (Wimberly et al., 1997). The protein contains a single α-helical domain comprising a bundle of six α-helices, α1-α2-α3-α4-α5-α6 (Fig. 2a). The N-terminal 11 residues are disordered in the structure, as are 5 internal loop residues immediately following α6. A distinguishing feature of the protein is an extended β-ribbon between helices α3 and α4. Helices α1-α2-α3-α4-α5 constitute the main body of the protein stabilized by a defined hydrophobic core, but α6 forms a bridge to the β-ribbon and appears to orient it with respect to the main body of the protein. Helix α6 is stabilized by a large number of salt bridges, perhaps necessitated by its exposed location. The β-ribbon has a defined twisted shape dictated by

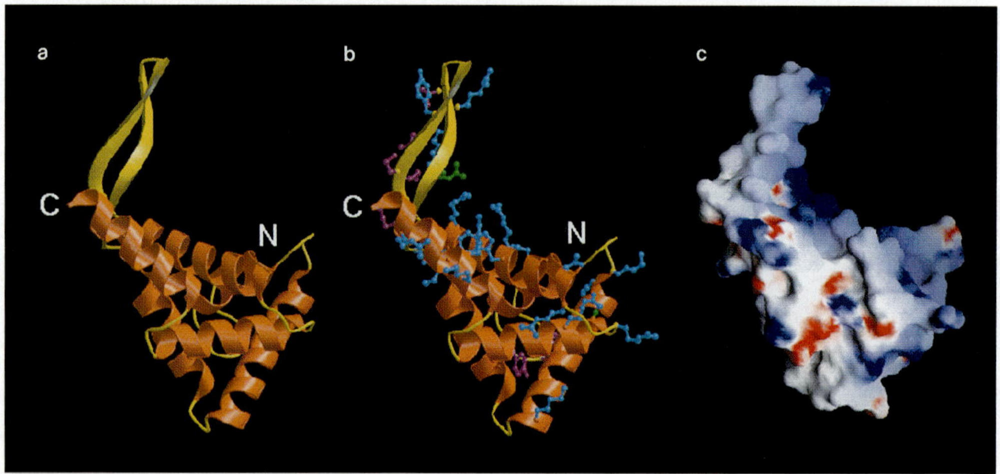

Figure 2. Structure of ribosomal protein S7 from *T. thermophilus*. (a) Cartoon showing locations of α-helices (orange), β-strands (yellow), and loops (yellow). The N and C termini are indicated. (b) Locations of functionally and biochemically important amino acids. Blue, putative RNA binding residues; magenta, conserved surface hydrophobic residues; green, residue cross-linked to RNA. (c) Surface charge potential as calculated by the program GRASP (Nicholls et al., 1991). Blue, positive; red, negative; white, hydrophobic.

three conserved proline residues. Also, the inbound strand contains mostly conserved hydrophobic residues, whereas the outbound strand contains mostly conserved basic residues. To date, S7 represents the largest single domain found in ribosomal proteins.

According to our standard criteria, S7 contains two putative RNA binding sites (Fig. 2b). The largest is a concave surface made up of the β-ribbon and portions of α1, α4, and α6. The shape and dimensions of this surface, which is highly electropositive due to a large number of conserved basic residues (Fig. 2c), approximately matches that of double-stranded (ds)RNA. Recently, a conserved basic residue within the β-ribbon (lysine 75 in *Escherichia coli*) was cross-linked to a nucleotide within the 3′ region of 16S rRNA, which supports this proposal (Urlaub et al., 1995, 1997). The N and C termini of S7 are also close to this concave surface, and both contain flexible conserved basic residues that probably become ordered when this region binds RNA. The second putative RNA binding surface, which is also supported by cross-linking data to the 3′ region of 16S rRNA (Urlaub et al., 1995, 1997), is at the opposite end of the molecule close to helix α4 (Fig. 2b). Finally, a conserved cluster of hydrophobic residues may represent a site of protein-protein interaction, and the best candidate is S9, which appears to be the closest neighboring protein (Capel et al., 1987).

Ribosomal Protein S8

S8 is a medium-size ribosomal protein with a molecular mass of 14.5 kDa. It is centrally located in the 30S subunit adjacent to proteins S2, S4, S5, S12, S15, and S17 (Capel et al., 1987; Stöffler and Stöffler-Meilicke, 1984; Lambert et al., 1983) and is therefore a useful reference point in the subunit for modeling purposes. Several early studies suggest that S8 directly interacts with S5 (Allen et al., 1979; Tindall and Aune, 1981). S8 has a crucial role in folding the central domain of 16S rRNA (Svensson et al., 1988), as evidenced by mutations that result in defective ribosome assembly (Geyl et al., 1977). The interaction of S8 with 16S rRNA has been studied by a number of groups with a variety of techniques (Oßwald et al., 1987; Wiener et al., 1988; Powers and Noller, 1995), and the protein has been shown to recognize and bind to a conserved structure within helix 21 (nucleotides 583 to 653 in *E. coli*) which contains a crucial small internal loop (Wu et al., 1994; Allmang et al., 1994). S8 is also an important regulatory protein controlling the expression of 10 ribosomal-protein genes in the *spc* operon by binding specifically to a region of the polycistronic mRNA (Cerretti et al., 1988).

The S8 gene was cloned from *B. stearothermophilus*, and the protein produced crystals that diffract to 1.9-Å resolution. The structure was determined by multiple isomorphous replacement techniques (Davies et al., 1996a). The structure comprises two α/β domains that are tightly associated through an interdomain hydrophobic core (Fig. 3a). In the N-terminal domain, two α-helices pack onto one surface of a three-stranded antiparallel β-sheet with an overall topology of α1-β1-α2-β2-β3. The C-terminal domain has a rather unusual pyramid-like structure with to-

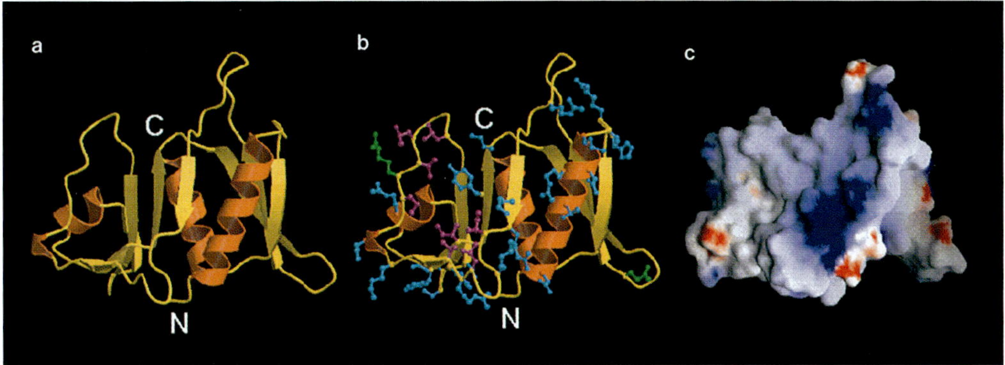

Figure 3. Structure of ribosomal protein S8 from *B. stearothermophilus*. (a) Cartoon showing locations of α-helices (orange), β-strands (yellow), and loops (yellow). The N and C termini are indicated. (b) Locations of functionally and biochemically important amino acids. Blue, putative RNA binding residues; magenta, conserved surface hydrophobic residues; green, residue cross-linked to RNA (bottom right) and ribosomal protein S5 (top left). (c) Surface charge potential as calculated by the program GRASP (Nicholls et al., 1991). Blue, positive; red, negative; white, hydrophobic.

pology β4-β5-β6-β7-α3-β8-β9. The three sides of the pyramid are formed by sheet β4-β9-β6-β7, sheet β5-β8-β6-β7, and helix α3. A conserved feature of the C-terminal domain is a large hydrophobic patch in the concave inner surface of the pyramid structure (Fig. 3b).

We have identified two putative RNA binding sites on S8 (Fig. 3b). The first is located at the top of the N-terminal domain at the C-terminal end of helix α1 and is flanked by regions of positive potential (Fig. 3c). The second site is on the lower surface of the C-terminal domain as viewed in Fig. 3b and is populated by a number of conserved exposed residues. An S8-RNA cross-link has been identified in the loop between β2 and β3 somewhat distant from the two sites (Urlaub et al., 1995, 1997), but an extended region of RNA bound below the protein could easily span site 2 and the cross-link. Finally, the curious hydrophobic patch on the C-terminal domain is flanked by several conserved basic residues (Fig. 3b) and has the hallmarks of a docking site for another protein-RNA complex. All three putative functional sites are on the same surface shown in Fig. 3b, which is generally highly conserved. In contrast, and like S4, the opposite surface contains no distinguishing features apart from being generally polar.

Ribosomal Protein S15

S15 is a small, highly basic protein with a molecular mass of 10.5 kDa. It is centrally located in the 30S subunit adjacent to proteins S6, S8, S17, and S18 (Capel et al., 1987; Stöffler and Stöffler-Meilicke, 1984; Lambert et al., 1983) and binds specifically to a well-defined region of 16S rRNA comprising helix 22 and the three-way helical junction of helices 21, 22, and 23 (Powers and Noller, 1995; Batey and Williamson, 1996a, 1996b; Serganov et al., 1996). Like S4, S7, and S8, S15 also controls the translation from its own operon by binding specifically to the mRNA (Portier et al., 1990).

The S15 gene was cloned from *B. stearothermophilus*, and crystals that diffract to 2.1Å were grown. The structure was determined by using multiwavelength anomalous dispersion phasing on crystals of the selenomethionyl protein (Clemons et al., 1998). The structure basically comprises a bundle of four α-helices (Fig. 4a). The structure of S15 has also been determined by NMR methods (Berglund et al., 1997), and although poorly resolved, it is clearly very similar to our structure. The core of the structure comprises a coiled-coil type of interaction between the hydrophobic faces of the two longer helices, α2 and α3. In the crystal, helix α4 packs against the coiled-coil at the bottom of the molecule as viewed in Fig. 4a, but the N-terminal helix, α1, associates with a neighbor in the crystal lattice. The NMR structure confirms that this is a crystal artifact, as suspected, and α1 actually joins α4 to form a four-helix bundle. As far as we know, this is the only example where the crystallization of a ribosomal protein has produced a major artifact. The NMR structure also shows that α1 has a rather tenuous connection to the rest of the protein, and the loop connecting α1 and α2 is also poorly defined in our electron density map. This flexibility may be required for functional reasons, for example, in RNA binding or protein-protein interactions.

The putative RNA binding residues of S15 are clustered at either end of the molecule on the same face (Fig. 4b). This entire face of the molecule is highly basic (Fig. 4c) and conserved, in contrast to

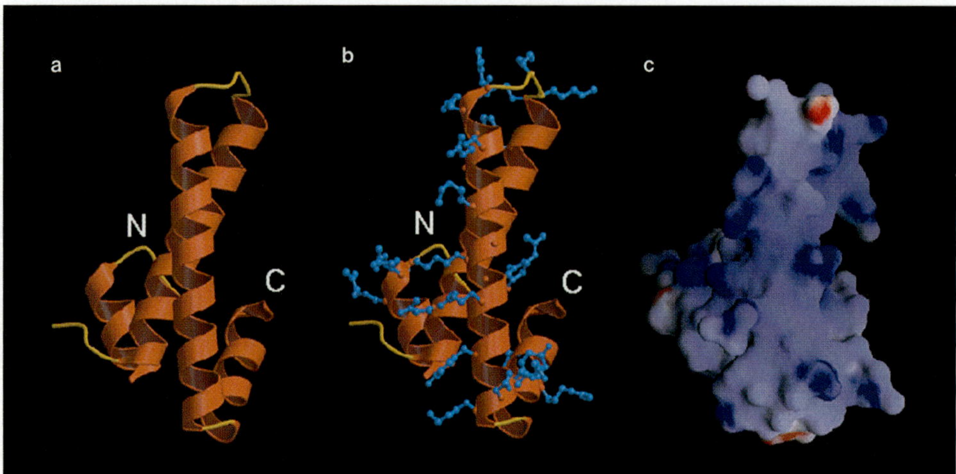

Figure 4. Structure of ribosomal protein S15 from *B. stearothermophilus*. (a) Cartoon showing locations of α-helices (orange) and loops (yellow). The N and C termini are indicated. Note that the N-terminal helix belongs to a neighbor in the unit cell, but its location on the molecule appears to be correct. (b) Locations of putative RNA binding residues (blue). (c) Surface charge potential as calculated by the program GRASP (Nicholls et al., 1991). Blue, positive; red, negative; white, hydrophobic.

the opposite face, which is featureless. The overall dumbbell-shaped architecture of the putative S15 RNA binding surface suggests that the protein may interact with two adjacent RNA sites.

RIBOSOMAL PROTEINS S5, S17, L6, AND L9 REVISITED

The four ribosomal proteins S5, S17, L6, and L9 have been studied and described earlier, but we have continued to improve their structures, and a number of interesting features have emerged which merit a brief description.

The S5 crystal structure was originally reported at 2.7 Å (Ramakrishnan and White, 1992), but this has now been extended to 2.2 Å (Davies et al., 1998b). The overall structure has not changed, but two important features are now apparent. First, the N-terminal domain has the exact topology of a typical dsRNA binding protein (dsRBD). This is important because the crystal structure of a dsRBD-dsRNA complex has now been reported (Ryter and Schultz, 1998), and this provides a possible model of how the N-terminal domain of S5 binds dsRNA. It also shows that two flexible regions almost certainly have defined functions in the complex. The basic loop between β1 and β2 becomes extended within the dsRNA minor groove, and the N terminus probably forms an α-helix that binds into the adjacent minor groove. The second feature is that the exposed β-sheet surface on the C-terminal domain, which has a convex shape, contains a distinct conserved hydrophobic patch.

An earlier NMR analysis of S17 by mainly homonuclear methods had revealed that the molecule comprises a five-stranded β-barrel with two extended and apparently mobile loop regions (Golden et al., 1993a). In order to determine the complete three-dimensional structure, the molecule was doubly labeled with ^{15}N and ^{13}C, and a complete set of heteronuclear experiments were performed to assign the maximal number of ^{1}H, ^{15}N, and ^{13}C resonances. These NMR data were extensively analyzed by using distance geometry, back-calculation, and simulated annealing, and the complete structure was determined (Jaishree et al., 1996). The structure confirmed the low-resolution model, but two extensive loop regions that were originally unresolved due to either mobility or disorder could not be visualized.

The structure of L6 was originally determined to 2.6 Å (Golden et al., 1993b), and this has been extended to 2.0 Å (Davies et al., 1998b). Again, the structure has not changed in any major way, but an additional putative RNA binding site was established at the tip of the N-terminal domain. We also characterized two gentamicin-resistant mutants of L6 and showed that both contain deletions at their C termini. These deletions compromise the major putative RNA binding site on the protein in common with similar functionally important mutations in other ribosomal proteins.

Finally, the structure of L9 was further investigated by NMR spectroscopy. L9 has two domains, but the smaller, N-terminal domain was poorly characterized in the original crystal structure due to flexibility and limited crystal contacts (Hoffman et al., 1994). To complete the refinement of the molecule, its NMR spectrum was assigned and a joint refinement procedure was performed with both crystal and NMR data (Hoffman et al., 1996). This is a rather unusual way of using the two methods in tandem to determine a single structure. The final model showed that the fold of the N-terminal domain as originally published was correct, but that a β-strand region of the sequence had been incorrectly placed in the electron density (shifted by 2 residues). The NMR data also confirmed that the unusual structure of L9, which contains two domains separated by an extended α-helix, is not a crystal artifact but actually exists in solution. Two papers fully describing the NMR analysis of L9 have been independently published (Lillemoen et al., 1997; Lillemoen and Hoffman, 1998).

BUILDING THE RIBOSOME—THE PROTEIN STRUCTURAL DATA

Steadily accumulating biochemical and biophysical data have increased our understanding of the structure and mechanism of the bacterial ribosome, and contemporary data have been used periodically to construct models of the subunits, which then form the basis for subsequent experimentation (Brimacombe et al., 1988; Stern et al., 1988; Mueller and Brimacombe, 1997a, 1997b; Mueller et al., 1997). These models contain many errors, ambiguities, and conflicts, but high-resolution structures of proteins and RNA fragments provide the means for improving these models and represent an independent way of testing their validity. More recently, X-ray-crystallographic images of complete ribosomal subunits have become sufficiently resolved to allow the visualization of ribosomal components, in particular, the individual proteins (see chapters 1, 2, and 13). Each protein has a unique shape at these resolutions, and they represent excellent places to begin interpreting the enormously complex electron density maps. Any information on how these proteins interact with rRNA, in particular, the precise regions they bind to, is also extremely useful for these interpretations. Here, we describe the pertinent information that we have identified for each of the protein structures that we have determined during the past 7 years. Protein L11, for which we have the structure of a complex with RNA, is a rather special case and is described separately in chapter 8.

The 30S Subunit

Our choice of proteins for structural analysis was partly based on the neutron map, which shows that S4, S5, S8, S15, and S17 form a cluster within the body of the subunit. This clustering is supported by the RNA footprinting data for these proteins, which show that they interact with adjacent regions of the 16S rRNA molecule. Thus, the prospects of building this important central region of the 30S subunit are excellent. S7 is somewhat distant from this cluster, organizing the 3′ domain of 16S rRNA that creates the head region. S7 will not be discussed here, but the structure fits the 16S rRNA model very well and has been reported elsewhere (Tanaka et al., 1998). The important architectural determinants of the remaining five proteins are summarized below.

S4 footprints and probably organizes a number of regions of 16S rRNA, but in the three-dimensional model of the 16S rRNA from Brimacombe's group, these regions form a cluster on the outside surface of the 30S subunit (Mueller and Brimacombe, 1997a). One entire face of S4 appears to interact with RNA, and it is reasonable to orient the protein with this face towards the RNA cluster and the opposite face towards the exterior. The latter face is polar and featureless, which fits this scenario well. S4 is known to form part of the accuracy center, which includes the 530 and 900 stem-loops. Since it is the α-helical domain 1 that is compromised in the *ram* mutants, the most likely orientation of the protein is with this domain facing, and possibly interacting with, these two regions of 16S rRNA.

S5 has a complicated environment that is extremely useful for modeling purposes. Mutations within S5 that produce a *ram* phenotype are within the C-terminal domain that should form part of the accuracy center close to S4. This is supported by site-directed hydroxyl radical probing experiments (Heilek and Noller, 1996) and is generally supported by many other independent data. Mutations within S5 that produce resistance to spectinomycin are within the N-terminal domain that appears to bind dsRNA. Spectinomycin binds to helix 34 within the 3′ domain of 16S rRNA that constitutes the head of the 30S subunit. It is therefore reasonable to assume that the N-terminal domain binds helix 34, and this is again supported by site-directed hydroxyl radical probing experiments (Heilek and Noller, 1996). It is also consistent with the observation that the N-terminal domain has a dsRBD motif. The picture that emerges is that S5 straddles the head and body regions of the 30S subunit and that it may mediate their relative movements during the translocation step with which S5 is known to be associated (Lodmell and Dahlberg, 1997). Finally, cross-linking studies show that S5 is adjacent to S8 (Allen et al., 1979), and the two proteins may directly interact (Tindall and Aune, 1981).

The structure of S8 also suggests that it has a complex environment. The protein is known to recognize the proximal region of helix 21, and the best candidate for the site of interaction on the protein is site 2 along the lower surface (Fig. 3b). It is the most extensive and highly conserved, and a region of dsRNA bound at this location would satisfy the observed cross-link and the homology of S8 to several DNA binding proteins (see below). The S5-S8 cross-link would be consistent with orienting the protein such that the C-terminal domain is facing the C-terminal domain of S5.

S15 has a very restricted footprint, and its RNA binding site has been well characterized as the three-way helical junction of helices 21, 22, and 23, and most of helix 22 (Batey and Williamson, 1996a, 1996b). One face of S15 appears to bind RNA, but its orientation cannot be determined. This complex region of RNA is likely to fold into a defined tertiary structure, and extensive modeling is not warranted.

Finally, S17 contacts two distant regions of 16S rRNA and appears to bring them together as part of the folding process. The protein footprints at helix 11 (Powers and Noller, 1995), but cross-linking (Greuer et al., 1987) and site-directed hydroxyl radical probing experiments agree that it contacts a secondary region on the distal half of helix 21. A suitable location does exist in Brimacombe's model of the 16S rRNA, adjacent to the binding site of S8. S17 is a member of the oligonucleotide-oligosaccharide binding-fold family of proteins, which have a common binding site that serves to orient the protein with respect to its RNA binding targets. This site on S17 is appropriately highly basic and contains two extended flexible loops that are highly conserved. The longer loop contains a mutation site for neamine resistance, and the smaller loop can harbor a mutation that leads to defective ribosome assembly. One possible scenario is that the longer loop is part of the major binding site for helix 11 and the smaller loop is the secondary site for helix 21 that may be compromised in the assembly mutation.

The 50S Subunit

L6 is located at the base of the L12 stalk adjacent to the central protuberance (Walleczek et al., 1988). It appears to be close to the 30S-50S interface and has been cross-linked to the end of helix 89 in domain V of 23S rRNA (Wower et al., 1981). This cross-link suggests that L6 is close to the peptidyltransferase site and/or the tRNA binding sites on the 50S subunit. Mutations in L6 that cause gentamicin resistance suggest that the protein is somehow involved in maintaining the ribosome's fidelity. Gentamicin causes translational inaccuracy, and mutations in L6 and the α-sarcin loop region of 23S rRNA are the only 50S alterations that can cause resistance. Indeed, the phenotypes of the two types of mutation are very similar (Melançon et al., 1992). We propose that these elements are close together or may interact directly within the 50S subunit and form part of the aminoacyl-tRNA·EF-Tu·GTP ternary complex binding site. To support this, L6 has been cross-linked to EF-G (Sköld, 1982) and has been shown to interact with the α-sarcin loop region (Uchiumi et al., 1999). As described earlier, we have shown that these L6 mutations result in C-terminal deletions that truncate the major putative RNA binding site of the protein. This suggests that the C-terminal domain of L6 is mediating the important interactions with the 23S rRNA.

L9 has been localized at, or very close to, the lateral protuberance adjacent to protein L1 (Walleczek et al., 1988). It is a primary RNA binding protein in the 50S subunit (Roth and Nierhaus, 1980) and appears to have an essential function in the ribosome (Dabbs, 1986; Nag et al., 1991). The structure of L9 immediately suggests that it has the role of an RNA scaffolding protein with its two separated but rigid RNA binding domains orienting two regions of 23S rRNA. Toeprinting experiments suggest that the two RNA sites are at opposite ends of helix 76, just below the L1 binding site (Adamski et al., 1996). This suggests the intriguing possibility that L9 acts as a support for the L1 stalk that is clearly visible on recent images of the 50S subunit (Stark et al., 1995; Frank et al., 1995; Ban et al., 1998).

L14 is located between the peptidyltransferase and GTPase centers, close to the subunit interface (Walleczek et al., 1988). This location between two important functional centers makes L14 a useful structural probe and landmark within the 50S subunit. The L14 structure suggests that the protein binds two regions of 23S rRNA, and this is consistent with the role of L14 in subunit assembly (Herold and Nierhaus, 1987). Protein-RNA cross-linking data support the close interaction of the two components and indicate that L14 may bind to helix 61 (Urlaub et al., 1995; Oßwald et al., 1990). Protein-protein cross-linking data suggest that the conserved hydrophobic patch on the L14 β-barrel (Davies et al., 1996b) may interact with ribosomal protein L19 (Walleczek et al., 1989).

STRUCTURAL HOMOLOGIES OF RIBOSOMAL PROTEINS

The new protein structures described above have further convinced us that ribosomal proteins may have been structural prototypes for many modern protein families. These structural homologies have turned out to be an unexpected bonus of the research program and have gained increasing significance in recent years. It is becoming apparent that there may be a relatively limited number of unique protein folds, and the 18 known ribosomal protein structures have now contributed 15 independent folding motifs. Those within L14, the N-terminal domain of L11, and the C-terminal domains of S8 and L9 remain unique at the time of writing. The two domains of S5 are a good illustration of how important this contribution has been. Although unique when first determined (Ramakrishnan and White, 1992), the N-terminal domain is now recognized as a dsRBD motif (Bycroft et al., 1995), and the C-terminal domain has been found in several proteins, including ribonuclease P (Stams et al., 1998). As described above, the dsRBD homology has allowed us to build an important model of an S5-dsRNA complex. Our recent findings are summarized below.

Both domains of S4 show similarities to well-characterized DNA binding motifs; the α-helical domain 1 is homologous to the DNA binding substructure within the Tet repressor, and the α/β inserted domain 2 has a fold almost identical to the E-26-specific DNA binding motif. However, the putative RNA binding sites within S4 do not overlap with the DNA binding regions of the two homologous motifs. The S7 helix ribbon motif is homologous to the DNA binding and bending protein HU, which suggests that S7 binds dsRNA and may bend it during ribosome assembly. HU is a highly interlocked dimeric protein, and it was difficult to imagine how it might have evolved. S7 now provides a possible answer, since its arrangement of α-helices is similar to that of the HU dimer (White et al., 1999). As for S8, the N-terminal domain folding topology is found in several other known protein structures, including DNase I, DNA methyltransferase, and IF3, but the structure of the C-terminal domain is unique among known protein structures. Finally, although S15 is a very small protein, it does contain an example of a coiled-coil structure.

REFERENCES

Adamski, F. M., J. F. Atkins, and R. F. Gesteland. 1996. Ribosomal protein L9 interactions with 23 S rRNA: the use of a translational bypass assay to study the effect of amino acid substitutions. *J. Mol. Biol.* **261:**357–371.

Alexander, R. W., P. Muralikrishna, and B. S. Cooperman. 1994. Ribosomal components neighboring the conserved 518–533 loop of 16S rRNA in 30S subunits. *Biochemistry* **33:**12109–12118.

Allen, G., R. Capasso, and C. Gualerzi. 1979. Identification of the amino acid residues of proteins S5 and S8 adjacent to each other in the 30S ribosomal subunit of *Escherichia coli*. *J. Biol. Chem.* **254:**9800–9806.

Allen, P. N., and H. F. Noller. 1989. Mutations in ribosomal proteins S4 and S12 influence the higher order structure of 16S ribosomal RNA. *J. Mol. Biol.* **208:**457–468.

Allmang, C., M. Mougel, E. Westhof, B. Ehresmann, and C. Ehresmann. 1994. Role of conserved nucleotides in building the 16S rRNA binding site of *E. coli* ribosomal protein S8. *Nucleic Acids Res.* **22:**3708–3714.

Ban, N., B. Freeborn, P. Nissen, P. Penczek, R. A. Grassucci, R. Sweet, J. Frank, P. B. Moore, and T. A. Steitz. 1998. A 9 Å resolution X-ray crystallographic map of the large ribosomal subunit. *Cell* **93:**1105–1115.

Batey, R., and J. Williamson. 1996a. Interaction of the *Bacillus stearothermophilus* ribosomal protein S15 with 16S rRNA. I. Defining the minimal RNA site. *J. Mol. Biol.* **261:**536–549.

Batey, R. T., and J. R. Williamson. 1996b. Interaction of the *Bacillus stearothermophilus* ribosomal protein S15 with 16S rRNA. II. Specificity determinants of RNA-protein recognition. *J. Mol. Biol.* **261:**550–567.

Berglund, H., A. Rak, A. Serganov, M. Garber, and T. Härd. 1997. Solution structure of the ribosomal RNA binding protein S15 from *Thermus thermophilus*. *Nat. Struct. Biol.* **4:**20–23.

Brimacombe, R., J. Atmadja, W. Stiege, and D. Schüler. 1988. A detailed model of the three-dimensional structure of *E. coli* 16S ribosomal RNA *in situ* in the 30S subunit. *J. Mol. Biol.* **199:**115–136.

Bycroft, M., S. Grunert, A. G. Murzin, M. Proctor, and D. St. Johnston. 1995. NMR solution structure of a double-stranded RNA-binding domain from *Drosophila* staufen protein reveals homology to the N-terminal domain of ribosomal protein S5. *EMBO J.* **14:**3563–3571.

Capel, M. S., D. M. Engelman, B. R. Freeborn, N. Kjeldgaard, J. A. Langer, V. Ramakrishnan, D. G. Schindler, D. K. Schneider, B. P. Schoenborn, I.-Y. Sillers, S. Yabuki, and P. B. Moore. 1987. A complete mapping of the proteins in the small ribosomal subunit of *Escherichia coli*. *Science* **238:**1403–1406.

Cerretti, D. P., L. C. Mattheakis, K. R. Kearney, L. Vu, and M. Nomura. 1988. Translational regulation of the spc operon in *Escherichia coli*. Identification and structural analysis of the target site for S8 repressor protein. *J. Mol. Biol.* **204:**309–329.

Changchien, L.-M., and G. R. Craven. 1976. The function of the N-terminal region of ribosomal protein S4. *J. Mol. Biol.* **108:**381–401.

Clemons, W. M., Jr., C. Davies, S. W. White, and V. Ramakrishnan. 1998. Conformational variability of the N-terminal helix in the structure of ribosomal protein S15. *Structure* **6:**429–438.

Dabbs, E. R. 1986. Mutant studies on the prokaryotic ribosome, p. 733–748. *In* B. Hardesty and G. Kramer (ed.), *Structure, Function and Genetics of Ribosomes*. Springer-Verlag, New York, N.Y.

Davies, C., V. Ramakrishnan, and S. W. White. 1996a. Structural evidence for specific S8-RNA and S8-protein interactions within the 30S ribosomal subunit: ribosomal protein S8 from *Bacillus stearothermophilus* at 1.9 Å resolution. *Structure* **4:**1093–1104.

Davies, C., S. W. White, and V. Ramakrishnan. 1996b. The crystal structure of ribosomal protein L14 reveals an important organizational component of the translational apparatus. *Structure* **4:**55–66.

Davies, C., R. G. Gerstner, D. E. Draper, V. Ramakrishnan, and S. W. White. 1998a. The crystal structure of ribosomal protein S4 reveals a two-domain molecule with an extensive RNA-binding surface: one domain shows structural homology to the ETS DNA-binding motif. *EMBO J.* **17:**4545–4558.

Davies, C., D. E. Bussiere, B. L. Golden, S. J. Porter, V. Ramakrishnan, and S. W. White. 1998b. Ribosomal proteins S5 and

L6: high-resolution crystal structures and roles in protein synthesis and antibiotic resistance. *J. Mol. Biol.* **279:**873–888.

Daya-Grosjean, L., R. A. Garrett, O. Pongs, G. Stöffler, and H. G. Wittmann. 1972. Properties of the interaction of ribosomal protein S4 and 16S RNA in *Escherichia coli* revertants from streptomycin dependence to independence. *Mol. Gen. Genet.* **119:**277–286.

Dontsova, O. A., K. V. Rosen, S. L. Bogdanova, E. A. Skripkin, A. M. Kopylov, and A. A. Bogdanov. 1992. Identification of the *Escherichia coli* 30S ribosomal subunit protein neighboring messenger RNA during initiation of translation. *Biochimie* **74:**363–371.

Döring, T., P. Mitchell, M. Oßwald, D. Bochkariov, and R. Brimacombe. 1994. The decoding region of 16S RNA: a cross-linking study of the ribosomal A, P and E sites using tRNA derivatized at position 32 in the anticodon loop. *EMBO J.* **13:** 2677–2685.

Dragon, F., and L. Brakier-Gingras. 1993. Interaction of *Escherichia coli* ribosomal protein S7 with 16S rRNA. *Nucleic Acids Res.* **21:**1199–1203.

Dragon, F., C. Payant, and L. Brakier-Gingras. 1994. Mutational and structural analysis of the RNA binding site for *Escherichia coli* ribosomal protein S7. *J. Mol. Biol.* **244:**74–85.

Frank, J. 1998. How the ribosome works. *Am. Sci.* **86:**428–439.

Frank, J., J. Zhu, P. Penczek, Y. Li, S. Srivistava, A. Verschoor, M. Radermacher, R. Grassucci, R. K. Lata, and R. K. Agrawal. 1995. A model of protein synthesis based on cryo-electron microscopy of the *E. coli* ribosome. *Nature* **376:**441–444.

Geyl, D., A. Bock, and H. G. Wittmann. 1977. Cold-sensitive growth of a mutant of *Escherichia coli* with an altered ribosomal protein S8: analysis of revertants. *Mol. Gen. Genet.* **152:**331–336.

Golden, B. L., D. W. Hoffman, V. Ramakrishnan, and S. W. White. 1993a. Ribosomal protein S17: characterization of the three-dimensional structure by ^{1}H- and ^{15}N-NMR. *Biochemistry* **32:**12812–12820.

Golden, B. L., V. Ramakrishnan, and S. W. White. 1993b. Ribosomal protein L6: structural evidence of gene duplication from a primitive RNA-binding protein. *EMBO J.* **12:**4901–4908.

Green, M., and C. G. Kurland. 1971. Mutant ribosomal protein with defective RNA binding site. *Nat. New Biol.* **234:**273–275.

Greuer, B., M. Oßwald, R. Brimacombe, and G. Stöffler. 1987. RNA-protein cross-linking in Escherichia coli 30S ribosomal subunits; determination of sites on 16S RNA that are cross-linked to proteins S3, S4, S7, S9, S10, S11, S17, S18 and S21 by treatment with bis-(2-chloroethyl)-methylamine. *Nucleic Acids Res.* **15:**3241–3255.

Heilek, G. M., and H. F. Noller. 1996. Site-directed hydroxyl radical probing of the rRNA neighborhood of ribosomal protein S5. *Science* **272:**1659–1662.

Heilek, G. M., R. Marusak, C. F. Meares, and H. F. Noller. 1995. Directed hydroxyl radical mapping of 16S rRNA using Fe(II) tethered to ribosomal protein S4. *Proc. Natl. Acad. Sci. USA* **92:** 1113–1116.

Herold, M., and K. N. Nierhaus. 1987. Incorporation of six additional proteins to complete the assembly map of the 50S subunit from *Escherichia coli* ribosomes. *J. Biol. Chem.* **262:**8826–8833.

Hoffman, D. W., C. Davies, S. E. Gerchman, J. H. Kycia, S. J. Porter, S. W. White, and V. Ramakrishnan. 1994. Crystal structure of prokaryotic ribosomal protein L9: a bi-lobed RNA-binding protein. *EMBO J.* **13:**205–212.

Hoffman, D. W., C. S. Cameron, C. Davies, S. W. White, and V. Ramakrishnan. 1996. Ribosomal protein L9: a structure determination by the combined use of X-ray crystallography and NMR spectroscopy. *J. Mol. Biol.* **264:**1058–1071.

Jaishree, T. N., V. Ramakrishnan, and S. W. White. 1996. Solution structure of prokaryotic ribosomal protein S17 by high-resolution NMR spectroscopy. *Biochemistry* **35:**2845–2853.

Kurland, C. G., F. Jörgensen, A. Richter, M. Ehrenberg, N. Bilgin, and A.-M. Rojas. 1990. Through the accuracy window, p. 513–526. *In* W. E. Hill, A. Dahlberg, R. A. Garrett, P. B. Moore, D. Schlessinger, and J. R. Warner (ed.), *The Ribosome: Structure, Function and Evolution.* American Society for Microbiology, Washington, D.C.

Lambert, J. M., G. Boileau, J. A. Cover, and R. R. Traut. 1983. Cross-links between ribosomal proteins of 30S subunits in 70S tight couples and in 30S subunits. *Biochemistry* **22:**3913–3920.

Lillemoen, J., and D. W. Hoffman. 1998. An investigation of the dynamics of ribosomal protein L9 using heteronuclear NMR relaxation measurements. *J. Mol. Biol.* **281:**539–551.

Lillemoen, J., C. S. Cameron, and D. W. Hoffman. 1997. The stability and dynamics of ribosomal protein L9: investigations of a molecular strut by amide proton exchange and circular dichroism. *J. Mol. Biol.* **268:**482–493.

Lodmell, J. S., and A. E. Dahlberg. 1997. A conformational switch in *E. coli* 16S ribosomal RNA during decoding of messenger RNA. *Science* **277:**1262–1267.

Malhotra, A., P. Penczek, R. K. Agrawal, I. S. Gabashvili, R. A. Grassucci, R. Junemann, N. Burkhardt, K. H. Nierhaus, and J. Frank. 1998. *Escherichia coli* 70S ribosome at 15 Å resolution by cryo-electron microscopy-localization of fmet-tRNA$^{f}_{met}$ and fitting of L1 protein. *J. Mol. Biol.* **280:**103–116.

Markus, M. A., R. B. Gerstner, D. E. Draper, and D. A. Torchia. 1998. The solution structure of ribosomal protein S4Δ41 reveals two subdomains and a positively charged surface that may interact with RNA. *EMBO J.* **17:**4559–4571.

Melançon, P., W. E. Tapprich, and L. Brakier-Gingras. 1992. Single-base mutations at position 2661 of *Escherichia coli* 23S rRNA increase efficiency of translational proofreading. *J. Bacteriol.* **174:**7896–7901.

Mueller, F., and R. Brimacombe. 1997a. A new model for the three-dimensional folding of *Escherichia coli* 16S ribosomal RNA. I. Fitting the RNA to a 3D electron microscopic map at 20 Å. *J. Mol. Biol.* **271:**524–544.

Mueller, F., and R. Brimacombe. 1997b. A new model for the three-dimensional folding of *Escherichia coli* 16S ribosomal RNA. II. The RNA-protein interaction data. *J. Mol. Biol.* **271:** 545–565.

Mueller, F., H. Stark, M. van Heel, J. Rinke-Appel, and R. Brimacombe. 1997. A new model for the three-dimensional folding of *Escherichia coli* 16S ribosomal RNA. III. The topography of the functional centre. *J. Mol. Biol.* **271:**566–587.

Muralikrishna, P., and B. S. Cooperman. 1994. A photolabile oligodeoxyribonucleotide probe of the decoding site in the small subunit of the *Escherichia coli* ribosome: identification of neighbouring ribosomal components. *Biochemistry* **33:**1392–1398.

Nag, B., S. S. Akella, P. A. Cann, D. W. Tewari, D. G. Glitz, and R. R. Traut. 1991. Monoclonal antibodies to *Escherichia coli* ribosomal proteins L9 and L10. Effects on ribosome function and localization of L9 on the surface of the 50 S ribosomal subunit. *J. Biol. Chem.* **266:**22129–22135.

Nicholls, A., K. A. Sharp, and B. Honig. 1991. Protein folding and association: insights from the interfacial and thermodynamic properties of hydrocarbons. *Proteins* **11:**281–296.

Noller, H. F., V. Hoffarth, and L. Zimniak. 1992. Unusual resistance of peptidyl transferase to protein extraction procedures. *Science* **256:**1416–1419.

Nowotny, V., and K. H. Nierhaus. 1988. Assembly of the 30S subunit from *Escherichia coli* ribosomes occurs via two assembly domains which are initiated by S4 and S7. *Biochemistry* **27:** 7051–7055.

Oßwald, M., B. Greuer, R. Brimacombe, G. Stöffler, H. Bäumert, and H. Fasold. 1987. RNA-protein cross-linking in *Escherichia coli* 30S ribosomal subunits: determination of sites on 16S RNA that are cross-linked to proteins S3, S4, S5, S7, S8, S9, S11, S19 and S21 by treatment with methyl *p*-azidophenyl acetimidate. *Nucleic Acids Res.* **15:**3221–3240.

Oßwald, M., B. Greuer, and R. Brimacombe. 1990. Localization of a series of RNA-protein cross-link sites in the 23S and 5S ribosomal RNA from *Escherichia coli*, induced by treatment of 50S subunits with three different bifunctional reagents. *Nucleic Acids Res.* **18:**6755–6760.

Portier, C., L. Dondon, and M. Grunberg-Manago. 1990. Translational autocontrol of the *Escherichia coli* ribosomal protein S15. *J. Mol. Biol.* **211:**407–414.

Powers, T., and H. F. Noller. 1991. A functional pseudoknot in 16S ribosomal RNA. *EMBO J.* **10:**2203–2214.

Powers, T., and H. F. Noller. 1995. Hydroxyl radical footprinting of ribosomal proteins on 16S rRNA. *RNA* **1:**194–209.

Ramakrishnan, V., and S. E. Gerchman. 1991. Cloning, sequencing and overexpression of genes for ribosomal proteins from *Bacillus stearothermophilus*. *J. Biol. Chem.* **266:**880–885.

Ramakrishnan, V., and S. W. White. 1992. The structure of ribosomal protein S5 reveals sites of interaction with 16S rRNA. *Nature* **358:**768–771.

Ramakrishnan, V., and S. W. White. 1998. Ribosomal protein structures: insights into the architecture, mechanism and evolution of the ribosome. *Trends Biochem. Sci.* **23:**208–212.

Roth, H. E., and K. H. Nierhaus. 1980. Assembly map of the 50-S subunit from *Escherichia coli* ribosomes, covering the proteins present in the first reconstitution intermediate particle. *Eur. J. Biochem.* **103:**95–98.

Ryter, J. M., and S. C. Schultz. 1998. Molecular basis of double-stranded RNA-protein interactions: structure of a dsRNA-binding domain complexed with dsRNA. *EMBO J.* **17:**7505–7513.

Saito, K., and M. Nomura. 1994. Post-transcriptional regulation of the str operon in *Escherichia coli*. Structural and mutational analysis of the target site for translational repressor S7. *J. Mol. Biol.* **235:**125–139.

Serganov, A. A., B. Masquida, E. Westhof, C. Cachia, C. Portier, M. Garber, B. Ehresmann, and C. Ehresmann. 1996. The 16S rRNA binding site of *Thermus thermophilus* ribosomal protein S15: comparison with *Escherichia coli* S15, minimum site and structure. *RNA* **2:**1124–1138.

Sköld, S.-E. 1982. Chemical cross-linking of elongation factor G to both subunits of the 70S ribosomes from *Escherichia coli*. *Eur. J. Biochem.* **127:**225–229.

Spedding, G., and D. E. Draper. 1993. Allosteric mechanism for translational repression in the *Escherichia coli* α operon. *Proc. Natl. Acad. Sci. USA* **90:**4399–4403.

Stams, T., S. Niranjanakumari, C. A. Fierke, and D. W. Christianson. 1998. Ribonuclease P protein structure: evolutionary origins in the translational apparatus. *Science* **280:**7552–7555.

Stark, H., F. Mueller, E. V. Orlova, M. Schatz, P. Dube, T. Erdemir, F. Zemlin, R. Brimacombe, and M. van Heel. 1995. The 70S *Escherichia coli* ribosome at 23 Å resolution: fitting the ribosomal RNA. *Structure* **3:**815–821.

Stark, H., E. V. Orlova, J. Rinke-Appel, N. Junke, F. Mueller, M. Rodnina, W. Wintermeyer, R. Brimacombe, and M. van Heel. 1997a. Arrangement of tRNAs in pre- and posttranslocational ribosomes revealed by electron cryomicroscopy. *Cell* **88:**19–28.

Stark, H., M. V. Rodnina, J. Rinke-Appel, R. Brimacombe, W. Wintermeyer, and M. van Heel. 1997b. Visualization of elongation factor Tu on the *Escherichia coli* ribosome. *Nature* **389:**403–406.

Stern, S., B. Weiser, and H. F. Noller. 1988. Model for the three-dimensional folding of 16S ribosomal RNA. *J. Mol. Biol.* **204:**447–481.

Stöffler, G., and M. Stöffler-Meilicke. 1984. Immunoelectron microscopy of ribosomes. *Annu. Rev. Biophys. Bioeng.* **13:**303–330.

Svensson, P., L.-M. Changchien, G. R. Craven, and H. F. Noller. 1988. Interaction of ribosomal proteins, S6, S8, S15 and S18 with the central domain of 16S ribosomal RNA. *J. Mol. Biol.* **200:**301–308.

Sylvers, L. A., A. M. Kopylov, J. Wower, S. S. Hixson, and R. A Zimmermann. 1992. Photochemical cross-linking of the anticodon loop of yeast $tRNA^{Phe}$ to 30S-subunit protein S7 at the ribosomal A and P sites. *Biochimie* **74:**381–389.

Tanaka, I., A. Nakagawa, H. Hosaka, S. Wakatsuki, F. Mueller, and R. Brimacombe. 1998. Matching the crystallographic structure of ribosomal protein S7 to a three-dimensional model of the 16S ribosomal RNA. *RNA* **4:**542–550.

Tindall, S. H., and K. C. Aune. 1981. Assessment by sedimentation equilibrium analysis of a heterologous macromolecular interaction in the presence of self-association: interaction of S5 with S8. *Biochemistry* **20:**4861–4866.

Uchiumi, T., N. Sato, A. Wada, and A. Hachimori. 1999. Interaction of the sarcin/ricin domain of 23 S ribosomal RNA with proteins L3 and L6. *J. Biol. Chem.* **274:**681–686.

Urlaub, H., V. Kruft, O. Bischof, E. C. Muller, and B. Wittmann-Liebold. 1995. Protein-rRNA binding features and their structural and functional implications in ribosomes as determined by cross-linking studies. *EMBO J.* **14:**4578–4588.

Urlaub, H., B. Thiede, E. C. Muller, R. Brimacombe, and B. Wittmann-Liebold. 1997. Identification and sequence analysis of contact sites between ribosomal proteins and rRNA in *Escherichia coli* 30S subunits by a new approach using matrix-assisted laser desorption/ionization-mass spectrometry combined with N-terminal microsequencing. *J. Biol. Chem.* **272:**14547–14555.

Walleczek, J., D. Schüler, M. Stöffler-Meilicke, R. Brimacombe, and G. Stöffler. 1988. A model for the spatial arrangement of the proteins in the large subunit of the *Escherichia coli* ribosome. *EMBO J.* **7:**3571–3576.

Walleczek, J., B. Redl, M. Stöffler-Meilicke, and G. Stöffler. 1989. Protein-protein cross-linking of the 50S ribosomal subunit of *Escherichia coli* using 2-iminothiolane. *J. Biol. Chem.* **264:**4231–4237.

White, S. W., K. S. Wilson, K. Appelt, and I. Tanaka. 1999. The high-resolution structure of DNA-binding protein HU from *Bacillus stearothermophilus*. *Acta Crystallogr.* **D55:**801–809.

Wiener, L., D. Schüler, and R. Brimacombe. 1988. Protein binding sites on *Escherichia coli* 16S ribosomal RNA; RNA regions that are protected by proteins S7, S9 and S19, and by proteins S8, S15 and S17. *Nucleic Acids Res.* **16:**1233–1250.

Wimberly, B. T., S. W. White, and V. Ramakrishnan. 1997. The structure of ribosomal protein S7 reveals a β-hairpin motif that binds double-stranded nucleic acids. *Structure* **5:**1187–1198.

Wower, I., J. Wower, M. Meineke, and R. Brimacombe. 1981. The use of 2-iminothiolane as an RNA-protein crosslinking agent in *Escherichia coli* ribosomes, and the localization on 23S RNA of sites crosslinked to proteins L4, L6, L21, L23, L27 and L29. *Nucleic Acids Res.* **9:**4285–4302.

Wu, H., L. Jiang, and R. A. Zimmermann. 1994. The binding site for ribosomal protein S8 in 16S rRNA and spc mRNA from *Escherichia coli*: minimum structural requirements and the effects of single bulged bases on S8-RNA interaction. *Nucleic Acids Res.* **22:**1687–1695.

The Ribosome: Structure, Function, Antibiotics, and Cellular Interactions
Edited by R. A. Garrett, S. R. Douthwaite, A. Liljas, A. T. Matheson, P. B. Moore, and H. F. Noller

Chapter 9

Structure and Evolution of the 23S rRNA Binding Domain of Protein L2

ISAO TANAKA, ATSUSHI NAKAGAWA, TAKASHI NAKASHIMA, MASAE TANIGUCHI, HARUMI HOSAKA, and MAKOTO KIMURA

During the last year or two, significant progress has been made in the determination of three-dimensional structures of ribosomal proteins from eubacteria either by X-ray crystallography or nuclear magnetic resonance methods (for a review see Ramakrishnan and White, 1998). These structures have suggested possible RNA or protein interaction sites within molecules and also shed light on how antibiotics bind and inhibit the functioning of ribosomal proteins. Furthermore, these studies facilitate the structure analysis of ribosomes or ribosomal subunits by superimposing the atomic coordinates on the electron densities obtained at an early stage of the analysis.

The crystal structure of protein S7, a protein located at the decoding center in the small subunits (Capel et al., 1987), was determined (Hosaka et al., 1998; Wimberly et al., 1998) and successfully fitted to the three-dimensional model of 16S rRNA by using two cross-linking sites between 16S rRNA and S7 (Tanaka et al., 1998). Very recently, Wimberly et al. (1999) have determined the crystal structure of a complex composed of ribosomal protein L11 and an rRNA fragment, which is an essential part of an important functional center (GTPase center) on the 50S subunits. These results provide the framework for understanding the role of the proteins involved in the two functional (decoding and GTPase) centers in ribosomes during protein biosynthesis.

The tertiary structure analyses of ribosomal proteins have also revealed that many ribosomal proteins show structural similarities to families of DNA and RNA binding proteins by sharing ribonucleoprotein (RNP), double-stranded (ds) RNA binding domain, K homology domain (KH domain), and helix-turn-helix motifs (Draper and Reynaldo, 1999). These findings have opened the door to understanding the evolutionary aspect of ribosomal proteins, which was not clear from sequence comparison alone.

We recently crystallized a 23S rRNA binding domain of the protein L2 from the *Bacillus stearothermophilus* ribosome (*Bst*L2-RBD) (Nakashima et al., 1998) and determined its structure (Nakagawa et al., 1999) in the expectation that this will ultimately be of help in establishing the three-dimensional structure of the peptidyltransferase center as well as ribosomes or ribosomal subunits. In this chapter we focus on the crystal structure of the RNA binding domain of *Bst*L2 and discuss its structure from functional and evolutionary points of view.

FEATURES OF PROTEIN L2

Ribosomal protein L2 is the largest protein in the large subunits, with about 280 amino acid residues, and is highly conserved among three phylogenetic kingdoms (Wittmann-Liebold et al., 1990). Biochemical studies during the last 2 decades, including chemical modification, site-directed mutagenesis, in vitro reconstitution, and affinity labeling studies, have established that protein L2 is an important constituent of the peptidyltransferase center of the 50S ribosomal subunits (Cooperman et al., 1995; Schulze and Nierhaus, 1982; and references therein). A recent study of *Thermus aquaticus* ribosomes eventually identified two ribosomal proteins, L2 and L3, as essential protein components of the peptidyltransferase center (Khaitovich et al., 1999).

Isao Tanaka, Atsushi Nakagawa, Takashi Nakashima, Masae Taniguchi, and Harumi Hosaka ■ Division of Biological Sciences, Graduate School of Science, Hokkaido University, Sapporo 060-0810, Japan. Makoto Kimura ■ Laboratory of Biochemistry, Faculty of Agriculture, Kyushu University, Fukuoka 812-8512, Japan.

In addition, L2 is known to be a primary 23S rRNA binding protein: it binds directly and specifically to a highly conserved stem-loop structure (helix 66) encompassing nucleotides from 1794 to 1825 in *Escherichia coli* 23S rRNA (Egebjerg et al., 1991). This localization was further confirmed by footprinting analysis of the L2-23S rRNA complex from *Bacillus caldolyticus* (Watanabe, unpublished); the *B. caldolyticus* protein L2 protects nucleotides which correspond to positions 1794 to 1825 in the *E. coli* 23S rRNA, from RNase digestion. Furthermore, it was reported that the nucleotides consisting of the target stem-loop structure in *E. coli* 23S rRNA are highly protected from chemical modifications when the protein L2 is bound to 23S rRNA (Beauclerk and Cundliffe, 1988). These facts suggest that L2 interacts with domain IV in the 23S rRNA covering the short stem-loop structure.

In a previous study, we localized the RBD in the center of *Bst*L2 between amino acids 60 and 201 by proteinase protection analysis of the *Bst*L2-23S rRNA complex (Watanabe and Kimura, 1985). It was found in the study that the fragment remained tightly bound to the RNA. By virtue of this property, the region from 60 to 201 in *Bst*L2 was referred to as the *Bst*L2-RBD. It was, however, shown by a filter binding assay that the *Bst*L2-RBD is necessary but not sufficient for 23S rRNA binding and that the C-terminal region is also required (Nakashima et al., unpublished). Subsequent deletion analysis of the C-terminal region of *Bst*L2 revealed that an additional 40 amino acid residues from the *Bst*L2-RBD, that is, up to position 240 in *Bst*L2, are indispensable for efficient binding to 23S RNA (Kimura et al., unpublished).

To investigate amino acid residues involved in 23S rRNA binding, we generated a number of *Bst*L2 point mutations in which all basic and aromatic residues within the *Bst*L2-RBD were replaced in turn with Gln and Ala, respectively, and analyzed their binding activities to a 23S rRNA fragment containing the target stem-loop structure. Site-directed mutagenesis of Arg86 or Arg155 significantly diminished RNA binding affinity, and in addition, Arg68 and Lys70 mutations caused partial loss of RNA binding. This analysis revealed that Arg86 and Arg155, and presumably Arg68 and Lys70, in *Bst*L2 may act as positively charged recognition groups for the negatively charged phosphate backbone of the 23S rRNA (Harada et al., 1998). As for aromatic residues, the mutants F66A, Y95A, and Y102A, in which Phe66, Tyr95, and Tyr102, respectively, were replaced by alanine, show almost no binding affinity to the 23S rRNA fragment (Harada et al., 1998). Replacement of these aromatic residues seemed to cause more drastic effects on binding affinity than did replacement of the basic residues; the circular-dichroism spectra of these mutants were slightly but significantly changed from that of the wild-type *Bst*L2. It is therefore likely that the mutations of the aromatic residues could impair RNA binding indirectly, probably by perturbing the structure of the *Bst*L2-RBD. It was thus suggested that Phe66, Tyr95, and Tyr102 may stabilize the *Bst*L2-23S rRNA interaction through intramolecular interactions (Harada et al., 1998).

CRYSTAL STRUCTURE OF THE *Bst*L2-RBD

In an attempt to elucidate the tertiary structure of protein *Bst*L2 by X-ray crystallography, a substantial effort has been made to crystallize the protein. While the intact protein (275 residues) still refuses to crystallize, the recombinant protein corresponding to the *Bst*L2-RBD has been crystallized (Nakashima et al., 1998) and the structure was solved at 2.3-Å resolution by multiwavelength anomalous diffraction on a selenomethionine derivative of the crystal (Nakagawa et al., 1999). Figure 1 is a stereo view of the overall structure of the *Bst*L2-RBD. The molecule has an all-beta structure consisting of two domains of approximately the same size. The amino-terminal domain (Arg62 to Pro130) has a five-stranded β-sheet (β1, 75 to 84; β2, 88 to 96; β3, 100 to 105; β4, 114 to 119; and β5, 127 to 131) and is folded into an open β-barrel structure with its open side facing the carboxyl-terminal domain. The carboxyl-terminal domain (Leu131 to Val194) is also folded into a five-stranded β-barrel (β6, 139 to 142; β7, 161 to 168; β8, 170 to 175; β9, 180 to 182; and β10, 188 to 192). The C-terminal β-barrel is characterized by a pair of short, antiparallel β-sheets facing each other. A long loop between β6 and β7 caps the top of the barrel.

These two domains are connected by a short 3_{10} helix (131 to 134), and they are arranged to create a putative RNA binding site between them. The barrel axes of the two domains are approximately perpendicular to each other. The assignment of the β-strands to two domains is somewhat artificial, because these two domains are not completely separated but are linked by β-sheet-type hydrogen bonds between β5 and β10; the C terminus of the peptide bond runs back into the interface region between the two domains and makes interdomain hydrogen bonds with N-terminal β5 while closing the C-terminal barrel structure by hydrogen bonds between β10 and β6. Thus, all the β-strands in the molecule are connected by main-chain hydrogen bonds, as illustrated in the topological diagram in Fig. 2. Both domains have extensive hydrophobic cores. The interface re-

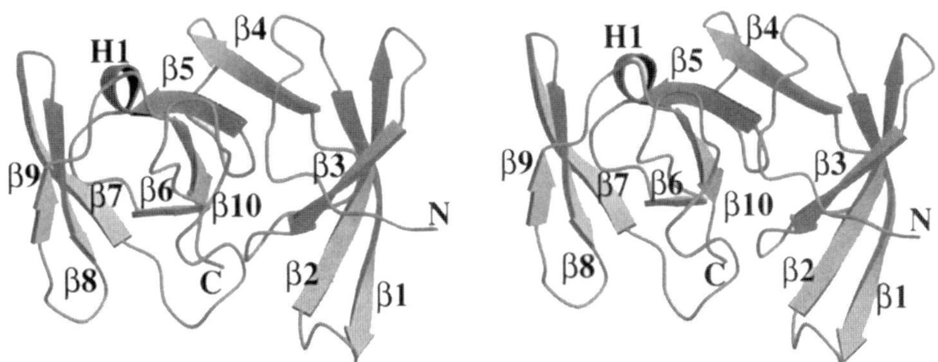

Figure 1. Stereo view of *Bst*L2-RBD. The molecule consists of two subdomains of approximately equal size. The N-terminal domain has an OB fold homologous to the S1 domain, and the C-terminal domain has an SH3-like-barrel motif. The fifth β-strand and a 3_{10} helix (H1) connect two subdomains.

gion between the two domains is less hydrophobic, suggesting that the *Bst*L2-RBD behaves as a two-domain molecule connected by a hinge.

POSSIBLE BINDING MECHANISM

Figure 3 shows the side chains of amino acid residues that were identified by mutagenesis as essential for 23S rRNA binding. An examination of these residues on the determined tertiary structure of the *Bst*L2-RBD shows that all except Arg155 are located in the N-terminal domain. Arg86 is located on the loop between the first (β1) and second (β2) strands of the N-terminal domain, while Arg155 lies at the bulge region of the β7 in the C-terminal domain. These two residues are close together in the tertiary structure and located at the gate of the cleft which is formed by the N-terminal and C-terminal domains of the *Bst*L2-RBD. On the other hand, Arg68 and Lys70 are located on the N-terminal long extended structure preceding the first strand (β1), and the side chain

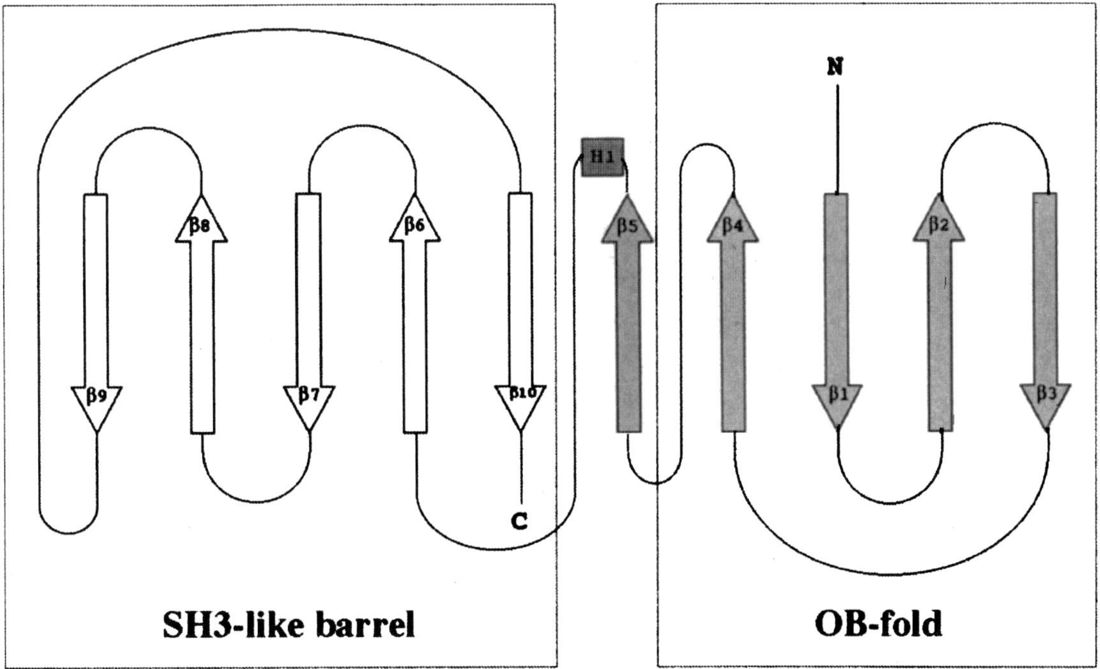

Figure 2. Topological diagram showing secondary structure of *Bst*L2-RBD. The molecule consists of 10 β-strands (indicated by arrows) and a short 3_{10} helix (indicated by a rectangle). Unlike in the ordinal OB fold, β-strands 4 and 5 are linked by a crossover connection, leaving the fifth β-strand at the interface between the N- and C-terminal domains. The C-terminal β10 makes a hydrogen bond to the β5, and thus all the β-strands in the molecule are connected by hydrogen bonds.

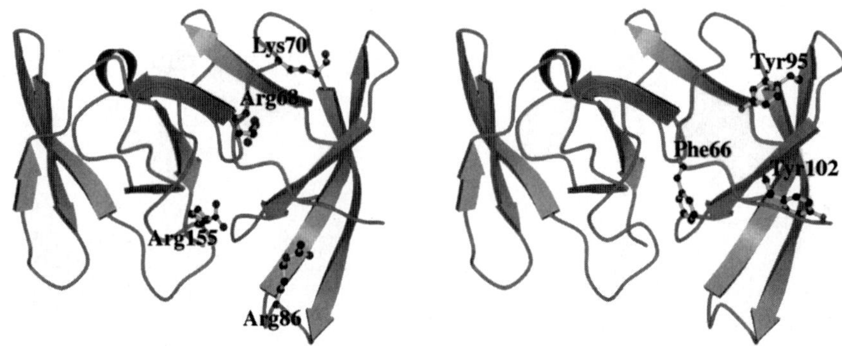

Figure 3. Amino acid residues identified by mutagenesis experiments as essential for 23S rRNA binding. (Left) Positively charged residues important for 23S rRNA binding. Most of the residues that affect 23S rRNA binding are located in the N-terminal domain. The most important residues (Arg86 and Arg155) are located in the protruding loops of the N- and C-terminal domains, respectively. The side chains of these residues face the gate of the interface between two subdomains, a putative RNA binding site. (Right) Aromatic residues important for 23S rRNA binding. Phe66 and Tyr95 are at the interface between two subdomains, and Tyr102 is exposed to the molecular surface at the N-terminal extension of the molecule.

of Arg68 protrudes into the cleft between the N- and C-terminal domains, while that of Lys70 protrudes into the solvent. Three aromatic residues (Phe66, Tyr95, and Tyr102) were in the protruding loops of the N-terminal domain. Phe66 is exposed to the molecular surface, and its phenolic side chain seems to adopt a conformation parallel to the side chain of Arg155. Tyr95 is located at the end of $\beta2$ and protrudes its side chain into the cleft. By contrast to Phe66 and Tyr95, Tyr102 extends its side chain towards the solvent along with the N-terminal extended structure.

The distribution of these basic and aromatic residues suggests that the *Bst*L2-RBD may have a large contact surface for the target RNA in the cleft formed by the interface region of the N- and C-terminal domains, where the RNA stem-loop structure might be wrapped by both domains, thereby being protected from RNase digestion and chemical modification. In this putative interaction, Phe66 and Tyr95 are probably involved in stabilizing the conformation of each domain, and Tyr102 may be involved in interaction with the N-terminal protruding loop of the *Bst*L2-RBD. As described above, the C-terminal extension (up to position 240) from the *Bst*L2-RBD is required for binding to 23S rRNA, possibly to fit the target RNA into the RNA binding site. It is likely that interaction of the C-terminal extension with the target RNA causes a conformational change in the RNA such that it binds to the *Bst*L2-RBD. The C terminus of the *Bst*L2-RBD is located on the molecular surface between the N- and C-terminal domains. This finding further supports the idea that the short target stem-loop binds in the cleft between the N- and C-terminal domains of the *Bst*L2-RBD.

EVOLUTIONARY ASPECTS

To date, the three-dimensional structures of over a dozen ribosomal proteins have been determined. Comparison of their structures with those of other known proteins in the Protein Data Bank (Abola et al., 1997) revealed that many ribosomal proteins share structural motifs, such as RNP, dsRNA binding domain, KH domain, and helix-turn-helix motifs, with RNA or DNA binding proteins (Draper and Reynaldo, 1999). Since the ribosome is an organelle which has a very ancient origin, these observations have led to the speculation that ribosomal proteins are evolutionary ancestors of more recently evolved RNA or DNA binding proteins. It is well known that ribosomal protein L2 is a primary 23S rRNA binding protein and one of the most important constituents in the peptidyltransferase center. It can thus be rationally expected that among the 50 ribosomal proteins, the origin of protein L2 is particularly old. Therefore, the molecular architecture of L2 has been of considerable interest.

An automatic structure alignment by all-against-all comparison of structures in the Protein Data Bank with the DALI server (Holm and Sander, 1993) detected that the N-terminal domain (positions 60 to 130 in *Bst*L2) shows a topology identical to that of the oligonucleotide-oligosaccharide binding (OB) fold (Murzin, 1993), whereas the C-terminal domain (positions 131 to 201 in *Bst*L2) shows a topology identical to that of the Src homology 3 (SH3)-like barrel in the SCOP database (Murzin et al., 1995), as shown in Fig. 4.

The most-conserved features of the OB fold are in strands 1 and 4 of the β-barrel: (i) strand 1 is in the middle of the β-sheet and is long and strongly

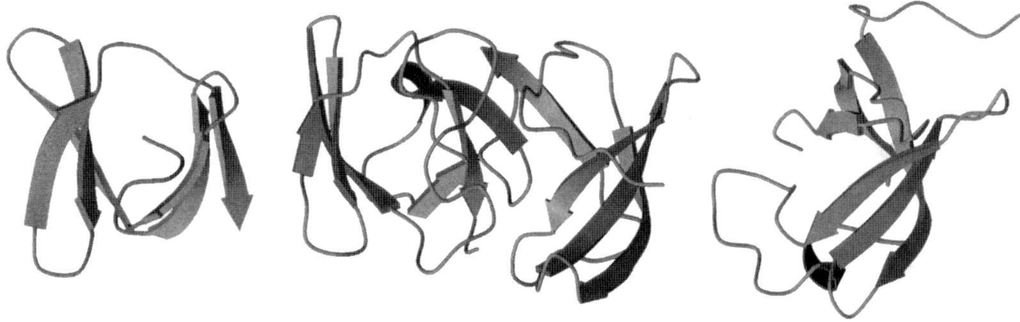

Figure 4. Comparison of the fold of N- and C-terminal subdomains of *Bst*L2-RBD with those of other proteins having similar motifs. (Left) The SH3-like-barrel domain of the biotin biosynthetic operon repressor, BirA, C-terminal domain (1BIA). (Middle) *Bst*L2-RBD. The N-terminal half of *Bst*L2-RBD is an OB fold, and the C-terminal half is an SH3-like barrel. The fifth strand ($\beta 5$) is at the linker region between the OB fold and the SH3-like barrel. (Right) The OB fold of the S1 domain of the polynucleotide phosphorylase of *E. coli* (1SRO).

coiled, (ii) the coiling of strand 1 is promoted by a conserved small residue (glycine or sometimes alanine) in the first half of the strand and by a conserved β-bulge in the second half, and (iii) the α_L conformation (usually at glycine) at the beginning of strand 4 facilitates the turn of the protein chain across the barrel end to allow the formation of the hydrogen bonds to the residues in the bulge (Murzin, 1993). In the *Bst*L2-RBD structure, all three of these features were conserved (Fig. 4). In a typical OB fold, however, $\beta 4$ and $\beta 5$ strands that are connected by a short loop form an antiparallel β-sheet, and the main-chain hydrogen bonds between $\beta 5$ and $\beta 3$ close the β-barrel. In the *Bst*L2-RBD structure, $\beta 4$ and $\beta 5$ strands have a crossover connection, thus forming a parallel β-sheet structure. Because of this connection, the $\beta 5$ strand is not involved in closing the β-barrel structure but is involved in the linker region between the N- and C-terminal domains. Furthermore, the N-terminal domain of *Bst*L2-RBD lacks the α-helix, which caps the barrel end between $\beta 3$ and $\beta 4$ in the classical OB fold (Holm and Sander, 1993). Instead, a short loop structure caps the β-barrel in the N-terminal domain of *Bst*L2-RBD. In spite of these disagreements with a typical OB fold, the overall topology of the N-terminal domain conserves the features of the OB-fold structure, as mentioned above.

The OB fold is found in a variety of proteins having properties of DNA and RNA binding, and among ribosomal proteins, it is found in S17 (Golden et al., 1993) and the internal repetitive domains of *E. coli* ribosomal protein S1 (S1 domain). The S1 domain was originally identified in *E. coli* ribosomal protein S1 (Subramanian, 1983). Although the three-dimensional structure of S1 is still not known, polynucleotide phosphorylase of *E. coli*, which shows sequence similarity to the S1 domain, has been analyzed by nuclear magnetic resonance (Bycroft et al., 1997); the results showed that the S1 domain has an entirely OB-fold motif. When the sequence of the N-terminal domain of *Bst*L2-RBD is aligned according to its structure with the sequences of S17 and S1 repetitive domains, *Bst*L2-RBD shares only a few residues with the corresponding domains. Similar observations that the structural resemblance is not reflected in an obvious homology at the sequence level have been noted in some cases for other ribosomal proteins, such as S5 and S6: their structures have a topology similar to those of the dsRNA binding and RNP domains, respectively (Ramakrishnan and White, 1998). Nevertheless, the consensus amino acids in the OB fold, such as Gly and hydrophobic residues, which are thought to play a structural role for antiparallel β-strands in the OB fold, are well conserved in the *Bst*L2-RBD. It is thus suggested that the N-terminal domain of the *Bst*L2-RBD, ribosomal protein S17, and the repeating domain of S1 have similar structures and thus may have evolved from a common ancestral RNA binding protein.

On the other hand, the C-terminal domain of *Bst*L2-RBD was found to have similarity to proteins having an SH3-like-barrel motif in the SCOP classification (Murzin et al., 1995). No SH3-like barrel has been reported for ribosomal proteins to date. Figure 4 (left) shows a protein that has an SH3-like barrel. It is BirA, the repressor of the *E. coli* biotin biosynthetic operon (Wilson et al., 1992). The SH3-like-barrel motif is found in several other proteins, including the N-terminal domain of the proto-oncogene product c-Crk (Wu et al., 1995) and the DNA binding domain of human immunodeficiency virus, type 1 integrase (Lodi et al., 1995). The five-stranded β-barrel motif designated SH3-like was first reported for an R67 plasmid-encoded dihydrofolate reductase from trimethoprim-resistant bacteria (Narayana et al., 1995) and has subsequently been found in SH3 domains. The SH3-like barrel is often described as a β sandwich comprising two sets of three-stranded antiparallel β-sheets.

Recent structure analysis of a eukaryotic or archaebacterial translation initiation factor (IF-5A) revealed that this protein has, like *Bst*L2-RBD, OB-fold and SH3-like-barrel domains joined with a flexible hinge (Kim et al., 1998; Peat et al., 1998). It is of interest to note that the N-terminal domain of IF-5A is an SH3-like barrel and the C-terminal domain is an OB fold, the reverse of the situation in L2 (Fig. 5). This finding supports the idea that the proteins involved in translation processes have evolved from a pool of a relatively small number of repertoires of ancient nucleic acid binding proteins by some genetic events, such as gene duplication or gene shuffling. Structural comparison of *Bst*L2-RBD and IF-5A, however, shows a significant difference in domain construction. While the OB fold and SH3-like-barrel domains in IF-5A are linked by a long linker and seem to act as two individual functional domains, these two domains in L2 are related more intimately, as described previously. Thus, the genetic event required for the advent of ribosomal protein L2 (or its RBD) is not merely gene fusion but rather a more complex process; it requires an insertion between β4 and β5 to make the crossover connection possible, which in turn allows the formation of interdomain hydrogen bonds between β5 and β10 such that the two domains are nearly fused into one and function as an RBD. In this context, it may be worthwhile to note that the longest insertion of amino residues is observed between β4 and β5 in the case of mitochondrial L2 of *Saccharomyces cerevisiae* (Nakagawa et al., 1999), which further supports this idea.

CONCLUSIONS

Structure analysis of the *Bst*L2-RBD demonstrates that even the largest protein component in the ribosome consists of domains of only ~70 amino acid residues. This domain construction is a common observation for all larger ribosomal proteins (L1, L6, L9, S5, and S8) and possibly reflects early genetic events of protein evolution. The N- and C-terminal domains of the *Bst*L2-RBD are OB-fold and SH3-like-barrel domains, respectively. Both structure motifs are often found in RNA or DNA binding proteins, and their RNA binding sites have been predicted to be regions constructed by loops protruding outward from the barrel axis. In the *Bst*L2-RBD structure, the corresponding regions are formed by L_{12} (the loop between β1 and β2), L_{34}, and L_{67}. Since residues Arg86 and Arg155, which are essential for 23S rRNA

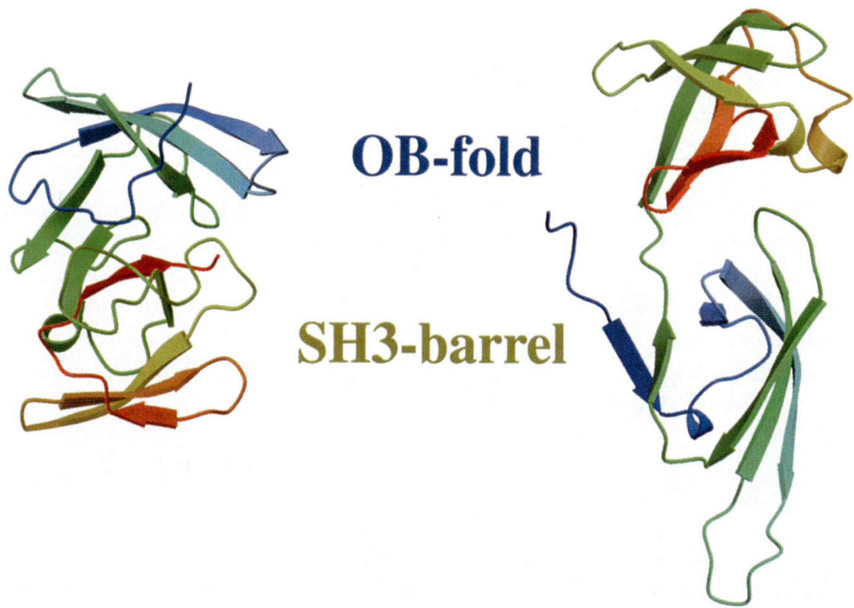

Figure 5. Comparison of molecules that have OB-fold and SH3-like-barrel domains. (Left) *Bst*L2-RBD. (Right) Eukaryotic translation initiation factor IF-5A (2EIF). The N-terminal domain of IF-5A is an SH3-like barrel, and the C-terminal domain is an OB fold, the reverse of the situation in the *Bst*L2-RBD. The molecules are drawn in rainbow colors to show the chain connectivity clearly.

binding, are located on L_{12} and L_{67}, respectively, it is likely that the *Bst*L2-RBD interacts with its target RNA in a way similar to those of other proteins containing either OB-fold or SH3-like-barrel domains. However, the essential Arg residues in the *Bst*L2-RBD are not conserved in other OB-fold and SH3-like-barrel proteins and, in addition, the *Bst*L2-RBD appears to bind the target RNA on the cleft formed by the N- and C-terminal domains, which probably confers on protein L2 a specific recognition for the target stem-loop structure.

Although it has been a long-standing question whether peptidyltransferase activity is due to the proteins or the RNA (Noller et al., 1992), it is well established that 23S rRNA plays a crucial role in the catalytic function of the peptidyltransferase. Recent studies of *T. aquaticus* ribosomes, however, demonstrated that peptidyltransferase activity is never attributed solely to the 23S rRNA, and they reduce the number of possible essential macromolecular components of the peptidyltransferase center to 23S rRNA and ribosomal proteins L2 and L3 (Khaitovich et al., 1999). It can thus be speculated that ancient rRNA worked as a ribozyme to create proteins in the so-called "RNA world" and that during evolution, the ancient rRNA binding proteins, probably prototypes of ribosomal proteins L2 and L3, were incorporated into the ancient ribosome to promote proper folding of 23S rRNA, and they have evolved, being directly or indirectly involved in the peptidyltransferase activity.

REFERENCES

Abola, E. E., J. L. Sussman, J. Prilusky, and N. O. Manning. 1997. Protein data bank archives of three-dimensional macromolecular structures. *Methods Enzymol.* **276:**556–571.

Beauclerk, A. A., and E. Cundliffe. 1988. The binding site for ribosomal protein L2 within 23S ribosomal RNA for *Escherichia coli*. *EMBO J.* **7:**3589–3594.

Bycroft, M., T. J. P. Hubbard, M. Proctor, S. M. Freund, and A. G. Murzin. 1997. The solution structure of the S1 RNA binding domain: a member of an ancient nucleic acid-binding fold. *Cell* **88:**235–242.

Capel, M. S., D. M. Engelman, B. R. Freeborn, M. Kjeldgaard, J. A. Langer, V. Ramakrishnan, D. G. Schindler, D. K. Schneider, B. P. Schoenborn, and I. Y. Sillers. 1987. A complete mapping of the proteins in the small ribosomal subunit of *Escherichia coli*. *Science* **238:**1403–1406.

Cooperman, B. S., T. Wooten, D. P. Romero, and R. R. Traut. 1995. Histidine 229 in protein L2 is apparently essential for 50S peptidyl transferase activity. *Biochem. Cell Biol.* **73:**1087–1094.

Draper, D. E., and L. P. Reynaldo. 1999. RNA binding strategies of ribosomal proteins. *Nucleic Acids Res.* **27:**381–388.

Egebjerg, J., J. Christiansen, and R. A. Garrett. 1991. Attachment sites of primary binding proteins L1, L2 and L23 on 23S ribosomal RNA of *Escherichia coli*. *J. Mol. Biol.* **222:**251–264.

Golden, B. L., D. W. Hoffman, V. Ramakrishnan, and S. W. White. 1993. Ribosomal protein S17: characterization of the three-dimensional structure by ^{1}H and ^{15}N NMR. *Biochemistry* **32:**12812–12820.

Harada, N., K. Maemura, N. Yamasaki, and M. Kimura. 1998. Identification by site-directed mutagenesis of amino acid residues in ribosomal protein L2 that are essential for binding to 23S ribosomal RNA. *Biochim. Biophys. Acta* **1429:**176–186.

Holm, L., and C. Sander. 1993. Protein structure comparison by alignment of distance matrices. *J. Mol. Biol.* **233:**123–138.

Hosaka, H., A. Nakagawa, I. Tanaka, N. Harada, K. Sano, M. Kimura, M. Yao, and S. Wakatsuki. 1998. Ribosomal protein S7: a new RNA-binding motif with structural similarities to a DNA architectural factor. *Structure* **5:**1199–1208.

Khaitovich, P., A. S. Mankin, R. Green, L. Lancaster, and H. Noller. 1999. Characterization of functionally active subribosomal particles from *Thermus aquaticus*. *Proc. Natl. Acad. Sci. USA* **96:**85–90.

Kim, K. K., L.-W. Hung, H. Yokota, R. Kim, and S. H. Kim. 1998. Crystal structures of eukaryotic translation initiation factor 5A from *Methanococcus jannaschii* at 1.8 Å resolution. *Proc. Natl. Acad. Sci. USA* **95:**10419–10424.

Kimura, M., M. Noda, M. Nakashima, and I. Tanaka. Unpublished data.

Lodi, P. J., J. A. Ernst, J. Kuszewski, A. B. Hickman, A. Engelman, R. Craigie, G. M. Clore, and A. M. Gronenborn. 1995. Solution structure of the DNA binding domain of HIV-1 integrase. *Biochemistry* **34:**9826–9833.

Murzin, A. G. 1993. OB (oligonucleotide/oligosaccharide binding)-fold: common structural and functional solution for non-homologous sequences. *EMBO J.* **12:**861–867.

Murzin, A. G., S. E. Brenner, T. Hubbard, and C. Chothia. 1995. SCOP: a structural classification of proteins database for the investigation of sequences and structures. *J. Mol. Biol.* **247:**536–540.

Nakagawa, A., T. Nakashima, M. Taniguchi, H. Hosaka, M. Kimura, and I. Tanaka. 1999. The three-dimensional structure of the RNA-binding domain of ribosomal protein L2: a protein at the peptidyl transferase center of the ribosome. *EMBO J.* **18:** 1459–1467.

Nakashima, T., M. Kimura, A. Nakagawa, and I. Tanaka. 1998. Crystallization and preliminary X-ray crystallographic study of a 23S rRNA binding domain of the ribosomal protein L2 from *Bacillus stearothermophilus*. *J. Struct. Biol.* **124:**99–101.

Nakashima, T., I. Tanaka, and M. Kimura. Unpublished data.

Narayana, N., D. A. Matthews, E. E. Howell, and X. Nguyen-huu. 1995. A plasmid-encoded dihydrofolate reductase from trimethoprim-resistant bacteria has a novel D2-symmetric active site. *Nat. Struct. Biol.* **2:**1018–1025.

Noller, H. F., V. Hoffarth, and L. Zimniak. 1992. Unusual resistance of peptidyl transferase to protein extraction procedures. *Science* **256:**1416–1419.

Peat, T. S., J. Newman, G. S. Waldo, J. Berendzen, and T. C. Terwilliger. 1998. Structure of translation initiation factor 5A from *Pyrobaculum aerophilum* at 1.75Å resolution. *Structure* **6:** 1207–1214.

Ramakrishnan, V., and W. S. White. 1998. Ribosomal protein structures: insights into the architecture, machinery and evolution of the ribosome. *Trends Biochem. Sci.* **23:**208–212.

Schulze, H., and K. H. Nierhaus. 1982. Minimal set of ribosomal components for reconstitution of the peptidyl transferase activity. *EMBO J.* **1:**609–613.

Subramanian, A. R. 1983. Structure and functions of ribosomal protein S1. *Prog. Nucleic Acid Res. Mol. Biol.* **28:**101–142.

Tanaka, I., A. Nakagawa, H. Hosaka, S. Wakatsuki, F. Mueller, and R. Brimacombe. 1998. Matching the crystallographic structure of ribosomal protein S7 to a three-dimensional model of the 16S ribosomal RNA. *RNA* **4:**542–550.

Watanabe, K. Unpublished results.

Watanabe, K., and M. Kimura. 1985. Location of the binding region for 23S ribosomal RNA on ribosomal protein L2 from *Bacillus stearothermophilus. Eur. J. Biochem.* **153:**299–304.

Wilson, K. P., L. M. Shewchuk, R. G. Brennan, A. J. Otsuka, and B. W. Matthews. 1992. *Escherichia coli* biotin holoenzyme synthetase/bio repressor crystal structure delineates the biotin- and DNA-binding domains. *Proc. Natl. Acad. Sci. USA* **89:**9257–9261.

Wimberly, B. T., S. W. White, and V. Ramakrishnan. 1998. The structure of ribosomal protein S7 at 1.9 Å resolution reveals a β-hairpin motif that binds double-stranded nucleic acids. *Structure* **5:**1187–1198.

Wimberly, B. T., R. Guymon, J. P. McCutcheon, S. W. White, and V. Ramakrishnan. 1999. A detailed view of a ribosomal active site: the structure of the L11-RNA complex. *Cell* **97:**491–502.

Wittmann-Liebold, B., A. K. E. Kopke, E. Arndt, W. Kromer, T. Hatakeyama, and H.G. Wittmann. 1990. Sequence comparison and evolution of ribosomal proteins and their genes, p. 598–616. *In* W. E. Hill, A. Dahlberg, R. A. Garrett, P. B. Moore, D. Schlessinger, and J. R. Warner (ed.), *The Ribosome: Structure, Function, and Evolution.* American Society for Microbiology, Washington, D.C.

Wu, X., B. Knudsen, S. M. Feller, J. Zheng, A. Sali, D. Cowburn, H. Hanafusa, and J. Kuriyan. 1995. Structural basis for the specific interaction of lysine-containing proline-rich peptides with the N-terminal SH3 domain of c-Crk. *Structure* **3:**215–226.

The Ribosome: Structure, Function, Antibiotics, and Cellular Interactions
Edited by R. A. Garrett, S. R. Douthwaite, A. Liljas, A. T. Matheson, P. B. Moore, and H. F. Noller

Chapter 10

How Ribosomal Proteins and rRNA Recognize One Another

ROBERT A. ZIMMERMANN, IRINA ALIMOV, K. UMA, HERREN WU, IWONA WOWER, EDWARD P. NIKONOWICZ, DENIS DRYGIN, PEINING DONG, and LIHONG JIANG

The assembly and structural integrity of the ribosome are guaranteed by specific interactions among its component proteins and RNAs (Draper, 1996). In *Escherichia coli* and other bacteria, many of the proteins that associate with rRNA also regulate the translation of ribosomal protein operons via interactions with the corresponding mRNAs (Lindahl and Zengel, 1986). Thus, an understanding of the structural features that underlie these specific, noncovalent interactions is important not only for determining how several dozen unique ribosomal components assemble into functional ribonucleoprotein complexes but also for comprehending the mechanisms of translational control. In a broader sense, this information will contribute to our knowledge of the conformational states accessible to RNA molecules, the patterns of RNA folding, and the mechanism of protein-RNA interaction. Although there appear to be few common structural motifs shared by the protein binding sites within the 16S and 23S rRNAs, a number of these sites are characterized by regular helical segments interrupted by irregularities, such as noncanonical base pairs, base triples, single bulged nucleotides, and small internal loops. In all likelihood, these features define the specific three-dimensional surfaces that are recognized by the ribosomal proteins (Draper, 1995).

Several of the protein binding sites in *E. coli* rRNA, including those for S4, S7, S8, and S15 in 16S rRNA and proteins L1 and L11 in 23S rRNA, have been investigated by a wide variety of methods (e.g., Sapag et al., 1990; Dragon et al., 1994; Wu et al., 1994; Serganov et al., 1996; Egebjerg et al., 1991; Ryan et al., 1991), and the first crystal structure of a ribosomal protein-rRNA complex has recently been reported (Conn et al., 1999; Wimberly et al., 1999). The results discussed in this chapter focus on our studies of the interactions between ribosomal proteins S8 and L1 and their binding sites in the 16S and 23S rRNAs, respectively, which take place during ribosomal subunit assembly. At the same time, S8 and L1 also regulate the translation of the operons that encode them through similar interactions with the *spc* and L11-L1 mRNAs (Gregory et al., 1988; Thomas and Nomura, 1987). We have defined rRNA fragments of 30 to 35 nucleotides that contain all of the structural information necessary for specific recognition of, and association with, S8 and L1. The binding sites are simple to prepare, they can be readily mutagenized, and they are amenable to modification by the incorporation of photolabile nucleotides or nucleotide analogs. Owing to their relatively small size, they are also suitable for high-resolution structural analysis. The three-dimensional structure of the S8 binding site, for instance, has recently been characterized by nuclear magnetic resonance (NMR) (Kalurachchi et al., 1997; Kalurachchi and Nikonowicz, 1998). In addition, crystal structures are now available for protein S8 from *Bacillus stearothermophilus* as well as for proteins S8 and L1 from *Thermus thermophilus* (Davies et al., 1996; Nevskaya et al., 1998; Nikonov et al., 1996).

Robert A. Zimmermann, Irina Alimov, K. Uma, Herren Wu, Iwona Wower, Denis Drygin, Peining Dong, and Lihong Jiang ■ Department of Biochemistry and Molecular Biology and Program in Molecular and Cellular Biology, University of Massachusetts, Amherst, MA 01003. **Edward P. Nikonowicz** ■ Department of Biochemistry and Cell Biology, Rice University, Houston, TX 77251.

THE INTERACTION OF RIBOSOMAL PROTEIN S8 WITH THE 16S rRNA

Conservation of Sequence and Secondary Structure in the S8 Binding Site

Protein S8 interacts with a portion of helix 21 within *E. coli* 16S rRNA that spans positions 588 to 604 and 634 to 651 (Fig. 1). This region consists of two duplex segments interrupted by a "core" element of irregular structure that encompasses nucleotides 595 to 598 and 640 to 644. The core is highly conserved in prokaryotic and chloroplast 16S rRNAs and contains determinants that are essential for S8 binding. The imperfect helix containing nucleotides 588 to 604 and 634 to 651 corresponds to the minimal site that supports specific S8 interaction with approximately the same affinity as intact helix 21. Deletion analysis with rRNA fragments prepared by in vitro transcription or chemical synthesis showed that further truncation of the helical segments led at first to moderate and then to severe decreases in the association constant for complex formation (Wu et al., 1994; Uma and Zimmermann, unpublished). It has been suggested on the basis of chemical protection studies that the S8 binding site includes helix 25 as well as other portions of the central domain (Powers and Noller, 1995; Svensson et al., 1988). The role of these nucleotides remains unclear, however, as nei-

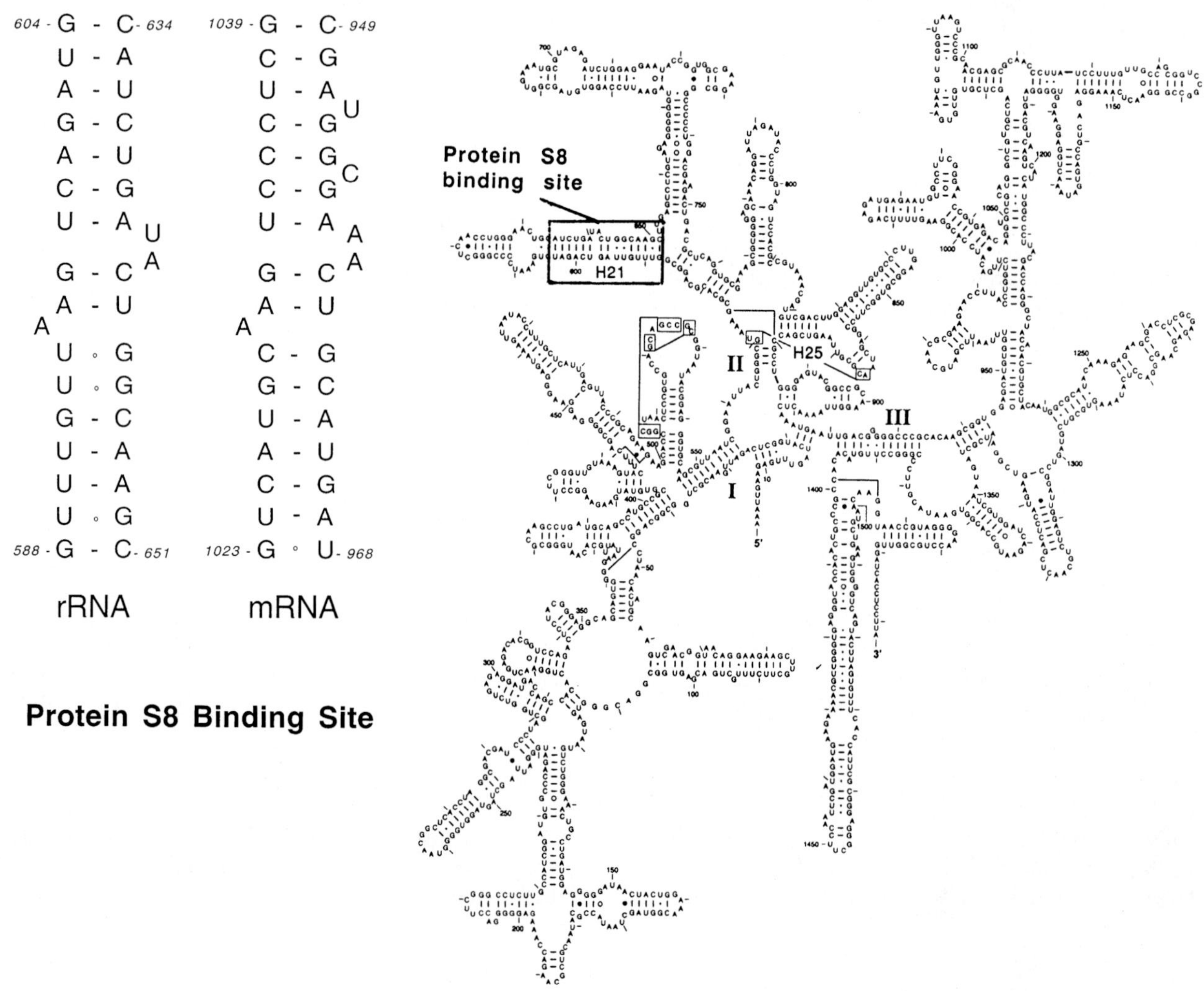

Figure 1. Binding site for protein S8. (Left) Structures of the binding site for ribosomal protein S8 in 16S rRNA (left) and *spc* mRNA (right) from *E. coli*. (Right) Position of the S8 binding site in the secondary structure of the 16S rRNA. Helices 21 and 25 are denoted as H21 and H25, respectively.

ther helix 25 nor other segments of the central domain have been tested for their abilities to interact independently with S8 or to stimulate its binding to helix 21. Thus, although they might form part of the primary S8 binding site, these nucleotides could also be sterically shielded from chemical attack in intact 16S rRNA or provide a set of secondary contacts that facilitate the folding of the 16S rRNA in the course of 30S subunit assembly.

As shown in Fig. 1, the site in *spc* mRNA through which S8 effects translational regulation is strikingly similar to that in 16S rRNA in both sequence and secondary structure, showing that the protein associates with structurally analogous sites in both RNAs (Gregory et al., 1988; Wu et al., 1994). However, the apparent association constant for the S8-mRNA interaction is roughly fivefold less than that for the S8-rRNA interaction. We suspected that this difference might be due to the influence of the two bulged bases located adjacent to the conserved core in the duplex portion of the mRNA binding site. Accordingly, we evaluated the role of these residues in S8-RNA interaction by synthesizing rRNA fragments with bulged bases inserted at equivalent positions and mRNA fragments from which the bulged bases had been deleted and testing them for their capacities to bind S8 (Wu et al., 1994). The results demonstrated that the difference in the affinity of S8 for the two sites can in fact be attributed to the bulged bases. We do not yet know if this is a general phenomenon, but we conclude that, in this case at least, single-base bulges in proximity to essential recognition features can modulate the strength of protein-RNA interaction.

The conservation of nucleotides within the conserved core of the S8 binding site—whether in rRNA or in mRNA—is remarkable (Fig. 2), and from evidence to be presented below, it is clearly related to the capacity of both RNAs to associate with S8. The G597-C643 base pair and the unpaired A residue at position 642 are essentially invariant, exhibiting a level of conservation in excess of 99% of all known prokaryotic and chloroplast 16S rRNAs (Gutell, personal communication). Furthermore, although A595, A596, and U644 are conserved at only the 80 to 90% level, they covary with G, C, and G, respectively, in 10% of the 16S rRNA inventory. Though not phylogenetically conserved, the identity of nucleotide 641 is highly constrained. While A or U occur at this position 93% of the time, G is found in only 6% of the cases and C is excluded entirely. Among the other base pairs in the S8 binding site, U598-A640 is conserved in more than 90% of prokaryotic 16S rRNAs and is required for S8 binding. On the other hand, G588-C651 and G604-C634, although nearly invariant, do not contribute to S8-RNA interaction, at least in vitro (Wu et al., 1994; Uma and Zimmermann, unpublished). Characterization of the S8 binding site by NMR has defined how these nucleotides maintain a structure that is competent to associate with S8, while site-directed mutagenesis and nucleotide analog substitution have helped to delineate the roles they play in S8-RNA interaction.

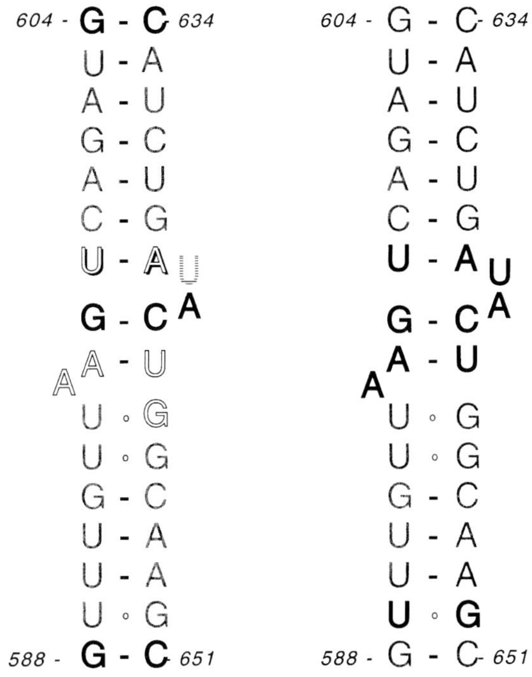

Figure 2. Phylogenetic conservation and mutagenesis in the binding site for protein S8. (Left) Conservation of nucleotides in bacterial and chloroplast 16S rRNAs; boldface, >95%; shadow, 90 to 95%; outline, 80 to 90%; hatched, constrained to A or U in over 93% of prokaryotic 16S rRNAs. (Right) Nucleotides which, when mutated, sharply decrease S8-RNA affinity (boldface).

Structural Analysis of the S8 Binding Site by NMR

Because of its small size and relative simplicity, the binding site for protein S8 is an attractive target for structural studies. We have investigated the solution structure of the S8 binding site both in the free state and in the presence of protein S8 by NMR spectroscopy (Kalurachchi et al., 1997; see also Kalurachchi and Nikonowicz, 1998). In this work, the integrity of the two helical segments was verified and the presence of the G597-C643 and A596-U644 base pairs within the conserved core—which had been predicted from comparative analysis—was confirmed. In addition, we found that the A596-U644 base pair was associated with A595 in an A·(A-U) base triple (Fig. 3a). In this case, A595 is nearly coplanar with the A596-U644 base pair and interacts with

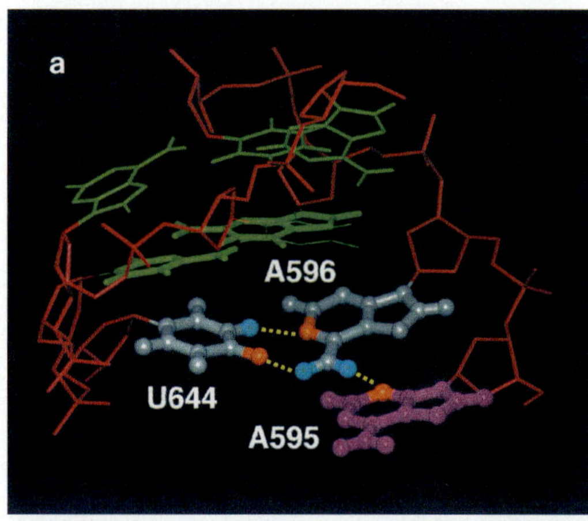

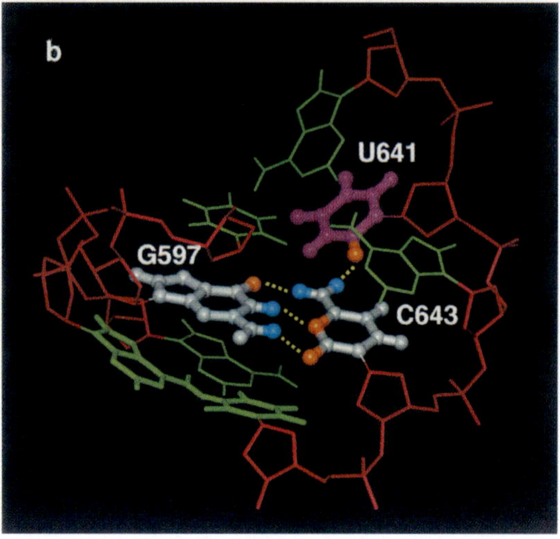

Figure 3. Base triples in the binding site for protein S8. (a) A595·(A596-U644). (b) U641·(G597-C643).

A596 via a hydrogen bond between its N-3 atom and one of the amino protons of A595. A significant proportion of the calculated structures place the C-2 carbonyl oxygen of U641 within hydrogen-bonding distance of the exocyclic amino group of C643, suggesting that a second base triple, U641·(G597-C643), occurs within the binding site core (Fig. 3b). The two triples confer considerable stability upon the core, as they are positioned so as to stack upon one another (Kalurachchi and Nikonowicz, 1998). Strong evidence for the stacked configuration has been provided by a short-range cross-link to A595 in the first triple from a s^4U residue at position 641 in the second (Zimmermann, unpublished).

The NMR data suggest that S8-RNA interaction is accomplished without significant changes in RNA conformation. Nonetheless, S8 binding promotes the formation of the U598-A640 base pair and stabilizes the G597-C643 and A596-U644 base pairs (Kalurachchi and Nikonowicz, 1998). In addition, the protein induced a small but characteristic change in the resonance of G650 at the base of the lower helical segment. Several functionalities within the core, including the C-6 amino groups of A595 and A642 and the C-4 carbonyl oxygen of U641, are exposed to the solvent and are thus potential recognition features for the protein.

Mutagenesis within the Core and Lower Helix of the S8 Binding Site

Site-directed-mutagenesis experiments have shown that the identities of most of the nucleotides in the conserved core (nucleotides 595 to 598 and 640 to 644) are critical for optimal S8-RNA interaction, whereas almost all of the base pairs in the duplex segments can be replaced with only minimal effects on protein binding as long as the helical elements are maintained (Fig. 2) (Gregory et al. 1988; Mougel et al., 1993; Wu et al., 1994; Allmang et al., 1994; Uma and Zimmermann, unpublished). Investigation of the S8 binding site by NMR revealed that the structure of the core is complex, consisting of two base triples in addition to base pairs and unpaired nucleotides (Kalurachchi et al., 1997; Kalurachchi and Nikonowicz, 1998). These results formed the basis for a more detailed mutational analysis of the S8 binding site.

NMR studies of the S8 binding site confirmed the presence of the A596-U644 base pair predicted from phylogenetic comparisons and showed that this pair forms an A·(A-U) triple with A595 (Fig. 3a). Positions corresponding to 595, 596, and 644 are occupied by A, A, and U in over 80% of several thousand known prokaryotic 16S rRNA sequences, but they covary with G, C, and G about 10% of the time, suggesting the possibility that an alternative G·(C-G) triple may form in these cases (Gutell, personal communication). To determine how S8-RNA interaction is affected by the identity of the bases at positions 595, 596, and 644, we carried out a systematic mutational analysis of the three positions with appropriate RNA transcripts and, in particular, we asked whether introduction of a G·(C-G) triple in place of A·(A-U) affected S8 binding. The substitution of G, C, and G at positions 595, 596, and 644, respectively, led to a fourfold decrease in S8-RNA affinity, but this indicated that the protein can nonetheless recognize a binding site that contains the alternative base triple. Other sequences that would lead to triple combinations such as G·(A-U), C·(A-U), A·(C-G), and C·(C-G), that contained base pairs other than A-U and C-G, or that lacked A595 were deleterious (Uma and Zimmermann, unpublished), although the substitution of U at position 595 was reported to have little effect on S8 binding (Mougel et al., 1993).

The second base triple identified by NMR, U641·(G597-C643), may contribute to S8-RNA interaction largely through stabilization of the binding-site core (Fig. 3b). Although the G597-C643 base pair is almost universally conserved in prokaryotic 16S rRNAs and is therefore presumed to be essential, the position corresponding to 641 is usually occupied by either A or U and occasionally by G, but never by C. Consistent with the phylogenetic analysis, reversal of the 597-643 base pair to C-G, substitution of C643 with U, or replacement of U641 by C severely reduces or abolishes affinity for S8, whereas a U-to-A exchange at position 641 has little or no effect (Gregory et al., 1988; Mougel et al., 1993). Interestingly, although nucleotides 641, 597, and 643 are not found to covary in naturally occurring 16S rRNAs owing to the invariance of the G597-C643 base pair (Gutell, personal communication), it has been reported that artificially selected variants of the S8 binding site in which positions 641, 597, and 643 are occupied by C, U, and G or A, C, and G, respectively, have the same affinity for the protein as the corresponding RNA fragment containing the U641·(G597-C643) triple characteristic of the *E. coli* binding site (Moine et al., 1997). It is not yet known how these triples are accommodated in the structure or whether they are in fact isomorphous with the wild-type binding site.

In another series of experiments, we observed that *E. coli* S8 is incapable of associating with the 16S rRNA from the archaeon *Thermoplasma acidophilum* (Thurlow and Zimmermann, unpublished). In this case, the positions equivalent to 641, 597, and 643 are occupied by G, A, and C, respectively, a combination of bases that is unique among known prokaryotic 16S rRNA sequences. To further examine the interdependence of these positions in S8-RNA association, we introduced specific nucleotides characteristic of the *Thermoplasma* binding site into that of *E. coli* and measured the affinity of the resulting RNA fragments for *E. coli* S8 (Uma and Zimmermann, unpublished). Although G is rare at position 641, its presence in a small minority of prokaryotic 16S rRNAs suggests that it is compatible with recognition by at least some S8 proteins. In the *E. coli* context, however, the U-to-G mutation at position 641 reduced S8 binding to an undetectable level. In addition, a fragment in which G597 was replaced by A, disrupting the highly conserved G597-C643 base pair, was unable to interact with S8. Similarly, the G641/A597 double mutant, which mimics the disposition of bases in the corresponding region of *Thermoplasma* 16S rRNA, does not support the binding of *E. coli* S8. The analog of protein S8 in *Thermoplasma* must therefore have evolved to recognize the noncanonical G·(A·C) opposition in addition to the features of the binding site that it shares with other prokaryotic 16S rRNAs.

Both chemical footprinting and mutagenesis have implicated the G588-C651 and U589·G650 base pairs in S8-RNA interaction (Mougel et al., 1987; Wu et al., 1994; Moine et al., 1997). In addition, NMR analysis has shown that the N-3 resonance of G650 is shifted upfield upon S8 binding whereas that of G588 shows no change (Kalurachchi et al., 1997). While the G588-C651 base pair is essentially invariant, there is considerable diversity in the nucleotide combinations that form the 589-650 base pair (Gutell, personal communication). To determine how changes in these base pairs affect S8 binding, a number of RNA variants were synthesized and tested for their abilities to associate with the protein. Base pair alterations at positions 588 and 651 had relatively little effect on S8 binding, whereas all of the replacements at positions 589 and 650 led to roughly equivalent decreases in the affinity of the RNA for S8. Surprisingly, the C589·C650 mismatch had about the same effect on the interaction as the canonical U589-A650 and C589-G650 substitutions. Deletion of either one of these base pairs led to a 5- to 10-fold reduction in S8 binding, while truncation of the lower stem by three or more base pairs reduced protein interaction to barely detectable levels. Taken together, these results show that the invariant G588-C651 base pair is dispensable for S8 binding but support the notion that the S8-RNA complex is stabilized by contacts with the phosphodiester backbone in the vicinity of the U589·G650 base pair in addition to the highly specific interactions mediated by the conserved core (Mougel et al., 1987; Kalurachchi et al., 1997; Moine et al., 1997; Uma and Zimmermann, unpublished).

Functional-Group Analysis in the S8 Binding Site: Atomic Mutagenesis

Conventional mutagenesis—the exchange of one of the four naturally occurring nucleotides for another—can disrupt the secondary and tertiary structure of an RNA molecule, often in unforeseen ways. This difficulty can be addressed through the use of a technique known as nucleotide analog substitution or atomic mutagenesis, which permits the assessment of the influence that individual nucleotide functional groups exert on RNA-RNA or RNA-ligand interactions (Grasby and Gait, 1994). Nucleotide analogs can be distributed randomly throughout an RNA by transcription in vitro or placed at specific positions in the RNA chain by chemical synthesis. In this manner, the effects of adding, exchanging, or eliminating specific atoms or chemical groups within the base, ribose, or phosphate moieties of A, G, C, or U can be tested. This approach is limited only by the availability of the appropriate nucleoside triphosphate or phosphoramidite precursors. Using this methodology, we have probed all of the nucleotide positions within the conserved core of the S8 binding site, as well as the U598·G650 base pair. The results are summarized in Fig. 4.

While S8-RNA interaction was in most cases unaffected when the core nucleotides were replaced with their 2′-deoxyribonucleotide counterparts, replacement of the invariant residues A642 and C643 (>99% conserved) with dA and dC, respectively, resulted in a more-than-10-fold drop in affinity for S8. The effects of the deoxynucleotides could result from a perturbation in the structure of the ribose phosphate backbone owing to their preference for the 2′-*endo* ribose conformation as opposed to the 3′-*endo* ribose conformation characteristic of ribonucleotides, or it could stem from the loss of hydrogen bonds to the protein. When A642 and C643 were replaced by their 2′-O-methyl analogues, which have a strong preference for the 3′-*endo* conformation but interfere with hydrogen bond formation, no improvement in binding occurred. Thus, it is likely that the 2′-OH groups of these two nucleotides are involved in direct interactions with protein S8. While no change in S8 binding was noted when U641 was replaced by dU, the substitution of dA, 2′-O-methylA, or any other 2′-deoxy analogue at this position led to a sharp decrease in the binding constant despite the fact the RNA molecules containing A at position 641 possess normal affinity for the protein. This suggests that the environments of U641 and A641 within the respective variants are not identical and that the presence of the latter nucleotide brings the backbone in closer contact with S8 upon complex formation. A slight increase in affinity was noted when G650 was replaced by dG, perhaps due to a subtle "improvement" in backbone configuration.

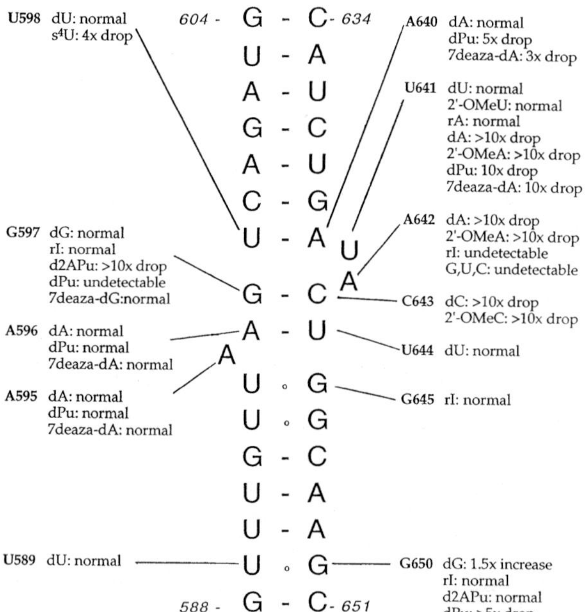

Figure 4. Effects of nucleotide analogue substitutions on S8-RNA interaction. Abbreviations are: r, ribo; d, 2′-deoxy; 2′OMe, 2′-O-methyl; s^{4}U, 4-thiouridine; Pu, purine; I, inosine; 2APu, 2-aminopurine; 7deaza, 7-deaza.

The key role of A642 in S8-RNA interaction is further underscored by the effects of base modifications at this position, as complex formation is undetectable when A642 is replaced by U, C, G, or I. These results make the exocyclic amino group of A642 a likely candidate for specific contact with the protein. A642 is the only core nucleotide that is not involved in base-base interactions, and although it stacks on C643, it is not rigidly positioned within the helical structure. It is thus in a unique position to interact with the protein. Some of the other substitutions, such as that which results from the exchange of A640 for deoxypurine, can be attributed to the disruption of base pairs in which the relevant nucleotides normally participate. The absence of such an effect at position 596, where the introduction of a purine causes no change in the affinity of S8 for the RNA, may be attributable to favorable stacking interactions between U594, A595, and G597 or the persistence of the 2′-*endo* ribose pucker preferred by the deoxynucleotide. The influence of substitutions at position 597 appears more complicated. In this case, replacement of the universally conserved G (>99% conserved) with I produces no effect, while 2-amino-deoxypurine leads to a 10- to 20-fold decrease in binding and deoxypurine renders the RNA molecule unable to form a detectable complex. This behavior is not due to the presence of the 2′-deoxyribose, as dG at this position does not affect binding. Although destabilization of the G597-C643 base pair per se may be a factor here, the exocyclic amino group of C643 may not be constrained to a position in which it can participate in a hydrogen bond with U641 to form the U641·(G597-C643) triple when 2-aminopurine or purine occupies this position. Alternatively, the absence of the C-6 carbonyl moiety of G597 may alter the properties of a divalent cation coordination site proposed on the basis of the NMR studies (Kalurachchi and Nikonowicz, 1998). The decrease in affinity noted when U598 is replaced by s^{4}U may also result from an alteration of a potential coordination site at the C-4 carbonyl of this residue. Several modifications at position 650, including the replacement of G by inosine, 2-aminopurine, and 7-deazaguanosine, had little influence on S8 binding, although the introduction of deoxypurine led to a fivefold decrease in affinity. The latter effect is probably due to a perturbation of the U589·G650 base pair. Together with the slight increase in affinity observed in the variant containing dG650 and the mutagenesis data, these results suggest that any interaction of protein S8 with the base of helix 21 is likely to be a nonspecific contact with the phosphodiester backbone.

Features of Ribosomal Protein S8 Involved in Interaction with 16S rRNA

The similarity of the sites to which S8 binds in rRNA and mRNA suggests that the protein makes use of a common RNA binding domain for both interactions. In order to delineate the structural features of S8 that are involved in its interaction with RNA, we have subjected the *rpsH* gene, which encodes protein S8, to random and site-directed mutagenesis, and prepared several truncated versions of the protein as well. In one series of experiments, the *rpsH* gene was cloned into a plasmid vector under control of the *lac* promoter. Upon induction, overexpression of protein S8 inhibits translation of *spc* operon mRNA via S8-mRNA interaction, leading to deficiencies in the supply of a number of ribosomal proteins, impairment of ribosome assembly, and a sharp increase in doubling time. We exploited this system to screen for mutants of protein S8 which, when overproduced, fail to exhibit growth rate inhibition (Wower et al., 1992). Such variants were presumed to be defective in the ability to bind *spc* mRNA and, by inference, 16S rRNA. In this way, we isolated over 40 mutants with single-amino-acid substitutions at 35 positions within the 129-residue protein.

We have purified roughly half of the mutant proteins and characterized them with respect to their RNA binding activities in vitro. Concurrently, the stabilities of many of the same mutants were measured in vivo by pulse-chase labeling and immunoprecipitation, as was their incorporation into 30S ribosomal subunits. Most of the S8 mutants tested thus far display a moderate to severe decrease in affinity for their RNA binding sites (Fig. 5). The deletion of as few as 8 amino acids from the C terminus completely abolishes the ability of the protein to bind RNA (Table 1) (Uma et al., 1995), while point mutations at positions 78, 80, 105, 106, 108, 121, and 126 within the C-terminal half sharply curtail complex formation (Fig. 5) (Zimmermann et al., unpublished). Several of these proteins are not incorporated into 30S subunits in vivo despite the fact that they are metabolically stable. Surprisingly, the removal of 19, 51, 69, or 82 residues from the N terminus of S8 does not abolish its affinity for RNA (Table 1), although the truncated proteins progressively lose the ability to discriminate between specific and nonspecific binding sites as the deletion becomes longer (Zimmermann et al., unpublished). We therefore conclude that the main determinants of RNA binding reside in the C-terminal portion of S8 but that structural features within the N terminus influence the specificity of the interaction. Since the circular dichroism spectra of most of the point mutants examined thus far

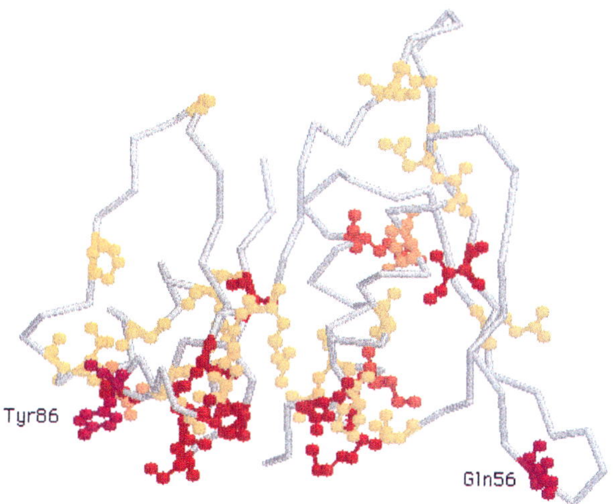

Figure 5. Structure of protein S8 showing amino acids implicated in RNA binding. Over 40 single-site mutants of protein S8 were isolated in a genetic screen designed to identify variants defective in RNA binding (Wower et al., 1992). Altered amino acids, depicted in ball-and-stick mode, are displayed on the backbone structure of protein S8 from *B. stearothermophilus* according to the crystallographic structure determined by Davies et al. (1996). The effects of the amino acid replacements on S8-RNA interaction are indicated as follows: red, strongly deleterious; red-orange, mildly deleterious; orange, no effect; yellow, identified in screen but not yet tested. Gln56 and Tyr86, in magenta, correspond to Lys55 and Tyr85 in *E. coli* S8 (see the text).

Table 1. Relative affinities of truncated variants of protein S8 for the S8 binding site

Amino acid residues	Relative affinity
Intact protein	1.00
C-terminal truncations	
1–100	<0.01
1–120	<0.01
N-terminal truncations	
20–129	0.20
52–129	0.08
70–129	0.10

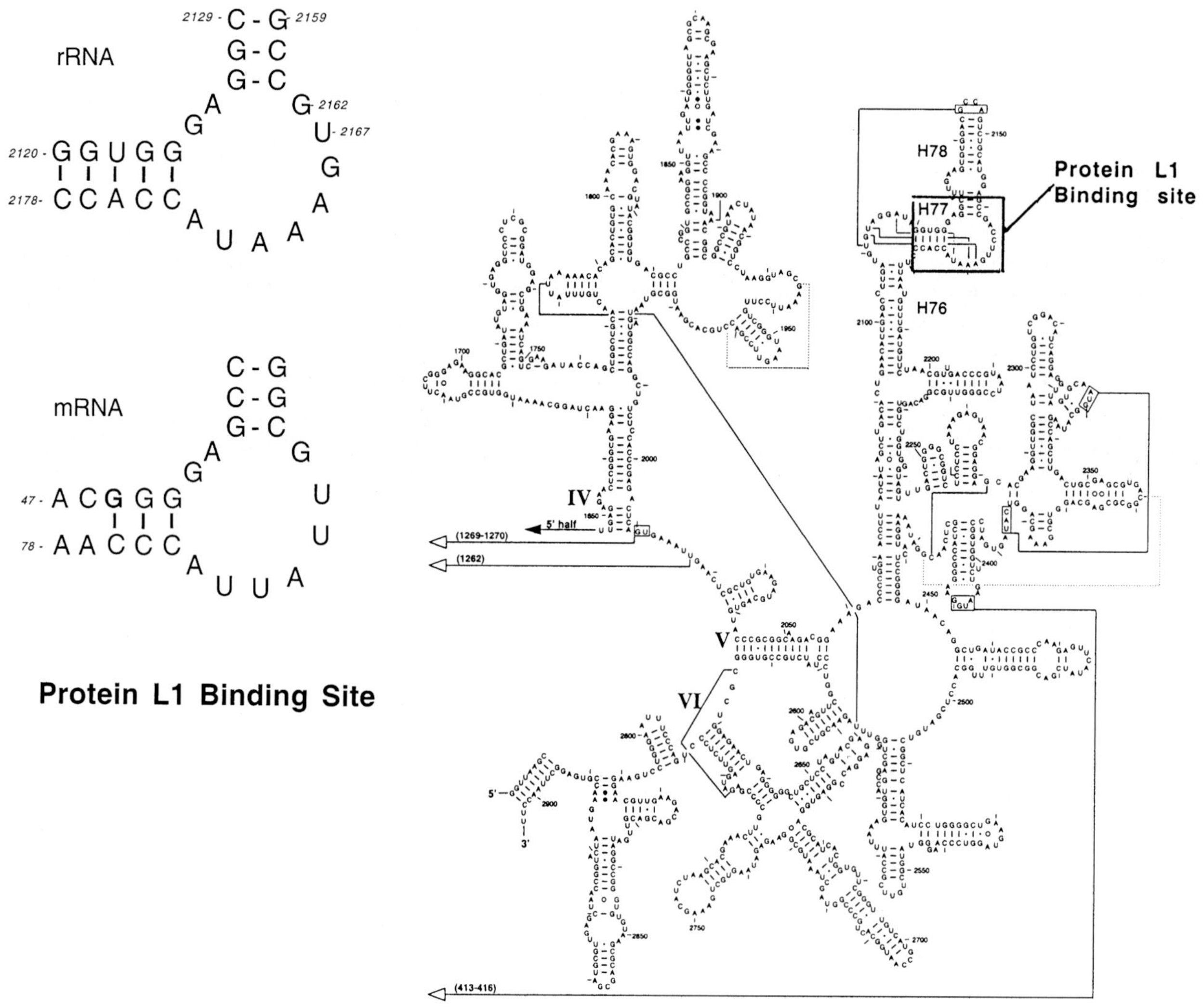

***E. coli* 23S rRNA 3'-half**

Figure 6. Binding site for protein L1. (Left) Structures of the binding site for ribosomal protein L1 in 23S rRNA (top) and L1-L11 mRNA (bottom) from *E. coli.* (Right) Position of the L1 binding site in the secondary structure of the 3′ half of the 23S rRNA. Helices 76, 77, and 78 are denoted as H76, H77, and H78, respectively.

are very similar to that of the wild-type protein, the altered amino acids are more likely to lead—either directly or indirectly—to the impairment of RNA binding than to major defects in the folding of the polypeptide chain. When mapped onto the crystallographic structure of S8 from *B. stearothermophilus*, mutations that lead to defects in RNA binding are found to cluster on one face of the protein (Fig. 5). It has been proposed that the conserved Tyr residue at position 85 of *E. coli* S8 (Tyr86 in *B. stearothermophilus* S8) contacts the conserved core of the binding site in the vicinity of nucleotide A642, while Lys55 (corresponding to Gln56 in *B. stearothermophilus* S8), which has been cross-linked to positions 651 to 654 of the RNA (Urlaub et al., 1997), contacts the base of the helix in the neighborhood of the U589·G650 base pair (Kalurachchi and Nikonowicz, 1998).

THE INTERACTION OF RIBOSOMAL PROTEIN L1 WITH THE 23S rRNA

Conservation of Sequence and Secondary Structure in the L1 Binding Site

Protein L1 protects a sequence encompassing nucleotides 2093 to 2196 of *E. coli* 23S rRNA from

RNase digestion. This region spans helices 76 to 79 and the nominally single-stranded loops that interconnect them. Homologous segments in a number of bacterial, archaeal, and eukaryotic large-subunit rRNAs are also protected from nuclease attack by L1, indicating that the L1-rRNA interaction has been unusually well conserved throughout evolution (e.g., Gourse et al., 1981). We have used a variety of approaches to identify the structural features within the nuclease-protected region that are essential for complex formation. Through the use of fragment selection techniques, deletion analysis, and L1 binding experiments involving the use of specifically tailored in vitro transcripts, we have concluded that the structural features essential for interaction with the protein occur within a region consisting of nucleotides 2120 to 2129 and 2159 to 2178. As in the S8 binding site, the two sequences are partially complementary, forming a 5-bp helix (nucleotides 2120 to 2124 and 2174 to 2178) and a 3-bp helix (nucleotides 2127 to 2129 and 2159 to 2161), which enclose an internal loop. Moreover, we have determined that nucleotides 2163 to 2166 can be deleted from the loop with little effect on protein-RNA association (Fig. 6).

In refined secondary-structure models of the 23S rRNA, the internal loop between the two helices is portrayed as single stranded, although it has been proposed on the basis of covariance analysis that several of its component nucleotides participate in tertiary interactions with other unpaired segments (Fig. 6). These tertiary interactions cannot play a significant role in L1 binding, however, as they involve nucleotides that lie outside the minimized binding site. The structure of the internal loop therefore remains enigmatic and, while preliminary NMR analysis provides evidence for all of the base pairs in the two helical regions, there are no resonances that would signal the presence of additional structured nucleotides (Nikonowicz, unpublished). It is of course possible that the loop becomes structured only when protein L1 is bound, an eventuality that is now under study.

Among the most conserved features of the L1 binding site are, surprisingly, the two short helices that enclose the internal loop (Gutell, personal communication). Three of the base pairs in the 5-bp helix are nearly invariant in bacterial 23S rRNAs and exhibit a level of conservation in excess of 90% when all three phylogenetic domains–*Bacteria, Archaea*, and *Eucarya*—are taken into account. Similarly, the first and third base pairs of the 3-bp helix are the same in over 95% of the bacterial 23S rRNAs, and the former is conserved in more than 95% of archaeal and eucaryal large-subunit rRNAs as well. In addition, several nucleotides of the internal loop, notably A2126, U2167, G2168, A2171, and A2173, are conserved at the level of 95% or over in all of the major phylogenetic domains. This level of conservation suggests that evolution has placed a high premium on the ability of large-subunit rRNAs to associate specifically with L1 or L1-like proteins, though, at least in *E. coli*, L1 is not an essential assembly protein. As in the case of S8, L1 has been found to be a translational repressor of the operon in which it is encoded both in bacteria and in archaea (Hanner et al., 1994). The site with which L1 interacts in the leader sequence of the *E. coli* L11-L1 operon shares many of the primary and secondary structure features characteristic of the rRNA binding site (Fig. 6).

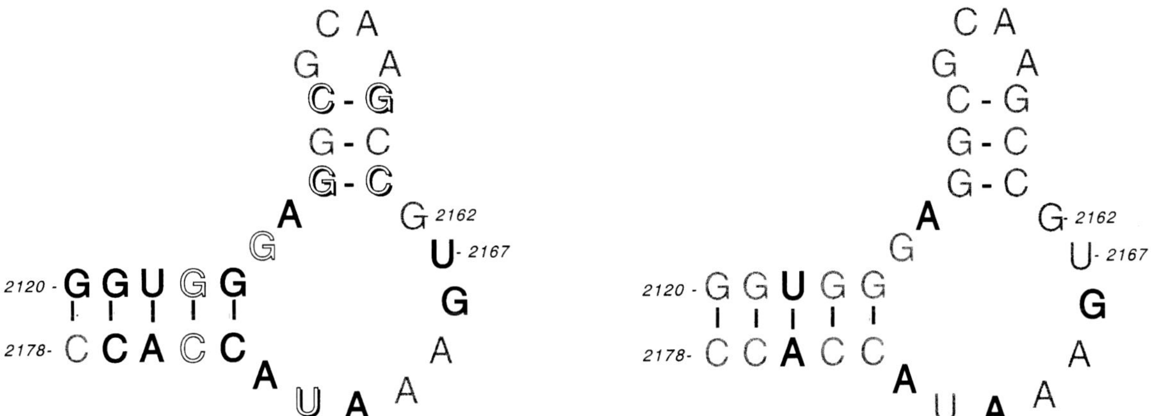

Figure 7. Phylogenetic conservation and mutagenesis in the binding site for protein L1. (Left) Conservation of nucleotides in bacterial and chloroplast 16S rRNAs; boldface, >95%; shadow, 90 to 95%; outline, 80 to 90%. (Right) Nucleotides which, when mutated, sharply decrease L1-RNA affinity (boldface).

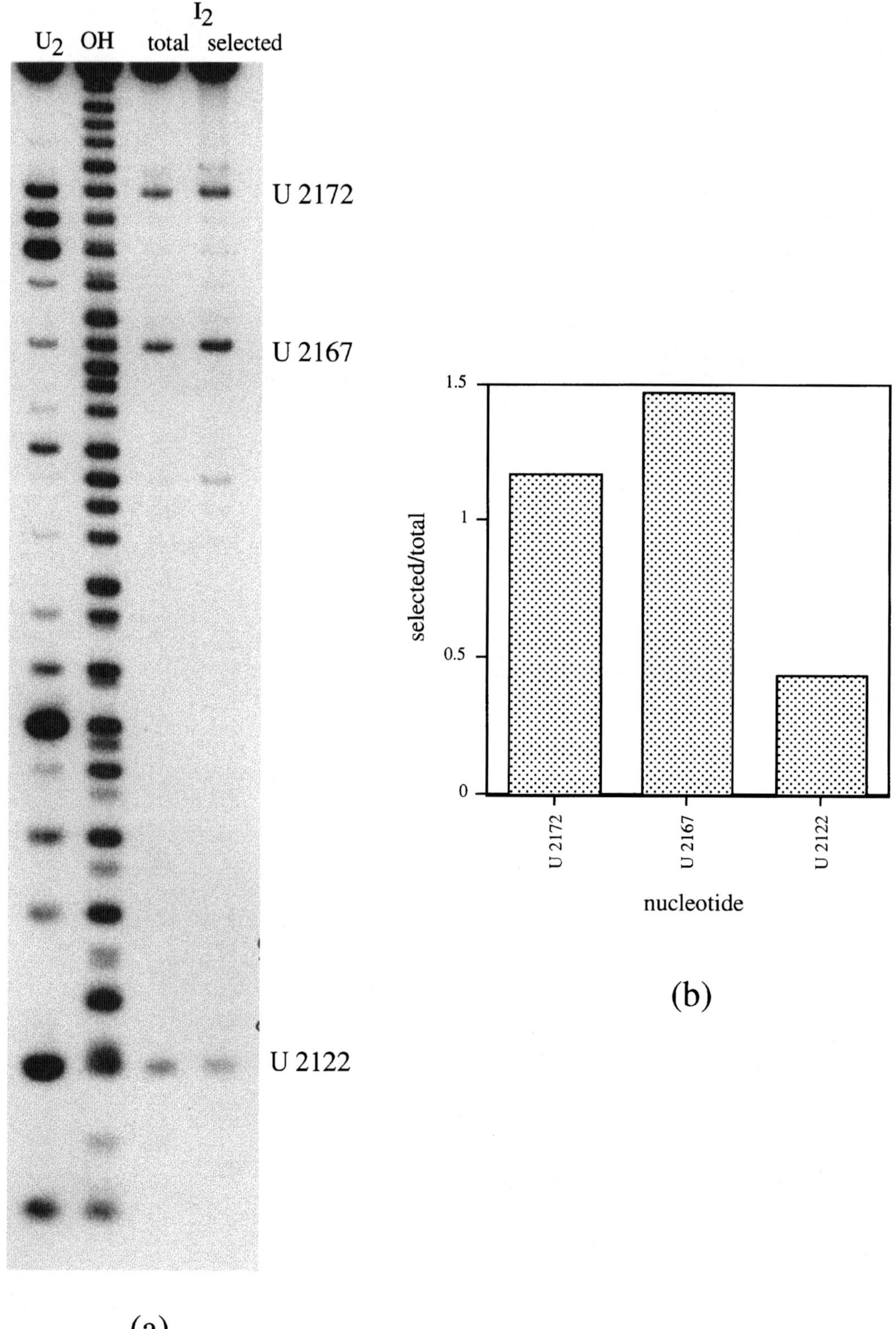

Figure 8. Modification-interference analysis of L1 binding site partially substituted with rUαS. (a) Autoradiogram of rUαS-substituted RNA after RNase U_2, alkali, and iodine treatment, followed by separation on a 15% polyacrylamide gel. Lane U_2, RNase U_2 digestion; lane OH, partial alkaline hydrolysis; lane I_2 total, iodine cleavage of unfractionated RNA; lane I_2 selected, iodine cleavage of L1-bound RNA. (b) Relative intensities of bands corresponding to individual U residues after iodine cleavage, as measured by phosphorimaging. A value near 1.0 indicates no enrichment of rU over rUαS in the bound fraction. Reduced values indicate a preference for the unmodified nucleotide.

Mutagenesis within the Helical and Nonhelical Portions of the L1 Binding Site

To gain further insight into the roles of individual bases and base pairs in L1-rRNA interaction, we have characterized the L1 binding site by site-directed mutagenesis (Dong et al., unpublished). Mutations that disrupt regular Watson-Crick base pairing in the 5-bp helix lead to a sharp decrease in the association constant. This structure is unlikely to mediate base-specific recognition by L1, however, as replacement of the central U2122-A2176 base pair with U·G or C-G has little or no influence on protein binding. The 3-bp helix is also essential for L1 binding; when the helix is deleted, the resulting RNA fragment is unable to interact with the protein at detectable levels. Some, but not all, base replacements at the conserved positions 2126, 2168, 2171, and 2173 severely reduce the affinity of the corresponding rRNA fragments for L1 (Fig. 7). Although these positions are assigned to single-stranded segments in current secondary-structure models of the 23S rRNA, it is likely that they acquire a more specific conformation when complexed with the protein. In this connection, nondenaturing polyacrylamide gel electrophoresis revealed that most of the mutant rRNA transcripts that exhibit defects in L1 binding also migrate more slowly than their wild-type counterparts. Conformational differences in several of these transcripts were detected through an analysis of their susceptibilities to chemical modification.

Role of the Phosphodiester Backbone in L1-RNA Interaction

As part of an effort to characterize the parts that various nucleotide functional groups play in L1-RNA interaction, we have recently focused our attention on the role of nonbridging phosphate oxygens and ribose 2′-OH groups in the phosphodiester backbone (Drygin and Zimmermann, unpublished). RNA molecules corresponding to the minimized L1 binding site in which either U, C, G, or A was partially replaced by its ribonucleoside-α-phosphorothioate (rNαS) or deoxyribo-(dN) analogue were synthesized by T7 transcription in vitro and analyzed for their ability to bind protein L1. Functional groups important for L1-RNA interaction were then identified by a modification-interference approach in which molecules capable of binding the protein were selected from a population of randomly substituted molecules by retention on nitrocellulose membranes. The bound population was then compared with the unfractionated population by its pattern of cleavage when treated with iodine, in the case of RNA containing ribonucleoside-α-phosphorothioates, or alkali, in the case of the deoxyribonucleotide analogues. Transcripts containing ribouridine-α-phosphorothioute (rUαS) exhibited a four- to fivefold decrease in affinity for L1, and analysis of the bound fraction revealed that those molecules containing an α-thiophosphate between positions 2121 and 2122 in the 5-bp helix were strongly selected against (Fig. 8). Replacement of a nonbridging oxygen with sulfur can interfere with the ability of the RNA to interact with the protein either directly, by impairing its capacity to form a hydrogen bond, or indirectly, by eliminating a specific coordination site for Mg^{2+}. The latter possibility was confirmed in the case of the nonbridging phosphate between positions 2121 and 2122 by the observation that Mn^{2+} ions, which have a higher affinity for sulfur than Mg^{2+}, could reverse the inhibition of L1 binding. A similar effect was noted in the case of an α-thiophosphate substitution between positions 2175 and 2176 on the opposite side of the 5-bp helix. Further experiments showed that the presence of deoxynucleotides at multiple positions within helix 77, including position 2122, provoked a sharp decrease in affinity for protein L1. Here again, the absence of specific 2′-OH groups could exert either direct or indirect effects on L1 binding. It is interesting to note that L1 binding is affected by changes in the phosphodiester backbone on opposite sides of the helix, as it has been proposed on the basis of the L1 crystal structure that the two domains of the protein may surround the RNA (Nikonov et al., 1996). If this occurs in the vicinity of the 5-bp helix, it could explain why changes in chemical groups on opposite surfaces of the RNA duplex influence L1 binding. It would also explain why disruption of the U2122-A2176 base pair obstructs L1-RNA interaction.

This work was supported by grant MCB-9818051 from the National Science Foundation and grant GM 22807 from the National Institutes of Health.

REFERENCES

Allmang, C., M. Mougel, E. Westhof, B. Ehresmann, and C. Ehresmann. 1994. Role of conserved nucleotides in building the 16S rRNA binding site of *E. coli* ribosomal protein S8. *Nucleic Acids Res.* **22:**3708–3714.

Conn, G. L., D. E. Draper, E. E. Lattman, and A. G. Gittis. 1999. Crystal structure of a conserved ribosomal protein-RNA complex. *Science* **284:**1171–1174.

Davies, C., V. Ramakrishnan, and S. W. White. 1996. Structural evidence for specific S8-RNA and S8-protein interactions within the 30S ribosomal subunit: ribosomal protein S8 from *Bacillus stearothermophilus* at 1.9 Å resolution. *Structure* **4:**1093–1104.

Dong, P., L. Jiang, K. Harington, P. B. C. Cahill, and R. A. Zimmermann. Unpublished data.

Dragon, F., C. Payant, and L. Brakier-Gingras. 1994. Mutational and structural analysis of the RNA binding site for *Escherichia coli* ribosomal protein S7. *J. Mol. Biol.* **244:**74–85.

Draper, D. E. 1995. Protein-RNA recognition. *Annu. Rev. Biochem.* **64:**593–620.

Draper, D. E. 1996. Ribosomal protein-RNA interactions, p. 171–197. *In* R. A. Zimmermann and A. E. Dahlberg (ed.), *Ribosomal RNA: Structure, Evolution, Processing and Function in Protein Biosynthesis.* CRC Press, Boca Raton, Fla.

Drygin, D., and R. A. Zimmermann. Unpublished data.

Egebjerg, J., J. Christiansen, and R. A. Garrett. 1991. Attachment sites of primary binding proteins L1, L2 and L23 on 23S ribosomal RNA of *Escherichia coli. J. Mol. Biol.* **222:**251–264.

Gourse, R. L., D. L. Thurlow, S. A. Gerbi, and R. A. Zimmermann. 1981. Specific binding of a prokaryotic ribosomal protein to a eukaryotic ribosomal RNA: implications for evolution and autoregulation. *Proc. Natl. Acad. Sci. USA* **78:**2722–2726.

Grasby, J. A., and M. J. Gait. 1994. Synthetic oligoribonucleotides carrying site-specific modifications for RNA structure-function analysis. *Biochimie* **76:**1223–1234.

Gregory, R. J., P. B. F. Cahill, D. L. Thurlow, and R. A. Zimmermann. 1988. The interaction of *Escherichia coli* ribosomal protein S8 with its binding sites in ribosomal RNA and messenger RNA. *J. Mol. Biol.* **204:**295–307.

Gutell, R. R. Personal communication.

Hanner, M., C. Mayer, C. Kohrer, G. Golderer, P. Grobner, and W. Piendl. 1994. Autogenous translational regulation of the ribosomal MvaL1 operon in the archaebacterium *Methanococcus vannielii, J. Bacteriol.* **176:**409–418.

Kalurachchi, K., and E. P. Nikonowicz., 1998. NMR structure determination of the binding site for ribosomal protein S8 from *Escherichia coli* 16S rRNA. *J. Mol. Biol.* **280:**639–654.

Kalurachchi, K., K. Uma, R. A. Zimmermann, and E. P. Nikonowicz. 1997. Structural features of the binding site for ribosomal protein S8 in *Escherichia coli* 16S rRNA defined using NMR spectroscopy. *Proc. Natl. Acad. Sci. USA* **94:**2139–2144.

Lindahl, L., and J. Zengel. 1986. Ribosomal genes in *Escherichia coli. Annu. Rev. Genet.* **20:**297.

Moine, H., C. Cachia, E. Westhof, B. Ehresmann, and C. Ehresmann. 1997. The RNA binding site of S8 ribosomal protein of *Escherichia coli*: selex and hydroxyl radical probing studies. *RNA* **3:**255–268.

Mougel, M., F. Eyermann, E. Westhof, P. Romby, A. Expert-Bezançon, J.-P. Ebel, B. Ehresmann, and C. Ehresmann. 1987. Binding of *Escherichia coli* ribosomal protein S8 to 16S rRNA. A model of the interaction and the tertiary structure of the RNA binding site. *J. Mol. Biol.* **198:**91–107.

Mougel, M., C. Allmang, F. Eyermann, C. Cachia, B. Ehresmann, and C. Ehresmann. 1993. Minimal 16S rRNA binding site and the role of conserved nucleotides in *Escherichia coli* ribosomal protein S8 recognition. *Eur. J. Biochem.* **215:**787–792.

Nevskaya, N., S. Tishchenko, A. Nikulin, S. al-Karadaghi, A. Liljas, B. Ehresmann, C. Ehresmann, M. Garber, and S. Nikonov. 1998. Crystal structure of ribosomal protein S8 from *Thermus thermophilus* reveals a high degree of structural conservation of a specific protein binding site. *J. Mol. Biol.* **279:**233–244.

Nikonov, S., N. Nevskaya, I. Eliseikina, N. Fomenkova, A. Nikulin, N. Ossina, M. Garber, B.-H. Jonsson, C. Briand, S. al-Karadaghi, A. Svensson, A. Ævarsson, and A. Liljas. 1996. Crystal structure of the RNA binding ribosomal protein L1 from *Thermus thermophilus. EMBO J.* **15:**1350–1359.

Nikonowicz, E. P. Unpublished data.

Powers, T., and H. F. Noller. 1995. Hydroxyl radical footprinting of ribosomal proteins on 16S rRNA. *RNA* **1:**194–209.

Ryan, P. C., M. Lu, and D. E. Draper. 1991. Recognition of the highly conserved GTPase center of 23S ribosomal RNA by ribosomal protein L11 and the antibiotic thiostrepton. *J. Mol. Biol.* **221:**1257–1268.

Sapag, A., J. V. Vartikar, and D. E. Draper. 1990. Dissection of the 16S rRNA binding site for ribosomal protein S4. *Biochim. Biophys. Acta* **1050:**34–37.

Serganov, A. A., B. Masquida, W. Westhof, C. Cachia, C. Portier, M. Garber, B. Ehresmann, and C. Ehresmann. 1996. The 16S rRNA binding site of *Thermus thermophilus* ribosomal protein S15: comparison with *Escherichia coli* S15, minimum site and structure. *RNA* **2:**1124–1138.

Svensson, P., L. M. Changchien, G. R. Craven, and H. F. Noller. 1988. Interaction of ribosomal proteins S6, S8, S15 and S18 with the central domain of 16S ribosomal RNA. *J. Mol. Biol.* **200:**301–308.

Thomas, M. S., and M. Nomura. 1987. Translational regulation of the L11 ribosomal protein operon of *Escherichia coli*: mutations that define the target site for repression by L1. *Nucleic Acids Res.* **15:**3085–3096.

Thurlow, D. L., and R. A. Zimmermann. Unpublished data.

Uma, K., and R. A. Zimmermann. Unpublished data.

Uma, K., E. P. Nikonowicz, K. Kalurachchi, H. Wu, I. Wower, and R. A. Zimmermann. 1995. Structural characterization of *Escherichia coli* ribosomal protein S8 and its binding site in 16S ribosomal RNA. *Nucleic Acids Symp. Ser.* **33:**8–10.

Urlaub, H., B. Thiede, E. C. Muller, R. Brimacombe, and B. Wittmann-Liebold. 1997. Identification and sequence analysis of contact sites between ribosomal proteins and rRNA in *Escherichia coli* 30S subunits by a new approach using matrix-assisted laser desorption/ionization-mass spectrometry combined with N-terminal microsequencing. *J. Biol. Chem.* **272:** 14547–14555.

Wimberly, B. T., R. Guymon, J. P. McCutcheon, S. W. White, and V. Ramakrishnan. 1999. A detailed view of a ribosomal active site: the structure of the L11-RNA complex. *Cell* **97:**491–502.

Wower, I., M. P. Kowaleski, L. E. Sears, and R. A. Zimmermann. 1992. Mutagenesis of ribosomal protein S8 from *Escherichia coli*: defects in the regulation of the *spc* operon. *J. Bacteriol.* **174:** 1213–1221.

Wu, H., L. Jiang, and R. A. Zimmmerann. 1994. The binding site for ribosomal protein S8 in 16S rRNA and *spc* mRNA from *Escherichia coli*: minimum structural requirements and the effects of single bulged bases on S8-RNA interaction. *Nucleic Acids Res.* **22:**1687–1695.

Zimmermann, R. A. Unpublished data.

Zimmermann, R. A, H. Wu, I. K. Wower, and K. Uma. Unpublished data.

The Ribosome: Structure, Function, Antibiotics, and Cellular Interactions
Edited by R. A. Garrett, S. R. Douthwaite, A. Liljas, A. T. Matheson, P. B. Moore, and H. F. Noller

Chapter 11

RNA Tertiary Structure and Protein Recognition in an L11-RNA Complex

DAVID E. DRAPER, GRAEME L. CONN, APOSTOLOS G. GITTIS, DEBRAJ GUHATHAKURTA, EATON E. LATTMAN, and LUIS REYNALDO

The importance of the L11-rRNA complex to ribosome function was first suggested by observations that L11 was necessary for high-affinity ribosome binding of thiostrepton (Highland et al., 1975), an antibiotic known to inhibit elongation factor-dependent ribosome activities (Cundliffe, 1979). Work in Eric Cundliffe's laboratory characterized a small rRNA fragment that formed a tight ternary complex with L11 and thiostrepton (Thompson et al., 1979; Schmidt et al., 1981) and identified A1067 within this RNA as the methylation site that confers resistance to thiostrepton (Cundliffe and Thompson, 1979). The protein and rRNA sequence appeared to be interchangeable between different organisms (Stark et al., 1980; Beauclerk et al., 1985; El-Baradi et al., 1987; Musters et al., 1991; Thompson et al., 1993), suggestive of a highly conserved and functionally important center. Several observations about the varying effects of L11, thiostrepton, and micrococcin (a thiazole antibiotic related to thiostrepton) on ribosome activity were initially explained by a proposal that the rRNA domain undergoes a series of conformational changes during translocation, which are promoted or inhibited in different ways by L11 or antibiotics (Cundliffe, 1986). More recently, studies of L11 point mutations conferring antibiotic resistance have prompted a hypothesis that L11 itself undergoes conformational changes which are inhibited by antibiotics (Porse et al., 1998, 1999).

Some time ago, we initiated physical studies of L11 and its target RNA with the expectation that structural and thermodynamic information about this complex would be essential for understanding its function. Several important points became clear during these studies: (i) the 58-nucleotide (nt) recognition site for L11 contains an extensive and (in some circumstances) very stable tertiary structure (Laing and Draper, 1994; Lu and Draper, 1994); (ii) the entire 58-nt RNA is required to form the binding sites for both L11 and thiostrepton (Ryan and Draper, 1989; Draper et al., 1995; Xing and Draper, 1995); and (iii) the protein folds into two domains, with the C-terminal domain entirely responsible for RNA binding and the N-terminal domain engaged in cooperative interactions with thiostrepton (Xing and Draper, 1996). During the course of these studies, it was found that some naturally occurring sequence variants of the rRNA domain had unusually stable tertiary structures (Lu and Draper, 1994) and that protein solubility was enhanced by elimination of the N-terminal domain, which tends to be poorly folded. Well-behaved protein and RNA fragments made structural studies possible, starting with study of the structure of the protein C-terminal domain (L11-C76) by nuclear magnetic resonance (NMR) methods (Xing et al., 1997) and further NMR studies of the protein-RNA complex that defined the RNA binding surface of L11 (Hinck et al., 1997). Solution of the RNA structure by NMR methods proved extremely difficult, but fortunately the L11–C76–58-nt RNA complex crystallized, and its structure at 2.8-Å resolution is now available (Conn et al., 1999). This structure provides our first atomic-resolution view of RNA tertiary folding and protein-RNA interactions within the ribosome.

The goal of this review is to relate the intricate architecture of the L11-RNA complex to previous studies that delineated crucial features of the RNA

David E. Draper, Graeme L. Conn, and Debraj GuhaThakurta ■ Department of Chemistry, Johns Hopkins University, Baltimore, MD 21218. **Apostolos G. Gittis and Eaton E. Lattman** ■ Department of Biophysics, Johns Hopkins University, Baltimore, MD 21218. **Luis Reynaldo** ■ Third Wave Technologies, Inc., 502 South Rosa Rd., Madison, WI 53719.

tertiary structure and protein-RNA interface. In describing the structure, it is interesting to note how conservation and variation of different nucleotides and amino acids serve as a guide to critical features of the complex, and we use the extreme conservation of some bases to speculate about functional surfaces of the rRNA domain. Lastly, we discuss the possibility that the functional role of L11-C76 is to promote a correct RNA tertiary fold.

OVERVIEW OF RNA TERTIARY FOLD

Figure 1 shows the RNA secondary structure and views of the folded molecule. The RNA secondary structure is formally a four-helix junction. As in DNA four-helix junctions, the four closing base pairs form two coaxially stacked units. The U1082-A1086 "lone pair" stacks onto the adjoining helix A. Helix B is closed by a G1056-A1103 base pair, which stacks with the adjacent helix C. These two highly irregular units are joined together by numerous tertiary hydrogen bonds that create sets of four or five coplanar bases across the width of the molecule. These planes of bases generate an unusually large interior of stacked bases enclosed by backbone strands. In this section we summarize the tertiary interactions within these planes of bases, starting from the bottom of the molecule as drawn in Fig. 1B.

The 3 bases closed by the U1082-A1086 base pair form a U-turn structure (see the discussion of RNA motifs below), which places A1084 and A1085 in the minor groove of helix B. A1084 makes a single hydrogen bond to the sugar of U1103, while A1085 is the center of a hydrogen bond network (Fig. 2A). The hydrogen bond to U1083 is part of the U-turn structure, and the other four hydrogen bonds line the minor groove of the G1055-C1104 base pair. An

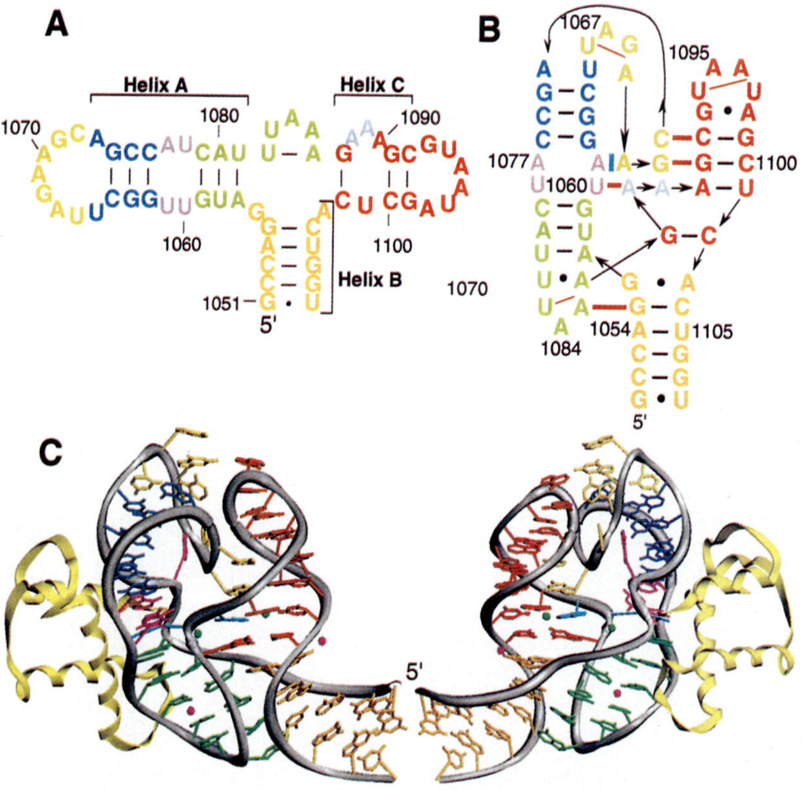

Figure 1. Folding of the 1051-to-1108 domain of large-subunit rRNA. (A) Phylogenetically conserved secondary structure, as drawn by Gutell et al. (1992b). The *E. coli* sequence and numbering are shown. (B) Representation of rRNA showing tertiary base-base hydrogen bonds (horizontal red bars) and base stacking (vertical blue bar) and approximate stacking of the bases as seen in the crystal structure. Thin red diagonal lines indicate U imino-phosphate hydrogen bonds characteristic of the U-turn motif. Thin black lines join consecutive nucleotides, with arrows indicating the 5′ → 3′ direction of the backbone. The crystallized sequence contains a U1061 → A mutation that stabilizes the tertiary fold. (C) Representation of the L11-RNA complex structure as determined by crystallography (Conn et al., 1998); the two views are rotated about the vertical axis by 180°. The bases are color coded to correspond to panels A and B. Two Mg ions (green spheres) and two $Os(NH_3)_6$ ions (magenta spheres) are located within the structure. The L11-C76 backbone is shown as a yellow ribbon.

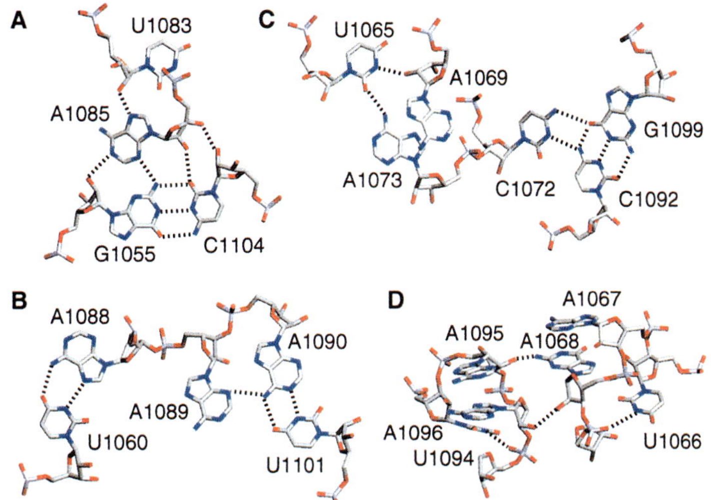

Figure 2. Tertiary interactions within the 1051-to-1108 RNA. (A) A1085 forms a minor-groove triple with G1055-C1104; all three of these bases are invariant. (B) A1088-A1089 sidestep connects a Hoogsteen pair (U1060) and a base triple (A1090-U1101). (C) The C1072(C1092-G1099) base triple and its linkage to a noncanonical A1065-C1073 pair. (D) Hydrogen bonding between the U1066 and U1094 U-turns. Dashed lines indicate potential hydrogen bonds.

A1085 → U mutation was previously found to strongly destabilize the RNA tertiary structure while leaving the secondary structure unperturbed (Lu and Draper, 1994); the hydrogen bonding and stacking interactions of this base must be an important component of the tertiary structure. A reversal of the G1055-C1104 pair to C-G likewise destabilized the tertiary structure (Lu and Draper, 1994).

The next layer of stacked bases contains two noncanonical pairings. U1082-A1086 is a reverse Watson-Crick pair (U-A O-2—HN-6, NH-3—N-1) with the adenine in a *syn* conformation. A U-A pair is found at this position in all bacteria, while C1082-G1086 is found in all eukaryotes. The identical conservation pattern is seen at the Levitt pair in tRNA (Levitt, 1969), which is either G15-C48 (e.g., *Saccharomyces cerevisiae* $tRNA^{Phe}$ [Westhof and Sundaralingham, 1986]) or A15-U48 ($tRNA^{Asp}$ [Westhof et al., 1985]). The Levitt pair makes a reverse Watson-Crick pair identical to U1082-A1086 but does not require a *syn* base conformation, as the two strands are parallel. The G-C Levitt pair is similar (G-C HN-1—O-2, HN-2—N-3); presumably the eukaryotic C1082-G1086 uses the same hydrogen bonding. Disruption of the 1082-1086 pair by an A1086 → U mutation destabilizes the tertiary structure, but a U1082 → A mutation surprisingly restores both the stability of the RNA and its ability to bind L11 (Ryan and Draper, 1991; Lu and Draper, 1994). To preserve the relative orientation of the two glycosidic bonds in an A1082-U1086 pair, the U would have to adopt a *syn* conformation, which is sterically unfavorable. A1082-U1086 has not been found in nature, despite its apparent stability.

Next to the 1082-1086 pair is G1056-A1103, which is a standard "edge-to-edge" or "sheared" pair, as first observed in DNA (Li et al., 1991) and subsequently found in many RNAs (Huang et al., 1996). An A-A pair can form the same structure lacking one hydrogen bond (cf. A113-A207 of the P4-P6 domain [Cate et al., 1996a]), and A1056 occurs commonly in rRNA sequences. The noncanonical U1082-A1086 and G1056-A1103 pairs are joined by hydrogen bonds from the 2′-OH of A1086 to both A1103 N-1 and G1056 2′-OH.

In going from A1086 to G1087, the backbone makes a sharp twist to point the two bases in opposite directions for pairing in different helices. This twist places the G1087 phosphate towards the interior of the RNA, where it forms hydrogen bonds to N6 of A1057. An A1057-U1081 pair is present in most bacteria but changes to C1057-G1081 in most eukaryotes. N-4 of C1057 might be able to preserve the same phosphate-hydrogen bond.

The most unusual plane of bases involves the 1088-1089 bulge loop (Fig. 2B). A1090 and U1101 form a Watson-Crick pair, though these positions are a U·U mismatch in many rRNA sequences. A1090 also forms a hydrogen bond with A1089 in an "A-A sidestep" to form an unusual base triple. An A-A sidestep was seen in a somewhat different context ("A platform") in the crystal structure of the P4-P6 domain of group I introns (Cate et al., 1996b). In rRNA sequences with a U1090-U1101 mismatch, position 1089 is a G and preserves a similar sidestep structure (Wimberly et al., 1999). A twist of the backbone between A1088 and A1090 points A1088 away from the base triple structure and intercalates it into the major groove of helix A, where it forms a Hoogsteen pair with U1060. A1088 is in the *syn* conformation, and U1060 has been rotated by 180° within the helix. The Hoogsteen pairing therefore exposes the major-groove edge of U1060 and the Watson-Crick acceptor and donor positions of A1088 in the minor groove of helix A. Mutation of either U1060 or A1088 disrupts the RNA tertiary structure (Lu and Draper, 1994), and as discussed below, this pair is also critical for protein recognition.

The A1089(A1090-U1101) base triple is the bottom of three stacked base triples; the other two are derived from two bases of the helix A hairpin loop placed in the major groove of helix C. The G1071(G1091-C1100) triple is conserved in nearly all rRNA sequences, but the C1072(C1092-G1099) triple (Fig. 2C) becomes U(U-A) in nearly all eukaryotes. The existence of this triple was predicted from comparative sequence analysis and confirmed by compensatory mutations (Conn et al., 1998). A C1072 → U mutation was particularly destabilizing for the tertiary structure. Comparative sequence analysis also showed a correlation of the 1065-1073 pair with the base triple: Y-A is generally present with C1072, and A-Y is present with U1072. Melting studies of 1065-1073 mutations with the two different base triples confirms the influence of this base pair on the stability of the base triple (Conn et al., 1998). U1065 and A1073 pair by a single noncanonical hydrogen bond (U O-2—A NH-6), and the imino proton of U1065 hydrogen bonds to the 2′-OH of A1069. An A1065-C1073 pair (as in most eukaryotes) cannot isosterically reproduce either of these hydrogen bonds. As suggested by Fig. 2C, any movement of position 1065 would also cause a repositioning of the 1072(1092-1099) triple and potentially affect its stability.

Placement of G1071-C1072 in the helix C major groove requires an unusual conformation for the helix A hairpin loop: a U-turn at the 5′ end of the loop directs the rest of the sequence alongside helix A. Helix C is also capped by a U-turn structure, as deduced previously by NMR studies of this hairpin (Huang et al., 1996). These two U-turns are adjacent to each other in the folded RNA and are held together by two hydrogen bonds (Fig. 2D). One of the helix A hairpin loop nucleotides, A1070, is turned out into the cleft between the two hairpin loops, where it stacks with A1061, which has been bulged out from helix A. 1061 is U in *Escherichia coli* rRNA, and its replacement with A increases the free energy of folding the RNA tertiary structure by ~4.5 kcal/mol (Lu and Draper, 1994). This increment is too large to be due solely to the change from a U-A stack to an A-A stack. It is likely that the U1061 → A mutation destabilizes base pairing of 1061 with A1077 and thereby lowers the cost of disrupting helix A to insert A1088.

RNA MOTIFS

Relatively few RNA structures that have noncanonical interactions have been determined at atomic resolution, and of these only tRNA (Jack et al., 1976; Holbrook et al., 1978) and the P4-P6 domain of group I intron (Cate et al., 1996a) have extensive tertiary structure. Despite this limited glimpse into the universe of RNA folding, a number of themes recur, which suggests that the number of RNA building blocks is limited (Conn and Draper, 1998).

A frequent motif is the U-turn, which was first described in the anticodon and TψC loops of tRNA (Quigley and Rich, 1976) and has been seen subsequently in the hammerhead ribozyme (Pley et al., 1994; Scott et al., 1995). The motif consists of three nucleotides, UNR, which reverse the direction of the backbone by a 120° rotation of a P-O bond between the U and N nucleotides; approximately A-form conformation is otherwise maintained for all three nucleotides. The U imino proton forms a hydrogen bond to the phosphate 3′ to R, and the 2′-OH of the U nucleotide forms a hydrogen bond to N-7 of R. The second two nucleotides of the motif are left in a stacked, helical conformation appropriate for Watson-Crick base pairing, as in the tRNA anticodon. The U-turn occurs three times in the L11 binding RNA. The U1083 turn positions the following two As in the minor groove of helix B, while the U1066 and U1094 turns each direct two stacked purines into solvent.

A second motif seen in this rRNA fragment is the A-A sidestep. In the A platform motif, an A-A sidestep in an internal loop stacks on the 3′ side with a noncanonical base pair and on the 5′ side with part of a base triple (Cate et al., 1996b). In the L11 binding RNA, the sidestep is not inserted into a helix in the same way as an A platform but is itself part of a base triple. In an aptamer binding theophylline, a nearly identical base triple is seen with an A-C sidestep and a C-G base pair (Zimmermann et al., 1997), and many ribosomal RNAs substitute a G-U sidestep and a U·U pair (Wimberly et al., 1999).

A third motif is the backbone S-turn that reverses the direction of the backbone at U1060. An identical S-turn was seen in the Rev-RRE complex, where the flipped nucleotide is part of a G·G mismatch (Battiste et al., 1996), and in one A of a sheared A-A pair in the sarcin-ricin loop (Correll et al., 1998). In all three RNAs a bulged base 3′ to the reversed nucleotide is present; the extra backbone segment allows resumption of standard helix conformations.

PROTEIN-RNA CONTACTS

From NMR studies of the free L11 RNA binding domain (L11-C76), it was known that the protein folds into three α-helices that are superimposable on the α-helices of the homeodomain class of DNA binding proteins (Xing et al., 1997). Further NMR studies also showed that L11-C76 uses the same surface, centered on α-helix 3, to bind RNA that homeodomains use to bind DNA (Hinck et al., 1997). The two proteins interact with nucleic acid helices differently, in that homeodomains place α-helix 3 in the DNA major groove while L11-C76 puts α-helix 3 against a distorted and flat RNA minor groove.

L11-RNA contacts are summarized in Fig. 3. α-helix 3 makes four hydrogen bonds to bases in RNA helix A, three of which are from backbone carbonyls. Two of these, Thr66 hydroxyl-U1060 and Gly65 carbonyl-A1088, are particularly interesting, as the U1060-A1088 pair is a universally conserved feature of the RNA tertiary structure. The Gly65 α carbon lies close enough to C1079 that insertion of a methyl group would push Gly65 out of hydrogen-bonding distance, and in fact a Gly65 → Ala mutation weakens RNA binding affinity substantially (Xing et al., 1997). A Thr66 → Val mutation, which substitutes a methyl group for the hydroxyl, is similarly deleterious. Over 50 L11 sequences are available, and they all show either Thr or Ser at 66 and only Gly at 65. L11 recognition of this unusual tertiary base pair is thus an extraordinarily well-conserved interaction.

At the C terminus of α-helix 3, Ser69 forms hydrogen bonds to G1062, but the interaction is not as highly conserved as that of Gly65 or Thr66. We previously noticed that mutations of Ser69 to several different amino acids weakened binding 5- to 20-fold, yet two of the substitutions tested, Gln and Asn, occurred naturally. Asn69 is present in all plastid sequences and in their close relative *Synechocystis*. Most plastid and *Synechocystis* ribosomes also carry U-A or C-U for the base pair at 1062-1076; these variants have not been seen in other parts of the phylogenetic tree. L11 protein carrying Ser69 discriminates against RNAs in which U1062-A1076 has been substituted for G1062-C1076; conversely, L11 with Asn69 discriminates in favor of the plastid-like

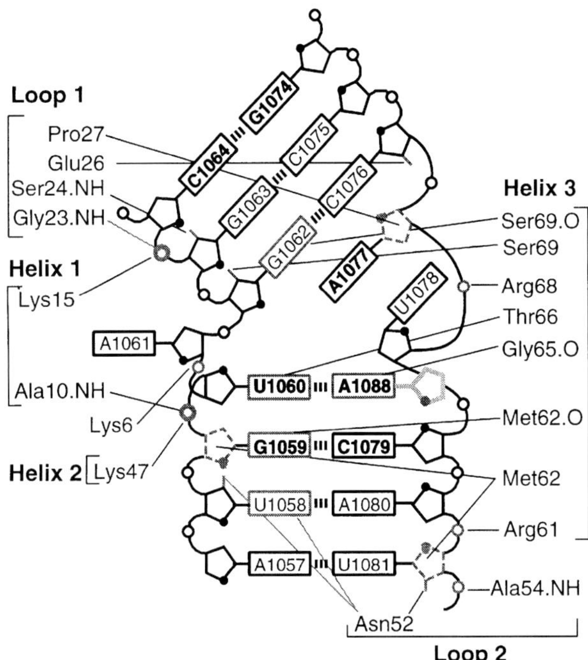

Figure 3. Schematic of protein contacts with RNA. The bases shown in boldface are conserved in >97% of rRNA sequences from the three main phylogenetic domains. The base, 2′-OH, and phosphate that form hydrogen bonds to protein are shown in gray. The dashed-line sugars make nonpolar contacts with protein. The A1088 sugar is in gray as a reminder that it is not contiguous with the RNA sequence shown but is intercalated from elsewhere in the RNA. Protein contacts are from the indicated amino acid side chain unless specified as backbone carbonyl (.O) or backbone amide (.NH).

U1062-A1076 substitution (GuhaThakurta and Draper, 1999). This protein-RNA covariation is most simply interpreted as a direct contact of Ser69 with G1062, but an indirect mechanism could not be ruled out on the basis of mutagenesis and binding studies. It was therefore interesting to find the suggestion of a direct interaction confirmed by the crystal structure. Ser69 makes two hydrogen bonds to G1062: the carbonyl bonds to N-2 of the base, and the hydroxyl bonds to 2′-OH. The G N-2 is replaced by pyrimidine O-2 in plastid sequences; it is not yet known how Asn at position 69 recognizes this substitution. It is also not yet understood how those proteins with Gln at 69 have compensated for the weakened binding affinity (GuhaThakurta and Draper, 1999).

α-helix 3 is flanked at either end by loops. One of these, loop 1, is disordered in the free protein but becomes as rigid as the rest of the protein when the complex forms (Markus et al., 1997). This loop makes several hydrogen bonds and nonpolar contacts with the backbone on either side of the helix. At the N terminus of α-helix 3, loop 2 uses Asn52 to make three hydrogen bonds, two with backbone 2′-OH and one with a base, U1058. This amino acid is not well conserved, varying to Thr, Ser, Leu, Met, and others; U1058-A1080 also becomes G-C in many rRNA sequences. What accommodations take place to preserve loop 2-RNA contacts in organisms with variant sequences is not yet known.

As in most nucleic acid binding proteins, L11 partially neutralizes the negative charge of the RNA backbone with basic residues. One way to experimentally determine which residues interact electrostatically is to measure the salt dependence of the binding constant when a basic residue is replaced by a neutral one. As monovalent cations compete with basic residues for electrostatic interactions with phosphates, the degree to which binding affinity is reduced by increasing KCl concentration reflects the electrostatic component of the binding free energy. A total of five Lys and Arg residues form salt bridges with phosphates on both strands of the helix. All of these residues were mutated to either Ala or Met and found to reduce the magnitude of the salt dependence of RNA binding (GuhaThakurta and Draper, in press). Mutation of some other basic residues that are away from the RNA binding surface did not affect the salt dependence of binding. Besides the salt bridge contacts shown in Fig. 3A, one additional residue in loop 1, Arg29, was found to contribute to electrostatic interactions. The electrostatically important basic residues line the perimeter of the protein-RNA contact surface and are conserved as basic residues in 90 to 100% of the available bacterial L11 sequences. Except for position 61, archaeal and eukaryotic sequences have different patterns of conserved basic residues and presumably place basic residues in different positions within the electrostatic field.

FUNCTIONAL SIGNIFICANCE OF CONSERVED NUCLEOTIDES

Covariation analysis has been an extremely powerful method for predicting rRNA secondary structure and providing clues to tertiary interactions (Gutell et al., 1992a). It is also informative to ask why some rRNA nucleotides never (or rarely) vary: are they components of unique tertiary structures, or are they targets for proteins or factors which place additional constraints on their variability? An exceptionally large fraction of the nucleotides within the L11 binding RNA, 23 of 58, are conserved at a very high level among the three main phylogenetic domains (*Bacteria, Archaea,* and *Eucarya*). With the RNA tertiary structure and contacts with L11 now known in detail, it is possible to ask which of the conserved residues are constrained by tertiary or protein contacts and which might be poised for interactions elsewhere in the ribosome. Figure 4 summarizes the highly conserved positions within the RNA and groups these into several classes depending on how well their conservation can be rationalized. Members of each of these groups are discussed in this section.

Eight of the invariant bases are involved in three sets of base-base tertiary hydrogen bonding (Fig. 4). These nucleotides may be conserved either because isosteric substitutions of other nucleotides are not possible or because concerted changes of several nucleotides are unlikely to arise. Any change in the U1060-A1088 tertiary pair would also require a compensating protein mutation(s).

Mutations in several residues involved in noncanonical interactions are probably also constrained by the folded RNA structure (Fig. 4). Gautheret et al. (1994) have shown that a sheared G-A pair can link to a Watson-Crick pair 5′ to the G, but it has a short P(G)-O3′(A) distance that prevents linkage to a standard base pair on the 3′ side. G1056-A1103 fits this pattern. In going from G1056 to A1057, the RNA backbone twists to place A1057 in helix B, and the displacement of G1056 from the position of a standard base pair helps to make this connection correctly. An A-A sheared pair has the same conformation; thus, the conservation pattern A/G1056 and A1103 is dictated by the junction conformation. The sheared G1093-A1098 pair, which closes a hairpin loop, may be conserved for similar reasons, though it is not as clear why an adenine at 1093 could not be tolerated. Lastly, U1094 is part of a U-

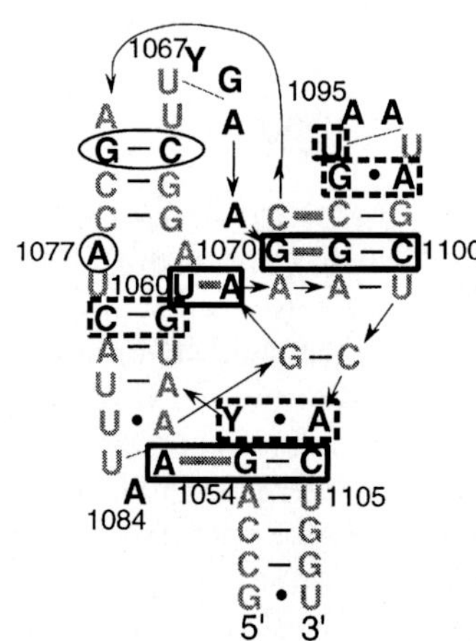

Figure 4. Highly conserved residues in the 1051-to-1108 rRNA domain. Residues conserved in greater than 97% of sequences from the three main phylogenetic domains are shown in boldface (Conn et al., 1998). Solid boxes, bases involved in base-base tertiary bonding; dashed boxes, bases involved in noncanonical interactions; circles, highly conserved residues; several other highly conserved bases unlikely to be required for proper folding are boldfaced and uncircled.

turn structure, though it is not obvious why the other two U-turns are tolerant of substitutions at a 5 to 15% level.

Protein recognition may prevent variants of the G1059-C1079 pair from occurring, as a carbonyl from α-helix 3 forms hydrogen bonds to the guanine.

Weak cases can be made for why several other residues are highly conserved (Fig. 4). A1077 is potentially hydrogen bonded to A1088 2′-OH (via N-1) and to A1089 phosphate (via N-6). However, A1088 2′-OH is also within hydrogen-bonding distance of the other A1089 phosphate oxygen, and cannot be donating a hydrogen bond to both groups simultaneously. It is not obvious whether a G at 1077 might be able to stack and form hydrogen bonds nearly as well as A; only a U substitution has been tried and was found to be destabilizing (Lu and Draper, 1994).

The C1064 amino group is close to A1070 phosphate oxygen (3.6 Å), but this does not seem sufficient to explain why C1064-G1075 is highly conserved. Substitution of a G-C for C1064-G1075 does affect L11 and thiostrepton binding to some degree (Ryan et al., 1991). It may be that C1064-G1075 is highly conserved because tertiary folding requires an extraordinarily precise helix B structure, or perhaps substitutions at these positions promote alternative conformations of the RNA. In support of these possibilities, we have previously noted that mutation of G1062-C1076 to a U-A pair destabilizes RNA tertiary structure, for no reason that can be discerned from the structure (GuhaThakurta and Draper, 1999).

There are several highly conserved positions which are very unlikely to be required for proper folding and also are suggestively displayed on the surface of the RNA (Fig. 4). The helix A and helix C hairpin loops contain several such bases. The bases at the tips of the U-turns, A1095 and A1067 (which is always a purine), are not constrained by the structure to be any particular base and are highly exposed. G1068 and A1069 are stacked underneath A1067, lining the cleft between the two U-turn loops. A1070 protrudes from this cleft and stacks with A1061. Among 1068 to 1070, the only real constraints on the bases are that 1068 be a purine (to form the U-turn) and that 1070 stack well. Likewise, A1096 is stacked underneath A1095 and is only constrained by the U-turn to be a purine.

These two clusters of highly conserved bases with edges and (for A1067, A1095, and A1070) faces exposed to solvent seem likely sites for interactions with other parts of the ribosome. An obvious candidate is elongation factor G, which binds this domain of rRNA (Munishkin and Wool, 1997) and protects A1067 and A1069 from reaction with dimethylsulfate (Moazed et al., 1988). Another ligand that binds to this rRNA domain is thiostrepton. Its binding site most likely lies within the same two clusters of conserved bases, as 2′-O-methylation of A1067 reduces thiostrepton binding affinity and bases at positions 1069 to 1070, 1095, and 1098 are protected from chemical reaction by the antibiotic (Egebjerg et al., 1989). Of course, a conserved surface has not evolved to aid binding of an antibiotic, but thiostrepton may well compete for an interaction normally occurring during the ribosome cycle.

STABILIZATION OF RNA TERTIARY STRUCTURE BY L11

In melting studies of the 58-nt RNA, it was proposed that the lowest-temperature melting transition is due to unfolding of a set of tertiary interactions that link the three helical elements (Laing and Draper, 1994). The unusual sensitivity of this unfolding step to the presence of Mg^{2+} (Laing et al., 1994; Bukhman and Draper, 1997) and to mutations in "loop" nucleotides (Lu and Draper, 1994) supported this interpretation. When a thermophilic homologue of L11 became available, it was possible to add the protein in melting experiments and ask whether it could stabilize tertiary structure. The results are very striking: no structure within the 58-nt RNA domain may unfold while L11-C76 is bound to the RNA (Fig. 5A).

It is easy to see from the L11-C76-RNA structure why the protein stabilizes the entire 58-nt RNA domain, as the correct conformation of the protein target site depends on RNA tertiary interactions. Two tertiary contacts are particularly relevant to L11 binding. First, the U1060-A1088 tertiary base pair is directly bonded to L11-C76 α-helix 3. As this pair links helix A and helix C, its stabilization is an important reason why the tertiary structure is unable to fold when L11 is bound. Second, L11 stabilizes the backbone S-turn at U1060, which in turn bulges nucleotide 1061 out of the helix; stacking of 1061 with A1070 stabilizes the unusual conformation of the helix A hairpin loop that is needed to form the G1071 and C1072 base triples. This is a more indirect effect of L11 on the tertiary structure but is also likely to be important. Stacking and hydrogen bonding of bases in the tertiary core of the RNA stitch together all the helical units of the molecule and make the tertiary structure unfold as a single cooperative unit. Stabilization of any one tertiary contact thus helps to stabilize the whole unit.

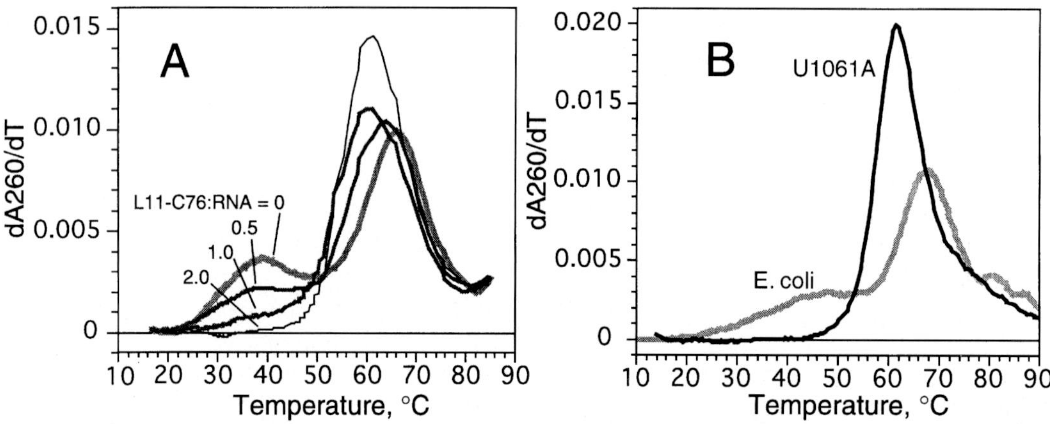

Figure 5. Stability of the 1051-to-1108 rRNA domain tertiary structure. The data are first-derivative melting curves taken in 5 mM $MgCl_2$, 100 mM KCl (A) or NH_4Cl (B), and 10 mM MOPS (morpholinepropanesulfonic acid) (pH 7.2). (A) Stabilization of rRNA structure by L11-C76 protein. GACG RNA, a variant in which the tertiary structure unfolds in a distinct first transition from the rest of the structure (Bukhman and Draper, 1997), was used in these experiments. The molar ratio of protein to RNA is indicated; the RNA concentration was 0.82 μM. Note that at the higher protein concentration, the RNA shows no hyperchromicity until >45°C. (B) Melting of either wild-type *E. coli* (gray) or a U1061A variant (black) RNA fragment. Data are from Lu and Draper (1994). Tertiary structure unfolds first in a broad transition in the *E. coli* rRNA fragment, but this tertiary structure is stabilized to much higher temperatures and melts simultaneously with some of the secondary structure in the variant.

Does the rRNA require L11 to fold its tertiary structure correctly? The tertiary structure of the *E. coli* sequence of the 1051-to-1108 rRNA fragment unfolds in a very broad, low-temperature transition in melting experiments (Fig. 5B). The tertiary structure is certainly not fully formed at physiological temperature and salt conditions, and hydroxyl radical reactivities of the RNA suggest that only a fraction of the molecules are correctly folded even at low temperatures (Conn, unpublished). *E. coli* rRNA, at any rate, appears to have a difficult time finding the correct tertiary fold. However, two simple mutations substantially stabilize tertiary structure. A change of the 1089-to-1090 A-A sidestep to G-U, which occurs in a large fraction of all rRNA sequences, increases the melting temperature (T_m) of the tertiary structure by ~20°C, though the range remains broad (Lu and Draper, 1994). Even more dramatic is the U1061-to-A mutation, which was used for NMR and crystallographic studies (Fig. 5B); the effective T_m of the tertiary structure increases by ~25°C, and the transition simultaneously sharpens (the ΔH of unfolding doubles). Strong protection of a number of nucleotides from reaction with hydroxyl radicals suggests that most of the RNA molecules have the same tertiary fold seen in the L11-RNA crystal structure (Conn, unpublished).

The U1061A variant is relatively rare, found only in some archaeal sequences. If ribosome function requires a folded rRNA tertiary structure, why has there not been more selective pressure for this single mutation? One possibility is that the rRNA must unfold during the ribosome cycle and that the structure should not be too stable. At the present time, there is no evidence that the rRNA unfolds or that L11 transiently dissociates during translation, so it is difficult to argue in favor of this possibility. We suggest a simpler explanation, based on NMR studies. The imino-proton region of the U1061A variant showed that this RNA is not folded into a unique conformation: there are more imino-proton peaks than there are Gs and Us in the RNA, and many of the protons have an intensity that is 10 to 15% that of the more prominent imino peaks in the spectrum (Huang, 1996). Thus, alternative conformations are available to this RNA, despite the fact that experiments were conducted well below the tertiary structure T_m. The "extra" imino protons disappear upon adding protein, and the remaining imino peaks become fairly uniform in intensity, indicating that bound protein drives the RNA into a single conformation.

These observations argue that the C-terminal domain of L11 is necessary for the rRNA to maintain a correctly folded conformation, even under conditions that strongly favor formation of the RNA tertiary structure, and we suggest that L11-C76 has evolved to insure correct folding of this rRNA domain. In the last decade rRNA has taken center stage as the functional component of ribosomes, and it has been suggested that the primary role of ribosomal proteins is to promote RNA folding. This suggestion agrees with the properties of L11-C76. It would be reckless to extrapolate this argument to all ribosomal proteins (note that the N-terminal domain of L11 undoubtedly has a specific function, which may not involve RNA recognition), but one can imagine that many other ribosomal proteins trap rRNA in tightly folded tertiary structures.

This work was supported by NIH grant R37 GM29048, a Wellcome Trust (United Kingdom) International Prize Traveling Fellowship (G.L.C.), and an NIH postdoctoral fellowship (L.R.).

REFERENCES

Battiste, J. L., H. Mao, N. S. Rao, R. Tan, D. R. Muhandiram, L. E. Kay, A. D. Frankel, and J. R. Williamson. 1996. α helix-RNA major groove recognition in an HIV-1 Rev peptide-RRE RNA complex. *Science* **273:**1547–1551.

Beauclerk, A. A. D., H. Hummel, D. J. Holmes, A. Böck, and E. Cundliffe. 1985. Studies of the GTPase domain of archaebacterial ribosomes. *Eur. J. Biochem.* **151:**245–255.

Bukhman, Y. V., and D. E. Draper. 1997. Affinities and selectivities of divalent cation binding sites within an RNA tertiary structure. *J. Mol. Biol.* **274:**1020–1031.

Cate, J. H., A. R. Gooding, E. Podell, K. Zhou, B. L. Golden, C. E. Kundrot, T. R. Cech, and J. A. Doudna. 1996a. Crystal structure of a group I ribozyme domain: principles of RNA packing. *Science* **273:**1678–1685.

Cate, J. H., A. R. Gooding, E. Podell, K. Zhou, B. L. Golden, A. A. Szewczak, C. E. Kundrot, T. R. Cech, and J. A. Doudna. 1996b. RNA tertiary structure mediation by adenosine platforms. *Science* **273:**1696–1699.

Conn, G. L. Unpublished data.

Conn, G. L., and D. E. Draper. 1998. RNA structure. *Curr. Opin. Struct. Biol.* **8:**278–285.

Conn, G. L., R. R. Gutell, and D. E. Draper. 1998. A functional ribosomal RNA tertiary structure involves a base triple interaction. *Biochemistry* **37:**11980–11988.

Conn, G. L., D. E. Draper, E. E. Lattman, and A. G. Gittis. 1999. Crystal structure of a conserved ribosomal protein-RNA complex. *Science* **284:**1171–1174.

Correll, C. C., A. Munishkin, Y. L. Chan, Z. Ren, I. G. Wool, and T. A. Steitz. 1998. Crystal structure of the ribosomal RNA domain essential for binding elongation factors. *Proc. Natl. Acad. Sci. USA* **95:**13436–13441.

Cundliffe, E. 1979. Antibiotics and prokaryotic ribosomes: action, interaction, and resistance, p. 555–581. *In* G. Chambliss, G. R. Craven, J. Davies, K. Davis, L. Kahan, and M. Nomura (ed.), *Ribosomes: Structures, Function, and Genetics.* University Park Press, Baltimore, Md.

Cundliffe, E. 1986. Involvement of specific portions of ribosomal RNA in defined ribosomal functions: a study utilizing antibiotics, p. 586–604. *In* B. Hardesty and G. Kramer (ed.), *Structure, Function, and Genetics of Ribosomes.* Springer-Verlag, New York, N.Y.

Cundliffe, E., and J. Thompson. 1979. Ribose methylation and resistance to thiostrepton. *Nature* **278:**859–861.

Draper, D. E., Y. Xing, and L. Laing. 1995. Thermodynamics of RNA unfolding: stabilization of a ribosomal RNA tertiary structure by thiostrepton and ammonium ion. *J. Mol. Biol.* **249:**231–238.

Egebjerg, J., S. Douthwaite, and R. A. Garrett. 1989. Antibiotic interactions at the GTPase-associated centre within *Escherichia coli* 23S rRNA. *EMBO J.* **8:**607–611.

El-Baradi, T. T. A. L., V. H. C. F. de Regt, S. W. C. Einerhand, J. Teixido, R. J. Planta, J. P. G. Ballesta, and H. A. Raué. 1987. Ribosomal proteins EL11 from *Escherichia coli* and L15 from *Saccharomyces cerevisiae* bind to the same site in both yeast 26 S and mouse 28 S rRNA. *J. Mol. Biol.* **195:**909–917.

Gautheret, D., D. Konings, and R. R. Gutell. 1994. A major family of motifs involving G·A mismatches in ribosomal RNA. *J. Mol. Biol.* **242:**1–8.

GuhaThakurta, D., and D. E. Draper. 1999. Protein-RNA sequence covariation in a ribosomal protein-rRNA complex. *Biochemistry* **38:**3633–3640.

GuhaThakurta, D., and D. E. Draper. Contributions of basic residues to ribosomal protein L11 recognition of RNA. *J. Mol. Biol.*, in press.

Gutell, R. R., A. Power, G. Z. Hertz, E. J. Putz, and G. D. Stormo. 1992a. Identifying constraints on the higher-order structure of RNA: continued development and application of comparative sequence analysis methods. *Nucleic Acids Res.* **20:**5785–5795.

Gutell, R. R., M. N. Schnare, and M. W. Gray. 1992b. A compilation of large subunit (23S and 23S-like) ribosomal RNA structures. *Nucleic Acids Res.* **20**(Suppl.)**:**2095–2109.

Highland, J. H., G. A. Howard, E. Ochsner, R. Hasenbank, J. Gordon, and G. Stoffler. 1975. Identification of a ribosomal protein necessary for thiostrepton binding to Escherichia coli ribosomes. *J. Biol. Chem.* **250:**1141–1145.

Hinck, A. P., M. A. Markus, S. Huang, S. Gresiek, I. Kustonovich, D. E. Draper, and D. A. Torchia. 1997. The RNA binding domain of ribosomal protein L11: three-dimensional structure of the RNA-bound form of the protein and its interaction with 23S rRNA. *J. Mol. Biol.* **274:**101–113.

Holbrook, S. R., J. L. Sussman, R. W. Warrant, and S.-H. Kim. 1978. Crystal structure of yeast phenylalanine transfer RNA. II. Structural features and functional implications. *J. Mol. Biol.* **123:**631–660.

Huang, S. 1996. Ph.D. thesis. Johns Hopkins University, Baltimore, Md.

Huang, S., Y. X. Wang, and D. E. Draper. 1996. Structure of a hexanucleotide RNA hairpin loop conserved in ribosomal RNAs. *J. Mol. Biol.* **258:**308–321.

Jack, A., J. E. Lander, and A. Klug. 1976. Crystallographic refinement of yeast phenylalanine transfer RNA at 2.5 Å resolution. *J. Mol. Biol.* **108:**619–649.

Laing, L. G., and D. E. Draper. 1994. Thermodynamics of RNA folding in a highly conserved ribosomal RNA domain. *J. Mol. Biol.* **237:**560–576.

Laing, L. G., T. C. Gluick, and D. E. Draper. 1994. Stabilization of RNA structure by Mg^{2+} ion: specific and non-specific effects. *J. Mol. Biol.* **237:**577–587.

Levitt, M. 1969. Detailed molecular model for transfer ribonucleic acid. *Nature* **224:**759–763.

Li, Y., G. Zon, and W. D. Wilson. 1991. NMR and molecular modeling evidence for a new type of G-A mismatch base pair in a purine-rich DNA duplex. *Proc. Natl. Acad. Sci. USA* **88:**26–30.

Lu, M., and D. E. Draper. 1994. Bases defining an ammonium and magnesium ion-dependent tertiary structure within the large subunit ribosomal RNA. *J. Mol. Biol.* **244:**572–585.

Markus, M., A. Hinck, S. Huang, D. E. Draper, and D. A. Torchia. 1997. High resolution structure of ribosomal protein L11-C76, a helical protein with a flexible loop that becomes structured upon binding RNA. *Nat. Struct. Biol.* **4:**70–77.

Moazed, D., J. M. Robertson, and H. F. Noller. 1988. Interaction of elongation factors EF-G and EF-Tu with a conserved loop in 23S RNA. *Nature* **334:**362–364.

Munishkin, A., and I. G. Wool. 1997. The ribosome-in-pieces: binding of elongation factor EF-G to oligoribonucleotides that mimic the sarcin/ricin and thiostrepton domains of 23S ribosomal RNA. *Proc. Natl. Acad. Sci. USA* **94:**12280–12284.

Musters, W., P. M. Gonçalves, K. Boon, H. A. Raué, H. van Heerikhuizen, and R. J. Planta. 1991. The conserved GTPase center and variable region V9 from *Saccharomyces cerevisiae* 26S rRNA can be replaced by their equivalents from other prokaryotes or eukaryotes without detectable loss of ribosomal function. *Proc. Natl. Acad. Sci. USA* **88:**1469–1473.

Pley, H. W., K. M. Flaherty, and D. B. McKay. 1994. Three-dimensional structure of a hammerhead ribozyme. *Nature* **372:** 68–74.

Porse, B. T., I. Leviev, A. S. Mankin, and R. A. Garrett. 1998. The antibiotic thiostrepton inhibits a functional transition within protein L11 at the ribosomal GTPase centre. *J. Mol. Biol.* **276:** 391–404.

Porse, B. T., E. Cundliffe, and R. A. Garrett. 1999. The antibiotic micrococcin acts on protein L11 at the ribosomal GTPase centre. *J. Mol. Biol.* **287:**33–45.

Quigley, G. J., and A. Rich. 1976. Structural domains of transfer RNA molecules. *Science* **194:**796–806.

Ryan, P. C., and D. E. Draper. 1989. Thermodynamics of protein-RNA recognition in a highly conserved region of the large subunit ribosomal RNA. *Biochemistry* **28:**9949–9956.

Ryan, P. C., and D. E. Draper. 1991. Detection of a key tertiary interaction in the highly conserved GTPase center of large subunit ribosomal RNA. *Proc. Natl. Acad. Sci. USA* **88:**6308–6312.

Ryan, P. C., M. Lu, and D. E. Draper. 1991. Recognition of the highly conserved GTPase center of 23 S ribosomal RNA by ribosomal protein L11 and the antibiotic thiostrepton. *J. Mol. Biol.* **221:**1257–1268.

Schmidt, F. J., J. Thompson, K. Lee, J. Dijk, and E. Cundliffe. 1981. The binding site for ribosomal protein L11 within 23 S ribosomal RNA of *Escherichia coli*. *J. Biol. Chem.* **256:**12301–12305.

Scott, W. G., J. T. Finch, and A. Klug. 1995. The crystal structure of an All-RNA hammerhead ribozyme: a proposed mechanism for RNA catalytic cleavage. *Cell* **81:**991–1002.

Stark, M. J. R., E. Cundliffe, J. Dijk, and G. Stöffler. 1980. Functional homology between *E. coli* ribosomal protein L11 and *B. megaterium* protein BM-L11. *Mol. Gen. Genet.* **180:**11–15.

Thompson, J., E. Cundliffe, and M. Stark. 1979. Binding of thiostrepton to a complex of 23-S rRNA with ribosomal protein L11. *Eur. J. Biochem.* **98:**261–265.

Thompson, J., W. Musters, E. Cundliffe, and A. E. Dahlberg. 1993. Replacement of the L11 binding region within *E. coli* 23S ribosomal RNA with its homologue from yeast: *in vivo* and *in vitro* analysis of hybrid ribosomes altered in the GTPase center. *Eur. J. Biochem.* **12:**1499–1504.

Westhof, E., and M. Sundaralingham. 1986. Restrained refinement of the monoclinic form of yeast phenylalanine transfer RNA. Temperature factors and dynamics, coordinated waters, and base-pair propeller twist angles. *Biochemistry* **25:**4868–4878.

Westhof, E., P. Dumas, and D. Moras. 1985. Crystallographic refinement of yeast aspartic acid transfer RNA. *J. Mol. Biol.* **184:** 119–145.

Wimberly, B. T., R. Guymon, J. P. McCutcheon, S. W. White, and V. Ramakrishnan. 1999. A detailed view of a ribosomal active site: the structure of the L11-RNA complex. *Cell* **97:**491–502.

Xing, Y., and D. E. Draper. 1995. Stabilization of ribosomal RNA tertiary structure by ribosomal protein L11. *J. Mol. Biol.* **246:** 319–331.

Xing, Y., and D. E. Draper. 1996. Cooperative interactions of RNA and thiostrepton antibiotic with two domains of ribosomal protein L11. *Biochemistry* **35:**1581–1588.

Xing, Y., D. GuhaThakurta, and D. E. Draper. 1997. The RNA binding domain of ribosomal protein L11 is structurally similar to homeodomains. *Nat. Struct. Biol.* **4:**24–27.

Zimmermann, G. R., R. D. Jenison, C. L. Wick, J. P. Simorre, and A. Pardi. 1997. Interlocking structural motifs mediate molecular discrimination by a theophylline-binding RNA. *Nat. Struct. Biol.* **4:**644–649.

The Ribosome: Structure, Function, Antibiotics, and Cellular Interactions
Edited by R. A. Garrett, S. R. Douthwaite, A. Liljas, A. T. Matheson, P. B. Moore, and H. F. Noller

Chapter 12

Structure of the Yeast Ribosomal Stalk

JUAN P. G. BALLESTA, ESTHER GUARINOS, JESUS ZURDO, PILAR PARADA, GRETEL NUSSPAUMER, VASSILIKI S. LALIOTI, JORGE PEREZ-FERNANDEZ, and MIGUEL REMACHA

The stalk of the large ribosomal subunit is involved in the interaction of the elongation factors with the ribosome, as biochemical data have clearly indicated (Liljas et al., 1986; Möller and Maassen, 1986) and as has been directly confirmed by electron microscopy (Agrawal et al., 1998; Stark et al., 1997). Moreover, the stalk might be directly involved in the translocation process, since mutations in one of its component proteins suppress the inhibition caused by some translocation inhibitors (Gomez-Lorenzo and Garcia-Bustos, 1998).

The stalk is an elongated and highly flexible lateral protuberance which in bacterial 50S ribosomal subunits is composed of two dimers of the 12-kDa acidic ribosomal protein L12, forming a very stable complex with protein L10 (Gudkov et al., 1978). This pentameric complex binds through the L10 N-terminal domain to the highly conserved GTPase-related region in domain II of 23S rRNA, partially overlapping the binding site of protein L11, which is also part of the stalk base (Egebjerg et al., 1990). The globular C-terminal domain of protein L12 (Leijonmarck and Liljas, 1987), which is linked to the elongated amino end through an unstructured hinge region (Bocharov et al., 1996), forms the external tip of the stalk (Marquis et al., 1981).

The eukaryotic ribosomal-stalk elements are functionally equivalent to the above-described bacterial components, but while some of them are highly conserved, others have evolved notably. The most conserved element is the rRNA GTPase-related region, which can be exchanged between bacterial and eukaryotic ribosomes with few negative functional effects (Musters et al., 1991; Thompson et al., 1993). In contrast, the eukaryotic 12-kDa acidic proteins (P1, P2, and P3), in spite of some apparent overall similarity, show a secondary and tertiary structure clearly different from that of their bacterial L12 counterpart (Zurdo et al., 1997). In protein P0, the equivalent of bacterial L10, both situations are found simultaneously. Thus, while the P0 domain involved in the interaction with proteins P1 and P2 seems to have evolved rapidly, the N-terminal domain, responsible for binding to the rRNA, is particularly conserved. In this way, a chimerical protein carrying the rRNA binding domain from human P0 and the rest of the protein from *Saccharomyces cerevisiae* is fully functional in yeast, while the complete human P0, although binding to the yeast ribosome, is defective in binding the yeast P1 and P2 proteins (Rodriguez-Gabriel et al., in press). A similar situation may occur in the case of protein L12, previously called L15, in *S. cerevisiae.* This protein, which shows immunological cross-reactivity with its bacterial counterpart, protein L11 (Juan-Vidales et al., 1981), also binds to the same highly conserved rRNA region, and it seems to contain an equally conserved RNA interaction domain (Xing and Draper, 1995; Xing et al., 1997). In contrast, the part of the protein involved in the interaction with other ribosomal components has evolved faster. Thus, while the absence of bacterial L11 does not affect the interaction of bacterial L12 with the ribosome (Stöffler et al., 1980), the absence of eukaryotic L12 reduces the acidic P1 and P2 proteins in the stalk (Briones et al., 1998), confirming the existence of interactions among them (Saenz-Robles et al., 1988).

In general, the eukaryotic stalk is considerably less stable and notably more complex than the bacterial one. Thus, changes have been reported in the

Juan P. G. Ballesta, Esther Guarinos, Jesus Zurdo, Pilar Parada, Gretel Nusspaumer, Vassiliki S. Lalioti, Jorge Perez-Fernandez, and Miguel Remacha ■ Centro de Biología Molecular "Severo Ochoa," CSIC and UAM, Canto Blanco, 28049 Madrid, Spain.

amount of 12-kDa acid protein found in the ribosome, depending on the metabolic state of the cells (Saenz-Robles et al., 1990). Moreover, the low stability is underscored by the capacity of the proteins to exchange with the free proteins in a cytoplasmic pool of these ribosomal components (Scharf and Nover, 1987; Tsurugi and Ogata, 1985; Zinker and Warner, 1976).

The larger size of protein P0, as well as the evolution of the 12-kDa acidic proteins from one to various protein families, confirms the complexity of the eukaryotic stalk (Ballesta and Remacha, 1996). Protein P0 contains an acidic-protein-like C-terminal extension, missing in bacterial L10 (Shimmin et al., 1989), which allows it to function as a minimal stalk, suppressing the essential character that the acidic proteins have in the bacterial ribosome (Remacha et al., 1995; Santos and Ballesta, 1995). In this way, the eukaryotic acidic proteins may act as ribosome modulators, which has led to the proposal that they are involved in a possible translation-regulatory mechanism (Ballesta and Remacha, 1996). Moreover, the number of acidic proteins has increased from only two forms of one polypeptide (L12 and its N-terminal acetylated form) to various types of protein families, P1, P2, and P3, which, in addition, can be phosphorylated. The number of families, as well as the members in each of them, varies with the species. Thus, mammals have only two families (P1 and P2) with one protein each while some lower eukaryotes have two members in each of the P1 and P2 families (Beltrame and Bianchi, 1990; Newton et al., 1990), and protozoa have more than two (Schijman et al., 1995). Plants seem to have a third protein family called P3 (Szick et al., 1998).

The mammalian stalk, like its bacterial homologue (Gudkov et al., 1978), has been shown by cross-linking studies to be formed by a P0-$(P1)_2$-$(P2)_2$ pentameric complex (Uchiumi and Kominami, 1997), which apparently confirms the dimeric state of the eukaryotic acidic proteins in the ribosome, as is the case for the bacterial L12 (Möller et al., 1972; Koteliansky et al., 1978; Oleinikov et al., 1993). Nevertheless, the presence of two different types of acidic proteins in the eukaryotic stalk increases its complexity and must have functional consequences which are presently unknown. The situation is even more complex in the case of species having several proteins in each family. There are indications that the four components present in *S. cerevisiae*, namely, proteins P1α, P1β, P2α, and P2β, are present as dimers when they are in solution (Juan-Vidales et al., 1984; Zurdo et al., 1997). If these proteins are also bound as dimers in the yeast ribosome and there is one of each protein per ribosome, the yeast stalk would contain eight copies of these proteins instead of four as in mammalian species. However, an estimate of the acidic proteins in the *S. cerevisiae* ribosome indicated the presence of around four copies per ribosome (Saenz-Robles et al., 1990), indicating that they cannot be in the particle as dimers. Alternatively, the stalk composition of the yeast ribosomes can be heterogeneous, with different subpopulations which contain two dimers of one protein of each type, i.e., $(P1\alpha)_2$-$(P2\alpha)_2$, $(P1\alpha)_2$-$(P2\beta)_2$, $(P1\beta)_2$-$(P2\alpha)_2$, $(P1\beta)_2$-$(P2\beta)_2$ (Fig. 1).

Apart from the obvious structural differences between these two alternative models for the yeast stalk structure, which it is important to clarify, each of them has notably different functional and, more importantly, regulatory implications. If the two proteins of the same family are not functionally identical, as is probably the case (Remacha et al., 1990), ribosomes carrying different stalk components can also be functionally heterogeneous, and consequently, the relative sizes of the different subpopulations would have metabolical consequences; moreover, changes in the subpopulation ratios can obviously have a relevant regulatory potential. It is, therefore, very important to distinguish between these two models of stalk structure in cells having more than one acidic protein of the P1 and P2 types.

DO ACIDIC PROTEINS EXIST AS DIMERS IN THE RIBOSOME?

Establishing the association state of the acidic proteins in the ribosome would help to resolve which of the two stalk models, the homogeneous or the heterogeneous, is correct. Unfortunately, the almost identical molecular weights of the four yeast proteins make it difficult to approach this question by using a general cross-linking reagent, since it would be very difficult to distinguish between homodimers and heterodimers. To overcome this problem, a specific cross-linking has been attempted by forming S-S bridges between cysteine residues, as has been done in bacterial systems (Oleinikov et al., 1993). Cysteines are absent in the four proteins; therefore, if one of them is mutated to introduce one or more Cys residues, only the mutated polypeptide will be able to form S-S bridges upon oxidation, thus stabilizing a possible dimer state of the protein. Cross-linking with other SH-containing proteins is possible but not with the other acidic proteins that do not carry Cys groups. As is shown in Fig. 2, a mutated protein, P1α, carrying a Cys in position 62 is fully able to form dimers upon oxidation with *o*-phenenthroline-Cu^{2+} when it is free in solution; however, dimers are not

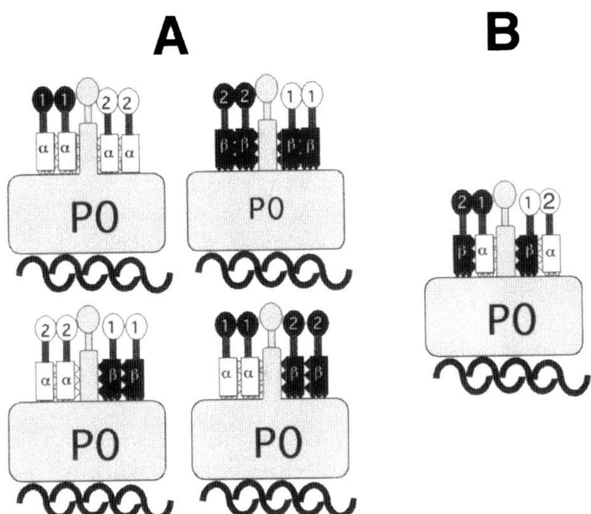

Figure 1. Models of acidic protein distribution in the yeast ribosomal stalk. (A) Heterogeneous model. Ribosomes carry one dimer of a P1 and another of a P2 protein type. Four associations of the four acidic proteins are possible: P1α-P2α, P1α-P2β, P1β-P2α, P1β-P2β. (B) Homogeneous model. All ribosomes carry one monomer of the four acidic proteins simultaneously.

detected when it is bound to the ribosome. Similarly, no dimers are found when a mutated P2α with Cys in position 28 is bound to the ribosome (not shown). Although a change in conformation upon binding that hinders the formation of the dimer cannot be excluded, these results strongly support the idea that proteins P1α and P2α, and probably the other acidic proteins, are present as monomers in the ribosome.

AFFINITY CHROMATOGRAPHY OF RIBOSOMES

The highly probable presence of the acidic proteins in the ribosome as monomers, together with the existence of one specific binding site for each 12-kDa protein in the particles, favors the homogeneous rather than the heterogeneous ribosomal-stalk model in yeast. To directly prove this notion, an attempt was made to purify ribosomes based on the presence of one of the stalk proteins. Afterwards, the purified particles were checked to find out whether they also contained the remaining three components. Protein P2α was tagged at the C terminus with a histidine tail, and the tagged protein, P2α-His, was expressed in *S. cerevisiae* D4, a strain lacking the wild-type P2α. The ribosomes from the transformed strain were fractionated by eluting them through an Ni^{2+} (Ni-nitrilotriacetic acid) column. More than 95% of the ribosome population was retained in the column and was eluted afterward with imidazol buffers (Fig. 3A). Under similar conditions, control ribosomes without protein P2α-His were not retained at all in the column (Fig. 3B). An analysis of the eluted ribosomes by isoelectrofocusing showed the presence of the tagged protein together with the other three acidic proteins (Fig. 4).

The fact that practically all ribosomes are retained in the column indicates that they must contain protein P2α-His, and the presence of each one of the four acidic proteins in all of the ribosomes is not easily compatible with the heterogeneous stalk model.

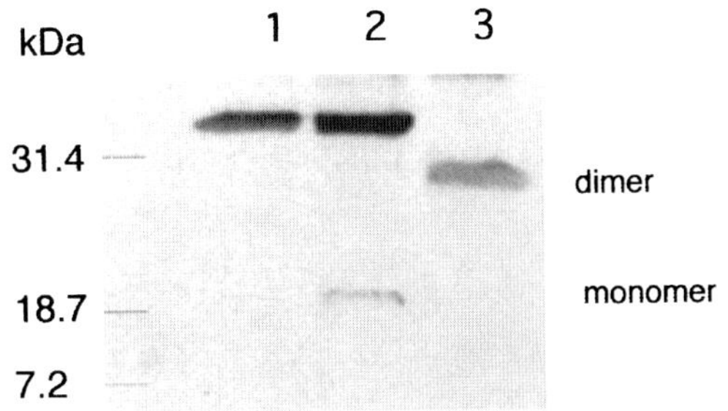

Figure 2. Cross-linking of P1α-cys62. *S. cerevisiae* D7, with a disrupted P1α gene, was transformed with plasmid pFL37-P1α-cys62 carrying a P1α protein in which a unique cysteine was introduced in position 62. Ribosomes (20 μg) from the transformed strain were treated in 10 μl of buffer with o-phenanthroline-Cu^{2+} (Kobashi, 1968; Oleinikov et al., 1993) to oxidize the cysteine SH groups, and the ribosomes were resolved by sodium dodecyl sulfate-polyacrylamide gel electrophoresis (lane 2). Similarly, a preparation of purified recombinant P1α-cys62 protein (5 μg) was treated in the same way and resolved in the same gel (lane 3). As a control, ribosomes from untransformed *S. cerevisiae* D7 were treated in the same way (lane 1). Proteins were blotted to nitrocellulose membranes and detected with P1α-specific antibodies.

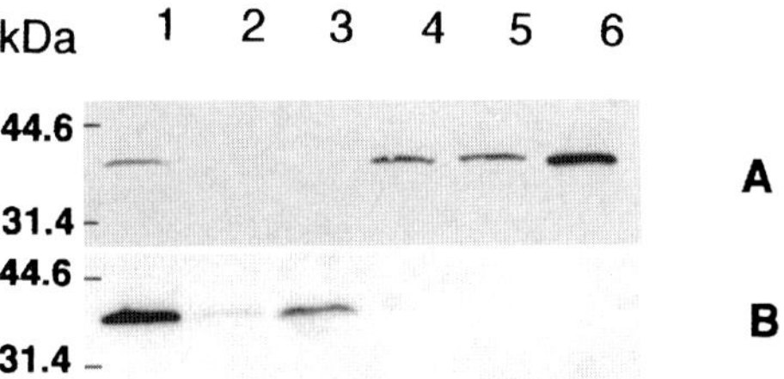

Figure 3. Affinity chromatography of *S. cerevisiae* ribosomes in Ni^{2+} (Ni-nitrilotriacetic acid) columns. (A) A P2α-HisT protein was constructed by fusing a six-His tail to the C-terminal end of protein P2α in plasmid pFL392α-His6. The tagged protein was expressed in *S. cerevisiae* D4 lacking P2α, and a ribosome preparation from the transformed strain was loaded into a 1-ml Ni^{2+} affinity column. The column was washed with 10 ml of 10 mM Tris-HCl (pH 7.4), 10 mM $MgCl_2$, and 100 mM KCl buffer containing increasing concentrations of imidazol. Aliquots of the eluted fractions were resolved in sodium dodecyl sulfate-polyacrylamide gels, and the presence of ribosomes in the fractions was checked by estimating the stalk protein P0 by Western blotting with a specific monoclonal antibody. (B) Ribosomes from a control strain carrying a wild-type P2α protein were obtained and processed as for panel A. Lanes: 1, excluded volume; 2 to 6, fractions eluted with buffer containing 0, 10, 40, 70, and 100 mM imidazol.

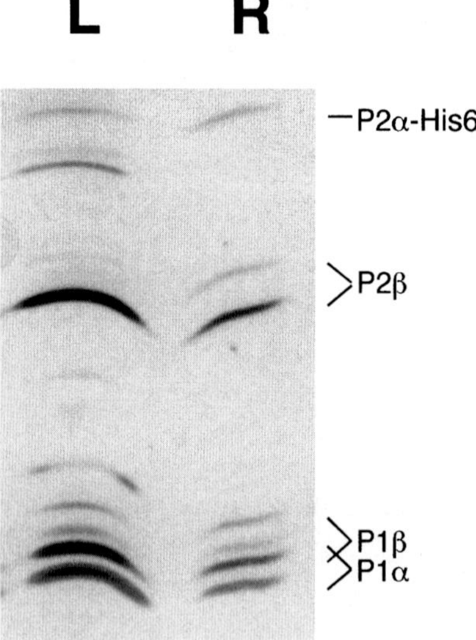

Figure 4. Isoelectrofocusing of ribosomes from the affinity chromatography columns. An aliquot of the loaded sample (L) and of the 100 mM imidazol-eluted fraction (R) were resolved in polyacrylamide gels in a 2.5 to 5.0 pH gradient as previously described (Rodriguez-Gabriel et al., 1998). The gels were silver stained. The positions of the different acidic proteins are marked. The upper band corresponds in each case to the phosphorylated form of the protein.

Table 1. Amino acid sequence similarity of yeast ribosomal acidic proteins[a]

Protein	Similarity						
	P1α	P1β		P2α		P2β	
		C	A	C	A	C	A
P1α	100	77.8	76.5	57.7	44.6	61.8	48.2
P1β		100	100	69.7	50.9	58.5	54.5
P2α				100	100	80.1	75.3
P2β						100	100

[a] Comparison carried out with the Genetics Computer Group Bestfit program. The scores were obtained by comparing either the whole amino acid sequence (C) or the amino-terminal domain, including the first 74 amino acids and excluding the conserved hinge and carboxyl regions (A).

Moreover, this conclusion is further confirmed by the fact that the three remaining acidic proteins, P1α, P1β, and P2β, are found in proportions similar to those in the control ribosomes. The experimental data, therefore, strongly indicate that the ribosomal stalk of yeast growing in exponential phase is made up of the four acidic proteins in addition to protein P0.

RELATIONSHIPS AMONG THE 12-kDa ACIDIC PROTEINS IN THE STALK

The experimental data convincingly indicate that the yeast ribosomal stalk does not contain dimers of any type of P1 and P2, as has been reported in mammals, but rather that one monomer of each of the four 12-kDa proteins, P1α, P1β, P2α, and P2β, is present in the particle. The overall structures of the stalks are similar in both cases, containing four 12-kDa acidic proteins bound to protein P0, but obviously the yeast ribosome shows a higher degree of complexity, a fact that probably has some still-obscure functional meaning.

The four *S. cerevisiae* 12-kDa ribosomal-stalk proteins share the same highly conserved C-terminal peptide, EESDDDMGFGLFD, which is also present in protein P0, but show substantial differences in the rest of the amino acid sequence (Newton et al., 1990). The overall sequence similarity in the two proteins of the same type is close to 80%, while in the proteins of different types it is around 50%, and these differences increase if the carboxyl end is excluded (Table 1). One of the first questions that arises regarding the yeast stalk is that of the relationships among the four acidic proteins.

STALK COMPOSITION IN ACIDIC-PROTEIN-DISRUPTED *S. CEREVISIAE* STRAINS

In order to understand the functional and structural relationships among the four proteins, disruptant strains lacking each of them were obtained. Analysis of the ribosomes from these strains clearly indicated that the absence of one protein does not result in an increase in the other protein of the same family, in spite of an excess of these proteins in the cytoplasm (Fig. 5). These results strongly suggest that the two proteins of the same family are not structurally, and probably not functionally, equivalent, no matter how close their sequence similarity is, and that there are specific binding sites for each one in the ribosome. Moreover, in some cases, the absence of one protein results in a clear reduction in the amount

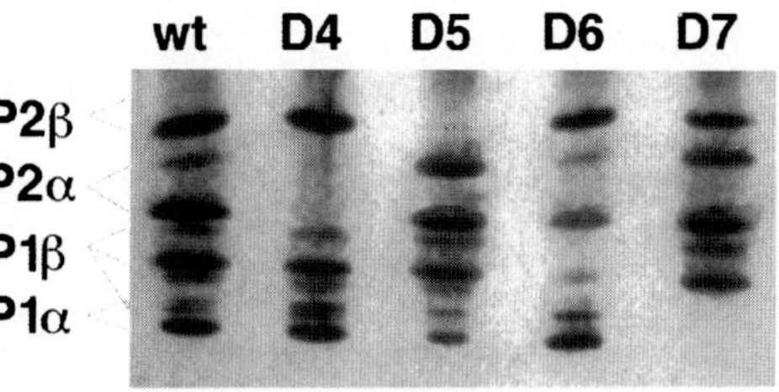

Figure 5. Isoelectrofocusing of ribosomes from *S. cerevisiae* disrupted strains lacking acidic proteins. Ribosomes (0.5 mg) from *S. cerevisiae* D4, D5, D6, and D7 lacking proteins P2α, P2β, P1β, and P1α, respectively, were resolved as for Fig. 4. wt, wild type.

of the other component. Thus, in the strain D5 lacking P2β, protein P1α is notably reduced; similarly, a reduction in P2α was detected upon disruption of the P1β in strain D6 (Fig. 5). It seems, therefore, that the interactions among the four proteins are not identical but that some preferential association in the ribosome must exist between P1α and P2β on the one hand and P1β and P2α on the other. This association is important for the binding of the proteins to P0 to form the stalk and suggests that they might bind as P1-P2 pairs.

In fact, there is previous evidence pointing to the existence of these two preferential protein pairs, P1α-P2β and P1β-P2α. This possibility is supported by results obtained from yeast mutants lacking protein L12. The absence of this protein results in the concomitant lack of proteins P1α and P2β but not of P1β and P2α (Briones et al., 1998). On the contrary, when the yeast P0 protein is replaced in the cell by the homologous protein from other eukaryotic species, like *Rattus norvegicus*, *Homo sapiens*, or *Dictyostelium discoideum*, proteins P1β and P2α are preferentially lost from the ribosome (Rodriguez-Gabriel et al., unpublished).

P0 PROTEIN CHIMERAS

The carboxyl domain of protein P0 is able to provide the minimal stalk structure necessary for ribosome function (Remacha et al., 1995; Santos and Ballesta, 1995). It seems, therefore, that in spite of the rather low amino acid sequence homology with P1-P2, except for the last 12 amino acids, this protein domain is apparently playing the role of these proteins. To study the functional similarities of this P0 domain and the 12-kDa acidic proteins in more detail, four chimerical constructs in the vector pFL39 were made, P0-1α, P0-1β, P0-2α, and P0-2β, in which the last 115 amino acids of P0 were replaced by the complete amino acid sequence of the P1 and P2 proteins. The four protein chimeras were transformed into P0 conditional null strains either carrying the four P1-P2 proteins (*S. cerevisiae* W303dGP0) or totally lacking the P1α and P1β proteins (*S. cerevisiae* D67dGP0) (Santos and Ballesta, 1994). When the proteins were expressed in *S. cerevisiae* W303dGP0, they were only able to partially complement the absence of the wild-type protein because the ribosomes carrying the chimeras were missing some of the 12-kDa proteins. As summarized in Table 2, the results confirmed the existence of specific associations between proteins P1α and P2β on one hand and proteins P1β and P2α on the other, in such a way that only the partner of the protein fused to P0 was detected in the ribosome. When the protein chimeras were expressed in *S. cerevisiae* D67dGP0, stimulation of the cell growth rate is detected in the case of P0-1α and P0-1β, since in these two cases the P2β and P2α proteins present in the cell are able to bind to the ribosome, increasing the particle efficiency (Table 2). In contrast, the absence of P1α and P1β from this strain prevents the formation of a complex with P0-2β and P0-2α, respectively, precluding the stimulatory effect on cell growth found in the previous cases. However, the fact that cells are viable, even in this case, indicates that P0 chimeras are able to substitute for the wild-type P0 in the absence of any 12-kDa bound protein.

Table 2. Effect of P0 protein chimera expression on *S. cerevisiae* strains[a]

Strain	P0 protein in the stalk	12-kDa protein(s) in the stalk	Growth rate (doubling time) (min)
W303dGP0	P0	P1α, P1β, P2α, P2β	100
W303dGP0	P0-1α	P2β	125
W303dGP0	P0-1β	P2α	123
W303dGP0	P0-2α	P1β	160
W303dGP0	P0-2β	P1α	160
D67dGP0	P0	None	170
D67dGP0	P0-1α	P2β	150
D67dGP0	P0-1β	P2α	136
D67dGP0	P0-2α	None	212
D67dGP0	P0-2β	None	215

[a] *S. cerevisiae* W303dGP0 and D67dGP0 (Santos and Ballesta, 1995) were transformed in galactose medium with plasmids pFL39-P0-1α, pFL39-P0-1β, pFL39-P0-2α, and pFL39-P0-2β. The transformants were shifted to glucose medium (yeast extract-peptone-dextrose) to estimate their growth rates. Ribosomes were obtained from the glucose-grown cells, and the acidic proteins in the particles were resolved by isoelectrofocusing as for Fig. 4.

These results clearly confirm that the mutual interactions of the four acidic proteins are not equivalent in the ribosomal stalk. There is a real preferential formation of two pairs, P1α-P2β and P1β-P2α, whose relative functional roles are still unknown but are not the same. Thus, the disruption of the two proteins of the P1α-P2β couple causes greater damage to cell growth than the equivalent disruption of the P1β-P2α couple (Remacha et al., 1992, 1995).

INTERACTION BETWEEN STALK COMPONENTS IN THE CELL DETECTED BY THE TWO-HYBRID METHOD

The previous results indicate that to fully understand the stalk structure-function relationships it is important to clearly establish the different associations and interactions that might exist among the stalk components. The extensive use of the two-hybrid technique (Chien et al., 1991) has convincingly shown it to be a very useful method for studying in vivo protein-protein interactions, and it has also been used to explore the protein associations in the yeast ribosomal stalk. The four acidic protein genes as well as the P0 gene have been fused to the binding and activation domains of the GAL4 gene in the pGBT9 and pGAD424 vectors and tested for β-galactosidase expression. Proteins P1α and P1β, when fused to the binding domain of GAL4, gave a false-positive response, inducing the expression of β-galactosidase in the absence of any cotransforming plasmid. All the other constructs, however, were negative when tested alone and therefore could be used in protein-protein interaction tests. The results are summarized in Table 3.

First, it must be noted that a strong interaction between the same protein fused to both GAL4 domains was not detected in any of the acidic proteins, indicating that in these conditions dimerization seems not to take place. This agrees with the absence of acidic protein dimers detected by cross-linking in the yeast ribosome, as mentioned previously, but in addition, it suggests that the acidic proteins are also unable to form dimers in the cytoplasm. Since the purified proteins seem to be able to dimerize in solution, these results suggest their interaction with other proteins that probably hinder the self-association, as was previously proposed based on the proteinase-sensitive structure of the free proteins (Zurdo et al., 1997).

Protein P1α interacts more strongly with protein P2β, confirming the preferred association of these two proteins and indicating that it can take place not only in the ribosome but also in the cytoplasm and in the nucleus. A similar relationship between P1β and P2α was not, however, confirmed by this method. This P1 protein seems to interact similarly in the cytoplasm with both P2 proteins. It is possible that in solution the specificity of this protein for P2α is not as strict as in the ribosome particle.

Table 3. Interaction between the different yeast ribosomal-stalk components estimated by the two-hybrid system[a]

pGBT9-transformed cells containing:	Induced β-galactosidase activity (U) in pGAD-transformed cells containing:					
	P0	P1α	P1β	P2α	P2β	Vector
P0	0	24	19	0.4	0.6	0
P2α	1.7	2.3	6.9	2.7	1.0	0
P2β	4.7	6.5	8.0	2.4	3.5	0
Vector	0	0	0	0	0	0

[a] *S. cerevisiae* cells transformed with pGAD and pGBT9 plasmids containing either the indicated proteins or the empty vector were grown to an A_{550} of 0.4. The cells were broken with glass beads, and the β-galactosidase activity in the extracts was estimated according to the method of Miller (1972).

Protein P0, on the other hand, interacts with the P1-type but not with the P2-type proteins. The results are quite clear, showing an evident structural difference between the P1 and P2 proteins in the formation of the stalk. They suggest that, if the acidic proteins bind to the stalk as couples, the interaction takes place mainly through the P1 protein partner.

ASSEMBLY OF THE YEAST RIBOSOMAL STALK

The data available on the assembly of the yeast stalk are not abundant. An analysis of two *S. cerevisiae* strains carrying a disruption of the genes encoding the two proteins of the same type, either P1 or P2, has shown that their ribosomes do not carry any 12-kDa acidic protein (Remacha et al., 1992). Since the acidic proteins whose genes were still intact did not show up in the ribosome, it was concluded that the assembly of the stalk requires the presence of one protein of each type. To explore the assembly process in more detail, an in vitro reconstitution approach was used.

Recombinant P1α and P2β proteins were purified and used to reconstitute ribosomes from *S. cerevisiae* D4567, which lacks the four 12-kDa acidic proteins. Incubation of P1α and P2β with the acidic-protein-deficient ribosomes results in the incorporation of both polypeptides into the particles in a highly specific way, since the binding resists washing through sucrose gradients (Fig. 6), and the activity of the reconstituted particle increases (Fig. 7). As expected from the in vivo data, when only P2β is in-

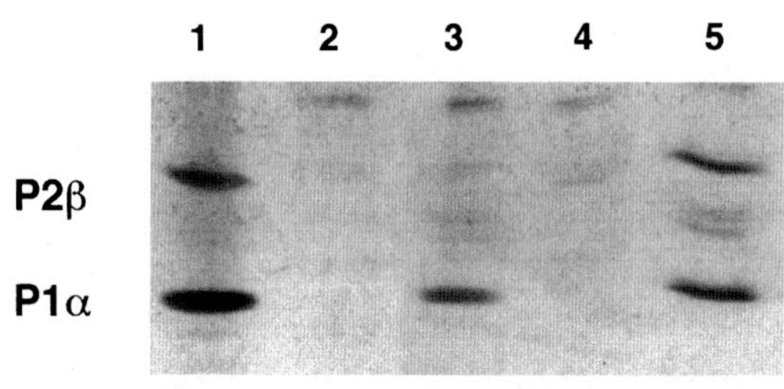

Figure 6. Isoelectrofocusing of ribosomes reconstituted with recombinant acidic proteins. Ribosomes from *S. cerevisiae* D4567 lacking the four acidic proteins were incubated at 30°C in 10 mM Tris-HCl (pH 7.4), 10 mM $MgCl_2$, and 100 mM KCl with a fourfold molar excess of recombinant protein P1α (lane 3), recombinant protein P2β (lane 4), and recombinant protein P1α plus P2β (lane 5). The reconstituted ribosomes were centrifuged through two layers of 20 and 40% sucrose in 10 mM Tris-HCl (pH 7.4), 100 mM $MgCl_2$, and 500 mM KCl buffer. Ribosomes incubated in the absence of proteins (lane 2) and a mixture of both free recombinant proteins (lane 1) were included as controls. Isoelectrofocusing was carried out as for Fig. 4.

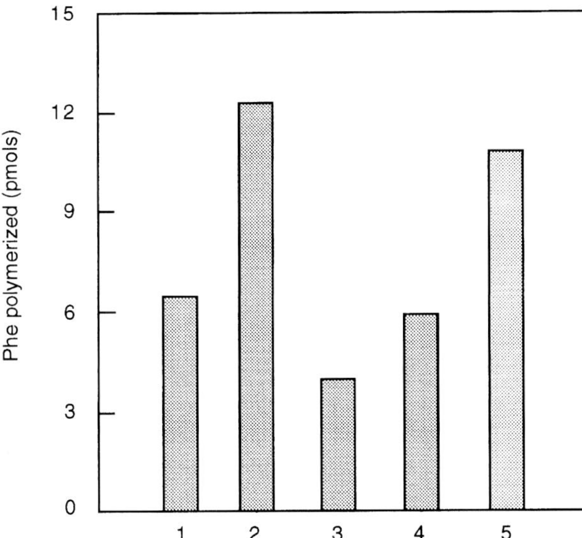

Figure 7. Activity of *S. cerevisiae* D4567 ribosomes reconstituted with recombinant proteins. Ribosomes reconstituted as indicated in the legend to Fig. 6 were tested in a poly(U)-dependent polymerization assay (Sanchez-Madrid et al., 1979). Bars: 1, D4567 ribosomes; 2, D4567 ribosomes plus SP fraction; 3, D4567 ribosomes plus P1α; 4, D4567 ribosomes plus P2β; 5, D4567 ribosomes plus P1α plus P2β. The SP fraction corresponds to an NH_4Cl-ethanol wash of wild-type ribosomes containing the four acidic proteins (Sanchez-Madrid et al., 1979). Control ribosomes (sample 2) polymerized 10.1 pmol of Phe per pmol of ribosomes.

cluded in the experiment, the protein does not bind and is not found associated with the particles. However, protein P1α was able to bind to the ribosome in the absence of protein P2β and is detected in the washed particles (Fig. 6), although this binding does not have functional consequences, since the activity of the particles is not affected (Fig. 7). These results clearly support the capacity of protein P1α for direct binding to the ribosome in the absence of P2 proteins, in agreement with the data from the two-hybrid experiments previously mentioned, indicating the existence of a direct interaction between the P1 proteins and protein P0. In contrast, the P2 proteins seem to require the presence of the P1 protein for assembly into the stalk. However, if this is the case, while the absence of P2 proteins in the ribosomes from strain D67 is expected, P1 proteins should be detected in the D45 ribosomes. However, as previously indicated, this is not so.

This discrepancy between the in vitro and in vivo data could indicate that the stalk assembly process is not the same in the cell as in the test tube. However, it has to be taken into account that while in *S. cerevisiae* D67 the P2-type proteins accumulate in the cytoplasm in large amounts, an important reduction in the amount of the P1 proteins was detected in strain D45 (Bermejo et al., 1994). These results, apart from revealing an interesting effect of the P2 proteins on the accumulation of the P1 polypeptides, suggest that the absence of these acidic proteins from the D45 ribosome could be due to the reduction in the protein available in the cell.

To test this possibility, overexpression of P1 proteins in the D45 strain was attempted by transforming strain D45 with the multicopy plasmids, YEp13-P1β and YEp356-P1α, carrying the corresponding genes under their own promoters. No increase in the amount of protein was detected in the strain transformed with the plasmid carrying P1α, indicating the existence of a very efficient control for overexpression of this protein in the absence of P2 proteins. In contrast, in the case of P1β, the amount of this protein in the cytoplasm of the transformed cells was similar to that detected in the wild-type cells (Table 4). Ribosomes from this P1β-overexpressing strain were obtained and analyzed by isoelectrofocusing. Protein P1β was indeed clearly detected in unwashed ribosomes, although the band is washed off by centrifuging the particles through a high-salt sucrose gradient (Fig. 8A). On the contrary, when ribosomes from strain D67, which, as previously indicated (Remacha et al., 1992), have a large amount of free protein P2β in the cytoplasm, are analyzed under similar conditions, this protein is not found in the unwashed particles (Fig. 8B). Taken together, these data indicate that protein P1β, but not P2β, is able to interact in the cell with the ribosome in the absence of proteins of the opposite type. The interaction is, however, weaker than that detected in vitro for P1α, which resists high-salt gradient centrifugation.

CONCLUSIONS

The existence of a larger number of ribosomal-stalk acidic proteins in a number of eukaryotes, like yeast, protozoa, and plants, is a puzzling fact that seems to imply that this structure is more complex in some species (Ballesta and Remacha, 1996). A first

Table 4. Estimation of protein P1α in S30 extracts from *S. cerevisiae* strains[a]

Strain	P1α(μg)/mg of S30 fraction
W303	0.5
D45	0.01
D45-pFLP1β[b]	0.6

[a] Exponentially growing cells were broken with glass beads and centrifuged at 30,000 × *g* to obtain an S30 extract. Protein in the extracts was estimated by enzyme-linked immunosorbent assay with P1α-specific monoclonal antibodies (Vilella et al., 1991), with purified recombinant protein as a standard.
[b] *S. cerevisiae* D45 transformed with the YEp13-P1β plasmid.

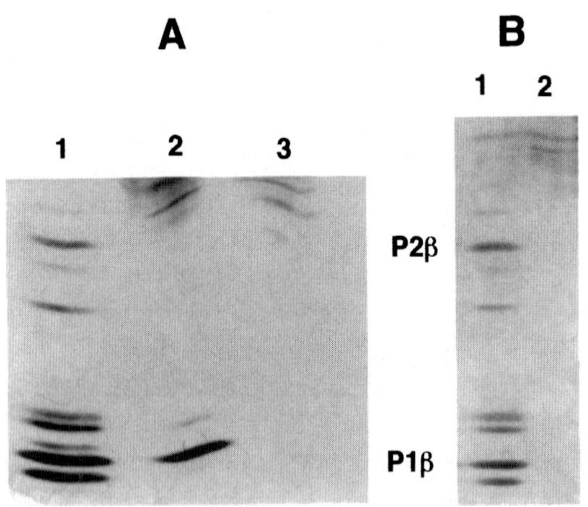

Figure 8. Analysis of ribosomes from *S. cerevisiae* D45 overexpressing protein P1β. (A) Ribosomes from the P2α-P2β-deficient *S. cerevisiae* D45 strain transformed with the multicopy plasmid pP1b were analyzed before (lane 2) and after (lane 3) centrifugation through a sucrose gradient as for Fig. 6. Ribosomes from *S. cerevisiae* W303 (lane 1) were included as a control. (B) Ribosomes from *S. cerevisiae* D67 (lane 2), lacking proteins P1α and P1β and containing an accumulation protein P2β, in the cytoplasm (Remacha et al., 1992) were analyzed as for panel A. Lane 1, control ribosomes as in panel A.

step towards finding out the functional consequences of this fact is understanding its structural repercussions. So far, this has been attempted only in the case of *S. cerevisiae*.

The stalk structure, which can be deduced from the previously discussed and still limited experimental yeast data, seems to differ in some important aspects from the general model present in other organisms, which is based on two acidic protein dimers bound to a larger RNA binding protein. The yeast stalk is also made of four different acidic proteins, P1α, P1β, P2α, and P2β, and the RNA binding protein P0. However, although the acidic proteins seem able to dimerize in solution (Zurdo et al., 1997), they are found as monomers in the ribosome. The four monomers, however, do not function independently in the ribosome; they interact, forming two protein pairs which might be considered equivalents of the protein dimers in the standard stalk model. These couples are not formed, however, by proteins of the same type, as might be expected, but rather by one protein of the P1 type and another of the P2 type. They form two heterodimers, namely, P1α-P2β and P1β-P2α. It has to be kept in mind, however, that these two protein couples are not totally independent in the stalk structure, and there must be some interaction between them. These interactions must be responsible for maintaining part of one of the couple partners bound to the ribosome when the other has been eliminated in the single-disrupted (Fig. 5) and in some double-disrupted strains (Remacha et al., 1992, 1995). These results suggest that a distribution of the four acidic proteins in the stalk structure as shown in Fig. 9 is quite probable.

The existence of specific associations among proteins of different types indicates that evolution increased not only the complexity of the yeast stalk but also its asymmetry. The functional consequences of higher stalk asymmetry are presently unknown. It must be noted, however, that the two protein couples are not functionally equivalent, P1α-P2β being more relevant to the ribosome function than P1β-P2α (Remacha et al., 1992). Similarly, the two protein pairs affect the ribosomal resistance to some antifungal compound inhibitors of translocation differently (Gomez-Lorenzo and Garcia-Bustos, 1998).

The stalk asymmetry probably does not affect the basic function of the stalk in the interaction of the elongation factors with the ribosome, in which the highly conserved carboxyl ends of the acidic proteins play the most relevant role. The differences among the various acidic proteins are found mainly in the amino-terminal domain (Table 1), involved in the interaction with other stalk components (Payo et al., 1995). Changes in this protein domain obviously may affect the stability of the stalk. Since a turnover of the acidic proteins seems to be a relevant functional feature of the eukaryotic stalk, the possibility that asymmetry has some effect on this process is an interesting hypothesis that may have significant functional consequences and is worth exploring in detail. It is known that the acidic protein content of the yeast stalk can change with the metabolic state of the cells (Saenz-Robles et al., 1990). No information on similar changes in other eukaryotic organisms is available, but it would be interesting to know whether the existence of a simpler structure has an effect on stability and turnover of the stalk.

In addition, the data on the yeast stalk open up the possibility that in the higher-eukaryotic stalk the

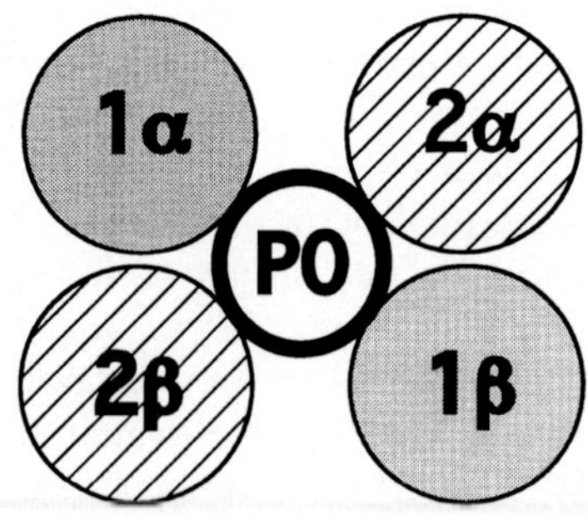

Figure 9. Probable distribution of the four acidic proteins in the yeast ribosomal stalk.

acidic proteins might not be present as real dimers. In fact, there is experimental evidence showing that the *Ceratitis capitata* P1 protein can apparently bind to the ribosome as a monomer when expressed in yeast (Gagou et al., in press). Although the existence of a close association between the two proteins of the same type has been shown by cross-linking studies in *Artemia salina* ribosomes (Uchiumi et al., 1987), the possibility that the two polypeptides are relatively independent in the ribosome cannot be ruled out, since a zero-distance cross-linking agent was not used.

The other set of reported data concerns the assembly of the yeast stalk. The experimental evidence indicates that the in vitro assembly of the yeast ribosomal stalk involves a number of consecutive binding steps which are apparently initiated by the interaction of the P1 proteins and is followed by the binding of the P2 polypeptides. The interaction of the P1 proteins, however, does not have a functional effect, indicating that the presence of the P2 proteins is required for the correct conformation of the structure.

However, the in vivo data raise the question of whether this assembly process is also working in the cell. The failure to detect P1 proteins in the ribosomes from *S. cerevisiae* D45 seems to be indeed due to a reduction in the amount of these proteins detected in this strain, which is related to the high sensitivity of P1 proteins to degradation in the absence of P2 proteins (Nusspaumer et al., unpublished). Thus, overexpression of P1β up to levels in the cytoplasm similar to those found in the wild-type cells allows detection of the protein in the ribosome. The binding is nevertheless much weaker than that found in vitro, and the protein is easily washed off the particle. However, it seems to be specific, since an interaction of protein P2β under similar conditions is not detected, suggesting a preferential interaction of the P1 proteins with the ribosome, also in the cell. There are, however, clear differences between the in vivo and in vitro stalk assembly mechanisms. This is not at all surprising, since as indicated above, P2 proteins protect P1 proteins from degradation in the cytoplasm (Nusspaumer et al., unpublished), indicating the existence of an association between them before their assembly into the stalk. It is therefore highly probable that the two protein pairs, P1α-P2β and P1β-P2α, are formed earlier in the cytoplasm and bind to the ribosome as pairs. Moreover, it is doubtful that the acidic proteins, due to their unstructured conformation, are free in the cytoplasm; rather, they are probably associated with other proteins which protect them from degradation (Zurdo et al., 1997). If so, an involvement of these unknown proteins in the stalk assembly process cannot be excluded. Presently, efforts are being made to identify these proteins by different approaches, such as affinity columns and immunoprecipitation.

It is important to clarify the stalk assembly mechanism in order to fully understand the possible stalk regulatory function (Ballesta and Remacha, 1996), since it must be a key step in the exchange between the proteins in the ribosome and in the cytoplasmic pool that has been reported in yeast (Zinker and Warner, 1976), mammals (Tsurugi and Ogata, 1985), and plants (Scharf and Nover, 1987). In the same way, it will be necessary to know the process responsible for the release of acidic proteins from the ribosome that results in the accumulation of protein-deficent particles in stationary phase (Saenz-Robles et al., 1990).

We thank M. C. Fernandez Moyano for expert technical assistance.

This work was partially supported by grant PB94-0032 from the Dirección General de Política Científica (Spain) and by an institutional grant to the Centro de Biología Molecular from the Fundación Ramón Areces (Madrid, Spain).

REFERENCES

Agrawal, R. K., P. Penzek, R. A. Grassucci, and J. Frank. 1998. Visualization of elongation factor G on the *Escherichia coli* 70S ribosome: the mechanism of translocation. *Proc. Natl. Acad. Sci. USA* **95:**6134–6138.

Ballesta, J. P. G., and M. Remacha. 1996. The large ribosomal subunit stalk as a regulatory element of the eukaryotic translational machinery. *Prog. Nucleic Acid Res. Mol. Biol.* **55:**157–193.

Beltrame, M., and M. E. Bianchi. 1990. A gene family for acidic ribosomal proteins in *Schizosaccharomyces pombe*: two essential and two nonessential genes. *Mol. Cell. Biol.* **10:**2341–2348.

Bermejo, B., M. Remacha, B. L. Ortiz-Reyes, C. Santos, and J. P. G. Ballesta. 1994. Importance of the 5′ untranslated region of the mRNA for acidic ribosomal phosphoproteins on gene expression and protein accumulation. *J. Biol. Chem.* **269:**3968–3975.

Bocharov, E. V., A. T. Gudkov, and A. S. Arseniev. 1996. Topology of the secondary structure elements of ribosomal protein L7/L12 from *E. coli* in solution. *FEBS Lett.* **379:**291–294.

Briones, E., C. Briones, M. Remacha, and J. P. G. Ballesta. 1998. The GTPase center protein L12 is required for correct ribosomal stalk assembly but not for *Saccharomyces cerevisiae* viability. *J. Biol. Chem.* **273:**31956–31961.

Chien, C.-T., P. L. Bartel, R. Sternglatz, and S. Fields. 1991. The two-hybrid system: a method to identify and clone genes for proteins that interact with a protein of interest. *Proc. Natl. Acad. Sci. USA* **88:**9578–9582.

Egebjerg, J., S. D. Douthwaite, A. Liljas, and R. A. Garrett. 1990. Characterization of the binding sites of protein L11 and the $L10(L12)_4$ pentameric complex in the GTPase domain of 23S ribosomal RNA from *E. coli*. *J. Mol. Biol.* **213:**275–288.

Gagou, M. E., M. A. Rodriguez-Gabriel, J. P. G. Ballesta, and S. Konyanou. The ribosomal P proteins of the medfly *Ceratitis capitata* form a heterogeneous stalk structure interacting with the endogenous P proteins in conditional P0-null strains of the yeast *Saccharomyces cerevisiae*. *Nucleic Acids Res.*, in press.

Gomez-Lorenzo, M. G., and J. F. Garcia-Bustos. 1998. Ribosomal P-protein stalk function is targeted by sordarin antifungals. *J. Biol. Chem.* **273:**25041–25044.

Gudkov, A. T., L. G. Tumanova, Y. S. Vanyaminov, and N. M. Khechinashvilli. 1978. Stoichiometry and properties of the complex between ribosomal proteins L7 and L10 in solution. *FEBS Lett.* **93:**215–218.

Juan-Vidales, F., F. Sanchez-Madrid, and J. P. G. Ballesta. 1981. Characterization of two acidic proteins of *Saccharomyces cerevisiae* ribosome. *Biochem. Biophys. Res. Commun.* **98:**717–726.

Juan-Vidales, F., M. T. Saenz-Robles, and J. P. G. Ballesta. 1984. Acidic proteins of the large ribosomal subunit in *Saccharomyces cerevisiae*. Effect of phosphorylation. *Biochemistry* **23:**390–396.

Kobashi, K. 1968. Catalytic oxidation of sulfhydryl groups by *o*-phanantroline copper complex. *Biochim. Biophys. Acta* **158:** 239–245.

Koteliansky, V. E., S. P. Domogatsky, and A. T. Gudkov. 1978. Dimer state of protein L7/12 and EFG dependent reactions on ribosomes. *Eur. J. Biochem.* **90:**319–323.

Leijonmarck, M., and A. Liljas. 1987. Structure of the C-terminal domain of the ribosomal protein L7/L12 from *E. coli* at 1.7Å. *J. Mol. Biol.* **195:**555–580.

Liljas, A., S. Thirup, and A. T. Matheson. 1986. Evolutionary aspects of ribosome-factor interactions. *Chemica Scripta* **26B:** 109–119.

Marquis, D. M., S. R. Fahnestock, E. Henderson, D. Woo, D. Schwinge, M. Clark, and J. A. Lake. 1981. The L7/L12 stalk, a conserved feature of the prokaryotic ribosome is attached to the large subunit through its N terminus. *J. Mol. Biol.* **150:**121–132.

Miller, J. H. 1972. *Experiments in Molecular Genetics*. Cold Spring Harbor Laboratory, Cold Spring Harbor, N.Y.

Möller, W., and J. A. Maassen. 1986. On the structure, function and dynamics of L7/12 from *Escherichia coli* ribosomes, p. 309–325. *In* B. Hardesty and G. Kramer (ed.), *Structure, Function and Genetics of Ribosomes*, Springer-Verlag, New York, N.Y.

Möller, W., A. Groene, C. Terhorst, and R. Amons. 1972. 50S ribosomal proteins. Purification and partial characterization of two acidic proteins, A1 and A2, isolated from 50S ribosomes from *Escherichia coli*. *Eur. J. Biochem.* **25:**5–12.

Musters, W., P. M. Gonzalves, K. Boon, H. A. Raué, H. van Heerikhuizen, and R. J. Planta. 1991. The conserved GTPase center and variable V9 from *Saccharomyces cerevisiae* 26S rRNA can be replaced by their equivalent from other prokaryotes or eukaryotes without detectable loss of ribosomal function. *Proc. Natl. Acad. Sci. USA* **88:**1469–1473.

Newton, C. H., L. C. Shimmin, J. Yee, and P. P. Dennis. 1990. A family of genes encode the multiple forms of the *Saccharomyces cerevisiae* ribosomal proteins equivalent to the *Escherichia coli* L12 protein and a single form of the L10-equivalent ribosomal protein. *J. Bacteriol.* **172:**579–588.

Nusspaumer, G., M. Remacha, and J. P. G. Ballesta. Unpublished data.

Oleinikov, A. V., G. G. Jokhadze, and R. R. Traut. 1993. Escherichia coli ribosomal protein L7/L12 dimers remain fully active after interchain crosslinking of the C-terminal domains in two orientations. *Proc. Natl. Acad. Sci. USA* **90:**9828–9831.

Payo, J. M., H. Santana-Roman, M. Remacha, J. P. G. Ballesta, and S. Zinker. 1995. The eukaryotic phosphoproteins interact with the ribosome through their amino terminal domain. *Biochemistry* **34:**7941–7948.

Remacha, M., C. Santos, and J. P. G. Ballesta. 1990. Disruption of single-copy genes encoding acidic ribosomal proteins in *Saccharomyces cerevisiae*. *Mol. Cell. Biol.* **10:**2182–2190.

Remacha, M., C. Santos, B. Bermejo, T. Naranda, and J. P. G. Ballesta. 1992. Stable binding of the eukaryotic acidic phosphoproteins to the ribosome is not an absolute requirement for *in vivo* protein synthesis. *J. Biol. Chem.* **267:**12061–12067.

Remacha, M., A. Jimenez-Diaz, B. Bermejo, M. A. Rodriguez-Gabriel, E. Guarinos, and J. P. G. Ballesta. 1995. Ribosomal acidic phosphoproteins P1 and P2 are not required for cell viability but regulate the pattern of protein expression in *Saccharomyces cerevisiae*. *Mol. Cell. Biol.* **15:**4754–4762.

Rodriguez-Gabriel, M. A., M. Remacha, and J. P. G. Ballesta. 1998. Phosphorylation of ribosomal protein P0 is not essential for ribosome function but affects translation. *Biochemistry* **37:** 16620–16626.

Rodriguez-Gabriel, M. A., M. Remacha, and J. P. G. Ballesta. The RNA interacting domain but not the protein interacting domain is highly conserved in ribosomal protein P0. *J. Biol. Chem.*, in press.

Saenz-Robles, M. T., M. D. Vilella, G. Pucciarelli, F. Polo, M. Remacha, B. L. Ortiz, F. Vidales, and J. P. G. Ballesta. 1988. Ribosomal protein interactions in yeast. Protein L15 forms a complex with the acidic proteins. *Eur. J. Biochem.* **177:**531–537.

Saenz-Robles, M. T., M. Remacha, M. D. Vilella, S. Zinker, and J. P. G. Ballesta. 1990. The acidic ribosomal proteins as regulators of the eukaryotic ribosomal activity. *Biochim. Biophys. Acta* **1050:**51–55.

Sanchez-Madrid, F., R. Reyes, P. Conde, and J. P. G. Ballesta. 1979. Acidic ribosomal proteins from eukaryotic cells. Effect on ribosomal functions. *Eur. J. Biochem.* **98:**409–416.

Santos, C., and J. P. G. Ballesta. 1994. Ribosomal phosphoprotein P0, contrary to phosphoproteins P1 and P2, is required for *S. cerevisiae* viability and ribosome assembly. *J. Biol. Chem.* **269:** 15689–15696.

Santos, C., and J. P. G. Ballesta. 1995. The highly conserved protein P0 carboxyl end is essential for ribosome activity only in the absence of proteins P1 and P2. *J. Biol. Chem.* **270:**20608–20614.

Scharf, K.-D., and L. Nover. 1987. Control of ribosome biosynthesis in plant cell cultures under heat shock conditions. II. Ribosomal proteins. *Biochim. Biophys. Acta* **909:**44–57.

Schijman, A. G., M. P. Vazquez, C. Ben Dov, S. Ghio, H. Lorenzi, and M. J. Levin. 1995. Cloning and sequence analysis of TcP2 beta cDNA variants of *Trypanosoma cruzi*. *Biochim. Biophys. Acta* **1264:**15–18.

Shimmin, L. C., G. Ramirez, A. T. Matheson, and P. P. Dennis. 1989. Sequence alignment and evolutionary comparison of the L10 equivalent and L12 equivalent ribosomal proteins from archaebacteria, eubacteria, and eucaryotes. *J. Mol. Evol.* **29:**448–462.

Stark, H., M. V. Rodnina, A. J. Rinke, R. Brimacombe, W. Wintermeyer, and M. van Heel. 1997. Visualization of elongation factor Tu on the *Escherichia coli* ribosome. *Nature* **389:**403–406.

Stöffler, G., E. Cundliffe, M. Stöffler-Meilicke, and E. R. Dabbs. 1980. Mutants of *Escherichia coli* lacking ribosomal protein L11. *J. Biol. Chem.* **255:**10517–10522.

Szick, K., M. Springer, and J. Bailey-Serres. 1998. Evolutionary analyses of the 12-kDa acidic ribosomal P-proteins reveal a distinct protein of higher plant ribosomes. *Proc. Natl. Acad. Sci. USA* **95:**2378–2383.

Thompson, J., W. Muster, E. Cundliffe, and A. E. Dahlberg. 1993. Replacement of the L11 binding region within E. coli 23S ribosomal RNA with its homologue from yeast: *in vivo* and *in vitro* analysis of hybrid ribosomes altered in the GTPase centre. *EMBO J.* **12:**1499–1504.

Tsurugi, K., and K. Ogata. 1985. Evidence for the exchangeability of acidic ribosomal proteins on cytoplasmic ribosomes in regenerating rat liver. *J. Biochem.* **98:**1427–1431.

Uchiumi, T., and R. Kominami. 1997. Binding of mammalian ribosomal protein complex P0-P1-P2 and protein L12 to the GTPase-associated domain of 28 S ribosomal RNA and effect on the accessibility to anti-28 S RNA autoantibody. *J. Biol. Chem.* **272:**3302–3308.

Uchiumi, T., A. J. Wahba, and R. R. Traut. 1987. Topography and stoichiometry of acidic proteins in large ribosomal subunits from *Artemia salina* as determined by crosslinking. *Proc. Natl. Acad. Sci. USA* **84:**5580–5584.

Vilella, M. D., M. Remacha, B. L. Ortiz, E. Mendez, and J. P. G. Ballesta. 1991. Characterization of the yeast acidic ribosomal phosphoproteins using monoclonal antibodies. Proteins L44/L45 and L44′ have different functional roles. *Eur. J. Biochem.* **196:**407–414.

Xing, Y., and D. E. Draper. 1995. Stabilization of a ribosomal RNA tertiary structure by ribosomal protein L11. *J. Mol. Biol.* **249:**319–331.

Xing, Y., D. GuhaThakurta, and D. E. Draper. 1997. The RNA binding domain of ribosomal protein L11 is structurally similar to homeodomains. *Nat. Struct. Biol.* **4:**24–27.

Zinker, S., and J. R. Warner. 1976. The ribosomal proteins of *Saccharomyces cerevisiae*. Phosphorylated and exchangeable proteins. *J. Biol. Chem.* **251:**1799–1807.

Zurdo, J., J. M. Sanz, C. Gonzalez, M. Rico, and J. P. G. Ballesta. 1997. The exchangeable yeast ribosomal acidic protein YP2beta shows characteristics of a partly folded state under physiological conditions. *Biochemistry* **36:**9625–9635.

IV. RIBOSOMAL MODELING

IV. RIBOSOMAL MODELING

Building ribosomal models has gone thorugh many stages over the past decades during which the macromolecular complexity of the ribosome has served as an ideal system for testing and developing many new biochemical and physical methods for defining macromolecular interactions or neighborhoods. Chemical cross-linking of proteins and RNA, in particular, has served to introduce constraints on three-dimensional folding of the rRNA and the relative positioning of the proteins. Combining all of these data with neutron-scattering models of protein locations, cryo-electron microscopy models, and, very recently, X-ray diffraction models has generated detailed and fairly exact ribosomal structures.

The Ribosome: Structure, Function, Antibiotics, and Cellular Interactions
Edited by R. A. Garrett, S. R. Douthwaite, A. Liljas, A. T. Matheson, P. B. Moore, and H. F. Noller

Chapter 13

Studies on the Structure and Function of Ribosomes by Combined Use of Chemical Probing and X-Ray Crystallography

HARRY F. NOLLER, JAMIE CATE, ANNE DALLAS, GLORIA CULVER, THOMAS N. EARNEST, RACHEL GREEN, LOVISA HOLMBERG, SIMPSON JOSEPH, LAURA LANCASTER, KATE LIEBERMAN, CHUCK MERRYMAN, LISA NEWCOMB, RAYMOND SAMAHA, UWE VON AHSEN, MARAT YUSUPOV, GULNARA YUSUPOVA, and KEVIN WILSON

The ribosome is among the most wonderfully complex molecular objects in all of biology. As it is the "enzyme" that catalyzes the process of protein synthesis, it is an intriguing fact that about two-thirds of its mass is composed of RNA. It should not be so surprising that it has a molecular mass of 2.5 MDa or more, as enzymes are typically about 2 orders of magnitude larger than their substrates, which for the ribosome is the 25-kDa tRNA. The study of the process of protein synthesis, which encompasses translation of the genetic code, has been intensively under way for about 4 decades. For most of this time, studies of the ribosome have been carried out with only a vague notion of its overall three-dimensional structure. However, with the advent of new structural information, from cryo-electron microscopy (EM) reconstruction methods (Malhotra et al., 1998; Stark et al., 1997) and, within the last year, X-ray crystallography of ribosomal subunits and whole ribosomes (Ban et al., 1998, 1999; Clemons et al., 1999; Cate et al., 1999) (see chapters 1 and 3), the veil is being lifted. Knowledge of the molecular structure of the ribosome will have a profound and permanent effect on how we think about the mechanism of translation from now on. This chapter summarizes recent studies of the structure and function of ribosomes by a combination of biochemical and X-ray-crystallographic approaches, focusing primarily on work from this laboratory. We apologize to any of our colleagues in the ribosome field whose contributions may not have been cited because of limitations of space.

In view of the molecular and mechanistic complexity of protein synthesis, a useful simplification is to ask, what are the fundamental mechanisms that the ribosome carries out? The three central paradoxes that the ribosome must confront are: (i) accurate selection of the correct aminoacyl-tRNA, based on codon-anticodon pairing (which, in the absence of the ribosome, is insufficient to ensure accurate discrimination); (ii) catalysis of peptide bond formation in an aqueous solvent (using the same enzymatic activity that is responsible for hydrolysis of peptidyl-tRNA during translational termination); and (iii) rapid translocation of tRNAs and their associated mRNA from one ribosomal binding site to another (which seemingly requires disruption of binding between mRNA, tRNA, and the ribosome yet must avoid disruption of base pairing between mRNA codons and tRNA anticodons). Most of the processes carried out by the ribosome can be considered to be related to, or based on, these three fundamental mechanisms.

THE ELONGATION CYCLE

For each amino acid added to the growing polypeptide chain, the ribosome executes one iteration of the elongation cycle. Studies in this laboratory led to the proposal of the hybrid-states model, shown in Fig. 1 (Moazed and Noller, 1989a). The main feature

Harry F. Noller, Jamie Cate, Anne Dallas, Gloria Culver, Rachel Green, Lovisa Holmberg, Simpson Joseph, Laura Lancaster, Kate Lieberman, Chuck Merryman, Lisa Newcomb, Raymond Samaha, Uwe von Ahsen, Marat Yusupov, Gulnara Yusupova, and Kevin Wilson ■ Center for Molecular Biology of RNA, Sinsheimer Laboratories, University of California—Santa Cruz, Santa Cruz, CA 95064. Thomas N. Earnest ■ Macromolecular Crystallography Facility, Advanced Light Source, Lawrence Berkeley National Laboratory, Berkeley, CA 94720.

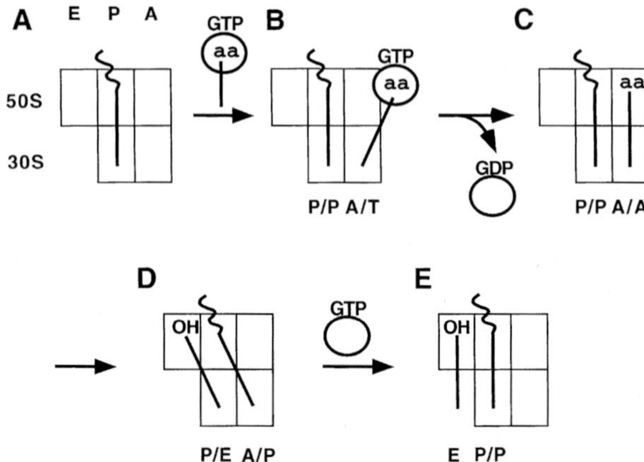

Figure 1. Hybrid-states model for the translocation cycle (Moazed and Noller, 1989a). The ribosomal binding sites (A, P, and E) for tRNA on the 30S and 50S subunits are schematized as rectangles. tRNAs are shown as sticks, with a squiggle and "aa" representing the peptidyl and aminoacyl moieties, respectively. The elongation factors EF-Tu and EF-G are represented as circles. The different binding states for tRNA (P/P, A/T, A/A, P/E, A/P, and E) are indicated at the bottom.

of this model was the independent interaction of the two extremities of tRNA (the anticodon and acceptor ends) with the two (30S and 50S) ribosomal subunits. The anticodon ends interact with mRNA codons in the 30S subunit, and the acceptor ends interact with the peptidyltransferase (PT) (and other sites) in the 50S subunit. In vitro chemical-probing experiments showed that the acceptor and anticodon ends can move independently (Fig. 1), suggesting that translocation is accomplished in (at least) two separate movements of a tRNA: first, the acceptor end moves relative to the 50S subunit, followed by EF-G-dependent movement of the anticodon end and its associated mRNA codon. A consequence is that tRNAs can exist in hybrid states in which, for example, the anticodon is bound to the 30S A site while the acceptor end is bound in the 50S P site (the A/P state).

One prediction of the hybrid-states model is that the acceptor end of the peptidyl-tRNA is always bound to the 50S P site, in agreement with the "displacement" model of Hardesty and coworkers (Odom and Hardesty, 1987). The model provides a rationale for the existence of the E site, as a binding site for the free acceptor end of deacylated tRNA, which could help to provide the thermodynamic driving force for the first, spontaneous step of translocation. It also suggests that the mechanism of tRNA movement might be related in some way to the molecular interactions between the 30S and 50S subunits.

More recently, new studies have provided evidence for additional states in the elongation cycle (Fig. 2). An initial interaction between the ribosome and the EF-Tu·tRNA·GTP ternary complex (Fig. 2B) has been observed kinetically by Wintermeyer and colleagues (Rodnina et al., 1996), in which the complex binds nonspecifically, presumably exclusively through its EF-Tu moiety, to the ribosome prior to codon-anticodon interaction. A puzzling result of the original hybrid-states experiments was that acetyl-

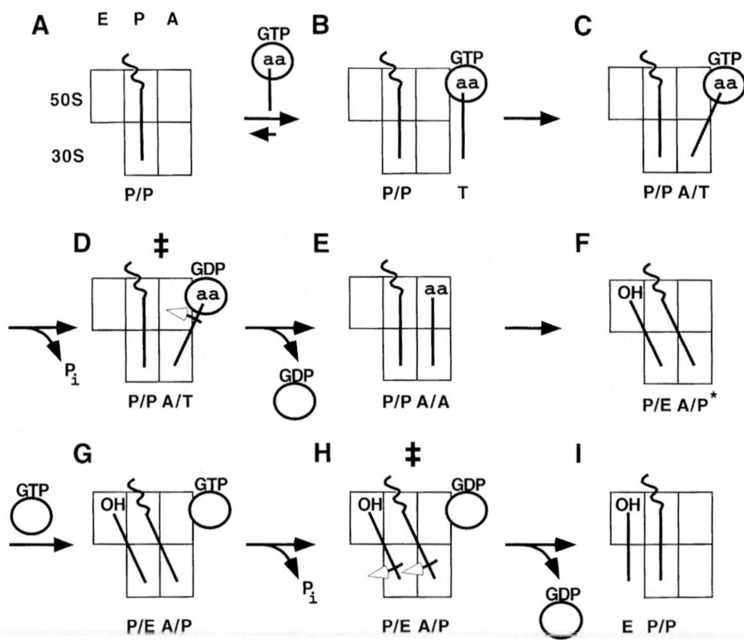

Figure 2. Extended hybrid-states model (Wilson and Noller, 1998b). Additional steps in the translocational cycle have been added, based on more recent studies described in the text.

Phe-tRNA bound in the A/P state was unreactive toward puromycin in the PT reaction, even though the acetyl-Phe acceptor end of the tRNA appeared to be bound to the PT P site by chemical footprinting and other criteria (although at least one detail—the anomalous reactivity of A2602 of 23S rRNA toward dimethyl sulfate—suggested that it was not identical to the "active" state of P-site binding). Wintermeyer and coworkers (Borowski et al., 1996) have shown that use of a deletion mutant of EF-G results in an A/P hybrid state that is reactive toward puromycin; accordingly, we refer to this newly observed state as the true A/P state (in keeping with a functional definition for the 50S P site), and redefine the previously observed, unreactive state as the A/P* state (Fig. 2F to G).

Another revision of the elongation mechanism concerns the role of GTP hydrolysis. Using pre-steady-state kinetics measurements, Wintermeyer and coworkers have shown that, contrary to what was previously believed, GTP hydrolysis precedes tRNA movement (Fig. 2H to I) and so could provide energy that is in some way directly coupled to the translocation event (Rodnina et al., 1997). Finally, translocation is known to occur spontaneously under appropriate solution conditions in vitro (Gavrilova and Spirin, 1974), suggesting that EF-G and GTP simply catalyze a reaction that is carried out by the ribosome itself. Accordingly, Spirin (1969) has proposed that the tRNAs are bound to the ribosome in "locked" states and need to pass through an "unlocked" (presumably high-energy) state (Fig. 2H) to move during translocation. It has been suggested that, in an analogous way, the EF-Tu ternary complex uses GTP hydrolysis to pass aminoacyl-tRNA through an unlocked state (Fig. 2D) in delivering it to the ribosomal A site (Wilson and Noller, 1998b). Figure 2 schematically summarizes these results in terms of an extended hybrid-states model for the elongation cycle. It would not be surprising to find even more states of tRNA-ribosome interaction in the coming years.

MAPPING THE LOCATION OF EF-G IN THE RIBOSOME BY DIRECTED HYDROXYL RADICAL PROBING

Ribosomologists have traditionally used a wide variety of experimental strategies to study the many molecular interactions of the translational machinery. In recent years, we have developed the method of directed hydroxyl radical probing to map the RNA environments around individual proteins (Heilek et al., 1995; Heilek and Noller, 1996a, 1996b). This approach complements the chemical-footprinting method (Stern et al., 1988a; 1989; Powers and Noller, 1995), which provides information about direct contact between proteins and RNA or protein-induced conformational changes in the RNA. Advantages of the directed-probing method are the large amount of information typically obtained and the fairly straightforward interpretation of the data. Fe(II) is tethered to specific positions on the surface of a protein via the EDTA linker bromacetamido benzyl ethylenediaminetetraacetic acid (BABE) (DeRiemer et al., 1981), which can be attached to single cysteine residues that are introduced by site-directed mutagenesis (Heilek and Noller, 1996a; Culver and Noller, 1999). Availability of a crystal or nuclear magnetic resonance solution structure for the protein greatly facilitates choosing tethering sites that are well distributed over the surface of the protein and, most importantly, provides a strong basis for interpreting the probing results in terms of the three-dimensional locations of the different RNA targets. Hydroxyl radicals can be initiated under mild conditions after formation of the protein-RNA complex. The radicals diffuse from the tethering site and cleave the backbone of nearby elements of RNA, which are identified by primer extension with reverse transcriptase. The approximate maximum distances between tethering site and target can be calibrated with the cleavage intensities.

This approach was used to map the location of EF-G on the ribosome in its posttranslocational state

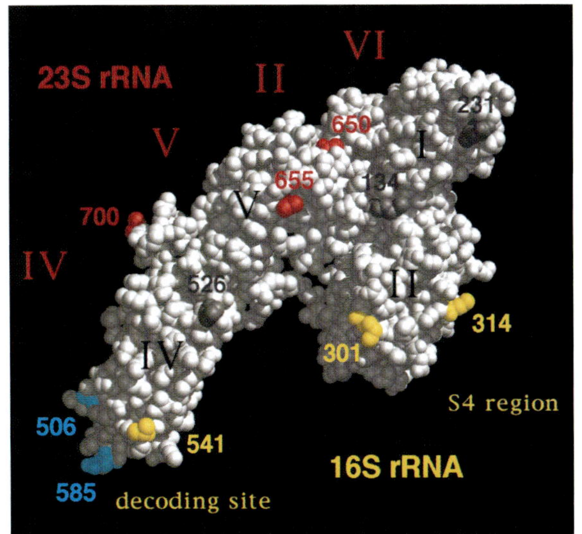

Figure 3. Crystal structure of EF-G·GDP (Czworkowski et al., 1994), showing several of the positions of tethering Fe(II) used for directed hydroxyl radical probing of rRNA in the ribosome (Wilson and Noller, 1998a). The tethering positions shown in yellow target 16S rRNA, those in red target 23S rRNA, and those in blue target both 16S and 23S rRNAs. Roman numerals in red indicate targeted domains of 23S rRNA; black roman numerals indicate domains of EF-G.

(Wilson and Noller, 1998a). First, the three natural cysteine residues in *Escherichia coli* EF-G were mutated to provide a fully active, Cys-free triple-mutant version of EF-G. Single cysteine mutations were then introduced at 19 positions distributed over the surface of the protein. EF-G was then bound in the presence of fusidic acid and GTP to ribosomes containing mRNA and tRNA bound to the P site, and hydroxyl radicals were initiated. Figure 3 shows the crystal structure of EF-G·GDP (Czworkowski et al., 1994) with the positions of the Fe(II) tethering sites and their respective RNA target regions. Cleavage was observed in specific positions of 16S and 23S, but not 5S, rRNA. Some of the tethering sites were specific for 16S rRNA, indicating that they face the 30S subunit, while others were specific for 23S rRNA, indicating proximity to the 50S subunit. Yet another class of tethering sites targeted both 16S and 23S rRNAs, suggesting that this part of EF-G resides in a region of close approach of the 30S and 50S subunits. Many functionally implicated sites were targeted, including the decoding site and P-tRNA binding region of 16S rRNA and the sarcin and thiostrepton loops of 23S rRNA. The extensive probing data allowed us to model the position and orientation of EF-G in the ribosome, based on our earlier model for the folding of 16S rRNA (Fig. 4). This docking arrangement is in striking agreement with an independent cryo-EM study of the EF-G–ribosome complex, in which the position of EF-G in the ribosome was directly observed at low resolution (Agrawal et al., 1998).

Based on this result, we also docked the EF-Tu–tRNA–GTP ternary complex on the model, using the extensive structural homology between EF-G and EF-Tu. This allowed us to predict the position and orientation of the EF-Tu-bound tRNA in the ribosome. The tRNA was predicted to be far from its orientation in the PT site, in a position that is also in agreement with independent cryo-EM studies (Stark et al., 1997). This arrangment of the tRNA bound in the A/T state explains the absence of a 23S A-site footprint for the ternary complex (Moazed and Noller, 1989b) and is consistent with its nonreactivity in the PT reaction. The predicted orientations of tRNA in the A/A, P/P, and A/T states suggested a model for the movement of tRNA during the ribosomal translocation reaction (Wilson and Noller, 1998b). The model proposes that certain rotational movements are inherent mechanical properties of the ribosome, which are coupled in two different ways to tRNA movement by the elongation factors EF-Tu and EF-G. This model provides a unified view of the actions of EF-G and EF-Tu, rationalizing their common molecular features and shared ribosomal binding site, and is consistent with the observation that both translocation and aminoacyl-tRNA binding can proceed in the absence of factors and GTP.

THE X-RAY STRUCTURE OF THE 70S RIBOSOME AT 7.8-Å RESOLUTION

Three-dimensional crystals of ribosomes and ribosomal subunits were obtained more than a decade ago (Yonath et al., 1983; Trakhanov et al., 1989), but only recently has X-ray crystallography begun to provide detailed information about the large-scale structure of the ribosome. A 9-Å crystallographic map of the 50S subunit of *Haloarcula marismortui* was

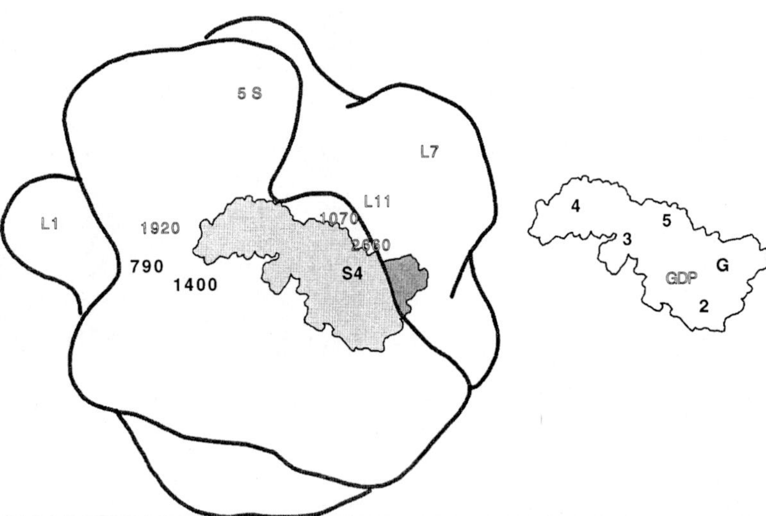

Figure 4. Predicted position of EF-G in the ribosome, based on directed hydroxyl radical probing results (Wilson and Noller, 1998a). The locations of different ribosomal features are as indicated.

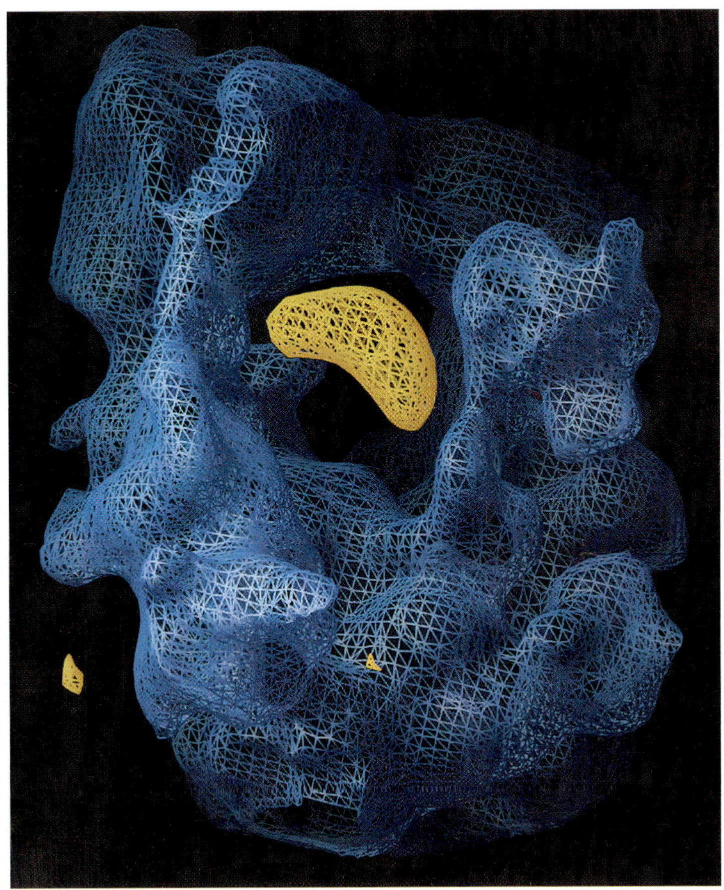

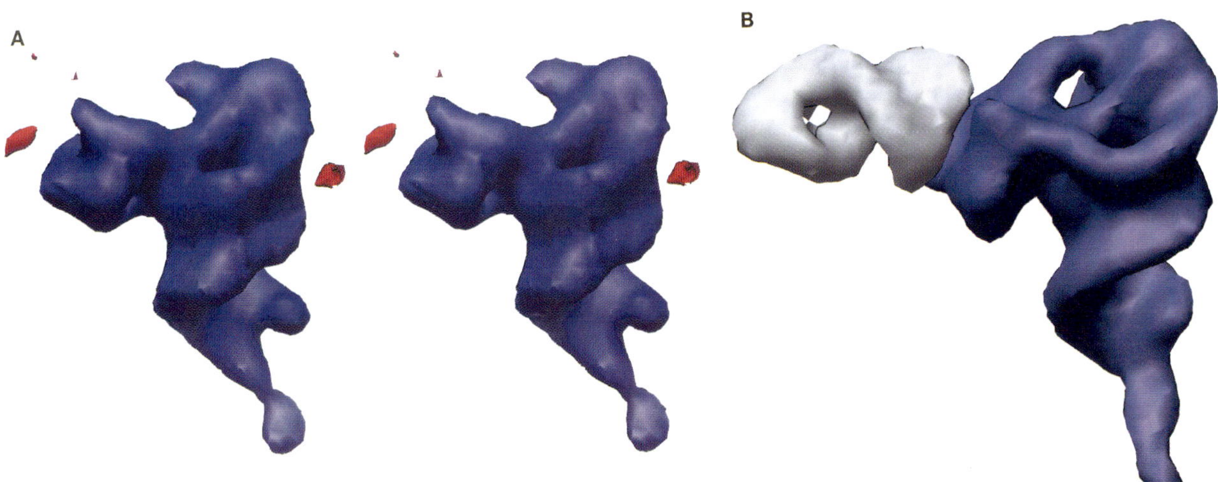

Figure 5 (top). Fourier difference map (yellow) of ribosome cocrystals containing A-site tRNA versus crystals with a vacant A site, at 25-Å resolution (Cate et al., 1999), superimposed on electron density of the complete 70S ribosome (blue), viewed from the L12 side of the ribosome. The 30S subunit is on the left, and the 50S subunit is on the right. The absence of visible calculated negative density (red) is indicative of the low noise level of the difference map at this resolution.

Figure 6 (bottom). (A) Stereo view of an 8.8-Å Fourier difference map of a ribosome cocrystal containing a full-length P-site tRNA versus a cocrystal containing only a P-site-bound ASL. Positive density is shown in blue, and negative density is shown in red. (B) Calculated electron density for $tRNA^{Phe}$ at 8.8Å, showing the ASL region in gray and the rest of the tRNA in blue.

solved within the past year (Ban et al., 1998), providing the best description so far of the structure of the large subunit, including views of clearly recognizable RNA helices. In this laboratory, we have obtained crystals of functional complexes of *Thermus thermophilus* 70S ribosomes containing mRNA and tRNA or a tRNA anticodon stem-loop (ASL) analogue. These crystals belong to the I422 tetragonal space group, have unit cell dimensions of 508 by 508 by 809 Å with a single 70S ribosome per asymmetric unit, and diffract to at least 5-Å resolution.

Initial low-resolution phases were obtained by molecular replacement, with the 25-Å-resolution *E. coli* 70S ribosome cryo-EM map (Penczek et al., 1994) as a search model. The quality of the low-resolution phases was confirmed by computing a Fourier difference map of crystals of 70S ribosomes containing a tRNA bound to the A site and ribosomes with a vacant A site. The difference map clearly reveals an L-shaped zone of positive density (Fig. 5) positioned in the cavity between the 30S and 50S subunits, consistent with current knowledge of the location of the ribosomal A site. Only small patches of negative density are observed, indicating a high signal-to-noise ratio for the map and minimal rearrangement of the ribosomal structure in response to A-site tRNA binding.

The low-resolution phase information was then used to locate heavy-atom clusters in Fourier difference maps, and multiwavelength anomalous dispersion (MAD) phasing was used to calculate an intermediate-resolution electron density map. The phases obtained in this way were then used to locate conventional single heavy-atom sites, again in Fourier difference maps. After phase refinement with MLPHARE (Otwinowski, 1991) and density modification with the CNS program (Brünger et al., 1998), an 8.8-Å electron density map was calculated. A Fourier difference map of ribosome crystals containing a full-length P-site tRNA and crystals containing only an ASL bound to the P site (Fig. 6A) clearly shows the expected density of a full-length tRNA missing precisely the 6-bp ASL, corresponding to nucleotides 26 through 44. The deep, narrow major grooves and shallow, broad minor grooves of the tRNA are clearly defined, and even the single-stranded CCA tail is visible. Figure 6B shows an electron density map for $tRNA^{Phe}$ calculated from the known crystal structure by using Fourier terms only to 8.8 Å, for comparison. The portion of the tRNA excluding the ASL is shown. The close agreement between the observed and predicted electron density maps provides strong evidence for the quality of the phases and for the absence of model bias, since no tRNA was present in the molecular-replacement search model.

A four-wavelength MAD dataset was collected on an ASL-mRNA-70S ribosome cocrystal, again by using conventional heavy atoms, providing useful phase information to 7.8-Å resolution (Table 1). An electron density map of the 70S ribosome calculated with these phases is shown in Fig. 7. The overall features resemble previous lower-resolution images of the complete ribosome obtained by cryo-EM reconstruction methods (Agrawal et al., 1998; Stark et al., 1997), but with a significantly higher level of detail. Many double-helical RNA features are recognizable, including a ~100-Å-long helix positioned vertically on the 50S-subunit face of the 30S subunit, identifiable as the penultimate stem in the 3′-minor domain of 16S rRNA (Fig. 8). The penultimate stem is protected at several discrete positions by the 50S subunit by hydroxyl radical footprinting (Merryman et al., 1999a) (see Fig. 12); the 3′-staggered protection patterns are characteristic of protection of the minor-groove face of an RNA helix. In the X-ray map, bridges from elements originating in the 50S subunit can be seen to contact the minor groove of this helix at positions corresponding closely to those predicted from the footprinting studies. Some additional features of the X-ray map are discussed below.

Table 1. MAD phasing of 70S ribosome structure

Parameter	Value				
Data set	λ1	λ2	λ3	λ4	N2
High-resolution limit (Å)	7.5	7.5	7.5	7.5	8.8
Rsym (%)[a]	7.3	7.2	8.9	7.9	8.4
Mean I/σ(I) at 7.8 Å	3.4	3.0	3.1	2.4	
No. of reflections					
Unique	121,730	118,423	124,437	119,051	41,520
Observational redundancy	4.0	3.4	4.4	3.4	5.9
Completeness (%)	95.9	93.1	97.7	93.4	99.8
MAD phasing figure of merit (at 7.8 Å)				0.50 (0.32)	
Phasing figure of merit after density modification (at 7.8 Å)				0.92 (0.80)	

[a] Rsym = Σ | *I* − ⟨I⟩ |/ΣI.

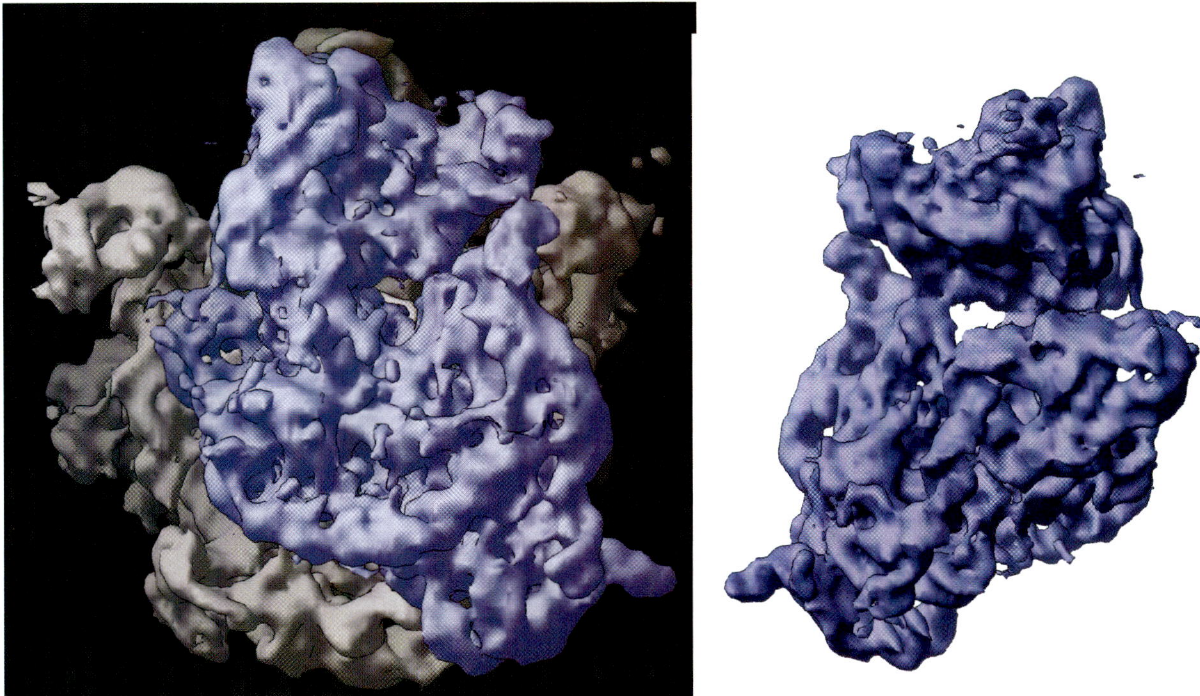

Figure 7 (left). Electron density map of the complete 70S ribosome from *T. thermophilus* at 7.8-Å resolution (Cate et al., 1999). The 30S subunit (blue) is in the foreground, with the head at the top and the platform to the left. The 50S subunit (gray) is behind, with its L1 ridge protruding at the left and the L12 stalk region to the right.

Figure 8 (right). View of the 30S subunit excised from the electron density map of the 70S ribosome, viewed from the 50S subunit. The head is at the top, the platform is at the right, and body is to the left. The penultimate stem can be seen as an ~100-Å-long helix extending from just below the middle of the head, angling slightly to the left, to the bottom of the subunit.

PROTEIN-RNA NEIGHBORHOODS

Of continuing interest is the way in which ribosomal proteins interact with rRNA and their respective roles in the structure, function, and assembly of ribosomes. The rich background of biochemical studies, including nuclease protection, cross-linking, and footprinting experiments, has been further extended in this laboratory by directed hydroxyl radical probing studies, using the approach described above

Table 2. Directed hydroxyl radical probing of rRNA from specific positions on ribosomal proteins and translation factors

Protein	No. of tethering positions	rRNA target(s)	Reference(s)
S4	1	16S	Heilek et al., 1995
S5	3	16S, 23S	Heilek and Noller, 1996a; Culver et al., 1998
S8	8	16S	Lancaster et al., unpublished
S13	1	16S	Heilek and Noller, 1996b
S15	3	16S, 23S	Culver and Noller, unpublished
S17	1	16S	Heilek and Noller, unpublished
S20	4	16S	Culver and Noller, 1998
L12	5	23S	Holmberg and Noller, unpublished
L9	10	23S	Lieberman and Noller, submitted for publication
L11	5	16S, 23S	Holmberg and Noller, 1999; unpublished
L15	3	23S	Lieberman and Noller, 1998
EF-G	18	16S, 23S	Wilson and Noller, 1998a
IF-3	13	16S	Dallas and Noller, unpublished
RF-1	9	16S, 23S	Wilson et al., unpublished

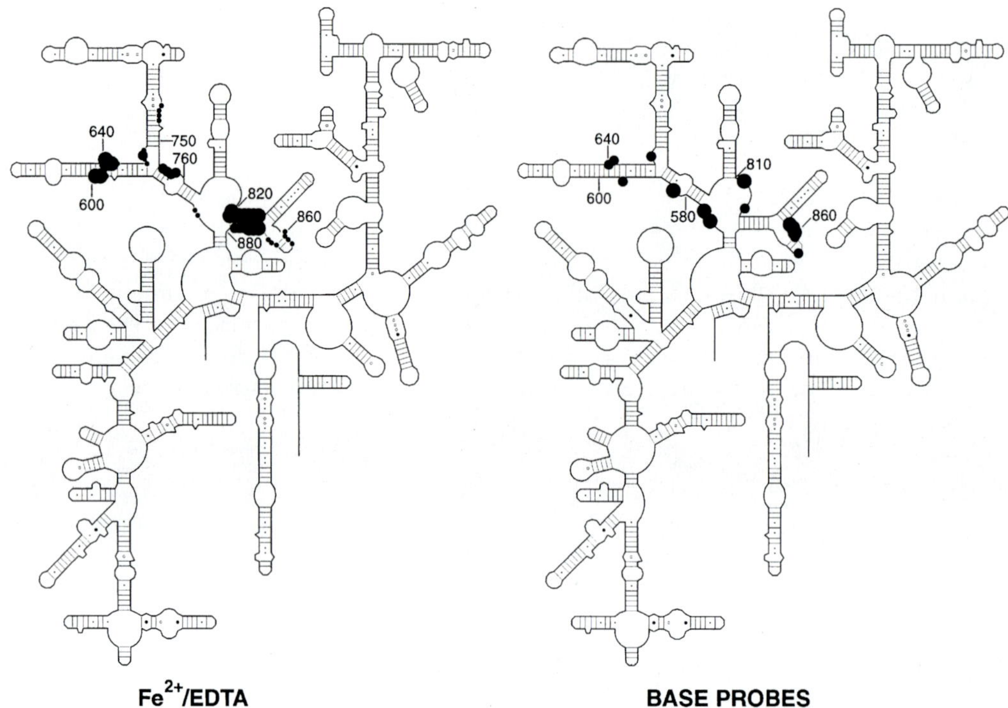

Figure 9. Chemical footprints of ribosomal protein S8 on 16S rRNA, using hydroxyl radicals generated by free Fe(II)-EDTA (Powers and Noller, 1995) and base-specific probes (Svensson et al., 1988). The relative strengths of protection are indicated by the sizes of the solid circles.

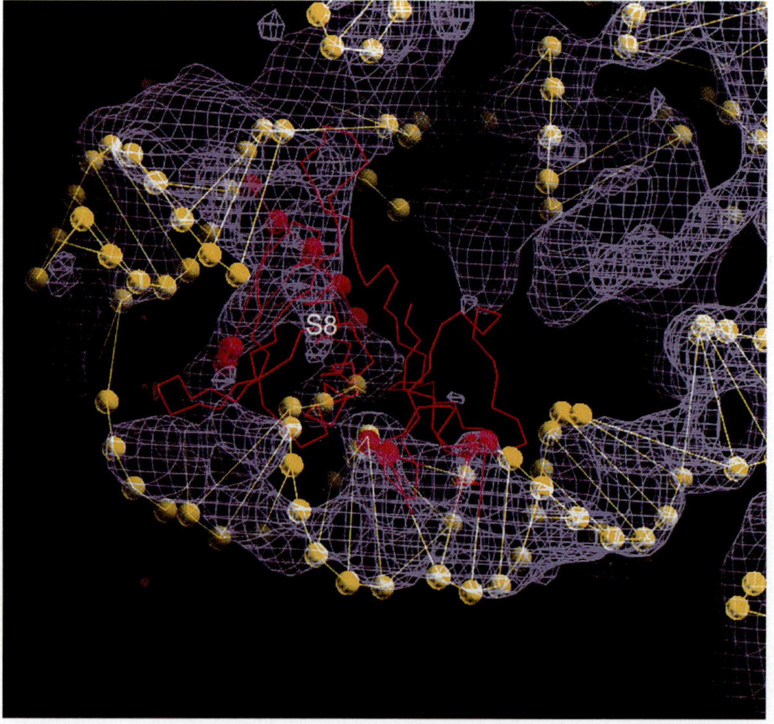

Figure 10. The S8 binding region of the 30S ribosomal subunit. A pseudoatom model for 16S rRNA (yellow) is superimposed on the 7.8-Å electron density map. Nucleotides protected from hydroxyl radicals by S8 (Powers and Noller, 1995) are shown in red. Density corresponding to protein S8 is indicated. The 620 stem runs from left to right at the bottom, and the 820 and 840 helices run from right to left at the upper left.

for mapping elongation factor EF-G. So far, directed probing has been applied to the study of the RNA environments of more than a dozen ribosomal proteins and translational factors (Table 2). Some of these data have provided constraints for modeling the three-dimensional folding of 16S rRNA (Heilek and Noller, 1996; Fink et al., 1996). These directed-probing data can now be used to help interpret the low-resolution electron density maps obtained by X-ray-crystallographic studies of ribosomes and ribosomal subunits.

A case in point is ribosomal protein S8, for which crystal structures have been determined for both the *Bacillus stearothermophilus* (Davies et al., 1996) and *T. thermophilus* (Nevskaya et al., 1998) proteins. Early nuclease protection studies localized the binding site for S8 to the 600/640 region of the long 620 stem in the central domain of 16S rRNA (reviewed by Zimmermann, 1980). Chemical footprinting studies with both base-specific probes and hydroxyl radicals generated from free Fe(II)-EDTA (Svensson et al., 1988; Powers and Noller, 1995), as well as protein-RNA cross-linking studies (Osswald et al., 1987; Rinke-Appel et al., 1995; Wower and Brimacombe, 1983) suggested a more extensive set of contacts between S8 and 16S rRNA, including the middle of the 620 stem, the 820 stem, and the three-helix junction at the base of the 620 stem (Fig. 9). Based on our previous modeling studies, certain distinctive features on the solvent face of the 30S subunit were recognizable in the electron density map as likely features of the S8 binding site (Fig. 10). The long rod of helical density at the bottom corresponds to the 620 stem, with its apex at the right-hand side and the three-way junction behind at the left. The 820 stem is in the upper middle, with the 840 stem extending to the upper left. The 16S rRNA is displayed as a pseudoatom model, using a single pseudoatom at the P position for each helical nucleotide (Stern et al., 1988). Nucleotides protected from hydroxyl radical attack by protein S8 are shown. In addition to the helical RNA density, protein S8 is identifiable as an irregular blob wedged between the 620 and 820 stems, merging with the protected positions of the RNA.

The arrangement of the S8-16S rRNA interaction has been investigated extensively by directed hydroxyl radical probing (Lancaster et al., unpublished). Based on the crystal structure of *T. thermophilus* S8 and phylogenetic sequence comparisons of more than 30 S8 sequences, six nonconserved surface positions were chosen for mutagenesis to cysteine. Each of the Fe(II)-derivatized mutant proteins was reconstituted into 30S subunits and used to form 70S ribosomes, which were purified by sucrose gradient sedimentation and probed. Each tethering site generated a characteristic cleavage pattern in 16S rRNA (Fig. 11). The proximities of the tethering sites in S8 to their respective cleavage targets in the RNA model provide strong, detailed evidence for the proposed location of S8 and its associated elements of 16S rRNA. Further support comes from a cross-link between positions 653 of 16S rRNA and Lys 55 of S8 (Urlaub et al., 1997), which are within 5 Å of each other in the model, and a protein-protein cross-link between Lys 93 of S8 and Lys 166 of protein S5 (Allen et al., 1979), which can be placed directly adjacent to that face of S8, based on previous chemical probing and neutron diffraction studies (Capel et al., 1987; Heilek and Noller, 1996a).

S8 is a primary binding protein whose position in the assembly map of the 30S subunit (Held et al., 1974) is independent of the coordinated assembly of the other central-domain-associated proteins S6, S11, S15, S18, and S21, whose assembly is sequentially coordinated. Our findings suggest that its role is to organize the back of the central domain, which makes a connection to the body of the subunit via interactions of protein S16 both with the end of the 620 stem and to elements of the 5′ domain of 16S rRNA (Stern et al., 1989; Powers and Noller, 1995). The other central-domain proteins appear to be located on the interface side, forming part of the structure of the platform and cleft, explaining their independence from S8 during assembly. The apparent interaction of S8 with protein S5 would make a further connection to the central region of the 30S subunit, near the convergence of the three major domains of 16S rRNA. This suggests a role for S8 in stabilization of the junction between the central and 5′ domains of 16S rRNA in the structure of the 30S subunit

MOLECULAR INTERACTIONS AT THE SUBUNIT INTERFACE

Most, if not all, of the business of translation takes place in the cavity between the two ribosomal subunits. Molecular interactions between the subunits may coordinate functional events taking place on the different subunits and thus are likely to be of functional as well as structural importance. Early nuclease and chemical-footprinting studies showed that specific elements of 16S and 23S rRNA are protected upon association of 30S and 50S ribosomal subunits (Santer and Shane, 1977; Chapman and Noller, 1977; Herr and Noller, 1978). Modification-interference studies with kethoxal showed that modification of a set of guanines in 16S rRNA interferes with subunit association (Herr et al., 1979). More recently, systematic hydroxyl radical and base-specific footprinting studies (Merryman et al., 1999a, 1999b)

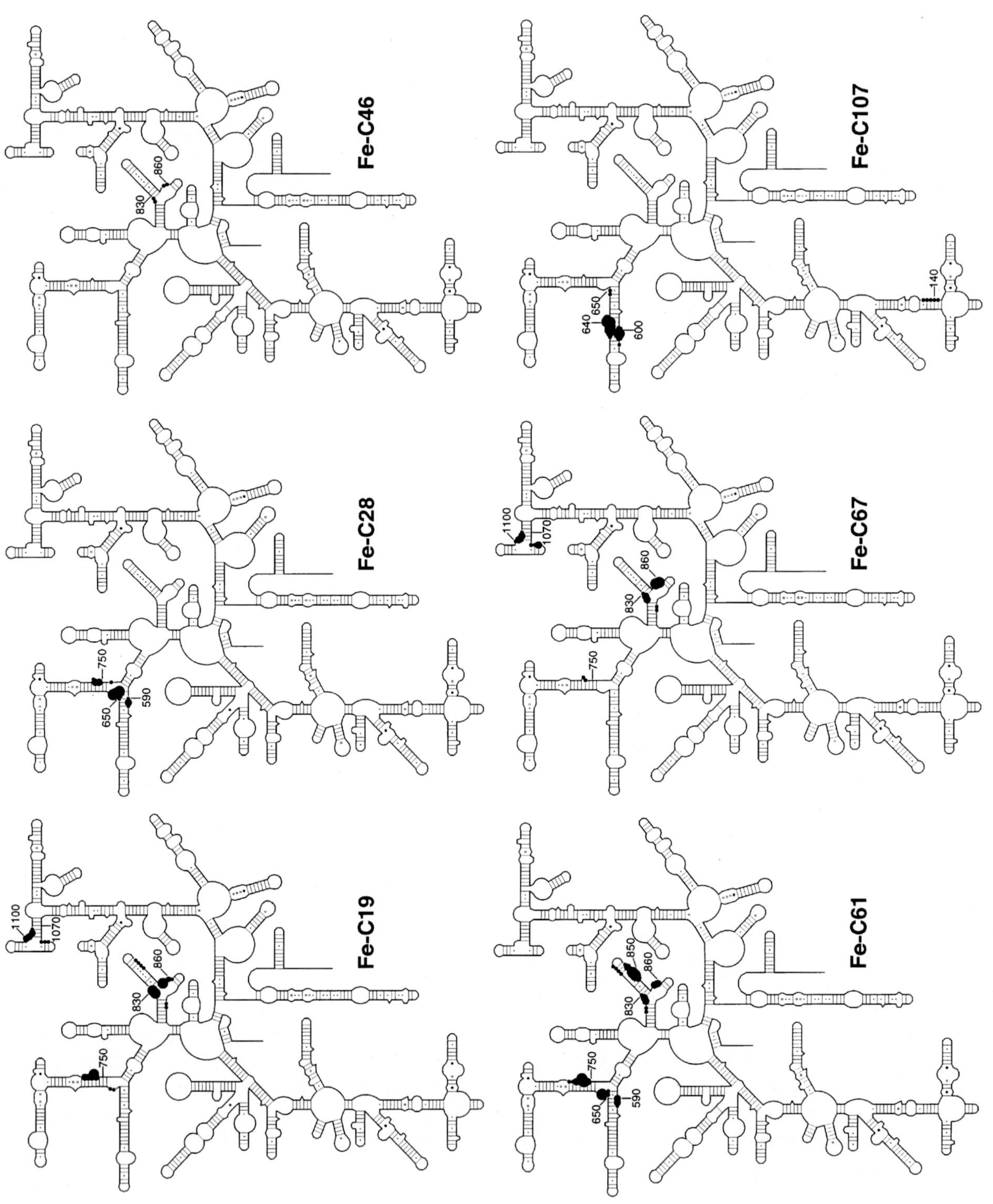
Fe-C46
830
860
Fe-C107
640
650
600
140
Fe-C28
750
650
590
Fe-C67
1100
1070
830
860
750
Fe-C19
1100
1070
830
860
750
Fe-C61
850
860
830
750
650
590

Figure 11. Directed hydroxyl radical probing of 16S rRNA from six different positions on the surface of protein S8 (Lancaster et al., unpubished). The relative strengths of cleavage are indicated by the sizes of the solid circles.

have localized the likely subunit interface contact sites to a few discrete regions in the central and 3′ major domains of 16S rRNA (Fig. 12A) and domains II and IV of 23S rRNA (Fig. 12B). The hydroxyl radical data, in particular, show an interesting tendency for protection of hairpin stem-loops in 23S rRNA, in contrast to apparent protection of the minor-groove faces of internal helical positions in 16S rRNA. These results suggested that some of the subunit interface contacts might be formed by RNA-RNA loop-receptor interactions. As noted above, the sites of subunit protection in the penultimate stem of 16S rRNA indeed appear to be due to 50S contacts with its minor groove at several positions.

Initiation factor IF-3 causes dissociation of 70S ribosomes into 30S and 50S subunits. Earlier studies, using base-specific probes, showed a correlation between bases protected by IF-3 and those protected by subunit association (Muralikrishna and Wickstrom, 1989; Moazed et al., 1995). More recent experiments, with hydroxyl radical footprinting, provide a more comprehensive picture of the potential contact surface between 16S rRNA and IF-3 (Dallas and Noller, unpublished). These studies (Fig. 13) show an extensive overlap between the footprint of IF-3 and that of the 50S subunit, supporting the idea that the dissociation activity of IF-3 is due to competition with the 50S subunit for overlapping binding sites on the 30s subunit.

These RNA chemical-probing studies left unaddressed the possibility that protein-RNA (or protein-protein) interactions might play a role in subunit association. A striking double-helical bridge, reaching from the 50S subunit to the bottom of the platform region of the 30S subunit, was revealed in the electron density map (Fig. 14A). Based on preliminary modeling of 16S rRNA, this region of the 30S subunit was placed in the vicinity of the S15 binding site. Directed-probing experiments, using Fe(II) tethered to S15, showed specific cleavage of the 715 loop of 23S rRNA from positions 12 and 46 (Fig. 15); interestingly, 30S subunits containing S15 derivatized at position 70 with Fe(II)-BABE failed to associate with 50S subunits. When 30S subunits lacking S15 were associated with 50S subunits, protection of the 715 loop by 30S subunits was abolished, although the other 23S protections were unaffected. More remarkably, pure S15 specifically protected the 715 loop when bound to isolated 50S subunits; again, the other sites in 23S rRNA were unaffected. These results provide convincing evidence that the observed intersubunit bridge connects the 715 loop of 23S rRNA with protein S15.

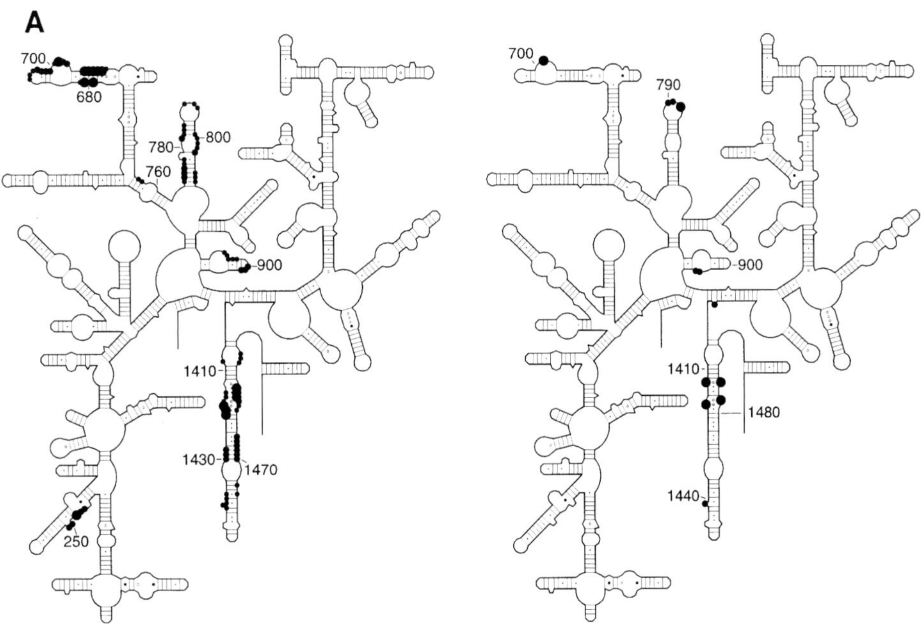

Figure 12. Protection of 16S (A) and 23S (B) rRNAs from hydroxyl radical and base-specific probes (Merryman et al., 1999a, 1999b). The relative strengths of protection are indicated by the sizes of the solid circles. (*Continued on next page.*)

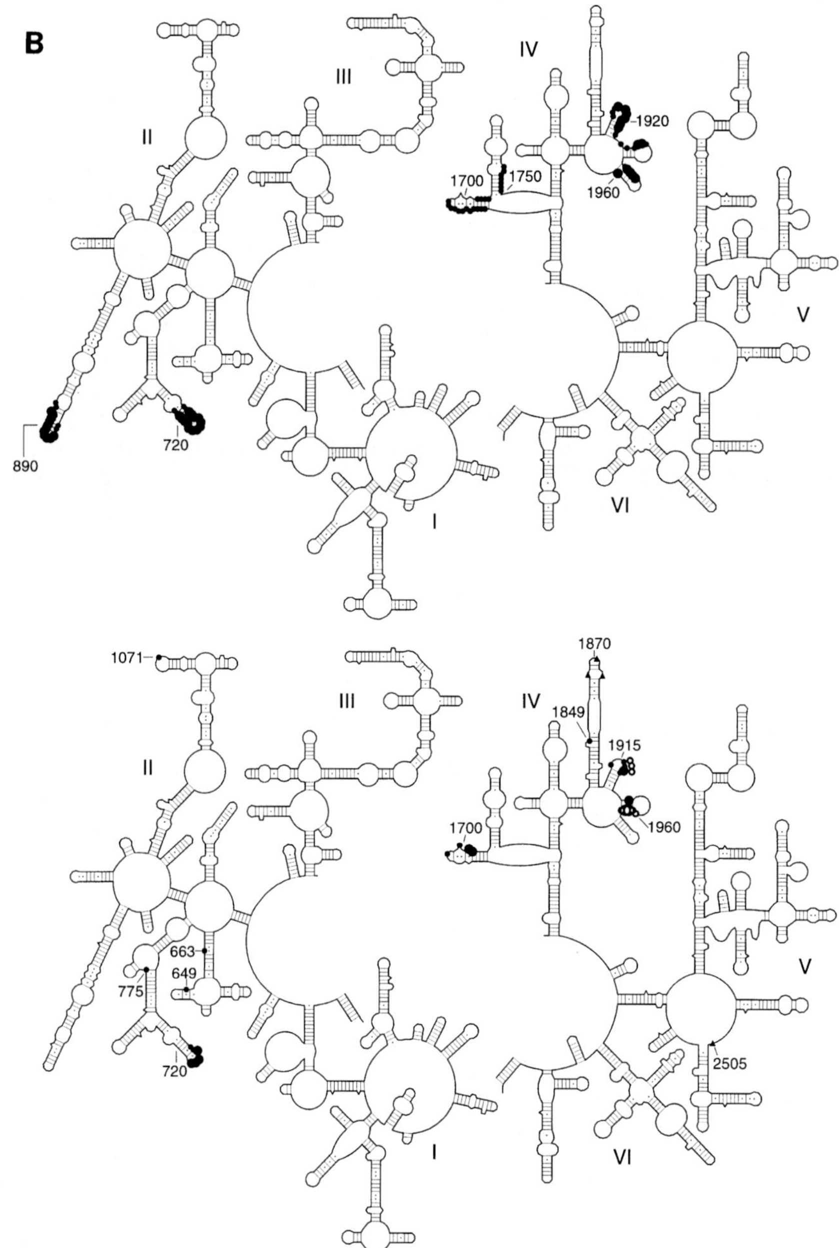

Figure 12.—Continued.

The nucleotide sequence of the 715 loop is identical to that of another RNA loop of considerable biological interest, the U2 snRNA loop IIA (Fig. 16), whose solution structure has been determined by nuclear magnetic resonance spectroscopy (Stallings and Moore, 1997). Figure 14B shows the close fit of the loop IIA structure to the helical bridge in the 7.8-Å electron density map of the ribosome, providing further evidence for the identification of this feature as the 715 loop of 23S rRNA. Shown are A715 and A716, the two most conserved bases in the loop, which are centrally located in the region of contact with the 30S subunit. The precise features of S15 contacted by the 715 loop have yet to be identified.

MOLECULAR INTERACTIONS BETWEEN tRNA, mRNA, AND THE 30S SUBUNIT P SITE

In addition to its interactions with mRNA, tRNA also interacts with the ribosome itself. The 30S subunit P site must not only help to secure the peptidyl-tRNA but must also maintain the correct translational reading frame by stabilizing mRNA-tRNA interac-

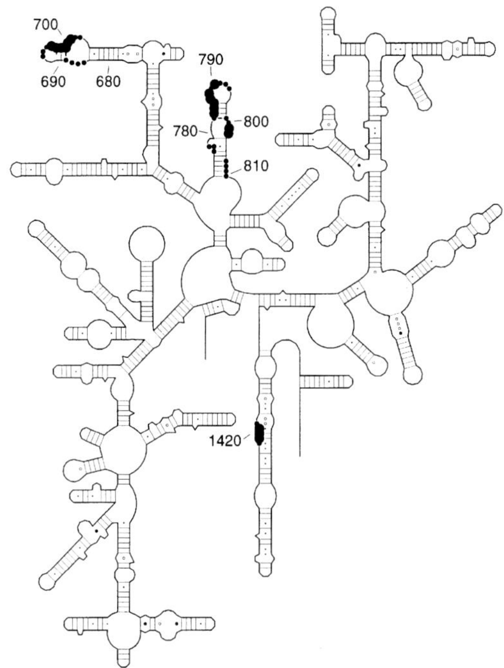

Figure 13. Protection of 16S rRNA in 30S ribosomal subunits from hydroxyl radical probing by initiation factor IF-3 (Dallas and Noller, unpublished). The relative strengths of protection are indicated by the sizes of the solid circles.

tion. Early studies (Noller and Chaires, 1972) showed that modification of a few guanines in the 16S rRNA of the 30S subunit at their N-1 and N-2 positions by kethoxal caused loss of its ability to bind tRNA to the 30S P site [although binding of poly(U) mRNA was unaffected]. Prior binding of tRNA to the P site protected the subunits against inactivation, suggesting that inactivation was caused by modification of the binding site itself rather than by generalized damage to the 30S structure. Reconstitution experiments with combinations of RNA and protein from modified and unmodified subunits showed conclusively that 16S rRNA, not the ribosomal proteins, was the target of inactivation. Later, the tRNA-protected guanines (and other bases) were identified and all but one were found to be conserved in ribosomes throughout the phylogenetic spectrum (Moazed and Noller, 1986, 1990) (Fig. 17). It was further shown that only the ASL of P-tRNA was required for the protection, consistent with the earlier finding that the binding constants for the interaction of ASL and full-length tRNA with the 30S subunit are virtually identical (Rose et al., 1983). These findings pointed to a possible role for 16S rRNA in binding the ASL of tRNA to the 30S subunit P site. Further evidence came from modification-interference experiments, in which active 30S subunits were separated from inactivated ones in a partially chemically modified population, using biotin-derivatized tRNA to capture the active subunits (von Ahsen and Noller, 1995). Modification of a subset of four of the tRNA-protected bases (G926, G966, and G1338 by kethoxal at N-1 and N-2 and G1401 by dimethyl sulfate at N-7) was found to interfere with P-site binding (Fig. 17). Cross-linking (Rinke-Appel et al., 1995; Döring et al., 1994; Osswald et al., 1995) and directed-probing (Joseph et al., 1997) experiments have provided extensive evidence for the physical proximity of several of these RNA elements to the P-site ASL.

In a general approach to mapping the intersubunit cavity of the ribosome, a set of ASL analogues, ranging from 4 to 33 bp in length, was engineered (Joseph et al., 1997). Fe(II) was tethered to the 5′ end of each of the ASLs via a 5′-phosphorothioate group to allow localized generation of hydroxyl radicals from the ends of the different ASLs inside the ribosome. Different cleavage patterns in 16S, 23S, and 5S rRNA were found for ASLs of different lengths and depending on whether they were bound to the A or P sites of the ribosome. A 10- to 12-nucleotide periodicity in the cleavage patterns was often observed, no doubt reflecting the 10- to 12-nucleotide periodicity of the RNA helix, as the ASLs extended into ribosomal space. Based on calibration experiments, the maximum distances between tethering sites and RNA targets were estimated for each cleavage event, and the values for a given target were combined to calculate the allowed volume of space for the location of the RNA target. Clouds for allowed positions of some of the rRNA targets are shown in Fig. 18.

We can now directly visualize the P-site ASL bound to its codon in the ribosome, since its position can be localized unambiguously by using the Fourier difference map of the full-length P-tRNA. The close fit of the ASL model (taken from the crystal structure of free $tRNA^{Phe}$) with the 7.8-Å-resolution ribosome cocrystal map is evident (Fig. 19); the major and minor grooves of the ASL helix are readily distinguishable. The codon was modeled with the tRNA Asp crystal structure, in which two tRNAs pair with each other via anticodon-anticodon triplet interactions; in our model, the tRNA Asp anticodon was replaced with the UUU codon that is present in the P site of our cocrystal; density corresponding to the codon can be seen at the predicted position (Fig. 19A). There are three regions (a to c) where electron density from the 30S ribosomal subunit merges with that of the ASL stem and three regions (d to f) where it merges with that of the codon-anticodon interaction. Contacts with the stem appear to be via the sugar-phosphate backbone, on the minor-groove side of the backbone ridge. Interaction e with the mRNA codon appears also to be a backbone contact, but interaction f has the appearance of a possible coaxial stacking

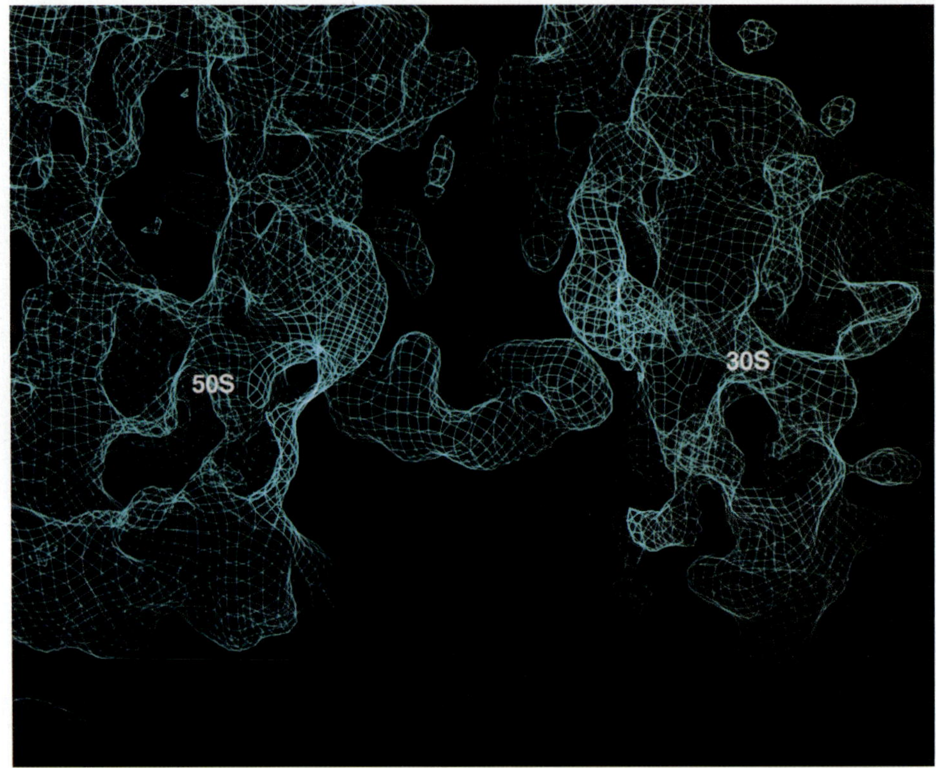

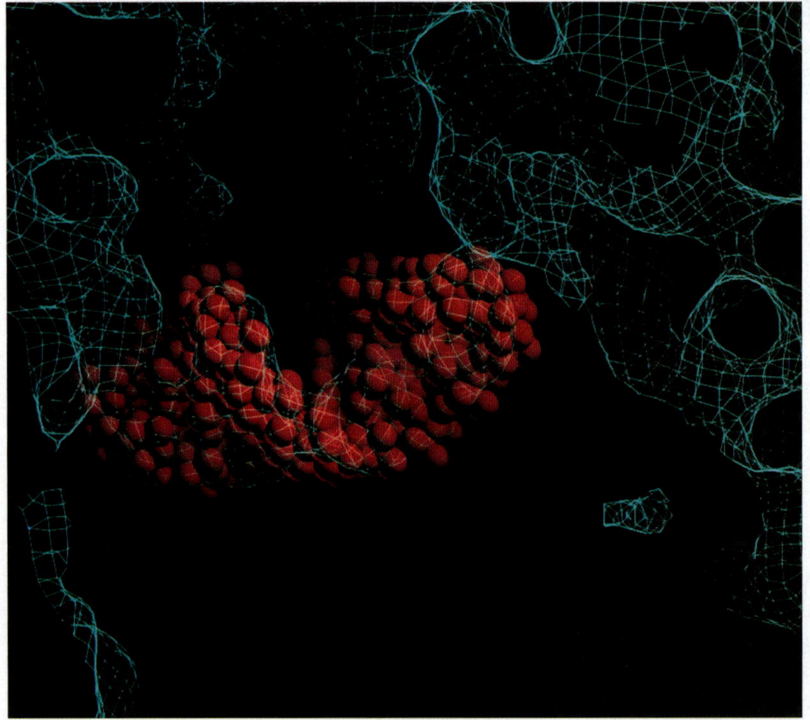

Figure 14. (A) View of the subunit interface region of the 7.8-Å electron density map of the 70S ribosome, showing an RNA helical bridge extending from the 50S subunit to the bottom of the platform of the 30S subunit. (B) Fitting the solution structure of the U2 snRNA loop IIA (Stallings and Moore, 1997) to the electron density of the intersubunit bridge. The loop region is at the right, in contact with the 30S subunit.

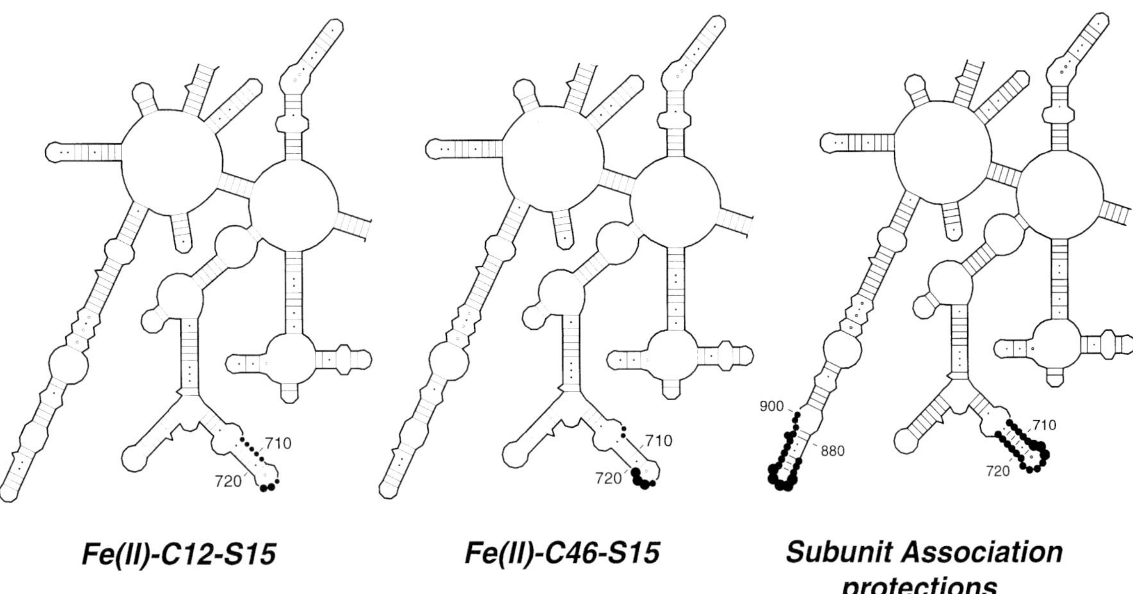

Figure 15. (Left and center) Cleavage targets in 23S rRNA after directed hydroxyl radical probing from positions 12 and 46 of protein S15 (Culver and Noller, unpublished). (Right) Protection of 23S rRNA from free hydroxyl radicals by association of 30S and 50S subunits (Merryman et al., 1999b). The relative strengths of cleavage and protection are indicated by the sizes of the solid circles.

interaction at the wobble pair end of the codon-anticodon helix. It is not yet possible to make unambiguous identifications of the elements of the 30S subunit that participate in these interactions.

The overall arrangment of the P-site ribosome-ASL interactions resembles an elaborate clamp, in which contacts a, b, and c serve to rigidly position the ASL stem while contacts d and e appear to hold the first 2 bp of the codon-anticodon triplet together, interacting with the codon and anticodon backbones, respectively, from opposite directions. A possible role for contact f may be to stabilize the weakest element of the codon-anticodon triplet, the wobble pair, by coaxial stacking. None of the contacts, with the possible exception of f, appear to interact with the bases of either the ASL or the codon; this allows exactly the same interactions to be made between the ribosomal P site and all possible tRNA-codon combinations, independent of their respective mRNA and tRNA sequences. Most, if not all, of the contacts appear to be to the sugar-phosphate backbone, which is studded with hydrogen bond acceptors and donors; indeed, one interesting possibility would be to find hydrogen-bonded contacts between phosphate oxygens of the ASL-codon complex and the N-1 positions of guanines identified in the chemical-probing experiments described above.

MOLECULAR INTERACTIONS BETWEEN THE ACCEPTOR END OF tRNA AND THE PT CENTER

One of the most fundamental molecular processes in biology is catalysis of peptide bond forma-

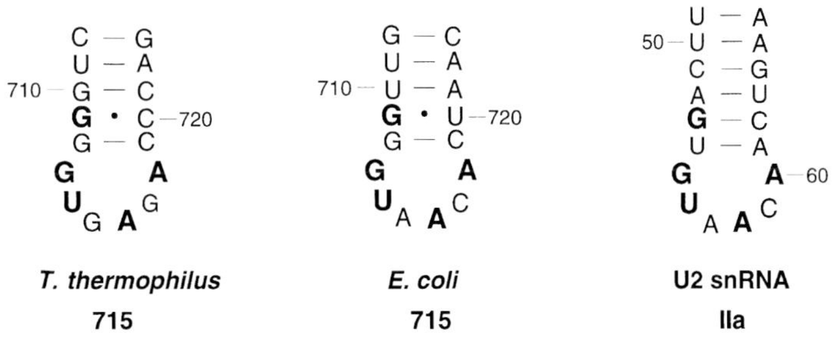

C — G
U — A
710 — G — C
G • C — 720
G — C
G **A**
U G **A** G

T. thermophilus
715

G — C
U — A
710 — U — A
G • U — 720
G — C
G **A**
U A **A** C

E. coli
715

U — A
50 — U — A
C — G
A — U
G — C
U — A
G **A** — 60
U A **A** C

U2 snRNA
IIa

Figure 16. Similarity between sequences of the 715 loop of 23S rRNA and the U2 snRNA loop IIA. Nucleotides in bold are conserved in all three loop structures.

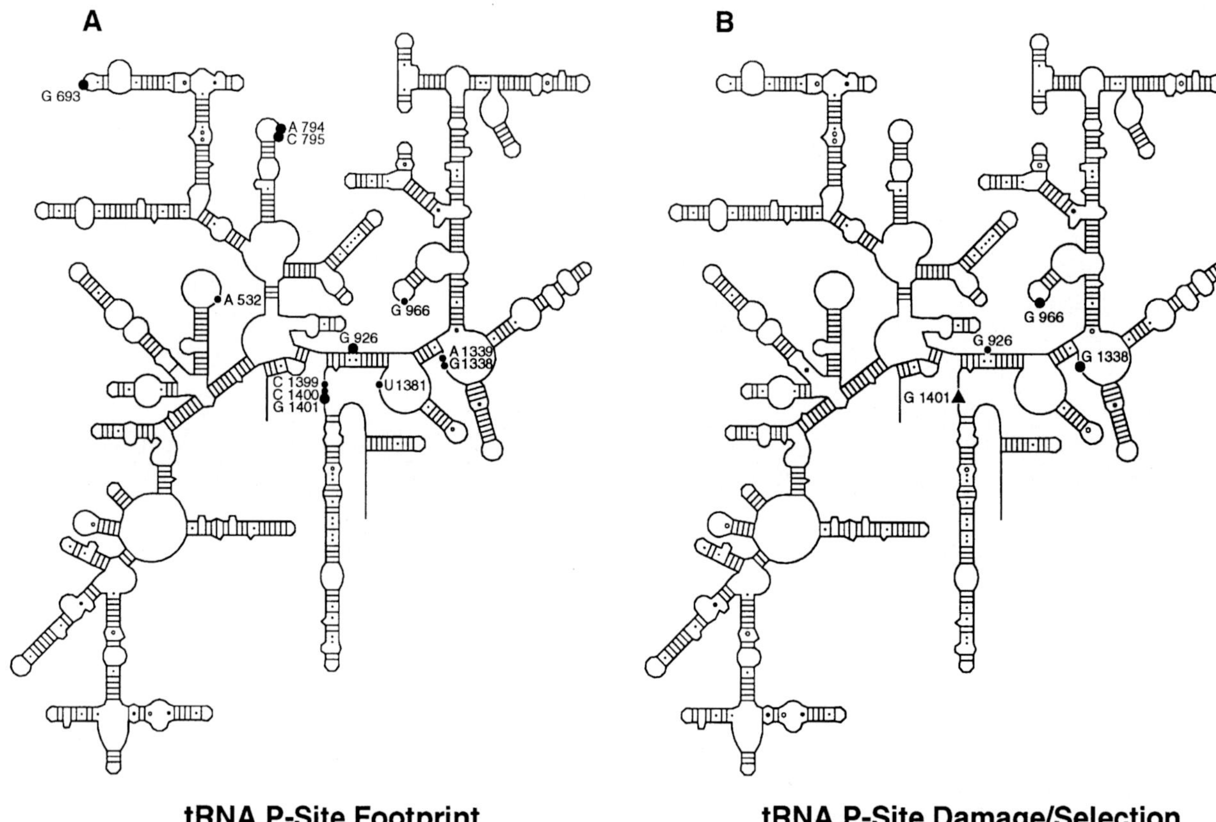

Figure 17. Nucleotides in 16S rRNA protected from base-specific probes by P-site tRNA (Moazed and Noller, 1986, 1990) (A) and those whose modification interferes with P-site tRNA binding (von Ahsen and Noller, 1995) (B). Solid circles and triangle are (A) protections or (B) selected against for tRNA binding. The triangle indicates N7 position.

tion—the PT reaction. Although PT has been known to be an integral part of the structure of the 50S ribosomal subunit for more than 3 decades (Maden et al., 1968), it has yet to be shown whether it is made of RNA or protein (or both). A steadily mounting tide of evidence seems to favor the RNA possibility, although a notable lack of passion among most contemporary ribosomologists for the protein possibility may be at least partly responsible for this favoritism. At present count, the only macromolecular species that have not been excluded as being essential for the PT reaction are 23S rRNA and proteins L2 and L3 (Khaitovich et al., 1999). The main obstacle is the difficulty in separating 23S rRNA from a group of core proteins which appear to be essential for maintaining its three-dimensional folding (Khaitovich et al., 1999). The final answer to this question may have to come from a high-resolution X-ray structure of the 50S subunit, when it may be solved by inspection.

Meanwhile, it is becoming ever clearer that the CCA-terminal moieties of aminoacyl- and peptidyl-tRNAs, which can be considered to be the substrates of this enzymatic reaction, are bound to the ribosome in a site that is richly populated by universally conserved elements of 23S rRNA. The first indications came from cross-linking studies with peptidyl-tRNA analogues that were derivatized with reactive groups on their aminoacyl moieties; although proteins L2 and L27 were frequently identified, by the mid-1970s cross-links to 23S rRNA began to be found. A particularly convincing finding was the cross-link obtained by Barta et al. (1984) using Phe-tRNA derivatized with a photochemically reactive benzophenone group. This reagent cross-linked 23S rRNA with an unusually high (ca. 50%) yield, almost exclusively at position 2451 in the universally conserved central loop of domain V. Most importantly, it was shown that the covalently cross-linked tRNA retained

Figure 18. Calculated allowed positions of targets of directed hydroxyl radical probing from Fe(II) tethered to ribosome-bound ASLs ranging in length from 4 to 33 bp (Joseph et al., 1997). Colored clouds are shown for nucleotide targets in 16S rRNA (A and B), 23S rRNA (C and D), and 5S rRNA (E), relative to P-site tRNA (left) and A-site tRNA (right).

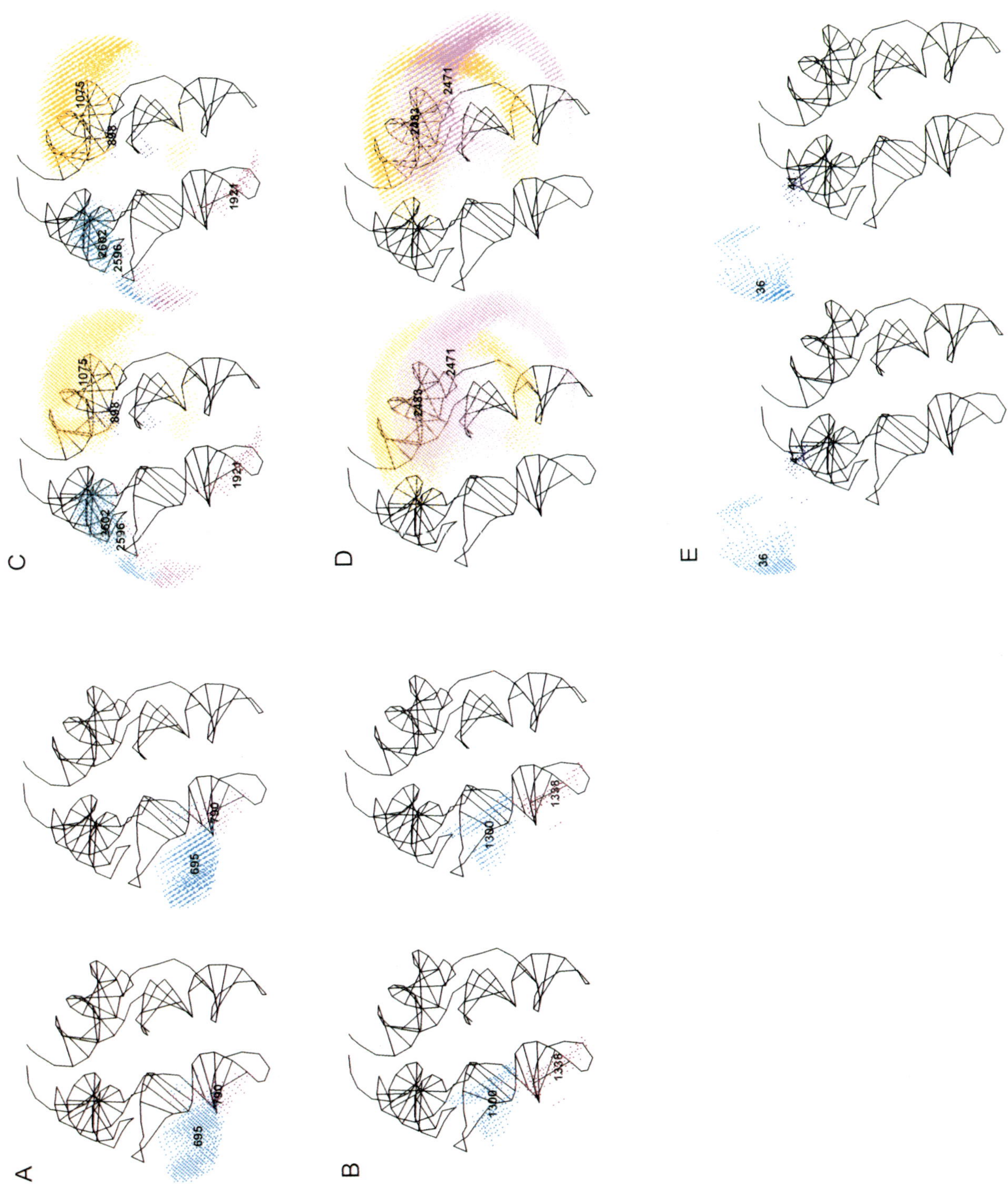
A
B
C
D
E
695
730
1300
1338
1075
2602
2596
2471
36

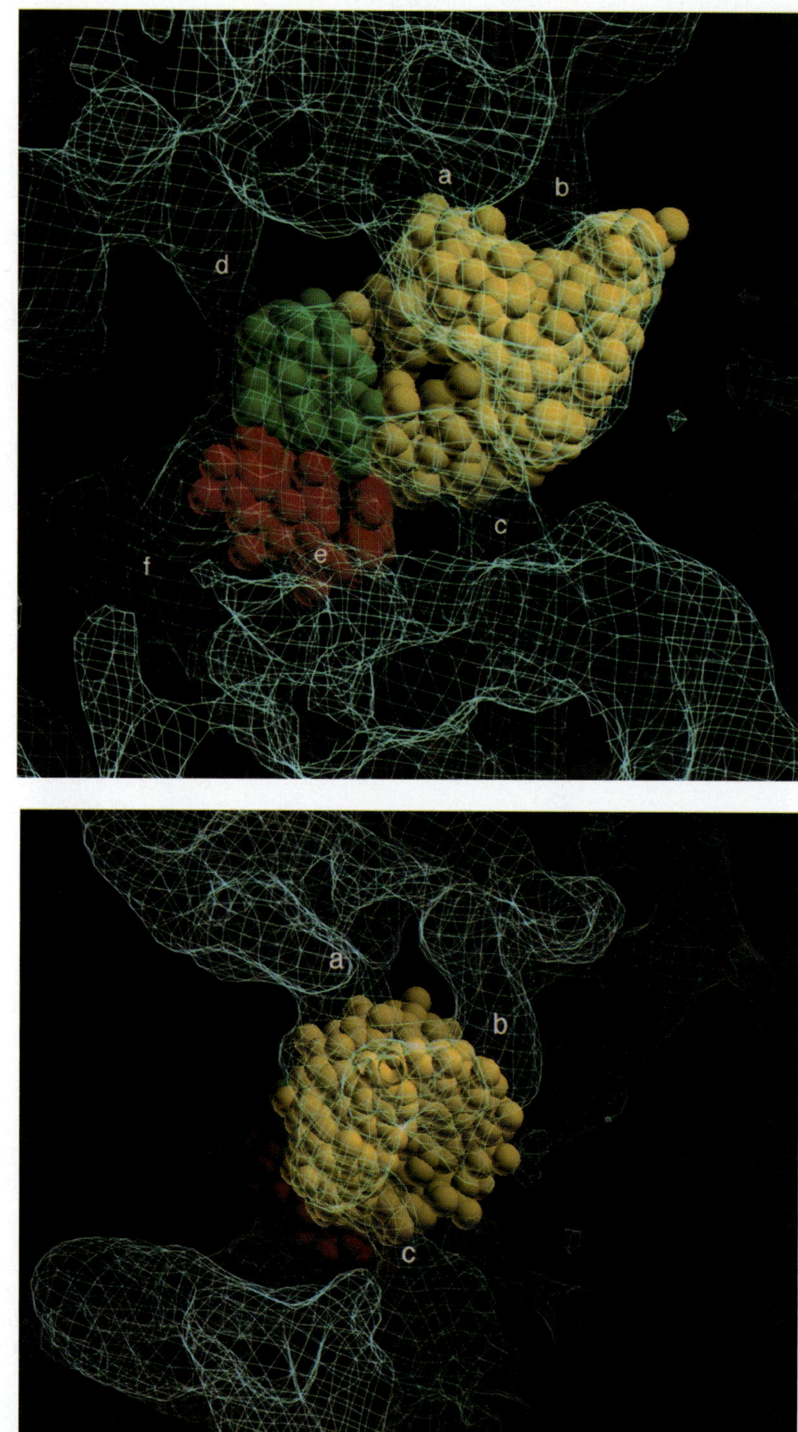

Figure 19. (A) View of the P-site ASL bound to the ribosome in the 7.8-Å electron density map, from the right-hand side (subunit interface at the right). The ASL is shown in yellow, with its anticodon in green; the P-site codon is red. (B) Axial view of the P-site ASL, viewed from the 50S subunit. a, b, and c, bridges to the ASL stem from the 30S subunit; d and e, bridges to the anticodon and codon, respectively; f, an apparent stacking interaction between the 30S subunit and the wobble base pair.

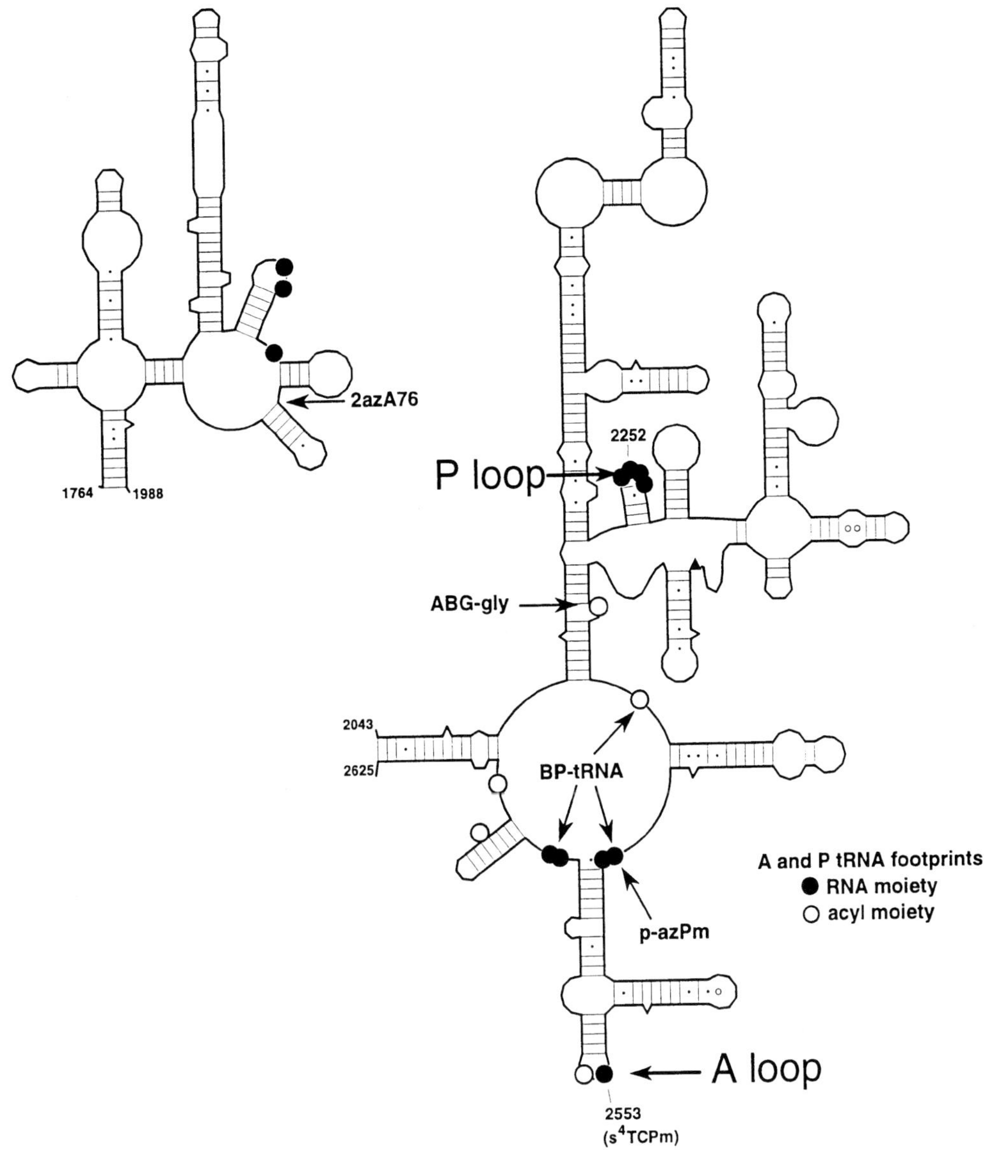

Figure 20. Elements of 23S rRNA associated with the PT function of the ribosome. Closed and open circles indicate bases protected by the acceptor ends of A- and P-site tRNAs (Moazed and Noller, 1989b, 1991). Several sites cross-linked by the acceptor ends of A- and P-site tRNAs are indicated by small arrows (Wower et al., 1989; Mitchell et al., 1993; Hall et al., 1988; Steiner et al., 1988). Features that base-pair with the C74 of P-tRNA (Samaha et al., 1995) and C75 of A-tRNA (Green et al., 1998; Kim and Green, in press) are indicated as the P loop and A loop, respectively.

its activity in the PT reaction, showing that 23S rRNA was indeed in close proximity to the catalytic center.

More recently, it was established that a Watson-Crick base pair is formed between C74, in the conserved CCA terminus of peptidyl-tRNA, and G2252, also in domain V of 23S rRNA (Samaha et al., 1995). G2252 had been shown to be protected by binding of peptidyl-tRNA or by the CCA end of peptidyl-tRNA to the 50S subunit P site; it had also been shown that protection of G2252 from kethoxal attack depended on the presence of both Cs in the CCA sequence of P-tRNA (Moazed and Noller, 1989b, 1991). Formation of the G2252-C74 base pair was shown to be critical not only for binding of the CCA end of the tRNA substrate to the P site but also for PT activity. This was the first demonstration that 23S rRNA is directly involved in a translational function and was the second example of a universally conserved element of 23S rRNA (the "P loop" [Fig. 20]) proven to be located at the PT center. No Watson-Crick interactions involving the other bases in the P-tRNA CCA sequence have been found so far.

Figure 21. Fourier difference map of P-tRNA (yellow) superimposed on 7.8-Å electron density map of the *T. thermophilus* 70S ribosome in the PT region of the 50S subunit (green), showing the pinching of the CCA tail of the tRNA by features of the 50S subunit.

Using a novel synthetic A-site substrate, s4TpCpPuro (4-thio-thymidylyl-cytidylyl-puromycin), a high-yield photochemical cross-link was formed with the 50S subunit, exclusively with 23S rRNA, and at a single position, G2553–again, a universally conserved base and again in domain V (Green et al., 1998). The covalently cross-linked substrate was found to be fully reactive in the PT reaction, placing yet another conserved element of 23S rRNA—the "A loop" (Fig. 20) at the catalytic center, this time in the A site. In vitro suppression experiments by Kim and Green (in press) have very recently obtained evidence for a restricted base-base interaction between C75 of A-tRNA and G2553 of 23S rRNA. Thus, the CCA ends of both the P-tRNA and A-tRNA substrates interact directly with 23S rRNA, and the aminoacyl moiety of P-tRNA must also be in close proximity to 23S rRNA. It can reasonably be asked, in view of these findings, whether there remains sufficient space in the PT site for protein.

Again using the Fourier difference map of $tRNA_f^{Met}$ bound to the P site as a guide, we have located the position of its CCA end in the 50S subunit (Fig. 21). The crystal structure of the free tRNA conforms closely to the difference density up until the CCA tail, which in the ribosome cocrystal map continues in a straight line parallel to the axis of the acceptor arm rather than following the quasihelical path of the free-tRNA structure. The CCA tail lies along the side of a channel that continues all the way through the 50S subunit, corresponding to the presumed polypeptide exit channel observed previously (Yonath et al., 1987). There is a pronounced, almost spherical blob of density corresponding to the expected position of C74 of the bound tRNA, suggesting that this element of the P-tRNA CCA tail is especially well ordered. The CCA tail is pinched by a constriction in the PT site, at the density predicted to correspond to C74. Thus, the CCA end of P-tRNA appears to be clamped between two structural elements of the 50S subunit, one of which is likely to be the P loop of 23S rRNA. Further studies will be needed for definitive identification of these functionally important ribosomal features.

FUTURE DIRECTIONS

Ribosomology is presently entering a new era in which investigators will be able to see the molecular details of the translational apparatus at increasingly higher resolution. The transition from a "black box" era into the brave new world of concrete molecular interactions promises to be a rich and exciting one. However, there will soon be a further stage (perhaps we are already there) at which we will recognize that even a high-resolution structure of the ribosome will only represent a still from a movie. Ultimately, we will want the movie.

We thank Chuck Wilson for his help, advice, and encouragement during the early stages of the ribosome crystallography project and Django Sussman and Jay Nix for help with data collection. We also thank Claude Meares and members of his group for their generous advice on synthesis and the use of BABE.

These studies were supported by grants (to H.F.N.) from the NIH, NSF, HFSP, and (to the Center for Molecular Biology of RNA) the Lucille P. Markey Charitable Trust.

REFERENCES

Agrawal, R. K., P. Penczek, R. A. Grassucci, and J. Frank. 1998. Visualization of elongation factor G on the Escherichia coli 70S ribosome: the mechanism of translocation. *Proc. Natl. Acad. Sci. USA* 95:6134–6138.

Allen, G., R. Capasso, and C. Gualerzi. 1979. Identification of the amino acid residues of proteins S5 and S8 adjacent to each other in the 30 S ribosomal subunit of Escherichia coli. *J. Biol. Chem.* 254:9800–9806.

Ban, N., B. Freeborn, P. Nissen, P. Penczek, R. A. Grassucci, R. Sweet, J. Frank, P. B. Moore, and T. A. Steitz. 1998. A 9Å resolution X-ray crystallographic map of the large ribosomal subunit. *Cell* 93:1105–1115.

Ban, N., P. Nissen, J. Hansen, M.Capel, P. B. Moore, and T. A. Steitz. 1999. Placement of protein and RNA structures into a 5

Å-resolution map of the 50S ribosomal subunit. *Nature* **400:** 841–847.

Barta, A, G. Steiner, J. Brosius, H. F. Noller, and E. Kuechler. 1984. Identification of a site on 23S ribosomal RNA located at the peptidyl transferase center. *Proc. Natl. Acad. Sci. USA* **81:** 3607–3611.

Borowski, C., M. V. Rodnina, and W. Wintermeyer. 1996. Truncated elongation factor G lacking the G domain promotes translocation of the 3′ end but not of the anticodon domain of peptidyl-tRNA. *Proc. Natl. Acad. Sci. USA* **93:**4202–4206.

Brünger, A. T., P. D. Adams, G. M. Clore, W. L. DeLano, P. Gros, R. W. Grosse-Kunstleve, J. S. Jiang, J. Kuszewski, M. Nilges, N. S. Pannu, R. J. Read, L. M. Rice, T. Simonson, and G. L. Warren. 1998. Crystallography and NMR system: a new software suite for macromolecular structure determination. *Acta Crystallogr.* **D54:**905–921.

Capel, M. S., D. M. Engelman, B. R. Freeborn, M. Kjeldgaard, J. A. Langer, V. Ramakrishnan, D. G. Schindler, D. K. Schneider, B. P. Schoenborn, I. Y. Sillers, S. Yabuki, and P. B. Moore, 1987. A complete mapping of the proteins in the small ribosomal subunit of Escherichia coli. *Science* **238:**1403–1406.

Cate, J. H., M. M. Yusupov, G. Z. Yusupova, T. N. Earnest, and H. F. Noller. 1999. X-ray crystal structures of 70S ribosome functional complexes. *Science* **285:**2095–2104.

Chapman, N. M., and H. F. Noller. 1977. Protection of specific sites in 16S RNA from chemical modification by association of 30S and 50S ribosomes. *J. Mol. Biol.* **109:**131–149.

Clemons, W. M., Jr., J. L. May, B. T. Wimberly, J. P. McCutcheon, M. S. Capel, and V. Ramakrishnan. 1999. Structure of a bacterial 30S ribosomal subunit at 5.5 Å resolution. *Nature* **400:** 833–840.

Culver, G. M., and H. F. Noller. 1998. Directed hydroxyl radical probing of 16S ribosomal RNA in ribosomes containing Fe(II) tethered to ribosomal protein S20. *RNA* **4:**1471–1480.

Culver, G. M., and H. F. Noller. 1999. Efficient reconstitution of functional *Escherichia coli* 30S ribosomal subunits from a complete set of recombinant small subunit ribosomal proteins. *RNA* **5:**832–834.

Culver, G. M., G. M. Heilek, and H. F. Noller. 1998. Probing the rRNA environment of ribosomal protein S5 across the subunit interface and inside the 30 S subunit using tethered Fe(II). *J. Mol. Biol.* **286:**355–364.

Czworkowski, J., J. Wang, T. A. Steitz, and P. B. Moore. 1994. The crystal structure of elongation factor G complexed with GDP, at 2.7 Å resolution. *EMBO J.* **13:**3661–3668.

Dallas, A., and H. Noller. Unpublished data.

Davies, C., V. Ramakrishnan, and S. W. White. 1996. Structural evidence for specific S8-RNA and S8-protein interactions within the 30S ribosomal subunit: ribosomal protein S8 from Bacillus stearothermophilus at 1.9 Å resolution. *Structure* **4:**1093–1104.

DeRiemer, L. H., C. H. Meares, D. A. Goodwin, and C. J. Diamanti. 1981. BLEOTA II: synthesis of a new tumor-visualizing derivative of Co(III)-bleomycin. *J. Labelled Compd. Radiopharm.* **18:**1517–1534.

Döring, T., P. Mitchell, M. Osswald, D. Bochkariov, and R. Brimacombe. 1994. The decoding region of 16S RNA; a cross-linking study of the ribosomal A, P and E sites using tRNA derivatized at position 32 in the anticodon loop. *EMBO J.* **13:** 2677–2685.

Fink, D. L., R. O. Chen, H. F. Noller, and R. B. Altman. 1996. Computational methods for defining the allowed conformational space of 16S rRNA based on chemical footprinting data. *RNA* **2:**851–866.

Gavrilova, L. P., and A. S. Spirin. 1974. "Nonenzymatic" translation. *Methods Enzymol.* **30:**452–462.

Green, R., C. Switzer, and H. F. Noller. 1998. Ribosome-catalyzed peptide-bond formation with an A-site substrate covalently linked to 23S rRNA. *Science* **280:**286–289.

Hall, C. C., D. Johnson, and B. S. Cooperman. 1988. [3H]-p-azidopuromycin photoaffinity labeling of Escherichia coli ribosomes: evidence for site-specific interaction at U-2504 and G-2502 in domain V of 23S ribosomal RNA. *Biochemistry* **27:** 3983–3990.

Heilek G., and H. F. Noller. 1996a. Site-directed hydroxyl radical probing of the rRNA neighborhood of ribosomal protein S5. *Science* **272:**1659–1662.

Heilek, G. M., and H. F. Noller. 1996b. Directed hydroxyl radical probing of the rRNA neighborhood of ribosomal protein S13 using tethered Fe(II). *RNA* **2:**597–602.

Heilek, G., R. Marusek, C. F. Meares, and H. F. Noller. 1995. Directed hydroxyl radical probing of 16S rRNA using Fe(II) tethered to ribosomal protein S4. *Proc. Natl. Acad. Sci. USA* **92:** 1113–1116.

Held, W. A., B. Ballou, S. Mizushima, and M. Nomura. 1974. Assembly mapping of 30S ribosomal proteins from Escherichia coli. Further studies. *J. Biol. Chem.* **249:**3103–3111.

Herr, W., and H. F. Noller. 1978. Nucleotide sequences of accessible regions of 23S RNA in 50S ribosomal subunits. *Biochemistry* **17:**307–315.

Herr, W., N. M. Chapman, and H. F. Noller. 1979. Mechanism of ribosomal subunit association: discrimination of specific sites in 16S RNA essential for association activity. *J. Mol. Biol.* **130:** 433–449.

Holmberg, L., and H. F. Noller. 1999. Mapping the ribosomal RNA neighborhood of protein L11 by directed hydroxyl radical probing. *J. Mol. Biol.* **289:**223–233.

Holmberg, L., and H. F. Noller. Unpublished data.

Joseph, S., B. Weiser, and H. F. Noller. 1997. Mapping the inside of the ribosome using an RNA helical ruler. *Science* **270:**1093–1098.

Khaitovich, P., A. S. Mankin, R. Green, L. Lancaster, and H. F. Noller. 1999. Characterization of functionally active subribosomal particles from Thermus aquaticus. *Proc. Natl. Acad. Sci. USA* **96:**85–90.

Kim, D. F., and R. Green. Base-pairing between 23S rRNA and tRNA in the ribosomal A site. *Mol. Cell*, in press.

Lancaster, L., G. Culver, and H. F. Noller. Unpublished data.

Lieberman, K. R., and H. F. Noller. 1998. Ribosomal protein L15 as a probe of 50 S ribosomal subunit structure. *J. Mol. Biol.* **284:** 1367–1378.

Lieberman, K. R., and H. F. Noller. The 23S rRNA environment of ribosomal protein L9 in the 50S ribosomal subunit, submitted for publication.

Maden, B. E., R. R. Traut, and R. E. Monro. 1968. Ribosome-catalysed peptidyl transfer: the polyphenylalanine system. *J. Mol. Biol.* **35:**333–345.

Malhotra, A., P. Penczek, R. K. Agrawal, I. S. Gabashvili, R. A. Grassucci, R. Junemann, N. Burkhardt, K. H. Nierhaus, and J. Frank. 1998. Escherichia coli 70 S ribosome at 15 Å resolution by cryo-electron microscopy: localization of fMet-$tRNA_f^{Met}$ and fitting of L1 protein. *J. Mol. Biol.* **280:**103–116.

Merryman, C., D. Moazed, J. McWhirter, and H. F. Noller. 1999a. Nucleotides in 16S rRNA protected by association of 30S and 50S ribosomal subunits. *J. Mol. Biol.* **285:**97–103.

Merryman, C., D. Moazed, G. Daubresse, and H. F. Noller. 1999b. Nucleotides in 23S rRNA protected by association of 30S and 50S ribosomal subunits. *J. Mol. Biol.* **285:**105–113.

Mitchell, P., K. Stade, M. Osswald, and R. Brimacombe. 1993. Site-directed cross-linking studies on the E. coli tRNA-ribosome complex: determination of sites labelled with an aromatic azide attached to the variable loop or aminoacyl group of tRNA. *Nucleic Acids Res.* **21:**887–896.

Moazed, D., and H. F. Noller. 1986. Transfer RNA shields specific nucleotides in 16S ribosomal RNA from attack by chemical probes. *Cell* **47:**985–994.

Moazed, D., and H. F. Noller. 1989a. Intermediate states in the movement of transfer RNA in the ribosome. *Nature* **342:**142–148.

Moazed, D., and H. F. Noller. 1989b. Interaction of tRNA with 23S rRNA in the ribosomal A, P, and E sites. *Cell* **57:**585–597.

Moazed, D., and H. F. Noller. 1990. Binding of tRNA to the ribosomal A and P sites protects two distinct sets of nucleotides in 16S rRNA. *J. Mol. Biol.* **211:**135–145.

Moazed, D., and H. F. Noller. 1991. Sites of interaction of the CCA end of peptidyl-tRNA with 23S rRNA. *Proc. Natl. Acad. Sci. USA* **88:**3725–3728.

Moazed, D., R. Samaha, C. Gualerzi, and H. F. Noller. 1995. Specific protection of 16S rRNA by translational initiation factors. *J. Mol. Biol.* **248:**207–210.

Muralikrishna, P., and E. Wickstrom. 1989. Escherichia coli initiation factor 3 protein binding to 30S ribosomal subunits alters the accessibility of nucleotides within the conserved central region of 16S rRNA. *Biochemistry* **28:**7505–7510.

Nevskaya, N., S. Tishchenko, A. Nikulin, S. Al-Karadaghi, A. Liljas, B. Ehresmann, C. Ehresmann, M. Garber, and S. Nikonov. 1998. Crystal structure of ribosomal protein S8 from Thermus thermophilus reveals a high degree of structural conservation of a specific RNA binding site. *J. Mol. Biol.* **279:**233–244.

Noller, H. F., and J. B. Chaires. 1972. Functional modification of 16S ribosomal RNA by kethoxal. *Proc. Natl. Acad. Sci. USA* **69:** 3115–3118.

Odom, O. W., and B. Hardesty. 1987. An apparent conformational change in tRNA(Phe) that is associated with the peptidyl transferase reaction. *Biochimie* **69:**925–938.

Osswald, M., B. Greuer, R. Brimacombe, G. Stöffler, H. Baumert, and H. Fasold. 1987. RNA-protein cross-linking in Escherichia coli 30S ribosomal subunits; determination of sites on 16S RNA that are cross-linked to proteins S3, S4, S5, S7, S8, S9, S11, S13, S19 and S21 by treatment with methyl p-azidophenyl acetimidate. *Nucleic Acids Res.* **15:**3221–3240.

Osswald, M., T. Döring, and R. Brimacombe. 1995. The ribosomal neighbourhood of the central fold of tRNA: cross-links from position 47 of tRNA located at the A, P or E site. *Nucleic Acids Res.* **23:**4635–4641.

Otwinowski, Z. 1991. MLPHARE, a maximum-likelihood phase refinement program. *In* W. Wolf, P. R. Evans, and A. G. W. Leslie (ed.), *Isomorphous Replacement and Anomalous Scattering.* SERC Daresbury Laboratory, Warrington, United Kingdom.

Penczek, P. A., R. A. Grassucci, and J. Frank. 1994. The ribosome at improved resolution: new techniques for merging and orientation refinement in 3D cryo-electron microscopy of biological particles. *Ultramicroscopy* **53:**251–270.

Powers, T., and H. F. Noller. 1995. Hydroxyl radical footprinting of ribosomal proteins on 16S rRNA. *RNA* **1:**194–209.

Rinke-Appel, J., N. Jünke, M. Osswald, and R. Brimacombe. 1995. The ribosomal environment of tRNA: crosslinks to rRNA from positions 8 and 20:1 in the central fold of tRNA located at the A, P, or E site. *RNA* **1:**1018–1028.

Rodnina, M. V., T. Pape, R. Fricke, L. Kuhn, and W. Wintermeyer. 1996. Initial binding of the elongation factor Tu.GTP.aminoacyl-tRNA complex preceding codon recognition on the ribosome. *J. Biol. Chem.* **271:**646–652.

Rodnina, M. V., A. Savelsbergh, V. I., Katunin, and W. Wintermeyer. 1997. Hydrolysis of GTP by elongation factor G drives tRNA movement on the ribosome. *Nature* **385:**37–41.

Rose, S. J., P. T. Lowary, and O. C. Uhlenbeck. 1983. Binding of yeast $tRNA^{Phe}$ anticodon arm to Escherichia coli 30 S ribosomes. *J. Mol. Biol.* **167:**103–117.

Samaha, R. R., R. Green, and H. F. Noller. 1995. A base pair between tRNA and 23S rRNA in the peptidyl transferase centre. *Nature* **377:**309–314.

Santer, M., and S. Shane. 1977. Area of 16S ribonucleic acid at or near the interface between 30S and 50S ribosomes of Escherichia coli. *J. Bacteriol.* **130:**900–910.

Spirin, A. S. 1969. A model of the functioning ribosome: locking and unlocking of the ribosome subparticles. *Cold Spring Harbor Symp. Quant. Biol.* **34:**197–207.

Stallings, S. C., and P. B. Moore. 1997. The structure of an essential splicing element: stem loop Iia from yeast U2 snRNA. *Structure* **5:**1173–1185.

Stark, H., M. V. Rodnina, J. Rinke-Appel, R. Brimacombe, W. Wintermeyer, and M. van Heel. 1997. Visualization of elongation factor Tu on the Escherichia coli ribosome. *Nature* **389:** 403–406.

Steiner, G., E. Kuechler, and A. Barta. 1988. Photo-affinity labelling at the peptidyl transferase centre reveals two different positions for the A- and P-sites in domain V of 23S rRNA. *EMBO J* **7:**3949–3955.

Stern, S., D. Moazed, and H. F. Noller. 1988a. Structural analysis of RNA structure using chemical and enzymatic probing monitored by primer extension. *Methods Enzymol.* **164:**481–489.

Stern, S., B. Weiser, and H. F. Noller. 1988b. Model for the three-dimensional folding of 16S ribosomal RNA. *J. Mol. Biol.* **204:** 447–481.

Stern, S., T. Powers, L.-M. Changchien, and H. F. Noller. 1989. RNA-protein interactions in 30S ribosomal subunits: folding and function of 16S rRNA. *Science* **244:**783–790.

Svensson, P., L.-M. Changchien, G. R. Craven, and H. F. Noller. 1988. Interaction of ribosomal proteins S6, S8, S15 and S18 with the central domain of 16S ribosomal RNA. *J. Mol. Biol.* **200:**301–308.

Trakhanov, S., M. Yusupov, V. Shirokov, M. Garber, A. Mitschler, M. Ruff, J. C. Thierry, and D. Moras. 1989. Preliminary X-ray investigation of 70 S ribosome crystals from Thermus thermophilus. *J. Mol. Biol.* **209:**327–328.

Urlaub, H., B. Thiede, E. C. Muller, R. Brimacombe, and B. Wittmann-Liebold. 1997. Identification and sequence analysis of contact sites between ribosomal proteins and rRNA in Escherichia coli 30S subunits by a new approach using matrix-assisted laser desorption/ionization-mass spectrometry combined with N-terminal microsequencing. *J. Biol. Chem.* **272:** 14547–14555.

von Ahsen, U., and H. F. Noller. 1995. Identification of bases in 16S rRNA essential for tRNA binding at the 30S ribosomal P site. *Science* **267:**234–237.

Wilson, K., and H. F. Noller. 1998a. Mapping the position of translational elongation factor EF-G in the ribosome by directed hydroxyl radical probing. *Cell* **92:**131–139.

Wilson, K., and H. F. Noller. 1998b. Molecular movement inside the translational engine. *Cell* **92:**337–349.

Wilson, K., K. Ito, Y. Nakamura, and H. F. Noller. Unpublished data.

Wower, I., and R. Brimacombe. 1983. The localization of multiple sites on 16S RNA which are cross-linked to proteins S7 and S8 in Escherichia coli 30S ribosomal subunits by treatment with 2-iminothiolane. *Nucleic Acids Res.* **11:**1419–1437.

Wower, J., S. S. Hixson, and R. A. Zimmermann. 1989. Labeling the peptidyl transferase center of the Escherichia coli ribosome with photoreactive tRNA(Phe) derivatives containing azidoadenosine at the 3′ end of the acceptor arm: a model of the tRNA-ribosome complex. *Proc. Natl. Acad. Sci. USA* **86:**5232–5236.

Yonath, A. G., B. Tesche, S. Lorenz, J. Mussig, V. A. Erdmann, and H. G. Wittmann. 1983. Several crystal forms of the Bacillus stearothermophilus 50S ribosomal particles. *FEBS Lett.* **154:**15–20.

Yonath, A., K. R. Leonard, and H. G. Wittmann. 1987. A tunnel in the large ribosomal subunit revealed by three-dimensional image reconstruction. *Science* **236:**813–816.

Zimmermann, R. A. 1980. Interactions among protein and RNA components of the ribosome, p. 135–170. *In* G. Chambliss, G. R. Craven, J. Davies, K. Davis, L. Kahan, and M. Nomura (ed.), *Ribosomes: Structure, Function and Genetics.* University Park Press, Baltimore, Md.

The Ribosome: Structure, Function, Antibiotics, and Cellular Interactions
Edited by R. A. Garrett, S. R. Douthwaite, A. Liljas, A. T. Matheson, P. B. Moore, and H. F. Noller

Chapter 14

Three-Dimensional Organization of the Bacterial Ribosome and Its Subunits: Transition from Low-Resolution Models to High-Resolution Structures

RICHARD BRIMACOMBE, BARBARA GREUER, FLORIAN MUELLER, MONIKA OSSWALD, JUTTA RINKE-APPEL, and INGOLF SOMMER

The structure of the bacterial ribosome, in particular that of *Escherichia coli*, has been studied over a period of several decades with the help of a wide variety of biochemical and biophysical techniques. In addition to the determination of the primary amino acid and nucleotide sequences of the many ribosomal-protein and rRNA components, and the derivation of the secondary structures of the rRNA molecules, considerable effort has been devoted to the investigation of the three-dimensional (3-D) arrangement of the ribosomal constituents. Here the approaches that have been applied include immunoelectron-microscopic (IEM), neutron-scattering, or cross-linking techniques to determine the spatial distribution of the ribosomal proteins; the analysis of interactions between the proteins and rRNA by footprinting and cross-linking methods; and the study of neighborhoods within or between the rRNA molecules by intra- and inter-rRNA cross-linking (see Brimacombe, 1995, and Green and Noller, 1997, for reviews). Similar methodologies have also been used to study interactions between the ribosomal components and functional ligands such as tRNA or mRNA. The results from all these approaches have been combined in different ways by several research groups (e.g., Nagano et al., 1988; Brimacombe et al., 1988; Stern et al., 1988; Malhotra and Harvey, 1994) in order to derive crude 3-D models for the 16S or 23S rRNA molecules in relation to the ribosomal proteins. Such models are essentially only "cartoons" of the ribosome, but nevertheless, until recently they have represented the best (and indeed the only) approximations that we have had to the fine structure of the ribosome.

The situation has changed dramatically with the rapid advances that have been made during the last 4 or 5 years in the fields of both cryo-electron microscopy (cryo-EM) (Stark et al., 1995, 1997a, 1997b; Agrawal et al., 1996; Malhotra et al., 1998) and X-ray crystallography (Ban et al., 1998; Yonath and Françeschi, 1998). As a result, electron density maps of ribosomal complexes or ribosomal subunits with increasingly high quality are becoming available, and it is to be expected that these studies will soon have the potential of offering us the first view of a ribosome at atomic resolution. However, this does not imply that an atomic solution of the ribosome structure will automatically follow, as is frequently assumed. The huge size, complexity, flexibility, and total lack of symmetry of the ribosome are such that the fitting of the atomic structures of the ribosomal proteins and rRNA to the electron density represents an immensely difficult problem, which will only be resolvable if all the available information from other sources is fully exploited.

We have for some time been of the opinion that the only effective way to solve the ribosome structure problem is to combine the cartoon molecular-modeling approach with the electron density maps obtained by cryo-EM (or X-ray crystallography). The cryo-EM reconstructions on the one hand provide a direct physical framework which defines the 3-D ribosomal structure, whereas the biochemically derived models on the other formulate constraints, or neigh-

Richard Brimacombe, Barbara Greuer, Florian Mueller, Monika Osswald, Jutta Rinke-Appel, and Ingolf Sommer ■ Max-Planck-Institut für Molekulare Genetik, AG-Ribosomen, Ihnestrasse 73, 14195 Berlin, Germany.

borhoods, between or within the individual ribosomal components which must be accommodated in that 3-D structure. We anticipate that the combination of these two approaches will yield structures for the rRNA which can be improved by a process of successive approximation, ultimately leading to a structure of the ribosome at atomic resolution. Contributions to this process of stepwise improvement would be expected from four different directions, namely, (i) the derivation of cryo-EM reconstructions (or X-ray structures) at ever-higher resolution, (ii) a growing pool of biochemical information, (iii) an increasing ability to differentiate (both structurally and biochemically) between ribosomes in different functional states, and (iv) the availability of X-ray or nuclear magnetic resonance (NMR) structures of individual ribosomal proteins, small segments of rRNA, or rRNA-protein complexes.

In line with this strategy, we have published a 3-D model of the *E. coli* 16S rRNA (Mueller and Brimacombe, 1997a, 1997b; Mueller et al., 1997) which was fitted to a cryo-EM reconstruction at 20-Å resolution of the 70S ribosome carrying tRNAs at the ribosomal A and P sites (Stark et al., 1997a). Subsequently, the crystal structure of ribosomal protein S7 was added to the model (Tanaka et al., 1998). Since then, we have continued the process by fitting all three rRNA molecules (5S, 16S, and 23S) to an improved cryo-EM reconstruction at 13-Å resolution (Stark et al., submitted) of 70S ribosomes carrying a P-site tRNA and a kirromycin-stalled EF-Tu–tRNA complex (cf. Stark et al., 1997b). The available X-ray and NMR structures of proteins from both the 30S and 50S subunits have been incorporated, as well as the NMR structures of several rRNA fragments. The resulting model of the 70S ribosome is being prepared for publication as a series of papers, and there is no space here to describe this model in detail. Rather, our purpose is to summarize briefly how the 70S structure was derived and to describe some of its principal features, with a view to convincing the reader that we are indeed involved in an ongoing transition from a "low-resolution model" to a "high-resolution structure" of the ribosome.

FITTING THE rRNA TO A CRYO-EM RECONSTRUCTION

The majority of the cartoon models of the 16S rRNA that have been published (e.g., Stern et al., 1988; Malhotra and Harvey, 1994), including our own earlier efforts (Brimacombe et al., 1988), relied heavily on the neutron-scattering map of Capel et al. (1988) to fit the rRNA helices to the mass centers of the 30S subunit proteins via RNA-protein footprinting (Powers and Noller, 1995) or cross-linking (Brimacombe, 1991) data. In the case of the 50S subunit, there is no complete neutron-scattering map for the proteins, nor is there a comprehensive set of RNA-protein footprint data corresponding to that for the 30S subunit (Powers and Noller, 1995), although partial data sets are available (May et al., 1992; Østergaard et al., 1998, and references therein). This is undoubtedly the reason why less attention has been paid to developing models for the 23S rRNA. However, for reasons that have been discussed in detail elsewhere (Brimacombe, 1995), we are now of the opinion that combining the neutron map with RNA-protein footprinting data can be misleading when used as a "primary constraint" in model-building studies and that these data are more appropriate for testing the quality of a model that has already been built (Mueller and Brimacombe, 1997b) (see below).

Accordingly, in our newer models we rely more on inter- and intra-RNA cross-linking data as primary constraints on the tertiary folding of the rRNA. The inter-RNA cross-links include cross-links between tRNA and 16S or 23S rRNA (Rinke-Appel et al., 1995), between mRNA and 16S rRNA (Sergiev et al., 1997), between 5S rRNA and 23S rRNA (Osswald and Brimacombe, 1999), and between 16S and 23S rRNA (Mitchell et al., 1992). With the exception of the 16S-23S rRNA cross-links, all of these data are from site-directed cross-linking experiments. On the other hand, apart from some recent data (Baranov et al., 1998; Mundus and Wollenzien, 1998), the intra-rRNA cross-links are largely from older experiments involving in situ cross-linking of the 16S or 23S rRNA within 30S or 50S subunits (Brimacombe, 1995). The latter data sets are, however, extensive, and moreover, many of the individual cross-links have been corroborated by data from completely independent sources, such as phylogenetic covariance of pairs of nucleotides in the 16S or 23S rRNA (Gutell et al., 1994). It is also worth noting that the older cross-linking data define neighborhoods between nucleotides with much greater precision than some of the techniques that have been more recently introduced, such as the identification of Fe(II)-induced chain scissions from tethered EDTA (Joseph et al., 1997) or cross-links from photolabile oligodeoxynucleotide probes (Wang et al., 1999).

The cryo-EM reconstructions of 70S ribosomal complexes carrying tRNA are ideal substrates for combining electron density maps with biochemical models, because the tRNA molecules are directly visible in these reconstructions. The tRNA positions in turn define the location of the mRNA in the vicinity of the A- and P-site codon-anticodon interactions. Thus, the position of the tRNA-mRNA complex within the 70S EM reconstruction provides a focal point for constructing the models for both 16S and 23S rRNAs with the help of the inter- and intra-RNA cross-linking data just described. As already mentioned above, the cryo-EM reconstruction that we are currently using is a refined version (at 13-Å resolution) of the *E. coli* 70S kirromycin-stalled complex published at 18 Å by Stark et al. (1997b). Not only are the tRNA molecules clearly visible at this resolution, but the reconstruction also shows a highly detailed fine structure—particularly in the 50S subunit—which imposes rigorous constraints on the possible modes of 3-D folding of the rRNA molecules.

The fitting of the 16S rRNA to the 30S moiety of the 13-Å reconstruction represents a refinement of the 16S model at 20 Å that we have already published (Mueller and Brimacombe, 1997a). This refinement entailed making a number of changes to the structure, which will be described below. In contrast, in the case of the 23S rRNA (together with the 5S rRNA), the 13-Å reconstruction has for the first time enabled us to derive a complete rRNA model fitted to the electron density of the 50S subunit. The secondary structure of the 23S rRNA is very compact, and there are in fact considerably more inter- and intra-RNA cross-linking data involving the 23S rRNA than there are for the 16S. These two facts served to facilitate the fitting process and effectively offset the greater size of the 23S molecule. Nevertheless, several rounds of refinement were necessary before we were satisfied that the entire 23S rRNA, as well as the 5S rRNA, were optimally incorporated into the EM density. In the final structure, which is described in detail elsewhere (Mueller et al., submitted), there was no case where a helix or a single strand of the rRNA had to cross a gap or canyon in the EM density. Furthermore, the cross-link sites were all at positions in the structure which appeared plausibly accessible for a cross-linking agent. The model was built as before (Mueller and Brimacombe, 1997a) with the program ERNA-3D, and energy minimization procedures were also applied to remove Van der Waals conflicts and to optimize the positions of single strands involved in cross-links.

Two views of the rRNA structures are illustrated in Fig. 1, in which only the helical elements of the rRNA secondary structures are displayed, in the form of cylinders. In Fig. 1 (top) the two subunits have been artificially separated for clarity, with the 16S rRNA on the left and the 23S and 5S rRNAs on the right. The cylinders are color coded according to

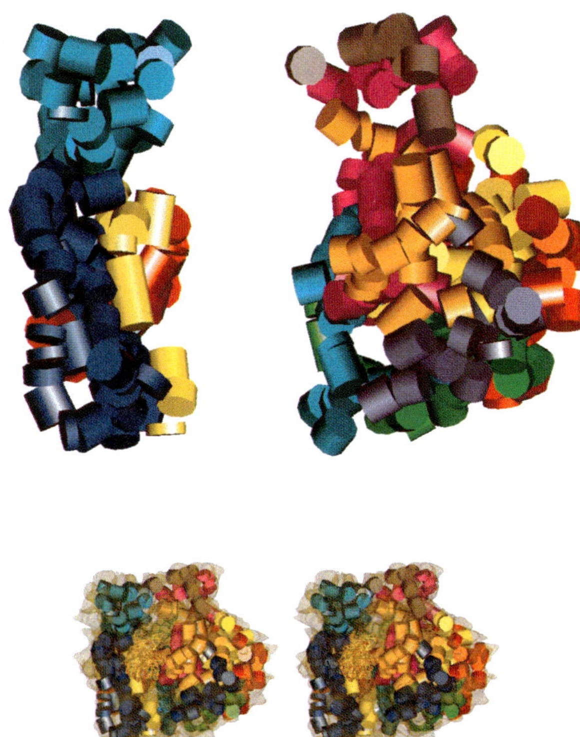

Figure 1. Arrangement of rRNA in the 70S ribosome. The views are from the L7/12 side of the ribosome, and in the upper diagram the 16S rRNA (modeled into the 30S subunit) has been separated from the 23S and 5S rRNAs (modeled into the 50S subunit); the 16S rRNA is on the left. Only the double-helical elements of the rRNA molecules are shown, color coded according to secondary-structure domains. In the 16S rRNA, the 5′ domain is blue, the central domain is red, the 3′ domain is turquoise, and the 3′ minor domain is yellow. In the 23S rRNA, domain I is red, domain IIa is yellow, domain IIb is orange, domain III is green, domain IV is turquoise, domain V is magenta, and domain VI is blue-gray. The 5S rRNA is brown. In the lower (stereo) diagram, the same representation of the 16S, 23S, and 5S molecules is shown in situ within a semitransparent contour of the kirromycin-stalled 70S ribosome at 13 Å. The atomic structures of the P-site tRNA and of the EF-Tu–tRNA ternary complex are included as backbone tube models; the P-site tRNA is green, and the EF-Tu and tRNA moieties of the ternary complex are orange and yellow, respectively.

the secondary structural domains, as in Mueller and Brimacombe (1997a) for the 16S rRNA and as in Mueller et al. (submitted) (see above) for the 23S and 5S rRNAs. Fig. 1 (bottom) is a stereo view of all three rRNA molecules, displayed within a semitransparent rendering of the 70S EM contour. The P-site tRNA and the EF-Tu–tRNA complex (Nissen et al., 1995), both of which are present in the EM reconstruction (cf. Stark et al., 1997b), are also included in the figure as backbone tube models. In the fitted structure, empty density is available at positions in both the 30S and 50S subunits where ribosomal proteins could be expected, and the X-ray or NMR structures of the corresponding proteins were incorporated at the appropriate positions; this is the subject of the following sections.

FITTING 30S RIBOSOMAL PROTEIN STRUCTURES TO THE EM DENSITY

The refined structure for the 16S rRNA at 13-Å resolution provides a good example of the process of successive approximation to a high-resolution structure outlined above. In our published model at 20 Å (Mueller and Brimacombe, 1997b), the neutron-scattering map of Capel et al. (1988) was placed empirically in the EM density, and the locations of RNA-protein footprint and cross-link sites in the 16S rRNA model were compared with the neutron positions of the corresponding proteins. The crystal structure of protein S7 (Hosaka et al., 1997; Wimberly et al., 1997) could be incorporated satisfactorily at this stage (Tanaka et al., 1998), but with other crystal structures there were problems. In particular, protein S8 appeared to lie in a "hole" in the EM density, and while we could place it close to its binding site (Allmang et al., 1994) and footprint sites (Powers and Noller, 1995) in helix 21 and close to its footprint sites in helix 25 (Powers and Noller, 1995), as well as satisfying the cross-link between residue 55 of the protein (in *E. coli* sequence numbering) and nucleotide 653 at the helix 21-helix 22 junction of the 16S rRNA (Urlaub et al., 1997), we were unable to satisfy the cross-link (Allen et al., 1979) from residue 93 of S8 to the C terminus of the neighboring protein, S5 (cf. Capel et al., 1988).

In the EM reconstruction at 13 Å it was immediately clear that our location for helix 21 (Mueller and Brimacombe, 1997a) must be incorrect, as it would not fit to the EM density at the higher resolution. The problem was resolved by reversing the chirality of the three-way junction between helices 20, 21, and 22. This caused a "chain reaction" in the central domain of the 16S rRNA which resulted not only in a satisfactory fit for protein S8 in all respects (see Fig. 3), but also improved the fit of the atomic structure of S17 (Jaishree et al., 1996); this protein has a cross-link site in helix 21, which has been localized to residue 29 of the protein (in *E. coli* numbering) and nucleotide 632 of the 16S rRNA (Urlaub et al., 1997). A number of other features of the central domain of the 16S model were also spontaneously improved by this rearrangement, such as the fit to the cross-linking data of Mundus and Wollenzien (1998).

The protein structures that have now been incorporated into the 16S model at 13-Å resolution are those of S4 (Davies et al., 1998), S5 (Ramakrishnan and White, 1992), S6 (Lindahl et al., 1994), S7 (Hosaka et al., 1997; Wimberly et al., 1997), S8 (Davies et al., 1996a; Nevskaya et al., 1998), S15 (Berglund et al., 1997; Clemons et al., 1998), and S17 (Jaishree et al., 1996). The fit of S7 (see Fig. 7) has been slightly adjusted in relation to its previously described position (Tanaka et al., 1998), and the fits of S4, S5, S8, S15, and S17 to the refined 16S rRNA model will be described in detail elsewhere. In contrast, the fit of S6 is only tentative, due to the lack of RNA-protein interaction data for this protein; it has only been footprinted on the 16S rRNA in combination with S18 (Powers and Noller, 1995).

After placing all of the atomic structures of the 30S proteins in the EM density, we were able to optimize the orientation of the neutron map in relation to the 30S contour by refitting it onto the positions of these atomic structures. The result is shown in Fig. 2, from which it can be seen that there is good agreement between the positions of the mass centers of the proteins concerned and their locations in the EM

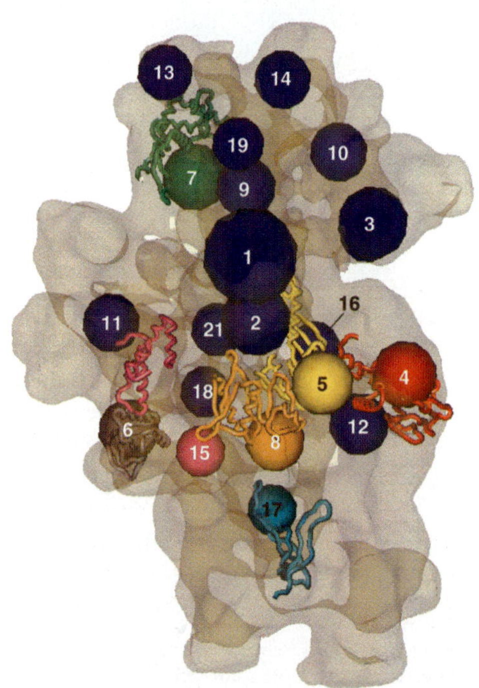

Figure 2. Arrangement of proteins in the 30S subunit. The neutron map of Capel et al. (1988) is shown superimposed (see the text) on a transparent silhouette of the 30S subunit (excised from the 13-Å reconstruction of the 70S ribosome). The X-ray or NMR structures of proteins that have been modeled into the subunit via the 16S rRNA structure are shown as backbone tube models, and the corresponding protein spheres in the neutron map are rendered in the same colors. Protein numbers are shown for simplicity without the prefix S. The view is from the solvent side of the 30S subunit.

density. Figure 3 shows a close-up view of proteins S8 and S5, together with the elements of the 16S rRNA with which they interact (cf. Mueller and Brimacombe, 1997b). Protein S8 lies in good contact with its binding site in helix 21 and with its footprint sites in helices 21 and 25 of the 16S rRNA. At the same time, the cross-links to both helix 21 and protein S5 mentioned above are clearly satisfied. Protein S5 (Fig. 3) is placed within a well-defined element of EM density so that it lies close to its cross-link site on the 16S rRNA (Brimacombe, 1991) and so that the C terminus of the protein faces S8 in accordance with the cross-link to the latter. Furthermore, the protein is oriented in such a way that sites within S5 that are implicated in resistance to the antibiotics spectinomycin and streptomycin (Piepersberg et al., 1975) are located close to the main group of sites on the 16S rRNA that are similarly implicated in interaction with these same antibiotics, both by mutational and footprinting studies (see Brimacombe, 1992, for references). Protein S4 (not shown in Fig. 3, but see Fig. 2) can also be placed so that sites in S4 that are implicated in streptomycin resistance (Björkman et al., 1999) lie close to the corresponding sites in S5.

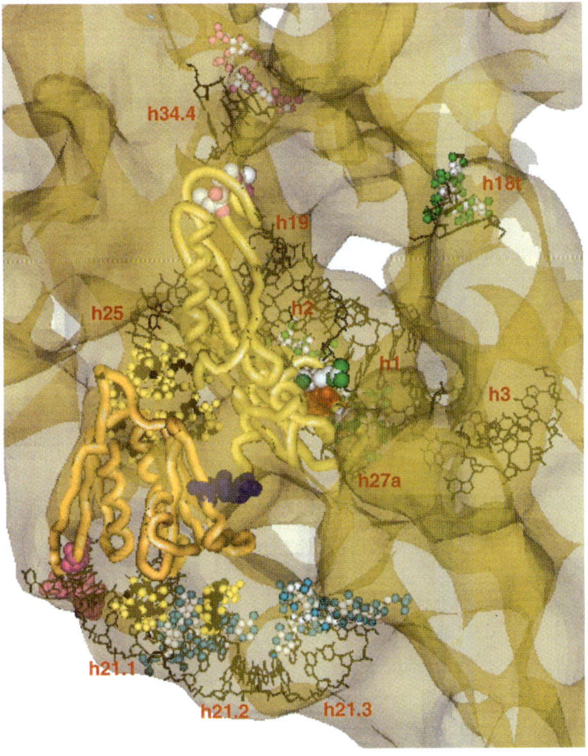

Figure 3. Close-up view of proteins S5 and S8 in the semitransparent 30S subunit, together with appropriate elements of the 16S rRNA. The proteins are shown as backbone tube models (S5 is yellow; S8 is orange) as in Fig. 2, and the view is approximately the same as in that figure. Helical elements of the rRNA are numbered as in Mueller and Brimacombe (1997a). Individual nucleotides and amino acids are color coded as follows. The magenta nucleotide in helix 21 and the corresponding magenta amino acid in S8 are those that are cross-linked together. Other cross-link sites from S8 in helix 21 are turquoise and white. The brown and yellow nucleotides in helices 21 and 25 are footprint sites for S8. The purple amino acid in S8 is cross-linked to the purple amino acid at the C terminus of S5. The red nucleotide (between helices 3 and 19) is the cross-link site from S5. The magenta and white nucleotides and amino acids (in S5) are spectinomycin sites, and the green and white nucleotides and amino acids are the corresponding streptomycin sites. See the text for references.

However, in order for protein S5 to bridge the gap between the spectinomycin sites (in helix 34) and the streptomycin sites (in the neighborhood of helix 2) as well as the S5-16S rRNA cross-link site (in the neighborhood of helix 3), it was necessary to open up the two domains of the crystal structure of the protein (Ramakrishnan and White, 1992). While this domain separation also gives a better fit to the EM density (cf. Fig. 3), it disrupts a number of hydrophobic contacts in the crystal structure, which is clearly undesirable and raises the question of how far the RNA-protein interactions in the ribosome represent induced fits between the RNA and protein moieties. This will require a case-by-case inspection at higher resolution (cf. the L11-rRNA complex [see below]), and for this reason we have not so far tried to minimize Van der Waals contacts between the fitted protein structures and their contact sites on the rRNA. For the purposes of this chapter, the point to be made is that the atomic structures of the proteins have been fitted into elements of the EM density in a manner which is dictated primarily by the rRNA model but which at the same time both satisfies the biochemical data relating to the proteins and "feeds back" into the neutron map, as shown in Fig. 2.

FITTING X-RAY AND NMR STRUCTURES TO THE EM DENSITY OF THE 50S SUBUNIT

The 50S subunit proteins for which X-ray or NMR structures are available include L1 (Nikonov et al., 1996), the RNA-binding domain of L2 (Nakagawa et al., 1999), L6 (Golden et al., 1993), the C-terminal domain of L7/12 (Leijonmarck and Liljas, 1987), L9 (Hoffman et al., 1996), the RNA-binding domain of L11 both in isolated form and bound to a 23S rRNA fragment (Markus et al., 1997; Hinck et al., 1997), L14 (Davies et al., 1996b), L22 (Unge et al., 1998), L25 (Stoldt et al., 1998), and L30 (Wilson et al., 1986). Of these, proteins L1, L2, L9, L11, L14, and L25 have been fitted to the 23S-5S rRNA model, and the fits will be described in detail elsewhere. The positions of the proteins in the structure are in good agreement with their positions in the model of Walleczek et al. (1988), which was derived from IEM determinations combined with interprotein cross-link analyses. L6 and the L7/12 fragment have not yet been incorporated into the model, as there are several EM density elements which could plausibly accommodate these proteins (and there are of course four copies of L7/12). L22 and L30 have also not yet been fitted to the rRNA model, due to the lack of RNA-protein interaction data for these proteins.

In addition to the protein structures, NMR and X-ray structures (Szewczak and Moore, 1995; Correll et al., 1998) have been determined for the α-sarcin loop (helix 95 in our nomenclature) and for parts of the 5S rRNA (Dallas and Moore, 1997; Correll et al., 1997). The NMR structures of the sarcin loop and of helices IV and V of the 5S molecule have been incorporated into the model, and Fig. 4 shows a view of the L7/12 side of the 50S subunit, including these helices (helix 95 and helices IV and V), together with protein L25, which is known to interact with 5S helices IV and V (Huber and Wool, 1984) and which was fitted on the basis of preliminary NMR data obtained from a 5S rRNA-L25 complex (in collaboration with M. Stoldt and M. Görlach). Figure 4 also shows the crystal structure of EF-Tu from the EF-Tu–tRNA complex (Nissen et al., 1995), the RNA binding domain of protein L11, and the binding site for L11 on the 23S rRNA (Schmidt et al., 1981), which comprises nucleotides 1051 to 1108 (helices 42.4, 43, and 44).

The arrangement of L11 and its binding site in the model (Fig. 4) is of particular interest, because an X-ray structure is now available for the L11-rRNA complex (Wimberly et al., 1999). A comparison of our model structure with the X-ray structure (in collaboration with S. White and V. Ramakrishnan) showed that the basic arrangement of the helices concerned (42.4, 43, and 44) is correct in our model, although—not surprisingly—the detailed arrangement of the single-stranded rRNA regions was wrong. Moreover, we had placed the protein on the correct side of the structure, and—most important for our purposes—the connection to the rest of the 23 S rRNA model (to the remainder of helix 42 and helix 41) could be made almost perfectly from the X-ray structure. This comparison with the X-ray structure of the L11-rRNA complex represents the most searching test to which our rRNA models have so far been subjected.

The proteins and rRNA regions illustrated in Fig. 4, from L25 and 5S rRNA at the top to the sarcin loop at the bottom, show that a substantial area on this side of the 50S subunit has now been filled with structures at atomic or near-atomic resolution. The position of the sarcin loop deserves some comment, as it does not appear to be in direct contact with EF-Tu, although a contact would be expected from the footprinting data relating to the factor (Moazed et al., 1988). It is possible that in the kirromycin-stalled complex the EF-Tu is slightly displaced from its normal position, and in this context it should be noted that in the EM reconstruction at 20-Å resolution made by Agrawal et al. (1998) of a ribosomal complex carrying EF-G, the factor was located lower down in the structure, at a position which would bring it into contact with helix 95 in our model. It is also noteworthy (Fig. 4) that the end of the hairpin loop of helix 95 (where the sarcin-ricin nucleotide is located) lies very close to the 30S subunit.

THE POSITIONS OF tRNAs IN THE 70S RIBOSOME

As already noted above, the cryo-EM reconstructions of 70S ribosomal complexes allow the direct visualization of tRNA molecules, and at the current levels of resolution (Malhotra et al., 1998; Stark et al., submitted) the X-ray structure of a tRNA molecule (Kim et al., 1974) can be fitted very precisely to the corresponding density elements in the EM reconstructions. In earlier reconstructions there was some controversy as to the positioning of the A- and P-site tRNAs (Agrawal et al., 1996; Stark et al., 1997a), but now this situation has been clarified in favor of the version of Stark et al. (1997a) (cf. Agrawal et al., 1999). In the kirromycin-stalled 70S ribosomal complex (Stark et al., 1997b), the P site is occupied by fMet-$tRNA_f^{Met}$, whereas the EF-Tu–tRNA ternary complex is located at the "pre-A site"; the positions of these ligands in the cryo-EM reconstruction at 13 Å are illustrated in Fig. 5. The figure shows the 50S subunit, electronically excised from the 70S ribosome and viewed from the interface side of the subunit. The subunit is rendered as an opaque structure, and the density corresponding to the P-site tRNA and to the EF-Tu–tRNA complex has also been removed to enable the ligands to be seen in their fitted positions, displayed as backbone tube models. It should be noted that the angle between the EF-Tu molecule and the tRNA in the ternary complex was slightly modified from that in the X-ray structure (Nissen et al., 1995) to give a better fit to the EM density.

Figure 5 also includes an A-site and an E-site tRNA molecule, taken from the 20-Å reconstruction of Stark et al. (1997a). To position these tRNAs, the coordinates of the complexes of two tRNA molecules at the A and P sites, or P and E sites, respectively, from the 20-Å reconstruction were superimposed on the 13-Å reconstruction, so that in each case the

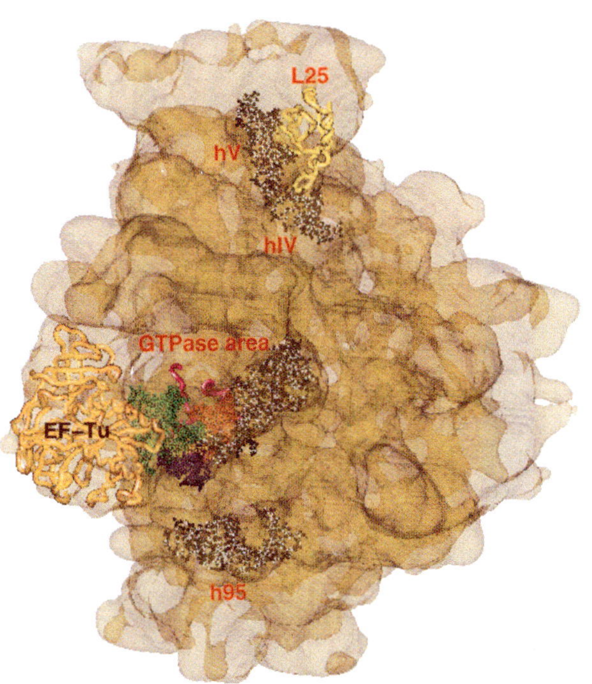

Figure 4. Ribosomal components in the 50S subunit. Shown is a view from the L7/12 side of the semitransparent subunit. The atomic structures of EF-Tu, L25, and the RNA binding domain of L11 are displayed as backbone tube models colored orange, yellow, and magenta, respectively. The NMR structures of helices IV and V of the 5S rRNA, and of helix 95 of the 23S rRNA (the sarcin loop), are in black and white, and the modeled structure of helices 42, 43, and 44 of the 23S rRNA (the "GTPase area") is also included. Helix 42.4 is red-brown, helix 43 is green, helix 44 is purple, and the connecting strand between 43 and 44 (barely visible) is orange, as in the illustrations of the X-ray structure of Wimberly et al. (1999). The remainder of helix 42 (i.e., 42.1 to 42.3) is black and white. The helix numbering is as in Mueller et al. (1999, submitted). See the text for further references.

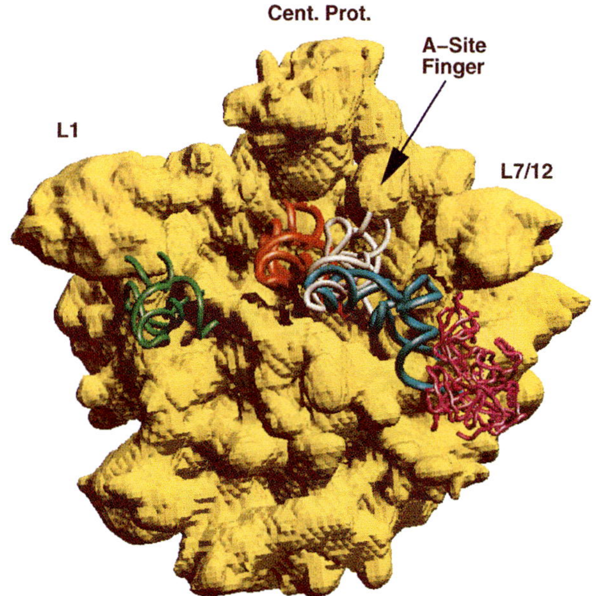

Figure 5. Positions of tRNA sites on the 50S subunit. The view is of the interface side of the 50S subunit, rendered as an opaque structure with prominent features (the L1 and L7/12 protuberances, the central protuberance [Cent. Prot.], and the A-site finger) marked. Density corresponding to the P-site tRNA and to the EF-Tu–tRNA complex has been removed. The tRNA molecules at the different sites and EF-Tu are displayed as backbone tube models, color coded as follows. EF-Tu is magenta, and tRNA in the EF-Tu–tRNA complex is blue. The A-site tRNA is white, the P-site tRNA is red, and the E site (corresponding approximately to the E2 site of Agrawal et al. [1999]) is green. See the text for further details.

P-site molecule of the former reconstruction coincided with the P site of the latter. This gives a reasonably precise positioning of the A-site tRNA in the 13-Å reconstruction (Fig. 5), although in contrast, that of the E site is only approximate, due both to differences in the 13-Å and 20-Å maps in the area of the L1 protuberance and to the relative uncertainty of the E-site tRNA placement in the 20-Å reconstruction (Stark et al., 1997a). The A- and E-site tRNAs are depicted with thinner backbone tubes in Fig. 5 to indicate that their locations are indirect. The position of the E site corresponds approximately to that of the "E2" site observed by Agrawal et al. (1999), the E site as defined by these authors being located in the gap between the P and E sites of Fig. 5.

As the tRNA traverses the ribosome, moving from one site to the next, it sweeps a broad arc from the L7/12 side to the L1 side of the 50S subunit (Fig. 5). In the pre-A-site position, the CCA end of the tRNA lies well to the right of its position in the A site proper, although it is noteworthy that the position of the anticodon loop is coincident in both of these sites; this is as would be expected, since the pre-A-site tRNA should be able to "read" the A-site codon when it is brought to the ribosome in the form of the ternary complex with EF-Tu. The movement from pre-A to A site is thus a rotation of the tRNA about its anticodon loop, bringing the CCA end into a position close to the CCA end of the already-present P-site tRNA. The EM density in the 13-Å reconstruction suggests that the 3′ terminus of the P-site tRNA is turned upwards slightly. This enables it to make contact with helix 80 of the 23S rRNA and allows the base pair to form between C74 of the tRNA and G2252 in helix 80, as demonstrated by Samaha et al. (1995) (Fig. 6).

The CCA ends of the tRNAs at the P and A sites (Fig. 5) lie in the interface canyon, at the entrance to the peptide tunnel (see Fig. 8). At the same time, their anticodon loops are close together, so that simulta-

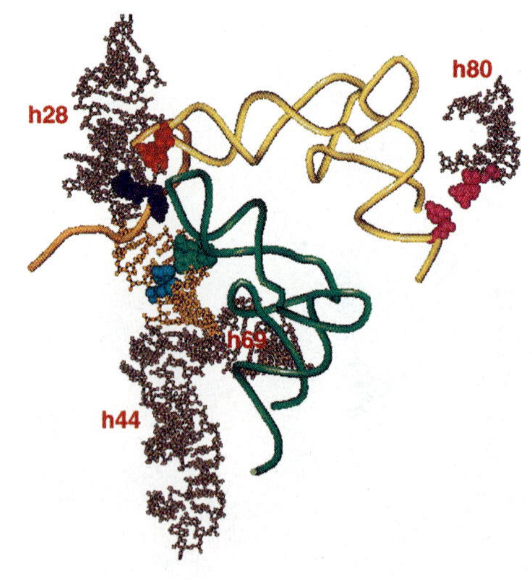

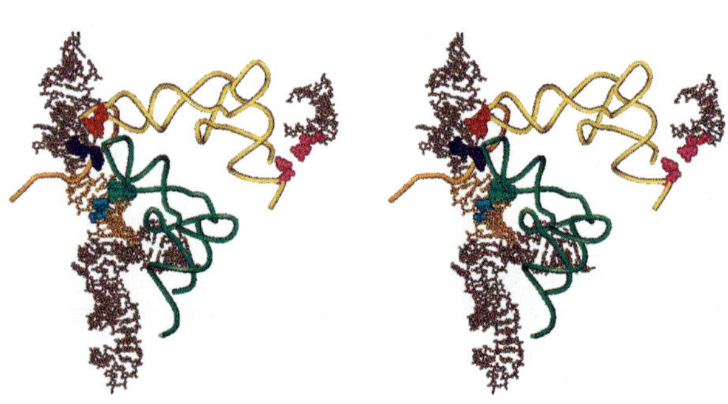

Figure 6. Decoding area of the 16S rRNA, together with the relative locations of elements of the 23S rRNA. Helix 28 and part of 44 of the 16S rRNA are shown, with the nucleotides described in the NMR structure of Fourmy et al. (1996) highlighted in orange. Helices 69 and 80 are included from the 23S rRNA. The P-site tRNA and mRNA are displayed as backbone tube models (yellow and orange, respectively), in positions adapted from the model of VanLoock et al. (1999), and the pre-A-site tRNA (in the EF-Tu–tRNA ternary complex) is similarly shown in green. Positions 34 of the P-site tRNA and C1400 of the 16S rRNA (which are cross-linked) are shown as red CPK nucleotides. Similarly, positions +4 of the mRNA and C1402 of the 16S rRNA are purple. Nucleotides 1492 and 1493 (footprint sites for A-site tRNA) are green, and nucleotide 1408 (footprint site for A-site tRNA and cross-link site to helix 69 in 23S rRNA) is turquoise. The CCA end of the P-site tRNA has been turned upwards (in accordance with the EM density), and the potential base pair from C74 to G2252 in helix 80 is indicated by the magenta CPK nucleotides. The lower part of the figure shows a stereo pair of the same structural elements.

neous codon-anticodon interaction can occur at the two sites. In the E site as observed by Agrawal et al. (1999), which as just noted was not seen in the reconstructions of Stark et al. (1997a, 1997b), codon-anticodon interaction is also possible, in accordance with the experiments of Gnirke et al. (1989). In contrast, at the E2 site (or the E site of Fig. 5) the anticodon loop of the tRNA molecule is clearly no longer in a location where it could interact with the corresponding codon on the mRNA.

An interesting feature of the tRNA locations as depicted in Fig. 5 is that at each site the "elbow" region of the tRNA appears to interact with a specific feature of the 50S subunit. In the pre-A site, the tRNA elbow is close to the L7/12 protuberance, whereas in the A site proper it contacts the "A-site finger" (Stark et al., 1997a, 1997b). Our cross-linking (Rinke-Appel et al., 1995; Osswald and Brimacombe, 1999) and modeling (Mueller et al., submitted) studies indicate that this A-site finger corresponds to helix 38 of the 23S rRNA. Similarly, at the P site the elbow of the tRNA contacts the underside of the central protuberance of the 50S subunit and is cross-linked to sites in helices 84 and 85 within this region (Rinke-Appel et al., 1995). At the E (or E2) site, the contact appears to be to the L1 mushroom itself, although here the cross-linking data (Rinke-Appel et al., 1995) are not so clear-cut.

THE DECODING SITE ON 16S rRNA AND THE PATH OF mRNA IN THE 30S SUBUNIT

The position of the P-site tRNA is particularly well defined in the EM density at 13 Å, and as already mentioned the tRNA at this site is precisely "anchored" to the 23S rRNA by virtue of the C74-G2252 base pair (Samaha et al., 1995). There is a similar anchor to the P-site tRNA in the 30S subunit, provided by the well-known cross-link (Prince et al., 1982) from position 34 in the anticodon loop to nucleotide C1400 in the 16S rRNA. An NMR structure has been published for an RNA oligomer corresponding to the A-site moiety of the 16S decoding area (Fourmy et al., 1996), and VanLoock et al. (1999) have extrapolated this structure into a model for the whole decoding area covering both P and A sites. Their model encompasses nucleotides 1399 to 1412 and 1488 to 1504 of the 16S rRNA and thus extends from approximately the beginning of helix 28 (Mueller and Brimacombe, 1997a) into the long penultimate helix 44. The positions of both of these helices are firmly constrained by the EM density, helix 28 forming the covalent connection between the head and body of the 30S subunit and helix 44 running down to the bottom of the subunit on the interface side (cf. Mueller and Brimacombe, 1997a). Since the P-site tRNA is also firmly constrained, we asked the question, how well can the model of VanLoock et al. (1999) be incorporated into the revised 16S rRNA model at 13 Å?

To fit the VanLoock model, we superimposed it on the 16S rRNA via the position of the P-site tRNA in the two respective structures and then made minor rotations of the rRNA moiety around the tRNA-C1400 cross-link site so as to dock it onto helices 28 and 44 in our model. The result is illustrated in Fig. 6. VanLoock et al. (1999) had presented convincing arguments that the tRNA-mRNA complex is bound to the major groove of the rRNA, and this concept is maintained (Fig. 6) as well as receiving additional support from the cross-link (Sergiev et al., 1997) between position +4 of the mRNA and nucleotide 1402 of the 16S rRNA. On the other hand, the rotations that we had to make resulted in the loss of the hydrogen-bonded interaction proposed by VanLoock et al. (1999) between position +5 of the mRNA and nucleotides 1492 and 1493. Nonetheless, there is still a reasonable proximity between these bases and the anticodon loop of the pre-A-site tRNA, as located in the 13-Å reconstruction; nucleotides 1492 and 1493, together with 1408, form part of the footprint of A-site-bound tRNA (Moazed and Noller, 1990).

Figure 6 also includes helices 80 and 69 of the 23S rRNA, the former to show the potential base pairing between C74 of the tRNA and G2252 in the 23S rRNA (Samaha et al., 1995) (see above) and the latter to demonstrate the proximity of the loop end of helix 69 to nucleotide 1408 of the 16S rRNA in accordance with the 16S-23S rRNA cross-linking data of Mitchell et al. (1992). The 16S and 23S rRNA models are thus connected to one another at the subunit interface (cf. Fig. 1) both via the P-site tRNA interactions and by the 16S-23S rRNA cross-link.

One of the more intriguing predictions to emerge from our previous model for the 16S rRNA at 20-Å resolution (Mueller and Brimacombe, 1997a) was that the incoming mRNA chain passes through a "hole" at the head-body junction of the 30S subunit (Mueller et al., 1997). A similar prediction had been made previously by Lata et al. (1996). The hole is formed as a result of a second—noncovalent—contact between the head and the body of the subunit (in addition to the covalent contact made via helix 28, as just mentioned), and it was suggested that this hole should open and close in a manner analogous to the ring or clamp structures that have been observed with both DNA and RNA polymerases (e.g., Herendeen and Kelly, 1996). Evidence in support of this idea comes from EM reconstructions of isolated 30S subunits (Gabashvili et al., 1999), in which the hole does indeed appear to be open.

Our prediction that the mRNA passes through the hole was based on the locations in the 16S rRNA model (Mueller and Brimacombe, 1997a) of a series of highly specific cross-links between downstream sites on the mRNA and nucleotides in the 16S rRNA (Sergiev et al., 1997). In the revised 16S rRNA model at 13 Å, the downstream mRNA-16S rRNA contacts have been modeled more precisely (e.g., that from position +4 of the mRNA to nucleotide 1402 [Fig. 6]), and the mRNA path through the hole is more clearly defined than before. This is illustrated in Fig. 7, which shows the mRNA and P-site tRNA in the 30S subunit, together with the refined position for protein S7 (cf. Tanaka et al., 1998) (see above). The positions in the more flexible upstream region of the mRNA that become cross-linked to S7 (cf. Stade et al., 1989; Rinke-Appel et al., 1994) are marked, and as we have previously noted (Tanaka et al., 1998), the model makes the strong prediction that these cross-links are either to the β-sheet region or to the C terminus of S7, both of which are located at the lower apex of the protein, as depicted in Fig. 7. This prediction has now been experimentally confirmed (Greuer et al., in press), with the cross-link site at the C terminus of the protein.

THE PATH OF THE NASCENT PEPTIDE THROUGH THE 50S SUBUNIT

The final topic to be discussed is the path followed by the growing peptide chain after it leaves the peptidyltransferase center in the 50S subunit. The cryo-EM reconstructions of the 50S subunit show that it contains many canyons, cavities, and tunnels, and the most prominent of these leads from the peptidyltransferase center (as defined by the position of

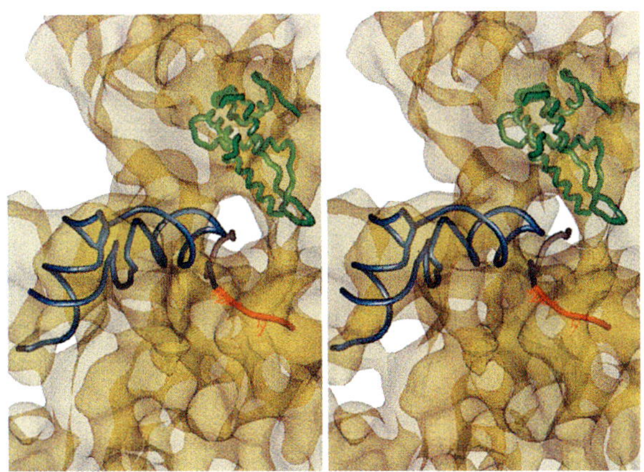

Figure 7. Path of mRNA through the 30S subunit. A close-up stereo view of the interface side of the (semitransparent) 30S subunit is shown. The P-site tRNA and protein S7 are displayed as backbone tube models, in blue-gray and green, respectively. The mRNA is displayed as a backbone tube model, with the upstream sequence red, the P-site codon blue-gray and the downstream region brown. Nucleotides at positions −1, −3, and −4 of the mRNA, which can be cross-linked to S7, are shown as red wireframe structures.

the CCA ends of the A- and P-site tRNA molecules in the interface canyon [Fig. 5]) through the subunit, to emerge at a point close to that proposed as the peptide exit site by Bernabeu and Lake (1982) on the basis of IEM studies. This tunnel exit has been demonstrated by cryo-EM to be in line with the doughnut-shaped Sec61 complex in yeast (Beckmann et al., 1997), which in turn is involved in the translocation of proteins across the endoplasmic reticulum membrane.

Direct evidence for the path followed by the growing peptide chain relative to the 23S rRNA has been obtained in our laboratory in a series of cross-linking experiments in which peptides of varying lengths carrying a photoreactive agent at their N termini were synthesized by in situ translation on the ribosome of suitable mRNA fragments and then subjected to photoactivation (Stade et al., 1994, 1995). The analysis was extended to peptide families derived from several proteins (Choi and Brimacombe, 1998). The results showed that, with increasing length, the N termini of the peptides became cross-linked to sites first within domain V of the 23S rRNA (close to the peptidyltransferase region), then to a site in domain IV, then domain II, then domain III, and finally domain I.

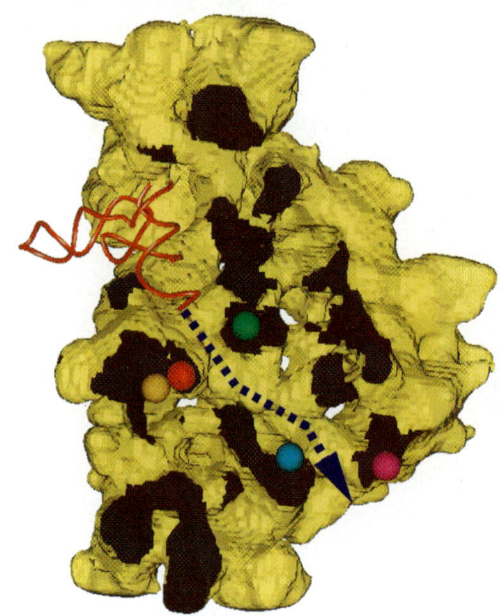

Figure 8. Path of the nascent peptide through the 50S subunit. The 50S subunit, viewed from the L7/12 side, is displayed as an opaque structure with density elements cut away (leaving the dark brown "scars") to reveal the peptide tunnel. The EM density corresponding to the P-site tRNA has also been removed, and the P-site tRNA is shown as a red backbone tube model. The arrow indicates the path followed by the growing peptide chain, with the colored balls showing the sites on 23S rRNA that are cross-linked from the N termini of peptides with increasing minimum lengths, color coded as follows: red, position 2609 (3 amino acids); orange, position 1781 (4 amino acids); green, position 750 (6 amino acids); turquoise, position 1614 (12 or 25 amino acids; see Choi and Brimacombe, 1998); magenta, position 91 (30 amino acids).

The locations of these sites in the 23S rRNA model are illustrated in Fig. 8, in which the EM density of the 50S subunit has been cut away to optimally reveal the putative peptide tunnel. (The tunnel is not straight and cannot be seen optimally from a planar section through the subunit as in the publication of Beckmann et al. [1997]). It is important to note that the peptide cross-link sites on 23S in Fig. 8 were not used as primary constraints during the model-building process. Nevertheless, it can be seen that all of the sites lie within the tunnel, those from the progressively longer peptides lying farther from the peptidyltransferase center and closer to the exit site. A further observation in our cross-linking studies (Stade et al., 1995; Choi and Brimacombe, 1998) was that the N termini of the longer peptides were able to "reach back" and become simultaneously cross-linked to the sites closer to the peptidyltransferase center found with the shorter peptides. This indicates that the growing peptide chain must be very flexible, which would be expected if cotranslational folding is to occur, as suggested by the experiments of Kudlicki et al. (1995) or Makeyev et al. (1996). The broad width of the peptide tunnel (Fig. 8) is consistent with this concept.

FUTURE PROSPECTS

The current versions of the 3-D structures that we have derived for the rRNA molecules are still far from perfect, and as stated at the beginning of this chapter, we anticipate that they will be progressively improved as more data become available and as the resolution of the electron density maps increases. It is to be expected that the highest resolutions will now come from X-ray crystallography (cf. Ban et al., 1998) rather than from the cryo-EM reconstructions. However, the crystals for X-ray analysis are being obtained from halophilic or thermophilic bacteria, whereas the EM reconstructions are primarily from *E. coli* ribosomes, with the attendant advantages of all the biochemical knowledge and experience that have been accumulated for this organism. We would predict that in the next phase of development an extrapolation of our combined "molecular modeling plus density fitting" approach into high-resolution crystallographic maps of the ribosomal subunits will yield the most accurate models of the 3-D structure of the rRNA. On the other hand, the continued development of the cryo-EM techniques will be better suited to improving our understanding of the ribosomal function by following conformational changes in the ribosomes or by observing the positions of functional ligands at different stages of the protein biosynthetic cycle. The cryo-EM method is also more appropriate for experimental testing of the structures, by making reconstructions of appropriately modified ribosomes or ribosomal subunits.

ADDENDUM IN PROOF

At the time of going to press, a new reconstruction of the *E. coli* 50S subunit at 7.5-Å resolution has become available (R. Matadeen, A. Patwardhan, B. Gowen, E. V. Orlova, T. Pape, M. Cuff, F. Mueller, R. Brimacombe, and M. van Heel. The *E. coli* large ribosomal subunit at 7.5 Å resolution. *Structure*, in press). A number of improvements have been made to the 23S rRNA model on the basis of this new reconstruction (F. Mueller, I. Sommer, P. Baranov, R. Matadeen, M. Stoldt, M. Görlach, M. van Heel, and R. Brimacombe, unpublished data).

REFERENCES

Agrawal, R. K., P. Penczek, R. A. Grassucci, Y. Li, A. Leith, K. H. Nierhaus, and J. Frank. 1996. Direct visualization of A-, P- and E-site transfer RNAs in the *E. coli* ribosome. *Science* **271:** 1000–1002.

Agrawal, R. K., P. Penczek, R. A. Grassucci, and J. Frank. 1998. Visualization of elongation factor G on the *E. coli* 70S ribosome: the mechanism of translocation. *Proc. Natl. Acad. Sci. USA* **95:** 6134–6138.

Agrawal, R. K., P. Penczek, R. A. Grassucci, N. Burkhardt, K. H. Nierhaus, and J. Frank. 1999. Effect of buffer conditions on the position of tRNA on the 70S ribosome as visualized by cryoelectron microscopy. *J. Biol. Chem.* **274:**8723–8729.

Allen, G., R. Capasso, and C. Gualerzi. 1979. Identification of the amino acid residues of proteins S5 and S8 adjacent to each other in the 30S ribosomal subunit of *E. coli*. *J. Biol. Chem.* **254:**9800–9806.

Allmang, C., M. Mougel, E. Westhof, B. Ehresmann, and C. Ehresmann. 1994. Role of conserved nucleotides in building the 16S rRNA binding site of *E. coli* ribosomal protein S8. *Nucleic Acids Res.* **22:**3708–3714.

Ban, N., B. Freeborn, P. Nissen, P. Penczek, R. A. Grassucci, R. Sweet, J. Frank, P. B. Moore, and T. A. Steitz. 1998. A 9 Å resolution X-ray crystallographic map of the large ribosomal subunit. *Cell* **93:**1105–1115.

Baranov, P. V., O. L. Gurvich, A. A. Bogdanov, R. Brimacombe, and O. A. Dontsova. 1998. New features of 23S ribosomal RNA folding: the long helix 41-42 makes a "U-turn" inside the ribosome. *RNA* **4:**658–668.

Beckmann, R., D. Bubeck, R. Grassucci, P. Penczek, A. Verschoor, G. Blobel, and J. Frank. 1997. Alignment of conduits for the nascent polypeptide chain in the ribosome-Sec61 complex. *Science* **278:**2123–2126.

Berglund, H., A. Rak, A. Serganov, M. Garber, and T. Härd. 1997. Solution structure of the ribosomal RNA binding protein S15 from *Thermus thermophilus*. *Nat. Struct. Biol.* **4:**20–23.

Bernabeu, C., and J. A. Lake. 1982. Nascent polypeptide chains emerge from the exit domain of the large ribosomal subunit: immune mapping of the nascent chain. *Proc. Natl. Acad. Sci. USA* **79:**3111–3115.

Björkman, J., P. Samuelsson, D. I. Andersson, and D. Hughes. 1999. Novel ribosomal mutations affecting translational accuracy, antibiotic resistance and virulence of *Salmonella typhimurium*. *Mol. Microbiol.* **31:**53–58.

Brimacombe, R. 1991. RNA-protein interactions in the *E. coli* ribosome. *Biochimie* **73:**927–936.

Brimacombe, R. 1992. Structure-function correlations (and discrepancies) in the 16S ribosomal RNA from *E. coli*. *Biochimie* **74:**319–326.

Brimacombe, R. 1995. The structure of ribosomal RNA: a three-dimensional jigsaw-puzzle. *Eur. J. Biochem.* **230:**365–383.

Brimacombe, R., J. Atmadja, W. Stiege, and D. Schüler. 1988. A detailed model for the three-dimensional structure of E. coli 16S ribosomal RNA *in situ* in the 30S subunit. *J. Mol. Biol.* **199:** 115–136.

Capel, M. S., M. Kjeldgaard, D. M. Engelman, and P. B. Moore. 1988. Positions of S2, S13, S16, S17, S19 and S21 in the 30S ribosomal subunit of *E. coli*. *J. Mol. Biol.* **200:**65–87.

Choi, K. M., and R. Brimacombe. 1998. The path of the growing peptide chain through the 23S rRNA in the 50S ribosomal subunit: a comparative cross-linking study with three different peptide families. *Nucleic Acids Res.* **26:**887–895.

Clemons, W. M., C. Davies, S. W. White, and V. Ramakrishnan. 1998. Conformational variability of the N-terminal helix in the structure of ribosomal protein S15. *Structure* **6:**429–438.

Correll, C. C., B. Freeborn, P. B. Moore, and T. A. Steitz. 1997. Metals, motifs and recognition in the crystal structure of a 5S rRNA domain. *Cell* **91:**705–712.

Correll, C. C., A. Munishkin, Y. Chan, Z. Ren. I. G. Wool, and T. A. Steitz. 1998. Crystal structure of the ribosomal RNA loop essential for binding both elongation factors. *Proc. Natl. Acad. Sci. USA* **95:**13436–13441.

Dallas, A., and P. B. Moore. 1997. The loop E-loop D region of 5S rRNA: the solution structure reveals an unusual loop that may be important for binding ribosomal proteins. *Structure* **5:** 1639–1653.

Davies, C., V. Ramakrishnan, and S. W. White. 1996a. Structural evidence for specific S8-RNA and S8-protein interactions within the 30S ribosomal subunit: ribosomal protein S8 from *Bacillus stearothermophilus* at 1.9 Å resolution. *Structure* **4:**1093–1104.

Davies, C., S. W. White, and V. Ramakrishnan. 1996b. The crystal structure of ribosomal protein L14 reveals an important organizational component of the translational apparatus. *Structure* **4:** 55–66.

Davies, C., R. B. Gerstner, D. E. Draper, V. Ramakrishnan, and S. W. White. 1998. The crystal structure of ribosomal protein S4 reveals a two-domain molecule with an extensive RNA-binding surface: one domain shows structural homology to the ETS DNA-binding motif. *EMBO J.* **17:**4545–4558.

Fourmy, D., M. I. Recht, S. C. Blanchard, and J. D. Puglisi. 1996. Structure of the A site of *E. coli* 16S ribosomal RNA complexed with an aminoglycoside antibiotic. *Science* **274:**1367–1371.

Gabashvili, I. S., R. K. Agrawal, R. Grassucci, and J. Frank. 1999. Structure and structural variations of the *E. coli* 30S ribosomal subunit as revealed by three-dimensional cryo-electron microscopy. *J. Mol. Biol.* **286:**1285–1291.

Gnirke, A., U. Geigenmüller, H. J. Rheinberger, and K. H. Nierhaus. 1989. The allosteric three-site model for the ribosomal elongation cycle: analysis with a heteropolymeric mRNA. *J. Biol. Chem.* **264:**7291–7301.

Golden, B. L., V. Ramakrishnan, and S. W. White. 1993. Ribosomal protein L6: structural evidence of gene duplication from a primitive RNA binding protein. *EMBO J.* **12:**4901–4908.

Green, R., and H. F. Noller. 1997. Ribosomes and translation. *Annu. Rev. Biochem.* **66:**679–716.

Greuer, B., B. Thiede, and R. Brimacombe. The cross-link from the upstream region of mRNA to ribosomal protein S7 is located in the C-terminal peptide: experimental verification of a prediction from modeling studies. *RNA*, in press.

Gutell, R. R., N. Larsen, and C. R. Woese. 1994. Lessons from an evolving rRNA: 16S and 23S rRNA structures from a comparative perspective. *Microbiol. Rev.* **58:**10–26.

Herendeen, D. R., and T. J. Kelly. 1996. DNA polymerase III; running rings around the fork. *Cell* **84:**5–8.

Hinck, A. P., M. A. Markus, S. Huang, S. Grzesiek, I. Kustonovich, D. E. Draper, and D. A. Torchia. 1997. The RNA binding domain of ribosomal protein L11: three-dimensional structure of the RNA-bound form of the protein and its interaction with 23S rRNA. *J. Mol. Biol.* **274:**101–113.

Hoffman, D. W., C. S. Cameron, C. Davies, S. W. White, and V. Ramakrishnan. 1996. Ribosomal protein L9: a structure determination by the combined use of X-ray crystallography and NMR spectroscopy. *J. Mol. Biol.* **264:**1058–1071.

Hosaka, H., A. Nakagawa, I. Tanaka, N. Harada, K. Sano, M. Kimura, M. Yao, and S. Wakatsuki. 1997. Ribosomal protein S7: a new RNA-binding motif with structural similarities to a DNA architectural factor. *Structure* **5:**1199–1208.

Huber, P. W., and I. G. Wool. 1984. Nuclease protection analysis of ribonucleoprotein complexes: use of the cytotoxic ribonuclease α-sarcin to determine the binding sites for *E. coli* ribosomal proteins L5, L18 and L25 on 5S rRNA. *Proc. Natl. Acad. Sci. USA* 81:322–326.

Jaishree, T. N., V. Ramakrishnan, and S. W. White. 1996. Solution structure of prokaryotic ribosomal protein S17 by high-resolution NMR spectroscopy. *Biochemistry* 35:2845–2853.

Joseph, S., B. Weiser, and H. F. Noller. 1997. Mapping the inside of the ribosome with an RNA helical ruler. *Science* **278:**1093–1098.

Kim, S. H., F. L. Suddath, G. J. Quigley, A. McPherson, J. L. Sussman, A. H. J. Wang, N. C. Seeman, and A. Rich. 1974. Three-dimensional tertiary structure of yeast phenylalanine transfer RNA. *Science* **185:**435–440.

Kudlicki, W., J. Chirgwin, G. Kramer, and B. Hardesty. 1995. Folding of an enzyme into an active conformation while bound as peptidyl tRNA to the ribosome. *Biochemistry* **34:**14284–14287.

Lata, K. R., R. K. Agrawal, P. Penczek, R. Grassucci, J. Zhu, and J. Frank. 1996. Three-dimensional reconstruction of the *E. coli* 30S ribosomal subunit in ice. *J. Mol. Biol.* **262:**43–52.

Leijonmarck, M., and A. Liljas. 1987. Structure of the C-terminal domain of the ribosomal protein L7/L12 from *E. coli* at 1.7 Å. *J. Mol. Biol.* **195:**555–579.

Lindahl, M., L. A. Svensson, A. Liljas, S. E. Sedelnikova, I. A. Eliseikina, N. P. Fomenkova, N. Nevskaya, S. V. Nikonov, M. B. Garber, T. A. Muranova, A. I. Rykonova, and R. Amons. 1994. Crystal structure of the ribosomal protein S6 from *Thermus thermophilus. EMBO J.* **13:**1249–1254.

Makeyev, E. V., V. A. Kolb, and A. S. Spirin. 1996. Enzymatic activity of the ribosome-bound nascent peptide. *FEBS Lett.* **378:** 166–170.

Malhotra, A., and S. C. Harvey. 1994. A quantitative model of the *E. coli* 16S RNA in the 30S ribosomal subunit. *J. Mol. Biol.* **240:**308–340.

Malhotra, A., P. Penczek, R. K Agrawal, I. S. Gabashvili, R. A. Grassucci, R. Jünemann, N. Burkhardt, K. H. Nierhaus, and J. Frank. 1998. *E. coli* 70S ribosome at 15 Å resolution by cryo-electron microscopy: localization of fMet-$tRNA_f^{Met}$ and fitting of L1 protein. *J. Mol. Biol.* **280:**103–116.

Markus, M. A., A. P. Hinck, S. Huang, D. E. Draper, and D. A. Torchia. 1997. High resolution solution structure of ribosomal protein L11-C76, a helical protein with a flexible loop that becomes structured upon binding to RNA. *Nat. Struct. Biol.* **4:**70–77.

May, R. P., V. Nowotny, P. Nowotny, H. Voss, and K. H. Nierhaus. 1992. Interprotein distances within the large subunit from *E. coli* ribosomes. *EMBO J.* **11:**373–378.

Mitchell, P., M. Osswald, and R. Brimacombe. 1992. Identification of intermolecular cross links at the subunit interface of the *E. coli* ribosome. *Biochemistry* **31:**3004–3011.

Moazed, D., and H. F. Noller. 1990. Binding of tRNA to the ribosomal A and P sites protects two distinct sets of nucleotides in 16S rRNA. *J. Mol. Biol.* **211:**135–145.

Moazed, D., J. M. Robertson, and H. F. Noller. 1988. Interaction of elongation factors EF-G and EF-Tu with a conserved loop in 23S RNA. *Nature* **334:**362–364.

Mueller, F., and R. Brimacombe. 1997a. A new model for the three-dimensional folding of *E. coli* 16S ribosomal RNA. I. Fitting the RNA to a 3D electron microscopic map at 20 Å. *J. Mol. Biol.* **271:**524–544.

Mueller, F., and R. Brimacombe. 1997b. A new model for the three-dimensional folding of *E. coli* 16S ribosomal RNA. II. The RNA-protein interaction data. *J. Mol. Biol.* **271:**545–565.

Mueller, F., H. Stark, M. van Heel, J. Rinke-Appel, and R. Brimacombe. 1997. A new model for the three-dimensional folding of *E. coli* 16S ribosomal RNA. III. The topography of the functional centre. *J. Mol. Biol.* **271:**566–587.

Mueller, F., I. Sommer, P. Baranov, H. Stark, M. van Heel, M. Rodnina, W. Wintermeyer, and R. Brimacombe. Submitted for publication.

Mundus, D., and P. Wollenzien. 1998. Neighbourhood of 16S rRNA nucleotides U788/U789 in the 30S ribosomal subunit determined by site-directed cross-linking. *RNA* **4:**1373–1385.

Nagano, K., M. Harel, and M. Takezawa. 1988. Prediction of three-dimensional structure of *E. coli* ribosomal RNA. *J. Theor. Biol.* **134:**199–256.

Nakagawa, A., T. Nakashima, M. Taniguchi, H. Hosaka, M. Kumura, and I. Tanaka. 1999. The three-dimensional structure of the RNA-binding domain of ribosomal protein L2; a protein at the peptidyl transferase center of the ribosome. *EMBO J.* **18:** 1459–1467.

Nevskaya, N., S. Tishchenko, A. Nikulin, S. Al-Karadaghi, A. Liljas, B. Ehresmann, C. Ehresmann, M. Garber, and S. Nikonov. 1998. Crystal structure of ribosomal protein S8 from Thermus thermophilus reveals a high degree of structural conservation of a specific RNA binding site. *J. Mol. Biol.* **279:**233–244.

Nikonov, S., N. Nevskaya, I. Eliseikina, N. Fomenkova, A. Nikulin, N. Ossina, M. Garber, B. H. Jonsson, C. Briand, S. Al-Karadaghi, A. Svensson, A. Ævarsson, and A. Liljas. 1996. Crystal structure of the RNA binding ribosomal protein L1 from *Thermus thermophilus. EMBO J.* **15:**1350–1359.

Nissen, P., M. Kjeldgaard, S. Thirup, G. Polekhina, L. Reshetnikova, B. F. C. Clark, and J. Nyborg. 1995. Crystal structure of the ternary complex of Phe-$tRNA^{Phe}$, EF-Tu and a GTP analog. *Science* **270:**1464–1472.

Osswald, M., and R. Brimacombe. 1999. The environment of 5S rRNA in the ribosome: cross-links to 23S rRNA from sites within helices II and III of the 5S molecule. *Nucleic Acids Res.* **27:**2283–2290.

Østergaard, P., H. Phan, L. B. Johansen, J. Egebjerg, L. Østergaard, B. Porse, and R. A. Garrett. 1998. Assembly of proteins and 5S rRNA to transcripts of the major structural domains of 23S rRNA. *J. Mol. Biol.* **284:**227–240.

Piepersberg, W., A. Böck, M. Yaguchi, and H. G. Wittmann. 1975. Genetic position and amino acid replacement of several mutations in ribosomal protein S5 from *E. coli. Mol. Gen. Genet.* **143:**43–52.

Powers, T., and H. F. Noller. 1995. Hydroxyl radical foot-printing of ribosomal proteins on 16S rRNA. *RNA* **1:**194–209.

Prince, J. B., B. H. Taylor, D. L. Thurlow, J. Ofengand, and R. A. Zimmermann. 1982. Covalent cross-linking of $tRNA_1^{Val}$ to 16S RNA at the ribosomal P site; identification of cross-linked residues. *Proc. Natl. Acad. Sci. USA* **79:**5450–5454.

Ramakrishnan, V., and S. W. White. 1992. The structure of ribosomal protein S5 reveals sites of interaction with 16S rRNA. *Nature* 358:768–771.

Rinke-Appel, J., N. Jünke, R. Brimacombe, I. Lavrik, S. Dokudovskaya, O. Dontsova, and A. Bogdanov. 1994. Contacts between 16S ribosomal RNA and mRNA, within the spacer region separating the AUG initiator codon and the Shine-Dalgarno sequence; a site-directed cross-linking study. *Nucleic Acids Res.* **22:** 3018–3025.

Rinke-Appel, J., N. Jünke, M. Osswald, and R. Brimacombe. 1995. The ribosomal environment of tRNA: cross-links to rRNA from positions 8 and 20:1 in the central fold of tRNA located at the A, P or E site. *RNA* **1:**1018–1028.

Samaha, R. R., R. Green, and H. F. Noller. 1995. A base pair between tRNA and 23S rRNA in the peptidyl transferase centre of the ribosome. *Nature* **377:**309–314.

Schmidt, F. J., J. Thompson, K. Lee, J. Dijk, and E. Cundliffe. 1981. The binding site for ribosomal protein L11 within 23S ribosomal RNA of *E. coli*. *J. Biol. Chem.* **256:**12301–12305.

Sergiev, P. V., I. N. Lavrik, V. A. Wlasoff, S. S. Dokudovskaya, O. A. Dontsova, A. A. Bogdanov, and R. Brimacombe. 1997. The path of mRNA through the bacterial ribosome: a site-directed cross-linking study using new photoreactive derivatives of guanosine and uridine. *RNA* **3:**464–475.

Stade, K., J. Rinke-Appel, and R. Brimacombe. 1989. Site-directed cross-linking of mRNA analogues to the E. coli ribosome; identification of 30S ribosomal components that can be cross-linked to mRNA at various points 5′ with respect to the decoding site. *Nucleic Acids Res.* **17:**9889–9908.

Stade, K., S. Riens, D. Bochkariov, and R. Brimacombe. 1994. Contacts between the growing peptide chain and the 23S RNA in the 50S ribosomal subunit. *Nucleic Acids Res.* **22:**1394–1399.

Stade, K., N. Jünke, and R. Brimacombe. 1995. Mapping the path of the nascent peptide chain through the 23S RNA in the 50S ribosomal subunit. *Nucleic Acids Res.* **23:**2371–2380.

Stark, H., F. Mueller, E. V. Orlova, M. Schatz, P. Dube, T. Erdemir, F. Zemlin, R. Brimacombe, and M. van Heel. 1995. The 70s E. coli ribosome at 23 Å resolution; fitting the ribosomal RNA. *Structure* **3:**815–821.

Stark, H., E. V. Orlova, J. Rinke-Appel, N. Jünke, F. Mueller, M. Rodnina, W. Wintermeyer, R. Brimacombe, and M. van Heel. 1997a. Arrangement of tRNAs in pre- and post-translocational ribosomes revealed by electron cryomicroscopy. *Cell* **88:**19–28.

Stark, H., M. Rodnina, J. Rinke-Appel, R. Brimacombe, W. Wintermeyer, and M. van Heel. 1997b. Visualization of elongation factor EF-Tu on the *E. coli* ribosome. *Nature* **389:**403–406.

Stark, H., M. Rodnina, F. Zemlin, W. Wintermeyer, and M. van Heel. Submitted for publication.

Stern, S., B. Weiser, and H. F. Noller. 1988. Model for the three-dimensional folding of 16S ribosomal RNA. *J. Mol. Biol.* **204:** 447–481.

Stoldt, M., J. Wöhnert, M. Görlach, and L. Brown. 1998. The NMR structure of *E. coli* ribosomal protein L25 shows homology to general stress proteins and glutaminyl-tRNA synthetases. *EMBO J.* **17:**6377–6384.

Szewczak, A. A., and P. B. Moore. 1995. The sarcin/ricin loop, a modular RNA. *J. Mol. Biol.* **247:**81–98.

Tanaka, I., A. Nakagawa, H. Hosaka, S. Wakatsuki, F. Mueller, and R. Brimacombe. 1998. Matching the crystallographic structure of ribosomal protein S7 to a three-dimensional model of the 16S ribosomal RNA. *RNA* **4:**542–550.

Unge, J., A. Åberg, S. Al-Karadaghi, A. Nikulin, S. Nikonov, N. L. Davydova, N. Nevskaya, M. Garber, and A. Liljas. 1998. The crystal structure of ribosomal protein L22 from *Thermus thermophilus*: insights into the mechanism of erythromycin resistance. *Structure* **6:**1577–1586.

Urlaub, H., B. Thiede, E. C. Müller, R. Brimacombe, and B. Wittmann-Liebold. 1997. Identification and sequence analysis of contact sites between ribosomal proteins and rRNA in *E. coli* 30S subunits by a new approach using matrix-assisted laser desorption/ionization mass spectrometry combined with N-terminal microsequencing. *J. Biol. Chem.* **272:**14547–14555.

VanLoock, M. S., T. R. Easterwood, and S. C. Harvey. 1999. Major groove binding of the tRNA/mRNA complex to the 16S ribosomal RNA decoding site. *J. Mol. Biol.* **285:**2069–2078.

Walleczek, J., D. Schüler, M. Stöffler-Meilicke, R. Brimacombe, and G. Stöffler. 1988. A model for the spatial arrangement of the proteins in the large subunit of the *E. coli* ribosome. *EMBO J.* **7:**3571 3576.

Wang, R., R. W. Alexander, M. VanLoock, S. Vladimirov, Y. Bukhtiyarov, S. C. Harvey, and B. S. Cooperman. 1999. Three-dimensional placement of the conserved 530 loop of 16S rRNA and of its neighbouring components in the 30S subunit. *J. Mol. Biol.* **286:**521–540.

Wilson, K. S., K. Appelt, J. Badger, I. Tanaka, and S. W. White. 1986. Crystal structure of a prokaryotic ribosomal protein. *Proc. Natl. Acad. Sci. USA* **83:**7251–7255.

Wimberly, B. T., S. W. White, and V. Ramakrishnan. 1997. The structure of ribosomal protein S7 at 1.9 Å resolution reveals a β-hairpin motif that binds double-stranded nucleic acids. *Structure* **5:**1187–1198.

Wimberly, B. T., R. Guymon, J. P. McCutcheon, S. White, and V. Ramakrishnan. 1999. A detailed view of a ribosomal active site: the structure of the GTPase center at 2.6 Å resolution. *Cell* **97:** 491–502.

Yonath, A., and F. Françeschi. 1998. Functional universality and evolutionary diversity: insights from the structure of the ribosome. *Structure* **6:**679–684.

The Ribosome: Structure, Function, Antibiotics, and Cellular Interactions
Edited by R. A. Garrett, S. R. Douthwaite, A. Liljas, A. T. Matheson, P. B. Moore, and H. F. Noller

Chapter 15

Functional Interpretation of the Cryo-Electron Microscopy Map of the 30S Ribosomal Subunit from *Escherichia coli*

MARGARET S. VanLOOCK, ARUN MALHOTRA, DAVID A. CASE, RAJENDRA K. AGRAWAL, PAWEL PENCZEK, THOMAS R. EASTERWOOD, JOACHIM FRANK, and STEPHEN C. HARVEY

Over the past 3 decades understanding the influence of structure on the function of the ribosome has been of particular interest. Consequently, an enormous amount of both biochemical and direct structural data has been collected. This wealth of information has been used by structural biologists to develop a series of both low-resolution and atomic-resolution models for either all or part of the ribosome (Brimacombe, 1988; Hubbard and Hearst, 1991; Malhotra and Harvey, 1994; Masquida et al., 1997; Mueller and Brimacombe, 1997a, 1997b; Mueller et al., 1997; Stern and Weiser, 1988; Wang et al., 1999). Recently, the ever-improving cryo-electron microscopy maps of the 70S ribosome have increased the overall accuracy of the modeling studies (Malhotra and Harvey, 1994; Mueller and Brimacombe, 1997a, 1997b; Mueller et al., 1997). These electron density maps have identified the binding sites for many of the ribosomal components, including the A- and P-site tRNAs (Agrawal et al., 1996; Malhotra et al., 1998; Agrawal et al., 1999a, 1999b; Stark et al., 1997a), EF-G (Agrawal et al., 1999a, 1999b) (see chapter 6), EF-Tu (Stark et al., 1997b), and the large-subunit protein L1 (Malhotra et al., 1998). In addition, the mRNA binding channel in the 30S subunit (Frank et al., 1995) and the polypeptide exit channel in the 50S subunit have been identified (Beckman et al., 1997; Frank et al., 1995; Yonath and Leonard, 1987). Together, this information will allow structural biologists to develop new models that not only give details of the structure of the ribosome but also add insight into how the structure influences ribosomal function.

To date, ribosome models can be divided into two general classes: those built by hand (Brimacombe, 1988; Mueller and Brimacombe, 1997a, 1997b; Mueller et al., 1997; Stern and Weiser, 1988) and those built with computer-automated algorithms (Hubbard and Hearst, 1991; Malhotra and Harvey, 1994; Wang et al., 1999). There are advantages and disadvantages to both approaches. In the past, we have focused our efforts on developing low-resolution models for the 30S ribosomal subunit by using the molecular-mechanics software package Yammp (Malhotra et al., 1994; Tan and Harvey, 1993). In this approach, molecular-mechanics protocols were developed that generated models compatible with the large body of low-resolution data for the structure of the ribosome. These data were generated by a variety of experimental techniques, including chemical and enzymatic probing, chemical and UV cross-linking, immuno-electron microscopy, and neutron diffraction. The folding algorithm used to generate our 30S subunit models is similar to that used when building and refining models based on nuclear overhauser effect spectroscopy and correlated-spectroscopy data from nuclear magnetic resonance (NMR). The computer is given a series of weighted distances derived from the experimental data, which it uses to interactively fold the RNA into conformations that best agree with the constraints. However, unlike NMR or X-ray crystallography, we used low-

Margaret S. VanLoock, Thomas R. Easterwood, and Stephen C. Harvey ■ Department of Biochemistry and Molecular Genetics, University of Alabama at Birmingham, Birmingham, AL 35294-0005. **Arun Malhotra, Rajendra K. Agrawal, and Pawel Penczek** ■ Wadsworth Center, P.O. Box 509, Albany, NY 12201-0509. **David A. Case** ■ Department of Molecular Biology, The Scripps Institute, 10550 N. Torrey Pines Rd., La Jolla, CA 92037. **Joachim Frank** ■ Howard Hughes Medical Institute, P.O. Box 509, Albany, NY 12201-0509.

resolution data to generate our 30S subunit model, thus making an atomic model inappropriate (Malhotra and Harvey, 1994). In particular, we used a pseudoatom representation in which each nucleotide was represented by one pseudoatom and each protein was treated as one large pseudoatom. The protein pseudoatoms were fixed onto the positions determined by neutron diffraction (Capel et al., 1988) and were used as a scaffold on which to fold the RNA. A series of models were generated representing a range of possible structures that conform to the experimental data. Not only did this approach explore a range of conformations, providing quantitative estimates of uncertainties in the positions of the various regions of RNA, but it also highlighted conflicts in the experimental data.

The approach we used to generate our previous models was sufficient to provide information concerning the global conformation of the 30S subunit, but it did not offer insight into how the RNA and proteins within the subunit contributed to ribosomal function. Information of this kind can only be derived from structures with atomic resolution in those regions of the ribosome that are functionally significant. Therefore, we are now focusing our efforts on developing multiscale-modeling protocols that will allow us to build second-generation models of the 30S ribosomal subunit with atomic detail in some regions and lower-resolution representations in others. As these models progress, we are beginning to develop hypotheses concerning structure-function relationships in ribosomes.

This chapter provides our view of three important structure-function relationships. First, we briefly review our recently published model of interactions between the 16S rRNA decoding site and the mRNA-tRNA complex (VanLoock et al., 1999). Second, we discuss the functional implications of our placement of that model in the electron density map. Finally, we identify regions that we believe compose the "gate" that closes to form the channel in which the mRNA is located.

The second-generation models incorporate atomic-resolution data for those components whose structures have been determined by crystallography or NMR, and for selected components we have modeled at atomic detail because of their functional significance. The models also incorporate the wealth of biochemical data available for the 30S subunit to ensure proper folding and the cryo-electron microscopy maps of the intact 70S ribosome (Malhotra et al., 1998). These electron density maps are currently reported at 15-Å resolution, thus allowing some pieces of the RNA to be identified. In particular, the position of the P-site tRNA has been determined by using the cryo-electron density map of the 70S ribosome from *Escherichia coli* complexed with both mRNA and fMet-$tRNA_f^{Met}$ positioned in the P site (Malhotra et al., 1998). The A-site tRNA binding site has also been located by using an electron density map of a pretranslocation complex (Agrawal et al., 1999b).

We used these tRNA positions to dock our atomic model of the 16S rRNA decoding site (Fig. 1) (VanLoock et al., 1999) in the electron density map. In this model, the mRNA-tRNA complex binds in the major groove of the decoding-site RNA such that interactions between the complex and A1492, A1493, and C1400 could be maintained (Fig. 2). A1492 and A1493 in the A-site region of the decoding-site RNA interact directly with the A-site codon in a sequence-independent manner. This interaction suggests a mechanism for ribosomal proofreading and may explain how aminoglycoside antibiotics that bind to this region of the 16S rRNA disrupt this proofreading. In particular, ribosomal fidelity requires that the A-site region of the decoding-site RNA distinguish cognate from noncognate tRNAs. According to our model, when cognate tRNAs bind to the A-site codon, the necessary hydrogen bonds between A1492-A1493 and the mRNA are formed. However, when noncognate tRNAs attempt to bind, the orientation of A1492 and A1493 is distorted such that the proper interactions cannot be made with the mRNA. When aminoglycoside binds to the A-site region, A1492 and A1493 are pushed out into the minor groove of the decoding-site RNA, thus making hydrogen bonds between the mRNA and A1492 or A1493 impossible. This of course decreases translational fidelity. In addition, the aminoglycosides may offer many nonspecific hydrogen-bonding donors and acceptors that can interact with the mRNA-tRNA complex directly. This would lock the mRNA-tRNA complex into the A-site region of the decoding site, thus explaining the decreased off rate of A-site tRNA in the presence of aminoglycoside.

Superimposition of the P-site tRNA from our decoding-site model onto the P-site tRNA that had been manually placed into the electron density map tests the hypothesis that mRNA lies in the channel of the small subunit, as earlier proposed (Frank et al., 1995). Figure 3 shows that this superimposition exercise places the mRNA right in the center of the channel and that the walls of the channel represent the decoding site of the 16S rRNA. This is an exciting result, since the modeling of the mRNA–tRNA–decoding-site RNA complex was done without prior reference to the electron density map.

It is also worth noting that the mRNA in the P site lies in a region of relatively high density, while the density in the region of the A-site mRNA is rel-

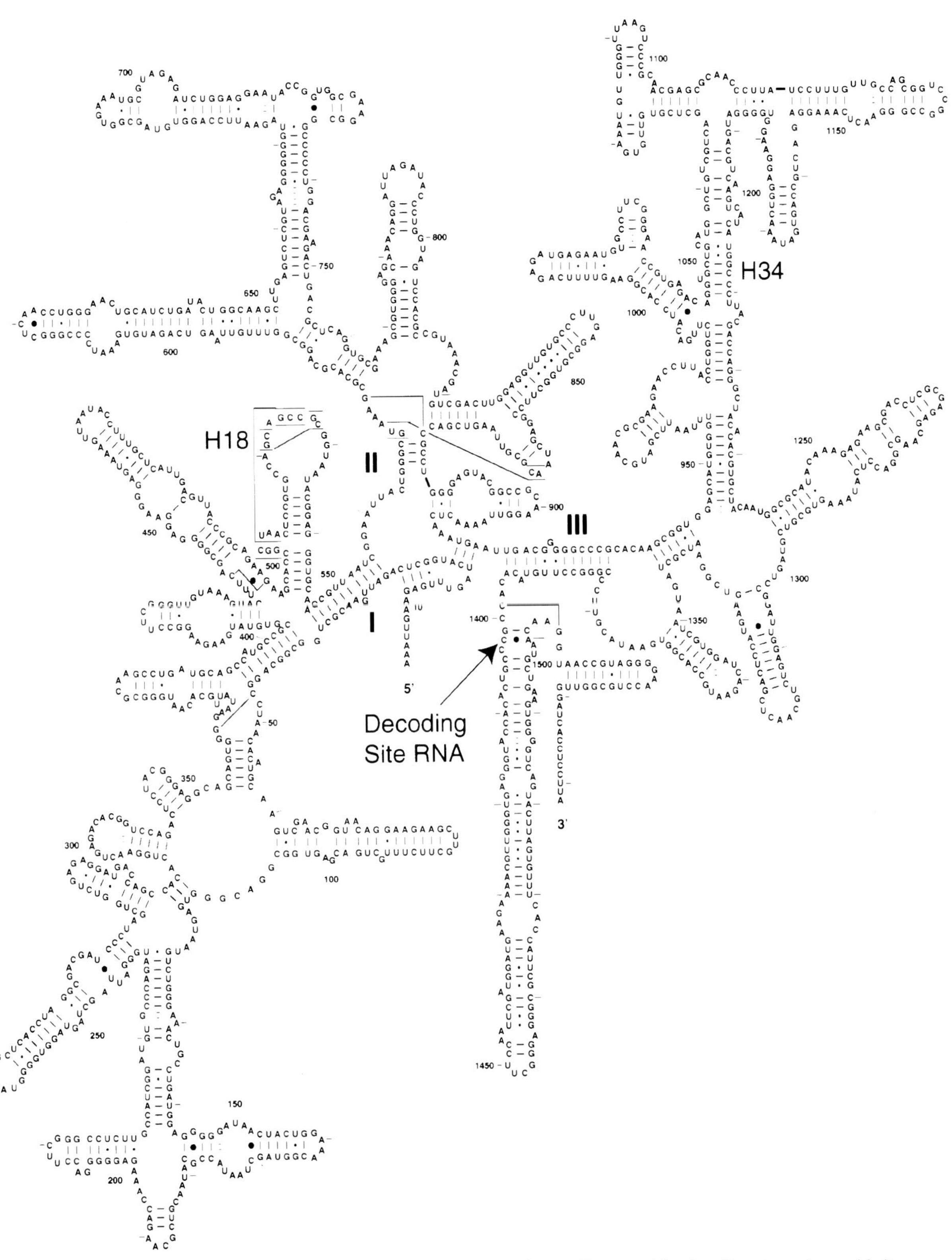

Figure 1. Secondary-structure diagram of the 16S rRNA from *E. coli* (Gutell, 1994). The decoding site region and helices 18 and 34 have been labeled.

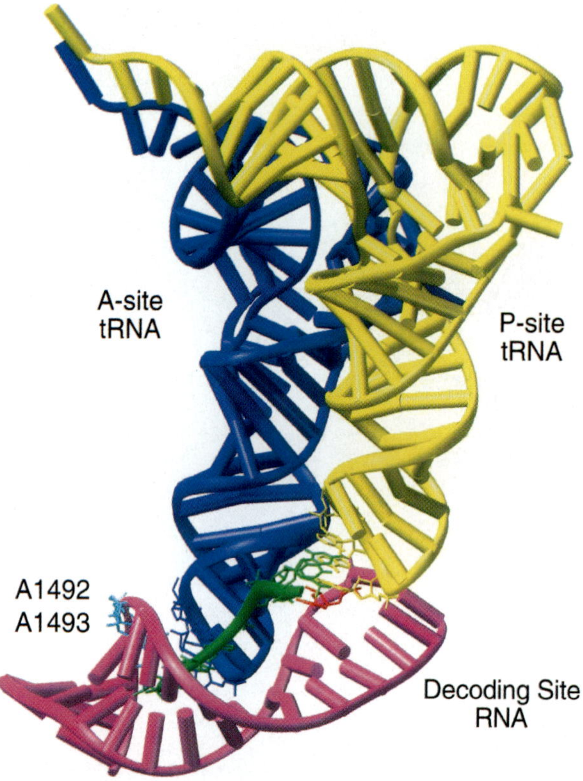

Figure 2. Ribbon diagram (Carson, 1987) of the 16S rRNA decoding-site model (VanLoock et al., 1999) with the mRNA-tRNA complex bound in the major groove of the 16S rRNA. A1492 and A1493 (cyan) form sequence-independent interactions with the A-site codon of the mRNA (green). C1400 (red) stacks directly beneath the wobble base of the P-site tRNA (yellow), as was indicated by a UV-induced cross-link between these two residues (Prince et al., 1982). The A-site tRNA is shown in blue.

atively low (Fig. 3 and 4). This difference suggests a dynamic proofreading model: the electron density map used in Fig. 3 and 4 was derived from ribosomes with fMet-$tRNA_f^{Met}$ bound at the P site and no tRNA bound to the A site. Under these circumstances, we believe that the mRNA-16S rRNA conformation is dynamic at the A site, due to the absence of tRNA, thus explaining the lack of density in the A-site region. Binding of the cognate tRNA to the A site produces the single well-defined conformation shown in Fig. 2, which is the signal for the ribosome to proceed with transpeptidation and translocation. We therefore predict that the density in the channel will rise (the channel will appear to be narrower) when tRNA is bound at the A site.

Another interesting aspect of this model is that the axis of the decoding-site RNA lies perpendicular to the long axis of the 30S ribosomal subunit. This aligns the decoding site RNA parallel to the direction of movement of the anticodon loops during translocation. We believe that the decoding-site RNA drives the mRNA through the channel by a back and forth ratchet-like motion.

Once the decoding-site RNA was positioned, we then used distance geometry to position the remainder of the RNA and to generate a series of starting models for both manual and automated refinement. In the starting models, the A- and P-site tRNAs and the decoding-site RNA were locked into position and the ribosomal proteins were held firmly at positions specified by the neutron diffraction map (Capel et al., 1988). A large list of approximately 250 biochemical and structural constraints between the 16S rRNA, A- and P-site tRNAs, mRNA, and ribosomal proteins was included in the refinement protocol. In addition, a newly developed surface function (Tan and Harvey, unpublished) was included to ensure that the rRNA was positioned in the regions of highest electron density during model refinement. After the initial round of refinement, the structures were improved in an iterative process of manual and automated refinement, analogous to that used in refining structures from X-ray crystallography. This process is ongoing and will continue until a series of models have been generated that satisfy the biochemical, structural, and electron density constraints.

In the small ribosomal subunit there is a channel through the body in 70S ribosomes with bound mRNA (Fig. 4) (Frank et al., 1995). One wall of the channel corresponds to the neck region of the subunit, while the other is a thinner wall that is disrupted in the free form of the 30S subunit (Lata et al., 1996). Since the latter opens to permit mRNA binding and then closes, we call it the gate. The gate has been identified as either a set of tertiary interactions between regions of the 16S rRNA or a set of interactions between the rRNA and ribosomal proteins (Mueller et al., 1997). We have placed a series of helices from domains I and II of the 16S rRNA into the electron density map and have identified the helices that lie on either side of the gate. Our distance geometry models have suggested that helix 18, which lies on the body side of the gate, interacts with helix 34 or 35, which lie on the head side of the gate. In support of these findings, the 30S subunit model of Mueller et al. (1997) places helix 34 across from helix 18, while the model developed from the Santa Cruz-Stanford collaboration has helix 35 close to helix 18 (coordinates obtained from http://smi-web.stanford.edu/projects/helix/ribo3dmodels/index.html). We believe that the gate is a tertiary interaction formed between helix 34 (1050 region) and helix 18 (530 stem-loop) (Fig. 1). This may resemble a tetraloop-tetraloop receptor interaction. The details

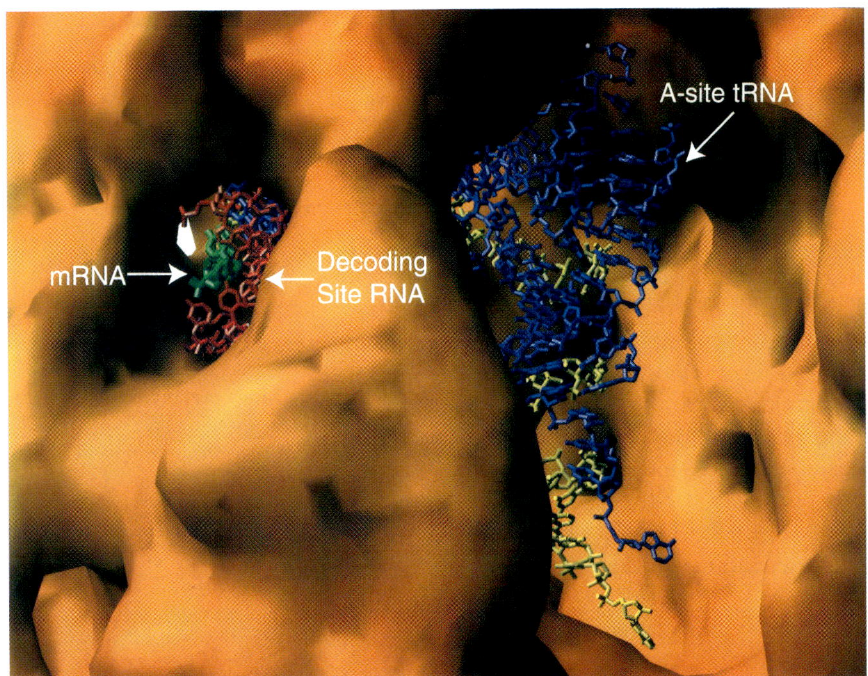

Figure 3. Ribbon image showing the position of the 16S rRNA decoding-site model (VanLoock et al., 1999) docked into the 15-Å cryo-electron microscopy map of the intact 70S ribosome (Malhotra et al., 1998). The mRNA (green) lies in the center of the channel that is surrounded by the decoding-site RNA (red). This cryo-electron microscopy reconstruction has an fMet-$tRNA_f^{Met}$ (yellow) bound to the P site and no tRNA bound to the A site. However, the A-site tRNA (blue) from the decoding-site model was included in this docking exercise. The P-site tRNA (yellow) is also visible in this image. The reason for this is threefold: first, the threshold of the electron microscopy map is high, thus eliminating regions of low electron density; second, the P site is not fully occupied by fMet-$tRNA_f^{Met}$; and third, there is motion in the acceptor stem. All three of these factors together decrease the electron density in the region of the P-site tRNA, thus making it visible in this figure.

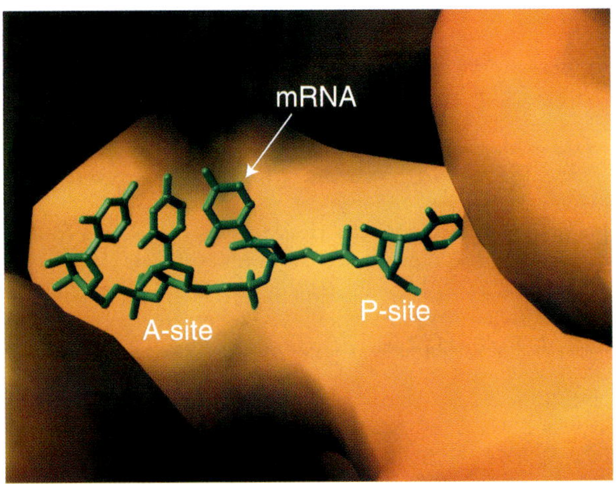

Figure 4. The mRNA within the channel, viewed from the direction of the large subunit. At this density threshold, the mRNA in the P site is buried in the density, reflecting the well-defined structure of the tRNA, mRNA, and 16S RNA in the P site. A lack of tRNA in the A site produces a dynamic, poorly defined structure for the A-site mRNA-rRNA complex. This is reflected by the lower density in the A-site part of the channel and the visibility of the A-site mRNA in this model.

of the tertiary interactions are still preliminary, but we believe that nucleotides 530 to 535 adopt either a GNRA tetraloop or a U-turn conformation that allows it to dock in the minor groove of the lower stem in helix 34 (nucleotides 1047 to 1053 and 1205 to 1210).

A variety of both structural and biochemical data support the interaction between these two regions of the 16S rRNA. In particular, comparative-analysis studies have shown that G529, G530, U531, A532, and A533 and all but C1049, G1207, and C1210 in helix 34 are over 95% conserved (Gutell, 1994). Therefore, all of the bases necessary for the tertiary interactions have been conserved throughout evolution, thus highlighting their importance for ribosome function. Cross-linking studies between mRNA and the 16S rRNA also suggest that these two regions of the 16S rRNA are close together in the ribosomal subunit (Rinke-Appel et al., 1993; Sergiev et al., 1997). The +6 and +8 nucleotides in an mRNA with 4-thiouridine modifications have been shown to cross-link to helix 34, and the +11 and +12 residues in the same modified mRNA cross-link to nucleotides 532 and 530 in helix 18 (Rinke-Appel et al., 1993; Sergiev et al., 1997). In order for these cross-links to occur, both regions of the 16S rRNA must be close together in the ribosomal subunit.

We have already suggested that the gate must open and close in order to accommodate the translation of circular mRNAs (Lata et al., 1996). Gabashvili et al. (1999) have revealed the dynamic relationship between the head and the body by using cryo-electron microscopy on free 30S subunits. They showed that the head and neck of the 30S subunit display "nodding" motions, and these motions are presumably similar to those required to close the gate when mRNA is bound to intact ribosomes. These results support the dynamic nature of the interaction between the head and the body in that the ribosome can alternate between the closed and open forms of the gate. The gate is opened and closed as the proposed interaction between helix 34 and helix 18 is broken and reformed, depending on the state of the ribosome.

Multiscale modeling permits investigation of structure-function relationships at atomic detail within the low-resolution framework of those regions that are less well defined and less functionally significant. This allows one to focus on those regions of the system that are most critical for function while paying less attention to those regions that are less critical. It should be noted that the structure-function models we put forward are not definitive. Only X-ray crystallography can provide definitive details of the atomic structure adopted by systems like the ribosome. The purpose of our efforts is to provide a framework for examining relationships between structure and function in the ribosome and to offer suggestions concerning experimental tests for these hypotheses.

This work was supported by grants from the National Institutes of Health: P41 RR 12255 (S.C.H. and D.A.C.), R01 GM 53827-04 (S.C.H.), R37 GM 29169 (J.F.), and R01 GM 55440 (J.F.).

REFERENCES

Agrawal, R. K., P. Penczek, R. A. Grassucci, Y. Li, A. Leith, K. H. Nierhaus, and J. Frank. 1996. Direct visualization of A-, P-, and E-site transfer RNAs in the Escherichia coli ribosome. *Science* **271**:1000–1002.

Agrawal, R. K., A. B. Heagle, P. Penczek, R. Grassucci, and J. Frank. 1999a. EF-G dependent GTP hydrolysis-induces translocation accompanied by large conformational changes in the 70S ribosome. *Nat. Struct. Biol.* **6**:643–647.

Agrawal, R. K., P. Penczek, R. A. Grassucci, N. Burkhardt, K. H. Nierhaus, and J. Frank. 1999b. Effect of buffer conditions on the position of tRNA on the 70S ribosome as visualized by cryoelectron microscopy. *J. Biol. Chem.* **274**:8723–8729.

Beckman, R., D. Bubeck, R. A. Grassucci, P. Penczek, A. Verschoor, G. Blobel, and J. Frank. 1997. Alignment of conduits for the nascent polypeptide chain in the ribosome-Sec61 complex. *Science* **278**:2123–2126.

Brimacombe, R. 1988. The emerging three-dimensional structure and function of 16S ribosomal RNA. *Biochemistry* **28**:4207–4214.

Capel, M. S., M. Kjeldgaard, D. M. Engelman, and P. B. Moore. 1988. Positions of S2, S13, S16, S17, S19 and S21 in the 30 S ribosomal subunit of Escherichia coli. *J. Mol. Biol.* **200**:65–87.

Carson, M. L. 1987. Ribbon models of macromolecules. *J. Mol. Graph.* **5**:103–106.

Frank, J., J. Zhu, P. Penczek, Y. Li, S. Srivastava, A. Verschoor, M. Radermacher, R. Grassucci, K. R. Lata, and R. K. Agrawal. 1995. A model of protein synthesis based on cryo-electron microscopy of the E. coli ribosome. *Nature* **376**:441–444.

Gabashvili, I. S., R. K. Agrawal, R. Grassucci, and J. Frank. 1999. Structure and structural variations of the Escherichia coli 30 S ribosomal subunit as revealed by three-dimensional cryoelectron microscopy. *J. Mol. Biol.* **286**:1285–1291.

Gutell, R. R. 1994. Collection of small subunit (16S- and 16S-like) ribosomal RNA structures: 1994. *Nucleic Acids Res.* **22**:3502–3507.

Hubbard, J. M., and J. E. Hearst. 1991. Computer modeling 16S ribosomal RNA. *J. Mol. Biol.* **221**:889–907.

Lata, K. R., R. K. Agrawal, P. Penczek, R. Grassucci, J. Zhu, and J. Frank. 1996. Three dimensional reconstruction of the Escherichia coli 30 S ribosomal subunit in ice. *J. Mol. Biol.* **262**:43–52.

Malhotra, A., and S. C. Harvey. 1994. A quantitative model of the *Escherichia coli* 16S RNA in the 30S ribosomal subunit. *J. Mol. Biol.* **240**:308–340.

Malhotra, A., R. K.-Z. Tan, and S. C. Harvey. 1994. Modeling large RNAs and ribonucleoprotein particles using molecular mechanics techniques. *Biophys. J.* **66**:1777–1795.

Malhotra, A., P. Penczek, R. Agrawal, I. S. Gabashvili, R. A. Grassucci, R. Junemann, N. Burkhardt, K. H. Nierhaus, and J. Frank. 1998. Escherichia coli 70 S ribosome at 15A resolution by cryo-election microscopy: localization of fMet-tRNAfMet and fitting of L1 protein. *J. Mol. Biol.* **280**:103–116.

Masquida, B., B. Felden, and E. Westhof. 1997. Context dependent RNA-RNA recognition in a three-dimensional model of the 16S rRNA core. *Bioorg. Med. Chem.* **5**:1021–1035.

Mueller, F., and R. Brimacombe. 1997a. A new model for the three-dimensional folding of Escherichia coli 16 S ribosomal RNA. I. Fitting the RNA to a 3D electron microscopic map at 20Å. *J. Mol. Biol.* **271**:524–544.

Mueller, F., and R. Brimacombe. 1997b. A new model for the three-dimensonal folding of Escherichia coli 16 S ribosomal RNA. II. The RNA-protein interaction data. *J. Mol. Biol.* **271**: 545–565.

Mueller, F., H. Stark, M. van Heel, J. Rinke-Appel, and R. Brimacombe. 1997. A new model for the three-dimensional folding of Escherichi coli 16 S ribosomal RNA. III. The topography of the functional centre. *J. Mol. Biol.* **271**:566–587.

Prince, J. B., B. H. Taylor, D. L. Thurlow, J. Ofengand, and R. A. Zimmermann. 1982. Covalent crosslinking of $tRNA^{Val}$ to 16S RNA at the ribosomal P site: identification of crosslinked residues. *Proc. Natl. Acad. Sci. USA* **79**:5450–5454.

Rinke-Appel, J., N. Junke, R. Brimacombe, S. Dokudovskaya, O. Dontsova, and A. Bogdanov. 1993. Site-directed cross-linking of mRNA analogues to 16 S ribosomal RNA; a complete scan of cross-links from all positions between +1 and +16 on the mRNA, downstream from the decoding site. *Nucleic Acids Res.* **21**:2853–2859.

Sergiev, P. V., I. N. Lavrik, and V. A. Wlasoff. 1997. The path of mRNA through the bacterial ribosome: a site-directed crosslinking study using new photoreactive derivatives of guanosine and uridine. *RNA* **3**:464–475.

Stark, H., E. V. Orlova, J. R. Appel, N. Junke, F. Mueller, M. Rodnina, W. Wintermeyer, and R. Brimacombe. 1997a. Arrangement of tRNAs in pre- and posttranslocational ribosomes revealed by electron cryomicroscopy. *Cell* **88:**19–28.

Stark, H., M. V. Rodnina, J. R. Appel, and R. Brimacombe. 1997b. Visualization of elongation factor Tu on the Escherichia coli ribosome. *Nature* **389:**403.

Stern, S., and B. Weiser. 1988. Model for the three-dimensional folding of 16 S ribosomal RNA. *J. Mol. Biol.* **204:**447–481.

Tan, R. K.-Z., and S. C. Harvey. 1993. Yammp: development of a molecular mechanics program using the modular programming method. *J. Comp. Chem.* **14:**455–470.

Tan, R. K.-Z., and S. C. Harvey. Unpublished data.

VanLoock, M. S., T. R. Easterwood, and S. C. Harvey. 1999. Binding of the tRNA/mRNA complex to the 16S ribosomal RNA decoding site. *J. Mol. Biol.* **285:**2069–2078.

Wang, R., R. W. Alexander, M. VanLoock, S. Vladimirov, Y. Buktiyarov, S. C. Harvey, and B. S. Cooperman. 1999. Three-dimensional placement of the conserved 530 loop of 16 S rRNA and of its neighboring components in the 30 S subunit. *J. Mol. Biol.* **286:**521–540.

Yonath, A., and K. R. Leonard. 1987. A tunnel in the large ribosomal subunit revealed by three dimensional image reconstruction. *Science* **236:**813–816.

V. rRNA MODIFICATION AND PROTEIN SIGNALS

V. *r*RNA MODIFICATION AND PROTEIN SIGNALS

rRNA modifications tend to be concentrated in rRNA regions known to be functionally important, and they are generally considered to be required for tuning the various ribosomal functions. Moreover, they tend to vary in nature and position from organism to organism, perhaps reflecting the different prevailing conditions in vivo. Their importance is underlined by the fact that their synthesis is quite expensive for the cell. The first two chapters in this section concentrate on the mechanisms of synthesis of the most common modifications, pseudouridines and methylations. The final chapter summarizes evidence for small sequence-specific signals within the proteins which determine their locations in the cell.

The Ribosome: Structure, Function, Antibiotics, and Cellular Interactions
Edited by R. A. Garrett, S. R. Douthwaite, A. Liljas, A. T. Matheson, P. B. Moore, and H. F. Noller

Chapter 16

Bacterial, Archaeal, and Organellar rRNA Pseudouridines and Methylated Nucleosides and Their Enzymes

JAMES OFENGAND and KENNETH E. RUDD

rRNAs contain a number of different modified nucleosides, mainly, but not exclusively, pseudouridine (Ψ, 5-ribosyl-uracil) and nucleosides methylated on either the base or the 2′-hydroxyl of the ribose (Rozenski et al., 1999). Of these, Ψ is by far the most prevalent, although the sum of all the methylated nucleosides frequently is close to the number of Ψ. In this review, we focus on the rRNAs of bacteria, archaea, and organelles, including the last because of their presumed evolutionary relationship to bacterial organisms. Modified nucleosides in eukaryotic rRNA are discussed in chapter 17.

Ψ and Ψ synthases will be emphasized in this review for the simple reason that considerably more has been learned in the past several years about how Ψ is made and what it does than about the methylated nucleosides and their associated methyltransferases. The pseudouridines of ribosomal RNA have been reviewed previously (Ofengand et al., 1995; Ofengand and Fournier, 1998).

PSEUDOURIDINE

General Considerations

Ψ is formed from U by enzymatic cleavage of the *N*-glycosyl bond, rotation (apparently while still on the enzyme surface) of the uracil ring about one of the two axes shown in Fig. 1 so as to bring C-5 to the location previously occupied by N-1, and reformation of the glycosyl link as a C-C bond. In this process, the C-5 H is released to water. This forms the basis of a quantitative assay for Ψ formation when the substrate contains [5-^{3}H]uridine (Cortese et al., 1974). Ψ formation takes place with selected uridines only after they have been incorporated into a polynucleotide chain. Therefore, the first major question is how particular uridine residues are selected for isomerization. A partial answer to this question is now available (see below).

The second major question, still largely unresolved, is why make Ψ at all. Ψ is found only in RNAs whose tertiary structure is integral to their function, such as rRNA, tRNA, and snRNA, and the most obvious structural effect of Ψ formation is the creation of a new H bond donor located where C-5 used to be. These facts suggest that Ψ may be a stabilizer of important intramolecular or intermolecular tertiary structural interactions. However, no direct evidence exists to support this hypothesis. Nevertheless, it has recently been shown that at least in two cases, lack of a Ψ synthase which makes specific Ψ residues in *Escherichia coli* rRNA compromises the cell's ability to grow, especially when in competition with wild-type cells (Raychaudhuri et al., 1998, 1999). A molecular explanation for these defects is not yet available.

Occurrence of Ψ in rRNA

The number of Ψ residues in the small-subunit (SSU) and large-subunit (LSU) rRNAs of bacteria, archaea, and organelles range from 0 to 9 for those organisms and organelles studied, accounting for up to 2% of the uridines and ca. 0.3% of the sum of all nucleotides (Table 1). A distinction between SSU RNA and LSU RNA is that while there are two examples with no Ψ in their SSU RNA, all organisms so far have at least one Ψ in their LSU RNA. It is also noteworthy that while *E. coli* has 10 Ψ and *Sul-*

James Ofengand and Kenneth E. Rudd ■ Department of Biochemistry and Molecular Biology, University of Miami School of Medicine, Miami, FL 33101.

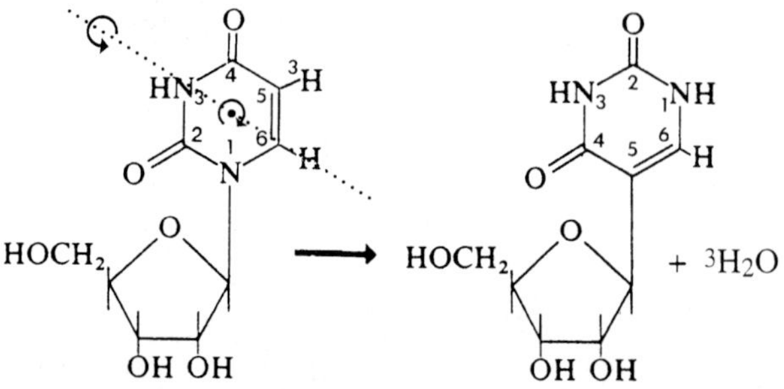

Figure 1. Conversion of uridine (left) to pseudouridine (right). The reaction results in the release of the C-5 H of uridine (shown as ^3H) to water. Two axes are shown about which the uracil ring could rotate so as to place C-5 at the position previously occupied by N-1.

folobus solfataricus has 8 to 9, they are distributed between the two ribosomal subunits very differently. It is too early to tell if this is a singular occurrence or if it reflects a difference between bacteria and archaea. The locations of the *E. coli* residues are shown in Fig. 2 and 3. The Ψ sites in *S. solfataricus* are not known. Eukaryotes contain many more Ψ, up to 44 in the cytoplasmic SSU RNA and up to 57 in LSU RNA (Ofengand and Fournier, 1998). Most of the data in Table 1 and for eukaryotes were obtained by Ψ sequencing, so in most cases the exact locations are known. The striking result of these studies was that the LSU RNA Ψ of all the organisms studied were localized at three distinct regions, all three of which were at or near the peptidyltransferase center of the ribosome (Ofengand and Bakin, 1997) (Fig. 3). This was so despite the up-to-10-fold variation in the absolute number of Ψ residues. In contrast, there was no such localization apparent in SSU RNA (reviewed in Ofengand and Fournier, 1998). The correlation between the positions of Ψ and methylated nucleosides in LSU RNA has been previously noted (Bakin et al., 1994b; Ofengand and Bakin, 1997), as well as the lack of any correlation in SSU RNA (Bakin and Ofengand, 1995; Maden, 1990).

Identification and Properties of the Ψ Synthases of *E. coli*

Identification

With a single exception in *Bacillus subtilis*, all of the available information on the rRNA Ψ synthases comes from *E. coli* (Table 2). As shown in Table 2 and in Fig. 2 and 3, five synthases specific for 9 of the 10 Ψ in *E. coli* rRNA have been identified. A synthase for the unassigned Ψ2605 has been identified in *B. subtilis*, but its counterpart in *E. coli* has

Table 1. Number of pseudouridine residues in LSU and SSU ribosomal RNAs of bacteria, archaea, and organelles

Organism or organelle	No. of Ψ residues detected	%U	% Total nucleotides	Analysis method[a]	% RNA sequenced	Reference(s)
SSU cytoplasm						
Escherichia coli	1	0.3	0.06	Seq.	100	Bakin et al., 1994a
Bacillus subtilis	1	0.3	0.06	Seq.	15	Niu and Ofengand, 1999
Halobacter halobium	0			Seq.	70	Bakin and Ofengand, 1995
Halobacterium volcanii	0			Tot.		McCloskey, personal communication
Sulfolobus solfataricus	4–5	1.8–2.2	0.28–0.35	Tot.		Noon et al., 1998
LSU cytoplasm						
Escherichia coli	9[b]	1.5	0.31	Seq.	99	Bakin and Ofengand, 1993
Bacillus subtilis	5[b]	0.9	0.17	Seq.	56	Ofengand and Bakin, 1997
Halobacter halobium	4	0.7	0.14	Seq.	54	Ofengand and Bakin, 1997
Sulfolobus solfataricus	4	0.8	0.13	Tot.		Noon et al., 1998
LSU mitochondria						
Saccharomyces cerevisiae	1	0.1	0.03	Seq.	99	Bakin et al., 1994b
Mus musculus	1	0.2	0.06	Seq.	17	Ofengand and Bakin, 1997
Homo sapiens	1	0.3	0.06	Seq.	41	Ofengand and Bakin, 1997
Trypanosoma brucei	6	0.8	0.35	Seq.	98	Ofengand and Bakin, 1997
LSU chloroplasts						
Zea mays	4[b]	0.7	0.14	Seq.	29	Ofengand and Bakin, 1997

[a] Seq., by Ψ sequencing, locations are known; Tot., by total Ψ analysis, locations are unknown.
[b] Includes the N_3-methyl Ψ in *E. coli* (Kowalak et al., 1996) and the unidentified modified U found at the same site in *B. subtilis* and *Z. mays*.

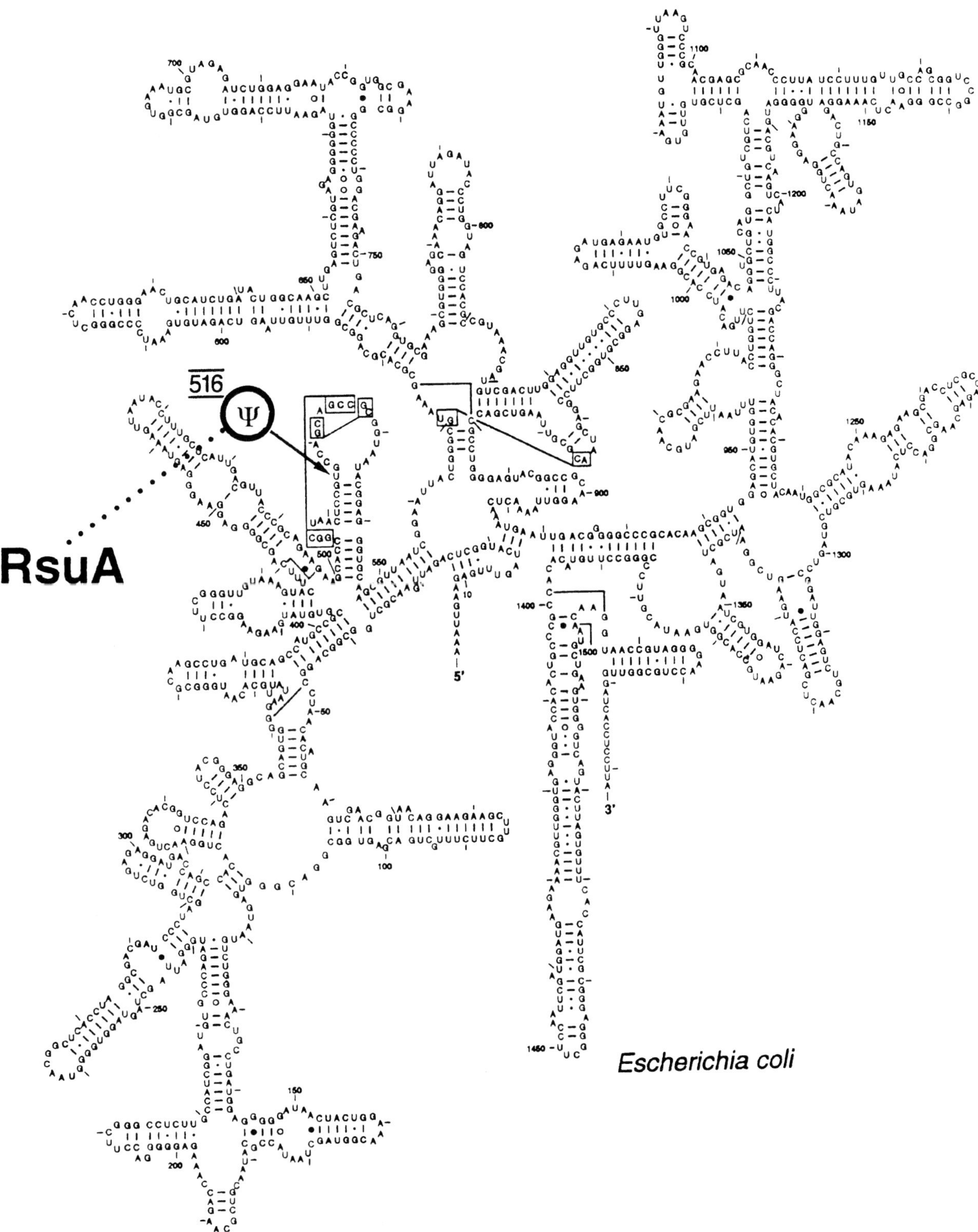

Figure 2. Secondary structure of *E. coli* 16S RNA showing the site of the single Ψ along with the designation of the synthase which forms it.

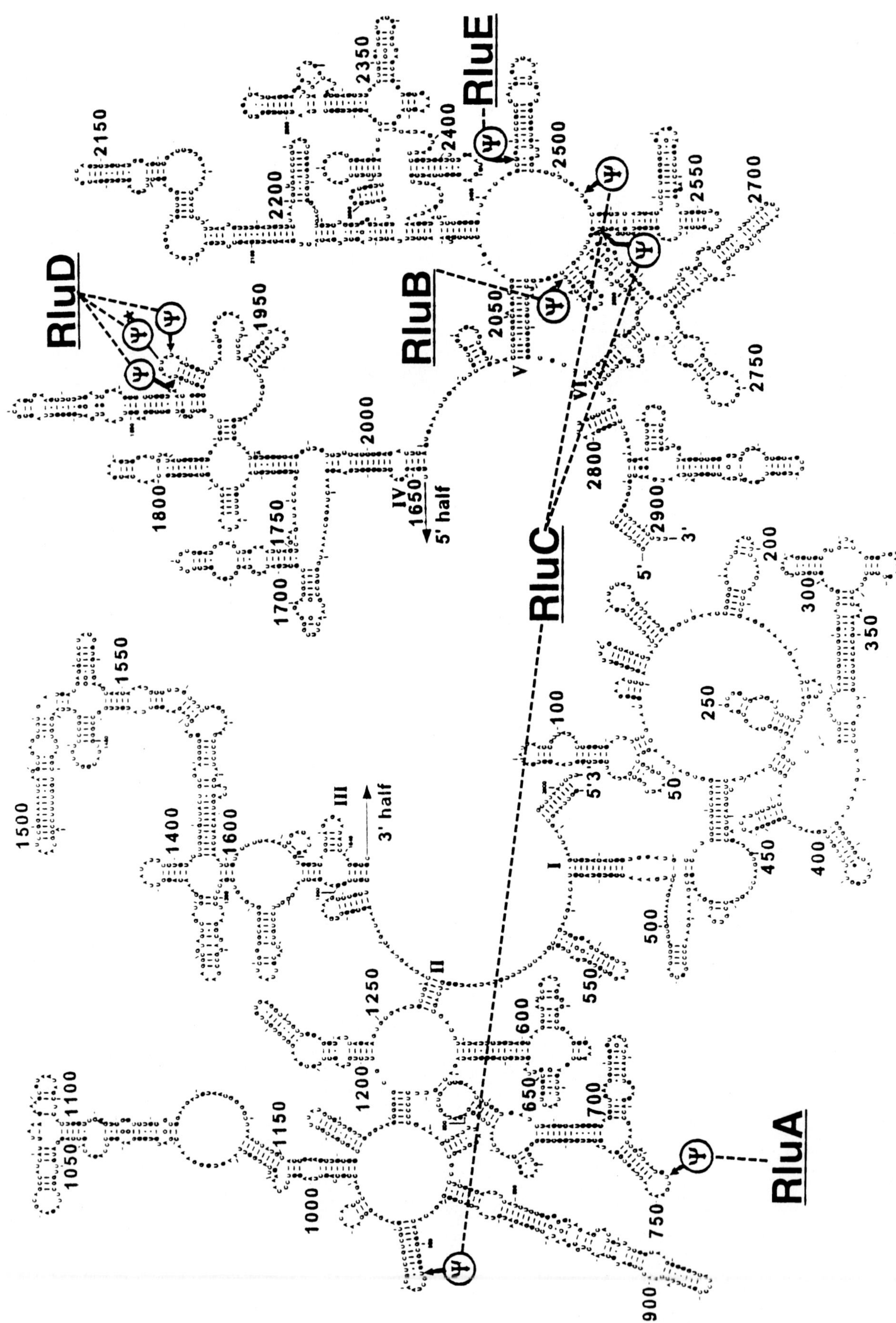
RluA
RluB
RluC
RluD
RluE
5' half
3' half
I
II
III
IV
V
VI
5'
3'
50
100
200
250
300
350
400
450
500
550
600
650
700
750
900
1000
1050
1100
1150
1200
1250
1400
1500
1550
1600
1650
1700
1750
1800
1950
2000
2050
2150
2200
2350
2400
2500
2550
2700
2750
2800
2900

not yet been found. Assignment of Ψ to specific synthases was accomplished by two complementary methods. Overexpression of the putative synthase and purification, usually by means of an N-terminal His-tag sequence, allowed in vitro quantitation of Ψ formation when [5-^{3}H]uridine-labeled substrate was used. This assay also was used to assess RNA substrate specificity, that is, whether tRNA, SSU RNA, or LSU RNA was the substrate. Subsequent Ψ sequence analysis (Bakin and Ofengand, 1993, 1998) showed the site(s) of formation in the given RNA. The problem with this approach is that, unknowingly, an inappropriate substrate may be used. For example, RsuA does not react with intact SSU RNA or several of its fragments, nor does it react well with complete 30S particles. The preferred substrate is an intermediate RNP particle (Wrzesinski et al., 1995a). In vitro activity of RsuA was only detected because, fortuitously, the 30S particles used initially were sufficiently unfolded to be reactive. On the other hand, deletion of the *rsuA* gene showed clearly that it coded for the synthase responsible for SSU RNA Ψ516 (Conrad et al., 1999).

Deletion of the synthase gene and subsequent RNA analysis for Ψ can also be misleading if another synthase shares the specificity for forming a particular Ψ. In that event, both synthase genes need to be deleted to see an effect. So far, 8 of the 10 Ψ of *E. coli* rRNA have been analyzed by deletion of synthase genes without any evidence for such shared specificity. The only two remaining are Ψ2605, for which no synthase has yet been identified in *E. coli*, and Ψ2457, which has been assigned to RluE only by the overexpression method.

Multiple specificity

Three instances of multiple-site specificity have been found so far. RluA was the first example of "dual specificity." It forms only Ψ746 in LSU RNA, but it also makes Ψ32 in tRNA (Wrzesinski et al., 1995b). Moreover, it is the only protein in *E. coli* able to make both Ψ, since gene deletion causes the loss of both Ψ and replacement of the gene on a plasmid restores them both (Raychaudhuri et al., 1999). A similar example has recently been described. In this case, a *Saccharomyces cerevisiae* tRNA Ψ synthase which recognizes eight sites in tRNA (Motorin et al., 1998) also specifically forms Ψ44 in yeast U2 snRNA (Massenet et al., 1999).

RluC and RluD are two examples of multiple-site specificity in the same RNA molecule. Deletion of RluC resulted in the loss of Ψ955, Ψ2504, and Ψ2580, and replacement of the gene on a plasmid restored all three Ψ. Moreover, reaction of overexpressed RluC with 23S RNA in vitro also resulted in formation of the same three Ψ. The surprising aspect is that U955 is far from either U2504 or U2580 and the three sites share virtually no common elements of either primary or secondary structure. The only discernible common feature is that each of the U residues is followed by a G. However, there are many other UG sequences in 23S RNA that are not recognized by RluC. This aspect needs further work.

The three Ψ made by RluD, Ψ1911, Ψ1915, and Ψ1917, cluster in a small loop which contains no other U residues. Therefore, it may be that RluD is specifically bound to the stem-loop structure and converts all U residues within a certain range. There are no other U residues nearby; the closest is 6 residues away from U1917. Moreover, if RluD has a recognition specificity for UA analogous to the putative UG specificity of RluC, the nearest other UA sequences would be at U1898 and U1926, which may be too far removed from the site of action. The isomerization of several nearby U residues by one enzyme has a precedent in TruA, which is able to recognize and isomerize U residues at positions 38, 39, and 40 in tRNA. However, in that case, no more than two Ψ are formed in any one tRNA and the maximum spacing is 1 residue, whereas with RluD, the spacing is 3 between Ψ1911 and Ψ1915, and three Ψ are formed. The deletion analysis identification of the Ψ residues formed by RluC and RluD has been confirmed by Huang et al. (1998a). For unknown reasons, these authors were unable to also confirm the in vitro specificity described above and in Table 2.

Other properties

The Ψ synthases of *E. coli* range from ~220 to ~330 amino acids in length, with genes distributed throughout the chromosome (Table 2). Their amino acid sequences place them in four related families, as indicated. The four families are named for the four original members and do not necessarily indicate the RNA substrate for the synthase. This issue is discussed in more detail below.

Figure 3. Secondary structure of *E. coli* 23S RNA showing the locations of the nine Ψ in this RNA along with the assignments of the five synthases which form them.

Table 2. Properties of known bacterial rRNA and tRNA pseudouridine synthases

Organism	Name	Previous gene name(s)	SWISSPROT file name	No. of amino acids	Gene locus (kb)[a]	Family[b]	Substrate (site[s])	How identified[c]	Reference(s)
E. coli	RsuA	*yejD*	P33918	231	2277.8	RsuA	16S rRNA (516)	Overexp. Deletion	Wrzesinski et al., 1995a Conrad et al., 1999
E. coli	RluA	*yabO*	P39219	219	59.7	RluA	23S rRNA (746) tRNA (32)	Overexp. Deletion	Wrzesinski et al., 1995b Raychaudhuri et al., 1999
B. subtilis	RluB	*ypuL*	P35159	229	2421.4	RsuA	23S rRNA (2633[d])	Overexp.	Niu and Ofengand, 1999
E. coli	RluC	*yceC*	P23851	319	1144.2	RluA	23S rRNA (955, 2504, 2580)	Deletion Overexp.	Conrad et al., 1998 Conrad and Ofengand, unpublished
E. coli	RluD	*yfiI, sfhB*	P33643	326	2733.1	RluA	23S rRNA (1911, 1915, 1917)	Deletion; overexp.	Raychaudhuri et al., 1998
E. coli	RluE	*ymfC*	P75966	217 (207)[e]	1193.5	RsuA	23S rRNA (2457)	Overexp.	Conrad et al., unpublished
E. coli	TruA	*hisT*	P07649	270	2432.8	TruA	tRNA (38–40)	Deletion; overexp.	Arps et al., 1985; Marvel et al., 1985; Kammen et al., 1988
E. coli	TruB	*yhbA*	P09171	314	3309.5	TruB	tRNA (55)	Overexp. Deletion	Nurse et al., 1995 Englund et al., unpublished

[a] Positions for *E. coli* taken from Rudd (1998); the *B. subtilis* value is from the Colibri website (www.pasteur.fr/Bio/SubtiList/).
[b] According to amino acid sequence homology (Koonin, 1996; Conrad et al., 1999).
[c] Overexp., overexpression of the protein from a cloned gene, in vitro formation with an RNA or RNP substrate, and sequence analysis of the modified RNA; Deletion, deletion of the synthase gene and sequence analysis of the RNA from the deletion strain.
[d] Equivalent to position 2605 in *E. coli*.
[e] Values obtained with the next downstream initiator AUG. The true start site is uncertain.

Other ORFs

In addition to the known synthases listed in Table 2, three other *E. coli* open reading frames (ORFs) were identified by Koonin (1996) as possessing Ψ synthase motifs (see below). They are known as *yciL*, *yjbC*, and *yqcB*. Their properties are similar to those in Table 2. Interestingly, the Ψ sites in *E. coli* remaining to be assigned to a synthase also number three, namely, LSU RNA Ψ2605, $tRNA^{Glu}$ Ψ13, and $tRNA^{Asp}$ and $tRNA^{Ile}$ Ψ65. The other three tRNA Ψ sites, 32, 38 to 40, and 55, have already been assigned to synthases, as indicated in Table 2. The one-to-one correspondence between known sites and putative ORFs gives confidence that the assignments will soon be forthcoming and that there are not likely to be any other Ψ synthases in *E. coli* with differing sequence motifs.

Functional Effects of Deletion of *E. coli* Ψ Synthases

The availability of deletion mutants for four rRNA synthases and three tRNA synthases (RluA has both functions) made it possible to examine the effect of the absence of selected Ψ residues from SSU RNA, LSU RNA, and tRNA on cell growth (Table 3). Exponential-phase growth rates were the same as those of the wild-type for five of the six synthases studied, even at different temperatures and in rich and poor media. Only RluD-minus cells had a lower growth rate. Although the growth rate was only reduced by half, this has a major effect on the ability of the cells to propagate. When wild-type and RluD-minus cells were grown on Luria-Bertani plates for 19 h, the wild-type cells produced healthy colonies but the mutant cells produced only tiny, pinpoint colonies (Fig. 4). In fact, from the doubling times of 29 and 58 min, respectively, one can calculate that the relative numbers of cells in such colonies grown for 19 h should be 820,000 to 1. This effect was specifically due to the lack of RluD, because when the *rluD* gene was supplied on a plasmid, normal growth was restored (Fig. 4, lower right panel) and the exponential growth rate also returned to normal (Raychaudhuri et al., 1998). The stem-loop containing these three Ψ is near the decoding center of the 30S subunit, as

Table 3. Effect of deletion and conserved aspartate mutation of *E. coli* pseudouridine synthases on cell growth and synthase activity in vivo and in vitro

Synthase	Substrate (site[s])	Growth properties of the synthase deletion strain		Site of aspartate mutation	Mutant synthase activity (%)		Reference(s)
		Expon. ph.[a]	Competition[b]		In vivo	In vitro	
RsuA	16S RNA (516)	No effect	42	D102T D102N	<1 <1	NT NT	Conrad et al., 1999; Raychaudhuri et al., 1999
RluA	23S RNA (746) tRNA (32)	No effect	0.1	D64T D64N	<1 <1	<2 <2	Raychaudhuri et al., 1999; Ramamurthy et al., 1999
RluC	23S RNA (955, 2504, 2580)	No effect	21	D144T D144N	NT NT	NT NT	Conrad et al., 1998; Raychaudhuri et al., 1999
RluD	23S RNA (1911, 1915, 1917)	50% of wild type	NT[d]	D139T D139N	NT NT	<5 <1	Raychaudhuri et al., 1998; Raychaudhuri and Ofengand, unpublished
TruA	tRNA (38–40)	No effect[c]	NT	D60X[e]	NT	<0.01	Tsui et al., 1991; Huang et al., 1998b
TruB	tRNA (55)	No effect	NT	D48	NT	<0.1	Englund et al., unpublished; Ramamurthy et al., 1999

[a] Expon. ph., exponential-phase growth in Luria-Bertani or M-9 medium at 24, 37, or 42°C. TruB and TruA were only grown at 37°C.
[b] Growth in competition with wild-type cells by the 24-h growth cycle method described in the text. The values are the percentages of the original mutant culture remaining after six growth cycles.
[c] A three- to four-fold decrease in growth rate was only observed in minimal medium without uracil. This is thought to be due to the *rph* mutation in the test strain (Jensen, 1993).
[d] NT, not tested.
[e] X, represents A, E, K, N, or S.

shown by cross-linking (Mitchell et al., 1992) and mutation studies (O'Connor and Dahlberg, 1995). Therefore, it is possible that the severe growth defect observed in the absence of these three Ψ residues is caused by an effect on codon recognition.

The strong effect of RluD deletion prompted a closer examination of the growth rates of strains carrying deletions of RsuA, RluA, and RluC. Recognizing that in a natural setting, a mutant which arises spontaneously has to compete with wild-type cells for survival in a milieu where nutrients are periodically supplied and exhausted, a competition experiment was designed such that a series of cycles of initiation of growth, exponential growth, approach to stationary phase, and stationary phase was experienced by a mixed culture of wild-type and mutant cells. This was achieved by growing a mixed culture to stationary phase, dilution, and regrowth to stationary phase for a number of cycles (Raychaudhuri et al., 1999). The results are shown in Table 3. RluA-minus cells did not compete well at all, with only 0.1% of the initial number of cells present after six cycles of growth. The effects of the absence of RluC or RsuA were much less marked but still significant. The strong effect of RluA deletion might be due to the loss of Ψ from both LSU RNA and tRNA. Nevertheless, the loss of three Ψ from LSU RNA in the RluC-minus strain had much less of an effect. The rate of loss of RluA-minus cells as a function of the number of cycles of growth did not depend on the dilution used between cycles. The same decay rate was found with dilutions which varied by twofold the number of required doublings to reach stationary phase (Raychaudhuri et al., 1999), suggesting that the selection against RluA-deficient cells did not occur during exponential-phase growth, in agreement with the short-term rate measurements cited above. Exactly how the loss of these Ψ from rRNA (and tRNA) negatively affects cell physiology is not known.

These growth defects could also reflect the loss of the synthase more directly. Should these synthases have other essential functions in the cell in addition to their role in Ψ formation, the presence or absence of Ψ may not be related to the growth defects. There

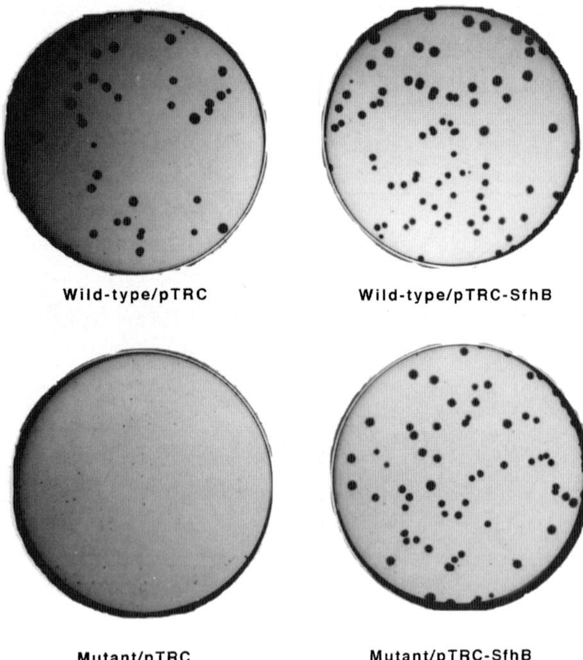

Figure 4. Diminished growth of *E. coli* lacking a functional RluD. The *rluD* gene was disrupted and inactivated by a mini-Tn*10*::*cat* insertion two-thirds of the way along the gene (Raychaudhuri et al, 1998). The defect was rescued by supplying an intact gene in *trans* on a plasmid. Wild-type/pTRC, MG1655 ($rluD^+$) containing the plasmid vector pTrc; Wild-type/pTRC-SfhB, same, except the pTrc vector contained the *rluD* gene; Mutant/pTRC, MG1655 (*rluD* minus) containing the plasmid vector pTrc; Mutant/pTRC-SfhB, same, except the pTrc vector contained the *rluD* gene. (Reprinted from Raychaudhuri et al., 1998, with the permission of Cambridge University Press.)

is precedent for such a situation. The *E. coli* RUMT enzyme, which forms m^5U54 in tRNA, is essential, but its methylation activity is dispensable (Persson et al., 1992). This problem could be addressed by use of the Asp mutants described in the next section.

A Functionally Essential Aspartate Residue Found in a Conserved Sequence Motif

Since all Ψ synthases catalyze the same reaction, it is reasonable to assume that the catalytic centers are constructed in similar ways and thus that there should be amino acid sequence motifs common to all Ψ synthases. Using the sequences of the four Ψ synthases known at the time, RsuA, RluA, TruB, and TruA, Koonin (1996) identified three conserved sequence motifs and defined four distinct families, one for each of the original Ψ synthases. None of the three motifs were identified in TruA, nor was motif III identified in the TruB family (but see below). However, a motif II-like region is highly conserved within the TruA family (Huang et al., 1998b) and all Ψ synthase family members (including the TruA family) have an invariant Asp residue in motif II (Fig. 5), despite the fact that extensive motif-based database searches did not provide statistical support for a homologous relationship between TruA and the other Ψ synthases (Conrad et al., 1999; but see below). Site-directed mutagenesis of the invariant Asp residue in TruA demonstrated that this residue is essential for in vitro activity, even when Asp was replaced by the similar amino acids Asn and Glu (Huang et al., 1998b) (Table 3). Mutations in the invariant Asp of RsuA (Conrad et al., 1999), RluA (Raychaudhuri et al., 1999; Ramamurthy et al., 1999), RluD (Raychaudhuri and Ofengand, unpublished), and TruB (Ramamurthy et al., 1999) also produced inactive enzymes (Table 3). RsuA and RluA were tested in vitro and in vivo, while RluD and TruB were analyzed only in vitro. Mutant RluA was also inactive in tRNA Ψ32 formation (Raychaudhuri et al., 1999). Although these results fall short of demonstrating a role for the COOH of Asp at the catalytic center, they do show that Asp in this particular sequence context is an essential residue for Ψ formation activity in several, and possibly all, Ψ synthases.

Ψ Synthases Identified by Sequence Homology

In addition to the 4 known Ψ synthases, Koonin (1996) identified 31 putative ones by searching for similarity among genomic sequences available at that time. The RluA and RsuA families were identified as being homologous (also shown by Gustafsson et al. [1996]), and the possibility that TruB was also a member of a Ψ synthase superfamily was suggested. This work was recently reexamined and expanded, since many new genome sequences have now become available (Conrad et al., 1999). A total of 111 members were identified for the RluA, RsuA, and TruB families, and 33 members were identified in the TruA family. The additional sequences and use of the PSI-BLAST reiterative motif search tool produced significant statistical support for an RsuA-RluA-TruB superfamily. An updated version of the compilation by Conrad et al. (1999) is shown in Fig. 5. It consists of 27 RsuA, 61 RluA, 44 TruB, and 43 TruA family members. Within each family, the conserved motif II region sequences are grouped as bacterial, archaeal, and eukaryotic, in that order. Subtle variations in the motif II sequences that allow numerous subgroups to be identified are discussed in detail by Conrad et al. (1999). Since the TruA family includes some TruA orthologues that are quite distantly related to each other, we label TruA a second Ψ synthase superfamily. Not depicted are a large number of putative Ψ synthases available only as unfinished genome sequences not yet deposited in the DNA sequence databases.

Although the TruA (Ψ synthase superfamily II) members are bona fide Ψ synthases, their amino acid sequences have not previously been shown to have statistically significant similarity to the RsuA-RluA-TruB Ψ synthase superfamily I. However, a preliminary analysis with the MACAW multiple-alignment program (Schuler et al., 1991) has identified all three conserved motifs in all four *E. coli* Ψ synthase families in similar positions and with high statistical significance ($P < 10^{-10}$), suggesting that there may actually be only one Ψ synthase superfamily (Rudd and Ofengand, unpublished) in agreement with the Asp mutational studies described above.

Number and Classification of Putative Ψ Synthases in Complete Genomes

A complete set of all the Ψ synthases in a given organism can be obtained from the data in Fig. 5 when the complete genome sequence is available. This is the case for the organisms listed in Table 4. These should be the total number of Ψ synthases in those organisms unless there are sequences so divergent from those in superfamilies I and II that they are unrecognizable by the search method used. There is no evidence that such synthases exist, but the limited extent of our knowledge does not allow a definitive assessment of this possibility. It is also important to note that the family designations based on amino acid sequence homology do not necessarily define the specificities of the member synthases. Thus, although the original RsuA makes SSU RNA Ψ516, two other RsuA family members, namely, RluB (*B. subtilis*) and RluE, make LSU RNA Ψ (Table 2). On the other hand, all RluA family members so far make only Ψ in LSU RNA. TruB family members so far make only Ψ55 in tRNA, although it must be admitted that only two examples, in *E. coli* (Nurse et al., 1995) and in yeast (Becker et al., 1997), are known. As indicated in Table 4, a TruB homologue related to yeast Cbf5, the putative eukaryotic rRNA Ψ synthase (see chapter 17), is also found in this family. TruA members are not necessarily restricted to positions 38 to 40 in tRNA, as for the classic *E. coli* TruA (Table 2), although so far they all do act on tRNA. Pus1p of yeast is responsible for eight sites in tRNA, but not positions 38 to 40 (Motorin et al., 1998), plus one site in U2 snRNA (Massenet et al., 1999); Pus2p does not yet have a defined site or RNA specificity (Simos et al., 1996); while Pus3p is the classic TruA homologue (Lecointe et al., 1998). In mice (combined with *Homo sapiens* in Table 4), the TruA member is an analogue of Pus1p (Chen and Patton, 1999). The specificity of the single TruA synthase in all of the bacterial and archaeal organisms in Table 4 is not known. Conceivably, they do not all have a classic TruA specificity.

The data in Table 4 support the importance of Ψ in the life cycle of a cell. *Mycoplasma genitalium*, with only 467 protein-coding ORFs, nevertheless has three synthases, corresponding to a use of 0.64% of the entire genome for Ψ biosynthesis. A similar-percentage use, but with higher absolute values, is found for five other organisms. *M. genitalium*, *Mycoplasma pneumoniae*, and *Helicobacter pylori* lack a TruB synthase. Either they do not have Ψ55 in their tRNA, a synthase from another family is the Ψ55 synthase, or one of the other synthases has an expanded specificity to encompass U55. If the latter, the TruA synthase is the most likely candidate. Note also that neither *M. genitalium* nor *M. pneumoniae* has any RsuA family members. Either one of the RluA members can make the necessary Ψ (although so far no RluA-class members are known to make Ψ in SSU RNA), or Ψ is lacking in the SSU RNA. Also noteworthy is the fact that there is never more than one member of the TruB and TruA families in members of the domains *Bacteria* and *Archaea*, despite the expansion in the RsuA and RluA families of *Bacteria*.

The archaeal organisms listed in Table 4 have identical synthase distributions, one each in the TruB and TruA families but none for making Ψ in rRNA. Unfortunately, there are no data on the presence of Ψ in the rRNAs of these organisms, although other archaeal species have from four to eight Ψ in their rRNAs (Table 1). It would be striking if the rRNAs of the listed organisms were devoid of Ψ. However, it is more likely that one or both of the existing synthases are able to make Ψ in rRNA either instead of or in addition to forming Ψ in tRNA. In this regard, note that *Methanobacterium thermoautotrophicum* has a Ψ54Ψ55 sequence in at least two of its tRNAs (Gu et al., 1984) and so should need a TruB-like activity. Alternatively, these archaea could use a guide RNA system for their rRNA Ψ, like eukaryotes. In that event, one or the other of the two synthases, most likely the TruB synthase, would have to assume the role of the enzyme working with the guide RNAs in addition to its role in forming Ψ55 and probably Ψ54. These archaea could also make use of synthase(s) with sequences too divergent from those of the four families described here to be detected.

The eukaryotes lack RsuA-like synthases. Since the mitochondrial ribosomes of yeast and human each contain one Ψ in the LSU RNA (Ofengand and Bakin, 1997), one of the RluA synthases in yeast and the only one in *Caenorhabditis elegans* and human may be for a mitochondrial LSU RNA. The single

```
                    Ψ synthase superfamily I

                 RsuA family
Db|Access|Gene_Organism          Motif
                              $GRLD +T G$$$$
                              A       S
sp|P33918|RSUA_ECOLI          AGRLDIDTTGLVLM
sp|P75966|RLUE_ECOLI          AGRLDRDSEGLLVL
sp|P37765|YCIL_ECOLI          VGRLDVNTCGLLLF
sp|P32684|YJBC_ECOLI          IGRLDKDSQGLIFL
sp|P45124|RSUA_HAEIN          AGRLDVDTTGLVLL
sp|P44827|RLUE_HAEIN          AGRLDRDSEGLLIL
sp|P45104|YCIL_HAEIN          VGRLDINTSGLLLF
sp|O32068|YTZF/BACSU          AGRLDKDTEGFLLL
sp|P35159|RLUB_BACSU          IGRLDYDTSGLLLL
sp|P72581|Y612_SYNY3          VGRLDQDSEGLLLL
sp|Q55578|Y361_SYNY3          VGRLDRNSTGALLL
sp|O66829|Y554_AQUAE          AGRLDVDAEGLLLI
sp|O67444|YE64_AQUAE          VGRLDYNTEGLLIL
SANG|RSUX/CAMJE               VGRLDYASEGLLLL
sp|P55986|YE59_HELPY          VGRLDFASEGVLLL
gb|4980760|RSUA/THEMA         VGRLDKDAEGLLII
TIGR|RSUA/DEIRA               VGRLDKDSEGLLLL
tr|O83472|RSUX/TREPA          IGRLDVRSEGALLF
sp|O51155|Y129_BORBU          IGRLDFKSSGLLLF
em|e1342837|RSUX/RICPR        IGRLDLNSEGLLLL
tr|O84728|YJBC/CHLTR          VGRLDKETSGLILV
gb|4377181|RSUA/CHLPN         VGRLDKETSGLILV
sp|P42395|YCIL_BUCAP          VGRLDINTKGLLLF
sp|O05668|YRSU_MYCLE          VGRLDADTEGLILL
sp|O33210|YRSU_MYCTU          VGRLDADTEGLMLL

gb|3845303|RSUX/PLAFA         VGRLDRNTSGVLLL
tr|O80967|RSUX/ARATH          VGRLDVATTGLIVV

                 RluA family
                              $HRLD++TSG$$$$

sp|P39219|RLUA_ECOLI          VHRLDMATSGVIVV
sp|P33643|RLUD_ECOLI          VHRLDKDTTGLMVV
sp|P23851|RLUC_ECOLI          VHRLDRDTSGVLLV
sp|Q46918|YQCB_ECOLI          AHRLDRPTSGVLLM
sp|P44782|RLUA_HAEIN          VHRLDMATSGIIVF
sp|P44445|RLUD_HAEIN          VHRLDKDTTGLMVV
sp|P44433|RLUC_HAEIN          VHRLDRDTSGILLI
sp|P44197|YQCB_HAEIN          IHRLDRPTSGVLLF
sp|P43930|Y042_HAEIN          VHRLDKVTSGLLIL
sp|O25441|Y745_HELPY          VHRLDKDTSGGIVI
sp|O25610|Y956_HELPY          LHRLDKETSGVVLL
sp|O25114|Y347_HELPY          AHRLDYETSGLVLA
tr|O83359|RLUX/TREPA          VHRLDKDTSGVLLT
tr|O83189|YQCB/TREPA          LHRLDRGTEGLIAF
tr|O83259|RLUX/TREPA          LHRLDKDTAGVLLF

             RluA family (con't)
Db|Access|Gene_Organism          Motif
                              $HRLD++TSG$$$$

sp|Q45480|YLYB_BACSU          VHRIDKDTSGLLMV
sp|P54604|YHCT_BACSU          VHRLDQDTSGAIVF
sp|O31613|YJBO_BACSU          VTRLDRDTSGIMLV
SANG|RLUX/CAMJE               AHRLDKETSGLILI
SANG|RLUY/CAMJE               VHRLDKDTSGAILI
SANG|RLUZ/CAMJE               LNRLDKETSGVILL
TIGR|RLUX/DEIRA               VHRLDKDTSGVIVV
TIGR|RLUY/DEIRA               LHRLGRGTSGLVLF
TIGR|RLUZ/DEIRA               PHRLDRETSGAQLL
sp|Q10786|Y04P_MYCTU          VHRLDVGTSGVMVV
tr|O07166|YQCB/MYCTU          AHRLDRLTAGVLLF
sp|P74346|YG29_SYNY3          VHRLDKDTTGAMVV
sp|P72970|YF92_SYNY3          LHRLGTGTSGLLLL
sp|P47451|Y209_MYCGE          VHRLDRDTSGAIVV
sp|P47610|Y370_MYCGE          AHRIDRNTSGIVIG
sp|P75230|Y370_MYCPN          AHRIDRNTCGLVIG
sp|P75485|Y209_MYCPN          VHRLDRDTSGVIML
em|1343115|RLUD/RICPR         VHRLDKDTTGLMVV
em|e1342564|RLUX/RICPR        VHRLDKETSGLLLI
sp|P70870|RLUD_BORBU          VHRLDKDTSGVLIC
tr|O51755|RLUX/BORBU          VHRLDRNTSGIIIF
gb|4980971|RLUD/THEMA         VHRLDKETSGVIVV
gb|4981478|RLUC/THEMA         VHRLDKETSGLLVV
tr|O84665|RLUD/CHLTR          VHRLDKDTSGLLIT
tr|O84108|RLUC/CHLTR          VHRLDRDTSGCILF
sp|P50513|RLUD_ZYMMO          VHRIDKDTSGLLVV
sp|O66114|YLP4_ZYMMO          VHRLDRDTSGCLLL
gb|4155455|RLUX/HELP2         LHRLDKETSGVILL
sp|O67638|YH58_AQUAE          VHRLDKETAGVMVI
sp|P33640|RLUD_PSEAE          VHRLDKDTTGLMVV
gb|4377023|RLUA/CHLPN         VHRLDKDTSGLIIT
gb|4539134|RLUD/MYCLE         VHRLDVGTSGVMLV
sp|O50310|YBC5_CHLVI          VHRLDKDTSGLIII
sp|Q45826|YMDA_CHLAU          VHRLDRDTSGLLVI
tr|P70863|RLUX/BARBA          VHRLDRETSGLLVV
sp|Q47417|YQCB_ERWCA          VHRLDRPTSGVLLL

gb|3834304|RLUX/ARATH         VHRLDRETSGLLVM
sp|Q12069|YD36_YEAST          CNRLDRLTSGLMFL
sp|P53294|YG3X_YEAST          CYRLDKITSGLLIL
sp|Q12362|RIB2_YEAST          CNRLDKPTSGLMFL
tr|Q06244|Y165C/YEAST         VHRLDHCVTGGMLI
sp|Q09709|YA32_SCHPO          CNRLDRLTSGLMFF
tr|Q25346|RLUX/LEIMA          CHNLDTETSGCVVL
tr|Q25257|RLUX/LEIDO          VHRLDAETSGCLLI
sp|O16686|YK27_CAEEL          LHRLDRATSGVLLF
gb|AC006246|RLUX/DROME        IHRLDRLTSGLLLF
```

Figure 5. Alignment of 132 motif II sequences of the Ψ synthase superfamily I and 43 TruB motif II-like sequences. The 14 amino acids of motif II are aligned and grouped according to families. The motif consensus shown above the sequences summarizes the sequence patterns but does not indicate all pattern variations. Motif conventions: single capital letters, invariant or nearly invariant; vertical letter pairs indicate the two most common amino acids at that position; $, usually one of ILVM; +, charged. The *B. subtilis* YTZF/BASCU protein sequence is a reconstruction of a probable frameshift mutation that fuses two adjacent protein sequences, YTZF_BACSU (O32068) and YTZG_BACSU (O32069). There is also a human orthologue to the YD36_YEAST subgroup of the RluA family (tr, Q92939) which is not listed because it is a C-terminal partial sequence with the motif II region still unsequenced. Two TruA sequences were omitted from the alignment. The conserved motif in the TruA homologue YQN3_CAEEL (sp, Q09524) appears to have two gaps and therefore was not aligned. A TruA homologue found in the database of expressed sequence tags, db EST, supposedly from *Arabidopsis thaliana* (gb, N37304), is very similar to the *E. coli* sequence, and we suspect it represents a bacterial contaminant in that particular EST library. The database record accession numbers are taken from a variety of databases abbreviated as follows: sp, SWISS-PROT; pi, PIR; tr, TREMBL, em, EMBL; gb, GenBank; dd, DDBJ. The citations to the original sequence papers can be found within the database records. Locus names were taken from SWISS-PROT or are provisional designations (if Gene and Organism are connected by a backslash instead of an underline). When no orthologue was obvious and no other name was available, a generic family name was given to the protein as a temporary identifier, e.g., RLUX or RSUX. The SWISS-PROT organism codes used are as follows: ACICA, *Acinetobacter calcoaceticus*; AQUAE, *Aquifex aeolicus*; ARATH, *Arabidopsis thaliana*; ARCFU, *Archaeoglobus fulgidus*; ASPFU, *Aspergillus fumigatus*; BACHD, *Bacillus halodurans*; BACSP, *Bacillus* sp. strain KSM-64; BACSU, *Bacillus subtilis*; BARBA, *Bartonella bacilliformis*; BUCAP, *Buchnera aphidicola*; BORBU, *Borrelia burgdorferi*; CHLTR, *Chlamydia trachomatis*; CHLPN, *Chlamydia pneumoniae*; CAEEL, *Caenorhabditis elegans;* CAMJE, *Campylobacter jejuni*; CANAL, *Candida albicans*; CHLAU, *Chloroflexus aurantiacus*; CHLVI, *Chlorobium vibrioforme*; CRYPV, *Cryptosporidium parvum*; DEIRA, *Deinococcus radiodurans*; DROME, *Drosophila melanogaster*; ECOLI, *Escherichia coli*; EMENI, *Emericella nidulans*; ERWCA, *Erwinia carotovora*; HAEIN, *Haemophilus influenzae*; HELPY, *Helicobacter pylori*; HUMAN, *Homo sapiens*; KLULA, *Kluyveromyces lactis*; LACLA, *Lactococcus lactis*; LEIDO, *Leishmania donovani*; LEIMA, *Leishmania major*; METJA, *Methannococcus jannaschii*; METTH, *Methanobacterium thermoautotrophicum*;

Ψ synthase superfamily I

TruB family

```
Db|Access|Gene_Organism        Motif
                               GTLDPKATG$L$$
                                   $VS   P
sp|P09171|TRUB_ECOLI          TGALDPLATGMLPI
sp|O34273|TRUB_YEREN          TGALDPLATGMLPI
sp|P45141|TRUC_HAEIN          TGALDPLATGMLPI
pi|C57253|TRUB/ACICA          TGALDPLATGLLPI
sp|P72154|TRUB_PSEAE          TGSLDPLATGVLPL
sp|P74696|TRUB_SYNY3          GGTLDPLAEGVLPL
SANG|TRUB/CAMJE               SGTLDPFAKGVLIV
tr|O32785|TRUB/LACLA          GGTLDPQVTGVLPV
sp|O51743|TRUB_BORBU          AGTLDKFASGILVC
em|e1342797|TRUB/RICPR        AGTLDVEAEGILPL
sp|O83859|TRUB_TREPA          TGTLDRFADGLLLL
dd|d1038085|TRUB/SYNP7        GGTLDPAVTGVLPI
gb|4981390|TRUB/THEMA         GGTLDPFACGVLII
tr|O84096|TRUB/CHLTR          AGTLDPFATGVMVM
gb|4376596|TRUB/CHLPN         AGTLDPFATGVMVM
TIGR|TRUB/DEIRA               TGTLDPLATGVVVL
sp|O66922|TRUB_AQUAE          TGTLDPIATGLLII
sp|P32732|TRUB_BACSU          TGTLDPEVSGVLPI
sp|O33335|TRUB_MYCTU          AGTLDPMATGVLVI
gb|4455683|TRUB/MYCLE         AGTLDPMATGVLVL
gb|4464271|TRUB/STRCO         AGTLDPMATGVLVL

sp|Q57612|TRUB_METJA          GGTLDPKVTGVLPV
sp|O26140|TRUB_METTH          GGTLDPKVTGVLPL
sp|O30001|TRUB_ARCFU          AGTLDPRVTGVLPI
sp|O59357|TRUB_PYRHO          GGTLDPKVSGVLPV
GENO|TRUB/PYRAB               GGTLDPKVSGVLPV
UCG|TRUB/PYRFU                GGTLDPKVSGVLPV

tr|O59721|TRUB/SCHPO          GGTLDPLASGVLVV
em|1319405|TRUB2/SCHPO        GGTLDPLASGVLVV
sp|O14007|CBF5_SCHPO          SGTLDPKVTGCLII
sp|P33322|CBF5_YEAST          SGTLDPKVTGCLIV
sp|P40567|PUS4_YEAST          GGTLDPLASGVLVI
sp|O13473|CBF5_KLULA          SGTLDPKVTGCLIV
sp|O43101|CBF5_CANAL          SGTLDPKVTGCLIV
sp|O43102|CBF5_ASPFU          SGTLDPKVTGCLIV
gb|AA532324|CBF5/CRYPV        SGTLDPKVTGCLLV
sp|O17919|NO50_CAEEL          SGTLDPKVSGCLIV
sp|O43100|CBF5_EMENI          SGTLDPKVTGCLIV
sp|O44081|NO60_DROME          SGTLDPKVTGCLIV
sp|P040615|DKC1_RAT           SGTLDPKVTGCLIV
gb|AI120090|TRUB/MOUSE        GGTLDSAARGVLVV
em|Z45640|TRUB2/HUMAN         XXXLDAQASGVLVL
em|HS1186784|TRUB/HUMAN       GGTLDSARRGVLVV
sp|O60832|DKC1_HUMAN          SGTLDPKVTGCLIV
```

Ψ synthase superfamily II

TruA family

```
Db|Access|Gene_Organism        Motif
                               AGRTDKGVHA GQ$
                               S    A      N
sp|P07649|TRUA_ECOLI          AGRTDAGVHGTGQV
sp|P45291|TRUA_HAEIN          AGRTDSGVSGTGQV
tr|O87016|TRUA/PSEAE          AGRTDAAVHASGQV
em|e1343126|TRUA/RICPR        SGRTDAGVHAIGQV
SANG|TRUA/CAMJE               ASRTDKGVHASYAV
sp|P56144|TRUA_HELPY          AGRTDKGVHANNQV
sp|P70973|TRUA_BACSU          SGRTDSGVHAAGQV
gb|4512436|TRUA/BACHD         SGRTDTGVHARGQI
sp|Q45557|TRUA_BACSP          SGRTDAGVHALGQV
sp|Q50291|TRUA_MYCPN          SGRTDKGVHAINQT
sp|P47428|TRUA_MYCGE          SGRTDKGVHAINQT
gb|4539110|TRUA/MYCLE         AGRTDTGVHATGQV
sp|O06322|TRUA_MYCTU          AGRTDAGVHASGQV
tr|O86776|TRUA/STRCO          AGRTDAGVHARGQV
sp|O66953|TRUA_AQUAE          CCRTDSGVHALDYI
sp|P70830|TRUA_BORBU          SGRTDKGVHAKKQI
TIGR|TRUA/DEIRA               AGRTDAGVHAEAMP
tr|O84469|TRUA/CHLTR          SGRTDAGVHAQGQI
gb|4376873|TRUA/CHLPN         SGRTDAGVHAYGQV
sp|O24712|TRUA_SYNP6          AGRTDTGVHRAAQV
sp|P73295|TRUA_SYNY3          AGRTDAGVHAAAQV
gb|4982143|TRUA/THEMA         AGRTDTGVHANGQL
sp|O83802|TRUA_TREPA          SGRTDSGVHAVGQA

sp|Q59069|TRUA_METJA          GGRTDKGVSALGNF
sp|O58941|TRUA_PYRHO          ASRTDKGVSALGNV
GENO|TRUA/PYRAB               ASRTDRGVSALGNV
UCG|TRUA/PYRFU                ASRTDRGVSARGNV
sp|O26928|TRUA_METTH          AGRTDRGVHALGNF
sp|O28544|TRUA_ARCFU          AGRTDAGVHAYGQV

sp|Q12211|PUS1_YEAST          AARTDKGVHAGGNL
sp|P53167|PUS2_YEAST          AARTDKGVHAMLNL
sp|P31115|PUS3_YEAST          GGRTDKGVSAMNQV
em|e1359943|TRUA/SCHPO        AARTDKGVHAAGNV
em|e1339974|TRUA2/SCHPO       AARTDKGVHTLRNL
em|e1349968|TRUA/CAEEL        AARTDRAVSAARQM
tr|O65241|TRUA/ARATH          AGRTDKGVSALNQV
tr|O04502|TRUA2/ARATH         GVLQDAGVHALSNV
em|Z83321|TRUA3/ARATH         SARTDKGVSAVGQV
sp|O22928|PUSH_ARATH          SSRTDKGVHSLATS
gb|AA696025|TRUA/DROME        SSRTDAGVHALHST
gb|W48211|TRUA/MOUSE          SSRTDAGVHALSNA
gb|4455035|PUS1/HUMAN         CARTDKGVSAAGQV
gb|4455033|PUS1/MOUSE         CARTDKGVSAAGQV
```

Figure 5. *Continued.* MOUSE, *Mus musculus*; MYCGE, *Mycoplasma genitalium*; MYCLE, *Mycobacterium leprae*; MYCPN, *Mycoplasma pneumoniae*; MYCTU, *Mycobacterium tuberculosis*; PLAFA, *Plasmodium falciparum*; PSEAE, *Pseudomonas aeruginosa*; PYRAB, *Pyrococcus abyssi*; PYRFU, *Pyrococcus furiosus*; PYRHO, *Pyrococcus horikoshii*; RAT, *Rattus norvegicus*; RICPR, *Rickettsia prowazekii*; SCHPO, *Schizosaccharomyces pombe*; STRCO, *Streptomyces coelicolor*; SYNP6, *Synechococcus* sp. strain PCC 6301; SYNP7, *Synechococcus* sp. strain PCC 7942; SYNY3, *Synechocystis* sp. strain PCC 6803; THEMA, *Thermotoga maritima*; TREPA, *Treponema pallidum*; YEAST, *Saccharomyces cerevisiae;* YEREN, *Yersinia enterocolitica*; ZYMMO, *Zymomonas mobilis*. HELP2 is a second strain (J99) of *H. pylori* that has been recently sequenced. The motif II sequence from this strain was included, since it differed from the HELPY version. Other motif II sequences from different strains of the same species were identical and have been omitted from this compilation. The complete genome sequences for *C. jejuni* (SANG), *D. radiodurans* (TIGR), *P. abyssi* (GENO), and *P. furiosus* (UGC) were taken from the following private databases indicated in parentheses: SANG, sequence data produced by the *C. jejuni* Sequencing Group at the Sanger Centre (they can be obtained from ftp://ftp.sanger.ac.uk/pub/pathogens.cj/); TIGR, preliminary sequence data obtained from The Institute for Genome Research website at http://www.tigr.org; sequencing of *D. radiodurans* was accomplished with support from the U.S. Department of Energy; UGC, the Utah Genome Center, Department of Human Genetics, University of Utah; GENO, the GENOSCOPE website (www.genoscope.cns.fr).

Cbf5-like synthase in all three eukaryotes probably acts in conjunction with guide RNAs to form all of the rRNA Ψ. In *C. elegans*, this leaves two TruA-like synthases for all tRNA modifications. Assuming that Ψ55 is present in the tRNA of this organism, one of the TruA synthases must act like a TruB enzyme. Although the human genome is incomplete, there are already more than enough synthases. In fact, there is an extra TruB-class synthase whose function is unknown. Yeast has even more excess synthases of the RluA class. Even if one is used for mitochondrial ribosomes (YD36 shows the characteristics of a mitochondrial enzyme), there are still three others of unknown function. The two TruB enzymes are ac-

Table 4. Number and family distribution of pseudouridine synthases in sequenced genomes[a]

Organism	Family				Sum	No. of ORFs[b]	% ORFs devoted to Ψ synthases
	RsuA	RluA	TruB	TruA			
Bacteria							
Escherichia coli	4	4	1	1	10	4,034	0.25
Haemophilus influenzae	3	5	1	1	10	1,709	0.59
Rickettsia prowazekii	1	2	1	1	5	834	0.60
Campylobacter jejuni	1	3	1	1	6	~1,500[c]	0.40
Helicobacter pylori	1	3	0	1	5	1,566	0.32
Bacillus subtilis	2	3	1	1	7	4,100	0.17
Mycoplasma genitalium	0	2	0	1	3	467	0.64
Mycoplasma pneumoniae	0	2	0	1	3	677	0.44
Mycobacterium tuberculosis	1	2	1	1	5	3,918	0.13
Aquifex aeolicus	2	1	1	1	5	1,522	0.33
Borrelia burgdorferi	1	2	1	1	5	850	0.59
Treponema pallidum	1	3	1	1	6	1,031	0.58
Chlamydia pneumoniae	1	1	1	1	4	1,052	0.38
Chlamydia trachomatis	1	2	1	1	5	894	0.56
Deinococcus radiodurans	1	3	1	1	6	~2,900[c]	0.21
Synechocystis sp.	2	2	1	1	6	3,169	0.19
Thermotoga maritima	1	2	1	1	5	1,877	0.27
Archaea							
Methanococcus jannaschii	0	0	1	1	2	1,715	0.12
Methanobacterium	0	0	1	1	2	1,869	0.11
thermoautotrophicum	0	0	1	1	2	2,407	0.08
Archaeoglobus fulgidus	0	0	1	1	2	1,765	0.11
Pyrococcus abyssii	0	0	1	1	2	~1,700[c]	0.12
Pyrococcus furiosus	0	0	1	1	2	1,979	0.10
Pyrococcus horikoshii							
Eucarya							
Saccharomyces cerevisiae	0	4	2 (1)	3	9	5,885[d]	0.14
Caenorhabditis elegans	0	1	1 (1)	2	4	19,099[e]	0.02
Homo sapiens (incomplete genome)	0	1	3 (1)	1			

[a] Data compiled from Fig. 5. The numbers in parentheses under TruB are the number of ORFs with homology to Cbf5.
[b] ORFs predicted to encode proteins (NCBI Genomes [www.ncbi.nlm.nih.gov/PMGifs/Genome/org.html]). The *E. coli* value is from EcoGene 11 (Rudd, 1998; bmb.med.miami.edu/EcoGene).
[c] Estimated from genome length.
[d] Goffeau et al., 1996.
[e] *C. elegans* Sequencing Consortium, 1998.

counted for as the classic TruB (Becker et al., 1997) and Cbf5, the putative rRNA enzyme, and two of the three TruA family members have defined sites in tRNA (Lecointe et al., 1998; Motorin et al., 1998). Possible functions for the four synthases of undefined specificity may be in Ψ formation in snRNA. Another possibility is that yeast retains elements of both rRNA Ψ formation systems, using specific synthases for certain critical sites, and a complete set of guide RNAs as well.

METHYLATED AND OTHER MODIFIED NUCLEOSIDES

Kind, Number, and Location

The wide variety of base-methylated and other more esoterically modified nucleosides found in rRNA are listed in the excellent compilation of Rozenski et al. (1999). Unfortunately, rather little information is available on how many of each of these residues are present in rRNA and in which subunit, and even less information is available on their locations. In *E. coli*, the locations of the 10 methylated nucleosides in the SSU RNA and the 14 methylated nucleosides (one is $m^3\Psi$) plus one dihydrouridine in the LSU RNA are known (Rozenski et al., 1999). The locations of the four modified nucleosides in the SSU RNA of *Halobacterium volcanii* are also known (Gupta et al., 1983). Their identities, while not determined, can be reasonably deduced from alignment of the *H. volcanii* sequence with those of *E. coli* and *Xenopus laevis*. Quantitative data are also available for another archaeon, *Sulfolobus solfataricus*, but without information on the location of the modified residues (Noon et al., 1998).

Methyltransferases

In this section we note briefly those modifying enzymes which have been characterized by cloning and whose sites of action have been identified. They are all methyltransferases which use S-adenosylmethionine as a methyl donor. The first to be identified was the *E. coli* methyltransferase that forms two m^6_2A residues at the 3′ end of SSU RNA, the product of the *ksgA* gene (Van Buul and van Knippenberg, 1985). Inactivation of this gene makes the cell resistant to kasugamycin but otherwise has little effect (Van Knippenberg, 1986). More recently, two *E. coli* enzymes, RsmB and RsmC, which make m^5C967 and m^2G1207, respectively, have been cloned (Tscherne et al., 1999a, 1999b). m^5C967 is at the site of the hypermodified nucleoside m^1acp^3U in eukaryotes (Maden, 1990) and is in a small loop implicated in tRNA P-site binding (Döring et al., 1994; von Ahsen and Noller, 1995). Nevertheless, deletion of RsmB did not alter the growth rate in either rich or minimal medium (Gu et al., 1999). The *rsmC* gene has not yet been deleted. The enzyme which forms m^1G745 in *E. coli* LSU RNA has also been cloned (Gustafsson and Persson, 1998). When inactivated, the cells grow 40% more slowly in rich medium and show other aberrant effects. The *Saccharopolyspora erythraea* methyltransferase which forms m^6_2A2058 (*E. coli* numbering) in LSU RNA as a means of protecting itself from the erythromycin it manufactures has been cloned and characterized (Kovalic et al., 1994; Vester and Douthwaite, 1994). Lastly, 2′-O-methylation of G2251 (*E. coli* numbering) in yeast mitochondrial LSU RNA was shown to be catalyzed by the PET56 gene product, whose loss by inactivation of the gene leads to a defect in LSU subunit assembly (Sirum-Connolly and Mason, 1993; Mason, 1998).

This work was supported by NIH grant GM58879 (J.O.) and by a Markey Foundation grant to the Department of Biochemistry and Molecular Biology, University of Miami School of Medicine (K.E.R.)

REFERENCES

Arps, P. J., C. C. Marvel, B. C. Rubin, D. A. Tolan, E. E. Penhoet, and M. E. Winkler. 1985. Structural features of the *hisT* operon of *Escherichia coli* K-12. *Nucleic Acids Res.* **13:**5297–5315.

Bakin, A., and J. Ofengand. 1993. Four newly located pseudouridylate residues in *Escherichia coli* 23S ribosomal RNA are all at the peptidyl transferase center: analysis by the application of a new sequencing technique. *Biochemistry* **32:**9754–9762.

Bakin, A., and J. Ofengand. 1995. Mapping of the thirteen pseudouridine residues in *Saccharomyces cerevisiae* small subunit ribosomal RNA to nucleotide resolution. *Nucleic Acids Res.* **23:** 3290–3294.

Bakin, A., and J. Ofengand. 1998. Mapping of pseudouridine residues in RNA to nucleotide resolution, p. 297–309. *In* R. Martin (ed.), *Methods in Molecular Biology*, vol. 77: *Protein Synthesis: Methods and Protocols.* Humana Press, Inc., Totowa, NJ.

Bakin, A., J. A. Kowalak, J. A. McCloskey, and J. Ofengand. 1994a. The single pseudouridine residue in *Escherichia coli* 16S RNA is located at position 516. *Nucleic Acids Res.* **22:**3681–3684.

Bakin, A., B. G. Lane, and J. Ofengand. 1994b. Clustering of pseudouridine residues around the peptidyl transferase center of yeast cytoplasmic and mitochondrial ribosomes. *Biochemistry* **33:**13475–13483.

Becker, H. F., Y. Motorin, R. J. Planta, and H. Grosjean. 1997. The yeast gene YNL292w encodes a pseudouridine synthase (Pus4) catalyzing the formation of Ψ55 in both mitochondrial and cytoplasmic tRNAs. *Nucleic Acids Res.* **25:**4493–4499.

***C. elegans* Sequencing Consortium.** 1998. Genome sequence of the nematode *C. elegans*: a platform for investigating biology. *Science* **282:**2012–2018.

Chen, J., and J. R. Patton. 1999. Cloning and characterization of a mammalian pseudouridine synthase. *RNA* **5:**409–419.

Conrad, J., and J. Ofengand. Unpublished data.

Conrad, J., D. Sun, N. Englund, and J. Ofengand. 1998. The *rluC* gene of *Escherichia coli* codes for a pseudouridine synthase which is solely responsible for synthesis of pseudouridine at positions 955, 2504, and 2580 in 23S ribosomal RNA. *J. Biol. Chem.* **273:**18562–18566.

Conrad, J., L. Niu, K. Rudd, B. G. Lane, and J. Ofengand. 1999. 16S ribosomal RNA pseudouridine synthase RsuA of *Escherichia coli*: deletion, mutation of the conserved Asp102 residue, and sequence comparison among all other pseudouridine synthases. *RNA* **5:**751–763.

Conrad, J., C. Alabiad, and J. Ofengand. Unpublished data.

Cortese R., H. O. Kammen, S. J. Spengler, and B. N. Ames. 1974. Biosynthesis of pseudouridine in transfer ribonucleic acid. *J. Biol. Chem.* **249:**1103–1108.

Döring, T., P. Mitchell, M. Osswald, D. Bochkariov, and R. Brimacombe. 1994. The decoding region of 16S RNA; a cross-linking study of the ribosomal A, P and E sites using tRNA derivatized at position 32 in the anticodon loop. *EMBO J.* **13:** 2677–2685.

Englund, N., L. Niu, B. Lane, and J. Ofengand. Unpublished data.

Goffeau, A., B. G. Barrell, H. Bussey, R. W. Davis, B. Dujon, H. Feldmann, F. Galibert, J. D. Hoheisel, C. Jacq, M. Johnston, E. J. Louis, H. W. Mewes, Y. Murakami, P. Philippsen, H. Tettelin, and S. G. Oliver. 1996. Life with 6000 genes. *Science* **274:** 546–567.

Gu, X.-R., K. Nicoghosian, and R. J. Cedergren. 1984. Glycine and asparagine tRNA sequences from the archaebacterium, *Methanobacterium thermoautotrophicum*. *FEBS Lett.* **176:**462–466.

Gu, X. R., C. Gustafsson, J. Ku, M. Yu, and D. V. Santi. 1999. Identification of the 16S rRNA m5C967 methyltransferase from *Escherichia coli*. *Biochemistry* **38:**4053–4057.

Gupta, R., J. M. Lanter, and C. R. Woese. 1983. Sequence of the 16S ribosomal RNA from *Halobacterium volcanii*, an archaebacterium. *Science* **221:**656–659.

Gustafsson, C. and B. C. Persson. 1998. Identification of the *rrmA* gene encoding the 23S rRNA m1G745 methyltransferase in *Escherichia coli* and characterization of an m1G745-deficient mutant. *J. Bacteriol.* **180:**359–365.

Gustafsson, C., R. Reid, P. J. Greene, and D. V. Santi. 1996. Identification of new RNA modifying enzymes by iterative genome search using known modifying enzymes as probes. *Nucleic Acids Res.* **24:**3756–3762.

Huang, L., J. Ku, M. Pookanjanatavip, X. Gu, D. Wang, P. J. Greene, and D. V. Santi. 1998a. Identification of two *Escherichia* *coli* pseudouridine synthases that show multisite specificity for 23S RNA. *Biochemistry* 37:15951–15957.

Huang, L., M. Pookanjanatavip, X. Gu, and D. V. Santi. 1998b. A conserved aspartate of tRNA pseudouridine synthase is essential for activity and a probable nucleophilic catalyst. *Biochemistry* 37:344–351.

Jensen, K. J. 1993. The *Escherichia coli* K-12 "wild types" W3110 and MG1655 have an *rph* frameshift mutation that leads to pyrimidine starvation due to low *pyrE* expression levels. *J. Bacteriol.* 175:3401–3407.

Kammen, H. O., C. C. Marvel, L. Hardy, and E. E. Penhoet. 1988. Purification, structure, and properties of *Escherichia coli* tRNA pseudouridine synthase I. *J. Biol. Chem.* 263:2255–2263.

Koonin, E. V. 1996. Pseudouridine synthases: four families of enzymes containing a putative uridine-binding motif also conserved in dUTPases and dCTP deaminases. *Nucleic Acids Res.* 24:2411–2415.

Kovalic, D., R. B. Giannattasio, H. J. Jin, and B. Weisblum. 1994. 23S rRNA domain V, a fragment that can be specifically methylated in vitro by the ErmSF (TlrA) methyltransferase. *J. Bacteriol.* 176:6992–6998.

Kowalak, J. A., E. Bruenger, T. Hashizume, J. M. Peltier, J. Ofengand, and J. A. McCloskey. 1996. Structural characterization of 3-methylpseudouridine in Domain IV from *E. coli* 23S ribosomal RNA. *Nucleic Acids Res.* 24:688–693.

Lecointe, F., G. Simos, A. Sauer, E. C. Hurt, Y. Motorin, and H. Grosjean. 1998. Characterization of yeast protein Deg1 as pseudouridine synthase (Pus3) catalyzing the formation of Ψ38 and Ψ39 in tRNA anticodon loop. *J. Biol. Chem.* 273: 1316–1323.

Maden, B. E. H. 1990. The numerous modified nucleotides in eukaryotic ribosomal RNA. *Prog. Nucleic Acids Res. Mol. Biol.* 39:241–300.

Marvel, C. C., P. J. Arps, B. C. Rubin, H. O. Kammen, E. E. Penhoet, and M. E. Winkler. 1985. *hisT* is part of a multigene operon in *Escherichia coli* K-12. *J. Bacteriol.* 161:60–71.

Mason, T. L. 1998. Functional aspects of the three modified nucleotides in yeast mitochondrial large-subunit rRNA, p. 273–280. *In* H. Grosjean and R. Benne (ed.), *Modification and Editing of RNA.* ASM Press, Washington, D.C.

Massenet, S., Y. Motorin, D. L. J. Lafontaine, E. C. Hurt, H. Grosjean, and C. Branlant. 1999. Pseudouridine mapping in the *Saccharomyces cerevisae* spliceosomal UsnRNAs reveals that pseudouridine synthase Pus1p exhibits a dual substrate specificity for U2 snRNA and tRNA. *Mol. Cell. Biol.* 19:2142–2154.

McCloskey, J. A. Personal communication.

Mitchell, P., M. Osswald, and R. Brimacombe. 1992. Identification of intermolecular RNA cross-links at the subunit interface of the *Escherichia coli* ribosome. *Biochemistry* 31:3004–3011.

Motorin, Y., G. Keith, C. Simon, D. Foiret, G. Simos, E. Hurt, and H. Grosjean. 1998. The yeast tRNA:pseudouridine synthase Pus1p displays a multisite substrate specificity. *RNA* 4:856–869.

Niu, L., and J. Ofengand. 1999. Cloning and characterization of the 23S RNA pseudouridine 2633 synthase from *Bacillus subtilis. Biochemistry* 38:629–635.

Noon, K. R., E. Bruenger, and J. A. McCloskey. 1998. Posttranscriptional modifications in 16S and 23S rRNAs of the archaeal hyperthermophile *Sulfolobus solfataricus. J. Bacteriol.* 180: 2883–2888.

Nurse, K., J. Wrzesinski, A. Bakin, B. G. Lane, and J. Ofengand. 1995. Purification, cloning, and properties of the tRNA Ψ55 synthase from *Escherichia coli. RNA* 1:102–112.

O'Connor, M., and A. E. Dahlberg. 1995. The involvement of two distinct regions of 23 S ribosomal RNA in tRNA selection. *J. Mol. Biol.* 254:838–847.

Ofengand, J., and A. Bakin. 1997. Mapping to nucleotide resolution of pseudouridine residues in large subunit ribosomal RNAs from representative eukaryotes, prokaryotes, archaebacteria, mitochondria, and chloroplasts. *J. Mol. Biol.* 266:246–268.

Ofengand, J., and M. Fournier. 1998. The pseudouridine residues of rRNA: number, location, biosynthesis, and function, p. 229–253. *In* H. Grosjean and R. Benne (ed.), *Modification and Editing of RNA.* ASM Press, Washington, D.C.

Ofengand, J., A. Bakin, J. Wrzesinski, K. Nurse, and B. G. Lane. 1995. The pseudouridine residues of ribosomal RNA. *Biochem. Cell Biol.* 73:915–924.

Persson, B. C., C. Gustafsson, D. E. Berg, and G. R. Björk. 1992. The gene for a tRNA modifying enzyme, m5U54-methyltransferase, is essential for viability in *Escherichia coli. Proc. Natl. Acad. Sci. USA* 89:3995–3998.

Ramamurthy, V., S. L. Swann, J. L. Paulson, C. J. Spedaliere, and E. G. Mueller. 1999. Critical aspartic acid residues in pseudouridine synthases. *J. Biol. Chem.* 274:22225–22230.

Raychaudhuri, S., J. Conrad, B. G. Hall, and J. Ofengand. 1998. A pseudouridine synthase required for the formation of two universally conserved pseudouridines in ribosomal RNA is essential for normal growth of *Escherichia coli. RNA* 4:1407–1417.

Raychaudhuri, S., L. Niu, J. Conrad, and J. Ofengand. 1999. Functional effect of deletion and mutation of the *Escherichia coli* ribosomal RNA and tRNA pseudouridine synthase RluA. *J. Biol. Chem.* 274:18880–18886.

Raychaudhuri, S., and J. Ofengand. Unpublished results.

Rozenski, J., P. F. Crain, and J. A. McCloskey. 1999. The RNA modification database: 1999 update. *Nucleic Acids Res.* 27:196–197.

Rudd, K. E. 1998. Linkage map of *Escherichia coli* K-12, edition 10: the physical map. *Microbiol. Mol. Biol. Rev.* 62:985–1019.

Rudd, K. E., and J. Ofengand. Unpublished results.

Schuler, G. D., S. F. Altschul, and D. J. Lipman. 1991. A workbench for multiple alignment construction and analysis. *Proteins* 9:180–190.

Simos, G., H. Tekotte, H. Grosjean, A. Segref, K. Sharma, D. Tollervey, and E. C. Hurt. 1996. Nuclear pore proteins are involved in the biogenesis of functional tRNA. *EMBO J.* 15:2270–2284.

Sirum-Connolly, K., and T. L. Mason. 1993. Functional requirement of a site-specific ribose methylation in ribosomal RNA. *Science* 262:1886–1889.

Tscherne, J. S., K. Nurse, P. Popienick, H. Michel, M. Sochacki, and J. Ofengand. 1999a. Purification, cloning, and characterization of the 16S RNA m5C967 methyltransferase from *Escherichia coli. Biochemistry* 38:1884–1892.

Tscherne, J. S., K. Nurse, P. Popienick, and J. Ofengand. 1999b. Purification, cloning, and characterization of the 16 S RNA m2G1207 methyltransferase from *Escherichia coli. J. Biol. Chem.* 274:924–929.

Tsui, H.-C. T., P. J. Arps, D. M. Connolly, and M. E. Winkler. 1991. Absence of *hisT*-mediated tRNA pseudouridylation results in a uracil requirement that interferes with *Escherichia coli* K-12 cell division. *J. Bacteriol.* 173:7395–7400.

Van Buul, C. P. J. J., and P. H. van Knippenberg. 1985. Nucleotide sequence of the *ksgA* gene of *Escherichia coli*: comparison of the methyltransferases effecting dimethylation of adenosine in ribosomal RNA. *Gene* 38:65–72.

Van Knippenberg, P. H. 1986. Structural and functional aspects of the N^6,N^6 dimethyladenosines in 16S ribosomal RNA, p. 412–424. *In* B. Hardesty and G. Kramer (ed.), *Structure, Function, and Genetics of Ribosomes.* Springer-Verlag, Berlin, Germany.

Vester, B., and S. Douthwaite. 1994. Domain V of 23S rRNA contains all the structural elements necessary for recognition by the ErmE methyltransferase. *J. Bacteriol.* **176:**6999–7004.

von Ahsen, U., and H. F. Noller. 1995. Identification of bases in 16S rRNA essential for tRNA binding at the 30S ribosomal P site. *Science* **267:**234–237.

Wrzesinski, J., A. Bakin, K. Nurse, B. G. Lane, and J. Ofengand. 1995a. Purification, cloning, and properties of the 16S RNA Ψ516 synthase from *Escherichia coli. Biochemistry* **34:**8904–8913.

Wrzesinski, J., K. Nurse, A. Bakin, B. G. Lane, and J. Ofengand. 1995b. A dual-specificity pseudouridine synthase: purification and cloning of a synthase from *Escherichia coli* which is specific for both Ψ746 in 23S RNA and for Ψ32 in $tRNA^{Phe}$. *RNA* **1:** 437–448.

The Ribosome: Structure, Function, Antibiotics, and Cellular Interactions
Edited by R. A. Garrett, S. R. Douthwaite, A. Liljas, A. T. Matheson, P. B. Moore, and H. F. Noller

Chapter 17

Nucleotide Modifications of Eukaryotic rRNAs: the World of Small Nucleolar RNA Guides Revisited

JEAN-PIERRE BACHELLERIE, JÉRÔME CAVAILLÉ, and LIANG-HU QU

The biogenesis of eukaryotic ribosomes in the nucleolus involves the processing of a long rRNA primary transcript which removes extended spacer regions from the precursor and produces stoichiometric amounts of small- and large-subunit (LSU) mature rRNAs. Before its cleavage by endo- and exonucleases, the nascent pre-rRNA undergoes a pattern of nucleoside modifications considerably more complex than that for *Escherichia coli* rRNA, essentially because of the presence of two prevalent types of modification, 2′-O-ribose methylations and pseudouridines. Each of these modifications is found at about 50 to 100 sites per eukaryotic ribosome (~100 sites in a vertebrate ribosome, but only ~50 sites in the yeast *Saccharomyces cerevisiae*), in contrast to the *E. coli* ribosome, which contains only four ribose-methylated nucleotides and 10 pseudouridines (Maden, 1990; Bachellerie and Cavaillé, 1998; Ofengand and Fournier, 1998). In eukaryotes, ribose methylations and pseudouridines both occur on the nascent rRNA primary transcript in the nucleolus, but they are not absolutely essential for its processing and their role remains elusive. Remarkably, ribose methylations and pseudouridines are not only restricted to the mature RNA regions of the rRNA precursor but are located exclusively within the most conserved, functionally important domains of mature RNAs, which have clear counterparts in prokaryotic rRNAs, especially in structural elements contributing to the peptidyltransfer region and its vicinity (Brimacombe et al., 1993). Moreover, their precise locations within the rRNA sequences are largely (but not perfectly) conserved among distantly related eukaryotic organisms, indicating that they play some important role.

Breakthroughs in the field of small nucleolar RNAs (snoRNAs) have illuminated the process of selection of the nucleotides to be ribose methylated and pseudouridylated in eukaryotic rRNA, opening new prospects for elucidating the role of these modifications in the assembly and function of the eukaryotic ribosome. A decade ago, only a few snoRNAs were known, and those with an established role were involved in pre-rRNA cleavage. Over the last few years the nucleolus has been shown to contain a population of small RNAs of unexpected complexity. The first novel snoRNAs immediately generated a deep interest because of their bizarre mode of biosynthesis: they were encoded in introns and produced by processing of the pre-mRNA intron (Leverette et al., 1992; Sollner-Webb, 1993; Maxwell and Fournier, 1995; Bachellerie et al., 1995). But these were only the tip of the iceberg, and more than 100 novel snoRNAs are presently known. All of these snoRNAs, which are all intronic in vertebrates but mostly nonintronic in yeast, belong to two distinct families, box C/D antisense snoRNAs and H/ACA snoRNAs. These two families have been found to perform unanticipated functions in ribosome biogenesis, i.e., to guide the ribose methylations and pseudouridylations, respectively, of eukaryotic rRNAs.

This chapter summarizes our present knowledge of snoRNA guide function, which in both cases involves formation of a specific RNA duplex at each rRNA modification site, and focuses on recent progress and issues in this area. It also incorporates new developments regarding snoRNA biogenesis, which, despite the diversity and peculiarity of snoRNA genomic organization and modes of expression among

Jean-Pierre Bachellerie and Jérôme Cavaillé ■ Laboratoire de Biologie Moléculaire Eucaryote du C.N.R.S., Université Paul-Sabatier, 118 route de Narbonne, 31062 Toulouse Cédex, France. **Liang-Hu Qu** ■ Biotechnology Research Center, Zhongshan University, Guangzhou 510 275, China.

eukaryotes, appears to be systematically coordinated with that of components of the translational apparatus, not only at the transcriptional level but also at the RNA-processing level. For more detailed insights into the studies which have established the guide function, the reader is encouraged to consult previous reviews on the topic (Ofengand and Fournier, 1998; Bachellerie and Cavaillé, 1997, 1998). The reader is also referred to comprehensive review articles dealing with patterns of rRNA nucleotide modifications and their functional significance (Maden, 1990; Lane et al., 1995; Ofengand et al., 1995) and the processing of pre-rRNAs (Sollner-Webb et al., 1995; Tollervey, 1996) in eukaryotes.

TWO MAJOR FAMILIES OF snoRNAs

All of the newly discovered snoRNAs fall into two major families, based on structural features.

The first family has been termed box C/D antisense snoRNAs because of the presence of two short conserved sequence motifs, box C (5′ PuUGAUGA 3′) and box D (5′ CUGA 3′), always located only a few nucleotides away from the 5′ and 3′ end, respectively, and because of a 10- to 20-nucleotide (nt)-long complementarity to rRNA, called the antisense element (Bachellerie et al., 1995). The antisense elements, all different among different box C/D snoRNA species, are phylogenetically conserved among homologous snoRNAs, even among distantly related eukaryotes, corresponding to the most conserved portion of the snoRNA sequence, in addition to the box motifs. Antisense elements always match highly conserved sequences in eukaryotic mature rRNAs (Bachellerie and Cavaillé, 1997, 1998). Moreover, they always have the same location within snoRNA sequences (Fig. 1a), immediately upstream from either the box D motif or another CUGA (box D′) in the 5′ half of the snoRNA molecule (Tycowski, 1996b). Some box C/D snoRNAs, such as U24, U32, U36, U45, and U50, contain antisense elements at these two positions (Bachellerie and Cavaillé, 1997). A C′ box carrying up to two deviations from the consensus C box has been identified in the central region of all box C/D antisense snoRNAs and found to be important for methylation guide function (Kiss-Laszlo et al., 1998).

Most box C/D antisense snoRNAs contain a characteristic 5′-3′-terminal 4- to 5-bp stem-box structure bringing the two box motifs into close proximity (Fig. 1a). This structure is the key signal for both exonucleolytic processing and nucleolar localization (Cavaillé and Bachellerie, 1996; Caffarelli et al., 1996; Samarsky et al., 1998). For snoRNAs devoid of a 5′-3′-terminal stem, the juxtapositioning of the box C and D motifs appears to be mediated by an internal stem flanking the boxes (Watkins et al., 1996). In the initial stages of the characterization of this family the fact that all box C/D antisense snoRNAs interact with nucleolar protein fibrillarin (Nop1p in *S. cerevisiae*) has been extensively used. This protein is required for pre-rRNA processing and methylation and for ribosome assembly in yeast (Tollervey et al., 1991, 1993). As detailed below, a few additional box C/D snoRNP proteins have been subsequently identified.

The modification guide function was first established for the box C/D family, essentially because of their long complementarities to rRNA which, together with the box motifs, provided the basis for the detection of a large number of new snoRNAs through computer searches of genomic databases (Qu et al., 1994, 1995; Nicoloso et al., 1996). Following the identification of this function, scores of novel methylation guide snoRNAs have been detected, most of them by further genome searches, in vertebrates (Smith and Steitz, 1998) and especially in yeast. These studies include, in addition to work carried out in our laboratory (Qu et al., 1999), a major contribution based on the utilization of an algorithm incorporating all the structural requirements eventually identified for members of this family (Lowe and Eddy, 1999). Of the 55 ribose methylations in *S. cerevisiae* rRNA, 51 have been assigned a single cognate snoRNA guide: only two remain unassigned, and two others (LSU-Um2918 and LSU-Gm2919) might have redundant cognate snoRNAs (Lowe and Eddy, 1999). Intriguingly, Gm2919 in the invariant UmGmψ sequence of eukaryotic LSU rRNA is unique among all rRNA ribose methylations: it is methylated at a late stage of LSU rRNA processing, and its formation is selectively depressed during depletion of the yeast nucleolar protein Nop2 (Maden, 1990; Hong et al., 1997).

The second family is called H/ACA snoRNAs, because of an ACA trinucleotide at a fixed position 3 nt away from the 3′ end and a related H motif (box ANANNA) in the hinge region linking the two long stem structures shared by all of these snoRNAs, as depicted in Fig. 1b (Balakin et al., 1996; Ganot et al., 1997a). H/ACA snoRNAs also contain complementarities to rRNA, located within one (or both) of the large internal loops of their long stems (Fig. 1b). However, these complementarities were much harder to discover because they are shorter (4 to 8 nt) and in two separate pieces, as detailed below.

H/ACA snoRNAs have a mode of formation analogous to that of box C/D antisense snoRNAs, with their exonucleolytic processing also exclusively

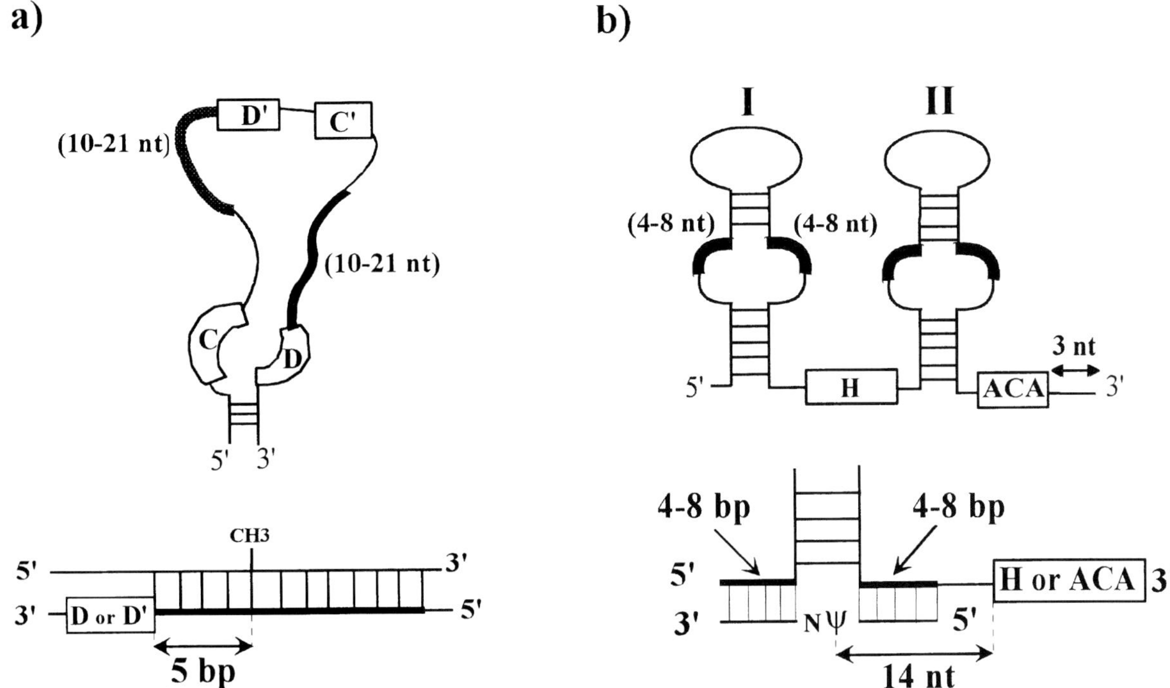

Figure 1. Generic structure of the two major snoRNA families and corresponding guide RNA duplexes at rRNA modification sites. (a) Box C/D antisense snoRNAs. (Top) Locations of the two box motifs (and less conserved copies C′ and D′) and of the antisense elements (thick lines), present either as a single copy or as a pair. (Bottom) Canonical structure of the RNA duplex formed at each ribose-methylated site in rRNA. (b) H/ACA snoRNAs. (Top) Schematized conserved secondary structure. The total number of nucleotides within each of the long helical domains I and II may vary substantially among the different snoRNAs, with the variations affecting both the stem and loop regions. The pair of 4- to 8-nt-long complementarities to rRNA present within the long internal loop in domain I or II (or in both) is depicted by a thicker line. (Bottom) Canonical structure of the RNA duplex formed at each pseudouridylation site in rRNA.

directed by mature snoRNA sequences (Balakin et al., 1996; Ganot et al., 1997a). In addition to their critical role in the function of pseudouridylation guide snoRNAs, the H and ACA boxes are also essential for snoRNA processing and accumulation, similar to boxes C and D in methylation guide snoRNAs, and their integrity is required for exonucleolytic formation of the correct 5′ and 3′ ends of the snoRNA, respectively (Bortolin et al., 1999). All H/ACA snoRNAs specifically associate with nucleolar protein Gar1, required for pseudouridylation of *S. cerevisiae* rRNAs (Bousquet-Antonelli et al., 1997), and a few additional common proteins (see below).

BOX C/D ANTISENSE snoRNAs GUIDE rRNA RIBOSE METHYLATIONS

The comparison of a large collection of these snoRNAs provided the key to the function of this snoRNA family (Kiss-Laszlo et al., 1996; Nicoloso et al., 1996). All antisense elements in box C/D antisense snoRNAs were found to span different sites of ribose methylation in rRNA. Moreover, in each corresponding RNA duplex, the methylated site always appeared at the same position (Fig. 1a, bottom), i.e., paired to the fifth nucleotide upstream from box D, suggesting that each ribose methylation was directed by a cognate snoRNA guide. The hypothesis has been experimentally tested by gene disruptions in yeast: as reported for U24 and several other yeast box C/D antisense snoRNAs (Kiss-Laszlo et al., 1996; Qu et al., 1999; Lowe and Eddy, 1999), the snoRNA gene disruption completely and selectively abolishes the cognate rRNA methylation. Likewise, the same effect has been observed after selective snoRNA depletion in *Xenopus* oocytes (Tycowski et al., 1996a). The key role of the rRNA-snoRNA duplex at each rRNA ribose methylation site has been confirmed by a different approach, by targeting the ribose methylation to novel nucleotide positions, not only in endogenous rRNA but also in ectopically expressed nonribosomal RNAs, by expression of a snoRNA with the tailored antisense element in transfected mammalian cells (Cavaillé et al., 1996). This study clearly showed that formation of the long duplex involving the antisense element and vicinal box D is the sole determinant of the site of ribose methylation and that the precise nucleotide position to be methylated in the duplex is determined by its spacing from box D. Moreover, the integrity of box D was found to be essential for the reaction.

Dissecting the Reaction

Further analyses performed with yeast U24 have shown that box C′ also functions in concert with the D′ box and plays a crucial role in the efficiency of the reaction (Kiss-Laszlo et al., 1998). This suggests a similar importance for box C in the methyltransfer reaction. No in vitro system faithfully reproducing the snoRNA-directed methylation has been reported. A preliminary analysis of features of the RNA duplex important for the reaction has been performed in vivo (Cavaillé and Bachellerie, 1998) by cotransfection with constructs expressing an artificial intron-encoded RNA guide and, through RNA polymerase I transcription, an RNA substrate containing the appropriate complementarity to the guide snoRNA (Fig. 2a and b). The thermodynamic stability of the intermolecular duplex is an important parameter for the targeted methylation. The minimum size of a functional duplex is 10 bp when the duplex has a 70% G+C content, but it reaches 15 to 16 bp for a duplex with a low G+C content. Moreover, some helix irregularities are tolerated within the duplex, such as a single mispairing anywhere in the duplex, even at the target site. Even a bulge nucleotide is not detrimental, except at a few positions (Fig. 2c) where the bulge completely blocks the reaction. On the snoRNA strand of the duplex, the five forbidden positions are clustered around the target site, whereas the presence of a bulge on the substrate RNA strand of the duplex is inhibitory at only the two positions immediately downstream from the target nucleotide. While the significance of these data in terms of the inner workings of the reaction remains elusive, the observations have initiated further genome searches for putative guide snoRNAs forming slightly irregular duplexes with rRNA, resulting in the positive identification of a few in *S. cerevisiae* (Cavaillé and Bachellerie, 1998).

Given the critical role of peculiar 2′ hydroxyl groups in RNA function, methylation guide snoRNAs could serve as highly selective tools for altering gene expression at the posttranscriptional level when directed at nonribosomal-RNA targets. However, the efficiency of the reaction directed at ectopically expressed RNA polymerase II transcripts is considerably decreased compared to that directed to RNA polymerase I transcripts (Cavaillé et al., 1996). To identify the reason for this discrepancy, a different kind of RNA polymerase II transcript has recently been tested, this time engineered so as to be addressed to the nucleolus by taking advantage of the nucleolar localization signal of methylation guide snoRNAs, i.e., their 5′-3′-terminal stem-box C/D structure (Cavaillé and Bachellerie, 1996; Samarsky et al., 1998; Lange et al., 1998). In these experiments (Fig. 3), the sequence targeted for methylation was inserted in an intron and flanked by sequence elements able to form the characteristic snoRNA stem-box structure, which is also the *cis*-acting signal for exonucleolytic processing of intronic box C/D snoRNAs. The RNA polymerase II transcript is therefore processed, as expected, at the borders of the stem-box structure. After coexpression of a tailored guide snoRNA carrying an antisense element matching the substrate RNA sequence, the RNA polymerase II transcript is very efficiently methylated at the selected target site. Thus, the snoRNA-guided reaction is not dependent upon RNA polymerase I transcription of the substrate per se but rather on its nucleolar localization.

New Frontiers for Methylation Guides

Recent reports indicate that rRNA is not the only natural substrate for methylation guide snoRNAs. Thus, two different ribose methylations in vertebrate U6 snRNA, Am47 and Cm77, are directed by two members of the box C/D antisense snoRNA family, one of which also directs a 28S rRNA ribose methylation through an overlapping antisense element (Tycowski et al., 1998). A putative box C/D antisense snoRNA that should direct the ribose methylation of A53 in human U6 has also been detected (Kiss-Laszlo et al., 1998). These developments might have far-reaching implications in the field of mRNA biogenesis in eukaryotes, particularly since the crucial importance of snRNA nucleotide modifications for pre-mRNA splicing has just been established in the case of U2 snRNA (Yu et al., 1998). They should also open a new era in the search for novel natural RNA targets for methylation guide snoRNAs, reinforcing the potential interest of the snoRNA-directed reaction as a tool, especially since the nucleolus contains, at least transiently, several important RNAs in addition to rRNA (see Pederson, 1998, for a review), such as RNase P (Jacobson et al., 1997), signal recognition particle RNA (Jacobson and Pederson, 1998), precursors to tRNAs (Bertrand et al., 1998), and telomerase RNA (Mitchell et al., 1999). Indeed, even pre-mRNAs were detected in the nucleolus in earlier studies (Bond and Wolde, 1993).

Archaebacterial rRNAs have recently been found to contain a large number of ribose methylations, amounting to 67 per ribosome in *Sulfolobus solfataricus* (Noon et al., 1998), i.e., a number very similar

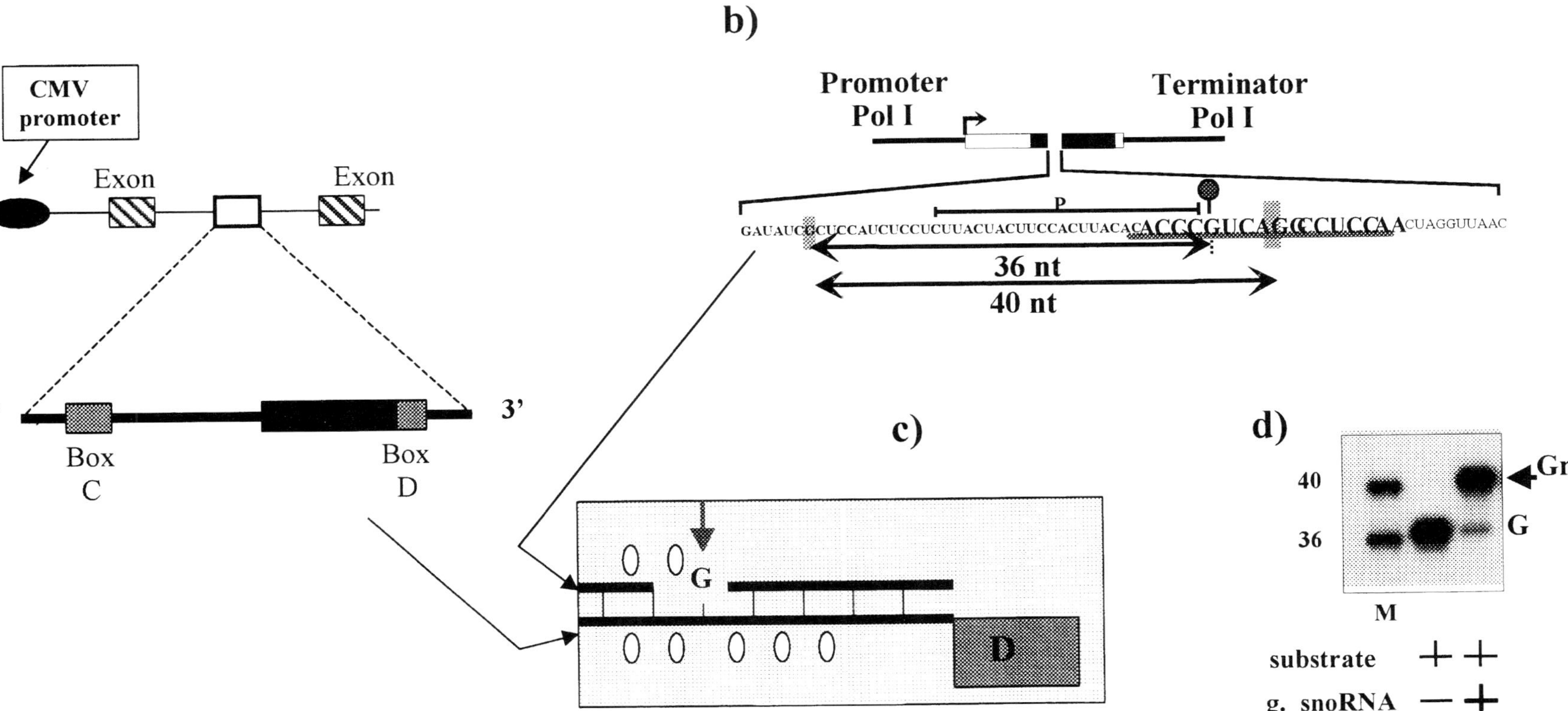

Figure 2. In vivo dissection of the methylation guide RNA duplex by cotransfection of artificial guide snoRNA and RNA substrate in mammalian cells. (a) Construct expressing an artificial intron-encoded guide snoRNA under the control of the strong cytomegalovirus (CMV) promoter. (b) rRNA minigene construct expressing, under the control of RNA polymerase I, an RNA substrate (sequence shown) carrying the appropriate complementarity (underlined) to the artificial guide snoRNA. The detection of ribose methylation was based on the selection of a guanosine as the target nucleotide in the RNA substrate. The guanosine was placed in a sequence devoid of any proximal upstream guanosine. Two long oligonucleotides (horizontal arrows) generated by RNase T1 digestion reflect the degree of ribose methylation at the targeted guanosine (below circle; the two proximal upstream and downstream guanosines are shaded). P, oligonucleotide P (see below). (c) Schematized structure of the guide RNA duplex. Positions where a 1-nt bulge is strongly inhibitory for the targeted methylation are denoted by ovals. (d) Northern assay of the ribose methylation at the targeted guanosine. T1 RNase digests of total RNA from cells cotransfected with the guide (g.) snoRNA and its cognate RNA substrate were probed with oligonucleotide P, shown in panel b. +, present; −, absent.

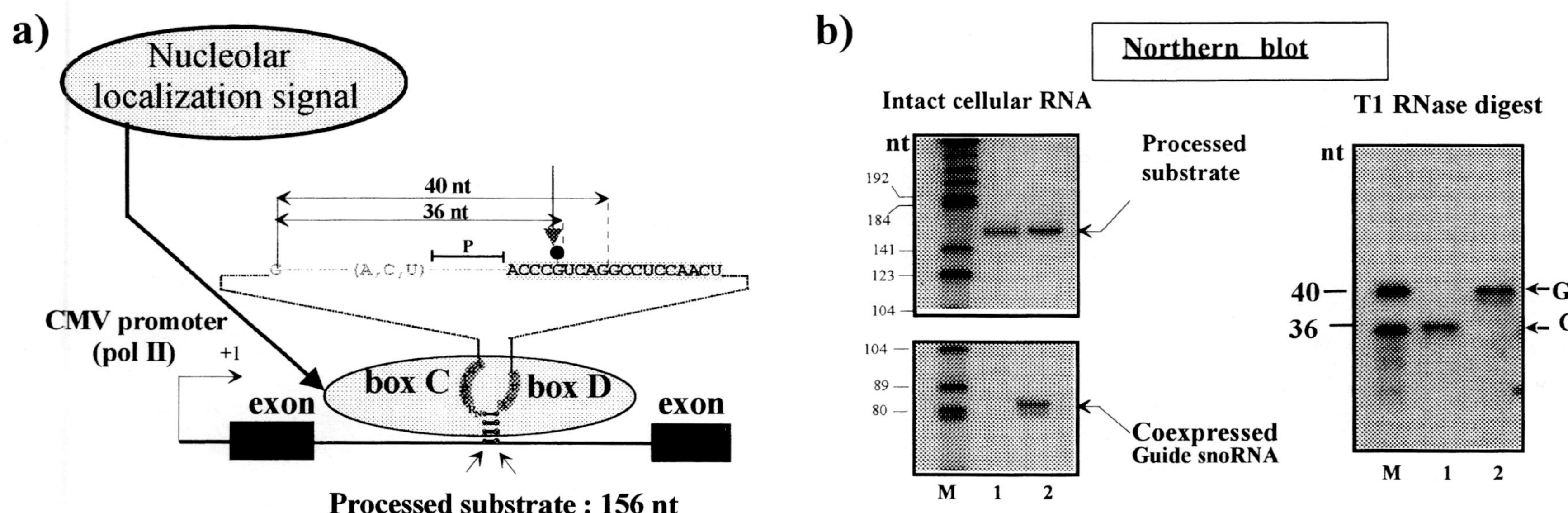

Figure 3. Efficient methylation of an RNA polymerase II transcript addressed to the nucleolus. (a) Structure of the construct expressing the RNA polymerase II transcript, showing the guanosine targeted for methylation (vertical arrow) in the sequence (shaded) complementary to the antisense element of the cotransfected artificial guide snoRNA (not shown). The RNA substrate sequence was inserted in an intron, flanked by sequence elements able to form the typical stem-box terminal structure of box C/D snoRNAs. P, oligonucleotide P. (b) The level of targeted-ribose methylation was determined after RNase T1 digestion by Northern blotting with the oligonucleotide probe (P) delineated in panel a. Accumulation of the RNA substrate processed at the borders of the stem-box structure and expression of the cognate artificial guide snoRNA were both assessed on intact cellular RNA. Lanes 1, transfection with the RNA substrate alone. Lanes 2, cotransfection of the RNA substrate and the guide snoRNA.

to that of eukaryotes and quite different from the very simple eubacterial pattern. While the location of the *S. solfataricus* rRNA ribose-methylated nucleotides has not yet been reported, the very large number points to the potential presence of methylation guide RNAs in archaebacteria. In line with this notion, archaebacteria have unambiguous homologues of two box C/D antisense snoRNP proteins, fibrillarin (Amiri, 1994) and Nop58p (Lafontaine and Tollervey, 1998, 1999). Precise mapping of the *S. solfataricus* rRNA methylation pattern combined with searches of archaebacterial genomes for potential RNA guides may therefore illuminate the origin of eukaryotic methylation guide snoRNAs.

Methylase(s) and Other Protein Factors

In *E. coli*, the four rRNA ribose methylations are thought to each involve a distinct methylase (Gustafsson et al., 1996). In contrast, a single methylase is thought to catalyze the reaction at all eukaryotic rRNA sites—and possibly also on other RNA substrates—since formation of a canonical guide RNA duplex at the target site is sufficient to direct the reaction. The search for this common methylase, hampered by the lack of an in vitro system for the reaction, has been unsuccessful so far. Nucleolar protein fibrillarin (yeast Nop1p), which binds to all box C/D snoRNAs, was initially considered as an unlikely common methylase on the basis of several pieces of indirect evidence (Bachellerie and Cavaillé, 1997, 1998). However, a search of the yeast genome for *S*-adenosylmethionine-dependent methylases has recently shown that fibrillarin contains marginally significant matches to methylase consensus sequence motifs (Niewmierzycka and Clarke, 1999). This revives the issue, especially since one of the *nop1.3* mutations inhibiting ribose methylation of yeast rRNAs (Tollervey et al., 1993) corresponds to an alteration of one of these consensus amino acid sequence motifs. In another search of the *S. cerevisiae* genome for uncharacterized open reading frames exhibiting similarities to prokaryotic RNA 2′-O-ribose methylases, a single good match has been detected, corresponding to an enzyme not involved in rRNA methylation but catalyzing the site-specific formation of Gm18 in yeast tRNAs (Cavaillé et al., 1999).

In addition to Nop1p, two other common protein components of box C/D snoRNPs in *S. cerevisiae* have been identified, Nop56p and Nop58p (Nop5) (Gautier et al., 1997; Wu et al., 1998; Lafontaine and Tollervey, 1999). The two proteins show 45% identity, are both essential, and localize to the nucleolus. They exhibit a stoichiometric association, and both associate with Nop1p in complexes. In vertebrates, a few novel snoRNP proteins have also been detected, after cross-linking to a box C/D snoRNA or utilization of a tagged RNA (Caffarelli et al., 1998; Watkins et al., 1998a). The characterization of these proteins, the binding of which is dependent on the 5′-3′-terminal stem and vicinal conserved boxes C and D, should provide further insight into the mechanisms of snoRNA-guided methylation.

H/ACA snoRNAs: THE GUIDES FOR rRNA PSEUDOURIDYLATIONS

Following the discovery of methylation guide snoRNAs and the telling observation that protein Gar1, which binds to all H/ACA snoRNAs, was required for pseudouridylation of *S. cerevisiae* rRNAs (Bousquet-Antonelli et al., 1997), the second large snoRNA family immediately became a candidate for guiding rRNA pseudouridylations. The notion has been rapidly confirmed, and detailed models have been proposed and tested (Ni et al., 1997; Ganot et al., 1997b).

In a first stage, the *cis*-acting elements directing site-specific pseudouridylation of rRNA were delineated (Ganot et al. 1997b), taking advantage of the experimental approach previously utilized for establishing the methylation guide function of box C/D antisense snoRNAs, i.e., expression of severely truncated rRNA substrates by rRNA minigenes in transfected mammalian cells (Cavaillé et al., 1996). These elements were found to be restricted within the 6 or 7 nt upstream and downstream from the uridine to be isomerized (Ganot et al., 1997b). The systematic analysis of the base-pairing potential between, on one hand, the 14- to 15-nt sequences spanning known rRNA pseudouridylation sites and, on the other hand, all available H/ACA snoRNA sequences then provided the basis for a model which has been experimentally confirmed (Ganot et al., 1997b). Each H/ACA snoRNA is able to base-pair with a particular site of pseudouridylation in rRNA, not through the formation of a single long duplex, as for methylation guide snoRNAs, but through two shorter stems (involving a minimum of 10 bp) which precisely flank the target nucleotide, thus leaving this uridine readily accessible for isomerization (Fig. 1b, bottom). Remarkably, the two 4- to 8-nt-long rRNA complementarities have the same location in all H/ACA snoRNAs, always within the large internal loop of one of its two long helical domains and invariably abutting the apex-proximal stem (Fig. 1b, bottom). Finally, in the bipartite RNA duplex, the target uridine is at a conserved distance, 14 nt, upstream from the ACA or related H motif, suggesting a role for this motif in target site selection similar to that of box D in methylation guide snoRNAs. As observed for methylation guides, a few "double" pseudouridylation guides have been detected, such as snoRNAs E2, U65, and U69 in vertebrates and snR5, snR34, snR44, and snR189 in *S. cerevisiae*, which all exhibit a canonical "pseudouridylation pocket" in each of their long stem structures (Ganot et al., 1997b). A potential triple guide has even been identified in yeast: snR49, which can specify Ψ990 in 25S rRNA and Ψ211 in 18S rRNA with its 5′ pseudouridylation pocket and Ψ302 in 18S rRNA with its 3′ pseudouridylation pocket (Ganot et al., 1997b; Chetouani, 1999). As expected, gene disruptions in yeast have shown that each snoRNA is selectively required for formation of the predicted pseudouridine in rRNA (Ganot et al., 1997b; Ni et al., 1997).

snoRNA Features and Protein Factors Involved in Pseudouridylation Guide Function

In the absence of an in vitro system, the function of this second snoRNA family has also been dissected in vivo, revealing properties very reminiscent of those of methylation guides, including the key importance of the two H and ACA box motifs and the possibility of targeting the modification in vivo to novel nucleotides by using snoRNAs with the appropriate target site bipartite complementarity (Bortolin et al., 1999). Moreover, RNA helices flanking a pseudouridylation pocket are essential for the activity of the guide snoRNA. Interestingly, the integrity of both helical domains is required for pseudouridylation directed by either the 5′ or 3′ pseudouridylation pocket, indicating that the two long helical domains act in a highly cooperative way, even when the H/ACA snoRNA contains a single pseudouridylation pocket (Bortolin et al., 1999). Unexpectedly, the extra domain exhibited by mammalian telomerase RNAs compared to ciliate telomerase RNAs strongly resembles an H/ACA snoRNA, but no modification target sequence has been predicted so far for this snoRNA-like domain (Mitchell et al., 1999).

Presently, protein components of H/ACA snoRNPs are better characterized than those associated with methylation guides. Gar1p binds directly and specifically to H/ACA snoRNAs through the cooperative action of two distinct, highly conserved amino acid domains (Bagni and Lapeyre, 1998). The putative common modification enzyme, Cbf5p, has been identified in *S. cerevisiae*, although its enzymatic activity has not been tested in vitro (Lafontaine et al., 1998b). It exhibits sequence similarity with *truB*, the pseudouridine synthase catalyzing formation of Ψ55 in *E. coli* tRNAs. Remarkably, mutations in the human homologue of Cbf5, dyskeratin, cause dyskeratosis, an inherited X-linked disorder with a spectrum of clinical manifestations, which thus appears as a "ribosomapathy" (Heiss et al., 1998). The *Drosophila melanogaster* homologue of Cbf5p has also been identified recently. It is encoded by the *minifly* (*mfl*) gene essential for *Drosophila* viability and fertility (Giordano et al., 1999). *mfl* partial loss-of-function mutations cause pleiotropic defects, such as extreme reduction in body size, developmental delay, and reduced female fertility. Remarkably, *mfl* also hosts an intron-encoded H/ACA snoRNA, which directs an 18S rRNA pseudouridylation and could represent the functional equivalent of human snoRNA U70. Two other essential *S. cerevisiae* proteins, Nh2p and Nop10p, identified by isolation of affinity-purified complexes containing epitope-tagged Gar1p protein, appear to represent core components of H/ACA snoRNAs (Henras et al., 1998). Finally, electron microscopy observations of purified snoRNP core particles have revealed a highly symmetric bipartite structure and predicted that the particle consists of a snoRNA and two copies each of the Gar1, Nhp2, and Cbf5 proteins, in line with the notion that the two major structural domains of H/ACA snoRNAs have striking structural and functional similarities (Watkins et al., 1998b).

BIOSYNTHESIS OF GUIDE snoRNAs IS COORDINATED WITH THAT OF COMPONENTS OF THE TRANSLATIONAL MACHINERY

The genomic organizations and modes of expression of both families of guide snoRNAs exhibit great diversity, with, in addition to intronic snoRNAs, snoRNAs encoded by independent genes and processed by a splicing-independent mechanism (Fig. 4). Independent genes for modification guide snoRNAs have not been detected so far in vertebrates but are prevalent in *S. cerevisiae*, which also contains a few intron-encoded specimens (Qu et al., 1995; Ooi et al., 1998; Samarsky and Fournier, 1999; Lowe and Eddy, 1999). Remarkably, several independent genes are polycistronic, corresponding to the juxtaposition of different guide snoRNA coding regions, in both yeast and plants. In *S. cerevisiae*, the most complex snoRNA polycistronic cluster encodes seven distinct methylation guides (Qu et al., 1999). In maize, a snoRNA polycistron includes both methylation and pseudouridylation guides (Leader et al., 1997; Shaw et al., 1998). As for intron-encoded snoRNAs, they are present as a single unit per intron, in line with the observation that almost all intronic

A) Intron-encoded snoRNAs

exon snoRNA exon
1 2 pre-mRNA
1 2
splicing
lariat debranching
Major pathway

exon snoRNA exon
1 2 pre-mRNA
Endonucleolytic cleavage of the intron
Minor pathway

intron processed by 5'-3' (Rat1p)
and 3'-5' exonucleases

B) snoRNAs processed from independent genes

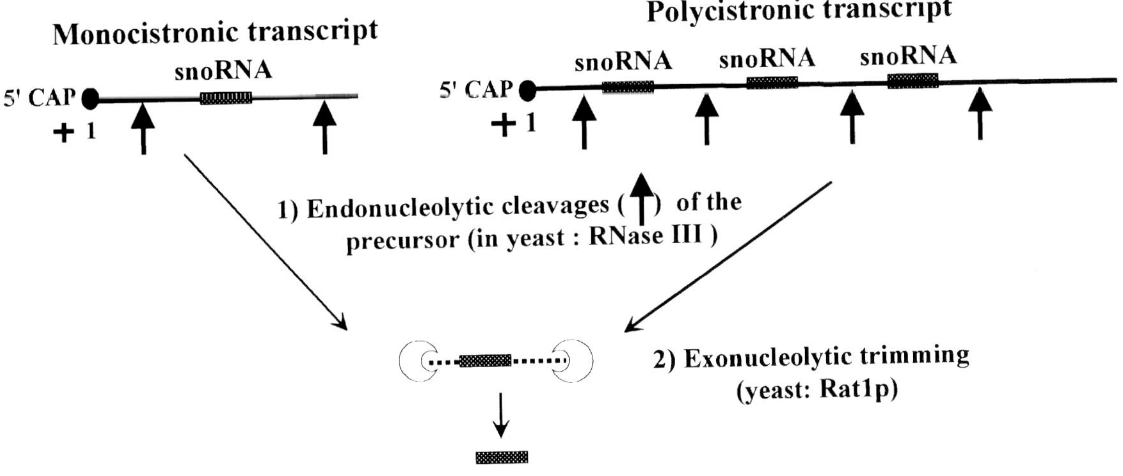

Figure 4. Diversity of guide snoRNA biosynthetic pathways.

snoRNAs are exclusively produced by 5′ → 3′ and 3′ → 5′ exonucleolytic processing of the debranched lariat (Fig. 4a), but intronic snoRNA host genes frequently exhibit more than a single snoRNA-containing intron.

Coordination at the Transcription Level

Remarkably, the vast majority of intronic snoRNAs, not only in vertebrates but also in yeast, are hosted by genes involved in the formation of the translational machinery, i.e., ribosomal proteins (S3, S8, L1a, L5, L7a, and L13a), nucleolar proteins (nucleolin and also hsc70, which becomes nucleolar following heat shock), or protein synthesis factors ($eIF4A_{I}$, $eIF4A_{II}$, EF-2, and EF-1β). This genomic organization and biosynthesis through processing of the pre-mRNA intron (Fig. 4) provides the basis for coordination, at the transcriptional level, with the production of components of the translational machinery. However, several exceptions to this pattern have been detected in vertebrates, with a few host genes not obviously related to ribosomal assembly or function (such as ATP synthase β) or even not coding for a protein (Tycowski et al., 1996b; Bortolin and Kiss, 1998; Pelczar and Filipowicz, 1998; Smith and Steitz, 1998). For this last category of genes, termed UHG (Tycowski et al., 1996b), the intronic snoRNA sequences are much more conserved phylogenetically than the exon sequences, suggesting that the only functional part of the transcript is the modification guide snoRNA. All vertebrate snoRNA host genes, including those without apparent ribosome-related functions and those without protein-coding potential, have recently been found to belong to the 5′-terminal oligopyrimidine class of vertebrate genes (Smith and Steitz, 1998). Each of them displays a long pyrimidine-rich sequence around its transcription start site, which seems likely to mediate transcriptional regulation of all snoRNA-containing genes.

As for the yeast independent snoRNA genes, whether they are monocistronic or polycistronic, their promoter regions generally contain binding sites for global regulatory factor Rap1, which plays an essential role in the transcription activation of yeast ribosomal-protein genes, again pointing to a coordinated transcriptional control (Qu et al., 1999).

Coordination at the RNA-Processing Level

In *S. cerevisiae*, processing of both families of modification guide snoRNAs involves enzymes which also process pre-rRNA. The two 5′ → 3′ exonucleases Xrn1 and Rat1, which degrade pre-rRNA spacers, are also required for the 5′-terminal processing of both intronic and nonintronic snoRNAs (Petfalski et al., 1998; Qu et al., 1999). Yeast Rrp6p, a 3′ → 5′ exonuclease required for 5.8S rRNA biosynthesis, is also involved in 3′-end formation of the box C/D antisense snoRNAs U14, U18, and U24 in *S. cerevisiae* (Tollervey, personal communication).

RNase III also appears to be a key player in snoRNA processing. Thus, in the case of the polycistronic unit encoding seven distinct methylation guide snoRNAs, a 1.4-kb-long polycistronic precursor accumulates after disruption of the RNase III gene, and the production of the seven mature snoRNAs is abolished. RNase III cuts have been mapped within the polycistronic transcript, using yeast cells lacking a functional exonuclease Rat1, in which the products of RNase III endonucleolytic cleavage accumulate. The RNase III cuts are sufficient to free each of the snoRNAs from the primary transcript, and pairs of cuts are found next to each other on opposite strands of long double-helical stems in the secondary structure predicted for the polycistronic precursor, in line with known properties of RNase III. The processing of several other yeast snoRNAs transcribed from independent genes, belonging to either the methylation or the pseudouridylation guide family, has been shown independently to involve RNase III, with the cuts again distributed pairwise, on opposite sides of RNA duplexes (Chanfreau et al., 1998a, 1998b). Interestingly, a remarkably conserved feature is detected in the immediate vicinity of all cleavage sites, including, in addition to a wide range of snoRNA precursors, precursors to U2 and U5 snRNAs and the 3′ external transcribed spacer of yeast pre-rRNA. A potential tetraloop with the consensus sequence AGNN is always located 14 to 17 bp from the pair of RNase III cleavage sites and seems likely to act as an important structural determinant of the cleavage site (Chanfreau et al., 1998a).

CONCLUSIONS

Progress has been made in understanding how the sites of ribose methylations and pseudouridines are selected within the sequences of eukaryotic rRNAs. However, we still do not know the biological role of either type of modification, and there is no firm evidence so far for an essential function of any of these modifications. Even the simultaneous suppression of seven different rRNA ribose methylations in yeast did not result in any growth delay under the culture conditions tested (Qu et al., 1999). Paradoxically, the strong conservation of the complex patterns of rRNA ribose methylations and pseudouridines in eukaryotes points to some important, although not essential, function. The identification of elaborate repertoires of modification guide snoRNAs, the modes of formation of which allow for multiple coordinations with the synthesis of the translational apparatus, despite the diversities and peculiarities of their genomic organization in eukaryotes, reinforces this notion. Taken together, the specific RNA duplexes formed by both families of guide snoRNAs involve more than 2.5 kb of rRNA sequences. Binding of the entire set of modification guide snoRNAs onto the elongating pre-rRNA transcript in eukaryotes, therefore, should dramatically constrain the folding of the rRNA precursor and affect early stages of ribosomal protein assembly. Box C/D and H/ACA snoRNAs could represent intrinsic RNA chaperones, and the sole role of some nucleotide modifications might be to signal completion of the chaperone function and trigger disruption of the snoRNA-rRNA duplex (Bachellerie et al., 1995; Steitz and Tycowski, 1995; Bachellerie and Cavaillé, 1997, 1998; Ofengand and Fournier, 1998), reminiscent of the quality control mechanism proposed for the yeast 18S rRNA dimethylase Dim1p in ribosome synthesis (Lafontaine et al., 1998a). Thus, box C/D snoRNA U14, which contains an antisense element guiding the formation of Cm462 in 18S rRNA, also carries a phylogenetically conserved 14-nt-long rRNA complementarity not involved in rRNA ribose methylation but essential for rRNA processing, and its function can be rescued by an RNA helicase protein (Liang et al., 1997). In any case, the tightly coordinated actions of both families of guide snoRNAs seem likely to involve a complex interplay with RNA helicases in the nucleolus (de la Cruz et al., 1999). Clearly, we still have a long way to go in evaluating the full implications of guide snoRNA-rRNA interactions, and many more challenges and exciting developments lie ahead.

We acknowledge our colleagues of the L.B.M.E., Toulouse, M. Nicoloso, B. Michot, M. Caizergues-Ferrer, Y. Henry, A. Henras, T. Kiss, P. Ganot, and M. L. Bortolin, who have contributed many aspects of the work reviewed in this manuscript. Our work was supported by the C.N.R.S. (Programme Physique et Chimie du Vivant 1997), the Ministère de l'Education Nationale, de l'Enseignement Supérieur et de la Recherche (MENESR, grant ACC-SV1), Université Paul Sabatier, Toulouse, and the Association Franco-Chinoise pour la Recherche Scientifique et Technique (Programme de Recherches Avancées 1998).

REFERENCES

Amiri, K. A. 1994. Fibrillarin-like proteins occur in the domain *Archaea. J. Bacteriol.* **176**:2124–2127.

Bachellerie, J. P., and J. Cavaillé. 1997. Guiding ribose methylation of rRNA. *Trends Biochem. Sci.* **22**:257–261.

Bachellerie, J. P., and J. Cavaillé. 1998. Small nucleolar RNAs guide the ribose methylations of eukaryotic rRNAs, p. 255–272. *In* H. Grosjean and R. Benne (ed.), *Modification and Editing of RNA: the Alteration of RNA Structure and Function.* ASM Press, Washington, D.C.

Bachellerie, J. P., B. Michot, M. Nicoloso, A. Balakin, J. Ni, and M. J. Fournier. 1995. Antisense snoRNAs: a family of nucleolar RNAs with long complementarities to rRNA. *Trends Biochem. Sci.* **20**:261–264.

Bagni, C., and B. Lapeyre. 1998. Gar1p binds to the small nucleolar RNAs snR10 and snR30 *in vitro* through a nontypical RNA binding element. *J. Biol. Chem.* **273**:10868–10873.

Balakin, A. G., L. Smith, and M. J. Fournier. 1996. The RNA world of the nucleolus: two major families of small nucleolar RNAs defined by different box elements with related functions. *Cell* **86**:823–834.

Bertrand, E., F. Houser-Scott, A. Kendall, R. H. Singer, and D. R. Engelke. 1998. Nucleolar localization of early tRNA processing. *Genes Dev.* **12**:2463–2468.

Bond, V. C., and B. Wold. 1993. Nucleolar localization of *myc* transcripts. *Mol. Cell. Biol.* **13**:3221–3230.

Bortolin, M. L., and T. Kiss. 1998. Human U19 intron-encoded snoRNA is processed from a long primary transcript that possesses little potential for protein coding. *RNA* **4**:445–454.

Bortolin, M. L., P. Ganot, and T. Kiss. 1999. Elements essential for accumulation and function of small nucleolar RNAs directing site-specific pseudouridylation of ribosomal RNAs. *EMBO J.* **18**: 457–469.

Bousquet-Antonelli, C., Y. Henry, J.-P. Gélugne, M. Caizergues-Ferrer, and T. Kiss. 1997. A small nucleolar RNP protein is required for pseudouridylation of eukaryotic ribosomal RNAs. *EMBO J.* **15**:4770–4776.

Brimacombe, R., P. Mitchell, M. Osswald, K. Stade, and D. Bochkariov. 1993. Clustering of modified nucleotides at the functional center of bacterial ribosomal RNA. *FASEB J.* **7**:161–167.

Caffarelli, E., A. Fatica, S. Prislei, E. De Gregorio, P. Fragapane, and I. Bozzoni. 1996. Processing of the intron-encoded U16 and U18 snoRNAs: the conserved C and D boxes control both the processing and the stability of the mature snoRNA. *EMBO J.* **15**: 1121–1131.

Caffarelli, E., M. Losito, C. Giorgi, A. Fatica, and I. Bozzoni. 1998. In vivo identification of nuclear factors interacting with the conserved elements of box C/D small nucleolar RNAs. *Mol. Cell. Biol.* **18**:1023–1028.

Cavaillé, J., and J. P. Bachellerie. 1996. Processing of fibrillarin-associated snoRNAs from pre-mRNA introns: an exonucleolytic process exclusively directed by the common stem-box terminal structure. *Biochimie* (Paris) **78**:443–456.

Cavaillé, J., and J. P. Bachellerie. 1998. SnoRNA-guided ribose methylation of rRNA: structural features of the guide RNA duplex influencing the extent of the reaction. *Nucleic Acids Res.* **26**:1576–1587.

Cavaillé, J., M. Nicoloso, and J. P. Bachellerie. 1996. Targeted ribose methylation of RNA *in vivo* directed by tailored antisense RNA guides. *Nature* **383**:732–735.

Cavaillé, J., F. Chetouani, and J. P. Bachellerie. 1999. The yeast *Saccharomyces cerevisiae* YDL112w ORF encodes the putative 2′-O-ribose methyltransferase catalyzing the formation of Gm18 in tRNAs. *RNA* **5**:66–81.

Chanfreau, G., P. Legrain, and A. Jacquier. 1998a. Yeast RNase III as a key processing enzyme in small nucleolar RNAs metabolism. *J. Mol. Biol.* **284**:975–988.

Chanfreau, G., G. Rotondo, P. Legrain, and A. Jacquier. 1998b. Processing of a dicistronic small nucleolar RNA precursor by the RNA endonuclease Rnt1. *EMBO J.* **17**:3726–3737.

Chetouani, F. 1999. Développement d'un logiciel d'édition/analyse de la structure secondaire des ARNs. Recherche informatique de motifs ARNs structurés et d'enzymes de modifications de nucléotides. PhD thesis. Université Paul-Sabatier, Toulouse, France.

de la Cruz, J., D. Kressler, and P. Linder. 1999. Unwinding RNA in *Sacchaaromyces cerevisiae:* DEAD-box proteins and related families. *Trends Biochem. Sci.* **24**:192–198.

Ganot, P., M. Caizergues-Ferrer, and T. Kiss. 1997a. The family of box ACA snoRNAs is defined by an evolutionarily conserved secondary structure and ubiquitous sequence elements essential for RNA accumulation. *Genes Dev.* **11**:946–972.

Ganot, P., M. L. Bortolin, and T. Kiss. 1997b. Site-specific pseudouridine formation in eukaryotic pre-rRNAs is guided by small nucleolar RNAs. *Cell* **89**:799–809.

Gautier, T., T. Berges, D. Tollervey, and E. Hurt. 1997. Nucleolar KKE/D repeat proteins Nop56p and Nop58p interact with Nop1p and are required for ribosome biogenesis. *Mol. Cell. Biol.* **17**:7088–7098.

Giordano, E., I. Peluso, S. Senger, and M. Furia. 1999. *minifly*, a *Drosophila* gene required for ribosome biogenesis. *J. Cell. Biol.* **144**:1123–1133.

Gustafsson, C., R. Reid, P. J. Greene, and D. V. Santi. 1996. Identification of new RNA modifying enzymes by iterative genome search using known modifying enzymes as probes. *Nucleic Acids Res.* **24**:3756–3762.

Heiss, N. S., S. W. Knight, T. J. Vulliamy, S. M. Klauck, S. Wiemann, P. J. Mason, A. Poustka, and I. Dokal. 1998. X-linked dyskeratosis congenita is caused by mutations in a highly conserved gene with putative nucleolar functions. *Nat. Genet.* **19**: 32–38.

Henras, A., Y. Henry, C. Bousquet-Antonelli, J. Noaillac-Depeyre, J. P. Gelugne, and M. Caizergues-Ferrer. 1998. Nhp2p and Nop10p are essential for the function of H/ACA snoRNPs. *EMBO J.* **17**:7078–7090.

Hong, B., J. S. Brockenbrough, P. Wu, and J. P. Aris. 1997. Nop2 is required for pre-rRNA processing and 60S ribosome subunit in yeast. *Mol. Cell. Biol.* **17**:378–388.

Jacobson, M. R., and T. Pederson. 1998. Localization of signal recognition particle RNA in the nucleolus of mammalian cells. *Proc. Natl. Acad. Sci. USA* **95:**7981–7986.

Jacobson, M. R., L. G. Cao, K. Taneja, R. H. Singer, Y. L. Wang, and T. Pederson. 1997. Nuclear domains of the RNA subunit of RNase P. *J. Cell Sci.* **110:**829–837.

Kiss-Laszlo, Z., Y. Henry, J. P. Bachellerie, M. Caizergues-Ferrer, and T. Kiss. 1996. Site-specific ribose methylation of preribosomal RNA: a novel function for small nucleolar RNAs. *Cell* **85:** 1077–1088.

Kiss-Laszlo, Z., Y. Henry, and T. Kiss. 1998. Sequence and structural elements of methylation guide snoRNAs essential for site-specific ribose methylation of pre-rRNA. *EMBO J.* **17:**797–807.

Lafontaine, D. L., and D. Tollervey. 1998. Birth of the snoRNPs: the evolution of the modification-guide snoRNAs. *Trends Biochem. Sci.* **23:**383–388.

Lafontaine, D. L., and D. Tollervey. 1999. Nop58p is a common component of the box C+D snoRNPs that is required for snoRNA stability. *RNA* **5:**455–467.

Lafontaine, D. L., T. Preiss, and D. Tollervey. 1998a. Yeast 18S rRNA dimethylase Dim1p: a quality control mechanism in ribosome synthesis? *Mol. Cell. Biol.* **18:**2360–2370.

Lafontaine D. L. J., C. Bousquet-Antonelli, Y. Henry, M. Caizergues-Ferrer, and D. Tollervey. 1998b. The box H + ACA snoRNAs carry Cbf5p, the putative rRNA pseudouridine synthase. *Genes Dev.* **12:**527–537.

Lane, B. G., J. Ofengand, and M. W. Gray. 1995. Pseudouridine and O2′-methylated nucleosides. Significance of their selective occurrence in rRNA domains that function in ribosome-catalyzed synthesis of the peptide bonds in proteins. *Biochimie* (Paris) **77:**7–15.

Lange, T. S., A. Borovjagin, E. S. Maxwell, and S. A. Gerbi. 1998. Conserved boxes C and D are essential nucleolar localization elements of U14 and U8 snoRNAs. *EMBO J.* **17:**3176–3187.

Leader, D. J., G. P. Clark, J. Watters, A. F. Beven, P. J. Shaw, and J. W. S. Brown. 1997. Clusters of multiple different small nucleolar RNA genes in plants are expressed as and processed from polycistronic pre-snoRNAs. *EMBO J.* **16:**5742–5751.

Leverette, R. D., M. T. Andrews, and E. S. Maxwell. 1992. Mouse U14 snRNA is a processed intron of the cognate hsc70 heat-shock pre-messenger RNA. *Cell* **71:**1215–1221.

Liang, W. -Q., J. A. Clark, and M. J. Fournier. 1997. The rRNA-processing function of the yeast U14 small nucleolar RNA can be rescued by a conserved RNA-helicase-like protein. *Mol. Cell. Biol.* **17:**4124–4132.

Lowe, T. M., and S. R. Eddy. 1999. A computational screen for methylation guide snoRNAs in yeast. *Science* **283:**1168–1171.

Maden, B. E. H. 1990. The numerous modified nucleotides in eukaryotic ribosomal RNA. *Prog. Nucleic Acid Res. Mol. Biol.* **39:**241–301.

Maxwell, E. S., and M. J. Fournier. 1995. The small nucleolar RNAs. *Annu. Rev. Biochem.* **35:**897–934.

Mitchell, J. R., J. Cheng, and K. Collins. 1999. A box H/ACA small nucleolar RNA-like domain at the human telomerase RNA 3′ end. *Mol. Cell. Biol.* **19:**567–576.

Ni, J., A. L. Tien, and M. J. Fournier. 1997. Small nucleolar RNAs direct site-specific synthesis of pseudouridine in ribosomal RNA. *Cell* **89:**565–573.

Nicoloso, M., L.-H. Qu, B. Michot, and J. P. Bachellerie. 1996. Intron-encoded, antisense small nucleolar RNAs : the characterization of nine novel species points to their direct role as guides for the 2′-O-ribose methylation of rRNAs. *J. Mol. Biol.* **260:** 178–195.

Niewmierzycka, A., and S. Clarke. 1999. S-Adenosylmethionine-dependent methylation in *Saccharomyces cerevisiae*. Identification of a novel protein arginine methyltransferase. *J. Biol. Chem.* **274:**814–824.

Noon, K. R., E. Bruenger, and J. A. McCloskey. 1998. Posttranscriptional modifications in 16S and 23S rRNAs of the archaeal hyperthermophile *Sulfolobus solfataricus*. *J. Bacteriol.* **180:** 2883–2888.

Ofengand, J., and M. J. Fournier. 1998. The pseudouridine residues of rRNA: number, location, biosynthesis and function, p. 229–253. *In* H. Grosjean and R. Benne (ed.), *Modification and Editing of RNA: the Alteration of RNA Structure and Function.* ASM Press, Washington, D.C.

Ofengand, J., A. Bakin, J. Wrzesinski, K. Nurse, and B. G. Lane. 1995. The pseudouridine residues of ribosomal RNA. *Biochem. Cell Biol.* **73:**915–924.

Ooi, S. L., D. A. Samarsky, M. J. Fournier, and J. D. Boeke. 1998. Intronic snoRNA biosynthesis in *Saccharomyces cerevisiae* depends on the lariat-debranching enzyme: intron length effect and activity of a precursor snoRNA. *RNA* **4:**1096–1110.

Pederson, T. 1998. The plurifunctional nucleolus. *Nucleic Acids Res.* **26:**3871–3876.

Pelczar, P., and W. Filipowicz. 1998. The host gene for intronic U17 small nucleolar RNAs in mammals has no protein-coding potential and is a member of the 5′-terminal oligopyrimidine gene family. *Mol. Cell. Biol.* **18:**4509–4518.

Petfalski, E., T. Dandekar, Y. Henry, and D. Tollervey. 1998. Processing of the precursors to small nucleolar RNAs and rRNAs requires common components. *Mol. Cell. Biol.* **18:**1181–1189.

Qu, L. H., M. Nicoloso, B. Michot, M. C. Azum, M. Caizergues-Ferrer, M. H. Renalier, and J. P. Bachellerie. 1994. U21, a novel small nucleolar RNA with a 13 nt. complementarity to 28S rRNA, is encoded in an intron of ribosomal protein L5 gene in chicken and mammals. *Nucleic Acids Res.* **22:**4073–4081.

Qu, L. H., Y. Henry, M. Nicoloso, B. Michot, M. C. Azum, M. H. Renalier, M. Caizergues-Ferrer, and J. P. Bachellerie. 1995. U24, a novel intron-encoded small nucleolar RNA with two 12 nt. long, phylogenetically conserved complementarities to 28S rRNA. *Nucleic Acids Res.* **23:**2669–2676.

Qu, L. H., A. Henras, Y. J. Lu, H. Zhou, W. X. Zhou, Y. Q. Zhu, J. Zhao, Y. Henry, M. Caizergues-Ferrer, and J. P. Bachellerie. 1999. Seven novel methylation guide small nucleolar RNAs are processed from a common polycistronic transcript by Rat1p and RNase III in yeast. *Mol. Cell. Biol.* **19:**1144–1158.

Samarsky, D. A., and M. J. Fournier. 1999. A comprehensive database for the small nucleolar RNAs from *Saccharomyces cerevisiae*. *Nucleic Acids Res.* **27:**161–164.

Samarsky, D. A., M. J. Fournier, R. H. Singer, and E. Bertrand. 1998. The snoRNA box C/D motif directs nucleolar targeting and also couples snoRNA synthesis and localization. *EMBO J.* **17:**3747–3757.

Shaw, P. J., A. F. Beven, D. J. Leader, and J. W. Brown. 1998. Localization and processing from a polycistronic precursor of novel snoRNAs in maize. *J. Cell. Sci.* **111:**2121–2128.

Smith, C. M., and J. A. Steitz. 1998. Classification of gas5 as a multi-small-nucleolar-RNA (snoRNA) host gene and a member of the 5′-terminal oligopyrimidine gene family reveals common features of snoRNA host genes. *Mol. Cell. Biol.* **18:**6897–6909.

Sollner-Webb, B. 1993. Novel intron-encoded small nucleolar RNAs. *Cell* **75:**403–405.

Sollner-Webb, B., K. Tyc, and J. A. Steitz. 1995. Ribosomal RNA processing in eukaryotes, p. 469–490. *In* R. A. Zimmermann and A. E. Dahlberg (ed.), *Ribosomal RNA. Structure, Evolution, Processing, and Function in Protein Biosynthesis.* Telford, Caldwell, N.J.

Steitz, J. A., and K. T. Tycowski. 1995. Small RNA chaperones for ribosome biogenesis. *Science* **270:**1626–1627.

Tollervey, D. 1996. *trans*-acting factors in ribosome biogenesis. *Exp. Cell Res.* **229:**226–232.

Tollervey, D. Personal communication.

Tollervey, D., H. Lehtonen, M. Carmo-Fonseca, and E. C. Hurt. 1991. The small nucleolar RNP protein NOP1 (fibrillarin) is required for pre-rRNA processing in yeast. *EMBO J.* **10:**573–583.

Tollervey, D., H. Lehtonen, R. Jansen, H. Kern, and E. C. Hurt. 1993. Temperature-sensitive mutations demonstrate roles for yeast fibrillarin in pre-rRNA processing, pre-rRNA methylation, and ribosome assembly. *Cell* **72:**443–457.

Tycowski, K. T., C. M. Smith, M.-D. Shu, and J. A. Steitz. 1996a. A small nucleolar RNA requirement for site-specific ribose methylation of rRNA in *Xenopus. Proc. Natl. Acad. Sci. USA* **93:** 14480–14485.

Tycowski, K. T., M.-D. Shu, and J. A. Steitz. 1996b. A mammalian gene with introns instead of exons generating stable RNA products. *Nature* (London) **379:**464–466.

Tycowski, K. T., Z. H. You, P. J. Graham, and J. A. Steitz. 1998. Modification of U6 spliceosomal RNA is guided by other small RNAs. *Mol. Cell* **2:**629–638.

Watkins, N. J., R. D. Leverette, L. Xia, M. T. Andrews, and E. S. Maxwell. 1996. Elements essential for processing intronic U14 snoRNA are located at the termini of the mature snoRNA sequence and include conserved nucleotide boxes C and D. *RNA* **2:**118–133.

Watkins, N. J., D. R. Newman, J. F. Kuhn, and E. S. Maxwell. 1998a. In vitro assembly of the mouse U14 snoRNP core complex and identification of a 65-kDa box C/D-binding protein. *RNA* **4:**582–593.

Watkins, N. J., A. Gottschalk, G. Neubauer, B. Kastner, P. Fabrizio, M. Mann, and R. Luhrmann. 1998b. Cbf5p, a potential pseudouridine synthase, and Nhp2p, a putative RNA-binding protein, are present together with Gar1p in all H BOX/ACA-motif snoRNPs and constitute a common bipartite structure. *RNA* **4:1549–1568.**

Wu, P., J. S. Brockenbrough, A. C. Metcalfe, S. Chen, and J. P. Aris. 1998. Nop5p is a small nucleolar ribonucleoprotein component required for pre-18 S rRNA processing in yeast. *J. Biol. Chem.* **273:**16453–16463.

Yu, Y. T., M. D. Shu, and J. A. Steitz. 1998. Modifications of U2 snRNA are required for snRNP assembly and pre-mRNA splicing. *EMBO J.* **17:**5783–5795.

The Ribosome: Structure, Function, Antibiotics, and Cellular Interactions
Edited by R. A. Garrett, S. R. Douthwaite, A. Liljas, A. T. Matheson, P. B. Moore, and H. F. Noller

Chapter 18

Nuclear Import of Ribosomal Proteins: Evidence for a Novel Type of Nuclear Localization Signal

ROGIER STUGER, ANTONIUS C. J. TIMMERS, HENDRIK A. RAUÉ, and JAN VAN 'T RIET

The formation of eukaryotic ribosomes is a highly complex process which requires the coordinated expression of a large set of ribosomal genes, transcribed by three different RNA polymerases, to ensure production of equimolar amounts of the four rRNAs and the approximately 80 ribosomal protein (r-protein) species under all growth conditions (for a recent review, see Planta, 1997).

A further level of complexity is added to eukaryotic ribosome biogenesis by the fact that it involves different cellular compartments. Transcription of the rRNA and r-protein genes takes place in the nucleolus and nucleoplasm, respectively. The r-protein mRNAs have to be exported to the cytoplasm. After translation, the r-proteins must be imported into the nucle(ol)us, where they have to be present in equimolar amounts to be assembled with the rRNAs into ribosomal subunits. The subunits are then exported from the nucleus to take up their function in the cytosol. Thus, ribosome biogenesis in eukaryotic cells involves massive transport of macromolecules and macromolecular complexes in both directions across the nuclear envelope (reviewed in Scheer and Weisenberger, 1994). This transport not only concerns the ribosomes themselves (or their components) but also a large number of accessory factors, varying from constituents of the transcription, translation, and splicing machinery to pre-rRNA-processing and ribosome assembly factors (Tollervey, 1996).

In recent years considerable insight has been gained into the mechanisms of nucleocytoplasmic transport (see Mattaj and Englmeier, 1998; Weis, 1998; and Wen et al., 1995 for recent reviews). Interestingly, a number of strong indications were found that nuclear import of r-proteins uses a specialized import pathway different from that used by the majority of karyophilic proteins. This suggests that the nuclear localization signals or sequences (NLSs) of r-proteins may be structurally distinct from the classical NLSs present in the latter class of proteins (Chelsky et al., 1989; Dingwall and Laskey, 1991). In this chapter we review present knowledge of the mechanism and signals responsible for the nuclear import of proteins, in particular, r-proteins. A database search of the complete set of yeast r-proteins for putative NLSs by using a set of criteria derived from the comparison of all experimentally identified signals in yeast r-proteins indicates that the large majority of these proteins may indeed possess a novel type of NLS, characterized by the presence of a sequence motif consisting of three basic residues within a 4- to 7-amino acid-long sequence, of which the first, the last, or both are helix-breaking residues.

NUCLEOCYTOPLASMIC TRANSPORT

Transport from cytoplasm to nucleus and vice versa is channeled through the nuclear pore complexes (NPCs). Such NPCs are complicated, proteinaceous, three-dimensional structures spanning the nuclear envelope (for reviews, see Fahrenkrog et al., 1998; Goldberg and Allen, 1995; Panté and Aebi, 1996). Although the NPC is freely permeable to small (macro)molecules (with masses below ~40 kDa) (Roberts et al., 1987), translocation through the NPC

Rogier Stuger ■ Department of Molecular Cell Physiology, Vrije Universiteit, De Boelelaan 1087, 1081 HV Amsterdam, The Netherlands. **Antonius C. J. Timmers** ■ Laboratoire de Biologie Moléculaire des Relations Plantes-Microorganismes, CNRS-INRA, BP 27, 31326 Castanet-Tolosan Cedex, France. **Hendrik A. Raué and Jan van 't Riet** ■ Faculty of Science, Division of Chemistry, Department of Biochemistry and Molecular Biology, Institute for Molecular Biological Sciences, BioCentrum Amsterdam, Vrije Universiteit, de Boelelaan 1083, 1081 HV Amsterdam, The Netherlands.

is energy dependent (reviewed by Dingwall, 1991) and carrier mediated, depending on the presence of an NLS within the primary structure of the nuclear protein, even if its mass is far below 40 kDa (Breeuwer and Goldfarb, 1990; Schaap et al., 1991). Similarly, nuclear export depends upon the presence of nuclear export signals (Fischer et al., 1995; Michael et al., 1995; Wen et al., 1995). Several soluble factors that are required for import and export of nuclear proteins have been identified (Görlich and Mattaj, 1996; Melchior and Gerace, 1995). In the original model for nuclear import, an NLS receptor (Görlich et al., 1995), also named karyopherin-α or importin α, binds its karyophilic transport ligand before it interacts with importin β_1, also named karyopherin-β or p97 (Adam and Adam, 1994; Moore and Blobel, 1994). With the aid of the transport factors p10 (also called pp15 or NTF2) (Nehrbass and Blobel, 1996; Paschal and Gerace, 1995) and Ran-GDP, this complex then binds to nucleoporins at the cytoplasmic face of the NPC. After translocation of the complex into the nucleus, it is dissociated by the exchange of GDP for GTP through the action of the RanGEF protein RCC1. The transport factors are recycled into the cytoplasm, using receptors that also belong to the growing family of importin β-like proteins (see Mattaj and Englmeier, 1998; Weis, 1998; and Wozniak et al., 1998, for recent reviews).

The importin α adaptor recognizes the classical mono- and bipartite NLSs, of which the ones present in the simian virus 40 large T antigen (PKKKRKV) (Kalderon et al., 1984) and in nucleoplasmin (Dingwall and Laskey, 1991) are the respective prototypes. Importin α appears to have two binding sites for the monopartite signal. The same sites are also responsible for binding the two parts of the bipartite NLS (Conti et al., 1998). Importin α itself contains a bipartite signal in its N-terminal region, the so-called IBB domain, that is recognized by the importin β receptor (Görlich and Mattaj, 1996). Recently, it was found that the IBB domain can also interact with the NLS binding domain of the same importin α molecule (Kobe, 1999). This interaction may take place in the nucleoplasm after disruption of the nuclear cargo-importin α-importin β complex by Ran-GTP, thereby preventing reassembly of the complex. In the cytoplasm this self-inhibitory effect is prevented by the high affinity of the IBB domain for the IBB binding domain of importin β, thus exposing the NLS binding domain of importin α for interaction with the nuclear cargo.

NLSs

As indicated above, two major classes of NLSs have been identified in eukaryotic proteins. The classical, monopartite NLS consists of a short stretch of 4 to 7 amino acids in which basic and hydrophobic residues predominate. Chelsky et al. (1989) have proposed the sequence KR/KXR/K (X being any amino acid) as a consensus sequence for such monopartite NLSs. However, not every sequence matching this consensus shows nuclear targeting activity (e.g., Schmidt et al., 1995). Dang and Lee (1989) observed that many monopartite NLSs contain a helix-breaking residue, either glycine or proline, which may reflect the requirement for a "random" structure of the sequence for it to be able to function as an NLS. An additional type of NLS having the sequence KIPIK, and thus similar but not identical to the monopartite type, was identified in the yeast transcriptional repressor Matα2 (Hall et al., 1984).

The bipartite NLS is characterized by the presence of a basic dipeptide separated from another sequence rich in basic amino acids by about 10 arbitrary residues (Dingwall and Laskey, 1991). It should be noted that the downstream portion of the bipartite NLS conforms to the Chelsky consensus sequence.

In the past few years a number of NLSs that cannot be grouped into either of the two classes described above have been discovered. These include NLSs of karyophilic proteins, such as the M9 domain of the hnRNP A1 protein (Aitchison et al., 1996; Pollard et al., 1996), as well as the complex signals of U snRNPs (Fischer et al., 1991). Nuclear import of these components has been found to involve adaptors and/or receptors different from the ones required for transport of components containing a classical mono- or bipartite NLS, although they clearly belong to the same family as importin β (Huber et al., 1998; Wozniak et al., 1998). In yeast the importin β family presently consists of 13 members, most of which have a vertebrate homologue (reviewed in Wozniak et al., 1998). In many cases, these proteins interact directly with the NLS-bearing protein without the intervention of an importin α-like adaptor.

In yeast, NLSs vary considerably in length and composition. Many show only weak similarity to the consensus sequences described above, although in most cases they are composed of basic and hydrophobic residues (Osborne and Silver, 1993). Nevertheless, the machinery for nuclear protein import appears to be functionally conserved between yeast and higher eukaryotes, since, regardless of their source, NLSs are functional in either type of cell (Jeeninga et al., 1996; Schaap et al., 1991; Wagner and Hall, 1993).

NUCLEAR IMPORT OF r-PROTEINS

Nuclear import of r-proteins has been most extensively investigated in yeast. So far, the NLS sequences of six yeast r-proteins (L3, L25, L28, S17, S22, and S25, according to the new nomenclature of Mager et al., 1997) have been identified experimentally by assessing the ability of various r-protein fragments to stimulate nuclear import of a reporter protein that normally resides in the cytoplasm. A caveat to this approach is that a region identified as having nuclear targeting activity might not be a bona fide NLS, i.e., a sequence recognized by a nuclear import receptor, but might be required for the association of the protein with another NLS-bearing protein ("piggy-back" import) (Booher et al., 1989). Although the existence of such "preassembly" complexes of r-proteins has been suggested (Fried, 1993), it is at present entirely hypothetical.

The first notable property shared by the yeast r-protein NLSs is their occurrence close to the N terminus of the protein. In the cases of both L3 (Moreland et al., 1985) and L25 (Schaap et al., 1991), the NLSs are located within a region of the r-protein that does not have a counterpart in its prokaryotic homologue (Fried, 1993; Rutgers et al., 1990). Such extensions, therefore, may have evolved to accommodate this eukaryote-specific functional signal.

A second notable feature is the presence of multiple NLSs in three of the six yeast r-proteins so far investigated. In L25 and L28 these two NLSs are located in the immediate neighborhood of each other and, since they enhance each other's activity (Underwood and Fried, 1990; Schaap et al., 1991), might together be considered a degenerate bipartite NLS (Fried, 1993). However, each of the sequences on its own acts as an NLS, as shown by mutational analysis within the context of the r-protein itself and the ability of the individual sequences to stimulate nuclear import of a reporter protein (Underwood and Fried, 1990; Schaap et al., 1991). In S25 the two NLSs are separated by about 50 residues. Moreover, the N-terminal signal is considerably more potent than the C-terminal one (Timmers et al., 1999).

Both the multiplicity and the localization close to the N terminus might increase the probability of interaction of the r-protein NLSs with the receptors and thus ensure rapid translocation of the r-protein through the nuclear pores. Nuclear import of r-proteins has to be highly efficient, since no pools of free r-proteins are present in the cytosol under any circumstances. Moreover, in yeast cells unassembled r-proteins are rapidly degraded in the cytosol (Warner et al., 1985; El Baradi et al., 1986; Maicas et al., 1988).

The yeast r-protein NLSs shown in Table 1 can be divided into three groups. Those present in L3 and the N-terminal region of S25 correspond to the classical bipartite consensus (but see below). The C-terminal, weak NLS of S25 has features in common with the import signal identified in the yeast Matα2 transcription factor (KIPIK) (Hall et al., 1984). It should be noted that the efficiency of this signal is significantly improved when adjacent sequences are present (Timmers et al., 1999). The NLSs of the yeast r-proteins S17, S22, L25, and L28, apart from a predominance of basic and hydrophobic amino acids, show only weak similarity to the classical NLSs. On the other hand, they do display mutual similarity in that they all contain a motif consisting of three basic residues clustered within a 4- to 7-amino-acid-long sequence that is either N- or C-terminally flanked by a helix-breaking Pro or Gly residue. It should be noted that the same motif is also found as part of the (supposedly) bipartite NLSs of proteins L3 and S25 (Table 1).

The existence of a specific, functionally distinct type of NLS in yeast r-proteins is supported by the recent observation that these proteins use a nuclear import pathway different from the one employed by karyophilic proteins that carry a classical mono- or bipartite NLS. The first evidence was reported by Nehrbass et al. (1993), who found that a mutation in the yeast nuclear pore protein Nsp1 did not affect nuclear import of r-protein L25 while it prevented the import of nonribosomal proteins. Many yeast r-proteins, in particular, L25, were shown to be recognized by the importin β homologues Kap123p (also called Yrb4p and yer110c), Kap121p (also called Pse1p and ymr308c), and Sxm1p (or ydr395w) but not by importin α or importin β (Rosenblum et al., 1997; Rout et al., 1997; Schlenstedt et al., 1997). A similar situation exists in mammalian cells, where the nuclear import of rat r-proteins S7, L5, and L23a could be achieved by four different vertebrate importins, namely, transportin; RanBP5, the homologue of yeast Pse1p; and Ran BP7; as well as importin β itself, but without the aid of importin α (Jäkel and Görlich, 1998).

A CONSENSUS NLS FOR YEAST r-PROTEINS

In order to obtain further support for the existence of a functionally distinct NLS (YRP-NLS) in yeast r-proteins, we performed a computer search of the complete set of yeast r-protein sequences present in the Swiss-Prot database by using the consensus sequence $(K/R)_3X_{1-4}$ either preceded or followed by a helix-breaking Gly or Pro. No acidic residues, either within or adjacent to the consensus, were allowed. The consensus is based upon the common features of

Table 1. NLSs identified experimentally in yeast r-proteins[a]

Protein		NLS	Reference
L3		1-SHRKYEAPRHGHL**GFLPRKRA**-21	Moreland et al., 1985
L25	NLS1	1-MAPSAKATAA**KKAVVKG**-17	Schaap et al., 1991
	NLS2	18-TN**GKKALKVR**TSATFRLPKTLKLAR-41	
L28	NLS1	6-**KHRKHPG**-12	Underwood and Fried, 1990
	NLS2	23-**KTRKHRG**-29	
S17		2-**GRVRTK**-7	Gritz et al., 1985
S22		20-**GKRQVLIRP**-28	Timmers et al., 1999
S25		11-AKAAAALAG**GKKSK**KKWSKKSMKDRA-36	Timmers et al., 1999
		87-GII*KPISK*H-95	

[a] The minimal sequences identified as able to direct a reporter protein efficiently into the nucleus are shown. The numbers indicate the position of the first and last residue in the complete sequence of the protein. Underlining indicates sequences conforming to the classical bipartite consensus. The sequences in boldface match the YRP consensus discussed in the text. The sequences corresponding to the Matα-2-like NLS (Hall et al., 1984) are in italics.

the r-protein NLSs discussed above, the results of the detailed mutational analysis of the L28 and S22 NLSs (Timmers et al., 1999; Underwood and Fried, 1990), and the observation that the C-terminal region of L25 contains two sequences that are very similar to its NLSs except for the absence of the flanking glycines. These sequences had no nuclear targeting activity (Schaap et al., 1991). Additional searches were carried out to score for the presence of potential bipartite and Matα2-like sequences with a strategy to identify peptide sequences containing ambiguous residues (Vodkin et al., 1996).

As shown in Table 2, 61 of the 78 individual yeast r-proteins (78%) contain one or more sequences matching the proposed YRP-NLS consensus. The emergence of these putative NLSs is not simply a consequence of the abundance of basic residues in r-proteins: of a total of 283 basic stretches, only 112 were identified by our search as containing a match to the consensus. Furthermore, for S22, S25, L25, and L28, the only matches to the YRP-NLS consensus found correspond to the experimentally identified signals. Only for S17 did we find an additional match, which is located in a region of the protein not tested for nuclear targeting activity (Gritz et al., 1985).

About half of the putative YRP-NLSs identified by our search might also be considered to be part of a putative bipartite NLS. However, in this case the abundance of basic residues may well lead to a high percentage of false positives. Nevertheless, 35 yeast r-proteins contain matches to the YRP-NLS that are not part of a bipartite consensus, and about half of these r-proteins do not have an identifiable match to the bipartite consensus at all. Only eight yeast r-proteins (10%) do not possess a sequence matching the YRP-NLS criteria but do contain a region corresponding to the bipartite consensus. Finally, a Matα2-like sequence was found in six of the yeast r-proteins, but all of these proteins also contain matches to the YRP-NLS and/or bipartite consensus.

These data clearly support the existence of a novel type of NLS in at least the majority of the yeast r-proteins that could be responsible for directing the protein to the specialized import pathway. The r-proteins that failed to give a match to the YRP-NLS consensus might use the standard importin α-importin β pathway, since they do contain a putative bipartite NLS.

No putative NLSs were found in the acidic r-proteins P1 and P2, which are known to assemble in the cytoplasm (reviewed in Ballesta et al., 1999). Acidic r-protein P0, on the other hand, which is the functional homologue of the rRNA-binding *Escherichia coli* protein L10 (Shimmin et al., 1989), does contain a YRP-like sequence. We were also unable to find matches to any of the NLS consensus sequences in a further seven nonacidic yeast r-proteins. This does not necessarily mean that these r-proteins fail to be imported into the nucleus, however. They may contain an NLS that diverges too much from our search criteria, or they could be imported by a piggyback mechanism (Booher et al., 1989) after associating with another r-protein(s) in the cytoplasm. Interestingly, a large portion of the set (proteins S27, S28, L12, and L30) was classified as "late-assembling" proteins on the basis of kinetic labeling studies carried out in our laboratory (Kruiswijk et al., 1978).

NLSs OF MAMMALIAN r-PROTEINS

Mammalian r-proteins also use a specialized nuclear import pathway not involving importin α. The situation is even more complex than in yeast, however, since four different receptors have been identified, all belonging to the importin β family:

Table 2. Putative NLSs identified by computer search of the complete set of yeast r-protein sequences[a]

YRP	Residues	Type of NLS[b]			Sequence with (putative) NLSs
S1	3–18	Y*	B		**GKNKRLSKGKKGQKKR**
	150–165		B		KRHSYAQSSHIRAIRK
S2	9–27	Y*	B		KRGGFG**GRNRGRPNRRG**PR
S3	63–77		B		RRINELTLLVQKRFK
S4	4–10	Y			**PKKHLKR**
	123–127	Y			**GKVKK**
S5	74–80	Y			**GRYANKR**
	100–111	Y*			**GRNNGKKLKAVR**
S6	87–99	Y	B		RRDGER**KRKSVRG**
	135–142	Y			**PKRANNIR**
	181–188	Y			**PQRLQRKR**
	214–230		B		KRLSERKAEKAEIRKRR
S7	95–115	Y	B		RRILPKPSRTS**RQVQKRP**RSR
	136–140	Y			**GKRVR**
S8	9–25		B		KRSATGAKRAQFRKKRK
	36–55		B	M	*KIGAK*RIHSVRTRGGNKKYR
S9	38–91		B		KREIYRISFQLSKIRR
	53–70		B		RRAARDLLTRDEKDPKRL
	77–91				RRLVRVGVLSEDKKK
	169–174	Y			**GRVARR**
S10	3–20	Y*	B		**GKNKRLSKGKKGQKKRVV**
	150–165		B		KRHSYAQSSHIRAIRK
S11	22–29	Y			**PKVKTSKR**
	86–104		B		RRAYLHYIPKYNRYEKRHK
S12		No matches			
S13	38–43	Y			**KYARKG**
	92–108		B		KKAVSVRKHLERNRKDK
S14	126–135	Y*			**RKKGGRRGRR**
S15	42–61	Y	B		**RRRFARG**MTSKPAGFMKKLR
S16	25–29			M	*KGLIK*
	126–132	Y			**KKFGGKG**
S17	2–7	Y			**GRVRTK**
	32–49		B		KRLCDEIATIQSKRLRNK
	59–64	Y			**KRIQKG**
S18	117–124	Y			**KKIRAHRG**
S19	101–122	Y	B		RKVLQALEKIGIVEIS**PKGGRR**
S20	52–59	Y			**KKGPVRLP**
	68–89		B		RKTPNGEGSKTWETYEMRIHKR
S21	3–20	Y*	B		KKRASN**GRNKKGRGHVRP**
S22	20–28	Y			**GKRQVLIRP**
S23	2–8	Y			**GKGKPRG**
	13–32		B*		RKLRVHRRNNRWAENNYKKR
	69–82		B		RKCVRVQLIKNGKK
	108–115	Y			**GRKGKAKG**
S24	107–131	Y	B		RKQKKNRDKKIFGT**GKRLAKK**VARR
S25	20–33	Y	B		**GKKSKKK**WSKKSMK
	90–94			M	KPISK
S26	3–15	Y	B		KKRASN**GRNKKGR**
S27		No matches			
S28		No matches			
S29	10–16	Y			**PRRYGKG**
S30	26–54	Y	B*		**PKKPKGRAYKR**LLYTRRFVNVTLVNGKRR
S31	1–23	Y	B*		**GKKRKKK**VYTT**PKKIKHK**HKKVK
P0	54–61	Y			**GKNTMVRR**
P1		No matches			
P2		No matches			
L1	91–106		B		KKLNKNKKLIKKLSKK
L2	5–10	Y			**RNQRKG**
	240–247	Y			**RRTGLLRG**
L3	3–29	Y	B		RKYEAPRHGHL**GFLPRKRA**ASIRARVK
	114–127		B		KRRFYKNWYKSKKK
	235–247		B		KKLPRKTHRGLRK

Continued on next page

Table 2. —*Continued.*

YRP	Residues	Type of NLS[b]	Sequence with (putative) NLSs
L4	184–203	Y* B	KKLRA**GKGKYRNRRW**TQRRG
L5	21–35	B	RRRREGKTDYYQRKR
	258–273	B	KKFTKEQYAAESKKYR
L6A	21–26	Y	**RKAARP**
	44–50	Y	**GRFRGKR**
L6B	17–26	Y	**PKKTRKAVRP**
	44–50	Y	**GRFRGKR**
L7	93–100	Y M	**_KIPPK_PRK**
	144–151	Y	**RQLVYKRG**
	153–161	Y	**GKINKQRVP**
	216–220	Y	**PRKFK**
L8	41–62	Y B	**PKRNLSR**YVKWPEYVRVQRQKK
	170–184	B	KKMGVPYAIVKGKAR
	229–250	B	KKHWGGGILGNKAQAKMDKRAK
L9	85–91	Y	**GYKYKMR**
L10	198–204	Y	**KFLSKKG**
L11	139–145	Y	**RRKRCKG**
L12		No matches	
L13	15–36	Y B	RKHWQERVKVHFDQA**GKKVSRR**
	177–199	B	KKFRGIREKRAREKAEAEAEKKK
L14	18–26	Y	**GRVVLIKKG**
	108–128	B	RRAALTDFERFQVMVLRKQKR
L15	48–77	Y* B	RRLGYKAKQGFVIY**RVRVRRGNRKRPVPKG**
	106–113	Y	**GRRAANLR**
	167–203	Y B*	**GKKSRG**INKGHKFNNTKA**GRRKTWKR**QNTLSLWRYRK
L16	113–133	B	KKKRVVVPQALRVLRLKPGRK
	158–171	B	KRKVSSAEYYAKKR
L17	122–127	Y	**PKQRRR**
	162–180	Y B	KKVVRLTSRQR**GRIAAQKR**
L18	170–183	Y B	RKFERAR**GRRRSKG**
L19	7–20	B	KRLAASVVGVGKRK
	51–63	B	KKAVTVHSKSRTR
	69–99	Y B	KREGRHSGY**GKRKGTR**EARLPSQVVWIRRLR
	161–172	B	RRLKNRAARDRR
L20	163–170	Y	**KTFSYKRP**
L21		No matches	
L22	41–48	Y	**PKRNLSRY**
	170–184	B	KKMGVPYAIVKGKAR
L23	63–88	B*	KKGKPELRKKVMPAIVVRQAKSWRRR
L24	43–77	Y B*	RKNPRRIAWTVLF**RKHHKKG**ITEEVAKKRSRKTVK
	93–127	B	RRSLKPEVRKANREEKLKANKEKKKAEKAARKAEK
L25	10–26	Y* B	**KKAVVKG**TN**GKKALKVR**
L26	50–64	Y B	RRDDEVLVV**RGSKKG**
L27	16–28	Y*	**GRYAGKKVVIVKP**
	55–73	Y B M	KKHGAKKVA**KRTKI_KP_**_FIK_
L28	8–28	Y* B	**RKHRG**HVSAGK**GRIGKHRKHP**
	110–114	M	*KILGK*
L29	12–40	Y B*	RKAH**RNGIKKP**KTYKYPYLKGVDPKFRR
L30		No matches	
L31	17–30	B	KRLHGVYFKKRAPRAVKEIKKFAK
	60–78	B	KRGVKGVEYRLRLRIYRKR
L32	7–12	Y	**PKIVKK**
	15–48	Y B*	KKFKRHHSDRYHRVAENW**RKQKG**IDSVV**RRRFRG**
L33		No matches	
L34	36–42	Y	**KKLATRP**
	102–121	B	KKVVKEQTEAAKKSEKKSKK
L35	42–49	Y	**PKIKTVRK**
	72–107	Y B*	**GKKYQPKDLRAKK**TRALRRALTKFEASQVTEKQRKK
L36	23–30	Y	**PKISYKKG**
	54–85	B	RRLIDLIRNSGEKRARKVAKKRLGSFTRAKAK
L37	9–24	Y B	**GKRHNK**SHTLCNRCGRR
	53–74	B	KRRHTTGTGRMRYLKHVSRRFK

Continued on next page

Table 2. —*Continued.*

YRP	Residues	Type of NLS[b]		Sequence with (putative) NLSs
L38	16–46		B*	RRADVKTATVKINKKLNKAGKPFRQTKFKVR
	60–64	Y		**GKAKK**
L39	16–21	Y		**KKQNRP**
L40	21–52	Y	B	RKCYARLP**PRATNCRK**RKCGHTNQLR**PKKKLK**
L41	6–25		B	RKKRTRRLKRKRRKVRARSK
L42	4–8	Y		**PKTRK**
	13–18	Y		**GKTCRK**
	85–99		B	KRCKHFELGGEKKQK
L43	5–27		B	KKVGITGKYGVRYGSSLRRQVKK
	26–49	Y	B	KKLEIQQHARYDCSFC**GKKTVKRG**

[a] The complete sequences of all *S. cerevisiae* r-proteins were searched for the presence of the three types of NLSs (YRP-like, bipartite, and Matα2-like). The regions containing YRP-like sequences are shown in boldface. The underlined sequences have been shown experimentally to act as NLSs (cf. Table 1).
[b] Y, YRP-like; B, bipartite; M, Matα2-like. An asterisk indicates that the sequence in question contains multiple matches to the consensus.

transportin, which is also involved in nuclear import of hnRNP proteins; importin *β* itself; RanBP5, the homologue of the yeast r-protein importin Pse1p; and RanBP7. Each of these receptors was shown to be able to promote nuclear import of at least three different r-proteins (S7, L5, and L23a) directly, without the help of importin *α*. The r-protein S7, but not L5 or L23a, is also able to use the classical importin *α*-importin *β* pathway (Jäkel and Görlich, 1998).

Regions containing NLSs have been experimentally identified so far in five different mammalian r-proteins, including S7 and L23a (Table 3). In almost all cases the signals were found to be quite complex. They consist either of rather long stretches of amino acids (S7 and L23a) or of multiple sequences (S6 and L7a). In the latter case the individual sequences do have nuclear targeting activity by themselves, but more efficient import is observed when all of them are present (Annilo et al., 1998; Russo et al., 1997; Schmidt et al., 1995). Only L31 appears to possess a "simple" NLS (Quaye et al., 1996).

The NLS of L23a has been subjected to detailed functional characterization. This NLS encompasses residues 32 to 74, a sequence that is able to bind each of the four import receptors mentioned above but does not appear to contain separate binding sites for these receptors. Interestingly, as shown in Fig. 1, this NLS corresponds exactly to the C-terminal portion of the eukaryotic extension of its yeast homologue, L25, which contains the NLSs (Rutgers et al., 1990, 1991; Schaap et al., 1991). We have previously shown that rat L23a can be incorporated into functional yeast 60S subunits (Jeeninga et al., 1996). Presumably, therefore, L23a can use the r-protein-specific import pathway of yeast cells. While the NLS of L23a clearly is not identical to NLS2 of L25, it contains a large number of overlapping YRP motifs (Fig. 1) that could possibly allow its interaction with the yeast r-protein-specific import receptors. However, the complete region of L23a from position 32 to 74 is required for nuclear import in HeLa cells (Jäkel and Görlich, 1998). The simple

Table 3. NLSs identified experimentally in mammalian r-proteins[a]

Protein[b]	NLS	Reference
hS6	165-E**GKKPR**TKAPK-175	Schmidt et al., 1995
	182-**PRVLQHKR**RRIALKKQ-197	
	215-RMKEAKEKRQEQIAKRRRLSSLRA-239	
hS7	98-**RRILPKPTRKSR**TKN**KQKRP**R-118	Annilo et al., 1998
hL7a	23-EAKKVVNPLFEKRPKNTFGIGQDIQPKRDL-51	Russo et al., 1997
	52-TRFVKWPRYIRLQRQRAILY**KRLKVP**PAINQFTQALDRQTATQLLKLAH-100	
	101-KYRPETKQEKKQRLLARAEKKAAGKGDVPTKRPPVLRAGVNTVTTLVENKKAQLVV IAHDDPIELVVFLPALCRKM-220	
hL23a	31-**GVHSHKKKKIRTSPTFRRPKTLRLRRQPKYPRKSAPR**RNKLDHYA-74	Jäkel and Görlich, 1998
rL31	81-**PYRIRVR**SLRKR-92	Quaye et al., 1996

[a] The minimal sequences identified as able to direct a reporter protein to the nucleus are underlined. The regions within these sequences matching either the mono- or bipartite consensus but not experimentally tested for NLS activity are doubly underlined. The sequences containing one or more matches to the YRP consensus are shown in boldface. The numbers indicate the positions of the first and last residues in the complete sequence of the protein.
[b] The species of origin is indicated by the prefix: h, human; r, rat.

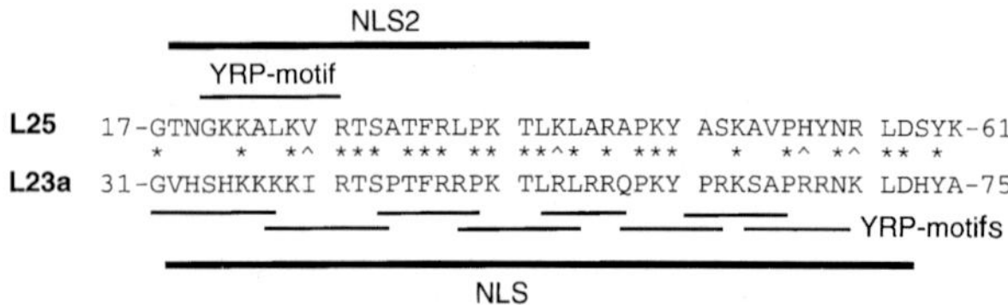

Figure 1. Comparison of the NLS of human r-protein L23a with the corresponding region in its L25 functional homologue from *S. cerevisiae*. *, identical residues; ^, similar residues. The sequences identified experimentally as having NLS activity are indicated by thick lines. YRP-like motifs are indicated by thin lines.

NLS identified in mammalian L31, on the other hand, might conform to the YRP consensus because the sequence RLSRKR, when fused to the β-galactosidase reporter, is preceded by a proline residue (Quaye et al., 1996). Similarly, YRP motifs that overlap with the NLSs are present in S6 and L7a (Table 3). Mutation of Pro^{117} in the C-terminal YRP motif significantly impaired nuclear import of S7 (Annilo et al., 1998). We have also searched the complete set of rat r-protein sequences and found that sequences matching the YRP consensus are present in 87% of these proteins. A similar analysis of the *E. coli* r-proteins, on the other hand, revealed YRP-like motifs in only 42% of the 55 species. Thus, such motifs occur considerably more frequently in eukaryotic r-proteins than in their prokaryotic counterparts.

NUCLEOLAR LOCALIZATION OF r-PROTEINS

Most of the assembly of ribosomal subunits in eukaryotic cells takes place in the nucleolus (for a review see Raué and Planta, 1991, and Mélèse and Xue, 1995). Shortly after their synthesis newly formed r-proteins accumulate rapidly in this subcompartment of the nucleus (Warner, 1979). However, the manner in which r-proteins (and other nucleolar proteins) find their way into the nucleolus is still obscure.

Sequences required for nucleolar accumulation have been identified in a number of r-proteins of mammalian origin (Schmidt et al., 1995; Quaye et al., 1996; Russo et al., 1997; Annilo et al., 1998), as well as the yeast r-protein S25. In the latter case the fragment containing the NLS of the r-protein (cf. Table 1) was also necessary and sufficient to direct the tagged r-protein to the nucleolus (Timmers et al., 1999). A similar colocalization of nuclear and nucleolar targeting activity was seen in mammalian r-protein L31, although the two activities could be separated by mutation of specific residues (Quaye et al., 1996). In mammalian r-proteins S6 (Schmidt et al., 1995), S7 (Annilo et al., 1998), and L7 (Russo et al., 1997), the nucleolar targeting activity is more complex in nature, requiring the cooperation of multiple functional domains that partly overlap with the sequences showing nuclear targeting activity.

Because nucleolar accumulation does not require passage through a membrane, the precise function of sequences that direct a protein to the nucleolus is still debated. In a number of cases, evidence has been presented that nucleolar accumulation is based upon retention of the protein by its binding to another nucleolar component(s), either protein or nucleic acid, rather than by directed transport through the action of a nucleolar targeting sequence (Schmidt-Zachmann and Nigg, 1993; Xue and Mélèse, 1994). For instance, yeast proteins like Nsr1p that have RNA binding as well as acidic (phosphorylated) domains may bring together rRNA and the (basic) r-proteins (Xue and Mélèse, 1994). In mammalian cells nucleolin may play a similar role (Bouvet et al., 1998).

CONCLUSION

In conclusion, it is now well established that most, if not all, eukaryotic r-proteins are imported into the nucleus via multiple specialized pathways, distinct from the one mediated by the importin α-importin β dimer (Rout et al., 1997; Schlenstedt et al., 1997; Jäkel and Görlich, 1998). The redundancy may be necessary to satisfy the massive requirement for import of r-proteins, particularly during rapid growth, when several hundred r-protein molecules have to be imported per NPC per minute. It may also act as a "fail-safe" mechanism to ensure the supply of r-proteins in case one of the pathways is inactivated.

The import pathways for r-proteins are simpler than the one involving the classical mono- and bipartite NLSs because they involve interaction of the r-protein only with an importin β-like receptor that delivers its cargo directly to the NPC. The domain of this receptor recognizing the r-protein differs from the domain that interacts with the bipartite signal of importin α (Jäkel and Görlich, 1998). The existence of a functionally distinct NLS implied by these observations is supported by the computer analysis of the yeast r-protein sequence database discussed in this chapter. This novel type of NLS appears to be characterized by the presence of a $[G/P](K/R)_3X_{1-4}[G/P]$ consensus motif in which there is a preference for X being hydrophobic and acidic amino acids are excluded from positions within or adjacent to the consensus motif. However, in some cases, notably in L25

(Schaap et al., 1991), the presence of adjacent sequences that themselves have no nuclear targeting activity considerably improves the activity of the region containing the YRP motif. In other yeast r-proteins, such as L28 (Underwood and Fried, 1990) and S22 (Timmers et al., 1999), the YRP motif by itself appears to have strong NLS activity (cf. Table 1).

Although matches to this consensus motif are found in many mammalian r-proteins, their NLSs appear to be more complex in nature, often distributed over multiple regions, each of which contributes part of the activity.

The work carried out in our laboratory was supported in part by the Netherlands Foundation for Chemical Science (NWO-CW), by a "STIMULANS" grant from the Netherlands Organization for Scientific Research (NWO) to A.C.J.T., and a grant to R.S. from the Royal Netherlands Chemical Society (KNCV) and the Netherlands Ministry of Economic Affairs.

REFERENCES

Adam, E. J. H., and S. A. Adam. 1994. Identification of cytosolic factors required for nuclear location: sequence-mediated binding to the nuclear envelope. *J. Cell. Biol.* **125:**547–555.

Aitchison, J. D., G. Blobel, and M. P. Rout. 1996. Kap104p: a karyopherin involved in the nuclear transport of messenger RNA binding proteins. *Science* **274:**624–627.

Annilo, T., A. Karis, S. Hoth, T. Rikk, J. Kruppa, and A. Metspalu. 1998. Nuclear import and nucleolar accumulation of the human ribosomal protein S7 depends on both a minimal nuclear localization sequence and an adjacent basic region. *Biochem. Biophys. Res. Commun.* **249:**759–766.

Ballesta, J. P. G., M. A. Rodriguez-Gabriel, G. Bou, E. Briones, R. Zambrano, and M. Remacha. 1999. Phosphorylation of the yeast ribosomal stalk. Functional effects and enzymes involved in the process. *FEMS Microbiol. Lett.* **23:**537–550.

Booher, R. N., C. E. Alfa, J. S. Hyams, and D. H. Beach. 1989. The fission yeast cdc2/cdc13/suc1 protein kinase: regulation of catalytic activity and nuclear localization. *Cell* **58:**485–497.

Bouvet, P., J. J. Diaz, K. Kindbeiter, J. J. Madjar, and F. Amalric. 1998. Nucleolin interacts with several ribosomal proteins through its RGG domain. *J. Biol. Chem.* **273:**19025–19029.

Breeuwer, M., and D. S. Goldfarb. 1990. Facilitated nuclear transport of Histone H1 and other small nucleophilic proteins. *Cell* **60:**999–1008.

Chelsky, D., R. Ralph, and G. Jonak. 1989. Sequence requirements for synthetic peptide-mediated translocation to the nucleus. *Mol. Cell. Biol.* **9:**2487–2492.

Conti, E., M. Uy, L. Leighton, G. Blobel, and J. Kuriyan. 1998. Crystallographic analysis of the recognition of a nuclear localization signal by the nuclear import factor karyopherin alpha. *Cell* **94:**193–204.

Dang, C. V., and W. M. F. Lee. 1989. Nuclear and nucleolar targeting sequences of c-*erb*-A, c-*myb*, N-*myc*, p53, Hsp70 and HIV *tat* proteins. *J. Biol. Chem.* **264:**18019–18023.

Dingwall, C. 1991. Transport across the nuclear envelope: enigmas and explanations. *Bioessays* **13:**213–217.

Dingwall, C. M., and R. A. Laskey. 1991. Nuclear targeting sequences—a consensus? *Trends Biochem. Sci.* **16:**478–481.

El Baradi, T. T. A. L., C. A. F. M. Van der Sande, W. H. Mager, H. A. Raué, and R. J. Planta. 1986. The cellular level of yeast ribosomal protein L25 is controlled principally by rapid degradation of excess protein. *Curr. Genet.* **10:**733–739.

Fahrenkrog, B., E. C. Hurt, U. Aebi, and N. Pante. 1998. Molecular architecture of the yeast nuclear pore complex—localization of Nsp1p subcomplexes. *J. Cell Biol.* **143:**577–588.

Fischer, U., E. Darzynkiewicz, S. M. Tahara, N. A. Dathan, R. Lührmann, and I. W. Mattaj. 1991. Diversity in the signals required for nuclear accumulation of U snRNPs and variety in the pathways of nuclear transport. *J. Cell Biol.* **113:**705–714.

Fischer, U., J. Huber, W. C. Boelens, I. W. Mattaj, and R. Lührmann. 1995. The HIV-1 Rev activation domain is a nuclear export signal that accesses an export pathway used by specific cellular RNAs. *Cell* **82:**475–483.

Fried, H. M. 1993. Nucleocytoplasmic transport in ribosome biogenesis, p. 257-267. *In* A. J. P. Brown, M. F. Tuite, and J. E. G. McCarthy (ed.), *Protein Synthesis and Targeting in Yeast.* Springer-Verlag, Berlin, Germany.

Goldberg, M. W., and T. D. Allen. 1995. Structural and functional organization of the nuclear envelope. *Curr. Opin. Cell Biol.* **7:** 301–309.

Görlich, D., and I. W. Mattaj. 1996. Nucleocytoplasmic transport. *Science* **271:**1513–1518.

Görlich, D., F. Vogel, A. D. Mills, E. Hartmann, and R. A. Laskey. 1995. Distinct functions for the two importin subunits in nuclear protein import. *Nature* **377:**246–248.

Gritz, L., N. Abovich, J. L. Teem, and M. Rosbash. 1985. Post-transcriptional regulation and assembly into ribosomes of a *Saccharomyces cerevisiae* ribosomal protein–β-galactosidase fusion. *Mol. Cell. Biol.* **5:**3436–3442.

Hall, M. N., L. Hereford, and I. L. Herskowitz. 1984. Targeting of *Escherichia coli* β-galactosidase to the nucleus in yeast. *Cell* **36:**1057–1065.

Huber, J., U. Cronshagen, M. Kadokura, C. Marshallsay, T. Wada, M. Sekine, and R. Lührmann. 1998. Snurportin1, an M(3)G-Cap-specific nuclear import receptor with a novel domain structure. *EMBO J.* **17:**4114–4126.

Jäkel, S., and D. Görlich. 1998. Importin beta, Transportin, Ranbp5 and Ranbp7 mediate nuclear import of ribosomal proteins in mammalian cells. *EMBO J.* **17:**4491–4502.

Jeeninga, R. E., J. Venema, and H. A. Raué. 1996. Rat RL23a ribosomal protein efficiently competes with its *Saccharomyces cerevisiae* L25 homologue for assembly into 60 S subunits. *J. Mol. Biol.* **263:**648–656.

Kalderon, D., E. L. Roberts, W. D. Richardson, and A. E. Smith. 1984. A short amino acid sequence able to specify nuclear location. *Cell* **39:**499–509.

Kobe, B. 1999. Autoinhibition by an internal nuclear localization signal revealed by the crystal structure of mammalian importin alpha. *Nat. Struct. Biol.* **6:**388–397.

Kruiswijk, T., R. J. Planta, and J. M. Krop. 1978. The course of assembly of ribosomal subunits in yeast. *Biochim. Biophys. Acta* **517:**378–389.

Mager, W. H., R. J. Planta, J. P. G. Ballesta, J. C. Lee, K. Mizuta, K. Suzuki, J. R. Warner, and J. Woolford. 1997. A new nomenclature for the cytoplasmic ribosomal proteins of *Saccharomyces cerevisiae*. *Nucleic Acids Res.* **25:**4872–4875.

Maicas, E., F. G. Pluthero, and J. D. Friesen. 1988. The accumulation of three yeast ribosomal proteins under conditions of excess mRNA is determined primarily by fast protein decay. *Mol. Cell. Biol.* **8:**169–175.

Mattaj, I. W., and L. Englmeier. 1998. Nucleocytoplasmic transport—the soluble phase. *Annu. Rev. Biochem.* **67:**265–306.

Melchior, F., and L. Gerace. 1995. Mechanisms of nuclear protein import. *Curr. Opin. Cell. Biol.* **7:**310–318.

Mélèse, T., and Z. Xue. 1995. The nucleolus: an organelle formed by the act of building a ribosome. *Curr. Opin. Cell Biol.* **7:**319–324.

Michael, W. M., M. Y. Choi, and G. Dreyfuss. 1995. A nuclear export signal in hnRNP A1: a signal-mediated, temperature-dependent nuclear protein export pathway. *Cell* **83:**415–422.

Moore, M. S., and G. Blobel. 1994. Purification of a Ran-interacting protein that is required for protein import into the nucleus. *Proc. Natl. Acad. Sci. USA* **91:**10212–10216.

Moreland, R. B., H. G. Nam, L. M. Hereford, and H. M. Fried. 1985. Identification of a nuclear localization signal of a yeast ribosomal protein. *Proc. Natl. Acad. Sci. USA* **82:**6561–6565.

Nehrbass, U., and G. Blobel. 1996. Role of the nuclear transport factor p10 in nuclear import. *Science* **272:**120–122.

Nehrbass, U., E. Fabre, S. Dihlmann, W. Herth, and E. C. Hurt. 1993. Analysis of nucleocytoplasmic transport in a thermosensitive mutant of nuclear pore protein NSP1. *Eur. J. Cell Biol.* **62:**1–12.

Osborne, M. A., and P. A. Silver. 1993. Nucleocytoplasmic transport in the yeast *Saccharomyces cerevisiae*. *Annu. Rev. Biochem.* **62:**219–254.

Panté, N., and U. Aebi. 1996. Molecular dissection of the nuclear pore complex. *Crit. Rev. Biochem. Mol. Biol.* **31:**153–199.

Paschal, B. M., and L. Gerace. 1995. Identification of NTF2, a cytosolic factor for nuclear import that interacts with nuclear pore complex protein p62. *J. Cell Biol.* **129:**925–937.

Planta, R. J. 1997. Regulation of ribosome synthesis in yeast. *Yeast* **13:**1505–1518.

Pollard, V. W., W. M. Michael, S. Nakielny, M. C. Siomi, F. Wang, and G. Dreyfuss. 1996. A novel receptor-mediated nuclear protein import pathway. *Cell* **86:**985–994.

Quaye, I. K. E., S. Toku, and T. Tanaka. 1996. Sequence requirement for nucleolar localisation of rat ribosomal protein L31. *Eur. J. Cell Biol.* **69:**151–155.

Raué, H. A., and R. J. Planta. 1991. Ribosome biogenesis in yeast. *Prog. Nucleic Acid Res. Mol. Biol.* **41:**89–129.

Roberts, B. L., W. D. Richardson, and A. E. Smith. 1987. The effect of protein context on nuclear location signal function. *Cell* **50:**465–475.

Rosenblum, J. S., L. F. Pemberton, and G. Blobel. 1997. A nuclear import pathway for a protein involved in tRNA maturation. *J. Cell Biol.* **139:**1655–1661.

Rout, M. P., G. Blobel, and J. D. Aitchison. 1997. A distinct nuclear import pathway used by ribosomal proteins. *Cell* **89:**715–725.

Russo, G., G. Ricciardelli, and C. Pietropaolo. 1997. Different domains cooperate to target the human ribosomal L7a protein to the nucleus and to the nucleoli. *J. Biol. Chem.* **272:**5229–5235.

Rutgers, C. A., P. J. Schaap, J. Van 't Riet, C. L. Woldringh, and H. A. Raué. 1990. *In vivo* and *in vitro* analysis of structure-function relationships in ribosomal protein L25 from *Saccharomyces cerevisiae*. *Biochim. Biophys. Acta* **1050:**74–79.

Rutgers, C. A., J. M. J. Rientjes, J. Van 't Riet, and H. A. Raué. 1991. rRNA binding domain of yeast ribosomal protein L25. Identification of its borders and a key leucine residue. *J. Mol. Biol.* **218:**375–385.

Schaap, P. J., J. Van 't Riet, C. L. Woldringh, and H. A. Raué. 1991. Identification and functional analysis of the nuclear localization signals of ribosomal protein L25 from *Saccharomyces cerevisiae*. *J. Mol. Biol.* **221:**225–237.

Scheer, U., and D. Weisenberger. 1994. The nucleolus. *Curr. Opin. Cell Biol.* **6:**354–359.

Schlenstedt, G., E. Smirnova, R. Deane, J. Solsbacher, U. Kutay, D. Gorlich, H. Ponstingl, and F. R. Bischoff. 1997. Yrb4p, a yeast Ran-GTP-binding protein involved in import of ribosomal protein L25 into the nucleus. *EMBO J.* **16:**6237–6249.

Schmidt, C., E. Lipsius, and J. Kruppa. 1995. Nuclear and nucleolar targeting of human ribosomal protein S6. *Mol. Biol. Cell* **6:**1875–1885.

Schmidt-Zachmann, M. S., and E. A. Nigg. 1993. Protein localization to the nucleolus—a search for targeting domains in nucleolin. *J. Cell Sci.* **105:**799–806.

Shimmin, L. C., G. Ramirez, A. T. Matheson, and P. P. Dennis. 1989. Sequence alignment and evolutionary comparison of the L10 equivalent and L12 equivalent ribosomal proteins from archaebacteria, eubacteria, and eucaryotes. *J. Mol. Evol.* **29:**448–462.

Timmers, A. C. J., R. Stuger, P. J. Schaap, J. Van 't Riet, and H. A. Raué. 1999. Nuclear and nucleolar localization of *Saccharomyces cerevisiae* ribosomal proteins S22 and S25. *FEBS Lett.* **452:**335–340.

Tollervey, D. 1996. Trans-acting factors in ribosome synthesis. *Exp. Cell Res.* **229:**226–232.

Underwood, M., and H. M. Fried. 1990. Characterization of nuclear localizing sequences derived from yeast ribosomal protein L29. *EMBO J.* **9:**91–99.

Vodkin, M. H., R. J. Novak, and G. L. McLaughlin. 1996. Database searches with multiple oligopeptides containing ambiguous residues. *BioTechniques* **21:**1116–1117.

Wagner, P., and M. N. Hall. 1993. Nuclear protein transport is functionally conserved between yeast and higher eukaryotes. *FEBS Lett.* **321:**261–266.

Warner, J. R. 1979. Distribution of newly formed ribosomal proteins in HeLa cell fractions. *J. Cell Biol.* **80:**767–772.

Warner, J. R., G. Mitra, F. Schwindiger, M. Studeny, and H. M. Fried. 1985. *Saccharomyces cerevisiae* coordinates accumulation of yeast ribosomal proteins by modulating mRNA splicing, translational initiation, and protein turnover. *Mol. Cell. Biol.* **5:**1512–1521.

Weis, K. 1998. Importins and exportins—how to get in and out of the nucleus. *Trends Biochem. Sci.* **23:**235.

Wen, W., J. L. Meinkoth, R. Y. Tsien, and S. S. Taylor. 1995. Identification of a signal for rapid export of proteins from the nucleus. *Cell* **82:**463–473.

Wozniak, R. W., M. P. Rout, and J. D. Aitchison. 1998. Karyopherins and kissing cousins. *Trends Cell Biol.* **8:**184–188.

Xue, Z., and T. Mélèse. 1994. Nucleolar proteins that bind NLSs: a role in nuclear import or ribosome biogenesis? *Trends Cell Biol.* **4:**414–417.

VI. PROBING OF FUNCTIONAL SITES

VI. PROBING OF FUNCTIONAL SITES

Probing of functional sites has been one of the most creative areas of ribosome research. A large variety of biochemical and genetic approaches have been developed and applied to characterizing the main functional sites on the ribosome, which include the peptidyltransferase center, where peptide bond formation occurs, as well as the GTPase-associated site, the decoding site, and the peptide channel. This section provides a broad overview of the methods used and the functional sites under study.

The Ribosome: Structure, Function, Antibiotics, and Cellular Interactions
Edited by R. A. Garrett, S. R. Douthwaite, A. Liljas, A. T. Matheson, P. B. Moore, and H. F. Noller

Chapter 19

Probing Ribosomal Structure and Function: Analyses with rRNA and Protein Mutants

MICHAEL O'CONNOR, MARK BAYFIELD, STEVEN T. GREGORY, WYAN-CHING MIMI LEE, J. STEPHEN LODMELL, ANUJ MANKAD, JILL R. THOMPSON, ANTON VILA-SANJURJO, CATHERINE L. SQUIRES, and ALBERT E. DAHLBERG

The past few years have witnessed enormous advances in our understanding of the structure of the ribosome and of its individual components. Advances in cryo-electron microscopy have permitted visualization of the bacterial ribosome in far greater detail than has ever previously been possible, while crystallographic analyses have yielded high-resolution structures of most of the initiation and elongation factors and many ribosomal proteins (Ramakrishnan and White, 1998; Agrawal and Frank, 1999). Together with functional studies, these varied approaches have provided new insights into the organization and function of the protein synthetic apparatus. The ability to produce and analyze altered variants of ribosomal components also continues to be a powerful tool in the analysis of ribosome structure and function. The recent development of new genetic systems for the construction of rRNA and ribosomal protein mutants in *Escherichia coli* has facilitated both structural and functional analyses of these molecules. Mutations in ribosomal proteins resulting in antibiotic resistance or alterations in the accuracy of translation can now be interpreted in structural terms (Davies et al., 1998), while the consequences for ribosome structure of alterations in rRNA have been visualized directly for the first time by cryo-electron-microscopic techniques (see chapter 5). In this chapter, we review the recent developments that our laboratory has made in this area.

EXPRESSION AND ANALYSIS OF rRNA MUTATIONS IN *E. COLI*: PROBLEMS AND SOLUTIONS

The study of rRNA mutations in *E. coli* has been hampered by the presence of multiple copies of rRNA cistrons on the chromosome. Until recently, genetic analysis of *E. coli* rRNA has been carried out with a plasmid copy of a selected *rrn* operon, and plasmid-encoded mutant rRNA has been coexpressed with wild-type, chromosomally encoded rRNA. Cells carrying such plasmids thus inevitably contain a mixture of wild-type and mutant rRNAs. An elaboration of this system has been the inclusion of antibiotic resistance mutations in the plasmid-encoded 16S and 23S rRNAs and subsequent analysis of the ribosomes in the presence of the relevant antibiotic (Powers and Noller, 1991; Porse and Garrett, 1995; Saarma et al., 1998). However, this approach is not without its problems, since antibiotics invariably target functional sites on the ribosome and the resistance mutations are rarely, if ever, completely without effect on ribosome function. An alternative approach has been the development of systems for genetic analysis of rRNA in halophilic bacteria or mycobacteria that naturally carry a single *rrn* operon (Mankin, 1997; Sander et al., 1997). In yeast, deletion of the ribosomal DNA repeats on chromosome XII and expression of 18S and 25S rRNA from a multicopy plasmid

Michael O'Connor, Mark Bayfield, Steven T. Gregory, Wyan-Ching Mimi Lee, J. Stephen Lodmell, Anuj Mankad, Jill R. Thompson, Anton Vila-Sanjurjo, and Albert E. Dahlberg ■ J. W. Wilson Laboratory, Department of Molecular and Cellular Biology and Biochemistry, Brown University, Providence, RI 02912. **Catherine L. Squires** ■ Department of Molecular Biology and Microbiology, Tufts University School of Medicine, Boston, MA 02111.

have allowed the analysis of rRNA mutations in the absence of any chromosomally encoded wild-type rRNA (Liu and Liebman, 1996). Despite the development of these systems, the *E. coli* ribosome remains the best characterized in terms of genetics, structure, and function, and results obtained with other systems are interpreted in terms of our understanding of the corresponding element of the *E. coli* ribosome. The recent generation of a strain of *E. coli* in which all seven chromosomal *rrn* operons have been deleted and in which total rRNA is transcribed from a plasmid-associated *rrn* operon has successfully circumvented many of the earlier problems associated with analysis of heterogeneous ribosome populations (Asai et al., 1999). While rRNA mutations that confer dominant-lethal phenotypes or produce nonfunctional ribosomes cannot be analyzed under conditions where they constitute the total pool of ribosomes, this new system now allows for the biochemical and structural study of functional rRNA mutants without any of the complications associated with analysis of mixed populations of ribosomes. Genetic analysis, including the generation of recessive mutations, is now greatly simplified in this strain. Below, we describe some of the recent applications of this strain and its potential for further analysis of rRNA function.

ANTIBIOTIC RESISTANCE MUTATIONS IN rRNA

Antibiotic inhibitors of protein synthesis cause inhibition by binding to specific sites on the ribosome and interfering with discrete steps of translation. As such, they remain among the best probes of ribosome function. Elucidation of the mechanisms of action of antibiotics, together with the determination of their ribosome binding sites and mechanisms of antibiotic resistance, has been invaluable in indicating the regions of the ribosome which must interact during translation and has provided insights into the dynamics of translation. Our laboratory has focused primarily on the isolation of mutations in rRNA, and recently, we isolated kasugamycin resistance mutations in the rRNA, using the *E. coli* strain in which all of the rRNA is encoded on a multicopy plasmid carrying a single *rrn* operon (strain Δ7 prrn) (Asai et al., 1999; Vila-Sanjurjo et al., 1999). The kasugamycin-resistant rRNA mutants were isolated simply by plating Δ7 prrn cells onto rich, solid medium containing kasugamycin. Three mutations in 16S rRNA, at positions A794, G926, and A1519, were identified, and site-directed mutagenesis showed that virtually any base change at these residues produced a viable resistance phenotype in the Δ7 prrn strain (Fig. 1).

Antibiotics target functionally important sites on the ribosome; thus, the isolation of kasugamycin resistance mutations at three universally conserved bases in 16S rRNA confirms the previous suggestion that these bases are important for ribosome function (Gutell, 1994). A1519 is one of the two bases (A1518 and A1519) modified by the KsgA methyltransferase (Sparling, 1970; Helser et al., 1971, 1972) the lack of which confers resistance to kasugamycin. Hence, it is not surprising that a base change at 1519 may mimic the effect of a mutation in the *ksgA* gene encoding KsgA methyltransferase. In fact, any mutation at A1519 confers at least as much resistance to the drug as do *ksgA* mutations, without causing major changes in growth rates compared to those of wild-type cells. In contrast, all the mutations at the adjacent A1518 result in extremely deleterious phenotypes and do not increase kasugamycin resistance over wild-type levels. Interestingly, expression of the A1519C base change in the *ksgA19* background produces a dramatically increased doubling time, whereas the same *ksgA19* mutation partially rescues the remarkably slow growth phenotype caused by the A1518C base change (Vila-Sanjurjo et al., 1999). These observations suggest that the effects imposed

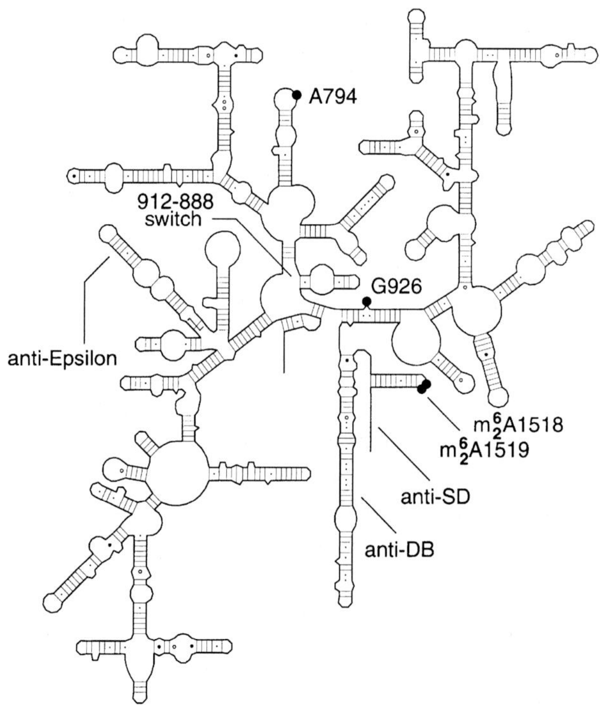

Figure 1. Secondary structure of *E. coli* 16S rRNA (Gutell, 1996) showing the locations of bases involved in kasugamycin resistance, the 912-888 switch, and proposed sites of mRNA-rRNA interaction.

on the structure of the dimethyl A stem-loop by the introduction of mutations at either A1518 or A1519 are modulated by the state of methylation at the unchanged adenosine. This is in agreement with results obtained with reconstituted ribosomes, which showed that the methylation of either adenosine by the KsgA methylase is unaffected by the identity of the base at the adjacent residue (Cunningham et al., 1990).

Both A794 and G926 are protected by P-site-bound tRNA in chemical-probing studies (Moazed and Noller, 1986, 1990). Moreover, modification interference experiments indicated that G926 is an essential nucleotide for binding tRNA at the ribosomal P site and forms a zero-length cross-link to the second base of the P-site codon on the mRNA (von Ahsen and Noller, 1995; Sergiev et al., 1997). At the same time, kasugamycin is known to inhibit initiation of protein synthesis, a process that starts with the binding of tRNA at the ribosomal P site. Thus, it is conceivable that the A794G and G926A mutations confer resistance to kasugamycin by rendering one of the macromolecular interactions occurring during the initiation of protein synthesis insensitive to the effects of the antibiotic, although a decreased affinity for the drug in the mutants may also explain the resistance phenotype. Despite the universal conservation and functional importance of residues A794 and G926, we have shown that virtually any base change at these positions produces functional ribosomes in vivo. This indicates that the results of the chemical-probing and chemical-interference studies are probably the consequence of conformational changes induced by tRNA binding at the ribosomal P site rather than an effect of a direct interaction between the tRNA and residues A794 and G926.

Some of the mutations that confer kasugamycin resistance in the Δ7 prrn strain are recessive and are sensitive to the antibiotic when introduced into a wild-type *E. coli* strain carrying seven intact chromosomal *rrn* operons. This shows that coexpression of mutant rRNA in the presence of wild-type ribosomes may, in some cases, prevent the detection of the resistance phenotype. A similar situation exists with rRNA mutations conferring resistance to streptomycin. Both rRNA and ribosomal protein S12 mutations conferring resistance to this drug are recessive (Cundliffe, 1981). Streptomycin-resistant rRNA mutations have been readily obtained in mycobacteria and in organelles, each of which contains a single locus encoding rRNA. However, in *E. coli* strains containing intact chromosomal *rrn* operons, expression of the streptomycin resistance phenotype has previously been achieved only by combining the relevant rRNA mutations with the spectinomycin-resistant C1192U mutation and assessing streptomycin resistance in the presence of spectinomycin, so that only ribosomes containing plasmid-encoded rRNA are able to function (Powers and Noller, 1991; Lodmell and Dahlberg, 1997). However, we have observed that introduction of the 16S rRNA A523C mutation (Melancon et al., 1988) into the Δ7 prrn strain allows expression of streptomycin resistance without the requirement for further manipulation of rRNA expression (O'Connor and Dahlberg, unpublished). Similarly, we have previously reported that expression of paromomycin resistance associated with the C1491U 16S rRNA mutation in *E. coli* strains containing intact chromosomal *rrn* operons requires the presence of a streptomycin-resistant form of ribosomal protein S12 (O'Connor et al., 1991). More recently, we have observed that this rRNA mutation confers paromomycin resistance in the Δ7 prrn strain in the absence of any *rpsL* mutation (O'Connor et al., unpublished). Thus, the behavior of antibiotic resistance mutations in the Δ7 prrn strain of *E. coli* more closely resembles that observed in organisms that contain single *rrn* genes.

The successful isolation of three kasugamycin resistance mutations underscores the potential usefulness of the Δ7 prrn strain for the isolation of mutations in rRNA genes. Future applications of this strain are not limited to the isolation of new mutants resistant to drugs that target the ribosome but also allow the development or improvement of any genetic scheme that pursues the isolation of mutations in rRNA under conditions where all of the ribosomes carry the specified mutation. These may include the isolation of second-site suppressors of deleterious rRNA mutations and the isolation of rRNA mutants that complement alterations in other (non-rRNA) components of the protein synthesis machinery.

ENHANCERS OF TRANSLATION: THE DOWNSTREAM BOX

The demonstration that the 3′ end of bacterial 16S rRNA base-pairs with the Shine-Dalgarno (SD) sequence upstream of AUG initiation codons was among the first experiments that suggested an important functional role for the RNA component of the ribosome (Shine and Dalgarno, 1975; Steitz and Jakes, 1975). The elegance and simplicity of this mechanism of initiation has encouraged searches for similar base-pairing schemes in archeal and eukaryal mRNAs and has also led to the proposal of mRNA-rRNA base-pairing schemes involving other regions of 16S rRNA in bacteria. While the sequence and position of the SD sequence in the initiation region of mRNAs is of prime importance in determining translational efficiency, mutagenesis studies of specific mRNAs and analyses of mRNAs lacking recognizable SD sequences have indicated that other sequences surrounding the initiation codon contribute to the efficiency of the initiation signal. Certain sequences, loosely termed enhancers, have been shown to be responsible for efficient initiation in their native contexts and to enhance expression of reporter genes. The best studied of these bacterial enhancers is the downstream box (DB), so-called because of its invariant location downstream of the initiation codon. This element was first described by Sprengart and coworkers (1990) in the 0.3 mRNA of phage T7, and the partial complementarity of the DB to nucleotides 1469 to 1483 of 16S rRNA led to the proposal that enhancement of translation was due to base-pairing of the DB to the complementary region of 16S rRNA, the so-called anti-DB. Similar sequences bearing partial complementarity to the 1469-to-1483 region were subsequently observed in the initiation regions of *E. coli rpoH, lysU, glnV,* and *cspA* and the T7 gene 10 mRNAs, as well as the leaderless *c*I, *vph*, and *dnaX* mRNAs of bacteriophage λ, *Streptomyces vinaceus*, and *Caulobacter crescentus*, respectively (Faxen et al., 1991; Nagai et al., 1991; Shean and Gottesman, 1992; Ito et al., 1993; Sprengart et al., 1996; Wu and Janssen, 1996; Mitta et al., 1997; Winzeler and Shapiro, 1997). In addition, several mutagenesis studies of these various DB elements showed that, in general, increases in complementarity to the anti-DB increased expression of the relevant reporter gene while decreases in mRNA-rRNA complementarity decreased expression. These experimental data have led to the widespread acceptance of the DB–anti-DB base-pairing model for enhancer action and the application of similar base-pairing models to explain the activity of other enhancers.

While the mutagenesis studies involving manipulation of mRNA sequences lend considerable support to the proposed base-pairing interaction, the limited biochemical analyses of DB-ribosome interactions cast doubt on this model. Chemical footprinting of the T4 gene 32-derived mRNA, with probes sensitive to base-pairing interactions, showed that while the SD region of this mRNA was protected from modification by bound ribosomes in a manner consistent with rRNA-mRNA pairing, no ribosome-dependent protection of the DB region of this mRNA was observed (Huttenhofer and Noller, 1994). A similar result was subsequently obtained with the DB-containing λ *c*I mRNA (Resch et al., 1996). We have recently tested the proposed DB–anti-DB base-pairing model by constructing mutations in the proposed anti-DB region of 16S rRNA and examining their effects on expression of DB-containing mRNAs (O'Connor et al., 1999). Nucleotides 1470 to 1481 form part of a conserved helical structure in 16S rRNA (helix 44 [Fig. 2]), and our first mutants introduced four base mismatches on each side of the helix. Disruption of base pairing within helix 44 decreased the formation of 70S ribosomes by mutant subunits, and strains expressing these mutants had increased doubling times and displayed elevated levels of misreading. Restoration of base pairing through the introduction of compensatory base pairs ameliorated most of these effects, suggesting that maintenance of base pairing within this region of 16S rRNA is critical for ribosome function (Firpo and Dahlberg, 1998). Based on these findings, a more radical rRNA mutant was constructed that reversed all 12 bp within this helix (anti-DB flip mutant [Fig. 2]). This mutant was viable, and using the Δ7 prrn strain described above, we have been able to produce strains of *E. coli* that express either mutant or wild-type rRNAs exclusively. Analysis of the expression of the DB-containing *glnS, rpoH, lysU,* λ *c*I, and *vph* mRNAs in strains expressing either anti-DB flip mutant or wild-type rRNAs showed that the expression of all of these mRNAs was unaltered by radical alterations in the proposed anti-DB sequence. The results of these genetic experiments led us to conclude that enhancement of translation by the DB does not involve pairing of the DB to the 1491-to-1483 region of 16S rRNA and that the proposed anti-DB does not exist. Furthermore, the lack of protection of the DB region of the λ *c*I and T4 gene 32 mRNAs in initiation complexes from modification by base-specific chemical

	wild type		anti-DB flip
1419	G-U	1481	U-G
	U-A		A-U
	G-C		C-G
	G-U		U-G
	G-U		U-G
	U-A		A-U
	U-G		G-U
	G-U		U-G
	C-G		G-C
	A-U		U-A
	A-U		U-A
1430	A-U	1470	U-A

Figure 2. Secondary structures of wild-type and mutant anti-DB regions of helix 44 of 16S rRNA.

probes indicates that neither the anti-DB region nor any other region of rRNA is involved in base-pairing interactions with the DB.

The DB has also been proposed to be involved in binding tmRNA to the ribosome (Muto et al., 1998), and a region of tmRNA including the first 5 bases of the tag sequence is partially complementary to the anti-DB. Disruption of the *ssrA* gene encoding tmRNA produces rather subtle phenotypes, among which is the inability of *ssrA* mutant strains to propagate hybrid λ-P22 phages (Withey and Friedman, 1999). We have taken advantage of this phenotype to investigate whether disruption of the proposed DB-rRNA base-pairing interaction affects tmRNA function. Cells expressing either wild-type or the anti-DB flip mutant rRNA were tested for the ability to plate the λ-P22 hybrid phages $\lambda imm^{P22}hy25$ and $\lambda imm^{P22}dis$ (Withey and Friedman, 1999). Both hybrid phages plated equally well on each host strain, indicating that alteration of the proposed tmRNA-rRNA base-pairing interaction did not affect tmRNA function (O'Connor and Dahlberg, unpublished). The proposed tmRNA-rRNA base-pairing interaction has also been investigated by Zwieb and coworkers (1999) by comparative sequence analysis, and no covariation between the DB region of tmRNAs and the anti-DB region of 16S rRNA was observed. Together, these data indicate that tmRNA does not base-pair to the 1480 region of 16S rRNA.

While our data effectively disprove the base-pairing model for DB enhancer action, the demonstrated effects of the DB on translation remain unaffected by these experiments, and other models must be invoked to explain the mechanism by which the DB enhances translation. The distribution of codons near the initiation codon has been proposed to influence the rate of initiation by affecting the rate of ribosome clearance from the initiation region (Goldman et al., 1995; Irwin et al., 1995). We have tested the possibility that the DB enhances translation indirectly through the choice of codons that constitute this mRNA sequence. Derivatives of an *rpoH-lacZ* fusion were constructed containing the DB in zero, +1, and +2 reading frames, relative to the zero-frame AUG initiation codon. However, all three constructs supported identical levels of β-galactosidase activity, suggesting that, at least in the *rpoH* mRNA, the activity of the enhancer was not dependent on the choice of codons that constitute the DB.

Analysis of the factors influencing expression of the leaderless λ *c*I mRNA led to the isolation of mutations in the *rpsB* locus encoding ribosomal protein S2 that increased λ *c*I translation (Shean and Gottesman, 1992). We have recently shown that this same *rpsB* mutation increased expression of all of the DB-containing *lacZ* constructs examined so far, while expression of a wild-type *lacZ* gene (not containing a DB) remained unchanged. A similar effect on expression of the DB-containing *cspA* mRNA by mutations in S2 was recently reported by Etchegaray and Inouye (1999). If the effect of S2 is indeed limited to DB-containing mRNAs, this then raises the exciting possibility that S2 is involved in ribosomal recognition of the DB region of mRNAs, either directly or through its influence on other ribosomal components, including ribosomal protein S1, with which it interacts (Laughrea and Moore, 1978).

The acceptance of the base-pairing model of enhancer action has led to the conclusion that all DB elements enhance translation by the same mechanism and has prompted the identification of DB elements by 16S rRNA sequence complementarity alone. Since the DB–anti-DB base-pairing model is no longer tenable, systematic mutagenesis studies are now required to identify these enhancer elements and to define the sequence requirements for DB function.

THE EPSILON ENHANCER

A pyrimidine-rich sequence (UUUAACUUUAA), termed epsilon, has been found to contribute to the efficiency of translation of the T7gene 10 mRNA and of several reporter genes containing this sequence (Olins and Rangwala, 1989). The original report of the T7 gene 10 epsilon element noted its complementarity to the 460 region of *E. coli* 16S rRNA, and an mRNA-rRNA base-pairing scheme was invoked to explain the enhancer effect of this mRNA sequence (Fig. 3). Subsequent mutagenesis studies of epsilon-containing chloramphenicol acetyltransferase gene constructs showed that while the original T7 gene 10 epsilon element still required an upstream SD sequence for efficient initiation, extension of the mRNA-rRNA complementarity to 16 bases allowed effective initiation in the absence of any SD elements (Golshani et al., 1997). The 460 region of 16S rRNA proposed to base-pair with epsilon (the anti-epsilon sequence) forms a stable stem-loop that must presumably unwind to engage in mRNA pairing. Considerable sequence variation is observed in this region of bacterial 16S rRNAs, even in the rRNA from *Salmonella typhimurium*, which is closely related to *E. coli* (Fig. 3). Nevertheless, an *S. typhimurium rrn* operon can replace the resident *E. coli rrn* operon in a Δ7 prrn strain without apparent effect on ribosome function (Asai et al., 1999). An epsilon-like element and an adjacent hexapyrimidine sequence have each been shown to have negative effects on the expression of *recF-lacZ* fusions in *E. coli* (Sandler and

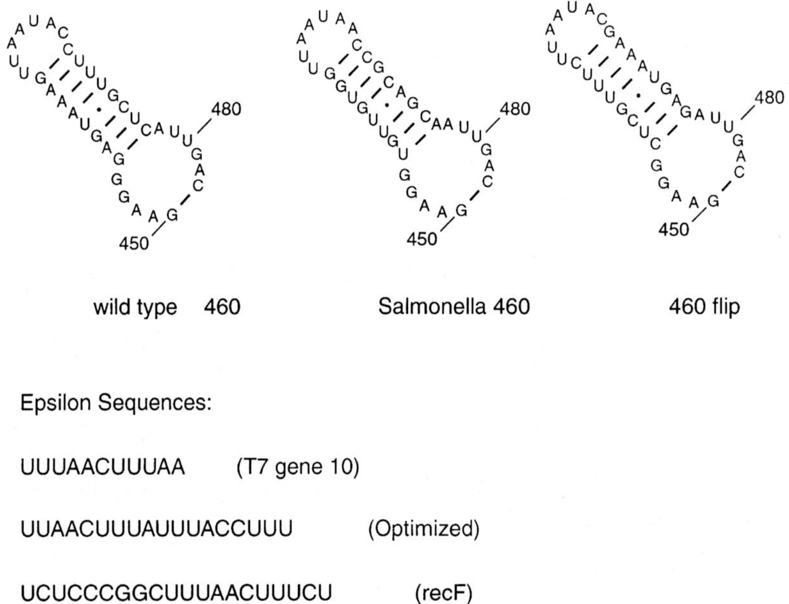

Figure 3. Secondary structures of the wild-type 460 stem-loop from *E. coli* (left) and *S. typhimurium* (middle) 16S rRNAs and a mutant in which all 8 bp of the *E. coli* stem-loop are reversed (right). The sequences of the original epsilon element from T7 gene 10 (Olins and Rangwala, 1989), an optimized epsilon element (Golshani et al., 1997), and the epsilon-containing regulatory region of the *E. coli recF* mRNA (Sandler and Clark, 1994) are shown beneath.

Clark, 1994). Each of these regions of the *recF* mRNA is complementary to the 460 stem-loop (Fig. 3). However, the sequence of the *Salmonella recF* gene is available (Sandler et al., 1992), and although it differs slightly from the *E. coli* sequence, no covariation between the *Salmonella recF* epsilon/hexapyrimidine elements and the 460 region of *Salmonella* rRNA is observed. Further doubt is cast upon the epsilon-rRNA base-pairing model by our recent finding that all 8 bp of the *E. coli* 460 stem-loop can be reversed (Fig. 3) without discernible effects on cell growth or expression of any of the *recF-lacZ* fusions constructed by Sandler and Clark (1994).

While the mechanism of translational enhancement by epsilon may not involve rRNA-mRNA base pairing, epsilon does have discernible effects on the efficiency of the initiation signal, and other explanations need to be considered. All epsilon sequences are pyrimidine rich; ribosomal protein S1 has a well-documented affinity for pyrimidine-rich sequences and has been shown to be involved in mRNA binding to 30S subunits (reviewed in Subramanian, 1983). Uridine-rich sequences are found upstream of the AUG codon of several *E. coli* mRNAs (Boni et al., 1991). In one instance, the *rnd* mRNA encoding RNase D, an uninterrupted stretch of eight uridines has been shown to be critical for *rnd* expression (Zhang and Deutscher, 1992). The role of such uridine-rich sequences has been investigated by Boni and colleagues (1991), who have shown that in binary complexes of ribosomes and mRNAs from the RNA phages Qβ and fr, ribosome-bound S1 cross-linked to uridine-rich regions upstream of the AUG start codons. This has led to the proposal that such uridine-rich sequences might serve as high-affinity binding sites for S1 during initiation of protein synthesis, thereby increasing the ribosome concentration near the initiation codon and enhancing translation efficiency. Thus, an alternative explanation for the enhancer activity of the epsilon element is that it provides a binding site for ribosome-bound S1. It is of interest that the longer 17-nucleotide epsilon sequence (UUAACUUUAUUUACCUU [Fig. 3]), which promotes efficient translation in the absence of an SD sequence, consists mostly of pyrimidines and so may constitute an efficient S1 affinity site that can bind ribosomes in the absence of any SD sequence.

A DYNAMIC SWITCH IN 16S rRNA

A long-held principle of ribosome function is that the transitions between different stages of translation are accompanied or even driven by conformational changes in the ribosome (reviewed in Wilson and Noller, 1998). Neutron-scattering experiments have indicated that a major structural change occurs on the 70S ribosome between pre- and post-translocation states (Spirin et al., 1987), while electron microscopy indicates that starvation conditions alter ribosome conformation (Ofverstedt et al., 1994). More recently, Dabrowski and colleagues (1998) have proposed that tRNAs are actively moved by a mobile ribosome "shuttle" during translocation. Identification of the individual ribosomal components involved in these conformational changes has, in general, proved elusive. We have used rRNA mutagenesis to identify a conformational rearrangement occurring during protein synthesis which involves the shift between alternating base-paired arrangements of nucleotides 910 to 912 with either 885 to 887 (912-885 conformation) or 888 to 890 (912-888 conformation) in 16S rRNA (Lodmell and Dahlberg, 1997). The switch affects tRNA binding to the ribosome and decoding of mRNA. It occurs in the central region of 16S rRNA, at the junction of the three major domains and in the region to which ribosomal proteins S5 and S12 bind. Site-directed mutagenesis was used to construct 16S rRNA mutants which favored either the 912-885 or 912-888 conformation and, in conjunction with structure-probing data, they were shown to be physiologically relevant structures. The switch is facilitated by ribosomal proteins S5 and S12, and changes occur in the 16S rRNA structure both locally and more distally at sites which previously had been shown to be cross-linked to mRNA. The switch may influence the positioning of mRNA for proper selection of tRNA at the ribosomal A site, since only one form has an enhanced affinity for tRNA. It produces a significant change in the conformation of the 30S subunit despite the fact that it involves a shift of only 3 nucleotides. This has been confirmed recently by cryo-electron-microscopic studies in collaboration with Joachim Frank and coworkers (see chapter 5).

The native state of the 30S subunit more closely resembles the 912-885 conformation, which is more error prone and has an enhanced affinity for tRNA. The second conformation (912-888) displays a hyperaccurate phenotype with a significantly reduced affinity for tRNA. This conformation may represent the state of the ribosome during proofreading, and in this restrictive (hyperaccurate) state, a noncognate tRNA would have a greater likelihood of diffusing away from the ribosome than would a cognate tRNA, thereby increasing the accuracy of decoding. It follows that the length of time that the ribosome spends in the 912-888 conformation influences the fidelity of translation. It is likely that the conformational-switch mechanism was developed very early in the evolution of rRNA, and comparative sequence analysis suggests that it is functioning in ribosomes of all species. While it is most likely involved in tRNA selection during decoding of mRNA, it is reasonable to expect that the switch mechanism might also occur during translocation.

More recently, our efforts have focused on determining the factors that tend to keep the native 30S ribosome in the 912-885 conformation, the state of high affinity for tRNA. The presence of an E-loop structure adjacent to the shift site, proposed by Leontis and Westhof (1998), favors the 912-885 conformation (Fig. 4). In preliminary experiments with a site-directed-mutagenesis approach, we have modified bases in the E loop and either enhanced or decreased the strength of this putative structure. Mutations which are predicted to stabilize the E-loop conformation increase the stability of the 912-885 structure, hence increasing the error-prone nature of the ribosome. These mutants are compatible with a ribosome containing a mutation in protein S12, which promotes increased accuracy. Similarly, mutations that decrease the stability of the E loop are compatible with error-prone S5 mutations. These results are consistent with the hypothesis and support the presence of the E loop in wild-type ribosomes. What precisely triggers the switch between the two conformational states in the 16S rRNA is currently under investigation. It is possible that disruption of the E-loop structure initiates the switch from the 912-885 to the 912-888 structure. Proper codon-anticodon base pairing by the ribosome could initiate GTP hydrolysis in the ternary complex, resulting in a conformational change which triggers a ribosomal protein (e.g., S5 or S12)-mediated relaxation of the E-loop structure and a switch from the 912-885 to the 912-888 structure.

23S rRNA TERTIARY STRUCTURE

Much of the study of ribosomes in the past 15 years has focused on the central role of RNA in all aspects of translation. While protein synthesis in the RNA world may indeed have been carried out by rRNA alone, all modern ribosomes contain both RNA and protein, and it seems unlikely that RNA-only translation systems could ever have approached the catalytic ability of modern ribosomes. Thus, in modern ribosomes, ribosomal proteins and rRNAs each play integral roles in ensuring rapid and accurate translation. Analysis of the structural changes in rRNA and ribosomal proteins brought about by changes in either component has the potential to provide much insight into their interaction and function during protein synthesis. We have taken advantage of the numerous ribosomal protein and rRNA mutations that are available to explore the three-dimensional organization of 23S rRNA in the 50S subunit by structure probing. The immediate goal of this effort is to detect RNA-RNA and protein-RNA interactions

a.

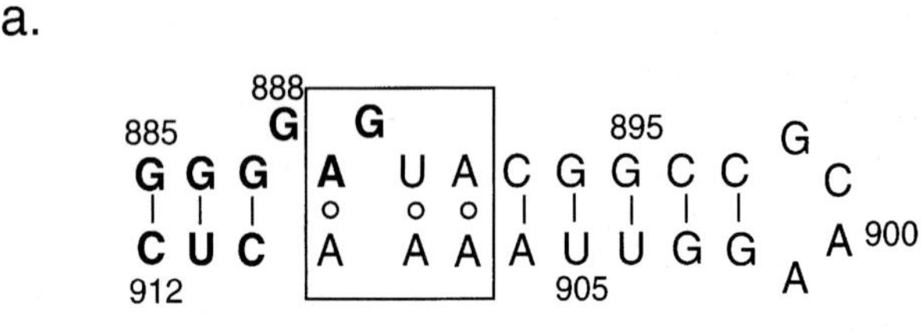

b.

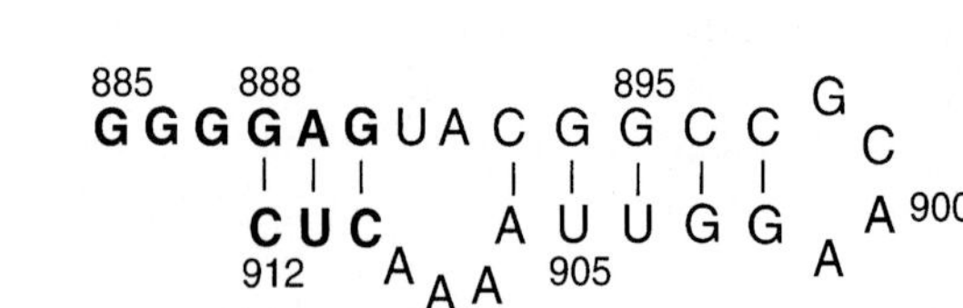

Figure 4. Two conformational states of *E. coli* helix 27 16S rRNA. (a) 912-885 *ram* conformation, with E loop shown in box. (b) 912-888 restrictive conformation, with disrupted E loop. Destabilization of the E loop could trigger the switch between the *ram* and restrictive structures. The bases directly involved in the triplet switch are shown in boldface.

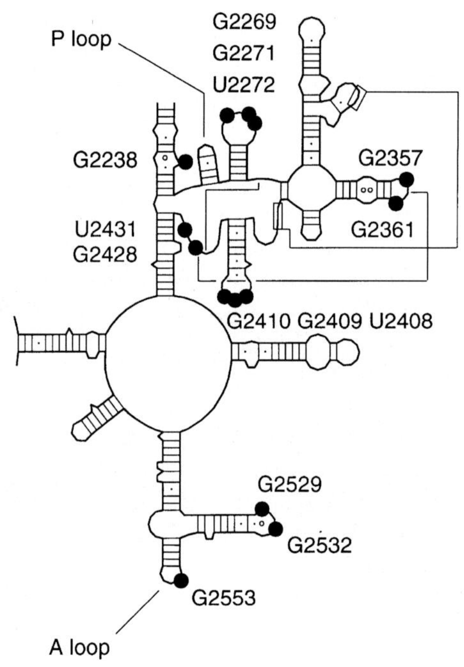

Figure 5. Effects of mutations in the P loop of 23S rRNA on reactivity to base-specific chemical probes. The positions marked by solid dots exhibit novel or enhanced reactivity in ribosomes bearing mutations at the universally conserved nucleotide G2251 or G2252. The locations of the P and A loops, which interact with P-site- and A-site-bound tRNAs, respectively, are indicated.

that contribute to the formation of catalytically active conformations of 23S rRNA. Ultimately, it is hoped that these experiments will reveal some underlying principles that govern the architecture and function of such large protein-RNA complexes.

In one study, introduction of mutations in the conserved P loop (nucleotides G2250 to C2254), an important element of the peptidyltransferase center, resulted in enhanced reactivity of a number of nucleotides in domain V to base-specific chemical probes (Fig. 5) (Gregory and Dahlberg, 1999a). This suggested the participation of the loop not only in peptide bond formation but in tertiary interactions as well. We had previously found that mutations in the P loop affect both peptidyltransferase activity (Lieberman and Dahlberg, 1994) and translational fidelity (Gregory et al., 1994), while G2252 had been identified by Samaha and coworkers (1995) as the site of a Watson-Crick base-pairing interaction with C74 of P-site-bound tRNA. The existence of these changes in reactivity correlated with the severity of the functional defects associated with each of the mutations, such that the most functionally defective mutations produced the most pronounced structural changes. Many of the affected positions could also be brought into close proximity by the existence of tertiary interactions detected by comparative sequence analysis (Gutell, 1996). Interestingly, long-range effects were also observed in and around the A loop (nucleotides U2552 to C2556), a site of interaction with A-site tRNA (Moazed and Noller, 1989). G2553, which is protected from kethoxal modification by the 3′-terminal adenosine of aminoacyl tRNA, exhibited enhanced reactivity in the P-loop mutants. This is consistent with a close physical proximity of the A and P loops and with the findings that mutations in either the A loop (O'Connor and Dahlberg, 1993) or the P loop (Gregory et al., 1994) affect tRNA selection. In addition, these results imply the participation of the P loop in RNA-RNA tertiary interactions or protein-RNA interactions which might function to facilitate the positioning of the P and A loops in proximity for peptide bond formation.

Mutations in ribosomal proteins which confer resistance to antibiotics often occur in putative RNA binding sites (Ramakrishnan et al., 1995; Davies et al., 1998). Structure probing of ribosomes bearing erythromycin resistance mutations in either ribosomal protein L22 or ribosomal protein L4 revealed a number of changes in reactivity located at positions dispersed throughout the 23S rRNA secondary-structure model (Gregory and Dahlberg, 1999b) (Fig. 6). Neither of these mutations produced changes in the central loop of domain V, the site of erythromycin resistance mutations and chemical footprints (reviewed in Moazed and Noller, 1987), suggesting that the ribosomal protein mutations confer erythromycin resistance in a manner distinct from that of the rRNA mutations.

These results also implied that some or all of the affected nucleotide residues are in close proximity to

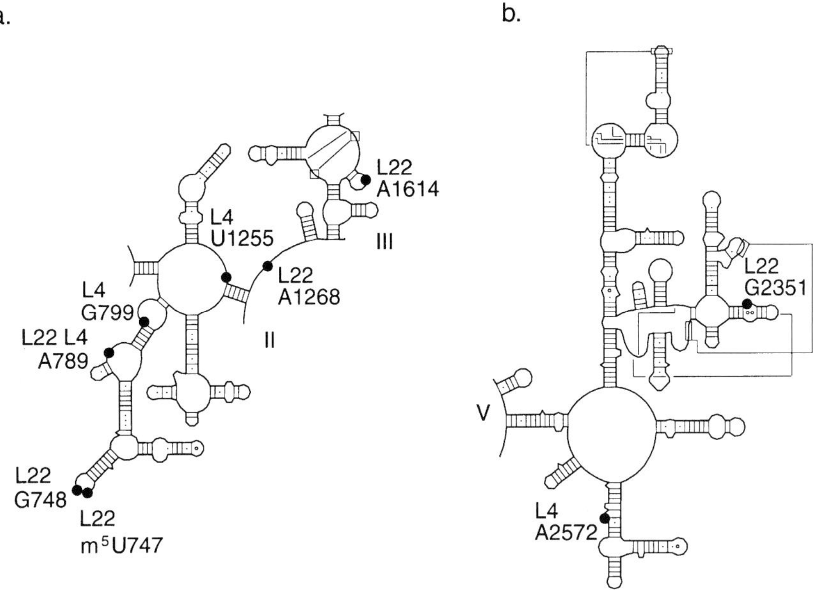

Figure 6. Effects of erythromycin resistance mutations in ribosomal proteins L22 and L4 (*eryB* and *eryA*, respectively) on the reactivity of 23S rRNA to base-specific chemical probes. Newly reactive positions are indicated by solid dots. The *eryB* and *eryA* mutations produce distinct patterns of reactivity across several secondary structure domains.

one another. One of the affected structures, the 750 loop, can be placed near the central loop by cross-linking and by antibiotic footprints (reviewed in Moazed and Noller, 1987), whereas another, A2572, is located in a helix emanating from the central loop of domain V. The remaining affected nucleotides are scattered throughout domains II, III, and V and provide evidence for a highly convoluted tertiary structure of 23S rRNA.

POSTTRANSCRIPTIONAL MODIFICATION OF 23S rRNA: THE ROLE OF D2449

Dihydrouridine (D) is one of the most common posttranscriptional modifications found in bacterial and eukaryal tRNAs, where it is believed to allow conformational flexibility in the loop regions of tRNAs that are involved in tertiary interactions (Dalluge et al., 1996). In contrast, there is but a single D modification, at position 2449, in *E. coli* 23S rRNA (Kowalak et al., 1995). The role of D modification at this invariant U residue was investigated by constructing and analyzing mutations at position 2449. Surprisingly, despite the lack of sequence variation at D2449, the D2449C mutation was viable and could displace the resident *rrnC* plasmid from the Δ7 prrn strain to provide all the rRNA requirements of the cell. Strains expressing the D2449C mutant rRNA exclusively had no detectable defect in growth rate, subunit assembly, or translational fidelity. In contrast, the D2449A mutation could not be expressed at high levels in any *E. coli* strain and could not displace the resident *rrnC* plasmid from a Δ7 prrn strain. This suggests that a pyrimidine may be required at this position. The ability of the D2449C mutant to supply all of the rRNA needs of the cell indicates that a D modification at this position is not required for assembly or any critical ribosome function. This is in agreement with the observation by Dalluge (1996) that while the rRNA modification profile of *Aeromonas hydrophila* closely resembles that found for *E. coli*, D was conspicuously absent.

CONCLUSIONS AND PERSPECTIVES

The development of an *E. coli* strain that permits the expression of pure populations of mutant rRNAs has greatly facilitated genetic, functional, and structural analyses of rRNA. This strain has now made possible the isolation of novel rRNA mutations and visualization of ribosomes containing mutant rRNAs. The application of rRNA mutagenesis has effectively disproved two widely accepted models of enhancer action. The question of how these enhancers operate now has to be readdressed, and we are forced to reconsider other rRNA-mRNA base-pairing models. Structure probing of ribosomal protein and rRNA mutants has provided new insights into the organization of the ribosome that can be incorporated into models of ribosome structure. Finally, while crystal or other high-resolution structures capture a static, "snapshot" view of the ribosome and its components, the challenge remains to identify and define the dynamic conformational changes that underlie ribosome function.

We are indebted to SunThorn Pond-Tor for his unstinting assistance and unfailing good humor. We thank J. Dalluge and James McCloskey for sharing their unpublished data with us and George Q. Pennabble for his prosaic, elegiac, and soporific comments on the manuscript.

Work in our laboratories was supported by grants GMS19756 (to A.E.D.) and GM24751 (to C.L.S) from the National Institutes of Health.

REFERENCES

Agrawal, R. K., and J. Frank. 1999. Structural studies of the translational apparatus. *Curr. Opin. Struct. Biol.* **9:**215–221.

Asai, T., D. Zaporojets, C. Squires, and C. L. Squires. 1999. An *Escherichia coli* strain with all chromosomal rRNA operons inactivated: complete exchange of rRNA genes between bacteria. *Proc. Natl. Acad. Sci. USA* **96:**1971–1976.

Boni, I. V., D. M. Isaeva, M. L. Musychenko, and N. V. Tzareva. 1991. Ribosome-messenger recognition: mRNA target sites for ribosomal protein S1. *Nucleic Acids Res.* **19:**155–162.

Cundliffe, E. 1981. Antibiotic inhibitors of ribosome function, p. 439–442. *In* E. F. Gale, E. Cundliffe, P. E. Reynolds, M. H. Richmond, and M. J. Waring (ed.), *The Molecular Basis of Antibiotic Action,* 2nd ed. John Wiley and Sons, New York, N.Y.

Cunningham, P. R., C. J. Weitzmann, K. Nurse, R. Masurel, P. H. Van Knippenberg, and J. Ofengand. 1990. Site-specific mutation of the conserved m6(2)Am6(2)A residues of *E. coli* 16S ribosomal RNA. Effects on ribosome function and activity of the ksgA methyltransferase. *Biochim. Biophys. Acta* **1050:**18–26.

Dabrowski, M., C. M. Spahn, M. A. Schafer, S. Patzke, and K. H. Nierhaus. 1998. Protection patterns of tRNAs do not change during ribosomal translocation. *J. Biol. Chem.* **273:**32793–32800.

Dalluge, J. J. 1996. Conformational flexibility and cold-adaptation of RNA: the role of dihydrouridine. Ph.D. dissertation. University of Utah, Salt Lake City, Utah.

Dalluge, J. J., T. Hashizume, A. E. Sopchik, J. A. McCloskey, and D. R. Davis. 1996. Conformational flexibility in RNA: the role of dihydrouridine. *Nucleic Acids Res.* **24:**1073–1079.

Davies, C., D. E. Bussiere, B. L. Golden, S. J. Porter, V. Ramakrishnan, and S. W. White. 1998. Ribosomal proteins S5 and L6: high-resolution crystal structures and roles in protein synthesis and antibiotic resistance. *J. Mol. Biol.* **279:**873–888.

Etchegaray, J. P., and M. Inouye. 1999. Translational enhancement by an element downstream of the initiation codon in *Escherichia coli. J. Biol. Chem.* **274:**10079–10085.

Faxen, M., J. Plumbridge, and L. A. Isaksson. 1991. Codon choice and potential complementarity between mRNA downstream of the initiation codon and bases 1471–1480 in 16S ribosomal RNA affects expression of *glnS. Nucleic Acids Res.* **19:**5247–5251.

Firpo, M. A., and A. E. Dahlberg. 1998. The importance of base pairing in the penultimate stem of *Escherichia coli* 16S rRNA for ribosomal subunit association. *Nucleic Acids Res.* **26:**2156–2160.

Goldman, E., A. H. Rosenberg, G. Zubay, and F. W. Studier. 1995. Consecutive low-usage leucine codons block translation only when near the 5′ end of a message in *Escherichia coli. J. Mol. Biol.* **245:**467–473.

Golshani, A., V. Golomehova, R. Mironova, I. G. Ivanov, and M. G. AbouHaidar. 1997. Does the epsilon sequence of phage T7 function as an initiator for the translation of CAT mRNA in *Escherichia coli*? *Biochem. Biophys. Res. Commun.* **236:**253–256.

Gregory, S. T., and A. E. Dahlberg. 1999a. Mutations in the conserved P loop perturb the conformation of two structural elements in the peptidyl transferase center of 23S ribosomal RNA. *J. Mol. Biol.* **285:**1475–1483.

Gregory, S. T., and A. E. Dahlberg. 1999b. Erythromycin resistance mutations in ribosomal proteins L22 and L4 perturb the higher order structure of 23S ribosomal RNA. *J. Mol. Biol.* **289:**827–834.

Gregory, S. T., K. R. Lieberman, and A. E. Dahlberg. 1994. Mutations in the peptidyl transferase region of *E. coli* 23S rRNA affecting translational accuracy. *Nucleic Acids Res.* **22:**257–261.

Gutell, R. R. 1994. Collection of small subunit (16S- and 16S-like) ribosomal RNA structures: 1994. *Nucleic Acids Res.* **22:**3502–3507.

Gutell, R. R. 1996. Comparative sequence analysis and the structure of 16S and 23S rRNA, p. 111–128. *In* R. A. Zimmermann and A. E. Dahlberg (ed.), *Ribosomal RNA: Structure, Evolution, Processing and Function in Protein Biosynthesis.* CRC Press, Boca Raton, Fla.

Helser, T. L., J. E. Davies, and J. E. Dahlberg. 1971. Change in methylation of 16S ribosomal RNA associated with mutation to kasugamycin resistance in *Escherichia coli. Nat. New Biol.* **233:**12–14.

Helser, T. L., J. E. Davies, and J. E. Dahlberg. 1972. Mechanism of kasugamycin resistance in *Escherichia coli. Nat. New Biol.* **235:**6–9.

Huttenhofer, A., and H. F. Noller. 1994. Footprinting mRNA-ribosome complexes with chemical probes. *EMBO J.* **13:**3892–3901.

Irwin, B., J. D. Heck, and G. W. Hatfield. 1995. Codon pair utilization biases influence translational elongation step times. *J. Biol. Chem.* **270:**22801–22806.

Ito K., K. Kawakami, and Y. Nakamura. 1993. Multiple control of *Escherichia coli* lysyl-tRNA synthetase expression involves a transcriptional repressor and a translational enhancer element. *Proc. Natl. Acad. Sci. USA* **90:**302–306.

Kowalak, J. A., E. Bruenger, and J. A. McCloskey. 1995. Posttranscriptional modification of the central loop of domain V in *Escherichia coli* 23 S ribosomal RNA. *J. Biol. Chem.* **270:**17758–17764.

Laughrea, M., and P. B. Moore. 1978. On the relationship between the binding of ribosomal protein S1 to the 30 S subunit of *Escherichia coli* and 3′ terminus of 16 S RNA. *J. Mol. Biol.* **121:**411–430.

Leontis, N. B., and E. Westhof. 1998. A common motif organizes the structure of multi-helix loops in 16S and 23S ribosomal RNAs. *J. Mol. Biol.* **283:**571–583.

Lieberman, K. R., and A. E. Dahlberg. 1994. The importance of conserved nucleotides of 23S ribosomal RNA and transfer RNA in ribosome catalyzed peptide bond formation. *J. Biol. Chem.* **269:**16163–16169.

Liu, R., and S. W. Liebman. 1996. A translational fidelity mutation in the universally conserved sarcin/ricin domain of 25S yeast ribosomal RNA. *RNA* **2:**254–263.

Lodmell, J. S., and A. E. Dahlberg. 1997. A conformational switch in *Escherichia coli* 16S ribosomal RNA during decoding of messenger RNA. *Science* **277:**1262–1267.

Mankin, A. S. 1997. Pactamycin resistance mutations in functional sites of 16S rRNA. *J. Mol. Biol.* **274:**8–15.

Melancon, P., C. Lemieux, and L. Brakier-Gingras. 1988. A mutation in the 530 loop of *Escherichia coli* 16S ribosomal RNA causes resistance to streptomycin. *Nucleic Acids Res.* **16**:9631–9639.

Mitta, M., L. Fang, and M. Inouye. 1997. Deletion analysis of *cspA* of *Escherichia coli*: requirement of the AT-rich UP element for *cspA* transcription and the downstream box in the coding region for its cold shock induction. *Mol. Microbiol.* **26**:321–335.

Moazed, D., and H. F. Noller. 1986. Transfer RNA shields specific nucleotides in 16S ribosomal RNA from attack by chemical probes. *Cell* **47**:985–994.

Moazed, D., and H. F. Noller. 1987. Chloramphenicol, erythromycin, carbomycin and vernamycin B protect overlapping sites in the peptidyl transferase region of 23S ribosomal RNA. *Biochimie* **69**:879–884.

Moazed, D., and H. F. Noller. 1989. Interaction of tRNA with 23S rRNA in the ribosomal A, P, and E sites. *Cell* **57**:585–597.

Moazed, D., and H. F. Noller. 1990. Binding of tRNA to the ribosomal A and P sites protects two distinct sets of nucleotides in 16 S rRNA. *J. Mol. Biol.* **211**:135–145.

Muto, A., C. Ushida, and H. Himeno. 1998. A bacterial RNA that functions as both a tRNA and an mRNA. *Trends Biochem. Sci.* **23**:25–29.

Nagai, H., H. Yuzawa, and T. Yura. 1991. Interplay of two *cis*-acting mRNA regions in translational control of sigma 32 synthesis during the heat shock response of *Escherichia coli*. *Proc. Natl. Acad. Sci. USA* **88**:10515–10519.

O'Connor, M., and A. E. Dahlberg. 1993. Mutations at U2555, a tRNA-protected base in 23S rRNA, affect translational fidelity. *Proc. Natl. Acad. Sci. USA* **90**:9214–9218.

O'Connor, M., and A. E. Dahlberg. Unpublished observations.

O'Connor, M., E. A. De Stasio, and A. E. Dahlberg. 1991. Interaction between 16S ribosomal RNA and ribosomal protein S12: differential effects of paromomycin and streptomycin. *Biochimie* **73**:1493–1500.

O'Connor, M., T. Asai, C. L. Squires, and A. E. Dahlberg. 1999. Enhancement of translation by the downstream box does not involve base pairing of mRNA with the penultimate stem sequence of 16S rRNA. *Proc. Natl. Acad. Sci. USA* **96**:8973–8978.

O'Connor, M., W.-C. M. Lee, and A. E. Dahlberg. Unpublished observations.

Ofverstedt, L. G., K. Zhang, S. Tapio, U. Skoglund, and L. A. Isaksson. 1994. Starvation *in vivo* for aminoacyl-tRNA increases the spatial separation between the two ribosomal subunits. *Cell* **79**:629–638.

Olins, P. O., and S. H. Rangwala. 1989. A novel sequence element derived from bacteriophage T7 mRNA acts as an enhancer of translation of the *lacZ* gene in *Escherichia coli*. *J. Biol. Chem.* **264**:16973–16976.

Porse, B. T., and R. A. Garrett. 1995. Mapping important nucleotides in the peptidyl transferase centre of 23 S rRNA using a random mutagenesis approach. *J. Mol. Biol.* **249**:1–10.

Powers, T., and H. F. Noller. 1991. A functional pseudoknot in 16S ribosomal RNA. *EMBO J.* **10**:2203–2214.

Ramakrishnan, V., and S. W. White. 1998. Ribosomal protein structures: insights into the architecture, machinery and evolution of the ribosome. *Trends Biochem. Sci.* **23**:208–212.

Ramakrishnan, V., C. Davies, S. Gerchman, B. L. Golden, D. W. Hoffmann, T. N. Jaishree, J. H. Kycia, S. Porter, and S. W. White. 1995. Structures of prokaryotic ribosomal proteins: implications for RNA binding and evolution. *Biochem. Cell Biol.* **73**:979–986.

Resch, A., K. Tedin, A. Grundling, A. Mundlein, and U. Blasi. 1996. Downstream box–anti-downstream box interactions are dispensable for translation initiation of leaderless mRNAs. *EMBO J.* **15**:4740–4748.

Saarma, U., C. M. Spahn, K. H. Nierhaus, and J. Remme. 1998. Mutational analysis of the donor substrate binding site of the ribosomal peptidyltransferase center. *RNA* **4**:189–194.

Samaha, R., R. Green, and H. F. Noller. 1995. A base pair between tRNA and 23S rRNA in the peptidyltransferase centre of the ribosome. *Nature* 377:309–314.

Sander, P., T. Prammananan, A. Meier, K. Frischkorn, and E. C. Bottger. 1997. The role of ribosomal RNAs in macrolide resistance. *Mol. Microbiol.* **26**:469–480.

Sandler, S. J., and A. J. Clark. 1994. Mutational analysis of sequences in the *recF* gene of *Escherichia coli* K-12 that affect expression. *J. Bacteriol.* **176**:4011–4016.

Sandler, S. J., B. Chackerian, J. T. Li, and A. J. Clark. 1992. Sequence and complementation analysis of recF genes from *Escherichia coli, Salmonella typhimurium, Pseudomonas putida* and *Bacillus subtilis*: evidence for an essential phosphate binding loop. *Nucleic Acids Res.* **20**:839–845.

Sergiev, P. V., I. N. Lavrik, V. A. Wlasoff, S. S. Dokudovskaya, O. A. Dontsova, A. A. Bogdanov, and R. Brimacombe. 1997. The path of mRNA through the bacterial ribosome: a site-directed crosslinking study using new photoreactive derivatives of guanosine and uridine. *RNA* **3**:464–475.

Shean, C. S., and M. E. Gottesman. 1992. Translation of the prophage lambda cl transcript. *Cell* **70**:513–522.

Shine, J., and L. Dalgarno. 1975. Determinant of cistron specificity in bacterial ribosomes. *Nature* **254**:34–38.

Sparling, P. F. 1970. Kasugamycin resistance: 30S ribosomal mutation with an unusual location on the *Escherichia coli* chromosome. *Science* **167**:56–58.

Spirin, A. S., V. I. Baranov, G. S. Polubesov, I. N. Serdyuk, and R. P. May. 1987. Translocation makes the ribosome less compact. *J. Mol. Biol.* **194**:119–126.

Sprengart, M. L., H. P. Fatscher, and E. Fuchs. 1990. The initiation of translation in *E. coli*: apparent base pairing between the 16S rRNA and downstream sequences of the mRNA. *Nucleic Acids Res.* **18**:1719–1723.

Sprengart, M. L., E. Fuchs, and A. G. Porter. 1996. The downstream box: an efficient and independent translation initiation signal in *Escherichia coli*. *EMBO J.* **15**:665–674.

Steitz, J. A., and K. Jakes. 1975. How ribosomes select initiator regions in mRNA: base pair formation between the 3′ terminus of 16S rRNA and the mRNA during initiation of protein synthesis in *Escherichia coli*. *Proc. Natl. Acad. Sci. USA* **72**:4734–4738.

Subramanian, A. R. 1983. Structure and functions of ribosomal protein S1. *Prog. Nucleic Acid Res. Mol. Biol.* **28**:101–142.

Vila-Sanjurjo, A., C. L. Squires, and A. E. Dahlberg. 1999. Isolation of kasugamycin resistant mutants in the 16S ribosomal RNA of *Escherichia coli*. *J. Mol. Biol.* **293**:1–8.

von Ahsen, U., and H. F. Noller. 1995. Identification of bases in 16S rRNA essential for tRNA binding at the 30S ribosomal P site. *Science* **267**:234–237.

Wilson, K. S., and H. F. Noller. 1998. Molecular movement inside the translational engine. *Cell* **92**:337–349.

Winzeler, E., and L. Shapiro. 1997. Translation of the leaderless *Caulobacter dnaX* mRNA. *J. Bacteriol.* **179**:3981–3988.

Withey, J., and D. Friedman. 1999. Analysis of the role of *trans*-translation in the requirement of tmRNA for λimm^{P22} growth in *Escherichia coli*. *J. Bacteriol.* **181**:2148–2157.

Wu, C. J., and G. R. Janssen. 1996. Translation of *vph* mRNA in *Streptomyces lividans* and *Escherichia coli* after removal of the 5′ untranslated leader. *Mol. Microbiol.* **22**:339–355.

Zhang, J., and M. P. Deutscher. 1992. A uridine-rich sequence required for translation of prokaryotic mRNA. *Proc. Natl. Acad. Sci. USA* **89**:2605–2609.

Zwieb, C., I. Wower, and J. Wower. 1999. Comparative sequence analysis of tmRNA. *Nucleic Acids Res.* **27**:2063–2071.

The Ribosome: Structure, Function, Antibiotics, and Cellular Interactions
Edited by R. A. Garrett, S. R. Douthwaite, A. Liljas, A. T. Matheson, P. B. Moore, and H. F. Noller

Chapter 20

Reconstitution of the 50S Subunit with In Vitro-Transcribed 23S rRNA: a New Tool for Studying Peptidyltransferase

PHILIPP KHAITOVICH and ALEXANDER S. MANKIN

More than 3 decades ago, peptidyltransferase activity was found to be associated with the large ribosomal subunit (Monro, 1967). After years of studies, we know that 23S rRNA plays an important role in peptidyltransferase activity, we know which ribosomal proteins or rRNA segments are located close to the peptidyltransferase center, and we also know that many ribosomal proteins, as well as 5S rRNA, are not essential for peptidyltransferase activity. However, the basic molecular mechanism of the reaction and the structural organization of the catalytic center remain obscure. Furthermore, it remains unknown whether the main catalyst of the peptidyltransferase reaction is RNA or protein.

RNA plays a critical role in all the steps of protein synthesis. The list of various RNA molecules that are involved in protein synthesis includes rRNA, tRNA, mRNA, and many other auxiliary but essential RNA molecules, like tmRNA, 4.5S (or 7S) RNA, M1 RNA of RNase P, small nucleolar RNAs, etc. Already in 1968, Crick had proposed that a primitive translational machine could consist predominantly or entirely of RNA (Crick, 1968). Nevertheless, most of the early studies of peptidyltransferase were focused primarily on a search for a conventional protein catalyst among more than 30 different proteins composing the large ribosomal subunit. All attempts to assign the catalytic activity to one isolated ribosomal protein or to any of their combinations were, however, unsuccessful (see Noller, 1993, for a review). As the research progressed, more and more ribosomal proteins were found to be dispensable for the peptidyltransferase activity (Hampl et al., 1981; Moore et al., 1975; Franceschi and Nierhaus, 1990; Dabbs, 1991). Meanwhile, a large amount of experimental data which pointed to the functional significance of rRNA was accumulated. Discovery of the first ribozymes and the growing popularity of the RNA World evolutionary doctrine turned the mainstream of peptidyltransferase research from ribosomal proteins to rRNA. The idea that the modern ribosome evolved from the original RNA-only protein-synthesizing machinery was so appealing, and made so much sense from the evolutionary standpoint, that it immediately found many proponents. A logical consequence of this hypothesis was that the modern ribosome remains fundamentally an RNA machine, and hence, the central functions, including catalysis of peptide bond formation, are still carried out by RNA. A lot of experimental data implicated large-ribosomal-subunit rRNA in functions of the peptidyltransferase center. Furthermore, the general possibility of RNA catalysis of the peptidyltransferase reaction was demonstrated by in vitro selection of RNA aptamers capable of joining two amino acids by a peptide bond (Zhang and Cech, 1997). It is therefore extremely tempting to think that 23S rRNA is directly responsible for the peptidyltransferase catalysis.

We feel, however, that to plan future research and for the sake of scientific accuracy, it is vitally important to separate "wishful thinking" from real knowledge. Therefore, in the following few paragraphs we attempt to summarize briefly some of the arguments regarding the nature of peptidyltransferase in order to demonstrate that it is not yet possible to conclude which ribosomal component is responsible for the catalysis. More detailed information about the ribosomal peptidyltransferase center can be found in several recent reviews (Wower et al., 1995; Garrett

Philipp Khaitovich and Alexander S. Mankin ■ Center for Pharmaceutical Biotechnology-m/c 870, University of Illinois, 900 S. Ashland Ave., Chicago, IL 60607.

and Rodriguez-Fonseca, 1996; Barta and Halama, 1996; Green and Noller, 1997).

PEPTIDYLTRANSFERASE: RNA VERSUS PROTEIN

The high degree of rRNA sequence conservation is usually perceived as an indication of the functional importance of the corresponding rRNA segment. Several nucleotide sequence segments of 23S rRNA, as well as a number of individual nucleotides, are invariant among all the studied organisms. Although the functional importance of some invariant positions has been experimentally demonstrated (see Garrett and Rodriguez-Fonseca, 1996, and Green and Noller, 1997, for reviews), conservation of some other positions may not necessarily reflect their direct involvement in ribosomal functions. Mutations of some of the conserved positions in 23S rRNA have very little effect on normal ribosomal functions (Triman et al., 1998). Hence, conservation per se does not prove functional importance of the nucleotide residue.

Generally amino acid sequences of ribosomal proteins are conserved less than nucleotide sequences of rRNA (Wittmann-Liebold et al., 1990). However, the lack of extensive sequence conservation may well be compatible with the functional importance of a protein—after all, only a few strategically positioned amino acids form the catalytic centers of most enzymes. On the other hand, some proteins, for example, L2, are quite well conserved (Cooperman et al., 1995). L2 proteins from eukarya, archaea, and bacteria exhibit about 30% identity. Antibodies recognizing the tertiary-structure elements of the *Escherichia coli* L2 protein cross-react with L2 homologues from a variety of species (Stöffler-Meilicke and Stöffler, 1990), demonstrating a high degree of conservation of the protein tertiary structure. Furthermore, eukaryotic and archaeal L2 proteins can serve as functional substitutes for the *E. coli* homologue (Uhlein et al., 1998). Thus, conservation of structure and function of some ribosomal proteins, for example, L2, is compatible with their possible significance for the key functions of the ribosome, such as peptidyltransferase activity.

A strong argument in support of rRNA involvement in peptidyltransferase activity came from antibiotic studies. Mutations conferring resistance to antibiotics, inhibitors of peptidyltransferase, were mapped within or near the central loop of domain V of 23S rRNA (annotated in Triman et al., 1998). Several peptidyltransferase inhibitors protect nucleotides within the central loop of domain V from chemical modification in RNA footprinting experiments (see Garrett and Rodriguez-Fonseca, 1996, for a review), and some drugs could be cross-linked to rRNA in the ribosome or even to isolated 23S rRNA (Porse et al., 1999).

On the other hand, similar experimental evidence suggests close contacts between peptidyltransferase-inhibiting antibiotics and ribosomal proteins. Affinity labeling has identified possible interaction of some ribosomal proteins with antibiotics that affect peptidyltransferase activity (Lazaro et al., 1991; Le Goffic et al., 1980). Mutations in ribosomal proteins conferred resistance to several peptidyltransferase inhibitors (Hummel et al., 1979; Baughman and Fahnestock, 1979; Böck et al., 1982) or to drugs with binding sites located in the immediate vicinity of peptidyltransferase (e.g., erythromycin) (Chittum and Champney, 1994). It is of course possible that antibiotic resistance mutations in ribosomal protein alter the conformation of rRNA (Gregory and Dahlberg, 1999). It is also possible, however, that some of the rRNA mutations affect the conformation of the neighboring proteins. Accordingly, antibiotic studies do not prove the exclusive role of rRNA in formation of the peptidyltransferase center.

Peptidyltransferase substrates, aminoacyl- and peptidyl-tRNAs, form tight contacts with rRNA, which has been demonstrated most clearly by cross-linking and footprinting experiments. Most of the cross-links involving the amino acceptor end of tRNA or its analogues were localized in domain V of 23S rRNA (reviewed in Barta and Halama, 1996, and Brimacombe et al., 1995). Cross-linking data are in general agreement with results of footprinting experiments where tRNA substrates bound to ribosomal A and P sites protect from chemical modification a set of nucleotides located primarily in domain V of 23S rRNA (Moazed and Noller, 1989). It should be mentioned, however, that not all of the protected residues are essential for tRNA binding. In damage selection experiments, modification of only 4 nucleotides (G2252, A2451, U2506, and U2585) of 12 protected by the P-site-bound tRNA interferes with binding of peptidyl-tRNA to the 50S subunit (Bocchetta et al., 1998). One of these 4 nucleotides, G2252, forms a base pair with C74 of the acceptor end of the P-site-bound tRNA (Samaha et al., 1995). The existence of this base pair is the strongest argument to date in favor of direct participation of rRNA in the activity of the peptidyltransferase center. Although G2252-C74 contact is clearly important, it can nevertheless be replaced by a weaker interaction, since tRNA with C74 mutated to U can still participate in translation in vivo (O'Connor et al., 1993).

Affinity labeling offers equally convincing proof of the close proximity of aminoacyl- and peptidyl-tRNAs in the peptidyltransferase center to several ribosomal proteins. Among the proteins identified in cross-linking experiments were L2, L11, L14, L15, L16, L18, L23, and L27 (for a review, see Wower et al., 1995). A number of these proteins, including L2, L15, L18, and L27, were mapped close to the peptidyltransferase center by immuno-electron microscopy (Walleczek et al., 1988). It was even shown that purified L2 and L16 exhibited substantial affinity for tRNA in solution (Maimets et al., 1984; Remme et al., 1985). Furthermore, binding of L2 increased the hydrolysis rate of peptidyl-tRNA, which could indicate participation of L2 in the peptidyltransferase reaction (Remme et al., 1985).

Strong support for rRNA involvement in peptidyltransferase function comes from mutational studies. Several point mutations located in the vicinity of the central loop of domain V of 23S rRNA abolished the peptidyltransferase activity (annotated in Triman et al., 1998). In turn, there are mutations in ribosomal proteins that significantly affect the peptidyltransferase function. For example, mutation of the conserved His229 in *E. coli* protein L2 and in its eukaryotic and archaeal homologues drastically reduced peptidyltransferase activity without affecting ribosome assembly (Cooperman et al., 1995; Uhlein et al., 1998). Though the proposed critical role of His299 was challenged by the finding that corresponding mutation in yeast mitochondria conferred only a conditionally lethal phenotype (Pan and Mason, 1997) and that this residue was not conserved in the L2 protein of *Mycoplasma capricolum* (Ohkubo et al., 1987), it remains possible that another amino acid residue of protein L2 or some other protein may prove to be essential for peptidyltransferase.

One of the arguments against protein-based peptidyltransferase is that none of the isolated ribosomal proteins, nor any of their mixtures, have ever been found to possess any detectable peptidyltransferase activity. The same is true, however, for the isolated rRNA. One can argue that removal of ribosomal proteins may perturb the tertiary structure of rRNA required for the catalysis. In an attempt to address this possibility, Noller and coworkers chose to use ribosomes from a thermophilic bacterium, *Thermus aquaticus*, instead of the commonly used ribosomes from mesophilic *E. coli*. Since *T. aquaticus* ribosomes are designed to function at high temperatures (the optimal growth temperature for *T. aquaticus* is 70°C), the tertiary structure of rRNA must be extremely robust and may withstand removal of ribosomal proteins without loss of the 23S rRNA structural integrity. Indeed, Noller and colleagues showed that *T. aquaticus* ribosomes did not lose the ability to catalyze the peptidyltransferase reaction after being subjected to extensive protein extraction procedures, which included treatment with proteinase K in the presence of sodium dodecyl sulfate and phenol extractions (Noller et al., 1992). This result confirmed the earlier observation of Bernabeu et al. (1979) that the peptidyltransferase activity of *E. coli* ribosomes is rather resistant to proteinase K treatment. The resistance of ribosomal peptidyltransferase to protein extraction was greeted with great enthusiasm not only by those working in the field of translation but also by many working on other aspects of RNA function, since this result was perceived as long-awaited proof of the catalytic activity of rRNA. Unfortunately, Noller's cautious note that "direct proof for the hypothesis that peptide bond formation is catalyzed solely by rRNA will require demonstration of activity with completely protein-free preparations, such as *in vitro* transcripts of 23S rRNA" was largely ignored. Later, the detailed analysis of the peptidyltransferase-active particles obtained by treatment of *T. aquaticus* large ribosomal subunits with proteinase K, sodium dodecyl sulfate, and phenol (termed KSP particles to reflect the corresponding treatments) showed that these protein extraction procedures failed to remove approximately one-fourth of the ribosomal proteins of *T. aquaticus* 50S subunits (Fig. 1). Removal of the remaining proteins by repeated phenol extractions led to unfolding of 23S rRNA and the loss of peptidyltransferase activity (Khaitovich et al., 1999a).

Ribosomal proteins associated with rRNA in the KSP particles remained intact in spite of proteinase treatment and were identified as homologues of *E. coli* proteins L2, L3, L13, L15, L17, L18, L21, and L22 (Khaitovich et al., 1999a) (Fig. 1). An additional weak spot (X) was frequently found when KSP proteins were separated by two-dimensional gels. Analysis of the spot X in the total 50S subunit protein preparation showed that it is composed of two proteins, L1 and L4. Therefore, the drastic decrease of intensity of this spot in KSP particle protein preparation may reflect the fact that only one of the proteins, most probably L1, is efficiently removed during protein extraction. Thus, we cannot exclude the presence of some amount of protein L4 in KSP particles. The previous partial disassembly and omission reconstitution experiments showed that many proteins in the large ribosomal subunit are dispensable for the peptidyltransferase activity. However, removal of L2, L3, or L4 protein resulted in loss of activity (discussed in Noller, 1993, and Khaitovich et al., 1999a). Thus, neither of these proteins can yet be excluded from the list of potential components required for catalysis of peptide bond formation. It should be kept

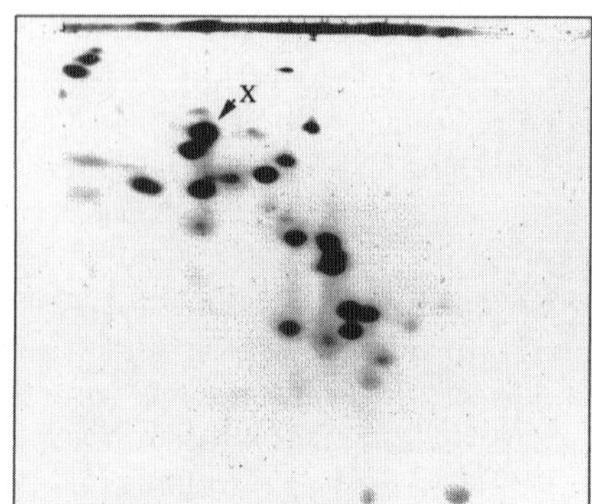

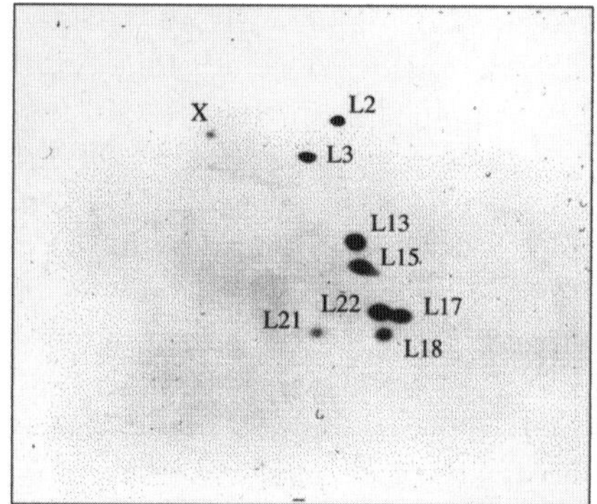

Figure 1. Two-dimensional gel electrophoresis of proteins from *T. aquaticus* 50S subunits and KSP particles. The identified proteins of KSP particles are labeled. Spot X in protein preparations from KSP particles was not sequenced; the corresponding spot in total protein of the 50S subunit contains unresolved proteins L1 and L4.

in mind, however, that the presence of L2, L3, possibly L4, and the other six ribosomal proteins in the catalytically active KSP particles does not prove their involvement in peptidyltransferase activity and may merely reflect their tight association with 23S rRNA in *T. aquaticus* large ribosomal subunits.

Although experiments with *T. aquaticus* 50S subunits failed to prove that peptidyltransferase is an RNA enzyme, they opened a spectacular possibility for preparation of subribosomal particles that lack approximately three-fourths of the large-ribosomal-subunit proteins and still possess virtually full peptidyltransferase activity.

Two other papers published in 1998 claimed to present the ultimate proof of catalytic activity of 23S rRNA (Nitta et al., 1998a, 1998b). It was reported that in vitro-transcribed protein-free *E. coli* 23S rRNA or even its separate domains could catalyze in vitro peptidyltransferase reaction between acetyl-Phe-tRNA and Phe-tRNA. Subsequent studies did not, however, confirm these claims. The main reaction product, originally described as acetyl-Phe-Phe dipeptide, was subsequently shown to be an alcohol ester of acetylphenylalanine resulting from alcoholysis of acetyl-Phe-tRNA. The appearance of this product did not depend on the nature of RNA used in the reaction and could be explained by alcohol contamination in RNA preparations (Khaitovich et al., 1999c).

To summarize the evidence presented in the previous paragraphs, we can conclude that although 23S rRNA undoubtedly plays a significant role in the formation of the catalytic center, there is to date no convincing evidence that rRNA is the actual catalyst of the reaction. The possibility of direct functional or structural roles for at least three ribosomal proteins, L2, L3, and L4, in the catalysis of peptide bond formation cannot be excluded yet. From an evolutionary prospective, the origin of protein synthesis machinery as an all-RNA system is extremely appealing. It would, however, be naïve to think that billions of years of coevolution of rRNA and proteins could not affect the composition and function of peptidyltransferase. It is obvious, therefore, that further studies are required to clarify the functions of rRNA in catalysis of peptide bond formation in the modern ribosome. These studies will likely employ the new experimental techniques discussed in the following sections of this chapter.

PEPTIDYLTRANSFERASE EXPLORATION: THINK IN VITRO

One of the main obstacles to understanding the molecular mechanisms of translation is the enormous complexity of the ribosome. Discovering the molecular mechanisms of peptidyltransfer catalysis and unraveling the possible enzymatic activity of rRNA require methods which would make it possible to maximally simplify the composition of the peptidyltransferase center and to assess the functions of its individual components. Any of these goals is difficult to achieve through in vivo experiments. The introduction of any alterations in rRNA is complicated by the presence of multiple copies of rRNA genes in most organisms. Expression of mutant rRNA from a multicopy plasmid results in a mixed population of ribosomes in the cell, with mutant ribosomes being "contaminated" by wild-type ribosomes containing chromosome-encoded rRNA. Several approaches have been proposed which allow the study of peptidyltransferase activity of the ribosomes in a mixed population (Saarma and Remme, 1992; Leviev et al., 1995). Unfortunately, neither one of these methods can completely eliminate the contribution of the wild-type ribosomes, which makes it extremely difficult to detect weak activities of the mutant ribosomes. Furthermore, neither those few organisms which contain single rRNA operons (Mankin, 1995) nor the recently engineered *E. coli* strain with deletion of all seven chromosomal rRNA operons (Asai et al., 1999) can be efficiently used for studying dominantly lethal or highly deleterious rRNA mutations, which are the most interesting in a search for the functionally essential nucleotides. Similarly, although some ribosomal-protein genes are not essential, it is unlikely that cells lacking any substantial number of ribosomal proteins will be viable (Dabbs, 1991).

The in vitro approach appears to be the best option for dissecting the composition of peptidyltransferase and the functions of its components. One of the most fruitful techniques used for the analysis of the peptidyltransferase center and of the ribosome in general has been in vitro reconstitution of ribosomal subunits. In 1970, Nomura and Erdman succeeded in in vitro assembly of *Bacillus stearothermophilus* large ribosomal subunits from rRNA and ribosomal proteins (Nomura and Erdmann, 1970). To reconstitute the functional 50S subunit, the rRNA and proteins had to be incubated together at 60°C for 45 min in the presence of 20 mM $MgCl_2$. Attempts to apply the same procedure to the reconstitution of *E. coli* 50S subunits, however, were unsuccessful. It was not until 1974 that Nierhaus and Dohme came up with the two-step incubation protocol for reconstitution of the functional *E. coli* 50S subunits (Nierhaus and Dohme, 1974). In their protocol, a mixture of large-subunit rRNA and proteins was incubated first at 44°C in a buffer with a low Mg^{2+} concentration (4 mM), followed by incubation at 50°C in the presence of 20 mM Mg^{2+}. The modified two-step protocol also turned out to be useful for reconstitution of *T. aquaticus* large ribosomal subunits, although the temperature of the second incubation step had to be increased to 60°C to account for the thermophilic nature of *T. aquaticus* ribosomes (Fig. 2A) (Khaitovich et al., 1999b). By analogy with *B. stearothermophilus* or *E. coli* systems, one- or two-step incubation protocols were developed for reconstitution of large ribosomal subunits from a number of other organisms (see Khaitovich et al., 1999b, for references).

In vitro reconstitution of large ribosomal subunits provided the tool for analysis of the contributions of individual proteins and rRNA molecules to peptidyltransferase activity: this could now be accomplished by omission of one or several components or by incorporating their mutated variations. The main conclusion of reconstitution experiments, performed mostly in the *E. coli* system, was that the presence of 23S rRNA and three ribosomal proteins, L2, L3, and L4, appears to be critical for reconstitution of peptidyltransferase-active particles. Omission of any of these components from reconstitution mixtures resulted in inactive particles (see Noller, 1993, for a discussion). To our knowledge, however, no successful attempts to reconstitute minimal peptidyltransferase from these four components alone were reported.

The interest in rRNA function in translation called for an easy and efficient way to manipulate rRNA structure. As discussed above, the in vivo techniques offer fairly limited opportunities for changing rRNA; therefore, it was important to be able to incorporate into the ribosome synthetic rRNA prepared either by chemical synthesis or by in vitro transcription of modified rRNA genes. The reconstitution of the functional small ribosomal subunit with the in vitro-transcribed 16S rRNA was pioneered by Ofengand's group (Krzyzosiak et al., 1987). Incorporation of the 16S rRNA transcript, lacking all 11 known posttranscriptional modifications, resulted in a slightly looser conformation of the central core of the reconstituted subunits; nevertheless, assembled 30S subunits were quite active in tRNA binding, subunit association, and in vitro translation, exhibiting 60 to 80% of the activity of subunits reconstituted with natural 16S rRNA (Moine et al., 1997; Krzyzosiak et al., 1987). Development of this procedure allowed analysis of the functions of a number of nucleotides in the decoding center and other functionally important regions of 16S rRNA (Moine et al., 1997; Melancon et al., 1990). With the large ribosomal subunit, the situation turned out to be more difficult. It was fairly easy to incorporate the 5S rRNA transcript into the in vitro-assembled subunits—this technique proved extremely useful for studies of 5S rRNA conformation and environment in the ribosome (Dontsova et al., 1994; Dokudovskaya et al., 1996; Sergiev et al., 1998). It was also possible to partly assemble individual domains of 23S rRNA with in vitro-prepared corresponding segments of 23S rRNA—this approach was extremely useful for elucidating sites of interaction of some ribosomal proteins and 5S rRNA with 23S rRNA (Østergaard et al., 1998; Draper and Reynaldo, 1999). In contrast, attempts to incorporate in vitro-transcribed full-length 23S rRNA into functional *E. coli* 50S subunits were unsuccessful (Weitzmann et al., 1990). Using an extremely sensitive peptidyltransferase assay, Green and Noller (1996) showed that activity of *E. coli* 50S subunits assembled with the 23S rRNA transcript was 5 orders of magnitude lower than that of the native 50S subunits.

The two most probable explanations for the failure to reconstitute catalytically active 50S subunits

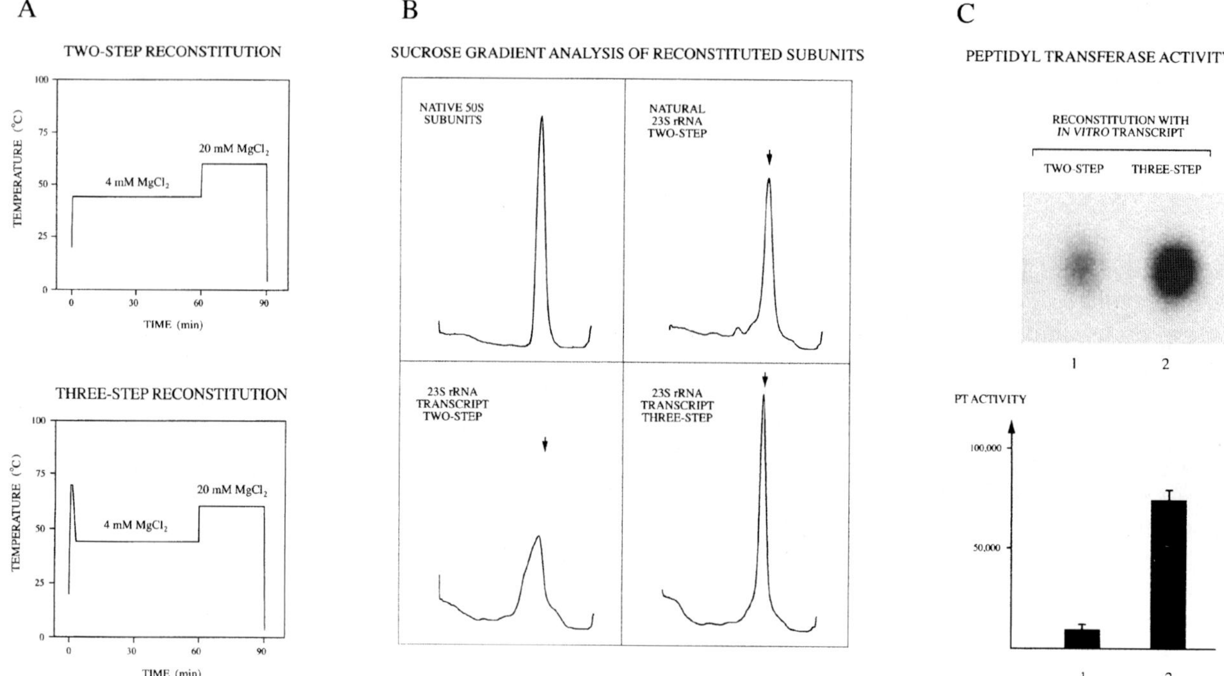

Figure 2. Reconstitution of *T. aquaticus* 50S subunits with natural and in vitro-transcribed 23S rRNA. (A) Temperature profiles for the two-step and three-step reconstitutions. (B) Sucrose gradient analysis of native 50S subunits or subunits reconstituted with natural or in vitro-transcribed 23S rRNA. The arrows indicate the positions of the peaks of native 50S subunits. (C) Peptidyltransferase activity of subunits reconstituted with in vitro-transcribed 23S rRNA in a two- or three-step reconstitution procedure. The spots shown correspond to the product of the peptidyltransferase reaction, [^{35}S]fMet-puromycin, resolved by paper electrophoresis (for details, see Khaitovich et al., 1999b). The histogram at the bottom of panel C shows the results of quantification of the [^{35}S]fMet-puromycin spot (total counts over a 10-h exposure).

with the in vitro-transcribed 23S rRNA were the lack of posttranscriptional modifications in the rRNA transcript or its incorrect folding. A significant number of modified nucleotides are present in all of the analyzed bacterial 23S rRNA molecules, as well as in their archaeal and eukaryotic counterparts (Rozenski et al., 1999). *E. coli* 23S rRNA contains 23 modified nucleotides (Rozenski et al., 1999); the lowest number of modifications was found in the large-ribosomal-subunit rRNA of yeast mitochondria, which contains only three posttranscriptionally modified nucleotides (Sirum-Connolly et al., 1995). The clustering of modified nucleotides in functionally significant regions of rRNA, including the central loop of domain V, suggested their importance for ribosome function (Brimacombe et al., 1993). The chimeric reconstitution procedure, in which fragments of natural 23S rRNA were combined with the remaining RNA fragments prepared by in vitro transcription, allowed identification of a region of 23S rRNA (positions 2445 to 2523) that could not be replaced by in vitro transcripts and thus appeared to contain modified nucleotides essential for assembly of functional *E. coli* 50S subunits (Green and Noller, 1996). However, although the indispensable natural rRNA segment included six posttranscriptional modifications, none of them was universally conserved and they were thus unlikely to play any major role in ribosomal function. Hence, modified nucleotides in *E. coli* 23S rRNA could be important for subunit assembly rather than for the catalytic activity itself. Indeed, the absence of posttranscriptional modifications in *E. coli* 23S rRNA prevented assembly of compact 50S subunits (Green and Noller, 1996).

RECONSTITUTION OF ACTIVE 50S SUBUNITS WITH IN VITRO-TRANSCRIBED rRNA

The question remained, however, whether dependence of the 50S subunit assembly on the presence of modified nucleotides in 23S rRNA is a general rule or merely an *E. coli*-specific phenomenon. *E. coli* 50S subunits require notoriously elaborate conditions for in vitro reconstitution. The necessity of the two-step incubation and an extremely strict Mg^{2+} requirement during the first incubation step suggest that *E. coli* 23S rRNA has to follow a very narrow folding pathway in order to assemble into active ribosomal subunits (Green and Noller,

1996). The assembly of 50S subunits of other organisms can be less restricted and may thus depend less on the presence of modified nucleotides in 23S rRNA. Thermophiles were a natural choice. The assembly of 23S rRNA into 50S subunits in thermophilic organisms must be facilitated by the thermodynamically favorable active subunit conformation required to sustain protein synthesis at elevated temperatures. Highly robust 50S subunits of *T. aquaticus* and "easily reconstitutable" 50S subunits of *B. stearothermophilus* were chosen in our own and Noller's laboratories as models for in vitro assembly of peptidyltransferase-active subunits with in vitro-transcribed 23S rRNA (Green and Noller, 1999; Khaitovich et al., 1999b).

Active 50S subunits of these thermophilic ribosomes could be reconstituted from natural (posttranscriptionally modified) rRNA and proteins by one-step (*B. stearothermophilus*) (Nomura and Erdmann, 1970) or two-step (*T. aquaticus*) (Khaitovich et al., 1999b) (Fig. 2) procedures. When 23S rRNA transcript was used instead of natural rRNA, the peptidyltransferase activity of reconstituted 50S subunits from both organisms was rather low, approximately 1% of the activity of natural subunits in the "fragment reaction" assay. Nevertheless, this activity was 3 orders of magnitude higher than that of *E. coli* 50S subunits assembled with 23S rRNA transcript, suggesting that we were on the right track. Optimization of buffer composition, the RNA/protein ratio, and protein-preparation procedures brought catalytic activity of *B. stearothermophilus* subunits reconstituted with the 23S rRNA transcript up to approximately 10% of the activity of native 50S subunits (Green and Noller, 1999). Such activity was already quite high. Similar adjustments of the reconstitution protocol did not, however, significantly increase the activity of reconstituted *T. aquaticus* subunits.

The relatively low activity of *T. aquaticus* subunits assembled with in vitro-transcribed 23S rRNA correlated with their "bad" sedimentation profile: reconstituted subunits containing 23S rRNA transcript sedimented as a broad asymmetric peak in which only the faster-sedimenting shoulder had peptidyltransferase activity (Fig. 2C). Hence, aberrant subunit assembly appeared to be one of the main problems, likely related to the improper folding of the rRNA transcript. Heat denaturing of the natural 23S rRNA prior to reconstitution reduced activity of the assembled subunits approximately 20-fold, while denaturing of the 23S rRNA transcript did not decrease the activity of the reconstituted subunits. Consequently, it appeared that in vitro-transcribed 23S rRNA was from the very beginning trapped in the stable "assembly-incompetent" conformation. In order to rescue the misfolded RNA, an additional "refolding" step was introduced into the reconstitution protocol (Fig. 2A). During the refolding step, the in vitro 23S rRNA transcript, mixed with 5S rRNA and ribosomal proteins in a buffer containing 4 mM Mg^{2+}, was incubated for 1 min at 70°C and then cooled down over 2 min to 44°C. This was followed by the standard two-step reconstitution procedure. Brief heating at low Mg concentration should partly unfold the rRNA transcript, while binding of ribosomal proteins may "lock" 23S rRNA in a conformation competent for the subsequent assembly of the 50S subunit. Introducing the refolding step improved dramatically the gradient profile of the assembled particles, which now sedimented as a sharp symmetric 50S peak (Fig. 2B). More importantly, subunits assembled with the 23S rRNA transcript in a three-step reconstitution procedure (including the refolding step) showed significantly improved peptidyltransferase activity, which, after purification of reconstituted subunits from the sucrose gradient, approached the activity of subunits assembled with natural 23S rRNA (23 and 25% of the activity of the intact 50S subunits, respectively) (Fig. 2C). *T. aquaticus* 50S subunits assembled with 23S rRNA transcript and combined with 30S subunits from either *T. aquaticus* or *E. coli* also exhibited substantial activity in a poly(U)-dependent cell-free translation system: approximately 30% of that of native 50S subunits (Khaitovich et al., 1999b).

Analysis of the protein composition of *T. aquaticus* subunits assembled with in vitro-transcribed 23S rRNA showed that while most of the 50S subunit proteins were present in the assembled particles, several proteins (as yet unidentified) were present in substoichiometric amounts. Incomplete incorporation of these proteins could partially explain why the activity of reconstituted particles did not quite reach the level of the native 50S subunits. In general, the quality of protein preparation was an essential factor for the activity of the assembled 50S subunits of *T. aquaticus* and *B. stearothermophilus*.

Activity of *T. aquaticus* 50S subunits reconstituted with either natural or in vitro-transcribed 23S rRNA depended critically on the presence of 5S rRNA. Omission of 5S resulted in virtually inactive material. This was in contrast to the *B. stearothermophilus* system, where omission of 5S rRNA reduced subunit activity only approximately twofold (Green and Noller, 1999). However, the nature of 5S rRNA was not critical in either system. Equally efficient reconstitution was achieved with either cognate 5S rRNA or 5S rRNA from *E. coli*. Furthermore, natural 5S rRNA could be replaced with its in vitro transcript without any loss of activity of the assembled subunits.

Reconstitution of functionally active large ribosomal subunits with rRNA lacking any modified nucleotides complemented the previous finding that the small ribosomal subunit can be reconstituted with modification-free 16S rRNA. We can now conclude that modified nucleotides are not essential for any of the crucial ribosome functions. This finding does not of course exclude the possible functional significance of posttranscriptional modifications, which may facilitate ribosome biogenesis and fine-tuning of translation in vivo (Sirum-Connolly et al., 1995; Sirum-Connolly and Mason, 1993; Gustafsson and Persson, 1998; Raychaudhuri et al., 1998).

Perhaps even more importantly, the ability to assemble catalytically active large ribosomal subunits with in vitro-transcribed 23S rRNA opens a wide range of opportunities for exploring the composition of the peptidyltransferase center and the functions of 23S rRNA in peptide bond formation.

CURRENT AND FUTURE APPLICATIONS

Virtually unlimited amounts of 23S rRNA can be prepared by in vitro transcription of the cloned gene or the PCR product by using RNA polymerase of the T7 bacteriophage. Once we can incorporate in vitro-transcribed 23S rRNA into functional 50S subunits, a whole new range of experimental approaches, which were difficult or even impossible to use before, becomes available.

Among other things, it is now possible:

- To study in vitro the effects of rRNA mutations in ribosomal populations containing exclusively mutant rRNA
- To obtain and analyze ribosomes containing lethal or highly deleterious rRNA mutations
- To use 23S rRNA virtually free from possible contamination with other ribosomal components (ribosomal proteins or 5S rRNA)
- To isolate specific rRNA mutations by in vitro selection

We will now illustrate with several examples the type of information about structure and function of ribosomal peptidyltransferase and the large ribosomal subunit in general that one may expect to gain by using subunit reconstitution with the in vitro rRNA transcript. Some of the projects mentioned are still in progress; others are only beginning and indicate future directions of research rather than present, finished studies.

Point Mutations in rRNA

One of the opportunities opened by incorporation of the in vitro-transcribed 23S rRNA into functional subunits is the possibility of introducing into rRNA any desired mutations and obtaining ribosome preparations containing exclusively mutant rRNA. The prototype of this approach, known as in vitro genetics, was used previously by Samaha et al. (1995) to demonstrate the existence of a Watson-Crick base pair between G2252 in *E. coli* 23S rRNA and C74 of the peptidyl-tRNA in the ribosomal P site. In order to prepare ribosomes with the G2252 mutations in 23S rRNA, these authors employed the chimeric reconstitution, in which in vitro-transcribed fragments of 23S rRNA were used in combination with the fragments of natural rRNA. In spite of a very low efficiency of chimeric reconstitution, it provided the critical proof of a complementary interaction between 23S rRNA and tRNA.

The adequacy of the *T. aquaticus* reconstitution system for in vitro genetic experiments was tested with two well-characterized point mutations, G2252A and A2058G, in 23S rRNA (to avoid possible confusion, we will use *E. coli* numbering of 23S rRNA positions throughout this chapter). The mutations were engineered in the cloned *T. aquaticus* 23S rRNA, and in vitro-transcribed mutant 23S rRNA was incorporated into 50S subunits (Khaitovich et al., 1999b). The substitution of other nucleotides for G2252 interferes with tRNA binding in the P site and thus with the peptidyltransferase function (Samaha et al., 1995). As expected, *T. aquaticus* 50S subunits reconstituted with in vitro-transcribed 23S rRNA carrying the G2252A mutation displayed dramatically reduced peptidyltransferase activity. The A2058G mutation renders *E. coli* ribosomes resistant to clindamycin, an antibiotic inhibitor of peptidyltransferase activity (Leviev et al., 1995). Indeed, *T. aquaticus* 50S subunits reconstituted with 23S rRNA containing the A2058G mutation showed markedly increased resistance to clindamycin compared to that of subunits reconstituted with the wild-type 23S rRNA transcript. These results demonstrated that the *T. aquaticus* reconstitution system can be used for *in vitro* analysis of 23S rRNA function.

The effect of mutations at several universally conserved positions, including U2506, U2584, and A2602, was studied in the *B. stearothermophilus* system (Green and Noller, 1999). These positions were suspected to be involved in donor substrate binding or the catalytic function of the peptidyltransferase center (Moazed and Noller, 1989; Porse et al., 1996; Bocchetta et al., 1998). The mutations of U2584 (to A or C) had little effect on peptidyltransferase activity of reconstituted subunits. The activities of subunits assembled with 23S rRNA containing U2506A and U2506C mutations were reduced approximately two-fold compared to that of subunits reconstituted with wild-type 23S rRNA, while the U2506G and A2602C mutations reduced activity approximately 20-fold and 10-fold, respectively. The advantage of the in vitro-reconstitution technique is that it made it possible to clearly detect low enzymatic activities of mutant subunits, something that was difficult to do accurately with the mixed ribosomal population prepared by in vivo expression of mutant rRNA (Porse et al., 1996). Accordingly, it was possible to conclude that none of the tested universally conserved nucleotides was essential for the catalysis of the peptidyltransferase reaction.

In the experiments with *T. aquaticus* 23S rRNA, we studied the effects of mutations at another universally conserved position, G2553, on peptidyltransferase activity and binding of the acceptor substrate. G2553 was shown to be protected from chemical modification by the 3′ end of the A-site-bound tRNA (Moazed and Noller, 1989). Chemical modification of U2555 located nearby interferes with the binding of puromycin, a structural analogue of the aminoacyl-tRNA 3′ end (Bocchetta et al., 1998). Furthermore, a puromycin derivative, 4-thio-dT-pCp-puromycin, could be efficiently cross-linked to G2553 (Green et al., 1998). All these data pointed to the possible interaction of G2553 with the universally conserved CCA end of the A-site-bound tRNA. In order to test this hypothesis, *T. aquaticus* 50S subunits were assembled with the in vitro-transcribed 23S rRNA containing all three possible nucleotide substitutions of G2553. Neither of the mutations affected assembly of 50S subunits as judged by the sucrose gradient profiles. However, the peptidyltransferase activity of 50S subunits containing the G2553C mutation, measured under the fragment reaction conditions with fMet-tRNA as a donor substrate and a puromycin-containing oligonucleotide, CpCp-puromycin (CC-Pm), as an acceptor, was significantly reduced (Fig. 3A).

To test whether the deleterious effect of the G2553C mutation can be compensated for by mutations in the acceptor substrate, the "mutant" puromycin-containing oligonucleotides, GC-Pm and CG-Pm, were used. In the reaction, catalyzed by wild-type 50S subunits, GC-Pm was as active as CC-Pm. However, when CG-Pm was used under the same reaction conditions, the peptidyltransferase activity was almost completely abolished (Fig. 3A). Remarkably, the activity was partially restored when mutant CG-Pm was used in combination with reconstituted ribosomes containing 23S rRNA with the G2553C mutation. The apparent compensatory effect of mutations in 23S rRNA and in the CCA end of the acceptor tRNA analogue was confirmed further in an experiment with authentic aminoacyl-tRNA. For this experiment, reconstituted wild-type or mutant (G2553C) *T. aquaticus* 50S subunits were combined with *E. coli* small ribosomal subunits and ribosomes were programmed with synthetic mRNA containing Met and Val codons. The peptidyltransferase reaction between fMet-tRNA as a donor and either wild-type Val-tRNA or its C75G mutant as an acceptor was monitored by formation of the fMet-Val dipeptide (Fig. 3B). The results of this experiment were qualitatively similar to those obtained with puromycin-containing oligonucleotides, although the amplitude of the effect was reduced, probably due to a stronger interaction of aminoacyl-tRNA with the 70S ribosome compared to the interaction of puromycin oligonucleotides with the isolated 50S subunit. These data are consistent with a direct interaction between G2553 in 23S rRNA and C75 of aminoacyl-tRNA in the ribosomal A site. Similar results have been independently obtained in R. Green's laboratory with the use of the *B. stearothermophilus* system (Green, personal communication).

rRNA Deletions

The large-ribosomal-subunit rRNA of animal mitochondria is much smaller than 23S-like rRNA of bacteria, archaea, and eukaryotes. Extended rRNA segments "missing" in mitochondrial rRNA should not be essential for the critical ribosome functions (Glotz et al., 1981; Mankin and Kopylov, 1981). Using mitochondrial RNA as a guideline, we are trying to reduce the complexity of the functional prokaryotic large ribosomal subunit by introducing large-scale deletions into 23S rRNA. In the first experiments in this direction, two entire domains (domains I and III), neither of which is conserved in the mitochondrial ribosome, were individually deleted for the *T. aquaticus* 23S rRNA. The deletions were engineered in the cloned 23S rRNA gene; subunits were reconstituted with the truncated 23S rRNA transcript, 5S rRNA, and large-ribosomal-subunit proteins, and their peptidyltransferase activities were tested under the fragment reaction conditions.

Although deletion of practically the entire domain I (positions 45 to 432) notably reduced the activity of reconstituted particles, subunits assembled with truncated 23S rRNA still displayed approximately 1% of the peptidyltransferase activity of subunits reconstituted with the wild-type 23S rRNA transcript. The clearly detectable activity of the sub-

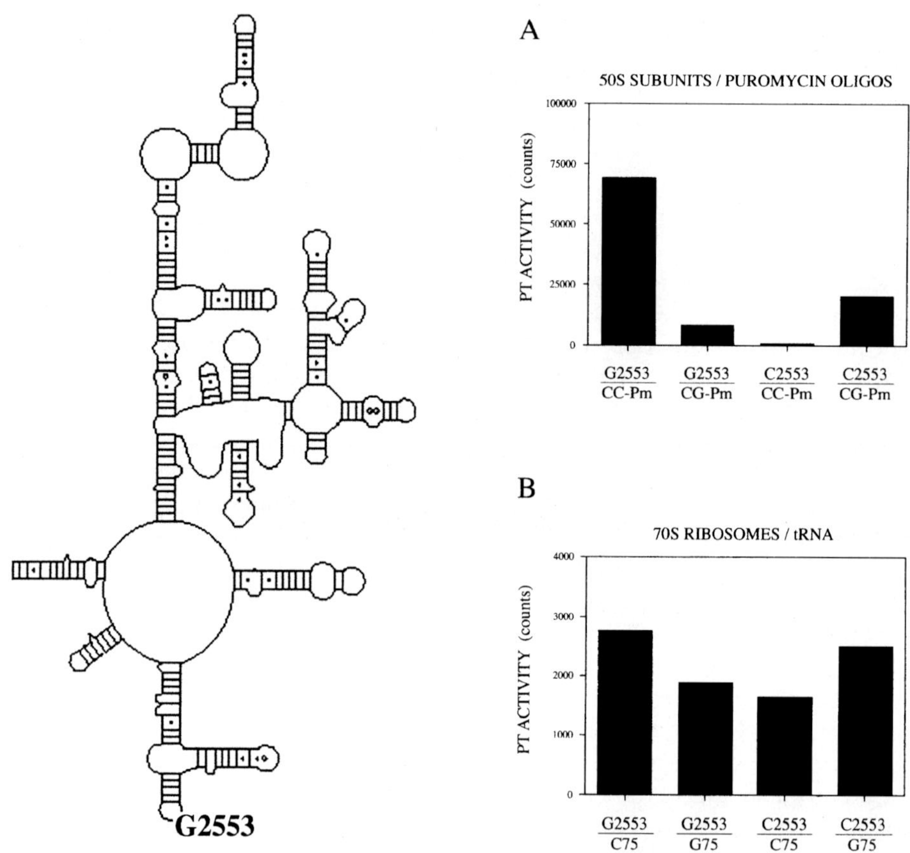

Figure 3. Compensatory effect of G2553C mutation in *T. aquaticus* 23S rRNA and C75G mutation in aminoacyl-tRNA or its analogues. The schematic secondary structure of domain V of 23S rRNA and the position of G2553 are shown on the left. (A) Peptidyltransferase reaction between [^{35}S]fMet-tRNA and puromycin-containing oligonucleotides, CC-Pm or CG-Pm, catalyzed by 50S subunits reconstituted with the wild-type (G2553) or mutant (C2553) 23S rRNA. The reaction product, [^{35}S]fMet-puromycin, was resolved by paper electrophoresis and quantified. (B) Effect of G2553C mutation in 23S rRNA on peptidyltransferase reaction between [^{35}S]fMet-tRNA and either wild-type (C75) or mutant (G75) Val-tRNA (Liu et al., 1998). The reaction product, [^{35}S]fMet-Val dipeptide, was resolved by thin-layer chromatography. In both experiments, the amount of radioactivity in the reaction product, [^{35}S]fMet-puromycin or [^{35}S]fMet-Val, was quantified after a 10-h exposure on a beta-scanner.

units assembled with truncated 23S rRNA confirmed the assumption that extended segments of 23S rRNA could be removed without abolishing the catalytic activity of 50S subunits. The binding site of the ribosomal protein L4, one of the three proteins that may be essential for peptidyltransferase activity, has been mapped within domain I of 23S rRNA (Østergaard et al., 1998). Therefore, although the protein composition of the particles assembled with the truncated 23S rRNA has yet to be analyzed, we expect that the particles may be lacking protein L4. This would confirm our earlier notion that the presence of protein L4 may not be critical for the catalysis of peptide bond formation (Khaitovich et al., 1999a).

Deletion of domain III resulted in a complete loss of peptidyltransferase activity. Sucrose gradient analysis showed that deletion of domain III interferes dramatically with the assembly of compact ribonucleoprotein particles.

In these initial experiments with truncated 23S rRNA, we used the reconstitution procedure developed previously for the assembly of subunits with the complete 23S rRNA transcript. We hope that the compactness and activity of the subunits assembled with truncated 23S rRNA can be improved by adjusting the reconstitution conditions and/or introducing a limited number of mutations into the truncated 23S rRNA, which can possibly compensate for the missing rRNA segments by establishing new or alternative intramolecular interactions.

Possible Role of 5S rRNA in Assembly of the Catalytic Center

The role of 5S rRNA in the ribosome remains a mystery. The conservation of its structure and its presence in ribosomes of practically all known organisms, with the possible exception of fungi and animal mitochondria (Lund and Dahlberg, 1998; Dallas et al., 1995), suggest the importance of 5S rRNA for the ribosome. Omission of 5S rRNA during reconstitution of *E. coli* 50S subunits did not, however, eliminate completely any of the functions, even though peptidyltransferase activity decreased to approximately 10% of its original level (Dohme and Nierhaus, 1976). Similarly, *B. stearothermophilus* 50S subunits reconstituted without 5S rRNA showed only a twofold reduction in catalytic activity (Green and Noller, 1999). By contrast, the omission of 5S rRNA during reconstitution of *T. aquaticus* 50S subunits with either natural or in vitro-transcribed 23S rRNA had a significantly more severe effect on the peptidyltransferase activity, which was reduced to the level of no more than 0.03% of that of native 50S subunits (Khaitovich and Mankin, 1999). This dramatic dependence of the peptidyltransferase activity of *T. aquaticus* 50S subunits on the presence of 5S rRNA during subunit reconstitution allowed us to test the possible function of 5S rRNA in the ribosome.

We found that the lack of 5S rRNA during subunit reconstitution could be partially compensated for by several antibiotics. The presence of several macrolide, ketolide, or streptogramin B-type drugs during subunit reconstitution can restore to some extent the activity of 50S subunits assembled without 5S rRNA (Fig. 4C). With the "best" drugs, for example, macrolide RU 69874, activity of the in vitro-assembled 5S rRNA-deficient 50S subunits of *T. aquaticus* could be increased more than 30-fold. Antibiotics apparently stimulated the assembly of catalytically active subunits rather than peptidyltransferase activity per se, because addition of the drugs after the completion of subunit reconstitution did not increase the activity of the assembled particles (Fig. 4B and C).

The binding site of the 5S rRNA-protein complex was mapped within the segment 2250 to 2410 of 23S rRNA near the central loop of domain V (Østergaard et al., 1998) (Fig. 4A). At the same time, cross-linking experiments showed the proximity of U89 of 5S rRNA to both the central loop of domain V (position C2475) and several regions in domain II (positions 958, 1022, and 1138) of 23S rRNA (Dokudovskaya et al., 1996; Sergiev et al., 1998). In their turn, antibiotics which stimulate assembly of functional 5S rRNA-deficient subunits also interact simultaneously with the central loop of domain V and the loop of helix 35 in domain II (Fig. 4) (Moazed and Noller, 1987; Xiong et al., 1999; Hansen et al., 1999). Because simultaneous interaction with domains II and V of 23S rRNA is the only apparent feature that small antibiotic molecules and 5S rRNA have in common, it is likely that such interaction is essential for assembly of catalytically active *T. aquaticus* 50S subunits.

The RNA component of the peptidyltransferase center is composed of RNA segments scattered in the 23S rRNA primary structure. While some rRNA sequences are brought together in the rRNA secondary structure, other elements are brought into proximity with the catalytic center in the tertiary structure of the 50S subunit. Segments of domain II appear to be related structurally and functionally to the peptidyltransferase. For instance, mutations in domain II affect ribosome sensitivity to the peptidyltransferase-targeted antibiotic amicetin, while the amicetin resistance mutation in domain V influences the sensitivity of the cell to thiostrepton, which binds at the GTPase center in domain II (Mankin et al., 1994). The segment 739 to 748 of helix 35 in domain II could be UV-cross-linked to sequence 2609 to 2618 in the central loop of domain V (Brimacombe et al., 1995), and covariation analysis suggests the existence of triple-base interaction involving Ψ746 in helix 35 and the G2057-C2611 base pair in the central loop of domain V (Gutell, 1996). In addition, production in vivo of a short RNA fragment complementary to the helix 35 loop affects translation termination, a ribosomal function associated with the activity of the peptidyltransferase center (Chernyaeva et al., 1999).

Considering the possible involvement of segments of domain II in the formation or even the activity of the ribosomal peptidyltransferase center, it is conceivable that proper interpositioning of domains II and V is essential for the assembly of the catalytically active large ribosomal subunit. Interaction of 5S rRNA with domains II and V can assist in aligning their functional segments in the ribosome catalytic center. Similar to antibiotics, the presence of 5S rRNA is known to be important during the last steps of in vitro subunit assembly, when most ribosomal proteins are already attached to the RNA and a compact particle (as well as probably individual ribonucleoprotein domains) is formed (Dohme and Nierhaus, 1976). Accordingly, 5S rRNA is likely to facilitate interpositioning of the preassembled ribonucleoprotein domains. This chaperone-like activity may reflect at least one of the functions 5S rRNA plays in the ribosomes.

Given the scope of this chapter, it is important to emphasize that due to the very low activity of 50S subunits reconstituted without 5S rRNA, these experiments were possible only with the use of the in vitro 23S rRNA transcript. All the preparations of natural 23S rRNA would almost inevitably be contaminated with small amounts of 5S rRNA, which would not allow measurement of either the activity

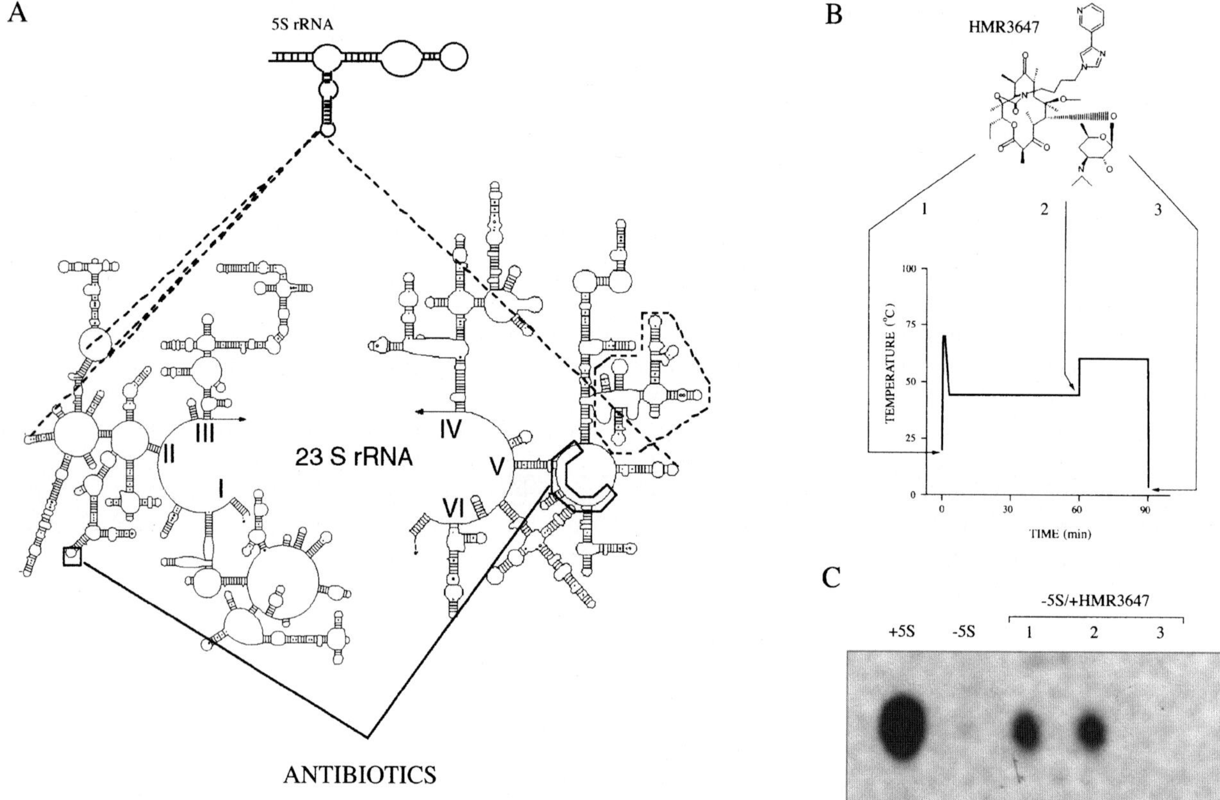

Figure 4. Effect of 5S rRNA and the ketolide antibiotic HMR3647 on the reconstitution of catalytically active *T. aquaticus* large ribosomal subunits. (A) Sites of interactions of 5S rRNA and some macrolide, ketolide, and streptogramin B-type antibiotics with 23S rRNA. The regions boxed by solid lines in domains II and V show the sites of occurrence of antibiotic resistance mutations and/or rRNA positions whose accessibilities to chemical modifications are affected by the antibiotics. The region in domain V boxed by a dashed line shows the locations of 23S rRNA positions protected from chemical modification upon binding of the 5S rRNA-protein complex. 5S rRNA-23S rRNA cross-links are shown by dashed lines (see the text for references). (B and C) Effect of the presence of a ketolide, HMR3647, during the reconstitution procedure on the peptidyltransferase activity of *T. aquaticus* 50S subunits assembled without 5S rRNA. The spots in panel C represent the product of the peptidyltransferase reaction, [^{35}S]fMet-puromycin. The drug was added before the first incubation step (lane 1), before the third incubation step (lane 2), or after completion of reconstitution (lane 3), as also shown in panel B. +, present; −, absent.

of the "truly" 5S rRNA-free subunits or the effects of antibiotics on subunit assembly.

Future Directions

The conditions for reconstitution of functional 50S subunits with in vitro-transcribed 23S rRNA have been developed only recently, and therefore, there has not been enough time to explore the full potential of this system. The examples discussed in this chapter outline some directions for research on the ribosomal peptidyltransferase center, such as minimization of the RNA component, investigation of the functions of individual nucleotides in peptidyltransferase activity, and studies of the contributions of other ribosomal components, for example, 5S rRNA, to the assembly or function of the catalytic center. A number of experiments which are just being started or are still being contemplated have not been discussed. Among them are experiments which should verify whether ribosomal RNA can catalyze formation of the peptide bond without the participation of ribosomal proteins. It is also clear that peptidyltransferase is not the only functional activity of the large ribosomal subunit that can be studied by using synthetic 23S rRNA incorporated into the functional subunits. Studies of possible intra- and intermolecular interactions involving 23S rRNA should be facilitated significantly by the new reconstitution techniques. These include RNA-RNA contacts required for maintaining large-ribosomal-subunit architecture, participation of 23S rRNA in the binding of translation factors, and rRNA interaction with the growing polypeptide chain, with ribosomal proteins, with the small ribosomal subunit, and with other cellular components.

Finally, the lack of posttranscriptional modifications in the 23S rRNA transcript makes it an excellent template for reverse transcription: the full-length cDNA copy of the entire 23S rRNA transcript can be easily produced, PCR amplified, and then used as a template for production of new generations of 23S rRNA transcripts. With the availability of the appropriate selection methods, and with the randomly mutagenized 23S rRNA gene as a starting pool, this amplification method would make possible in vitro selection of specific rRNA mutations. Mutations that improve the activity of the assembled subunit, compensate for the lack of specific rRNA segments or ribosomal proteins, or simply confer resistance to new antibiotics could be isolated by such a SELEX-like in vitro approach. The selection methods which would make such an approach possible are now being developed in several laboratories.

This work was supported by the National Institutes of Health (grant GM53762 from the National Institute of General Medical Sciences and grant TW00870 from the J. E. Fogarty International Center) and a research grant from Hoechst-Marion-Roussel.

We are indebted to our colleagues, Liqun Xiong, Patricia Kloss, Julianna Oh, Tobin Chettiath, and Tanel Tenson for their assistance. We express appreciation to Jack Horowitz for providing tRNA mutants. We also thank Marynel Ryan for help in preparing the manuscript.

REFERENCES

Asai, T., D. Zaporojets, C. Squires, and C. L. Squires. 1999. An *Escherichia coli* strain with all chromosomal rRNA operons inactivated: complete exchange of rRNA genes between bacteria. *Proc. Natl. Acad. Sci. USA* **96:**1971–1976.

Barta, A., and I. Halama. 1996. The elusive peptidyl transferase—RNA or protein? p. 35–54. *In* R. Green and R. Schroeder (ed.), *Ribosomal RNA and Group I Introns.* R. G. Landes Company, Austin, Tex.

Baughman, G. A., and S. R. Fahnestock. 1979. Chloramphenicol resistance mutation in *Escherichia coli* which maps in the major ribosomal protein gene cluster. *J. Bacteriol.* **137:**1315–1323.

Bernabeu, C., P. Conde, D. Vazquez, and J. P. Ballesta. 1979. Peptidyl transferase of bacterial ribosome: resistance to proteinase K. *Eur. J. Biochem.* **93:**527–533.

Bocchetta, M., L. Q. Xiong, and A. S. Mankin. 1998. 23S rRNA positions essential for tRNA binding in ribosomal functional sites. *Proc. Natl. Acad. Sci. USA* **95:**3525–3530.

Böck, A., F. Turnowsky, and G. Hogenauer. 1982. Tiamulin resistance mutations in *Escherichia coli. J. Bacteriol.* **151:**1253–1260.

Brimacombe, R., P. Mitchell, M. Osswald, K. Stade, and D. Bochkariov. 1993. Clustering of modified nucleotides at the functional center of bacterial ribosomal RNA. *FASEB J.* **7:**161–167.

Brimacombe, R., P. Mitchell, and F. Müller. 1995. The organization of rRNA, tRNA, and mRNA in the ribosome, p. 129–147. *In* R. A. Zimmermann and A. E. Dahlberg (ed.), *Ribosomal RNA: Structure, Evolution, Processing, and Function in Protein Biosynthesis.* CRC Press, Boca Raton, Fla.

Chernyaeva, N. S., E. J. Murgola, and A. S. Mankin. 1999. Suppression of nonsense mutations induced by expression of an RNA complementary to a conserved segment of 23S rRNA. *J. Bacteriol.* **181:**5257–5262.

Chittum, H. S., and W. S. Champney. 1994. Ribosomal protein gene sequence changes in erythromycin-resistant mutants of *Escherichia coli. J. Bacteriol.* **176:**6192–6198.

Cooperman, B. S., T. Wooten, D. P. Romero, and R. R. Traut. 1995. Histidine 229 in protein L2 is apparently essential for 50S peptidyl transferase activity. *Biochem. Cell Biol.* **73:**1087–1094.

Crick, F. H. C. 1968. The origin of the genetic code. *J. Mol. Biol.* **38:**367–379.

Dabbs, E. R. 1991. Mutants lacking individual ribosomal proteins as a tool to investigate ribosomal properties. *Biochimie* **73:**639–645.

Dallas, A., R. Rycyna, and P. Moore. 1995. A proposal for the conformation of loop E in *Escherichia coli* 5S rRNA. *Biochem. Cell Biol.* **73:**887–897.

Dohme, F., and K. H. Nierhaus. 1976. Role of 5S RNA in the assembly and function of the 50S subunit from *Escherichia coli. Proc. Natl. Acad. Sci. USA* **73:**2221–2225.

Dokudovskaya, S., O. Dontsova, O. Shpanchenko, A. Bogdanov, and R. Brimacombe. 1996. Loop IV of 5S ribosomal RNA has contacts both to domain II and to domain V of the 23S RNA. *RNA* **2:**146–152.

Dontsova, O., V. Tishkov, S. Dokudovskaya, A. Bogdanov, T. Döring, J. Rinke-Appel, S. Thamm, B. Greuer, and R. Brimacombe. 1994. Stem-loop IV of 5S rRNA lies close to the peptidyltransferase center. *Proc. Natl. Acad. Sci. USA* **91:**4125–4129.

Draper, D. E., and L. P. Reynaldo. 1999. RNA binding strategies of ribosomal proteins. *Nucleic Acids Res.* **27:**381–388.

Franceschi, F. J., and K. H. Nierhaus. 1990. Ribosomal proteins L15 and L16 are mere late assembly proteins of the large ribosomal subunit. Analysis of an *Escherichia coli* mutant lacking L15. *J. Biol. Chem.* **265:**16676–16682.

Garrett, R. A., and C. Rodriguez-Fonseca. 1996. The peptidyl transferase center, p. 327–355. *In* R. A. Zimmermann and A. E. Dahlberg (ed.), *Ribosomal RNA: Structure, Evolution, Processing, and Function in Protein Biosynthesis.* CRC Press, Boca Raton, Fla.

Glotz, C., C. Zwieb, R. Brimacombe, K. Edwards, and H. Kossel. 1981. Secondary structure of the large subunit ribosomal RNA from *Escherichia coli, Zea mays* chloroplast, and human and mouse mitochondrial ribosomes. *Nucleic Acids Res.* **9:**3287–3306.

Green, R. Personal communication.

Green, R., and H. F. Noller. 1996. In vitro complementation analysis localizes 23S rRNA postranscriptional modifications that are required for *Escherichia coli* 50S ribosomal subunit assembly and function. *RNA* **2:**1011–1021.

Green, R., and H. F. Noller. 1997. Ribosomes and translation. *Annu. Rev. Biochem.* **66:**679–716.

Green, R., and H. F. Noller. 1999. Reconstitution of functional 50S ribosomes from in vitro transcripts of *Bacillus stearothermophilus* 23S rRNA. *Biochemistry* **38:**1772–1779.

Green, R., C. Switzer, and H. F. Noller. 1998. Ribosome-catalyzed peptide-bond formation with an A-site substrate covalently linked to 23S rribosomal RNA. *Science* **280:**286–289.

Gregory, S. T., and A. E. Dahlberg. 1999. Erythromycin resistance mutations in ribosomal proteins L22 and L4 perturb the higher order structure of 23 S ribosomal RNA. *J. Mol. Biol.* **289:**827–834.

Gustafsson, C., and B. C. Persson. 1998. Identification of the *rrmA* gene encoding the 23S rRNA m^1G745 methyltransferase in

Escherichia coli and characterization of an m[1]G745-deficient mutant. *J. Bacteriol.* **180**:359–365.

Gutell, R. R. 1996. Comparative sequence analysis and the structure of 16S and 23S rRNA, p. 111–128. *In* R. A. Zimmermann and A. E. Dahlberg (ed.), *Ribosomal RNA: Structure, Evolution, Processing, and Function in Protein Biosynthesis.* CRC Press, Boca Raton, Fla.

Hampl, H., H. Schulze, and K. H. Nierhaus. 1981. Ribosomal components from *Escherichia coli* 50S subunits involved in the reconstitution of peptidyltransferase activity. *J. Biol. Chem.* **256**: 2284–2288.

Hansen, L. H., P. Mauvais, and S. Douthwaite. 1999. The macrolide-ketolide antibiotic binding site is formed by structures in domains II and V of 23S ribosomal RNA. *Mol. Microbiol.* **31**:623–632.

Hummel, H., W. Piepersberg, and A. Böck. 1979. Analysis of lincomycin resistance mutations in *Escherichia coli. Mol. Gen. Genet.* **169**:345–347.

Khaitovich, P., and A. S. Mankin. 1999. Effect of antibiotics on large ribosomal subunit assembly reveals possible function of 5S rRNA. *J. Mol. Biol.* **291**:1025–1034.

Khaitovich, P., A. S. Mankin, R. Green, L. Lancaster, and H. F. Noller. 1999a. Characterization of functionally active subribosomal particles from *Thermus aquaticus. Proc. Natl. Acad. Sci. USA* **96**:85–90.

Khaitovich, P., T. Tenson, P. Kloss, and A. S. Mankin. 1999b. Reconstitution of functionally active *Thermus aquaticus* large ribosomal subunity with in vitro-transcribed rRNA. *Biochemistry* **38**:1780–1788.

Khaitovich, P., T. Tenson, A. S. Mankin, and R. Green. 1999c. Peptidyl transferase activity catalyzed by protein-free 23S ribosomal RNA remains elusive. *RNA* **5**:605–608.

Krzyzosiak, W., R. Denman, K. Nurse, W. Hellmann, M. Boublik, C. W. Gehrke, P. F. Agris, and J. Ofengand. 1987. In vitro synthesis of 16S ribosomal RNA containing single base changes and assembly into a functional 30S ribosome. *Biochemistry* **26**:2353–2364.

Lazaro, E., L. A. G. M. van den Broek, A. S. Felix, H. C. J. Ottenheijm, and J. P. G. Ballesta. 1991. Chemical, biochemical and genetic endeavours characterizing the interaction of sparsomycin with the ribosome. *Biochimie* **73**:1137–1143.

Le Goffic, F., M. L. Capmau, L. Chausson, and D. Bonnet. 1980. Photo-induced affinity labeling of *Escherichia coli* ribosomes by chloramphenicol. *Eur. J. Biochem.* **106**:667–674.

Leviev, I., S. Levieva, and R. A. Garrett. 1995. Role for the highly conserved region of domain IV of 23S-like rRNA in subunit-subunit interactions at the peptidyl transferase centre. *Nucleic Acids Res.* **23**:1512–1517.

Liu, J. C. H., M. Liu, and J. Horowitz. 1998. Recognition of the universally conserved 3′-CCA end of tRNA by elongation factor EF-Tu. *RNA* **4**:639–646.

Lund, E., and J. E. Dahlberg. 1998. Proofreading and aminoacylation of tRNAs before export from the nucleus. *Science* **282**: 2082–2085.

Maimets, T., J. Remme, and R. Villems. 1984. Ribosomal protein L16 binds to the 3′ end of transfer RNA. *FEBS Lett.* **153**:53–56.

Mankin, A. S. 1995. Selection for spontaneous and engineered mutations in the rRNA genes in halophilic archaea, p. 209–216. *In* F. T. Robb, A. R. Place, K. R. Sowers, H. J. Schreier, S. DasSarma, and E. M. Fleischmann (ed.), *Archaea, a Laboratory Manual.* Cold Spring Harbor Laboratory Press, Cold Spring Harbor, N.Y.

Mankin, A. S., and A. M. Kopylov. 1981. A secondary structure of the mitochondrial 12S rRNA as an example of rRNA economy. *Biochem. Int.* **3**:587–593.

Mankin, A. S., I. Leviev, and R. A. Garrett. 1994. Cross-hypersensitivity effects of mutations in 23 S rRNA yield insight into aminoacyl-tRNA binding. *J. Mol. Biol.* **244**:151–157.

Melancon, P., D. Leclerc, N. Destroismaisons, and L. Brakier-Gingras. 1990. The anti-Shine-Dalgarno region in *Escherichia coli* 16S ribosomal RNA is not essential for the correct selection of translational starts. *Biochemistry* **29**:3402–3407.

Moazed, D., and H. F. Noller. 1987. Chloramphenicol, erythromycin, carbomycin and vernamycin B protect overlapping sites in the peptidyl transferase region of 23S ribosomal RNA. *Biochimie* **69**:879–884.

Moazed, D., and H. F. Noller, 1989. Interaction of tRNA with 23S rRNA in the ribosomal A, P, and E sites. *Cell* **57**:585–597.

Moine, H., K. Nurse, B. Ehresmann, C. Ehresmann, and J. Ofengand. 1997. Conformational analysis of *Escherichia coli* 30S ribosomes containing the single-base mutations G530U, U1498G, G1401C, and C1501G and the double-base mutation G1401C/C1501G. *Biochemistry* **36**:13700–13709.

Monro, R. E. 1967. Catalysis of peptide bond formation by 50 S ribosomal subunits from *Escherichia coli. J. Mol. Biol.* **26**:147–151.

Moore, V. G., R. E. Atchison, G. Thomas, M. Moran, and H. F. Noller. 1975. Identification of a ribosomal protein essential for peptidyl transferase activity. *Proc. Natl. Acad. Sci. USA* **72**:844–848.

Nierhaus, K. H., and F. Dohme. 1974. Total reconstitution of functionally active 50S ribosomal subunits from *Escherichia coli. Proc. Natl. Acad. Sci. USA* **71**:4713–4717.

Nitta, I., Y. Kamada, H. Noda, T. Ueda, and K. Watanabe. 1998a Reconstitution of peptide bond formation with *Escherichia coli* 23S ribosomal RNA domains. *Science* **281**:666–669.

Nitta, I., T. Ueda, and K. Watanabe. 1998b. Possible involvement of *Escherichia coli* 23S ribosomal RNA in peptide bond formation. *RNA* **4**:257–267.

Noller, H. F. 1993. Peptidyl transferase: protein, ribonucleoprotein, or RNA? *J. Bacteriol.* **175**:5297–5300.

Noller, H. F., V. Hoffarth, and L. Zimniak. 1992. Unusual resistance of peptidyl transferase to protein extraction procedures. *Science* **256**:1416–1419.

Nomura, M., and V. A. Erdmann. 1970. Reconstitution of 50S ribosomal subunits from dissociated molecular components. *Nature* **228**:744–748.

O'Connor, M., N. M. Willis, L. Bossi, R. F. Gesteland, and J. F. Atkins. 1993. Functional tRNAs with altered 3′ ends. *EMBO J.* **12**:2559–2566.

Ohkubo, S., A. Muto, Y. Kawauchi, F. Yamao, and S. Osawa. 1987. The ribosomal protein gene cluster of *Mycoplasma capricolum. Mol. Gen. Genet.* **210**:314–322.

Østergaard, P., H. Phan, L. B. Johansen, J. Egebjerg, L. Ostergaard, B. T. Porse, and R. A. Garrett. 1998. Assembly of proteins and 5 S rRNA to transcripts of the major structural domains of 23 S rRNA. *J. Mol. Biol.* **284**:227–240.

Pan, C., and T. L. Mason. 1997. Functional analysis of ribosomal protein L2 in yeast mitochondia. *J. Biol. Chem.* **272**:8165–8171.

Porse, B. T., H. P. Thi-Ngoc, and R. A. Garrett. 1996. The donor substrate site within the peptidyl transferase loop of 23 S rRNA and its putative interactions with the CCA-end of N-blocked aminoacyl-tRNA^{Phe}. *J. Mol. Biol.* **264**:472–483.

Porse, B. T., S. V. Kirillov, M. J. Awayez, and R. A. Garrett. 1999. UV-induced modifications in the peptidyl transferase loop of 23S rRNA dependent on binding of the streptogramin B antibiotic, pristinamycin IA. *RNA* **5**:585–595.

Raychaudhuri, S., J. Conrad, B. G. Hall, and J. Ofengand. 1998. A pseudouridine synthase required for the formation of two universally conserved pseudouridines in ribosomal RNA is essential for normal growth of *Escherichia coli. RNA* **4**:1407–1417.

Remme, J., E. Metspalu, T. Maimets, and R. Villems. 1985. The properties of the complex between ribosomal protein L2 and tRNA. *FEBS Lett.* **190:**275–278.

Rozenski, J., P. F. Crain, and J. A. McCloskey. 1999. The RNA modification database: 1999 update. *Nucleic Acids Res.* **27:**196–197.

Saarma, U., and J. Remme. 1992. Novel mutants of 23S rRNA: characterization of functional properties. *Nucleic Acids Res.* **20:** 3147–3152.

Samaha, R. R., R. Green, and H. F. Noller. 1995. A base pair between tRNA and 23S rRNA in the peptidyl transferase centre of the ribosome. *Nature* **377:**309–314.

Sergiev, P., S. Dokudovskaya, E. Romanova, A. Topin, A. Bogdanov, R. Brimacombe, and O. Dontsova. 1998. The environment of 5S rRNA in the ribosome: cross-links to the GTPase-associated area of 23S rRNA. *Nucleic Acids Res.* **26:** 2519–2525.

Sirum-Connolly, K., and T. L. Mason. 1993. Functional requirement of a site-specific ribose methylation in ribosomal RNA. *Science* **262:**1886–1889.

Sirum-Connolly, K., J. M. Peltier, P. F. Crain, J. A. McCloskey, and T. L. Mason. 1995. Implications of a functional large ribosomal RNA with only three modified nucleotides. *Biochimie* **77:**30–39.

Stöffler-Meilicke, M., and G. Stöffler. 1990. Topography of the ribosomal proteins from *Escherichia coli* within the intact subunits as determined by immunoelectron microscopy and protein-protein cross-linking, p. 123–133. *In* W. E. Hill, A. Dahlberg, R. A. Garrett, P. B. Moore, D. Schlessinger, and J. R. Warner (ed.), *The Ribosome: Structure, Function, and Evolution.* American Society for Microbiology, Washington, D.C.

Triman, K. L., A. Peister, and R. A. Goel. 1998. Expanded versions of the 16S and 23S ribosomal RNA mutation databases (16SMDBexp and 23SMDBexp). *Nucleic Acids Res.* **26:**280–284.

Uhlein, M., W. Weglohner, H. Urlaub, and B. Wittmann-Liebold. 1998. Functional implications of ribosomal protein L2 in protein biosynthesis as shown by in vivo replacement studies. *Biochem. J.* **331:**423–430.

Walleczek, J., D. Schüler, M. Stöffler-Meilicke, R. Brimacombe, and G. Stöffler. 1988. A model for the spatial arrangement of the proteins in the large subunit of the *Escherichia coli* ribosome. *EMBO J.* **7:**3571–3576.

Weitzmann, C. J., P. R. Cunningham, and J. Ofengand. 1990. Cloning, in vitro transcription, and biological activity of *Escherichia coli* 23S ribosomal RNA. *Nucleic Acids Res.* **18:**3515–3520.

Wittmann-Liebold, B., A. K. E. Köpke, E. Arndt, W. Krömer, T. Hatakeyama, and H.-G. Wittmann. 1990. Sequence comparison and evolution of ribosomal proteins and their genes, p. 598–616. *In* W. E. Hill, A Dahlberg, R. A. Garrett, P. B. Moore, D. Schlessinger, and J. R. Warner (ed.), *The Ribosome: Structure, Function, and Evolution.* American Society for Microbiology, Washington, D.C.

Wower, J., I. K. Wower, S. V. Kirillov, K. V. Rosen, S. S. Hixson, and R. A. Zimmermann. 1995. Peptidyl transferase and beyond. *Biochem. Cell Biol.* **73:**1041–1047.

Xiong, L., S. Shah, P. Mauvais, and A. S. Mankin. 1999. A ketolide resistance mutation in domain II of 23S rRNA reveals proximity of hairpin 35 to the peptidyl transferase centre. *Mol. Microbiol.* **31:**633–639.

Zhang, B., and T. R. Cech. 1997. Peptide bond formation by in vitro selected ribozymes. *Nature* **390:**96–100.

The Ribosome: Structure, Function, Antibiotics, and Cellular Interactions
Edited by R. A. Garrett, S. R. Douthwaite, A. Liljas, A. T. Matheson, P. B. Moore, and H. F. Noller

Chapter 21

Modern Site-Directed Cross-Linking Approaches: Implication for Ribosome Structure and Functions

ALEXEY A. BOGDANOV, PETR V. SERGIEV, INNA N. LAVRIK, OLGA V. SPANCHENKO, ANDREY A. LEONOV, and OLGA A. DONTSOVA

Recent impressive advances in cryo-electron microscopy (Stark et al., 1997; Agrawal et al., 1999) and X-ray crystallographic analysis of ribosomes (Ban et al., 1998) raise hopes that their structure will be solved at atomic resolution in the near future. The completion of this extremely important task, however, appears to be dependent to a great extent on biochemical data that provide reliable information on the mutual arrangement of specific ribosomal components at nucleotide and amino acid resolution. Moreover, protein biosynthesis is a dynamic process, and one can predict that with the appearance of a static atomic structure of the ribosome, the biochemical approaches that allow conformational alterations to be followed in macromolecular complexes in solution will dominate in ribosomology. The introduction of new efficient photoreagents and the development of appropriate techniques for site-directed incorporation of photoprobes into ribosomes and their ligands, as well as substantial improvement in the methods for the determination of exact positions of cross-linked monomers in both RNA and proteins, have made cross-linking data particularly important.

The purpose of this chapter is to review modern trends in the development of cross-linking approaches and their implication for studies of rRNA folding in the ribosome and the organization of ribosomal functional centers. Several excellent reviews of photoaffinity labeling of ribosomes have been written in the past (Cooperman et al., 1988; Barta et al., 1990; Wower et al., 1993). Also, the database of ribosomal cross-links is available (Baranov et al., 1998b). Here we emphasize more recent progress in this field. We will illustrate the results emanating from the ribosomal cross-linking studies with data taken from our work on the investigation of the environment of mRNA and 5S rRNA in the ribosome.

THE CHEMISTRY

The application of a combination of short-range (also often called zero-length) and long-range cross-linking reagents has proved to be a very fruitful strategy for investigation of the topography of RNA and proteins in the ribosome. In this section we will briefly describe the photoreactive compounds used in most cross-linking experiments with ribosomes. Several reviews can be consulted for more details concerning the photochemistry, chemical structure, and synthesis of these compounds (Favre, 1990; Wower et al., 1994; Fleming, 1995; Meisenheimer and Koch, 1997; Kotziba-Hibert et al., 1995).

Zero-Length Cross-Linking Reagents

RNA-RNA and RNA-protein photo-cross-linking by means of zero-length reagents gives particularly valuable information about intraribosomal interactions, since in this case ribosomal components can only be covalently bound if their nucleotide and/or amino acid residues are in direct van der Waals contact.

In principle, the most easily available zero-length cross-linking agent is UV light with a wavelength near 260 nm. However, even short irradiation time cannot exclude unpredictable damage to ribosomes irradiated at wavelengths of <300 nm. To avoid this prob-

Alexey A. Bogdanov, Petr V. Sergiev, Inna N. Lavrik, Olga V. Spanchenko, Andrey A. Leonov, and Olga A. Dontsova ■ Department of Chemistry and Belozersky Institute of Physico-Chemical Biology, Moscow State University, Moscow 119899, Russia.

lem, the application of photoreactive analogues of nucleotide bases was introduced. Zimmermann and coworkers used azidoadenosine as a site-specific photoaffinity label substituting for "normal" adenosine for tRNA photolabeling (reviewed in Wower et al., 1994) (Fig. 1). Upon photoactivation, azidonucleoside residues in RNA are able to cross-link nonspecifically to almost any group within a nucleotide or amino acid residue that appears to be in close contact with them. The serious drawback of azides is their instability in aqueous solution, which leads to a significant decrease in the yield of cross-linking.

Among the zero-length cross-linking reagents, 4-thiouridine (4-thioU) has been used most frequently (Fig. 1). It is a close analogue of uridine which retains an ability for Watson-Crick base-pairing, and being incorporated into an RNA molecule does not perturb its structure. In most cases the mechanism of the 4-thioU cross-linking reactions is unknown. However, in the case of cross-linking of the naturally occurring 4-thioU8 and C13 in tRNA one can suggest that overlapping of the thione group of 4-thioU with the C5-C6 double bond of its pyrimidine partner is necessary for photoaddition (Favre, 1990).

In some studies 4-thiothymidine (4-thio-dT) was efficiently substituted for 4-thioU. An interesting example is the covalent photochemical binding of 4-thio-dTpCp-puromycin to the A-site region of 23S rRNA in the 50S subunit (Green et al., 1998). This compound formed a single cross-linked product (with G2553 of 23S rRNA), and its puromycin moiety retained the ability to participate in the peptidyltransferase reaction.

We have found that the use of 6-thioguanosine (6-thioG) as a photoreactive probe as well as 4-thioU has provided a great deal of information on mRNA-16S rRNA interactions in the *Escherichia coli* ribosome (Sergiev et al., 1997). The application of 6-thioG as a photochemical probe has the same advantages as 4-thioU, except for the lower efficiency of 6-thioG triphosphate as a substrate in the reaction catalyzed by T7 RNA polymerase. The major drawback for both 4-thioU and 6-thioG is their inability to form RNA-RNA cross-links from double-stranded RNA.

Nevertheless, two other classes of photoactive modified nucleotide bases that can help to overcome this difficulty, namely, pyrimidines with 5-bromo- and 5-iodo substitutions, have been widely used for DNA- and RNA-protein zero-length cross-linking (Blatter et al., 1992; Meisenheimer and Koch, 1997, and references therein). 5-bromo- and 5-iodouridine can easily be incorporated into an RNA chain by means of the T7 RNA-polymerase reaction. They can be excited at wavelengths above 300 nm.

Long-Range Cross-Linking Photoreagents

The application of cross-linking reagents that contain a bridge connecting a photosensitive group with an RNA monomer allows the spatial proximity between ribosomal elements that are located at a definite distance from one another to be investigated.

Two types of photosensitive groups, arylazido and aryldiazirino, have been almost universally used in recent cross-linking studies of ribosomes. Irradiation of the reagents containing these groups with UV light induces the release of nitrogen, with concomitant formation of highly reactive nitrenes and carbenes, respectively. Trifluoromethylaryldiazirines (TFMAD) have a particularly promising photoaffinity potential, since they are very stable in the dark (Kotzyba-Hibert et al., 1995). In contrast to the relatively stable nitrenes, the lifetime of carbenes is less than 100 ps. Therefore, TFMAD-containing photoreagents can be used for scanning very fast conformational changes within the ribosome structure. Arylazido and aryldiazirino groups can be introduced into RNA by derivatization of both naturally occurring and synthetic modified nucleotides (Fig. 1B).

THE TECHNIQUES

In principle, both RNA and protein molecules modified with photosensitive reagents can be used for photoaffinity labeling of ribosomes. However, so far, only rRNA, tRNA, and mRNA derivatives have been applied to structural and functional studies of ribosomes, although mutant forms of ribosomal proteins and translational factors, suitable for site-specific modification with photoprobes, were recently obtained (Heilek and Noller, 1996; Wilson and Noller, 1998).

Incorporation of Photosensitive Analogues of Nucleotides into Ribosomes and Their Ligands

Two strategies for the incorporation of photoreagents into RNA molecules have been successfully used in ribosomology: (i) statistical (random) substitution of one of the four types of nucleotides in relatively short RNA molecules (tRNA, 5S rRNA, or synthetic mRNA) with photoreactive analogues and (ii) site-directed incorporation of a cross-linking photoreagent into the RNA molecule.

The statistical approach usually involves the random introduction of modified nucleotides into the RNA chain in the course of a T7 RNA polymerase reaction. After UV irradiation, the position of a cross-linked nucleotide in the RNA molecule that carries

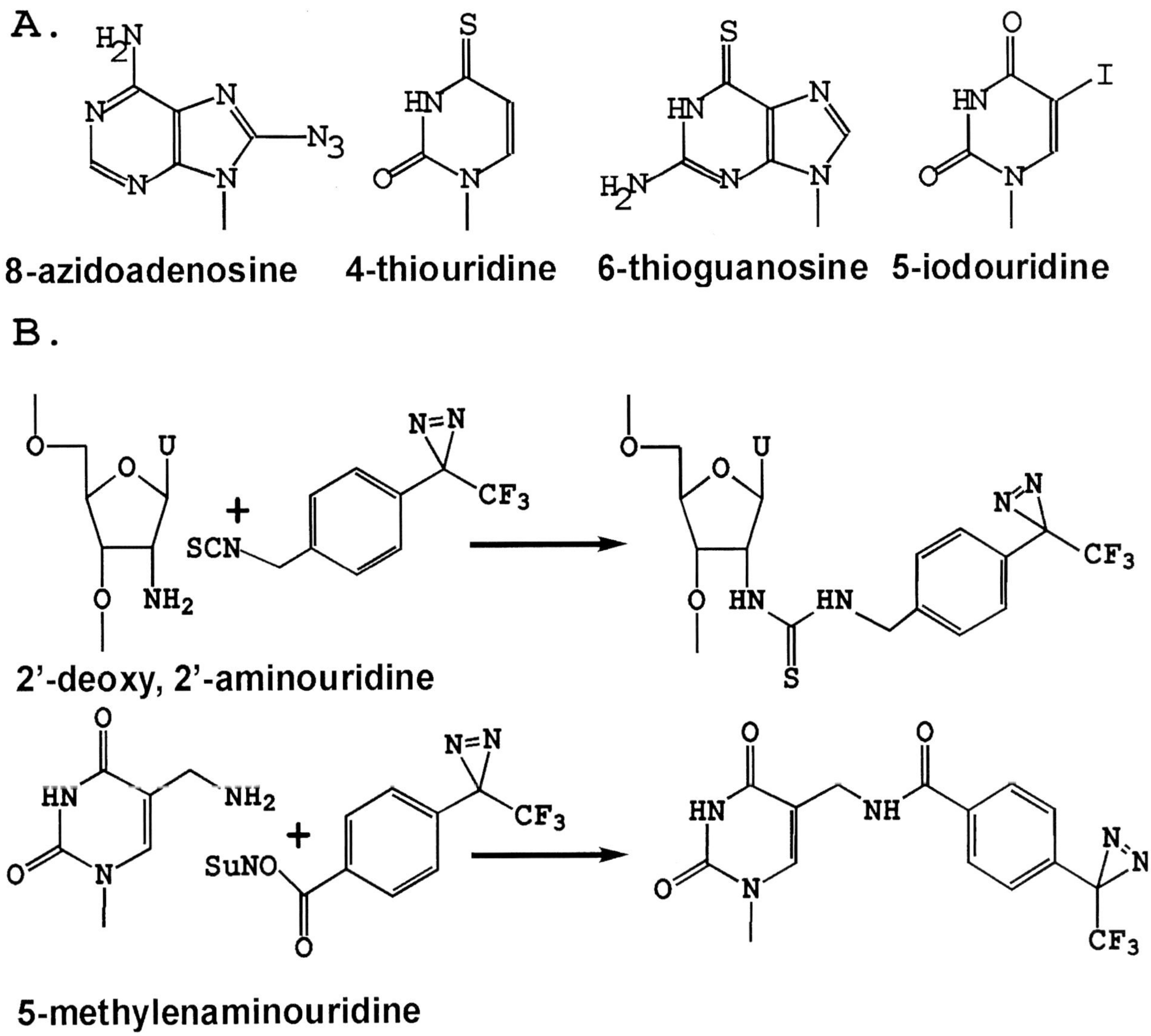

Figure 1. Examples of photoreagents used in ribosome studies. (A) Zero-length reagents. (B) Derivatization of modified uridine residues in the RNA chain with TFMAD.

the photoprobes can be localized at nucleotide resolution. The statistical approaches are relatively simple, but they are not universal. Firstly, the triphosphate of a modified nucleoside has to be a substrate for T7 RNA polymerase, and this is not always the case. Secondly, not every position in an rRNA molecule can have a modified residue substituted without disturbing the molecule's biological activity. Thirdly, in the case of long RNAs, too many cross-links would be formed and localization of the cross-linked nucleotides at mononucleotide resolution is not realistic. For this reason, special approaches for the incorporation of photoreagents into selected sites of an RNA molecule independent of size have been developed.

Photoprobes can be specifically attached to the 5′ or 3′ terminus of an RNA. For 3′-end labeling, the T4 RNA ligase reaction with the 3′,5′-diphosphate of the modified nucleoside [p(4-thio)Up, p(2-azido)Ap, or p(8-azido)Ap, for example] as a donor (Lingner and Keller, 1993) or the DNA polymerase reaction (Huang and Szostak, 1996) has been used.

It has been shown that the use of 4-thioUpG for initiation of the T7 RNA polymerase reaction results in site-specific incorporation of the photolabel at the 5′ terminus of an RNA transcript. This approach can be applied for the preparation of 5′-end-labeled RNA sequences that normally start with a G residue. Another possibility for introducing a photolabel to the 5′ ends of an RNA sequence starting with G is to initiate the T7 RNA polymerase transcription with guanosine monophosporothioate and then to modify this residue with an appropriate bifunctional photoreagent. This reaction, in combination with a circular-permutation technique, was recently used by Montpetit and coworkers to study RNA-RNA tertiary contacts in the 3′ major domain of 16S rRNA (Montpetit et al., 1998).

Several approaches for the site-directed incorporation of photoreactive nucleotide analogues into internal parts of the RNA have also been developed. They can be subdivided into methods for the insertion of photoprobes into (i) intact and (ii) fragmented RNA. Moore and Sharp (1992) proposed a universal approach for the incorporation of 4-thioU into any desired internal position of an RNA chain. The principle is to exploit the ability of T4 DNA ligase to join RNA segments with 3′ hydroxyl and 5′ monophosphate termini within a DNA-RNA heteroduplex. Although the Moore and Sharp technique has been widely used in molecular biology (see Sontheimer, 1994, for a review), its application to the study of the ribosome can be illustrated, so far, with only three examples (Wower et al., 1994; Juzumiene and Wollenzien, 1997; Osswald and Brimacombe, 1999).

One of the most promising innovations in ribosome cross-linking technology is the method recently proposed by Mondus and Wollenzien (1998). It allows the preparation of ribosomes quantitatively derivatized with a photolabel at selected internal sites in the rRNA. The method was successfully applied to the characterization of the environment of the U788-U789 residues of 16S rRNA in *E. coli* 30S subunits.

It has been shown in our laboratory that 30S subunits of *E. coli* ribosomes with their 16S rRNA specifically fragmented with RNase H in complexes with complementary oligodeoxyribonucleotides retain their abilities to associate with 50S subunits and, in some cases, their biological activities (Afonina et al., 1991). Recently, we have substantially improved the specificity of the RNase H cleavage of rRNA by using chimeric oligonucleotides. As a result, rRNA samples fragmented at a single selected phosphodiester bond are now available (Baranov et al., 1997). Both 16S and 23S rRNAs can then be ligated with p(4-thio)Up and used for the reconstitution of ribosomal subunits. In several cases cleavage of rRNA or the incorporation of an additional nucleotide residue causes disruption of ribosomal activity. Therefore, only those subunits that were able to form 70S ribosomes were used in cross-linking experiments (Baranov et al., 1997, 1998a). Identification of the cross-linked nucleotide residues has provided a great deal of useful information about the environment of the modified sites in the rRNA.

Localization of Cross-Linked Residues

For the analysis of cross-links to RNA molecules, three basic techniques are used. The preliminary determination of the cross-linked region of RNA is best made by an oligonucleotide-directed RNase H cleavage (Dontsova et al., 1991). The basis of the method is the excision of a short fragment of RNA (30 to 200 nucleotides) from the larger RNA molecule. If this fragment contains the cross-linked residue, and if the ligand originally carrying the photoreagent has been radioactively labeled, then the excised RNA fragment will be radioactive. Using a suitable set of oligonucleotide pairs, one can scan the whole RNA molecule.

After the excision of the cross-linked RNA fragment, the precise position of the cross-linked residue can be determined by reverse transcription. The reverse transcriptase stops the elongation of a primer at the nucleotide preceding the cross-linked residue (Bhangu and Wollenzien, 1992). It should be noted, however, that reverse transcriptase stops or pauses at a number of sites in the RNA that are not yet well characterized. Thus, the analysis of cross-link sites by reverse transcription alone, without a parallel determination of the cross-linked region by RNase H analysis, can lead to some false-positive results.

If a radioactive label has been inserted into the RNA molecule to be analyzed, the identification of the cross-linked residue can be done by a specific ribonuclease digestion. The resulting oligonucleotides are separated by thin-layer chromatography (Stiege et al., 1982; Dontsova et al., 1991). The absence or the reduced mobility of a certain oligonucleotide will indicate the cross-link position.

The determination of cross-link sites at single-residue precision is now possible for proteins as well. The application of mass spectrometry, particularly matrix-assisted laser desorption-ionization combined with N-terminal microsequencing, has simplified the analysis of the cross-linked amino acids and, most important, made it possible to operate with an extremely small amount of cross-linked material (Thiede et al., 1998).

The application of the new methods for cross-link analysis makes it possible to study RNA-RNA, RNA-protein, and protein-protein contacts inside the ribosome at the level of single nucleotides and/or amino acids. This resolution makes possible the detailed modeling of the tertiary structure of the rRNAs, and known protein structures can also be fitted into the model (Brimacombe, personal communication).

EXAMPLE I: THE *ESCHERICHIA COLI* RIBOSOME DECODING CENTER

The structure of the decoding center, as a part of the ribosome, will probably only be known in detail when an X-ray structure becomes available. However, the biochemical data will remain essential, providing the necessary constraints for fitting the rRNA secondary structure into the electron density maps.

Among the elements of the ribosome structure, the decoding center has been the classical object for cross-linking studies for many years (see Wower et al., 1993, and Bogdanov et al., 1993, for reviews). As might be expected, the majority of the cross-link sites in the 16S rRNA that have been ascribed to the decoding center belong to a set of a few secondary-structure elements, such as helices 18, 22 to 24, 28, 31, 34, and 43 to 45 (Fig. 2). The contributions of these helices to the structure and function of the decoding center correlate with the data obtained by complementary approaches, such as footprinting (Moazed and Noller, 1986) and site-directed mutagenesis (O'Connor et al., 1995). However, the interpretation of the cross-linking data is more straightforward in terms of the three-dimensional structure and can be used directly as a constraint for the modeling of the rRNA.

Binding Site of mRNA Region 5′ to the P Site

Apart from the well-known Shine-Dalgarno (SD) interaction (Shine and Dalgarno, 1975), docking the 5′ part of the mRNA at the stage of initiation, contacts of the 5′ part of the mRNA have been found with nucleotides 1527 to 1533, adjacent to the 16S rRNA anti-SD region (Rinke-Appel et al., 1994). The

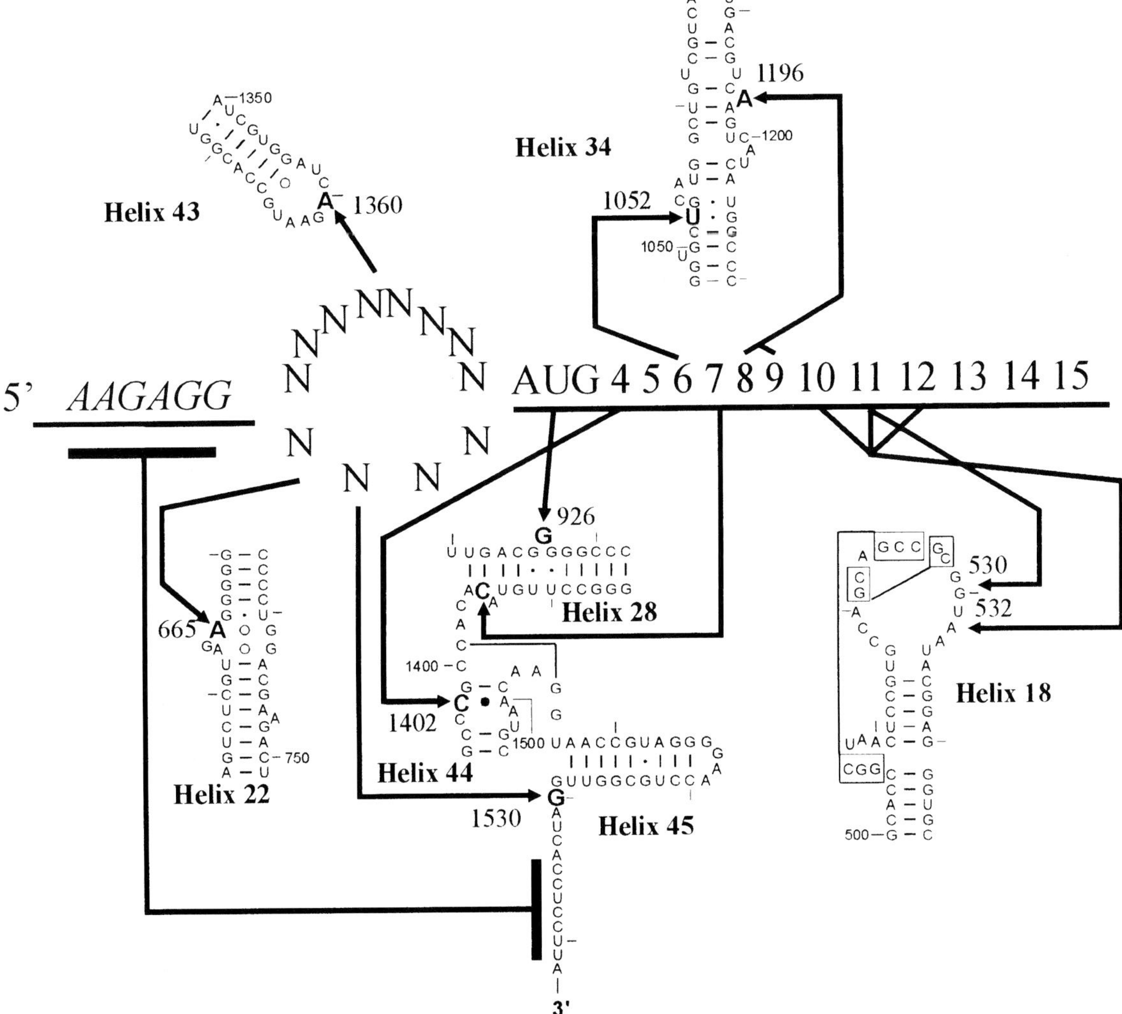

Figure 2. Summary of the cross-linking data on the environment of mRNA in the ribosome. Secondary-structural elements cross-linked to the mRNA are shown. The solid line in the middle corresponds to the mRNA. The position of the SD sequence is indicated by italic letters. Nucleotides within the variable spacer 5′ to the AUG P-site codon are denoted by Ns. The mRNA positions 3′ to the AUG P-site codon are shown by numbers. Position-specific cross-link sites are shown by arrows and boldface.

loop of the neighboring helix 45 has been cross-linked to the large ribosomal subunit (Mitchell et al., 1992). Based on these data, helix 45 most likely provides the bridge of electron density, joining the mRNA binding cleft at the side lobe of the 30S subunit with the 50S subunit (Mueller and Brimacombe, 1997). Contacts between the 5′ part of the mRNA and nucleotides A665 and A1360 of 16S rRNA have also been found in cross-linking studies (Rinke-Appel et al., 1994). The former belongs to helix 22 of 16S rRNA, which is known to be the protein S15 binding site. According to neutron-scattering and immuno-electron microscopy studies, S15 resides low down in the side lobe, away from the mRNA (Stoffler-Meilicke and Stoffler, 1990). The location of helix 22 in the rRNA models is thus dependent on the weight ascribed to the cross-linking data (as in Mueller and Brimacombe, 1997) versus the neutron-scattering data (Malhotra and Harvey, 1994).

Nucleotide 1360, which is also cross-linked to the 5′ part of mRNA, is part of helix 43 located in the head of the 30S subunit. Footprints of both proteins S7 and S19 were found in helix 43 (Powers et al., 1988). The location of the helix in the models is dependent on whether the cross-linking or footprinting data are used for modeling (see Mueller and Brimacombe, 1997, for a discussion).

The P-Site Part of the Decoding Center

Several lines of evidence point to helices 44 and 28 as the structure surrounding the codon-anticodon interactions in the P site. The well-known cross-link from residue 34 of the tRNA to C1400 of the 16S rRNA (Prince et al., 1982), together with footprinting data of tRNA bound to the P site (Moazed and Noller, 1986), proves that the 1400 region is in contact with the anticodon of the P-site-bound tRNA. Nucleotide C1402 was cross-linked to the first residue of the mRNA A-site codon (Rinke-Appel et al., 1993). However, the cross-link was only observed in the absence of A-site-bound tRNA, thus marking the 3′ border of the P site.

The bulged nucleotide G926, part of helix 28, was shown to be involved in tRNA binding to the P site (Moazed and Noller, 1986; von Ahsen and Noller, 1995). However, only the cross-link from a diazirine derivative of uridine located in the AUG codon of the mRNA showed that G926 interacts directly with the major groove of the P-site codon-anticodon duplex at its midpoint (Sergiev et al., 1997). Several other areas of 16S rRNA are known to contribute to the P site. The residues 966, 1338 to 1339, and 1381 interact with the P-site-bound tRNA (Moazed and Noller, 1986; von Ahsen and Noller, 1995). Several cross-links from different positions of tRNA at the A, P, and E sites are clustered around helix 43; however, not all of these data can be explained in terms of a unique structure (see Mueller et al., 1997, for a discussion).

Another cluster of tRNA cross-link sites is located in the central domain of the 16S rRNA in helices 23 and 24 (see Mueller et al., 1997, for a review). As in the case of the 1335 to 1348 region, cross-links from tRNAs in the A, P, and E sites to practically the same 16S rRNA residues were observed. The footprints of both the P-site bound tRNA (nucleotides 693 and 794 to 795) (Moazed and Noller, 1986) and the 50S subunit (702, 790 to 791, and 793) (Merryman et al., 1999) were found in this area. The tips of helices 23 and 24 are known to form a binding site for protein S11 (Stern et al., 1988; Powers and Noller, 1995), located on the edge of the side lobe (Stoffler-Meilicke and Stoffler, 1990). This location favors a contact with the E-site-bound tRNA (Wower et al., 1993) rather than the P or A site. In models of 16S rRNA tertiary structure, helices 23 and 24 are directed either upward from the S15 site to the S11 site on the tip of the side lobe (Malhotra and Harvey, 1994) or downward in the side lobe, thus allowing the neighboring helix 22 to contact the mRNA (Mueller and Brimacombe, 1997).

The A-Site Part of the Decoding Center

Two cross-links from the A-site codon of mRNA have been found. The first base of the codon was cross-linked to residue C1402 of 16S rRNA helix 44 (Rinke-Appel et al., 1993). Nucleotide C1402 is adjacent to the P-site-specific tRNA protections surrounding nucleotide C1400 and also close to the set of A-site-specific tRNA protections clustered around the 1407 to 1493 segment of helix 44 (Moazed and Noller, 1986). In the recent model of the helix 44-mRNA-tRNA complex, this area interacts directly with the codon-anticodon duplex at the A site (VanLoock et al., 1999). However, the direct experimental data do not restrict the interaction to only the anticodon itself but to the whole anticodon stem-loop.

The adjacent nucleotides 1409 to 1418 and 1483 to 1487 were shown to be proximal to the 50S subunit (Mitchell et al., 1992; Merryman et al., 1999). Thus, helix 44 could be a good candidate for a second bridge between the small and large ribosomal subunits. An alternative view is that elements of the 50S subunit, rather than the 30S subunit, form the bridge (Mueller and Brimacombe, 1997).

Nucleotide C1395 was cross-linked to the mRNA base adjacent to the last nucleotide of the A-site codon (Dontsova et al., 1992). It seems rather unlikely that the mRNA can make a loop covering the distance from nucleotide 1408 to 1395 by a single internucleotide bond. A probable explanation might be that the anticodon stem, rather than the anticodon loop, covers the protections of 16S rRNA residues 1405, 1408, and 1493 to 1495. This could also explain the discrepancy in the simultaneous docking of the aminoglycoside antibiotics (Fourmy et al., 1996) and the mRNA-tRNA duplex in the same major groove of helix 44 discussed by VanLoock et al. (1999).

The last nucleotide residue of the A-site codon forms a cross-link with U1052 of helix 34 of the 16S rRNA (Dontsova et al., 1992). Helix 34 is known to carry a cluster of mutations affecting translational accuracy and frameshifting (O'Connor et al., 1995). The site of interaction with the antibiotic spectinomycin (Moazed and Noller, 1987), a translocation inhibitor, and the binding site of protein S5 (Culver et al., 1999) are located on the opposite edge of helix 34. This helix is believed to belong to the head of the 30S subunit, and it is thus constrained to be "above" the mRNA. Its position, fixing the 3′ edge of the A site, could explain its function in maintaining the reading frame. At the same time, its role in translocation could be to unlock the mRNA.

Binding Site of mRNA Region 3′ to the A Site

According to cryo-electron microscopy data, the 3′ part of the mRNA bound to the ribosome passes through a tunnel in the neck of the small ribosomal subunit (Stark et al., 1997). The upper part of the tunnel, beginning at the 3′ edge of the A site, is most likely to be helix 34 of the 16S rRNA. This idea is supported by a cross-link from mRNA residues +8 and +9 to nucleotide A1196 in this helix (Sergiev et al., 1997). The region around nucleotide 1192, known to interact with protein S5 and forming, together with this protein, the spectinomycin binding site, is a good candidate for a clamp closing the tunnel and fixing the mRNA.

The lower part of the tunnel might be formed by the joint between helices 28 and 44. Nucleotide C1395 in the lower part of helix 28 was cross-linked to mRNA positions as far as residue +7. More distant from this point are the cross-links from helices 28 and 44 to residues +2 and +4 of the mRNA. mRNA nucleotides +10 to +12 were cross-linked to nucleotides 530 to 532 in helix 18 of the 16S rRNA (Dontsova et al., 1992; Rinke-Appel et al., 1993; Sergiev et al., 1997). The location of this helix was long discussed in the literature. The most contradictory data were the footprints of the tRNAs at the A and P sites (Moazed and Noller, 1986) as opposed to the footprints from protein S12 (Stern et al., 1988), known to be located rather distant from the decoding center (Capel et al., 1988). Possible explanations were the allosteric origin of the protections and imprecision of the protein positioning. The cross-linking data, in contrast to the footprinting results, are indicative of direct contacts. It was noted that the specific cross-links from the 3′ part of mRNA were formed only when a tRNA was present at the P site. These data favor a location of helix 18 closer to the solvent side of the tunnel and suggest that it is being protected in a tRNA-dependent manner by the mRNA region +10 to +13.

The extensive cross-linking study of the decoding center of the ribosome, together with other biochemical and electron-microscopic data, has enabled us to understand the spatial arrangement of the main elements of the decoding center.

EXAMPLE II: THE ENVIRONMENT OF 5S rRNA IN THE RIBOSOME

5S rRNA, a component of the large ribosomal subunit, plays an essential role in ribosome assembly and function. Without 5S rRNA, ribosomes have very low activity in peptide synthesis in vitro (Spanchenko et al., 1998, and references therein). Until recently, the role that 5S rRNA might play in the ribosome remained unknown. Since 5S rRNA was discovered, numerous attempts to solve this problem were undertaken without substantial success (see Moore, 1996, and Bogdanov et al., 1996, for reviews). However, the use of modern cross-linking approaches for elucidation of the environment of 5S rRNA in the *E. coli* 50S subunit with respect to 23S rRNA has been a turning point in our understanding of both the location of the 5S rRNA molecule in the ribosome and its possible functional role.

Initially, 5S rRNA randomly labeled with 4-thioU was incorporated into 70S ribosomes, with retention of a reasonable level of biological activity. After UV irradiation and localization of the positions of the cross-links, it was found that the residue U89, located in the D loop of 5S rRNA, forms covalent bonds with two highly conserved positions in 23S rRNA—namely, C2475 in helix 89 (domain V) and A960 in helix 39 (domain II) (Dontsova et al., 1994; Dokudovskaya et al., 1996). The overall cross-linking yield was very high (>50%), suggesting a rather firm fixation of the 4-thioU residue incorporated into position 89 of 5S rRNA relative to the C2475 and A960 residues of 23S rRNA. The conformation of U89 in the 5S rRNA structure favors these internucleotide

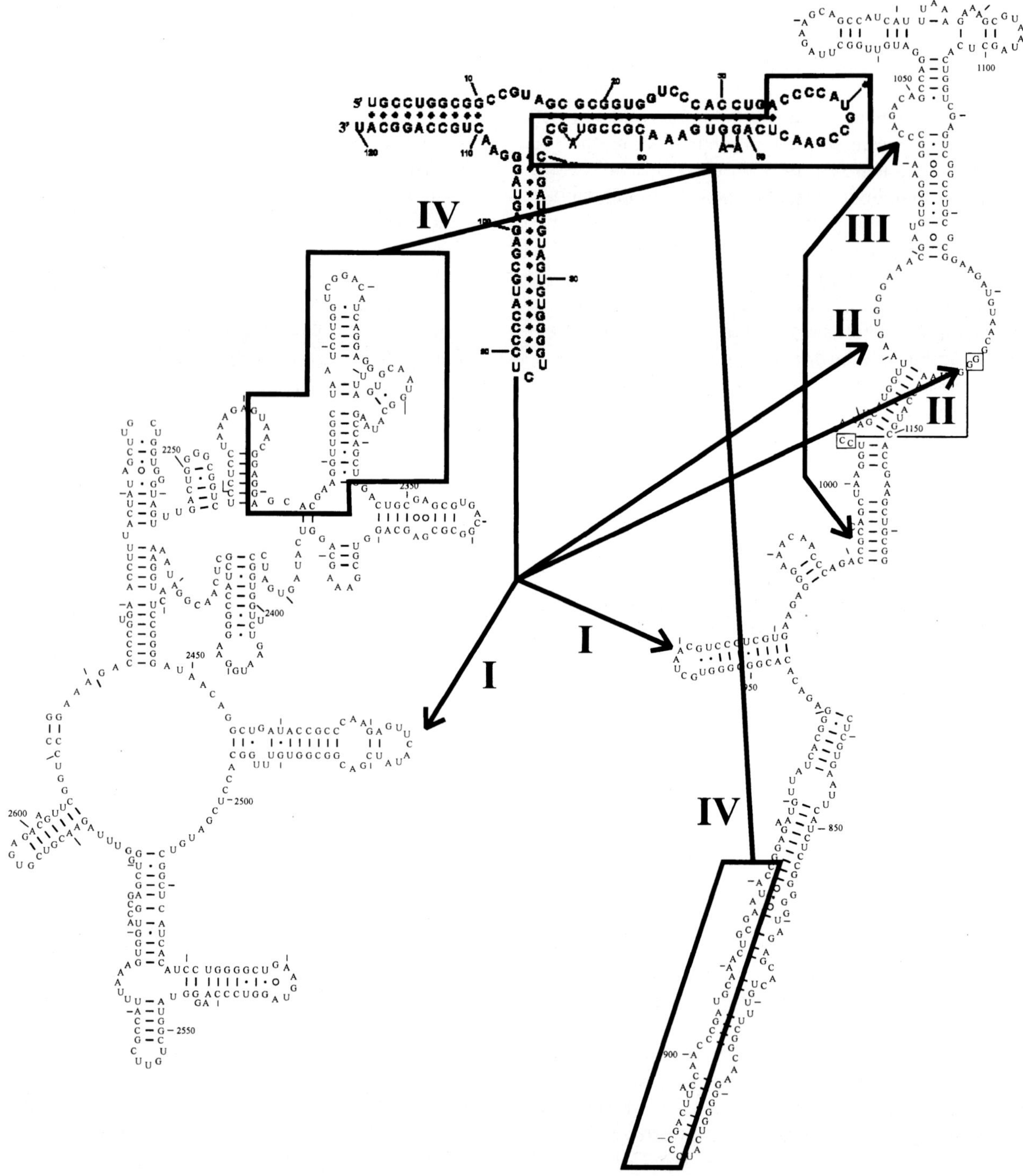

Figure 3. A network of interactions between the 5S rRNA and domains II and V of 23S rRNA established by cross-linking data. I. Cross-links formed with 4-thioU randomly incorporated into the 5S rRNA (Dontsova et al., 1994; Dokudovskaya et al., 1996). II. Cross-links formed by the diazirine derivative of 5-methyleneaminouridine (Sergiev et al., 1998). III. Cross-link formed by 4-thiouridine attached to residue 1045 of the 23S rRNA (Baranov et al., 1998a). IV. Cross-links formed by APAB, attached to a sequence within the 5S rRNA (Osswald and Brimacombe, 1999). The cross-linked rRNA regions are shown as boxes.

contacts; it is known from the nuclear magnetic resonance structure of the D loop of 5S rRNA that the uracil base in this nucleotide residue is unstacked and completely accessible for interaction with external nucleotide residues (Dallas and Moore, 1997). In a parallel site-directed-mutagenesis study we have shown that residue U89 is very important for an efficient incorporation of 5S rRNA into the 50S subunit (Zvereva et al., 1998). In addition, chemical probing of the 5S rRNA sugar-phosphate backbone in various ribosomal states revealed a conformational change of the RNA region neighboring U89 during the association of 50S and 30S subunits (Shpanchenko et al., 1998).

Further cross-linking studies revealed the close proximity of residue A960 to another important functional site of the 50S subunit, namely, its GTPase-associated center. By using long-range photoaffinity reagents, we found that the 5S rRNA residue U89 can be cross-linked not only to the A960 loop of 23S rRNA but also to G1138 (Sergiev et al., 1998), which is a component of the functionally important pseudoknot involving nucleotides 1137 to 1138 and 1005 to 1006 in helix 41. Its disruption abolishes the ability of the 50S subunit to associate with the 30S subunit (Rosendahl et al., 1995). It was shown in a parallel study that helices 41 and 42 can be cross-linked via a 4-thioU residue attached to helix 42 (Baranov et al., 1998a). The network of cross-links presented in Fig. 3 thus demonstrates that the D loop of 5S rRNA is located in the neighborhood of helices 43 and 44, the well-known GTPase-associated region of 23S rRNA. These topographical data allow us to suggest that 5S rRNA in the ribosome is part of a passageway ("informational channel") that serves for signal transmission from the peptidyltransferase center to the GTPase-associated region, the binding site of EF-G. This concept has recently obtained very strong support from a site-directed-mutagenesis study; it was shown that mutations at position A960 of 23S rRNA caused structural rearrangements both in the 5S rRNA D loop and in three compact regions of the peptidyltransferase center which are involved in tRNA binding (Sergiev et al., unpublished).

FURTHER PROSPECTS

The data reviewed in this chapter show that site-specific photo-cross-linking is a powerful tool for the identification of interacting elements of the translational machinery and for the characterization of functional centers in the ribosome. Furthemore, it has provided invaluable information concerning constraints on ribosome structure which have already been used in the interpretation of crystallographic and electron microscopy data. It is to be expected that the existing techniques will be substantially improved by the development of new cross-linking reagents and by the application to ribosomology of approaches for site-directed incorporation of photoaffinity probes into RNA that have been created in other areas of molecular biology. Photoreagents that allow cross-linking experiments to be carried out on the millisecond and nanosecond time scale are particularly desirable. Such compounds offer the possibility of investigating the structural and functional dynamics of the ribosome. The successful application of aryldiazirines to topographical studies of mRNA (Sergiev et al., 1997) and 5S rRNA (Sergiev et al., 1998) can be considered a first step in this direction.

We are very thankful to Richard Brimacombe for many years' standing collaboration in ribosome cross-linking study, for constant support, and for critical reading of the manuscript. We thank Roger Garrett and Stephen Douthwaite for organizing the meeting.

This work was supported by grants from the Volkswagen Foundation (I 72 338) and from the Russian Foundation for Basic Research.

REFERENCES

Afonina, E., N. Chichkova, S. Bogdanova, and A. Bogdanov. 1991. 30S ribosomal subunits with fragmented 16S RNA: a new approach for structure and function study of ribosomes. *Biochimie* 73:777–787.

Agrawal, R. K., P. Penczek, R. A. Grassucci, N. Burkhardt, K. H. Nierhaus, and J. Frank. 1999. Effect of buffer conditions on the position of tRNA on the 70 S ribosome as visualized by cryo-electron microscopy. *J. Biol. Chem.* **274**:8723–8729.

Ban, N., B. Freeborn, P. Nissen, P. Penczek, R. A. Grassucci, R. Sweet, J. Frank, P. B. Moore, and T. A. Steitz. 1998. A 9A resolution X-ray crystallographic map of the large ribosomal subunit. *Cell* **93**:1105–1115.

Baranov, P. V., S. S. Dokudovskaya, T. S. Oretskaya, O. A. Dontsova, A. A. Bogdanov, and R. Brimacombe. 1997. A new technique for the characterization of long-range tertiary contacts in large RNA molecules: insertion of a photolabel at a selected position in 16S rRNA within the Escherichia coli ribosome. *Nucleic Acids Res.* **25**:2266–2273.

Baranov, P., O. Gurvich, A. Bogdanov, R. Brimacombe, and O. Dontsova. 1998a. New features of 23S ribosomal RNA folding: the long helix 41-42 makes a "U-turn" inside the ribosome. *RNA* **6**:658–668.

Baranov, P. V., P. V. Sergiev, O. A. Dontsova, A. A. Bogdanov, and R. Brimacombe. 1998b. The database of ribosomal cross links (DRC). *Nucleic Acids Res.* **26**:187–189.

Barta, A., E. Kuechler, and G. Steiner. 1990. Photoaffinity labeling of the peptidyltransferase region, p. 358–365. *In* W. E. Hill, A. Dahlberg, R. A. Garrett, P. B. Moore, D. Schlessinger, and J. R. Warner (ed.), *The Ribosome: Structure, Function and Evolution.* American Society for Microbiology, Washington, D.C.

Bhangu, R., and P. Wollenzien. 1992. The mRNA binding track in the Escherichia coli ribosome for mRNAs of different sequences. *Biochemistry* **31**:5937–5944.

Blatter, E. E., Y. W. Ebright, and R. H. Ebright. 1992. Identification of an amino-acid base contact in the GCN4-DNA complex by bromouracil-mediated photocrosslinking. *Nature* **359:** 650–652.

Bogdanov, A. A., O. A. Dontsova, and R. Brimacombe. 1993. Messenger RNA path through the procaryotic ribosome, p. 421–432. *In* K. H. Nierhaus, F. Franceschi, A. R. Subramanian, V. A. Erdmann, and B. Wittmann-Liebold (ed.), *The Translational Apparatus: Structure, Function, Regulation, Evolution.* Plenum Press, New York, N.Y.

Bogdanov, A. A., O. A. Dontsova, S. S. Dokudovskaya, and I. N. Lavrik. 1996. Structure and function of 5S rRNA in the ribosome. *Biochem. Cell Biol.* **73:**869–876.

Brimacombe, R. Personal communication.

Capel, M. S., M. Kjeldgaard, D. M. Engelman, and P. B. Moore. 1988. Positions of S2, S13, S16, S17, S19 and S21 in the 30 S ribosomal subunit of Escherichia coli. *J. Mol. Biol.* **200:**65–87.

Cooperman, B. S., C. J. Weitzmann, and M. A. Buck. 1988. Affinity labeling of ribosomes. *Methods Enzymol.* **164:**341–361.

Culver, G. M., G. M. Heilek, and H. F. Noller. 1999. Probing the rRNA environment of ribosomal protein S5 across the subunit interface and inside the 30 S subunit using tethered Fe(II). *J. Mol. Biol.* **286:**355–364.

Dallas, A., and P. B. Moore. 1997. The loopE-loopD region of Escherichia coli 5S rRNA: the solution structure reveals an unusual loop that may be important for binding ribosomal proteins. *Structure* **5:**1639–1653.

Dokudovskaya, S., O. Dontsova, O. Shpanchenko, A. Bogdanov, and R. Brimacombe. 1996. Loop IV of 5S ribosomal RNA has contact both to domain II and to domain V of the 23S RNA. *RNA* **2:**146–152.

Dontsova, O., A. Kopylov, and R. Brimacombe. 1991. The location of mRNA in the ribosomal 30S initiation complex; site-directed cross-linking of mRNA analogues carrying several photo-reactive labels simultaneously on either side of the AUG start codon. *EMBO J.* **10:**2613–2620.

Dontsova, O., S. Dokudovskaya, A. Kopylov, A. Bogdanov, J. Rinke-Appel, N. Junke, and R. Brimacombe. 1992. Three widely separated positions in the 16S RNA lie close to the ribosomal decoding region; a site-directed cross-linking study with mRNA analogues. *EMBO J.* **11:**3105–3116.

Dontsova, O., V. Tishkov, S. Dokudovskaya, A. Bogdanov, T. Doring, J. Rinke-Appel, S. Thamm, B. Greuer, and R. Brimacombe. 1994. Stem-loop of 5S rRNA lies in close contact to the peptidyltransferase center. *Proc. Natl. Acad. Sci. USA* **91:**4125–4129.

Favre, A. 1990. 4-Thiouridine as an intrinsic photoaffinity probe of nucleic acid structure and interactions. p. 379–425. *In* H. Morrison (ed.), *Bioorganic Photochemistry.* John Wiley & Sons, New York, N.Y.

Fleming, A. S. 1995. Chemical reagents in photoaffinity labeling. *Tetrahedron* **46:**12479–12520.

Fourmy, D., M. I. Recht, S. C. Blanchard, and J. D. Puglisi. 1996. Structure of the A site of Escherichia coli 16S ribosomal RNA complexed with an aminoglycoside antibiotic. *Science* **274:** 1367–1371.

Green, R., C. Switzer, and H. F. Noller. 1998. Ribosome-catalyzed peptide-bond formation with an A-site substrate covalently linked to 23S ribosomal RNA. *Science* **280:**286–289.

Heilek, G. M., and H. F. Noller. 1996. Site-directed hydroxyl radical probing of the rRNA neighborhood of ribosomal protein S5. *Science* **272:**1659–1662.

Huang, Z., and J. W. Szostak. 1996. A simple method for 3′-labeling of RNA. *Nucleic Acids Res.* **24:**4360–4361.

Juzumiene, D., and P. Wollenzien. 1997. Determination of the 16S ribosomal RNA folded structure with site-directed photoreactive reagents in the 5′ and central pseudoknot region. *Nucleic Acid Symp. Ser.* **36:**168–170.

Kotziba-Hibert, F., I. Kapfer, and M. Goeldner. 1995. Recent trends in photoaffinity labeling. *Angew. Chem. Int. Ed.* **34:**1296–1312.

Lingner, J., and W. Keller. 1993. 3′-end labeling of RNA with recombinant yeast poly(A) polymerase. *Nucleic Acids Res.* **21:** 2917–2920.

Malhotra, A., and S. C. Harvey. 1994. A quantitative model of the Escherichia coli 16 S RNA in the 30 S ribosomal subunit. *J. Mol. Biol.* **240:**308–340.

Meisenheimer, K. M., and T. H. Koch. 1997. Photocrosslinking of nucleic acids to associated proteins. *Crit. Rev. Biochem. Mol. Biol.* **32:**101–140.

Merryman, C., D. Moazed, J. McWhirter, and H. F. Noller. 1999. Nucleotides in 16S rRNA protected by the association of 30S and 50S ribosomal subunits. *J. Mol. Biol.* **285:**97–105.

Mitchell P., M. Osswald, and R. Brimacombe. 1992. Identification of intermolecular RNA cross-links at the subunit interface of the Escherichia coli ribosome. *Biochemistry* **31:**3004–3011.

Moazed, D., and H. F. Noller. 1986. Transfer RNA shields specific nucleotides in 16S ribosomal RNA from attack by chemical probes. *Cell* **47:**985–994.

Moazed, D., and H. F. Noller. 1987. Interaction of antibiotics with functional sites in 16S ribosomal RNA. *Nature* **327:**389–394.

Mondus, D., and P. Wollenzien. 1998. Neighborhood of 16S RNA nucleotides U788/U789 in the 30S ribosomal subunit determined by site-directed cross-linking. *RNA* **4:**1373–1385.

Montpetit, A., C. Payant, J. M. Nolan, and L. Brakier-Gingras. 1998. Analysis of the conformation of the 3′ major domain of Escherichia coli 16S ribosomal RNA using site-directed photoaffinity crosslinking. *RNA* **4:**1455–1466.

Moore, M. J., and P. A. Sharp. 1992. Site-specific modification of pre-mRNA: the 2′-hydroxyl groups at the splice sites. *Science* **262:**992–997.

Moore, P. B. 1996. The structure and function of 5S ribosomal RNA, p. 199–236. *In* A. E. Dahlberg and R. A. Zimmermann (ed.), *Ribosomal RNA: Structure, Evolution, Processing and Function in Protein Biosynthesis.* CRC Press, Boca Raton, Fla.

Mueller, F., and R. Brimacombe. 1997. A new model for the three-dimensional folding of *Escherichia coli* 16 S ribosomal RNA. I. Fitting the RNA to a 3D electron microscopic map at 20 A. *J. Mol. Biol.* **271:**524–544.

Mueller, F., H. Stark, M. van Heel, J. Rinke-Appel, and R. Brimacombe. 1997. A new model for the three-dimensional folding of Escherichia coli 16S ribosomal RNA. III. The topography of the functional centre. *J. Mol. Biol.* **271:**566–587.

O'Connor, M., C. A. Brunelli, M. A. Firpo, S. T. Gregory, K. R. Lieberman, J. S. Lodmell, H. Moine, D. I. Van Ryk, and A. E. Dahlberg. 1995. Genetic probes of ribosomal RNA function. *Biochem. Cell Biol.* **73:**859–868.

Osswald, M., and R. Brimacombe. 1999. The environment of 5S rRNA in the ribosome: cross-links to 23S RNA from sites within helices II and III of the 5S molecule. *Nucleic Acids Res.* **27:**2283–2290.

Powers, T., and H. F. Noller. 1995. Hydroxyl radical footprinting of ribosomal proteins on 16S rRNA. *RNA* **1:**194–209.

Powers, T., L. M. Changchien, G. R. Craven, and H. F. Noller. 1988. Probing the assembly of the 3′ major domain of 16 S ribosomal RNA. Quaternary interactions involving ribosomal proteins S7, S9 and S19. *J. Mol. Biol.* **200:**309–319.

Prince, J. B., B. H. Taylor, D. L. Thurlow, J. Ofengand, and R. A. Zimmermann. 1982. Covalent cross-linking of $tRNA_1^{Val}$ to 16S RNA at the ribosomal P site: identification of cross-linked residues. *Proc. Natl. Acad. Sci. USA* **79:**5450–5454.

Rinke-Appel, J., N. Junke, R. Brimacombe, S. Dokudovskaya, O. Dontsova, and A. Bogdanov. 1993. Site-directed cross-linking of mRNA analogues to 16S ribosomal RNA; a complete scan of cross-links from all positions between +1 and +16 on the mRNA, downstream from the decoding site. *Nucleic Acids Res.* **21:**2853–2859.

Rinke-Appel, J., N. Junke, R. Brimacombe, I. Lavrik, S. Dokudovskaya, O. Dontsova, and A. Bogdanov. 1994. Contacts between 16S ribosomal RNA and mRNA, within the spacer region separating the AUG initiator codon and the Shine-Dalgarno sequence; a site-directed cross-linking study. *Nucleic Acids Res.* **22:** 3018–3025.

Rosendahl, G., L. H. Hansen, and S. Douthwaite. 1995. Pseudoknot in domain II of 23S rRNA is essential for ribosome function. *J. Mol. Biol.* **23:**2396–2403.

Sergiev, P. V., I. N. Lavrik, V. A. Wlasoff, S. S. Dokudovskaya, O. A. Dontsova, A. A. Bogdanov, and R. Brimacombe. 1997. The path of mRNA through the bacterial ribosome: a site-directed crosslinking study using new photoreactive derivatives of guanosine and uridine. *RNA* **3:**464–475.

Sergiev, P. V., S. S. Dokudovskaya, E. Romanova, A. Topin, A. Bogdanov, R. Brimacombe, and O. Dontsova. 1998. The environment of 5S rRNA in the ribosome: cross-links to the GTPase-associated area of 23S rRNA. *Nucleic Acids Res.* **26:**2519–2525.

Sergiev, P. V., A. A. Bogdanov, A. E. Dahlberg, and O. A. Dontsova. Unpublished data.

Shine, J., and L. Dalgarno. 1975. Terminal-sequence analysis of bacterial ribosomal RNA. Correlation between the 3′-terminal-polypyrimidine sequence of 16-S RNA and translational specificity of the ribosome. *Eur. J. Biochem.* **57:**221–230.

Shpanchenko, O. V., O. A. Dontsova, A. A. Bogdanov, and K. H. Nierhaus. 1998. Structure of 5S rRNA within the Escherichia coli ribosome: iodine-induced cleavage patterns of phosphorothioate derivatives. *RNA* **4:**1154–1164.

Sontheimer, E. J. 1994. Site-specific cross-linking with 4-thiouridine. *Mol. Biol. Rep.* **20:**35–44.

Stark, H., E. V. Orlova, J. Rinke-Appel, N. Junke, F. Mueller, M. Rodnina, W. Wintermeyer, R. Brimacombe, and M. van Heel. 1997. Arrangement of tRNAs in pre- and posttranslocational ribosomes revealed by electron cryomicroscopy. *Cell* **10:**19–28.

Stern, S., T. Powers, L. M. Changchien, and H. F. Noller. 1988. Interaction of ribosomal proteins S5, S6, S11, S12, S18 and S21 with 16 S rRNA. *J. Mol. Biol.* **201:**683–695.

Stiege, W., C. Zwieb, and R. Brimacombe. 1982. Precise localisation of three intra-RNA cross-links in 23S RNA and one in 5S RNA, induced by treatment of Escherichia coli 50S ribosomal subunits with bis-(2-chloroethyl)-methylamine. *Nucleic Acids Res.* **10:**7211–7229.

Stoffler-Meilicke, M., and G. Stoffler. 1990. Topography of the ribosomal proteins from Escherichia coli within the intact subunits as determined by immunoelectron microscopy and protein-protein cross-linking, p. 123-133. *In* W. E. Hill, A. E. Dahlberg, R. A. Garrett, P. B. Moore, D. Schlessinger, and J. R. Warner (ed.), *The Ribosome: Structure, Function, and Evolution.* American Society for Microbiology, Washington, D.C.

Thiede, B., H. Urlaub, H. Neubauer, G. Grelle, and B. Wittmann-Liebold. 1998. Precise determination of RNA-protein contact sites in the 50S ribosomal subunit of Escherichia coli. *Biochem. J.* **334:**39–42.

VanLoock, M. S., T. R. Easterwood, and S. C. Harvey. 1999. Major groove binding of the tRNA/mRNA complex to the 16 S ribosomal RNA decoding site. *J. Mol. Biol.* **285:**2069–2078.

von Ahsen, U., and H. F. Noller. 1995. Identification of bases in 16S rRNA essential for tRNA binding at the 30S ribosomal P site. *Science* **267:**234–237.

Wilson, K. S., and H. F. Noller. 1998. Mapping the position of translational elongation factor EF-G in the ribosome by directed hydroxyl probing. *Cell* **92:**131–139.

Wower, J., L. A. Sylvers, K. V. Rosen, S. S. Hixon, and R. A. Zimmermann. 1993. A model of the tRNA binding sites on the Escherichia coli ribosome, p. 455–464. *In* K. H. Nierhaus, F. Franceschi, A. R. Subramanian, V. A. Erdmann, and B. Wittmann-Liebold (ed.), *The Translational Apparatus: Structure, Function, Regulation, Evolution.* Plenum Press, New York, N.Y.

Wower, J., K. V. Rosen, S. S. Hixon, and R. A. Zimmermann. 1994. Recombinant photoreactive tRNA molecules as probes for cross-linking studies. *Biochimie* **76:**1235–1246.

Zvereva, M. I., O. V. Shpanchenko, O. A. Dontsova, K. H. Nierhaus, and A. A. Bogdanov. 1998. Effect of point mutations at position 89 of the E. coli 5S rRNA on the assembly and activity of the large ribosomal subunit. *FEBS Lett.* **421:**249–251.

The Ribosome: Structure, Function, Antibiotics, and Cellular Interactions
Edited by R. A. Garrett, S. R. Douthwaite, A. Liljas, A. T. Matheson, P. B. Moore, and H. F. Noller

Chapter 22

Chemical Cleavage as a Probe of Ribosomal Structure

WALTER E. HILL, GREGORY W. MUTH, JAMES M. BULLARD, SCOTT P. HENNELLY, JING YUAN, WENDY T. GRACE, DOUGLAS J. BUCKLIN, MICHAEL A. VAN WAES, and CHARLES M. THOMPSON

An axiom embraced by investigators in many fields, including ribosomes, is that to understand function, one must know the structure. In the case of the ribosome, we are approaching a time when the structures of the ribosome and ribosomal subunits will be known at a few angstroms resolution. Yet much more is needed to understand the structural details that relate to the function of the ribosome in protein biosynthesis.

It is essential to continue to probe details of ribosome structure and function, using all the tools available. The crystallographic evidence will provide solid pillars upon which to attach the structural cords which occur between the relevant functional states. Detailed analyses, using cross-linking, site-specific mutagenesis, chemical modification, fluorescence, hydroxyl radical probing, and chemical cleavage, will be essential to fill in the functional and structural gaps between the crystallographic monoliths.

In this chapter we will deal with the cleavage of RNA as an approach to the study of ribosome structure. Chemical nucleases, such as EDTA-Fe(II) and phenanthroline-Cu(II), have been widely used to cleave both RNA and DNA, and more recently, they have been used to study ribosomes.

BACKGROUND

The discovery in 1979, by David Sigman and coworkers, that phenanthroline could cleave the phosphodiester backbone of DNA opened an entirely new approach to the analysis of nucleic acids (Sigman et al., 1979, 1993b; Sigman, 1990). The discovery of other nucleolytic reagents with various specificities and binding characteristics has greatly enlarged the potential and efficacy of this approach (Hertzberg and Dervan, 1982; Pogozelski and Tullius, 1998; Burrows and Muller, 1998). By using such synthetic nucleolytic reagents, it is possible to identify and monitor structural characteristics of nucleic acids (Pope and Sigman, 1984; Dreyer and Dervan, 1985; Dervan, 1986; Chen and Sigman, 1988; Sluka et al., 1990; Oakley and Dervan, 1990; Sigman et al., 1993a; Chen et al., 1993a; Han and Dervan, 1994). These chemical nucleases can also be used as footprinting reagents, with potentially higher resolution than biological nucleases (Van Dyke and Dervan, 1983; Kuwabara and Sigman, 1987; Huttenhofer and Noller, 1992, 1994; Pearson et al., 1994; Powers and Noller, 1995; Wilson and Noller, 1998). In addition, by covalently attaching these complexes to probes, or ligands such as mRNA, tRNA, proteins, and factors, they can be used as proximity probes with high affinity for particular nucleic acid structures (Strobel and Dervan, 1990; Han and Dervan, 1994; Bullard et al., 1995, 1998; Gallagher et al., 1996; Heilek and Noller, 1996a, 1996b; Bucklin et al., 1997; Lieberman and Noller, 1998; Culver et al., 1999; Newcomb and Noller, 1999; Muth et al., 1999a, b).

The rRNA structure of ribosomes has been probed mainly with two cleavage reagents. EDTA-Fe(II) and derivatives have been used by the Noller laboratory, while our laboratory has been utilizing derivatives of phenanthroline-Cu(II). The EDTA-Fe(II) approach has been developed extensively by Peter Dervan and coworkers since they discovered its nucleolytic characteristics in 1982 (Hertzberg and

Walter E. Hill, James M. Bullard, Scott P. Hennelly, Jing Yuan, Wendy T. Grace, Douglas J. Bucklin, and Michael A. Van Waes ■ Division of Biological Sciences, The University of Montana, Missoula, MT 59812. Gregory W. Muth and Charles M. Thompson ■ Department of Chemistry, The University of Montana, Missoula, MT 59812.

Dervan, 1982, 1984; Van Dyke and Dervan, 1983; Strobel and Dervan, 1990; Han and Dervan, 1994; Baliga et al., 1995). Noller's laboratory, using both untethered EDTA-Fe(II) and an EDTA compound (*p*-bromoacetamidobenzyl-EDTA·Fe) which can be tethered, has looked at the proximity of nucleic acid neighborhoods to a number of ribosomal proteins, translation factors, and other ligands, as well as potential subunit interaction sites and specific sites within the rRNA itself (Huttenhofer and Noller, 1992; Noller et al., 1995; Powers and Noller, 1995; Heilek and Noller, 1996b; Wilson and Noller, 1998; Merryman et al., 1999). Our studies, with tethered phenanthroline-Cu(II), have focused primarily on rRNA sites proximal to mRNA, tRNA, and rRNA regions targeted by oligonucleotide probes (Bullard et al., 1995, 1998; Hill et al., 1995; Bucklin et al., 1997; Muth et al., 1999a).

The use of different cleavage reagents is warranted due to the quite different mechanisms of cleavage utilized by the two compounds. Fe(II)-EDTA, in conjunction with H_2O_2, produces a hydroxyl radical (HO·), which is capable of diffusing up to 60 Å from the site of formation (Que et al., 1980). These radicals primarily attack the ribose (or deoxyribose) sugar, abstracting a hydrogen atom and promoting direct strand scission (Pogozelski and Tullius, 1998). The bases can also be attacked, leading to oxidized nucleobases (base radicals and radical cations), which in turn can either revert by further reduction or, under defined conditions, promote cleavage of the nucleic acid (Burrows and Muller, 1998). In either case, the damage is done by the diffusible hydroxyl radical and takes place in a non-base-specific manner within the diffusion sphere.

In the case of phenanthroline-Cu(II), the mechanism is much less clear. Prevailing evidence strongly indicates that cleavage occurs via a hydroxyl radical intermediate (Meijler et al., 1997; Chen and Greenberg, 1998; Burrows and Muller, 1998). In addition, there is an absolute requirement for H_2O_2, as is true in the case of EDTA-Fe(II) (Chen et al., 1993b; Perrin et al., 1996). However, with phenanthroline-Cu(II), there is considerable evidence that, at least in the absence of added H_2O_2, the hydroxyl radical remains bound to the copper (Johnson and Nazhat, 1987). Thus, there is no diffusion of the reactive species, making the cleavage distance from the copper essentially the distance from the copper to the bound HO·. The requirement for H_2O_2 is readily satisfied by endogenous H_2O_2, which is generated by the redox reaction of copper and dissolved oxygen (Burkitt, 1994). Catalase, which efficiently scavenges and destroys H_2O_2, completely quenches the phenanthroline-Cu(II) reaction (Que et al., 1980). There is an additional requirement that the phenanthroline-Cu(II) be placed quite close to the hydrogen atom which is to be abstracted. This requirement is at once stringent and beneficial, allowing the cleavage reaction to be "tuned" to reveal subtle interactions.

Phenanthroline-Cu(II) will tend to form a bis complex when it is in the free form, coordinated around the copper in a square planar [Cu(II)] or tetrahedral [Cu(I)] manner (Fig. 1A). This species, in its untethered form, can readily dock in the minor groove of B-form DNA, allowing the bound hydroxyl radical to abstract the necessary sugar hydrogens and induce cleavage. Since RNA can only take the A form, which has a much shallower minor groove, docking of phenanthroline-Cu(II) does not occur (Fig. 1B). The result is that although double-stranded B-form DNA is labile to scission by free phenanthroline-Cu(II), double-stranded RNA is not. A cleavage reaction which produces a freely diffusible hydroxyl radical, such as with EDTA-Fe(II), should equally cleave both structures, which it does. Such is not the case with phenanthroline-Cu(II). In addition, it is not ordinarily possible for phenanthroline-Cu(II) to dock even with single-stranded RNA regions in the A form. It appears that for the partial intercalation to occur, the bases must be strained and splayed apart to some degree (Fig. 1C).

When the phenanthroline-Cu(II) complex is delivered to a specific site via a tether to a biomolecule, it may be found as a mono adduct, with a single phenanthroline coordinated to the Cu(II), which in itself is a highly reactive species (Fig. 1D). It is still necessary, as with the bis complex, for this species to dock or partially intercalate in order for cleavage to occur (Fig. 1B and C). Upon docking, with the addition of a reducing agent such as mercaptopropionic acid, cleavage is induced by hydrogen atom abstraction from a neighboring ribose moiety, probably on C_1' (Meijler et al., 1997; Chen and Greenberg, 1998; Pogozelski and Tullius, 1998). This abstraction leads to the release of the base, followed by β elimination and scission of the phosphodiester bond.

From the foregoing discussion it can be seen that phenanthroline-Cu(II) cleavage is quite discriminating, selecting sites that would contain single-stranded, splayed bases. In addition, if the phenanthroline-Cu(II) is tethered, such possible sites would have to lie within the length of the tether from the ligand to the phenanthroline-Cu(II). On the other hand, EDTA-Fe(II) induces a diffusible hydroxyl radical attack, which is more robust and not site selective but has a much broader range. This latter approach holds much more promise for footprinting and similar techniques, while the tethered phenanthroline holds more promise for identifying distances to proximal rRNAs.

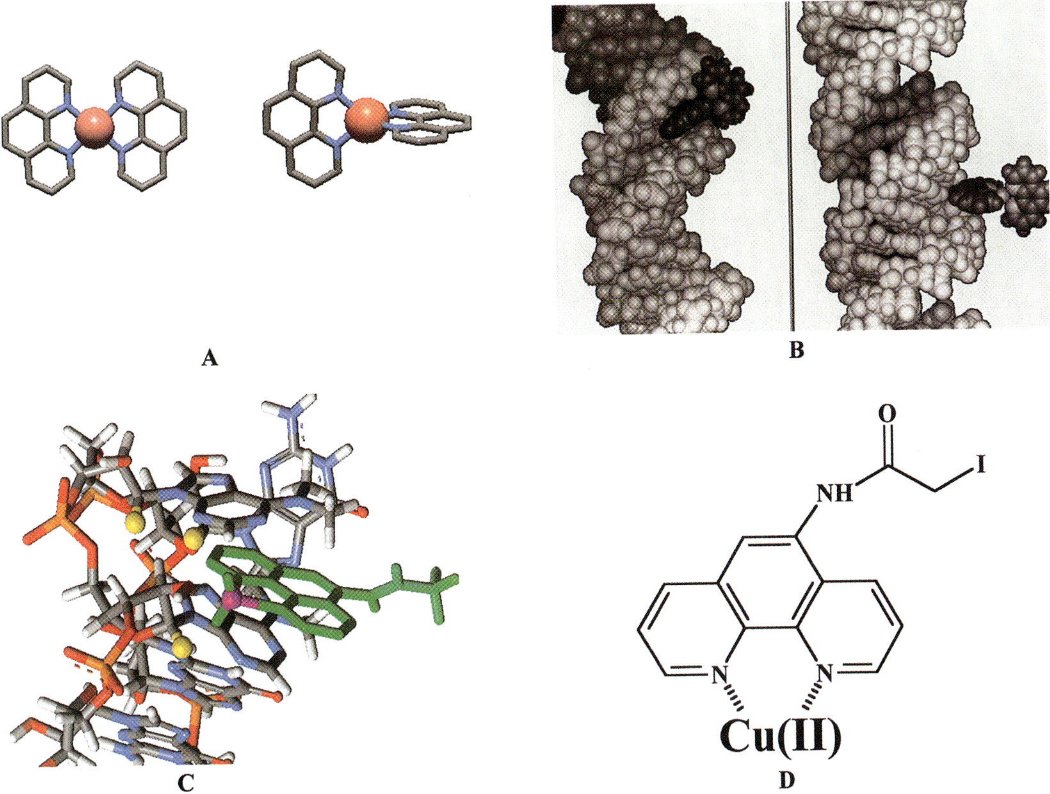

Figure 1. (A) Diagram showing the structure of phenanthroline coordinated with Cu(II) (left) in a square planar conformation and Cu(I) (right) in a tetrahedral conformation. (B) Space-filling diagram showing tetrahedral phenanthroline-Cu(I) docking in the minor groove of B-form DNA (left) and A-form RNA (right). Note that due to the shallow minor groove in the RNA, phenanthroline cannot dock. (C) Diagram portraying RNA hairpin loop with phenanthroline-Cu(II) partially intercalated. The three larger spheres represent proximal C_1' carbon atoms from which a hydrogen may be abstracted. (D) Structure of 5-iodoacetamido-1,10-orthophenanthroline (IoP).

It should be noted that if exogenous H_2O_2 is added to the reaction mixture, a diffusible cleavage species occurs with tethered phenanthroline-Cu(II). This is likely due to oxidative scission of the tether coupling phenanthroline-Cu(II) to the nucleotide, or perhaps even of the nucleotide itself in the presence of excess H_2O_2. This would then allow the phenanthroline-Cu(II) complex to act as a diffusible species in the surrounding region.

STRUCTURAL STUDIES WITH PHENANTHROLINE

We have used 5-iodoacetamido 1,10-orthophenanthroline (IoP) (Fig. 1D) conjugated with thiol groups on 4-thiouracil or with a phosphorothioate in the phosphodiester backbone of the RNA or DNA for the majority of our studies. We have tethered the phenanthroline-Cu(II) to tRNA, mRNA, and, recently, to short DNA oligomers targeted to various sites on the rRNA. We discuss each of these studies below.

Untethered Phenanthroline

Although, as noted above, untethered phenanthroline is not as robust a footprinting agent as EDTA-Fe(II), phenanthroline is an excellent determinant of strained structure. In unpublished results, we have identified all of the sites on the 23S and 16S rRNA which were cleaved by phenanthroline when untethered phenanthroline-Cu(II) was added to ribosomes or ribosomal subunits and cleavage was induced with mercaptopropionic acid. However, the pseudoknot region found in domain I of 23S rRNA, especially the loops found near nucleotides 65 and 90, showed exceptionally robust cleavage (Fig. 2). The results of this study suggested that these particular loops were strained and the bases were splayed (Muth et al., 1999a) (Fig. 1C). Upon removal of the protein, or upon heating of the rRNA, structure was

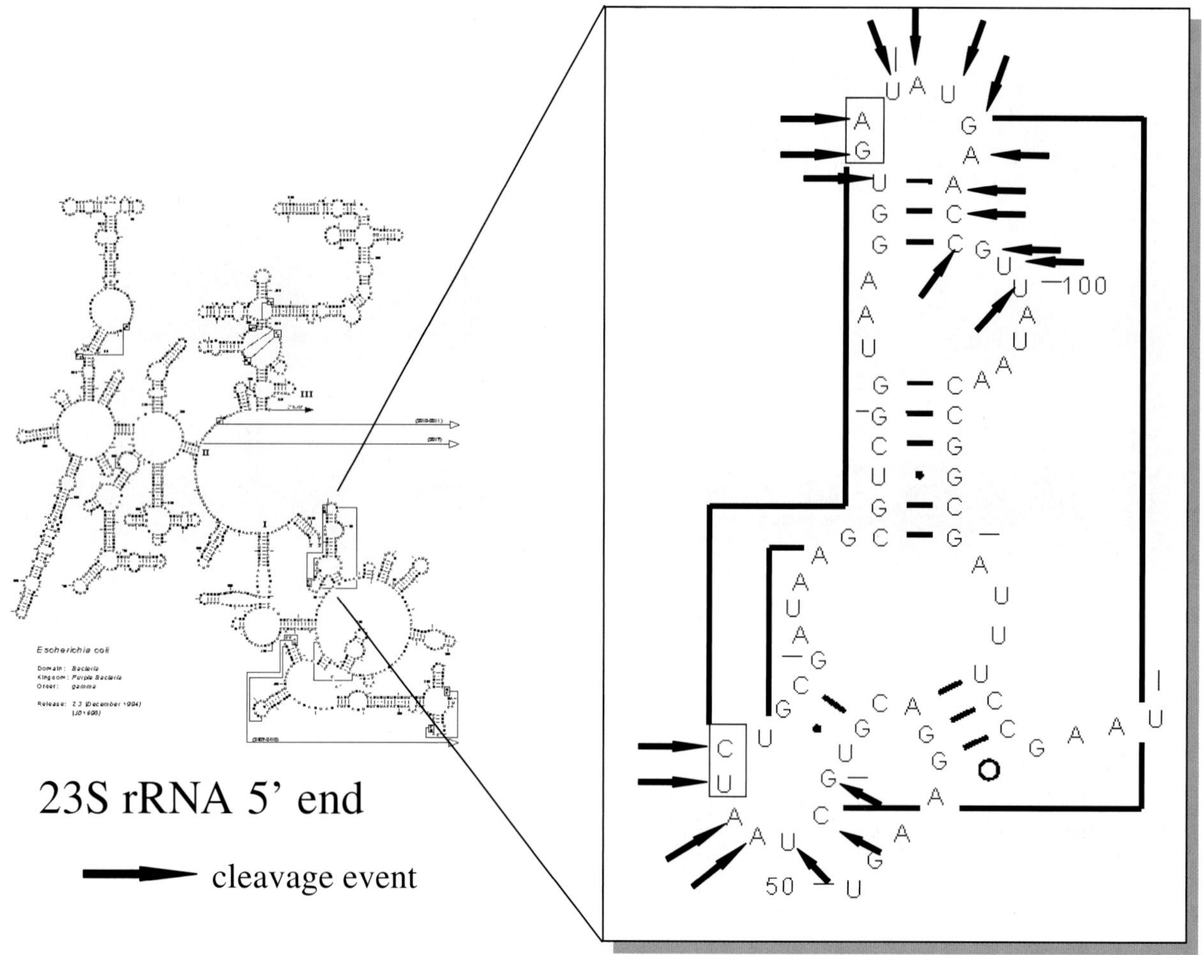

Figure 2. Domains I, II, and III of 23S rRNA with a portion of domain I enlarged to show phenanthroline-Cu(II) cleavages in the region of the pseudoknot.

lost and the cleavage pattern was attenuated significantly or disappeared entirely. These results, as well as others, confirm the use of phenanthroline-Cu(II) as a structural determinant.

tRNA

In early studies, we synthesized tRNA from a plasmid kindly given to us by Olke Uhlenbeck, using 4-thiouracil triphosphate as a substrate. This placed thiol groups throughout the tRNA transcript at 1 or more of the 18 uracil positions. Phenanthroline-Cu(II) was then conjugated to these modified transcripts, the conjugated transcripts were bound to the 50S ribosomal subunits on the P-P or P-E sites (Moazed and Noller, 1989), and cleavage was induced. On the 23S rRNA, 118 cleavages occurred, with various intensities, compared to those obtained in control experiments (Bullard et al., 1995). These cleavages occurred in all domains, but the great majority occurred in domains I, II, and V. There was an intense concentration of cleavages in the peptidyltransferase center. In almost all cases, there was extrinsic evidence that the cleaved regions were folded proximal to the peptidyltransferase region. In all cases the cleavages were attenuated when competed with deacylated native tRNA.

Similar experiments with 30S ribosomal subunits produced 21 cleavages at 9 sites on 16S rRNA within the active 30S ribosomal subunits alone and 23 cleavages at 12 sites when the 30S ribosomal subunits were in 70S ribosomes (Bullard et al., 1998). There were significant differences between the intensities and locations of cleavages in free 30S ribosomal subunits and those found in the 70S ribosomes. Again, there was other evidence which suggested the cleaved regions might be folded together in 16S rRNA in the 30S ribosomal subunit.

These experiments provide a general positioning of the deacylated tRNA on the ribosomes, presumably at the P-E or possibly P-P sites (Moazed and Noller, 1989). Although it is not possible to identify which cleavage emanated from a specific site on the tRNA, the distance from each cleavage site to the tRNA must lie within about 15 Å, as determined from computer modeling. Thus, we have a general knowledge of those regions of rRNA which are proximal to the tRNA.

Additional experiments with mini- and microhelix analogues to the aminoacceptor stem of tRNA have also been carried out in order to more closely identify those regions of rRNA proximal to this portion of the tRNA (Bullard et al., in press). These analogues have been shown to bind the 50S ribosomal subunits and 70S ribosomes, and they are readily competed by deacylated tRNA. Four uracils are present at positions equivalent to positions 50, 51, and 66 and near position 8 in native $tRNA^{Phe}$, all of which are located near the TΨC loop, close to the elbow of tRNA. Phenanthroline-Cu(II) was attached to these modified uracils, the microhelix was bound, and cleavage was induced. The result was 34 cleavages in 14 regions of 23S rRNA, mainly in domain V near the peptidyltransferase region. All of the cleavages but one had been noted previously when full-length tRNA was used (Bullard et al., 1995).

To further refine the results obtained from tRNA, we used $tRNA^{Phe}$ from *Escherichia coli*, which has a naturally occurring 4-thiouracil at position 8. To this a phenanthroline-Cu(II) was conjugated, the tRNA was bound, and cleavage was induced. When deacylated tRNA-oP was bound in either the P or the E site, cleavages occurred at positions 2112 and 2113 and 2169 and 2170 in 23S rRNA (Fig. 3). Acylated tRNA also produced these cleavages. No cleavages were observed when acylated tRNA-oP was bound to the A site.

More recently, efforts have been made to increase the specificity of the previous study by inserting a single phosphorothioate group between adjacent nucleotides in transcribed tRNA. This is an ongoing effort, but we have been successful in inserting a phosphorothioate group between nucleotides 55 and 56 in transcribed $tRNA^{Phe}$.

mRNA

mRNA has been modified with 4-thiouracil to allow conjugation of phenanthroline-Cu(II) to the modified base. When placed at position +5 on the mRNA, cleavage occurred at nucleotides 528 to 532, 1053 to 1055, 1196, and 1396 and 1397 of 16S rRNA when the message was bound to tight-couple 70S ribosomes (Bucklin et al., 1997) (Fig. 4). These results provide additional evidence that these three regions of 16S rRNA must lie within 15 Å of the second nucleotide in the second codon of mRNA. Moreover, these results suggest that these three regions are all proximal to the decoding region of the 30S ribosomal subunit. Further evidence for this proximity has come from a number of studies (Rinke-Appel et al., 1991, 1993; Dontsova et al., 1991, 1992; Alexander et al., 1994; Muth et al., in press).

Short DNA Oligomers Complementary to Targeted Regions of rRNA

Previously, using short DNA oligomers targeted to specific regions of the rRNA in situ, we have iden-

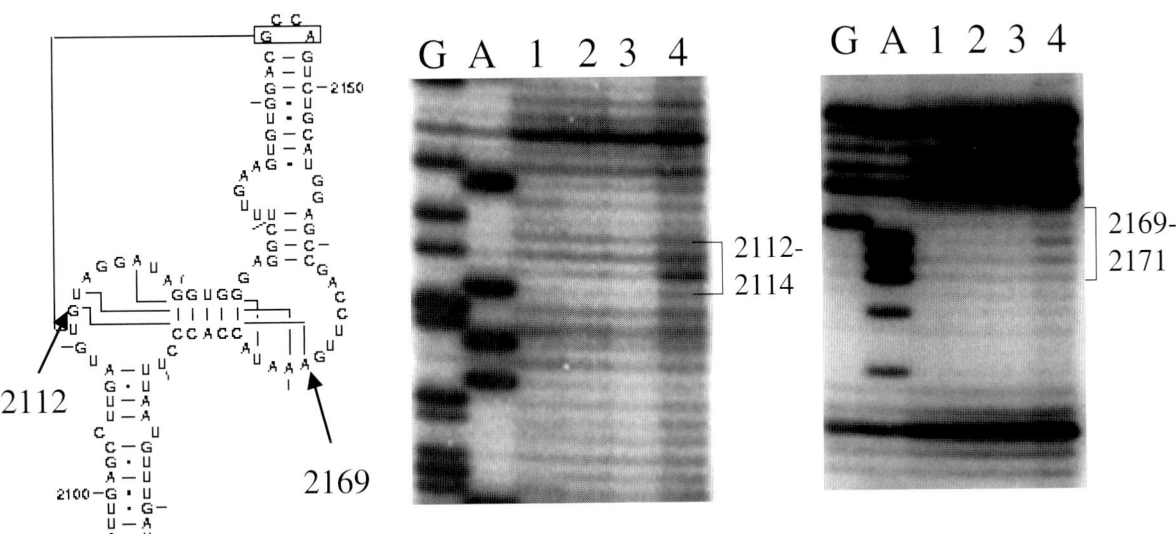

Figure 3. Cleavage of 23S rRNA domain V emanating from phenanthroline-Cu(II) attached to position 8 of $tRNA^{Phe}$ bound to the P or E site on 70S ribosomes.

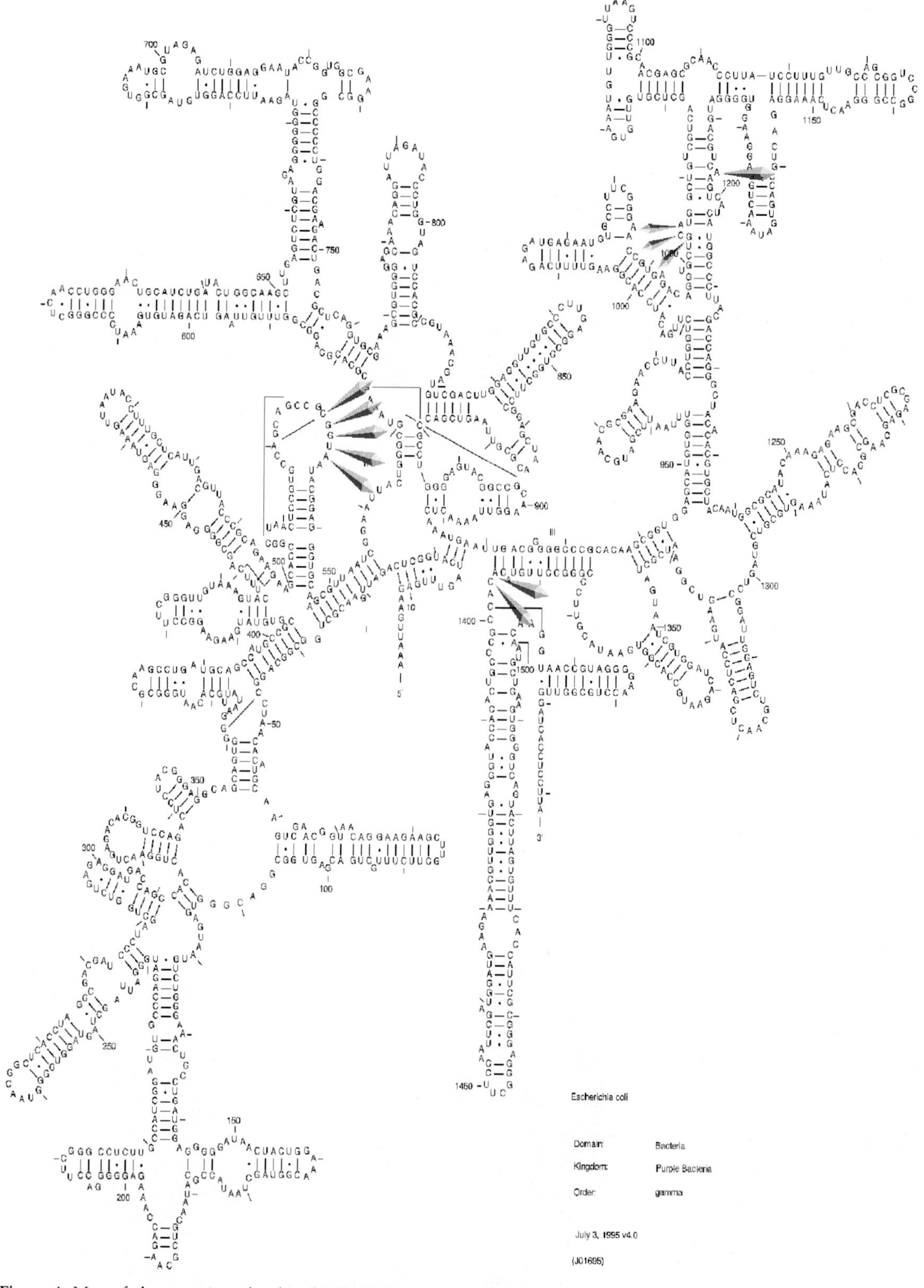

Figure 4. Map of cleavages (arowheads) of 16S rRNA emanating from phenanthroline-Cu(II) conjugated to position +5 of mRNA bound to the 30S ribosomal subunit.

tified several regions of both 16S and 23S rRNA which are quite accessible to such probes (Hill et al., 1986, 1990; Hill and Tassanakajohn, 1987). More recently, we have attached a phosphorothioate group to the 5′ and 3′ ends of such probes, to which a phenanthroline-Cu(II) is conjugated. By binding this to the rRNA in situ and inducing cleavage, we can determine the neighborhood of the targeted region. The distance from the tethered site to the cleavage point is no greater than 15 Å, as determined by computer modeling. However, in annealing the probe to rRNA, an A-form helix would be created, with the 5′ phenanthroline-Cu(II) moiety emerging from the side opposite the 3′ base-paired nucleotide of the target rRNA sequence (Fig. 5). A priori, it cannot be known to which side of the rRNA the phenanthroline-Cu(II) may point, but a distinct advantage of the phosphorothioate is that it is positioned on the outside of the helix.

Our initial studies using this approach have involved regions of 16S rRNA, namely, the 790 region and the decoding region near nucleotide 1400. Using phenanthroline covalently attached to a DNA oligomer complementary to nucleotides 787 to 795, we found that nucleotides 582 to 584, 693 to 694, 787 to 790, and 795 to 797 were cleaved and must lie within about 15 Å of the tethered site at the 5′ end of the DNA oligomer, which is adjacent to nucleotide 795 of 16S rRNA (Muth et al., 1999b). These results corroborate other studies which have shown that the 790 loop is proximal to the 693 loop (Atmadja et al., 1986; Malhotra and Harvey, 1994; Mundus and Wollenzien, 1998). However, our results give, for the first time, the proximity of nucleotides 582 and 583 to this region (Fig. 6). We can now place these two regions within 15 Å of each other, which will provide

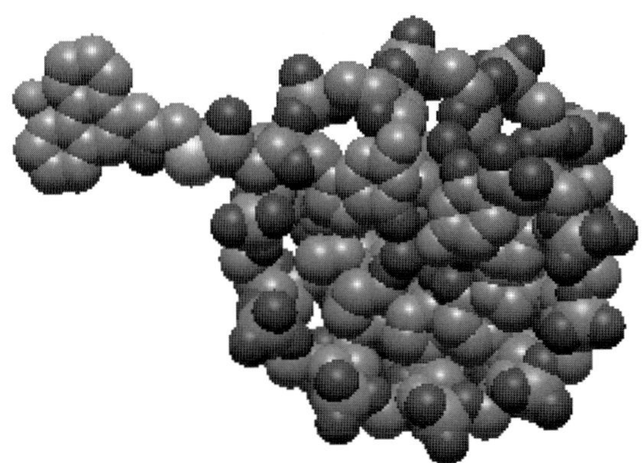

Figure 5. Spaced-filling diagram of phenanthroline-Cu(II) conjugated to a short DNA oligomer which is hybridized to RNA. The view is along the helical axis. Note that the phenanthroline can be positioned at any one of the 11 positions around the helix occupied by a nucleotide.

a determinant for the region, which now has a 15-Å deviation of uncertainty in the model published by Malhotra and Harvey (1994).

Perhaps the most interesting region we have studied is the decoding region. By placing a conjugated probe complementary to nucleotides 1396 to 1403 of 16S rRNA on both active and inactive subunits (Zamir et al., 1971), we were able to show structural changes between the two conformations. The probe was conjugated to a phenanthroline-Cu(II) through a thiol phosphate at the 5′ end of the probe, adjacent to nucleotide 1403 of 16S rRNA. We were able to show that nucleotides 923 to 929, 1190 to 1192, and 1391 to 1396 are within approximately 15 Å of this site in the inactive subunit (Fig. 6). However, in the active subunit, cleavage at all three sites disappears (Muth et al., in press). There is corroborating evidence from cross-linking studies that the 925 and 1395 regions are proximal to the decoding region (Muralikrishna and Cooperman, 1994; Alexander et al., 1994; Mundus and Wollenzien, 1998). Moazed et al. (1986) also showed a change in chemical protection as activation occurred. Evidence from our phenanthroline cleavage experiments suggests that either this region is more base paired as activation occurs or there is net movement of the region relative to position 1403 as activation occurs. The fact that the region around bulged G926 does not remain cleaved in the active conformation suggests that increased base pairing of this region is not causing the attenuation of cleavage. Therefore, relative movement between these two regions is most probable.

The placement of the spectinomycin-sensitive 1192 region (Sigmund et al., 1984) near the decoding site is not without precedent (Rinke-Appel et al., 1991, 1993; Bhangu and Wollenzien, 1992; Juzumiene and Wollenzien, 1997). In a recent model of the 30S ribosomal subunit, Mueller and Brimacombe positioned helix 28 (containing the 920 region), which connects the head of the subunit to the body, close behind helix 34 (where 1192 is positioned), with the mRNA and anticodon loops of tRNA seen in near proximity to these regions and with C1400 of the decoding region on the right side of the neck at the base of the cleft (Mueller and Brimacombe, 1997). The evidence from our study shows that nucleotides 1192 and 1193 must be within about 15 Å of the tethered site of the phenanthroline-Cu(II) complex in the inactive subunit, but upon activation, this region apparently moves relative to the decoding site. This is borne out, to a degree, by the 37-Å model derived from cryo-electron microscopy provided by Lata and coworkers (1996) showing the 30S ribosomal subunit in both inactive and active forms. The phenanthro-

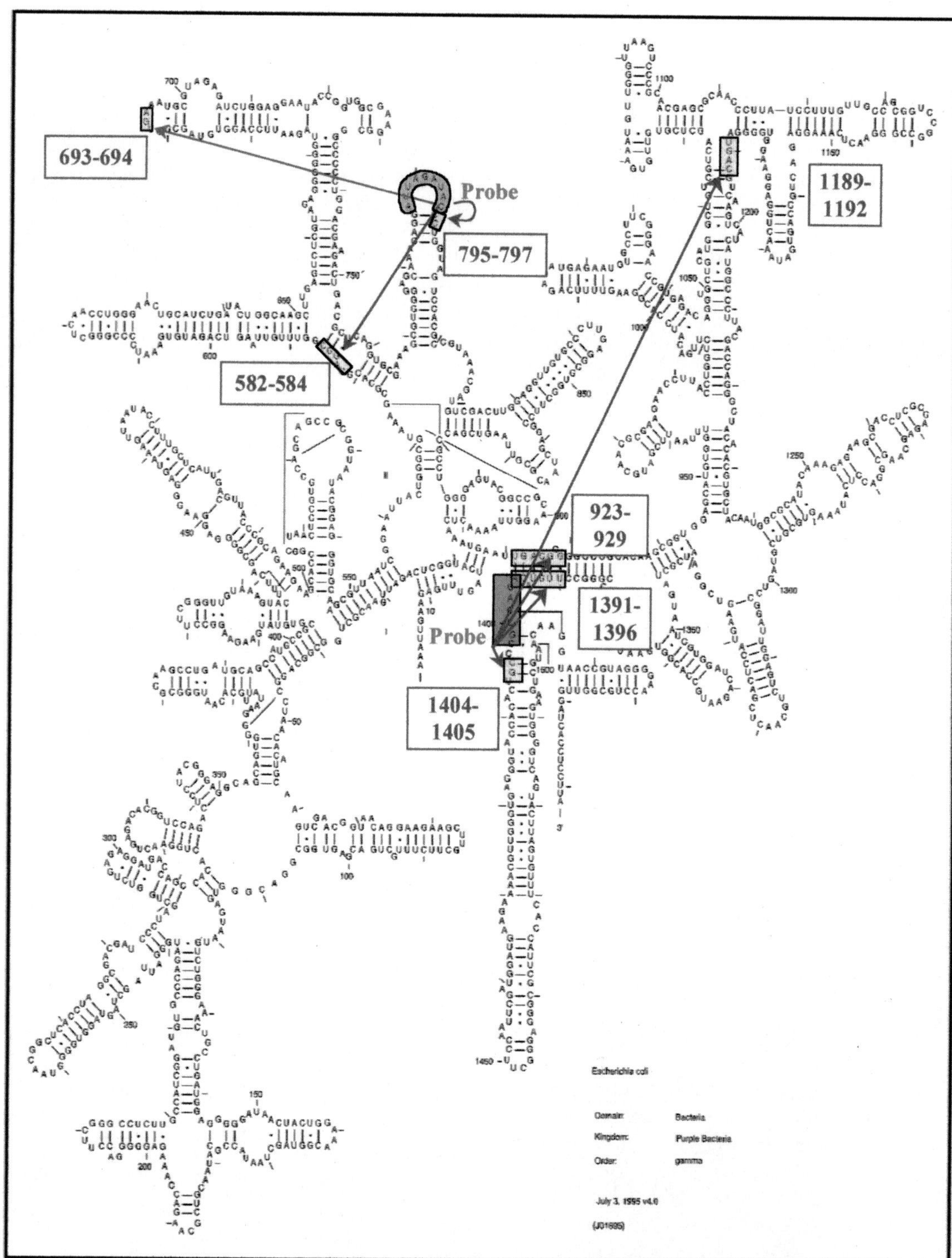

Figure 6. Map of cleavages emanating from DNA oligomers complementary to 16S nucleotides 787 to 795 and 1396 to 1403, to which phenanthroline-Cu(II) was conjugated at the 5′ ends. Cleavages from 1396 to 1403 occurred only when the 30S ribosomal subunit was in the inactive conformation (Zamir et al., 1971). Upon activation, all cleavages from the 1396-to-1403 oligomer except those occurring at nucleotides 1404 to 1405 were markedly attenuated or disappeared entirely.

line-Cu(II) cleavage results shed additional light on such movement.

There are some limitations to this approach, since the presence of a DNA oligomer, even a short one, may cause deformation of rRNA structure. However, evidence provided by assessing free phenanthroline cleavage of the rRNA both with and without the probe being present has indicated that the structural deformation must not be great. We recognize that it is possible that the terminal or penultimate nucleotides "dangle," but there is no evidence that they do.

Because of the fastidious nature of the phenanthroline-Cu(II) cleavage, it can readily be used as a proximity probe to identify movement of the rRNA. By comparing the cleavage patterns before and after an event such as activation or translocation, movement can be discovered. By using tethers of different lengths, the amount of movement may be quantified as well.

Cleavage with Phenanthroline-Co(II)

In addition to the studies with phenanthroline-Cu(II), we have also used cobalt. Results from early studies show that cobalt has properties very different from those of copper. Cobalt is an extremely poor cleavage agent, unless exogenous H_2O_2 is added. When H_2O_2 is added, phenanthroline-Co(II) becomes a very robust cleavage agent, showing patterns similar to those of phenanthroline-Cu(II), except that there is a much larger region cleaved (Fig. 7). Since there is still a very robust central intensity at sites shown to be proximal to the probe by phenanthroline-Cu(II) studies, the envelope of additional cleavages in the nearby regions suggests that a diffusible species emanates from this reaction. We believe, but have not yet proved, that this species is due to cleavage of the tether or perhaps the conjugated nucleotide, in which case the nucleotide–tether–phenanthroline-Co(II) complex itself would be diffusing. Whatever the case, the distances from the tethered phenanthroline conjugate would be roughly the same, and this becomes a different kind of proximity probe, with a much larger sphere of cleavage. Thus, with the addition of exogenous peroxide, phenanthroline-Co(II) provides diffusible cleavage emanating from the site of initial tethering of the conjugate, giving much the same pattern as phenanthroline-Cu(II) in the presence of exogenous peroxide. From these studies it is clear that other metals can be used for cleavage purposes under defined conditions.

The Peptidyltransferase Region

Recently, we have begun studies with a probe complementary to nucleotides 2580 to 2588 in the peptidyltransferase region of domain V in 23S rRNA. This probe is conjugated to phenanthroline-Cu(II) on the 3′ end, across from nucleotide 2580. Although this region was not assayed in our previous probing studies, the probe binds well to it and gives a strong RNase H cleavage, which disappears with the mismatch probe. When cleavage is induced, we see cleavage at nucleotides 2575 and 2576, proximal to the 3′ end of the probe, which would be expected. In addition, we see cleavage at nucleotides 2660 to 2665 (the α-sarcin region), 2750 to 2754, and 2775 to 2778 (Fig. 8). There may be others as well. These results are exciting, since for the first time, the proximity of the α-sarcin loop and a portion of the peptidyltransferase region is suggested.

The results obtained from cleavage directed by targeted DNA oligomers are summarized in Table 1.

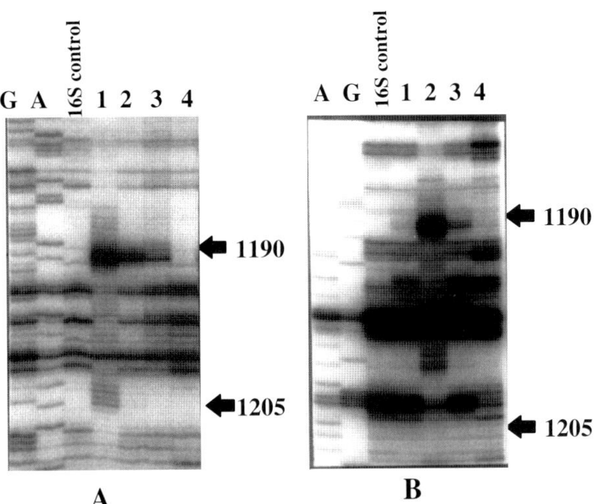

Figure 7. Cleavages of 16S rRNA emanating from a DNA oligomer complementary to 16S nucleotides 1396 to 1403, to which a phenanthroline-Co(II) [or Cu(II)] was conjugated at the 5′ end. These cleavages were done for 10 min with inactive 30S subunits. (Left) Lane 1, probe with conjugated phenanthroline-Cu(II) and H_2O_2; lane 2, probe with conjugated phenanthroline-Cu(II) without H_2O_2; lane 3, mismatch probe with conjugated phenanthroline-Co(II) and H_2O_2; lane 4, free phenanthroline-Co(II) and H_2O_2. (Right) Lane 1, probe with conjugated phenanthroline-Co(II) without H_2O_2; lane 2, probe with conjugated phenanthroline-Co(II) and H_2O_2; lane 3, mismatch probe with conjugated phenanthroline-Co(II) and H_2O_2; lane 4, free phenanthroline-Co(II) and H_2O_2.

THE FUTURE

It is clear that chemical cleavage will play a substantial role in the future in unraveling the biological structure and function of various macromolecules and macromolecular complexes. Many new approaches are being developed which will have direct

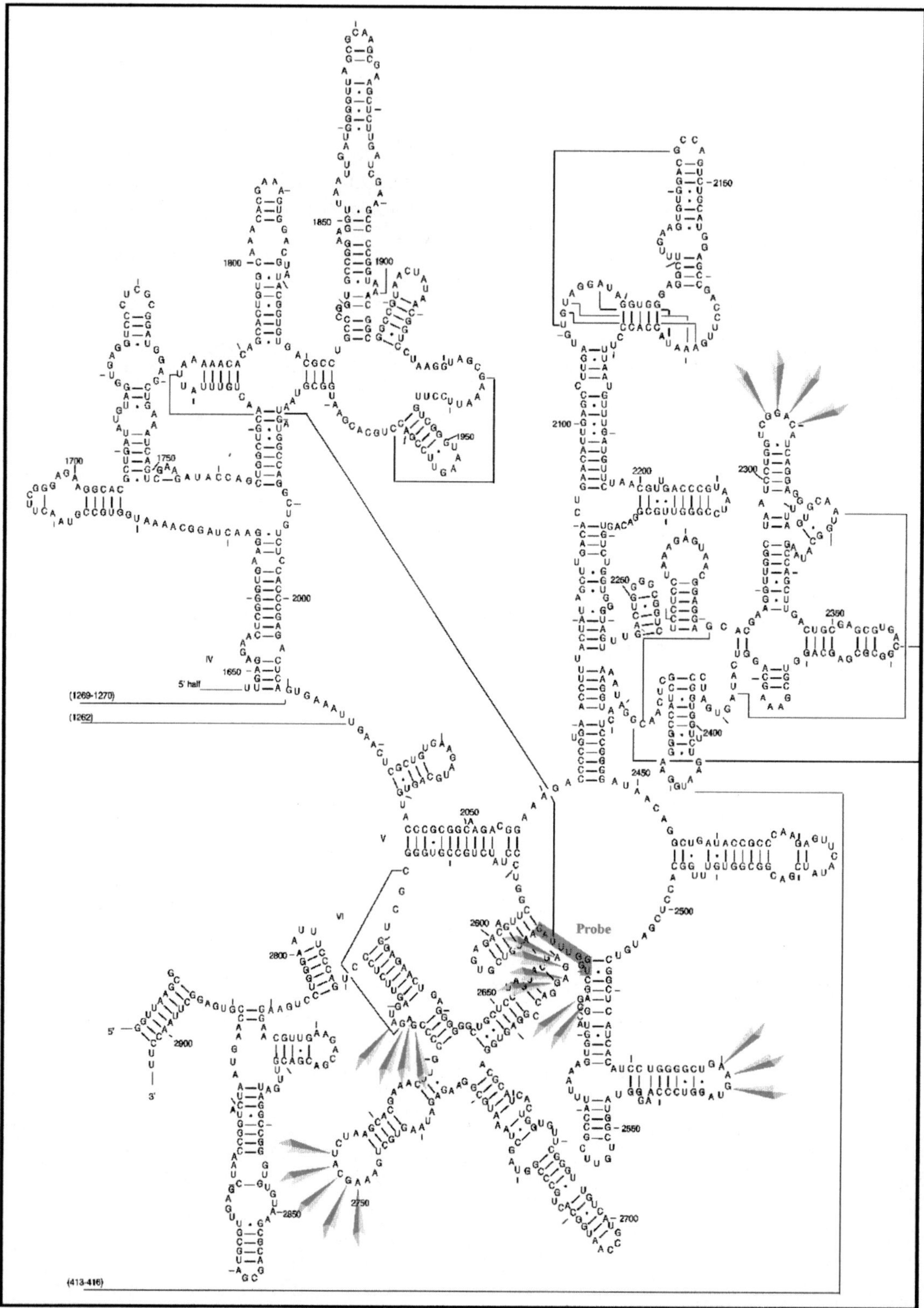

Figure 8. Map of cleavages (arrowheads) emanating from a DNA oligomer complementary to 23S nucleotides 2580 to 2588, to which phenanthroline-Cu(II) was conjugated at the 3′ end.

Table 1. Results of cleavage by targeted DNA oligomers

RNA source	DNA oligomer complementary to nucleotides[a]	Nucleotides cleaved
16S	787–795*	582–584
		693–694
		787–790
		795–797
16S	1397–1403*	923–929
		1190–1192
		1391–1396
		1404–1405
23S	2580*–2588	2308–2310
		2530–2533
		2660–2665 (the α-sarcin region)
		2575–2576
		2750–2754
		2775–2778

[a]*, position of tethered phenanthroline.

application to the ribosome. For instance, chromium derivatives have been shown to be able cleavage reagents. Cr(V) has been shown to cleave DNA readily, and it is directed toward the deoxyribose moieties (Sugden and Wetterhahn, 1997), whereas Cr(III) apparently attacks the nucleotide bases more readily (Sugden et al., 1992). Various nickel compounds have been used to probe RNA and DNA structure (Ross and Burrows, 1996; Burrows et al., 1996; Burrows and Rokita, 1996). Cobalt shows similar qualities as well (Zheng et al., 1998). These have been developed to be both base and structure specific and show considerable promise as tools for structure analysis.

Not only can different metals be used with phenanthroline for cleavage, but the length and position of the tether can be adjusted as well. We have succeeded in synthesizing several variations of the tether, which not only provide four- and six-carbon lengths but have also placed a phenyl group in the tether to make it more rigid. Using these, we should be able to bracket distances to a cleavage site, providing even more structural detail.

CONCLUSIONS

Phenanthroline cleavage provides a powerful method to identify regions of rRNA proximal to a single site on rRNA, tRNA, mRNA, protein, or translation factors. The results of such experiments provide primary evidence that two regions lie within a few angstroms of each other. These results can be used to identify regions of rRNA that may go through conformational changes during the translational processes. Such results are essential if we are to utilize the phenomenal results emanating from cryo-electron microscopy and crystallography to develop functional models of the ribosomal translation process.

The considerable arsenal of chemistries and tethers available for chemical cleavage will provide considerable latitude in designing experiments to probe the detailed structure of the dynamic ribosome in the future. Only through such studies will the dynamic behavior of the ribosome be delineated.

We specially thank Martha Rice for her expert technical assistance and William Knight for his help in computer modeling.

This project is supported, in part, by NIH Grant GM 35717.

REFERENCES

Alexander, R. W., P. Muralikrishna, and B. S. Cooperman. 1994. Ribosomal components surrounding the conserved 518–533 loop of 16S rRNA in 30S subunits. *Biochemistry* **33**:12109–12118.

Atmadja, J., W. Stiege, M. Zobawa, B. Greuer, M. Osswald, and R. Brimacombe. 1986. The tertiary folding of Escherichia coli 16S RNA, as studied by in situ intra-RNA cross-linking of 30S ribosomal subunits with bis-(2-chloroethyl)-methylamine. *Nucleic Acids Res.* **14**:659–674.

Baliga, R., J. W. Singleton, and P. B. Dervan. 1995. RecA oligonucleotide filaments bind in the minor groove of double-stranded DNA. *Proc. Natl. Acad. Sci. USA* **92**:10393-10397.

Bhangu, R., and P. Wollenzien. 1992. The mRNA binding track in the *Escherichia coli* ribosome for mRNAs of different sequences. *Biochemistry* **31**:5937–5944.

Bucklin, D. J., M. A. van Waes, J. M. Bullard, and W. E. Hill. 1997. Cleavage of 16S rRNA within the ribosome by mRNA modified in the A-site codon with phenanthroline-Cu(II). *Biochemistry* **36**:7951–7957.

Bullard, J. M., M. A. van Waes, D. J. Bucklin, and W. E. Hill. 1995. Regions of 23S ribosomal RNA proximal to transfer RNA bound at the P and E sites. *J. Mol. Biol.* **252**:572–582.

Bullard, J. M., M. A. van Waes, D. J. Bucklin, M. J. Rice, and W. E. Hill. 1998. Regions of 16S ribosomal RNA proximal to transfer RNA bound at the P-site of Escherichia coli ribosomes. *Biochemistry* **37**:1350–1356.

Bullard, J. M., S. A. Martinus, and W. E. Hill. Regions of 23S ribosomal RNA proximal to a transfer RNA acceptor stem microhelix bound at the E site of *Escherichia coli* ribosomes. *Nucleic Acids Res.*, in press.

Burkitt, M. J. 1994. Copper-DNA adducts. *Methods Enzymol.* **234**:66–79.

Burrows, C. J., and J. G. Muller. 1998. Oxidative nucleobase modifications leading to strand scission. *Chem. Rev.* **98**:1109–1151.

Burrows, C. J., and S. E. Rokita. 1996. Nickel complexes as probes of guanine sites in nucleic acid folding. *Met. Ions Biol. Syst.* **33**:537–560.

Burrows, C. J., J. G. Muller, G. T. Poulter, and S. E. Rokita. 1996. Nickel-catalyzed oxidations: from hydrocarbons to DNA. *Acta Chem. Scand.* **50**:337–344.

Chen, C. B., M. B. Gorin, and D. S. Sigman. 1993a. Sequence-specific scission of DNA by the chemical nuclease activity of 1,10-phenanthroline-copper(I) targeted by RNA. *Proc. Natl. Acad. Sci. USA* **90**:4206–4210.

Chen, C. H., A. Mazumder, J. F. Constant, and D. S. Sigman. 1993b. Nuclease activity of 1,10-phenanthroline-copper. New conjugates with low molecular weight targeting ligands. *Bioconjug. Chem.* 4:69–77.

Chen, C.-H. B., and D. S. Sigman. 1988. Sequence specific scission of RNA by 1,10 phenanthroline-copper linked to deoxyoligonucleotides. *J. Am. Chem. Soc.* **110**:6570–6572.

Chen, T., and M. M. Greenberg. 1998. Model studies indicate that copper phenanthroline induces direct strand breaks via beta-elimination of the 2′-deoxyribonolactone intermediate observed in enediyne mediated DNA damage. *J. Am. Chem. Soc.* **120**: 3815–3816.

Culver, G. M., G. M. Heilek, and H. F. Noller. 1999. Probing the rRNA environment of ribosomal protein S5 across the subunit interface and inside the 30 S subunit using tethered Fe(II). *J. Mol. Biol.* **286**:355–364.

Dervan, P. B. 1986. Design of sequence-specific DNA-binding molecules. *Science* **232**:464–471.

Dontsova, O., A. Kopylov, and R. Brimacombe. 1991. The location of mRNA in the ribosomal 30S initiation complex: site-directed cross-linking of mRNA analogues carrying several photo-reactive labels simultaneously on either side of the AUG start codon. *EMBO J.* **10**:2613–2620.

Dontsova, O., S. Dokudovskaya, A. Kopylov, A. Bogdanov, J. Rinke-Appel, N. Jünke, and R. Brimacombe. 1992. Three widely separated positions in the 16S RNA lie in or close to the ribosomal decoding region: a site-directed cross-linking study with mRNA analogues. *EMBO J.* **11**:3105–3116.

Dreyer, G. B., and P. B. Dervan. 1985. Sequence-specific cleavage of single-stranded DNA: oligodeoxynucleotide-EDTA X Fe(II). *Proc. Natl. Acad. Sci. USA* **82**:968–972.

Gallagher, J., C. H. Chen, C. Q. Pan, D. M. Perrin, Y. M. Cho, and D. S. Sigman. 1996. Optimizing the targeted chemical nuclease activity of 1,10-phenanthroline-copper by ligand modification. *Bioconjug. Chem.* **7**:413–420.

Han, H., and P. B. Dervan. 1994. Visualization of RNA tertiary structure by RNA-EDTA·Fe(II) autocleavage: analysis of tRNA(Phe) with uridine-EDTA·Fe(II) at position 47. *Proc. Natl. Acad. Sci. USA* **91**:4955–4959.

Heilek, G. M., and H. F. Noller. 1996a. Directed hydroxyl radical probing of the rRNA neighborhood of ribosomal protein S13 using tethered Fe(II). *RNA* **2**:597–602.

Heilek, G. M., and H. F. Noller. 1996b. Site-directed hydroxyl radical probing of the rRNA neighborhood of ribosomal protein S5. *Science* **272**:1659–1662.

Hertzberg, R. P., and P. B. Dervan. 1982. Cleavage of double-helical DNA by (methidiumpropyl-EDTA)-iron(II). *J. Am. Chem. Soc.* **104**:313–315.

Hertzberg, R. P., and P. B. Dervan. 1984. Cleavage of DNA with methidiumpropyl-EDTA-iron(II): reaction conditions and product analyses. *Biochemistry* **23**:3934–3945.

Hill, W., W. Tapprich, and A. Tassanakajohn. 1986. Probing ribosomal structure and function, p. 233–252. *In* B. Hardesty, (ed.), *The Structure, Function and Genetics of Ribosomes.* Springer-Verlag, New York, N.Y.

Hill, W. E., and A. Tassanakajohn. 1987. Probing ribosome structure using short oligodeoxyribonucleotides: the question of resolution. *Biochimie* **69**:1071–1080.

Hill, W. E., J. W. Weller, T. Gluick, C. Merryman, R. Marconi, and W. E. Tapprich. 1990. Probing the function and structure of the ribosome using short, complementary DNA oligomers, p. 253–261. *In* W. E. Hill, A. E. Dahlberg, R. A. Garrett, P. B. Moore, D. Schlessinger, and J. R. Warner (ed.), *The Ribosome: Structure, Function, and Evolution.* American Society for Microbiology, Washington, D.C.

Hill, W. E., D. J. Bucklin, J. M. Bullard, A. L. Galbraith, N. V. Jammi, C. C. Rettberg, B. S. Sawyer, and M. A. van Waes. 1995. Identification of ribosome-ligand interactions using cleavage reagents. *Biochem. Cell Biol.* **73**:1033–1039.

Huttenhofer, A., and H. F. Noller. 1992. Hydroxyl radical cleavage of tRNA in the ribosomal P site. *Proc. Natl. Acad. Sci. USA* **89**:7851–7855.

Huttenhofer, A., and H. F. Noller, 1994. Footprinting mRNA-ribosome complexes with chemical probes. *EMBO J.* **13**:3892–3901.

Johnson, G. R. A., and N. B. Nazhat. 1987. Kinetics and mechanism of the reaction of the bis(1,10-phenanthroline)copper(I) ion with hydrogen peroxide in aqueous solution. *J. Am. Chem. Soc.* **109**:1990–1994.

Juzumiene, D., and P. Wollenzien. 1997. Determination of the 16S ribosomal RNA folded structure with site-directed photoreactive reagents in the 5′ and central pseudoknot region. *Nucleic Acids Symp. Ser.* **36**:168–170.

Kuwabara, M. D., and D. S. Sigman. 1987. Footprinting DNA-protein complexes in situ following gel retardation assays using 1,10-phenanthroline-copper ion: Escherichia coli RNA polymerase-lac promoter complexes. *Biochemistry* **26**:7234–7238.

Lata, K. R., R. K. Agrawal, P. Penczek, R. Grassucci, J. Zhu, and J. Frank. 1996. Three-dimensional reconstruction of the Escherichia coli 30S ribosomal subunit on ice. *J. Mol. Biol.* **262**:43–52.

Lieberman, K. R., and H. F. Noller. 1998. Ribosomal protein L15 as a probe of 50 S ribosomal subunit structure. *J. Mol. Biol.* **284**: 1367–1378.

Malhotra, A., and S. C. Harvey. 1994. A quantitative model of the Escherichia coli 16S RNA in the 30S ribosomal subunit. *J. Mol. Biol.* **240**:308–340.

Meijler, M. M., O. Zelenko, and D. S. Sigman. 1997. Chemical mechanism of DNA scission by (1,10-phenanthroline)copper. Carbonyl oxygen of 5-methylenefuranone is derived from water. *J. Am. Chem. Soc.* **119**:1135–1136.

Merryman, C., D. Moazed, J. McWhirter, and H. F. Noller. 1999. Nucleotides in 16S rRNA protected by the association of 30S and 50S ribosomal subunits. *J. Mol. Biol.* **285**:97–105.

Moazed, D., and H. F. Noller. 1989. Intermediate states in the movement of transfer RNA in the ribosome. *Nature* **342**:142–148.

Moazed, D., B. Van Stolk, S. Douthwaite, and H. F. Noller. 1986. Interconversion of active and inactive 30S ribosomal subunits is accompanied by a conformational change in the decoding region of 16S RNA. *J. Mol. Biol.* **191**:483–493.

Mueller, F., and R. Brimacombe. 1997. A new model for the three-dimensional folding of Escherichia coli 16S ribosomal RNA. I. Fitting the RNA to a 3D electron microscopic map at 20 A. *J. Mol. Biol.* **271**:524–544.

Mundus, D., and P. Wollenzien. 1998. Neighborhood of 16S rRNA nucleotides U788/U789 in the 30S ribosomal subunit determined by site-directed crosslinking. *RNA* **4**:1373–1385.

Muralikrishna, P., and B. S. Cooperman. 1994. A photolabile oligodeoxyribonucleotide probe of the decoding site in the small subunit of the Escherichia coli ribosome: identification of neighboring ribosomal components. *Biochemistry* **33**:1392–1398.

Muth, G. W., C. M. Thompson, and W. E. Hill. 1999a. Cleavage of a 23S rRNA pseudoknot by phenanthroline-Cu(II). *Nucleic Acids Res.* **27**:1906–1911.

Muth, G. W., S. Hennelley, and W. E. Hill. 1999b. Positions in the 30S ribosomal subunit proximal to the 790 loop as determined by phenanthroline cleavage. *RNA* **5**:856–864.

Muth, G. W., S. P. Hennelley, and W. E. Hill. Conformational changes of the rRNA within 30S ribosomal subunits as elucidated by phenanthroline cleavage. *Biochemistry*, in press.

Newcomb, L. F., and H. F. Noller. 1999. Directed hydroxyl radical probing of 16S ribosomal RNA in 70S ribosomes from internal positions of the RNA. *Biochemistry* **38**:945–951.

Noller, H. F., R. Green, G. Heilek, V. Hoffarth, A. Huttenhofer, S. Joseph, I. Lee, K. Lieberman, A. Mankin, and C. Merryman. 1995. Structure and function of ribosomal RNA. *Biochem. Cell Biol.* **73**:997–1009. (Erratum, **74**:417, 1996.)

Oakley, M. G., and P. B. Dervan. 1990. Structural motif of the GCN4 DNA binding domain characterized by affinity cleaving. *Science* **248**:847–850.

Pearson, L., C. B. Chen, R. P. Gaynor, and D. S. Sigman. 1994. Footprinting RNA-protein complexes following gel retardation assays: application to the R-17-procoat-RNA and tat-TAR interactions. *Nucleic Acids Res.* **22**:2255–2263.

Perrin, D. M., A. Mazumder, and D. S. Sigman. 1996. Oxidative chemical nucleases. *Prog. Nucleic Acid Res. Mol. Biol.* **52**:123–151.

Pogozelski, W. K., and T. D. Tullius. 1998. Oxidative strand scission of nucleic acids: routes initiated by hydrogen abstraction from the sugar moiety. *Chem. Rev.* **98**:1089–1107.

Pope, L. E., and D. S. Sigman. 1984. Secondary structure specificity of the nuclease activity of the 1,10- phenanthroline-copper complex. *Proc. Natl. Acad. Sci. USA* **81**:3–7.

Powers, T., and H. F. Noller. 1995. Hydroxyl radical footprinting of ribosomal proteins on 16S rRNA. *RNA* **1**:194–209.

Que, B. G., K. M. Downey, and A. G. So. 1980. Degradation of deoxyribonucleic acid by a 1,10-phenanthroline-copper complex: the role of hydroxyl radicals. *Biochemistry* **19**:5987–5991.

Rinke-Appel, J., N. Junke, K. Stade, and R. Brimacombe. 1991. The path of mRNA through the Escherichia coli ribosome; site-directed cross-linking of mRNA analogues carrying a photo-reactive label at various points 3′ to the decoding site. *EMBO J.* **10**:2195–2202.

Rinke-Appel, J., N. Jünke, R. Brimacombe, S. Dokudovskaya, O. Dontsova, and A. Bogdanov. 1993. Site-directed cross-linking of mRNA analogues to 16S ribosomal RNA: a complete scan of cross-links from all positions between '+1' and '+16' on the mRNA, downstream from the decoding site. *Nucleic Acids Res.* **21**:2853–2859.

Ross, S. A., and C. J. Burrows. 1996. Cytosine-specific chemical probing of DNA using bromide and monoperoxysulfate. *Nucleic Acids Res.* **24**:5062–5063.

Sigman, D. S. 1990. Chemical nucleases. *Biochemistry* **29**:9097–9105.

Sigman, D. S., D. R. Graham, V. D'Aurora, and A. M. Stern. 1979. Oxygen-dependent cleavage of DNA by 1,10-phenanthroline-cuprous complex. *J. Biol. Chem.* **254**:12269–12272.

Sigman, D. S., C. H. Chen, and M. B. Gorin. 1993a. Sequence-specific scission of DNA by RNAs linked to a chemical nuclease. *Nature* **363**:474–475.

Sigman, D. S., A. Mazumder, and D. M. Perrin. 1993b. Chemical nucleases. *Chem. Rev.* **93**:2295–2316.

Sigmund, C. D., M. Ettayebi, and E. A. Morgan. 1984. Antibiotic resistance mutations in 16 S and 23 S ribosomal RNA genes of Escherichia coli. *Nucleic Acids Res.* **12**:4653–4663.

Sluka, J. P., S. J. Horvath, A. C. Glasgow, M. I. Simon, and P. B. Dervan. 1990. Importance of minor-groove contacts for recognition of DNA by the binding domain of Hin recombinase. *Biochemistry* **29**:6551–6561.

Strobel, S. A., and P. B. Dervan. 1990. Site-specific cleavage of a yeast chromosome by oligonucleotide-directed triple-helix formation. *Science* **249**:73–75.

Sugden, K. D., R. D. Geer, and S. J. Rogers. 1992. Oxygen radical-mediated DNA damage by redox-active Cr(III) complexes. *Biochemistry* **31**:11626–11631.

Sugden, K. D., and K. E. Wetterhahn. 1997. Direct and hydrogen peroxide-induced chromium(V) oxidation of deoxyribose in single-stranded and double-stranded calf thymus DNA. *Chem. Res. Toxicol.* **10**:1397–1406.

Van Dyke, M. W., and P. B. Dervan. 1983. Footprinting with MPE·Fe(II). Complementary-strand analyses of distamycin- and actinomycin-binding sites on heterogeneous DNA. *Cold Spring Harb. Symp. Quant. Biol.* **47**:347–353.

Wilson, K. S., and H. F. Noller. 1998. Mapping the position of translational elongation factor EF-G in the ribosome by directed hydroxyl radical probing. *Cell* **92**:131–139.

Zamir, A., R. Miskin, and D. Elson. 1971. Inactivation and reactivation of ribosomal subunit amino acyl-transfer RNA binding activity of the 30S subunit of E. coli. *J. Mol. Biol.* **60**:347–364.

Zheng, P., C. J. Burrows, and S. E. Rokita, 1998. Nickel- and cobalt-dependent reagents identify structural features of RNA that are not detected by dimethyl sulfate or RNase T1. *Biochemistry* **37**:2207–2214.

The Ribosome: Structure, Function, Antibiotics, and Cellular Interactions
Edited by R. A. Garrett, S. R. Douthwaite, A. Liljas, A. T. Matheson, P. B. Moore, and H. F. Noller

Chapter 23

Applying Photolabile Derivatives of Oligonucleotides To Probe the Peptidyltransferase Center

BARRY S. COOPERMAN, SERGUEI N. VLADIMIROV, YURI BUKHTIYAROV, ZHANNA DRUZINA, RUO WANG, and HYUK-SOO SEO

It is increasingly clear that rRNA, rather than being a mere framework for the assembly of ribosomes, is directly involved in ribosome function (Green and Noller, 1997). It is therefore quite important to identify ribosomal components in the vicinity of the rRNA sequences having particular significance for ribosome structure and function. Here we describe an approach we have been using (Muralikrishna and Cooperman, 1991, 1994, 1995; Alexander et al., 1994; Alexander and Cooperman, 1998; Muralikrishna et al., 1997; Wang et al., 1999; Bukhtiyarov et al., 1999; Cooperman et al., in press; Vladimirov et al., in press) to form defined photo-cross-links from targeted RNA sites within the ribosome, with an emphasis on those that are important functionally. In this approach, radioactive, photolabile derivatives of oligonucleotides (PHONTs) having sequences complementary to rRNA sequences are bound to their targeted sequences in intact ribosomal subunits and, on photolysis, form cross-links with neighboring ribosomal components. Such cross-links can be used, along with related data of others, as constraints for the construction and testing of proposed models of ribosome structure, as well as to describe conformational changes within the dynamic ribosome (Wilson and Noller, 1998).

The PHONT approach derives from earlier work of Bogdanov and colleagues (Skripkin et al., 1979; Mankin et al., 1981), who first introduced oligoDNAs that are complementary to rRNA sequences to probe the structure of rRNA, both of the native molecule and within ribosomal subunits. More recently, Hill and coworkers (1990) demonstrated that single-stranded regions of rRNA form stable complexes with their complementary oligoDNAs, as evidenced by filter binding assays and by the demonstration that treatment of the complex with RNase H cleaves rRNA at the appropriate position. They and others have exploited this approach in determining the functional importance of targeted rRNA sequences in both 16S and 23S rRNAs (Marconi and Hill, 1988, 1989; Marconi et al., 1990; Weller and Hill, 1992, 1994; Lodmell et al., 1993; Agrawal and Burma, 1996; Meyer et al., 1996; Azad et al., 1998). A further application has been to attach electron microscopic (EM) markers to complementary oligoDNAs and to visualize complexes of such oligoDNAs bound to ribosome subunits, thus obtaining a three-dimensional localization of the target rRNA sequence (Lasater et al., 1989, 1990; Oakes and Lake, 1990; Oakes et al., 1990; McWilliams and Glitz, 1991).

The PHONT approach offers several advantages. First, it allows targeting of sequences of particular functional or structural significance throughout the ribosome structure. Second, the cross-links formed provide a defined upper-limit distance for the separation of the linked components within the ribosome, given by the length of the tether. As this length can be varied, the approach can be used to identify components both immediately neighboring the target site and somewhat farther away. Further, placement of the photolabile group at different positions within a series of PHONT molecules targeting the same rRNA sequence allows exploration of a considerable volume of ribosome space surrounding the target sequence (Fig. 1) (Wang et al., 1999). Third,

Barry S. Cooperman, Serguei N. Vladimirov, Zhanna Druzina, and Hyuk-Soo Seo ■ Department of Chemistry, University of Pennsylvania, Philadelphia, PA 19104-6323. **Yuri Bukhtiyarov** ■ DuPont Pharmaceuticals Co., Experimental Station E400/3410, Wilmington, DE 19880. **Ruo Wang** ■ Schering-Plough Research Institute, 2015 Galloping Hill Rd., K-15-3-3545, Kenilworth, NJ 07033-0539.

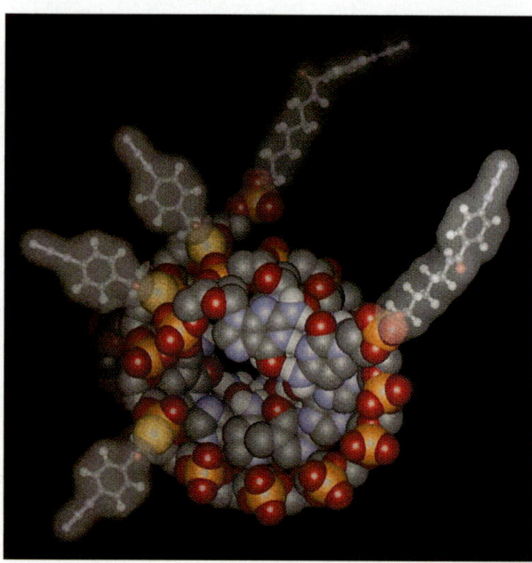

Figure 1. Space-filling model of heteroduplex between PHONTs and nt 517 to 527 in 16S rRNA. The tethered aryl azides correspond to five different PHONTs targeting this sequence (Wang et al., 1999).

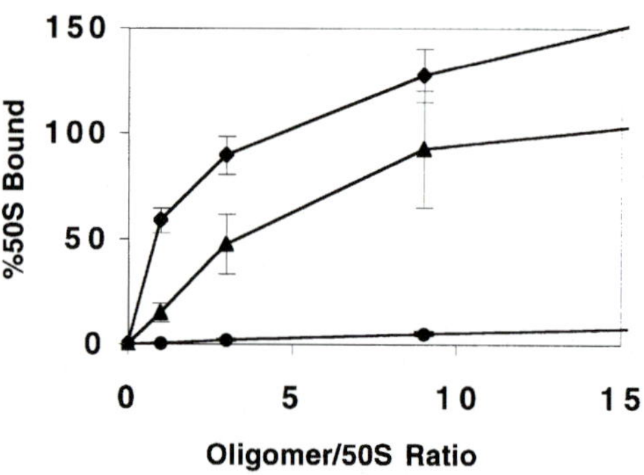

Figure 2. Noncovalent binding of oligonucleotides to 50S subunits. ♦, 2′-OMe-RNA-p*2612-2604; ●, cDNA-p*2612-2604; ▲, 2′-OMe-RNA-p*2258-2253/52(S)-2248. The last oligonucleotide is the precursor to PHONT 5 (Fig. 3), in which a nonbridging oxygen on the phosphoryl group connecting nucleotides complementary to G2253 and G2252 is replaced by a sulfur. The error bars indicate average deviations. Asterisks denote ^{32}P-labeled material.

sample preparation, utilizing readily synthesized PHONTs, intact ribosomes, and, in some cases, ribosomal ligands, is straightforward, an important attribute given the large number of constraints for construction and testing of structural models and for describing conformational changes.

Below, we first describe the PHONT approach in general before presenting the results of recent applications of the approach to the study of the peptidyltransferase center (PTC).

THE PHONT APPROACH

PHONT Design

We identify four principal elements in PHONT design: first, the backbone structure; second, the placement of the photolabile group within the oligonucleotide sequence; third, the length and flexibility of the tether linking the photolabile group with the oligonucleotide backbone; and fourth, the introduction of radioactivity. As is evident in the descriptions that follow, PHONT design continues to evolve.

Backbone structure

We synthesize oligonucleotides by using standard phosphoramidite chemistry. In most of our published work we have used oligoDNAs as PHONTs. More recently we have switched to 2′-OMe oligoRNAs because they bind more tightly to their target sites (Inoue et al., 1987). The results shown in Fig. 2 demonstrate this effect and are typical in showing two-phase binding, with high-affinity (K_d, approximately micromolar) sites of stoichiometry, 0.1 to 0.9 per subunit, and an indeterminate number of low-affinity sites. However, 2′-OMe oligoRNAs do not generate an RNase H-cleavable site on binding to the target site, and such a site is useful both for demonstrating target site binding and for identifying labeled sequences by RNase H analysis (see below). Therefore, a further refinement is to use a chimeric PHONT, made up of four consecutive DNA residues, to generate an RNase H site (Inoue et al., 1988), with the remainder of the residues being 2′-OMe RNAs. This allows retention of the RNase H site while binding to its target with high affinity. Such chimeric PHONTs are now in general use in our laboratory.

Placement of photolability within the oligonucleotide sequence

We introduce photolability by (i) placing a phenylazide or non-sequence complementary s^{4}U at either the 3′ or 5′ terminus of the PHONT, either by acylation of an amino group placed synthetically at either terminus with *N*-hydroxysuccinimidyl-4-azidobenzoate (HSAB) or by direct oligonucleotide synthesis (s^{4}U) or (ii) replacing a phosphoryl with a thiophosphoryl group at any position within the oligonucleotide and alkylating the sulfur with *p*-azidophenacyl bromide (APAB). These methods generate the PHONTs shown in Fig. 3.

Most of our early work employed the former method, because of the ease with which PHONTs having variable tether lengths could be synthesized. However, we now consider thiophosphoryl derivatization to be generally superior for two reasons. First,

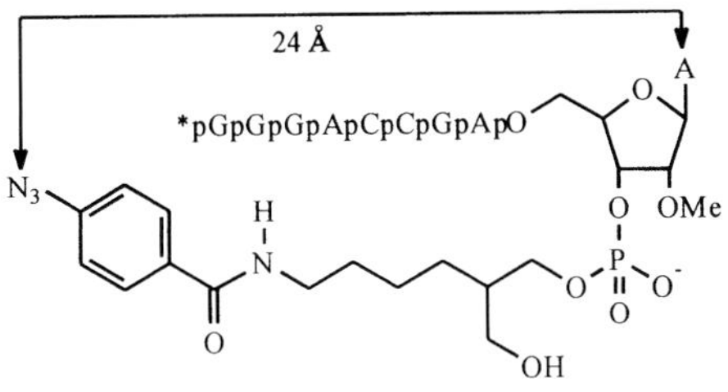
24 Å
*pGpGpGpApCpCpGpApO
A
N3
H
N
O
OMe
OH

1 *p2612-2604-3'-ABA

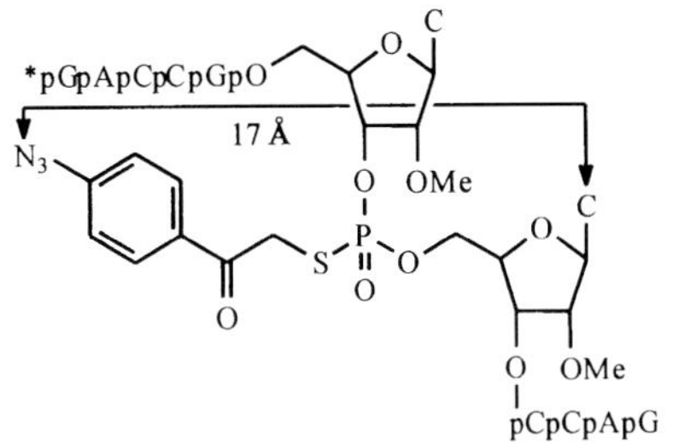
*pGpApCpCpGpO
C
17 Å
N3
OMe
S
C
OMe
pCpCpApG

5 *p2258-2252(SAz)-2248

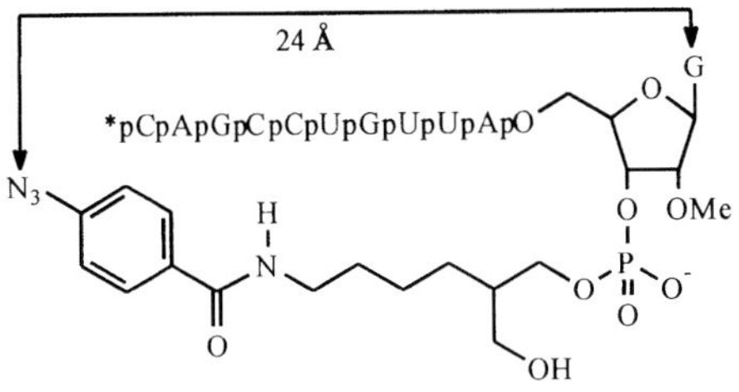
24 Å
*pCpApGpCpCpUpGpUpUpApO
G
N3
H
N
OMe
OH

2 *p2458-2448-3'-ABA

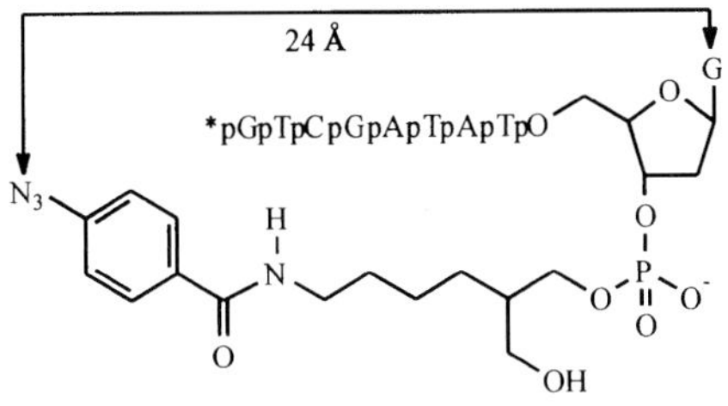
24 Å
*pGpTpCpGpApTpApTpO
G
N3
H
N
OH

6 *p2483-2475-3'-ABA

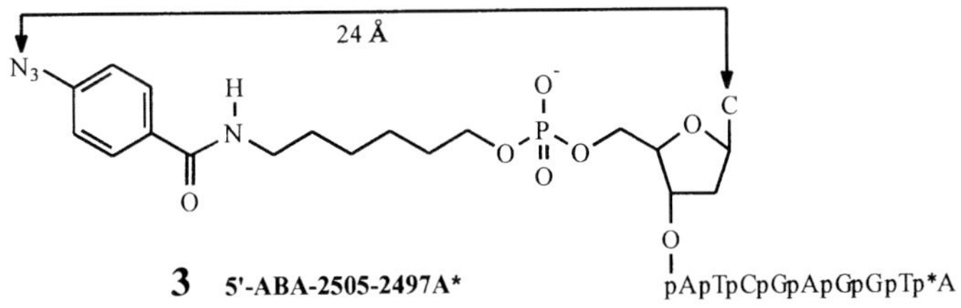
24 Å
N3
H
N
C
pApTpCpGpApGpGpTp*A

3 5'-ABA-2505-2497A*

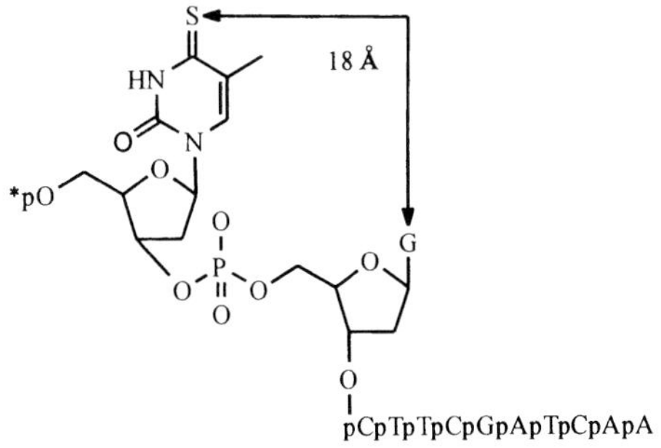
S
18 Å
HN
N
*pO
G
pCpTpTpCpGpApTpCpApA

7 *p-S4T-1892-1882

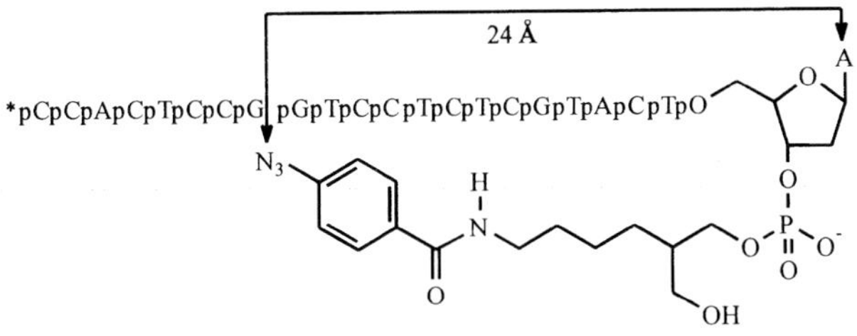
24 Å
*pCpCpApCpTpCpCpG
pGpTpCpCpTpCpTpCpGpTpApCpTpO
A
N3
H
N
OH

4 *p2674-2653-3'-ABA

Figure 3. Structures of PHONTs **1** to **7**. The distances between the photogenerated nitrenes (or 4-thio position) and the bases complementary to rRNA nucleotides are indicated. ABA, azidobenzoylamide; SAz, *S*-*p*-azidophenacyl derivative of a thiophosphate; *p and A*, ^{32}P-labeled phosphoryl and adenyl groups placed at the 5′ and 3′ positions, respectively.

it allows us to perform photo-cross-linking experiments from any phosphoryl group within the PHONT molecule. Second, PHONTs derivatized at internal positions are not subject to the uncertainty that the effective tether length could be larger than that calculated if there is some unraveling of the heteroduplex at the terminal positions.

Many other options exist for introducing photolability at internal PHONT positions. Phosphoramidite synthons are available for the introduction of s^4U, s^2dT, Br^5dT, and Br^5dC. In addition, synthons for Br^8dA and Br^8dG can be used to introduce azido^{8}G, or azido^{8}A, via displacement of Br-8 by an azide ion. PHONTs incorporating such nucleotides ("zero-length cross-linkers") are useful for studies in which the goal is to identify only the closest ribosomal component neighboring the complemented base in rRNA. Additionally, reagents are available that utilize purine or pyrimidine phosphoramidite synthons modified with leaving groups that are displaceable by nucleophiles after oligonucleotide synthesis has been completed and for which Watson-Crick base pairing is preserved. These synthons have been called "convertible" nucleosides (MacMillan and Verdine, 1991). For example, following oligoDNA synthesis employing the O^6-phenyl derivative of dI, the O^6-phenyl group can be displaced with linear diamines of the type $H_2N(CH_2)_nNH_2$. Condensation of the resulting free amine with a photolabile reagent such as HSAB permits the introduction of photolability at a variable distance from the resulting adenosine residue.

Tether length and flexibility

There is a clear trade-off between using shorter versus longer tethers and between more rigid versus flexible tethers. PHONTs utilizing zero-length cross-linkers provide tighter constraints for model building and testing. On the other hand, a PHONT (e.g., 3 [Fig. 3]) in which azide is attached to the backbone not only by a long and relatively flexible tether but at a terminus as well can access a wide range of neighboring components, both those that are quite far away (up to 25 to 30 Å from the complemented rRNA nucleotide if unraveling occurs) and those that are considerably closer (because the photolabile group could bend back toward the oligonucleotide backbone). The cost of such versatility is the generation of looser constraints. In between these extremes is, for example, PHONT 5, in which the tether is of intermediate size (9 Å from the PHONT phosphate backbone to the nitrene), is attached at an interior position, and is rather rigid (clearly, no doubling back is possible).

Which tether length to employ depends on the application. Experiments directed toward structure modeling through identification of ribosomal components that neighbor a targeted RNA sequence are best performed with zero-length and intermediate tethers, in order to identify neighboring ribosomal components over a significant but limited distance range. In principle, all three tether classes can be used to study conformational change. However, given the current state of ignorance about the detailed nature of conformational change in the ribosome, more information should be forthcoming from the use of intermediate and long-range tethers, which allow exploration of a wider range of ribosomal space. In any case, it makes good sense to use more than one tether class for a target site that binds PHONTs tightly, with high stoichiometry and minimal subunit perturbation (see below), to ensure that the maximum useful information is extracted.

Placement of radioactivity

Radioactivity may be conveniently introduced into either the 5′ end, with γ-^{32}P-ATP and T4 polynucleotide kinase, or the 3′ end, with α-^{32}P-ATP and deoxynucleotidyl transferase. Generally, we have introduced radioactivity and photolability at separate positions in a PHONT. However, introduction at the same position offers the advantage in identifying labeled proteins that subjecting labeled proteins to complete hydrolysis by S1 nuclease would result in modified, radioactive proteins differing from unmodified proteins only by the incorporation of a single nucleotide. Such modified proteins should show little change from unmodified protein in migration on polyacrylamide gel electrophoresis (PAGE) analysis and elution on reverse-phase high-performance liquid chromatography (RP-HPLC) analysis, thereby simplifying identification. Simultaneous introduction of radioactivity and photolability is possible with ^{3}H-HSAB or ^{14}C-APAB, both of which are commercially available.

Availability of Sites within the Ribosome for PHONT Binding

Our recent work and that of others suggests that many rRNA sequences are available for binding by a PHONT coming from solution. Most of the PHONTs we have used are 9 to 11 nucleotides (nt) long and target single-stranded sequences. Such sequences in general present the most attractive targets for our approach, since they are the best studied both from a functional point of view and with regard to their availability for binding within subunits. However, we have also shown that successful experiments can be performed with a longer, 22-nt PHONT (PHONT 4 [Fig. 3]) targeting the proximal stem in addition to the loop of the α-sarcin stem-loop region. Thus, it is feasible to consider both loop and stem-loop regions as potential targets for the PHONT approach. In addition, Tranque et al. (1998), in testing the hypothesis that mRNAs might interact with ribosomes by complementary duplex formation, have recently found that the large majority of mRNA-derived PHONTs tested formed target-specific cross-links with 18S rRNA within the mouse 40S subunit. In this work, s^4U was the source of photolability. More than half of these 18S rRNA targets correspond to sequences within 16S rRNA, suggesting that these sequences would be available for PHONT binding to 30S subunits as well.

Identification of Labeled Ribosomal Components

RNA

Localization of photoincorporation sites within 16S or 23S rRNA is accomplished in a two-step process, a partial localization employing RNase H digestion followed by identification of specific labeled nucleotides by using pauses or halts in reverse transcriptase primer extension. The first step, which monitors ^{32}P-labeled-PHONT radioactivity, can be carried out even for cross-links formed in very low yield. The second step, which measures pauses or halts in reverse transcription of photolabeled rRNA, requires significant labeling stoichiometry (≥0.2 to 0.5%) to be successful.

RNase H. Sites are first localized to limited sequences by hybridizing ^{32}P-labeled-PHONT rRNA with pairs of oligoDNAs complementary to two different sequences, digestion with RNase H, and, following PAGE separation of fragments, determination of radioactivity in the resulting RNA fragments by autoradiography (Brimacombe et al., 1990). Interestingly, oligoDNA probe-labeled rRNA is itself a substrate for RNase H. This is because the specific heteroduplex between photolyzed, covalently incorporated oligoDNA probe and rRNA is typically re-formed under the conditions used for RNase H digestion and provides a site for RNase H cleavage (see above).

Because the RNase H step allows easy quantification of the extent of labeling, it is particularly useful for testing the effects of variables such as light fluence, added complementary oligonucleotide as a competitor, and PHONT concentration. Sample data are presented in Fig. 4. This step is typically carried

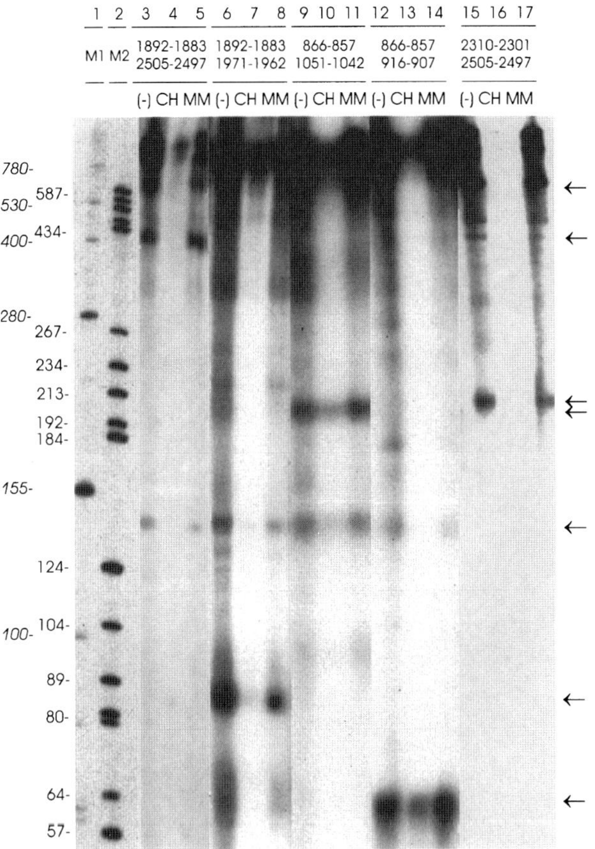

Figure 4. RNase H digestion of 50S rRNA labeled by PHONTs 1 and 2. Labeled 50S RNA isolated from 50S subunits photolyzed in the presence of PHONT 1 or 2 was incubated with two equivalents of the indicated cDNA probe(s) and digested with RNase H. The cleavage products were subjected to urea-PAGE and visualized by autoradiography. Photolabeling experiments were carried out either in the absence of competitor 2′-OMe-oligoRNA [lanes (-)] or in the presence of a 10-fold excess (over the 50S subunit) of 2′-OMe-oligoRNA either complementary to nt 2604 to 2612 (PHONT 1 target) or 2448 to 2458 (PHONT 2 target) (lanes CH) or containing mismatches (lanes MM) to these two targets. Lane 1, RNA size markers. Lane 2, DNA size markers, with sizes (in nucleotides) indicated to the left of the gel. RNA sizes are italicized. Lanes 3 to 14, 23S rRNA labeled with photoprobe 1, digested with cDNAs 1892 to 1883 and 2505 to 2497 (lanes 3 to 5); cDNAs 1892 to 1883 and 1971 to 1962 (lanes 6 to 8); cDNAs 866 to 857 and 1051 to 1042 (lanes 9 to 11); cDNAs 866 to 857 and 916 to 907 (lanes 12 to 14). Lanes 15 to 17, 23S rRNA labeled with photoprobe 2, digested with cDNAs 2505 to 2497 and 2310 to 2301. The arrows point to specifically labeled RNase H fragments.

out at ribosomes in large excess over PHONT, in order to maximize the fraction of photoincorporation arising from PHONTs bound to the high-affinity target site (Fig. 2).

A drawback of the use of oligoDNAs to generate RNase H sites is that, because RNase H can cut each RNA-DNA heteroduplex at several positions, fragment formation by cutting at two sites generates a family of labeled fragments, typically spreading over 10 to 20 nt, and limiting localization to 50 nt or so. In a recent improvement, we have generated RNase H cleavage sites with chimeric oligonucleotides containing four consecutive DNA residues, with the remainder of the residues being 2′-OMe RNAs, rather than with oligoDNAs. This results in RNase H cuts at single positions within each of the heteroduplexes (Inoue et al., 1988), giving rise to a single labeled fragment and making it feasible to localize labeled RNase H fragments to ~10 nt.

Reverse transcriptase primer extension. In the second step of localization, PHONT-labeled rRNA is hybridized with single-stranded oligoDNA primer complementary to the 3′ side of each region of rRNA found to be labeled in the first step. This heteroduplex is then used as a substrate for reverse transcriptase (Barta et al., 1984), and the exact position (or positions) of labeling is identified with a sequencing gel as the position(s) following that (or those) for which a halt or pause is observed (Fig. 5). As controls, analyses are also carried out of samples from unmodified subunits, from subunits irradiated in the absence of PHONTs, and from samples incubated with PHONTs but not irradiated. Other controls are discussed below.

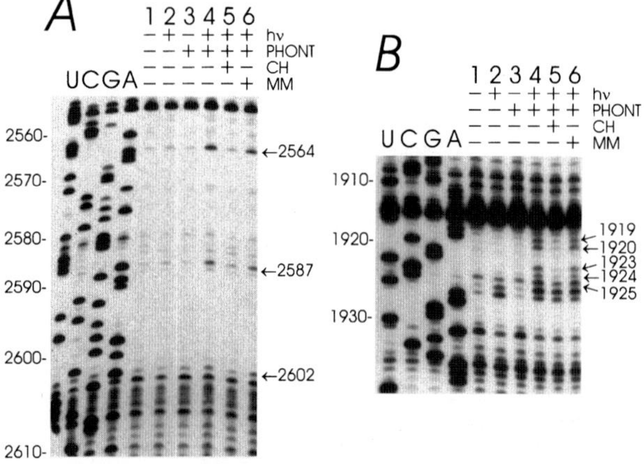

Figure 5. Reverse transcriptase analyses of PHONT 1-labeled 23S RNA extracted from PHONT 1-labeled 50S subunits. Lanes U, C, G, and A, sequencing products generated from control (nonphotolyzed) 23S rRNA in the presence of ddATP, ddGTP, ddCTP, and ddTTP, respectively. Lanes 1 to 3, control experiments for rRNA isolated from samples with (+) or without (−) photolysis (hν) as indicated. Lanes 4 to 6, samples photolyzed with PHONT 1 in the presence or absence of complementary (CH) or mismatched (MM) 2′-OMe-oligoRNAs as indicated. The arrows point to nucleotides at which pauses or stops induced by photoincorporation of PHONT 1 are observed. (A) With primer complementary to 23S rRNA nt 2639 to 2623. (B) With primer complementary to 23S rRNA nt 1983 to 1965.

Labeled proteins

We identify proteins labeled with ^{32}P-labeled PHONTs by RP-HPLC, sodium dodecyl sulfate (SDS)-PAGE, and, as needed, specific immunochemical analyses. RP-HPLC analysis with multistep acetonitrile-water gradients (Kerlavage et al., 1983; Cooperman et al., 1988) (Fig. 6A) permits rapid analysis of incorporation into proteins. We have found it to be generally true that labeled protein coelutes with unmodified protein, or only slightly differently. Thus, labeled proteins that are well resolved can be identified unambiguously by this analysis. However, in more crowded regions of the chromatogram, even small changes in elution volume on protein modification can introduce ambiguity into the identification process. Typically, SDS-PAGE analysis of labeled peaks seen by RP-HPLC (Fig. 6B), in which labeled protein migrates with an apparent molecular mass equal to the sum of the masses of protein and PHONT, permits unambiguous identification of labeled protein.

In some cases, further analysis is needed. At present, agarose antibody affinity chromatography analysis (Gulle et al., 1988), graciously carried out in the laboratory of Richard Brimacombe (Max Planck Institute, Berlin, Germany), provides the needed additional analysis. To minimize background problems, the samples subjected to such analysis are first purified, either by PAGE (by gel elution) or by RP-HPLC. Immunochemical analyses are performed with antibodies to proteins migrating or eluting in the general vicinity of the labeled protein. However, this approach demands a considerable effort for the maintenance of active stocks of some 55 antibodies, and we are in the process of replacing it with matrix-assisted laser desorption ionization mass spectral analysis.

The analysis of labeled proteins can also be extended to the amino acid level. Such analysis, by localizing the cross-link to a limited structural domain of the labeled protein, permits modeling to be carried out at a higher resolution level. For labeled proteins of known three-dimensional structure, such localization makes possible attempts to fit regions of resolved electron density, as determined by both EM and X-ray crystallography, to specific RNA-protein contacts. Such an attempt has been made in describing S7 binding to 16 S rRNA (Tanaka et al., 1998).

The Significance of Photo-Cross-Linking Results

We have identified three critical questions in evaluating the significance of photo-cross-links by the PHONT approach. First, do the cross-links arise

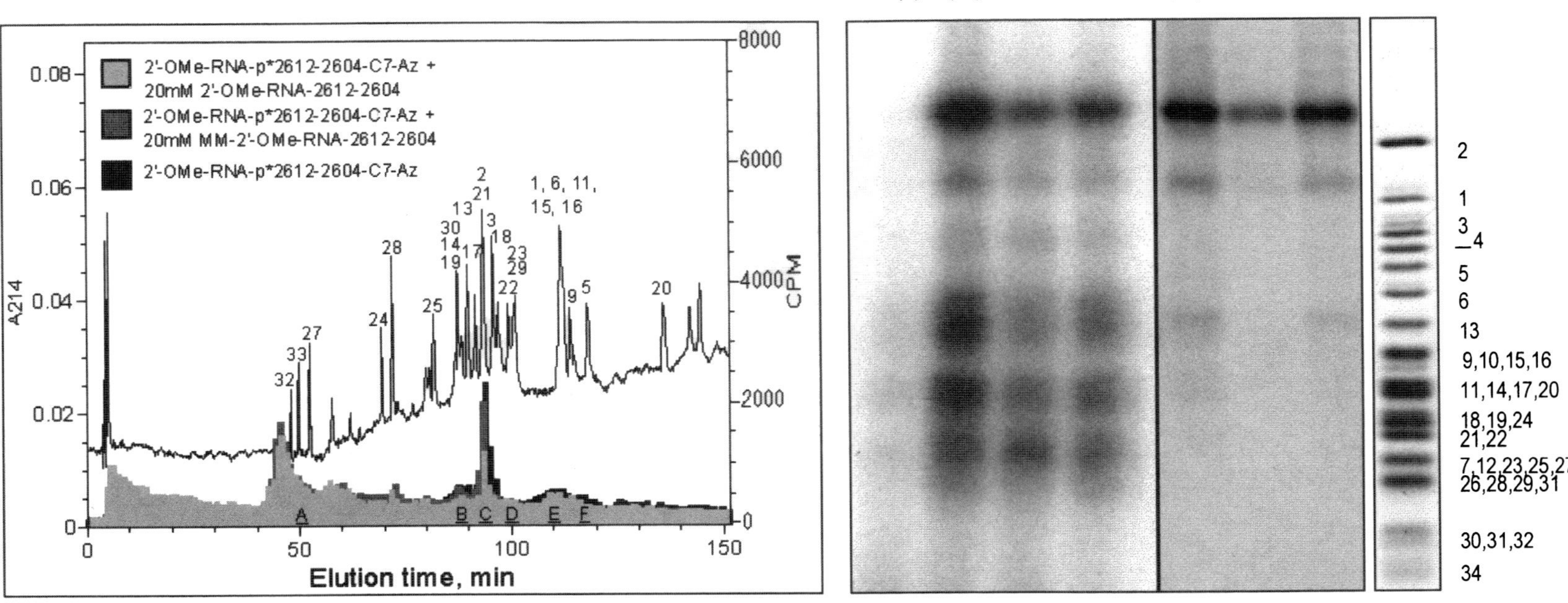

Figure 6. Analyses of PHONT 1-labeled TP50. (A) RP-HPLC. TP50 was extracted from 50S subunits labeled with PHONT 1 (2′-OMe-RNA-p*2612-2604-C7-Az) in the absence or presence of either competitive complementary 2′-OMe–oligoRNA (2′-OMe-RNA-2612-2604) or MM-2′-OMe-RNA-2612-2604). (B) SDS-PAGE and autoradiographic analyses of RP-HPLC peak C in panel A. The lefthand four lanes represent TP50; the right-hand three lanes represent peak C. Lane (−), 50S subunits were incubated with prephotolyzed 1 with no further photolysis. Lanes (+), subunits photolyzed with 1. Lanes CH and MM, subunits photolyzed with 1 in the presence of 10-fold excess (over 50S subunits) of either 2′-OMe-RNA-2612-2604 (lanes CH) or MM-2′-OMe-RNA-2612-2604 (lanes MM). The far righthand lane displays TP50 stained with Coomassie blue.

from the PHONT bound to its target site? Second, do the cross-links arise from the incorporated photolabile group? Third, does PHONT binding induce a long-range distortion in the ribosome?

Cross-linking from the target site

The validity of our approach rests on the assumption that the cross-links used as constraints derive from PHONTs bound to their respective target sites. We have used three criteria to demonstrate that PHONTs bind noncovalently to the target sites: (i) noncovalent binding of the probe to the target subunit; (ii) RNase H cleavage at the target site on addition of the parent unmodified cDNA (or chimeric PHONT) to the ribosomal subunit; and (iii) PHONT photoincorporation at or adjacent to its target site.

Our principal criterion for showing that photoincorporation proceeds from PHONT binding to the target site is the observation of a large decrease in photo-cross-link yield when photolysis is conducted in the presence of added excess complementary unlabeled parent oligonucleotide (cDNA or 2′-OMe-oligoRNA) contrasted with little effect on yield in the presence of an excess of a similar oligonucleotide containing 2 to 4 mismatches to the target at internal positions (Fig. 4 to 6). Such mismatches severely weaken binding to the target site but not necessarily to secondary sites (e.g., to proteins). In addition, we also check for possible cross-linking from sites that, by chance, have substantial sequence overlap (e.g., 7 or 8 of 10 nt) with the targeted site. Such sites are found by sequence scanning, using the freely available program AMPLIFY. The criterion is that there is little, if any, decrease in the yield of photo-cross-link for photolysis conducted in the presence of an oligonucleotide complementary to a "fortuitous" site. The set of oligonucleotides used to test the target site specificity of, for example, PHONT 5 is shown in Fig. 7.

Photochemistry of the cross-link

We use low-pressure Hg 300-nm-wavelength lamps, which have appreciable (>15% of peak) intensity from 270 to 330 nm, for photolysis. Azide- or s^{4}U-derived radicals are quenched by rapid addition of 2-mercaptoethanol after photolysis. To insure that the distance constraint is accurately reflected in a given cross-link, we verify that cross-linking occurs from reaction of the reactive intermediate (the aryl nitrene for most of our work to date) generated on photolysis. This can be done by demonstrating that photolysis with the oligonucleotide lacking the phenylazido group does not yield the cross-link (Wang et al., 1999). However, this procedure does not rule out the possibility (admittedly remote) that the photoincorporation results from the structural perturbation caused by introducing the phenylazide group rather than from the nitrene. Thus, more recently we have shown that an oligonucleotide derivatized with a phenyl group in place of the azidophenyl group does not give photoincorporation (Bukhtiyarov et al., 1999), and this is now our procedure of choice. The same logic is applied to s^{4}U, by replacement with U.

TARGET: 23 S rRNA 2248-58 — 5′-CUGGG GCGGUC-3′
PHONT 5: 2258-2253/52(SAz)-2248 — 3′-GACCC*CGCCAG-5′
CHASE: 2258-48 — 3′-GACCC CGCCAG-5′
MISMATCH: — 3′-GACC$\underline{A}$ $\underline{ACA}$CAG-5′
FORTUITOUS MATCH (1): 2531-21 — 3′-GACCC CG$\underline{A}$C$\underline{UU}$-5′
FORTUITOUS MATCH (2): 2646-36 — 3′-GAC$\underline{U}$C C$\underline{C}$CC$\underline{C}$G-5′

Figure 7. Oligonucleotides testing the target site specificity of PHONT 5 photoincorporation. The asterisk indicates the position of aryl azide attachment. Underlined nucleotides are not complementary to the target site.

Long-range subunit structure distortion induced by PHONT binding

Distortion arising from PHONT binding is a potential problem for the use of cross-linking results to consider three-dimensional models of ribosome structure. In order to assess how widespread such changes might be, we employ a chemical footprinting approach (Moazed et al., 1986) to determine which rRNA bases have their chemical reactivities toward dimethyl sulfate or kethoxal altered on PHONT binding. The generalization that emerges from these studies is that such binding does not lead to major alterations in ribosome structure. This is not unexpected. PHONTs that are found to bind tightly are unlikely to cause major structural change, since the energy required to induce such a change would result in overall weak binding. The reactivity changes that are seen tend to be at or near the targeted site in the primary sequence. In addition, we see some clusters of changes that occur with PHONTs bound at different target sites, implying that these changes occur in conformationally labile regions of rRNA that are sensitive to complementary oligonucleotide binding (Vladimirov et al., in press; Bukhtiyarov et al., 1999).

Modeling

In collaboration with Steve Harvey and Margaret vanLoock, we apply the YAMMP protocol (Malhotra et al., 1994) to construct three-dimensional models of ribosomal subunits or portions of subunits. This protocol uses molecular mechanics to convert empirically determined topographical constraints into a three-dimensional model. These constraints include cross-links, such as those we identify in our work, as well as electron density and X-ray diffraction data. An important attribute of the YAMMP protocol is that constraints may be individually weighted. Thus, as the level of resolution of a cross-link is increased (i.e., from a stretch of RNA sequence or ribosomal protein to a single nucleotide or amino acid) its weight in determining the overall structure is also increased. In addition, the level of detail of the model can be varied within the overall structure, depending on the information available for different parts of the structure, through the use of pseudoatoms of different sizes.

PHONT STUDIES OF THE PTC

Several regions of 23S rRNA domain V have been specifically implicated in PT (Garrett and Rodriguez-Fonseca, 1996). Below, we summarize work with PHONTs **1** to **6** (Fig. 3) targeting several functionally important regions within domain V, as well as the nearby α-sarcin stem-loop region.

Significance of the Target Sites

PHONTs 1 to 3

PHONTs **1** to **3** target the central loop, which brings together nucleotides from a large section of the 3′ half of 23S rRNA, from G2057 to C2611 (Garrett and Rodriguez-Fonseca, 1996). Mutations of nucleotides within the central loop give rise either to losses in PT activity (Porse and Garrett, 1995) or to resistance to antibiotic inhibitors of PT activity (Cundliffe, 1990). In addition, sites within the loop can be photo-cross-linked to both photolabile derivatives of aminoacyl tRNAs (Barta and Halama, 1996) and puromycin (Hall et al., 1988), and they are protected from chemical modification by A-site- and P-site-bound tRNAs and by antibiotic inhibitors of PT activity (Moazed and Noller, 1987; Rodriguez-Fonseca et al., 1995).

PHONT 4

PHONT **4** targets the α-sarcin stem-loop region in domain VI, stretching from C2646 to G2674 (Leffers et al., 1988). It contains the longest universally conserved sequence among rRNAs, corresponding in 23S rRNA to nt 2654 to 2665 (Szewczak et al., 1995). Both elongation factors EF-Tu and EF-G interact with this region, as shown by the inhibition of factor-dependent steps of the elongation cycle caused by α-sarcin cleavage of the α-sarcin loop or depurination of A2660 by the ricin (Fernandez-Puentes and Vazquez, 1977; Leffers et al., 1988), by footprinting (Hausner et al., 1987), and by EF-G protection of the ribosome from α-sarcin inactivation (Moazed et al., 1988). The α-sarcin loop is also important for translational accuracy (Miller and Bodley, 1991; Tapprich and Dahlberg, 1990; Tapio and Isaksson, 1991; Melançon et al., 1992).

PHONT 5

PHONT **5** targets the so-called P loop (nt 2246 to 2258) that is a vital part of the PTC. It contains three invariant guanosine residues, G2251 to G2253, of which two, G2252 and G2253, are protected from kethoxal modification by P-site-bound tRNA (Moazed and Noller, 1989). In addition, mutational analysis has shown that G2252 forms a Watson-Crick pair with C74 of the P-site-bound tRNA (Samaha et al., 1995a, 1995b), that mutations at G2251 abolish both binding of wild-type tRNA fragments and PT activity, and that there is evidence for direct or indirect interaction of the 2250 loop with U2585, which falls within the central loop of domain V (Green et al., 1997).

PHONT 6

PHONT **6** targets nt 2475 to 2483, which form part of the C2475 loop connected by a stem to the central loop of domain V. Binding of complementary oligoDNAs to the C2475 loop is inhibited in the presence of poly(U) (Hill et al., 1990), suggesting that the C2475 loop may be at or near the 30S-50S interface. Furthermore, U89 in 5S rRNA can be photo-cross-linked to two positions in this loop (Steiner et al., 1988).

Site-Specific Photoincorporation

PHONTs **1** to **6** each labeled nucleotides within or immediately adjacent to its target site. In addition, PHONTs **1** to **3**, and **5** labeled several other nucleotide sites. All PHONTs except **5** gave site-specific photoincorporation into one or more proteins. These results are summarized in Table 1.

Comparison of Cross-Linking Results with Related Results of Others

Cross-links from the central loop of domain V (PHONTs 1 to 3)

The cross-links **1**-2586, **3**-2454, **3**-2583, and **3**-2584 are consistent with the standard model (Gutell

Table 1. Site-specific cross-links from PHONTs for 23S rRNA

PHONT[a]	23S rRNA nucleotide[b]	Tether length (Å)[c]	Cross-link[d]
1	2604	25	886; 1918–1921; 1923–1924; 2563; 2586; L2+++
2	2448	25	L2++; L3++; (L17/18/21)++; (L9/15/16)++
3	2505	24	2454; 2583–2584; L3+++
4	2653	23	L2+++; L15+++; L16++; L27++; L9+; L1+; L5+; L17+; L24+
5	2252–2253	17	2358; 2388 ± 12; 2405–2407; 2411; 2430
6	2475	23	L1+++; L13++; L16++; L32++; L33++

[a] Sources: **1** and **2**, Vladimirov et al., in press; **3**, Muralikrishna and Cooperman, 1991; **4**, Muralikrishna et al., 1997; **5**, Bukhtiyarov et al., 1999; **6**, Muralikrishna and Cooperman, 1995.
[b] Complemented by PHONT nucleotide to which azide is attached.
[c] To nearest base.
[d] Number of plus signs provides a qualitative sense of relative stoichiometry of cross-linked proteins.

et al., 1994) for the secondary structure of 23S rRNA, which brings these nucleotides together within the central loop of domain V. Results from studies employing tRNA or tRNA analogues and photo- or electrophilic cross-linking, chemical footprinting, and induced chemical cleavage experiments, directly support the proximal relationships defined by the cross-links in Table 1. An important reference location is the 3′ end of P-site-bound tRNA, from A_{73} through the attached amino acid. Experiments utilizing this location show proximity between central loop positions A2451, C2452, A2503, U2506, and A2602 (Noller et al., 1995; Barta and Halama, 1996) and domain IV positions A1916, A1918, and C1926 (Wower et al., 1989; Noller et al., 1995), as well as with protein L2 (Pellegrini and Cantor, 1977; Johnson and Cantor, 1980; Graifer et al., 1989; Wower et al., 1995), consistent with cross-links **1**-1918/24, **2**-L2, and **1**-L2. The low PT activity resulting from U1926C mutation (Porse et al., 1995) is also consistent with the cross-link **1**-1918/24. Similarly, a photolabile derivative at position 47 of A-site-bound Phe-tRNA links C2452 with nucleotides falling within sequences 865 to 910 and 1892 to 1945 (Osswald et al., 1995), consistent with cross-links **1**-886 and **1**-1918/24. A more precise overlap with cross-link **1**-886 comes from a study employing RNA helices of various lengths attached to a P-site-bound anticodon loop (so-called ASLs). These are derivatized with Fe(II) and induce chemical cleavages at varying distances from the anticodon loop binding site. Using the same size ASLs (4 to 12 bp), cleavages are induced at or close to A2602 in the central loop and at or close to G880 and C890 (Joseph et al., 1997). Placement of L2 and L3 at or near the central loop by cross-links **1**-L2, **2**-L2, **2**-L3, and **3**-L3 is consistent with studies demonstrating that of all ribosomal proteins, only L2 and L3 could be essential for PT activity (Tate et al., 1987; Khaitovich et al., 1999). We and others have provided evidence for a direct role of L2 in such activity (Cooperman et al., 1995; Uhlein et al., 1998). Finally, cross-link **1**-2563 is consistent with both cross-linking and mutant studies showing the importance of helix 92 (Fig. 8) for PT activity (Green et al., 1997, 1998).

Cross-links from neighbors of the central loop of domain V (PHONTs 4 to 6)

PHONT **4** labels proteins L2, L15, L16, and L27 (major or intermediate labeled proteins) and L1, L5, L9, L17, and L24 (minor labeled proteins). Since each of the most heavily labeled proteins, L2, L15, L16, and L27, is located at or near the PTC (Cooperman et al., 1990), a clear result of our work is to place U2653 close to this center. This is consistent with work (Joseph and Noller, 1996) showing that Fe(II)-EDTA attached to the 5′ end of A-site-bound tRNA induced hydroxyl radical cleavages within both the α-sarcin loop and the 2550 loop in domain V, which has been implicated in PT activity. However, the minor labeling of proteins L1 and L9 suggests a placement for the α-sarcin stem-loop region between the PTC and the L1 projection. Such placement is different from the prevailing view placing it between the PTC and the L12 projection, based on the binding sites for EF-G and EF-Tu, which have been localized by immuno-EM and more recent cryo-EM studies (Stark et al., 1997; Agrawal et al., 1998) near the L12 stalk.

The reasons for this discrepancy are unclear. We cannot at present exclude the possibility that **4**, in addition to forming a heteroduplex with 23S rRNA along its entire length, also forms stable partial heteroduplexes. One example would be to nt 2660 to 2674. In such a complex, the azide attached to the adenyl residue in **4** that is complementary to U2653 could label a 50S protein that is quite far from the sarcin loop stem region. Alternatively, it is possible

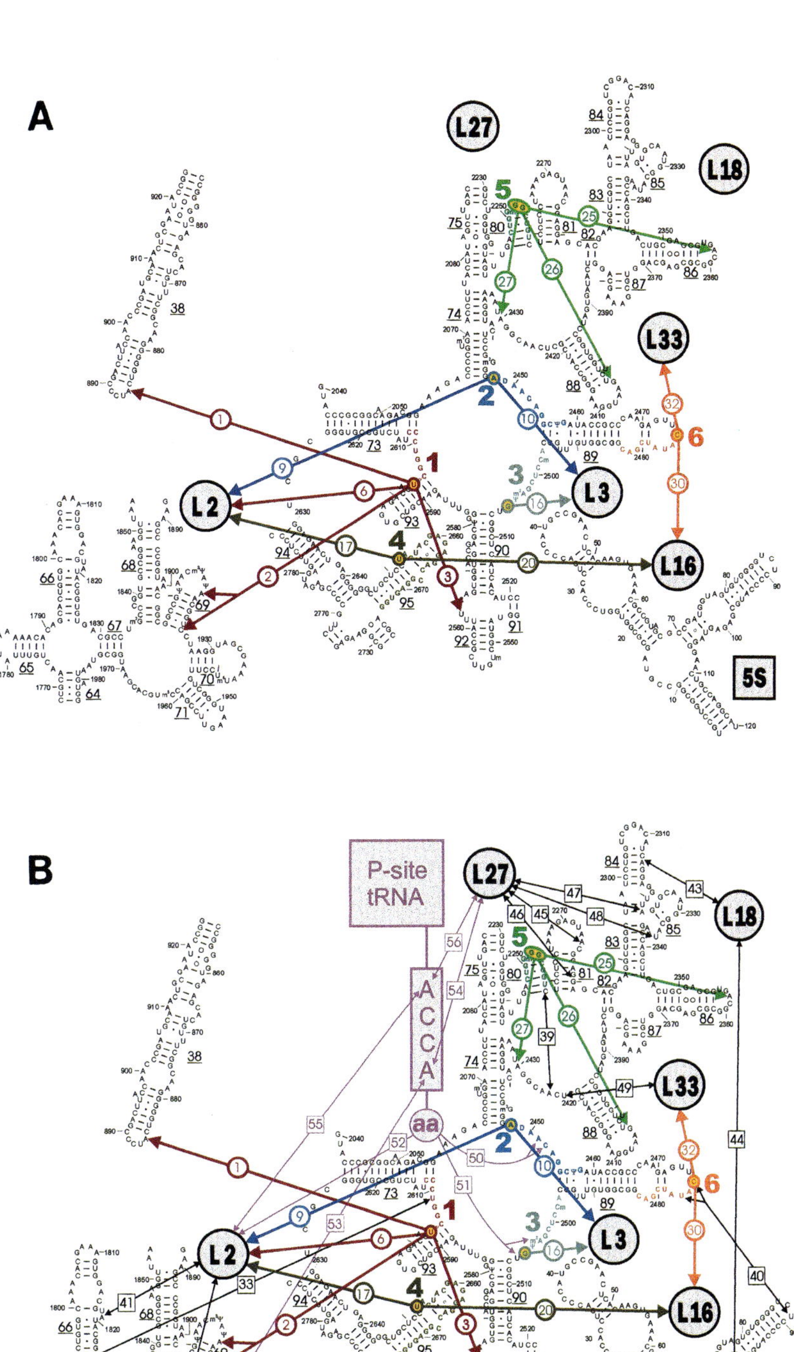

Figure 8. Summary of cross-linking results useful for construction of a peptidyltransferase model. (A) The circled cross-links are from Table 1. The large boldface numbers refer to target sites for PHONTs 1 to 6. (B) All relevant cross-links. The circled cross-links are as described in panel A. The boxed cross-links involve a variety of other approaches, mostly direct photolysis or via introduction of a photolabile group. The helices (underlined) are numbered as in Brimacombe, 1995.

that we are detecting conformational changes within the 50S subunit, e.g., a movement of L1 toward the L12 projection. Further experiments are planned to resolve this ambiguity.

PHONT 5 forms cross-links to helices 86 (A2358-A2359) and 88 (G2405-G2411 and A2430). The 8-2430 cross-link is consistent with a direct cross-link between nt 2258 and 2425 reported by Stiege et al. (1986). This cross-link, as well as 5-2405/7, 5-2411, and 5-2430, fits well with the observation (Gregory and Dahlberg, 1998) that mutations at positions 2251 to 2253 altered the chemical reactivity of nt G2357, G2361, U2408 to U2410, G2428, and U2431. On the other hand, we did not see any cross-links to the 2585 region in 23S rRNA, although Green et al. (1997) have shown that mutations of G2251 perturb the chemical reactivity of bases 2584 to 2586. Work is under way to determine whether other PHONTs targeting the 2250 loop will cross-link to nt 2584 to 2586 or other 50S components.

The five proteins cross-linked to PHONT 6 are linked to one another by other cross-linking studies. Thus, L1 forms cross-links with L33 (Walleczek et al., 1989a, 1989b; Redl et al., 1989), L32, and L13 (Traut et al., 1986), and L1, L16, and L33 each form cross-links to the 3′-end P-site-bound tRNAs (Wower et al., 1993). In addition, U89 in 5S rRNA forms cross-links with C2475 and U2477 (Dontsova et al., 1994). These results suggest a placement for the 2475 loop near the PTC, between the central protuberance and the L1 projection.

Toward a Three-Dimensional Model of the PTC

We are currently applying the YAMMP approach to develop a three-dimensional model of the PTC based on cross-linking results, which we consider to provide the clearest set of constraints for model construction. The comparison of such a model with electron density maps emerging from crystallographic and EM studies should allow for clear placement of the PTC region within the ribosome. Examination of the pertinent literature (Traut et al., 1986; Cooperman et al., 1990; Dontsova et al., 1994; Brimacombe, 1995; Noller et al., 1995; Garrett and Rodriguez-Fonseca, 1996; Green and Noller, 1997; Baranov et al., 1998) leads to the choices of 50S components indicated in Fig. 8 to include in modeling the PTC region. The cross-links we are currently using include those from our own work (Table 1) and those obtained by others (Fig. 8). Work now in progress will increase this number.

This approach can obviously be extended to other domains. Of these, domain IV, which interacts with the PTC in a number of locations (Fig. 8), is clearly of interest. For example, PHONT 7, targeted toward the 1882-to-1892 region, forms a cross-link to nt 1094 and 1095 in the thiostrepton binding site (Seo and Cooperman, unpublished).

THE FUTURE OF PHONT STUDIES

It is clear that the PHONT approach can be applied to any RNA-containing molecule, including, in addition to ribosomes, particles such as spliceosomes, RNAse P, and ribozymes. However, for ribosomes themselves it is worth considering what the value of PHONT or other cross-linking studies will be, given the recent dramatic advances in both cryo-EM and X-ray crystallographic investigation of ribosome structure, both already published (Yonath and Franceschi, 1998; Moore, 1998; Frank, 1998; Ban et al., 1998) and reported at the Helsingør Ribosome Conference (see chapters 1 to 6 and 13).

As a result of these advances, structural models for the 30S and 50S subunits, as well as for some complexes of the 70S ribosome, are now available at 5- to 8-Å resolution. In these structures, the relative spatial relationships of the great majority of ribosomal components with respect to each other within the ribosome will soon be clear, although some poorly defined areas may remain. Less certain, but still likely, is that X-ray structures will be forthcoming for 30S, 50S, and/or 70S particles at resolutions of 3.0 to 3.5 Å within the next several years, although here problems remain to be solved that could prove very time-consuming. Moreover, taking into account the conformational flexibility of the ribosome, it is not at all clear, and is probably unlikely, that all or even most functionally important conformations of ribosome structure will be amenable to determination by X-ray crystallography. While the resolution available from cryo-EM (currently 9 to 11 Å) is unlikely to match that obtainable from the best crystal structures, cryo-EM enjoys the distinct advantage of being readily applicable to studying changes in structure as the functional state of the ribosome varies, providing information quite complementary to that obtained by cross-linking studies (see, for example, Mueller and Brimacombe, 1997).

As a consequence of the rather remarkable progress in structural studies, PHONT studies will change focus, from generating photo-cross-links to aid in the construction of a static ribosome structure to using cross-links as a way of defining conformational changes occurring in different functional states of the dynamic ribosome—for example, those obtained by functional ligand binding (e.g., mRNA, tRNA, and protein factors), by mutation of rRNA (Lodmell and

Dahlberg, 1997), and/or by antibiotic binding. Here, the intrinsic ability of the PHONT approach to sample available conformations in solution from functionally significant targeted sequences is particularly important. Most interesting will be those cross-links not fitting the static structures, since they will constitute a priori evidence for conformational change bringing a portion (or portions) of the ribosome into closer proximity to the target site than is allowed by the X-ray structures. Such cross-links and relevant cryo-EM data can then be used to generate a three-dimensional model of the ribosome in a particular functional state, using the coordinates of a crystallographically known ribosome state for initial positions, from which refinement can be carried out by application of the YAMMP protocol.

This work was supported by a grant (GM 53146) from the National Institutes of Health.

REFERENCES

Agrawal, R. K., and D. P. Burma. 1996. Sites of ribosomal RNAs involved in the subunit association of tight and loose couple ribosomes. *J. Biol. Chem.* **271:**21285–21291.

Agrawal R. K., P. Penczek, R. A. Grassucci, and J. Frank. 1998. Visualization of elongation factor G on the Escherichia coli 70 S ribosome: the mechanism of translocation. *Proc. Natl. Acad. Sci. USA* **95:**6134–6138.

Alexander, R. W., and B. S. Cooperman. 1998. Ribosomal proteins neighboring 23 S rRNA nucleotides 803–811 within the 50 S subunit. *Biochemistry* **37:**1714–1721.

Alexander, R. W., P. Muralikrishna, and B. S. Cooperman. 1994. Ribosomal components neighboring the conserved 518-533 loop of 16 S ribosomal RNA in 30 S subunits. *Biochemistry* **33:** 12109–12118.

Azad, A. A., P. Failla, and J. Hanna. 1998. Inhibition of ribosomal subunit association and protein synthesis by oligonucleotides corresponding to defined regions of 18S rRNA and 5S rRNA. *Biochem. Biophys. Res. Commun.* **248:**51–56.

Ban, N., B. Freeborn, P. Nissen, P. Penczek, R.A. Grassucci, R. Sweet, J. Frank, P. B. Moore, and T. A. Steitz. 1998. A 9-Å resolution x-ray crystallographic map of the large ribosomal subunit. *Cell* **93:**1105–1115.

Baranov, P. V., P. V. Sergiev, O. A. Dontsova, A. A. Bogdanov, and R. Brimacombe. 1998. The database of ribosomal cross links (DRC). *Nucleic Acids Res.* **26:**187–189.

Barta, A., and I. Halama. 1996. The elusive peptidyl transferase—RNA or protein? p. 35–54. *In* R. Green and R. Schroeder (ed.), *Ribosomal RNA and Group I Introns.* R. G. Landes Company, Austin, Tex.

Barta, A., G. Steiner, J. Brosius, H. F. Noller, and E. Kuechler. 1984. Identification of a site on 23 S ribosomal RNA located at the peptidyl transferase center. *Proc. Natl. Acad. Sci. USA* **81:** 3607.

Brimacombe, R. 1995. The structure of ribosomal RNA: a three-dimensional jigsaw puzzle. *Eur. J. Biochem.* **230:**365–383.

Brimacombe, R., B. Greuer, P. Mitchell, M. Osswald, J. Rinke-Appel, D. Schüler, and K. Stade. 1990. Three-dimensional structure and function of *Escherichia coli* 16S and 23S rRNA as studied by cross-linking techniques, p. 93–106. *In* W. E. Hill, A. Dahlberg, R. A. Garrett, P. B. Moore, D. Schlessinger, and J. R. Warner (ed.), *The Ribosome: Structure, Function, and Evolution.* American Society for Microbiology, Washington, D.C.

Bukhtiyarov, Y., Z. Druzina, and B. S. Cooperman. 1999. Identification of 23 S rRNA nucleotides neighboring the P-loop in the *E. coli* 50 S subunit. *Nucleic Acids Res.* **27:**4376–4384.

Cooperman, B. S. 1988. Affinity labeling of ribosomes. *Methods Enzymol.* **164:**341–361.

Cooperman, B. S., C. J. Weitzmann, and M. A. Buck. 1988. Reverse phase high performance liquid chromatography of ribosomal proteins. *Methods Enzymol.* **164:**523–532.

Cooperman, B. S., C. J. Weitzmann, and C. L. Fernandez. 1990. Antibiotic probes of *E. coli* ribosomal peptidyltransferase, p. 491–501. *In* W. E. Hill, A. Dahlberg, R. A. Garrett, P. B. Moore, D. Schlessinger, and J. R. Warner (ed.), *The Ribosome: Structure, Function, and Evolution.* American Society for Microbiology, Washington, D.C.

Cooperman, B. S., T. Wooten, D. P. Romero, and R. R. Traut. 1995. Histidine 229 in protein L2 is apparently essential for 50 S peptidyl transferase activity. *Biochem. Cell Biol.* **73:**1087–1094.

Cooperman, B. S., R. W. Alexander, Y. Bukhtiyarov, S. N. Vladimirov, Z. Druzina, R. Wang, and N. Zuño. Photolabile derivatives of oligonucleotides (PHONTS) as probes of ribosomal structure. *Methods Enzymol.,* in press.

Cundliffe, E. 1990. Recognition sites for antibiotics within rRNA, p. 479–490. *In* W. E. Hill, A. Dahlberg, R. A. Garrett, P. B. Moore, D. Schlessinger, and J. R. Warner (ed.), *The Ribosome: Structure, Function, and Evolution.* American Society for Microbiology, Washington, D.C.

Dontsova, O., V. Tishkov, S. Dokudovskaya, A. Bogdanov, T. Döring, J. Rinke-Appel, S. Thamm, B. Greuer, and R. Brimacombe. 1994. Stem-loop IV of 5S rRNA lies close to the peptidyltransferase center. *Proc. Natl. Acad. Sci. USA* **91:**4125–4129.

Fernandez-Puentes, C., and D. Vazquez. 1977. Effects of some proteins that inactivate the eukaryotic ribosome. *FEBS Lett.* **78:** 143–146.

Frank, J. 1998. The ribosome-structure and functional ligand-binding experiments using cryo-electron microscopy. *J. Struct. Biol.* **124:**142–150.

Garrett, R. A., and C. Rodriguez-Fonseca. 1996. The peptidyl transferase center, p. 327–355. *In* R. A. Zimmermann and A. Dahlberg (ed.), *Ribosomal RNA: Structure, Evolution, Processing, and Function.* CRC Press, Boca Raton, Fla.

Graifer, D. M., G. T. Babkina, N. B. Matasova, S. N. Vladimirov, G. G. Karpova, and V. V. Vlassov. 1989. Structural arrangement of transfer-RNA binding-sites on Escherichia coli ribosomes, as revealed from data on affinity labeling with photoactivatable transfer-RNA derivatives. *Biochim. Biophys. Acta* **1008:**146–156.

Green, R., and H. F. Noller. 1997. Ribosomes and translation. *Annu. Rev. Biochem.* **66:**679–716.

Green, R., R. R. Samaha, and H. F. Noller. 1997. Mutations at nucleotides G2251 and U2585 of the 23 S rRNA perturb the transferase centre of the ribosome. *J. Mol. Biol.* **266:**40–50.

Green, R., C. Switzer, and H. F. Noller. 1998. Ribosome-catalyzed peptide-bond formation with an A-site substrate covalently linked to 23 S ribosomal RNA. *Science* **280:**286–289.

Gregory, S. T., and A. E. Dahlberg. 1998. Mutations in the conserved P loop perturb the conformation of two structural elements in the peptidyl transferase center of 23 S ribosomal RNA. *J. Mol. Biol.* **285:**1475–1483.

Gulle, H., E. Hoppe, M. Osswald, B. Greuer, R. Brimacombe, and G. Stoffler. 1988. RNA-protein cross-linking in Escherichia coli 50 S ribosomal subunits; determination of sites on 23 S RNA that are cross-linked to proteins L2, L4, L24 and L27 by treatment with 2-iminothiolane. *Nucleic Acids Res.* **16:**815–832.

Gutell, R. R., N. Larsen, and C. R. Woese. 1994. Lessons from an evolving rRNA: 16 S and 23 S rRNA structures from a comparative perspective. *Microbiol. Rev.* **58:**10–26.

Hall, C. C., D. Johnson, and B. S. Cooperman. 1988. [^{3}H]-p-azidopuromycin photoaffinity labeling of *E. coli* ribosomes: evidence for site-specific interaction at U-2504 and G-2502 in domain V of 23 S ribosomal RNA. *Biochemistry* **27:**3983–3990.

Hausner, T.-P., J. Atmadja, and K. Nierhaus. 1987. Evidence that the G2661 region of 23 S rRNA is located at the ribosomal binding sites of both elongation factors. *Biochimie* **69:**911–923.

Hill, W. E., J. Weller, T. Gluick, C. Merryman, R. T. Marconi, A. Tassanakajohn, and W. E. Tapprich. 1990. Probing ribosome structure and function by using short complementary DNA oligomers, p. 93–106. *In* W. E. Hill, A. Dahlberg, R. A. Garrett, P. B. Moore, D. Schlessinger, and J. R. Warner (ed.), *The Ribosome: Structure, Function, and Evolution.* American Society for Microbiology, Washington, D.C.

Inoue, H., Y. Hayase, A. Imura, S. Iwai, K. Miura, and E. Ohtsuka. 1987. Synthesis and hybridization studies on two complementary non(2′-O-methyl)ribonucleotides. *Nucleic Acids Res.* **15:**6131–6148.

Inoue, H., Y. Hayase, S. Iwai, and E. Ohtsuka. 1988. Sequence-specific cleavage of RNA using chimeric DNA splints and RNase H. *Nucleic Acids Res. Symp. Ser.* **19:**135–138.

Johnson, A. E., and C. R. Cantor. 1980. Elongation factor-dependent affinity labeling of E. coli ribosomes. *J. Mol. Biol.* **138:**273–297.

Joseph, S., and H. F. Noller. 1996. Mapping the rRNA neighborhood of the acceptor end of tRNA in the ribosome. *EMBO J.* **15:**910–916.

Joseph, S., B. Weiser, and H. F. Noller. 1997. Mapping the inside of the ribosome with an RNA helical ruler. *Science* **278:**1093–1098.

Kerlavage, A. R., T. Hasan, and B. S. Cooperman. 1983. Reverse-phase high performance liquid chromatography of *Escherichia coli* ribosomal proteins: standardization of 70S, 50S, and 30S protein chromatograms. Functional activity of purified proteins. *J. Biol. Chem.* **258:**6313–6318.

Khaitovich, P., A. S. Mankin, R. Green, L. Lancaster, and H. F. Noller. 1999. Characterization of functionally active subribosomal particles from Thermus aquaticus. *Proc. Natl. Acad. Sci. USA* **96:**85–90.

Lasater, L. S., P. A. Cann, and D. G. Glitz. 1989. Localization of the site of cleavage of ribosomal RNA by colicin-E3—placement on the small ribosomal subunit by electron microscopy of antibody-complementary oligodeoxynucleotide complexes. *J. Biol. Chem.* **264:**21798–21805.

Lasater, L. S., L. Montesano-Roditis, P. A. Cann, and D. G. Glitz. 1990. Localization of an oligodeoxynucleotide complementing 16 S ribosomal RNA residues 520-531 on the small subunit of *E. coli* ribosomes—electron microscopy of ribosome-cDNA-antibody complexes. *Nucleic Acids Res.* **18:**477–485.

Leffers, H., J. Egebjerg, A. Andersen, T. Christensen, and R. A. Garrett. 1988. Domain VI of *E. coli* 23 S ribosomal RNA: structure, assembly and function. *J. Mol. Biol.* **204:**507–522.

Lodmell, J. S., and A. Dahlberg. 1997. A conformational switch in Escherichia coli 16 S ribosomal RNA during decoding of messenger RNA. *Science* **277:**1262–1267.

Lodmell, J. S., W. E. Tapprich, and W. E. Hill. 1993. Evidence for a conformational change in the exit site of the E. coli ribosome upon tRNA binding. *Biochemistry* **32:**4067–4072.

MacMillan, A. M., and G. L. Verdine. 1991. Engineering tethered DNA molecules by the convertible nucleotide approach. *Tetrahedron* **47:**2603–2616.

Malhotra, A., R. K.-Z. Tan, and S. C. Harvey. 1994. Modeling large RNAs and ribonucleoprotein particles using molecular mechanics techniques. *Biophys. J.* **66:**1777–1795.

Mankin, A. S., E. A. Skripkin, N. V. Chichkova, A. M. Kopylov, and A. A. Bogdanov. 1981. An enzymatic approach for localization of oligodeoxyribonucleotide binding sites on RNA. *FEBS Lett.* **131:**253–256.

Marconi, R. T., and W. E. Hill. 1988. Identification of defined sequences in domain V of *E. coli* 23 S rRNA in the 50 S subunit accessible for hybridization with complementary oligodeoxyribonucleotides. *Nucleic Acids Res.* **16:**1603–1615.

Marconi, R. T., and W. E. Hill. 1989. Evidence for a tRNA/rRNA interaction site within the peptidyltransferase center of the *E. coli* ribosome. *Biochemistry* **28:**893–899.

Marconi, R. T., J. S. Lodmell, and W. E. Hill. 1990. Identification of a rRNA/chloramphenicol interaction site within the peptidyltransferase center of the 50 S subunit of the *E. coli* ribosome. *J. Biol. Chem.* **265:**7894–7899.

McWilliams, R. A., and D. G. Glitz. 1991. Localization of a segment of 16 S RNA on the surface of the small ribosomal subunit by immune electron microscopy of complementary oligodeoxynucleotides. *Biochimie* **73:**911–918.

Melançon, P., W. E. Tapprich, and L. Brakier-Gingras. 1992. Single-base mutations at position 2661 of *Escherichia coli* 23 S rRNA increase efficiency of translational proofreading. *J. Bacteriol.* **174:**7896–7901.

Meyer, H. A., F. Triana-Alonso, C. M. Spahn, T. Twardowski, A. Sobkiewicz, and K. H. Nierhaus. 1996. Effects of antisense DNA against the alpha-sarcin stem-loop structure of the ribosomal 23 S rRNA. *Nucleic Acids Res.* **24:**3996–4002.

Miller, S. P., and J. W. Bodley. 1991. Alpha-sarcin cleavage of ribosomal-RNA is inhibited by the binding of elongation factor-G or thiostrepton to the ribosome. *Nucleic Acids Res.* **19:**1657–1660.

Moazed, D., and H. F. Noller. 1987. Interaction of antibiotics with functional sites in 16 S ribosomal RNA. *Nature* **327:**389–394.

Moazed, D., and H. F. Noller. 1989. Interaction of tRNA with 23 S rRNA in the ribosomal A, P, and E sites. *Cell* **57:**585–597.

Moazed, D., S. Stern, and H. F. Noller. 1986. Rapid chemical probing of conformation in 16 S ribosomal RNA and 30 S ribosomal subunits using primer extension. *J. Mol. Biol.* **187:**399–416.

Moazed, D., J. M. Robertson, and H. F. Noller. 1988. Interaction of elongation factors EF-G and EF-Tu with a conserved loop in 23 S rRNA. *Nature* **334:**362–364.

Moore, P. B. 1998. The three-dimensional structure of the ribosome and its components. *Annu. Rev. Biophys. Biomol. Struct.* **27:**35–58.

Mueller, F., and R. Brimacombe. 1997. A new model for the three-dimensional folding of Escherichia coli 16 S ribosomal RNA. I. Fitting the RNA to a 3D electron microscopic map at 20 Å. *J. Mol. Biol.* **271:**524–544.

Muralikrishna, P., and B. S. Cooperman. 1991. A photolabile oligodeoxyribonucleotide probe of the peptidyl transferase center: identification of neighboring ribosomal components. *Biochemistry* **30:**5421–5428.

Muralikrishna, P., and B. S. Cooperman. 1994. A photolabile oligodeoxyribonucleotide probe of the decoding site in the small subunit of the *Escherichia coli* ribosome: identification of neighboring ribosomal components. *Biochemistry* **33:**1392–1398.

Muralikrishna, P., and B. S. Cooperman. 1995. Ribosomal proteins neighboring the 2475 loop in *Escherichia coli* 50 S subunits. *Biochemistry* **34:**115–121.

Muralikrishna, P., R. W. Alexander, and B. S. Cooperman. 1997. Placement of the a-sarcin loop within the 50 S subunit: evidence derived using a photolabile deoxyoligonucleotide probe. *Nucleic Acids Res.* **25:**4562–4569.

Noller, H. F., R. Green, G. Heilek, V. Hoffarth, A. Huttenhofer, S. Joseph, I. Lee, K. Lieberman, A. Mankin, C. Merryman, T.

Powers, E. V. Puglisi, R. R. Samaha, and B. Weiser. 1995. Structure and function of ribosomal RNA. *Biochem. Cell Biol.* **73:** 997–1009.

Oakes, M. I., and J. A. Lake. 1990. DNA-hybridization electron microscopy—localization of 5 regions of 16-S rRNA on the surface of 30-S ribosomal subunits. *J. Mol. Biol.* **211:**897–906.

Oakes, M. I., L. Kahan, and J. A. Lake. 1990. DNA-hybridization electron microscopy—tertiary structure of 16-S rRNA. *J. Mol. Biol.* **211:**907–918.

Osswald, M., T. Döring, and R. Brimacombe. 1995. The ribosomal neighborhood of the central fold of tRNA: cross-links from position 47 to tRNA located at the A, P or E site. *Nucleic Acids Res.* **23:**4635–4641.

Pelligrini, M., and C. R. Cantor. 1977. Affinity labeling of ribosomes, p. 203–244. *In* H. Weissbach and S. Pestka (ed.), *Molecular Mechanisms of Protein Synthesis.* Academic Press, New York, N.Y.

Porse, B. T., and R. A. Garrett. 1995. Mapping important nucleotides in the peptidyl transferase center of 23 S rRNA using a random mutagenesis approach. *J. Mol. Biol.* **249:**1–10.

Porse, B. T., C. Rodriguez-Fonseca, I. Leviev, and R. A. Garrett. 1995. Antibiotic inhibition of the movement of tRNA substrates through a peptidyl transferase cavity. *Biochem. Cell Biol.* **73:** 877–885.

Redl, B., J. Walleczek, M. Stöffler-Meilicke, and G. Stöffler. 1989. Immunoblotting analysis of protein-protein cross-links within the 50 S ribosomal subunit of E. coli. *Eur. J. Biochem.* **181:**351–356.

Rodriguez-Fonseca, C., R. Amils, and R. A. Garrett. 1995. Fine structure of the peptidyl transferase centre on 23 S-like rRNAs deduced from chemical probing of antibiotic-ribosome complexes. *J. Mol. Biol.* **247:**224–235.

Samaha, R. R., R. Green, and H. F. Noller. 1995a. A base pair between tRNA and 23 S rRNA in the peptidyl transferase centre of the ribosome. *Nature* **377:**309–314.

Samaha, R. R., R. Green, and H. F. Noller. 1995b. A base-pair between transfer-RNA and 238-ribosomal-RNA in the peptidyl transferase center of the ribosome. *Nature* **378:**419.

Seo, H. S., and B. S. Cooperman. Unpublished data.

Skripkin, E. A., A. M. Kopylov, A. A. Bogdanov, S. V. Vinogradov, and Y. A. Berlin. 1979. rRNA topography in ribosomes. IV. The accessibility of the 5′-end region of 16 S tRNA. *Mol. Biol. Rep.* **5:**221–224.

Stark, H., M. V. Rodnina, J. Rinke-Appel, R. Brimacombe, W. Wintermeyer, and M. van Heel. 1997. Visualization of elongation factor Tu on the E. coli ribosome. *Nature* **389:**403–406.

Steiner, G., E. Kuechler, and A. Barta. 1988. Photo-affinity labeling at the peptidyl transferase centre reveals two different positions of the A- and P-sites in domain V of 23 S rRNA. *EMBO J.* **7:**3949–3955.

Stiege, W., J. Atmadja, M. Zobawa, and R. Brimacombe. 1986. Investigation of the tertiary folding of E. coli ribosomal RNA by intra-RNA crosslinking in vivo. *J. Mol. Biol.* **191:**135–138.

Szewczak, A. A., and P. B. Moore. 1995. The sarcin/ricin loop, a modular RNA. *J. Mol. Biol.* **247:**81–98.

Tanaka, I., A. Nakagawa, H. Hosaka, S. Wakatsuki, F. Mueller, and R. Brimacombe. 1998. Matching the crystallographic structure of ribosomal protein S7 to a three-dimensional model of the 16 S ribosomal RNA. *RNA* **4:**542–550.

Tapio, S., and L. A. Isaksson. 1991. Base 2661 in Escherichia coli 23 S rRNA influences the binding of elongation factor Tu during protein synthesis in vivo. *Eur. J. Biochem.* **202:**981–984.

Tapprich, W. E., and A. Dahlberg. 1990. A single base mutation at position 2661 in *E. coli* 23 S ribosomal RNA affects the binding of ternary complex to the ribosome. *EMBO J.* **9:**2649–2655.

Tate, W. P., V. G. Sumpter, C. N. A. Trotman, M. Herold, and K. H. Nierhaus. 1987. The peptidyltransferase centre of the *E. coli* ribosome. *Eur. J. Biochem.* **165:**403–408.

Tranque, P., M. C. Hu, G. M. Edelman, and V. P. Mauro. 1998. rRNA complementarity within mRNAs: a possible basis for mRNA-ribosome interactions and translational control. *Proc. Natl. Acad. Sci. USA* **95:**12238–12243.

Traut, R. R., D. S. Tewari, A. Sommer, G. R. Gavino, H. M. Olson, and D. G. Glitz. 1986. Protein topography of ribosomal functional domains: effects of monoclonal antibodies to different epitopes in E. coli protein L7/L12 on ribosome function and structure, p. 286–308. *In* B. Hardesty and G. Kramer (ed.), *Structure, Function, and Genetics of Ribosomes.* Springer, New York, N.Y.

Uhlein, M., W. Weglöhner, H. Urlaub, and B. Wittmann-Liebold. 1998. Functional implications of ribosomal protein L2 in protein biosynthesis as shown by in vivo replacement studies. *Biochem. J.* **331:**423–430.

Vladimirov, S. N., Z. Druzina, R. Wang, and B. S. Cooperman. Identification of 50 S components neighboring 23 S rRNA nucleotides A2448 and U2604 within the peptidyl transferase center of *E. coli* ribosomes. *Biochemistry,* in press.

Walleczek, J., B. Redl, M. Stoffler-Meilicke, and G. Stoffler. 1989a. Protein-protein cross-linking of the 50 S ribosomal subunit of *E. coli* using 2-iminothiolane. *J. Biol. Chem.* **264:**4231–4237.

Walleczek, J., T. Martin, B. Redl, M. Stoffler-Meilicke, and G. Stoffler. 1989b. Comparative cross-linking study on the 50 S ribosomal subunit from *E. coli. Biochemistry* **28:**4099–4105.

Wang, R., R. W. Alexander, M. van Loock, S. Vladimirov, Y. Bukhtiyarov, S. C. Harvey, and B. S. Cooperman. 1999. Three-dimensional placement of the conserved 530 loop of 16 S rRNA and of its neighboring components in the 30 S subunit. *J. Mol. Biol.* **286:**521–540.

Weller, J., and W. E. Hill. 1992. Probing dynamic changes in rRNA conformation in the 30 S subunit of the *E. coli* ribosome. *Biochemistry* **31:**2748–2757.

Weller, J., and W. E. Hill. 1994. Probing the interactions of poly(U) and $tRNA^{PHE}$ with nucleotides 1530–1542 and 1390–1417 of 16 S rRNA of E. coli. *J. Biol. Chem.* **269:**19369–19374.

Wilson, K. S., and H. F. Noller. 1998. Molecular movement inside the translational engine. *Cell* **92:**337–349.

Wower, J., S. S. Hixson, and R. A. Zimmermann. 1989. Labeling the peptidyltransferase center of the E. coli ribosome with photoreactive tRNAPhe derivatives containing azidoadenosine at the 3′ end of the acceptor arm: a model of the tRNA-ribosome complex. *Proc. Natl. Acad. Sci. USA* **86:**5232–5236.

Wower, J., L. A. Sylvers, K. V. Rosen, S. S. Hixson, and R. A. Zimmermann. 1993. A model of the tRNA binding sites on the *E. coli* ribosome, p. 455–464. *In* K. H. Nierhaus, F. Franceschi, A. R. Subramanian, V. A. Erdmann, and B. Wittmann-Liebold (ed.), *The Translation Apparatus.* Plenum, New York, N.Y.

Wower, J., I. K. Wower, S. V. Kirillov, K. V. Rosen, S. S. Hixson, and R. A. Zimmermann. 1995. Peptidyl transferase and beyond. *Biochem. Cell Biol.* **73:**1041–1047.

Yonath, A., and F. Franceschi. 1998. Functional universality and evolutionary diversity: insights from the structure of the ribosome. *Structure* **6:**679–684.

The Ribosome: Structure, Function, Antibiotics, and Cellular Interactions
Edited by R. A. Garrett, S. R. Douthwaite, A. Liljas, A. T. Matheson, P. B. Moore, and H. F. Noller

Chapter 24

Folding of Nascent Peptides on Ribosomes

BOYD HARDESTY, GISELA KRAMER, TAMARA TSALKOVA, VASANTHI RAMACHANDIRAN, BRYAN McINTOSH, and DELBERT BROD

The *Escherichia coli* genome contains coding sequences for about 4,300 proteins (Ewalt et al., 1997). Most of these genes are expressed in growing cells, and their products may exist in multiple copies. In rapidly growing *E. coli*, the proteins are synthesized and folded into their active conformation within a cell generation time, which may be as short as 20 min. How is folding of these proteins accomplished? Many proteins do not refold efficiently from the denatured state without the assistance of molecular chaperones. However, the concentrations of most of the molecular chaperones shown to be effective in refolding are much lower than the concentration of nascent peptides. It has been calculated that the chaperonin GroEL could assist folding of only about 5% of newly synthesized protein in logarithmically growing *E. coli* (Lorimer, 1996). Ewalt et al. (1997) put this number at 10 to 15% based on in vivo studies. In either case, the data prompt the question of how cotranslational folding of the nascent peptide is accomplished at the required rate. This question leads directly to consideration of how the nascent peptide is moved from its point of synthesis in the peptidyltransferase center to the exit site on the distal surface of the large ribosomal subunit and the nature of interactions that must take place between the nascent peptide and components of the ribosome in this process.

In this review we will consider nascent peptides on *E. coli* ribosomes and how the *E. coli* ribosome may contribute to the folding of nascent proteins. The topics presented strongly reflect our own experiences and our frequently prejudiced ideas about the mechanisms that may be involved.

We restrict the term nascent protein to mean peptides that exist as peptidyl-tRNA bound to a ribosome in the A or P site. A nascent protein may be incomplete in that it requires further elongation or it may be a full-length polypeptide that has not been released from the ribosome by codon-dependent termination. Results from eukaryotic systems will be mentioned only if they appear to be relevant to results found in the bacterial system. More complicated situations in eukaryotes involving translocation across intracellular membranes and modifications of peptides that take place in the endoplasmic reticulum will not be discussed.

THE RIBOSOMAL TUNNEL

An open space, a tunnel, inside the 50S ribosomal subunit was first detected in the early analyses of crystalline arrays of *Bacillus stearothermophilus* ribosomes (Yonath et al., 1987). The recognition of a tunnel through the 50S subunit led to the suggestion that this might be the path followed by the nascent peptide through the ribosome. Immuno-electron microscopy originally placed the site at which the nascent protein emerges from the ribosome, the exit site, at a point on the 50S subunit distal to the peptidyltransferase center (Bernabeu and Lake, 1982). Thus, the tunnel might constitute the path followed by the nascent peptide between the peptidyltransferase center and the exit site. Recent results from a number of laboratories favor this conclusion. However, this was challenged by Ryabova and coworkers (1988), who tagged the amino group of Met-tRNA$_f^{Met}$ with 2,4-dinitrophenol and then used antibodies against 2,4-dinitrophenol and immuno-electron microscopy. These authors proposed that the nascent peptide follows an open channel to the cytoplasmic environ-

Boyd Hardesty, Gisela Kramer, Tamara Tsalkova, Vasanthi Ramachandiran, Bryan McIntosh, and Delbert Brod ■ Department of Chemistry and Biochemistry, The University of Texas at Austin, Austin, TX 78712-1167.

ment. However, recent results strongly favor the conclusion that the path of the nascent peptide is a tunnel rather than a channel on the surface of the large subunit. Refinements of the structure of the 50S subunit indicated several tunnels (Yonath and Berkovitch-Yellin, 1993; Eisenstein et al., 1994; Frank et al., 1995), with a narrow neck close to the peptidyltransferase center and a wider section that appeared to terminate at the distal surface of the subunit near the exit site. Recent image reconstruction from cryo-electron micrographs (Beckmann et al., 1997; Malhotra et al., 1998) shows a branched tunnel or tunnels within the 50S ribosomal subunit. It can be calculated from these images that the distance from the peptidyltransferase center to the exit site is about 71 Å with a width between 16 and 19 Å. These dimensions are important for consideration of the possible secondary structure of the section of the nascent peptide that might traverse the tunnel.

23S RNA AND PEPTIDE BOND FORMATION

Noller and coworkers reported peptide bond formation by the RNA portion of the 50S ribosomal subunit (reviewed in Green and Noller, 1997). Many other functions related to peptide bond formation have been associated with the 23S RNA. The peptidyl transferase center has been associated with nucleotides in domain V of 23S RNA by using nucleotide protection with antibiotics that inhibit peptide bond formation. For example, chloramphenicol binding protects A2058 in the central loop of domain V (Moazed and Noller, 1987). The ribosomal A site on 23S RNA has been mapped by cross-linking with a puromycin analogue (Green et al., 1998). Its reaction with a tRNA in the P site led to identification of positions on the 23S RNA protected by this tRNA. The entrance to the peptide tunnel was assumed to be located in proximity to the CCA end of P-site-bound tRNA (Frank, 1998).

Picking et al. (1991) used modified Ala-tRNA^{Ala} with an AAA anticodon in combination with poly(U) to bind *N*-coumarin–Ala-tRNA to ribosomes. When the fluorophore–Ala-tRNA was bound to the P site of the ribosome, the probe was held rigidly, as indicated by high fluorescence anisotropy, and in a hydrophobic environment, as indicated by the high quantum yield and shift in the fluorescence emission spectrum.

Other studies (Picking et al., 1992a) indicated that the fluorophore of *N*-acyl–aminoacyl-tRNA is bound to the ribosome in close proximity to RNA. This conclusion was reached from results involving fluorescence quenching by methyl viologen (1,1′ dimethyl-4, 4′ bipyridinium dichloride). Methyl viologen as a cation in solution has a strong affinity for the ionized phosphates of a nucleic acid. Fluorescence lifetime measurements indicated that the increase in fluorescence quenching observed when the tRNA was bound into the P site of the ribosome was due to static quenching by methyl viologen bound to RNA. The result led to the conclusion that the probe of *N*-acyl–aminoacyl-tRNA was bound in very close proximity to the RNA (probably within about 15 Å) and strongly supported the hypothesis that RNA is the primary functional component in the ribosomal synthesis of a peptide bond. Also, the results indicated that the quenching agent, methyl viologen, could penetrate the ribosome in the region of the peptidyltransferase center. The recent crystallographic map of the large ribosomal subunit (Ban et al., 1998) exhibits a forest of criss-crossing helices of the 23S RNA with large open spaces between them. In part, these spaces may be occupied by ribosomal proteins. A detailed understanding of the ribosome function considered below is critically dependent on the refinement of this map.

NASCENT POLYPEPTIDES

Proteins are synthesized vectorially from their N termini to their C termini on ribosomes. Most studies have led to the conclusion that the C-terminal part of a nascent peptide is shielded by the ribosome. This protected segment is at least 15 amino acids long and may extend to 60 or 70 amino acids for some proteins. Different biochemical, fluorescence, and immunological approaches have confirmed this finding during the last 30 years. These results prompt questions about how, where, and with what the nascent peptide interacts as it is extended within a ribosome and which factors may function in this process. Directly related questions involve the conformation of the nascent peptide as it is involved in these processes.

Picking et al. (1992b) used anticodon-modified forms of *N*-coumarin–aminoacyl-tRNA to study the poly(U)-directed synthesis of *N*-coumarin-polyalanine and *N*-coumarin-polyserine. Fluorescent lysyl-tRNA derivatized at the ε-amino group of lysine was used to initiate synthesis of polylysine from poly(A) in comparable experiments. All of the *N*-acyl–aminoacyl-tRNA species were held rigidly in what appeared to be a relatively hydrophobic environment when they were bound into the ribosomal P site. However, the fluorescence anisotropy and quantum yield changed in markedly different ways as the three types of homopolymeric peptides were elongated. The quantum yield and anisotropy declined as the polyalanine peptides were extended to an average length of 60 residues. This value corresponded closely to the average peptide length at which the N-terminal probe became accessible to anticoumarin-immunoglobulin G (IgG) and was suggested to be the length at which the probe emerged from the ribosome. It corresponds to a distance of 90 Å if the alanine polypeptide was in the conformation of an α-helix. Polyalanine has a high propensity to take an α-helical conformation. The corresponding length for recognition by Fab (M_r, ~50,000) derived from these antibodies was about 50 amino acids and about 40 amino acids for proteinase K (M_r, ~28,000).

The results with other nascent homopolymeric peptides were quite different. Anisotropy declined very rapidly as polylysine peptides were extended to a length of about 10 residues. A minimum value was reached at this length, indicating that nascent polylysine peptides, presumably positively charged due to the ionization of the ε-amino group, do not bind tightly to the ribosomes. This may be due to their conformation, which is likely to be largely unstructured. Other results indicated that short lysine peptides are easily aborted and lost from the ribosomes. Anisotropy changes during synthesis of *N*-coumarin-polyserine peptides were intermediate between those for polylysine and polyalanine. The results for polyphenylalanine (Picking et al., 1991) were strikingly different from those indicated above. The quantum yield, blue shift, and anisotropy of fluorescence were initially very high and tended to increase slightly as the peptides were extended. Lim and Spirin (1986) concluded from a detailed stereochemical analysis of the peptidyltransferase reaction that the nascent peptide was generated in an α-helical conformation. However, nascent peptides of different amino acid compositions appear to acquire different conformations as they are extended. Polyalanine and polyserine tend to form α-helices, whereas polylysine is charged, is very soluble, and tends to form a random coil. In sharp contrast, polyphenylalanine is highly insoluble in aqueous and most organic solvents. It appears to form a highly insoluble, amorphous glob near the peptidyltransferase center as it is formed.

The results indicate that the different conformations of these homopolymeric peptides can be accommodated by the ribosomes without blocking formation of peptide bonds and extension of the nascent peptides. It is unlikely that either nascent polylysine or polyphenylalanine peptides enter the tunnel and are pushed through the ribosome. This raises the possibility that nascent peptides of different proteins may follow different routes and be processed differently after they leave the peptidyltransferase center.

Contact sites between nascent peptides and 23S RNA have been mapped in cross-linking experiments by Brimacombe and coworkers. Most recently (Choi and Brimacombe, 1998), nascent peptides of *E. coli* OmpA protein and peptides from the phage T4 gene 60 were analyzed and compared to the previous results. A photoreactive diazirine moiety attached to the α-amino group of methionine in Met-tRNA$_f$ was used to synthesize peptides of defined length on *E. coli* ribosomes. After irradiation, the peptides that were formed were cross-linked about equally to 23S RNA and ribosomal proteins. The results for different lengths of nascent peptides showed progressive 23S RNA cross-linking first to nucleotides in domain V and then in domains II, III, and I—with some variations. Surprisingly, relatively long peptides also gave some cross-linking to domain V, which is thought to be associated with the peptidyltransferase center. The results were interpreted to show a continuous path of the nascent peptide through the tunnel inside the 50S ribosomal subunit that allows flexibility of the nascent peptide.

Cross-linking to both 30S proteins and 16S RNA also occurred in these experiments. The protein cross-linking was predominantly to S1, S2, S4, and, to a lesser extent, S3 (Choi et al., 1998). Cross-link sites of the 16S RNA were only partially localized, but one weak site to nucleotides in the 1385-to-1415 region was tentatively identified. These are very surprising and provocative results that, if substantiated, would require major modification of the generally accepted dogma that the nascent peptide is transported through the tunnel and folded in association with the 50S subunit. Clearly, answers to the two following questions are needed. What are the forces that direct the nascent peptide into and along the proper path in the first place? Do all nascent peptides of proteins follow the same path through the ribosome?

COTRANSLATIONAL FOLDING OF NASCENT PEPTIDES

Chantrenne (1961) appears to have been the first to rigorously state the hypothesis that proteins fold vectorially as they are synthesized from their N termini to their C termini. However, this has proven to be a difficult point to establish. The recent literature on this topic has been reviewed in Hardesty et al. (1999).

In some early studies, the presence of enzymatic activity with ribosomes on which the enzyme had been synthesized was taken as evidence for cotranslational folding. It now appears that many of these positive reports involved full-length enzyme that had been properly released but remained associated with the ribosome due to nonspecific interaction. We have found that nearly all enzyme proteins synthesized in the *E. coli* coupled transcription-translation system are enzymatically inactive when bound to the ribosome. However, enzymatically active nascent rhodanese (Kudlicki et al., 1995) and luciferase (Makeyev et al., 1996) were found if these proteins were extended at the C terminus by at least 23 and 26 amino acids, respectively. Apparently the C-terminal segment of wild-type enzyme is held within the ribosome, probably within the tunnel near the peptidyltransferase center, in such a way as to prevent folding of the polypeptide into the native conformation.

Nascent globin appears to constitute an exception to the principle that nascent proteins cannot fold into the native conformation. Komar and coworkers (1997) found that nascent (truncated) N-terminal globin fragments of 86 amino acids or longer were able to bind heme. These results demonstrated that nascent peptides can fold into their native conformations while they are bound to the ribosome as peptidyl-tRNAs.

Direct evidence for the folding of relatively short segments of nascent proteins on the ribosomes during their synthesis was obtained by determining the length of the nascent peptide at which its N terminus became accessible to specific antibodies that recognize an N-terminal probe (Tsalkova et al., 1998). Rhodanese, chloramphenicol acetyltransferase (CAT), and MS2 coat protein were synthesized in the bacterial cell-free system by using *N*-coumarin–Met-$tRNA_f$ to initiate the protein. After synthesis, the length that each of the three nascent peptide types must reach in order to be reactive with anti-coumarin IgG was determined.

The results indicated that this size was quite different for the three different nascent polypeptides: 4.5 kDa for the MS2 coat protein, 6 kDa for rhodanese, and 8.5 kDa for CAT. Since the C-terminal ends of the nascent peptides are bound as peptidyl-tRNA in the peptidyltransferase center and the N-terminal coumarin presumably emerges from the large ribosomal subunit at the same point for each of the nascent proteins, the results were interpreted to indicate that each of the nascent peptides has a different conformation and they have folded into these conformations in a protected cavity within the ribosome.

Whether additional factors are required for folding of newly synthesized proteins is not clear. Trigger factor, peptidyl-*cis/trans* prolyl isomerase, is bound to the large subunit of *E. coli* ribosomes and appears to have protein folding activity and to function cotranslationally (Stoller et al., 1995; Hesterkamp et al., 1996). DnaJ may interact with a nascent peptide (Kudlicki et al., 1996), but this may be a special situation in which the molecular chaperone functions to correct misfolding that has occurred in a small portion of the nascent peptides. A DnaK homologue, Hsc 70, appears to be involved in cotranslational folding of polypeptides larger than 20 kDa in eukaryotic cells in vivo (Thulasiraman et al., 1999).

PAUSE SITE PEPTIDES

Synthesis of most if not all proteins is accompanied by a unique pattern of ribosome-bound peptides of different lengths that appears to result from discontinuous synthesis. This is demonstrated in Fig. 1 for six different proteins synthesized in vitro by coupled transcription-translation with an *E. coli* extract. The plasmids containing the coding sequences of the proteins under the control of the T7 promoter were transcribed with T7 RNA polymerase and translated in the presence of [^{14}C]leucine. The autoradiogram derived from a sodium dodecyl

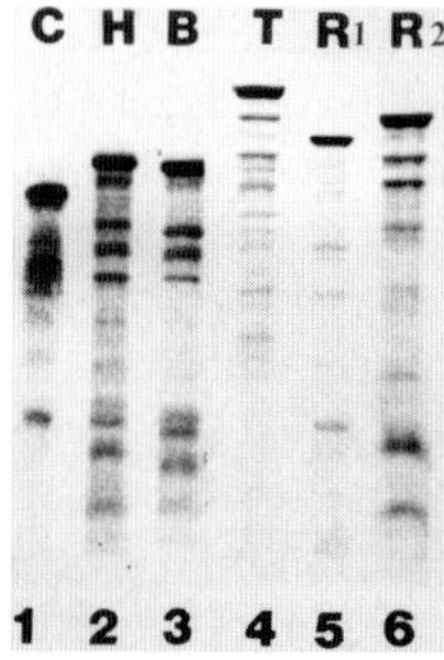

Figure 1. Translational pause sites differ for different proteins. Six different proteins were synthesized by coupled transcription-translation in the cell-free *E. coli* system with a small amount of the A19 S30 fraction (cf. Tsalkova et al., 1999). The coding sequences were in plasmids under the control of the T7 promoter. [^{14}C]leucine was the radioactive precursor. After coupled transcription-translation, an aliquot (15 μl) was withdrawn to determine the amount of polypeptides formed, another aliquot (15 μl) of the reaction mixtures was analyzed by polyacrylamide gel electrophoresis according to the method of Schägger and von Jagow (1987), and the radioactive bands were visualized by phosphorimaging. Lane 1, CAT (M_r, ~25,600; 13 leucines; 207 pmol of leucine incorporated); lane 2, hamster rhodanese (M_r, ~33,000; 28 leucines; 103 pmol of leucine incorporated); lane 3, bovine rhodanese (M_r, ~33,000; 25 leucines; 115 pmol of leucine incorporated); lane 4, trigger factor (M_r, ~58,000; 31 leucines; 73 pmol of leucine incorporated); lane 5, release factor 1 (M_r, ~40,500; 32 leucines; 50 pmol of leucine incorporated); lane 6, release factor 2 (M_r, ~41,200; 29 leucines; 61 pmol of leucine incorporated).

sulfate-polyacrylamide gel on which the resulting polypeptides were separated is shown. The pattern of the peptide bands is different for each of the six proteins. Their names and molecular masses, the number of leucine residues in their full-length amino acid sequences, and the total amount of leucine incorporated into polypeptides are given in the legend to Fig. 1. Note that the relative intensity of a specific peptide band is dependent on the number of leucine residues the peptide contained in relation to its relative abundance. This is in contrast to peptide labeled with radioactive cytidylic acid-puromycin (C-puro), as shown in Fig. 2.

CAUSE OF TRANSLATIONAL PAUSING

The mechanism by which the pause site peptides are generated is unknown. Suggested causes involve rarely used codons and small amounts of certain cognate tRNA species, truncation of mRNA, the secondary structure of the mRNA, slow steps in cotranslational folding, or the interaction of the nascent peptide with the ribosome (discussed and referenced in Tsalkova et al., 1999). Before considering how pause site peptides are generated, it must be established that they are nascent peptides as defined above. That is, they have not been prematurely aborted and released from the ribosomes and are not peptide fragments generated by proteolysis. Although premature release, particularly of short peptides, does occur, this does not account for the pattern of peptides that is observed. This conclusion is substantiated by reactions of pause site peptides with an analogue of puromycin. The results are shown in Fig. 2. After coupled transcription-translation, the reaction mixtures were incubated with ^{32}P-labeled C-puro, which reacts only with peptidyl-tRNA in the ribosomal P site. The use of puromycin derivatives has been described by Green et al. (1998). Two important features of C-puro in contrast to puromycin itself are its low K_m value, which is in the micromolar range, and the possibility of phosphorylating C-puro and thereby labeling the compound with radioactive phosphate.

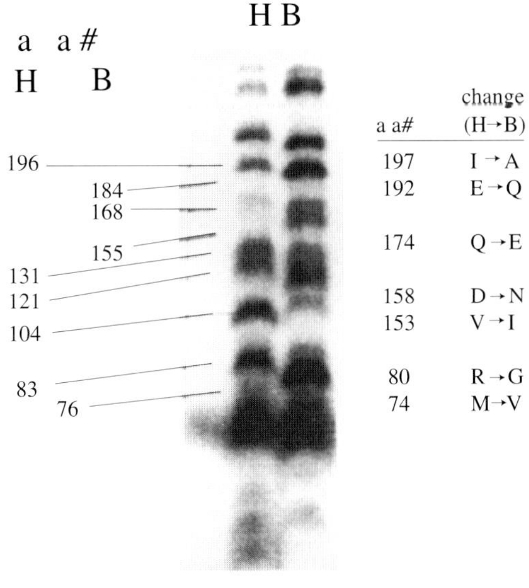

Figure 2. Puromycin reactivity of nascent peptides. Hamster (H) and bovine (B) rhodanese were synthesized as described in the legend to Fig. 1 except that 5 μl of S30 (*E. coli* MRE 600) and nonradioactive amino acids were used. After coupled transcription-translation, ^{32}P-labeled C-puro was added to give 8 μM, and the incubation continued for 10 min. Then one aliquot was used to determine the incorporation of C-puro into nascent peptides, and another aliquot was processed for polyacrylamide gel electrophoresis and phosphorimaging. The result is shown. Lane 1, no plasmid added; lane 2, hamster rhodanese; lane 3, bovine rhodanese. About 4.5 and 5 pmol of C-puro was incorporated into hamster and bovine polypeptides, respectively. The numbers on the left side of the gel indicate the numbers of amino acids (a a #) corresponding to the puromycin-labeled band in either lane 2 (H) or lane 3 (B). On the right side of the gel, amino acid positions in which a change in amino acid composition from H to B occurred are given.

The result, shown in Fig. 2, demonstrates that pause site peptides are reactive with C-puro and thus bound to the ribosome as nascent peptides. Also, considered with the similar pattern for peptides labeled at their N termini (not shown), the results lead to the conclusion that they are not proteolytic degradation products of nascent peptides. Beyond these important conclusions, Fig. 2 shows that the patterns for hamster and bovine rhodanese peptides are similar but not identical. The numbers at the left of the figure are the approximate positions in the hamster or bovine amino acid sequence at which the pause site occurs, as well as this can be deduced from molecular-weight markers run on the same gel. The positions and kinds of amino acid changes in the sequences of the two proteins are given on the right. There appears to be no correspondence between differences in amino acid sequences that might account for differences in the pause sites. No changes occur between amino acids 80 and 153, between amino acids 158 and 174, or between amino acids 174 and 192, yet numerous differences in the pause site patterns are discernible in all these regions. On this basis, it appears unlikely that the amino acid sequences of the nascent peptides are directly related to the size pattern of the pause site peptides. However, there are numerous differences in the nucleotide sequences of the corresponding segments of the coding region (Miller et al., 1991; Trevino et al., 1995) that might lead to differences in translation related to the mRNA.

THE EFFECTS OF N-TERMINAL PROBES

In some situations, the character of an N-terminal residue may influence the quantity of certain specific pause site peptides (Tsalkova et al., 1999). However, it does not appear to affect the position of the translational pause site on the mRNA and thus the size of the resulting pause site peptide. An accumulation of a heterogeneous band of relatively small peptides occurs during the synthesis of CAT with coumarin-labeled initiator tRNA. The hypothesis was that the addition of a hydrophobic residue to a relatively hydrophilic N terminus of a nascent protein may affect its initial folding and transport path through the ribosome, depending on its amino acid composition and conformation. In turn, this might affect the quantity of pause site peptides. This hypothesis was supported by the observation that no effect similar to that seen with CAT was observed during the synthesis of N (coumarin)-labeled rhodanese (see Fig. 4 and 5 in Tsalkova et al., 1999). Rhodanese has a relatively hydrophobic leader sequence.

A size-heterogeneous band of CAT peptides of 3,500 to 4,000 M_r (size range, 35 amino acids) accumulates transiently on the ribosomes even when they have fMet at their N termini. This apparent accumulation is markedly increased if the peptides are initiated with coumarin–Met-tRNA$_f$ or other fluorophore–Met-tRNA species (see below). The N-terminal group appears to have little effect on the size or relative abundance of the larger pause site peptides.

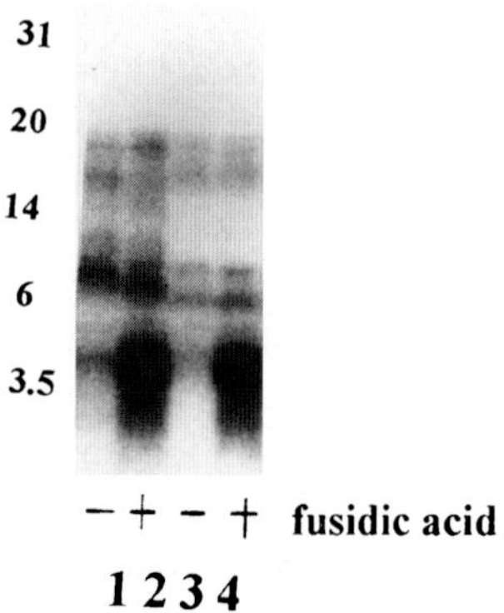

Figure 3. Fusidic acid keeps short CAT peptides in the ribosomal P site. CAT was synthesized for either 5 min (lanes 1 and 2) or 30 min (lanes 3 and 4) by coupled transcription-translation with an S30 fraction from MRE 600 and unlabeled amino acids. Then, either H_2O (lanes 1 and 3) or fusidic acid (lanes 2 and 4) was added and the samples were kept on ice for 3 min before the addition of ^{32}P-labeled C-puro. The samples were incubated and processed as described in the legend to Fig. 2. The numbers on the left refer to positions of molecular weight markers as follows: 31, carbonic anhydrase; 20, soybean trypsin inhibitor; 14, lysozyme; 6, aprotinin; 3.5, insulin β chain. +, present; −, absent.

The reactivity of the short CAT peptides with radioactive C-puro is relatively low but is greatly increased if fusidic acid is added to the reaction mixture immediately after peptide synthesis but before the reaction with C-puro is initiated. This important result, shown in Fig. 3, demonstrates that the peptides are present on the ribosomes as peptidyl-tRNA. Fusidic acid binds to EF-G and blocks the release of GDP from the factor so that EF-G–GDP remains bound to the ribosome after the hydrolysis of GTP (Miller and Bodley, 1991). Thus, it may function in these experiments by preventing the EF-G-assisted release of short peptides from the ribosome (Heurgué-Hamard et al., 1998).

OTHER N-TERMINAL PROBES

The results with *N*-coumarin–Met-tRNA$_f$ prompted the synthesis of other derivatives with different fluorescent probes. An objective was to begin to characterize the limits in terms of size and chemical character of probes that can be incorporated into a protein in addition to characterization of the effect of the N-terminal probes on cotranslational folding. *N*-Acyl derivatives of Met-tRNA$_f$ were synthesized with cascade yellow, eosin Y, or pyrene covalently linked to the α-amino group of methionine and tested for incorporation during protein synthesis (Ramachandiran et al., in press). The structures of the fluorophore–Met-tRNA$_f$ species are given in Fig. 4. They provide a series of fluorescent *N*-acyl derivatives that vary in size and chemical character compared to *N*-formyl methionine.

Total synthesis of CAT or rhodanese polypeptides under optimal conditions is somewhat reduced with the coumarin, cascade yellow, or pyrene derivative relative to synthesis with formyl-methionine, as shown in Table 1. However, synthesis is sharply reduced with the eosin derivative. Eosin, with its larger ring containing four bromine atoms, appears to be near the limit of the size and structure that can be incorporated into a protein.

When CAT polypeptides are initiated with the fluorophore–Met-tRNA$_f$ and synthesized with a reduced amount of the S30 fraction, small peptides, including the 3.5- to 4-kDa species, appear to accumulate on the ribosomes over time (Fig. 5). This is especially pronounced when cascade yellow or pyrene-Met forms the N terminus. Again, synthesis is

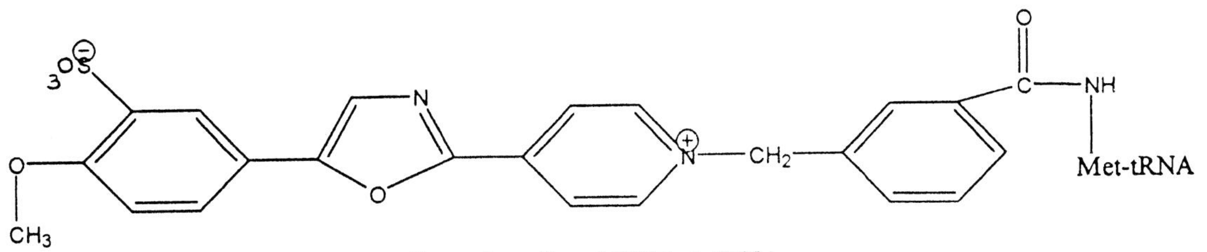

Cascade yellow-NH(Met)-tRNA

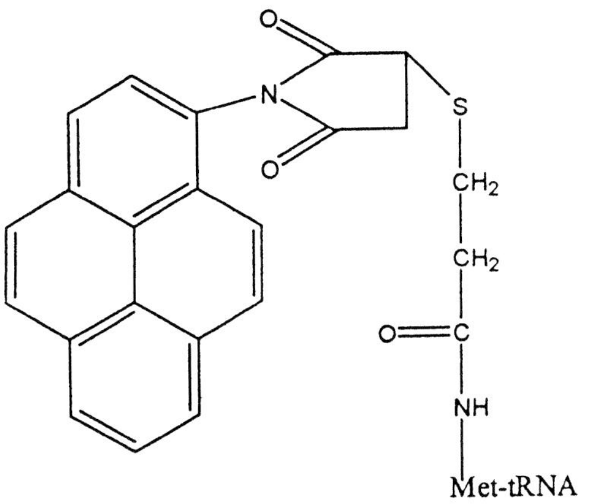

Pyrene-Succinimido-S-P-NH(Met)-tRNA

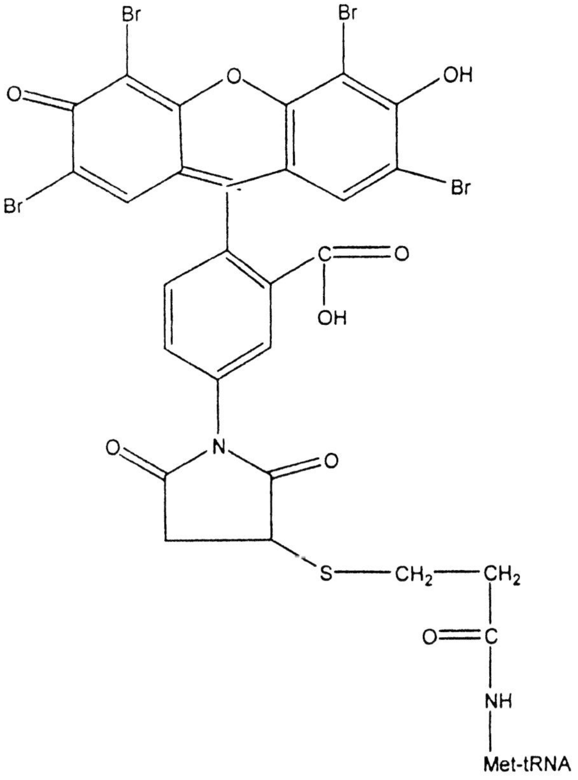

Eosin-Succinimido-S-P-NH (Met)-tRNA

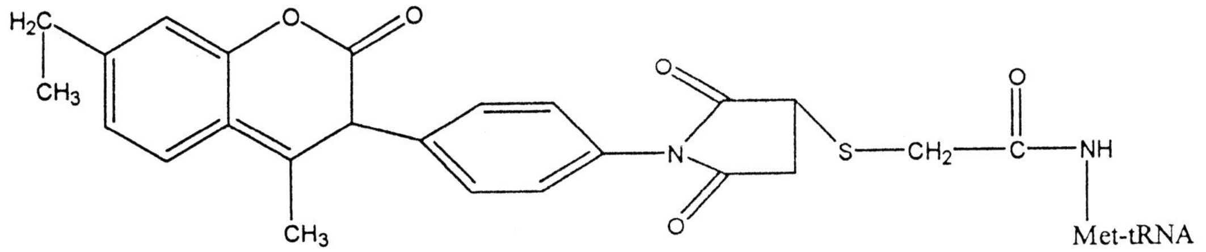

CPM-S-Ac-NH(Met)-tRNA

Figure 4. Structures of different fluorophore–Met-tRNA$_f$ species that were synthesized and used in coupled transcription-translation.

greatly reduced with eosin–Met-tRNA$_f$. However, the data presented in Fig. 6A indicate that eosin–Met-tRNA$_f$ can form a binary complex with initiation factor IF2. The binding constant for the interaction is similar to that of fMet-tRNA$_f$ and IF2. Scatchard plots of the interaction indicated a 1:1 ratio of the two components in the resulting complex (data not shown). This interaction was measured in the absence of Mg^{2+} ions, as the Met-tRNA$_f$ complex with IF2 is unstable in the presence of these divalent ions (Sundari et al., 1976). Binding of eosin–Met-tRNA$_f$ to salt-washed ribosomes in the absence or presence of IF2 is shown in Fig. 6B. This is carried out at an Mg^{2+} concentration similar to that used in coupled

Table 1. Incorporation of *N*-acyl-methionine into RHO and CAT[a]

N-acyl-Met type	Incorporation (pmol of *N*-acyl-Met)	
	RHO	CAT
fMet	2.4	2.5
Cascade yellow-Met	1.3	1.4
Pyrene-Met	1.2	1.3
Coumarin-Met	1.2	1.3
Eosin-Met	1.0	0.4

[a] RHO or CAT was synthesized by coupled transcription-translation under optimal conditions (with 5 μl of S30/30-μl reaction mixture) with CPM-SAc-[^{35}S]Met-tRNA$_f$, cascade yellow–[^{35}S]Met-tRNA$_f$, eosin-SP-[^{35}S]Met-tRNA$_f$ or fMet-tRNA$_f$ as the initiator tRNA. After 30 min of incubation at 37°C, an aliquot from the reaction mixture was precipitated by trichloroacetic acid and its radioactivity was determined. The data presented are calculated for a 30-μl reaction mixture. The values for incorporation of the [^{35}S]methionine derivatives in the absence of added plasmid were 0.08 ± 0.01 pmol. This value was subtracted.

transcription-translation. Other results show no large difference in tertiary complex formation between fMet-tRNA$_f$ or cascade yellow–Met-tRNA$_f$, IF2, and the ribosomes. Preliminary data show puromycin reactivity of eosin–Met-tRNA$_f$ bound to salt-washed ribosomes with IF2. Although this conclusion must be considered very tentative, it prompts questions about what part of the translational apparatus is not fully compatible with eosin. Might it be the inability of eosin to be transported through the tunnel?

SHIFT OF NASCENT PEPTIDES TO FULL-LENGTH PROTEIN

The data presented above indicate that the amount of nascent peptides on the ribosomes can be manipulated. In addition, we observed differences in different S30 preparations in how efficiently full-length product can be formed. We excluded the possibility that a poor performance is due to increased amounts of ribonuclease or lack of tRNA. Rather, we observed that the efficiency of a "poor" S30 preparation can be increased by adding back a certain protein fraction of a "good" S30. Figure 7 demonstrates this effect. Increasing amounts of the protein fraction were added to a reaction mixture in which either CAT or RHO was synthesized. Increased amounts of full-length protein were obtained, and the relative amount of short peptides decreased. These full-length polypeptides have enzymatic activity, i.e., they are not only formed but folded correctly into their native structures.

Preliminary characterization of the active fraction indicates that a protein is an integral component, as shown by heat inactivation of the stimulatory effect. Rhodanese enzymatic activity was determined after coupled transcription-translation in the absence of the factor and in its presence and in the presence of the activating fraction after it had been preincubated for 10 min at elevated temperatures. A rapid loss was observed at temperatures above 40°C. Also, the increased formation of full-length protein as measured by rhodanese enzymatic activity was abolished by pretreatment of the fraction with *N*-ethyl maleimide.

RENATURATION MEDIATED BY 50S SUBUNITS AND 23S RNA

Refolding from the denatured state of many (especially larger) proteins is assisted in vitro by chaperones and chaperone-like molecules (Ellis, 1997). An interesting observation has been that ribosomes, their 50S subunits, or 23S RNA can increase the rate

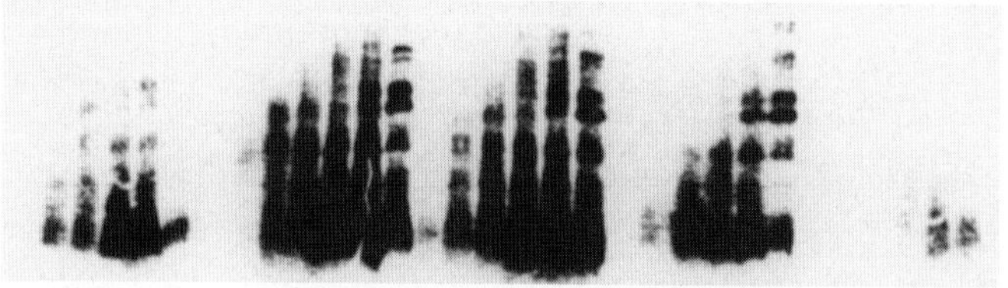

Figure 5. Time course of CAT synthesis with fluorophore–Met-tRNA$_f$. CAT was synthesized by coupled transcription-translation with a reduced amount of the S30 fraction (2.5 μl/30-μl assay). Protein synthesis was initiated with either f[^{35}S]Met-tRNA$_f$, cascade yellow–[^{35}S]Met-tRNA$_f$, pyrene–[^{35}S]Met-tRNA$_f$, coumarin–[^{35}S]Met-tRNA$_f$, or eosin–[^{35}S]Met-tRNA$_f$. The reaction mixtures were incubated, and at the indicated times aliquots were withdrawn and analyzed by sodium dodecyl sulfate-polyacrylamide gel electrophoresis and phosphorimaging. The results are shown.

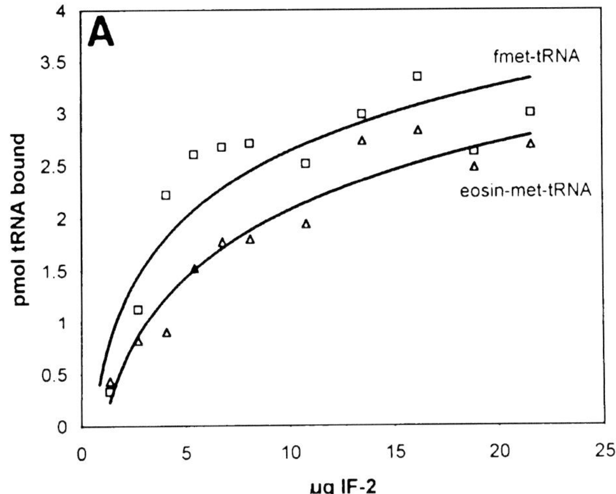

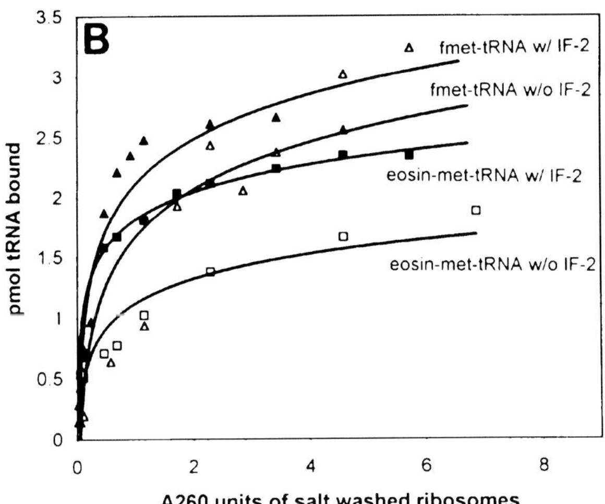

Figure 6. Interaction of eosin–Met-tRNA$_f$ with IF2 and its binding to salt-washed ribosomes. (A) Binding of f[^{35}S]Met-tRNA$_f$ and eosin–[^{35}S]Met-tRNA$_f$ to IF2 was determined by the Millipore filter binding assay in the absence of Mg^{2+} (Sundari et al., 1976). IF2 was isolated according to the method of Mortensen et al. (1991) from an *E. coli* strain transformed by a plasmid containing the IF2 sequence. (We thank U. RajBhandari for providing the plasmid.) (B) Binding of f[^{35}S]Met-tRNA$_f$ and eosin–[^{35}S]Met-tRNA$_f$ to salt-washed ribosomes was carried out in the presence of 18 mM Mg^{2+} in the absence (w/o) or presence (w/) of an excess of IF2.

and extent of the refolding reaction (Das et al., 1996; Kudlicki et al., 1997). Recently, Chattopadhyay et al. (1999) reported that a similar activity of the 50S ribosomal subunit may exist in vivo.

Ribosomes appear to be in two interconvertible states, or conformations, active and inactive, with respect to renaturation of a denatured protein. Inactive ribosomes were activated for renaturation by EF-G plus GTP or by α-sarcin cleavage of a specific tetraloop in their 23S RNAs. Renaturation was inhibited by both fusidic acid and thiostrepton.

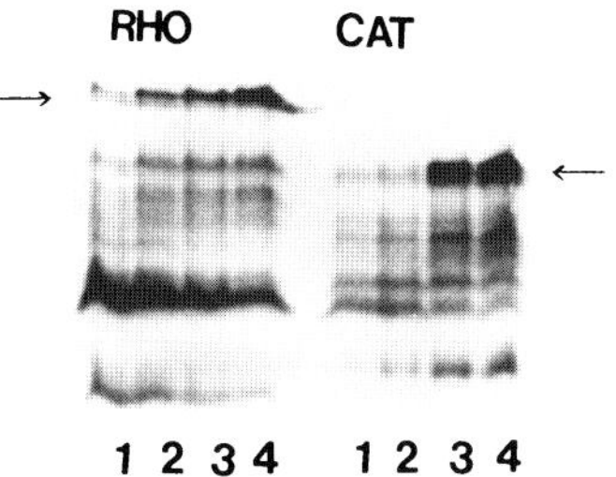

Figure 7. Increased synthesis of full-length protein in the S30 fraction from MRE 600. Both RHO and CAT were synthesized for 20 min in the presence of [^{14}C]Leu. The samples shown in lanes 2 to 4 received increasing amounts of a fraction derived from a different S30 before protein synthesis was started. This fraction was isolated by S300 chromatography and eluted well behind ribosomes but before ~100-kDa proteins. The arrows indicate the positions of the full-length proteins.

Inhibition of refolding activity by both antibiotics suggests that the activity is dependent on GTP-mediated cycling between two ribosomal conformations.

α-Sarcin cleaves a single phosphodiester bond at the 3′ side of G2661 of domain VI of 23S RNA. Hydrolysis abolishes the functions of both EF-Tu and EF-G that may promote or stabilize a specific conformation of the 50S ribosomal subunit (Nierhaus et al., 1992). We suggest that α-sarcin activation of the ribosome for renaturation is associated with increased flexibility of the 50S subunits resulting from a failure to be held in the conformation stabilized by either EF-G or EF-Tu.

HYPOTHESIS: THE RIBOSOME AS A CHAPERONE FOR FOLDING OF NASCENT PEPTIDES

The results considered above lead to the conclusion that most nascent peptides acquire secondary and specific tertiary structures as they are formed; that is, they undergo cotranslational folding. The nascent peptide is sheltered from external proteins the size of IgG and proteases as it is extended from the peptidyltransferase center. Beyond this region but within or on the ribosome the nascent peptides of at least some enzymes can fold into conformations related to their enzymatically active, native states. These regions of the ribosome are probably the tunnel and cavities near the exit site on the large ribosomal subunit. Thus, the hypothetical folding cavity appears not only to provide a place where folding can take place but also functions to shield the nascent protein from destructive factors and interaction with other proteins that might lead to misfolding and formation of nonspecific complexes. Upon release of the nascent peptide, the latter may lead to aggregation and inclusion body formation.

Two particularly perplexing but provocative problems are brought into focus by the results considered above. First, with what specific component of the peptidyltransferase center or adjacent region does the N terminus of the nascent peptide interact so that it is held rigidly in a hydrophobic environment? This interaction is likely to be with 23S RNA, although this remains to be firmly established. What structural feature of the 23S RNA could provide this interaction and effect?

The second, related question involves the mechanism by which the nascent peptide is pushed, pulled, or otherwise propelled through the ribosome and how this might affect the folding of the nascent peptide. This is likely to involve conformational changes in the structural components of the large ribosomal subunit.

Large ribosomal subunits and 23S RNA are able to promote renaturation of many proteins from their denatured state. Although renaturation and cotranslational folding are fundamentally different, the two processes are likely to involve similar mechanisms. Renaturation of denatured rhodanese by ribosomes or 23S RNA is promoted by EF-G plus GTP and by α-sarcin, respectively. We suggest that changes in the conformation of the large subunit or 23S RNA are involved as part of the reaction cycle of peptide elongation. Such changes in conformation would likely stretch and compress an interacting nascent or denatured peptide, thus facilitating unfolding and refolding of misfolded sections. Such sections may constitute folding traps in which the peptide is caught in a misfolded, semistable conformation. Misfolded regions are likely to have a relatively high proportion of exposed hydrophobic amino acid side chains that might interact with hydrophobic structure along the path followed by the nascent chain through the ribosome. The now unfolded section could then refold spontaneously into the stable conformation of the native state, in which the hydrophobic side chains are mostly buried within the protein and thus are no longer available for interaction with structural components of the ribosome. We envision such a GTP-dependent cycle within the ribosome to mechanistically resemble the ATP- and GroES-dependent processing of a denatured protein within the central cavity of the GroEL complex (Xu and Sigler, 1998).

The results considered above indicate that folding of a nascent peptide can take place in a sheltered region within the ribosome to yield a functional protein and that a hydrophobic moiety at the N terminus of a nascent peptide can interact with the ribosome. However, they do not establish that conformational changes in the ribosomes that are associated with reactions of the peptide elongation cycle facilitate folding of a nascent protein.

The research presented here was supported by grants from the National Institutes of Health (GM 53152) and from the Welch Foundation (F 1348).

We thank Barbara Jann for help in preparing the manuscript.

REFERENCES

Agrawal, R. K., P. Penczek, R. A. Grassucci, and J. Frank. 1998. Visualization of elongation factor G on the *Escherichia coli* ribosome: the mechanism of translocation. *Proc. Natl. Acad. Sci. USA* **95:**6134–6138.

Ban, N., B. Freeborn, P. Nissen, P. Penczek, R. A. Grassucci, R. Sweet, J. Frank, P. B. Moore, and T. A. Steitz. 1998. A 9 Å resolution x-ray crystallographic map of the large ribosomal subunit. *Cell* **93:**1105–1116.

Beckman, R., D. Bubeck, R. Grassucci, P. Penzcek, A. Verschoor, G. Blobel, and J. Frank. 1997. Alignments of conduits for the nascent polypeptide chain in the ribosome-Sec61 complex. *Science* **278:**2123–2126.

Bernabeu, C., and J. A. Lake. 1982. Nascent polypeptide chains emerge from the exit domain of the large ribosomal subunit: immune mapping of the nascent chain. *Proc. Natl. Acad. Sci. USA* **79:**3111–3115.

Chantrenne, H. 1961. The biosynthesis of proteins, p. 122–147. *In* P. Alexander and Z. Bacq (ed.), *Modern Trends in Physiological Science*, vol. 14. Pergamon Press, Elmsford, N.Y.

Chattopadhyay, S., S. Pal, D. Pal, D. Sarkar, C. Suparna, and C. Das Gupta. 1999. Protein folding in *Escherichia coli*: role of 23S ribosomal RNA. *Biochim. Biophys. Acta* **1429:**293–298.

Choi, K. M., and R. Brimacombe. 1998. The path of the growing peptide chain through the 23S rRNA in the 50S ribosomal subunit: a comparative cross-linking study with three different peptide families. *Nucleic Acids Res.* **26:**887–895.

Choi, K. M., J. F. Atkins, R. F. Gesteland, and R. Brimacombe. 1998. Flexibility of the nascent polypeptide chain within the ribosome. Contacts from the peptide N-terminus to the 30S subunit. *Eur. J. Biochem.* **255:**409–413.

Das, B., S. Chattopadhyay, A. K. Bera, and C. Dasgupta. 1996. In vitro protein folding by ribosomes from *Escherichia coli*, wheat germ and rat liver. The role of the 50S particle and its 23S rRNA. *Eur. J. Biochem.* **235:**613–621.

Eisenstein, M., B. Hardesty, O. W. Odom, W. Kudlicki, G. Kramer, T. Arad, F. Franceschi, and A. Yonath. 1994. Modeling and experimental study of the progression of nascent proteins in ribosomes, p. 213–246. *In* G. Pifat (ed.), *Supramolecular Structure and Function.* Ruder Boskovic Institute, Zagreb, Croatia.

Ellis, R. J. 1997. Do molecular chaperones have to be proteins? *Biochem. Biophys. Res. Commun.* **238:**687–692.

Ewalt, K. L., J. P. Hendrick, W. A. Houry, and F. U. Hartl. 1997. *In vivo* observation of polypeptide flux through the bacterial chaperonin system. *Cell* **90:**491–500.

Frank, J. 1998. The ribosome—structure and functional ligand-binding experiments using cryo-electron microscopy. *J. Struct. Biol.* **124:**142–150.

Frank, J., J. Zhu, P. Penczek, Y. Li, S. Srivastava, A. Verschoor, M. Radermacher, R. Grassucci, R. K. Lata, and R. K. Agrawal. 1995. A model of protein synthesis based on cryo-electron microscopy of the *E. coli* ribosome. *Nature* **376:**441–444.

Green, R., and H. F. Noller. 1997. Ribosomes and translation. *Annu. Rev. Biochem.* **66:**679–716.

Green, R., C. Switzer, and H. F. Noller. 1998. Ribosome-catalyzed peptide-bond formation with an A-site substrate covalently linked to 23S RNA. *Science* **280:**286–289.

Hardesty, B., T. Tsalkova, and G. Kramer. 1999. Co-translational folding. *Curr. Opin. Struct. Biol.* **9:**111–114.

Hesterkamp, T., S. Hauser, H. Lütcke, and B. Bukau. 1996. *Escherichia coli* trigger factor is a prolyl isomerase that associates with nascent peptide chains. *Proc. Natl. Acad. Sci. USA* **93:**4437–4441.

Heurgué-Hamard, V., R. Karimi, L. Mora, J. MacDougall, C. Leboeuf, G. Grentzmann, M. Ehrenberg, and R. H. Buckingham. 1998. Ribosome release factor RF4 and termination factor RF3 are involved in dissociation of peptidyl-tRNA from the ribosome. *EMBO J.* **17:**808–816.

Komar, A. A., A. Kommer, I. A. Krasheninnikov, and A. S. Spirin. 1997. Cotranslational folding of globin. *J. Biol. Chem.* **272:**10646–10651.

Kudlicki, W., J. Chirgwin, G. Kramer, and B. Hardesty. 1995. Folding of an enzyme into an active conformation while bound as peptidyl-tRNA to the ribosome. *Biochemistry* **34:**14284–14287.

Kudlicki, W., O. W. Odom, G. Kramer, and B. Hardesty. 1996. Binding of an N-terminal rhodanese peptide to DnaJ and to ribosomes. *J. Biol. Chem.* **271:**31160–31165.

Kudlicki, W., A. Coffman, G. Kramer, and B. Hardesty. 1997. Ribosomes and ribosomal RNA as chaperones. *Fold. Des.* **2:**101–108.

Lim, V. I., and A. S. Spirin. 1986. Stereochemical analysis of ribosomal transpeptidation. Conformation of nascent peptide. *J. Mol. Biol.* **188:**565–577.

Lorimer, G. H. 1996. A quantitative assessment of the role of chaperonin proteins in protein folding *in vivo*. *FASEB J.* **10:**5–9.

Makeyev, E. V., V. A Kolb, and A. S. Spirin. 1996. Enzymatic activity of the ribosome-bound nascent polypeptide. *FEBS Lett.* **376:**166–170.

Malhotra, A., P. Penczek, R. Agrawal, I. S. Gabashvili, R. A. Grassucci, R. A. Junemann, N. Burkhardt, K. Nierhaus, and J. Frank. 1998. *Escherichia coli* 70 S ribosome at 15Å resolution by cryo-electron microscopy: localization of fMet-tRNA$_f^{Met}$ and fitting of L1 protein. *J. Mol. Biol.* **280:**103–116.

Miller, D. M., R. Delgado, J. M. Chirgwin, S. C. Hardies, and P. M. Horowitz. 1991. Expression of cloned bovine adrenal rhodanese. *J. Biol. Chem.* **266:**4686–4691.

Miller, S. P., and J. W. Bodley. 1991. α-Sarcin cleavage of ribosomal RNA is inhibited by the binding of elongation factor G or thiostrepton to the ribosome. *Nucleic Acid Res.* **19:**1657–1660.

Moazed, D., and H. F. Noller. 1987. Chloramphenicol, erythromycin, carbomycin and vernamycin B protect overlapping sites in the peptidyl transferase region of 23S ribosomal RNA. *Biochimie* **69:**879–884.

Mortensen, K. K., N. R. Nyengaard, J. W. B. Hershey, S. Laalami, and H. U. Sperling-Petersen. 1991. Superexpression and fast purification of *E. coli* initiation factor IF-2. *Biochimie* **73:**983–989.

Nierhaus K. H., S. Schilling-Bartetzko, and T. Twardowski. 1992. The two main states of the elongating ribosome and the role of the α-sarcin stem loop structure of 23S RNA. *Biochimie* **74:**403–410.

Picking, W. D., O. W. Odom, T. Tsalkova, I. Serdyuk, and B. Hardesty. 1991. The conformation of nascent polylysine and polyphenylalanine peptides on ribosomes. *J. Biol. Chem.* **266:**1534–1542.

Picking, W. D., O. W. Odom, and B. Hardesty. 1992a. Evidence for RNA in the peptidyl transferase center of *Escherichia coli* ribosomes as indicated by fluorescence. *Biochemistry* **31:**12565–12570.

Picking, W. D., W. L. Picking, O. W. Odom, and B. Hardesty. 1992b. Fluorescence characterization of the environment encountered by nascent polyalanine and polyserine as they exit *Escherichia coli* ribosomes during translation. *Biochemistry* **31:**2368–2375.

Ramachandiran, V., C. Willms, G. Kramer, and B. Hardesty. Fluorophores at the N terminus of nascent chloramphenicol acetyltransferase peptides affect translation and movement through the ribosome. *J. Biol. Chem.*, in press.

Ryabova, L. A., O. M. Selivanova, V. I. Baranov, V. D. Vasiliev, and A. S. Spirin. 1988. Does the channel for nascent peptide exist inside the ribosome? *FEBS Lett.* **236:**255–260.

Schägger, M., and G. von Jagow. 1987. Tricine-sodium dodecyl sulfate-polyacrylamide gel separation of proteins in the range from 1 to 100 kDa. *Anal. Biochem.* **166:**368–379.

Stoller, G., K. P. Rücknagel, K. Nierhaus, F. X. Schmid, G. Fischer, and J. U. Rahfeld. 1995. Identification of the peptidyl-prolyl cis/trans isomerase bound to the *Escherichia coli* ribosome as the trigger factor. *EMBO J.* **14:**4939–4948.

Sundari, R., E. A. Stringer, L. H. Schulmann, and U. Maitra. 1976. Interaction of bacterial initiation factor 2 with initiator tRNA. *J. Biol. Chem.* **251:**3338–3345.

Thulasiraman, V., C. Yang, and J. Frydman. 1999. In vivo newly translated polypeptides are sequestered in a protected folding environment. *EMBO J.* **18:**85–95.

Trevino, R. J., J. Hunt, P. M. Horowitz, and J. M. Chirgwin. 1995. Chinese hamster rhodanese cDNA: activity of the expressed protein is not blocked by a C-terminal extension. *Protein Expr. Purif.* **6:**693–699.

Tsalkova, T., O. W. Odom, G. Kramer, and B. Hardesty. 1998. Different conformations of nascent peptides on ribosomes. *J. Mol. Biol.* **278:**713–723.

Tsalkova, T., G. Kramer, and B. Hardesty. 1999. The effect of a hydrophobic N-terminal probe on translational pausing of chloramphenicol acetyl transferase and rhodanese. *J. Mol. Biol.* **286:**71–81.

Xu, Z., and P. B. Sigler. 1998. GroEL/GroES: structure and function of a two-stroke folding machine. *J. Struct. Biol.* **124:**129–141.

Yonath, A., and Z. Berkovitch-Yellin. 1993. Hollows, voids, gaps and tunnels in the ribosome. *Curr. Opin. Struct. Biol.* **3:**175–181.

Yonath, A., K. R. Leonard, and H. G. Wittmann. 1987. A tunnel in the large ribosomal subunit revealed by three-dimensional image reconstruction. *Science* **236:**813–816.

VII. FACTOR-MEDIATED STEPS IN THE ELONGATION CYCLE

VII. FACTOR-MEDIATED STEPS IN THE ELONGATION CYCLE

During the past few years, crystal structures have been determined both for the ternary complex of aminoacyl-tRNA, elongation factor Tu, and GTP and for elongation factor G. This has led, in combination with kinetic, biochemical, and functional studies, to important new insights into the role of the factors and GTP hydrolysis in facilitating and modulating the movements of the tRNA-mRNA complexes through the subunit interface region of the ribosome.

The Ribosome: Structure, Function, Antibiotics, and Cellular Interactions
Edited by R. A. Garrett, S. R. Douthwaite, A. Liljas, A. T. Matheson, P. B. Moore, and H. F. Noller

Chapter 25

Mechanisms of Partial Reactions of the Elongation Cycle Catalyzed by Elongation Factors Tu and G

MARINA V. RODNINA, TILLMANN PAPE, ANDREAS SAVELSBERGH, DAGMAR MOHR, NATALIA B. MATASSOVA, and WOLFGANG WINTERMEYER

The elongation cycle of protein synthesis comprises three major steps. Starting with the ribosome bound to mRNA and carrying peptidyl-tRNA (or initiator tRNA) in the P site, in the first step aminoacyl-tRNA (aa-tRNA) binds to the A site in a codon-dependent manner. The reaction requires an elongation factor, (EF-Tu in bacteria), and GTP, which form a high-affinity ternary complex with aa-tRNA that binds to the ribosomal A site. Codon recognition triggers GTP hydrolysis, and subsequently the aa-tRNA is released from EF-Tu·GDP and is accommodated in the peptidyltransferase center to take part in rapid peptidyl transfer, the second step of the elongation cycle. The third major step of the cycle, translocation, is catalyzed by another elongation factor, EF-G in bacteria, which hydrolyzes GTP during the reaction. During translocation, peptidyl-tRNA is displaced from the A site to the P site while deacylated tRNA is transferred from the P site to the E site, from where it dissociates. Thus, translocation restores the initial state of the ribosome, which then enters the next elongation cycle unless a stop codon in the A site causes termination.

EF-Tu and EF-G belong to a subgroup of the GTPase superfamily and have homologies in sequence and structure. Their functional cycles have in common the fact that the intrinsically low GTPase activities of both factors are triggered on the ribosome. On the other hand, in line with the different functions of the two factors, the details of GTPase activation as well as the consequences of GTP hydrolysis are quite different. Recent years have seen major advances, in particular in structural and mechanistic work, that have greatly expanded our knowledge of the functional mechanisms of the elongation factors and the ribosome during elongation. While this chapter concentrates on pre-steady-state kinetic work, it will become evident how much the interpretation of kinetic results in molecular-mechanistic terms owes to structural information obtained from crystallography and cryo-electron microscopy.

A-SITE BINDING OF aa-tRNA

Our studies of EF-Tu function address two main issues: (i) the elucidation of the reaction pathway to identify intermediate steps of A-site binding and (ii) the quantitative evaluation of the pathway in order to understand specificity. In the following section, we compare the elemental rate constants of A-site binding determined for cognate, near-cognate, and noncognate ternary complexes. On the basis of a comprehensive set of rate constants, we discuss the mechanism through which the ribosome achieves high decoding fidelity.

Kinetic Pathway of A-Site Binding

A-site binding of aa-tRNA comprises several steps (Fig. 1), as has been established by experiments with either ternary complexes containing nonhydrolyzable and slowly hydrolyzable analogues of GTP, ribosomes in different functional states, or antibiotics. Initial binding of the ternary complex to the ribosome (Rodnina et al., 1996) is followed by codon recognition. Provided a codon is recognized, the complex of EF-Tu·GTP·aa-tRNA is stabilized in the A site by interactions of the tRNA with the mRNA

Marina V. Rodnina, Andreas Savelsbergh, Dagmar Mohr, Natalia B. Matassova, and Wolfgang Wintermeyer ■ Institute of Molecular Biology, University of Witten/Herdecke, 58448 Witten, Germany. **Tillmann Pape** ■ Department of Biochemistry, Imperial College of Science, Technology and Medicine, London SW7 2AY, United Kingdom.

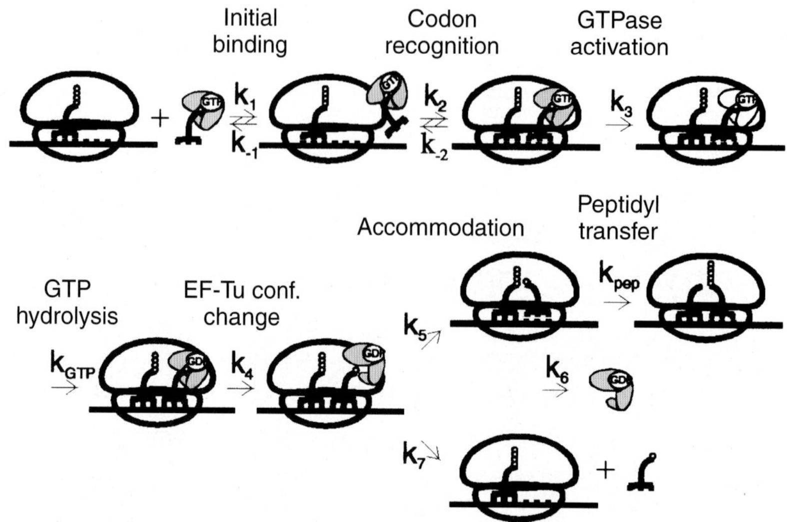

Figure 1. Mechanism of EF-Tu-dependent binding of aa-tRNA to the ribosomal A site. EF-Tu is depicted in three conformations: the GTP-bound form (gray), the transient GTPase-activated form on the ribosome (white), and the GDP-bound form (gray) that dissociates from the ribosome. The kinetic parameters are summarized in Table 1.

and, possibly, the ribosome. Codon-anticodon interaction generates an activation signal that is transmitted to the G domain of EF-Tu and leads to the formation of the activated GTPase state of the ribosome·EF-Tu·aa-tRNA complex (Rodnina et al., 1995a), which is followed by GTP hydrolysis. As a consequence, the conformation of EF-Tu switches from the GTP to the GDP form, which has a greatly reduced affinity for aa-tRNA (see chapters 27 and 28). Subsequently, aa-tRNA is released from EF-Tu·GDP, accommodates in the A site, and takes part in the peptidyltransferase reaction, while EF-Tu·GDP dissociates from the ribosome. In the following sections, the individual steps of the model are discussed.

Initial binding (k_1 and k_{-1})

The initial step in the interaction of EF-Tu·GTP·aa-tRNA with the ribosome is codon-independent binding (Fig. 1); the formation of the initial complex is fast, and the complex is labile (Pape et al., 1998; Rodnina et al., 1996) (Table 1). At about 10^8 M^{-1} s^{-1} (20°C), the value for the rate constant (k_1) is unusually high compared to typical second-order rate constants of 10^5 to 10^7 M^{-1} s^{-1} that have been reported for other systems. This indicates that the encounter of the ternary complex with the ribosome is nonrandom. One possible explanation is preorientation, for instance, by electrostatic forces, of the ternary complex upon approaching the ribosome so as to hit the A-site region more frequently than other regions of the ribosome. Another possible explanation is suggested by the observation that EF-Tu early in the sequence of A-site binding interacts with protein L12 (Stark et al., 1997b), four copies of which form the stalk of the 50S ribosomal subunit. A possible scenario is that the formation of the initial binding complex starts with the binding of EF-Tu to L12, most likely to the highly mobile C-terminal domain (Traut et al., 1995), thus introducing a favorable statistical factor. The role of EF-Tu in initial binding is to promote binding to the ribosome and to position aa-tRNA such that the anticodon is preoriented for codon recognition. The dissociation rate constant of the initial binding complex (k_{-1}) is 25 to 30 s^{-1} at 20°C and is likely to increase to about 50 to 70 s^{-1} at 37°C (Rodnina et al., 1996).

Codon recognition (k_2 and k_{-2})

The codon recognition complex is dominated by interactions of the tRNA anticodon with the codon exposed in the A site. The rate constant of codon recognition, k_2, is about 100 s^{-1} for cognate or near-cognate ternary complexes (Table 1). Dissociation rate constants, k_{-2}, of cognate and near-cognate codon recognition complexes have been measured previously (Thompson, 1988). Due to the strong dependence of k_{-2} on the Mg^{2+} concentration and temperature, values measured by different laboratories are difficult to compare. Cognate ternary complex has been reported to dissociate more than 3,000 times more slowly than near-cognate ternary complex (0.002 s^{-1} versus >6 s^{-1}, respectively, at 5°C and 5 mM Mg^{2+}). The values for k_{-2} in the cognate case strongly increase with temperature (0.002 s^{-1} versus 0.2 s^{-1} at 5 and 25°C, respectively, and 5 mM

Table 1. Elemental rate constants of cognate, near-cognate, and noncognate aa-tRNA binding to the A site[a]

Step	Symbol	Rate constant (s^{-1})		
		Value		
		Cognate	Near-cognate	Noncognate
Initial binding	k_1	110 ± 10[b]	110 ± 20[b]	60 ± 10[b]
	k_{-1}	25 ± 5	25 ± 5	25 ± 5
Codon recognition	k_2	100 ± 15	100 ± 20	
	k_{-2}	**0.2 ± 0.1**	**17 ± 8**	
GTPase activation and GTP hydrolysis[c]	k_3	**500 ± 100**	**50 ± 20**	**0.005**
GTP-GDP conformation change of EF-Tu	k_4	60 ± 20	70 ± 20	
aa-tRNA accommodation and peptide bond formation[c]	k_5	**7 ± 2**	**0.1 ± 0.03**	
Dissociation of EF-Tu	k_6	3 ± 1	2 ± 1	
aa-tRNA rejection	k_7	**<0.3**	**6 ± 1**	

[a] Kinetic steps and rate constants are defined in Fig. 1. Poly(U)-programmed ribosomes were used with ternary complexes of Phe-$tRNA^{Phe}$ (cognate), Leu-$tRNA^{Leu2}$ (GAG) (near cognate), or poly(A)-programmed ribosomes with Phe-$tRNA^{Phe}$ (noncognate). The rate constants that determine the discrimination are in boldface type. The data were taken from Pape et al., 1998, 1999, and Rodnina et al., 1996.
[b] $Micromolar^{-1}$ $second^{-1}$.
[c] Grouped for analysis, because the former reaction is rate limiting.

Mg^{2+} [Karim and Thompson, 1986]) and with decreasing Mg^{2+} concentration (0.2 and 2 s^{-1} at 10 and 5 mM Mg^{2+}, respectively, at 20°C [Pape et al., 1998]). An estimated value for k_{-2} for a near-cognate ternary complex is available from Thompson's group (greater than 6 s^{-1} at 5°C and 5 mM Mg^{2+} [Thompson and Dix, 1982]), which is similar to the value we measured (17 s^{-1} at 20°C and 10 mM Mg^{2+} [Pape et al., 1999]). It is likely, therefore, that the large stability differences between cognate and near-cognate codon recognition complexes observed at low temperature (5°C) do not reflect the difference under more physiological conditions.

The present values of the dissociation rate constants of the codon recognition complex are about 10 times smaller than those of the complexes formed from tRNAs with complementary anticodons (Grosjean et al., 1976). This suggests that binding interactions with the ribosome contribute up to a factor of 10 (5 to 6 kJ/mol) to the stabilization of the codon-anticodon complex in the A site. On the other hand, the 100-fold difference in k_{-2} for cognate and near-cognate ternary complexes on the ribosome is remarkably similar to the value that has been found by using the tRNA-tRNA model system in the absence of ribosomes (Grosjean et al., 1978). Therefore, the contribution of the ribosome to the stabilization of the mRNA·aa-tRNA complexes appears to be non-codon specific and is unlikely to improve the discrimination between cognate and near-cognate aa-tRNAs.

GTPase activation and GTP hydrolysis (k_3 and k_{GTP})

Rapid GTP hydrolysis in EF-Tu·GTP·aa-tRNA on the ribosome depends on codon recognition, as GTP hydrolysis in the noncognate initial binding complex is very slow (10^{-3} to 10^{-2} s^{-1}) (Rodnina et al., 1996) while GTP is hydrolyzed with a rate constant, k_3, of about 500 s^{-1} upon recognition of a cognate codon (Pape et al., 1998). The near-cognate codon recognition complex containing Leu-$tRNA^{Leu2}$ on poly(U), featuring a single mismatch in the first position of the codon, hydrolyzes GTP with a 10-times-smaller rate constant (Pape et al., 1999). This difference was overlooked previously, since rate constants of GTP hydrolysis were not determined with precision (25 s^{-1} and >4 s^{-1} for the cognate and near-cognate ternary complexes, respectively, at 5 mM Mg^{2+} and 20°C [Thompson and Dix, 1982; Thompson et al., 1986]). The GTPase rate constant depends on Mg^{2+} concentration and temperature. Reported values range from 25 s^{-1} (Eccleston et al., 1985; Thompson and Dix, 1982) or 55 s^{-1} (Pape et al., 1998) at 5 mM Mg^{2+} and 20°C to 100 s^{-1} in polyamine-containing buffer ("polymix") at 37°C (Bilgin et al., 1992).

In studies with the fully functional fluorescent derivative of GTP, mant-dGTP, a step was identified (GTPase activation) that precedes GTP hydrolysis in EF-Tu (Rodnina et al., 1995a). The step is likely to represent a conformational change of EF-Tu in the ternary complex and, possibly, the ribosome that aligns the catalytic groups in the G domain in their correct orientation for GTP cleavage. The kinetic analysis has revealed similar rate constants for the isomerization and GTP hydrolysis steps (Pape et al., 1998, 1999). This indicates that the isomerization step limits the rate of the subsequent chemistry step and suggests that GTP hydrolysis is triggered by an allosteric conformational change induced by codon-anticodon interaction.

Rate-limiting conformational changes preceding NTP cleavage have been demonstrated for several other enzymes, including different classes of GTPases as well as DNA and RNA polymerases. Studies with heterotrimeric G proteins, G_0 and G_α, have identified a substrate-induced conformational change that is important for the interaction of the G protein with its target; the correlation with GTP hydrolysis has not been established (Neubig et al., 1994; Remmers and Neubig, 1996; Remmers et al., 1994). Conflicting evidence has been reported for the proto-oncogene product Ras (Ahmadian et al., 1997; Neal et al., 1990; Rensland et al., 1991); most probably, the rate of GTP cleavage is not limited by a protein isomerization in that case. The most convincing results came from the studies of the mechanism of the ATP sulfurylase-GTPase, where the rate of GTP hydrolysis was shown to be preceded by, and strictly coupled to, a rate-limiting conformational change of the enzyme (Wei and Leyh, 1998). A rate-limiting rearrangement step preceding NTP cleavage was also reported for DNA polymerase (Johnson, 1992; Rittinger et al., 1995; Spence et al., 1995). It seems that isomerization steps that limit the rate of NTP cleavage are characteristic of enzyme systems that either couple the energy of NTP hydrolysis to subsequent chemical steps (ATP sulfurylase-GTPase and possibly EF-Tu) or catalyze reactions with particularly high demands for accuracy (DNA and RNA polymerases and EF-Tu).

Conformational change of EF-Tu (k_4) and dissociation of EF-Tu·GDP (k_6)

As a result of GTP hydrolysis and/or the release of P_i, the conformation of EF-Tu rearranges into that of the GDP-bound form and aa-tRNA dissociates from EF-Tu. The k_4 is about 60 s^{-1} (20°C), independent of whether the aa-tRNA is cognate or near cognate (Pape et al., 1998, 1999). The dissociation of aa-tRNA from EF-Tu probably takes place during the transition, although it has not been observed directly. At a rate of 60 s^{-1}, the rearrangement is fast enough not to limit the rate of the subsequent step, that is, accommodation or rejection of aa-tRNA. Thus, following the conformational transition of EF-Tu to the GDP-bound form and the release of the aa-tRNA, the reactions of EF-Tu·GDP and aa-tRNA on the ribosome are independent of each other. EF-Tu·GDP dissociates from the ribosome with a rate constant (k_6) of about 3 s^{-1}.

Accommodation or rejection of aa-tRNA and peptide bond formation (k_5, k_7, and k_{pep})

Following release from EF-Tu, the aminoacyl end of aa-tRNA moves into the peptidyltransferase center of the 50S A site (accommodation) to take part in peptide bond formation. For cognate aa-tRNA, the rate constant of accommodation is about 7 s^{-1} (k_5). After aa-tRNA is accommodated in the A site, peptide bond formation, which completes the sequence of A-site binding, takes place instantaneously ($k_{pep} > 100$). Peptide bond formation with cognate aa-tRNA is virtually quantitative, which implies that rejection is below the detection limit ($k_7 < 0.3$ s^{-1}) (Pape et al., 1998).

In contrast to cognate aa-tRNA, near-cognate Leu-$tRNA^{Leu2}$ dissociates from the ribosome to a considerable extent during the rearrangement that follows release from EF-Tu. The rate constant of rejection, k_7, is about 6 s^{-1}, while the rate constant of accommodation, k_5, is 0.1 s^{-1}. Thus, the rate constants of rejection and accommodation in the A site differ by >60-fold and about 70-fold, respectively, for near-cognate and cognate aa-tRNA. The small fraction of Leu-$tRNA^{Leu2}$ that is retained on the ribosome during accommodation is stably bound and competent for peptide bond formation, which again seems to be instantaneous (Pape et al., 1999).

The values of the rejection rate constants of Phe-$tRNA^{Phe}$ and Leu-$tRNA^{Leu2}$ determined here are similar to previously published values (<0.08 and 6 s^{-1}, respectively [Thompson and Dix, 1982; Thompson and Karim, 1982]). On the other hand, the rate constants of peptide bond formation were previously determined to be between 0.3 and 1.1 s^{-1} for cognate Phe-$tRNA^{Phe}$ (Eccleston et al., 1985; Thompson and Dix, 1982) and about 0.3 ± 0.15 s^{-1} for near-cognate Leu-$tRNA^{Leu2}$. Since the rate of dipeptide formation is limited by the accommodation in the A site and the accommodation step was not distinguished from peptide bond formation in the earlier experiments, the previous values represent the rate of accommodation. The large difference in the rate constants of accommodation (k_5) for cognate and near-cognate aa-tRNA, which according to our results amounts to a factor of about 70, was not reported previously.

Fidelity of aa-tRNA Selection on the Ribosome

The error frequency of *Escherichia coli* translation in vivo on internal codons was estimated to be in the range from 6×10^{-4} to 5×10^{-3} (average, 3×10^{-3}) (Kurland and Ehrenberg, 1987). Similar selectivity can be achieved in vitro at a low Mg^{2+} concentration (Kurland and Ehrenberg, 1984). The high fidelity of ribosomal decoding has long been puzzling, as model studies in aqueous solution show that the free energy difference of forming a correct base pair compared to an incorrect one is on the order of 10 kJ/mol, providing a maximum discrimination of about 100-fold in a single selection step (Uhlenbeck et al., 1981). Thus, proofreading mechanisms of various kinds have been proposed which have in common the fact that the discriminatory interaction is used more than once, the selection steps being separated by irreversible energy-consuming steps. General selection schemes, including proofreading, have been suggested by Hopfield (1974) and Ninio (1975).

On the basis of limited kinetic information, a two-stage selection mechanism comprising initial selection before and proofreading after GTP hydrolysis was put forward (Thompson, 1988). Based on measured rates of GTP hydrolysis and peptide bond formation, Thompson and colleagues proposed that the rate of GTP hydrolysis by EF-Tu is independent of the tRNA, thereby providing an internal kinetic standard for translational accuracy. According to the model, cognate and near-cognate ternary complexes, unlike noncognate ones, undergo GTP hydrolysis due to the long lifetime of their complex with the ribosome and therefore have to be discriminated by subsequent proofreading. However, the distinction of cognate from near-cognate and noncognate ternary complexes by the time they remain bound to the ribosome proved invalid (Rodnina et al., 1996). The conclusion that ternary complexes or aa-tRNAs are discriminated solely on the basis of different rejection rates in either initial selection or proofreading appears to have been premature, since at the time not all steps of A-site binding contributing to selection had been resolved and some rate constants had not been measured with sufficient precision. Moreover, the error frequency predicted from the rate constants reported by Thompson's group was significantly higher than that determined experimentally, and hence, the internal-standard model was unsatisfactory.

The comprehensive set of elemental rate constants of A-site binding determined for cognate (Pape et al., 1998), near-cognate (Pape et al., 1999), and noncognate (Rodnina et al., 1996) ternary complexes provides a kinetic model which consistently explains translational accuracy. Similar to previously suggested models, the present model features two selection steps, that is, initial selection before and proofreading after GTP hydrolysis. Also, the observed differences in the stabilities of the codon-anticodon complexes depending on the extent of the base pairing, before (k_{-2} [Fig. 1]) or after (k_7) GTP hydrolysis in EF-Tu, are consistent with earlier reports. Additionally, however, an important regulatory influence of codon recognition on two forward reactions was revealed by the present analysis. The GTPase rate constant, k_3, is 10-fold lower in the near-cognate complex (noncognate, 10^5-fold) than in the cognate complex, and the rate constant of aa-tRNA accommodation, k_5, is decreased about 60 times for near-cognate aa-tRNA (noncognate, not measurable). In both cases, rearrangements that limit the rates of the subsequent irreversible chemical steps (GTP hydrolysis and peptide bond formation) are affected. This suggests the existence of yet another selection mechanism in addition to discrimination by rejection, that is, discrimination by induced fit, which contributes to both initial selection and proofreading. The quantitative analysis shows, as discussed below, that only the combined contributions of both selection mechanisms fully account for the fidelity levels observed in vitro or in vivo.

How the ribosome senses structural differences between cognate and near-cognate codon-anticodon duplexes, and how these differences may affect the GTPase activation and accommodation steps, is not known. One attractive possibility is that the formation of the codon-anticodon duplex induces a conformational change in the decoding center of 16S rRNA and that the rate of this change depends upon structural details, such as potential hydrogen bonds, which are provided by the cognate, and less so by the near-cognate, codon-anticodon duplex. The induced conformational change would constitute the signal that is transmitted to the G domain of EF-Tu to trigger the GTPase. The transmission may be accomplished either by a series of subsequent conformational changes of the ribosome traveling across the subunit interface or by a movement (or distortion) of the tRNA molecule that is induced by the conformational change of the decoding center, or by both mechanisms.

The model could also explain how a cognate codon-anticodon duplex may promote the accommodation of the aa-tRNA in the A site more efficiently than a near cognate. In order to reach the peptidyltransferase center, the acceptor arm has to move into the 50S A site after it has been released from EF-Tu. According to the structural reconstruction of the codon recognition complex by cryo-electron microscopy (Stark et al., 1997b), the movement involves a rotation of the whole tRNA molecule pivoting around the anticodon. It is conceivable that conformational changes in the decoding center are instrumental in promoting the rotational movement of the tRNA. In this model, the rate of accommodation is determined by the rate of the conformational change in the decoding center, and the latter, in turn, is promoted by structural determinants provided by the cognate, and less so by the near-cognate, codon-anticodon duplex.

The present model of an induced-fit mechanism of aa-tRNA discrimination places the ribosome beside DNA polymerases and RNA polymerase, for which analogous models have been put forward. Extensive results obtained for DNA polymerases suggest that there is additional discrimination, apart from exonuclease proofreading, at the level of nucleotide incorporation in that correct nucleotides are incorporated much faster than incorrect ones (Johnson, 1992; Rittinger et al., 1995; Spence et al., 1995). The kinetic contribution to selection in these models is due to the acceleration of the forward reactions of the cognate substrate compared to those of the noncognate ones. Direct evidence suggesting that structural differences between correct and incorrect base pairs influence the catalytic center of the DNA polymerase by induced fit is provided by crystal structures (Doublié et al., 1998; Kiefer et al., 1998; Li et al., 1998). A similar mechanism appears to operate in *E. coli* RNA polymerase (Erie et al., 1993).

In these systems, induced fit leads to enhanced fidelity because rearrangements of the enzyme-substrate complexes, and not the chemical steps, are rate limiting and are accelerated for the correct substrates (Herschlag, 1988; Post and Ray, 1995). Otherwise, induced fit would not increase fidelity, since the structure of the transition state would be the same for every substrate. In the present case of the ribosome, the rate-limiting rearrangements that are accelerated by structural determinants of the correct codon-anticodon complexes are GTPase activation and A-site accommodation.

Rejection of noncognate aa-tRNAs

There is no estimate of the misincorporation frequency of noncognate amino acids in vivo, although clearly it must be lower than that of near-cognate amino acids, that is, below 10^{-3}. To predict the selectivity of an enzyme in an initial selection step, k_{cat} and K_m values have to be compared for correct and incorrect substrates. The relative usage of correct versus incorrect substrate is defined by the ratio of the respective k_{cat} and K_m values, weighted by the concentrations; the error frequency is roughly the reciprocal of that ratio. For the binding of EF-Tu·GTP·Phe-tRNA^{Phe} to poly(A)-programmed ribosomes, that is, when no base pairing is possible, the k_{cat}/K_m ratio is 0.012 μM^{-1} s^{-1} (10 mM Mg^{2+}) (Rodnina et al., 1996); the value for the cognate situation is 88 μM (Pape et al., 1999). Thus, the error frequency of initial selection of noncognate ternary complexes is 1.4×10^{-4}, and the actual error in initial selection is about 3×10^{-3}, assuming a 20-fold excess of noncognate over cognate ternary complexes. This value is likely to be even lower at Mg^{2+} concentrations closer to physiological ones. Thus, the bulk of noncognate ternary complexes can be discriminated in a single selection step with essentially no cost with respect to GTP hydrolysis.

Rejection of near-cognate aa-tRNAs

The frequency of misincorporation of a near-cognate amino acid in vivo is about 3×10^{-3}. The measurements performed in vitro under conditions of comparable selectivity show that the overall error frequency of translation is determined by the selectivity of A-site binding and that there are no additional selection steps (Pape et al., 1999). When the relative contributions of initial selection and proofreading have been measured, it appears that near-cognate ternary complexes are poorly discriminated during initial selection (up to 10:1) while efficient rejection (100:1) is achieved in the proofreading step. These values are probably representative of the contributions of initial selection and proofreading in vivo. In vitro, where the error frequency can be manipulated in the range from 10^{-4} to 10^{-2}, the efficiency of initial selection can vary drastically, from about 100:1 (overall error frequency 10^{-4}) (Bilgin and Ehrenberg, 1994) to almost none (error frequency, 10^{-2}) (Pape et al., 1999), while proofreading works at about 100:1 efficiency, more or less independent of the experimental conditions (Bilgin and Ehrenberg, 1994; Pape et al., 1999).

The error frequency measured in vitro at 10 mM Mg^{2+} is about 10^{-2}; there is essentially no initial selection under these conditions, and the observed fidelity is due to proofreading only. The reason is that rapid GTP hydrolysis at 10 mM Mg^{2+} overrides the lower stability of the near-cognate codon-anticodon complex. Lowering the Mg^{2+} concentration decreases the GTPase rate, thereby increasing the efficiency of near-cognate ternary complex rejection in initial selection.

Decoding errors induced by aminoglycosides

Many antibiotics induce translation errors by binding to specific sites in rRNA. For instance, the structurally related aminoglycosides neomycin, paromomycin, gentamicin, and kanamycin bind to the decoding region of 16S rRNA (Moazed and Noller, 1987) and induce a particular conformation of the RNA which may affect decoding (see chapter 34). While tRNA binding in the A site is stabilized about sixfold in the presence of aminoglycosides (Hornig et al., 1987; Karimi and Ehrenberg, 1994), this effect is too small to explain the aminoglycoside-induced increase of misreading observed in vitro and in vivo, indicating that additional effects must be involved.

The analysis of A-site binding of near-cognate aa-tRNA [Leu2 on poly(U)] has shown that the aminoglycoside paromomycin affects all steps that were characterized kinetically, except initial binding (Table 2). k_2, the rate constant of codon recognition, was reduced about three times; the same effect was observed with cognate ternary complex (unpublished data). The rate constant of GTPase activation, k_3, and with it, the rate of GTP hydrolysis, is increased more than 10-fold, to $>500\ s^{-1}$, up to the level observed in the cognate situation (Pape et al., 1998, 1999). The stability of codon-anticodon interaction is increased as the rate constants of ternary complex dissociation before, and aa-tRNA dissociation after, GTP hydrolysis, k_{-2} and k_7, are decreased five to six times. The rate constant for the conformational switch of EF-Tu from the GTP to the GDP conformation, k_4, is decreased about 10-fold; the same decrease was observed in the cognate situation (data not shown). The effect may indicate an influence of the 30S subunit on EF-Tu which is modulated by paromomycin binding to 30S; according to the mechanism shown in Fig. 1, such an effect would have no relevance for the selection. Functionally, the most important effect of paromomycin is that it increases the accommodation rate constant, k_5, by a factor of 10 and, at the same time, decreases the rejection rate constant, k_7, about sixfold, to make them about equal, 0.9 and 1 s^{-1}. Accordingly, about 50% of leucine is incorporated into dipeptide in the presence of paromomycin, about 30 times the level (1.5%) observed in the absence of antibiotic (Pape et al., in press).

The increased level of near-cognate amino acid misincorporation induced by paromomycin is due to (i) the failure of rejection and (ii) the acceleration of A-site accommodation of near-cognate aa-tRNA in the proofreading step. Under in vivo conditions, when there is initial selection, the effect of the antibiotic will be larger, because paromomycin, by accelerating GTP hydrolysis more than 10-fold, also abolishes initial selection.

To explain the effect of paromomycin, one may envisage two possibilities. The antibiotic may directly interact with the anticodon region of the aa-tRNA and/or the mRNA (VanLoock et al., 1999); alternatively, it may induce a conformation of 16S rRNA which favors interactions of the 16S rRNA with tRNA and/or mRNA (Fourmy et al., 1996). While both interpretations seem consistent with the observed stabilization by paromomycin of near-cognate aa-tRNA binding, the former model does not explain the threefold-slower codon recognition observed in the presence of paromomycin and is also difficult to reconcile with the acceleration observed for the forward reactions of GTPase activation and A-site accommodation. Rather, the present kinetic results suggest that the particular conformation of 16S rRNA induced by aminoglycoside binding is involved (see chapter 34). This seems to be a principal feature of the mechanism of decoding, as the steps of GTPase activation and A-site accommodation of aa-tRNA are much faster in the cognate situation than in the near-cognate one, suggesting an induced-fit mechanism of selection (Pape et al., 1999).

It is likely that the formation of the codon-anticodon complex induces a conformational transition of the decoding region of 16S rRNA by

Table 2. Effect of paromomycin on the elemental rate constants of A-site binding of near-cognate aa-tRNA[a]

Step	Symbol	Rate constant (s^{-1}) Value: Without paromomycin	Rate constant (s^{-1}) Value: With paromomycin
Initial binding	k_1	110 ± 20[b]	140 ± 20[b]
	k_{-1}	25 ± 5	25 ± 10
Codon recognition	k_2	**100 ± 20**	**37 ± 2**
	k_{-2}	**17 ± 8**	**3.5 ± 0.5**
GTPase activation and GTP hydrolysis[c]	k_3	**50 ± 20**	**>500**
GTP-GDP conformation change of EF-Tu	k_4	60 ± 20	6 ± 2
aa-tRNA accommodation and peptide bond formation[c]	k_5	**0.1 ± 0.03**	**1 ± 0.1**
Dissociation of EF-Tu·GDP	k_6	3 ± 0.5	ND[d]
aa-tRNA rejection	k_7	**6 ± 1**	**0.9 ± 0.2**

[a] Kinetic steps and rate constants are defined in Fig. 1. The rate constants that are changed in the presence of paromomycin are in boldface type. The data were taken from Pape et al., in press.
[b] Micromolar^{-1} second^{-1}.
[c] Grouped for analysis, because the former reaction is rate limiting.
[d] ND, not determined; does not contribute to selection, as aa-tRNA behaves independently of EF-Tu·GDP.

establishing structure-specific, sequence-independent interactions. Such contacts, which would increase the stability of the complex, could be formed between the A-site 16S rRNA and the sugar-phosphate backbone of the codon-anticodon complex or residues of the codon-anticodon base pairs that are in the same positions in all Watson-Crick base pairs (Doublié et al., 1998; Fourmy et al., 1996). One prediction of the model is that codon recognition should be impaired when the antibiotic is bound. As mentioned above, this is, in fact, observed (Table 2). The model assumes that the conformational transition of 16S rRNA induced by codon-anticodon complex formation is structurally coupled to the acceleration of GTPase activation and A-site accommodation of aa-tRNA; the mechanism of coupling is not clear at present, although there are indications that the tRNA is involved. The near-cognate codon-anticodon complex would be less efficient in promoting the transition of 16S rRNA to the active conformation, and binding of paromomycin to the decoding center is required to induce the active conformation of 16S rRNA.

TRANSLOCATION

Translocation is arguably the most complex step of elongation. During translocation, two tRNAs together with the mRNA move rapidly from their respective pretranslocation sites to their post-translocation sites. To allow the movement, tRNA-ribosome interactions have to be released temporarily, while codon-anticodon interaction is retained. The potential for the molecular movements involved in translocation resides in the structure of the ribosome, and GTP hydrolysis is not required to drive the reaction thermodynamically, since the spontaneous, EF-G-independent reaction takes place, albeit very slowly, under certain conditions in vitro (Spirin, 1985). To what extent this prevails under other conditions, for instance, when highly structured mRNAs are translated, has not been established yet. Under any conditions, there is a large activation barrier to be overcome for translocation to proceed at the rate necessary to sustain protein elongation in vivo (in *E. coli*, on average, 10 amino acids are incorporated per s). Molecular aspects of translocation have been reviewed recently (Wilson and Noller, 1998b). In the following discussion, we will concentrate on kinetic aspects and on the role of EF-G and GTP hydrolysis by EF-G in translocation catalysis.

While discussing individual steps of translocation, we will refer to our current working model, depicted in Fig. 2. In brief, the model comprises six steps, as follows. In the initial pretranslocation state, deacylated tRNA resides in the P site with the 3′ end oriented towards, or bound in, the E site (P/E hybrid state) and peptidyl-tRNA in the A site with the peptidyl end towards the P site (A/P* hybrid state [P* indicates that this state is not puromycin reactive]). Immediately following the binding of EF-G·GTP (step 1), the GTPase activity of EF-G is activated and GTP is hydrolyzed (step 2). GTP hydrolysis and/or subsequent release of P_i causes a conformational change of EF-G, which in turn induces the formation of the transition state of the ribosome (step 3). In the transition state, the movement of the tRNA-mRNA complex takes place (step 4); it comprises (i) the movement of the 3′ ends of the two tRNAs on the 50S subunit into their respective posttranslocation positions, i.e., the E site for the deacylated tRNA and the puromycin-reactive position of the peptidyl end of the peptidyl-tRNA in the P site (A/P state), and (ii) the movement of the anticodon arms of both tRNA molecules together with the mRNA to their immediate posttranslocation positions on the 30S subunit. In step 5, the ribosome returns to the ground state and EF-G assumes the GDP-bound conformation; the latter step is the one inhibited by fusidic acid binding to EF-G on the ribosome. Step 6 then comprises the dissociation of EF-G·GDP and deacylated tRNA (the order of dissociation is not known) to reach the final posttranslocation state of the ribosome with peptidyl-tRNA in the P site and a free A site.

Ribosome Binding of EF-G and GTP Hydrolysis

The initial binding of EF-G·GTP to the ribosome (Fig. 2, step 1) is rapid and readily reversible (Table 3); it does not depend on the presence of GTP and is also observed with GDP or without a nucleotide (Rodnina et al., 1997). Binding is followed by rapid GTP hydrolysis (step 2); as discussed below, ribosomal protein 12 has a major role in GTPase activation. The rate of single-round GTP hydrolysis is not influenced by the tRNA occupancy of the ribosome and is not affected by the antibiotic viomycin, an efficient inhibitor of translocation (Rodnina et al., 1997). Thus, GTP hydrolysis by EF-G on the ribosome does not depend upon subsequent translocation. The reverse is not true, as translocation is much slower (more than 50-fold) when GTP is replaced with non-hydrolyzable GTP analogues or GDP. GTP hydrolysis seems to be "uncoupled" from translocation either when there is no A-site tRNA to be translocated, i.e., when the ribosomes are vacant or in the posttranslocation state, or when translocation is inhibited. In these situations, a single round of GTP hydrolysis takes place rapidly, and more GTP is hydrolyzed in

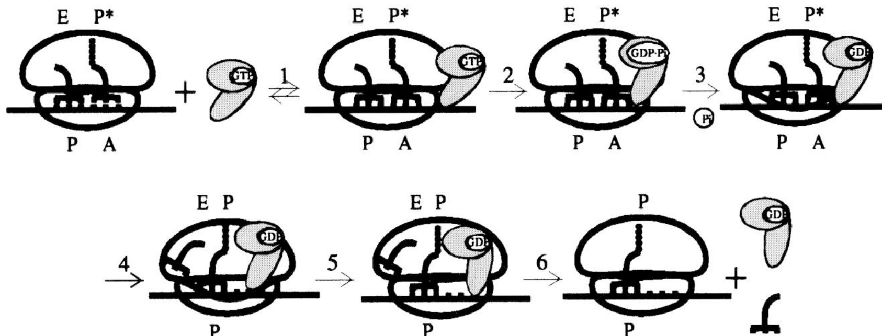

Figure 2. Reaction scheme of translocation as discussed in the text. EF-G is depicted in three conformations: the GTP-bound form, an intermediate GDP-bound form on the ribosome, and the GDP-bound form that dissociates from the ribosome. The transition state of the ribosome, formed in step 3, is symbolized by an altered conformation of the small ribosomal subunit. A, P (P*), and E denote the tRNA binding sites on the two subunits and are indicated when occupied. The kinetic parameters are summarized in Table 3.

the turnover reaction. In contrast, with pretranslocation ribosomes in the absence of inhibitors, GTP hydrolysis and subsequent rapid translocation are tightly coupled in a 1:1 stoichiometric fashion. This implies that, as a result of GTP hydrolysis, the ribosome–EF-G complex is stabilized such that the dissociation of EF-G prior to translocation is slower than translocation, hence minimizing unproductive turnover.

Phosphate Release Following GTP Hydrolysis

There is a delay of about 35 ms between GTP hydrolysis and subsequent translocation (Rodnina et al., 1997), indicating that a slow rearrangement of the pretranslocation complex precedes translocation. To address the question of whether GTP hydrolysis itself or subsequent release of the phosphate (P_i) initiates the subsequent steps, we have performed rapid kinetic measurements of P_i release from the complex after GTP hydrolysis. The liberation of P_i was monitored by the fluorescence change of fluorescence-labeled mutant phosphate binding protein from *E. coli* as described by Brune et al. (1994). It was found that the rate constant of P_i release was 30 s^{-1} (37°C) (Table 3), similar to the rate of translocation, 25 s^{-1}, measured by stopped flow under the same conditions, while GTP hydrolysis was much faster, about 120 s^{-1} (Rodnina et al., 1997). According to these results, P_i release and translocation are coupled reactions. Preliminary data suggest that P_i release precedes, and possibly limits the rate of, translocation.

Table 3. Kinetic parameters of translocation[a]

Step	Rate constant (s^{-1})	
	Symbol	Value
Initial binding	k_1	140 ± 50[b]
	k_{-1}	70 ± 20
GTP hydrolysis	k_2	170 ± 50
P_i release	k_3	30 ± 10
tRNA movement[c]	k_4	25 ± 5
Conformational change of EF-G[d]	k_5	5 ± 2
Dissociation of EF-G		ND[e]
Dissociation of tRNA		ND

[a] Kinetic steps are defined in Fig. 2. Rapid kinetic measurements were performed at 37°C with pretranslocation complexes carrying $tRNA^{fMet}$ in the P site and fMetPhe-$tRNA^{Phe}$ in the A site as described previously (Rodnina et al., 1997). P_i release was measured as described in the text.
[b] $Micromolar^{-1}$ $second^{-1}$.
[c] Apparent rate constant; may be limited by preceding step.
[d] Apparent rate constant.
[e] ND, not determined.

tRNA Movement

In the pre-translocation state, A- and P-site-bound tRNAs are arranged such that the two anticodons and acceptor ends, respectively, are close to each other and the planes of the molecules enclose an angle of about 60°. This arrangement was suggested on the basis of distance measurements by fluorescence resonance energy transfer (Johnson et al., 1982; Paulsen et al., 1983) and was confirmed by cryo-electron microscopy (Stark et al., 1997a). The detailed positions of the acceptor ends on the 50S ribosomal subunit depend on the functional state of the tRNA, as chemical footprinting on 23S rRNA revealed different footprinting patterns of deacylated tRNA in A, P, and E sites (Moazed and Noller, 1989a, 1989b). In ribosomal complexes carrying peptidyl-tRNA in the A site and deacylated tRNA in the P site, the A-site pattern on the 50S subunit was no longer observed, but rather the P-site and E-site patterns. This indicates that, upon forming peptidyl-tRNA in the A site, the acceptor end is moved towards, or into, the 50S P site and the acceptor end of deacylated tRNA is moved towards the E site to form A/P* and P/E "hybrid states."

Mobility of the acceptor end does not necessarily mean that the 3′ end of P-site-bound deacylated tRNA is bound in the E site exclusively. In fact, prebound E-site tRNA does not inhibit peptidyl transfer (Robertson and Wintermeyer, 1987) and is not competed out of the E site when the peptide is formed (unpublished data). Hence the observed P/E footprinting pattern may reflect one of several possible structural states of the 3′ end of P-site-bound tRNA. Therefore, the argument that the hybrid-state model necessitated the ejection of E-site-bound tRNA upon peptide bond formation in the next round of elongation (Nierhaus et al., 1997) is not conclusive.

The formation of the P/E state is important for translocation, since modifications at the 3′ end of the tRNA that impair the interaction with the E site (Lill et al., 1986) strongly reduce the rate of translocation (Lill et al., 1989). It has been shown that the 3′-terminal adenine 76 (Lill et al., 1986) and, to a lesser extent, cytosine 75 (unpublished data) of $tRNA^{Phe}$ are important for the E-site binding affinity, possibly by base-pairing interactions with 23S rRNA. The importance of 50S E-site interaction of the tRNA leaving the P site is corroborated by the observation that an isolated anticodon arm bound to the P site inhibits translocation (Joseph and Noller, 1998).

The movement of deacylated tRNA during translocation and subsequent dissociation from the ribosome takes place in several steps. From the P/E hybrid state, attained by peptidyl transfer, the tRNA is translocated to the E site to establish an intermediate E-site-bound state with the anticodon still bound to the mRNA. The latter intermediate state, E′, which was demonstrated by kinetic experiments (Paulsen and Wintermeyer, 1986; Robertson et al., 1986), rearranges to another state, E, in which the anticodon of the E-site-bound tRNA is about 35 Å away from the anticodon of the tRNA in the P site (Paulsen and Wintermeyer, 1986). The affinity of tRNA binding in the E site strongly depends on the concentrations of Mg^{2+} (Robertson and Wintermeyer, 1987) and polyamines (Gnirke et al., 1989; Semenkov et al., 1996). Under any conditions, however, the E-site-bound tRNA is bound in a kinetically labile fashion (Semenkov et al., 1996). An influence of aa-tRNA binding to the A site on E-site-bound tRNA, and vice versa, as reported by one group (Nierhaus, 1990; Spahn and Nierhaus, 1998), could not be confirmed by others (Semenkov et al., 1996, and references therein).

In keeping with the indirect data, different arrangements of deacylated tRNA in the E site have been revealed by cryo-electron microscopy (Spahn and Nierhaus, 1998; Stark et al., 1997a). The heterogeneity of tRNA arrangements in the E site reflects the existence of several states that the tRNA may assume after translocation, such as the two states discussed above. The extent to which the different states are populated is determined by the ionic conditions.

Early pre-steady-state kinetic experiments with translocation were performed with poly(U)-programmed ribosomes carrying deacylated $tRNA^{Phe}$ in the P site and AcPhe-$tRNA^{Phe}$ in the A site (Robertson and Wintermeyer, 1987). In that system, translocation was rather slow (about 1 s^{-1} at 20°C). Using AcPhePhe-$tRNA^{Phe}$ in the A site, a rate of about 10 s^{-1} was observed, indicating a strong influence of the peptidyl residue (Wintermeyer et al., 1986). In more recent experiments with mRNA-programmed ribosomes carrying $tRNA^{fMet}$ in the P site and fMetPhe-$tRNA^{Phe}$ in the A site, a translocation rate of 25 s^{-1} (10 mM Mg^{2+}; 37°C) was measured (Rodnina et al., 1997). This is 10 times the overall rate of elongation, about 2 s^{-1}, measured in that system. As pointed out above, the observed translocation rate seems to be limited by preceding steps, i.e., P_i release and/or a structural rearrangement of the ribosome, indicating that the intrinsic rate of tRNA-mRNA movement in translocation may be even faster.

Molecular Mechanics of EF-G

To study the functional role of EF-G domains in translocation, domain deletion mutants which were truncated from either the N or the C terminus were prepared. EF-G lacking the G domain exhibited partial translocation activity in that it brought about the transfer of the 3′ end of peptidyl-tRNA to the P site on the 50S ribosomal subunit into a puromycin-reactive state, but not translocation on the 30S subunit (Borowski et al., 1996). This behavior is consistent with the idea that the 50S phase of translocation, i.e., the formation of the puromycin-reactive A/P state of peptidyl-tRNA, may represent the first EF-G-dependent step of translocation.

Most interesting are the effects of deleting domain 4. The translocation activity of truncated EF-G lacking domain 4 was reduced about 1,000-fold compared to that of native EF-G, while the activity in GTP hydrolysis was unaffected (Rodnina et al., 1997). The activity was restricted to a single round, indicating tight binding of the mutant factor to the ribosome after GTP hydrolysis. The data suggest that domain 4 is essential for coupling the conformational change of EF-G induced by GTP hydrolysis to the rearrangement of the ribosome required for rapid tRNA translocation and for subsequent release of the factor.

Unlike EF-G, the mutant lacking domain 4 does not dissociate from the ribosome following GTP hydrolysis and translocation, indicating that the presence of domain 4 is required not only for rapid translocation but also for the release of EF-G (Rodnina et al., 1997). This striking result can be explained by assuming that, early in the sequence of conformational changes induced by GTP hydrolysis, EF-G establishes additional interactions with the ribosome to form a kinetically more stable complex, thus enabling the factor to force the ribosome into a structure which favors translocation. A movement of domain 4 into the A site, as discussed above, may then serve to resolve the tight interaction and allow the release of the factor. There are two ways in which this could happen. A passive movement may induce a structural change in the body of the EF-G molecule that lowers the affinity for the ribosome and facilitates dissociation. Alternatively, an active movement of domain 4 may change the conformation of the ribosome towards low affinity for EF-G. Mutants lacking domain 4 would lack either activity and would, therefore, remain bound in the kinetically stable complex.

Ribosome Structure and Translocation

On the basis of structural information from cryo-electron microscopy (Agrawal et al., 1998; Stark et al., submitted), a structural model of EF-G function in translocation has been derived. In the state prior to translocation, which was stabilized by thiostrepton, the factor is positioned with its body bound to protein L12, while domain 4 is oriented towards the 30S subunit and binds in the region where protein S4 is located (Stark et al., submitted). The latter interaction appears to induce substantial structural changes in the 30S subunit, in particular around the neck and in the region where head and body are associated. We hypothesize that this structure of the ribosome represents, or is related to, the transition state of translocation. After translocation, EF-G is found in an entirely different position, so that domain 4 reaches into the decoding region (Agrawal et al., 1998; Stark et al., submitted; Wilson and Noller, 1998a) and the ribosome is present in the ground state structure. The movement of domain 4 into the 30S A site may be functional in that it closes the site and, by inhibiting the backward movement of the tRNAs, helps to stabilize the posttranslocation state while the ribosome is in the transition state conformation.

The conformational change of the ribosome with low probability may occur spontaneously, resulting in slow spontaneous translocation. A potential role for EF-G in translocation catalysis may be (i) to promote the conformational change of the ribosome required for the tRNA-mRNA movement or (ii) to interact with the A-site tRNA and actively displace the tRNA-mRNA complex to the P site. The latter possibility seems less likely, since a tRNA analogue comprising a minimal 15-nucleotide-long anticodon stem-loop structure is a substrate for EF-G-dependent translocation (Joseph and Noller, 1998). It is known that the anticodon stem-loop interacts exclusively with the small ribosomal subunit (Moazed and Noller, 1986; Rose et al., 1983) and binds to the decoding region of 16S rRNA (Purohit and Stern, 1994), so that a direct contact with EF-G seems unlikely. In fact, in the pretranslocation complex studied by cryo-electron microscopy, the only contact of EF-G with the 30S subunit is the interaction of domain 4 in the S4 region, and there is no contact with tRNA. Thus, tRNA movement is more likely to be promoted indirectly by a structural change of the ribosome that is induced by EF-G rather than by a direct interaction which pushes the tRNA out of the A site.

Involvement of rRNA in Translocation

The site of tRNA-mRNA interaction is the decoding region of 16S rRNA, and the anticodon arms of both tRNAs interact with 16S rRNA. In the transition state of translocation, these interactions have to be released in order to allow tRNA-mRNA movement in translocation. There are numerous observations that relate structural changes of the decoding center, or their inhibition, to translocation. For instance, certain mutations in protein S5 affect the fidelity of decoding and others confer resistance to spectinomycin, an inhibitor of EF-G-dependent translocation (Brink et al., 1994). Aminoglycoside antibiotics that bind to the decoding region on 16S rRNA and lower the fidelity also inhibit translocation (Davies and Davis, 1968). The antibiotic viomycin, which specifically inhibits translocation (Modolell and Vazquez, 1977), also seems to bind to the decoding center (Moazed and Noller, 1987).

The important question is how the interaction of EF-G with thc ribosome may affect the decoding center. We suggest that the structural changes in the small ribosomal subunit related to translocation originate at the site of interaction of domain 4 of EF-G in the pretranslocation complex, which is in the vicinity of the binding region of protein S4. S4 binds to the rRNA junction of five helical elements (S4 junction [Heilek et al., 1995; Powers and Noller, 1995]), one of which includes helix 18 of 16S rRNA, a region known to contain a pseudoknot and to be crucially important for ribosome function (Powers and Noller, 1991; O'Connor et al., 1995). Motion at the S4 binding region may switch the pseudoknot structure. The 530 loop seems to be located close to helix 34 (see chapters 1, 13, and 14) and is structurally coupled to the decoding region. A switch in the 530 pseudoknot structure may induce a rearrangement in the 16S rRNA decoding region towards the conformation that has low affinity for the codon-anticodon complexes. Destabilization of the 16S rRNA–codon-anticodon complex, caused by mutation or antibiotic binding, may reduce the activation energy barrier of translocation, thus facilitating movement.

Comparably less extensive conformational changes related to EF-G binding and translocation were observed in the large ribosomal subunit (Stark et al., submitted). The interactions of EF-G with the α-sarcin stem-loop and the thiostrepton binding region of 23S rRNA are well documented, although stabilization of the ribosome–EF-G complex with fusidic acid is required to observe the respective footprints (Moazed et al., 1988; Munishkin and Wool, 1997; Wilson and Noller, 1998a). According to cryo-electron microscopy, EF-G binds to the region of the large ribosomal subunit that is supposed to comprise the α-sarcin loop and the thiostrepton binding region in the complex stabilized by fusidic acid (Agrawal et al., 1998); the interaction is not observed in the posttranslocation complex stabilized by thiostrepton (Stark et al., submitted). Thus, it appears that EF-G interactions with these structural elements of 23S rRNA are late events in translocation and may be involved in the release of EF-G from the ribosome. However, conformational transitions of the thiostrepton binding region seem to also be important for earlier steps in translocation. This is suggested by experiments with the antibiotic thiostrepton, which binds specifically to 23S rRNA in the region around residue 1070.

As discussed in more detail below, binding of thiostrepton to the 1070 region of 23S rRNA interferes with translocation, as it strongly inhibits P_i release, translocation, and subsequent turnover of EF-G; in contrast, EF-G binding and GTP hydrolysis are not affected (Rodnina et al., 1999). The tertiary structure of the thiostrepton binding region is relatively unstable, suggesting that it may undergo conformational transitions during elongation. The region is the binding site of protein L11, which stabilizes the tertiary structure of the region and increases the affinity of thiostrepton binding (see chapters 1 and 11). Mutations in L11 that confer resistance to thiostrepton do not affect the binding of the antibiotic, indicating that the role of L11 is to modulate the structural flexibility of the rRNA region to which it is bound. These findings suggest that conformational transitions in that region of 23S rRNA are important for EF-G function on the ribosome and that thiostrepton may interfere with these transitions by stabilizing a conformation which may prevail in the complex with L11 (Draper and Xing, 1995; Porse et al., 1998). Other functionally related regions of the ribosome, such as the L10-$(L12)_2$ complex forming the stalk of the 50S subunit and the α-sarcin loop, may be involved as well.

Interestingly, a very similar inhibition pattern, that is, a strong inhibition of translocation and EF-G turnover, was observed with truncated EF-G that lacked domain 4 (Rodnina et al., 1997). This parallel suggests that structural rearrangements involved in both translocation and subsequent dissociation of EF-G are coupled rearrangements of the factor and the ribosome.

Mechanochemical Function of EF-G

Classic GTPases function as molecular switches. The active, GTP-bound form is formed by GDP-GTP exchange, catalyzed by exchange factors, and is inactivated by GTP hydrolysis triggered by GTPase-activating proteins. Traditionally, EF-G was considered to follow basically the same scheme, although, for instance, it exchanges GDP for GTP spontaneously and the affinities of ribosome binding of EF-G·GTP and EF-G·GDP are similar (Baca et al., 1976). Nevertheless, EF-G was thought to bind to the ribosome in the GTP-bound form and, by binding, to catalyze translocation; subsequently, GTP would be hydrolyzed and the affinity of the factor to the ribosome would be switched from high to low, thereby enabling the factor to dissociate from the ribosome. The picture emerging from the kinetic and mutational analyses, however, is inconsistent with the switch model.

Association and dissociation rate constants of the initial complex of EF-G and the ribosome are the same with GTP, nonhydrolyzable caged GTP, or GDP (1.5×10^8 M^{-1} s^{-1} and 50 to 100 s^{-1}, respectively, at 37°C [Rodnina et al., 1997]). K_m values for EF-G, as determined in steady-state translocation experiments, are similar with caged GTP and GDP (0.3 and 1 μM, respectively [unpublished results]). Finally, the kinetic analysis shows that GTP is hydrolyzed immediately following the binding and that the tRNA-mRNA movement takes place subsequent to, and is accelerated by, GTP hydrolysis. These features are not consistent with a switch model, which predicts that EF-G·GTP is the active form of the factor; rather, the active form of EF-G on the ribosome seems to be the form in which GDP·P_i (or GDP alone) is bound.

Elongation factors consist of several domains, and the homology to other GTPases, like Ras and the G_α subunit of heterotrimeric G proteins, is restricted to the G domain. The conformational switch of Ras and G_α brought about by exchanging GDP for GTP (activation) and GTP hydrolysis (inactivation) changes their affinities to their respective downstream effector proteins, Raf and adenylate cyclase; that is, it regulates intermolecular interactions. By contrast, in EF-Tu the conformational switch of the G domain affects the interactions of the G domains with neighboring domains within the molecule, leading to a dramatic change of the overall architecture of the molecule. The effect of GTP hydrolysis on the structure of EF-G is not known, since only the structures of EF-G·GDP and of nucleotide-free EF-G were determined (see chapters 29 and 45). By analogy to the well-characterized structural change of EF-Tu brought about by GTP hydrolysis, it is likely that the loss of the γ-phosphate introduces a conformational change in the G domain of EF-G. This change may affect the interactions of the G domain with neighboring domains (2 and 5), thereby changing the relative arrangement of the domains. Although the effect on the overall structure of EF-G is probably less extensive than in the case of EF-Tu (Czworkowski and Moore, 1997), the presumed rearrangement may nevertheless suffice to induce rapid translocation. It is appealing to assume that a movement of domain 4 is involved which may affect structural elements of the ribosome it is bound to.

This scenario bears interesting parallels to systems where chemical energy is transformed into directed molecular movement. The classic, ATP-driven motor proteins, such as myosin and kinesin, resemble G proteins in that the affinities of the motors to their protein partners depend on the nucleotide that occupies the active site (Vale, 1996). The motor is active in the ADP-bound form. For instance, both ATP- and ADP·P_i forms of myosin are weakly bound to actin. P_i release and concomitant protein structural rearrangements allow myosin·ADP to bind 10^4-fold more tightly to the actin filament and enable the motor to produce force ("power stroke") (Taylor, 1992). Thus, the ATP-to-ADP transition converts the motor into its active state. It has been pointed out (Sablin et al., 1996; Vale, 1996) that there are structural similarities between motor proteins and several GTPases, as well as potential analogies in the switch mechanisms induced by NTP hydrolysis. While the structural similarities are confined to the nucleotide binding domains, there are striking parallels in the functional cycles of EF-G and, for instance, myosin. Binding of EF-G·GTP to the ribosome is weak and readily reversible, and GTP hydrolysis is required to promote rapid tRNA movement (Rodnina et al., 1997). The kinetic data suggest that the energy released by GTP hydrolysis is stored in the complex, probably by retaining P_i in the active site of EF-G, and that translocation is induced by the release of P_i. While this model is consistent with the available experimental evidence, there remain many open questions. What seems clear, though, is that the GTPase switch model is unsuitable to describe the function of EF-G. Rather, EF-G seems to work as a mechanochemical device to promote molecular rearrangements on, and of, the ribosome at the expense of chemical energy derived from GTP hydrolysis.

ACTIVATION OF GTP HYDROLYSIS IN EF-Tu AND EF-G

The GTPase activities of EF-Tu and EF-G intrinsically are very low and are strongly enhanced on the ribosome. The molecular mechanisms of GTPase activation and GTP hydrolysis are not known. Mechanisms of GTPase activation were elucidated for heterotrimeric G proteins and for the small GTPases Ras and Rho (for a review, see Sprang, 1997). Upon interaction with their respective targets, both RasGAP and RhoGAP (GTPase activating protein) supply a catalytic arginine residue to the active site of Ras (the "arginine finger"), thereby stabilizing the GTPase transition state and increasing the rate of GTP hydrolysis more than 1,000-fold. Also, in G_α proteins, an arginine residue is crucial for the stabilization of the GTPase transition state; in this case, however, the catalytic residue belongs to the G_α subunit. Furthermore, the structure of the active center of the Ras-RasGAP and Rho-RhoGAP complexes appeared to be remarkably similar to the putative GTPase transition state structure of G_α, suggesting a similar stereochemistry of the GTPase reaction, regardless of whether the catalytic arginine is provided in *cis* or in *trans*.

Both elongation factors contain a conserved arginine residue at the position of the catalytic arginine in G_α; however, in *Thermus thermophilus* EF-Tu, replacement of this arginine (position 59) with threonine did not significantly affect the GTPase activity (Zeidler et al., 1995). Furthermore, the characteristic of the GTPase activity of the two elongation factors (very low intrinsic activity; about 10^5-fold activation by the ribosome) resembles that of Ras-like proteins, rather than that of heterotrimeric G proteins. It is not known which component of the ribosome is responsible for GTPase activation, although ribosomal protein L12 has been implicated very early in factor function on the ribosome (Kischa et al., 1971; Traut et al., 1995), as ribosomes depleted of L12 are not able to interact with elongation factors and the activity is restored by the addition of L12. Cryo-electron microscopic reconstructions of ribosome complexes with EF-Tu (Stark et al., 1997b) and EF-G (Agrawal et al., 1998; Stark et al., submitted) suggest there is direct contact between L12 and the G domains of both factors.

Recently, we have studied the ability of isolated L12 protein to stimulate GTP hydrolysis by either EF-Tu or EF-G (Savelsbergh et al., in press). Strong stimulation of GTP hydrolysis upon addition of L12 was observed for EF-G at a K_m of about 40 μM. The low affinity is probably the reason why the activating interaction has not been detected in previous experiments with isolated L12 (Donner et al., 1978; Sander et al., 1980). The k_{cat} of the reaction is 0.3 s^{-1}, that is, much higher than the undetectably low intrinsic GTPase of EF-G but still about 500 times lower than the rate constant of GTP hydrolysis by EF-G on the ribosome (Rodnina et al., 1997). No GTPase activity was induced in EF-Tu (EF-Tu·GTP or EF-Tu·GTP·aa-tRNA complex), even at very high concentrations of L12, although the complex is formed under these conditions (K_d = 10 μM). This is consistent with the earlier finding that EF-Tu in the ternary complex is refractory to GTPase stimulation unless a cognate, or near-cognate, codon is recognized on the ribosome (Pape et al., 1999; Rodnina et al., 1995b).

Alignments of L12 sequences revealed a unique, highly conserved arginine (position 74 in *E. coli*) in the C-terminal domain. This suggested that L12 may function similarly to RasGAP and RhoGAP, namely, by providing a catalytic arginine in *trans* to the catalytic centers of the elongation factors. To test this hypothesis, arginine 74 in *E. coli* L12 was replaced with lysine or methionine. Neither mutation significantly affected the stimulatory effect, suggesting that arginine 74 of L12 is not essential and that L12 accelerates GTP hydrolysis by inducing or stabilizing the transition state of EF-G, similar to the activation mechanism found for some heterotrimeric G proteins (De Vries and Farquhar, 1999). Support for this contention comes from the recent observation that arginine 29, which is located close to the nucleotide binding site in the G domain of EF-G, is essential for GTP hydrolysis (unpublished data).

Contacts of EF-G with 23S rRNA were repeatedly invoked in GTPase activation, in particular, the loops around residues 1070 and 1100 in domain II, which take part in EF-G binding (Moazed et al., 1988; Munishkin and Wool, 1997; Wilson and Noller, 1998a). The 1070 loop is also the binding site of protein L11 (see chapters 1 and 11) as well as of an antibiotic, thiostrepton, which is generally thought to inhibit ribosome-induced GTP hydrolysis by EF-G (Gale et al., 1981); hence the designation "GTPase-associated region." However, recent single-round experiments have shown that thiostrepton has no effect on GTP hydrolysis by EF-G, whereas it strongly inhibits the steps following GTP hydrolysis, that is, P_i release, translocation, and dissociation of EF-G·GDP from the ribosome (Rodnina et al., 1999). This explains why very little GTP hydrolysis was observed in previous experiments with thiostrepton which were conducted under multiple-turnover conditions. Thus, there is no evidence suggesting a direct involvement of the L11 binding region of 23S rRNA in GTPase activation.

Another region of 23S rRNA presumed to interact with EF-G (and EF-Tu) is the α-sarcin stem-loop in domain VI (positions 2646 to 2674), although chemical footprints indicative of the interaction (A2660 and A2662) were observed only when fusidic acid was present to stabilize the EF-G–ribosome complex (Moazed et al., 1988). The isolated α-sarcin stem-loop was shown to bind to EF-G in solution (Munishkin and Wool, 1997). In the fusidic acid-stabilized complex of EF-G with the ribosome, the α-sarcin stem is close to position 196 in the G domain of EF-G, which lies just above the GTP binding site, while the α-sarcin loop region is in the vicinity of position 650 in domain 5 of the factor (Wilson and Noller, 1998a). While the arrangement of EF-G relative to the α-sarcin stem-loop would be consistent with an influence on the GTPase site, we did not observe the chemical footprints indicative of the interaction when thiostrepton was present in the complex (Rodnina et al., 1999). This suggests that, in the latter complex, EF-G does not interact with the α-sarcin region, while the GTPase activity is unaffected. We conclude that the contact of EF-G with the α-sarcin region does not contribute to the GTPase activation, but is rather a later event in the functional cycle.

REFERENCES

Agrawal, R. K., P. Penczek, R. A. Grassucci, and J. Frank. 1998. Visualization of elongation factor G on the *Escherichia coli* 70S ribosome: the mechanism of translocation. *Proc. Natl. Acad. Sci. USA* 95:6134–6138.

Ahmadian, M. R., P. Stege, K. Scheffzek, and A. Wittinghofer. 1997. Confirmation of the arginine-finger hypothesis for the GAP-stimulated GTP-hydrolysis reaction of Ras. *Nat. Struct. Biol.* 4:686–689.

Baca, O. G., M. S. Rohrbach, and J. W. Bodley. 1976. Equilibrium measurements of the interactions of guanine nucleotides with *Escherichia coli* elongation factor G and the ribosome. *Biochemistry* 15:4570–4574.

Bilgin, N., and M. Ehrenberg. 1994. Mutations in 23 S ribosomal RNA perturb transfer RNA selection and can lead to streptomycin dependence. *J. Mol. Biol.* 235:813–824.

Bilgin, N., F. Claesens, H. Pahverk, and M. Ehrenberg. 1992. Kinetic properties of *Escherichia coli* ribosomes with altered forms of S12. *J. Mol. Biol.* **224:**1011–1027.

Borowski, C., M. V. Rodnina, and W. Wintermeyer. 1996. Truncated elongation factor G lacking the G domain promotes translocation of the 3′ end but not of the anticodon domain of peptidyl-tRNA. *Proc. Natl. Acad. Sci. USA* **93:**4202–4206.

Brink, M. F., G. Brink, M. P. Verbeet, and H. A. de Boer. 1994. Spectinomycin interacts specifically with the residues G1064 and C1192 in 16S rRNA, thereby potentially freezing this molecule into an inactive conformation. *Nucleic Acids Res.* **22:**325–331.

Brune, M., J. L. Hunter, J. E. Corrie, and M. R. Webb. 1994. Direct, real-time measurement of rapid inorganic phosphate release using a novel fluorescent probe and its application to actomyosin subfragment 1 ATPase. *Biochemistry* **33:**8262–8271.

Czworkowski, J., and P. B. Moore. 1997. The conformational properties of elongation factor G and the mechanism of translocation. *Biochemistry* **36:**10327–10334.

Davies, J., and B. D. Davis. 1968. Misreading of ribonucleic acid code words induced by aminoglycoside antibiotics. The effect of drug concentration. *J. Biol. Chem.* **243:**3312–3316.

De Vries, L., and M. G. Farquhar. 1999. RGS proteins: more than just GAPs for heterotrimeric G proteins. *Trends Cell. Biol.* **9:** 138–144.

Donner, D., R. Villems, A. Liljas, and C. G. Kurland. 1978. Guanosinetriphosphatase activity dependent on elongation factor Tu and ribosomal protein L7/L12. *Proc. Natl. Acad. Sci. USA* **75:** 3192–3195.

Doublié, S., S. Tabor, A. M. Long, C. C. Richardson, and T. Ellenberger. 1998. Crystal structure of bacteriophage T7 DNA replication complex at 2.2 Å. *Nature* **391:**251–259.

Draper, D. E., and Y. Xing. 1995. Protein recognition of a ribosomal RNA tertiary structure. *Nucleic Acids Symp. Ser.* **33:**5–7.

Eccleston, J. F., D. B. Dix, and R. C. Thompson. 1985. The rate of cleavage of GTP on the binding of Phe-tRNA·elongation factor Tu·GTP to poly(U)-programmed ribosomes of *Escherichia coli*. *J. Biol. Chem.* **260:**16237–16241.

Erie, D. A., O. Hajiseyedjavadi, M. C. Young, and P. H. von Hippel. 1993. Multiple RNA polymerase conformations and GreA: control of the fidelity of transcription. *Science* **262:**867–873.

Fourmy, D., M. I. Recht, S. C. Blanchard, and J. D. Puglisi. 1996. Structure of the A site of *Escherichia coli* 16S RNA complexed with an aminoglycoside antibiotic. *Science* **274:**1367–1371.

Gale, E. F., E. Cundliffe, P. E. Reynolds, M. H. Richmond, and M. Waring. 1981. *Antibiotic Inhibitors of Ribosomal Function*, p. 402–457. Wiley, London, England.

Gnirke, A., U. Geigenmuller, H. J. Rheinberger, and K. H. Nierhaus. 1989. The allosteric three-site model for the ribosomal elongation cycle. Analysis with a heteropolymeric mRNA. *J. Biol. Chem.* **264:**7291–7301.

Grosjean, H., D. G. Soll, and D. M. Crothers. 1976. Studies of the complex between transfer RNAs with complementary anticodons. I. Origins of enhanced affinity between complementary triplets. *J. Mol. Biol.* **103:**499–519.

Grosjean, H. J., S. de Henau, and D. M. Crothers. 1978. On the physical basis for ambiguity in genetic coding interactions. *Proc. Natl. Acad. Sci. USA* **75:**610–614.

Heilek, G. M., R. Marusak, C. F. Meares, and H. F. Noller. 1995. Directed hydroxyl radical probing of 16S rRNA using Fe(II) tethered to ribosomal protein S4. *Proc. Natl. Acad. Sci. USA* **92:** 1113–1116.

Herschlag, D. 1988. The role of induced fit and conformational changes of enzymes in specificity and catalysis. *Bioorg. Chem.* **16:**62–96.

Hopfield, J. J. 1974. Kinetic proofreading: a new mechanism for reducing errors in biosynthetic processes requiring high specificity. *Proc. Natl. Acad. Sci. USA* **71:**4135–4139.

Hornig, H., P. Woolley, and R. Lührmann. 1987. Decoding at the ribosomal A site: antibiotics, misreading and energy of aminoacyl-tRNA binding. *Biochimie* **69:**803–813.

Johnson, A. E., H. J. Adkins, E. A. Matthews, and C. R. Cantor. 1982. Distance moved by transfer RNA during translocation from the A site to the P site on the ribosome. *J. Mol. Biol.* **156:** 113–140.

Johnson, K. A. 1992. Conformational coupling in DNA polymerase fidelity. *Annu. Rev. Biochem.* **62:**685–713.

Joseph, S., and H. F. Noller. 1998. EF-G-catalyzed translocation of anticodon stem-loop analogs of transfer RNA in the ribosome. *EMBO J.* **17:**3478–3483.

Karim, A. M., and R. C. Thompson. 1986. Guanosine 5′-O-(3-thiotriphosphate) as an analog of GTP in protein biosynthesis. *J. Biol. Chem.* **261:**3238–3243.

Karimi, R., and M. Ehrenberg. 1994. Dissociation rate of cognate peptidyl-tRNA from the A-site of hyper-accurate and error-prone ribosomes. *Eur. J. Biochem.* **226:**355–360.

Kiefer, J. R., C. Mao, J. C. Braman, and L. S. Beese. 1998. Visualizing DNA replication in a catalytically active Bacillus DNA polymerase crystal. *Nature* **391:**304–307.

Kischa, K., W. Möller, and G. Stöffler. 1971. Reconstitution of a GTPase activity by a 50S ribosomal protein from *E. coli*. *Nat. New. Biol.* **233:**62–63.

Kurland, C. G., and M. Ehrenberg. 1984. Optimization of translational accuracy. *Prog. Nucleic Acids Res. Mol. Biol.* **31:**191–219.

Kurland, C. G., and M. Ehrenberg. 1987. Growth-optimizing accuracy of gene expression. *Annu. Rev. Biophys. Biophys. Chem.* **16:**291–318.

Li, Y., S. Korolev, and G. Waksman. 1998. Crystal structures of open and closed forms of binary and ternary complexes of the large fragment of *Thermus aquaticus* DNA polymerase I: structural basis for nucleotide incorporation. *EMBO J.* **17:**7514–7525.

Lill, R., J. M. Robertson, and W. Wintermeyer. 1986. Affinities of tRNA binding sites of ribosomes from *Escherichia coli*. *Biochemistry* **25:**3245–3255.

Lill, R., J. M. Robertson, and W. Wintermeyer. 1989. Binding of the 3′ terminus of tRNA to 23S rRNA in the ribosomal exit site actively promotes translocation. *EMBO J.* **8:**3933–3938.

Moazed, D., and H. F. Noller. 1986. Transfer RNA shields specific nucleotides in 16S ribosomal RNA from attack by chemical probes. *Cell* **47:**985–994.

Moazed, D., and H. F. Noller. 1987. Interaction of antibiotics with functional sites in 16S ribosomal RNA. *Nature* **327:**389–394.

Moazed, D., and H. F. Noller. 1989a. Interaction of tRNA with 23S rRNA in the ribosomal A, P, and E sites. *Cell* **57:**585–597.

Moazed, D., and H. F. Noller. 1989b. Intermediate states in the movement of transfer RNA in the ribosome. *Nature* **342:**142–148.

Moazed, D., J. M. Robertson, and H. F. Noller. 1988. Interaction of elongation factors EF-G and EF-Tu with a conserved loop in 23S RNA. *Nature* **334:**362–364.

Modolell, J., and D. Vazquez. 1977. The inhibition of ribosomal translocation by viomycin. *Eur. J. Biochem.* **81:**491–497.

Munishkin, A., and I. G. Wool. 1997. The ribosome-in-pieces: binding of elongation factor EF-G to oligoribonucleotides that mimic the sarcin/ricin and thiostrepton domains of 23S ribosomal RNA. *Proc. Natl. Acad. Sci. USA* **94:**12280–12284.

Neal, S. E., J. F. Eccleston, and M. R. Webb. 1990. Hydrolysis of GTP by p21N-ras, the N-ras protooncogene product, is accompanied by a conformational change in the wild-type protein: use of a single fluorescent probe at the catalytic site. *Proc. Natl. Acad. Sci. USA* **87:**3562–3565.

Neubig, R. R., M. P. Connolly, and A. E. Remmers. 1994. Rapid kinetics of G protein subunit association: a rate-limiting conformational change? *FEBS Lett.* **355**:251–253.

Nierhaus, K. H. 1990. The allosteric three-site model for the ribosomal elongation cycle: features and future. *Biochemistry* **29**: 4997–5008.

Nierhaus, K. H., R. Junemann, and C. M. T. Spahn. 1997. Are the current three-site models valid descriptions of the ribosomal elongation cycle? *Proc. Natl. Acad. Sci. USA* **94**:10499–10500.

Ninio, J. 1975. Kinetic amplification of enzyme discrimination. *Biochimie* **57**:587–595.

O'Connor, M., C. A. Brunelli, M. A. Firpo, S. T. Gregory, K. R. Lieberman, J. S. Lodmell, H. Moine, D. I. Van Ryk, and A. Dahlberg. 1995. Genetic probes of ribosomal RNA functions. *Biochem. Cell Biol.* **73**:859–868.

Pape, T., W. Wintermeyer, and M. V. Rodnina. 1998. Complete kinetic mechanism of elongation factor Tu-dependent binding of aminoacyl-tRNA to the A site of the *E. coli* ribosome. *EMBO J.* **17**:7490–7497.

Pape, T., W. Wintermeyer, and M. V. Rodnina. 1999. Induced fit in initial selection and proofreading of aminoacyl-tRNA on the ribosome. *EMBO J.* **18**:3800–3807.

Pape, T., W. Wintermeyer, and M. V. Rodnina. Conformational switch in the decoding region of 16S rRNA during aminoacyl-tRNA selection on the ribosome. *Nat. Struct. Biol.*, in press.

Paulsen, H., and W. Wintermeyer. 1986. tRNA topography during translocation: steady-state and kinetic fluorescence energy-transfer studies. *Biochemistry* **25**:2749–2756.

Paulsen, H., J. M. Robertson, and W. Wintermeyer. 1983. Topological arrangement of two transfer RNAs on the ribosome. Fluorescence energy transfer measurements between A and P site-bound $tRNA^{Phe}$. *J. Mol. Biol.* **167**:411–426.

Porse, B. T., I. Leviev, A. S. Mankin, and R. A. Garrett. 1998. The antibiotic thiostrepton inhibits a functional transition within protein L11 at the ribosomal GTPase centre. *J. Mol. Biol.* **276**: 391–404.

Post, C. B., and W. J. Ray. 1995. Reexamination of induced fit as a determinant of substrate specificity in enzymatic reactions. *Biochemistry* **34**:15881–15890.

Powers, T., and H. F. Noller. 1991. A functional pseudoknot in 16S ribosomal RNA. *EMBO J.* **10**:2203–2214.

Powers, T., and H. F. Noller. 1995. Hydroxyl radical footprinting of ribosomal proteins on 16S rRNA. *RNA* **1**:194–209.

Purohit, P., and S. Stern. 1994. Interactions of a small RNA with antibiotic and RNA ligands of the 30S subunit. *Nature* **370**:659–662.

Remmers, A. E., and R. R. Neubig. 1996. Partial G protein activation by fluorescent guanine nucleotide analogs. Evidence for a triphosphate-bound but inactive state. *J. Biol. Chem.* **271**: 4791–4797.

Remmers, A. E., R. Posner, and R. R. Neubig. 1994. Fluorescent guanine nucleotide analogs and G protein activation. *J. Biol. Chem.* **269**:13771–13778.

Rensland, H., A. Lautwein, A. Wittinghofer, and R. S. Goody. 1991. Is there a rate-limiting step before GTP cleavage by H-ras p21? *Biochemistry* **30**:11181–11185.

Rittinger, K., G. Divita, and R. S. Goody. 1995. Human immunodeficiency virus reverse transcriptase substrate-induced conformational changes and the mechanism of inhibition by nonnucleoside inhibitors. *Proc. Natl. Acad. Sci. USA* **92**:8046–8049.

Robertson, J. M., and W. Wintermeyer. 1987. Mechanism of ribosomal translocation. tRNA binds transiently to an exit site before leaving the ribosome during translocation. *J. Mol. Biol.* **196**:525–540.

Robertson, J. M., H. Paulsen, and W. Wintermeyer. 1986. Pre-steady-state kinetics of ribosomal translocation. *J. Mol. Biol.* **192**:351–360.

Rodnina, M. V., R. Fricke, L. Kuhn, and W. Wintermeyer. 1995a. Codon-dependent conformational change of elongation factor Tu preceding GTP hydrolysis on the ribosome. *EMBO J.* **14**: 2613–2619.

Rodnina, M. V., T. Pape, R. Fricke, and W. Wintermeyer. 1995b. Elongation factor Tu, a GTPase triggered by codon recognition on the ribosome: mechanism and GTP consumption. *Biochem. Cell Biol.* **73**:1221–1227.

Rodnina, M. V., T. Pape, R. Fricke, L. Kuhn, and W. Wintermeyer. 1996. Initial binding of the elongation factor Tu·GTP·aminoacyl-tRNA complex preceding codon recognition on the ribosome. *J. Biol. Chem.* **271**:646–652.

Rodnina, M. V., A. Savelsbergh, V. I. Katunin, and W. Wintermeyer. 1997. Hydrolysis of GTP by elongation factor G drives tRNA movement on the ribosome. *Nature* **385**:37–41.

Rodnina, M. V., A. Savelsbergh, N. B. Matassova, V. I. Katunin, Y. P. Semenkov, and W. Wintermeyer. 1999. Thiostrepton inhibits turnover but not GTPase of elongation factor G on the ribosome. *Proc. Natl. Acad. Sci. USA* **96**:9586–9590.

Rose, S. J., P. T. Lowary, and O. C. Uhlenbeck. 1983. Binding of yeast tRNAPhe anticodon arm to *Escherichia coli* 30 S ribosomes. *J. Mol. Biol.* **167**:103–117.

Sablin, E. P., F. J. Kull, R. Cooke, R. D. Vale, and R. J. Fletterick. 1996. Crystal structure of the motor domain of the kinesin-related motor ncd. *Nature* **380**:555–559.

Sander, G., R. Ivell, J. B. Crechet, and A. Parmeggiani. 1980. Interaction of elongation factor Tu with the ribosome. A study using the antibiotic kirromycin. *Biochemistry* **19**:865–870.

Savelsbergh, A., D. Mohr, W. Wintermeyer, and M. V. Rodnina. Ribosomal protein L7/12 stimulates GTP hydrolysis on elongation factor G by an RGS-type mechanism. *J. Biol. Chem.*, in press.

Semenkov, Y. P., M. V. Rodnina, and W. Wintermeyer. 1996. The "allosteric three-site model" of elongation cannot be confirmed in a well-defined ribosome system from *Escherischia coli. Proc. Natl. Acad. Sci. USA* **93**:12183–12188.

Spahn, C. M. T., and K. H. Nierhaus. 1998. Models of the elongation cycle: an evaluation. *Biol. Chem.* **379**:753–772.

Spence, R. A., W. M. Kati, K. S. Anderson, and K. A. Johnson. 1995. Mechanism of inhibition of HIV-1 reverse transcriptase by nonnucleoside inhibitors. *Science* **267**:988–992.

Spirin, A. S. 1985. Ribosomal translocation: facts and models. Prog. *Nucleic Acid. Res. Mol. Biol.* **32**:75–114.

Sprang, S. R. 1997. G protein mechanisms: insights from structural analysis. *Annu. Rev. Biochem.* **66**:639–678.

Stark, H., E. V. Orlova, J. Rinke-Appel, N. Jünke, F. Mueller, M. V. Rodnina, W. Wintermeyer, R. Brimacombe, and M. van Heel. 1997a. Arrangement of tRNAs in pre- and posttranslocational ribosomes revealed by electron cryomicroscopy. *Cell* **88**: 19–28.

Stark, H., M. V. Rodnina, J. Rinke-Appel, R. Brimacombe, W. Wintermeyer, and M. van Heel. 1997b. Visualization of elongation factor Tu on the *Escherichia coli* ribosome. *Nature* **389**: 403–406.

Stark, H., M. V. Rodnina, M. van Heel, and W. Wintermeyer. Large-scale movement of elongation factor G and extensive conformational changes of the ribosome during translocation. Submitted for publication.

Taylor, E. W. 1992. Mechanism and energetics of actomyosin ATPase, p. 1281–1293. *In* H. A. Fozzard (ed.), *The Heart and Cardiovascular System*, Raven Press, Ltd., New York, N.Y.

Thompson, R. C. 1988. EF-Tu provides an internal kinetic standard for translational accuracy. *Trends Biochem. Sci.* **13**:91–93.

Thompson, R. C., and D. B. Dix. 1982. Accuracy of protein biosynthesis. A kinetic study of the reaction of poly(U)-programmed ribosomes with a leucyl-tRNA$_2$-elongation factor Tu-GTP complex. *J. Biol. Chem.* **257**:6677–6682.

Thompson, R. C., and A. M. Karim. 1982. The accuracy of protein biosynthesis is limited by its speed: high fidelity selection by ribosomes of aminoacyl-tRNA ternary complexes containing GTPγS. *Proc. Natl. Acad. Sci. USA* **79**:4922–4926.

Thompson, R. C., D. B. Dix, and A. M. Karim. 1986. The reaction of ribosomes with elongation factor Tu·GTP complexes. Aminoacyl-tRNA-independent reactions in the elongation cycle determine the accuracy of protein synthesis. *J. Biol. Chem.* **261**: 4868–4874.

Traut, R. R., D. Dey, D. E. Bochkariov, A. V. Oleinikov, G. G. Jokhadze, B. Hamman, and D. Jameson. 1995. Location and domain structure of *Escherichia coli* ribosomal protein L7/L12: Site specific cysteine cross-linking and attachment of fluorescent probes. *Biochem.Cell Biol.* **73**:949–958.

Uhlenbeck, O. C., F. H. Martin, and P. Doty. 1981. Self-complementary nucleotides: effects of helix defects and guanylic acid-cytidilyc acid base pairs. *J. Mol. Biol.* **57**:217–229.

Vale, R. D. 1996. Switches, latches and amplifiers: common themes of G proteins and molecular motors. *J. Cell Biol.* **135**: 291–302.

VanLoock, M. S., T. R. Easterwood, and S. C. Harvey. 1999. Major groove binding of the tRNA/mRNA complex to the 16S ribosomal RNA decoding center. *J. Mol. Biol.* **285**:2069–2078.

Wei, J., and T. S. Leyh. 1998. Conformational change rate-limits GTP hydrolysis: the mechanism of the ATP sulfurylase-GTPase. *Biochemistry* **37**:17163–17169.

Wilson, K. S., and H. F. Noller. 1998a. Mapping the position of translational elongation factor EF-G in the ribosome by directed hydroxyl radical probing. *Cell* **92**:131–139.

Wilson, K. S., and H. F. Noller. 1998b. Molecular movement inside the translational engine. *Cell* **92**:337–349.

Wintermeyer, W., R. Lill, H. Paulsen, and J. M. Robertson. 1986. Mechanism of ribosomal translocation, p. 523–540. *In* B. Hardesty and G. Kramer (ed.), *Structure, Function, and Genetics of Ribosomes*. Springer-Verlag, New York, N.Y.

Zeidler, W., C. Egle, S. Ribeiro, A. Wagner, V. Katunin, R. Kreutzer, M. Rodnina, W. Wintermeyer, and M. Sprinzl. 1995. Site-directed mutagenesis of Thermus thermophilus elongation factor Tu. Replacement of His85, Asp81 and Arg300. *Eur. J. Biochem.* **2295**:596–604.

The Ribosome: Structure, Function, Antibiotics, and Cellular Interactions
Edited by R. A. Garrett, S. R. Douthwaite, A. Liljas, A. T. Matheson, P. B. Moore, and H. F. Noller

Chapter 26

Ribosomal Elongation Cycle

KNUD H. NIERHAUS, CHRISTIAN SPAHN, NILS BURKHARDT, MARYLENA DABROWSKI, GUNDO DIEDRICH, EDDA EINFELDT, DETLEV KAMP, VITER MARQUEZ, SEBASTIAN PATZKE, MARKUS A. SCHÄFER, ULRICH STELZL, GREGOR BLAHA, REGINE WILLUMEIT, and HEINRICH B. STUHRMANN

After the detection of the third tRNA binding site on *Escherichia coli* ribosomes, the E site specific for deacylated tRNA (Grajevskaja et al., 1982; Kirillov et al., 1983; Lill et al., 1984; Rheinberger and Nierhaus, 1980; Rheinberger et al., 1981), a third site was also found in ribosomes from organisms of all kingdoms (domains; for a review, see Burkhardt et al., 1998). Today three tRNA binding sites are accepted as a universal feature of ribosomes.

A controversy still exists concerning the importance of the E site for the translational process and the movement of tRNAs on the ribosome during translocation. These different views extend to the models that are currently used for the description of the elongation cycle and thus are of fundamental importance for our understanding of the principles of protein synthesis. Therefore, a test of these different views is of general interest.

Protection patterns of the tRNA phosphate groups in the various sites and the identification of the tRNA locations on the ribosome allow discrimination between the different views of the E site. Furthermore, the data also have an impact on the models of the elongation cycle. Here we give a brief review of the reactions of the elongation cycle and discuss recent data that clarify and explain some points of the divergent aspects of the current models of the elongation cycle. We will see that the location and thus the features of the deacylated tRNA in either the P or E site are extremely sensitive to the buffer conditions applied. These data explain the discrepancies of the current models of the elongation cycle and the controversy about the features and importance of the E site.

DIFFERENT VIEWS OF THE E SITE

Table 1 summarizes the two radically different views of the E site. The view of the "Nierhaus group" is characterized by a reciprocal coupling of the A and E sites, a tight binding of the deacylated tRNA at the E site, and codon-anticodon interaction at this site. The view of the "Wintermeyer group" denies all three features. Both groups have presented a wealth of data supporting their respective views. The data of the Nierhaus group have been reviewed (Nierhaus, 1990; Rheinberger et al., 1990; a summary of the data collected in the eighties can be found in the form of tables and figures in Rheinberger, 1991) as well as those of the Wintermeyer group (Wintermeyer et al., 1990).

We will see that the striking differences can be traced back to differences in the buffer systems used

Knud H. Nierhaus, Marylena Dabrowski, Edda Einfeldt, Detlev Kamp, Viter Marquez, Sebastian Patzke, Markus A. Schäfer, and Ulrich Stelzl ■ Max-Planck-Institut für Molekulare Genetik, AG Ribosomen, Ihnestraße 73, D-14195 Berlin, Germany. **Christian Spahn** ■ Max-Planck-Institut für Molekulare Genetik, AG Ribosomen, Ihnestraße 73, D-14195 Berlin, Germany, and Wadsworth Center, New York State Department of Health, Empire State Plaza, P.O. Box 509, Albany, NY 12201-0509. **Nils Burkhardt and Gundo Diedrich** ■ Max-Planck-Institut für Molekulare Genetik, AG Ribosomen, Ihnestraße 73, D-14195 Berlin, and BAYER AG-Wuppertal, Gebäude 405, Abteilung MST, D-42096 Wuppertal, Germany. **Gregor Blaha** ■ Max-Planck-Institut für Molekulare Genetik, AG Ribosomen, Ihnestraße 73, D-14195 Berlin, and GKSS Research Center, Max-Planck-Straße, 21502 Geesthacht, Germany. **Regine Willumeit** ■ GKSS Research Center, Max-Planck-Straße, 21502 Geesthacht, Germany. **Heinrich B. Stuhrmann** ■ GKSS Research Center, Max-Planck-Straße, 21502 Geesthacht, Germany, and Institut de Biologie Structurale, 41 Avenue des Martyrs, F-38027 Grenoble, CEDEX 1, France.

Table 1. Two views of the E site

Feature	View	
	Nierhaus group	Wintermeyer group
A- and E-site coupling	**A and E sites are reciprocally coupled.** If the E site is occupied, the A site has a low affinity for tRNAs and vice versa.	Such a **coupling does not exist.**
E-site binding of tRNA	The E site specifically binds deacylated tRNA in a **tight** manner.	Only **weak** binding: the tRNA is released from the E site as soon as it is translocated to this site.
E-site codon-anticodon interaction	Both tRNAs present on PRE ribosomes at the A and P sites and on POST ribosomes at the P and E sites undergo codon-anticodon interaction simultaneously. Occupation of the E site **depends on codon-anticodon interaction.**	Occupation of the E site **does not depend on codon-anticodon interaction.**

by the two groups. The Nierhaus group uses a polyamine buffer, the essential components of which are very similar to the corresponding concentrations found in vivo, whereas the Wintermeyer group and others use a conventional buffer system that deviates significantly from the physiological values (Table 2). A second important but less significant factor is the harvesting point of the *E. coli* cells used as the source for the ribosomes. Ribosomes isolated from early-log-phase cells (harvesting point, 0.5 A_{560} units in a 100-liter fermentor) tend to show the features observed in the polyamine buffer system under conventional conditions as well.

THE CURRENT MODELS OF THE ELONGATION CYCLE

The elongation cycle represents a series of reactions during which a peptidyl-tRNA is prolonged by one aminoacyl-tRNA. A-site occupation, peptide bond formation, and translocation are the three basic reactions. In the course of two elongation cycles a tRNA sequentially occupies the A site, then the P site, and eventually the E site. A tRNA enters a ribosome as an aminoacyl-tRNA at the A site complexed with the elongation factor EF-Tu (EF-1 in archaea and eukarya) and GTP. After GTP hydrolysis EF-Tu·GDP dissociates. The aminoacyl residue receives the peptidyl residue from the peptidyl-tRNA at the adjacent P site via peptide bond formation. The peptidyl-tRNA, prolonged by one aminoacyl residue, is then translocated to the P site. When the peptidyl-tRNA at the P site has lost its peptidyl residue in the next elongation cycle, the subsequent translocation reaction moves the deacylated tRNA from the P site to the E site.

One of the prevailing current models of the elongation cycle, the hybrid-site model (Moazed and Noller, 1989), contains the Wintermeyer view of the E site, although it does not essentially depend on that view. The essence of this model (Fig. 1A) is a creeping movement of the tRNAs through the ribosome. The model states that after peptide bond formation the tRNA portion on the large subunit moves to the next tRNA binding site, viz., the peptidyl-tRNA part on the large subunit moves from the A to the P site and the deacylated tRNA moves from the P to the E site. The peptidyl-tRNA is now present in the hybrid site A/P with the anticodon stem-loop region still at the A site and the remaining region of the tRNA at the P site. Correspondingly, the deacylated tRNA is in the P/E hybrid site. This hybrid-site state after peptide bond formation is the diagnostic feature of the model: a peptidyl-tRNA is never at the A site but rather is at the A/P hybrid site, and a deacylated tRNA is never at the P site but is at the P/E hybrid site in the frame of the model.

Table 2. Concentrations of ions and polyamines important for ribosomal functions

System	Concn (mM)				
	Mg^{2+}	K^+, NH_4^{+a}	Polyamine		
			Spermidine	Spermine	Putrescine
Conventional buffer	7–20	100	None	None	None
Polyamine buffer	3–6	150	2	0.05	None[b]
In vivo[c]	~4	~150	1–4	~0.03	20[c]

[a] K^+ and NH_4^+ are more or less equivalent in in vitro systems of protein synthesis due to their similar ionic radii.
[b] Putrescine has no effect on protein synthesis in vitro; additions of cadaverine and agmatine, which are also present in significant amounts in vivo, also have no effect on protein synthesis (Lewicki and Nierhaus, unpublished).
[c] Lusk et al., 1968; Tabor and Tabor, 1985; Kamekura et al., 1987.

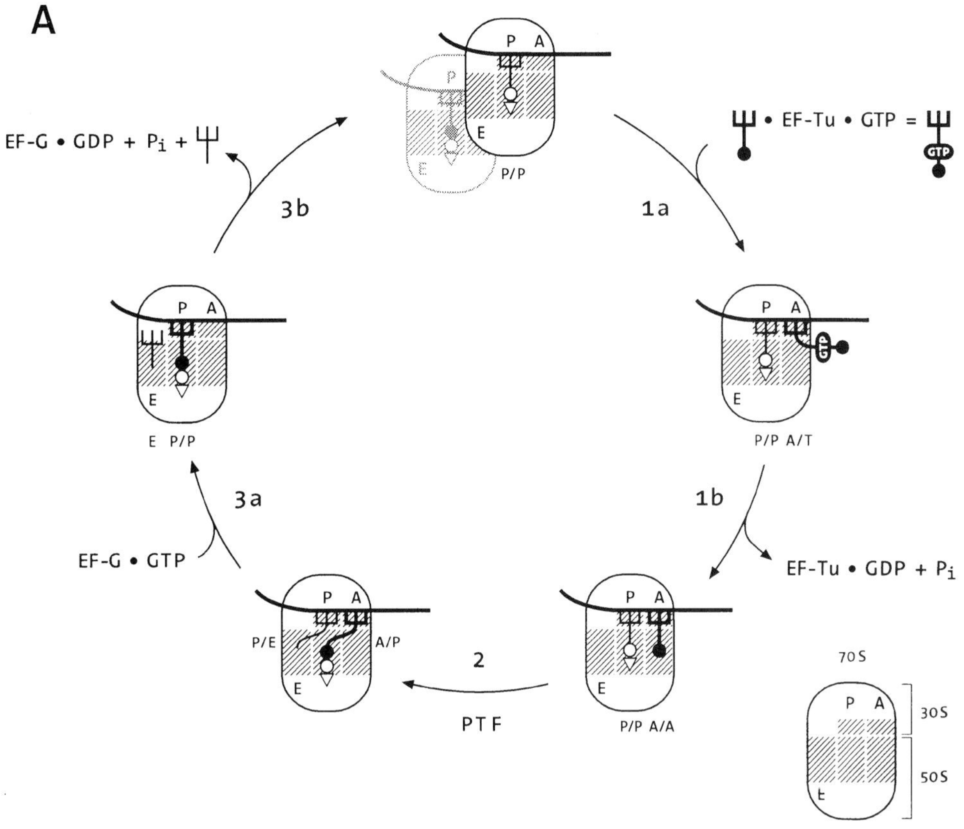

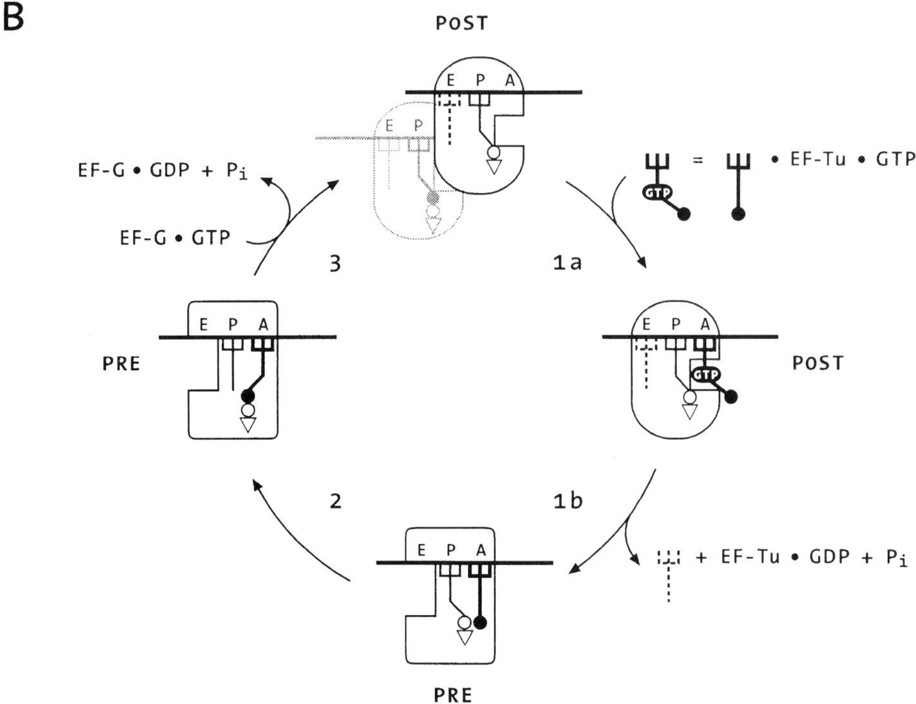

Figure 1. Models of the elongation cycle. (A) Hybrid-site model. (B) Allosteric three-site model. For explanations, see the text.

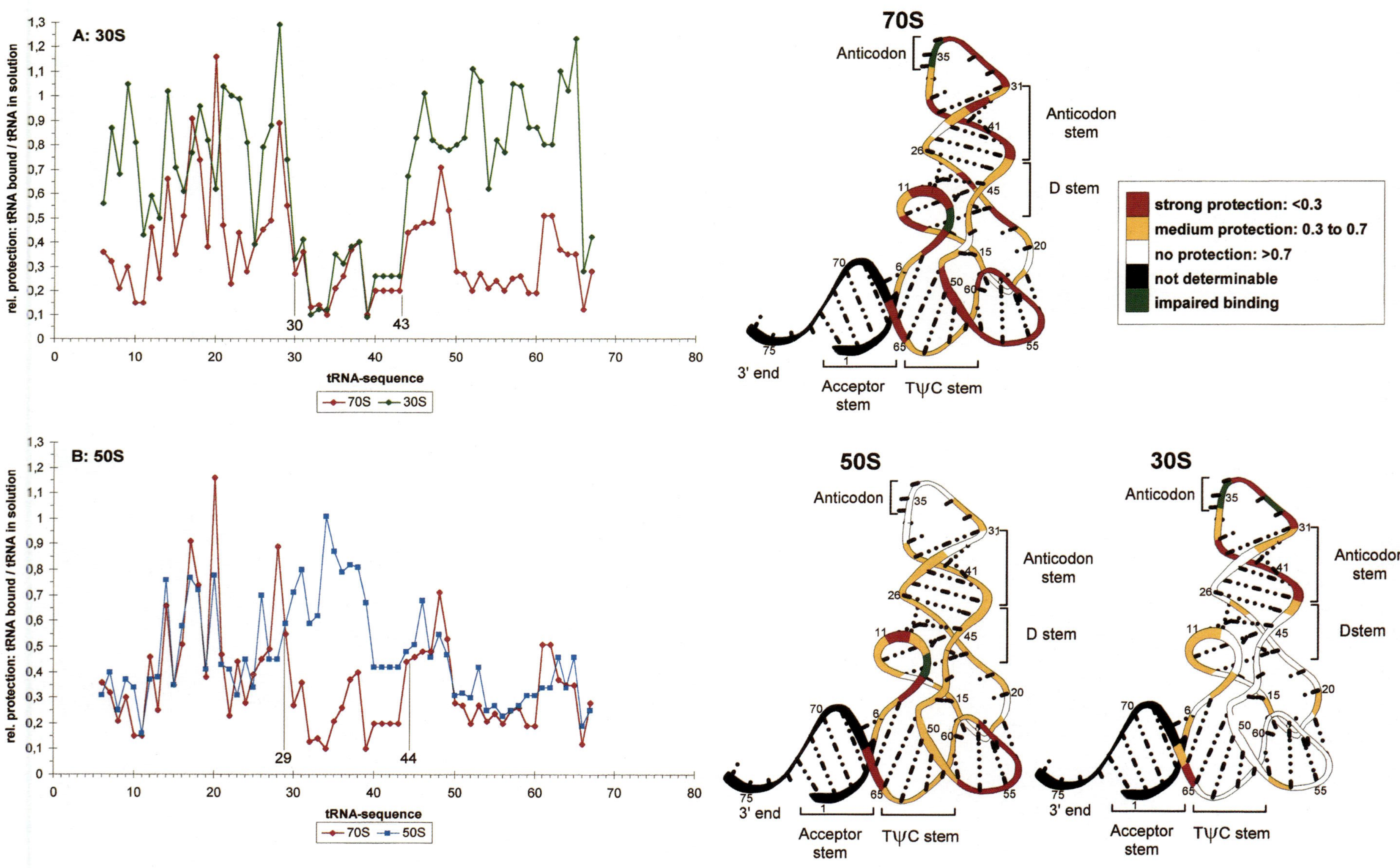
A: 30S
rel. protection: tRNA bound / tRNA in solution
tRNA-sequence
70S
30S
30
43
B: 50S
rel. protection: tRNA bound / tRNA in solution
tRNA-sequence
70S
50S
29
44
70S
Anticodon
Anticodon stem
D stem
Acceptor stem
TψC stem
3' end
strong protection: <0.3
medium protection: 0.3 to 0.7
no protection: >0.7
not determinable
impaired binding
50S
Anticodon
Anticodon stem
D stem
Acceptor stem
TψC stem
3' end
30S
Anticodon
Anticodon stem
Dstem
Acceptor stem
TψC stem
3' end

The model is based essentially on three observations. (i) The simple peptidyl-tRNA analogue AcPhe-tRNA protects some bases on the 16S rRNA and the 23S rRNA against modifying reagents; the bases are diagnostic features of the P site in both subunits. After puromycin reaction the resulting deacylated tRNA shows an "E-site" pattern on the 23S rRNA and a "P-site" pattern on the 16S rRNA. The conclusion was that the deacylated tRNA is in the P /E hybrid site. (ii) An AcPhe-tRNA at the classical A site was found with the pattern of the A/P hybrid site rather than with that of a mere A site. (iii) After translocation the 16S rRNA protections of the peptidyl-tRNA changed from those of the A site to those of the P site, whereas the 23S rRNA pattern remained unchanged. All these experiments were done under conventional buffer conditions, and it is also notable that 12 of 16 protections of the 23S rRNA used for assignment to the P or E site depend on an intact CCA 3′ end of the tRNA (Noller et al., 1990). The last 4 nucleotides, NCCA-3′, are single stranded and allow a movement of the terminal A of up to 40 Å if the tRNA body is fixed. It follows that the CCA end is not a reliable reporter structure for the position of the tRNA.

The second model for the elongation cycle is the allosteric three-site model that is based on a functional analysis of the importance of the E site. The main features are listed in Table 1. A consequence of the reciprocal linkage of the A and E sites of this model is that statistically two tRNAs are on the ribosome, at the A and P sites before translocation (PRE state) and at the P and E sites after (POST state). Only during the decoding process, a substep of A-site occupation, are three tRNAs on the ribosome (Fig. 1B). The reciprocal linkage was first shown in 1986 with *E. coli* ribosomes exploiting the differential binding behavior of the A site at 0 and 37°C, depending on the occupation state of the E site. The kinetics of the A-site occupation were identical with the kinetics of the E-site release (Rheinberger and Nierhaus, 1986). This coupling of the A and E sites could be confirmed in the presence of a heteropolymeric mRNA that displays different codons at the three ribosomal tRNA binding sites (Gnirke et al., 1989). In agreement with this view it was demonstrated that yeast 80S ribosomes bound deacylated tRNA at the E site so tightly that the A site could not be occupied by a ternary complex aminoacyl tRNA·EF-1·GTP. A third elongation factor, EF-3, specific for higher fungi, was required that "opened" the E site with the help of ATP hydrolysis, thus enabling the A-site occupation and the concomitant E-site release (Triana-Alonso et al., 1995). It was thought that the reciprocal linkage between the A and E states was caused by an allosteric effect in the sense of a negative cooperativity. Evidence for codon-anticodon interaction at the E site was gathered in in vitro studies by applying various experimental strategies (for reviews, see Nierhaus, 1990, and Rheinberger et al., 1990) and was also found in vivo (Horsfield et al., 1995; see Nierhaus et al., 1997, for a discussion).

The recently suggested α-ε model (Dabrowski et al., 1998; Nierhaus et al., 1995) is in fact a revised allosteric three-site model that keeps all the features of the latter but explains the reciprocal linkage between the A and E sites radically differently (see below). For a detailed evaluation of the three models, see Spahn and Nierhaus, 1998.

CONTACT PATTERNS OF DEACYLATED tRNA WITH THE RIBOSOMAL SUBUNITS AND THE P/E HYBRID SITE

The tRNA regions in contact with either ribosomal subunit were explored by the phosphorothioate technique (Schatz et al., 1991). This method monitors the accessibility of the tRNA phosphate groups when they are bound to a ribosomal subunit. The tRNA was transcribed from its gene in vitro, and during transcription one of the four nucleotides was introduced as a phosphorothioate derivative. The result was that a nonbridging oxygen atom of a phosphate residue was replaced by sulfur in about four randomly distributed positions. A tRNA modified in this way was almost fully active in various biological functions, such as chargeability, by the phenylalanyl-tRNA synthetase, ternary complex formation, and poly(Phe) synthesis (Dabrowski et al., 1995). The point is that the addition of iodine (I_2), a small and

Figure 2. Relative accessibilities of the phosphate groups of a phosphorothioated tRNA on either subunit and a 70S ribosome. The relative accessibility of a phosphate position means the accessibility relative to that of the corresponding position of a deacylated tRNA in solution. (Left) Accessibility graphs of tRNAs bound to ribosomal subunits or 70S ribosomes, always in the presence of poly(U). The *y* axes indicate the relative accessibilities (1, full accessibility; 0, full protection); the *x* axis gives the nucleotide position. (A) Comparison of the accessibility pattern of a tRNA bound to the P site of a 70S ribosome (red) with that of a tRNA bound to 30S subunits (green). (B) Same as graph A except that the pattern obtained with 50S subunits is shown (blue). (Right) The same patterns projected onto the three-dimensional model of tRNA with a color code for the accessibility of a phosphate group. (Taken from Schäfer et al., unpublished.)

rather inert molecule, can induce with low efficiency (around 5%) a cleavage at the position of the modified phosphate as long as a ribosomal component does not prevent access. If the phosphorothiated tRNA is ^{32}P labeled at its 5′ end, a sequencing gel allows the quantitative assessment of each phosphate group, which can be compared with the accessibility pattern of a tRNA in solution. The accessibilities of 65 to 70 positions of a tRNA were assessed; for technical reasons the very ends of a tRNA could not be resolved on one sequencing gel.

Figure 2A, left, shows the pattern of relative accessibilities of the phosphate groups of a tRNA bound to the 30S subunit compared to that of tRNA present at the P site of a 70S ribosome. It is clear that the sequence 30 to 43 is protected as in the 70S ribosome, whereas at lower and higher numbers the 30S positions are much less protected than in the 70S ribosome. With 50S subunits the reverse pattern is obtained: a similar protection pattern below 30 and above 43 but no correlation from 30 to 43.

These results allow at least two conclusions. (i) The patterns obtained in the presence of 30S and 50S subunits add together to form a pattern of a deacylated tRNA at the 70S P site. In isolated 50S subunits the only site that can be occupied is one that becomes the E site in 70S ribosomes (a prospective E site [Gnirke and Nierhaus, 1986; Kirillov et al., 1983]), and in 30S submits, only the prospective P site is available (Gnirke and Nierhaus, 1986). It follows that this result is consistent with the hybrid-site model stating that a deacylated tRNA is never found at the P site but rather at the P/E hybrid site. (ii) Only the anticodon loop and a part of the anticodon stem (30 ± 1 to 43 ± 1) contact the 30S subunit in isolated 30S subunits as well as within the 70S ribosome. The remaining 83% of the tRNA is in contact with the 50S subunit, again both in isolated subunits and within the 70S ribosomes (Fig. 3).

The fact that most of a tRNA is located on the 50S subunit means that a deacylated tRNA in the P/E hybrid site occupies a position significantly different from the P-site location. In fact, a P/E hybrid site with a deacylated tRNA has been observed by cryo-electron microscopy (cryo-EM) when a conventional buffer system was applied. In contrast, the majority of the deacylated tRNA was found in the classical P site under polyamine buffer conditions. A tRNA in the P site overlaps with a tRNA at the P/E site such that the anticodon regions are at the same place; i.e., both positions cannot be occupied at the same time (Fig. 4) (Agrawal et al., 1999). In native polysomes the P/E site is not observed but rather the classical P-site position is seen (Agrawal et al., 1998), although about 40% of ribosomes in polysomes are present in the PRE state and 60% in the POST (Remme et al., 1989). It follows that the hybrid-site P/E is not a stable feature of ribosomes, as suggested by both native polysomes and functional complexes analyzed under polyamine buffer conditions.

If the P site rather than the P/E hybrid site is occupied by a deacylated tRNA in PRE ribosomes, the adjacent peptidyl-tRNA must be present in the A site rather than in the A/P hybrid site, since occupations of the P site (deacylated tRNA) and of the A/P hybrid site (peptidyl-tRNA) are mutually exclusive. The diagnostic feature of the hybrid-site model, viz., location of the peptidyl-tRNA in the A/P site and the adjacent deacylated tRNA in the P/E site, cannot be confirmed by the data described in this section.

CONTACT PATTERNS OF tRNAs IN THE RIBOSOMAL PRE AND POST STATES

When the phosphorothioate method was applied with the tRNAs of either elongation state, viz., the PRE and the POST states, two main results were obtained: (i) the tRNAs present on the elongating ribosome have strikingly different protection patterns (Dabrowski et al., 1995), and (ii) neither tRNA changes its protection pattern during the translocation reaction (Fig. 5) (Dabrowski et al., 1998). The conclusion was that the microtopographies of the two tRNAs do not change and therefore a movable ribosomal domain exists that binds both tRNAs firmly and keeps them bound during and after translocation. In other words, the movable domain displays the two tRNAs at the A and P sites and shifts them towards the P and E sites. At the A site only one of the patterns is observed (α pattern), whereas at the E site the other appears (ε pattern), and therefore the movable domain was called the α-ε domain and the resulting model of the elongation cycle was called the α-ε model (Fig. 6). This model has all the features of the allosteric three-site model but explains the reciprocal linkage of the A and E sites not by an allosteric interaction but rather by the fact that the α-ε domain moves from the A plus P sites to the P plus E sites in the course of translocation, thus leaving the A site in a low-affinity state. The decoding center is thought not to participate in the α-ε domain, since decoding occurs at the A site in the POST state when the tRNAs are displayed at the P and E sites (Nierhaus et al., 1995).

Whereas the allosteric three-site model could still be reconciled with the hybrid-site model, the α-ε model cannot. The essence of the hybrid-site model is that the tRNAs are moved in a creeping way through the ribosome, starting on the 50S subunit

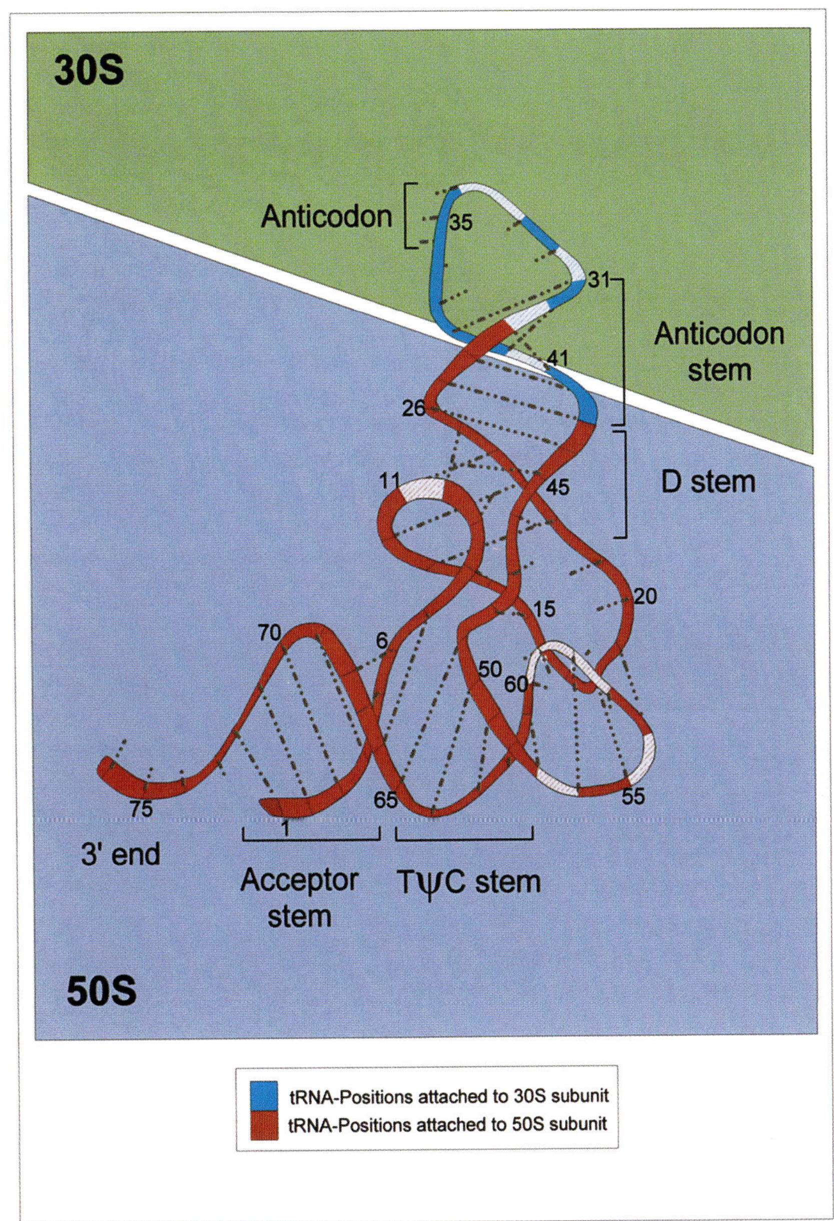

Figure 3. Contact border of a deacylated tRNA at the 70S P site separating the tRNA regions contacting the small and the large ribosomal subunit. (Taken from Schäfer et al., unpublished.)

after peptide bond formation. This step is followed by a movement on the 30S subunit during translocation (Fig. 1B). In contrast, the α-ε model demands a tRNA movement in toto and not in parts.

But what about the demonstration that the protection patterns of 16S rRNA change from the A-site pattern to the P-site pattern during translocation (Moazed and Noller, 1990)? Six base protections of 16S rRNA against modifying reagents were found to be of diagnostic importance for the A site (G529, G530, A1492, and A1493 [strongly] and A1408 and G1494), four of which are located in the decoding center (Purohit and Stern, 1994).

Of the 65 phosphate groups of an AcPhe-tRNA that could be analyzed by the phosphorothioate technique, 57 were almost identical at the A and P sites; the eight changes that were observed were not clustered at the anticodon stem-loop structure but rather were scattered over the whole tRNA (Fig. 5A), not supporting the idea that the tRNA movement is restricted to the 30S subunit. Since the decoding center, as mentioned above, is not movable and is located at the A site, a translocation of the anticodon stem-loop structure out of the decoding center is expected to change the base protections of 16S rRNA in this center. Therefore, the two sets of data, the accessibility

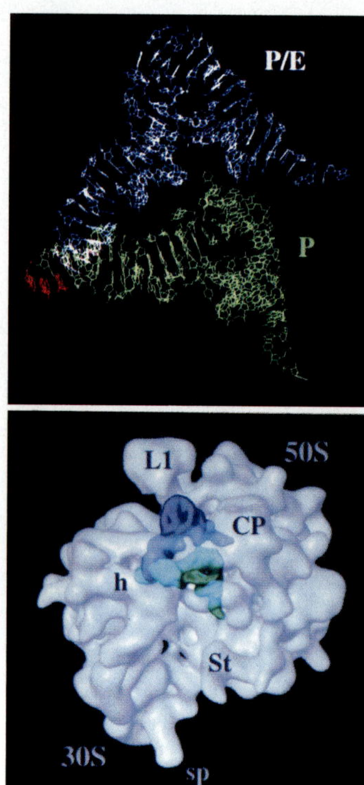

Figure 4. Deacylated tRNAs in the P and P/E sites as observed in polyamine and conventional buffer systems, respectively (Agrawal et al., 1999). (Top) Mutual arrangement of tRNAs in the two sites. The anticodon regions are highlighted in red and overlap; therefore, both positions cannot be occupied simultaneously in the same ribosome. (Bottom) tRNAs in the two positions seen within the ribosome. Landmarks of the small 30S subunit: h, head; sp, spore. Landmarks of the large 50S subunit: L1, L1 protuberance; CP, central protuberance; St, L12 stalk. (Adapted from Agrawal et al., 1999.)

of phosphate groups of the tRNA and the protection pattern of the 16S rRNA bases, can be reconciled.

THE E SITE AND THE E2 POSITION

If a movable α-ε domain carries the two tRNAs from the A and P sites to the P and E sites, respectively, during translocation and keeps them tightly bound during this reaction, one may expect that the mutual arrangement of the tRNAs is the same before and after translocation. The other models do not necessarily suggest this.

We tested this expectation by means of neutron scattering. This technique exploits the sharply different scattering at the atom nuclei of normal hydrogen (protons) and at those of heavy hydrogen (deuterons). We applied the new technique of proton spin contrast variation (Knop et al., 1992; Stuhrmann et al., 1995) that increases the signal-to-noise-ratio by more than 10-fold over that of conventional techniques and thus allows the determination of protein locations and ligand positions within the 70S ribosome that were hitherto not accessible by neutron scattering. The data processing is a fitting procedure where by means of models an intensity spectrum is calculated and compared to the measured one. The advantage of the method is that the results are not blurred by conformational changes; if, for example, the ribosome is deuterated and the bound mRNA-$tRNA_2$ complex is protonated, the structure of the latter can be measured with little interference of ribosomal conformational changes between various functional states.

E. coli cells grow happily in D_2O instead of H_2O, thus replacing all the protons with deuterons. In this way deuterated 70S ribosomes were prepared and specific PRE and POST complexes carrying two protonated tRNAs were constructed and analyzed by neutron scattering. The following results were obtained. (i) The mass center of gravity of the mRNA-$tRNA_2$ complex moved by 12 ± 4 Å, in good agreement with the length of a codon between 10 and 20 Å. (ii) No tRNA was lost during the translocation reaction (the protonated mass of the complex goes exponentially into the intensity and is therefore an extremely sensitive measure of any loss of the protonated mass [Wadzack et al., 1997]). (iii) The mutual arrangement of the tRNAs did not change during translocation. Both tRNAs move in a concerted fashion and rotate by about 20 ± 10° (Fig. 7). (iv) The ribosome does not change its conformation during translocation within a resolution of about 25 Å (Nierhaus et al., 1998). The lack of significant conformational change was also observed in cryo-EM analysis (Stark et al., 1997a).

The central finding, namely, that the mutual arrangement did not change during translocation, was obtained with different ribosomal models used for the fitting procedure. Since this finding is a model-invariant feature, it is considered to be a solid result (Nierhaus et al., 1998).

When the tRNA locations were determined in the first cryo-EM analysis, strikingly different results were obtained with the tRNA at the E site. All studies identified about the same A- and P-site locations of the tRNAs, but the mutual arrangement of the tRNAs seemed to change dramatically after translocation (compare the tRNAs in the A and P sites and in the P and "E2" sites in Fig. 8A). The tRNA thought to be present in the E site (E2 in Fig. 8A) clings with the inner elbow region at the L1 protuberance; other contacts with the ribosome seemingly do not exist. Since L1 is not an essential component for protein synthesis (mutants lacking L1 are viable [Dabbs, 1991]), this position is probably not even an authen-

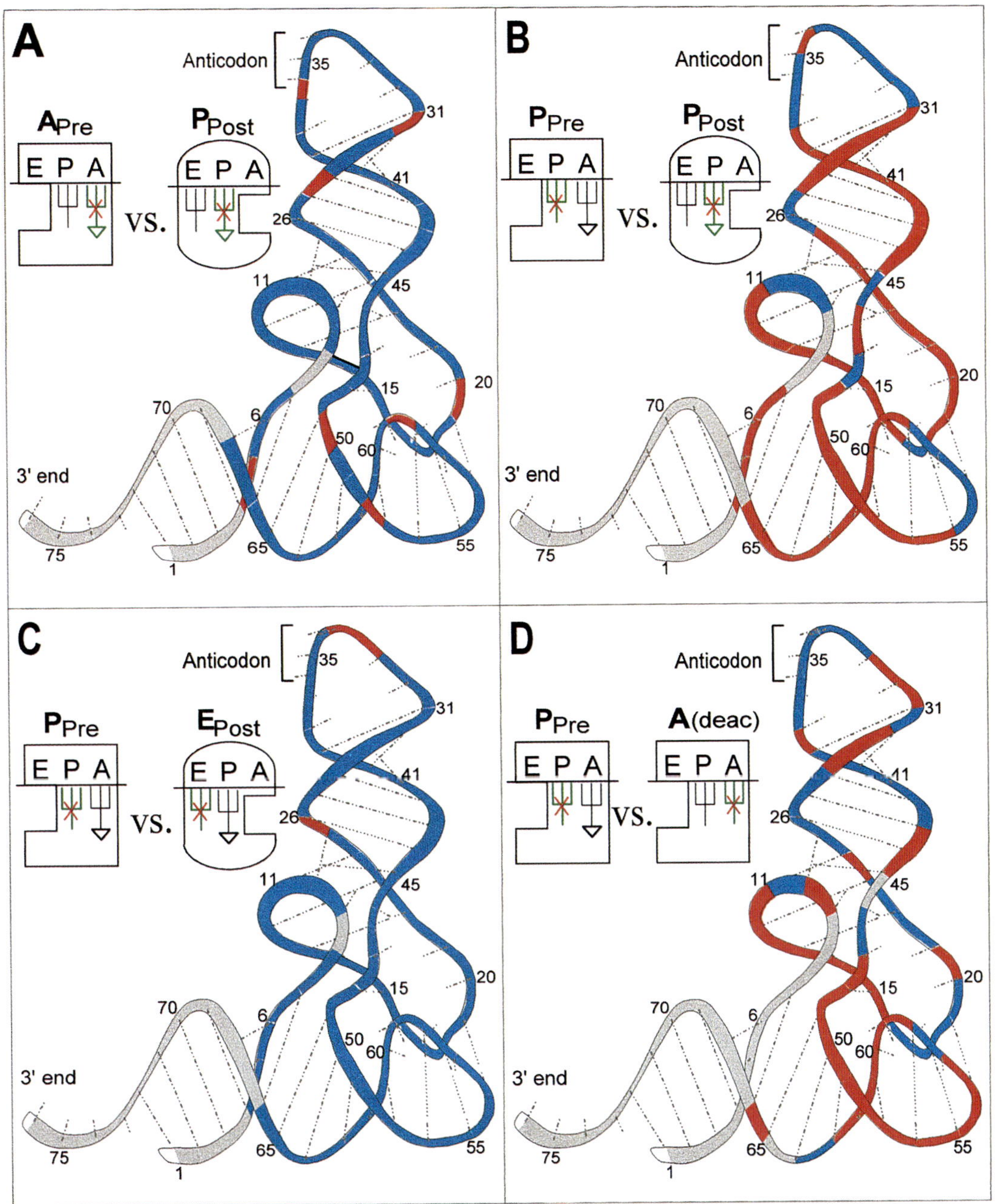

Figure 5. Difference patterns of thioated tRNAs. E, P, and A mark the respective tRNA binding sites. Red, positions where the relative intensities of the two states compared differed by at least a factor of 2; blue, no difference according to this criterion. (A) Differences in the protection patterns of AcPhe-tRNA before and after translocation. (B) Differences of $tRNA^{Phe}$ in the P_{Pre} site and AcPhe-tRNA in the P_{Post} site. (C) Differences in the protection patterns of $tRNA^{Phe}$ before and after translocation. (D) Differences in the protection patterns of deacylated $tRNA^{Phe}$ in the P_{Pre} site and in the A site [A(deac)]. (Adapted from Dabrowski et al., 1998.)

tic binding site for the tRNA. However, this position fulfils all the criteria ascribed to the E site by the Wintermeyer group (Table 1): (i) loose and weak binding causing an easy dissociation and (ii) no codon-anticodon interaction at the site, since the anticodon is about 50 Å from the anticodon of the adjacent tRNA at the P site.

The essential difference to the foregoing results is the conventional buffer system that was used for the preparation of the cryo-EM sample in contrast to the polyamine system applied in the other experiments. Therefore, we prepared a PRE and a POST state in vitro with an mRNA 46 nucleotides long with two defined codons, AUG-UUC, in the middle. The

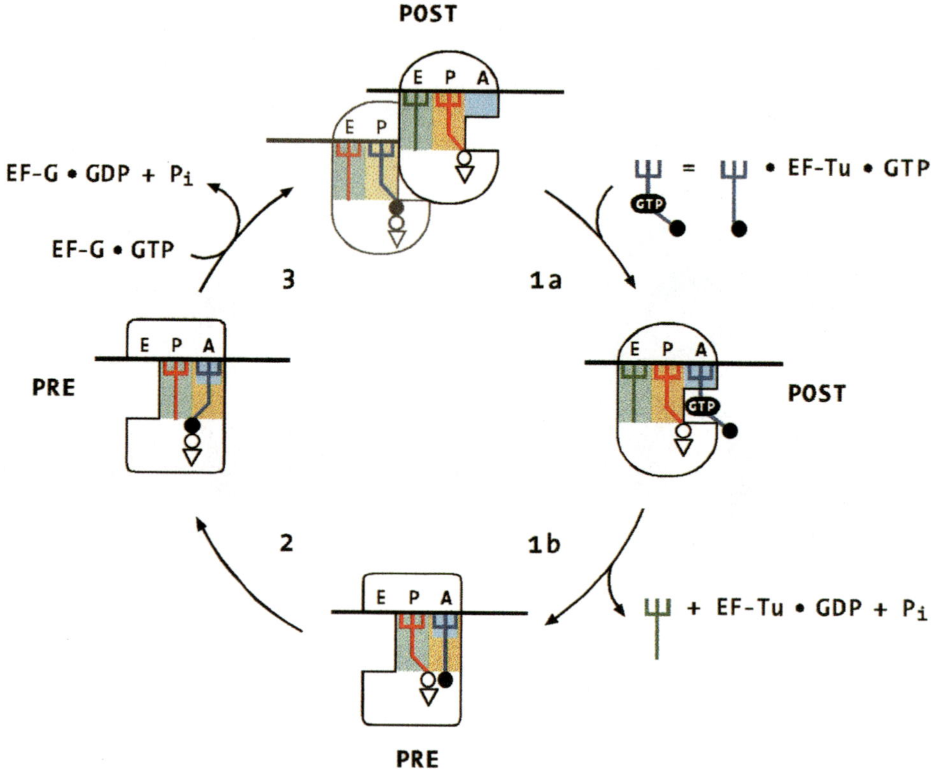

Figure 6. The α-ε model of the elongation cycle. E, P, and A mark the respective tRNA binding sites. The essential feature is a movable ribosomal α-ε domain that connects both subunits through the intersubunit space, binds both tRNAs of an elongating ribosome, and carries them from the A and P sites to the P and E sites, respectively, during translocation. The model keeps all the features of the allosteric three-site model (Fig. 1B) but explains the reciprocal linkage between the A and E sites by the fact that the α-ε domain moves out of the A site during translocation, leaving the decoding center alone at the A site, rather than by an allosteric coupling. Yellow and green, the two binding regions of the α-ε domain; blue, the decoding center at the A site.

position of the tRNA at the E site found in the POST state (Fig. 8B) differed sharply from that in the previous analysis (Fig. 8A), which we termed E2 in order to avoid confusion.

The E site now observed has all the features of a site in that it is embedded in ribosome structure (Fig. 8C and D) and thus is in agreement with the intense contact pattern all over the tRNA in the E site observed by the phosphothioate technique (Fig. 5C). The mutual arrangement of the tRNAs in P and E is almost identical to that in A and P, coinciding with the results from the neutron-scattering analysis and fulfilling the expectations of the α-ε model. Simultaneous codon-anticodon interaction of the tRNAs at the P and E sites of the POST ribosome is possible, as well as interaction of tRNAs at the A and P sites of PRE ribosomes. This "new" E site obviously agrees well with the features ascribed to the E site by our group (Table 1). Wintermeyer and coworkers inferred from fluorescence energy transfer measurements that after translocation the deacylated tRNA shifts from the P site to an "E′ site," where codon-anticodon interaction is still possible, to a site they called "E site" that does not allow codon-anticodon interaction and where the tRNA is bound only in a labile fashion (Paulsen and Wintermeyer, 1986). Obviously, their E′ and E sites are identical to the E and E2 sites, respectively, described here. The essential points are, however, that (i) two "E states" accumulating in the E2 position are found only under conventional buffer systems, whereas only one defined E site is observed in POST states under the conditions of the polyamine buffer and in native polysomes, and (ii) the features Wintermeyer and coworkers have ascribed to the E site are related to the E2 position rather than to the E site.

We have seen that five tRNA positions are detected by cryo-EM, viz., the A, P, and E sites, the hybrid site P/E, and the E2 position. A stable E site is not observed under conventional buffer conditions, in contrast to the polyamine system and to native polysomes. The locations P/E and E2 are seen with dea-

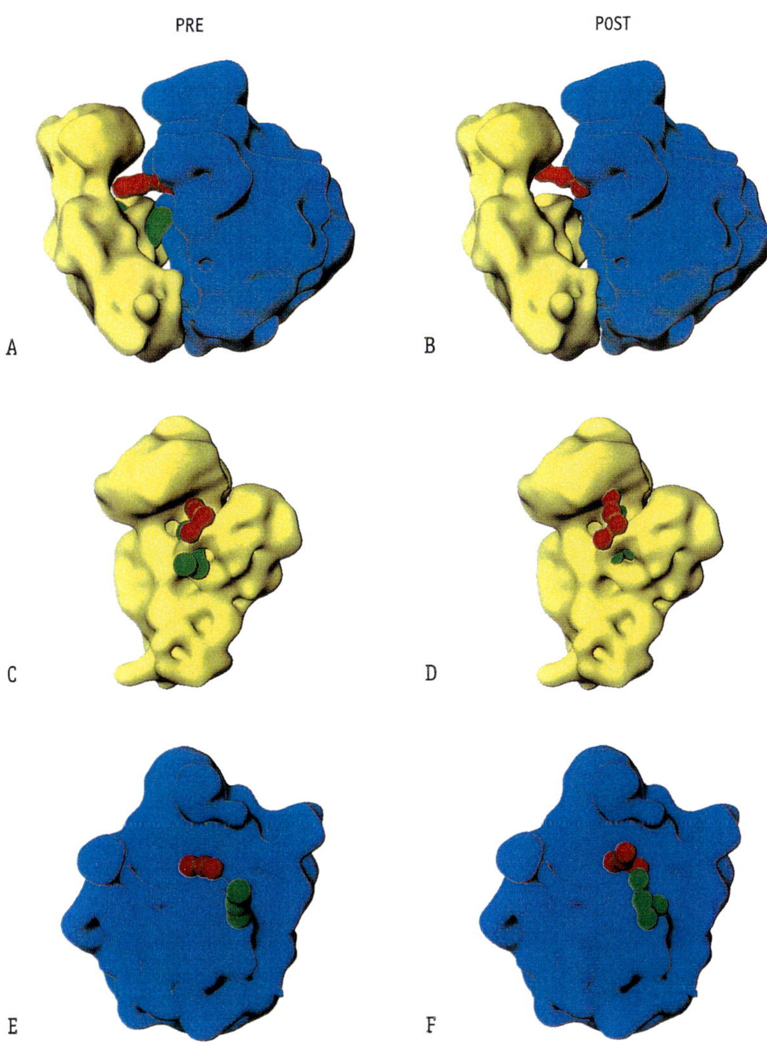

Figure 7. tRNA arrangement in PRE (left-hand panels) and POST (right-hand panels) states. The green and red tRNAs are thought to be at the A and P sites, respectively, of PRE states and at the P and E sites, respectively, of POST states. (A and B) 70S ribosome; (C and D) 30S subunit; (E and F) 50S subunit. (From Nierhaus et al., 1998.)

cylated tRNA under the conditions of conventional buffer systems that do not match the corresponding in vivo concentrations shown in Table 2, whereas these positions are not observed with elongation states in vitro under the polyamine conditions. Therefore, the P/E and E2 positions do not seem to belong to the features firmly established during the elongation cycle. The P/E location is a distinctive and essential element of the hybrid-site model. It follows that the assumption of a tRNA movement only on the 50S subunit after peptide bond formation is not supported by the data described here. Therefore, the α-ε model (in which tRNA moves in toto during translocation) rather than the hybrid-site model (in which tRNA moves on the 50S subunit after peptide bond formation and before translocation) seems to be an appropriate description of the elongation cycle. The distinct sensitivity of the P- and E-site locations for deacylated tRNA and probably also the A site for peptidyl-tRNA calls for the application of near-in vivo conditions for in vitro studies of functional complexes.

The α-ε model seems to reconcile the various observations more easily than the other models. It gives a satisfying explanation of the translocation movements of both tRNAs, whereas the occupation of the A site requires a shift of the α-ε conveyor from the E-P position back to the P-A position. It follows that the A-site occupation rather than the translocation reaction (as assumed hitherto) is the most puzzling reaction of the elongation cycle. This view is consistent with the observation that the A-site occu-

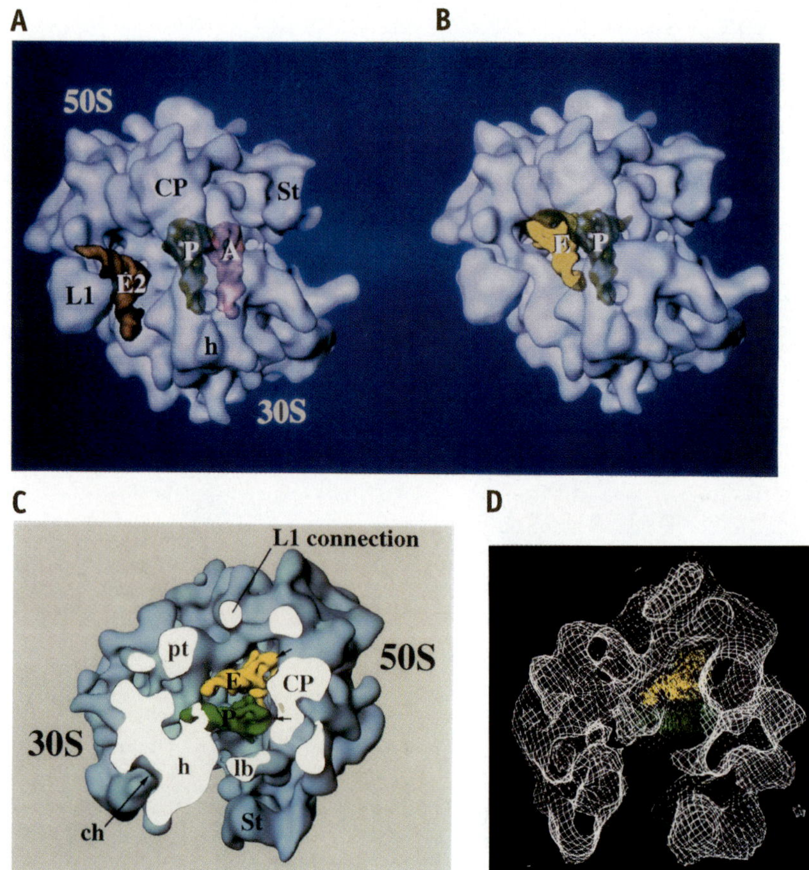

Figure 8. The tRNAs present on the elongating ribosome in the PRE and POST states. (A) tRNA position observed in poly(U)-programmed ribosomes saturated with deacylated tRNAs in the presence of the conventional buffer system. (B) tRNAs in the POST state in the presence of the polyamine system (adapted from Agrawal et al., 1998). (C) Same as panel B, but the top of the 70S ribosome has been cut off. (D) Same as panel B, but a slice of the ribosome is shown with the tRNAs kept at the P and E sites. A, P, and E mark the respective tRNA binding sites, and E2 marks the corresponding tRNA position. Landmarks of the small subunit: ch, tentative mRNA channel through the neck of the 30S subunit; h, head; pt, platform; 1b, part of bridge 1 connecting the subunits. Landmarks of the large subunit: CP, central protuberance; St, L12 stalk.

pation and not the translocation reaction is the rate-limiting step of protein synthesis (Bilgin et al., 1988; Schilling-Bartetzko et al., 1992a).

THE OCCUPATION OF THE A SITE

In the framework of the α-ε model, the A-site occupation is the most puzzling reaction, since the α-ε conveyor has to release the tRNAs present at the P and E sites of the POST state and swing back to the P and E sites. The result is that the E-site-bound deacylated tRNA loses its binding region and leaves the E site and the ribosome possibly via a transient presence at the E2 site, whereas the P-site-bound peptidyl-tRNA changes the binding region from the α to the ε region. This probably happens after decoding, the first step of A-site occupation. Therefore, two factors are mainly responsible for preventing a premature release of the peptidyl-tRNA from the ribosome during the conveyor shift: (i) the presence of the nascent peptide chain in the tunnel, in particular if the nascent chain is longer than a pentapeptide (erythromycin impairs the fixation of the growing chain in the tunnel and provokes a release preferentially of oligopeptides with a length of less than 5 amino acid residues [Tanaka and Teraoka, 1966]), and (ii) the fixation of the ternary complex by the decoding center at the adjacent downstream codon.

The A-site occupation separates into two steps, viz., the decoding step and the tight binding of the aminoacyl-tRNA. According to the α-ε model, the first step is to the low-affinity state of the A site that restricts the interaction of the ternary complex mainly to codon-anticodon interactions rather than allowing general contacts with the tRNA outside the anticodon or with the elongation factor EF-Tu. The important consequence is that the binding area is reduced to the anticodon region and thus the binding energy is more or less identical with the discrimination energy for the selection process. This feature allows an exploitation of the discrimination potential in a rather short time. Note that the exploitation of the discrimination potential requires the achievement of equilibrium, which is possible in a relatively short time if the binding energy is small but lasts very long if the discrimination energy is only a small fraction of the total binding energy.

On the contrary, if the binding included a strong interaction of the whole aminoacyl-tRNA and EF-Tu during the decoding step, the recognition of the codon-anticodon complex by the decoding center would involve a small fraction of the total binding area and thus the discrimination energy would be a small fraction of the total binding energy. The unfavorable consequence would be that all 40 different ternary complexes in *E. coli* would interfere significantly with the selection process, since EF-Tu, a nondiscriminatory element of the binding process, increases the affinity more than 2 orders of magnitude over that of an aminoacyl-tRNA outside a ternary complex (Schilling-Bartetzko et al., 1992b). In support of this view it has been shown that a high-affinity A site (E site not occupied) can bind even a noncognate ternary complex and that its amino acid is incorporated into the growing peptide chain, whereas a low-affinity A site (E site occupied) did not incorporate the amino acid from a noncognate ternary complex (Geigenmüller and Nierhaus, 1990). Amino acids from noncognate ternary complexes (about 90% of the various ternary complexes) are not detectably incorporated in vivo but rather amino acids derived from near-cognate ternary complexes (about four to six ternary complexes) are incorporated. The reciprocal linkage between the A and E sites explains this important feature of protein synthesis (for further discussion, Nierhaus, 1990). In agreement with this view, it has been demonstrated by means of cryo-EM that the ternary complex interacts with only the anticodon of the A site (Stark et al., 1997b; Agrawal et al., unpublished).

The question remains whether the selection of the cognate ternary complex from about six complexes requires a GTP-consuming proofreading mechanism in which EF-Tu is directly involved (a "proofreading factor" [Tapio and Kurland, 1986]). Selecting the correct cognate ternary complex from about six complexes is roughly as complex as the transcription process, where 1 of 4 nucleotides has to be selected. Without proofreading, transcription achieves an accuracy of better than 1 in 50,000 (this means less than one misincorporation during the incorporation of 50,000 nucleotides [Libby et al., 1989]). Discrimination of tRNA seems to happen via the recognition of the partial Watson-Crick structure rather than being determined by the stability of the codon-anticodon interaction (Potapov et al., 1995). Accordingly, this procedure should make available discrimination energy sufficient for the accuracy of about 1 in 3,000 observed in the ribosomal selection process (Bouadloun et al., 1983). In fact, it has been shown with the barely cleavable GTP analogue GTPγS that, in principle, the ribosome can select ternary complexes with a precision of better than 1 in 1,000 without needing a GTP cleavage. This precision was due to the dissociation rate (k_{-1}) of the ternary complexes (Thompson and Karim, 1982). It is conceivable that the first step, the decoding process, is fast compared with the second step, the high-affinity binding of the aminoacyl-tRNA to the A site, since the latter requires a conformational change seen in the movement of the α-ε conveyor from the P-E sites to the A-P site position. If so, the first step runs under equilibrium conditions even under steady-state protein synthesis. This mechanism would exploit the full discrimination potential of the dissociation rates (k_{-1}) of the cognate versus near-cognate ternary complexes during the initial binding of the decoding step (Nierhaus, 1990). Finally, the crystal structure of the ternary complex has demonstrated that EF-Tu is more than 50 Å from the anticodon (Nissen et al. 1995). This fact makes it prohibitively difficult for EF-Tu to regulate and control the decoding process. For all these reasons we consider it unlikely that an active proofreading mechanism like that in synthetases (Jakubowski, 1994) or in replicases (Echols and Goodman, 1991) exists in ribosomes at all.

However, a passive kinetic proofreading might exist in ribosomes. The basis could be the following. Assume a competition between the cognate ternary complex and a near-cognate ternary complex where the on rate is the same for both complexes but the off rate is 50 times higher for the near-cognate complex (although an off rate more than 1,000 times higher for the near-cognate complex has been observed [see preceding paragraph]; we assume 50 only to make the point). This would mean an accuracy of 1 in 50 for the initial binding step. After the decoding step, the second step of the A-site occupation is initiated. The decoding center fixes the codon-anticodon interaction and then allows the high-affinity binding of the EF-Tu moiety of the ternary complex, providing the energy for the backswing of the α-ε conveyor into the A-P position. For this movement the decoding center has to partially detach from the codon-anticodon complex, possibly allowing a dissociation of the near-cognate complex, again with a 50-times-larger off rate. This kinetic proofreading step improves the accuracy to up to 1 in 2,500 (50 times 50). The tRNA at the E site will be released, since this site has lost its binding capacity, and the EF-Tu-dependent GTPase will be triggered, resulting in the release of EF-Tu·GDP. The dissociation of EF-Tu·GDP enables a tight binding of the aminoacyl-tRNA at the A site and a docking of the aminoacyl residue in the A-site region of the peptidyltransferase (PTF) center. Now peptide bond formation can occur.

PEPTIDE-BOND FORMATION

The central enzymatic activity of the ribosome is the formation of the peptide bond that is formed at an active center on the large ribosomal subunit, the PTF center (Monro, 1967; Maden et al., 1968). The catalysis has two aspects, physical and chemical. (i) The largest effect of a catalytic acceleration of a reaction rate is due to the template effect, viz., the stereochemically optimal positioning of the two reactants that is often achieved by binding the transition state. (ii) A chemical component can be added to the reaction sequence if the active center is actively participating, being transiently linked to a reactant by a covalent bond. A famous example is the family of serine proteases, where a triad of essential amino acids is responsible for the chemical catalysis: nucleophilic seryl residue breaks the peptide bond via an attack on the carbonyl group of the peptide bond, thus forming an acyl-enzyme intermediate (seryl ester). A His-Asp system functions as a proton-accepting-and-donating system, displacing the seryl residue by a nucleophilic attack on, e.g., an amine in a second step. This method of cleaving a peptide bond is obviously so favorable that the active center has been developed independently at least twice during evolution, as exemplified by the bacterial subtilisin and the chymotrypsin of higher organisms. Since all enzymes catalyze both directions of a reaction, depending on the starting conditions, a mechanism for peptide bond formation has been suggested that follows the mechanism of the serine proteases by replacing the seryl ester with the ester bond of a peptidyl-tRNA. A number of observations are in agreement with such a mechanism (see Nierhaus et al., 1980, for a discussion; Rychlik and Cerna, 1980).

Different affinity-labeling approaches were applied to identify components at or near the PTF center (Cooperman, 1980; Sonenberg et al., 1973; Wower et al., 1995). As a topographical method, affinity labeling cannot directly identify the component actually involved in the enzymatic activity. Another approach is the single-omission test of the reconstitution technique, in which all but one of the ribosomal components were used for the reconstitution of a particle that was subsequently tested for peptide bond formation. A related strategy is the identification of the minimal set of ribosomal components that could reconstitute an active particle from *E. coli* components. These reconstitution strategies identified the components 23S rRNA, L2, L3, and L4 as possible candidates for this activity (Schulze and Nierhaus, 1982; Franceschi and Nierhaus, 1990). Recently this set of components has been further reduced to 23S rRNA, L2, and L3 by rigorously washing the 50S subunit of *Thermus aquaticus* ribosomes and analyzing the activities of peptide bond formation of the remaining cores (Khaitovich et al., 1999). A recent report that naked 23S rRNA is able to perform this activity (Nitta et al., 1998) could not be reproduced (Nitta et al., 1999). Therefore, the elusive activity of the large ribosomal subunit has to be sought within these three components. The most probable candidates are 23S rRNA and the L2 protein, since L3 is not present in all ribosomes, in contrast to L2, which belongs to the most conserved ribosomal proteins (Müller and Wittman-Liebold, 1997).

A direct involvement of 23S rRNA could be manifested for P-site substrates of the PTF center. The universally conserved G2252 forms a canonical base pair with the third-to-last nucleotide (C) of the universally conserved CCA 3′ ends of tRNAs (Samaha et al., 1995). A double mutation changing G2252-G2253 is still significantly active in peptide-bond formation (Lieberman and Dahlberg, 1994), and therefore, the defects of the PTF center hardly explain the dominant-lethal effect of the mutation in G2252. Rather, a mutation of this and neighboring nucleotides provokes a conformational change of the tertiary structure of domain V, also impairing tRNA binding and translational fidelity (Gregory and Dahlberg, 1999). The nucleotide G2252 is in domain V but not within the "PTF ring" of this domain. The ring comprises about 40 nucleotides, many of which are universally conserved, and it has been implicated in or near the PTF center by various experimental strategies (see Spahn et al., 1996a, for an overview). Another nucleotide that has been implicated in peptide bond formation is the universally conserved bulged G2581 just next to the ring. Mutation of this nucleotide has a dominant-lethal phenotype and exclusively abolishes peptide bond formation in systems specific for this activity of ribosomes but does not impair ribosome assembly, tRNA binding to P or A sites, and translocation, as demonstrated in a systematic functional analysis. Evidence that this nucleotide, possibly together with its adjacent Ψ2580 and G2582, fixes the second-to-last nucleotide of the

CCA 3′ ends of tRNAs was reported, but canonical base pairing could be excluded (Spahn et al., 1996b). These nucleotides, among a total of 18 positions, were also identified as important for peptide bond formation by a random-mutagenesis study of the "Southern" half of the PTF ring (Porse and Garrett, 1995).

A nucleotide neighboring to the A-site substrate of the PTF center could be identified on the 23S rRNA. A 4-thio-dT-pCp-puromycin was cross-linked to 50S subunits and thereafter was able to form a peptide bond with the P-site substrate CACCA-AcPhe in the presence of 33% methanol; the cross-linked nucleotide was G2553 (Green et al., 1998).

The protein L2 was also analyzed according to the "serine-protease" hypothesis outlined above. The highly conserved His229 was mutagenized, and indeed, a loss of PTF activity was observed (Cooperman et al., 1995). The conclusions from this reconstitution analysis were supported by an in vivo study in which archaebacterial as well as human L2 could be incorporated into active *E. coli* ribosomes. The incorporation worked equally well when the heterologous L2 carried a mutated His229 residue, but the corresponding 50S subunits were inactive in protein synthesis (Uhlein et al., 1998). In contrast, it was suggested that His229 of L2 is not essential in *Saccharomyces cerevisiae* mitochondria, because a His229Gln mutation behaves normally at 30°C. However, a conditionally lethal phenotype (strong respiratory deficiency and 40%-reduced protein synthesis at 18°C) is an indication of the importance of this histidine residue.

The apparent discrepancy was reconciled in a recent systematic mutagenesis study of L2 (Diedrich et al., submitted). Mutations of His229 specifically disturbed peptide bond formation but not other functions. Although peptide bond formation was abolished in systems specific for that activity, overall elongation assays, such as poly(Phe) synthesis, showed a significant activity of about 60%, possibly indicating that the template function of the ribosome still allows significant activity in the complete ribosomal system, although the chemical catalysis of peptide bond formation is abolished. In other words, low-efficiency protein synthesis would be possible without chemical catalysis involved in peptide bond formation. Two exceptions are known that do not have a His residue at position 229: (i) L2 of *Mycoplasma capricolum*, containing a Glu residue at the corresponding position, and (ii) the permissive situation with Gln229 in yeast mitochondria. If protein synthesis in these two systems is much less efficient than in *E. coli* and in almost all bacteria, archaebacteria, and the cytoplasmic systems of higher organisms, peptide bond formation might not require a chemical catalysis in addition to the template function. Furthermore, the mutation His229Ala hardly affects the binding of PTF substrates such as CACCA-AcPhe (P site) and CACCA-Phe (A site) (Diedrich et al., submitted). Since this mutation severely impairs peptide bond formation, these observations support the view that His229 is an active element of the PTF.

We thank Joachim Frank and Rajendra Agrawal for help and discussion.

REFERENCES

Agrawal, R., P. Penczek, A. Malhotra, R. Grassucci, I. Gabashvili, A. Heagle, S. Srivastava, M. Burkhardt, R. Jünemann, K. Nierhaus, and J. Frank. 1998. Binding positions of tRNAs in translating *Escherichia coli* ribosomes p. 717–718. *In Proceedings of the 14th International Congress of Electron Microscopy*.

Agrawal, R. K., P. Penczek, R. A. Grassucci, N. Burkhardt, K. H. Nierhaus, and J. Frank. 1999. Effect of buffer conditions on the position of tRNA on the 70S ribosome as visualized by cryo-electron microscopy. *J. Biol. Chem.* **274:**8723–8729.

Agraval, R. K., N. Burkhardt, K. H. Nierhaus, and J. Frank. Unpublished data.

Bilgin, N., L. A. Kirsebom, M. Ehrenberg, and C. G. Kurland. 1988. Mutations in ribosomal proteins L7/L12 perturb EF-G and EF-Tu functions. *Biochimie* **70:**611–618.

Bouadloun, F., D. Donner, and C. G. Kurland. 1983. Codon-specific missense errors *in vivo*. *EMBO J.* **2:**1351–1356.

Burkhardt, N., R. Jünemann, C. M. T. Spahn, and K. H. Nierhaus, 1998. Ribosomal tRNA binding sites: three-site models of translation. *Crit. Rev. Biochem. Mol. Biol.* **33:**95–149.

Cooperman, B. 1980. Photolabile antibiotics as probes of ribosomal structure and function. *Ann. N.Y. Acad. Sci.* **346:**302–323.

Cooperman, B. S., T. Wooten, D. P. Romero, and R. Traut. 1995. Histidine 229 in the protein L2 is apparently essential for 50S peptidyl transferase activity. *Biochem. Cell Biol.* **73:**1087–1094.

Dabbs, E. R. 1991. Mutants lacking individual ribosomal proteins as a tool to investigate ribosomal properties. *Biochimie* **73:**639–645.

Dabrowski, M., C. M. T. Spahn, and K. H. Nierhaus. 1995. Interaction of tRNAs with the ribosome at the A and P sites. *EMBO J.* **14:**4872–4882.

Dabrowski, M., C. M. T. Spahn, M. A. Schäfer, S. Patzke, and K. H. Nierhaus. 1998. Contact patterns of tRNAs do not change during ribosomal translocation. *J. Biol. Chem.* **273:**32793–32800.

Diedrich, G., C. Spahn, M. A. Schäfer, B. Cooperman, and K. H. Nierhaus. Ribosomal protein L2 is involved in the association of the ribosomal subunits, tRNA binding to A and P sites and peptidyl transferase. Submitted for publication.

Echols, H., and M. F. Goodman. 1991. Fidelity mechanisms in DNA replication. *Annu. Rev. Biochem.* **60:**477–511.

Franceschi, F., and K. H. Nierhaus. 1990. Ribosomal proteins L15 and L16 are mere late assembly proteins of the large ribosomal subunit. *J. Biol. Chem.* **265:**16676–16682.

Geigenmüller, U., and K. H. Nierhaus. 1990. Significance of the third tRNA binding site, the E site, on *E. coli* ribosomes for the accuracy of translation: an occupied E site prevents the binding of non-cognate aminoacyl-transfer RNA to the A site. *EMBO J.* **9:**4527–4533.

Gnirke, A., and K. H. Nierhaus. 1986. tRNA binding sites on the subunits of Escherichia coli ribosomes. *J. Biol. Chem.* **261:** 14506–14514.

Gnirke, A., U. Geigenmüller, H.-J. Rheinberger, and K. H. Nierhaus. 1989. The allosteric three-site model for the ribosomal elongation cycle. *J. Biol. Chem.* **264:**7291–7301.

Grajevskaja, R. A., Y. V. Ivanov, and E. S. Saminsky. 1982. 70S ribosomes of *Escherichia coli* have an additional site for deacylated tRNA. *Eur. J. Biochem.* **128:**47–52.

Green, R., E. Switzer, and H. F. Noller. 1998. Ribosome-catalyzed peptide-bond formation with an A-site substrate covalently lnked to 23S ribosomal RNA. *Science* **280:**286–289.

Gregory, S., and A. Dahlberg. 1999. Mutations in the conserved P loop perturb the conformation of two structural elements in the peptidyl transferase center of 23S ribosomal RNA. *J. Mol. Biol.* **285:**1475–1483.

Horsfield, J. A., D. N. Wilson, S. A. Mannering, F. M. Adamski, and W. P. Tate, 1995. Prokaryotic ribosomes recode the HIV-1 gag-pol-1 frameshift sequence by an E/P site post-translocation simultaneous slippage mechanism. *Nucleic Acids Res.* **23:**1487–1494.

Jakubowski, H. 1994. Energy cost of translational proofreading *in vivo*. The aminoacylation of transfer RNA in *Escherichia coli*. *Ann. N.Y. Acad. Sci.* **745:**4–20.

Kamekura, M., K. Hamana, and S. Matsuzaki. 1987. Polyamine contents and amino acid decarboxylation activities of extremely halophilic achaebacteria and some eubacteria. *FEMS Microbiol. Lett.* **43:**301–305.

Khaitovich, P., A. Mankin, R. Green, L. Lancaster, and H. Noller. 1999. Characterization of functionally active subribosomal particles from *Thermus aquaticus*. *Proc. Natl. Acad. Sci. USA* **96:** 85–90.

Kirillov, S. V., E. M. Makarov, and Y. P. Semenkov. 1983. Quantitative study of interaction of deacylated tRNA with E. coli ribosomes. Role of 50S subunits in formation of the E. site. *FEBS Lett.* **157:**91–94.

Knop, W., M. Hirai, H.-J. Schink, H. B. Stuhrmann, R. Wagner, J. Zhao, O. Schärpf, R. R. Crichton, M. Krumpolc, K. H. Nierhaus, A. Rijllart, and T. O. Niinikoski. 1992. A new polarized target for neutron scattering studies on biomolecules: first results from apoferritin and the deuterated 50S subunit of ribosomes. *J. Appl. Crystallogr.* **25:**155–165.

Lewicki, B., and K. H. Nierhaus. Unpublished data.

Libby, R. T., J. L. Nelson, J. M. Calvo, and J. A. Gallant. 1989. Transcriptional proofreading in *Escherichia coli*. *EMBO J.* **8:** 3153–3158.

Lieberman, K. R., and A. E. Dahlberg. 1994. The importance of conserved nucleotides of 23 S ribosomal RNA and transfer RNA in ribosome catalysed peptide bond formation. *J. Biol. Chem.* **269:**16163–16169.

Lill, R., J. M. Robertson, and W. Wintermeyer. 1984. tRNA binding sites of ribosomes from Escherichia coli. *Biochemistry* **23:** 6710–6717.

Lusk, J. E., R. J. P. Williams, and E. P. Kennedy. 1968. Magnesium and the growth of *Escherichia coli*. *J. Biol. Chem.* **243:**2618–2624.

Maden, B., R. Traut, R., and R. Monro. 1968. Ribosome-catalysed peptidyl transfer: the polyphenylalanine system. *J. Mol. Biol.* **35:** 333–345.

Moazed, D., and H. F. Noller. 1989. Intermediate states in the movement of transfer RNA in the ribosome. *Nature* **342:**142–148.

Moazed, D., and H. F. Noller. 1990. Binding of tRNA to the ribosomal A and P sites protects two distinct sets of nucleotides in the 16S rRNA. *J. Mol. Biol.* **211:**135–145.

Monro, R. 1967. Catalysis of peptide bond formation by 50S ribosomal subunits from *Escherichia coli*. *J. Mol. Biol.* **26:**147–151.

Müller, E. C., and B. Wittman-Liebold. 1997. Phylogenetic relationship of organisms obtained by ribosomal protein comparison. *Cell. Mol. Life Sci.* **53:**34–50.

Nierhaus, K. H. 1990. The allosteric three-site model for the ribosomal elongation cycle: features and future. *Biochemistry* **29:** 4997–5008.

Nierhaus, K. H., H. Schulze, and B. S. Cooperman. 1980. Molecular mechanisms of the ribosomal peptidyltransferase centre. *Biochem. Int.* **1:**185–192.

Nierhaus, K. H., D. Beyer, M. Dabrowski, M. A. Schäfer, C. M. T. Spahn, J. Wadzack, K.-U. Bittner, N. Burkhardt, G. Diedrich, R. Jünemann, D. Kamp, H. Voss, and H. B. Stuhrmann. 1995. The elongating ribosome: structural and functional aspects. *Biochem. Cell Biol.* **73:**1011–1021.

Nierhaus, K. H., R. Jünemann, and C. M. T. Spahn. 1997. Are the current three-site models valid descriptions of the ribosomal elongation cycle? *Proc. Natl. Acad. Sci. USA* **94:**10499–10500.

Nierhaus, K. H., J. Wadzack, N. Burkhardt, R. Jünemann, W. Meerwinck, R. Willumeit, and H. B. Stuhrmann. 1998. Structure of the elongating ribosome: arrangement of the two tRNAs before and after translocation. *Proc. Natl. Acad. Sci. USA* **95:** 945–950.

Nissen P., M. Kjeldgaard, S. Thirup, G. Polekhina, L. Reshetnikova, B. F. C. Clark, and J. Nyborg. 1995. Crystal structure of the ternary complex of Phe-tRNAPhe, EF-Tu, and a GTP analog. *Science* **270:**1464–1472.

Nitta, I., Y. Kamada, H. Noda, T. Ueada, and K. Watanabe. 1998. Reconstitution of peptide bond formation with *Escherichia coli* 23S ribosomal RNA domains. *Science* **281:**666–669.

Nitta, I., Y. Kamada, H. Noda, T. Ueada, and K. Watanabe. 1999. Peptide bond formation: retraction. *Science* **283:** 2019–2020.

Noller, H. F., D. Moazed, S. Stern, T. Powers, P. N. Allen, J. M. Robertson, B. Weiser, and K. Triman. 1990. Structure of rRNA and its functional interactions in translation, p. 73–92. *In* W. Hill, A. Dahlberg, R. A. Garrett, P. B. Moore, D. Schlessinger, and J. R. Warner (ed.), *The Ribosome: Structure, Function, and Evolution*. American Society for Microbiology, Washington, D.C.

Paulsen, H., and W. Wintermeyer. 1986. tRNA topography during translocation: steady-state and kinetic fluorescence energy-transfer studies. *Biochemistry* **25:**2749–2756.

Porse, B. T., and R. A. Garrett 1995. Mapping important nucleotides in the peptidyl transferase centre of 23 S rRNA using a random mutagenesis approach. *J. Mol. Biol.* **249:**1–10.

Potapov, A. P., F. J. Triana-Alonso, and K. H. Nierhaus. 1995. Ribosomal decoding processes at codons in the A or P sites depend differently on 2′-OH groups. *J. Biol. Chem.* **270:**17680–17684.

Purohit, P., and S. Stern. 1994. Interactions of a small RNA with antibiotic and RNA ligands of the 30S subunit. *Nature* **370:**659–662.

Remme, J., T. Margus, R. Villems, and K. H. Nierhaus. 1989. The third ribosomal tRNA-binding site, the E site, is occupied in native polysomes. *Eur. J. Biochem.* **183:**281–284.

Rheinberger, H. J. 1991. The function of the translating ribosome: allosteric three-site model of elongation. *Biochimie* **73:**1067–1088.

Rheinberger, H.-J., and K. H. Nierhaus. 1980. Simultaneous binding of the 3 tRNA molecules by the ribosome of E. coli. *Biochem. Int.* **1:**297–303.

Rheinberger, H.-J., and K. H. Nierhaus. 1986. Allosteric interactions between the ribosomal transfer RNA-binding sites A and E. *J. Biol. Chem.* **261:**9133–9139.

Rheinberger, H.-J., H. Sternbach, and K. H. Nierhaus. 1981. Three tRNA binding sites on E. coli ribosomes. *Proc. Natl. Acad. Sci. USA* **78:**5310–5314.

Rheinberger, H.-J., U. Geigenmüller, A. Gnirke, T. P. Hausner, J. Remme, H. Saruyam, and K. H. Nierhaus. 1990. Allosteric three-site model for the ribosomal elongation cycle, p. 318–330. *In* W. E. Hill, A. E. Dahlberg, R. A. Garrett, P. B. Moore, D. Schlessinger, and J. R. Warner (ed.), *The Ribosome: Structure, Function, and Evolution.* American Society for Microbiology, Washington, D.C.

Rychlik, I., and J. Cerna. 1980. Peptidyl transferase—involvement of histidine in substrate binding and peptide bond formation. *Biochem. Int.* 1:193–200.

Samaha, R. R., R. Green, and H. F. Noller. 1995. A base pair between tRNA and 23S rRNA in the peptidyl transferase center of the ribosome. *Nature* **377:**309–314.

Schäfer, M. A., S. Patzke, and K. H. Nierhaus. Unpublished data.

Schatz, D., R. Leberman, and F. Eckstein. 1991. Interaction of *Escherichia coli* $tRNA^{Ser}$ with its cognate aminoacyl-tRNA synthetase as determined by footprinting with phosphorothioate-containing tRNA transcripts. *Proc. Natl. Acad. Sci. USA* **88:** 6132–6136.

Schilling-Bartetzko, S., A. Bartetzko, and K. H. Nierhaus. 1992a. Kinetic and thermodynamic parameters for transfer RNA binding to the ribosome and for the translocation reaction. *J. Biol. Chem.* **267:**4703–4712.

Schilling-Bartetzko, S., F. Franceschi, H. Sternbach, and K. H. Nierhaus. 1992b. Apparent association constants of transfer RNAs for the ribosomal A-site, P-site, and E-site. *J. Biol. Chem.* **267:**4693–4702.

Schulze, H., and K. H. Nierhaus. 1982. Minimal set of ribosomal components for the reconstitution of the peptidyltransferase activity. *EMBO J.* 1:609–613.

Sonenberg, N., M. Wilchek, and A. Zamir. 1973. Mapping of *Escherichia coli* ribosomal components involved in peptidyl tranferase activity. *Proc. Natl. Acad. Sci. USA* **70:**1423–1426.

Spahn, C. M. T., and K. H. Nierhaus. 1998. Models of the elongation cycle: an evaluation. *Biol. Chem.* **379:**753–772.

Spahn, C. M. T., J. Remme, M. A. Schäfer, and K. H. Nierhaus. 1996a. Mutational analysis of two highly conserved UGG sequences of 23 S rRNA from *Escherichia coli. J. Biol. Chem.* **271:** 32849–32856.

Spahn, C. M. T., M. A. Schäfer, A. A. Krayevsky, and K. H. Nierhaus. 1996b. Conserved nucleotides of 23 S rRNA located at the ribosomal peptidyltransferase center. *J. Biol. Chem.* **271:** 32857–32862.

Stark, H., E. V. Orlova, J. Rinke-Appel, N. Junke, F. Mueller, M. Rodnina, W. Wintermeyer, R. Brimacombe, and M. vanHeel. 1997a. Arrangement of tRNAs in pre- and posttranslocational ribosomes revealed by electron cryomicroscopy. *Cell* **88:**19–28.

Stark, H., M. V. Rodnina, J. Rinke-Appel, R. Brimacombe, W. Wintermeyer, and M. van Heel. 1997b. Visualization of elongation factor Tu on the *Escherichia coli* ribosome. *Nature* **389:** 403–406.

Stuhrmann, H. B., N. Burkhardt, G. Diedrich, R. Jünemann, W. Meerwinck, M. Schmitt, J. Wadzack, R. Willumeit, J. Zhao, and K. H. Nierhaus. 1995. Proton- and deuteron spin targets in biological structure research. *Nucleic Instr. Methods Phys. Res.* **356:**124–132.

Tabor, C. W., and H. Tabor. 1985. Polyamines in microorganisms. *Microbiol. Rev.* **49:**81–99.

Tanaka, K., and H. Teraoka. 1966. Binding of erythromycin to *Escherichia coli* ribosomes. *Biochim. Biophys. Acta* **114:**204–206.

Tapio, S., and C. G. Kurland. 1986. Mutant EF-Tu increases missense error *in vitro. Mol. Gen. Genet.* **205:**186–188.

Thompson, R. C., and A. M. Karim. 1982. The accuracy of protein biosynthesis is limited by its speed: high fidelity selection by ribosomes of aminoacyl-tRNA ternary complexes containing GTP[γ S]. *Proc. Natl. Acad. Sci. USA* **79:**4922–4926.

Triana-Alonso, F. J., K. Chakraburtty, and K. H. Nierhaus. 1995. The elongation factor 3 unique in higher fungi and essential for protein biosynthesis is an E site factor. *J. Biol. Chem.* **270:** 20473–20478.

Ühlein, M., W. Weglöhner, H. Urlaub, and B. Wittmann-Liebold. 1998. Ribosomal protein L2 is essential for protein biosynthesis: replacement of E. coli L2 with the archaebacterial and human homologues in vivo. *Biochem. J.* 331:423–430.

Wadzack, J., N. Burkhardt, R. Jünemann, G. Diedrich, K. H. Nierhaus, J. Frank, P. Penczek, W. Meerwinck, M. Schmitt, R. Willumeit, and H. B. Stuhrmann. 1997. Direct localization of the tRNAs within the elongating ribosome by means of neutron scattering (proton-spin contrast-variation). *J. Mol. Biol.* **266:**343–356.

Wintermeyer, W., R. Lill, and J. M. Robertson. 1990. Role of the tRNA exit site in ribosomal translocation, p. 348–357. *In* W. Hill, A. Dahlberg, R. A. Garrett, P. B. Moore, D. Schlessinger, and J. R. Warner (ed.), *The Ribosome: Structure, Function, and Evolution.* American Society for Microbiology, Washington, D.C.

Wower, J., I. K. Wower, S. V. Kirillov, K. V. Rosen, S. S. Hixson, and R. A. Zimmermann. 1995. Peptidyl transferase and beyond. *Biochem. Cell Biol.* 73:1041–1047.

The Ribosome: Structure, Function, Antibiotics, and Cellular Interactions
Edited by R. A. Garrett, S. R. Douthwaite, A. Liljas, A. T. Matheson, P. B. Moore, and H. F. Noller

Chapter 27

Ternary Complex of EF-Tu and Its Action on the Ribosome

GREGERS R. ANDERSEN, VICTOR G. STEPANOV, MORTEN KJELDGAARD, SØREN S. THIRUP, and JENS NYBORG

The elongation phase of protein biosynthesis is the central function of the ribosome. Here the real synthesis of proteins in the correct sequence of amino acids is performed on the basis of genetic information transcribed into the mRNA. The basic functional details of the elongation phase are known from textbooks. However, it is fair to say that even today little is known about the detailed functional steps involved and even less about the structural details lying behind these steps. It is the general impression that the last decade has provided much more information on probable models for ribosomal RNAs (Mueller and Brimacombe, 1997a, 1997b; Mueller et al., 1997), as well as the crystal and solution structures of a number of ribosomal proteins (Liljas and Garber, 1995; Ramakrishnan and White, 1998). Much structural information on the tRNA synthetases has also been gathered (Cavarelli and Moras, 1993), to the point where one now begins to grasp the details of how both tRNAs and amino acids are recognized specifically. A detailed structure of both ribosomal subunits will most likely be available within a few years (Ban et al., 1998; Yonath and Franceschi, 1998). Nevertheless, the massive amount of work needed in order to understand in structural detail the ribosomal function during elongation is daunting.

It is generally accepted that the ribosome has three functional sites. One is the A site, where the aminoacyl-tRNA (aa-tRNA) is located after the correct decoding of mRNA by codon-anticodon recognition. This site is deep inside the ribosome and most likely puts the incoming amino acid close to the peptidyl transfer center of the ribosome. Peptide bond formation and peptidyl transfer could well happen right after the aa-tRNA is placed in the A site. Another site is the P site, where peptidyl-tRNA is located after translocation of tRNAs and mRNA. The third site is the E site, where deacylated tRNA is placed in order to be able to leave the ribosome. A possible fourth site is the testing or factor site, where elongation factors interact with the ribosome to test whether a codon-anticodon match of the ternary complex of EF-Tu can be made or where EF-G·GTP can induce the translocation of tRNAs and mRNA. It is known from previous work (Moazed and Noller, 1989) and from other chapters in this book that this overall textbook picture is much too simple. It is easy to predict that the elongation phase involves a number of functional and structural states of which we only have a glimpse at the moment. It is easy to foresee that understanding the structural transitions between such states will be of utmost importance in order to describe the ribosomal function in detail.

In this chapter we describe the advances made in the structural studies of elongation factor EF-Tu during the last decade. It will be shown that the structural transition between the active, aa-tRNA binding form and the inactive form of EF-Tu is surprisingly large. It will further be demonstrated that the structure of the ternary complex of EF-Tu is astonishingly similar to the structure of the inactive form of elongation factor EF-G. If this macromolecular mimicry tells us nothing else, it at least strongly suggests that EF-G when released from the ribosome leaves behind it an imprint on the ribosome ready to accept a ternary complex (Liljas, 1996). The function and structural results of EF-Tu have been described in a number of recent reviews (Krab and Parmeggiani, 1998; Liljas and Al-Karadaghi, 1997; Nyborg and Kjeldgaard, 1996; Nyborg and Liljas, 1998).

Gregers R. Andersen, Victor G. Stepanov, Morten Kjeldgaard, Søren S. Thirup, and Jens Nyborg ▪ Institute of Molecular and Structural Biology, University of Aarhus, Gustav Wieds Vej 10C, DK-8000 Aarhus C, Denmark.

CONFORMATIONAL CHANGE OF EF-Tu

Most of the various functional states of EF-Tu have been illustrated by structural results over the last few years. Crystal structures of the inactive EF-Tu·GDP have been obtained from *Escherichia coli* (Abel et al., 1996; Polekhina et al., 1996) and from *Thermus aquaticus* (Polekhina et al., 1996). A recent higher-resolution structure of EF-Tu from *E. coli* generally confirms the previous results (Song et al., 1999). Structures of the active EF-Tu·GDPNP, where GDPNP is a nonhydrolyzable analogue of GTP, have been determined from *T. thermophilus* (Berchtold et al., 1993) and from *T. aquaticus* (Kjeldgaard et al., 1993). A structural model of EF-Tu from *Bacillus stearothermophilus* has been put forward (Krásny et al., 1998).

Some of these structures are shown in Fig. 1. All of these studies show that EF-Tu is a three-domain structure, where domain 1, about 200 amino acid residues, is a nucleotide binding domain with a central six-stranded β-sheet surrounded by five to six α-helices. The β-sheet is mostly composed of parallel β-strands (one β-strand at one end of the sheet is antiparallel), and the nucleotide is bound near the C-terminal end of the sheet. The β-phosphate is held in place by a P loop and the positive end of a helical dipole, but the nucleotide is generally exposed to the solvent. All of these structural features can be found in many other nucleotide binding proteins, although the specific structural details and the consensus sequences found in loops around the nucleotide binding site are special to G proteins (Kjeldgaard et al., 1996). Domains 2 and 3, about 100 amino acid residues each, are both β-barrels, with the barrel axes almost perpendicular to each other. These are found in all known structures in the same relative orientation to each other and apparently function as one structural unit.

When the two structures of EF-Tu·GDP and EF-Tu·GDNP (Fig. 1) are compared, it is obvious that the two structural units of EF-Tu (domain 1 is one unit, and domains 2 and 3 together are the other) have vastly different orientations to each other. Domain 1 can be seen as rotating about 90° on the surface of domains 2 and 3. This alteration is accompanied by a relative translation. Such a large conformational change can only be accomplished if there is a transient dissociation and reassociation of the two structural units. Elongation factor EF-Ts catalyzes the nucleotide exchange such that EF-Tu·GDP is reactivated into EF-Tu·GTP. In the structures of the EF-Tu·EF-Ts complex, it is obvious that part of the function of EF-Ts is to separate the two structural units from each other (Kawashima et al., 1996; Wang et al., 1997). A similar structural change, but in the opposite direction, must happen on the ribosome during the stimulation of GTP hydrolysis. Is that structural change catalyzed by the ribosome?

On domain 1 there are two regions which are influenced by the state of the nucleotide. These are the so-called switch regions I and II. Switch region I of EF-Tu·GDP contains a β-hairpin which changes into a short α-helix in EF-Tu·GDPNP (Abel et al., 1996; Polekhina et al., 1996). Switch region II contains an α-helix in both forms of EF-Tu. This helix is found between two β-strands. However, the structural change is such that the helix shifts 4 amino acid residues along the sequence so that in EF-Tu·GDP there is a short loop between the end of the α-helix

Figure 1. Structures of EF-Tu. To the left is shown EF-Tu·GDP, in the middle EF-Tu in the EF-Tu·EF-Ts complex, and to the right EF-Tu·GDPNP. Domain 1 is yellow, domains 2 and 3 are green, and switch region I is red. The magnesium ion is shown as a gray ball. Note the differences in the relative orientations of the two parts of EF-Tu and of the structural change of switch region I as a function of the nature of the nucleotide. (The figure was produced with the program MOLSCRIPT [Kraulis, 1991].)

and the second β-strand while in EF-Tu·GDPNP there is a short loop between the first β-strand and the beginning of the α-helix. This results in a rotation of the helix axis of about 45° (Berchtold et al., 1993; Kjeldgaard et al., 1993).

Very recently we have finished the refinement of the structure of bovine mitochondrial EF-Tu·GDP at a resolution of 1.94 Å in a collaboration with Linda Spremulli, University of North Carolina (Andersen, unpublished). This structure is very similar to that of bacterial EF-Tu (Fig. 2). However, it has some features which resemble those of eukaryotic EF-1α. This is especially true for a 10-amino-acid C-terminal extension. The higher resolution allows a more accurate description of the whole structure but especially that of the nucleotide binding site.

NUCLEOTIDE EXCHANGE REACTION

The structure of the nucleotide exchange complex of EF-Tu·EF-Ts is now known from *E. coli* (Kawashima et al., 1996) and from *T. thermophilus* (Wang et al., 1997). Very recently, the solution structure of a fragment of the human elongation factor EF-1β has been determined (Pérez et al., 1999). This structure shows features which are similar to those of the bacterial elongation factor EF-Ts. Recently we have finished refinement of EF-Tu·EF-Ts from *E. coli* at 2.2-Å resolution (Thirup, unpublished).

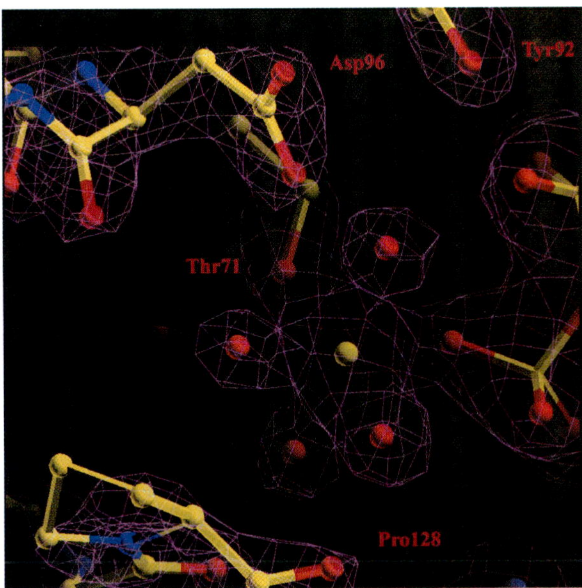

Figure 2. Details of nucleotide binding to mitochondrial EF-Tu·GDP. The electron density of the coordination of the Mg^{2+} ion of this 1.94-Å-resolution structure is shown, which indicates the quality of the structural model obtained. (The figure was produced with the program O [Jones et al., 1991].)

The folds of EF-Ts in the complexes from *E. coli* and *T. thermophilus* are similar (Fig. 3). However, the functional form of EF-Ts from *T. thermophilus* is a homodimer, while the functional form of EF-Ts from *E. coli* is a monomer. In the structure of *E. coli* EF-Ts, which has a longer sequence, a pseudo-structural symmetry can be recognized. In both functional forms there is a central β-sandwich, which holds together the two parts. Also, in both forms there is an N-terminal domain, which contacts domain 1 of EF-Tu. EF-Ts from *T. thermophilus*, because of its internal symmetry, exposes at the same time two possible interaction surfaces. Therefore, the complex in the crystal structure is seen as an EF-Tu·EF-Ts dimer.

Strangely enough the *E. coli* EF-Ts monomer exposes an α-helical hairpin near its C terminus, which interacts with a similar helical hairpin in another EF-Ts molecule (Kawashima et al., 1996). Thus, a dimer is also seen in the crystal structure of *E. coli* EF-Tu·EF-Ts but with an overall organization which is very different from the one seen for the *T. thermophilus* complex (Fig. 3).

One part of the function of EF-Ts is to separate the two structural units of EF-Tu to allow for the structural alteration of the switch regions of EF-Tu. Another is to interact with the loops of EF-Tu involved in nucleotide binding. Virtually every such loop is contacted, and its local structure is altered in order to influence the binding affinity of GDP or GTP. There is no reason to believe that the removal of Mg^{2+} alone is a central event in the nucleotide exchange reaction as claimed earlier (Kawashima et al., 1996).

TERNARY COMPLEX OF EF-Tu AND MACROMOLECULAR MIMICRY

Some years ago we determined the crystal structure of yeast Phe-tRNA in complex with *T. aquaticus* EF-Tu·GDPNP (Nissen et al., 1995). It has been shown that the structure is similar in solution (Bilgin et al., 1998). Although it is not a native biological complex, it was proposed that the structure represented the canonical structure of all ternary complexes. This is supported by the fact that all tRNAs have the same basic structure and that EF-Tu and EF-1α from various organisms have sequences with very little variation in length and many conserved amino acids. It was suggested at the time that the ternary complex could be a vehicle for obtaining the structures of some of the interesting tRNAs. We have obtained crystals for a number of such ternary complexes, but the structure determinations have

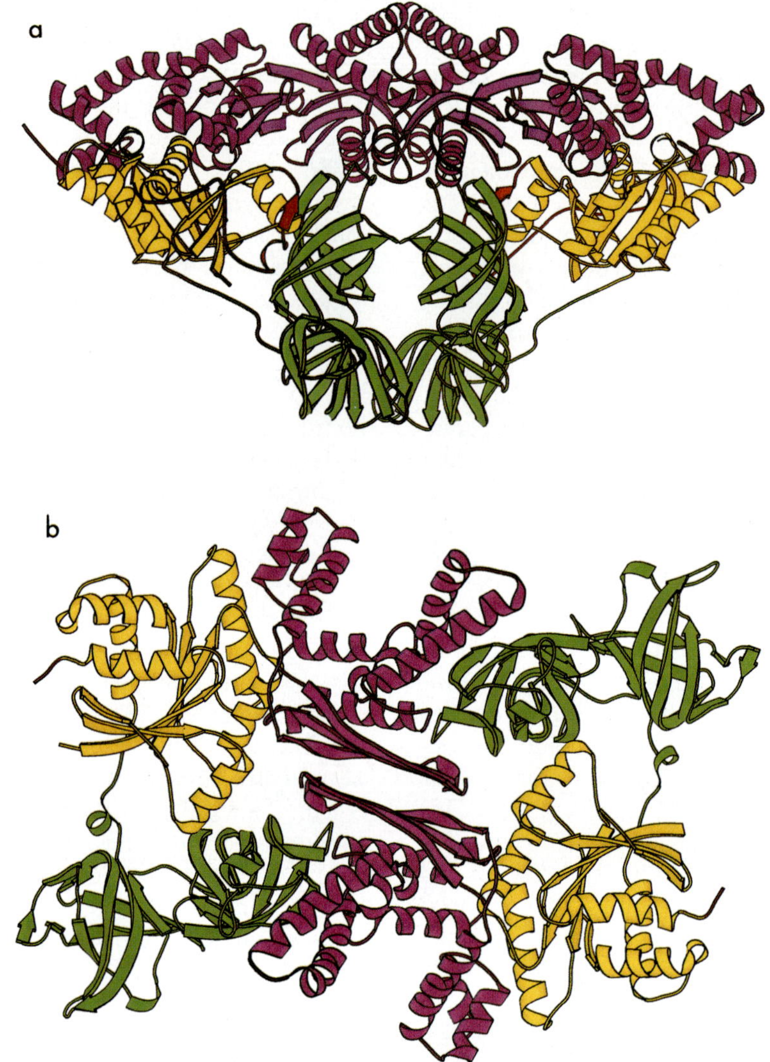

Figure 3. The EF-Tu·EF-Ts complexes from *E. coli* (a) and *T. thermophilus* (b). EF-Ts is shown in magenta, while the coloring of EF-Tu is as in Fig. 1. The *E. coli* structure is shown with a vertical pseudo-twofold symmetry axis in the plane of the paper. Note that the two molecules of EF-Tu are on the same side of the pseudodimer of EF-Ts. The *T. thermophilus* structure is shown with a twofold symmetry axis perpendicular to the plane of the paper. Here the two molecules of EF-Tu are on opposite sides of the true homodimer of EF-Ts. Note also that the homodimer of this complex is about half the size of the pseudodimer of EF-Ts in the *E. coli* structure. (The figure was produced with the program MOLSCRIPT [Kraulis, 1991].)

been somewhat hampered by the fact that tRNAs are difficult to isolate and purify on a large scale. Recently, the structure of the ternary complex of *E. coli* Cys-tRNA and *T. aquaticus* EF-Tu·GDPNP was determined (Nissen et al., 1999). Both structures have very similar overall shapes (Fig. 4).

The two parts of the protein-nucleic acid complex are very little distorted compared to the structures of the free components (Jack et al., 1976; Kjeldgaard et al., 1993). Only one side of the T-stem helix, the CCA end, and the 5′ phosphate of the tRNA are in direct contact with EF-Tu, while the part with the anticodon on the tip is pointing away from EF-Tu. This gives a very elongated complex with a maximum length of about 115 Å. The interaction between the surface of domain 3 and the T-stem helix of tRNA is nonspecific, and nonconserved residues of EF-Tu contact the backbone structure of tRNA. A comparison of the two known structures of ternary complexes indicates that this binding surface can accommodate small local variations of the tRNA structure. The 5′ phosphate is specifically recognized in a small binding pocket formed at the corner where all three domains of EF-Tu meet. A small number of conserved residues from the three domains are involved in this recognition. A shallow cleft between

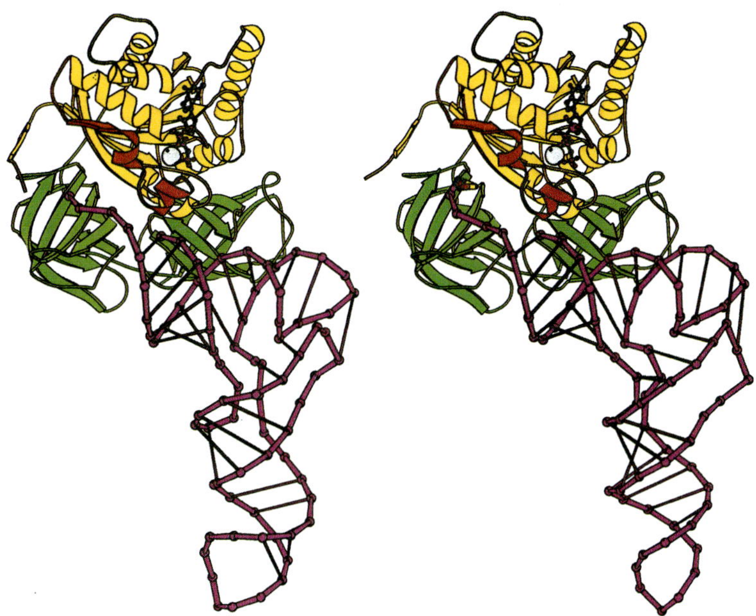

Figure 4. Ternary complexes of EF-Tu. Shown are the ternary complexes of *Saccharomyces cerevisiae* Phe-tRNA, *T. aquaticus* EF-Tu, and GDPNP (left) and of *E. coli* Cys-tRNA, *T. aquaticus* EF-Tu, and GDPNP (right). The coloring of EF-Tu is the same as in Fig. 1, and tRNAs are shown in magenta. Note the overall similarity of the structures of the two ternary complexes. However, in the structure of Cys-tRNA the angle between the two parts of the L-shape is slightly larger, and the major groove of the helix in contact with EF-Tu is a little wider. (The figure was produced with the program MOLSCRIPT [Kraulis, 1991].)

domains 1 and 2 binds the CCA end. Only the terminal A and the constant part of the amino acid ester are specifically recognized. On the surface of domain 2 is a pocket with conserved or semiconserved hydrophobic residues on one side and a conserved Glu making stacking interactions with the plane of the A base. The same Glu also makes a hydrogen bond to the 2′-OH of the ribose. The phosphate interacts with a conserved Arg, and the amino acid ester is held in place by backbone hydrogen bonds. The amino acid side chain is found in a pocket large enough to hold every amino acid.

When the structure of the ternary complex was compared to the structure of EF-G·GDP, it was immediately obvious that they have the same overall shape (Nissen et al., 1995). In EF-G, domains 1 and 2 are very similar to the corresponding domains of EF-Tu, although the relative orientation of the domains of the inactive EF-G·GDP is the same as that of the active EF-Tu·GDPNP. Domains 3, 4, and 5 of EF-G apparently mimic the shape of a tRNA molecule, with domain 4 mimicking the anticodon helix. If one assumes that translation in early evolutionary time was dominated by RNA, it is tempting to speculate that the first use of the structurally and functionally much more versatile proteins was to replace some of the RNA. The success of such a macromolecular mimicry may be reflected in the fact that such a replacement in elongation has survived evolutionary pressure. Much discussion has since been centered around other interpretations of this mimicry. However, with the unfortunate lack of structural information on EF-G·GTP, any further implications are hard to substantiate. It has not been our impression, although it was argued otherwise (Moore, 1995), that the conformational change of EF-G necessarily has to be as large as that for EF-Tu. This is mainly because the structural changes of the switch regions of EF-Tu are seen as the driving force for its large structural change, and domain 3 of EF-Tu, which is most influenced by this change, has no counterpart in EF-G. Moreover, there is no need for a nucleotide exchange factor for EF-G, indicating that there is no need for catalysis of a larger structural change.

It has been suggested that part of IF2 and the release factors RF1 and RF2 could also be mimicking tRNA (Ito et al., 1996; Nissen et al., 1995).

ELONGATION FACTOR INTERACTIONS WITH THE RIBOSOME

The modes by which EF-G and the ternary complex of EF-Tu interact with the ribosome have been elegantly demonstrated by cryo-electron microscopy (EM) reconstructions (Agrawal et al., 1998; Stark et al., 1997). Specific states of the ribosome have been trapped by the use of antibiotics, as in many biochemical investigations. The structure of the ternary complex on the ribosome is blocked with kirromycin. It has to be stressed that in these experiments the hydrolyzable GTP was used. Thus, it is entirely possible that the ribosome has induced GTP hydrolysis and EF-Tu is complexed to GDP but that the normal structural change and release of EF-Tu·GDP is blocked by kirromycin. The extra density seen compared to that of factor-free ribosomes is clearly very similar to the crystal structure of the ternary complex (Stark et al., 1997). However, there is a slight bending of the tRNA part. At the moment, we are refining the structure of a complex of kirromycin and the ternary complex of EF-Tu with GDPNP as a nucleotide (Kristensen et al., 1996). Although this refinement has not been completed, and the detailed interactions between kirromycin and EF-Tu have not been determined, it was clear at an early stage of refinement that kirromycin binds at a cleft between domains 1 and 3, most likely in contact with some of the residues known to give kirromycin resistance to EF-Tu (Abdulkarim et al., 1994; Kraal et al., 1995). Furthermore, the relative positions of the two domains are slightly altered, resulting in an overall bending of the tRNA very similar to the one seen in the cryo-EM reconstructions (Stark et al., 1997).

The cryo-EM reconstruction of EF-G·GTP inhibited by fusidic acid on the ribosome is very interesting (Agrawal et al., 1998). Also, in this case it could well be that GTP is hydrolyzed to GDP but that EF-G is kept by the antibiotic in a slightly altered GTP-like conformation. If this assumption holds, then this reconstruction provides a first low-resolution model for the structure of EF-G·GTP. When the extra density was compared to the known structures of EF-G·GDP, it was observed that the tRNA mimicking part of EF-G on the ribosome is bent such that EF-G obtains a more open conformation. Nevertheless, it was seen to occupy the same position as the anticodon helix of tRNA in the A site of the 30S subunit. Thus, the function of EF-G could be to physically force the tRNA out of the A site (Abel and Jurnak, 1996).

Such speculations inevitably bring into mind the question of why EF-Tu must undergo such a drastic conformational change. It is certainly not necessary in order to release the tRNA from the ternary complex. A small movement of domains 2 and 3 away from domain 1 should be sufficient for such a release. Likewise, it is hard to see why a smaller change would not disrupt the interactions between EF-Tu and the ribosome. Based on the results of cryo-EM reconstructions, these interactions are between domain 1 and the 50S subunit and between domain 2 and the 30S subunit (Agrawal et al., 1998; Stark et al., 1997). It is interesting that, based on sequence comparisons, IF2 and release factors seem to have domains similar to domains 1 and 2 of EF-Tu and EF-G (Ævarsson, 1995). This suggests a common initial binding to the ribosome and contact with its GTPase center for all the translation factors which are controlled by GTP. So is there an as-yet-unknown reason for the large conformational change of EF-Tu? One rather obvious but unsupported suggestion is that some of the energy of the GTPase reaction can be used through the conformational change to physically push aa-tRNA into the A site (or a proofreading site). In order for this to happen, EF-Tu must go through some structural states which are not known from the crystal structures.

GTPase REACTION

The mechanism of the GTPase reaction of EF-Tu has been exeedingly difficult to pin down (for a recent review, see Krab and Parmeggiani, 1998). Part of the reason for this is that EF-Tu is not a strong hydrolyzing enzyme on its own. It has been suspected for some time that a His residue in switch region II is involved, although mutagenesis studies showed very early that it was not an essential part of the mechanism (Cool and Parmeggiani, 1990). In other G proteins, such as ras p21 and the heterotrimeric G_α, a Gln is found at a similar position, and it has been shown to have a similar effect on the GTPase reaction when mutated. In elegant structural studies with AlF_4^- as a transition analogue of the reaction, it was shown that for G_α, most likely this Gln and an Arg from switch I are stabilizing the transition state (Coleman et al., 1994; Sondek et al., 1994). In ras p21 the corresponding residue of switch I is not an Arg, but as shown in the complex with its GAP, an "Arg finger" is provided by GAP and occupies a position which overlaps well with the Arg found internally in G_α (Scheffzek et al., 1997). The problem in comparing these results with EF-Tu is that, although the Arg of switch I is present in EF-Tu, it has been found that in all known structures it is far away from the γ-phosphate. Furthermore, the GTPase reaction of EF-Tu is delayed, as for ras p21, and is strongly induced by the ribosome. It is not known at present whether the ribosome provides an Arg finger or whether the interaction with the ribosome deforms the switch I region of EF-Tu to such an extent that its own Arg is brought into contact with the γ-phosphate. It has been shown very convincingly that the γ-phosphate itself activates the catalytic water (Schweins et al., 1995). Such an activation is likely to be the same for all G proteins.

PERSPECTIVES

Within the last 6 years we have obtained a rather complete structural picture of many of the important functional states of the elongation phase of bacterial protein biosynthesis. What can we expect from the future structural work in this field? Is it time to pack up and turn to other biologically interesting subjects? In our opinion we can easily find work to do for at least the next 6 years. One of the very obvious pieces of the puzzle still missing is the structure of EF-G·GTP. Why is this so difficult to obtain? One possible explanation is that although this structure does not have to be very different from the structure of EF-G·GDP, as mentioned above, it is sufficiently different to be more structurally flexible by having a looser interaction between domains 1 and 2. Is it a possibility that EF-Tu·GDP is so flexible that it can oscillate between two structural states, one which has now been observed several times and therefore is the most energetically favorable (Abel et al., 1996; Andersen, unpublished; Polekhina et al., 1996; Song et al., 1999) and another one which is more like the GTP form? This at least could explain why a weak ternary complex with EF-Tu·GDP can be observed (Pingoud et al., 1982).

The nucleotide exchange mechanism of EF-Ts is not well understood in structural terms. We know the structure of the nucleotide-free complex of EF-Tu·EF-Ts, but we do not know the structure of the intact free form of EF-Ts, although the structure of a fragment of EF-Ts from *T. thermophilus* has been determined (Jiang et al., 1996). Would it be possible somehow to trap EF-Tu·GDP·EF-Ts or EF-Tu·GDPNP·EF-Ts in a form that would allow their structures to be determined? Is the fact that EF-Tu·EF-Ts has been observed as two very different dimers in the crystals of the complexes from *E. coli* and *T. thermophilus* just a curious artifact, or does it have any biological relevance? Why is EF-Tu·EF-Ts from mitochondria more stable than the complexes from bacteria? The first structural information on eukaryotic elongation factor EF-1β indicates similarity to bacterial EF-Ts (Pérez et al., 1999). Are all the eukaryotic elongation factors similar to the bacterial ones?

Is the ternary complex of EF-Tu really a vehicle to obtain the structures of more tRNAs? If so it would be interesting to get the structures of noncanonical tRNAs. Among these are tRNAs with a long variable arm and the minimal tRNAs of mitochondria, some of which are missing the D arm. Very strangely many tRNAs from the mitochondria of a nematode are missing the T arm (Watanabe et al., 1994), which from the crystal structure is seen to be one of the interacting parts in the ternary complex. Is the macromolecular mimicry seen in the comparison of EF-G·GDP with the ternary complex just a strange coincidence or does it have real functional meaning? If so, will it be found also in initiation and release factors? The GTP hydrolysis reactions of the G proteins of translation are all highly stimulated during interaction with the ribosome. Do we really need to get a high-resolution structure of the complete ribosomal particle with a ternary complex in order to get structural information about this stimulation?

If all the structural biologists interested in the elongation phase of protein biosynthesis press on for the next 10 years or so, perhaps some of these very relevant biological questions will be answered, and in the same period many more will be raised.

We are extremely grateful to Ludmila S. Reshetnikova for her continued help with crystallization experiments relevant to many projects with translation factors. We are equally grateful for the huge efforts on crystallizations and structure determinations performed by Poul Nissen, Galina Polekhina, and Ole Kristensen when they were still members of our laboratory. Brian F. C. Clark has provided valuable support for our work even at times when progress was still in the future.

We acknowledge the continued support from the special Program for Biotechnological Research of the Danish Natural Science Research Council. Financial support for G.R.A. from the Carlsberg Foundation is gratefully acknowledged. Part of this work has been supported by the EU project on Protein Synthesis and Antibiotics under contract BIO4-CT97-2188.

REFERENCES

Abdulkarim, F., L. Liljas, and D. Hughes. 1994. Mutations to kirromycin resistance occur in the interface of domains I and III of EF-Tu·GTP. *FEBS Lett.* **352:**118–122.

Abel, K., and F. Jurnak. 1996. A complex profile of protein elongation: translating chemical energy into molecular movement. *Structure* **4:**229–238.

Abel, K., M. D. Yoder., R. Hilgenfeld, and F. Jurnak. 1996. An α to β conformational switch in EF-Tu. *Structure* **4:**1153–1159.

Ævarsson, A. 1995. Structure-based sequence alignment of elongation factors Tu and G with related GTPases involved in translation. *J. Mol. Evol.* **41:**1096–1104.

Agrawal, R. K., P. Penczek, R. A. Grassucci, and J. Frank. 1998. Visualization of the elongation factor G on the *Escherichia coli* 70S ribosome: the mechanism of translocation. *Proc. Natl. Acad. Sci. USA* **95:**6134–6138.

Andersen, G. R. 1999. Unpublished data.

Ban, N., B. Freeborn, P. Nissen, P. Penczek, R. A. Grassucci, R. Sweet, J. Frank, P. B. Moore, and T. A. Steitz. 1998. A 9 Å resolution X-ray crystallographic map of the large ribosomal subunit. *Cell* **93:**1105–1115.

Berchtold, H., L. Reshetnikova, C. O. A. Reiser, N. K. Schirmer, M. Sprinzl, and R. Hilgenfeld. 1993. Crystal structure of active elongation factor Tu reveals major domain rearrangements. *Nature* **365**:126–132.

Bilgin, N., M. Ehrenberg, C. Ebel, G. Zaccai, Z. Sayers, M. H. J. Koch, D. Svergun, C. Barberato, V. Volkov, P. Nissen, and J. Nyborg. 1998. The solution structure of the ternary complex between aminoacyl-tRNA, EF-Tu and GTP. *Biochemistry* **37**: 8163–8172.

Cavarelli, J., and D. Moras. 1993. Recognition of tRNAs by aminoacyl-tRNA synthetases. *FASEB J.* **7**:79–86.

Coleman, D. E., A. M. Berghuis, E. Lee, M. E. Linder, A. G. Gilman, and S. R. Sprang. 1994. Structures of active conformations of $G_i\alpha_1$ and the mechanism of GTP hydrolysis. *Science* **265**: 1405–1412.

Cool, R. H., and A. Parmeggiani. 1990. Substitution of His84 and the GTPase mechanism of elongation factor Tu. *Biochemistry* **30**:362–366.

Ito, K., K. Ebihara, M. Uno, and Y. Nakamura. 1996. Conserved motifs in prokaryotic and eukaryotic polypeptide release factors: tRNA-protein mimicry hypothesis. *Proc. Natl. Acad. Sci. USA* **93**:5443–5448.

Jack, A., J. E. Ladner, and A. Klug. 1976. Crystallographic refinement of yeast phenylalanine transfer RNA at 2.5Å resolution. *J. Mol. Biol.* **108**:619–649.

Jiang, Y., S. Nock, M. Nesper, M. Sprinzl, and P. B. Sigler. 1996. Structure and importance of the dimerization domain in elongation factor Ts from *Thermus thermophilus*. *Biochemistry* **35**: 10269–10278.

Jones, T. A., S. Cowan, J.-Y. Zou, and M. Kjeldgaard. 1991. Improved methods for building protein models in electron density maps and the location of errors in these models. *Acta Crystallogr. Sect. A* **47**:110–119.

Kawashima, T., C. Berthet-Colominas, M. Wulff, S. Cusack, and R. Leberman. 1996. The structure of the *Escherichia coli* EF-Tu:EF-Ts complex at 2.5 Å resolution. *Nature* **379**:511–518.

Kjeldgaard, M., P. Nissen, S. Thirup, and J. Nyborg. 1993. The crystal structure of elongation factor EF-Tu from *Thermus aquaticus* in the GTP conformation. *Structure* **1**:35–50.

Kjeldgaard, M., J. Nyborg, and B. F. C. Clark. 1996. The GTP-binding motif—variations on a theme. *FASEB J.* **10**:1347–1368.

Kraal, B., L. A. Zeef, J. R. Mesters, K. Boon, E. L. Vorstenbosch, L. Bosch, P. H. Anborgh, A. Parmeggiani, and R. Hilgenfeld. 1995. Antibiotic resistance mechanisms of mutant EF-Tu species in Escherichia coli. *Biochem. Cell Biol.* **73**:1167–1177.

Krab, I. M., and A. Parmeggiani. 1998. EF-Tu, a GTPase odyssey. *Biochim. Biophys. Acta* **1443**:1–22.

Krásny, L., J. R. Mesters, L. N. Tieleman, B. Kraal, V. Fucík, R. Hilgenfeld, and J. Jonák. 1998. Structure and expression of elongation factor Tu from *Bacillus stearothermophilus*. *J. Mol. Biol.* **283**:371–381.

Kraulis, P. J. 1991. MOLSCRIPT: a program to produce both detailed and schematic plots of protein structures. *J. Appl. Crystallogr.* **24**:946–950.

Kristensen, O., L. Reshetnikova, P. Nissen, G. Siboska, S. Thirup, and J. Nyborg. 1996. Isolation, crystallization and X-ray analysis of the quaternary complex of Phe-tRNA^{Phe}, EF-Tu, a GTP analog and kirromycin. *FEBS Lett.* **399**:59–62.

Liljas, A. 1996. Protein synthesis: imprinting through molecular mimicry. *Curr. Biol.* **6**:247–249.

Liljas, A., and S. Al-Karadaghi. 1997. Structural aspects of protein synthesis. *Nat. Struct. Biol.* **4**:767–771.

Liljas, A., and M. Garber. 1995. Ribosomal proteins and elongation factors. *Curr. Opin. Struct. Biol.* **5**:721–727.

Moazed, D., and H. F. Noller. 1989. Intermediate states in the movement of transfer RNA in the ribosome. *Nature* **342**:142–148.

Moore, P. B. 1995. Molecular mimicry in protein synthesis? *Science* **270**:1453–1454.

Mueller, F., and R. Brimacombe. 1997a. A new model for the three-dimensional folding of Escherichia coli 16 S ribosomal RNA. I. Fitting the RNA to a 3D electron microscopic map at 20 A. *J. Mol. Biol.* **271**:524–544.

Mueller, F., and R. Brimacombe. 1997b. A new model for the three-dimensional folding of Escherichia coli 16 S ribosomal RNA. II. The RNA-protein interaction data. *J. Mol. Biol* **271**: 545–565.

Mueller, F., H. Stark, M. van Heel, J. Rinke-Appel, and R. Brimacombe. 1997. A new model for the three-dimensional folding of Escherichia coli 16 S ribosomal RNA. III. The topography of the functional centre. *J. Mol. Biol.* **271**:566–587.

Nissen, P., M. Kjeldgaard, S. Thirup, G. Polekhina, L. Reshetnikova, B. F. C. Clark, and J. Nyborg. 1995. Crystal structure of the ternary complex of Phe-tRNA^{Phe}, EF-Tu, and a GTP analog. *Science* **270**:1464–1472.

Nissen, P., S. Thirup, M. Kjeldgaard, and J. Nyborg. 1999. The crystal structure of Cys-tRNA^{Cys}:EF-Tu:GDPNP reveals general and specific features in the ternary complex and in tRNA. *Structure* **7**:143–156.

Nyborg, J., and M. Kjeldgaard. 1996. Elongation in bacterial protein biosynthesis. *Curr. Opin. Biotechnol.* **7**:369–375.

Nyborg, J., and A. Liljas. 1998. Protein biosynthesis: structural studies of the elongation cycle. *FEBS Lett.* **430**:95–99.

Pérez, J. M. J., G. Siegal, J. Kriek, K. Hård, J. Dijk, G. W. Canters, and W. Möller. 1999. The solution structure of the guanine nucleotide exchange domain of elongation factor 1β reveals a striking resemblance to that of EF-Ts from *Escherichia coli*. *Structure* **7**:217–226.

Pingoud, A., W. Block, A. Wittinghofer, H. Wolf, and E. Fischer. 1982. The elongation factor Tu binds aminoacyl-tRNA in the presence of GDP. *J. Biol. Chem.* **257**:11261–11267.

Polekhina, G., S. Thirup, M. Kjeldgaard, P. Nissen, C. Lippmann, and J. Nyborg. 1996. Helix unwinding in the effector region of elongation factor EF-Tu:GDP. *Structure* **4**:1141–1151.

Ramakrishnan, V., and S. W. White. 1998. Ribosomal protein structures: insights into the architecture, machinery and evolution of the ribosome. *Trends Biochem. Sci.* **23**:208–212.

Scheffzek, K., M. R. Ahmadian, W. Kabsch, L. Wiesmüller, A. Lautwein, F. Schmitz, and A. Wittinghofer. 1997. The RasGAP complex: structural basis for the GTPase activation and its loss in oncogenic mutants. *Science* **277**:333–338.

Schweins, T., M. Geyer, K. Scheffzek, A. Warshel, H. R. Kalbitzer, and A. Wittinghofer. 1995. Substrate-assisted catalysis as a mechanism for GTP hydrolysis of *ras*-p21 and other GTP-binding proteins. *Nat. Struct. Biol.* **2**:36–44.

Sondek, J., D. G. Lambright, J. P. Noel, H. E. Hamm, and P. B. Sigler. 1994. GTPase mechanism of G-proteins from the 1.7 Å crystal structure of transducin α-GDP·AlF_4^-. *Nature* **372**:276–279.

Song, H., M. R. Parsons, S. Rowsell, G. Leonard, and S. E. V. Philips. 1999. Crystal structure of intact elongation factor EF-Tu from *Escherichia coli* in GDP conformation at 2.05 Å resolution. *J. Mol. Biol.* **285**:1245–1256.

Stark, H., M. V. Rodnina, J. Rinke-Appel, R. Brimacombe, W. Wintermeyer, and M. van Heel. 1997. Visualization of elongation factor Tu on the *Escherichia coli* ribosome. *Nature* **389**: 403–406.

Thirup, S. 1999. Unpublished data.

Wang, Y., Y. Jiang, M. Meyering-Voss, M. Sprinzl, and P. B. Sigler. 1997. Crystal structure of the EF-Tu:EF-Ts complex from *Thermus thermophilus*. *Nat. Struct. Biol.* **4**:650–656.

Watanabe, Y., H. Tsurui, T. Ueda, R. Furushima, S. Takamiya, K. Kita, K. Nishikawa, and K. Watanabe. 1994. Primary and higher order structures of nematode (*Ascaris suum*) mitochondrial tRNAs lacking either the T or D stem. *J. Biol. Chem.* **269**: 22902–22906.

Yonath, A., and F. Franceschi. 1998. Functional universality and evolutionary diversity: insights from the structure of the ribosome. *Structure* **6**:679–684.

The Ribosome: Structure, Function, Antibiotics, and Cellular Interactions
Edited by R. A. Garrett, S. R. Douthwaite, A. Liljas, A. T. Matheson, P. B. Moore, and H. F. Noller

Chapter 28

Insights into the GTPase Mechanism of EF-Tu from Structural Studies

ROLF HILGENFELD, JEROEN MESTERS, and TANIS HOGG

1999 marks the 35th anniversary of the discovery of elongation factor T by Fritz Lipmann. Several years after its initial discovery, it was shown that EF-T actually consists of two different proteins, the GTP binding protein EF-Tu and its nucleotide exchange factor, EF-Ts (for references see Lucas-Lenard and Lipmann, 1971). Ever since, EF-Tu has fascinated biochemists and, more recently, structural biologists because of its unique regulatory mechanisms. Even though a wealth of information on the protein has been collected over the years, some of the functions of EF-Tu remain elusive, and the protein still holds several surprises. Thus, few would have predicted that EF-Tu can act as a chaperone (Kudlicki et al., 1997; Caldas et al., 1998), or that it is the target of specific degradation upon bacteriophage infection, thereby resulting in programmed cell death (Georgiou et al., 1998).

The main role of EF-Tu, however, is clearly in the elongation phase of bacterial protein synthesis (for reviews see Abel and Jurnak, 1996, and Krab and Parmeggiani, 1998). The protein transports aminoacylated tRNA (aa-tRNA) molecules to the programmed ribosome and profoundly contributes to an accurate and fast translation of mRNAs into proteins. EF-Tu was the first protein found to be regulated by the binding and subsequent hydrolysis of GTP, making it a paradigm for the superfamily of regulatory GTPases (Hilgenfeld, 1995a). In its active form, bound to GTP·Mg^{2+}, EF-Tu forms a ternary complex with noninitiator aa-tRNA, without exhibiting much discrimination between different species of aa-tRNAs. Upon cognate recognition between the anticodon of the tRNA in the ternary complex and the codon of the mRNA in the ribosomal A site, GTP is hydrolyzed with high efficiency and the binary complex EF-Tu·GDP dissociates from the ribosome. This GTPase activity is several orders of magnitude faster than the intrinsic GTPase displayed by EF-Tu in the absence of ribosomes and aa-tRNA.

Crystallographic studies of EF-Tu have provided a wealth of structural information in the past few years. When we determined the three-dimensional structure at <1.7-Å resolution of *Thermus thermophilus* EF-Tu in complex with GppNHp, a nonhydrolyzable GTP analogue (Berchtold et al., 1993), it came as a surprise how different the mutual arrangement of the three domains of the molecule was from what had been seen previously in the GDP form of the *Escherichia coli* protein (Kjeldgaard and Nyborg, 1992). In its inactive, GDP-bound form, the EF-Tu molecule has an unusual central hole between the GTPase domain (G domain) and domain 2, whereas it is much more compact, with many contacts among all three domains, in the active GTP complex (Fig. 1). This observation was confirmed through the crystal structure of a highly similar EF-Tu from *Thermus aquaticus* at a resolution of 2.5 Å (Kjeldgaard et al., 1993).

The rather loose structure of the GDP complex, originally seen with an EF-Tu that had been proteolytically cleaved at at least two sites, was later validated by X-ray analysis of crystals of intact EF-Tu·GDP (Abel et al., 1996; Polekhina et al., 1996). These studies revealed that upon GTP hydrolysis, part of the presumable effector region of EF-Tu is transformed from an α-helix in the GTP form to a β-hairpin in the GDP form (Fig. 1). This conformational change is much larger than that of the corresponding switch I region in the small GTPase

Rolf Hilgenfeld, Jeroen Mesters, and Tanis Hogg ▪ Institute of Molecular Biotechnology, Beutenbergstr. 11, D-07745 Jena, Germany.

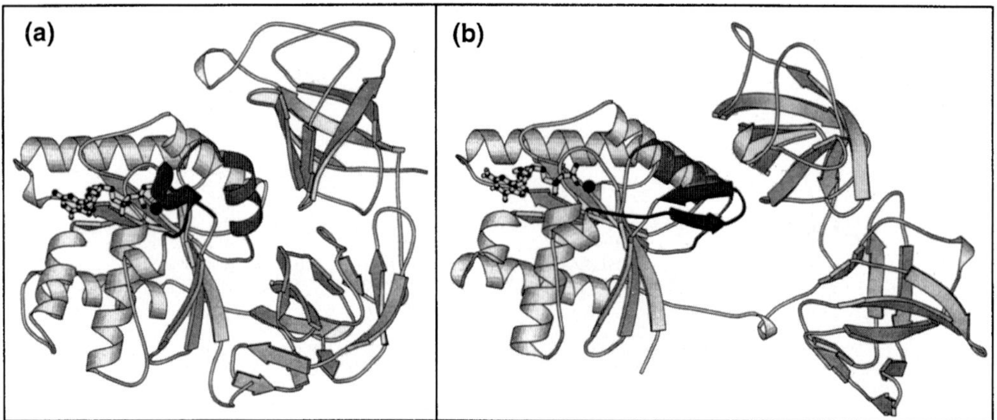

Figure 1. Conformational rearrangements in EF-Tu, as seen in the X-ray crystal structures of EF-Tu·Mg^{2+}·GppNHp (Berchtold et al., 1993) (a) and EF-Tu·Mg^{2+}·GDP (Abel et al., 1996) (b). Global rearrangements are effected by two mobile structural elements, the switch I (residues 40 to 62) and switch II (residues 80 to 100) regions, which are highlighted in dark and medium shading, respectively. The nucleotide and Mg^{2+} ion are rendered in diagrams for each structure.

Ras but was recently observed to be very similar in the Ras-related Ran (Vetter et al., 1999).

OPEN QUESTIONS CONCERNING THE GTPase MECHANISM

In spite of all the structural work, there are a number of open questions concerning the exact mechanism of GTP hydrolysis by EF-Tu and the dramatic rate enhancement of this reaction in the ternary EF-Tu complex with GTP and aa-tRNA on the ribosome. In particular, it is unclear whether the ribosome acts as a GTPase activator in a way similar to the GTPase-activating proteins (GAPs) of small Ras-like GTPases, i.e., by providing an arginine residue ("arginine finger") that stabilizes the transition state of hydrolysis about the γ-phosphorus atom (Scheffzek et al., 1997; Rittinger et al., 1997a, 1997b). Alternatively, ribosomal components could act similarly to the RGS (regulators of G protein signaling) proteins, which enhance the GTPase activity of the α subunit of heterotrimeric G proteins by interacting with their catalytic residues (Arg-178 and Gln-204 in $G_i\alpha$), thereby locking them in a position which is ideal for transition state stabilization (Tesmer et al., 1997).

While the mechanism of the ribosome-mediated GTPase reaction of EF-Tu is thus entirely unclear, we also know very little about the intrinsic GTP-hydrolyzing activity exhibited by the enzyme in the absence of the ribosome. Since we believe that an understanding of the mechanism of this slow GTPase will help to interpret the structural features of EF-Tu in the context of the ribosome, in particular with respect to identifying the interacting partners, we decided to work on structural aspects of the intrinsic GTPase and carried out crystallographic studies on EF-Tus that had key residues modified by site-directed mutagenesis.

The active GTPase site of EF-Tu as seen in its complex with the nonhydrolyzable GTP analogue, GppNHp, is depicted in Fig. 2. Similar to other GTPases, such as Ras and Gα subunits, a water molecule (wat 411) is found at a distance of 3.2 Å from the γ-phosphorus atom. This water is ideally placed to perform a nucleophilic attack on the γ-phosphate moiety, in line with the bond to the leaving group, GDP. Of course, since we used a GTP analogue that carries an imido group in place of the oxygen-bridging β- and γ-phosphates, the reaction does not occur in the present case but is frozen at the state of the Michaelis complex.

It is generally believed that the mechanism of substrate hydrolysis in the GTPases is largely associative, i.e., S_N2-like. On the other hand, there is an overall agreement that in enzyme-free aqueous solution, GTP hydrolysis occurs via a dissociative mechanism (Maegley et al., 1996). Undoubtedly, it is possible that an enzyme alters the mechanism of the reaction which it catalyzes, but in our view, good arguments should be presented to support such a case of deviation from the situation in enzyme-free solution.

The two mechanistic alternatives are displayed in Fig. 3. In a dissociative mechanism, the bond to the leaving group of the reaction, GDP, would be largely broken before much bond formation between the incoming nucleophile and the γ-phosphorus atom would have occurred. The transition state of this reaction would be trigonal planar, i.e., metaphosphate-like. In an associative mechanism, the bond to the

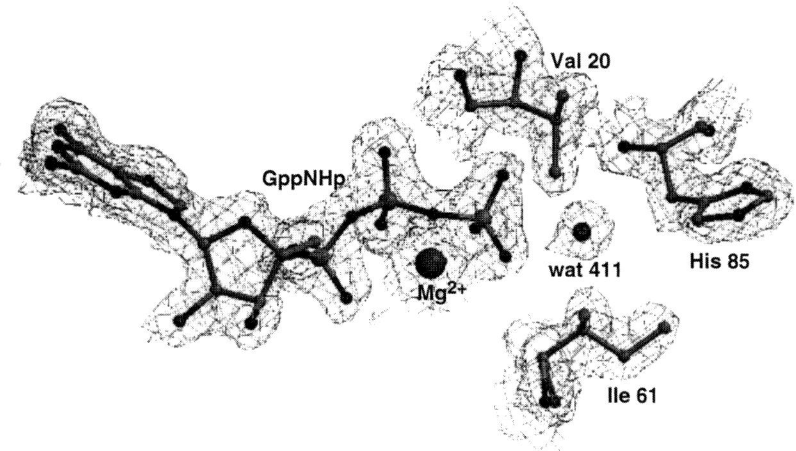

Figure 2. A $2F_o - F_c$ electron density map in the catalytic center of active EF-Tu, contoured at 1.5 σ above the mean. The nucleophilic water (wat) molecule (411) is shielded from His-85 and bulk solvent by the hydrophobic side chains of Val-20 and Ile-61, the so-called hydrophobic gate.

nucleophile would be formed before the bond to the leaving group is broken. This would result in a pentacovalent, trigonal-bipyramidal transition state. It is important to note that in an intermediate of the latter type, there would be an increase of negative charge compared to the ground state, whereas a dissociative mechanism would lead to a reduction of charge in the transition state (Fig. 3). As a consequence, different modes of transition state stabilization are required, depending on the ionic character of the intermediate. The identities of the residues stabilizing the transition state in EF-Tu are largely unclear, and so is the presence or absence of a general base activating the nucleophile.

HISTIDINE-85: GENERAL BASE OR RECOGNITION POINT FOR EF-Ts?

In an associative mechanism, it would be advantageous to have the attacking water molecule deprotonated by a general base in order to increase its nucleophilicity. It has been proposed many times that His-85 of *T. thermophilus* EF-Tu (His-84 in the *E. coli* protein; the *T. thermophilus* numbering scheme is used throughout this chapter) could fulfill this function. However, in the crystal structure of EF-Tu·GppNHp (Berchtold et al., 1993), the side chain of His-85 is more than 5.5 Å away from the γ-phosphate (Fig. 2). In principle, a simple rotation

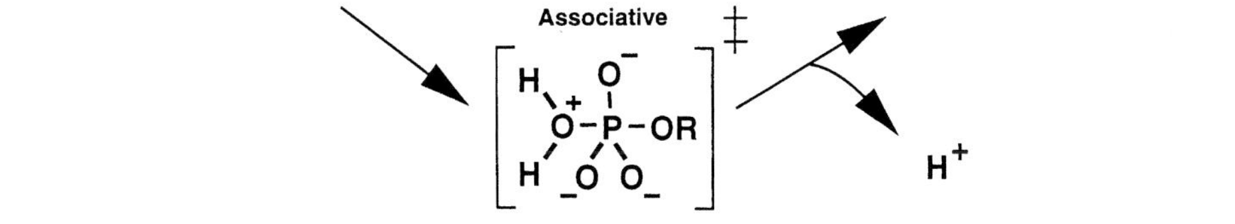

Dissociative

Associative

Figure 3. Alternative mechanistic pathway extremes for phosphoryl transfer reactions, including the GTP hydrolysis reaction in EF-Tu. A fully dissociative reaction (top) is characterized by a loss of negative charge on the transferred metaphosphate. The analogous phosphoryl moiety exhibits a net increase in negative charge in an associative transition state (bottom). GDP is denoted by "OR."

over the Cα-Cβ bond of this histidine could place its imidazole group next to the active water, from which it could abstract a proton. However, such a movement is prevented by the presence of the "hydrophobic gate," i.e., the side chains of Val-20 and Ile-61 (Fig. 2). These protect the active-site water molecule from an approach by His-85 and from bulk solvent. Neither Val-20 nor Ile-61 exhibits elevated atomic temperature factors in the crystal structure, so enhanced mobility of their side chains is unlikely. This was supported by extended molecular dynamics calculations over the unusually long simulation time of 1,000 ps, which failed to indicate any opening of the hydrophobic gate (Schütz and Hilgenfeld, unpublished). We therefore speculated that the largely hydrophobic environment of water no. 411 is sufficient to increase its nucleophilicity to a level where it can attack the γ-phosphate (Hilgenfeld, 1995b). Since one wing of the hydrophobic gate, Ile-61, is part of the presumed effector site of EF-Tu, i.e., its ribosome binding site, it is possible that upon binding of the factor to the ribosome, the gate is partly opened, allowing His-85 access to the active-site water. This could be one reason for the greatly accelerated GTP hydrolysis reaction on the ribosome (Berchtold et al., 1993). However, this should only occur after cognate codon-anticodon recognition, which means that a tRNA-dependent signal should be transmitted to the GTPase active center of EF-Tu.

His-85 of *T. thermophilus* EF-Tu and the equivalent His-84 in the *E. coli* protein have been replaced by site-directed mutagenesis. Replacement by glutamine reduced the intrinsic GTPase somewhat (Zeidler et al., 1995; Scarano et al., 1995), while exchange for alanine reduced it to 10% of the wild-type value (Scarano et al., 1995). Finally, replacement by leucine was reported to abolish the intrinsic GTPase activity altogether (Zeidler et al., 1995). The nucleotide exchange rate was found to be significantly increased in the latter mutant.

The H85L mutant appeared to offer an opportunity to crystallize the EF-Tu complex with the real substrate, GTP, rather than a nonhydrolyzable GTP analogue. However, when we determined the crystal structure of the presumed H85L·GTP complex, we were surprised to find that the electron density maps showed the presence of GDP instead of the expected GTP, indicating that the latter had been hydrolyzed by EF-Tu during crystallization or in the crystal. Furthermore, in agreement with the elevated nucleotide exchange rate measured for H85L EF-Tu, the occupancy of the GDP appears to be less than 50% and no electron density for Mg^{2+} can be identified. The absence of magnesium in a nucleotide complex has never before been observed for any GTP-GDP binding protein. Unfortunately, the whole switch II region in the H85L·GDP product complex appears to be disordered, since there is no electron density at all for the stretch including amino acids 84 to 97, which harbors the mutation site.

In order to obtain a structure corresponding to the GTP complex, we cocrystallized the H85L protein with GppNHp. Surprisingly, the crystals turned out to be isomorphous with those of the GDP complex, and indeed, the two complexes are isostructural. The only exception is the nucleotide binding site, which shows clear density for GppNHp and Mg^{2+}. The switch II region is again disordered, similarly to the GDP complex described above.

These results can be interpreted as a severe disturbance of the three-dimensional structure of EF-Tu in the neighborhood of the mutation site. Since His-85 is part of the switch II region, which in the wild-type protein undergoes specific conformational changes upon GTP hydrolysis, it is perhaps not surprising that replacement of a hydrophilic, solvent-accessible residue (His) by a hydrophobic one (Leu) leads to an ill-defined structure. In the GDP form of wild-type EF-Tu (Abel et al., 1996), His-85 is hydrogen-bonded to His-119 (Fig. 4). Upon binding of the nucleotide exchange factor EF-Ts, this interaction is broken by the insertion of a phenylalanine residue. The resulting conformational change is transmitted to the nucleotide binding site, where it leads to a weakening of the coordination of Mg^{2+} (Kawashima et al., 1996). In addition, by a flip of the His-19—Val-20 bond, the side chain of Val-20 is reoriented, leading to an opening of one wing of the hydrophobic gate and dissociation of the nucleotide (Wang et al., 1997). In the H85L mutant of EF-Tu, the leucine could easily play a role similar to that of the phenylalanine residue inserted by EF-Ts, i.e., disruption of the His-85...His-119 hydrogen bond and, ultimately, increased dissociation of the nucleotide. This idea is supported by only partial occupancy of the nucleotide binding site, the apparent absence of Mg^{2+} from this site, and the elevated nucleotide exchange rate measured for this mutant in solution (Zeidler et al., 1995). Thus, the H85L variant can be regarded as an EF-Tu with built-in nucleotide exchange factor properties. We are convinced that His-85 has nothing to do with the intrinsic GTPase activity but that it is conserved in the EF-Tu family because it serves as the initial recognition point for EF-Ts.

Since His-85 obviously does not function as a general base in the intrinsic GTP hydrolysis reaction of EF-Tu, the only remaining possibility is that the

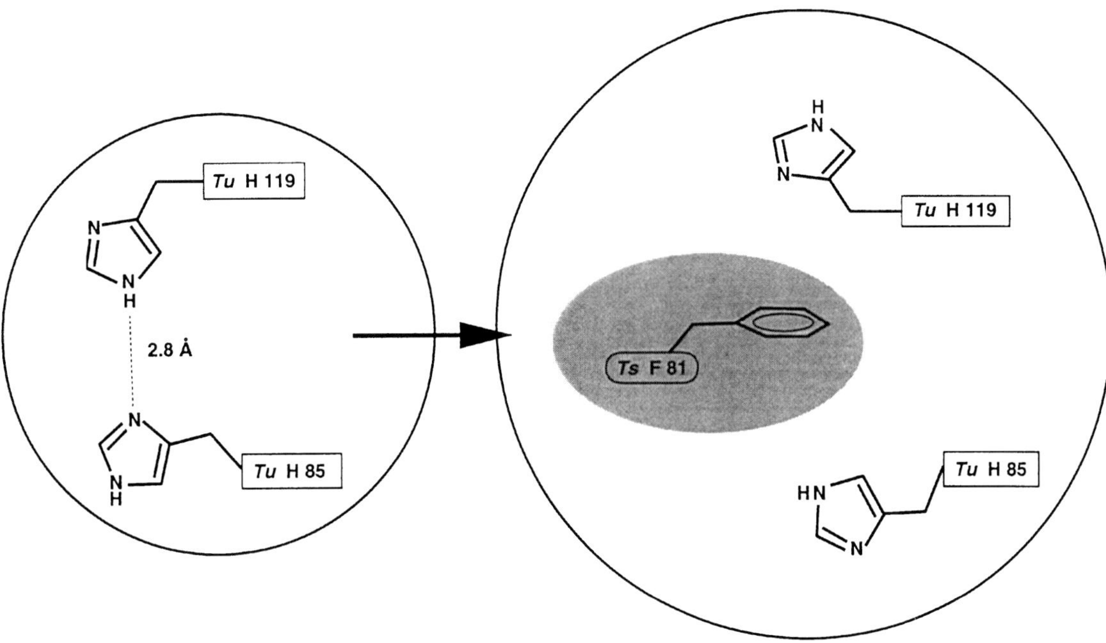

Figure 4. Schematic drawing illustrating the interaction between His-85 and His-119 in EF-Tu·GDP, which is interrupted by insertion of a phenylalanine from the nucleotide exchange factor in the EF-Tu·EF-Ts complex.

γ-phosphate itself abstracts a proton from the attacking water, as had been proposed by Schweins et al. (1995) for Ras. It should be noted, however, that the γ-phosphate in EF-Tu makes more hydrogen-bonding interactions with the protein than in Ras, so not all arguments used by Schweins et al. to support such a substrate-assisted mechanism can be directly applied to the situation in EF-Tu.

ARGININE-59: INVOLVED IN TRANSITION STATE STABILIZATION ON THE RIBOSOME?

The identities of side chains stabilizing the transition state of GTP hydrolysis in EF-Tu are as puzzling as the role played by His-85. In this case, it is instructive to briefly examine the structures of complexes of other GTPases with transition state analogues.

A good transition state analogue for α subunits of heterotrimeric G proteins is AlF_4^- or AlF_3, which, when complexed to the protein together with GDP, results in a conformation resembling the plausible transition state about the γ-phosphate. The planar aluminum fluoride moiety has a geometry (at least in the case of AlF_3) which can be expected for the trigonal base of the transition state of GTP hydrolysis. In $G_i\alpha$ as well as in transducin ($G_t\alpha$), the AlF_4^- moiety is hydrogen bonded to the side chains of Arg-178 (−174 in $G_t\alpha$) and Gln-204 (−200 in $G_t\alpha$) (Coleman et al., 1994; Sondek et al., 1994). While the latter is conserved in Ras-like small GTPases (Gln-61), the arginine is not present in these proteins, explaining why they do not normally bind aluminum fluorides (Kahn, 1991). However, in the complex with the GTPase-activating protein of Ras, Ras-GAP, an arginine is provided for GTPase transition state stabilization by the GAP (arginine finger). As a result, AlF_3 can be bound to Ras·GDP under these conditions (Scheffzek et al., 1997).

In the amino acid sequence of EF-Tu, the conserved glutamine residue of $G\alpha$ and Ras is replaced by His-85, which we discussed above and which is unlikely to stabilize the transition state. Furthermore, EF-Tu does not have an arginine anywhere close to the active GTPase site (see Fig. 2) and it is therefore not surprising that it does not bind GDP·AlF_x (Mesters et al., 1993). If one aligns the amino acid sequences of $G\alpha$ subunits and EF-Tu, there is an arginine in the same position as Arg-178 of $G_i\alpha$. However, in our crystal structure of EF-Tu·GppNHp, this Arg-59 is approximately 11 Å apart from the γ-phosphorus atom (Berchtold et al., 1993). Not only is this distance very large, but also there is one wing of the hydrophobic gate, the side chain of Ile-61, between Arg-59 and the GTP (see Fig. 6). Therefore, a major conformational rearrangement would be required to bring this arginine closer to the γ-phosphate. Our efforts to simulate such a movement by molecular modeling failed, but of course we cannot exclude the possibility that binding to the ribosome could induce a significant structural rearrangement of EF-Tu in this region.

To study the role of Arg-59 further, Sprinzl and coworkers have replaced it with threonine (Zeidler et al., 1996). Not unexpectedly, the intrinsic GTPase of the EF-Tu thus modified was found to be unchanged (or even slightly elevated), while the rate of the ribosome-enhanced GTP hydrolysis reaction was somewhat reduced. Therefore, we believe that Arg-59 is not an essential player in the GTP hydrolysis reaction, either in the isolated EF-Tu or in the ternary complex when bound to the ribosome.

In order to inspect the structural consequences of this mutation, we crystallized R59T EF-Tu (Wagner, 1996) and determined its structure at 1.8-Å resolution by X-ray crystallography. In the wild-type protein, the side chain of Arg-59 is involved in a salt bridge with Asp-87. Knudsen and Clark (1995) have shown that replacement of the latter does not affect the GTPase rate of EF-Tu. Thus, earlier suggestions (Kjeldgaard et al., 1993) that Asp-87 may be part of a distorted catalytic triad Asp-87...His-85...active-site water are certainly not correct.

The structure of the R59T mutant of EF-Tu is virtually superimposable on the wild-type molecule without significant deviations. The salt bridge between Arg-59 and Asp-87 is, of course, no longer present in the mutant structure, but this has no influence on the immediate environment. Even the removal in the R59T mutant of a water-mediated intermolecular contact between Arg-59 and the neighboring molecule in crystals of wild-type EF-Tu is not sufficient to induce a structural change. This is rewarding for the crystallographer, who is often faced with questions of whether crystal packing could lead to local structural artifacts.

TESTING THE IMPORTANCE OF THE HYDROPHOBIC GATE

In our interpretation, the hydrophobic gate protects the active-site water from being activated by His-85. So what would happen if one or both wings of the gate were partly removed? Would the histidine swing in and contact the active-site water, thereby enhancing the rate of GTP hydrolysis? To answer this question, Sprinzl and coworkers replaced Ile-61 by Ala and, in a separate experiment, Val-20 by Ser (Rütthard, 1999), and indeed, it was found that the intrinsic GTPase of these mutant proteins was increased by a factor of 1.5 to 2. We determined the crystal structures at 2.2- (I61A) and 2.0-Å (V20S) resolution. Surprisingly, the position of the His-85 side chain is unchanged in both mutants. The only real difference from the wild-type protein concerns the water structure around the γ-phosphate. While the active-site water 411 is in the same position as in the wild-type structure, we observe significant rearrangements of a chain of water molecules which starts at the γ-phosphate and involves waters 473, 499, and 583 before it enters bulk solvent. These waters adopt more ideal hydrogen-bonding distances in the mutants than in the wild-type structures (Fig. 5), leading us to suggest that they might be involved in protonation of the γ-phosphate before the nucleophilic water 411 (which is not part of this water chain) attacks the γ-phosphorus atom. The advantage of this "two-water mechanism" would be that the negative charge on the γ-phosphate would be reduced, thereby facilitating the nucleophilic attack of water 411. Removal of the Ile-61 wing of the hydrophobic gate permits a more ideal arrangement of the water chain, with stronger hydrogen bonds between the waters and between the γ-phosphate and the first water in the row (wat 473), thereby perhaps allowing a more efficient proton transfer to the γ-phosphate.

ASPARTATE-21: A GENERAL ACID IN A DISSOCIATIVE MECHANISM?

The chain of water molecules just discussed virtually wraps around the side chain of Asp-21, the fourth residue within the P-loop motif GX_4GKT, characteristic of GTPases (Fig. 5 and 6). We noticed during refinement of the hydrophobic-gate mutant structures that this portion of the P loop (to be more exact, residues 20 to 22) exhibits some degree of conformational heterogeneity. In order to find out whether this is a property of the mutants, we went back and refined the structural model of wild-type EF-Tu·GppNHp (Berchtold et al., 1993) further, at a resolution of 1.70 Å. And indeed, we found that in the native EF-Tu as well, there is conformational variability of the main chain around Asp-21, and in particular, of the side chain of this residue. Two independent conformations could be refined (Fig. 5). In one of them, the side chain of Asp-21 is hydrogen-bonded to waters 473 and 499 of the water chain discussed above. In the other, it accepts a hydrogen bond from the imido group bridging the β- and γ-phosphates in the GTP analogue, GppNHp. The latter situation can only occur in GTP itself if the bridging oxygen is protonated. Such protonation would occur at an early stage of the reaction in a dissociative but not in an associative mechanism.

Since the side chain of Asp-21 is in contact with the chain of water molecules connecting the γ-

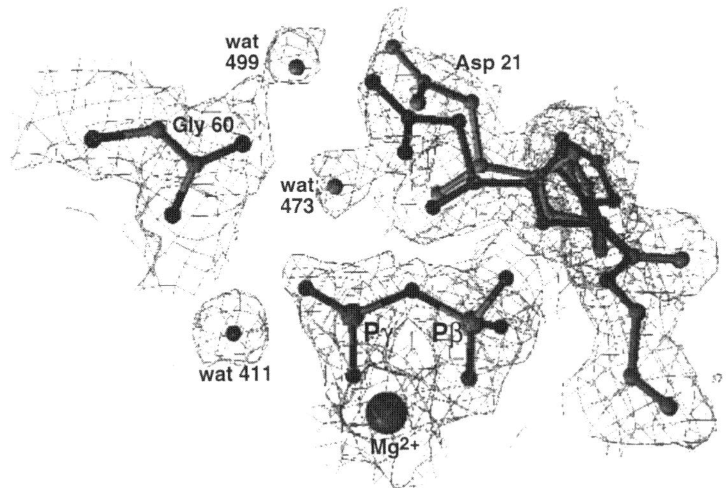

Figure 5. Conformational variability in the nucleotide binding pocket is observed for residues 20 to 22 of the phosphate binding loop (P loop) in the EF-Tu complex with GppNHp and manifested to the largest extent in the side chain of Asp-21. Waters 473 and 499 are only present in one of the two conformations of the P loop. Shown is a $2F_o - F_c$ electron density map contoured at 1.2 σ above the mean.

phosphate with bulk solvent in one of its two conformations and with the β,γ-bridging NH group of GppNHp in the other, we speculate that this residue may in fact abstract a proton from water 473 and transfer it directly to the bridging group in GTP (Fig. 7). It would thus be involved in protonation of the leaving group, most probably in a dissociative mechanism (where early protonation of the leaving group is required). In other words, Asp-21 would act as a general acid in the reaction. If this mechanism is true, replacement of the Asp-21 side chain should impair the GTPase reaction. While we do not yet have an Asp-21 mutant for EF-Tu, Martemyanov and Gudkov (personal communication) have prepared the corresponding D22A variant of elongation factor G (EF-G) on our suggestion, and indeed, their preliminary findings indicate that both the intrinsic and the ribosome-enhanced GTPase activities of this mutant are reduced. It should also be noted that Asp-21 is conserved among elongation factors Tu and G from almost all species. In Gα subunits, the corresponding residue is Glu, but an acidic residue occurs in only a

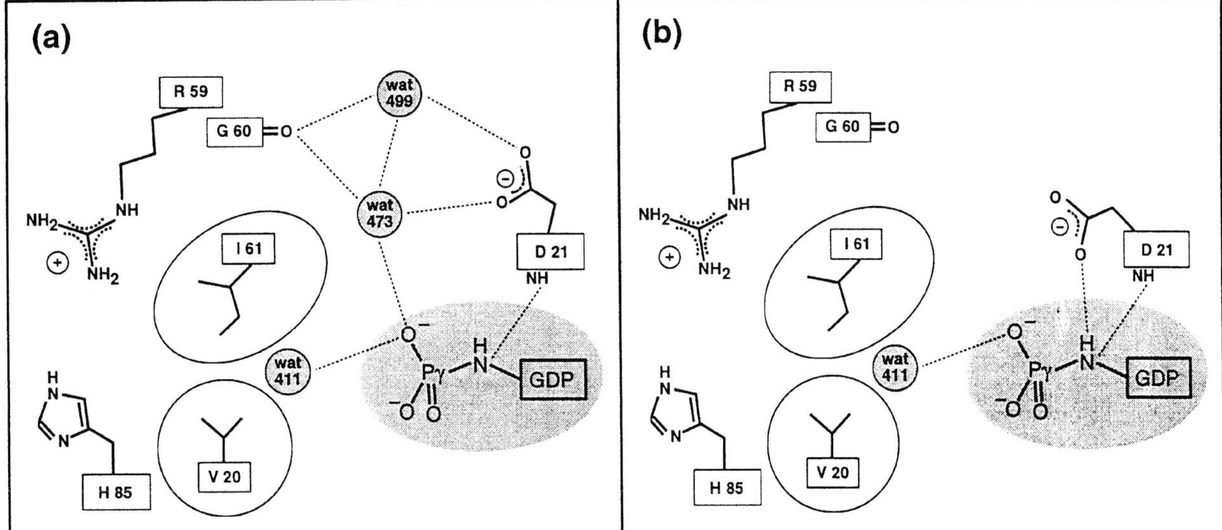

Figure 6. Schematic illustration of the alternative modes of interaction between Asp-21 and GppNHp. (a) In the open binding mode, the side chain of Asp-21 interacts with waters (wat) 473 and 499 of the water chain connecting the γ-phosphate to bulk solvent. (b) In the closed binding mode, the side chain of Asp-21 moves to accept a hydrogen bond from the β,γ-bridging group, thereby occluding formation of the water chain.

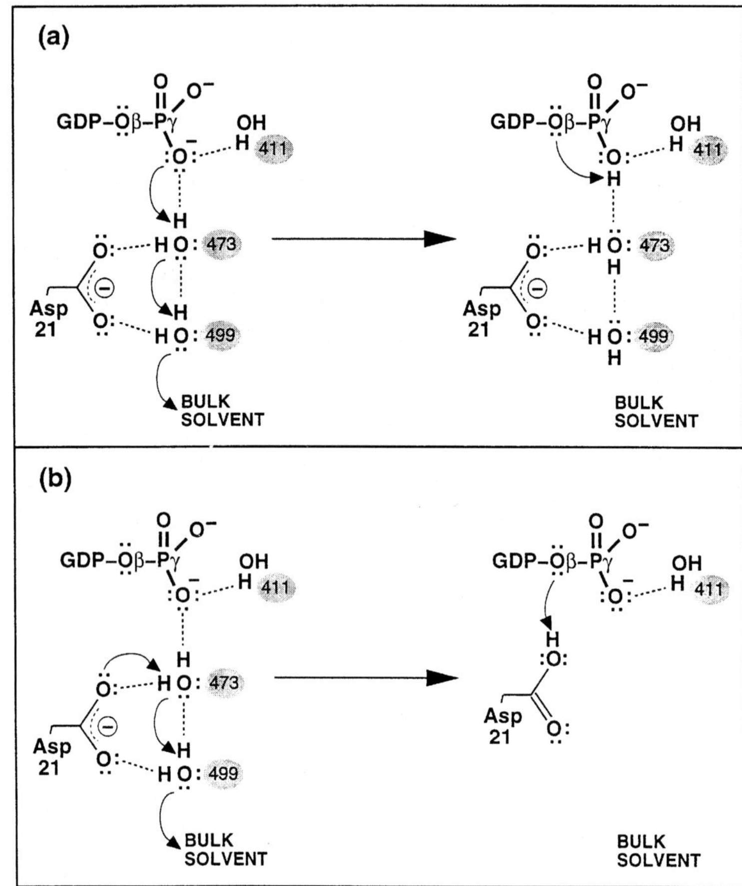

Figure 7. Possible pre-transition state pathways for dissociative GTP hydrolysis catalyzed by EF-Tu. (a) Two-water mechanism involving protonation of the γ-phosphate through the water chain prior to nucleophilic attack by the active-site water, 411. Very likely, the proton originating from the water chain is immediately transferred to the β,γ-bridging oxygen (Oβ), thereby facilitating the dissociation of the leaving group. (b) General-acid-catalyzed hydrolysis of GTP. Asp-21 transfers a proton from the proximal water molecule (473) of the water channel directly to the β,γ-bridging oxygen (Oβ) of GTP, which would be highly indicative of a dissociative mechanism.

few of the Ras-like small GTPases at this position (not in Ras itself). However, it is becoming increasingly apparent that there are significant differences between the members of the GTPase superfamily. Therefore, it is no longer helpful to expect that their mechanisms of catalysis should all be the same. In fact, such a view could prevent us from recognizing important features that are present in one class of the superfamily but not in another.

A largely dissociative GTPase mechanism involving general-acid catalysis by Asp-21 is attractive because it would be compatible with the observed absence of a general base and of an arginine or glutamine residue stabilizing the transition state. It is intriguing that similar catalytic elements, i.e., a general acid and a nucleophile, are also found in the tyrosine phosphatase from *Yersinia pestis*. In this enzyme, Cys-403 has been shown to perform an in-line attack on the phosphate group, and Asp-356 serves as a general acid/base (Fauman et al., 1996). Like EF-Tu, the phosphatase catalyzes the hydrolysis of a bond to a phosphate group, but in contrast to the former, this occurs through a two-step mechanism. In the first step, the thiolate group of Cys-403 will accept the PO_3 moiety, while the completely solvent-accessible Asp-356 serves to protonate the leaving group, i.e., the tyrosine side chain. In the second step, the phosphocysteine intermediate is hydrolyzed by a nucleophilic water molecule from which Asp-356 has abstracted a proton. Having crystal structures of several different phosphatase complexes at hand, Fauman et al. (1996) modeled the two extremes of the possible hydrolysis mechanism, i.e., fully associative or fully dissociative, and found that the results support a more dissociative character of the first step of the reaction.

DIFFERENCES BETWEEN GTP ANALOGUES

While GTP and GppNHp have hydrogen-bonding potential in their β,γ-bridging atoms, another widely used GTP analogue, GppCH$_2$p, carrying a methylene group between the two phosphates, does not. In order to see whether this induces any difference in binding, we also determined the crystal structure of wild-type EF-Tu in complex with this nonhydrolyzable GTP analogue, at a resolution of 1.80 Å. The approximate length of a P—N single bond (as in GppNHp) is 1.63 Å, while a P—C single bond is almost 0.2 Å longer (1.81 Å), leading us to expect interesting differences between these nucleotides (P—O, as in GTP, is intermediate at 1.73 Å). However, a "simulated-annealing omit" difference Fourier revealed that the positions of all three phosphorus atoms in the GppCH$_2$p complex are identical to their positions in the GppNHp structure. The only difference in the conformation of the nucleotide itself is the bond angle at the bridging group, which appears to be smaller in the methylene analogue.

The protein, however, reacts differently to the binding of GppCH$_2$p. Since the side chain of Asp-21 does not have a hydrogen-bonding partner in the β,γ-bridging group of the nucleotide, it does not adopt its second conformation, nor is this observed for the main chain between residues 20 and 22. This is the first report of differences in the binding between the two GTP analogues, GppNHp and GppCH$_2$p. Since we only noticed this fact after very careful inspection of difference maps and extensive refinement of the structures, it may well be that similar differences have been overlooked in other structures that were solved at lower resolution. However, the presence of a hydrogen-bonding aspartate (or possibly glutamate in the Gα subunits) in the GX$_4$GKT motif seems to be a cause of and prerequisite for the observed conformational heterogeneity.

So which nucleotide is the better analogue for GTP: GppNHp or GppCH$_2$p? From what we know now, GppNHp would be the better model if a dissociative mechanism requiring protonation of the leaving group early in the reaction were at work. In contrast, GppCH$_2$p might be the better model for an associative mechanism in which protonation of the bridging oxygen does not play a role until late in the reaction. What is needed to decide this question is the structure of a complex with a nucleotide which carries a β,γ-bridging oxygen atom, such as GTPγS, or GTP itself. The latter complex could possibly be generated by flash photolysis of caged GTP and immediate shock cooling thereafter.

FRACTIONAL BOND NUMBERS: ASSOCIATIVE VERSUS DISSOCIATIVE

As long as we do not have such structural information, can we rely on other features to discriminate between associative and dissociative mechanisms of GTP hydrolysis? Clearly, the true reaction mechanism of EF-Tu will be somewhere between these two extremes. Structural considerations allow a rough estimation of where on the continuous scale between a fully dissociative and a fully associative mechanism the actual case under study might lie. If the crystal structure of a complex with a transition state analogue is available, axial-bond distances between the incoming nucleophile and the atom attacked by it, as well as between the latter and the leaving group, can be estimated. These distances can be used to determine the fractional bond number, n, in the transition state by a formula derived by Pauling (1960): $D(n) = D(1) - 0.60 \log(n)$, as recently pointed out in a very useful discussion published by Mildvan (1997). In this formula, D(1) is the single-bond distance, 1.73 and 1.63 Å for P—O and P—N single bonds, respectively, and $D(n)$ is the observed bond length. This equation is applied to the distance between the Pγ atom (in the transition state) and the attacking water molecule. Using this approach, a true S$_N$2 reaction can be classified as 50% associative (the fractional bond number, n, in the transition state is 0.5 [Mildvan, 1997]).

Unfortunately, aluminum fluorides cannot be used as transition state analogues because they fail to bind to EF-Tu, as explained above. But it is also possible to obtain an estimate of the reaction coordinate distance between attacking nucleophiles and γ-phosphorus atoms before the reaction has begun (Mildvan, 1997). If this distance is $>$4.9 Å, a metaphosphate transition state would have enough space to place itself symmetrically between the two axial oxygens. Therefore, such a long distance would be indicative of a purely dissociative mechanism. The shorter this distance, the more associative character can be expected. In the present case of EF-Tu·GppNHp, the relevant distance is 3.2 Å, and the distance between the attacking and leaving axial ligands of the phosphorus is 4.8 Å. If we assume for the moment that a transition state could form in the case of GppNHp, its trigonal base would have a distance of 2.4 Å to each of the two axial ligands after having been moved into a symmetrical position midway between the two atoms. This would correspond to fractional-bond numbers of 0.076 and 0.052 to the two axial ligands [O(water 411) and N(β,γ-bridge), respectively], indicating that the mechanism is at least 7.6% associative. From this it follows that the transition state could have a more dissociative than associative character. Of course, we have to keep in mind that we are looking at a nonreactive analogue of GTP and that the situation might be different in the case of the real substrate. But we tend to believe that the possibility of a dissociative mechanism for GTP hydrolysis by EF-Tu should not be discarded (as is presently done by the majority of our colleagues in the field), because it would agree with the structural and biochemical observations better than the associative route. The facts supporting this view are the absence of a general base, the absence of a residue stabilizing a highly negative transition state, and the presence of a potential general acid.

CONCLUSIONS

Depending on the mechanism of GTP hydrolysis catalyzed by EF-Tu, different interactions of the ternary complex with the ribosome will have to be expected. In a largely associative mechanism, the transition state about the γ-phosphorus atom could be stabilized by hydrogen bonds donated by amino, guanidinium, or hydroxyl groups (glutamine, lysine, or arginine from ribosomal proteins or ribose hydroxyl groups from rRNA). On the other hand, acidic groups (Asp and Glu) can be expected to be in the neighborhood of a dissociative transition state, fulfilling a role as proton shuttles from water to the oxygen of the leaving group, GDP. For an understanding of the action of the enzyme on the ribosome, it is important to dissect the mechanism of its intrinsic GTPase reaction.

In this chapter, we discussed the two alternatives for the GTPase mechanism on the basis of high-resolution crystal structures of EF-Tu (from *T. thermophilus*) modified by site-directed mutagenesis. The replacement of His-85, long considered to be the general base, by leucine does not fully abolish the intrinsic GTPase activity, since GDP is found in the crystal after cocrystallization with GTP. In agreement with the elevated nucleotide exchange in this mutant, the GDP-binding site is only partly occupied, and no magnesium ion can be located. The mutation, located in the switch II region, apparently leads to a disturbance of the local three-dimensional structure. The complex of the same mutant with GppNHp adopts the GDP conformation as well, indicating that switch II is unable to function because of the mutation. Rather than playing a role in the GTPase reaction, His-85 could be important as a recognition point for the nucleotide exchange factor, EF-Ts.

In search of a residue stabilizing the transition state about the γ-phosphorus, Arg-59 has been replaced by threonine (Zeidler et al., 1996) without significant consequences for the three-dimensional structure. Since this residue is more than 11 Å away from the γ-phosphate, which it could only approach after a major conformational change that would move Ile-61 (one of the wings of the hydrophobic gate) out of the way, we cannot exclude the possibility that such a movement occurs on the ribosome.

The importance of the hydrophobic gate in protecting the nucleophilic water molecule from premature activation was tested by replacing the wing residues (V20S and I61A mutants [Rütthard, 1999]). Both mutants do show a somewhat elevated intrinsic GTPase, but instead of His-85 swinging in to activate the nucleophilic water, we observe the rearrangement of a chain of water molecules extending from bulk solvent to the γ-phosphate. We propose a two-water mechanism, in which the γ-phosphate is first protonated by proton transfer from this water chain before it is attacked by the active-site water molecule, wat 411 (which is not part of the water chain). This process may be supported by Asp-21, which exhibits two alternative conformations in the complex with GppNHp. In one of these, it accepts two hydrogen bonds from the water chain, whereas it is in contact with the atom bridging β- and γ-phosphates in the other. It is conceivable that this movement is used to transfer a proton directly from the water chain to the leaving group (GDP) of the reaction, supporting the concept of a dissociative mechanism. In contrast, such mobility of Asp-21 is not observed in the EF-Tu complex with GppCH$_2$p, in which the β,γ-bridging group does not have hydrogen-bonding potential. It is therefore possible that GppNHp is a good GTP analogue in the case of a dissociative mechanism (where early protonation of the bridging oxygen is essential), while GppCH$_2$p would be the better model for an associative mechanism (where such protonation does not play a role until late along the reaction coordinate). A decision between the two possibilities will have to await the determination of the structure of the EF-Tu complex with the real substrate, GTP, possibly augmented by theoretical calculations. Until then, the discussion of the GTPase mechanism is likely to continue.

We are grateful to Mathias Sprinzl and his group, in particular, Martina Meyering-Vos, Heike Rütthard, and Waltraud Zeidler, who prepared and crystallized most of the mutant proteins described in this chapter. The R59T variant was crystallized by Annett Wagner, who also carried out the early stages of crystallographic refinement.

Parts of the work described here were funded by the Deutsche Forschungsgemeinschaft, the Federal Ministry of Education and Research (BMBF), through DESY-HS, the European Commission, the Howard Hughes Medical Institute, and the Fonds der Chemischen Industrie.

REFERENCES

Abel, K., and F. Jurnak. 1996. A complex profile of protein elongation: translating chemical energy into molecular movement. *Structure* **4**:229–238.

Abel, K., M. D. Yoder, R. Hilgenfeld, and F. Jurnak. 1996. An α to β conformational switch in EF-Tu. *Structure* **4**:1153–1159.

Berchtold, H., L. Reshetnikova, C. O. A. Reiser, N. K. Schirmer, M. Sprinzl, and R. Hilgenfeld. 1993. Crystal structure of active elongation factor Tu reveals major domain rearrangements. *Nature* **365**:126–132.

Caldas, T. D., A. E. Yaagoubi, and G. Richarme. 1998. Chaperone properties of bacterial elongation factor EF-Tu. *J. Biol. Chem.* **273**:11478–11482.

Coleman, D. E., A. M. Berghuis, E. Lee, M. E. Linder, A. G. Gilman, and S. R. Sprang. 1994. Structures of active conformations of $G_i\alpha 1$ and the mechanism of GTP hydrolysis. *Science* **265**: 1405–1412.

Fauman, E. B., C. Yuvaniyama, H. L. Schubert, J. A. Stuckey, and M. A. Saper. 1996. The X-ray structures of *Yersinia* tyrosine phosphatase with bound tungstate and nitrate. *J. Biol. Chem.* **271**:18780–18788.

Georgiou, T., Y. N. Yu, S. Ekunwe, M. J. Buttner, A.-M. Zuurmond, B. Kraal, C. Kleanthous, and L. Snyder. 1998. Specific peptide-activated proteolytic cleavage of *Escherichia coli* elongation factor Tu. *Proc. Natl. Acad. Sci. USA* **95**:2891–2895.

Hilgenfeld, R. 1995a. Regulatory GTPases. *Curr. Opin. Struct. Biol.* **5**:810–817.

Hilgenfeld, R. 1995b. How do the GTPases really work? *Nat. Struct. Biol.* **2**:3–6.

Kahn, R. A. 1991. Fluoride is not an activator of the smaller (20–25 kDa) GTP-binding proteins. *J. Biol. Chem.* **266**:15595–15597.

Kawashima, T., C. Berthet-Colominas, M. Wulff, S. Cusack, and R. Leberman. 1996. The structure of the *Escherichia coli* EF-Tu·EF-Ts complex at 2.5 Å resolution. *Nature* **379**:511–518.

Kjeldgaard, M., and J. Nyborg. 1992. Refined structure of elongation factor Tu from *Escherichia coli*. *J. Mol. Biol.* **223**:721–742.

Kjeldgaard, M., P. Nissen, S. Thirup, and J. Nyborg. 1993. The crystal structure of elongation factor EF-Tu from *Thermus aquaticus* in the GTP conformation. *Structure* **1**:35–50.

Knudsen, C. R., and B. F. Clark. 1995. Site-directed mutagenesis of Arg58 and Asp86 of elongation factor Tu from *Escherichia coli*: effects on the GTPase reaction and aminoacyl-tRNA binding. *Protein Eng.* **8**:1267–1273.

Krab, I. M., and A. Parmeggiani. 1998. EF-Tu, a GTPase odyssey. *Biochim. Biophys. Acta* **1443**:1–22.

Kudlicki, W., A. Coffman, G. Kramer, and B. Hardesty. 1997. Renaturation of rhodanese by translation elongation factor Tu. *J. Biol. Chem.* **272**:32206–32210.

Lucas-Lenard, J., and F. Lipmann. 1971. Protein biosynthesis. *Annu. Rev. Biochem.* **40**:409–448.

Maegley, K. A., S. J. Admiraal, and D. Herschlag. 1996. Ras-catalyzed hydrolysis of GTP: a new perspective from model studies. *Proc. Natl. Acad. Sci. USA* **93**:8160–8166.

Martemyanov, K., and A. Gudkov. Personal communication.

Mesters, J. R., J. M. de Graaf, and B. Kraal. 1993. Divergent effects of fluoroaluminates on the peptide chain elongation factors EF-Tu and EF-G as members of the GTPase superfamily. *FEBS Lett.* **321**:149–152.

Mildvan, A. S. 1997. Mechanisms of signaling and related enzymes. *Protein Struct. Funct. Genet.* **29**:401–416.

Pauling, L. 1960. *The Nature of the Chemical Bond*, 3rd ed., p. 255–260. Cornell University Press, Ithaca, N.Y.

Polekhina, G., S. Thirup, M. Kjeldgaard, P. Nissen, C. Lippmann, and J. Nyborg. 1996. Helix unwinding in the effector region of elongation factor Tu·GDP. *Structure* **4**:1141–1151.

Rittinger, K., P. A. Walker, J. F. Ecclestone, K. Nurmahomed, D. Own, E. Laue, S. J. Gamblin, and S. K. Smerdon. 1997a. Crystal structure of the complex between Cdc42Hs·GMPPNP and p50rhoGAP. *Nature* **388**:693–697.

Rittinger, K., P. A. Walker, J. F. Ecclestone, S. K. Smerdon, and S. J. Gamblin. 1997b. Structure at 1.65 Å of RhoA and its GTPase-activating protein in complex with a transition-state analogue. *Nature* **389**:758–762.

Rütthard, H. 1999. Ph.D. thesis. University of Bayreuth, Beyreuth, Germany.

Scarano, G., I. M. Krab, V. Bocchini, and A. Parmeggiani. 1995. Relevance of histidine-84 in the elongation factor Tu GTPase activity and in poly(Phe) synthesis: its substitution by glutamine and alanine. *FEBS Lett.* **365**:214–218.

Scheffzek, K., M. R. Ahmadian, W. Kabsch, L. Wiesmüller, A. Lautwein, F. Schmitz, and A. Wittinghofer. 1997. The ras-rasGAP complex: structural basis for GTPase activation and its loss in oncogenic ras mutants. *Science* **277**:333–338.

Schütz, H., and R. Hilgenfeld. Unpublished data.

Schweins, T., M. Geyer, K. Scheffzek, A. Warshel, H. R. Kalbitzer, and A. Wittinghofer. 1995. Substrate-assisted catalysis as a mechanism for GTP hydrolysis of $p21^{ras}$ and other GTP-binding proteins. *Nat. Struct. Biol.* **2**:36–44.

Sondek, J., D. G. Lambright, J. P. Noel, H. E. Hamm, and P. B. Sigler. 1994. GTPase mechanism of G proteins from the 1.7Å crystal structure of transducin α-GDP-AlF_4. *Nature* **372**: 276–279.

Tesmer, J. J. G., D. M. Berman, A. G. Gilman, and S. R. Sprang. 1997. Structure of RGS4 bound to AlF_4^--activated $G_{i\alpha 1}$: stabilisation of the transition state for GTP hydrolysis. *Cell* **89**:251–261.

Vetter, I. R., C. Nowak, T. Nishimoto, J. Kuhlmann, and A. Wittinghofer. 1999. Structure of a Ran-binding domain complexed with Ran bound to a GTP analogue: implications for nuclear transport. *Nature* **398**:39–46.

Wagner, A. 1996. Ph.D. thesis. University of Bayreuth, Bayreuth, Germany.

Wang, Y., Y. Jiang, M. Meyering-Vos, M. Sprinzl, and P. B. Sigler. 1997. Crystal structure of the EF-Tu*EF-Ts complex from *Thermus thermophilus*. *Nat. Struct. Biol.* **4**:650–656.

Zeidler, W., C. Egle, S. Ribeiro, A. Wagner, V. Katunin, R. Kreutzer, M. Rodnina, W. Wintermeyer, and M. Sprinzl. 1995. Site-directed mutagenesis of *Thermus thermophilus* elongation factor Tu. Replacement of His 85, Asp 81 and Arg 300. *Eur. J. Biochem.* **229**:596–604.

Zeidler, W., N. K. Schirmer, C. Egle, S. Ribeiro, R. Kreutzer, and M. Sprinzl. 1996. Limited proteolysis and amino acid replacements in the effector region of *Thermus thermophilus* elongation factor Tu. *Eur. J. Biochem.* **239**:265–271.

The Ribosome: Structure, Function, Antibiotics, and Cellular Interactions
Edited by R. A. Garrett, S. R. Douthwaite, A. Liljas, A. T. Matheson, P. B. Moore, and H. F. Noller

Chapter 29

The States, Conformational Dynamics, and Fusidic Acid-Resistant Mutants of Elongation Factor G

ANDERS LILJAS, OLE KRISTENSEN, MARTIN LAURBERG, SALAM AL-KARADAGHI, ANATOLY GUDKOV, KIRILL MARTEMYANOV, DIARMAID HUGHES, and IVAN NAGAEV

Elongation factor G (EF-G) has an essential role in protein synthesis in catalyzing the translocation of the peptidyl-tRNA from the A site to the P site (Kaziro, 1978). It also has an extensive relationship to elongation factor Tu (EF-Tu): both factors are G proteins (Bourne et al., 1990), and there is a remarkable mimicry between the ternary complex of EF-Tu·tRNA·GDPNP and EF-G·GDP (Nissen et al., 1995). The large conformational changes of EF-Tu during its functional cycle (Berchtold et al., 1993; Kjeldgaard et al., 1993; Abel et al., 1996; Polekhina et al., 1996) raise the question of what conformational changes EF-G may go through during its functional cycle (Liljas et al., 1995; Czworkowski and Moore, 1997). The structural studies of EF-G have so far given little support for large conformational transitions. After the initial studies in which both the empty factor (Ævarsson et al., 1994) and its complex with GDP (Czworkowski et al., 1994; Al-Karadaghi et al., 1996) were found to have essentially the same structure, recent observations have identified a new, alternative conformation (Czworkowski and Moore, personal communication; Laurberg et al., unpublished). It is our aim in this review to summarize the current knowledge and try to propose a structural model for the function of EF-G. The large number of mutations associated with fusidic acid resistance are an essential ingredient in this analysis.

THE STRUCTURE OF EF-G

The crystallographic investigations of EF-G from *Thermus thermophilus* have revealed a highly extended protein organized into six domains. Two regions of the protein have not been clarified until recently. These are the effector loop (residues 37 to 68) and domain III. These regions may be more flexible than the rest of the protein and have a number of conformational states, leading to weak electron density which is difficult to interpret. The reason for the flexibility and its functional importance remain to be explored. Partial interpretations of domain III have been made for the elements of the structure that have been most visible in the electron density maps. Recent investigations of a mutant EF-G with a different crystal packing have led to a complete interpretation of domain III at relatively low resolution (Laurberg et al., unpublished). Details of the interpretation can be confirmed with the partially visible structural elements in the classical conformation but at higher resolution (see below).

THE FUNCTIONAL STATES OF EF-G

The functional cycle of EF-G can be described as a number of states both on and off the ribosome (Fig. 1). Additional states may exist. The affinity of EF-G for the ribosome is high in complex with GTP. Thus, there are GTP states for EF-G both on and off the ribosome. There are both a classical and a recent model for what happens after the factor has bound to the ribosome. According to the classical model, translocation precedes GTP hydrolysis, while according to observations by Rodnina et al. (1997), GTP hydrolysis is very rapid and precedes translocation.

Anders Liljas, Ole Kristensen, Martin Laurberg, and Salam Al-Karadaghi ■ Molecular Biophysics, Lund University, Box 124, SE-22100 Lund, Sweden. **Anatoly Gudkov and Kirill Martemyanov** ■ Institute of Protein Research, Russian Academy of Sciences, Pushchino, Moscow Region, Russia. **Diarmaid Hughes and Ivan Nagaev** ■ Department of Cell and Molecular Biology, BMC, Uppsala University, Box 596, SE-75124 Uppsala, Sweden.

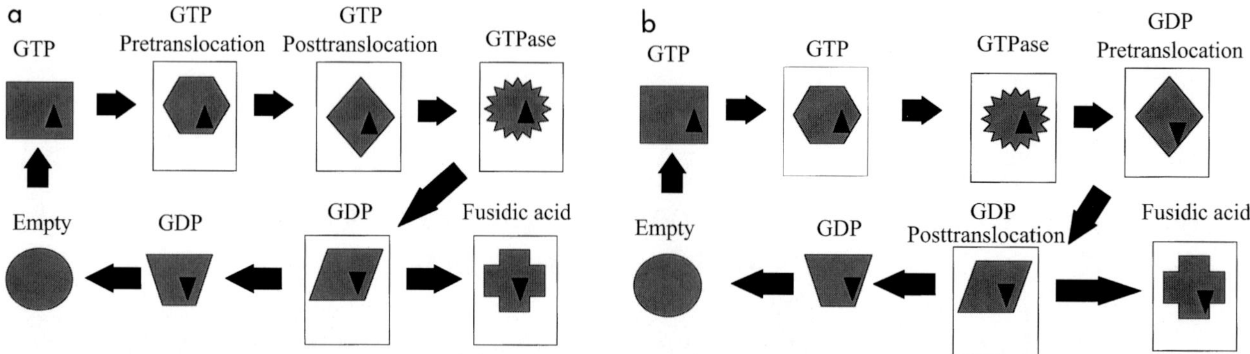

Figure 1. Representation of some of the possible functional states of EF-G according to the classical model (a) and the model of Rodnina et al. (1997) (b). The white boxes represent the ribosome, the gray symbols represent EF-G in different states, and the black triangles represent GTP (base down) or GDP (base up).

Thus, GTP hydrolysis occurs shortly after the factor has bound to the ribosome. Even though the model of Rodnina et al. (1997) is supported by interesting experiments, the translocation was followed indirectly from fluorescence-labeled tRNAs. Thus, additional observations would be valuable. In this model, the pretranslocation state with EF-G·GTP is followed by the GTPase state. This leads to the pretranslocation state with EF-G·GDP, followed by the posttranslocation state with EF-G·GDP bound to the ribosome. Subsequently, EF-G will dissociate from the ribosome and lose its bound GDP, which then is replaced by GTP.

Fusidic acid binds to EF-G·GTP on the ribosome. GTP hydrolysis occurs, but the EF-G·GDP complex cannot dissociate from the ribosome, which may be due to an inhibition of the transition to the EF-G·GDP conformation (Willie et al., 1975).

The different states of EF-G may not necessarily be associated with different conformations of EF-G, but to the extent that there are different conformations, they will be related to different states.

CONFORMATIONS OF EF-G

The structural studies of EF-G have revealed a few different conformations. The conformation that has been observed in numerous crystallographic studies is the one with a bound GDP molecule (Czworkowski et al., 1994; Al-Karadaghi et al., 1996). This conformation seems to be highly stable; we call it the GDP conformation. Czworkowski and Moore (personal communication) have also observed that EF-G with a bound GDPCP molecule retains this conformation. The conformation of the empty EF-G (Ævarsson et al., 1994) is slightly different from the GDP conformation. In particular the residues in the vicinity of the nucleotide in the G domain undergo interesting conformational changes (Al-Karadaghi et al., 1996). They involve the P loop, which is in close contact with the phosphates of the nucleotide, and the switch II loop, as well as helix C_G. However, in general the deviations from the GDP conformation are not very extensive. A different conformation with GDP that was obtained under slightly different crystallization conditions but with GDP bound is more bent (Czworkowski and Moore, personal communication). We call this the bent conformation of EF-G.

Agrawal et al. (1998) have used cryo-electron microscopy (cryo-EM) to study the structure of EF-G when it is bound to the ribosome in the presence of fusidic acid. The density of EF-G could be identified with difference methods and compared to the crystallographic GDP conformation. A significant rearrangement of EF-G was needed to fit the density. This conformation of EF-G is more open than those observed crystallographically. Thus, there are three main conformations of EF-G observed so far, the bent, the GDP, and the open conformations.

CRYSTALLOGRAPHIC ANALYSIS OF EF-G MUTANTS

A number of mutants of EF-G from *T. thermophilus* have been crystallized. Two of them are G16V and H573A. The crystallographic analysis of these mutants has added to the structural information about the factor (Laurberg et al., unpublished). Thus, the crystals of EF-G·GDP from the mutant H573A from the same experimental crystallization drop can have cell dimensions that differ in the *b* axis by as much as 20 Å. EF-G in crystals with the short *b* axis analyzed at 2.9-Å resolution has the bent conformation (Fig. 2). In this conformation domain III be-

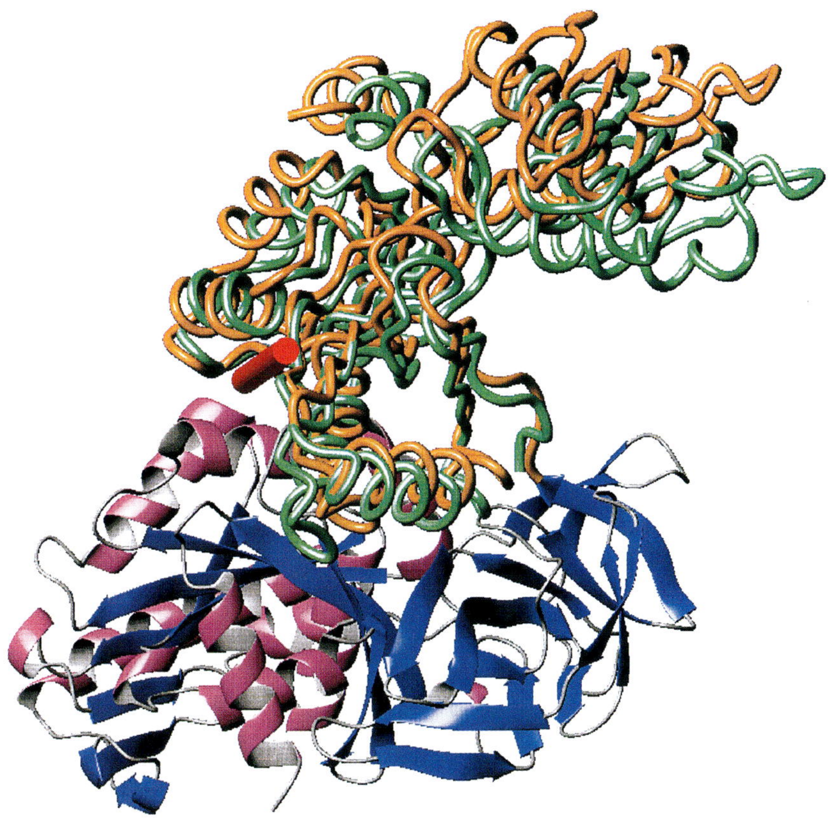

Figure 2. Comparison of the two main conformations of EF-G observed crystallographically. The broad ribbon represents domains G, G′, and II, which here are fixed. The narrow tube represents domains III to V. The GDP conformation is shown in orange, and the new bent conformation of mutant H573A is shown in green. The red bar indicates the axis around which the rotation of about 10° occurs.

comes more rigidly held between domains II and IV and can be completely interpreted. The fold of domain III is found to be identical to that of domain V (Fig. 3), as previously suggested by Ævarsson et al. (1994). This folding pattern has been observed repeatedly in proteins in the translational machinery (Liljas and Al-Karadaghi, 1997; Ramakrishnan and White, 1998) (see chapter 7).

EF-G domain III

EF-G domain V

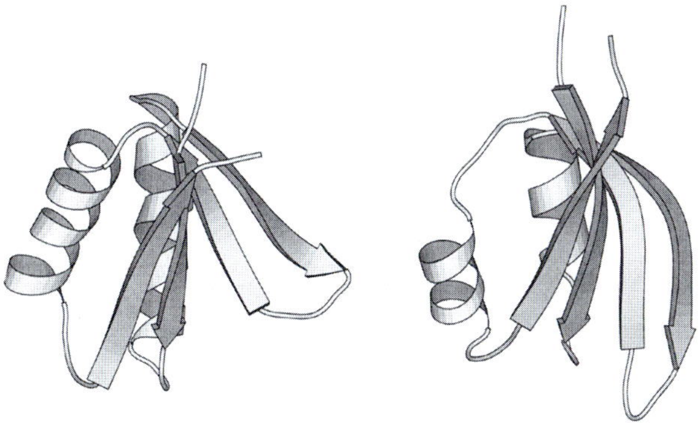

Figure 3. Comparison of the structures of domains III and V. The same fold is observed. This fold, the RNA recognition motif (RRM), has also been observed in many ribosomal proteins.

It is possible that the crystal contacts prevent wild-type (wt) EF-G from displaying the full range of conformational differences that may exist in solution. The bent conformation is obtained by a rotation of 9.4° of domains III, IV, and V in relation to domains G and II. The interdomain bending (Berendsen and Hayvard, 1998) involves residues 399 to 400, as well as 636 to 637 and 642 to 643. This should be compared with the molecular reorganization of EF-G that was needed to fit the crystal structure to the EM density (Agrawal et al., 1998). In this case domains III, IV, and V were rotated in the opposite direction by 10° in relation to domains G and II.

The mutant G16V is fusidic acid sensitive compared to wt *T. thermophilus* EF-G, which is relatively resistant to the antibiotic (Martemyanov et al., unpublished). The equivalent mutant in *Salmonella typhimurium* EF-G, G13V, reverses the fusidic acid-resistant phenotype of *fusA1* P413L (P407 in *T. thermophilus*) to almost wt sensitivity (Johanson et al., 1996). Crystals of the mutant G16V, grown in the presence of GDPNP and alkaline phosphatase, were studied at 2.5-Å resolution and found to have the GDP conformation. However, in this case no GDP is seen in the active site but mainly a phosphate ion equivalent to the position of the β-phosphate of GDP. An understanding of this phenomenon will require further experimentation. However, parts of the interpretation of domain III obtained from mutant H573A could be confirmed with the higher-resolution data from mutant G16V.

THE LOCATIONS OF FUSIDIC ACID-RESISTANT MUTATIONS IN EF-G

The interpretation of the structure of domain III in EF-G has allowed a complete analysis of new and old fusidic acid-resistant mutations in EF-G (Johanson et al., 1996; Nagaev et al., unpublished). An earlier analysis focused on fusidic acid-resistant mutations in the region between domains G and V, particularly in helix C_G, leading to the suggestion that this could be the binding site for fusidic acid (Johanson et al., 1996). The location of these mutations also corresponds to the location of kirromycin-resistant mutations in EF-Tu (Abdulkarim et al., 1994). However, many of the fusidic acid mutations in EF-G are located in domain III (Johanson et al., 1996; Nagaev et al., unpublished). In the structure, derived from the H573A mutant, these mutations are found lining a pocket on the surface between domains II and III (Fig. 4). It is tempting to suggest that the interface between domains II and III is the binding site for fusidic acid. The interface has hydrophobic patches that may fit the nature of fusidic acid. If fusidic acid were bound at this interface, it would lead to a more open conformation of EF-G, as observed by Agrawal et al. (1998). The vicinity of this interface to the switch II loop is worth noting. The previously suggested binding site for fusidic acid between domains G and V on EF-G would lead to a bent conformation, contrary to the observation by Agrawal et al. (1998).

Some fusidic acid-resistant mutants grow very slowly. Slow growth is associated with a reduced rate of exchange of GTP for GDP on EF-G off the ribosome (Johanson et al., 1996; Macvanin et al., submitted). Many revertants from slow growth are located between domains G and II, and most of these do not reduce the level of resistance to fusidic acid. The revertant mutations probably restore the balance between the GDP and GTP conformations of EF-G off the ribosome (Johanson et al., 1996). The structural aspects of nucleotide exchange of EF-G remain poorly understood.

Figure 4. Locations of fusidic acid resistant mutations in EF-G (indicated in pink). The six most resistant mutants are shown in red. The right panel is an enlargement of the central region of the left panel.

Figure 5. Superposition of EF-G (van der Waals surface) in complex with GDP and the ternary complex of EF-Tu with tRNA and GDPNP (atomic model). The ternary complex has been found to have a somewhat more open conformation.

COMPARISONS OF CONFORMATIONS OF EF-G AND EF-Tu

With the striking similarity of the overall shape of EF-G to the ternary complex of EF-Tu with tRNA and GDPNP, and with some conformational variation in EF-G available, a more detailed comparison may be done. In the case of the ternary complex, two states with different conformations have been observed. One was observed for the free ternary complex (Nissen et al., 1995, 1999). The fit of the crystallographic structure of EF-G·GDP to the free ternary complex is good but not perfect, due to a more open conformation of the ternary complex (Fig. 5). The other conformation is the one bound to the ribosome in complex with kirromycin (Stark et al., 1997). The fitting of the crystallographic ternary complex into the EM difference density is not perfect but is suggestive (Stark et al., 1997). If the kirromycin-resistant mutations in EF-Tu indicate the binding site of the antibiotic (Abdulkarim et al., 1994), the binding of kirromycin would move domain III away from the G domain. This would lead to a better fit of the ternary complex to the EM density (Stark et al., 1997). Such a conformation of the ternary complex would be close to the bent conformation of EF-G.

POSSIBLE CONFORMATIONAL CHANGES OF THE ELONGATION FACTORS DURING THEIR FUNCTIONAL CYCLES

A schematic illustration of the elongation factor conformations can be seen in Fig. 6. The ternary complex of EF-Tu with tRNA and GDPNP as ob-

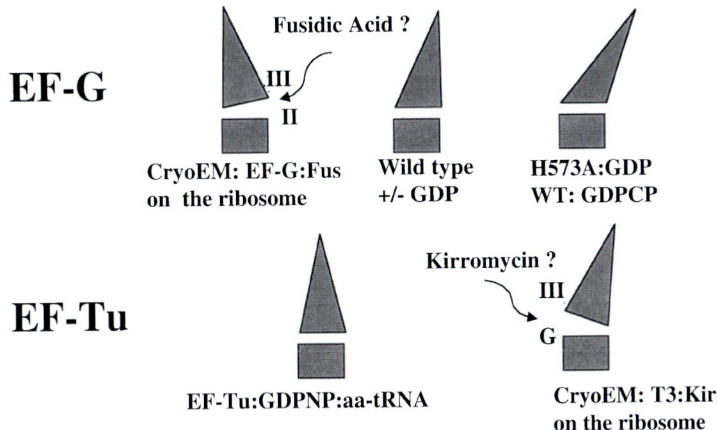

Figure 6. Comparison of the observed conformations of EF-G with those of the ternary complex of EF-Tu. Fus, fusidic acid; kirr, kirromycin; aa-tRNA, aminoacyl-tRNA.

served in the crystal structures is somewhat more open than the GDP conformation of EF-G (Fig. 5). The kirromycin complex of EF-Tu when bound to the ribosome (Stark et al., 1997) has a bent conformation comparable to the one observed for the H573A mutant of EF-G. The fusidic acid complex of EF-G when bound to the ribosome (Agrawal et al., 1998) has the most open structure. It binds with domain IV partly overlapping the A site tRNA close to the 30S subunit of the ribosome (Agrawal et al., 1998). It is evident that this is a primary site of interaction between the codon in the A site and the anticodon of the tRNA in the ternary complex during initial recognition. These observations can be related to the functional states.

It is not unlikely that the open conformation of EF-G, bound to the ribosome in complex with fusidic acid (Agrawal et al., 1998), is close to one in the normal elongation cycle. In such a case it must be a conformation that precedes the dissociation of EF-G·GDP. However, this complex, which dissociates from the ribosome, has a more bent conformation, the GDP conformation. Thus, in going from the open conformation to an intermediate GDP conformation, EF-G loses its affinity for the ribosome and vacates a space that would neatly fit the ternary complex, EF-Tu·tRNA·GTP, presumably in a conformation close to that in the crystal structures (Nissen et al., 1995, 1999).

The conformations that EF-Tu may go through to hydrolyze its GTP are not known. However, if the kirromycin-stalled ternary complex (Stark et al., 1997) has a conformation close to one in the normal elongation cycle, it should be a conformation preceding the dissociation of EF-Tu·GDP. When EF-G subsequently binds to pretranslocation ribosomes, it must have a conformation that does not overlap with the A-site tRNA, since translocation occurs after the factor has bound. Since the open fusidic acid conformation (Agrawal et al., 1998) overlaps with the A-site tRNA, the conformation of EF-G when binding should be more bent, perhaps like the one observed crystallographically (Czworkowski and Moore, personal communication; Laurberg et al., unpublished), but not so far identified as the GTP conformation. During one of the subsequent steps, EF-G adopts an open conformation like the one observed by cryo-EM. Since it overlaps the A-site tRNA, translocation must already have occurred, as is well known from studies of fusidic acid inhibition of protein synthesis. When EF-G has dissociated, it has the intermediate GDP conformation. Thus, during its functional cycle, EF-G may bind in a bent conformation and go through an open conformation before it dissociates in the intermediate GDP conformation.

SUMMARY

The conformational dynamics of the elongation factors is an important aspect of the elongation cycle. From the above discussion, the following hypothetical functional steps can be suggested. The ternary complex EF-Tu·tRNA·GTP binds in an open conformation (Nissen et al., 1995, 1999). If the match of the anticodon to the codon is of the cognate type, the GTPase activity of the factor will be induced (Rodnina et al., 1996). Before or after this GTP hydrolysis, the factor-tRNA complex adopts a more bent structure (Stark et al., 1997) and subsequently dissociates as the GDP complex, with its well-known conformation (Kjeldgaard and Nyborg, 1992).

As argued above, EF-G·GTP when binding to the ribosome probably must have a bent conformation to avoid overlap with the A-site tRNA. However, we do not yet know whether EF-G·GTP will adopt this conformation. Furthermore, upon translocation domains III and IV will move into this space. The structural details of this part of the elongation cycle remain unclear. However, after translocation but before the dissociation of EF-G·GDP with its well-known conformation, the complex may adopt the open conformation seen in the fusidic acid complex by Agrawal et al. (1998).

This work was supported by grants from the Swedish Natural Science Research Council, EC Biotechnology (BIO4-CT97-2188), INTAS (96-1562), and the Russian Academy of Sciences and Russian Foundation for Basic Research.

REFERENCES

Abdulkarim, F., L. Liljas, and D. Hughes. 1994. Mutations to kirromycin resistance occur in the interface of domains I and III of EF-Tu·GTP. *FEBS Lett.* **352:**118–122.

Abel, K., M. D. Yoder, R. Hilgenfeldt, and F. Jurnak. 1996. An α to β conformational switch in EF-Tu from *Escherichia coli. Structure* **4:**1153–1159.

Ævarsson, A., E. Brashnikov, M. Garber, J. Zheltonosova, Y. Chirgadze, S. Al-Karadaghi, L. A. Svensson, and A. Liljas. 1994. Three-dimensional structure of the ribosomal translocase: elongation G from *Thermus thermophilus. EMBO J.* **13:**3669–3677.

Agrawal, R. K., P. Penczek, R. A. Grassucci, and J. Frank. 1998. Visualization of elongation factor G on the *Escherichia coli* 70S ribosome: the mechanism of translocation. *Proc. Natl. Acad. Sci. USA* **95:**6134–6138.

Al-Karadaghi, S., A. Ævarsson, M. Garber, J. Zheltonosova, and A. Liljas. 1996. The structure of elongation factor G in complex with GDP: conformational flexibility and nucleotide exchange. *Structure* **4:**555–565.

Berchtold, H., L. Reshetnikova, C. O. Reiser, N. K. Schirmer, M. Sprinzl, and R. Hilgenfeldt. 1993. Crystal structure of active elongation factor Tu reveals major domain rearrangements. *Nature* **365:**126–132.

Berendsen, S., and H. J. C. Hayvard. 1998. Systematic analysis of domain motions in proteins from conformational change: new results on citrate synthetase and T4 lysozyme. *Proteins* **30:**144–154.

Bourne, H. R., D. A. Sanders, and F. McCormick. 1990. The GTPase superfamily: a conserved switch for diverse cell functions. *Nature* **248:**125–132.

Czworkowski, J., and P. B. Moore. 1997. The conformational properties of elongation factor G and the mechanism of translocation. *Biochemistry* **36:**10327–10334.

Czworkowski, J., and P. B. Moore. Personal communication.

Czworkowski, J., J. Wang, T. A. Steitz, and P. B. Moore. 1994. The crystal structure of elongation factor G complexed with GDP, at 2.7Å resolution. *EMBO J.* **13:**3661–3668.

Johanson, U., A. Ævarsson, A. Liljas, and D. Hughes. 1996. The dynamic structure of EF-G studied by fusidic acid resistance and internal revertants. *J. Mol. Biol.* **258:**420–432.

Kaziro, Y. 1978. The role of guanosine 5′-triphosphate in polypeptide chain elongation. *Biochim. Biophys. Acta* **505:**95–127.

Kjeldgaard, M., and J. Nyborg. 1992. Refined structure of elongation factor EF-Tu from Escherichia coli. *J. Mol. Biol.* **223:** 721–742.

Kjeldgaard, M., P. Nissen, S. Thirup, and J. Nyborg. 1993. The crystal structure of elongation factor EF-Tu from *Thermus aquaticus* in the GTP conformation. *Structure* **1:**35–50.

Kraulis, P. 1991. MOLSCRIPT: a program to produce both detailed and schematic plots of protein structures. *J. Appl. Crystallogr.* **24:**946–950.

Laurberg, M., O. Kristensen, S. Al-Karadaghi, K. Martemyanov, A. T. Gudkov, and A. Liljas. Unpublished data.

Liljas, A., and S. Al-Karadaghi. 1997. Structural aspects of protein synthesis. *Nat. Struct. Biol.* **4:**767–771.

Liljas, A., A. Ævarsson, S. Al-Karadaghi, M. Garber, J. Zheltonosova, and E. Brazhnikov. 1995. Crystallographic studies of elongation factor G. *Biochem. Cell Biol.* **73:**1209–1216.

Macvanin, M., U. Johanson, M. Ehrenberg, and D. Hughes. Fusidic acid resistant EF-G reduces the accumulation of (p)ppGpp. Submitted for publication.

Martemyanov, K. A., A. S.Yarunin, A. Liljas, and A. S. Gudkov. Unpublished data.

Nagaev, I., J. Björkman, D. I. Andersson, and D. Hughes. Unpublished data.

Nissen, P., M. Kjeldgaard, S. Thirup, L. Polekhina, L. Reshetnikova, B. F. C. Clark, and J. Nyborg. 1995. Crystal structure of the ternary complex of Phe-tRNA^{Phe}, EF-Tu, and a GTP analog. *Science* **270:**1464–1472.

Nissen, P., S. Thirup, M. Kjeldgaard, and J. Nyborg. 1999. The crystal structure of Cys-tRNACys–EF-Tu–GDPNP reveals general and specific features in the ternary complex and in tRNA. *Structure* **7:**143–156.

Polekhina, G., S. Thirup, M. Kjeldgaard, P. Nissen, C. Lippmann, and J. Nyborg. 1996. Helix unwinding in the effector region of elongation factor EF-Tu·GDP. *Structure* **4:**1141–1151.

Ramakrishnan, V., and S. W. White. 1998. Ribosomal protein structures: insights into the architecture, machinery and evolution of the ribosome. *Trends Biochem. Sci.* **23:**208–212.

Rodnina, M. V., T. Pape, R. Fricke, L. Kuhn, and W. Wintermeyer. 1996. Initial binding of the elongation factor Tu·GTP • aminoacyl-tRNA complex preceding codon recognition on the ribosome. *J. Biol. Chem.* **271:**646–652.

Rodnina, M. V., A. Savelsbergh, V. I. Katunin, and W. Wintermeyer. 1997. Hydrolysis of GTP by elongation factor G drives tRNA movement on the ribosome. *Nature* **385:**37–41.

Stark, H., M. V. Rodnina, J. Rinke-Appel, R. Brimacombe, W. Wintermeyer, and M. van Heel. 1997. Visualization of elongation factor Tu on the *Escherichia coli* ribosome. *Nature* **389:** 403–406.

Willie, G. R., N. Richman, W. O. Godtfredsen, and J. W. Bodley. 1975. Some characteristics and structural requirements for the interaction of 24,25-dihydrofusidic acid with ribosome elongation factor G complexes. *Biochemistry* **14:**1713–1718.

VIII. mRNA DECODING

VIII. mRNA DECODING

The mRNA decoding site on the 30S ribosomal subunit is a critical functional site where the codon-anticodon interaction between mRNA and tRNAs is regulated by the ribosome. The ribosomal site constitutes a highly conserved and highly modified region near the 3′ end of 16S rRNA, which is also a target site for the aminoglycoside antibiotics. In this region, the ribosome has the important role of maintaining the reading frame of the mRNA, and for some mRNAs it helps to modulate highly specific changes in the reading frame.

The Ribosome: Structure, Function, Antibiotics, and Cellular Interactions
Edited by R. A. Garrett, S. R. Douthwaite, A. Liljas, A. T. Matheson, P. B. Moore, and H. F. Noller

Chapter 30

Poking a Hole in the Sanctity of the Triplet Code: Inferences for Framing

JOHN F. ATKINS, ALAN J. HERR, CHRISTIAN MASSIRE, MICHAEL O'CONNOR, IVAYLO IVANOV, and RAYMOND F. GESTELAND

As our knowledge of the triplet code approaches its 40th birthday, the timing is opportune to reassess its readout. Once the reading frame is set at initiation, how does standard progressive triplet reading occur? What is the significance, for our perception of framing, of the exceptions where nontriplet decoding events efficiently occur at particular sequences? What determines and maintains step size? Is it just codon-anticodon base pairing, which is of course stabilized by the anticodon being part of a tRNA and by the ribosome? Or is there a triplet-counting mechanism inherent in the ribosome?

Ribosomes can change reading frame at particular sites—clearly there is mRNA sequence specificity to these sites. Does an inherent codon-base counting mechanism in the ribosome alter its behavior on sensing these sequences? The conventional view is that this is not the case but rather that sensing is done by tRNAs on the ribosome and that features of the tRNAs and their codon-anticodon interactions lead to the change in reading frame. There is no direct evidence for a counting mechanism in the ribosome, but there is the concern that we see frameshifts through the tRNAs that recognize these specific sequences and that a counting mechanism might underlie the tRNA-based interpretations.

It is universally accepted that tRNAs play a role in mRNA movement. A key question is whether that role influences framing. As will be seen below, there are great difficulties in proving that anticodon size determines step size in any particular instance, but the richness of data on the detailed properties of specific tRNAs leads us to conclude that tRNAs do play a role in framing.

In many cases the cognate tRNA first reads in a standard fashion and then detaches and re-pairs in a new frame—both events can be triplet pairing and still contribute to the change in framing. In addition, options of alternate anticodon presentation may play a crucial role. Also, as emphasized by Farabaugh and Björk (1999), in experiments where frameshifting is being studied, it is sometimes difficult to be sure which tRNA is doing the decoding—the mutant tRNA or a near-cognate tRNA. (Here we take cognate to mean the pair of tRNA and codon that has been evolutionarily selected, not just those pairs that use standard Watson-Crick or wobble pairing. We use the term near-cognate to mean pairing involving a tRNA other than the cognate tRNA but using some of the standard codon-anticodon base pairs. This usage is different from that of Farabaugh and Björk [1999] who we think inappropriately relegate some standard codon-anticodon interactions to the near-cognate category).

HISTORICAL PERSPECTIVE

Under special in vitro conditions, some translocation of A- and P-site tRNAs to E and P sites can take place in the absence of the elongation factors Tu and G (Peska, 1969; Gavrilova and Spirin 1971) or even of mRNA (Belitsina et al., 1981). These and other studies suggest the possibility that the translo-

John F. Atkins, Alan J. Herr, Ivaylo Ivanov, and Raymond F. Gesteland ■ Department of Human Genetics, University of Utah, 15N 2030E Room 7410, Salt Lake City, UT 84112-5330. **Christian Massire** ■ Institut de Biologie Moléculaire et Cellulaire du CNRS UPR 9002, 15 rue Descartes, 67084 Strasbourg, France. **Michael O'Connor** ■ Department of Molecular and Cellular Biology and Biochemistry, Brown University, Providence, RI 02912.

cation machine acts directly on tRNA and movement of mRNA is passive, driven by its association with tRNA (for a review, see Wilson and Noller, 1998). If so, then shouldn't anticodon size determine codon size? The genetic approach has been to isolate tRNA mutants that partially compensate locally for the effects of a frameshift mutation. The first sequencing of such a tRNA (encoded by *sufD41*) that restored some in-frame reading to a +1 frameshift mutant by a quadruplet step size (Yourno, 1972) revealed an anticodon loop with an extra base (Riddle and Carbon, 1973). However, the generality of a simple correlation between what was surmised to be a quadruplet anticodon and a corresponding quadruplet codon was cast in doubt by numerous early studies (e.g., Gaber and Culbertson, 1984), including our finding that the classic tRNA encoded by *sufD41* could, in the appropriate context, also mediate equally efficient shifting to the −1 frame (Weiss et al., 1990a). (tRNA frameshift suppressors for −1 frameshift mutants were never found to have 6 bases in the anticodon loop.) Nevertheless, with the expanded anticodon loops, +1 frameshifting is more efficient if a fourth base pair can form (Gaber and Culbertson, 1984; Curran and Yarus, 1987; O'Connor et al., 1989; Tuohy et al., 1992; Moore et al., submitted), indicating that base pairing is at least influential in determining step size. (The first +1 suppressors acted at runs of repeat bases, but later studies with other sequences focused on different anticodon loop stacking possibilities for 8-base-membered anticodon loops [Bossi and Smith, 1984; Curran and Yarus, 1987]. Interconversion between two conformations of 7-base-membered anticodon loops under certain conditions has been described by Labuda et al. [1985] and Gollnick et al. [1987].)

Mutant tRNAs with a normal-size anticodon loop can mediate both doublet and triplet translation (O'Mahony et al., 1989a, 1989b; Pagel et al., 1992; Herr, unpublished) or quadruplet translation (Björk et al., 1989; Tucker et al., 1989; Hüttenhofer et al., 1990; Tuohy et al., 1992; Hagervall et al., 1993; Pagel and Murgola, 1996; Qian and Björk, 1997). Interestingly, one of the mutant tRNAs isolated in these studies had its sole mutation not in the anticodon loop but in the 5′ C, C74, of the universally conserved CCA sequence at the 3′ end of a gene for *Escherichia coli* $tRNA_1^{Val}$ (O'Connor et al., 1993). Since there are four identical genes for this tRNA and the C74 mutation in just one gave the phenotype, the mutant tRNA must directly mediate the effect. Subsequent work showed that C74 pairs with G2252 of 23S rRNA (Samaha et al., 1995; Green et al., 1998) and that mutants of G2252 also caused frameshifting (Gregory et al., 1994). How this mutant causes frameshifting is unknown. One secondary mutant which had no phenotype on its own, the allosuppressor *moc-1*, increased the growth rate of strains with the C74 $tRNA_1^{Val}$ mutant (O'Connor et al., 1993). *moc-1* is now known to be an allele of the *hrpA* gene (O'Connor and Atkins, unpublished results), which encodes a 146-kDa protein with similarities to a DEAH box containing RNA helicase (Moriya et al., 1995).

A lack of strict correspondence between anticodon size and codon size has also been inferred from studies of mutants of other translation components that cause frameshifting with the normal wild-type (WT) tRNA complement. Such mutants have been found in elongation factor Tu–EF-1α (Hughes et al., 1987; Vijgenboom and Bosch, 1989; Sandbaken and Culbertson, 1988; Tuohy et al., 1990; Dinman and Kinzy, 1997; Farabaugh and Vimaladithan, 1998) and rRNA (O'Connor et al., 1992, 1997; O'Connor and Dahlberg, 1993, 1995, 1996; Dinman and Wickner, 1995) but strikingly are not generally found in EF-G (two intensive studies reinforce this point despite the positive properties of two mutants among the many studied [Dahlfors and Kurland, 1990; MacVanin et al., submitted]). Interestingly, a restrictive EF-G mutant (Hou et al., 1994) acts as an antisuppressor of the frameshift causing properties of the tRNA position 74 substitutions but not of the *sufS* or $tRNA^{Val}$ hopping mutants described in this chapter (O'Connor, unpublished). WT error frameshifting, as its designation implies, also occurs with WT tRNAs. The dramatic effect on the efficiency of certain types of frameshifting that can be achieved by perturbing the balance of specific WT components, especially tRNAs, can at first glance be interpreted as indicating a lack of strict correspondence between codon and anticodon sizes, but as the analysis presented below suggests, the permissible inferences are often not straightforward. Nevertheless, one example from this category will be dealt with here. Attempts to clone genetically selected tRNA external suppressors for a −1 frameshift mutant, *trpE91*, in the *Salmonella typhimurium* anthranilate synthetase gene did not give tRNA mutants but instead resulted in the repeated isolation of two WT genes which when overexpressed caused −1 frameshifting close to the site of the frameshift mutation with resultant synthesis of sufficient full-length product to permit cell growth. One of these two is the protein-encoding gene ORF 99 at map location 64.56 on the *E. coli* chromosome (ECD060.00; CDS u28375), for which no function is known (O'Connor, unpublished). How the product of this gene influences frameshifting remains to be determined. The other WT gene encodes $tRNA_1^{Gly}$ (anticodon CCC), which is near cognate for the codon GGA adjacent to the codon where the frameshift mutation occurred. (The cognate tRNA for GGA is $tRNA_2^{Gly}$ with anticodon 3′CCU*5′, where * indicates an unknown modification of U.) Misreading of GGA by the WT near-cognate $tRNA_1^{Gly}$ causes frameshifting, but the ratio of $tRNA_1^{Gly}$ to $tRNA_2^{Gly}$ has to be increased from the normal for an effect to be detectable (O'Connor, 1998). Furthermore, extensive important work on frameshift mutant suppression in Björk's laboratory has also focused attention on frameshifting mediated by WT near-cognate tRNAs in the presence of mutant cognate tRNAs (Qian et al., 1998).

FARABAUGH AND BJÖRK MODEL

Qian et al. (1998) have suggested that it is not the mutant tRNAs in the frameshift suppressor mutant strains that directly mediate frameshifting. Rather, the defects in the mutant tRNAs debilitate these tRNAs and allow decoding by WT near-cognate tRNAs to become competitive. In this model decoding by near-cognate tRNAs is then the direct cause of the frameshifting. Farabaugh and Björk (1999) extended the model to include the possibility that instead of a normal near-cognate tRNA an undermodified tRNA or a cognate tRNA altered in some other way may also be prone to frameshifting (due to slippage in the P site). This model also proposed that in the absence of mutant tRNAs, near-cognate tRNAs cause frameshifting. However, it is hard to see how this model explains frameshifting at runs of U, since in *E. coli* there is only one $tRNA^{Phe}$. Weiss et al. (1990a) found Phe to be inserted at the +1 frameshift site UUU U (the re-pair codon is underlined) in a WT strain, and Fu and Parker (1994) analyzed a similar situation in the *argI* gene of *E. coli* (but see Qian, 1997); Condron et al. (1991a, 1991b) found Val Phe corresponding to the phage T7 gene *10* −1 shift site G GUU UUC; Brierley et al. (1992) found very high −1 frameshifting in reticulocyte lysates at U UUU UUU; Schwartz and Curran (1997) analyzed frameshifting at UUU U in detail, and Doyon et al. (1998) found efficient frameshifting at the same sequence in a human immunodeficiency virus mutant resistant to protease inhibitors. As described below, it is WT cognate tRNA which mediates the shift in frame that occurs in phage T4 gene *60* decoding.

The significance of the Farabaugh-Björk model relates to its bearing on another proposal that they made at the same time: that triplet decoding is inherently a function of the ribosome and that while ribosomes mediate mRNA movement via pairing with tRNA, step size is unrelated to the number of codon-anticodon base pairs (provided it is enough to hold the codon-anticodon complex together) and to the relative positioning of tRNA and mRNA in the ribosomal A site.

COMPLEXITY OF THE RELATIONSHIP BETWEEN ANTICODON PAIRING AND STEP SIZE

Our only window into these issues is through exceptions to standard framing. Here we look at a few of these exceptions, keeping in mind the question of whether tRNA can act to influence codon size in the A site by its alignment as an incoming tRNA or in the P site where its "final" position determines the boundary of the next A-site codon.

Example 1: $tRNA_2^{Gly}$ Mutants and Frameshifting

Some of the external suppressors for the −1 frameshift mutant, *trpE91*, of *S. typhimurium* have as their sole mutation a U*-to-C change in anticodon position 34 of $tRNA_2^{Gly}$, to give the anticodon CCC (U* is an unknown modification of U) (O'Mahony et al., 1989b). The mutant tRNA directly causes −1 frameshifting at G GGA. At C GGA a significant amount of frameshifting still occurs, presumably due to pairing with just the GG of the GGA (O'Mahony et al., 1989a; Herr et al., unpublished). There are three models. (i) C36 and C35 pair with the GG of the GGA without C34 occluding the third codon base, A. Then a 2-base step ensues. (ii) Anticodon bases C35 and C34 of the incoming $tRNA_2^{Gly}$ in the A site pair with the first two codon bases, GG, of the GGA, and anticodon base C36 does not pair with a codon base, with a resulting 2-base step. (iii) C36 and C35 pair with the GG of the GGA (as in model 2) but with occlusion (or nonstandard pairing) of the third codon base, A, by anticodon base 34C. Then a triplet step ensues. In the P site (or during translocation?), pairing of anticodon bases 36 and 35 to the GG is disrupted and re-pairing to the same mRNA bases then occurs via anticodon bases 35 and 34 with no occlusion of the codon base A, leading to a −1 frameshift event in the P site. Such an "anticodon rollover" was previously suggested as one of the possibilities for mutant valine tRNA suppressors of *trpE91* (O'Connor et al., 1989). Current evidence (Herr et al., unpublished) favors model 3, and so this case does not appear to provide evidence for anticodon pairing influencing step size. (As an aside, it needs to be pointed out that for several $tRNA_2^{Gly}$ mutants which have their sole mutations outside the

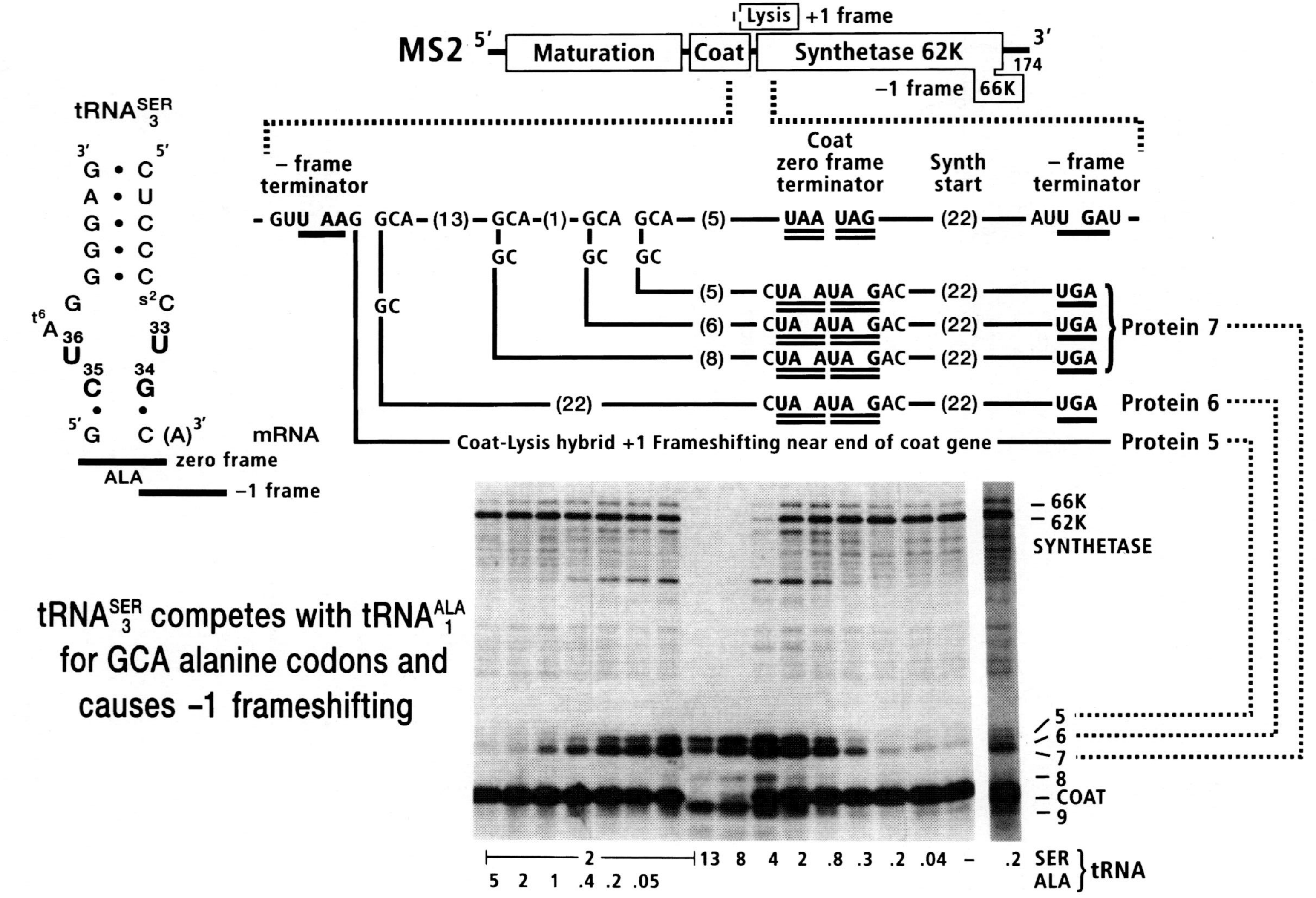
MS2 5′ Maturation Coat Synthetase 62K 3′
Lysis +1 frame
−1 frame 66K
174
tRNA$^{SER}_{3}$
3′ G • C 5′
A • U
G • C
G • C
G • C
G s^2C
t^6A 36
33 U
U 35
34 G
C
5′ G
C (A) 3′
mRNA
zero frame
ALA
−1 frame
− frame terminator
Coat zero frame terminator
Synth start
− frame terminator
- GUU AAG GCA−(13)−GCA−(1)−GCA GCA−(5)—UAA UAG—(22)—AUU GAU -
GC
GC GC GC
(5) — CUA AUA GAC— (22) —UGA
(6) — CUA AUA GAC— (22) —UGA
(8) — CUA AUA GAC— (22) —UGA
Protein 7
(22) — CUA AUA GAC— (22) —UGA
Protein 6
Coat-Lysis hybrid +1 Frameshifting near end of coat gene
Protein 5
− 66K
− 62K
SYNTHETASE
5
6
7
− 8
− COAT
− 9
2
13 8 4 2 .8 .3 .2 .04 − .2
5 2 1 .4 .2 .05
SER
ALA
tRNA
tRNA$^{SER}_{3}$ competes with tRNA$^{ALA}_{1}$ for GCA alanine codons and causes −1 frameshifting

anticodon arm [e.g., *sufS605*] it is the near-cognate $tRNA_1^{Gly}$ that directly mediates the frameshifting [Herr et al., unpublished], but for the U*34C mutant, *sufS601*, there is good evidence that it is the mutant $tRNA_2^{Gly}$ that directly mediates the frameshifting [Pagel et al., 1992].)

Frameshifting by the U*34C mutant of $tRNA_2^{Gly}$ sometimes occurs by dissociation and re-pairing to the overlapping GGG in the −1 frame in the sequence G GGA (O'Mahony et al., 1989a). This tRNA also mediates detachment and re-pairing to nonoverlapping codons (Herr et al., unpublished). With one of these nonoverlapping codons, CUG, only one codon base can be involved in Watson-Crick base pairing with the standard anticodon (which is relevant to an issue addressed later). (The involvement of $tRNA_2^{Gly}$ in T4 gene *60* bypassing is described below. Programmed frameshifting at the sequence G GGA occurs in lambda GT [Levin et al., 1993] and probably IS*407* and IS*869* frameshifting, but the tRNA involved has not yet been determined.)

In conclusion, the available evidence suggests that example 1 supports a role for tRNA influencing framing via its action in the P site.

Example 2: Decoding of a GCA Alanine Codon by $tRNA_3^{Ser}$

Two *E. coli* tRNAs can cause −1 frameshifting by reading specific codons for different amino acids. In *E. coli* extracts, AGC-decoding $tRNA_3^{Ser}$ (anticodon, 3′UCG5′) reads a GCA alanine codon to cause −1 frameshifting. This was only detected when the tRNA balance was changed by addition of the serine tRNA (Atkins et al., 1979) (Fig. 1). (A similar phenomenon was seen with $tRNA_3^{Thr}$ reading a CCA or CCG proline codon.) If $tRNA_3^{Ser}$ anticodon base U36 paired with the first codon base, G, then C:C apposition would be required for the second codon base and G-A for the third, which seems unlikely. The conclusion seems inescapable that anticodon bases 35 and 34, C and G, pair with codon bases G and C in the A site, which could simply lead to a −1 frameshift with a step size of 2, implying that anticodon size does influence step size. However, there are other alternatives. For instance, the A-site pairing might result in an initial step size of 3 and the real frameshift occurs in the P site. Perhaps a P-site shift could be facilitated by presentation of an alternate anticodon by the tRNA. Could there be a flux between standard anticodon loop stacking 2:5 (two on the 5′ side and five on the 3′ side) employed in standard decoding of serine codons AGU/C, with a novel 1:6 form that causes frameshifting? In the latter conformation, base U33 would stack below G34 (rather than below U32 [Fig. 2]), allowing transient pairing with codon base A. Unpairing with this A in the P site would allow the A to be the first base for the next tRNA entering the A site. Such a 1:6 anticodon stack may be possible despite the use of only a single base to connect the 3′ stack to the 5′ side of the anticodon stem. Indeed, the tight stacking of the six 3′-most nucleotides of the loop allows loop closure by the only remaining base, U32, thanks to the natural twist of the strand within regular RNA helices. This extensive 1:6 stacking makes the loop less widely open than in the 2:5 stacking, implying a slightly different position of the codon relative to the anticodon stem (Fig. 2). The 1:6 stacking is feasible at the expense of losing the noncanonical hydrogen bonds that stabilize the U-turn at base 33.

However, while a U is almost universally conserved at position 33, used for the ~180° turn of the phosphodiester backbone to the 3′ stack, U is not essential for the turn between the 5′ and 3′ stacks (Bare et al., 1983; Santos et al., 1997; von Ahsen et

Figure 1. Frameshifting near the end of the coat protein gene of the single-stranded RNA phage MS2 results in products larger than the coat protein. The first part of the lysis gene overlaps the end of the coat protein gene, and +1 frameshifting at an unknown site near the end of the coat protein gene yields a coat-lysis hybrid termed protein 5. Reading of the fourth-to-last GCA alanine codon in the coat protein gene by $tRNA_3^{Ser}$ causes a shift to the −1 frame, resulting in the synthesis of protein 6. Similar −1 frameshifting at any of the last three GCA codons yields protein 7, which, though it contains the same number of amino acids as protein 6, due to its different composition migrates faster on sodium dodecyl sulfate-polyacrylamide gel electrophoresis. Increasing the ratio of $tRNA_3^{Ser}$ (which normally reads AGC) to $tRNA_1^{Ala}$ (which normally reads GCA) (the numbers under the right-hand part of the autoradiogram give the amounts in micrograms per standard 12.5-μl reaction mixture) gives increasing levels of −1 frameshifting at GCA codons. The highest concentrations shown result in termination, to yield protein 9, at the last stop codon before the zero-frame terminator is reached. With elevated levels of $tRNA_3^{Ser}$ (2 μg), increasing the relative amount of $tRNA_1^{Ala}$ (the amounts in micrograms are given under the left-hand part of the autoradiogram) decreases the amount of frameshifting to the point where the only obvious product is the regular coat protein. The model for how stacking of 2 bases on the 5′ side and 5 bases on the 3′ side of the anticodon loop of $tRNA_3^{Ser}$ could mediate the shift to the −1 frame is indicated at the top left of the figure. With the normal balance of tRNAs in an *E. coli* extract (i.e., without tRNA addition), a −1 frameshift product, 66K, that is longer than the synthetase (the virus-encoded component of replicase) is detectable. This is due to an analogous type of noncognate decoding of a proline codon by $tRNA_3^{Thr}$.

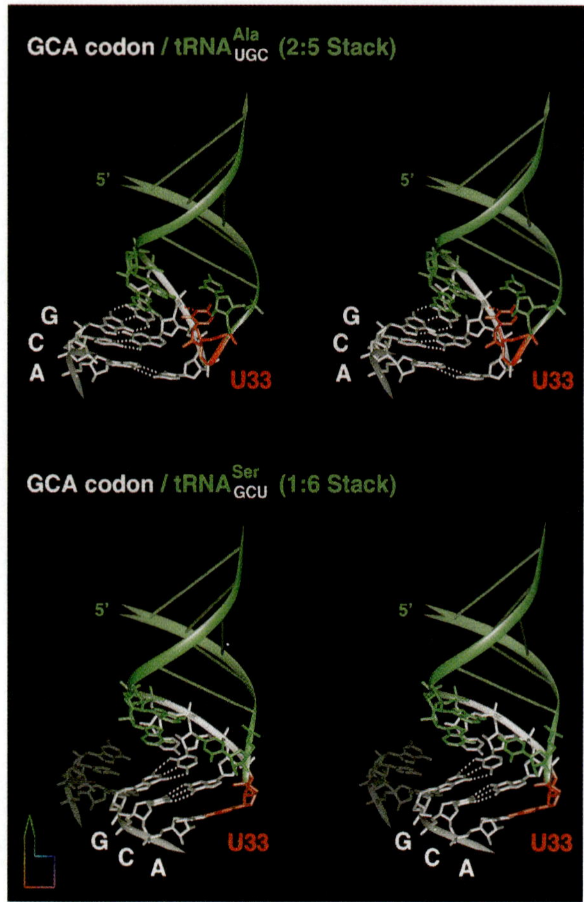

Figure 2. (Top) Stereographic representation of the three-dimensional modeling of the canonical interaction between a GCA codon (white) and the anticodon loop of $tRNA^{Ala}$. The anticodon loop displays the classical 2:5 stack. U33 (red) adopts a U-turn conformation. (Bottom) Stereographic representation of the three-dimensional modeling of the putative interaction between a GCA codon (white) and the anticodon loop of $tRNA^{GCU}$, involving a −1 frameshift. The anticodon loop displays an alternative 1:6 stack, with U33 flipped over to base-pair with codon base A. The trace of the GCA codon in the case of the canonical interaction is shown in gray. (These models were constructed with the MANIP modeling tool [Massire and Westhof, 1998]. The ribbon drawings were produced with DRAWNA software [Massire et al., 1994].)

al., 1997; Ashraf et al., 1999a; Ayer and Yarus, 1986). In $tRNA^{Phe}$ the standard anticodon stack is stabilized by the two noncanonical hydrogen bonds and stacking interactions (Quigley and Rich, 1976; Auffinger and Westhof, 1999). Irrespective of the generality of this feature, lack of some specialized features may contribute to instability of a 1:6 stack in the P site and a rollover of the anticodon in the P site, so that the third codon base, A, becomes available for pairing as the first codon base by the next tRNA, with a consequent −1 frameshift. If a 1:6 stack involving U33 pairing with the third codon base is the mechanism for this frameshift, then one might expect that the third codon base would need to be A, or perhaps G (unless more relaxed pairing rules are involved), and that the base at position 33 would need to be U when the third codon base is A. With A at tRNA position 33 and A as the third codon base (protein 6 in Dayhuff et al., 1986, Fig. 1), there is efficient −1 frameshifting (Bruce et al., 1986), and with U at position 33 and U as the third codon base weak frameshifting probably occurs (Dayhuff et al., 1986). It may be that a 1:6 stack with occlusion rather than pairing of the third codon base can be utilized in the A site. The possibility of a 1:6 stack is important to investigate because of the implication for the influence of anticodon structure and pairing on step size.

In conclusion, example 2 shows an influence of tRNA pairing on framing, but whether the effect is mediated at the A or P site is unresolved.

Example 3: a Role for Inosine in Frameshifting

The third candidate is the −1 frameshifting at A CGA AAG in decoding the *Bacillus subtilis* cytidine deaminase gene, *cdd* (Mejlhede et al., 1999). The $tRNA^{Lys}$ pairs first with AAG and then re-pairs with the underlined AAA in the −1 frame. However, the $tRNA^{Arg}$ must also be involved, since the identity of each base of its codon, CGA, is important for the frameshifting (the upstream A is not important). Unlike $tRNA^{Lys}$, the $tRNA^{Arg}$ is not likely to re-pair in the −1 frame, since no standard pairing is possible. The tRNA, $tRNA_2^{Arg}$, which normally decodes CGA, CGC, and CGU is the only tRNA in *B. subtilis* (or *E. coli*) that contains inosine—it is at position 34, the third codon base counterpart. There are four genes for $tRNA_2^{Arg}$, which is abundant compared to the sparse $tRNA_3^{Arg}$ (anticodon, 3′GCC5′; decodes CGG), which is encoded by a single gene, and the very sparse $tRNA_4^{Arg}$ (anticodon, 3′UC5-methylaminomethyluridine5′; decodes AGA and, much less efficiently, AGG, which is primarily decoded by $tRNA_5^{Arg}$), which is also encoded by only one gene (Dong et al., 1996). A-I apposition, as in CGA decoding by $tRNA_2^{Arg}$, is the only purine-purine combination used in decoding in *E. coli* and is at best very inefficient (see Curran, 1995). It seems most likely that it is $tRNA_2^{Arg}$ that is directly involved in the CGA AAG frameshifting, rather than $tRNA_3^{Arg}$ or $tRNA_4^{Arg}$ (it is known that Arg is the amino acid inserted).

The anticodon base inosine may simply facilitate the anticodon of the next tRNA re-pairing to the mRNA by permitting its base 36 to pair with the third codon base of the CGA. The A-I purine-purine apposition may also influence dissociation of the initial pairing of the next tRNA, via an effect on the pairing of the second ribosomal A-site base (Lim, 1995). (Decoding of a CGA immediately 5′ of a UAG [Curran, 1995] or a UGA [Björnsson et al., 1996] reduces readthrough of those codons by their respective suppressor tRNAs.) As the tRNA that decodes CGA is the only *E. coli* tRNA that contains inosine at position 34, it is not possible to test the importance, for the "inosine-mediated type" of frameshifting, of having the first two codon bases involved in G-C base pairs independent of their necessity of recruiting an inosine-containing tRNA. However, in certain eukaryotes (but not yeast), the sole tRNA that reads AUA contains inosine at position 34, but frameshifting tests with a codon pair such as AUA AAC have not yet been performed. (Where inosine occurs in mRNA, due to deamination of adenosine, its structural similarity of guanosine merely lacking the 2-amino group presumably ensures that tRNAs with A at the corresponding position are not competitive and that frameshifting at these sites is unlikely to be at a high level.)

At what stage in the ribosomal cycle in *B. subtilis* does the specific inosine-containing $tRNA^{Arg}$ involved act to mediate the frameshifting? One possibility is that when this tRNA is in the E site, it influences lysine tRNA anticodon re-pairing in the P site. Previously, an E/P-site posttranslocational slip was invoked (Horsfield et al., 1995) to explain a different case—frameshifting in *E. coli* on the human immunodeficiency virus shift sequence (see Lim, 1997). The alternative possibility is that when the $tRNA^{Arg}$ is in the P site, or during translocation, it permits $tRNA^{Lys}$ realignment from AAG to AAA. On the hybrid-sites model this would occur after peptidyl transfer, with the acceptor site of $tRNA^{Lys}$ in the P site (see Weiss et al., 1989; for a review, see Atkins and Gesteland, 1995). (One may be able to distinguish between these two alternatives by placing a stop, or very rare, codon directly 3′ of the AAG and determining whether the frameshifting efficiency changes and to what extent it is now dependent on the CG, since the stop codon itself will promote frameshifting at the A AAG.) At any rate, the codon utilized seems to be specified by the re-pairing of the $tRNA^{Lys}$ anticodon to mRNA (the anticodon of the sole lysine tRNA is 3′UUU*5′, where U* is 5-methylaminomethyl-2-thiouridine). The 2-thiouridine is important (see Kumar and Davis, 1997; Ashraf et al., 1999b), and the possible functioning of the different components of the position 34 modification have recently been reviewed (Curran, 1998). However, the anticodon loop of this tRNA is quite different from others that contain the same position 34 modification (Watanabe et al., 1994). Evidence has been provided that this is because of a cross-loop interaction with the *N*-6-threonylcarbamoyl A at position 37, resulting in a lack of the classical U-turn at position 33 (Agris et al., 1997), with a resultant 3:4 anticodon stack. Another possibility is rigidity in the structure of the anticodon loop. Only anticodon bases 36 and 35 may be paired to the zero-frame AAG. After the dissociation of this weak pairing, anticodon bases 36, 35, and 34 become involved in a stronger re-pairing to the mRNA via the −1 frame $\overline{\text{AAA}}$ in the sequence $\overline{\text{CGA A}}\text{AG}$ (see Tsuchihashi and Brown, 1992). Thus, in this case two tRNAs are involved at a hexanucleotide shift site—one re-pairs in a new frame aided by special properties of the other. This is distinct from commonly found −1 frameshifting on heptanucleotide sequences where two tRNAs simultaneously dissociate and re-pair in the −1 frame. This tandem shift mechanism can lead to high levels of frameshifting; it is no coincidence that the sequence most prone to −1 frameshifting in *E. coli* is $\overline{\text{A AA}}\ \overline{\text{A AA}}\text{G}$ (the codons involved in re-pairing are underlined) (Weiss et al., 1989). Initial weak pairing to AAG by $tRNA^{Lys}$ is utilized in the −1 programmed frameshifting in decoding *E. coli dnaX* (see Tsuchihashi and Brown, 1992), IS*911* (Rettberg et al., 1999, and references therein), and phage λ GT in the sequence G GGA AAG (Levin et al., 1993), and A AAG is also the site of −1 frameshifting in a base substitution revertant of a −1 frameshift mutant (Atkins et al., 1983). Also, peptidyl-$tRNA^{Lys}$ is most prone to complete dissociation from ribosomes (Heurgué-Hamard et al., 1996), presumably mostly from AAG codons.

In conclusion, example 3 shows that distinctive behavior of positions 34 of two adjacent tRNAs can influence framing, but the identities of their respective ribosomal sites have not been established.

Example 4: Out-of-Frame Pairing by an Incoming tRNA

Can a reading frame be reset by the positioning of an incoming tRNA by, for instance, skipping the first mRNA base in the ribosomal A site? If so, these would be cases where there is no re-pairing of tRNA in an alternate frame. Frameshifting of this type was not detected in an early study with shift sites immediately preceding a stop codon (Weiss et al., 1990a). However, with studies of the Ala Lys codons GCC $\text{A}\overline{\text{AG C}}$ in *E. coli* and starvation for lysine resulting in limiting aminoacyl-lysine-tRNA, Gallant and colleagues found the sequence Ala Ser due to reading of the underlined +1 frame codon by the cognate $tRNA^{Ser}_3$ (Peter et al., 1992). It has been implicitly assumed that re-pairing by the GCC-decoding $tRNA^{Ala}_2$ (anticodon, 3′CGG5′, with A at position 32 instead of the common pyrimidine [Mims et al., 1985]) does not occur on the overlapping +1 frame CCA (for a review, see Atkins and Gesteland, 1995), as only 1 bp could be formed. It seems likely that pairing of the cognate $tRNA_3^{Ser}$ to ribosomal A-site bases 2, 3, and 4 with the first A-site base being skipped is competitive with pairing to A-site bases 1, 2, and 3 by $tRNA^{Lys}$. (As introduced above, this lysine tRNA has a distinctive anticodon loop structure and pairs weakly with AAG but more strongly with AAA.) The simplest interpretation from this experiment is that the positioning and number of codon-anticodon bases influence the translocation step size.

In *Saccharomyces cerevisiae* Ty3 programmed frameshifting, there is out-of-frame pairing by an incoming tRNA. The frameshifting occurs at GCG AGU U, and the fact that AGU is a rare codon in yeast is important for the frameshifting (Farabaugh et al., 1993). Re-pairing of the GCG-decoding alanine tRNA with CGA is not possible without invoking non-Watson-Crick pairing. Rather, it seems that the incoming cognate aminoacyl-$tRNA^{Val}$ pairs with the second, third, and fourth bases in the A site. Only a small number of codons can substitute for GCG, including CGA (Vimaladithan and Farabaugh, 1994). It is likely that some special feature of the P-site wobble position pairing influences the ratio between productive pairing of an incoming tRNA cognate for bases 1, 2, and 3 of the A site or pairing by a different tRNA to 2, 3, and 4. Whether this base is considered part of the P site or A site is arbitrary until we know more about the sites. In *S. cerevisiae* the $tRNA^{Arg}$ that decodes CGA has the anticodon 3′GCI5′ (like *E. coli* [see above]). How the effect of the P-site I-A apposition in the case of CGA or the wobble pairing for the other tRNAs that are active influences skipping of the first A-site base is unclear. One proposed mechanism for skipping is occlusion, implying that an active tRNA physically blocks access to the first base of the next codon. However, it may be more appropriate to envisage a subtle effect on the positioning of the first A-site base—the extreme of this view is a complete bulging out of the mRNA base. Alternatively, maybe the angular alignment of the adjacent tRNAs on the ribosome is slightly different, allowing space for the anticodons to pair with codons 1 base apart. If so, the details would be a consequence of tRNA-tRNA interaction on ribosomes (Smith and Yarus, 1989; see Lim, 1997).

A second case of programmed frameshifting where skipping of the first A-site base has been inferred is with some mutants of the mammalian antizyme 1 frameshift site. However, whether this mechanism is operative with the WT sequence remains to be seen. With *Caenorhabditis elegans* antizyme, it is likely that detachment and re-pairing are involved (Ivanov et al., unpublished).

Summary: Examples 1 to 4

The question at the start of these four examples was whether tRNA can influence codon size in the A site by its alignment as an incoming tRNA or in the P site, with its "final" position determining the boundary of the next A-site codon. Example 1 is a candidate for the P site, where there is effectively a revision of step size which resets the frame. Example 2 could be A or P site. Example 3 illustrates either an A- and P- or P- and E-site effect of tRNA on framing. Prior to the important 1993 paper of Farabaugh et al. on Ty3 frameshifting, it was known that in addition to the importance of anticodon pairing to mRNA for mRNA movement in translocation, the nature of anticodon-codon pairing was important for realignment in the P site. As illustrated by example 4, we now know that the nature of the final P-site anticodon-codon pairing can influence the framing of the incoming tRNA, which redefines translocation. In example 4 the pairings of both the P-site and A-site tRNAs determine the frame. This again shows the importance of tRNA for framing but does not imply a direct effect on the translocation process itself. Whether there can be such a direct effect is unresolved. For instance, we do not know the translocation step size with mutant tRNAs, which directly mediate frameshifting and which have an extra anticodon loop base complementary to the fourth codon base. It is unknown whether four anticodon bases pair at any one time or whether a triplet anticodon initially pairs and then detaches and an alternate triplet anticodon that includes one "new" base re-pairs to mRNA (O'Connor et al., 1989). Such rollover could involve the type of different stacking conformations, 3:5 and 2:6, proposed earlier in this laboratory (Bossi and Smith, 1984). None of this should imply that step size determination is not multifaceted, with several important parameters. The tRNA data could also be interpreted to mean that tRNA-based mechanisms are able to override an intrinsic counting mechanism. All these examples have involved step sizes of 3 ±1. In contrast, we next look at a case where a normal tRNA leaves the imprint of a giant.

LONG-DISTANCE RE-PAIRING: SCANNING BY 70S RIBOSOMES CONTAINING PEPTIDYL-tRNA

During decoding of bacteriophage T4 gene *60*, ribosomes bypass 50 nucleotides in the mRNA coding sequence, between codons 46 and 47 (Fig. 3A). The peptidyl-tRNA$_2^{Gly}$ unpairs from a GGA codon to initiate scanning by the 70S peptidyl-tRNA ribosome complex and re-pairs with another GGA codon immediately preceding the "resume" codon (codon 47) (Weiss et al., 1990b; Herr et al., 1999). In several ways this mechanism is not unlike the unpairing and re-pairing in many types of frameshifting. However, it is strikingly different in that the peptidyl-tRNA completely escapes the local mRNA environment, freeing the mRNA to move over long distances rather than just ± 1 nucleotide. This bypassing gives us a different window into the frameshift problem.

One of the mRNA sequence elements that directs gene *60* bypass is a *cis*-acting signal corresponding to a stretch of amino acids translated from codons preceding the coding gap (Weiss et al., 1990b). One model for how the nascent peptide promotes bypassing is that it disrupts peptidyl-tRNA–mRNA pairing and switches the ribosome into a scanning mode (Herr, unpublished; Choi et al., 1998). Ribosomal protein L9 also plays a role (Herbst et al., 1994; Adamski et al., 1996); mutants partially suppress mutations in the nascent peptide, suggesting the possibility that L9 is the peptide's target. Not only do L9 mutants stimulate the 50-nucleotide gene *60* bypass, they also enhance hopping over stop codons, such as GGA UAA GGA (Herr, unpublished). This suggests that, like the nascent peptide signal, L9 mutants may destabilize peptidyl-tRNA–mRNA pairing directly and allow scanning. Recent work by Gallant and Lindsley (1998) has revealed interesting cases of long-distance bypassing stimulated by limitation of aminoacyl-tRNA for certain "hungry" codons (Gallant and Lindsley, 1998). Studies with L9 mutants in these situations have not yet been carried out.

L9 mutants have been found in one other genetic selection. The initial batch of external suppressors of the -1 frameshift mutant, *trpE91*, contained the *sufS* mutants of tRNA$_2^{Gly}$ discussed above in example 1. However, it also contained weaker suppressors (Riyasaty and Atkins, 1968) which mapped to the *supK* gene (Atkins and Ryce, 1974), which is now known to encode release factor 2 (Kawakami and Nakamura, 1990). An allosuppressor that increases the growth rate of the *trpE91*-containing strain with the mutant release factor 2 on selective media has a mutation in the gene for L9 (C. Johnston, unpublished). The effects of mutant L9 on different types of frameshifting are currently being assessed.

With gene *60*, the codon to which re-pairing of peptidyl-tRNA to mRNA occurs as a prelude to coding resumption is GGA, the same as the codon 47 nucleotides 5′, which was involved in the initial pairing. Six nucleotides 5′ of the "re-pairing" codon, the sequence GAG occurs, and this contributes to coding resumption. There is evidence that ribosomes traverse the coding gap via slippage of unpaired and unfolded mRNA, with the anticodon of peptidyl-tRNA within the ribosome scanning the mRNA for potential complementarity. With gene *60* there is no evidence that backward scanning can take place (the only case involving nonoverlapping codons where we have seen backward scanning, and then at a very low level, has been with constructs containing the antizyme 1 frameshift cassette expressed in *E. coli* [Matsufuji, unpublished]). In gene *60* scanning, it is likely that the GAG sequence acts as a mini-Shine-Dalgarno (SD) sequence to help re-pairing of the anticodon of peptidyl-tRNA (Adamski et al., unpublished). The SD-like interaction is unproved in this gene *60* case, which involves coding resumption at a considerable distance within the same mRNA. In the different case where tmRNA is involved, coding resumption occurs on a different mRNA (Keiler et al., 1996; Himeno et al., 1997) but without any hint of the involvement of an SD-like interaction. Williams et al. (1999) have intriguing evidence that a stop codon prior to the resume codon may influence coding resumption.

rRNA-mRNA INTERACTION

Studies of the programmed frameshifting involved in release factor 2 (Fig. 3B) (Weiss et al., 1987, 1988; Curran and Yarus, 1988), *dnaX* (Fig. 3C) (Larsen et al., 1994), *cdd* (Mejlhede et al., 1999), and IS*911* (Prère et al., unpublished; see Rettberg et al., 1999) decoding have shown that the anti-SD sequence near the 3′ end of 16S rRNA in 70S ribosomes scans mRNA for potential complementarity during translation.

The influence of the spacing of the internal SD sequence from the shift site on the directionality of the frameshifting is intriguing. An SD sequence spaced 3 nucleotides 5′ of an appropriate shift site stimulates $+1$ frameshifting, whereas when the spacing is 9 to 14, -1 frameshifting is stimulated (Larsen et al., 1994). A 5′ element is also important for the autoregulatory frameshifting in decoding mammalian antizyme, despite the absence of the possibility that an SD-type interaction is involved in this case (Matsufuji et al., 1995; Ivanov et al., 1998; Matsufuji et al., unpublished). Whether some alternative interaction with rRNA or an interaction with ribosomal proteins is involved is unknown, in contrast to the situation with the SD-like sequences just 5′ of the prokaryotic frameshift sites.

In an evolutionary sense, it is unknown whether the SD pairing by 30S subunits to identify the nearby

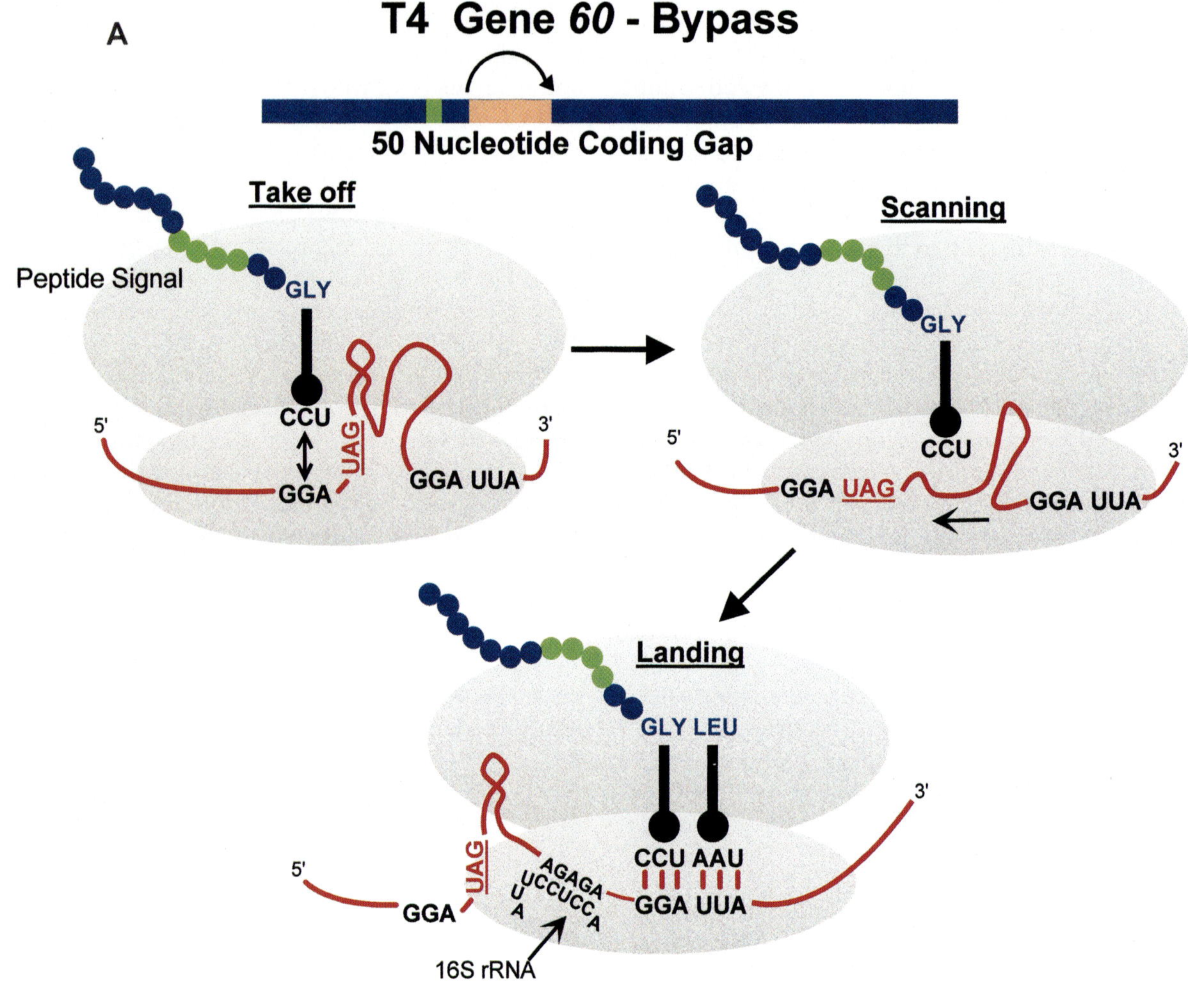

Figure 3. (A) Ribosomes bypass 50 nucleotides to decode phage T4 gene *60*. The signals important for bypass are matched "take-off" and "landing" sites (GGA), a stop codon immediately 3′ of the take-off site, a stem-loop at the beginning of the coding gap, and a critical region of the nascent peptide that acts within the ribosome. (B) An autoregulatory +1 frameshift at codon 26 is required to synthesize *E. coli* release factor (RF) 2, which mediates polypeptide chain release at UGA. tRNA Leu dissociates from pairing with its codon CUU to re-pair to the mRNA via the overlapping +1 frame codon UUU, which includes the first base of a zero-frame UGA stop codon. An SD sequence 3 bases 5′ of the shift site is important for the efficiency of the frameshifting. (C) Two-thirds of the way through the coding sequence of *E. coli dnaX*, 50% of the ribosomes shift to the −1 frame to synthesize the gamma product. Gamma and the product of standard decoding are present in a 1:1 ratio as subunits of DNA polymerase III. A "slippery" heptanucleotide shift site, a 5′ SD sequence sense by translating ribosomes, and a 3′ stem-loop are important for frameshifting.

AUG codons as initiators rather than as internal methionine codons followed from a role of temporarily tethering mRNA during translation. No trace of a remnant periodicity of internal SD sequences has been reported. Trifonov has made a specific proposal for mRNA sequence periodicity playing a role in framing via an interaction with the rRNA of translating ribosomes (Trifonov, 1987, 1992; Lagunez-Ortero and Trifonov, 1992). However, the experiments performed to test the model have not been positive (Weiss et al., 1990a; Curran and Gross, 1994).

PERSPECTIVE

Maybe selection balanced the need for accurate standard decoding with the opportunities provided by programmed frameshifting. It seems not to have been advantageous to achieve sacrosanct triplet decoding. Obviously selection operated to aid the utilization of programmed frameshifting with recoding signals present in mRNA, but what about selection at the tRNA level to optimize frameshifting? Are there cases where tRNAs which are optimal for triplet decoding are present in suboptimal concentrations or

B **Release Factor 2 Frameshift**

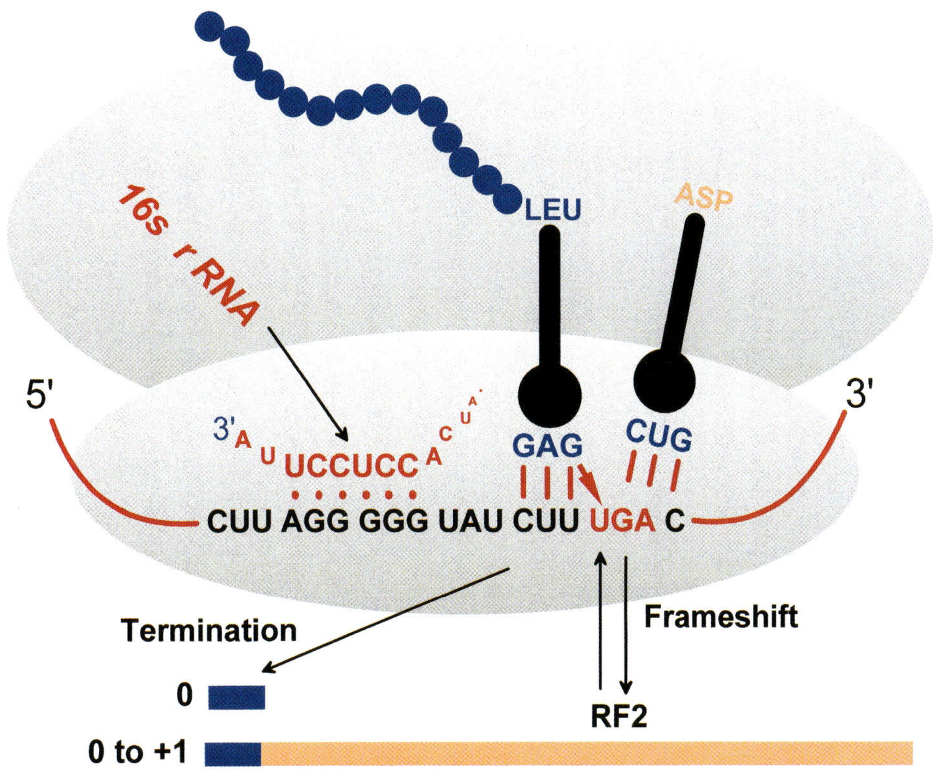

C ***dnaX*** **Frameshift - DNA Pol III Subunits**

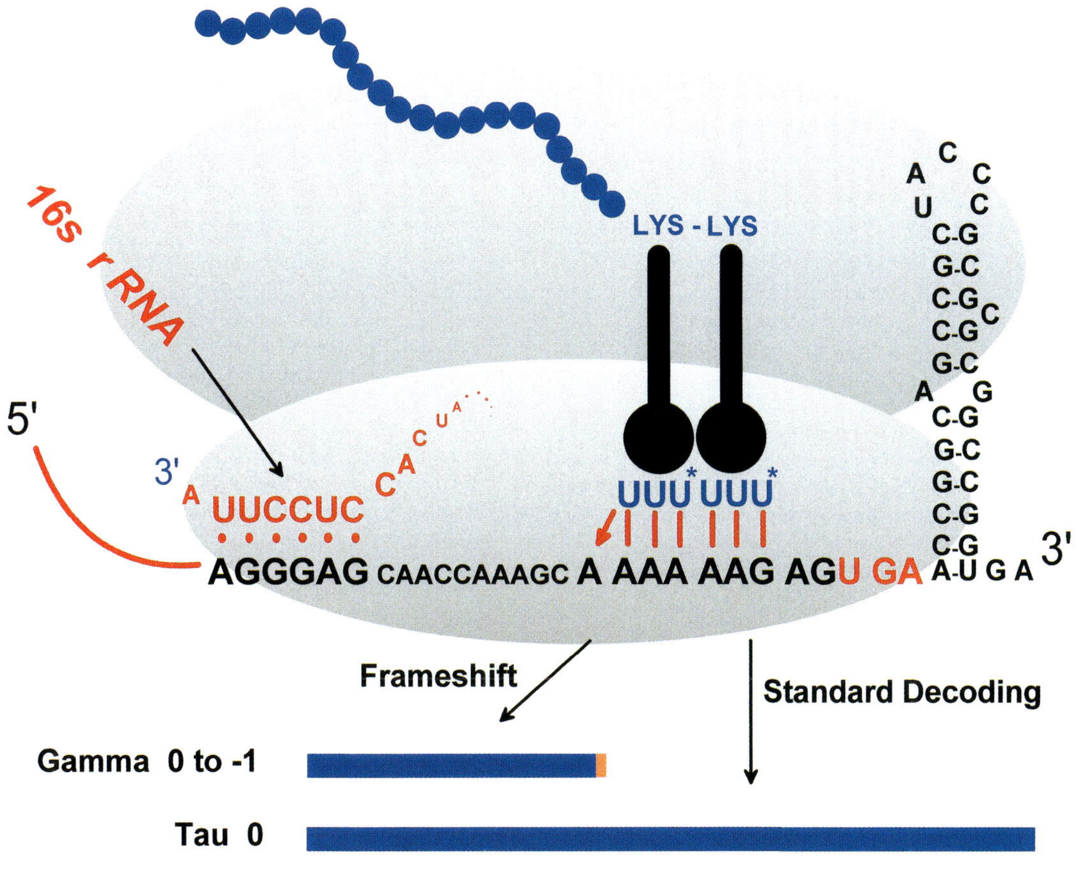

are missing as selection operated to balance the opportunity for frameshifting? Related to this, are there any small-genome organisms where one codon is used only for the dual purposes of permitting both standard decoding and frameshifting? Are there cases where tRNA modifications are added for the purpose of facilitating frameshifting? Are there special proteins whose function is to interact with specific stimulatory signals in mRNA for frameshifting or redefinition of stop codons and locally alter the decoding properties of the ribosome? In a sense, the special elongation factor SELB (Kromayer et al., 1996) does this for bacterial selenocysteine incorporation, but the answer for frameshifting is unknown. Only very limited answers to these and many other questions about frameshifting are known; there is much to discover.

Currently two rather different philosophical views about ribosomal frameshifting are being taken. One is to consider programmed frameshifting as amplified errors whose main interest is the insight they provide into translational errors (Farabaugh, 1997). The other is to value the richness of nature's exploitation of opportunities to generate high-efficiency frameshifting at particular sites for gene expression purposes. Certainly, framing errors (Atkins et al., 1972, 1983; Fox and Weiss-Brummer, 1980) are important for several reasons, including the possible amelioration of some genetic diseases due to frameshift mutant leakiness and immunological problems associated with epitopes derived from nonprimary open reading frames (see Mayrand and Green, 1998). When nature chooses to use non-Watson-Crick base pairs, should we judge her to be in error? We regard the phrase "programmed errors" as an oxymoron, because the sequences utilized in some cases of programmed frameshifting have been conserved for at least hundreds of millions of years.

We thank Al Dahlberg for his Brownian movements in sharing his ribosomal expertise and good humor with "nonribosomologists living in Utah," Senya Matsufuji for a great system and friendship, Glenn Björk and Jim McCloskey for imparting to us their appreciation of modified bases, Phil Farabaugh for his constructive and witty provocation, Ulfert Hornemann for helpful comments, and Owen White for a suggestion.

Our current work is supported by NIH grant GM48152 to J.F.A. and Department of Energy grant DEFG03-99ER62732 to R.F.G.

REFERENCES

Adamski, F. M., B. Moore, R. F. Gesteland, and J. F. Atkins. Unpublished data.

Adamski, F. M., J. F. Atkins, and R. F. Gesteland. 1996. Ribosomal protein L9 interactions with 23S rRNA: the use of a translational bypass assay to study the effect of amino acid substitutions. *J. Mol. Biol.* **261:**357–371.

Agris, P. F., R. Guenther, P. C. Ingram, M. M. Basti, J. W. Stuart, E. Sochacka, and A. Malkiewicz. 1997. Unconventional structure of $tRNA^{Lys}_{SUU}$ anticodon explains tRNA's role in bacterial and mammalian ribosomal frameshifting and primer selection by HIV-1. *RNA* **3:**420–428.

Ashraf, S. S., G. Ansari, R. Guenther, E. Sochacka, A. J. Malkiewicz, and P. F. Agris. 1999a. The uridine in "U-turn": contributions to tRNA-ribosomal binding. *RNA* **5:**503–511.

Ashraf, S. S., E. Sochacka, R. Cain, R. Guenther, A. Malkiewicz, and P. F. Agris. 1999b. Single atom modification (O→S) of tRNA confers ribosome binding. *RNA* **5:**188–194.

Atkins J. F., and R. F. Gesteland. 1995. Discontinuous triplet decoding with or without repairing by peptidyl tRNA, p. 471–490. *In* D. Söll and U. RajBhandary (ed.), *tRNA: Structure, Biosynthesis, and Function.* ASM Press, Washington, D.C.

Atkins, J. F., and S. Ryce. 1974. UGA and non-triplet suppressor reading of the genetic code. *Nature* **249:**527–530.

Atkins, J. F., D. Elseviers, and L. Gorini. 1972. Low activity of β-galactosidase in frameshift mutants of *Escherichia coli. Proc. Natl. Acad. Sci. USA* **69:**1192–1195.

Atkins, J. F., R. F. Gesteland, B. R. Reid, and C. W. Anderson. 1979. Normal tRNAs promote ribosomal frameshifting. *Cell* **18:**1119–1131.

Atkins, J. F., B. P. Nichols, and S. Thompson. 1983. The nucleotide sequence of the first externally suppressible −1 frameshift mutant, and of some nearby leaky frameshift mutants. *EMBO J.* **2:**1345–1350.

Aufinger, P., and E. Westhof. 1999. Singly and bifurcated hydrogen-bonded base-pairs in tRNA anticodon hairpins and ribozymes. *J. Mol. Biol.* **292:**467–483.

Ayer, D., and M. Yarus. 1986. The context effect does not require a fourth base pair. *Science* **231:**393–395.

Bare, L., A. G. Bruce, R. F. Gesteland, and O. C. Uhlenbeck. 1983. Uridine-33 in yeast tRNA is not essential for amber suppression. *Nature* **305:**554–556.

Belitsina, N. V., G. Z. Tnalina, and A. S. Spirin. 1981. Template-free ribosomal synthesis of polylysine from lysyl-tRNA. *FEBS Lett.* **131:**289–292.

Björk, G. R., P. M. Wikstrom, and A. S. Bystrom. 1989. Prevention of translational frameshifting by the modified nucleoside 1-methylguanosine. *Science* **244:**986–989.

Björnsson, A., S. Mottagui-Tabar, and L. A. Isaksson. 1996. Structure of the C-terminal end of the nascent peptide influences translation termination. *EMBO J.* **15:**1696–1704.

Bossi, L., and D. M. Smith. 1984. Suppressor *sufJ*: a novel type of tRNA mutant that induces translational frameshifting. *Proc. Natl. Acad. Sci. USA* **81:**6105–6109.

Brierley, I., A. J. Jenner, and S. C. Inglis. 1992. Mutational analysis of the "slippery sequence" component of a coronavirus ribosomal frameshifting signal. *J. Mol. Biol.* **227:**463–479.

Bruce, A. G., J. F. Atkins, and R. F. Gesteland. 1986. tRNA anticodon replacement experiments show that ribosomal frameshifting can be caused by doublet decoding. *Proc. Natl. Acad. Sci. USA* **83:**5062–5066.

Choi, K. M., J. F. Atkins, R. F. Gesteland, and R. Brimacombe. 1998. Flexibility of the nascent polypeptide chain within the ribosome: contacts from the peptide N terminus to a specific region of the 30S subunit. *Eur. J. Biochem.* **255:**409–413.

Condron, B. G., J. F. Atkins, and R. F. Gesteland. 1991a. Frameshifting in gene 10 of bacteriophage T7. *J. Bacteriol.* **173:**6998–7003.

Condron, B. G., R. F. Gesteland, and J. F. Atkins. 1991b. An analysis of sequences stimulating frameshifting in the decoding of gene 10 of bacteriophage T7. *Nucleic Acids Res.* **19:**5607–5612.

Curran, J. F. 1995. Decoding with an A:I wobble pair is inefficient. *Nucleic Acids Res.* **23:**683–688.

Curran, J. F. 1998. Modified nucleosides in translation, p. 493–516. *In* H. Grosjean and R. Benne (ed.), *Modification and Editing of RNA.* ASM Press, Washington, D.C.

Curran, J. F., and B. L. Gross. 1994. Evidence that GHN phase bias does not constitute a framing code. *J. Mol. Biol.* **235:**389–395.

Curran, J. F., and M. Yarus. 1987. Reading frame selection and transfer RNA anticodon loop stacking. *Science* **238:**1545–1550.

Curran, J. F., and M. Yarus. 1988. Use of tRNA suppressors to probe regulation of *Escherichia coli* release factor 2. *J. Mol. Biol.* **203:**75–83.

Dahlfors, A. A. R., and C. G. Kurland. 1990. Novel mutants of elongation factor G. *J. Mol. Biol.* **215:**549–557.

Dayhuff, T. J., J. F. Atkins, and R. F. Gesteland. 1986. Characterization of ribosomal frameshift events by protein sequence analysis. *J. Biol. Chem.* **261:**7491–7500.

Dinman, J. D., and T. G. Kinzy. 1997. Translational misreading: mutations in translation elongation factor 1α differentially affect programmed ribosomal frameshifting and drug sensitivity. *RNA* **3:**870–881.

Dinman, J. D., and R. B. Wickner. 1995. 5S rRNA is involved in fidelity of translational reading frame. *Genetics* **141:**95–105.

Dong, H., L. Nilsson, and C. G. Kurland. 1996. Co-variation of tRNA abundance and codon usage in Escherichia coli at different growth rates. *J. Mol. Biol.* **260:**649–663.

Doyon, L., C. Payant, L. Brakier-Gingras, and D. Lamarre. 1998. Novel Gag-Pol frameshift site in human immunodeficiency virus type 1 variants resistant to protease inhibitors. *J. Virol.* **72:**6146–6150.

Farabaugh, P. J. 1997. *Programmed Alternative Reading of the Genetic Code,* p. 1–208. R. G. Lanes Co., Austin, Tex.

Farabaugh, P. J., and G. R. Björk. 1999. How translational accuracy influences reading frame maintenance. *EMBO J.* **18:**1427–1434.

Farabaugh, P. J., and A. Vimaladithan. 1998. Effect of frameshift-inducing mutants of elongation factor 1α on programmed +1 frameshifting in yeast. *RNA* **4:**38–46.

Farabaugh, P. J., H. Zhao, and A. Vimaladithan. 1993. A novel programmed frameshift expresses the POL3 gene of retrotransposon Ty3 of yeast: frameshifting without tRNA slippage. *Cell* **74:**93–103.

Fox, T. D., and B. Weiss-Brummer. 1980. Leaky +1 and −1 frameshift mutations at the same site in a yeast mitochondrial gene. *Nature* **288:**60–63.

Fu, C., and J. Parker. 1994. A ribosomal frameshifting error during translation of the *argI* mRNA of *Escherichia coli. Mol. Gen. Genet.* **243:**434–441.

Gaber, R. F., and M. R. Culbertson. 1984. Codon recognition during frameshift suppression in *Saccharomyces cerevisiae. Mol. Cell. Biol.* **4:**2052–2061.

Gallant, J. A., and D. Lindsley. 1998. Ribosomes can slide over and beyond "hungry" codons, resuming protein chain elongation many nucleotides downstream. *Proc. Natl. Acad. Sci. USA* **95:**13771–13776.

Gavrilova, L. P., and A. S. Spirin. 1971. Stimulation of "non-enzymatic" translocation in ribosomes by p-chloromercuribenzoate. *FEBS Lett.* **17:**324–326.

Gollnick, P., C. C. Hardin, and J. Horowitz. 1987. ^{19}F nuclear magnetic resonance as a probe of anticodon structure in 5-fluorouracil-substituted *Escherichia coli* transfer RNA. *J. Mol. Biol.* **197:**571–584.

Green, R., C. Switzer, and H. F. Noller. 1998. Ribosome-catalyzed peptide-bond formation with an A-site substrate covalently linked to 23S ribosomal RNA. *Science* **280:**286–289.

Gregory, S. T., K. R. Lieberman, and A. E. Dahlberg. 1994. Mutations in the peptidyl transferase region of *E. coli* 23S rRNA affecting translational accuracy. *Nucleic Acids Res.* **22:**279–284.

Hagervall, T. G., T. M. F. Tuohy, J. F. Atkins, and G. R. Björk. 1993. Deficiency of 1-methylguanosine in tRNA from *Salmonella typhimurium* induces frameshifting by quadruplet translocation. *J. Mol. Biol.* **232:**756–765.

Herbst, K. L., L. M. Nichols, R. F. Gesteland, and R. B. Weiss. 1994. A mutation in ribosomal protein L9 affects ribosomal hopping during translation of gene 60 from bacteriophage T4. *Proc. Natl. Acad. Sci. USA* **91:**12525–12529.

Herr, A. J. Unpublished data.

Herr, A. J., J. F. Atkins, and R. F. Gesteland. 1999. Mutations which alter the elbow region of $tRNA_2^{Gly}$ reduce T4 gene 60 translational bypassing efficiency. *EMBO J.* **18:**2886–2896.

Herr, A. J., R. F. Gesteland, and J. F. Atkins. Unpublished data.

Heurgué-Hamard, V., L. Mora, G. Guarneros, and R. H. Buckingham. 1996. The growth defect in *Escherichia coli* deficient in peptidyl-tRNA hydrolase is due to starvation for $Lys\text{-}tRNA^{Lys}$. *EMBO J.* **15:**2826–2833.

Himeno, H., M. Sato, T. Tadaki, M. Fukushima, C. Ushida, and A. Muto. 1997. In vitro trans-translation mediated by alanine-charged 10Sa RNA. *J. Mol. Biol.* **268:**803–808.

Horsfield, J. A., D. N. Wilson, S. A. Mannering, F. M. Adamski, and W. P. Tate. 1995. Prokaryotic ribosomes recode the HIV-*gag-pol*-1 frameshift sequence by an E/P site post-translocation simultaneous slippage mechanism. *Nucleic Acids Res.* **23:**1487–1494.

Hou, Y., E. S. Yaskowiak, and P. E. March. 1994. Carboxyl-terminal amino acid residues in elongation factor G essential for ribosome association and translocation. *J. Bacteriol.* **176:**7038–7044.

Hughes, D., J. F. Atkins, and S. Thompson. 1987. Mutants of elongation factor Tu promote ribosomal frameshifting and nonsense readthrough. *EMBO J.* **6:**4235–4239.

Hüttenhofer, A., B. Weiss-Brummer, G. Dirheimer, and R. P. Martin. 1990. A novel type of +1 frameshift suppressor: a base substitution in the anticodon stem of a yeast mitochondrial serine-tRNA causes frameshift suppression. *EMBO J.* **9:**551–558.

Ivanov, I. P., R. F. Gesteland, and J. F. Atkins. 1998. A second mammalian antizyme: conservation of programmed ribosomal frameshifting. *Genomics* **52:**119–129.

Ivanov, I. P., S. Matsufuji, R. F. Gesteland, and J. F. Atkins. Unpublished data.

Johnston, C. Unpublished data.

Kawakami, K., and Y. Nakamura. 1990. Autogenous suppression of an opal mutation in the gene encoding peptide chain release factor 2. *Proc. Natl. Acad. Sci. USA* **87:**8432–8436.

Keiler, K. C., P. R. H. Waller, and R. T. Sauer. 1996. Role of a peptide tagging system in degradation of proteins synthesized from damaged messenger RNA. *Science* **271:**990–993.

Kromayer, M., R. Wilting, P. Tormay, and A. Böck. 1996. Domain structure of the prokaryotic selenocysteine-specific elongation factor SELB. *J. Mol. Biol.* **262:**413–420.

Kumar, R. K., and D. R. Davis. 1997. Synthesis and studies on the effect of 2-thiouridine and 4-thiouridine on sugar conformation and RNA duplex stability. *Nucleic Acids Res.* **25:**1272–1280.

Labuda, D., G. Stricker, H. Grosjean, and D. Pörschke. 1985. Mechanism of codon recognition by transfer RNA studies with oligonucleotides larger than triplets. *Nucleic Acids Res.* **13:** 3667–3683.

Lagunez-Otero, J., and E. N. Trifonov. 1992. mRNA periodical infrastructure complementary to the proof-reading site in the ribosome. *J. Biomol. Struct. Dyn.* **10:**455–464.

Larsen, B., N. M. Wills, and R. F. Gesteland. 1994. rRNA-mRNA base pairing stimulates a programmed −1 ribosomal frameshift. *J. Bacteriol.* **176:**6842–6851.

Levin, M. E., R. W. Hendrix, and S. R. Casjens. 1993. A programmed translational frameshift is required for the synthesis of a bacteriophage λ tail assembly protein. *J. Mol. Biol.* **234:**124–139.

Lim, V. I. 1995. Analysis of action of the wobble adenine on codon reading within the ribosome. *J. Mol. Biol.* **252:**277–282.

Lim, V. I. 1997. Analysis of interactions between the codon-anticodon duplexes within the ribosome: their role in translocation. *J. Mol. Biol.* **266:**877–890.

MacVanin, M., M. E. Johanson, and D. Hughes. Submitted for publication.

Massire, C., and E. Westhof. 1998. MANIP: an interactive tool for modeling RNA. *J. Mol. Graph. Model.* **16:**197–205.

Massire, C., C. Gaspin, and E. Westhof. 1994. DRAWNA: a program for drawing schematic views of nucleic acids. *J. Mol. Graph.* **12:**201–206.

Matsufuji, S., T. Matsufuji, Y. Miyazaki, Y. Murakami, J. F. Atkins, R. F. Gesteland, and S. Hayashi. 1995. Autoregulatory frameshifting in decoding mammalian ornithine decarboxylase antizyme. *Cell* **80:**51–60.

Matsufuji, S. Unpublished data.

Mayrand, S.-M., and W. R. Green. 1998. Non-traditionally derived CTL epitopes: exceptions that prove the rules? *Immunol. Today* **19:**551–556.

Mejlhede, N., J. F. Atkins, and J. Neuhard. 1999. Ribosomal −1 frameshifting during decoding of *Bacillus subtilis cdd* occurs at the sequence CGA AAG. *J. Bacteriol.* **181:**2930–2937.

Mims, B. H., N. E. Prather, and E. J. Murgola. 1985. Isolation and nucleotide sequence analysis of $RNA^{Ala}{}_{GGC}$ from *Escherichia coli* K-12. *J. Bacteriol.* **162:**837–839.

Moore, B., B. Persson, C. C. Nelson, R. F. Gesteland, and J. F. Atkins. Quadruplet codons: implications for code expansion and the specification of translation step size. Submitted for publication.

Moriya, H., H. Kasai, and K. Isono. 1995. Cloning and characterization of the hrpA gene in the terC region of *Escherichia coli* that is highly similar to the DEAH family RNA helicase genes of *Saccharomyces cerevisiae*. *Nucleic Acids Res.* **23:**595–598.

O'Connor, M. 1998. tRNA imbalance promotes −1 frameshifting via near-cognate decoding. *J. Mol. Biol.* **279:**727–736.

O'Connor, M. Unpublished data.

O'Connor, M., and J. F. Atkins. Unpublished results.

O'Connor, M., and A. E. Dahlberg. 1993. Mutations at U2555, a tRNA-protected base in 23S rRNA, affect translational fidelity. *Proc. Natl. Acad. Sci. USA* **90:**9214–9218.

O'Connor, M., and A. E. Dahlberg. 1995. The involvement of two distinct regions of 23S ribosomal RNA in tRNA selection. *J. Mol. Biol.* **254:**838–847.

O'Connor, M., and A. E. Dahlberg. 1996. The influence of base identity and base pairing on the function of the α-sarcin loop of 23S rRNA. *Nucleic Acids Res.* **24:**2701–2705.

O'Connor, M., R. F. Gesteland, and J. F. Atkins. 1989. tRNA hopping: enhancement by an expanded anticodon. *EMBO J.* **8:** 4315–4323.

O'Connor, M., H. U. Goringer, and A. E. Dahlberg. 1992. A ribosomal ambiguity mutation in the 530 loop of *E. coli*. *Nucleic Acids Res.* **20:**4221–4227.

O'Connor, M., N. M. Wills, L. Bossi, R. F. Gesteland, and J. F. Atkins. 1993. Functional tRNAs with altered 3′ ends. *EMBO J.* **12:**2559–2566.

O'Connor, M., C. L. Thomas, R. A. Zimmermann, and A. E. Dahlberg. 1997. Decoding fidelity at the ribosomal A and P sites: influence of mutations in three different regions of the decoding domain in 16S rRNA. *Nucleic Acids Res.* **25:**1185–1193.

O'Mahony, D. J., D. Hughes, S. Thompson, and J. F. Atkins. 1989a. Suppression of a −1 frameshift mutation by a recessive tRNA suppressor which causes doublet decoding. *J. Bacteriol.* **171:**3824–3830.

O'Mahony, D. J., B. H. Mims, S. Thompson, E. J. Murgola, and J. F. Atkins. 1989b. Glycine tRNA mutants with normal anticodon loop size cause −1 frameshifting. *Proc. Natl. Acad. Sci. USA* **86:**7979–7983.

Pagel, F. T., and E. J. Murgola. 1996. A base substitution in the amino acid acceptor stem of $tRNA^{Lys}$ causes both misacylation and altered decoding. *Gene Expr.* **6:**101–112.

Pagel, F. T., T. M. F. Tuohy, J. F. Atkins, and E. J. Murgola. 1992. Doublet translocation at GGA is mediated directly by mutant $tRNA_2^{Gly}$. *J. Bacteriol.* **174:**4179–4182.

Peska, S. 1969. Studies on the formation of transfer ribonucleic acid-ribosome complexes. *J. Biol. Chem.* **244:**1533–1539.

Peter, K., D. Lindsley, L. Peng, and J. A. Gallant. 1992. Context rules of rightward overlapping reading. *New Biol.* **4:**520–526.

Prère, M. F., C. Bertrand, R. F. Gesteland, J. F. Atkins, and O. Fayet. Unpublished data.

Qian, Q. 1997. Transfer RNA modification and translational frameshifting. Ph.D. thesis. Umea University, Umeå, Sweden.

Qian, Q., and G. R. Björk. 1997. Structural alterations far from the anticodon of the tRNA Pro GGG of *Salmonella typhimurium* induce +1 frameshifting at the peptidyl-site. *J. Mol. Biol.* **273:** 978–992.

Qian, Q., J.-N. Li, H. Zhao, T. G. Hagervall, P. J. Farabaugh, and G. R. Björk. 1998. A new model for phenotypic suppression of frameshift mutations by mutant tRNAs. *Mol. Cell* **1:**471–482.

Quigley, G. J., and A. Rich. 1976. Structural domains of transfer RNA molecules. *Science* **194:**796–806.

Rettberg, C. C., M. F. Prère, R. F. Gesteland, J. F. Atkins, and O. Fayet. 1999. A three-way junction and constituent stem-loops as the stimulator for programmed −1 frameshifting in bacterial insertion sequence IS911. *J. Mol. Biol.* **286:**1365–1378.

Riddle, D. L., and J. Carbon. 1973. Frameshift suppression: a nucleotide addition in the anticodon of a glycine transfer RNA. *Nat. New Biol.* **242:**230–234.

Riyasaty, S., and J. F. Atkins. 1968. External suppression of a frameshift mutant in *Salmonella*. *J. Mol. Biol.* **34:**541–557.

Samaha, R. R., R. Green, and H. F. Noller. 1995. A base pair between tRNA and 23S rRNA in the peptidyl transferase centre of the ribosome. *Nature* **377:**309–314.

Sandbaken, M. G., and M. R. Culbertson. 1988. Mutations in elongation factor EF-1α affect the frequency of frameshifting and amino acid misincorporation in the yeast *Saccharomyces cerevisiae*. *Genetics* **120:**923–934.

Santos, M. A. S., T. Ueda, K. Watanabe, and M. F. Tuite. 1997. The non-standard genetic code of *Candida* spp.: an evolving genetic code or a novel mechanism for adaptation? *Mol. Microbiol.* **26:**423–431.

Schwartz, R., and J. F. Curran. 1997. Analysis of frameshifting at UUU-pyrimidine runs. *Nucleic Acids Res.* **25:**2005–2011.

Smith, D., and M. Yarus. 1989. tRNA-tRNA interactions within cellular ribosomes. *Proc. Natl. Acad. Sci. USA* **86:**4397–4401.

Trifonov, E. N. 1987. Translation framing code and frame-monitoring mechanism as suggested by the analysis of mRNA and 16S rRNA nucleotide sequences. *J. Mol. Biol.* **194:**643–652.

Trifonov, E. N. 1992. Recognition of correct reading frame by the ribosome. *Biochimie* **74:**357–362.

Tsuchihashi, Z., and P. O. Brown. 1992. Sequence requirements for efficient translational frameshifting in the *Escherichia coli dnaX* gene and the role of an unstable interaction between $tRNA^{Lys}$ and an AAG lysine codon. *Genes Dev.* **6:**511–519.

Tucker, S. D., E. J. Murgola, and F. T. Pagel. 1989. Missense and nonsense suppressors can correct frameshift mutants. *Biochimie* 71:729–739.

Tuohy, T. M. F., S. Thompson, R. F. Gesteland, D. Hughes, and J. F. Atkins. 1990. The role of EF-Tu and other translation components in determining translocation step size. *Biochim. Biophys. Acta* **1050**:274–278. (Erratum, **1087**:347.)

Tuohy, T. M. F., S. Thompson, R. F. Gesteland, and J. F. Atkins. 1992. Seven, eight and nine-membered anticodon loop mutants of $tRNA_2^{Arg}$ which cause +1 frameshifting. Tolerance of DHU arm and other secondary mutations. *J. Mol. Biol.* **228**:1042–1054.

Vijgenboom, E., and L. Bosch. 1989. Translational frameshifts induced by mutant species of the polypeptide chain elongation factor Tu of *Escherichia coli*. *J. Biol. Chem.* **264**:13012–13017.

Vimaladithan, A., and P. J. Farabaugh. 1994. Special peptidyl-tRNA molecules promote translational frameshifting without slippage. *Mol. Cell. Biol.* **14**:8107–8116.

von Ahsen, U., R. Green, R. Schroeder, and H. F. Noller. 1997. Identification of 2′-hydroxyl groups required for interaction of a tRNA anticodon stem-loop region with the ribosome. *RNA* **3**: 49–56.

Watanabe, K., N. Hayashi, A. Oyama, K. Nishikawa, T. Ueda, and K. Miura. 1994. Unusual anticodon loop structure found in *E. coli* lysine tRNA. *Nucleic Acids Res.* **22**:79–87.

Weiss, R. B., D. M. Dunn, J. F. Atkins, and R. F. Gesteland. 1987. Slippery runs, shifty stops, backward steps, and forward hops: −2, −1, +1, +2, +5 and +6 ribosomal frameshifting. *Cold Spring Harbor Symp. Quant. Biol.* **52**:687–693.

Weiss, R. B., D. M. Dunn, A. E. Dahlberg, J. F. Atkins, and R. F. Gesteland. 1988. Reading frame switch caused by base-pair formation between the 3′ end of 16S rRNA and the mRNA during elongation of protein synthesis in *Escherichia coli*. *EMBO J.* **7**: 159–169.

Weiss, R. B., D. M. Dunn, M. Shuh, J. F. Atkins, and R. F. Gesteland. 1989. *E. coli* ribosomes re-phase on retroviral frameshift signals at rates ranging from 2 to 50 percent. *New Biol.* **1**:159–169.

Weiss, R. B., D. M. Dunn, J. F. Atkins, and R. F. Gesteland. 1990a. Ribosomal frameshifting from −2 to +50 nucleotides. *Prog. Nucleic Acids Res. Mol. Biol.* **39**:159–183.

Weiss, R. B., W. M. Huang, and D. M. Dunn. 1990b. A nascent peptide is required for ribosomal bypass of the coding gap in bacteriophage T4 gene 60. *Cell* **62**:117–126.

Williams, K. P., K. A. Martindale, and D. P. Bartel. 1999. Resuming translation on tmRNA: a unique mode of determining a reading frame. *EMBO J.* **18**:5423–5433.

Wilson, K. S., and H. F. Noller. 1998. Molecular movement inside the translational engine. *Cell* **92**:337–349.

Yourno, J. 1972. Externally suppressible +1 "glycine" frameshift: possible quadruplet isomers for glycine and proline. *Nat. New Biol.* **239**:219–221.

The Ribosome: Structure, Function, Antibiotics, and Cellular Interactions
Edited by R. A. Garrett, S. R. Douthwaite, A. Liljas, A. T. Matheson, P. B. Moore, and H. F. Noller

Chapter 31

The Unbearable Lightness of Peptidyl-tRNA

JONATHAN GALLANT, DALE LINDSLEY, and JUDY MASUCCI

PREHISTORY AND A SEEMINGLY PLAUSIBLE EARLY HYPOTHESIS

In the conventional picture of protein synthesis, the ribosome advances by 3 nucleotides along the mRNA in each translocation step—or, looked at differently, the mRNA with peptidyl-tRNA associated threads back through the ribosome by 3 nucleotides. In either case, triplet movement is assumed to be the natural state of affairs. Moreover, the picture assumes that mRNA and peptidyl-tRNA retain their point of association, at a specific triplet, while the two move together relative to the ribosome.

Both parts of the picture are challenged by our increasingly detailed knowledge of noncanonical ribosome movements. It has been clear for nearly 2 decades that the interaction of mRNA with ribosomes and with tRNAs was compatible with nontriplet movement. Atkins and coworkers showed that ordinary ribosomes translating in a cell-free system could be induced to shift reading frame—meaning a nontriplet movement—by simply adding an excess of certain tRNAs (Atkins et al., 1979; Bruce et al., 1986; Dayhuff et al., 1986). A similar result was demonstrated in whole cells by, in essence, subtracting one or another aminoacyl-tRNA through amino acid starvation or interference with an aminoacyl-tRNA synthase (Gallant and Foley, 1980; Weiss and Gallant, 1983, 1986). The same effect of aminoacyl-tRNA subtraction could also be demonstrated in Atkins' cell-free system (Weiss et al., 1984). It is worth emphasizing that the whole-cell and the cell-free experiments both demonstrated induced frameshifting in both directions, implying that ribosomes could wander by 1 nucleotide either left (5′-ward) or right (3′-ward) or perform more complicated movements.

Next, spontaneous ribosome frameshifting was found to occur at high frequency in the normal translation of specific sites in a variety of RNA sequences: rightward on the *Escherichia coli prfB* message for protein release factor II (Craigen and Caskey, 1986; Curran and Yarus, 1988) and the *Saccharomyces cerevisiae* Ty1 retrotransposon (Clare et al., 1988; Belcourt and Farabaugh, 1990) and leftward on the retroviruses Rous sarcoma virus (Jacks and Varmus, 1985; Jacks et al., 1988a), mouse mammary tumor virus (Hizi et al., 1987), human immunodeficiency virus type 1 (HIV-1) (Jacks et al., 1988b), and a bewildering variety of other retroviruses, animal viruses, plant viruses, and bacterial mobile elements (reviewed by Hatfield et al., 1992; Gramstat et al., 1994; Rohde et al., 1994; Brierly, 1995; Maia et al., 1996; and Farabaugh, 1996, 1997). A very few cases of programmed frameshifting in "normal" genes (such as the aforementioned *prfB*) have been reported, but most instances are found among sequences of parasitic or mobile elements, which presents an interesting ecological puzzle we will return to later.

In any case, the association of frameshifting with parasitic elements, including viruses of some notoriety, focused considerable interest on the subject. During the last dozen years or so, sequence rules governing programmed frameshifting in prokaryotic, viral, retroviral, and transposon systems have been worked out in great detail, including unexpected and profound interactions involving sequences and secondary structures quite distant from the site of frameshifting itself (see the aforementioned reviews). In this chapter, we will confine our attention to the "shifty" sequence in the immediate region of the frameshift event in relationship to certain limited aspects of ribosome behavior.

Jonathan Gallant, Dale Lindsley, and Judy Masucci ■ Genetics Department, University of Washington, Box 357360, Warehouse of the Muses, 1959 NE Pacific St.—J-205, Seattle, WA 98195-7360.

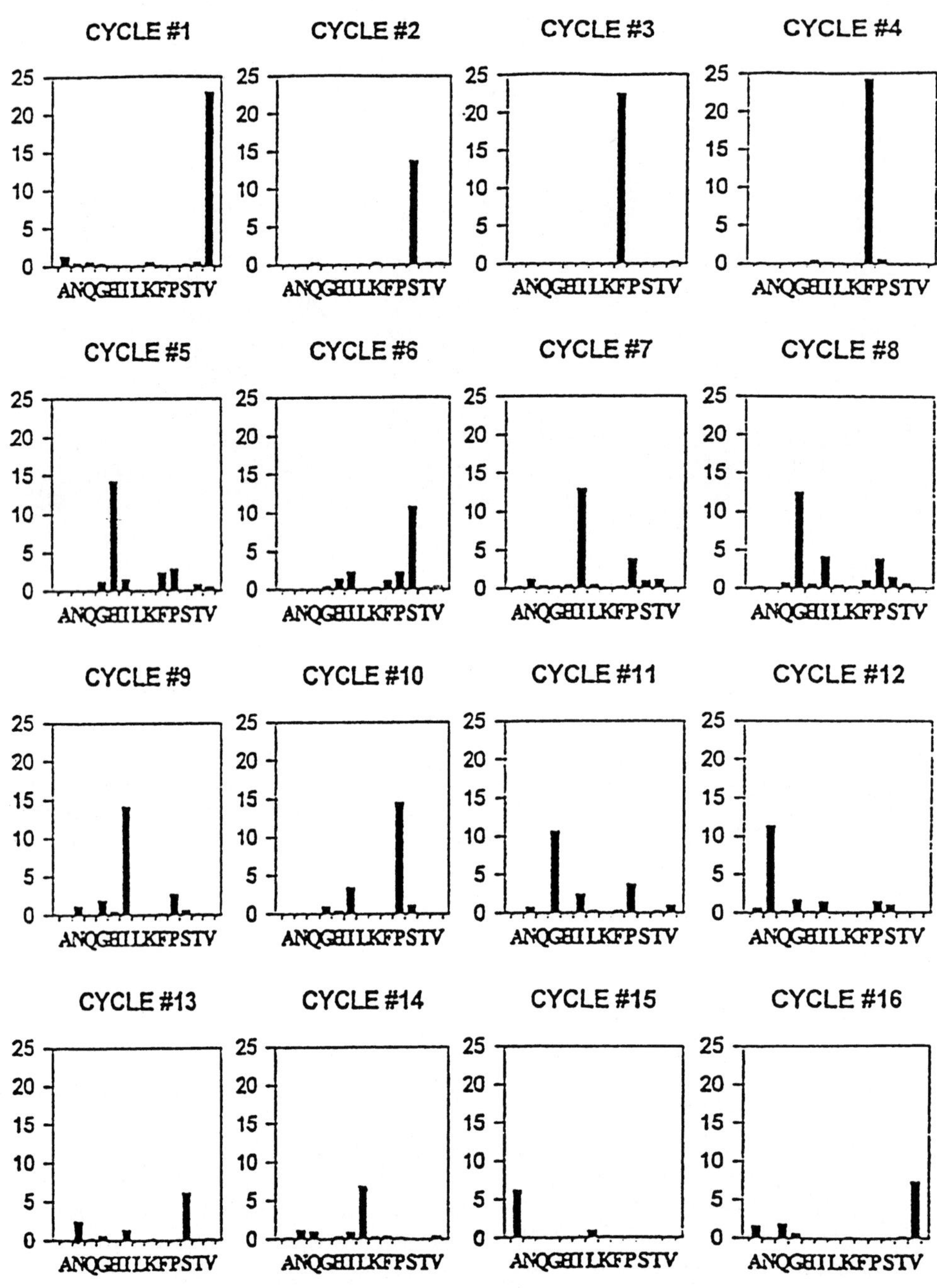


```
 1   2   3   4
val ser phe phe(ile)(val) ***
GUA AGC UUU UUC AUA GU AUA GGG AUC

                his ser ile gly ile
                 5   6   7   8   9


CCC GGG AAU UCA CUG GCC GUC
pro gly asn ser leu ala val
 10  11  12  13  14  15  16
```

At a very early stage, Kurland (1979) raised the idea that translocation and the codon-anticodon interaction were physically coupled. This postulate was further elaborated to suggest that nonstandard (which is to say erroneous) codon-anticodon associations in the ribosome's A site might distort the transition to the P site, so as to produce nontriplet translocation, a notion the authors termed "error coupling" (Kurland and Gallant, 1986; Kurland, 1992). This hypothesis was nicely consistent with our aforementioned finding that aminoacyl-tRNA limitation, which was known to stimulate missense errors, also stimulated ribosome frameshifting. The hypothesis seemed further vindicated, indirectly, by the finding that *relA* mutations, which exacerbate the occurrence of missense errors at hungry codons (reviewed by Parker, 1989), had precisely the same effect on starvation-promoted frameshift errors (Gallant and Foley, 1980; Weiss and Gallant, 1983).

These ideas lead to a simple, critical prediction: the amino acid present at the frameshift site should be an erroneous one, reflecting the nonstandard codon-anticodon interaction at the ribosome A site which provoked the frameshift error. Protein sequence data are now available to test this prediction. Leaving aside programmed frameshifts for the moment, let us first consider starvation-induced frameshifts

The clearest case is that analyzed by Barak et al. (1996a), who induced a high level of leftward frameshifting at the sequence U UUC AUA by limiting the isoleucine-tRNA encoded at the AUA triplet (Fig. 1). The protein sequence unambiguously demonstrates phenylalanine at the zero-frame UUC position, followed by histidine in place of isoleucine at the frameshift, and then the rest of the protein read in the left-shifted reading frame. This corresponds to correct reading of the UUC triplet by phenylalanyl-tRNA, and correct reading of the left-shifted C AU triplet by histidyl-tRNA, followed by decoding in the left-shifted reading frame from there on. No erroneous decoding is involved.

The same conclusion follows for a rightward frameshift provoked by the same limitation regime at a similar site (Fig. 2). Here, in right-shifted protein encoded by the sequence UUU AUA U, we find phenylalanine followed by tyrosine; these amino acids correspond to correct decoding of the UUU and then correct decoding of the right-shifted UA U triplet. Once again, only cognate reading is involved. Thus, the error-coupling hypothesis is evidently refuted for starvation-promoted shifts in both directions.

PEPTIDYL SLIPPPAGE

How do the frameshifting tRNAs (histidyl- and tyrosyl-tRNA, respectively, in the two cases mentioned above) enter the A site to read a cognate out-of-frame triplet which overlaps the hungry, in-frame AUA? It is most unlikely that either tRNA slips after first decoding the original-frame AUA triplet, for this would involve an initial mismatch at all three positions. Therefore, the most likely mechanism in both cases is that peptidyl-phenylalanyl-tRNA in the P site slips first, while the A site is empty due to isoleucyl-tRNA deficiency; this slip is then followed by cognate, out-of-frame decoding by the histidyl- or tyrosyl-tRNA. In the case shown in Fig. 1, the phe-tRNA in the P site slips leftward (from the UUC triplet to the fully cognate overlapping U UU), while in the case shown in Fig. 2, it slips rightward (from the UUU to the near-cognate UU A).

Starvation-induced frameshifting thus falls into the category of peptidyl-tRNA slippage. In the case of leftward frameshifting, we have reported mutational evidence that the frameshift depends upon the capacity of the P-site quadruplet to pair with the same tRNA in both the original and the shifted frame (Kolor et al., 1993; Barak et al., 1996a). This implies that peptidyl slippage is fundamental to the mechanism.

In fact, peptidyl slippage in the rightward direction is well established as the mechanism of programmed frameshifting on the slippage site of the *E. coli prfB* message (Craigen and Caskey, 1986). The shifty sequence here is CUU UGA C. In the protein, the leucine encoded by the zero-frame CUU is followed by aspartic acid encoded by the GA C in the rightward, or +1, frame, which is the reading frame decoded thereafter.

Figure 1. Amino acid sequence in the region of a leftward (−1) ribosome frameshift at a hungry codon. The *lacZ* protein was encoded by a left-frameshift reporter; the *lacZ* promoter was induced, and cells were subjected to partial inhibition by isoleucine-hydroxamate, which stimulated synthesis of frameshifted, active enzyme to about 10% that of a zero-frame control; the resulting β-galactosidase was purified and cleaved specifically, and the fragment with the indicated valine residue at the N terminus was submitted to Edman N-terminal sequencing. The PTH-amino acids released in successive cycles are shown in conventional, cycle-bar plot form. The messenger nucleotide sequence is shown below the cycle-bar plots. The amino acids encoded in the zero-frame are shown above the nucleotide sequence; those corresponding to the leftward (−1) frame after the shift (arrow) are shown below the nucleotide sequence. (From Barak et al., 1996a.)

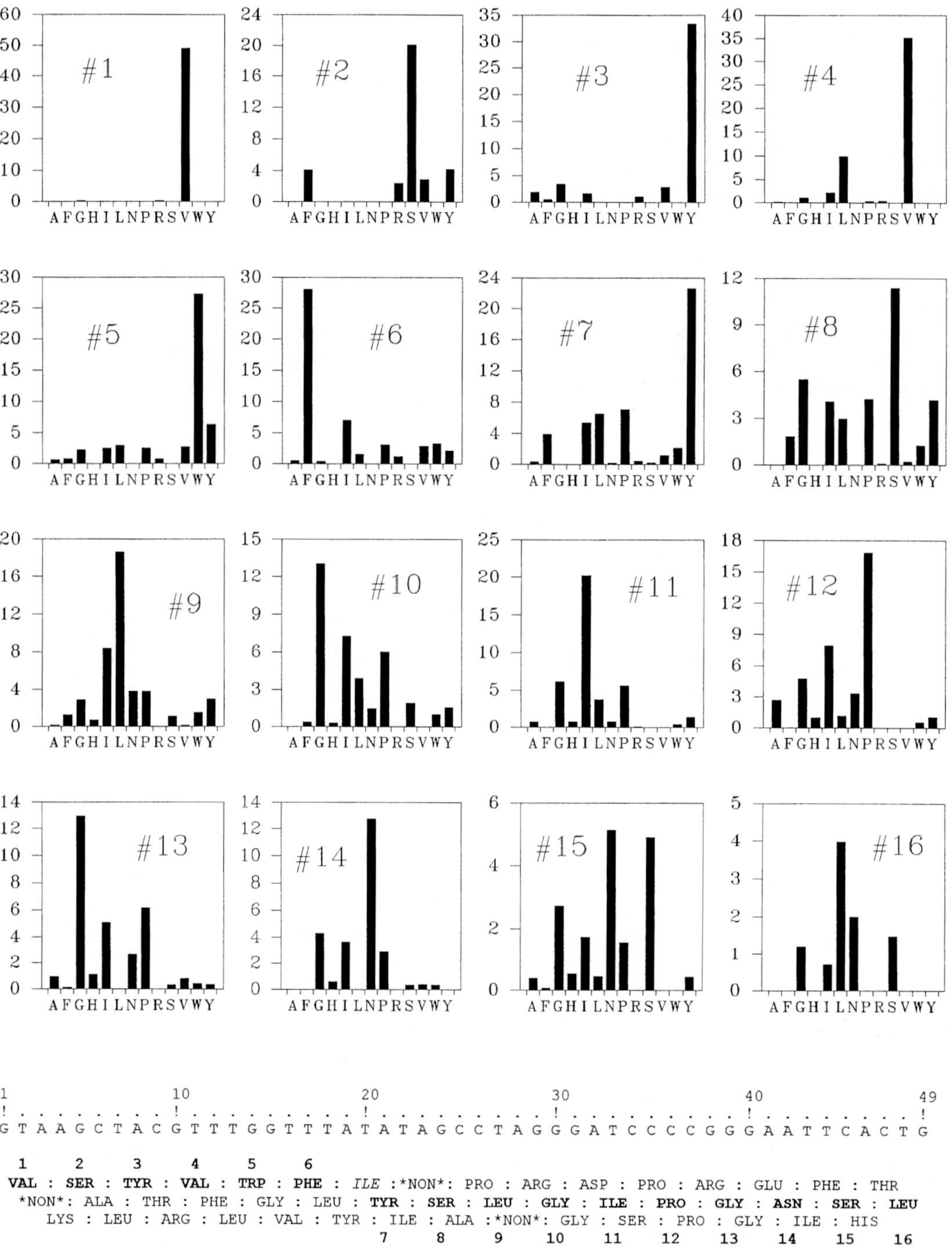

Figure 2. Amino acid sequence in the region of a rightward (+1) ribosome frameshift at a hungry codon. The *lacZ* protein was encoded by a right-frameshift reporter; cells were induced and inhibited by isoleucine-hydroxamate as for Fig. 1, stimulating enzyme synthesis 40-fold to 27% that of the parallel, zero-frame control; purification, cleavage, and sequencing were done as for Fig. 1. The nucleotide sequence is shown as the coding strand of the DNA; the predicted amino acid sequences in all three reading frames are shown below the nucleotide sequence, with the limiting isoleucine in the zero-frame position 7 in italics; the observed amino acid sequence, which shifts rightward at this position, is shown in boldface letters.

The ribosome frameshift requires that the codon after the CUU be a terminator, demonstrating that the CUU has already been decoded and must be in the P site when decoding slips rightward at the terminator. Thus, when the terminator triplet is in the A site (very likely causing a ribosome pause), the peptidyl-leucyl-tRNA in the P site slips rightwards and finds complementary pairing with the overlapping UU U codon by means of a wobble pair in the first position. A comprehensive analysis of the frameshifting proficiency of many triplets in the P site showed that this varied directly with the stability of the codon-anticodon association after slipping into the +1 frame (Weiss et al., 1987; Curran, 1993).

In prokaryotes, there is also strong evidence for peptidyl slippage in the case of programmed leftward frameshifting on the *insA-insB* sequence of IS*1* (Sekine and Ohtsubo, 1989). The shifty sequence here is A AAA AAC. Direct protein sequencing of the frameshift junction indicated (despite a rather complex signal) Lys-Lys rather than Lys-Asn (Sekine et al., 1992). This means that frameshifting occurred by leftward peptidyl slippage at the first in-frame lysine codon (AAA), before decoding of the following asparagine codon (AAC); as a result, the left-frame A AA codon is decoded instead of AAC, yielding the second lysine. Moreover, mutagenesis studies showed little effect of mutational change of the AAC triplet (Sekine and Ohtsubo, 1992), also implying that the frameshift occurs before the AAC triplet can be decoded.

Independent evidence of P-site leftward slippage has been reported by other groups for the HIV-1 frameshift site translated in bacteria (Horsfield et al., 1995) and for a shifty site in potato virus M (Gramstat et al., 1994). Moreover, Belcourt and Farabaugh (1990) have shown decisively that peptidyl slippage underlies rightward shifting in the Ty1 retrotransposon of yeast (see below).

It is worth emphasizing that peptidyl slippage should not be understood to be in contradiction to simultaneous slippage, the predominant mechanism of retroviral programmed frameshifting. On the contrary, both mechanisms have the common feature of depending upon peptidyl slippage, the former before the A site is filled and the latter after it is filled. Some sequences compatible only with slippage of the peptidyl-tRNA in the P site do not permit simultaneous slippage of the aminoacyl-tRNA in the A site. These include the shifts in either direction at a hungry AUA codon described above and the leftward frameshift in potato virus M, in which the shifty quadruplet (which could be either A AAA or U UUU) is followed by a termination triplet (Gramstat et al., 1994). In the latter case, peptidyl slippage occurs in the shifty quadruplet, but there is of course no aminoacyl-tRNA in the A site, which contains a terminator.

Other sequences, those at retroviral shift sites and ones like them in bacteria, fulfill the sequence requirements for simultaneous slippage of both tRNAs; at some of the latter sequences, the occurrence of both simultaneous slippage and peptidyl-tRNA-only slippage has been demonstrated by the detection of sequence heterogeneity at the frameshift site, with evidence of both of the expected different amino acids.

One example is the leftward shift in HIV-1, which occurs on the sequence U UUU UUA. Most of the frameshifting here occurs by the simultaneous mechanism: after a leucine-tRNA first decodes the cognate UUA codon, it slips back to the overlapping, near-cognate U UU codon, simultaneously with leftward slippage of the peptidyl-Phe-tRNA in the P site. The result, demonstrated in the transframe protein synthesized in vitro, is a protein which contains phenylalanine and then leucine at this position and, after that, the amino acid sequence corresponding to the leftward reading frame (Jacks et al., 1988b).

If slippage of peptidyl-Phe-tRNA alone were to occur when the A site containing the UUA codon was still empty, then the protein should contain a second phenylalanine residue corresponding to the left shifted U UU codon and then continue in the leftward reading frame. In fact, Jacks et al. (1988b) did detect a minority of in vitro product with precisely this sequence. This component was detected more decisively when a reporter carrying the HIV-1 shifty sequence was translated in *E. coli* cells, and it became the predominant product when the cells were subjected to leucine limitation so as to keep the A site programmed by UUA empty longer (Yelverton et al., 1994). Heterogeneity of exactly this kind was also detected by Weiss et al. (1989) in the transframe product of the mouse mammary tumor virus *gag-pro* frameshift sequence translated in *E. coli* cells.

Thus, there is abundant evidence for the occurrence of peptidyl-tRNA-only shifting, even on sequences where simultaneous shifting normally (?) occurs more often. The question mark in the last sentence alludes to the fact that virtually all studies of retroviral frameshifting have been done with heterologous translation systems, in vitro or whole cell, because of the technical difficulties posed by the natural system. Thus, we have virtually no information on the character of the transframe protein which (we suppose) is made in real life.

HOPPING AND SLIDING

We are left with the implication that noncognate interactions in the A site have little to do with frameshifting, while slippage of peptidyl-tRNA in the P site has everything to do with it. In fact, the mobility of peptidyl-tRNA is illustrated in much more dramatic fashion by two other related phenomena. More than a decade ago, Weiss and colleagues (1987) discovered that the ribosome·peptidyl-tRNA complex could apparently "hop" from the codon preceding a terminator to a synonymous codon if one was available +2, +5, or +6 nucleotides downstream. The phenomenon, unequivocally demonstrated by protein sequencing, occurred at very low frequency (in the vicinity of 1% or less). Later, a longer and much more efficient programmed hop was discovered in the T4 gene *60* topoisomerase (Huang et al., 1988). In this system, peptidyl-Gly-tRNA at a GGA codon abutting an in-frame UAG terminator apparently vaults over 50 nucleotides to a downstream GGA triplet and resumes translation there.

In a meticulous analysis, Weiss, Huang, and Dunn (1990) showed that this great leap forward depended upon several very specific sequence elements. It required the presence of (i) matched "takeoff" and "landing" triplets for the peptidyl-tRNA, (ii) a terminator triplet just after the takeoff site, (iii) a quite precisely defined hairpin loop in the mRNA starting just after the takeoff site, and (iv) a particular amino acid sequence encoded in the region between about −30 and −13 triplets upstream of the takeoff site, indicating that the peptide part of peptidyl-tRNA strongly influenced the phenomenon.

These complicated sequence determinants suggested a complicated and very special mechanism, and for some time the phenomenon seemed to be sui generis. Benhar et al. (1992) detected a truncated form of the *trpR* protein which they at first ascribed to frameshifting; then, after further characterization of the phenomenon with a *trpR-lacZ* reporter system, this group concluded that the aberrant protein represented a ribosomal hop from a *trpR* sequence into a *lacZ* sequence (Benhar and Engelberg-Kulka, 1993). The rules governing this putative hop, which seemingly did not require matching takeoff and landing site triplets, are made even more mysterious by two considerations. First, the original report involved only the *trpR* gene, without any of the *lacZ* sequence within which, according to the second report, the hop's landing site could be found. Second, the Utah frameshifting laboratory was unable to reproduce the evidence for hopping at all in a closely similar system (Wills et al., 1997). Pending clarification of these ambiguities, it is perhaps safest to avoid drawing any conclusions from the *trpR* translational phenomena.

On the other hand, something very like hopping can be induced at hungry codons, and seemingly without elaborate sequence constraints (Gallant and Lindsley, 1998). Ribosomes stalled for lack of isoleucine-tRNA at the same U UUC AUA sequence discussed above were shown to be capable of sliding various distances along the mRNA to reengage at the next phenylalanine codon downstream and to continue protein chain elongation there. The amino acid sequences of the protein products, which simply bypassed the sequences encoded from the beginning to the end of the slide, proved this gymnastic feat unequivocally (Gallant and Lindsley, 1998).

Slides of 7 to 16 nucleotides occurred with frequencies of about 20% and at lower but quite appreciable frequencies for longer distances up to at least 40 nucleotides. Figure 3 illustrates the induction of sliding over distances of 7 and 40 nucleotides in response to different degrees of limitation for isoleucine-tRNA. The frequency of successful slides depended on there being a landing site codon synonymous (UUU or UUC) with the takeoff site, that is, the codon preceding the hungry codon. Tests of a variety of constructs have recently shown that this behavior is not limited to phenylalanine codons but is displayed by several different takeoff-landing site pairs, with various efficiencies (Gallant, Lindsley, Bonthuis, and Heaton, unpublished). Kane et al.

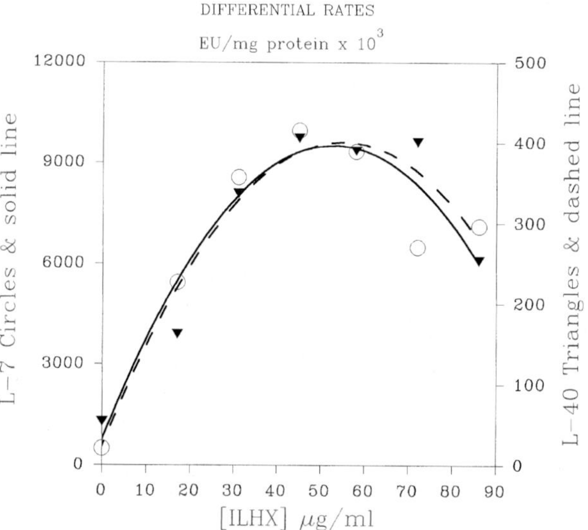

Figure 3. Efficiency of slides of 7 or 40 nucleotides induced at a hungry AUA codon. Constructs L-7 and L-40 (see Gallant and Lindsley, 1998) demand that the ribosome·peptidyl-tRNA complex slide downstream 7 or 40 nucleotides, respectively, from a hungry AUA codon to find a landing site beyond blocking terminators and thus produce β-galactosidase. The relevant region of the messenger sequence is, in schematic form, UUC AUA (7 or 40 nucleotides ending in UAA) UUU →[*lacZ* reading frame]. The differential rate of enzyme synthesis during one doubling of partial isoleucine-tRNA limitation is plotted on the *y* axis versus the concentration of isoleucine-hydroxamate (ILHX). Note the two different scales on the *y* axis for the slide of 7 (at left) and the slide of 40 (at right) nucleotides.

(1992), studying translation of a heterologous gene in *E. coli*, found evidence that peptidyl-leucyl-tRNA hopped 3 nucleotides from one UUG codon, over an AGG codon, to an adjacent in-frame UUG landing site. In this case, the bypassed codon was perhaps rendered hungry by a high level of the heterologous gene containing several AGG triplets, which are disfavored in *E. coli* genes and code for a very rare tRNA.

The phenomenon thus resembles the T4 gene *60* hop in its requirement for matching takeoff and landing sites. Perhaps the hungry codon plays a role similar to that of the terminator codon in the latter system. Otherwise, however, the movement of the ribosome·peptidyl-tRNA complex provoked at hungry codons seems quite general rather than characteristic of any special sequence. No special secondary structure was evident in the various bypassed regions, and the phenomenon was observed with two very different sequences upstream, neither of which bore any resemblance to that upstream of the T4 gene *60* hop.

Since no specific rules constrained the sequence between takeoff codon and landing site codon, and diverse takeoff-landing site pairs will support the resumption of protein chain synthesis, the ribosome departing the takeoff site must "scan" through the untranslated region for a triplet complementary to the anticodon of its peptidyl-tRNA. For this reason, we think the process is better described as "sliding" than as "hopping." The generality of sliding further refutes our old idea that the phenomena under discussion have anything whatever to do with mispairing at the A site or funny business during translocation.

The ribosome·peptidyl-tRNA complex is apparently able to disengage from the cognate codon in the P site after normal translocation, wander down the message for a remarkable distance, and then resume the ribosome cycle at the next triplet it encounters which is cognate to its anticodon. Here, the possibility that the process is initiated by a noncognate or near-cognate aminoacyl-tRNA in the A site seems ruled out of consideration by the topology of the event. How could the process be initiated by a noncognate interaction which occurs only after the complex has wandered 15 or 40 nucleotides down the message to find it?

Thus, the ribosome·peptidyl-tRNA complex is both more stable and more mobile than conventional pictures of ribosome dynamics had led us to suppose. The complex can evidently migrate along mRNA without translating while remaining intact to resume protein chain elongation at a new position. The implication is that peptidyl-tRNA slippage by a mere 1 nucleotide to the left or right, as occurs in ribosome frameshifting, is literally the least of its migratory tendencies.

A HYPOTHESIS ABOUT PEPTIDYL SLIPPAGE

Farabaugh and colleagues have suggested another kind of error-coupling hypothesis to account for the wanderlust of peptidyl-tRNA, based on their analysis of rightward frameshifting in the yeast Ty1 retrotransposon (Clare et al., 1988; Belcourt and Farabaugh, 1990; Farabaugh, 1997). The sequence at the site is CUU AGG C, and it is shifty because it encodes leucine tRNA in two different reading frames. The shift requires peptidyl-leucine tRNA in the P site to slip from the zero-frame CUU codon to the rightward UU A codon; this makes the following GG C (glycine) triplet the next one decoded, establishing the rightward reading frame. Yeast possesses a leucine-tRNA (anticodon UAG) which is unusual in that it is capable of decoding both CUU and UUA.

The initial zero-frame pairing with CUU, however, involves a mismatch in the third position and so qualifies as a near-cognate, rather than fully cognate, association. Thus, the movement of peptidyl-tRNA to the UUA triplet can be viewed as error coupling within a codon family. Farabaugh has proposed that near-cognate interactions of this sort in the peptidyl site may be a general feature underlying frameshifting (Farabaugh, 1997). If this mechanism were general, then a mismatch of peptidyl-tRNA in the P site would be evident in other cases of slippage at other frameshift sites.

One recently described case may fit this paradigm. O'Connor (1998) has shown that leftward frameshifting at CAG GGA ACC is stimulated by increased dosage (on a multicopy plasmid) of the *glyU* gene for $tRNA_1^{Gly}$, which normally reads the codon GGG. The author suggests that this tRNA is capable of decoding the GG doublet in the A site; after translocation, the next triplet will be A AC, corresponding to a leftward frameshift. An alternative possibility is that this tRNA misreads the GGA triplet by a mismatch in the third position, translocation occurs, and then peptidyl-$tRNA_1^{Gly}$ slips leftward to the cognate GGG triplet. This would be peptidyl-tRNA slippage in the leftward direction, due to a near-cognate interaction, and would reflect error coupling within a codon family. As such, it would be entirely analogous to the rightward slip at the Ty sequence.

Recent studies from Umeå, based on manipulation of the in vivo modification status of tRNAs, provide evidence that frameshift suppression by certain tRNAs occurs by virtue of near-cognate pairing in the

P site and subsequent peptidyl slippage (Qian et al., 1998). Moreover, the same group also showed that frameshift suppression is sensitive to the rate of decoding of the next in-frame codon. This elegantly demonstrates that the suppressing tRNA must already be in the P site when the frameshift event occurs and thus implies that peptidyl slippage is the mechanism.

All of the foregoing considerations suggest the following: (i) peptidyl slippage is a fundamental property of the ribosome·peptidyl-tRNA complex, (ii) slippage is by no means limited to a single nucleotide but can permit the complex to make great leaps forward (although perhaps not backward by more than 1 nucleotide), (iii) the probability of peptidyl slippage increases with the time the ribosome is constrained to pause at the next in-frame codon, and (iv) the probability of slippage can be enhanced when the codon-anticodon interaction in the P site is near cognate rather than cognate.

In the cases of starvation-promoted frameshifting, the idea of a near-cognate tRNA at the P site seems at first sight to be flatly contradicted by the protein sequence data. In both the leftward- and rightward-shifting cases discussed above, the amino acid preceding the frameshift is clearly phenylalanine, whose tRNA is perfectly cognate to the UUC or UUU triplet in the P site in the initial reading frame (Fig. 1 and 2). The same is true of rightward frameshifting at an entirely different sequence, where the protein sequencing data is admittedly not as decisive (Peter et al., 1992). And, finally, the same is true of the ribosome slides: the amino acid preceding the slide is clearly the correct one for the triplet at the takeoff-landing site pair, demonstrated for two such triplet pairs (see Fig. 2 and 5 in Gallant and Lindsley, 1998).

We believe that this apparent discrepancy does not really vitiate the model. First, if cognate peptidyl-tRNA exhibits some nonzero probability per unit time of slipping, even if this probability is very small, aminoacyl-tRNA limitation will necessarily increase the time during which such a slip may occur and thereby increase its overall probability of occurring. This simple relationship presumably explains the stimulation of frameshifting and sliding seen in *relA*+ cells.

We mentioned earlier that this stimulation is greater—for some, but not all, limitation regimes much greater—in *relA* mutant cells. An explanation of this feature is perhaps to be found in the arcane features of the "relaxed" response, that is, the physiology of *relA* mutants subjected to aminoacyl-tRNA limitation. The hallmark of this phenotype, first observed during the bronze age of molecular biology, is the overproduction of stable RNA, including all tRNA species, during amino acid limitation. Much later, it was shown that tRNAs accumulated under these conditions are peculiar in that they lack a number (perhaps many) of the base modifications characteristic of normal, mature tRNAs (Fournier et al., 1976; Kitchingman and Fournier, 1975, 1977).

Presumably, tRNA modification is coordinated with ongoing protein synthesis through a linkage which is broken when tRNAs accumulate faster than protein synthesis occurs—precisely the phenotype permitted by a *relA* defect. (Incidentally, the nature of this linkage is one of those interesting questions of bacterial physiology which has been allowed to drop through the cracks.)

In at least some of the starvation regimes we have investigated, the excess frameshifting in *relA* mutant cells is not observed right away but only after the cells have grown under conditions of aminoacyl-tRNA limitation for considerable time. Figure 4 shows an example. Note that in the *relA* mutant, isoleucine-tRNA limitation does not stimulate enzyme synthesis at first but only after about the third or fourth point, which corresponds to about half a doubling of protein, after which a very large increase over the control, uninhibited cells is evident. These slow kinetics suggest, at the very least, that the *relA* mutant excess is an indirect effect. More specifically, the slow kinetics are consistent with the hypothesis that the excess frameshifting in *relA* mutants depends upon the accumulation of new and presumably undermedicated tRNA molecules.

In fact, a variety of data demonstrate that certain tRNA modifications in the anticodon loop and stem affect the tendency of certain tRNA species to frameshift (reviewed by Björk, 1995, and Curran, 1998). Further experiments will, of course, be required to test the interpretation that this is the explanation of the *relA* effect and, if it is correct, to identify the particular undermedicated tRNA species, accumulating during the relaxed response, which do the mischief.

SPECULATIONS

In many cases of programmed leftward frameshifting, the event is partly (although not completely) dependent on a secondary structure, generally a pseudoknot, downstream of the frameshift site (ten Dam et al., 1990; Brierly, 1995; Farabaugh, 1997). The way in which these structures might influence frameshifting is not understood in detail, although a plausible hypothesis is that secondary structures stall the ribosome over the intrinsically shifty sequence at the frameshift site. One particularly mysterious feature is that the frameshift-enhancing pseudoknots or hair-

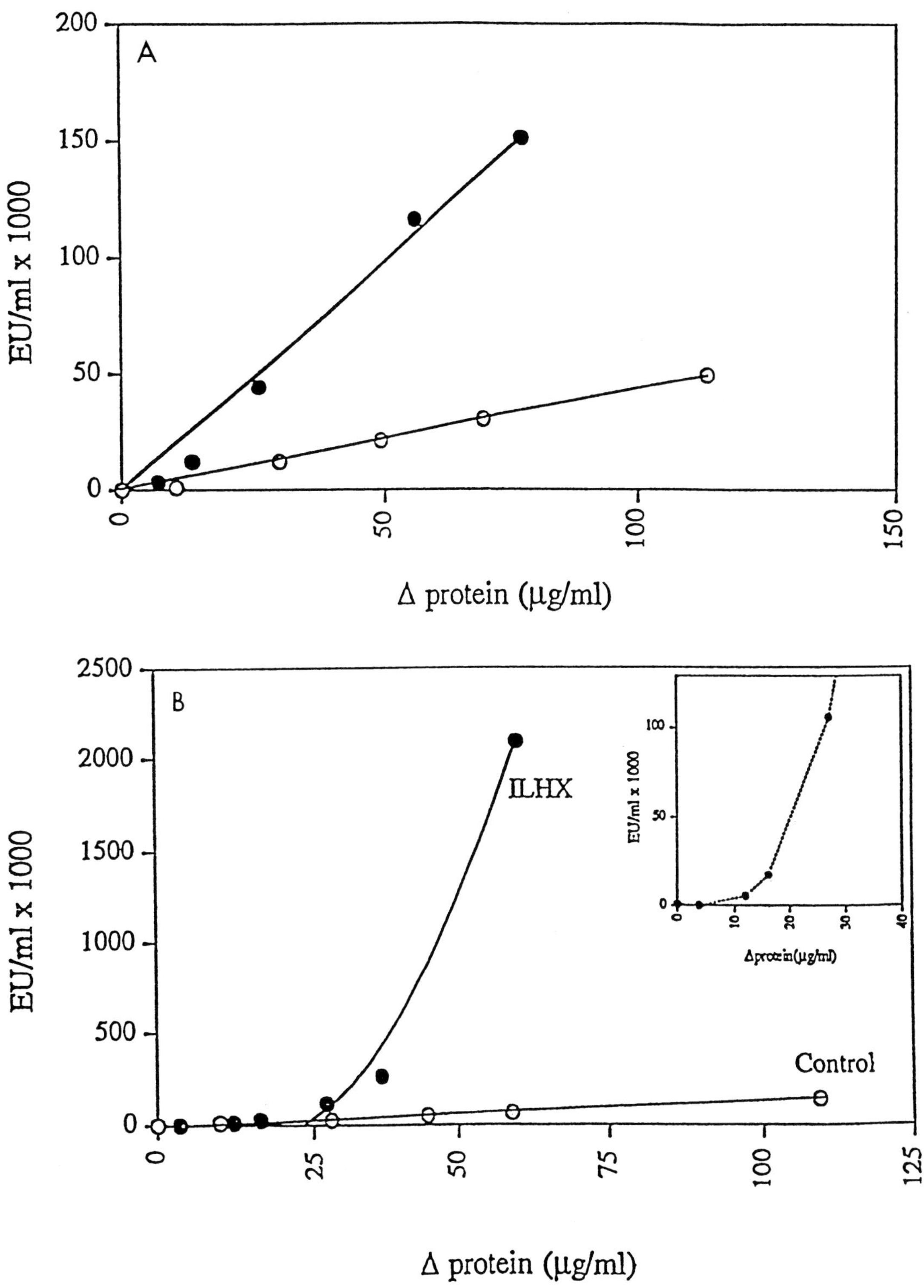

Figure 4. Kinetics of frameshift induction during isoleucine-tRNA limitation in a *relA*$^+$-*relA* mutant isogenic pair. The *lacZ* left-frameshift reporter is similar to those described by Barak et al. (1996a), with the following shifty sequence in the message: U UUU AUA. Log-phase cells were induced, and a subculture was subjected to isoleucine-hydroxamate (ILHX) inhibition (72 μg/ml) and sampled over a period of about two doublings. The data are shown in the form of a differential plot of enzyme synthesis versus increase in protein. Open circles, control, uninhibited subcultures; solid circles, ILHX-inhibited subcultures. (A) CP78 (*relA*$^+$); (B) CP79 (*relA* mutant). The inset shows the first four points of the ILHX-inhibited culture on an expanded scale. In the data for CP79 in ILHX, note the lag period and the subsequent enormous acceleration of the differential rate.

pins are not located a uniform distance downstream from the frameshift sites but at distances which vary from 1 to 10 nucleotides in different shifty sequences.

In the case of starvation-induced sliding, there is no evidence that secondary structures are involved at all (Gallant and Lindsley, 1998). In fact, the high efficiency of rightward (that is, 3′-ward) sliding suggests another possible role for the secondary structures in leftward frameshifting. Bear in mind that peptidyl-tRNA must dissociate from its zero-frame triplet on the message in order to frameshift, and we know that such an errant peptidyl-tRNA is likely to move rightward rather than leftward, go a long distance, and resume translation with good efficiency wherever it encounters a cognate codon. However, the sequences in the cases of programmed frameshifting under discussion are presumably optimized for the production of the single fusion protein corresponding to a leftward shift of 1 nucleotide rather than a long rightward slide.

Perhaps the presence of secondary structures serves to block rightward migration by the stalled ribosome·peptidyl-tRNA complex, leaving a 1-nucleotide leftward slip as its only remaining option to find an alternative complementary triplet. Inspection of a variety of programmed leftward-frameshift sequences in animal and plant virus RNAs reveals that there are potential landing sites downstream of the frameshift takeoff site; but in all cases they are either within the secondary structure or else within a few nucleotides of its end. Thus, a sliding ribosome could not reach them if the secondary structure bars the way. The heterogeneous placement of the secondary structures, in this view, relates to the triplets they function to mask rather than to the frameshift event itself. It will be interesting to test whether ribosome sliding is indeed impaired by secondary structures and, if so, by what kind and with what topological relationship to the slide.

Almost all the many cases of frameshifted reading have been found in retroviruses, retrotransposons, RNA viruses, and prokaryotic mobile elements. Why do these parasitic systems use something as eccentric as ribosome slippage in the expression of key genes in their traveling lifestyles, the reverse transcriptase or replicase in the former groups or the transposase in the latter? We suggest that the expression of these genes is keyed to some aspect of the host cells' general physiology so as to maximize expression when it is most selectively advantageous to the parasite. Perhaps ribosome movement is a very sensitive sensor of some feature of cellular energy metabolism that influences the usefulness of reverse transcriptase or transposase synthesis.

In this connection, we note that spontaneous frameshifting does not have the same probability in all growth phases. In balanced, exponential growth, frameshifting has a constant probability, as judged by a constant differential rate of enzyme synthesis encoded by frameshift reporters (demonstrated by Gallant and coworkers in innumerable unpublished experiments). However, when cells leave the log phase of growth, a large increase in the rate of frameshifting has been detected with several different reporters (Barak et al., 1996b; Schwartz and Curran, 1997; Wenthzel et al., 1998). Preliminary experiments indicate that this is true of sliding as well (Lindsley and Gallant, unpublished).

We have observed increased leftward frameshifting under the following conditions: in late log phase, as the cells slow down but before stationary phase proper; in stationary phase, due to glucose exhaustion; in log-phase cells in which glucose uptake is partly blocked by the addition of alpha-methylglucoside; and in cells shifted from aerobic to anaerobic growth (Gallant et al., unpublished). The simplest feature common to these various conditions is that they all reduce the rate of ATP regeneration; therefore, they presumably lower the levels of ATP and GTP and consequently the rate of ribosome movement, as we have confirmed for the case of simple stationary phase. They may also, like the relaxed response, affect the modification status of many different tRNAs.

However, very little is known yet about how these conditions enhance ribosome frameshifting. This subject surely warrants further study. It may provide a clue to the fascinating deeper puzzle of why so many parasitic and mobile elements have evolved a program which puts to use the otherwise accidental phenomenon of slipping by the ribosome·peptidyl-tRNA complex.

When first detected, ribosome frameshifting seemed an anomaly or a form of disturbed behavior by the translation system under abnormal stress. But then it turned out that many genetic elements (perhaps of the shadier sort) put it to use. Sliding, we have argued, reflects the same fundamental phenomenon as frameshifting. It also responds to the same physiological stimuli in *E. coli*. Perhaps it, too, occurs more frequently in the natural world than we suspect at present.

REFERENCES

Atkins, J., R. Gesteland, B. Reid, and C. Anderson. 1979. Normal tRNAs promote ribosomal frameshifting. *Cell* **18:**1119–1131.

Barak, Z., D. Lindsley, and J. Gallant. 1996a. On the mechanism of leftward frameshifting at several hungry codons. *J. Mol. Biol.* **256:**676–684.

Barak, Z., J. Gallant, D. Lindsley, B. Kwieciszewski, and D. Heidel. 1996b. Enhanced ribosome frameshifting in stationary phase cells. *J. Mol. Biol.* **263:**140–148.

Belcourt, M. F., and P. J. Farabaugh. 1990. Ribosomal frameshifting in the yeast transposon Ty: tRNAs induce slippage on a 7 nucleotide minimal site. *Cell* **62:**339–352.

Benhar, I., and H. Engelberg-Kulka. 1993. Frameshifting in the expression of the *E. coli trpR* gene occurs by the bypassing of a segment of its coding sequence. *Cell* **72:**121–130.

Benhar, I., C. Miller, and H. Engelberg-Kulka. 1992. Frameshifting in the expression of the *Escherichia coli trpR* gene. *Mol. Microbiol.* **6:**2777–2784.

Björk, G. R. 1995. Genetic dissection of synthesis and function of modified nucleosides in bacterial transfer RNA. *Prog. Nucleic Acid Res. Mol. Biol.* **50:**263–338.

Brierly, I. 1995. Ribosomal frameshifting on viral RNAs. *J. Gen. Virol.* **76:**1885–1892.

Bruce, A. G., J. F. Atkins, and R. F. Gesteland. 1986. Anticodon replacement experiments show that ribosomal frameshifting can be caused by doublet decoding. *Proc. Natl. Acad. Sci. USA* **83:** 5062–5066.

Clare, J. J., M. Belcourt, and P. J. Farabaugh. 1988. Efficient translational frameshifting occurs within a conserved sequence of the overlap between the two genes of a yeast Ty1 transposon. *Proc. Natl. Acad. Sci. USA* **85:**6816–6820.

Craigen, W. J., and C. T. Caskey. 1986. Expression of peptide chain release factor 2 requires high-efficiency frameshift. *Nature* **322:**273–275.

Curran, J. 1993. Analysis of effects of tRNA:message stability on frameshift frequency at the *Escherichia coli* RF2 programmed frameshift site. *Nucleic Acids Res.* **21:**1837–1843.

Curran, J. 1998. Modified nucleosides in translation, p. 493–516. *In* H. Grosjean and R. Benne (ed.), *Modification and Editing of RNA.* ASM Press, Washington, D.C.

Curran, J., and M. Yarus. 1988. Use of tRNA suppressors to probe regulation of *Escherichia coli* release factor 2. *J. Mol. Biol.* **203:** 75–83.

Dayhuff, T., J. F. Atkins, and R. Gesteland. 1986. Characterization of ribosomal frameshift events by protein sequence analysis. *J. Biol. Chem.* **261:**7491–7500.

Farabaugh, P. J. 1996. Programmed translational frameshifting. *Microbiol. Rev.* **60:**103–134.

Farabaugh, P. J. 1997. *Programmed Alternative Reading of the Genetic Code.* R. G. Landes, Austin, Tex.

Fournier, M. J., E. Webb, and G. R. Kitchingman. 1976. General and specific effects of amino acid starvation on the formation of undermodified Escherichia coli phenylalanine tRNA. *Biochem. Biophys. Acta* **454:**97–113.

Gallant, J., and D. Foley. 1980. On the causes and prevention of mistranslation, p. 615–638. *In* G. Chambliss, G. R. Craven, J. Davies, K. Davis, L. Kahan, and M. Nomura (ed.), *Ribosomes: Structure, Function, and Genetics.* University Park Press, Baltimore, Md.

Gallant, J., and D. Lindsley. 1998. Ribosomes can slide over and beyond "hungry" codons, resuming protein chain elongation many nucleotides downstream. *Proc. Natl. Acad. Sci. USA* **95:** 13771–13776.

Gallant, J., D. Lindsley, D. Heidel, and B. Kwieciszewski. Unpublished data.

Gallant, J., D. Lindsley, P. Bonthuis, and K. Heaton. Unpublished data.

Gramstat, A., D. Prufer, and W. Rohde. 1994. The nucleic acid-binding zinc finger protein of potato virus M is translated by internal initiation as well as by ribosomal frameshifting involving a shifty stop codon and a novel mechanism of P-site slippage. *Nucleic Acids Res.* **22:**3911–3917.

Hatfield, D. L., J. G. Levin, A. Rein, and S. Oroszlan. 1992. Translational suppression in retroviral gene expression. *Adv. Virus Res.* **41:**193–239.

Hizi, A., L. E. Henderson, T. D. Copeland, R. C. Sowder, C. V. Hixson, and S. Oroszlan. 1987. Characterization of mouse mammary tumor virus *gag-pol* gene products and the ribosomal frameshift by protein sequencing. *Proc. Natl. Acad. Sci. USA* **84:** 7041–7046.

Horsfield, J. A., D. N. Wilson, S. A. Mannering, F. M. Adamski, and W. P. Tate. 1995. Prokaryotic ribosomes recode the HIV-1 *gag-pol-1* frameshift sequence by an E/P site post-translocation simultaneous slippage mechanism. *Nucleic Acids Res.* **23:**1487–1494.

Huang, W. M., S.-Z. Ao, S. Casjens, R. Orlandi, R. Zeikus, R. Weiss, D. Winge, and N. Fang. 1988. A persistent untranslated sequence within bacteriophage T4 topoisomerase gene 60. *Science* **239:**1005–1012.

Jacks, T., and H. Varmus. 1985. Expression of Rous sarcoma virus pol gene by ribosomal frameshifting. *Science* **230:**1237–1242.

Jacks, T., H. D. Madshani, F. R. Masiarz, and H. E. Varmus. 1988a. Signals for ribosomal frameshifting in the Rous sarcoma virus *gag-pol* region. *Cell* **55:**447–458.

Jacks, T., M. D. Power, F. R. Masiarz, P. A. Luciw, P. J. Barr, and H. E. Varmus. 1988b. Characterization of ribosomal frameshifting in HIV-1 *gag-pol* shifting. *Nature* **331:**280–283.

Kane, J. F., B. N. Violand, D. F. Curran, N. R. Staten, K. L. Duffin, and G. Bogosian. 1992. Novel in-frame two codon translational hop during synthesis of bovine placental lactogen in a recombinant strain of *Escherichia coli. Nucleic Acids Res.* **20:** 6707–6712.

Kitchingman, G. R., and M. J. Fournier. 1975. Unbalanced growth and the production of unique transfer ribonucleic acids in relaxed-control Escherichia coli. *J. Bacteriol.* **124:**1382–1394.

Kitchingman, G. R., and M. J. Fournier. 1977. Modification-deficient transfer ribonucleic acids from relaxed control Escherichia coli: structures of the major undermodified phenylalanine and leucine transfer RNAs produced during leucine starvation. *Biochemistry* **16:**2213–2220.

Kolor, K., D. Lindsley, and J. Gallant. 1993. On the role of the P-site in leftward ribosome frameshifting at a hungry codon. *J. Mol. Biol.* **230:**1–5.

Kurland, C. G. 1979. Reading frame errors on ribosomes, p. 98–108. *In* J. Celis and J. D. Smith (ed.), *Nonsense Mutations and tRNA Suppressors.* Academic Press, London, United Kingdom.

Kurland, C. G. 1992. Translational accuracy and the fitness of bacteria. *Annu. Rev. Genet.* **26:**29–50.

Kurland, C. G., and J. Gallant. 1986. The secret life of the ribosome, p. 127–157. *In* T. B. L. Kirkwood, R. F. Rosenberger, and D. J. Galas (ed.), *Accuracy in Molecular Processes.* Chapman and Hall, London, United Kingdom.

Lindsley, D., and J. Gallant. Unpublished data.

Maia, I., K. Séron, A.-L. Haenni, and F. Bernardi. 1996. Gene expression from viral RNA genomes. *Plant Mol. Biol.* **32:**367–391.

O'Connor, M. 1998. tRNA imbalance promotes −1 frameshifting via near-cognate decoding. *J. Mol. Biol.* **279:**727–736.

Parker, J. 1989. Errors and alternatives in reading the universal genetic code. *Microbiol. Rev.* **53:**273–298.

Peter, K., D. Lindsley, L. Peng, and J. Gallant. 1992. Context rules of rightward overlapping reading. *New Biol.* **4:**1–7.

Qian, Q., J.-N. Li, H. Zhao, T. G. Hagervall, P. J. Farabaugh, and G. R. Björk. 1998. A new model for phenotypic suppression of frameshift mutations by mutant tRNAs. *Mol. Cell* **1:**471–482.

Rohde, W., A. Gramstat, J. Schmitz, E. Tacke, and D. Prüfer. 1994. Plant viruses as model systems for the study of noncanonical translation mechanisms in higher plants. *J. Gen. Virol.* 75:2141–2149.

Schwartz, R., and J. F. Curran. 1997. Analyses of frameshifting at UUU-pyrimidine sites. *Nucleic Acids Res.* **25**:2005–2011.

Sekine, U., and E. Ohtsubo. 1989. Frameshifting is required for the production of the transposase encoded by insertion sequence 1. *Proc. Natl. Acad. Sci. USA* **86**:4609–4613.

Sekine, Y., and E. Ohtsubo. 1992. DNA sequences required for translational frameshifting in production of the transposase encoded by IS1. *Mol. Gen. Genet.* **235**:325–332.

Sekine, Y., H. Nagasawa, and E. Ohtsubo. 1992. Identification of the site of translational frameshifting required for production of the transposase encoded by insertion sequence IS1. *Mol. Gen. Genet.* **235**:317–324.

ten Dam, E., C. Pleij, and L. Bosch. 1990. RNA pseudoknots: translational frameshifting and readthrough of viral RNAs. *Virus Genes* **4**:121–136.

Weiss, R., and J. Gallant. 1983. Mechanism of ribosome frameshifting during translation of the genetic code. *Nature* **302**:389–393.

Weiss, R., and J. Gallant. 1986. Frameshift suppression in aminoacyl-tRNA limited cells. *Genetics* **112**:727–739.

Weiss, R., J. Murphy, G. Wagner, and J. Gallant. 1984. The ribosome's frame of mind, p. 208–220. *In* B. F. C. Clark and H. U. Petersezn (ed.), *Gene Expression, Alfred Benzon Symposium 19.* Munksgaard, Copenhagen, Denmark.

Weiss, R., D. M. Dunn, J. F. Atkins, and R. F. Gesteland. 1987. Slippery runs, shifty stops, backward steps and forward hops: −2, −1, +5 and +6 ribosomal frameshifting. *Cold Spring Harbor Symp. Quant. Biol.* **52**:687–693.

Weiss, R., D. M. Dunn, M. Shuh, J. F. Atkins, and R. F. Gesteland. 1989. *E. coli* ribosomes re-phase on retroviral frameshift signals at rates ranging from 2 to 50 percent. *New Biol.* **1**:159–169.

Weiss, R., W. M. Huang, and D. M. Dunn. 1990. A nascent peptide is required for ribosomal bypass of the codon gap in bacteriophage T4 gene 60. *Cell* **62**:117–126.

Wenthzel, A.-M., M. Stancek, and L. A. Isaksson. 1998. Growth phase dependent stop codon readthrough and shift of translation reading frame in *Escherichia coli. FEBS Lett.* **421**:237–242.

Wills, N. M., J. A. Ingram, R. F. Gesteland, and J. F. Atkins. 1997. Reported translational bypass in a trp-R′-lacZ′ fusion is accounted for by unusual initiation and +1 frameshifting. *J. Mol. Biol.* **271**:491–498.

Yelverton, E., D. Lindsley, P. Yamauchi, and J. Gallant. 1994. The function of a ribosomal frameshifting signal from human immunodeficiency virus-1 in *Escherichia coli. Mol. Microbiol.* **11**: 303–313.

The Ribosome: Structure, Function, Antibiotics, and Cellular Interactions
Edited by R. A. Garrett, S. R. Douthwaite, A. Liljas, A. T. Matheson, P. B. Moore, and H. F. Noller

Chapter 32

tmRNA: a Case of Structural and Functional Mimicry on and off the Ribosome?

JACEK WOWER, CHRISTIAN ZWIEB, and IWONA K. WOWER

Protein synthesis in *Escherichia coli* takes place on the ribosome, a large macromolecular assembly of 54 ribosomal proteins and three rRNAs, 5S, 16S, and 23S. In concert, these components catalyze the formation of peptide bonds between aminoacyl and peptidyl derivatives of numerous tRNAs bound to the ribosomal A and P sites according to information encoded in mRNA. Evidence collected over the past 3 years indicates that this process includes tmRNA (previously called 10Sa RNA), a small stable RNA which binds to 70S ribosomes and can be aminoacylated with alanine (Komine et al., 1994, 1996). The first step towards understanding the function of tmRNA was made by Tu et al. (1995), who demonstrated that prematurely terminated murine interleukin-6, when expressed in *E. coli,* is tagged by an 11-amino-acid-long carboxy-terminal peptide recognized by the periplasmic protease Tsp. Tagging was prevented by disruption of a monocistronic *ssrA* gene, which encodes the 457-nucleotide precursor of tmRNA. Furthermore, the tag peptide code was present within the central part of the mature 10Sa RNA molecule. Thus, it was proposed that tagging of incompletely synthesized proteins occurs by a mechanism now called *trans*-translation. In this process, aminoacylated tmRNA binds to the A site of a ribosome which is stalled at the end of broken mRNA to accept an incomplete peptide from a P-site-bound peptidyl-tRNA. Upon translocation of the peptidyl-tmRNA, the ribosome is reprogrammed to bind aminoacyl-tRNAs as specified by tmRNA. This process results in the covalent attachment of the tag peptide. Finally, when the "stop" codon of tmRNA is reached, the tagged protein and the tmRNA are released from the ribosome. Presumably, several housekeeping proteases recognize the tag sequence and destroy the potentially harmful, incompletely synthesized polypeptide (Gottesman et al., 1998; Herman et al., 1998). (See Fig. 3, which provides an overview of tmRNA function but includes additional components as described below.)

The tmRNA-dependent *trans*-translational model (Muto et al., 1998) draws heavily on our current knowledge of bacterial protein synthesis (e.g., Wower and Zimmermann, 1991). The model proposes that a protein can be translated from more than one RNA molecule, as tmRNA combines the functions of tRNA and mRNA within a single molecule (hence the name, tmRNA).

Although portions of tmRNA are similar in structure to both tRNA and mRNA (Fig. 1), the degree to which tmRNA competes functionally with tRNA and mRNA on the ribosome remains unclear. Because the tmRNA molecule is considerably larger, one might postulate that conformational alterations must occur to allow the tag-coding segment of tmRNA to occupy the decoding site of the 30S ribosomal subunit and for the aminoacylated 3′ terminus to bind to the peptidyltransferase center of the 50S ribosomal subunit. These reactions are likely to be facilitated by known translational factors, but components not yet implicated in protein synthesis may also play a role. Given that many of the methods developed for investigating tRNA, mRNA, and ribosomes are applicable to the study of tmRNA, answers to the most pressing questions about the functional role of tmRNA in bacterial metabolism are likely to be forthcoming. However, since certain effects of *ssrA* gene inactivation are not fully explained by the current *trans*-translation model, the possibility that

Jacek Wower and Iwona K. Wower ■ Department of Animal and Dairy Sciences, Program in Cell and Molecular Biosciences, Auburn University, Auburn, AL 36849-5415. **Christian Zwieb** ■ Department of Molecular Biology, The University of Texas Health Science Center at Tyler, 11937 U.S. Highway 271, Tyler, TX 75708-3154.

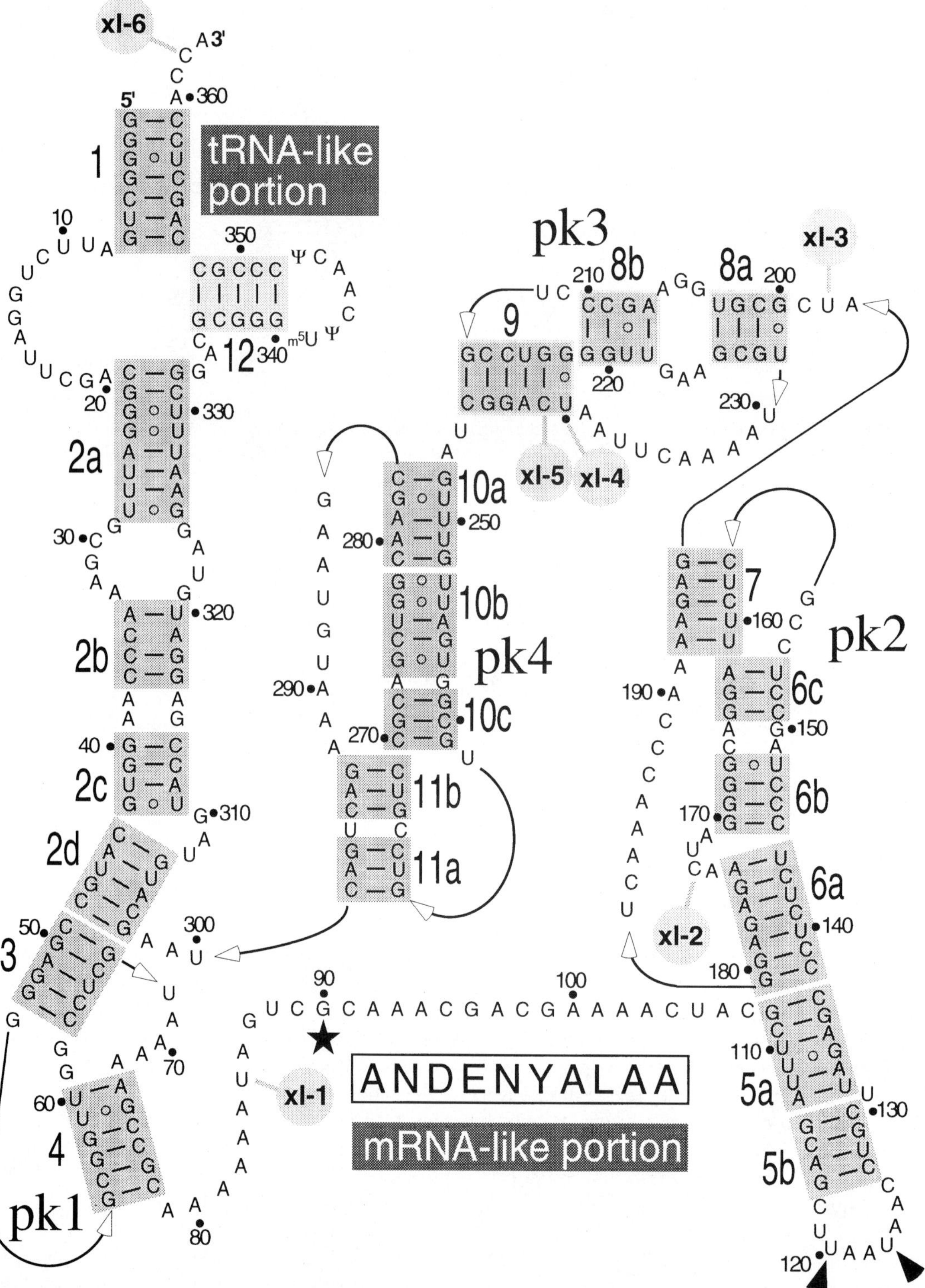

Figure 1. *E. coli* tmRNA secondary structure with base pairs supported by comparative sequence analysis (Fox and Woese, 1975; Larsen and Zwieb, 1991). Perfect base pairs are connected with lines; G·U pairs are connected with open circles. Helices are highlighted in gray and numbered from 1 to 12 from the 5′ end. Helical sections are given extensions with lowercase letters. The nucleosides are labeled with dots in increments of 10 and are numbered, if allowed by the available space. The 5′-to-3′ direction of the single RNA chain is indicated by lines with open arrowheads. The four pseudoknots are marked pk1 to pk4. The solid star marks the beginning of the tmRNA coding region. In-frame stop codons are marked by solid arrowheads. The tRNA-like structure and tmRNA-encoded tag peptide are shown at the top and the bottom of the tmRNA model, respectively. The cross-linked sites of covalent attachment of tmRNA to ribosomal protein S1 are indicated by "lollipops" denoted **xl-1** to **xl-6**.

tmRNA function is linked to processes other than protein tagging cannot be ruled out. For example, our finding that tmRNA interacts with ribosomal protein S1 provides a clue to how inactivation of the tmRNA gene might affect the growth of certain bacteriophages (Withey and Friedman, 1999).

tmRNA STRUCTURE

The tmRNA of *E. coli* was the first tmRNA discovered and was used initially to identify similar sequences in other organisms. These sequence comparisons, combined with biochemical analyses of the *E. coli* tmRNA, have given us considerable insight into the secondary structure of all known tmRNAs (Williams and Bartel, 1996; Felden et al., 1997; Zwieb et al., 1999a). In addition, three-dimensional models of several tmRNA sequences have been proposed recently (Zwieb et al., 1999b).

tmRNA SEQUENCES

New tmRNA sequences appear frequently at the tmRNA Website (http://sunflower.bio.indiana.edu/~kwilliam/tmRNA/home.html) (Williams, 1999), and as of 12 May 1999, tmRNAs from 64 species were listed. Sequences are also compiled in the tmRNA database (tmRDB) maintained at http://psyche.uthct.edu/dbs/tmRDB/tmRDB.html (Wower and Zwieb, 1999). As far as we can determine, all known tmRNAs are either of bacterial or plastid origin. tmRNAs have been found in 10 phyla of bacteria, thermophilic oxygen reducers, *Thermotogales*, green nonsulfur bacteria, *Flexibacter, Cytophaga, Bacteroides*, green sulfur bacteria, *Planctomycetes*, cyanobacteria, spirochetes, purple bacteria, and gram-positive bacteria. Chloroplast tmRNA sequences have been derived from organellar genomes of *Odontella sinensis*, *Porhyra purpurea*, *Guillardia theta*, *Thalassiosira weiiflogii*, and *Cyanophora paradoxa*. The recently sequenced genome of *Rickettsia prowazekii*, however, appears to lack a tmRNA homologue. We were unable to find structural tmRNA homologues in the mitochondrial genomes, and no homologues of tmRNA have been found yet in archaebacteria or eukaryotes.

SECONDARY STRUCTURE OF tmRNA

The common nomenclature of the tmRNA secondary structure describes 12 helices numbered consecutively from 1 to 12, starting at the 5′ end. Helices 1, 2a, and 12 form the tRNA-like structure. Other helices are part of four pseudoknotted regions (pk1 to pk4). The mRNA-like segment provides a coding sequence for the tag peptide and is flanked by pk1 and pk2. It should be noted that all our tmRNA secondary-structure models include only those base pairs that are supported by compensatory base changes of individual nucleotides by methods described by Fox and Woese (1975) and Larsen and Zwieb (1991). Despite this conservative approach, large portions of the tmRNA molecule could be determined to be base paired. These secondary structures were derived not only for *E. coli* (Fig. 1) but also for all other tmRNAs within the tmRDB and can be deduced directly from the aligned sequences (Zwieb et al., 1999a). Below, we compare and discuss the various tmRNA secondary-structure features in detail.

tRNA-Like Portion of tmRNA

The 3′ and 5′ termini are base paired to mimic the pairings found in alanyl-$tRNA^{Ala}$ (Komine et al., 1994; Ushida et al., 1994). This finding is compatible with the observation that the 3′ end of *E. coli* tmRNA can be aminoacylated with alanine. The similarity between tRNA and tmRNA is most pronounced in the aminoacyl acceptor stem. With an exception in *Alcaligenes eutrophus*, helix 1 of all tmRNAs consists of seven conserved or invariant base pairs and an $ACCA_{OH}$ 3′ terminus, as seen in canonical tRNAs. A common characteristic feature of the acceptor arm is the wobble base pair G3·U357 (numbered according to *E. coli* tmRNA [Fig. 1]), a major identity element which corresponds to G3·U73 in $tRNA^{Ala}$.

Similarity between tRNA and tmRNA is also observed with respect to helix 12 and the connecting loop, which are equivalent to the TΨC arm of $tRNA^{Ala}$. Both structures consist of 5 bp and a 7-nucleotide loop. An invariant G340-C348 base pair closes the loop, and five loop nucleotides (U341, U342, C343, A345, and U347) are identical. As in canonical tRNAs, positions 342 and 347 are pseudouridine (Ψ), and position 341 forms ribothymidine (T) (Felden et al., 1998). Chemical and enzymatic probing of the *E. coli* tmRNA (Felden et al., 1997) suggests that the conserved nucleotides T341 and A345 form a reverse-Hoogsteen base pair. This interaction is analogous to the T55-A58 base pair, which has been shown to stabilize the TΨC loops in tRNA.

At first sight (Fig. 1), the RNA segment between helices 1 and 2 differs from the DHU stem-loop region found in tRNA. In *E. coli* tmRNA, there are 13 nucleotides, but only 10 nucleotides may be present in other species (e.g., *Anabaena*). Comparative sequence analysis provides no support for any base pairing in this region. However, non-Watson-Crick interactions may occur. Furthermore, as most residues in this region are highly conserved, phylogenetic support for additional interactions (compensatory base changes) may become apparent only with more sequences. For example, in the DHU loop of canonical tRNAs, the invariant GG at positions 18 and 19 interacts with 55-ΨC-56 in the TΨC loop. Similarly, interaction between 13-GG-14 and 342-UC-343 might exist in tmRNA.

A typical anticodon arm of tRNA consists of 5 bp and a 7-nucleotide loop. This universal element is replaced in all tmRNA molecules by a much longer helix 2. In *E. coli*, helix 2 is composed of four sections, 2a, 2b, 2c, and 2d, with at least 16 Watson-Crick base pairs and three G·U wobble base pairs. Two nucleotides, C21 and G332, are invariant.

pk1

Formed by helices 3 and 4, pk1 is present in most tmRNAs. Helix 3 typically consists of 5 bp and displays a degree of conservation similar to that of helix 2d. Helix 4, however, varies in sequence and size. Formation of pk1 was verified by chemical and enzymatic probing and mutational analysis of the *E. coli* tmRNA in vitro (Felden et al., 1997, 1998; Hickerson et al., 1998). The susceptibility of A51 to chemical modification and nuclease cleavage at U65 suggests that these two nucleosides form an internal bulge in helix 3. The G54 residue may participate in long-range interactions, as it is protected from chemical modification. Alternatively, this protection of G54 may reflect a particular internal structure of pk1.

Changes of G61 and G62 to any other nucleotide affect the tagging process. The same two residues are sensitive to chemical modification, depending on the presence of magnesium ions. In the absence of magnesium ions, the G62 residue is protected. Felden et al. (1998) suggested that G61-G62 is a magnesium binding site similar to a GG in a domain of 5S rRNA (Correll et al., 1997).

mRNA-Like Tag-Encoding Segment

The first codon, known as the "resume" codon, is preceded by an adenosine-rich sequence which links it to pk1. The 3 nucleotides preceding the resume codon and the following codon are relatively conserved. Chemical probing of the *E. coli* tmRNA suggests that the resume codon is located in a hairpin formed by base pairing 88-UCGC-91 with 97-ACGA-100. However, this hairpin is not supported by comparative sequence analysis. In contrast, the stop codon is frequently located in the terminal loop of phylogenetically well-supported helix 5.

tmRNA-encoded tag peptides vary in length from 10 to 27 amino acid residues but are similar in their overall design (Fig. 2). Predictably, most tag peptides have alanine at the N terminus, but occasionally the resume codon appears to specify an aspartic acid, glycine, valine, phenylalanine, or serine residue. The second amino acid of the tag is usually an asparagine or, less frequently, a lysine residue. The most conserved part of the tag peptide is the nonpolar carboxyl terminus, typically containing either YALAA or FAVAA sequences (Fig. 2). In *E. coli*, Tu et al. (1995) showed that tagged proteins are degraded by the periplasmic protease Tsp (also known as Prc). Homologues of this enzyme have been found in both gram-positive and gram-negative bacteria (Anbudurai et al., 1994; Keiler et al., 1996). Recent studies identified three more proteases that degrade proteins tagged by tmRNA. One is the ATP-dependent, membrane-anchored zinc protease HflB (also called FtsH), the only essential protease in *E. coli* (Herman et al., 1998). Two others are the ATP-dependent cytosolic protease complexes ClpXP and ClpAP (Gottesman et al., 1998). These findings indicate that the tmRNA-dependent protein quality control system might have evolved to provide maximum degradation capacity by using many different proteases (Fig. 3). Since only five terminal amino acids of the tmRNA-encoded tag are needed for its recognition by all four proteases, the functional constraints imposed on the size and compositions of the tag-encoding region appear to be minimal.

pk2

Helices 6 and 7 constitute pk2. Helix 7 is relatively short, whereas helix 6 is composed of four segments in most known tmRNAs (Zwieb et al., 1999). In *E. coli* tmRNA, there are only three helical segments, denoted 6a, 6b, and 6c in Fig. 1. Probing the *E. coli* tmRNA with a square planar macrocyclic nickel (II) complex (NiCR) suggests that G150 and C166 may form a base pair and A149 and A165 may be bulged (Hickerson et al., 1998). However, this suggestion is not supported by phylogenetic analysis of the presently available sequences. Interestingly, helices 6a and 6d are absent in plastids and most gram-positive bacteria (e.g., *Mycoplasma pneumoniae*), respectively. In three purple bacteria (*Neisseria gonorrhoeae, Neisseria meningitis,* and *A. eutrophus*), helix 6 is modified to include a stabilizing GNRA

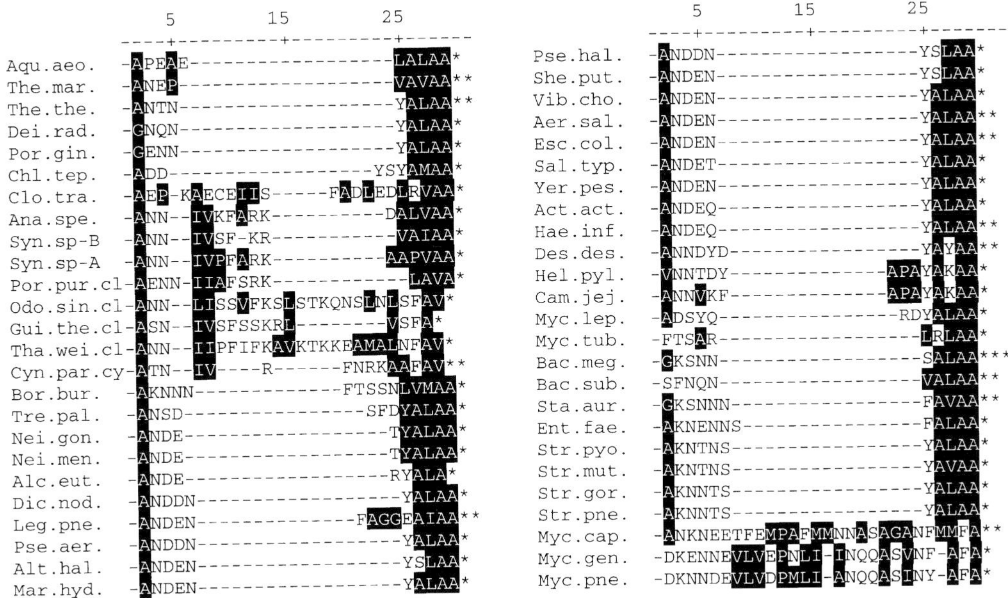

Figure 2. Alignment of tmRNA-encoded tag sequences as deduced from the tmRNA alignment (Zwieb et al., 1999a). The sequences are ordered phylogenetically with abbreviated species names, as in the tmRDB (http://psyche.uthct.edu/dbs/tmRDB). Nonpolar amino acids are shown in reverse print; stop codons are indicated by stars.

tetraloop, which has also been frequently observed in rRNA.

pk3

Typically, pk3 is formed by helices 8 and 9. In many species, including *E. coli*, helix 8 is divided into two segments, 8a and 8b. Although in the secondary-structure model of *E. coli* tmRNA these two segments are connected by an internal loop, chemical probing suggests that the residues of the loop region form the noncanonical base pairs G204-A225, G205-A224, and A206-G223 (Felden et al., 1997). According to Hickerson et al. (1998), G205-A224 and A206-G223 belong to the N7-N1 category of A-G base pairs, whereas G204-A225 may form an N1-N1 type A-G base pair, as proposed earlier by Heus and Pardi (1991) and Saenger (1984). The formation of noncanonical A-G base pairs is compatible with the suggestion that segments 8a and 8b are stacked to form a long helix 8, as it is realized, for example, in the tmRNA of *Synechocystis*.

pk4

In most tmRNA secondary structures, pk4 is formed by a long helix 10 and a shorter helix 11. As in the other tmRNA pseudoknots, the first loop is short, typically consisting of three nucleosides, whereas the second loop is extended and contains many adenosine residues. In *Synechocystis* spp., pk4 is divided into two smaller pseudoknots at the expense of the loops. In *E. coli* tmRNA, the formation of pk4 depends upon the presence of magnesium, as demonstrated by NiCR probing of full-length tmRNA and the isolated analogue of pk4 (Hickerson et al., 1998). This observation is consistent with earlier nuclear magnetic resonance (NMR) studies, which indicate that monovalent and divalent ions stabilize pseudoknots (Wyatt and Tinocco, 1993). Similar conclusions have been reached by testing the thermal stability of synthetic pseudoknots (Gluick et al., 1997).

THREE-DIMENSIONAL FOLDING OF tmRNA

Since tmRNA has resisted crystallization and is too large to be investigated by nuclear magnetic resonance techniques, we generated three-dimensional models by using comparative analysis assisted by the RNA modeling program ERNA-3D (Müller et al., 1995). This combined approach is justified, as a high degree of three-dimensional similarity is to be expected under the assumption that all tmRNAs perform the same function. Comparative analysis in three dimensions of 50 aligned tmRNA secondary-structure models provided support for the coaxial stacking of helices 2b and 2c, 2d and 3, and 5a and 5b and helices 1 and 12. Additional three-dimensional constraints were derived from computational comparative analysis of pseudoknots pk1 to

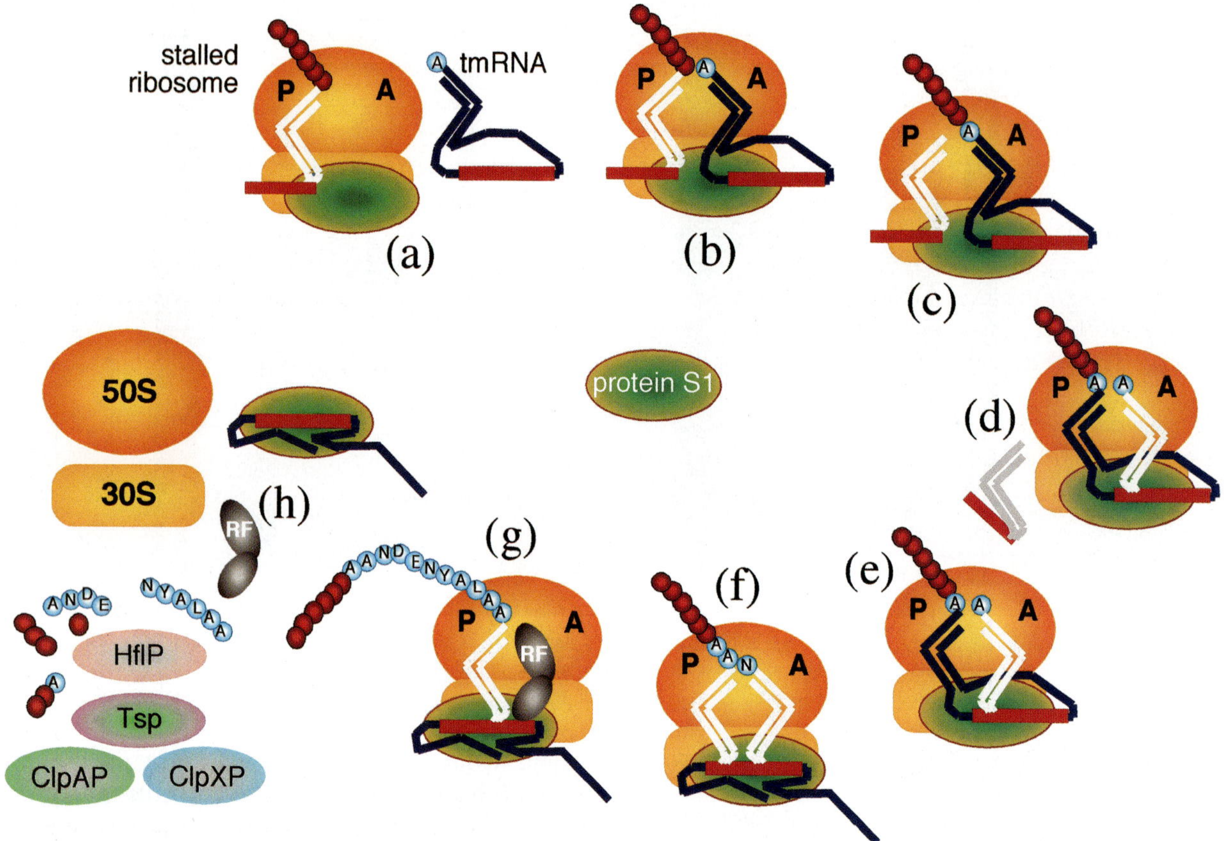

Figure 3. Mechanism of tmRNA in *trans*-translation. (a) The ribosome is stalled at the 3′ end of mRNA lacking a stop codon. The incomplete polypeptide chain remains bound to the P-site tRNA. (b) Alanyl-tmRNA binds the ribosomal A site assisted by ribosomal protein S1 and (c) accepts the polypeptide from the peptidyl-tRNA. (d) After the release of deacylated tRNA and damaged mRNA, peptidyl-tmRNA is translocated to the P site. (e) Aminoacyl-tRNA binds the resume codon in the A site. (f) After transpeptidylation, peptidyl-tRNA translocates to the P site and tmRNA adopts the mRNA-mode. (g) Releasing factors (RF) recognize the stop codon at the A site, and (h) the *trans*-translational complex dissociates. The tagged protein is degraded by proteases in the cytoplasm (ClpAP and ClpXP), cell membrane (HflP), or periplasm (Tsp).

pk4, which indicated coaxial stacking of helices 2d, 3, and 4; 6a, 6b, 6c, and 7; 8b and 9; and 10a, 10b, 10c, 11b, and 11a (Fig. 1). Nevertheless, the three-dimensional model of *E. coli* tmRNA assembled from these helical modules is still flexible in the tag coding region and between the individual pseudoknots, the longest stretch (242 Å) being between the 3′ adenosine at position 363 and the A122 residue of the stop codon (Zwieb et al., 1999b). The distance between the 3′ end of tmRNA and the first nucleoside of the resume codon is 192 Å, much longer than the 83 Å which separate the 3′ adenosine and the anticodon of canonical tRNA. In the future, additional three-dimensional constraints are expected to be obtained from forthcoming intra-tmRNA cross-linking studies, which aim to investigate the three-dimensional structure of tmRNA in solution or bound to ribosomes.

tmRNA INTERACTIONS

Interactions between tmRNA and ribosomes have been demonstrated experimentally both in vitro and in vivo (Komine et al., 1996; Tadaki et al., 1996). Irradiation of crude ribosomal preparations with UV light at 254 nm revealed that tmRNA cross-linked primarily to 16S rRNA and, to a lesser extent, to 23S rRNA. tmRNA binds exclusively to 70S ribosomes in vivo but appears to be absent in polysomes. Furthermore, when ribosomes are fractionated into subunits at low Mg^{2+} concentrations, interactions with tmRNA are no longer possible (Komine et al., 1996). These observations indicate that tmRNA binds to 70S ribosomes and may contact both subunits. The binding of tmRNA to ribosomes is much weaker than the analogous interaction with tRNA. This is understandable because tmRNA lacks a well-defined anticodon arm and thus may be unable to form tight complexes with the 30S ribosomal subunit. Alternatively, since mRNA-ribosome interactions are relatively weak, the tmRNA found in crude 70S ribosomes may exist in the "mRNA mode" of binding (Fig. 3).

To study ribosome tmRNA interactions, we applied cross-linking methods similar to those used successfully for investigating interactions between the ribosome, tRNA, and mRNA. In the past, these studies have led to the development of sophisticated models which represent different steps of the movement of tRNA across the ribosome (Moazed and Noller, 1989; Wower et al., 1989, 1993, 1995). Although, in the case of tmRNA, certain aspects of these models remain to be tested experimentally, it is reasonable to assume that they can be adopted to explain tmRNA functions.

To determine the interactions of tmRNA with ribosomes, complexes between tmRNA and ribosomal components were irradiated with UV light to induce cross-links. As described earlier (Wower et al., 1989), we used unmodified tmRNA irradiated with UV light at 254 nm as well as photoreactive tmRNAs containing 2-azidoadenosine ($2N_3A$) or 4-thiouridine (s^4U) modification irradiated with UV at 300 nm. Both unmodified tmRNA and photoreactive tmRNA derivatives were added to an in vitro translational system and, upon irradiation, formed covalent complexes with 70S ribosomes. Our preliminary analysis indicates that tmRNA cross-links exclusively to two segments of 16S rRNA which encompass nucleotides 1 to 450 and 1250 to 1543, respectively. Since the precise identification of these cross-linked sites is in progress, we focus below on a cytosolic protein of approximately 66 kDa which was cross-linked in the same experiments.

Immunological methods described earlier (Wower et al., 1993) identified the protein cross-linked to tmRNA as ribosomal protein S1. This identification is consistent with the molecular weight of the cross-linked complex and the partitioning of ribosomal protein S1 as free in the cytosol and associated with ribosomes, as described by Subramanian and van Duin (1977). In addition, we have demonstrated that ribosomes which lack protein S1 do not bind tmRNA. Furthermore, we showed that binding of tmRNA to free protein S1 impairs its association with ribosomes.

The covalently cross-linked sites of the tmRNA-protein S1 complexes were identified by combining RNase H fragmentation and primer extension as described previously (Wower et al., 1995). Most cross-linked sites with protein S1 were separated from each other in the secondary structure (see Fig. 1) and the current three-dimensional model of *E. coli* tmRNA. Interestingly, one cross-link (denoted xl-1 [Fig. 1]) was U85, a residue located in a single-stranded adenosine-rich region preceding the resume codon. Another cross-link (xl-2) was to C173 within pk2. Three other sites were U198 (xl-3), U240 (xl-4), and C241 (xl-5), located in or near pk3. The cross-links indicate an association between ribosomal S1 and tmRNA over a larger area which involves pk2 and pk3 and possibly other regions. The involvement of pseudoknots is supported by experiments of Ringquist et al. (1995), who used S1 to select pseudoknots from a pool of randomized RNA fragments.

Protein S1 was also labeled by a ribosome-bound photoreactive tmRNA derivative which was modified to contain $2N_3A$ at its 3′ terminus (xl-6 [Fig. 1]). Previously, analogous photoreactive tRNA probes were found to bind to either the ribosomal A site or the P site (Wower et al., 1989), and thus they are topographical markers for the peptidyltransferase center (Wower et al., 1998). It remains to be seen if a portion of tmRNA binds to the ribosome in an mRNA-like manner and how protein S1 might be involved in peptidyl transfer. Previously, S1 has been implicated in initiation and elongation (Linde et al., 1979; Potapov and Subramanian, 1992). For example, S1 may enhance both the efficiency and fidelity of elongation by adjusting the conformation of the mRNA and optimizing codon-anticodon interactions in the decoding site. The mechanism for binding tmRNA to ribosomes may be similar to one proposed by Odom et al. (1989), in which protein S1 attaches to the ribosome through its N-terminal domain and through its RNA binding domain "searches" the space around the ribosome for mRNA molecules which lack the strong Shine-Dalgarno sequences usually required for efficient binding of mRNA to ribosomes (Sorensen et al., 1998). Since tmRNA does not have a discernible Shine-Dalgarno sequence, one can speculate that protein S1 helps to direct tmRNA to the ribosome. Whether this process involves elongation factor Tu, needed for tRNA binding to the A site of the ribosome, is disputed. After binding to ribosomes and fulfilling its function as a tRNA mimic, it is possible that tmRNA undergoes a structural change in order to serve as an mRNA analogue (Fig. 3). Protein S1 may induce such changes in tmRNA by destabilizing one or more pseudoknots. It is also possible that protein S1 facilitates the departure of tmRNA from the ribosome, as the binding of S1 to the ribosome is relatively weak.

CONCLUSIONS AND OUTLOOK

Phylogenetic sequence analysis, combined with computer modeling and biochemical studies, has been a valuable tool for investigating the structure and function of tmRNA. The recent progress in understanding the three-dimensional structure of the ribosome, when combined with the identification of cross-linked sites, will help us to analyze tmRNA interactions with the ribosome at the level of single nucleotide and amino acid residues. Furthermore, genetic and in vivo studies are expected to contribute significantly to our understanding of the complex interactions that might occur among tmRNA, the ribosome, and its various ligands. The identification of protein S1 as a participant in *trans*-translation is an important step in the more detailed analysis of this interesting mechanism. Our task for the future will be to determine if tmRNA simply takes advantage of well-established ribosomal functions or if some aspects of tmRNA structure and function are unique and may be used to describe a ribosome that is more versatile than imagined previously.

This work was supported by National Science Foundation Grant MCB948732 and National Institutes of Health Grant GM58267 to J.W.

REFERENCES

Anbudurai, P. R., T. S. Mor, I. Ohad, S. V. Shestakov, and H. B. Pakrasi. 1994. The ctpA gene encodes the C-terminal processing protease for the D1 protein of the photosystem II reaction center complex. *Proc. Natl. Acad. Sci. USA* **91:**8082–8086.

Correll, C. C., B. Freeborn, P. B. Moore, and T. A. Steitz. 1997. Metals, motifs, and recognition in the crystal structure of a 5S rRNA domain. *Cell* **91:**705–712.

Felden, B., H. Himeno, A. Muto, J. P. McCutcheon, J. F. Atkins, and R. F. Gesteland. 1997. Probing the structure of the *Escherichia coli* 10Sa RNA (tmRNA). *RNA* **3:**89–103.

Felden, B., K. Hanawa, J. F. Atkins, H. Himeno, A. Muto, R. F. Gesteland, J. A. McCloskey, and P. F. Crain. 1998. Presence and location of modified nucleotides in *Escherichia coli* tmRNA: structural mimicry with tRNA acceptor branches. *EMBO J.* **17:** 3188–3196.

Fox, G. W., and C. R. Woese. 1975. 5S RNA secondary structure. *Nature* **256:**505–507.

Gluick, T. C., R. B. Gerstner, and D. E. Draper. 1997. Effects of Mg^{2+}, K^+, and H^+ on an equilibrium between alternative conformations of an RNA pseudoknot. *J. Mol. Biol.* **270:**451–463.

Gottesman, S., E. Roche, Y. Zhou, and R. T. Sauer. 1998. The ClpXP and ClpAP proteases degrade proteins with carboxy-terminal peptide tails added by the SsrA-tagging system. *Genes Dev.* **12:**1338–1347.

Herman, C., D. Thevenet, R. D'Ari, and P. Bouloc. 1998. Degradation of carboxy-terminal-tagged cytoplasmic proteins by the *Escherichia coli* protease HflB (FtsH). *Genes Dev.* **12:**1348–1355.

Heus, H. A., and A. Pardi. 1991. Structural features that give rise to the unusual stability of RNA hairpins containing GNRA loops. *Science* **253:**191–194.

Hickerson, R. P., C. D. Watkins-Sims, C. J. Burrows, J. F. Atkins, R. F. Gesteland, and B. Felden. 1998. A nickel complex cleaves uridine in folded RNA structures: application to *E. coli* tmRNA and related engineered molecules. *J. Mol. Biol.* **279:**577–587.

Keiler, K. C., P. R. Waller, and R. T. Sauer. 1996. Role of a peptide tagging system in degradation of proteins synthesized from damaged messenger RNA. *Science* **271:**990–993.

Komine, Y., M. Kitabatake, T. Yokogawa, K. Nishikawa, and H. Inokuchi. 1994. A tRNA-like structure is present in 10Sa RNA, a small stable RNA from *Escherichia coli. Proc. Natl. Acad. Sci. USA* **91:**9223–9227.

Komine, Y., M. Kitabatake, and H. Inokuchi. 1996. 10Sa RNA is associated with 70S ribosome particles in *Escherichia coli. J. Biochem.* (*Tokyo*) **119:**463–467.

Larsen, N., and C. Zwieb. 1991. SRP-RNA sequence alignment and secondary structure. *Nucleic Acids Res.* **19:**209–215.

Linde, R., N. Quoc Khan, R. Lipecky, and H. G. Gassen. 1979. On the function of the ribosomal protein S1 in the elongation cycle of bacterial protein synthesis. *Eur. J. Biochem.* **93:**565–572.

Moazed, D., and H. F. Noller. 1989. Intermediate states in the movement of transfer RNA in the ribosome. *Nature* **342:**142–148.

Müller, F., T. Döring, T. Erdemir, B. Greuer, N. Junke, M. Osswald, J. Rinke-Appel, K. Stade, S. Thamm, and R. Brimacombe. 1995. Getting closer to an understanding of the three-dimensional structure of ribosomal RNA. *Biochem. Cell Biol.* 73:767–773.

Muto, A., C. Ushida, and H. Himeno. 1998. A bacterial RNA that functions as both a tRNA and an mRNA. *Trends Biochem. Sci.* **23:**25–29.

Odom, O. W., H.-Y. Deng, A. R. Subramanian, and B. Hardesty. 1989. Relaxation time, interthiol distance, and mechanism of action of ribosomal protein S1. *Arch. Biochem. Biophys.* **230:** 178–193.

Potapov, A. P., and A. R. Subramanian. 1992. Effect of *E. coli* protein S1 on the fidelity of the translational elongation step: reading and misreading of poly(U) and poly(dT). *Biochem. J.* **27:** 745–753.

Ringquist, S., T. Jones, E. E. Snyder, T. Gibson, I. Boni, and L. Gold. 1995. High-affinity RNA ligands to *Escherichia coli* ribosomes and ribosomal protein S1: comparison of natural and unnatural binding sites. *Biochemistry* **34:**3640–3648.

Saenger, W. 1984. *Principles of Nucleic Acid Structure*. Springer-Verlag, New York, N.Y.

Sorensen, M. A., J. Fricke, and S. Pedersen. 1998. Ribosomal protein S1 is required for translation of most, if not all, natural mRNAs in *Escherichia coli in vivo. J. Mol. Biol.* **280:**561–569.

Subramanian, A. R., and J. van Duin. 1977. Exchange of individual ribosomal proteins between ribosomes as studied by heavy isotope-transfer experiments. *Mol. Gen. Genet.* **158:**1–9.

Tadaki, T., M. Fukushima, C. Ushida, H. Himeno, and A. Muto. 1996. Interaction of 10Sa RNA with ribosomes in *Escherichia coli. FEBS Lett.* **399:**223–226.

Tu, G. F., G. E. Reid, J.-G. Zhang, R. L. Moritz, and R. J. Simpson. 1995. C-terminal extension of truncated recombinant proteins in *Escherichia coli* with a 10Sa RNA decapeptide. *J. Biol. Chem.* **270:**9322–9326.

Ushida, C., H. Himeno, T. Watanabe, and A. Muto. 1994. tRNA-like structures in 10Sa RNAs of *Mycoplasma capricolum* and *Bacillus subtilis. Nucleic Acids Res.* **22:**3392–3396.

Williams, K. P. 1999. The tmRNA website. *Nucleic Acids Res.* **27:** 165–166.

Williams, K. P., and D. P. Bartel. 1996. Phylogenetic analysis of tmRNA secondary structure. *RNA* **2:**1306–1310.

Withey, J., and D. Friedman. 1999. Analysis of the role of *trans*-translation in the requirement of tmRNA for lambdaimmP22 growth in *Escherichia coli*. *J. Bacteriol.* **118**:2148–2157.

Wower, I. K., J. Wower, and R. A. Zimmermann. 1998. Ribosomal protein L27 participates in both 50S subunit assembly and the peptidyl transferase reaction. *J. Biol. Chem.* **273**:19847–19852.

Wower, J., and R. A. Zimmermann. 1991. A consonant model of the tRNA-ribosome complex during the elongation cycle of translation. *Biochimie* **73**:961–969.

Wower, J., and C. Zwieb. 1999. The tmRNA database (tmRDB). *Nucleic Acids Res.* **27**:167–168.

Wower, J., S. S. Hixson, and R. A. Zimmermann. 1989. Labeling the peptidyltransferase center of the *Escherichia coli* ribosome with photoreactive $tRNA^{Phe}$ derivatives containing azidoadenosine at the 3′ end of the acceptor arm: a model of the tRNA-ribosome complex. *Proc. Natl. Acad. Sci. USA* **86**:5232–5236.

Wower, J., P. Scheffer, L. A. Sylvers, W. Wintermeyer, and R. A. Zimmermann. 1993. Topography of the E site on the *Escherichia coli* ribosome. *EMBO J.* **12**:617–623.

Wower, J., I. K. Wower, S. V. Kirillov, K. V. Rosen, S. S. Hixson, and R. A. Zimmermann. 1995. Peptidyl transferase and beyond. *Biochem. Cell Biol.* **73**:1041–1047.

Wyatt, J. R., and J. I. Tinocco. 1993. RNA structural elements and RNA function, p. 465–493. *In* R. F. Gesteland and J. F. Atkins (ed.), *The RNA World*. Cold Spring Harbor Laboratory Press, Cold Spring Harbor, N.Y.

Zwieb, C., I. K. Wower, and J. Wower. 1999a. Comparative sequence analysis of tmRNA. *Nucleic Acids Res.* **27**:2063–2071.

Zwieb, C., F. Mueller, and J. Wower. 1999b. Comparative three-dimensional modeling of tmRNA. *Nucleic Acids Res. Symp.* **41**:200–204.

IX. FUNCTIONAL TARGET SITES FOR ANTIBIOTICS AND RIBOTOXINS

IX. FUNCTIONAL TARGET SITES FOR ANTIBIOTICS AND RIBOTOXINS

Many antibiotics that are widely used in clinical medicine and in animal feed act at different functional sites on the ribosome, where they either perturb or block protein biosynthesis. Of these, a large and structurally diverse group act within the peptidyltransferase center, where they interfere with the relative movements of the 3′ end of tRNA and the ribosome. Most of these antibiotics selectively, or preferentially, inactivate bacterial and archaeal ribosomes; other (universal) antibiotics inhibit all ribosomes. In addition, several ribotoxins, including α-sarcin and ricin, modify and functionally inactivate a region of the 23S rRNA involved in elongation factor binding in all ribosomes. Although the antibiotics and ribotoxins target rRNA-rich sites, there are indications that interactions with ribosomal proteins may also contribute to their inhibitory mechanisms. Considerable progress has been made over the past few years in characterizing the structures of their binding sites and in understanding their modes of action.

The Ribosome: Structure, Function, Antibiotics, and Cellular Interactions
Edited by R. A. Garrett, S. R. Douthwaite, A. Liljas, A. T. Matheson, P. B. Moore, and H. F. Noller

Chapter 33

Antibiotic Biosynthesis: Some Thoughts on "Why?" and "How?"

ERIC CUNDLIFFE

Other than for our enlightenment (Gale, 1966), it is a moot point why antibiotics are synthesized in Nature, although there are good models for how this is achieved. Until the advent of recombinant DNA technology, it was possible to argue that antibiotics are shunt metabolites, of no particular significance to producing organisms, arising from unbalanced growth. But all that has now changed. Antibiotics are synthesized by dedicated gene products; they are anything but accidents.

GENETICS OF ANTIBIOTIC PRODUCTION

The genetic complexity of low-*M*r metabolites can be immense, and many are produced only at considerable expense. Take, for example, the ribosome inhibitor tylosin, produced by *Streptomyces fradiae*. This macrolide antibiotic (*M*r, 916) consists of a polyketide lactone to which are added three deoxyhexose sugars (Fig. 1). As in other actinomycetes, the antibiotic-biosynthetic genes of *S. fradiae* are clustered within the genome (Fishman et al., 1987; reviewed in Baltz and Seno, 1988) and are adjacent to resistance determinants (Fig. 2). Including the latter (*tlrB*, *tlrC*, and *tlrD*, discussed in more detail below), the *tyl* cluster of *S. fradiae* contains at least 43 open reading frames. Five of these are megagenes (~41 kb in total) encoding the TylG polyketide synthase complex (~1.4 MDa), another 17 structural genes are required for biosynthesis and addition of the three sugars (Fig. 3), and somewhat unusually, the cluster also contains multiple regulatory genes (Bate et al., 1999). As presently defined, the *tyl* cluster occupies about 85 kb (about 1% of the total genome) and is flanked by the resistance determinants *tlrB* and *tlrC*. All the structural genes required for tylosin biosynthesis are probably present in the cluster, but additional regulatory genes might well be located elsewhere in the genome.

Antibiotic-biosynthetic gene clusters commonly include "pathway-specific" regulators. These typically encode transcriptional activators that turn on the structural genes for antibiotic production (Wietzorrek and Bibb, 1997). In turn, the pathway-specific regulators are typically controlled in "cascade" fashion by additional genes that exert pleiotropic control over multiple pathways (as in *Streptomyces coelicolor*, which produces four different antibiotics) or coregulate antibiotic production and sporulation (reviewed in Chater and Bibb, 1997). The pleiotropic regulatory genes, which are not usually found in antibiotic-biosynthetic clusters, collectively are not well defined, but a recurrent theme is the involvement of diffusible signaling factors, γ-butyrolactones. These resemble to some extent the homoserine lactones that are used for "quorum sensing" by gram-negative bacteria. Probably the most famous of the γ-butyrolactones is A-factor from *Streptomyces griseus*, the streptomycin producer (Fig. 4). This molecule is an inducer that turns on the streptomycin-biosynthetic gene cluster by derepressing a positive regulator whose product activates transcription of another regulator plus a resistance gene (reviewed in Horinouchi and Beppu, 1994). Since A-factor also regulates sporulation, it probably controls streptomycin production at a relatively high level in the hierarchical sense. Pathway-specific regulators function at a lower level and, as the name implies, specifically influence antibiotic production with no effects on morphological differentiation. Perhaps reflecting the complexity of the tylosin molecule, the *tyl* cluster is remarkable in containing two pathway-specific regulators (*tylS* and *tylT*), and the presence within the cluster of additional genes (*tylP* and *tylQ*), deduced to encode components involved in A-factor-

Eric Cundliffe ■ Department of Biochemistry, University of Leicester, Leicester LE1 7RH, United Kingdom.

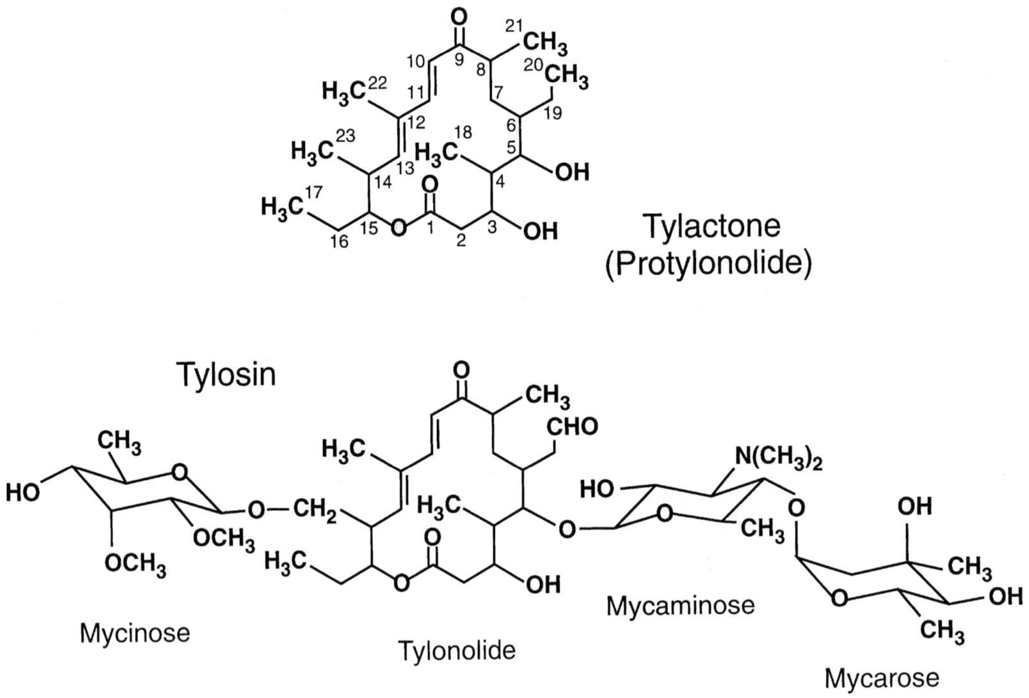

Figure 1. Structures of tylactone and tylosin.

mediated signal transduction, is also unprecedented (Bate et al., 1999). Since deletion of a fifth regulatory gene (*tylR*) knocks out not only polyketide metabolism but also synthesis of the tylosin sugars, *S. fradiae* evidently relies on interplay among multiple interactive components in order to sense conditions appropriate for tylosin production. This clearly implies some innate usefulness of the product; but what might that be?

WHAT ARE ANTIBIOTICS, WHAT KINDS OF ORGANISMS PRODUCE THEM, AND WHY?

Antibiotics are not just an isolated group of natural products with growth-inhibitory properties; they are part of a much larger family of compounds known as secondary metabolites (Vining, 1990). These are not essential for normal cell growth and reproduction but are typically produced in an idiosyncratic phase of metabolism that is separate from essential housekeeping activities. In microorganisms, for example, secondary metabolism commonly occurs mainly (or exclusively) after the phase of rapid growth is over, as the organism reacts to changes in nutrient levels and other environmental parameters. The nonessential nature of secondary metabolism is emphasized by the continued viability of nonproducing mutants. Not all secondary metabolites are antibiotics; some cause humans to hallucinate, others collectively exhibit a seemingly unlimited range of pharmacological properties, some are "merely" pigments, and others are quite unremarkable. To the extent that antibiotics are typical secondary metabolites, studies of their production and its regulation are interesting in a broader context. But the growth-inhibitory activities of antibiotics make them special, if only because organisms that produce them may have to take precautions for their own protection (Cundliffe, 1989).

Perhaps the greatest problem encountered in any discussion of antibiotics and their raison d'être is a tendency towards generalization in attempts to find unifying hypotheses. But even a casual survey of their extravagant range of chemical structures must surely raise the possibility that antibiotics might not all be produced for the same reason. Nevertheless, the kind of anthropocentric reasoning that suggested the title "antibiotic" in the first place seems to have led "naturally" to the competition hypothesis and suggestions of chemical weaponry. At its worst, neo-Darwinism encourages ex cathedra rationalization. Survival strategies lie at the heart of everything. Antibiotics are weapons, produced when times get hard (i.e., when nutrients are depleted), and confer a survival advantage in the wild. There is almost no need to discuss the subject—"it's obvious"! Of course, this argument requires not only that antibiotics actually be produced in the wild but that the levels of production be sufficient to influence the growth and survival of potential competitors.

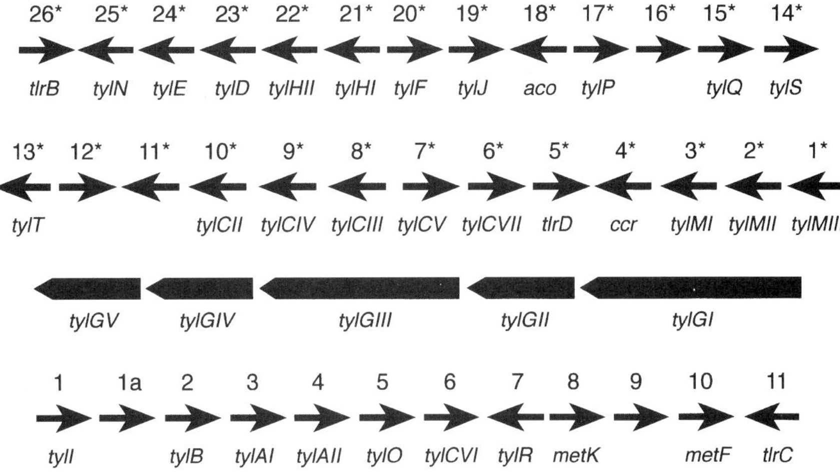

Figure 2. Tylosin-biosynthetic gene cluster of *S. fradiae* (not drawn to scale). The total size of the cluster is ~85 kb, of which the five *tylG* genes occupy ~41 kb. The data are taken from Merson-Davies and Cundliffe, 1994; Gandecha and Cundliffe, 1996; Gandecha et al., 1997; Wilson and Cundliffe, 1998, 1999; Bate and Cundliffe, 1999; and Bate et al., 1999, 2000. The *tylG* genes were sequenced at Lilly Research Laboratories (Indianapolis, Ind.) (GenBank accession no. U78289) but not formally published. Data relating to open reading frames 8 to 10 also originated at Lilly (DeHoff and Rosteck, personal communication).

This is probably a good point at which to digress and enquire what kinds of organisms actually produce antibiotics. Everyone is familiar with the Fleming parable, so we all know that fungi produce penicillin(s). Other microorganisms are also well known as prolific antibiotic producers, especially the bacteria known as actinomycetes. These are ubiquitous organisms, responsible for the earthy smell of soil, and members of the genus *Streptomyces* produce most of the currently known antibiotics, including most of the useful ones. However, it is less well appreciated that all organisms probably produce toxic metabolites. For example, animal cells and insects produce an extensive family of antimicrobial peptides, including the membrane-active "defensins" of human neutrophils (Lehrer and Ganz, 1996; Hancock and Chapple, 1999); somewhat similar compounds (magainins) are found in the skins of amphibians; many plant alkaloids are toxic; and as novel screening programs are developed, more and different organisms are being added to the list of antibiotic producers. Since the present discussion cannot adequately embrace such biodiversity, the arguments that follow relate mainly to microorganisms, in particular, the actinomycetes.

There is no doubt that some organisms do produce antibiotics in the wild, and in considerable amounts. Many plants produce bitter alkaloids (herbivore repellents), and ribosomologists familiar with the effects of narciclasine on the peptidyltransferase center of the 80S ribosome will not ingest the bulbs of King Alfred daffodils. More seriously, production of the potent carcinogen aflatoxin by *Aspergillus flavus* is responsible for the high incidence of liver carcinoma in East Africa. On the other hand, Julian Davies for years challenged anyone to demonstrate the production of any antibiotic by any actinomycete in the wild, or in the laboratory in unsupplemented soil. Nowadays, many antibiotics can be detected at levels well below those at which they display growth-inhibitory activity, and there are rumors that some have actually been found in natural soil. And that's the point; it has not been easy to detect chloramphenicol, tetracycline, aminoglycosides, thiostrepton, or you-name-it in the wild. Production of antibiotics in vanishingly small amounts hardly adds up to chemical warfare. And there is more.

Antibiotic-producing organisms are typically resistant to their own toxic metabolite(s), and some of the strategies employed for self-protection are discussed below and elsewhere (Cundliffe, 1989). The actinomycetes have been particularly well studied in this regard. Such resistance is usually quite specific for the avoidance of suicide; across-the-board resistance is not common. So, if I kill you and you kill me, who wins? And who are "you" and "me"? Who else is an actinomycete sure to find in its ecological niche? The original arguments relating to survival of the fittest considered more and less robust members of the same species, although that is not how the chemical warfarers like to argue in the present context. But the fact remains, if you are an actinomycete, the only organisms you are sure to find in your ecological niche are your siblings. Perhaps these antibiotic producers are not trying to kill dissimilar creatures, even if they occasionally do so. Maybe they

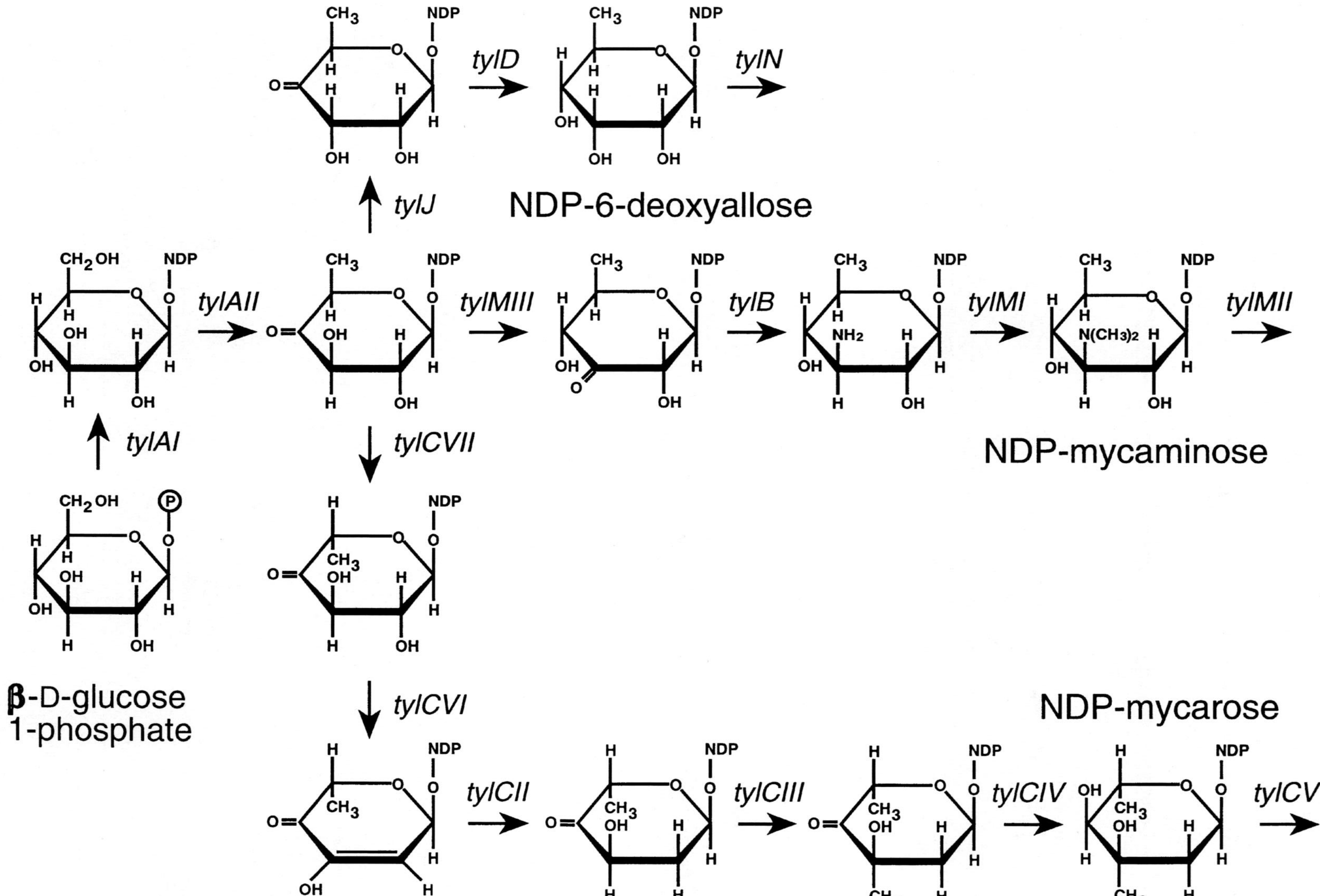

Figure 3. Biosynthesis of the tylosin sugars. During assembly of the tylosin molecule, mycaminose is the first sugar added to the polyketide ring, followed, in a preferred but not obligatory order, by deoxyallose and mycarose. After addition to the ring, the deoxyallose moiety is converted to mycinose via bis O-methylation, involving sequential action of the *tylE* and *tylF* products.

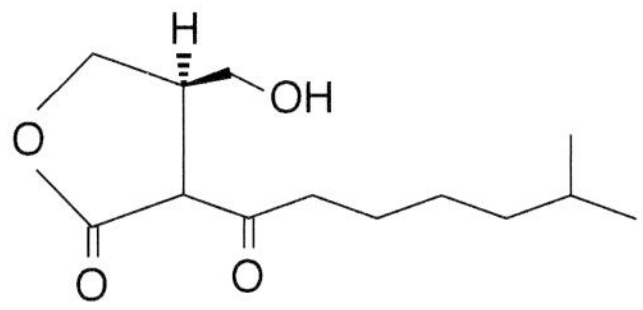

Figure 4. Structure of the γ-butyrolactone signal molecule, A-factor.

are communicating chemically with their own kind. In a similar vein, Julian Davies (again!) has repeatedly drawn attention to various intriguing effects exhibited by "antibiotics" at subinhibitory concentrations, including the promotion of gene transfer between bacteria (Davies, 1990). Later in this chapter the induction of gene expression by thiostrepton will also be discussed. In summary, the answer to the question of why organisms such as actinomycetes produce antibiotics is a simple one. The author doesn't know.

SELF-PROTECTION IN ANTIBIOTIC-PRODUCING ORGANISMS

The fungus that produces penicillin is not faced with any particular problem as a result of doing so, since fungal cell walls do not contain peptidoglycan. On the other hand, actinomycetes that produce antibacterial compounds pose an interesting challenge for themselves. Such strains can adopt either of two resistance strategies in order to avoid suicide. On the one hand, the targets (enzymes or other cellular components) to which the respective drugs would otherwise bind may be modified or redesigned de novo. Alternatively, intracellular drug levels can be maintained below toxic levels. Which of these options is appropriate for a given organism will, to some extent, depend on the antibiotic-biosynthetic pathway, since some drugs are produced as active molecules inside the cell whereas others are made and secreted as inert derivatives that are activated during or after export. Take, for example, *S. griseus*, the streptomycin producer. This organism produces and secretes 6-phosphorylstreptomycin, which is activated outside the mycelium by a specific phosphatase encoded within the streptomycin-biosynthetic (*str*) gene cluster (Mansouri and Piepersberg, 1991). Inside the mycelium, streptomycin 6-phosphotransferase [SPH(6)] acts on the 6-OH group of intermediates in the biosynthetic pathway and keeps them inactive (Walker and Skorvaga, 1973). Streptomycin itself is also a substrate for SPH(6), which can therefore inactivate any of the antibiotic that leaks back into the mycelium, although in order to avoid a futile and costly cycle of re- and dephosphorylation, influx of streptomycin into *S. griseus* is presumably limited. Antibiotic-inactivating enzymes in other producing organisms (Table 1) may similarly reveal the nature of the respective exported substrates (as in the oleandomycin producer, *Streptomyces antibioticus* [Vilches et al., 1992]), although the presence of multiple enzymes in various strains somewhat complicates the picture. In any event, one would not logically expect an antibiotic producer to defend itself exclusively by detoxifying its product.

Streptomycin is an inhibitor of protein synthesis that binds directly to ribosomes which, in *S. griseus*, remain fully sensitive to the drug at all times. Other actinomycetes, including those that produce aminoglycosides of the neomycin-paromomycin family, are also unable to desensitize their ribosomes and probably use a combination of drug modification and membrane permeability barriers to keep ribosomes and active drug molecules apart (Table 1). However, metabolic shielding of that kind may not suit the needs of other organisms in which active antibiotics are produced intracellularly. Thus, *Streptomyces azureus* renders its ribosomes insensitive to thiostrepton by methylating a specific nucleotide within 23S rRNA (Thompson et al., 1982). Quite apart from the obvious benefit to *S. azureus*, the thiostrepton resistance gene (*tsr*) is now used as the prime selectable marker on most of the vectors used for genetic manipulation of actinomycetes. Several other organisms that produce inhibitors of protein synthesis also employ ribosomal modification as a means of self-protection, and all examples documented thus far involve methylation of rRNA at single sites, each characteristic of a given resistance phenotype (Table

Table 1. Antibiotic-inactivating enzymes in organisms producing ribosome inhibitors[a]

Streptomyces sp.	Antibiotic(s) produced	Enzyme(s)[b]
S. griseus[c]	Streptomycin	SPH(6), SPH(3″)
S. fradiae[c]	Neomycin	APH(3′), AAC(3)
S. rimosus forma *paromomycinus*[c]	Paromomycin	APH(3′), AAC(3)
S. kanamyceticus	Kanamycin	AAC(6′)
S. tenebrarius	Tobramycin + apramycin	AAC(6′), AAC(2)
S. vinaceus[c]	Viomycin	VPH
S. alboniger[c]	Puromycin	PAC
S. antibioticus[c]	Oleandomycin	MGT
S. venezuelae[c]	Chloramphenicol	CPT

[a] For references and additional data, see Cundliffe, 1989, and Mosher et al., 1995.
[b] SPH(6), streptomycin 6-phosphotransferase; SPH(3″), streptomycin 3″-phosphotransferase; APH(3′), aminoglycoside 3′-phosphotransferase; AAC(3), aminoglycoside 3-*N*-acetyltransferase; AAC(6′), aminoglycoside 6′-*N*-acetyltransferase; AAC(2′), aminoglycoside 2′-*N*-acetyltransferase; VPH, viomycin phosphotransferase; PAC, puromycin *N*-acetyltransferase; MGT, macrolide glycosyltransferase; CPT, chloramphenicol phosphotransferase.
[c] Organism in which ribosomes are sensitive to the autogenous antibiotic(s).

Table 2. Resistance due to methylation of rRNA in antibiotic producers[a]

Organism[b]	Antibiotic(s) produced	Resistance gene(s)	rRNA and site[c]
S. azureus	Thiostrepton	*tsr*	23S; A1067
Sacc. erythraea[d]	Erythromycin	*ermE*	23S; A2058
S. fradiae	Tylosin	*tlrA, tlrD*	23S; A2058
S. thermotolerans	Carbomycin	*carB*	23S; A2058
S. caelestis	Celesticetin	*clr*	23S; A2058
S. pactum	Pactamycin	*pct*	16S; A964
S. kanamyceticus	Kanamycin	*kan*	16S; G1405
S. tenjimariensis	Istamycin	*kamA*	16S; A1408
S. tenebrarius	Tobramycin	*kgmB*	16S; G1405
	Apramycin	*kamB*	16S; A1408

[a] For references and additional data, see Cundliffe, 1989.
[b] *Sacc., Saccharopolyspora; S., Streptomyces.*
[c] Standard *E. coli* numbering system.
[d] Formerly *Streptomyces erythraeus.*

2). Interestingly, when *tsr* was expressed in *Streptomyces lividans* so that the host organism was immune to the ribosome-inhibitory action of thiostrepton, addition of the drug (at levels below those normally used to inhibit growth) triggered the appearance of several inducible proteins (Murakami et al., 1989). When these were characterized and the respective genes were isolated by reverse genetics, one of the latter (*tipA*) was found to possess a thiostrepton-inducible promoter and its product (the TipA protein) was shown to be a transcriptional activator that binds thiostrepton covalently and regulates its own production (Chiu et al., 1996). This in an organism that does not produce thiostrepton!

Ribosomes are not the only antibiotic targets that can be desensitized to autogenous drugs in the respective producing organisms (Table 3). Some strains simply produce resistant versions of the target enzymes, whereas others can make both sensitive and resistant isoforms. For example, *Streptomyces arenae* synthesizes a pentalenolactone-sensitive form of glyceraldehyde 3-phosphate dehydrogenase (GPDH) until pentalenolactone production is induced, whereupon drug-resistant GPDH activity appears. This does not involve modification of preexisting GPDH. Two distinct GPDH genes are present in *S. arenae*, and their expression is differentially regulated (Fröhlich et al., 1989). A conceptually similar situation is encountered in the novobiocin producer, *Streptomyces sphaeroides*, and in this case the mechanism of induction has been addressed.

Novobiocin is a coumarin antibiotic that acts on the B subunit of bacterial DNA gyrase, a tetrameric A_2B_2 enzyme. Gyrase is an essential DNA topoisomerase II enzyme that uses the energy of ATP hydrolysis to introduce negative supercoils into closed circular duplex DNA in vitro (for a review, see Reece and Maxwell, 1991). In contrast, DNA topoisomerase I acts in the opposite sense and removes supercoils. In intact bacteria, DNA superhelicity is affected by transcription and by the movement of DNA replication forks, and is dynamically controlled by the complementary activities of these and other DNA topoisomerases. When the novobiocin producer, *S. sphaeroides*, is grown in the absence of antibiotic, only novobiocin-sensitive DNA gyrase activity is detectable in mycelial extracts. However, when novobiocin is added to the growth medium (or during a novobiocin production fermentation), resistant DNA gyrase appears. This involves expression of a second *gyrB* gene (designated *gyrB*^{R}) encoding a resistant version of the gyrase B protein (Thiara and Cundliffe, 1988, 1989). When the *gyrB*^{R} promoter was transplanted into a reporter construct, it exhibited hair trigger activation in response to a reduction in DNA supercoiling. Although many promoters, especially those that control the production of DNA topoisomerases, probably respond to changes in DNA topology (Menzel and Gellert, 1983), the response of the *gyrB*^{R} promoter is extreme by any standards. This prompted the suggestion that during the early stages of novobiocin production in *S. sphaeroides*, the first appearance of the drug inhibits DNA gyrase activity and causes a transient reduction in genomic supercoiling, sufficient to turn on (sic!) expression of *gyrB*^{R}. The formation of drug-resistant gyrase then restores the relative levels of DNA topoisomerase I and II activities and allows continued growth in the presence of novobiocin (Thiara and Cundliffe, 1989).

Table 3. Antibiotic resistance due to altered properties of nonribosomal drug targets[a]

Organism	Antibiotic produced	Drug target
Streptomyces sphaeroides	Novobiocin	DNA gyrase
Amycolatopsis mediterranei[b]	Rifamycin	RNA polymerase
Streptomyces cinnamoneus	Kirrothricin	EF-Tu
Pseudomonas fluorescens	Pseudomonic acid	Ile-tRNA synthase
Cephalosporium caerulens	Cerulenin	Fatty acyl synthase
Streptomyces arenae	Pentalenolactone	GPDH

[a] For references and additional data, see Cundliffe, 1989.
[b] Formerly *Streptomyces*, then *Nocardia mediterranei.*

Although induction of resistance genes by antibiotics or their precursors may be fairly common in producing organisms, such genes may also be coregulated with those for antibiotic biosynthesis. As discussed earlier, in the streptomycin producer, *S. griseus*, *strR* (encoding a pathway-specific activator of the *str* gene cluster) and the resistance gene *aphD*, encoding SPH(6), are coexpressed as part of a regulatory cascade initiated by the γ-butyrolactone "hormone," A-factor (Distler et al., 1987). However, not all antibiotic resistance genes are inducible despite the widespread notion (propounded in several reviews) that producing organisms are typically sensitive to their products if challenged during the active-growth phase. To cite just one counterexample, the thiostrepton producer, *S. azureus*, is insensitive to inhibition by thiostrepton at all times. Indeed, two lines of reasoning demand that in some organisms resistance genes should be expressed constitutively. Firstly, not all antibiotics are produced exclusively during a distinct "idiophase" that is completely separated in time from primary metabolism, even though maximal drug production may occur as the cultures enter stationary phase. Secondly, development of resistance may take some time. For example, some of the enzymes that confer resistance by methylating rRNA do not act on mature ribosomes and presumably modify rRNA in vivo prior to its assembly into particles. Given the longevity of preexisting sensitive ribosomes, effective resistance levels might be attainable with these enzymes only if they are synthesized constitutively (as in the thiostrepton producer) or if they appear well ahead of drug production. In organisms where such enzymes are synthesized inducibly (e.g., in *S. fradiae*, the producer of tylosin), one finds additional resistance mechanisms.

Antibiotic-producing actinomycetes typically possess multiple resistance determinants (compare entries in Tables 1 and 2), and these are commonly clustered together with the antibiotic-biosynthetic genes. Thus, four resistance genes have been isolated from the tylosin producer. Two of these (*tlrA* and *tlrD*) are *erm*-type genes (for a review, see Weisblum, 1998) encoding methyltransferases that modify 23S rRNA and render ribosomes resistant to so-called MLS (for macrolides, lincomycin, and streptogramin type B) antibiotics, including tylosin. The TlrD protein (encoded within the *tyl* cluster) is produced constitutively and generates N^6-monomethylA at position 2058 (*Escherichia coli* numbering scheme) within 23S rRNA (Zalacain and Cundliffe, 1991). This confers moderate levels of resistance to macrolides. The TlrA protein is produced inducibly and introduces a second methyl group into A2058 (generating N^6,N^6-dimethylA). This increases the level of resistance to macrolides in general and tylosin in particular (Zalacain and Cundliffe, 1989). Expression of *tlrA* (which is not within the *tyl* cluster) is controlled via transcriptional attenuation and involves changes in the conformation of the mRNA leader sequence triggered by stalled ribosomes in the presence of inducing antibiotics (Kelemen et al., 1994). The induction specificity of *tlrA* is crucially dependent on the state of the ribosomes. Tylosin induces expression of *tlrA* in strains that express *tlrD* constitutively but not in *tlrD*-minus strains. In contrast, some tylosin precursors induce *tlrA* whether or not *tlrD* is present.

In addition to *tlrA* and *tlrD*, two other DNA fragments have been isolated from *S. fradiae* and shown to confer resistance to tylosin. One gene, *tlrB*, encodes a methyltransferase that confers resistance to tylosin in an ill-defined manner (Wilson and Cundliffe, 1999). The other, *tlrC*, is deduced to encode an ATP binding protein with a presumed role in tylosin efflux (Rosteck et al., 1991) and, on that basis, is probably not a resistance determinant in the strictest sense. These two genes flank the *tyl* cluster (Fig. 2), and their contributions to the production physiology of *S. fradiae* remain to be established.

ANTIBIOTICS AND RIBOSOMES

Ribosomologists embraced antibiotics once their potential as active site-directed inhibitors of the particle was realized in practice. Early work aimed at elucidating the modes of action of ribosome inhibitors contributed significantly to a superficial understanding of what ribosomes can do and which bits do what (for a review, see Gale et al., 1981). The actions of the various ribosome inhibitors were not identical. Studies with puromycin and chloramphenicol revealed that peptidyltransferase activity has to do with the larger ribosomal subunit, whereas tetracycline blocked aminoacyl-tRNA binding to the ribosome-mRNA complex from a site on the smaller subparticle. Thiostrepton targets the GTPase center on the 50S subunit. The early work, couched as it was in the framework of a mechanical two-site model that ultimately inhibited progress, held out the hope that studies of antibiotic binding sites in ribosomes would aid analysis of how the ribosomes work. A sea change occurred in the 1980s when rRNA became more amenable to study. In almost no time, chemical footprints produced by antibiotics were matched with sites of mutation or methylation associated with resistance (Cundliffe, 1990), and crucially, the inhibitory actions of antiribosomal drugs have been reinterpreted in terms of the hybrid-site model for ribosomal function (Kirillov et al., 1997). All this has been paralleled by high-resolution structural studies. Nuclear magnetic resonance analysis of antibiotics bound to rRNA fragments is now being complemented by crystallography of rRNA-protein complexes, soon to include antibiotic ligands (discussed elsewhere in this volume). The convergence of these various approaches will reveal in molecular detail how antibiotics block conformational transitions that underlie ribosomal function. The ribosome continues to be a focus for the application of ground-breaking methodology to biological systems, and enigmatic small molecules still contribute to our enlightenment.

Many workers have made important contributions to the analysis of antibiotic action against protein synthesis—none more so than the late David Vazquez, Julian Davies, and, latterly, Roger Garrett. Likewise, the contributions of David Hopwood and Arnold Demain to the genetics and physiology of antibiotic production have been incomparable.

REFERENCES

Baltz, R. H., and E. T. Seno. 1988. Genetics of *Streptomyces fradiae* and tylosin biosynthesis. *Annu. Rev. Microbiol.* **42:**547–574.

Bate, N., and E. Cundliffe. 1999. The mycinose-biosynthetic genes of *Streptomyces fradiae*, producer of tylosin. *J. Ind. Microbiol. Biotechnol.* **23:**118–122.

Bate, N., A. R. Butler, A. R. Gandecha, and E. Cundliffe. 1999. Multiple regulatory genes in the tylosin biosynthetic cluster of *Streptomyces fradiae*. *Chem. Biol.* **6:**617–624.

Bate, N., A. R. Butler, I. P. Smith, and E. Cundliffe. 2000. The mycarose-biosynthetic genes of *Streptomyces fradiae*, producer of tylosin. *Microbiology* **146:**139–146.

Chater, K. F., and M. J. Bibb. 1997. Regulation of bacterial antibiotic production, p. 59–105. *In* H. Kleinkauf and H. von Dören (ed.), *Biotechnology, Vol. 7. Products of Secondary Metabolism.* VCH, Weinheim, Germany.

Chiu, M. L., M. Folcher, P. Griffin, T. Holt, T. Klatt, and C. J. Thompson. 1996 Characterization of the covalent binding of thiostrepton to a thiostrepton-induced protein from *Streptomyces lividans*. *Biochemistry* **35:**2332–2341.

Cundliffe, E. 1989. How antibiotic-producing organisms avoid suicide. *Annu. Rev. Microbiol.* **43:**207–233.

Cundliffe, E. 1990. Recognition sites for antibiotics within rRNA, p. 479–490. *In* W. E. Hill, A. Dahlberg, R. A. Garrett, P. B. Moore, D. Schlessinger, and J. R. Warner (ed.), *The Ribosome: Structure, Function, and Evolution.* American Society for Microbiology, Washington, D.C.

Davies, J. 1990. What are antibiotics? Archaic functions for modern activities. *Mol. Microbiol.* **4:**1227–1232.

DeHoff, B. S., and P. R. Rosteck, Jr. Personal communication.

Distler, J., A. Ebert, K. Mansouri, K. Pissowotzki, M. Stockmann, and W. Piepersberg. 1987. Gene cluster for streptomycin biosynthesis in *Streptomyces griseus*: nucleotide sequence of three genes and analysis of transcriptional activity. *Nucleic Acids Res.* **15:**8041–8056.

Fishman, S. E., K. Cox, J. L. Larson, P. A. Reynolds, E. T. Seno, W.-K. Yeh, R. Van Frank, and C. L. Hershberger. 1987. Cloning genes for the biosynthesis of a macrolide antibiotic. *Proc. Natl. Acad. Sci. USA* **84:**8248–8252.

Fröhlich, K.-U., M. Wiedmann, F. Lottspeich, and D. Mecke. 1989. Substitution of a pentalenolactone-sensitive glyceraldehyde-3-phosphate dehydrogenase by a genetically distinct resistant isoform accompanies pentalenolactone production in *Streptomyces arenae*. *J. Bacteriol.* **171:**6696–6702.

Gale, E. F. 1966. The object of the exercise, p. 1–21. *In* B. A. Newton and P. E. Reynolds (ed.), *Biochemical Studies of Antimicrobial Drugs.* Cambridge University Press, Cambridge, United Kingdom.

Gale, E. F., E. Cundliffe, P. E. Reynolds, M. H. Richmond, and M. J. Waring. 1981. *The Molecular Basis of Antibiotic Action*, 2nd ed. John Wiley and Sons, London, United Kingdom.

Gandecha, A. R., and E. Cundliffe. 1996. Molecular analysis of *tlrD*, an MLS resistance determinant from the tylosin producer, *Streptomyces fradiae*. *Gene* **180:**173–176.

Gandecha, A. R., S. L. Large, and E. Cundliffe. 1997. Analysis of four tylosin biosynthetic genes from the *tylLM* region of the *Streptomyces fradiae* genome. *Gene* **184:**197–203.

Hancock, R. E. W., and D. S. Chapple. 1999. Peptide antibiotics. *Antimicrob. Agents Chemother.* **43:**1317–1323.

Horinouchi, S., and T. Beppu. 1994. A-factor as a microbial hormone that controls cellular differentiation and secondary metabolism in Streptomyces griseus. *Mol. Microbiol.* **12:**859–864.

Kelemen, G. H., M. Zalacain, E. Culebras, E. T. Seno, and E. Cundliffe. 1994. Transcriptional attenuation control of the tylosin resistance gene, *tlrA*, in *Streptomyces fradiae*. *Mol. Microbiol.* **14:**833–842.

Kirillov, S., B. T. Porse, B. Vester, P. Wolley, and R. A. Garrett. 1997. Movement of the 3′-end of tRNA through the peptidyl transferase centre and its inhibition by antibiotics. *FEBS Lett.* **406:**223–233.

Lehrer, R. I., and T. Ganz. 1996. Endogenous vertebrate antibiotics—defensins, protegrins and other cysteine-rich antimicrobial peptides. *Ann. N. Y. Acad. Sci.* **797:**228–239.

Mansouri, K., and W. Piepersberg. 1991. Genetics of streptomycin production in *Streptomyces griseus*: nucleotide sequence of five genes, *strFGHIK*, including a phosphatase gene. *Mol. Gen. Genet.* **228:**459–469.

Menzel, R., and M. Gellert. 1983. Regulation of the genes for *E. coli* DNA gyrase: homeostatic control of DNA supercoiling. *Cell* **34:**105–113.

Merson-Davies, L. A., and E. Cundliffe. 1994. Analysis of five tylosin biosynthetic genes from the *tylIBA* region of the *Streptomyces fradiae* genome. *Mol. Microbiol.* **13:**349–355.

Mosher, R. H., D. J. Camp, K. Yang, M. P. Brown, W. V. Shaw, and L. C. Vining. 1995. Inactivation of chloramphenicol by O-phosphorylation. A novel resistance mechanism in *Streptomyces venezuelae* ISP5230, a chloramphenicol producer. *J. Biol. Chem.* **270:**27000–27006.

Murakami, T., T. G. Holt, and C. J. Thompson. 1989. Thiostrepton-induced gene expression in *Streptomyces lividans*. *J. Bacteriol.* **171:**1459–1466.

Reece, R. J., and A. Maxwell. 1991. DNA gyrase: structure and function. *Crit. Rev. Biochem. Mol. Biol.* **26:**335–375.

Rosteck, P. R., Jr., P. A. Reynolds, and C. L. Hershberger. 1991. Homology between proteins controlling *Streptomyces fradiae* tylosin resistance and ATP-binding transport. *Gene* **102:**27–32.

Thiara, A. S., and E. Cundliffe. 1988. Cloning and characterization of a DNA gyrase B gene from *Streptomyces sphaeroides* that confers resistance to novobiocin. *EMBO J.* **7:**2255–2259.

Thiara, A. S., and E. Cundliffe. 1989. Interplay of novobiocin-resistant and -sensitive DNA gyrase activities in self-protection of the novobiocin producer, *Streptomyces sphaeroides*. *Gene* **81:** 65–72.

Thompson, J., F. J. Schmidt, and E. Cundliffe. 1982. Site of action of a ribosomal RNA methylase conferring resistance to thiostrepton. *J. Biol. Chem.* **257:**7915–7917.

Vilches, C., C. Hernandez, C. Mendez, and J. A. Salas. 1992. Role of glycosylation and deglycosylation in biosynthesis of and resistance to oleandomycin in the producer organism, *Streptomyces antibioticus*. *J. Bacteriol.* **174**:161–165.

Vining, L. C. 1990. Functions of secondary metabolites. *Annu. Rev. Microbiol.* **44**:395–427.

Walker, J. B., and M. Skorvaga. 1973. Phosphorylation of streptomycin and dihydrostreptomycin by *Streptomyces*. *J. Biol. Chem.* **248**:2435–2440.

Weisblum, B. 1998. Macrolide resistance. *Drug Res. Updates* **1**: 29–41.

Wietzorrek, A., and M. Bibb. 1997. A novel family of proteins that regulates antibiotic production in streptomycetes appears to contain an OmpR-like DNA-binding fold. *Mol. Microbiol.* **25**: 1177–1184.

Wilson, V. T. W., and E. Cundliffe. 1998. Characterization and targeted disruption of a glycosyltransferase gene in the tylosin producer, *Streptomyces fradiae*. *Gene* **214**:95–100.

Wilson, V. T. W., and E. Cundliffe. 1999. Molecular analysis of *tlrB*, an antibiotic-resistance gene from tylosin-producing *Streptomyces fradiae*, and discovery of a novel resistance mechanism. *J. Antibiot.* **52**:288–296.

Zalacain, M., and E. Cundliffe. 1989. Methylation of 23S rRNA caused by *tlrA* (*ermSF*), a tylosin resistance determinant from *Streptomyces fradiae*. *J. Bacteriol.* **171**:4254–4260.

Zalacain, M., and E. Cundliffe. 1991. Cloning of *tlrD*, a fourth resistance gene, from the tylosin producer, *Streptomyces fradiae*. *Gene* **97**:137–142.

The Ribosome: Structure, Function, Antibiotics, and Cellular Interactions
Edited by R. A. Garrett, S. R. Douthwaite, A. Liljas, A. T. Matheson, P. B. Moore, and H. F. Noller

Chapter 34

Aminoglycoside Antibiotics and Decoding

JOSEPH D. PUGLISI, SCOTT C. BLANCHARD, KAM D. DAHLQUIST, ROBERT G. EASON, DOMINIQUE FOURMY, STEPHEN R. LYNCH, MICHAEL I. RECHT, and SATOKO YOSHIZAWA

The ribosome is the target for many small-molecule drugs (Gale et al., 1981). These drugs interfere with distinct steps in translation and are powerful tools to dissect ribosome function. Aminoglycoside antibiotics bind directly to 16S rRNA in the 30S subunit of bacterial ribosomes (Moazed and Noller, 1987) and decrease the fidelity of translation (Davies et al., 1965; Davies and Davis, 1968). Aminoglycoside antibiotics remain important therapeutic agents and represent the archetype for RNA-targeted antibiotics. Here we present an overview of our investigations of how aminoglycoside antibiotics bind to rRNA and the insights these studies have provided into the decoding process.

We have investigated aminoglycosides that contain a 2-deoxystreptamine ring (Fig. 1) (Price et al., 1974; Gale et al., 1981). The members of this subclass of aminoglycosides target the same region of rRNA and interfere with the decoding process. Throughout this chapter, the term "aminoglycoside" will implicitly refer to this subclass. These aminoglycosides have common structural features, including the 2-deoxystreptamine ring (ring II) and chemical groups on the aminoglucose (ring I). These conserved elements are critical to direct specific interaction with rRNA (see below). The number of rings in the aminoglycosides can vary from two to five. Two general substitution patterns are observed for ring II: 4,5-disubstituted, which includes neomycin, paromomycin, ribostamycin, and neamine, and 4,6-disubstituted, which includes the gentamicins and kanamycins. The aminoglycosides have amino groups, many of which are protonated near neutral pH (Botto and Coxon, 1983). Therefore, there is a strong electrostatic component to RNA binding.

The 30S subunit is the site of the codon-anticodon interaction. Chemical probing and cross-linking experiments have mapped the sites of peptidyl (P-site)- and aminoacyl (A-site)-tRNAs on 16S rRNA (Green and Noller, 1997). The protection patterns and affinities of intact tRNAs for the P and A sites are reproduced by short anticodon stem-loops (Rose et al., 1983), indicating the local nature of tRNA-mRNA-rRNA contacts on the subunit. The P-site tRNA binds with higher affinity to the 30S subunit than A-site tRNA and can bind in the absence of mRNA. Consistent with these data, the P-site tRNA protects a wider range of nucleotides in 16S rRNA from chemical probes than A-site tRNA (Moazed and Noller, 1986, 1990). The P-site and A-site tRNAs are positioned in spatial proximity on adjacent codons. The region of 16S rRNA that closely contacts both tRNAs has been called the decoding region (Zimmermann et al., 1990; Cunningham et al., 1992, 1993) and consists of the 5′ and 3′ ends of the penultimate stem (helix 44) (Fig. 2a). Many nucleotides in the decoding region are highly or universally conserved (Gutell, 1994), consistent with their critical functional and structural roles in forming the tRNA binding sites. Mutations in many of these critical nucleotides yield ribosomes with impaired functions. It is within this conserved decoding region RNA that aminoglycoside antibiotics bind.

OLIGONUCLEOTIDE MODELS FOR AMINOGLYCOSIDE BINDING

Aminoglycosides that contain a 2-deoxystreptamine ring bind to the penultimate stem (helix

Joseph D. Puglisi, Scott C. Blanchard, Kam D. Dahlquist, Robert G. Eason, Dominique Fourmy, Stephen R. Lynch, Michael I. Recht, and Satoko Yoshizawa ■ Department of Structural Biology, Stanford University School of Medicine, Stanford, CA 94305-5421.

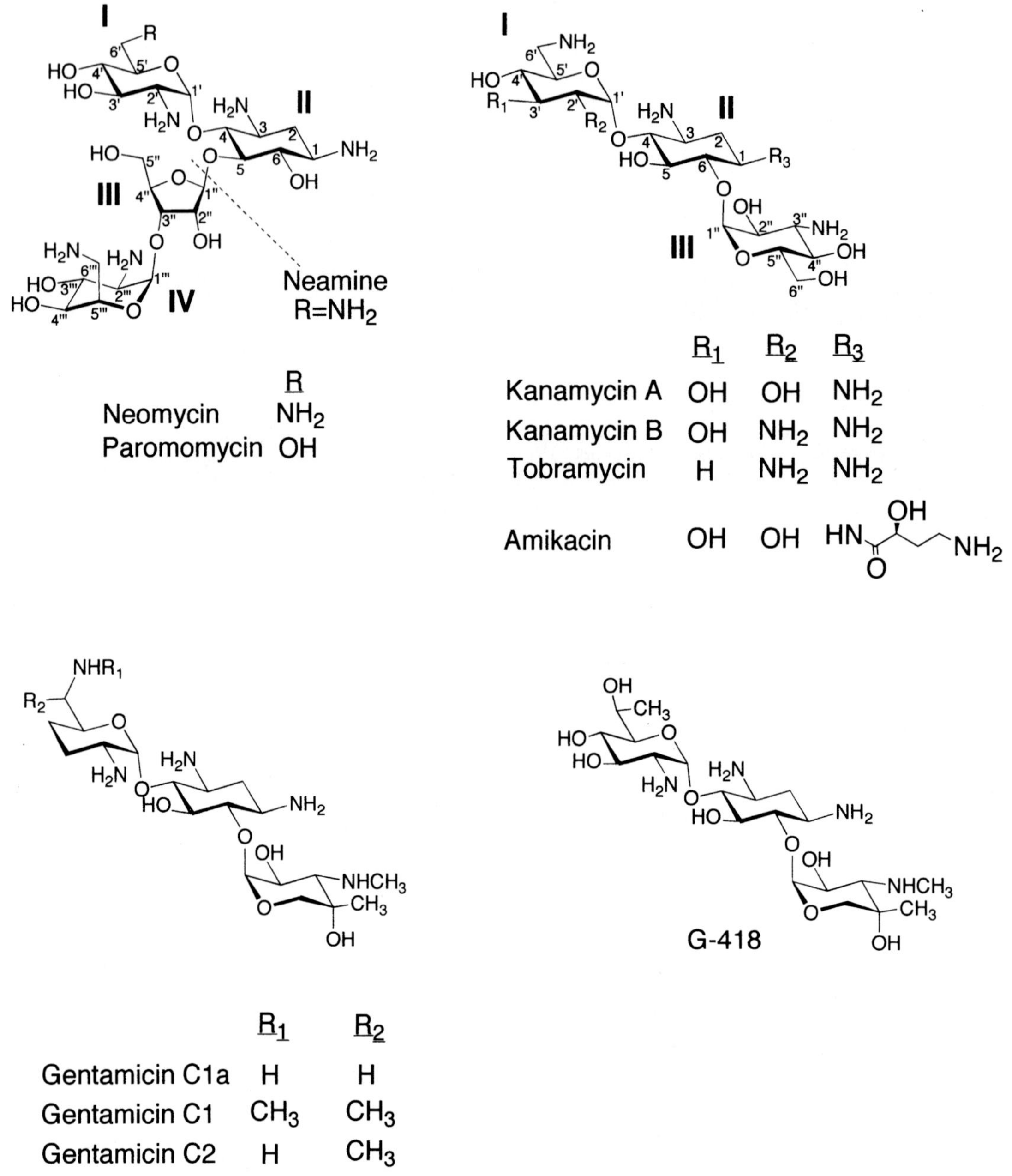

Neamine R=NH_2

	R
Neomycin	NH_2
Paromomycin	OH

	R_1	R_2	R_3
Kanamycin A	OH	OH	NH_2
Kanamycin B	OH	NH_2	NH_2
Tobramycin	H	NH_2	NH_2
Amikacin	OH	OH	HN–C(=O)–CH(OH)–CH$_2$–CH$_2$–NH_2

G-418

	R_1	R_2
Gentamicin C1a	H	H
Gentamicin C1	CH_3	CH_3
Gentamicin C2	H	CH_3

Figure 1. Aminoglycoside antibiotics containing a 2-deoxystreptamine ring.

44) in 16S rRNA in the region of A1408, A1492, and A1493, which is where A-site tRNAs bind (Fig. 2a) (Moazed and Noller, 1987). No ribosomal protein directly contacts this region of rRNA (Powers and Noller, 1995), so the aminoglycoside binding site is formed primarily by RNA. Aminoglycosides that contain streptamine, such as streptomycin, bind near the 900 region of rRNA (Moazed and Noller, 1987), and ribosomal protein S12 plays a role in the formation of the binding site for streptomycin (Davies et al., 1964; Bollen et al., 1969; Bonny et al., 1991). Mutations that yield streptomycin resistance occur in both rRNA and protein S12. In contrast, mutations that cause resistance to 2-deoxystreptamine antibiotics cluster mainly in helix 44 of 16S rRNA (DeStasio et al., 1989; DeStasio and Dahlberg, 1990; Alangaden et al., 1998; Prammananan et al., 1998); an important exception, a ribosomal protein L6 mutation (Kühberger et al., 1979), will be discussed later. The properties of the aminoglycoside antibiotic binding site suggested that it could be recreated within an RNA oligonucleotide.

Purohit and Stern showed that aminoglycosides bind specifically to a 44-nucleotide RNA that corre-

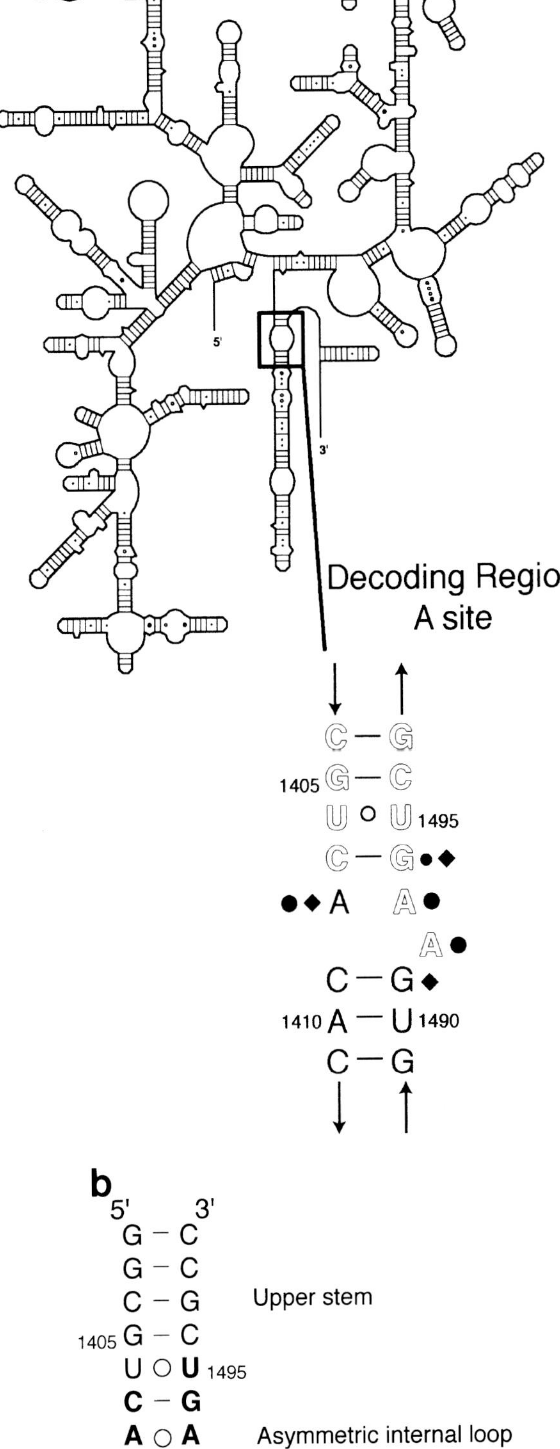

sponds to the 3′ and 5′ ends of helix 44 (Purohit and Stern, 1994). This critical experiment demonstrated the feasibility of using an RNA oligonucleotide to study aminoglycoside interaction with the ribosome. We have used a shorter construct, which corresponds to nucleotides 1403 to 1412 and 1488 to 1497 in helix 44 (Fig. 2b) (Recht et al., 1996). The short helix formed by C1404-G1497 and G1405-C1496, which is supported by phylogenetic covariations and mutational studies, was extended by two additional G-C base pairs, and the lower helix was capped with a UUCG tetraloop of known structure (Allain and Varani, 1995). This construct had the same chemical modification pattern as 16S rRNA in the 30S subunit; in particular, A1492 is more reactive to dimethyl sulfate (DMS) than A1493 at the N-1 position (Recht et al., 1996). The construct also recapitulated the aminoglycoside binding affinity and specificity of 30S subunits. The same nucleotides (A1408, G1494, and G1491) were protected in the oligonucleotide as were protected in the 30S subunit. Streptomycin and spermine did not yield protections on the oligonucleotide. Quantitative footprinting yielded a dissociation constant for paromomycin binding in the 200 nM range, consistent with the approximate K_d of aminoglycosides to 30S subunits (Fourmy et al., 1998b). These results validated the oligonucleotide construct for biochemical and structural studies of aminoglycoside binding to rRNA.

The nucleotides required for specific binding of aminoglycosides to rRNA have been determined with mutant versions of the oligonucleotides (Miyaguchi et al., 1996; Recht et al., 1996). The binding affinities were determined by chemical probing as a function of antibiotic concentration. The requirements for high-affinity aminoglycoside binding (Fig. 2b) are a C1407-G1494 base pair; U1495, A1408, and A1493; a bulged nucleotide at position 1492; and a base pair between positions 1409 and 1491. These results have been confirmed by using surface plasmon resonance to measure aminoglycoside affinities for the oligonucleotide (Hendrix et al., 1997; Wong et al., 1998).

Figure 2. (a) Secondary structure of *E. coli* 16S rRNA, highlighting the decoding region A site. Nucleotides conserved in greater than 95% of ribosomal sequences are shown in outline. Sites of protection from reaction with the chemical probe DMS in the presence of cognate A-site tRNA and mRNA are shown as solid circles. The size of the black dots indicates the relative level of chemical modification in the absence of ligand. Sites of protection from reaction with DMS upon addition of aminoglycoside antibiotics are shown as solid diamonds. (b) Secondary structure of the oligonucleotide construct that mimics the aminoglycoside binding properties of the 30S subunit. Nucleotides that are required for high-affinity binding of aminoglycosides are shown in boldface.

Mutations that disrupt aminoglycoside affinity for the oligonucleotide have also been incorporated into 30S subunits, and aminoglycoside binding to these mutant subunits has been monitored (Recht et al., 1999a, 1999b). A one-to-one correspondence exists between results on the oligonucleotide and results within the 30S subunit, further supporting the validity of the model oligonucleotide system.

STRUCTURE DETERMINATION BY NMR SPECTROSCOPY

Advances in nuclear magnetic resonance (NMR) spectroscopy have allowed structure determinations of RNA-ligand complexes in the 10- to 20-kDa range (Viani Puglisi and Puglisi, 1998). RNAs are typically labeled with ^{15}N and ^{13}C, which facilitates the critical resonance assignment procedure. Two- and three-dimensional NMR methods allow near-complete, unambiguous assignments of ^{1}H, ^{13}C, ^{15}N, and ^{31}P resonances. Once assignments are complete, interproton distance restraints and dihedral angle restraints can be extracted from the NMR data. Structures are calculated from the NMR restraints by using simulated-annealing molecular dynamics protocols. RNA-ligand complexes are particularly attractive targets for NMR structure determination (Puglisi and Williamson, 1999), as the ligand often stabilizes a particular RNA conformation and provides additional intermolecular distance restraints.

In the absence of antibiotic, the A-site oligonucleotide forms a structure with two A-form helical segments, separated by the asymmetric internal loop (Fourmy et al., 1998b). The internal loop is closed by a noncanonical U1406-U1495 base pair and a Watson-Crick C1407-G1494 pair. In the ensemble of calculated structures, A1408, A1492, and A1493 are more disordered than the rest of the molecule, suggesting increased dynamics in the internal loop (Fig. 3, top). A single-hydrogen-bond base pair between A1408 and A1493 forms. The high affinity of the aminoglycosides for the RNA oligonucleotide allowed formation of 1:1 complexes with several compounds (Recht et al., 1996; Fourmy et al., 1998a). Addition of the drug leads to stabilization of a distinct conformation for the internal loop (Fig. 3, bottom), as described below.

RNA complexes with either paromomycin (Fourmy et al., 1996) or gentamicin C1a (Yoshizawa et al., 1998) have been solved at high resolution by NMR, representing the two subclasses of aminoglycosides (4,5- and 4,6-disubstituted ring II). Paromomycin contains four rings and a 4,5-disubstituted ring II (Fig. 1). It differs from neomycin by a single NH_2-

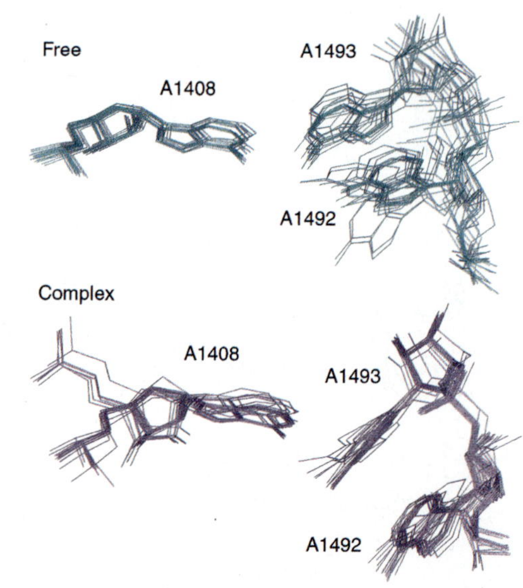

Figure 3. Comparison of the internal loops (A1408, A1492, and A1493) for the free (top) and paromomycin-bound (bottom) forms of the A-site oligonucleotide. The final 20 superimposed NMR structures for each state are shown.

to-OH change at the 6′ position of ring I. Gentamicin C1a contains three rings and a 4,6-disubstituted ring II (Fig. 1). Both drugs bind within the major groove of the RNA at the internal loop (Fig. 4). In both complexes the RNA conformations induced by drug binding are nearly identical; the root mean squared deviation for the core RNA residues (G1405 to A1410 and U1490 to C1496) between the two complexes is 1.48 Å (Yoshizawa et al., 1998). In both complexes, A1408 and A1493 form an A-A base pair in which the N-1 of A1408 forms a hydrogen bond with the N-6 of A1493 and the N-6 of A1408 forms a possible water-mediated hydrogen bond with the N-7 of A1493.

The A-site RNA undergoes a local conformational change upon binding of aminoglycosides. Compared to the free form, the bases A1492 and A1493 are displaced by ring I of the antibiotic towards the minor groove by 2.6 and 3.9 Å, respectively (Fig. 5). A bend in the RNA helix is apparent in the final structures of the complex, and we are currently confirming this bend by residual dipolar coupling measurements (Lynch and Puglisi, unpublished). The drug fits into a binding pocket formed by the RNA, and the complex is stabilized by specific drug-RNA contacts.

The chemical groups that are common among aminoglycoside antibiotics direct specific interaction with the RNA (Fig. 6). Rings I and II in the paromomycin and gentamicin C1a complexes are positioned in a similar manner with respect to the RNA. Ring II spans G1494 and U1495. The N-1 and N-3

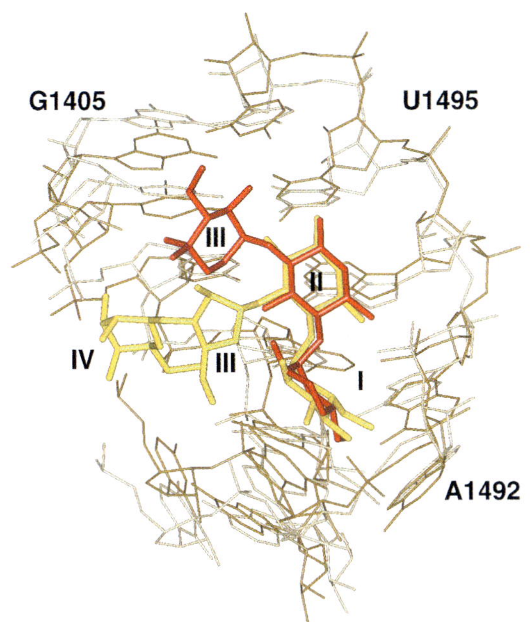

Figure 4. Comparison of the solution structures of the paromomycin- and gentamicin C1a-RNA oligonucleotide complexes. The drugs bind in the major groove of the RNA. Paromomycin is shown in yellow, and gentamicin C1a is in red. The RNA conformations are very similar. RNA in the paromomycin complex is brown, whereas the RNA in the gentamicin C1a complex is tan.

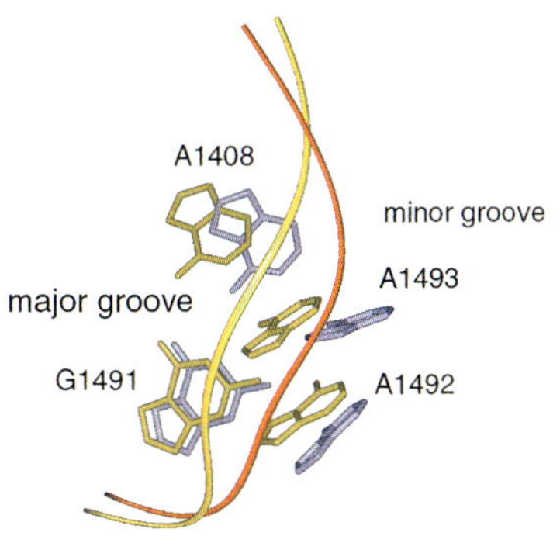

Figure 5. Binding of aminoglycoside antibiotics in the major groove of the A-site RNA oligonucleotide displaces A1492 and A1493 towards the minor groove. The phosphodiester backbone is represented by a yellow ribbon for the free-form RNA and an orange ribbon for the bound-form RNA. The bases A1408, G1491, A1492, and A1493 are shown in olive green for the free form and in lilac for the paromomycin complex.

amino groups of ring II form hydrogen bonds with the O-4 of U1495 and the N-7 of G1494, respectively. Ring I fits within the pocket formed by the A1408-A1493 pair, the bulged nucleotide A1492, and the C1409-G1491 pair on the bottom. The 6′ hydrogen bond donor (OH for paromomycin and NH_2 or NH_3^+ for gentamicin C1a) contacts the RNA phosphodiester backbone and helps to stabilize the conformation of the RNA pocket.

The additional rings on aminoglycosides contribute to RNA interaction. In the paromomycin-RNA complex, rings III and IV are directed towards the C1409-G1491 and A1410-U1490 base pairs (Fig. 4). These rings are more disordered in the ensemble of final structures, indicating increased conformational dynamics (Fourmy et al., 1996). Rings III and IV do not make specific hydrogen bonds with the RNA, but the amino groups on ring IV make favorable electrostatic contacts with the RNA backbone in the lower stem. Ring IV of neomycin can be removed to yield the active antibiotic ribostamycin, which binds to rRNA with decreased affinity (Fourmy et al., 1998b).

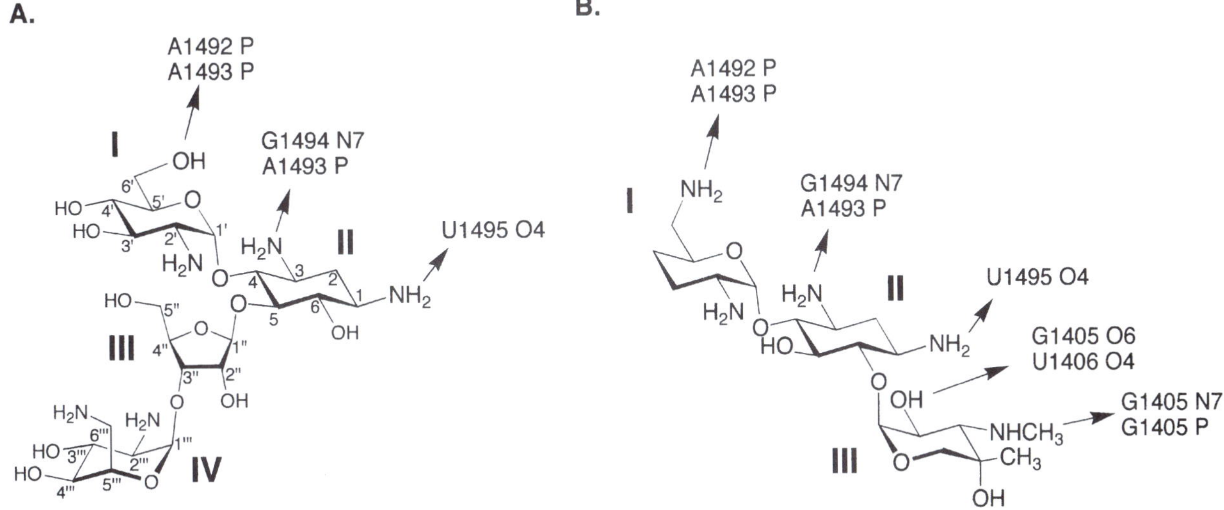

Figure 6. (A) Schematic showing intermolecular contacts between critical chemical groups in paromomycin and the A-site RNA. (B) Intermolecular contacts observed between gentamicin C1a and the A-site RNA.

In contrast, ring III of gentamicin C1a is directed towards U1406 and the G1405-C1496 base pair (Fig. 4) and a hydrogen bond is formed between the ring III 3″ and 2″ groups and the base moiety of G1405 (Fig. 6B). In strong support of this interaction, mutation of the G1405-C1496 pair to C1405-G1496 disrupts gentamicin C1a binding but has no effect on paromomycin binding (Yoshizawa et al., 1998). These results demonstrate the different functional roles of the additional rings in 4,5- and 4,6-disubstituted aminoglycosides.

The structural studies of aminoglycoside-rRNA complexes agree with biochemical and physical measurements of aminoglycoside binding affinities. The critical nucleotides for high-affinity drug binding to rRNA include U1495 and G1494, which are directly contacted by ring II of the aminoglycosides. Interestingly, the geometry of the U1406-U1495 pair is not required for ring II to contact the O-4 position of U1495; this base pair can be mutated to a Watson-Crick A1406-U1495 pair with little loss in paromomycin affinity (Recht et al., 1996, 1999a). This mutation does partially disrupt gentamicin C1a binding due to steric clash with ring III. The required A1408-A1493 pair allows formation of the correct binding pocket for ring I. A bulged nucleotide at position 1492 and a Watson-Crick pair at position 1409-1491 are also needed to form this binding pocket.

RESISTANCE AND SPECIFICITY FOR PROKARYOTIC RIBOSOMES

Mutations in rRNA can decrease aminoglycoside binding affinity and cause antibiotic resistance. The frequency of these mutations in clinical situations is limited by the multiple rRNA operons in many bacteria and the functional role of the decoding region in translation. Nonetheless, aminoglycoside resistance mutations have been discovered that change the 1409-1491 pair to a mispair (DeStasio et al., 1989; DeStasio and Dahlberg, 1990) and that change U1495 to C (Spangler and Blackburn, 1985) and A1408 to G (Alangaden et al., 1998; Prammananan et al., 1998). These mutations all decrease aminoglycoside antibiotic affinity for rRNA either by disrupting direct hydrogen bonding or by disrupting the conformation of the RNA binding pocket for the drug. RNA modification enzymes introduce more subtle chemical changes than mutations. Aminoglycoside-producing bacteria protect themselves from the effects of their own products by modifying their own ribosomes (Beauclerk and Cundliffe, 1987). The N-1 position of A1408 is a general target for a resistance methylase, consistent with the role of this position in forming the critical A1408-A1493 pair. Gentamicin-producing organisms protect themselves by methylation at the N-7 of G1405, which is contacted by ring III of the gentamicins.

The decoding region of 16S rRNA is highly conserved between prokaryotic and eukaryotic organisms (Fig. 7). Aminoglycosides interfere with eukaryotic translation in vitro only at concentrations 10- to 100-fold higher than those at which they affect prokaryotic organisms (Wilhelm et al., 1978). Clinically, aminoglycosides have serious toxicity problems that may be related to their interference with eukaryotic cytoplasmic or mitochondrial ribosomes (Prezant et al., 1993; Bacino et al., 1995; Guan et al., 1996). The major sequence difference between all prokaryotic ribosomes and all eukaryotic ribosomes in the aminoglycoside binding site is an A1408-to-G1408 change. Preliminary NMR data suggest that the G1408 mutation slightly changes the conformation of the RNA binding pocket, in which the base of G1408 is shifted towards the major groove in comparison to A1408 (Lynch and Puglisi, unpublished). In addition, most higher eukaryotic organisms have a mispair at the 1409-1491 position, which disrupts the bottom of the ring I binding pocket. As discussed above, these changes induce aminoglycoside resistance in various bacteria.

The critical role of A1408 in modulating aminoglycoside affinity to the ribosome was confirmed by mutagenesis of the *Escherichia coli* 30S subunit (Recht et al., 1999b). The change of A1408 to G in *E. coli* leads to high-level resistance to many aminoglycoside antibiotics. However, only low-level resistance was observed to G418 and paromomycin, which are the most effective aminoglycosides against eukaryotic organisms. The resistance observed with the G1408 mutant *E. coli* ribosomes mirrors the effects of aminoglycosides on translation in vitro (Wilhelm et al., 1978). Resistance is observed to aminoglycosides containing a 6′-NH_2 group, whereas little resistance is observed to those containing a 6′-OH group. The 6′ position is critical for aminoglycoside binding within the RNA pocket. For example, gentamicin C1, which contains an aminomethyl group at the 6′ position, binds with weaker affinity to rRNA than gentamicin C1a (Yoshizawa et al., 1998). The 6′ position is near the A1408-A1493 pair in the wild-type complex, but the details of how the geometry of the 1408-1493 pair correlates with the identity of the chemical group on the 6′ position of the aminoglycoside await structure determinations of the G1408 mutant oligonucleotide.

The major mechanism of aminoglycoside resistance is enzymatic modification of the drug (Davies

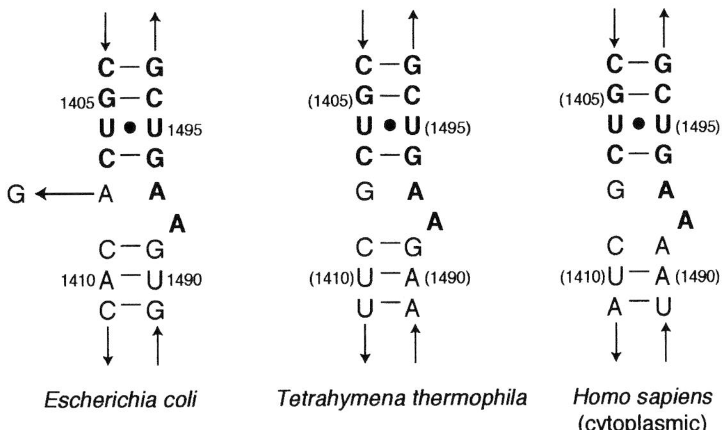

Figure 7. Secondary structures of the A-site decoding region in 16S-like rRNAs of three organisms: *E. coli*, *Tetrahymena thermophila*, and human (cytoplasmic). The G1408 mutation in *E. coli* is indicated. Highly conserved nucleotides are shown in boldface.

and Wright, 1997). A large number of aminoglycoside modification enzymes have been isolated. These enzymes are plasmid encoded and are readily transferred between bacterial strains. The activities of these modification enzymes, summarized in Fig. 8, include acetylation of amino groups, phosphorylation of hydroxyl groups, and adenylylation of hydroxyl groups (Shaw et al., 1993). The modified chemical groups on the aminoglycoside reside at the drug-RNA interface, and the modifications would introduce steric and electrostatic penalties to drug binding. Phosphorylation is the most prevalent resistance modification, as the negative charge introduced on ring I or III would cause electrostatic repulsion with the RNA.

The various resistance mechanisms underscore the need to understand the multiple forces that stabilize drug-RNA complexes. Effective antibiotics, such as the gentamicins, do not have OH groups at the 3′ and 4′ positions of ring I, eliminating these potential targets of modification enzymes. The removal of these hydrogen bond donors is compensated for by other physical forces, since gentamicin C1a binds tightly to rRNA. Likewise, acetylation of an amino group adds steric bulk and eliminates positive charge but maintains a hydrogen bond donor. Improved understanding of these contributions to drug binding will require further structure determinations coupled with more rigorous thermodynamic analyses.

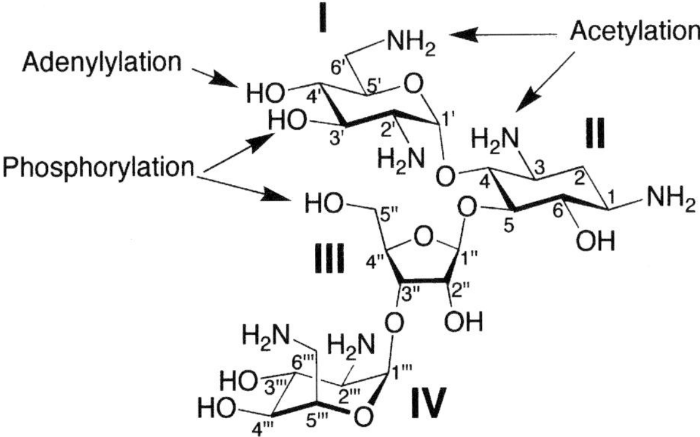

Figure 8. Sites of modification by resistance enzymes on neomycin.

Multiple structure determinations, coupled with many biochemical experiments and the vast literature on aminoglycosides, have revealed how aminoglycosides bind to their target RNAs. What remains less understood is how aminoglycoside binding to 30S ribosomal subunits causes misreading of the genetic code and why treated cells eventually die.

RIBOSOMAL DECODING AND INTERFERENCE BY AMINOGLYCOSIDES

Kinetic measurements have revealed the origins of high translational accuracy and the effects of aminoglycosides on the kinetic steps that establish this accuracy (Karimi and Ehrenberg, 1994; Pape et al., 1998). The error rate of translation is about 1 wrong amino acid for every 10^4 amino acids synthesized (Edelmann and Gallant, 1977). In the presence of aminoglycoside antibiotics, the error rate for prokaryotic ribosomes increases to about 1 error every 500 amino acids. Ribosomal discrimination of cognate from near-cognate tRNA is achieved by a series of steps in which near-cognate tRNAs are discarded from the ribosome more rapidly than cognate tRNAs, and cognate tRNAs induce specific and rapid conformational changes, which leads to efficient selection of the correct tRNAs. These mechanistic studies suggest ribosome recognition of the codon-anticodon interaction. In particular, there is coupling of codon-anticodon recognition to GTP hydrolysis by EF-Tu (Rodnina et al., 1995). Aminoglycosides perturb the selection process by decreasing the off rates of near-cognate tRNAs (Karimi and Ehrenberg, 1994) and increasing the forward rate constants for conformational signals. These results, coupled with our structure determinations for the decoding region, suggested a mechanism by which the ribosome recognizes the correct codon-anticodon pairing in the A site (Fourmy et al., 1996).

The ribosome recognizes 2′-OH groups within the A-site mRNA codon. Using a modification-interference approach, we delineated nucleotides in 16S rRNA that were required for A-site tRNA binding (Yoshizawa et al., 1999). The A-site tRNA was biotinylated, and competent 30S subunits for A-site tRNA binding were selected with streptavidin beads. These experiments indicated that the N-1 positions of A1492 and A1493, which are in the heart of the aminoglycoside binding pocket, are required for tRNA-mRNA binding in the A site. Further experiments indicated that changes of these universally conserved adenosines for G or C were lethal in *E. coli* and that these mutations also interfere with A-site tRNA interaction. However, the deleterious effects of the G1492 and G1493 mutations, which change the hydrogen bond acceptor at the N-1 of adenosine to a hydrogen bond donor, were eliminated by inclusion of 2′-F groups in the mRNA. The results strongly support interaction of A1492 and A1493, which are protected by A-site tRNA in the 30S subunit (Moazed and Noller, 1990), with 2′-OH groups in the mRNA backbone of the A-site codon (Patapov et al., 1995).

The proposed interactions between A1492 and A1493 and the mRNA backbone could provide the molecular mechanism for discrimination of cognate and near-cognate codon-anticodon pairs (Fig. 9) (Davies et al., 1964). Formation of a cognate codon-anticodon complex in the A site yields a short A-form helical duplex with at least two Watson-Crick base pairs. The conformation of the 2′-OH groups, as well as base chemical groups in the minor groove, may be recognized by the 30S subunit (A1492 and A1493). This recognition would then lead to conformational signaling in the 30S subunit, perhaps involving the 530 loop, the 900 region, and helix 44 of rRNA (Powers and Noller, 1994; Lodmell et al., 1995), which is transmitted to the 50S subunit. Near- or noncognate codon-anticodon pairs would present a different helical geometry, with base mispairs, that would distort the presentation of 2′-OH groups and base chemical groups in the minor groove. Disrupted interactions between the 30S subunit and the near-cognate codon-anticodon helix would lead to inefficient conformational signaling and increased off rates for the near-cognate tRNA.

Molecular contacts between A1492 and A1493 and mRNA during decoding may also explain the miscoding induced by aminoglycoside antibiotics. Aminoglycosides bind within the pocket formed by the A1408-A1493 pair and bulged nucleotide A1492. The binding of aminoglycosides displaces A1492 and A1493 towards the minor groove and stabilizes their conformation (Fourmy et al., 1998a). In the bound form, the N-1 positions of A1492 and A1493 are directed towards the minor groove, and the decoding region may contact the codon-anticodon helix in a minor groove-minor groove fashion, as observed in RNA crystal structures of the group I intron (Cate et al., 1996). The aminoglycoside-bound form of the decoding region may represent the conformation of rRNA that is formed upon contact with the codon-anticodon helix, and thus aminoglycosides could mimic the conformational signals normally induced by the correct tRNA-mRNA pairing. The initial conformational signal may be the local positions of A1492 and A1493 or global bending of helix 44. The role of conformational signaling in aminoglycoside-induced miscoding is supported by the presence of a gentamicin resistance mutation in the large-subunit protein L6 (Kühberger et al., 1979). This mutation is a C-terminal deletion of an RNA-binding domain of L6, which may help to structure the EF-Tu binding domain in the large-subunit (Davies et al., 1998).

How the effects of aminoglycosides on translation cause cell death remains a mystery. The general decrease in fidelity caused by aminoglycosides is probably not sufficient, as has been lucidly argued by Fast and Kurland (Fast et al., 1987; Kurland, 1992). There are ribosome mutations that decrease fidelity (*ram*), and bacterial strains containing these mutations are viable. In addition, other inhibitors of translation, such as chloramphenicol or erythromycin, are bacteriostatic, not bactericidal like the aminoglycosides (Gale et al., 1981). The most deleterious effect of aminoglycosides on protein synthesis may be the inhibition of translocation that comes with increased tRNA affinities in the A site. The early rounds of translation, between initiation and full elongation, may be particularly susceptible to the effects of aminoglycosides (Recht and Puglisi, unpublished). Much work remains to delineate the exact mode of action of aminoglycosides.

CONCLUSIONS

The work on aminoglycoside antibiotics has revealed the details of how aminoglycosides bind to the

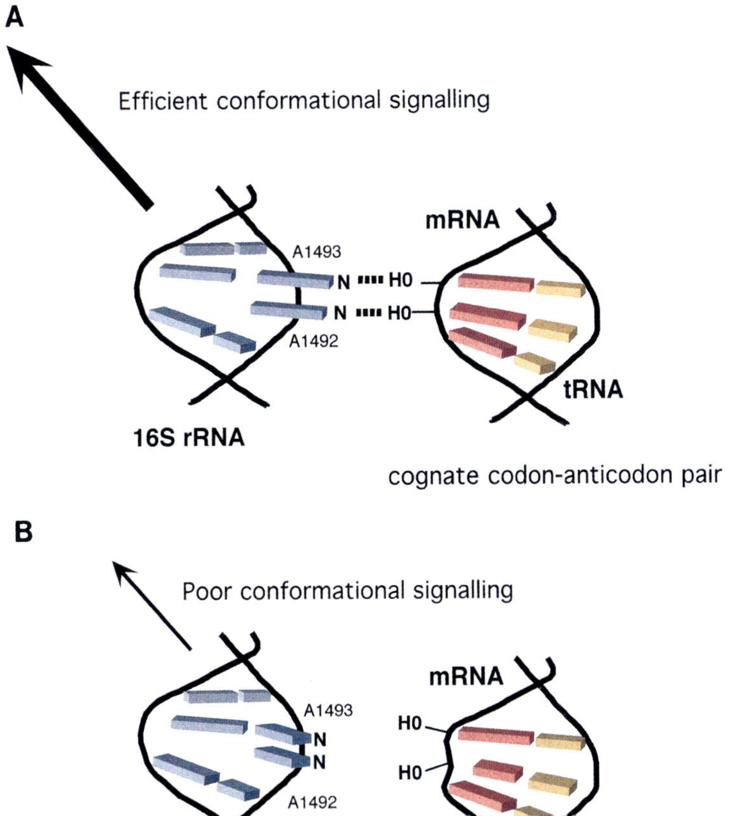

Figure 9. Molecular contacts between rRNA and mRNA in the A site may discriminate between cognate and near-cognate tRNAs. (A) A1492 and A1493 (N-1) positions interact with the phosphodiester backbone of the A-site mRNA, which forms an A-form helix with the cognate codon. These specific contacts drive conformational signaling between ribosomal subunits. (B) For near-cognate (or noncognate) tRNAs, the helix formed between the mRNA and tRNA is distorted by base mispairs. The distorted helix is poorly recognized by A1492 and A1493, and this results in poor conformational signaling during decoding.

ribosome, how resistance occurs, and how selectivity for prokaryotes is achieved. Only the combination of structure determination, biochemical, and biophysical approaches provides true insights into the workings of the ribosome. The use of a small oligonucleotide was essential for this work. The results with the oligonucleotide provide local features of drug-RNA interaction at high resolution. However, the global effects on ribosome structure and function can be observed only within intact ribosomes and subunits. The exact conformational changes involved in ribosome function and inhibition await the impending structures of the ribosome.

We thank H. Noller and the members of his laboratory for many important discussions.

This work was funded by NIH grant GM51266 and by grants from the Packard Foundation, the Sloan Foundation, the Lucille P. Markey Charitable Trust, and Rhone-Poulenc-Rorer.

REFERENCES

Alangaden, G. J., B. N. Kreiswirth, A. Aouad, M. Khetarpal, F. R. Igno, S. L. Moghazeh, E. K. Manavathu, and S. A. Lerner. 1998. Mechanism of resistance to amikacin and kanamycin in *Mycobacterium tuberculosis*. *Antimicrob. Agents Chemother.* **42:** 1295–1297.

Allain, F. H.-T., and G. Varani. 1995. Structure of the P1 helix from Group I self-splicing introns. *J. Mol. Biol.* **250:**333–353.

Bacino, C., T. R. Prezant, X. Bu, P. Fournier, and N. Fischel-Ghodsian. 1995. Susceptibility mutations in the mitochondrial small ribosomal RNA gene in aminoglycoside induced deafness. *Pharmacogenetics* **5:**165–172.

Beauclerk, A. A. D., and E. Cundliffe. 1987. Sites of action of two ribosomal RNA methylases responsible for resistance to aminoglycosides. *J. Mol. Biol.* **193:**661–671.

Bollen, A., T. Helser, T. Yamada, and J. Davies. 1969. Altered ribosomes in antibiotic-resistant mutants of E. coli. *Cold Spring Harbor Symp. Quant. Biol.* **34:**95–100.

Bonny, C., P. E. Montandon, S. Marc-Martin, and E. Stutz. 1991. Analysis of streptomycin-resistance of *Escherichia coli* mutants. *Biochim. Biophys. Acta* **1089:**213–219.

Botto, R. E., and B. Coxon. 1983. Nitrogen-15 nuclear magnetic resonance spectroscopy of neomycin B and related aminoglycosides. *J. Am. Chem. Soc.* **105:**1021–1028.

Cate, J. H., A. R. Gooding, E. Podell, K. Zhou, B. Golden, C. E. Kundrot, T. R. Cech, and J. A. Doudna. 1996. Crystal structure of a group I ribozyme domain: principles of RNA packing. *Science* **273:**1678–1685.

Cunningham, P. R., K. Nurse, A. Bakin, C. J. Weitzmann, M. Pflumm, and J. Ofengand. 1992. Interaction between the two conserved single-stranded regions at the decoding site of small subunit ribosomal RNA is essential for ribosome function. *Biochemistry* **31:**12012–12022.

Cunningham, P. R., K. Nurse, C. J. Weitzmann, and J. Ofengand. 1993. Functional effects of base changes which further define the decoding center of *Escherichia coli* 16S ribosomal RNA. Mutation of C1404, G1405, C1496, G1497, U1498. *Biochemistry* **32:**7172–7180.

Davies, C., D. E. Bussiere, B. L. Golden, S. J. Porter, V. Ramakrishnan, and S. W. White. 1998. Ribosomal proteins S5 and L6: high-resolution crystal structures and roles in protein synthesis and antibiotic resistance. *J. Mol. Biol.* **279:**873–888.

Davies, J., and B. D. Davis. 1968. Misreading of ribonucleic acid code words induced by aminoglycoside antibiotics: the effect of drug concentration. *J. Biol. Chem.* **243:**3312–3316.

Davies, J., and G. D. Wright. 1997. Bacterial resistance to aminoglycoside antibiotics. *Trends Microbiol.* **5:**234–240.

Davies, J., W. Gilbert, and L. Gorini. 1964. Streptomycin, suppression, and the code. *Proc. Natl. Acad. Sci. USA* **51:**883–890.

Davies, J., L. Gorini, and B. D. Davis. 1965. Misreading of RNA codewords induced by aminoglycoside antibiotics. *Mol. Pharmacol.* **1:**93–106.

DeStasio, E. A., and A. E. Dahlberg. 1990. Effects of mutagenesis of a conserved base-paired site near the decoding region of *Escherichia coli* 16S ribosomal RNA. *J. Mol. Biol.* **212:**127–133.

DeStasio, E. A., D. Moazed, H. F. Noller, and A. E. Dahlberg. 1989. Mutations in 16S ribosomal RNA disrupt antibiotic-RNA interactions. *EMBO J.* **8:**1213–1216.

Edelmann, P., and J. Gallant. 1977. Mistranslation in E. coli. *Cell* **10:**131–137.

Fast, R., T. H. Eberhard, T. Ruusala, and C. G. Kurland. 1987. Does streptomycin cause an error catastrophe? *Biochimie* **69:** 131–136.

Fourmy, D., M. I. Recht, S. C. Blanchard, and J. D. Puglisi. 1996. Structure of the A site of *E. coli* 16S ribosomal RNA complexed with an aminoglycoside antibiotic. *Science* **274:**1367–1371.

Fourmy, D., M. I. Recht, and J. D. Puglisi. 1998a. Binding of neomycin-class aminoglycoside antibiotics to the A site of 16S rRNA. *J. Mol. Biol.* **277:**347–362.

Fourmy, D., S. Yoshizawa, and J. D. Puglisi. 1998b. Paromomycin binding induces a local conformational change in the A site of 16S rRNA. *J. Mol. Biol.* **277:**333–345.

Gale, E. F., E. Cundliffe, P. E. Reynolds, M. H. Richmond, and M. J. Waring. 1981. *The Molecular Basis of Antibiotic Action.* John Wiley & Sons, London, United Kingdom.

Green, R., and H. F. Noller. 1997. Ribosomes and translation. *Annu. Rev. Biochem.* **66:**679–716.

Guan, M.-X., N. Fischel-Ghodsian, and G. Attardi. 1996. Biochemical evidence for nuclear gene involvement in phenotype of non-syndromic deafness associated with mitochondrial 12S rRNA mutation. *Hum. Mol. Genet.* **5:**963–971.

Gutell, R. R. 1994. Collection of small subunit (16S- and 16S-like) ribosomal RNA structures: 1994. *Nucleic Acids Res.* **22:**3502–3507.

Hendrix, M., E. S. Priestly, G. F. Joyce, and C.-H. Wong. 1997. Direct observation of aminoglycoside-RNA interactions by surface plasmon resonance. *J. Am. Chem. Soc.* **119:**3641–3648.

Karimi, R., and M. Ehrenberg. 1994. Dissociation rate of cognate peptidyl-tRNA from the A-site of hyper-accurate and error-prone ribosomes. *Eur. J. Biochem.* **226:**355–360.

Kühberger, R., W. Piepersberg, A. Petzet, P. Buckel, and A. Böck. 1979. Alteration of ribosomal protein L6 in gentamicin-resistant strains of Escherichia coli. Effects on fidelity of protein synthesis. *Biochemistry* **18:**187–193.

Kurland, C. G. 1992. Translational accuracy and the fitness of bacteria. *Annu. Rev. Genet.* **26:**29–50.

Lodmell, J. S., R. R. Gutell, and A. E. Dahlberg. 1995. Genetic and comparative analyses reveal an alternative secondary structure in the region of nt 912 of *Escherichia coli* 16S rRNA. *Proc. Natl. Acad. Sci. USA* **92:**10555–10559.

Lynch, S., and J. Puglisi. Unpublished data.

Miyaguchi, H., H. Narita, K. Sakamoto, and S. Yokoyama. 1996. An antibiotic-binding motif of an RNA fragment derived from the A-site-related region of *Esherichia coli* 16S rRNA. *Nucleic Acids Res.* **24:**3700–3706.

Moazed, D., and H. F. Noller. 1986. Transfer RNA shields specific nucleotides in 16S ribosomal RNA from attack by chemical probes. *Cell* **47:**985–994.

Moazed, D., and H. F. Noller. 1987. Interaction of antibiotics with functional sites in 16S ribosomal RNA. *Nature* **327:**389–394.

Moazed, D., and H. F. Noller. 1990. Binding of tRNA to the ribosomal A and P sites protects two distinct sets of nucleotides in 16 S rRNA. *J. Mol. Biol.* **211:**135–145.

Pape, T., W. Wintermeyer, and M. V. Rodnina. 1998. Complete kinetic mechanism of elongation factor Tu-dependent binding of aminoacyl-tRNA to the A site of the *E. coli* ribosome. *EMBO J.* **17:**7490–7497.

Patapov, A. P., F. J. Triana-Alonso, and K. Nierhaus. 1995. Ribosomal decoding processes at codons in the A or P sites depend differently on 2′-OH groups. *J. Biol. Chem.* **270:**17680–17684.

Powers, T., and H. F. Noller. 1994. The 530 loop of 16S rRNA: a signal to EF-Tu? *Trends in Genetics* **10:**27–31.

Powers, T., and H. F. Noller. 1995. Hydroxyl radical footprinting of ribosomal proteins on 16S rRNA. *RNA* **1:**194–209.

Prammananan, T., P. Sander, B. A. Brown, K. Frischkorn, G. O. Onyi, Y. Zhang, E. C. Bottger, and R. J. Wallace. 1998. A single 16S ribosomal RNA substitution is responsible for resistance to amikacin and other 2-deoxystreptamine aminoglycosides in *Mycobacterium abscessus* and *Mycobacterium chelonae. J. Infect. Dis.* **177:**1573–1581.

Prezant, T. R., J. V. Agapian, M. C. Bohlman, X. Bu, S. Oztas, W.-Q. Qui, K. S. Amos, G. A. Cortopassi, L. Jaber, J. I. Rotter, M. Shohat, and N. Fischel-Ghodsian. 1993. Mitochondrial ribosomal RNA mutation associated with both antibiotic-induced and non-syndromic deafness. *Nat. Genet.* **4:**289–294.

Price, K. E., J. C. Godfrey, and H. Kawaguchi. 1974. Aminoglycoside antibiotics containing 2-deoxystreptamine. *Adv. Appl. Microbiol.* **18:**191–307.

Puglisi, J. D., and J. R. Williamson. 1999. RNA interactions with small ligands and peptides, p. 403–425. *In* T. R. Cech and R. Gesteland (ed.), *The RNA World, 2nd ed.* Cold Spring Harbor Laboratory Press, Cold Spring Harbor, New York, N.Y.

Purohit, P., and S. Stern. 1994. Interactions of a small RNA with antibiotic and RNA ligands of the 30S subunit. *Nature* **370:**659–662.

Recht, M., and J. Puglisi. Unpublished data.

Recht, M. I., D. Fourmy, S. C. Blanchard, K. D. Dahlquist, and J. D. Puglisi. 1996. RNA sequence determinants for aminoglycoside binding to an A-site rRNA model oligonucleotide. *J. Mol. Biol.* **262:**421–436.

Recht, M. I., S. Douthwaite, K. D. Dahlquist, and J. D. Puglisi. 1999a. Effect of mutations in the A site of 16S rRNA on aminoglycoside antibiotic-ribosome interaction. *J. Mol. Biol.* **286:** 33–43.

Recht, M. I., S. Douthwaite, and J. D. Puglisi. 1999b. Basis for prokaryotic specificity of action of aminoglycoside antibiotics. *EMBO J.* **18:**3133–3138.

Rodnina, M. V., R. Fricke, L. Kuhn, and W. Wintermeyer. 1995. Codon-dependent conformational change of elongation factor Tu preceding GTP hydrolysis on the ribosome. *EMBO J.* **14:** 2613–2619.

Rose, S. J., P. T. Lowary, and O. C. Uhlenbeck. 1983. Binding of yeast $tRNA^{Phe}$ anticodon arm to *Escherichia coli* 30S ribosomes. *J. Mol. Biol.* **167:**103–117.

Shaw, K. J., P. N. Rather, R. S. Hare, and G. H. Miller. 1993. Molecular genetics of aminoglycoside resistance genes and familial relationships of the aminoglycoside-modifying enzymes. *Microbiol. Rev.* **57:**138–163.

Spangler, E. A., and E. H. Blackburn. 1985. The nucleotide sequence of the 17S ribosomal RNA gene of Tetrahymena thermophila and the identification of point mutations resulting in resistance to the antibiotics paromomycin and hygromycin. *J. Biol. Chem.* **260:**6334–6340.

Viani Puglisi, E., and J. D. Puglisi. 1998. Nuclear magnetic resonance spectroscopy of RNA, p. 117–146. *In* R. W. Simons and M. Grunberg-Manago (ed.), *RNA Structure and Function*. Cold Spring Harbor Laboratory Press, Cold Spring Harbor, New York, N.Y.

Wilhelm, J. M., J. J. Jessop, and S. E. Pettitt. 1978. Aminoglycoside antibiotics and eukaryotic protein synthesis: stimulation of errors in the translation of natural messengers in extracts of cultured human cells. *Biochemistry* **17:**1149–1153.

Wong, C.-H., M. Hendrix, E. S. Priestley, and W. A. Greenberg. 1998. Specificity of aminoglycoside antibiotics for the A-site of the decoding region of ribosomal RNA. *Chem. Biol.* **5:**397–406.

Yoshizawa, S., D. Fourmy, and J. D. Puglisi. 1998. Structural origins of gentamicin antibiotic action. *EMBO J.* **17:**6437–6448.

Yoshizawa, S., D. Fourmy, and J. D. Puglisi. 1999. Recognition of the codon-anticodon helix by ribosomal RNA. *Science* **285:** 1722–1725.

Zimmermann, R. A., C. L. Thomas, and J. Wower. 1990. Structure and function of rRNA in the decoding domain and in the peptidyl transferase center, p. 331–347. *In* W. E. Hill, A. Dahlberg, R. A. Garrett, P. B. Moore, D. Schlessinger, and J. R. Warner (ed.), *The Ribosome: Structure, Function, and Evolution*. American Society for Microbiology, Washington, D.C.

The Ribosome: Structure, Function, Antibiotics, and Cellular Interactions
Edited by R. A. Garrett, S. R. Douthwaite, A. Liljas, A. T. Matheson, P. B. Moore, and H. F. Noller

Chapter 35

Macrolide Resistance Conferred by Alterations in the Ribosome Target Site

STEPHEN DOUTHWAITE and BIRTE VESTER

The structural studies in this volume document the recent advances in mapping the vast array of unique molecular contours that are presented on the surface of the ribosome (see chapters 2, 8, and 13). Elucidation of the details of the ribosome's structural complexity at ever-improving resolution is enlightening and will continue to enlighten for many years to come. Not the least important application of this new knowledge will be to understand at atomic resolution the way in which many diverse groups of small ligands specifically interact with the ribosome. The majority of anti-infective drugs target specific sites on the bacterial ribosome (Gale et al., 1981; Vázquez, 1979), probably interacting with exposed rRNA structures, to exert their inhibitory effects (Cundliffe, 1990; Noller, 1991). To be an effective antibiotic, a ligand has to do more than merely find a specific niche on the ribosome: binding must also interfere with an essential ribosomal function. It is encouraging, therefore, that in regions which are known to function directly in protein synthesis, such as at the interface between the subunits, the rRNA components are richly represented.

Future ribosome structure work will undoubtedly indicate potential interaction sites for many new antibacterial drugs that await discovery or synthesis. However, the present state of affairs is less than rosy. Resistance to all the major groups of antibiotics has arisen hand in hand with the extensive use of these drugs in medicine and animal husbandry. Drug resistance is conferred by numerous mechanisms that collectively can be characterized as involving reduction in permeability of the bacterial envelope to the drugs, drug efflux, drug inactivation, or alterations in the drug target site. Target site alteration is a commonly encountered form of resistance to ribosome-specific drugs. This chapter concentrates on resistance to a subset of antibiotics, the 14-membered ring macrolides, and considers their fate as effective antibacterial agents now that resistance is widespread among bacterial pathogens.

MACROLIDE ANTIBIOTICS

Macrolides are produced in nature by many actinomycete strains (Gale et al., 1981; Vázquez, 1979) (see chapter 33). Clinically useful macrolides consist of a 14-, 15-, or 16-membered lactone ring, generally with two or more neutral and/or amino sugars substituted (Gale et al., 1981; Vázquez, 1979). One of the most commonly used drugs from this group is the 14-membered ring macrolide erythromycin A (Fig. 1).

The ribosome target site for macrolides lies within the 23S rRNA at the peptidyltransferase center of the 50S subunit (Cundliffe, 1990). Peptidyltransferase activity is associated with the central loop in 23S rRNA domain V, where macrolides make several contacts with the rRNA (Moazed and Noller, 1987). The binding site of erythromycin also includes hairpin 35 in domain II of the rRNA, where an additional drug interaction occurs (Hansen et al., 1999a; Xiong et al., 1999). Genetic and biochemical evidence suggests that in addition to hairpin 35 and the peptidyltransferase loop, other 23S rRNA sequences, including the loop of hairpin 72 (Douthwaite and Aagaard, 1993), participate in forming the drug binding pocket. Elucidation of this tertiary

Stephen Douthwaite ■ Department of Molecular Biology, Odense University, DK-5230 Odense M, Denmark. **Birte Vester** ■ RNA Regulation Centre, Department of Molecular Biology, University of Copenhagen, DK-1307 Copenhagen K, Denmark.

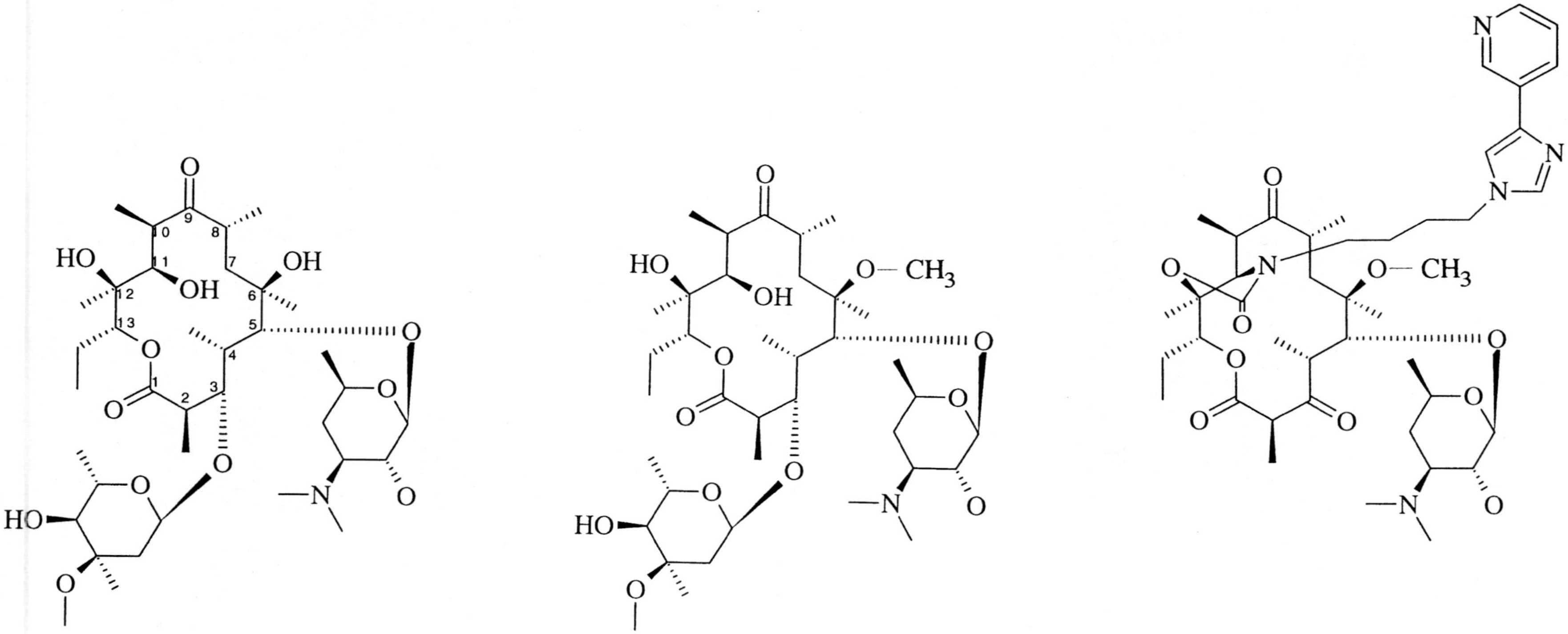

Erythromycin A

Clarithromycin

HMR 3647

Figure 1. Three types of macrolide antibiotics. The first-generation drug erythromycin A gave rise to a second generation of macrolides, including clarithromycin (6-O-methylerythromycin A). The third generation of drugs, the ketolides, include HMR 3647, which is presently undergoing clinical trials. The ketolides are characterized by a 3-keto group in place of the cladinose sugar residue. HMR 3647 also has an alkyl-aryl chain extending from a cyclic hydrazono-carbamate substitution at the 11/12 position of the lactone ring.

structure is now in the hands of the X-ray crystallographers.

After binding to the ribosome, erythromycin and the other 14-membered macrolides inhibit protein synthesis (Vannuffel and Cocito, 1996), probably by blocking elongation of the nascent peptide chain (Vázquez, 1979) and/or causing premature dissociation of the peptidyl-tRNA from the ribosome (Menninger, 1995). In addition to this, erythromycin and its derivatives have also been shown to reduce the cell's complement of 50S subunits by inhibiting the assembly of new subunits (Champney and Tober, 1999).

Shortly after the introduction of erythromycin in therapy in the 1950s, resistance to the drug became widespread in numerous pathogens (reviewed by Leclercq and Courvalin, 1991). More disturbing was the observation that the erythromycin-resistant strains were generally not only cross-resistant to all other macrolides but also to the chemically unrelated lincosamide and streptogramin B drugs. This phenotype was first observed in *Staphylococcus aureus* and came to be termed the MLS resistance phenotype (reviewed by Weisblum, 1995). In *S. aureus*, exposure to low concentrations of erythromycin induced expression of a methyltransferase enzyme (ErmC) which conferred drug resistance by methylation of the 23S rRNA (Lai and Weisblum, 1971). The site of Erm methylation was later localized to A2058 (*Escherichia coli* numbering) (Skinner et al., 1983), a pivotal nucleotide within the overlapping binding sites of the macrolide, lincosamide, and streptogramin B antibiotics at the peptidyltransferase loop of domain V of the rRNA (Moazed and Noller, 1987; Douthwaite, 1992; Rodriguez-Fonseca et al., 1995). Several dozen phylogenetically related *erm* methyltransferase genes have subsequently been identified (reviewed by Weisblum, 1995, and Jensen et al., 1999), all of which presumably methylate A2058.

Since the discovery of *erm* genes, another conceptually similar (but genetically different) means of resistance was identified. Mutation of A2058 in mitochondrial (Sor and Fukuhara, 1982) or bacterial rRNA (Sigmund et al., 1984) confers an MLS resistance phenotype similar to that of *erm*. Presumably, the mutation perturbs the site where the drugs interact and thereby interferes with their binding (Douthwaite and Aagaard, 1993) in a manner similar to methylation at this position (Goldman and Kadam, 1989). The degree of resistance that is conferred seems to depend on the proportion of ribosomes that contain a modification (or mutation) at position 2058. While a bacterium expressing an *erm* gene has the potential to modify all of its 23S rRNA molecules, the proportion of mutant ribosomes in a bacterium will depend on its relative expression of mutant and wild-type rRNA (*rrn*) operons.

rRNA MUTATIONS CAUSING RESISTANCE TO MACROLIDES

Single-nucleotide changes conferring macrolide resistance were originally identified after being created under laboratory conditions. These were first observed in the single *rrn* operon of the *Saccharomyces cerevisiae* mitochondrion (Sor and Fukuhara, 1982), and shortly afterwards similar phenotypes were obtained in *E. coli* by expression of mutant alleles from multiple-copy plasmids (e.g., Sigmund et al., 1984; Vester and Garrett, 1987). The mutations in these and later studies were located at or adjacent to A2058. One interesting exception to this is a mutation at 23S rRNA position 754 that confers low-level erythromycin resistance (Xiong et al., 1999). This mutation provides additional support for the idea that hairpin 35 and the peptidyltransferase loop (Fig. 2) are proximal in the rRNA tertiary structure.

Recent technical advances (not the least being reverse transcription-PCR methods) have made it possible to identify rRNA mutations that arise in pathogens during macrolide therapy. It is gratifying as well as disturbing to discover that the mutations selected for under clinical conditions are identical to those previously isolated in the laboratory. Pathogens that attain macrolide resistance by spontaneous rRNA mutations generally contain only one or two *rrn* operons. In organisms with multiple *rrn* operons (which includes most fast-growing bacteria, such as *E. coli*), mutation of a single operon would presumably be recessive, unless that operon is supplied in high copy numbers on a plasmid (Sigmund et al., 1984).

Helicobacter pylori is a human pathogen with two *rrn* operons (Wang and Taylor, 1998). *H. pylori* has recently been recognized as the probable etiologic agent in gastric cancer and ulcers and colonizes the stomach of every second adult (Covacci et al., 1999). The preferred treatment for aggressive infections is a drug combination including the second-generation macrolide clarithromycin (Fig. 1). Clarithromycin is an erythromycin derivative with improved acid stability and uptake properties (Goldman et al., 1994). In patients for whom *H. pylori* eradication failed, clarithromycin resistance was shown to have arisen during drug therapy (Debets-Ossenkopp et al., 1996) and could be traced to mutations at position A2058 or A2059 in the 23S rRNA (Versalovic et al., 1996). Mutations at position 2058 confer the MLS phenotype with high clarithromycin resistance (Wang and Taylor, 1998), with the G substitution at 2058 being

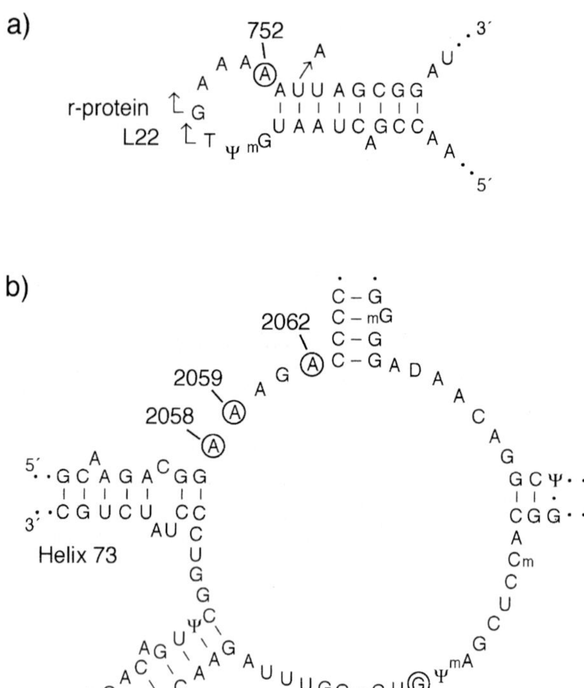

Figure 2. Hairpin 35 in domain II (a) and the peptidyltransferase center in domain V (b) of 23S rRNA. Nucleotides at which chemical footprint effects were observed upon erythromycin binding are circled (Moazed and Noller, 1987; Hansen et al., 1999a; Xiong et al., 1999). A mutation at position 754 conferring mild drug resistance is indicated (Xiong et al., 1999). A mutation in ribosomal protein L22, conferring erythromycin resistance, affects the accessibilities of G748 (the N7 position becomes more reactive to dimethyl sulfate) and T747 (becomes more reactive to carbodiimide) (Gregory and Dahlberg, 1999).

the most common in clinical isolates (Occhialini et al., 1997). Mutations at position 2059 confer a lower level of clarithromycin resistance and offer no resistance to streptogramins.

Erythromycin-resistant isolates of *Mycoplasma pneumoniae* display phenotypes similar to those of *H. pylori* for G2058 and G2059 mutants, and it was additionally shown that the G2059 mutant is the more resistant to 16-membered ring macrolides, such as tylosin and spiramycin (Lucier et al., 1995). This could be a peculiarity of the mutant *Mycoplasma* strains or it could reflect subtly different modes of interaction of 14- and 16-membered ring macrolides in the 2058 region of 23S rRNA (Moazed and Noller, 1987). The observation of similar phenotypes in resistant propionibacteria (Ross et al., 1997) suggests that the latter explanation is more likely, and this effect would therefore be expected to be valid for other species.

Pathogenic species of mycobacteria also develop resistance during clarithromycin treatment (Nash and Inderlied, 1995; Sander et al., 1997, and references therein). In *Mycobacterium intracellulare* and *Mycobacterium avium*, the rRNA mutations observed were limited to position 2058, with each of the three possible substitutions being isolated, whereas in the laboratory strain, *Mycobacterium smegmatis*, substitution at position 2059 was also found (Sander et al., 1997). The occurrence of MLS resistance mutations is not limited to erythromycin derivatives nor to human pathogens but can occur in any species with a low *rrn* copy number that is exposed to macrolides. *Brachyspira hyodysenteriae* possesses a single *rrn* operon and is the causative agent of swine dysentery. Seven isolates of *B. hyodysenteriae* showing resistance to tylosin (commonly used as a growth promoter in swine production) all exhibited single mutations at position 2058 (Karlsson et al., 1999). Surprisingly, no mutation was found at position 2059.

rRNA METHYLATION CONFERRING RESISTANCE TO MACROLIDES

When compared through the eye of the molecular biologist, methylation at A2058 in several ways appears to offer a more refined form of resistance than mutation. For some Erm methyltransferases, methylation can be induced only when required (Weisblum, 1995), the mechanism is not dependent on the number of *rrn* operons, and it does not fundamentally alter the identity of a nucleotide that has been totally conserved during bacterial evolution (Gutell et al., 1994). The mechanism does have one drawback in that it requires extra genetic information encoded by an *erm* gene and therefore cannot be acquired in the spontaneous manner of a mutation. Nature has attempted to overcome this by deploying *erm* genes on mobile genetic elements such as conjugative plasmids that are readily transferred between pathogens. The *erm* gene family is now the major cause of macrolide resistance in agricultural and clinical environments (Jensen et al., 1999).

Erm methyltransferases show sufficient sequence homology to suggest that they have a common ancestral origin and that they have retained the same function during evolution. This is reflected by the similar tertiary structures of ErmAM (Yu et al., 1997) and ErmC′ (Bussiere et al., 1998). The structure of the binding site for the methyl donor, S-adenosyl methionine, that is common to all Erm methyltransferases has recently been reported for ErmC′ (Schluckebier et al., 1999). All Erm methyltransferases target A2058, a nucleotide that occurs in a highly conserved and distinct structure in the rRNA (Gutell et al., 1994). The specificity of the Erm methyltransferases for A2058, together with their ability to methylate this nucleotide in phylogenetically diverse bacteria, suggests that they recognize a unique structural motif formed by features of this rRNA region.

THE Erm RECOGNITION MOTIF IN rRNA

We have used ErmE to investigate how this family of methyltransferases specifically recognize their A2058 target site on 23S rRNA. ErmE originates from the actinomycete *Saccharopolyspora erythraea*, the producer of erythromycin. ErmE dimethylates the N-6 position of A2058 to confer MLS resistance (Skinner et al., 1983). This enzyme methylates A2058 in diverse bacterial 23S rRNAs, but rRNA that is assembled in 50S subunits is no longer accessible as a substrate. RNA substrates that are specifically methylated by ErmE can be truncated to structures as small as 27 nucleotides, provided that they contain part of helix 73 and the single-stranded loop containing A2058 (Fig. 3a) (Vester et al., 1998). Slightly longer transcripts containing this region of 23S rRNA provide a good compromise between small size and the ability to be efficiently methylated by ErmE, and they formed the basis for studies to identify the Erm motif.

A negative in vitro selection procedure was developed to identify nucleotides that are essential for RNAs to function as effective substrates for the ErmE methyltransferase (Nielsen et al., 1999). This approach differed from previous RNA selection techniques in that pools of RNA molecules were screened for the loss rather than the acquisition of a desired characteristic. The starting material for the study was a 72-nucleotide RNA containing the 23S rRNA sequences previously identified as being required for efficient methylation by ErmE (Fig. 3b). Pools of degenerate RNAs were formed by doping nucleotide positions that extend over and beyond the putative Erm recognition motif within the 72-mer RNA. The RNAs were passed through a series of rounds of selection consisting of methylation by ErmE, amplification by reverse transcription-PCR, and T7 RNA polymerase transcription to form a new RNA pool. Dimethylation at adenosine blocks the path of reverse transcriptase. Thus, good substrates for ErmE cannot be amplified whereas poor (i.e., unmethylated) substrates pass through to subsequent rounds. Analyses of subclones after several rounds of methylation and selection revealed that 12 nucleotide positions are important for the methylation reaction. These nucleotides, corresponding to A2051 to A2060, C2611, and A2614 in 23S rRNA, presumably compose the RNA recognition motif for ErmE methyltransferase (Fig. 3b).

The use of small RNA aptamers can possibly give rise to artifactual structures not found in rRNA. Thus, the question remained whether the results from the negative selection study could be extrapolated to the intact 23S rRNA. The importance of the selected nucleotide positions was therefore tested by site-directed mutagenesis in a transcript containing the entire 23S rRNA domain V (Villsen et al., 1999). In vitro transcripts of domain V of *E. coli* 23S rRNA are methylated as efficiently as intact 23S rRNA by ErmE (Vester and Douthwaite, 1994), which is in accord with similar observations for ErmSF (Kovalic et al., 1994) and ErmC′ (Zhong et al., 1995). The mutagenesis study (summarized in Fig. 3c) largely supported the findings from the negative selection approach in that the irregular stem, with the bulge at position 2055, together with the identities of bases at positions 2056 to 2060, is of primary importance in the recognition motif. However, differences in the studies were evident at the periphery of the motif. The bulged A2051 was important for methylation of small RNA substrates (Vester et al., 1998) (Fig. 3a), but mutations there were of little consequence for methylation of domain V transcripts (Villsen et al., 1999) (Fig. 3c). The bulged A2051 is unimportant for methylation of similar, small RNA substrates by ErmC′ (Schluckebier et al., 1999). Although this could reflect differences in the enzymes studied, it remains clear that the data from the smallest RNA substrates should be interpreted with some caution.

ErmE shows remarkably high fidelity for its canonical site at A2058 under physiological conditions (Skinner et al., 1983; Vester and Douthwaite, 1994). However, upon relaxation of these conditions (by, for example, removal of magnesium ions) ErmE begins to methylate at other sites in rRNA with widely different degrees of reactivity (Hansen et al., 1999b). The canonical A2058 site is largely unaffected by magnesium ion depletion and remains the most reactive site in the rRNA. This suggests that methylation at the new sites results from changes in the RNA substrate rather than in the methyltransferase. Supporting this, chemical probing confirmed that the rRNA structure opens upon magnesium depletion, exposing potential new interaction sites to the enzyme (Hansen et al., 1999b). The new sites where ErmE begins to methylate show homology with the canonical A2058 site and have the consensus sequence aNNNcgGAHAg (where methylation occurs at the adenosine in boldface) (Fig. 3d). This consensus sequence contains the nucleotides that were also

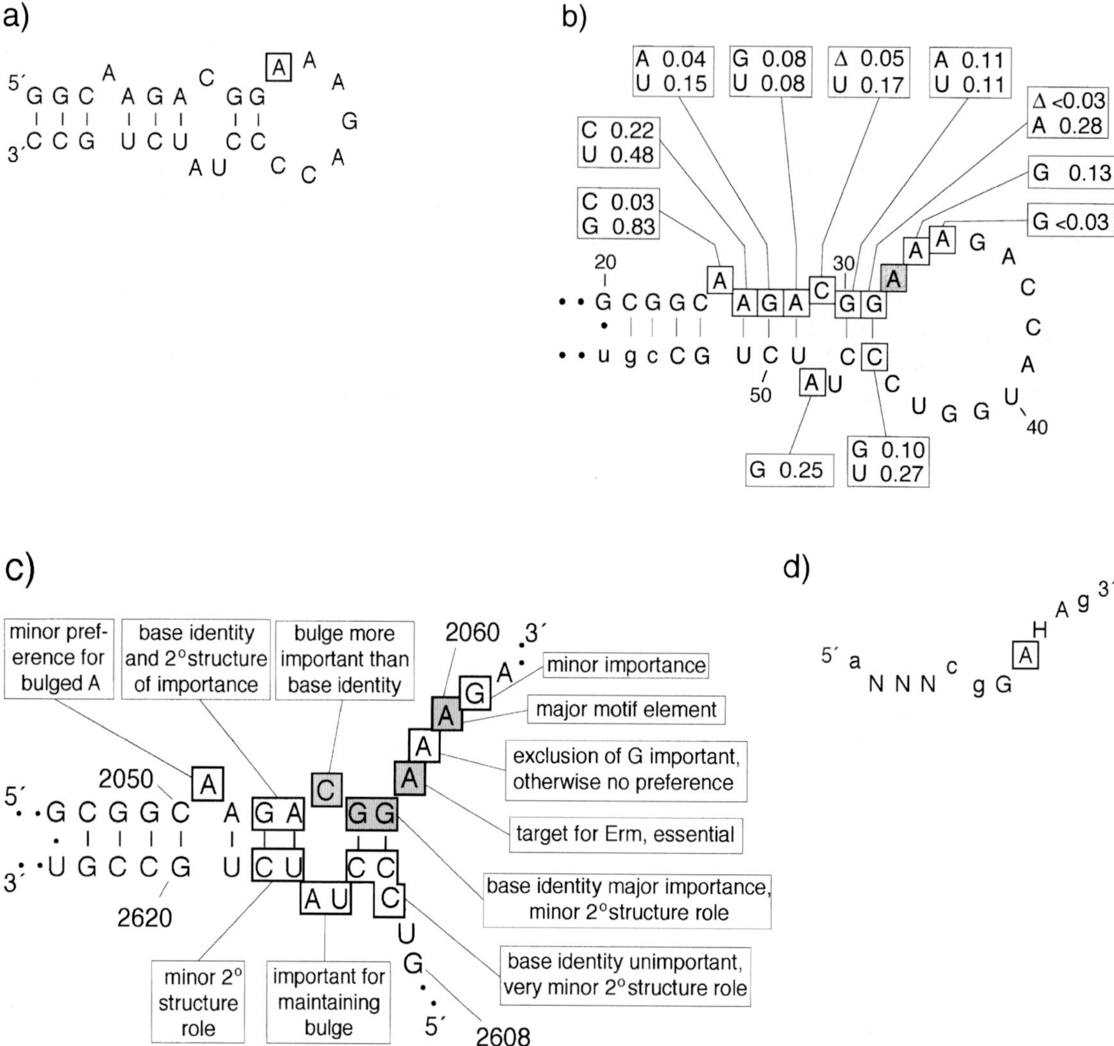

Figure 3. (a) Structure of a 27-mer RNA, the smallest substrate that could be methylated (albeit poorly) by ErmE (Vester et al., 1998). The structure corresponds to a portion of 23S rRNA helix 73 (Fig. 2b) and the single-stranded region adjacent to A2058 (boxed). (b) Part of the structure of the 72-mer RNA used in the negative in vitro selection study (Nielsen et al., 1999). The RNA was doped at 34 positions (uppercase nucleotides) and passed through several rounds of methylation and selection (see the text). Of 187 selected subclones, 43 had single-base substitutions that were limited to the 12 positions in boxes; Δ represents a single-nucleotide deletion. The relative effects of mutations at these positions on lowering the rate by ErmE methylation are indicated (wild type = 1). The lowercase nucleotides and the 5′ and 3′ sequences (not shown) were not doped and were used for amplifying and subcloning the selected RNA sequences. The shaded nucleotide is equivalent to A2058. (c) Summary of the results from a site-directed mutagenesis study on a domain V transcript of 23S rRNA and its methylation by ErmE (Villsen et al., 1999). The bases investigated by mutagenesis are boxed, and those most important for the Erm interaction are shaded. The nucleotides in the lower stem (from position 2611) are collectively essential for the methylation reaction (Vester et al., 1998), but the identities of individual bases here do not seem important provided their substitution does not unduly disrupt the stem structure (Villsen et al., 1999). (d) Consensus of the new sequences that are methylated by ErmE when its fidelity for A2058 is reduced by removal of magnesium ions (Hansen et al., 1999b). ErmE methylation occurs exclusively at adenosines (boxed), and these are preceded by a guanosine, equivalent to G2057; there is a high preference for the adenosine equivalent to A2060, and there are slight preferences for the nucleotides shown in lowercase. H, any nucleotide except G; N, any nucleotide.

shown by the selection and mutagenesis studies to represent the core of the motif that ErmE methyltransferase recognizes. The structure formed by these nucleotides is highly conserved throughout bacterial rRNAs, suggesting that this constitutes the motif that is recognized by all the Erm methyltransferases.

OTHER MACROLIDE RESISTANCE MECHANISMS

Bacteria are by no means totally dependent on changes in their rRNAs to acquire protection against macrolide antibiotics. Erythromycin can be removed from the bacterium by efflux mechanisms, which are prevalent in *Streptococcus* species, and alternatively, the drug can be inactivated by phosphorylation, glycosylation, or esterification (see Weisblum, 1995, and Jensen et al., 1999, for references). Recently it was shown that certain pentapeptide sequences confer mild resistance to erythromycin and its derivatives (Tripathi et al., 1998). The actual process of pentapeptide synthesis probably serves to eject the drug from the ribosome. The pentapeptides can be encoded by nucleotide sequences within 23S rRNA, which, after partial degradation, serve as mRNAs for other ribosomes.

Another mechanism that is more relevant to drug target site alterations involves mutations in ribosomal proteins L4 and L22 that confer erythromycin resistance (Wittmann et al., 1973). These ribosomal protein mutations have recently been shown to affect the conformations of nucleotides within domains II and V of 23S rRNA (Gregory and Dahlberg, 1999), although surprisingly, no changes were observed in the peptidyltransferase loop. Of particular interest are the effects of mutant L22 on the hairpin 35 loop, adjacent to a site where erythromycin was footprinted (Fig. 2). These observations would suggest that the domain II side of the drug binding site is disturbed in the L22 mutant ribosomes.

OVERCOMING MACROLIDE RESISTANCE

Naturally occurring macrolides have been derivatized in most conceivable ways to improve their acid stability, uptake, and resilience to modification and efflux and to improve ribosome binding, not least to MLS-resistant ribosomes. The third-generation macrolides, the ketolides (Fig. 1), are showing considerable promise in these respects. Removal of the 3-cladinose sugar enables the drugs to avoid tripping the expression of *erm* in inducible strains (Bonnefoy et al., 1997). In addition, the resulting 3-keto group in combination with an alkyl-aryl extension at the 11/12 of the lactone ring (Fig. 1) enables ketolides to bind to ribosomes with 10-fold-higher affinity than erythromycin (Hansen et al., 1999a). This is probably a direct consequence of better contact with the loop of hairpin 35 (Fig. 2). The improved interaction in domain II of 23S rRNA could explain the increased effectiveness of these drugs against MLS-resistant pathogens (e.g., Ednie et al., 1997).

Despite the improvements in the drugs, ribosomes methylated at A2058 in 23S rRNA are likely to remain less tractable to macrolide or ketolide inhibition than unmethylated ribosomes. One alternative, and highly ambitious, strategy is to attempt to prevent A2058 methylation, and thereby render the many redundant MLS antibiotics once again clinically effective. The details of the molecular contacts made between the Erm methyltransferases and the conserved rRNA motif await resolution by X-ray diffraction or NMR techniques. It can be hoped that determination of the three-dimensional structure of this motif will facilitate the design of compounds that will act as specific and effective competitive inhibitors of Erm methyltransferases.

REFERENCES

Bonnefoy, A., A. M. Girard, C. Agouridas, and J. F. Chantot. 1997. Ketolides lack inducibility properties of MLS(B) resistance phenotype. *J. Antimicrob. Chemother.* **40:**85–90.

Bussiere, D. E., S. W. Muchmore, C. G. Dealwis, G. Schluckebier, V. L. Nienaber, R. P. Edalji, K. A. Walter, U. S. Ladror, T. F. Holzman, and C. Abad-Zapatero. 1998. Crystal structure of ErmC′, an rRNA methyltransferase which mediates antibiotic resistance in bacteria. *Biochemistry* **37:**7103–7112.

Champney, W. S., and C. L. Tober. 1999. Superiority of 11,12 carbonate macrolide antibiotics as inhibitors of translation and 50S ribosomal subunit formation in *Staphylococcus aureus* cells. *Curr. Microbiol.* **38:**342–348.

Covacci, A., J. L. Telford, G. Del Giudice, J. Parsonnet, and R. Rappuoli. 1999. *Helicobacter pylori* virulence and genetic geography. *Science* **284:**1328–1333.

Cundliffe, E. 1990. Recognition sites for antibiotics in rRNA, p. 479–490. *In* W. Hill, A. Dahlberg, R. A. Garrett, P. B. Moore, D. Schlessinger, and J. Warner (ed.), *Ribosome: The Structure, Function, and Evolution.* American Society for Microbiology, Washington, D.C.

Debets-Ossenkopp, Y. J., M. Sparrius, J. G. Kusters, J. J. Kolkman, and C. M. J. E. Vandenbroucke-Grauls. 1996. Mechanism of clarithromycin resistance in clinical isolates of *Helicobacter pylori. FEMS Microbiol. Lett.* **142:**37–42.

Douthwaite, S. 1992. Interaction of the antibiotics clindamycin and lincomycin with *Escherichia coli* 23S ribosomal RNA. *Nucleic Acids Res.* **20:**4717–4720.

Douthwaite, S., and C. Aagaard. 1993. Erythromycin binding is reduced in ribosomes with conformational alterations in the 23S rRNA peptidyl transferase loop. *J. Mol. Biol.* **232:**725–731.

Ednie, L. M., S. K. Spangler, M. R. Jacobs, and P. C. Appelbaum. 1997. Susceptibilities of 228 penicillin- and erythromycin-susceptible and -resistant pneumococci to RU 64004 (HMR

3004), a new ketolide, compared with susceptibilities to 16 other agents. *Antimicrob. Agents Chemother.* **41**:1033–1036.

Gale, E. F., E. Cundliffe, P. E. Reynolds, M. H. Richmond, and M. J. Waring. 1981. *The Molecular Basis of Antibiotic Action.* John Wiley and Sons, London, United Kingdom.

Goldman, R. C., and S. K. Kadam. 1989. Binding of novel macrolide structures to macrolide-lincosamide-streptogramin B-resistant ribosomes inhibits protein synthesis and bacterial growth. *Antimicrob. Agents Chemother.* **33**:1058–1066.

Goldman, R. C., D. Zakula, R. Flamm, J. Beyer, and J. Capobianco. 1994. Tight binding of clarithromycin, its 14-(R)-hydroxy metabolite, and erythromycin to *Helicobacter pylori* ribosomes. *Antimicrob. Agents Chemother.* **38**:1496–1500.

Gregory, S. T., and A. E. Dahlberg. 1999. Erythromycin resistance mutations in ribosomal proteins L22 and L4 perturb the higher order structure of 23 S ribosomal RNA. *J. Mol. Biol.* **289**:827–834.

Gutell, R. R., N. Larsen, and C. R. Woese. 1994. Lessons from an evolving rRNA: 16S and 23S rRNA structures from a comparative perspective. *Microbiol. Rev.* **58**:10–26.

Hansen, L. H., P. Mauvais, and S. Douthwaite. 1999a. The macrolide-ketolide antibiotic binding site is formed by structures in domains II and V of 23S ribosomal RNA. *Mol. Microbiol.* **31**: 623–632.

Hansen, L. H., B. Vester, and S. Douthwaite. 1999b. Core sequence in the RNA motif recognized by the ErmE methyltransferase by reducing the enzyme stringency for its target. *RNA* **5**: 93–101.

Jensen, L. B., N. Frimodt-Moller, and F. M. Aarestrup. 1999. Presence of *erm* gene classes in gram-positive bacteria of animal and human origin in Denmark. *FEMS Microbiol. Lett.* **170**:151–158.

Karlsson, M., C. Fellstrom, M. U. Heldtander, K. E. Johansson, and A. Franklin. 1999. Genetic basis of macrolide and lincosamide resistance in *Brachyspira (Serpulina) hyodysenteriae*. *FEMS Microbiol. Lett.* **172**:255–260.

Kovalic, D., R. B. Giannattasio, J. Hyung-Jong, and B. Weisblum. 1994. 23S rRNA domain V, a fragment that can be specifically methylated in vitro by the ErmSF (TlrA) methyltransferase. *J. Bacteriol.* **176**:6992–6998.

Lai, C. J., and B. Weisblum. 1971. Altered methylation of ribosomal RNA in an erythromycin-resistant strain of Staphylococcus aureus. *Proc. Natl. Acad. Sci. USA* **68**:856–860.

Leclercq, R., and P. Courvalin. 1991. Bacterial resistance to macrolide, lincosamide and streptogramin antibiotics by target modification. *Antimicrob. Agents Chemother.* **35**:1267–1272.

Lucier, T., K. Heitzman, S.-K. Liu, and P.-C. Hu. 1995. Transition mutations in the 23S rRNA of erythromycin-resistant isolates of *Mycoplasma pneumoniae*. *Antimicrob. Agents Chemother.* **39**: 2770–2773.

Menninger, J. R. 1995. Mechanism of inhibition of protein synthesis by macrolide and lincosamide antibiotics. *J. Basic Clin. Physiol. Pharmacol.* **6**:229–250.

Moazed, D., and H. F. Noller. 1987. Chloramphenicol, erythromycin, carbomycin and vernamycin B protect overlapping sites in the peptidyl transferase region of 23S ribosomal RNA. *Biochimie* **69**:879–884.

Nash, K. A., and C. B. Inderlied. 1995. Genetic basis of macrolide resistance in *Mycobacterium avium* isolated from patients with disseminated disease. *Antimicrob. Agents Chemother.* **39**:2625–2630.

Nielsen, A. K., S. Douthwaite, and B. Vester. 1999. Negative *in vitro* selection identifies the rRNA recognition motif for ErmE methyltransferase. *RNA* **5**:1034–1041.

Noller, H. F. 1991. Ribosomal RNA and translation. *Annu. Rev. Biochem.* **60**:191–227.

Occhialini, A., M. Urdaci, F. Doucet-Populaire, C. M. Bébéar, H. Lamouliatte, and F. Mégraud. 1997. Macrolide resistance in *Helicobacter pylori:* rapid detection of point mutations and assays of macrolide binding to ribosomes. *Antimicrob. Agents Chemother.* **41**:2724–2728.

Rodriguez-Fonseca, C., R. Amils, and R. A. Garrett. 1995. Fine structure of the peptidyl transferase centre on 23S-like rRNAs deduced from chemical probing of antibiotic-ribosome complexes. *J. Mol. Biol.* **247**:224–235.

Ross, J. I., E. A. Eady, J. H. Cove, C. E. Jones, A. H. Ratyal, Y. W. Miller, S. Vyakrnam, and W. J. Cundliffe. 1997. Clinical resistance to erythromycin and clindamycin in cutaneous propionibacteria isolated from acne patients is associated with mutations in 23S rRNA. *Antimicrob. Agents Chemother.* **41**:1162–1165.

Sander, P., T. Prammananan, A. Meier, K. Frischkorn, and E. C. Böttger. 1997. The role of ribosomal RNAs in macrolide resistance. *Mol. Microbiol.* **26**:469–480.

Schluckebier, G., P. Zhong, K. D. Stewart, T. J. Kavanaugh, and C. Abad-Zapatero. 1999. The 2.2 Å structure of the rRNA methyltransferase ErmC′ and its complexes with cofactor and cofactor analogs: implications for the reaction mechanism. *J. Mol. Biol.* **289**:277–291.

Sigmund, C. D., M. Ettayebi, and E. A. Morgan. 1984. Antibiotic resistance mutations in 16S and 23S ribosomal RNA genes of *Escherichia coli. Nucleic Acids Res.* **12**:4653–4663.

Skinner, R. H., E. Cundliffe, and F. J. Schmidt. 1983. Site of action of a ribosomal RNA methylase responsible for resistance to erythromycin and other antibiotics. *J. Biol. Chem.* **258**:12702–12706.

Sor, F., and H. Fukuhara. 1982. Identification of two erthromycin resistant mutation sin the mitochondria: nature of the rib2 locus of the large ribosomal RNA gene. *Nucleic Acids Res.* **12**:8313–8318.

Tripathi, S., P. S. Kloss, and A. S. Mankin. 1998. Ketolide resistance conferred by short peptides. *J. Biol. Chem.* **273**:20073–20077.

Vannuffel, P., and C. Cocito. 1996. Mechanism of action of streptogramins and macrolides. *Drugs* **51**:20–30.

Vázquez, D. 1979. *Inhibitors of Protein Synthesis.* Springer-Verlag, New York, N.Y.

Versalovic, J., D. Shortridge, K. Kibler, M. V. Griffy, J. Beyer, R. K. Flamm, S. K. Tanaka, D. Y. Graham, and M. F. Go. 1996. Mutations in 23S rRNA are associated with clarithromycin resistance in *Helicobacter pylori. Antimicrob. Agents Chemother.* **40**:477–480.

Vester, B., and S. Douthwaite. 1994. Domain V of 23S rRNA contains all the structural elements necessary for recognition by the ErmE methyltransferase. *J. Bacteriol.* **176**:6999–7004.

Vester, B., and R. A. Garrett. 1987. A plasmid-coded and site-directed mutation in *Escherichia coli* 23S RNA that confers resistance to erythromycin: implications for the mechanism of action of erythromycin. *Biochimie* **69**:891–900.

Vester, B., A. K. Nielsen, L. H. Hansen, and S. Douthwaite. 1998. ErmE methyltransferase recognition elements in RNA substrates. *J. Mol. Biol.* **282**:255–264.

Villsen, I. D., B. Vester, and S. Douthwaite. 1999. ErmE methyltransferase recognizes features of the primary and secondary structure in a motif within domain V of 23S rRNA. *J. Mol. Biol.* **286**:365–374.

Wang, G., and D. E. Taylor. 1998. Site-specific mutations in the 23S rRNA gene of *Helicobacter pylori* confer two types of resistance to macrolide-lincosamide-streptogramin B antibiotics. *Antimicrob. Agents Chemother.* **42**:1952–1958.

Weisblum, B. 1995. Erythromycin resistance by ribosome modification. *Antimicrob. Agents Chemother.* **39**:577–585.

Wittmann, H. G., G. Stoffler, D. Apirion, L. Rosen, K. Tanaka, M. Tamaki, R. Takata, S. Dekio, and E. Otaka. 1973. Biochemical and genetic studies on two different types of erythromycin resistant mutants of *Escherichia coli* with altered ribosomal proteins. *Mol. Gen. Genet.* **127:**175–189.

Xiong, L., S. Shah, P. Mauvais, and A. S. Mankin. 1999. Ketolide resistance mutation in domain II of 23S rRNA reveals rRNA folding required for formation of macrolide-binding site. *Mol. Microbiol.* **31:**633–639.

Yu, L., A. M. Petros, A. Schnuchel, P. Zhong, J. M. Severin, K. Walter, T. F. Holzman, and S. W. Fesik. 1997. Solution structure of an rRNA methyltransferase (ErmAM) that confers macrolide-lincosamide-streptogramin antibiotic resistance. *Nat. Struct. Biol.* **4:**483–489.

Zhong, P., S. D. Pratt, R. P. Edalji, K. A. Walter, T. F. Holzman, A. G. Shivakumar, and L. Katz. 1995. Substrate requirements for ErmC′ methyltransferase activity. *J. Bacteriol.* **177:**4327–4332.

The Ribosome: Structure, Function, Antibiotics, and Cellular Interactions
Edited by R. A. Garrett, S. R. Douthwaite, A. Liljas, A. T. Matheson, P. B. Moore, and H. F. Noller

Chapter 36

Antibiotics and the Peptidyltransferase Center

BO T. PORSE, STANISLAV V. KIRILLOV, and ROGER A. GARRETT

The peptidyltransferase center is the ribosomal site where peptide bond formation occurs and the site where many antibiotics of diverse structures act. The 3′ termini of tRNAs anchored in the A or P site of the 30S subunit move through this center relative to the ribosome immediately before, during, and after peptide bond formation. At the current degree of resolution of the cryo-electron microscopy-based models, this center appears to constitute a cavity, which is consistent with the need to protect the ester linkage between the peptide and tRNA bond from water hydrolysis (Porse et al., 1995). The cavity is located on the interface side of the large subunit, and it leads into the peptide channel that passes through the body of the 50S subunit (see chapter 14).

By definition, the peptidyltransferase cavity must be able to accommodate any of the diverse hydrophobic and hydrophilic side chains of the amino acids, which suggests that it exhibits a high degree of structural complexity and/or conformational flexibility. It is this property which presumably accounts for its capacity to accommodate a wide range of antibiotics, many with apparently unrelated structures, that interfere in subtly different and barely understood ways with the relative movement of the 3′ termini of tRNAs and the ribosome.

An early model for the mechanisms of antibiotic inhibition at the peptidyltransferase center was based on molecular mimicry occurring between the antibiotics and the 3′ termini of either aminoacyl- or peptidyl-tRNAs. This hypothesis was given strong support by the crystal structure of puromycin, which not only mimics the 3′ end of aminoacyl-tRNA but can also act as an aminoacyl acceptor (see below). A major effort was made to demonstrate mimicry for a variety of the peptidyltransferase drugs (Harris and Symons, 1973a, 1973b), although many of the three-dimensional structures were only partially characterized and they were also known to be conformationally heterogeneous; chloramphenicol, for example, can assume at least four different conformers, only one of which is active (Harris and Symons, 1973a). Nevertheless, a credible case was made for some of the drugs resembling the 3′-terminal adenosine or the aminoacyl or peptidyl residues. Furthermore, for chloramphenicol it was suggested that, since it is produced by the same biosynthetic pathway as tyrosine and phenylalanine, it may have evolved to mimic the carboxyl terminus of the nascent peptide (Harris and Symons 1973a).

The molecular-mimicry hypothesis developed directly from the Watson two-site tRNA model, which proposed that there were two tRNA binding sites for the anticodon loop on the 30S subunit and two for the 3′-terminal end on the 50S subunit. Hence, any antibiotic that bound at the peptidyltransferase center was assumed to bind at one of these two physically distinct ribosomal sites and thereby sterically block tRNA binding (Harris and Symons, 1973a). Only in the past decade or so has it become clear that this model is not credible, firstly, because it does not deal with ribosomal movement, and secondly, because the 3′ termini of tRNAs are now known to pass through multiple states on the 50S subunit (see chapters 13 and 25). In fact, the Watson model has been a major conceptual liability for developing an understanding of the mechanisms of antibiotic action at the peptidyltransferase center.

Bo T. Porse, Stanislav V. Kirillov, and Roger A. Garrett ■ RNA Regulation Centre, Institute of Molecular Biology, Copenhagen University, DK-1307 Copenhagen K, Denmark.

STRUCTURE OF THE PEPTIDYLTRANSFERASE LOOP

Many lines of evidence, including mutational studies and rRNA footprinting and cross-linking studies with antibiotics and tRNAs, strongly suggest that the peptidyltransferase loop of 23S rRNA constitutes the main component of the peptidyltransferase cavity (Garrett and Rodriguez-Fonseca, 1995). Moreover, the few accessible bases in this region (Egebjerg et al., 1990) tend to show reactivity changes in the presence of bound antibiotics (Rodriguez-Fonseca et al., 1995) as well as constituting cross-linking sites for antibiotics or the 3′ termini of tRNAs (Porse et al., 1999b, 1999c; Kirillov et al., 1999; Wower et al., unpublished). Although these nucleotides are likely to be physically very close, little progress has been made in deducing additional higher-order structure in this region. The compensating base change approach that was so successful in predicting base pairing in the remainder of the rRNAs is generally precluded by the high level of nucleotide conservation. Nevertheless, limited insight derives from the large number of viable mutants carrying single-site mutations in this rRNA region (see below). Recent evidence from molecular modeling studies based on biochemical data and cryo-electron microscopy models tentatively places most of the nucleotides on the surface of the peptidyltransferase cavity in the 23S rRNA, where the 3′ termini of the tRNAs are juxtaposed during peptide bond formation (Müller et al., in press). The only ribosomal protein component which is consistently probed within the P′ site, by different approaches, is L27 (Wower et al., 1995), which also cross-links and assembles with domain V of 23S rRNA (Osswald et al., 1990; Østergaard et al., 1998).

STRUCTURAL AND FUNCTIONAL CONTRIBUTIONS OF THE OTHER rRNA DOMAINS

Although most of the structural effects that are caused by antibiotic binding to the peptidyltransferase center are confined to the peptidyltransferase loop of the 23S rRNA (Fig. 1), there are indications that domain V alone cannot assemble and generate the peptidyltransferase cavity. In particular, none of the peptidyltransferase drugs tested have yielded an rRNA footprint on a reconstituted domain V-protein complex (unpublished data), although a streptogramin B drug was recently shown to interact with a 23S rRNA fragment of the peptidyltransferase loop (Porse et al., 1999b). Moreover, other regions of the 50S subunit have also been implicated in the binding or mechanism of action of the peptidyltransferase drugs, which mainly belong to the MLS_B group (macrolides, lincosamides, and streptogramin B) and may act primarily by perturbing movement of the nascent peptide on peptide bond formation. Thus, mutations that cause resistance to erythromycin and lincomycin have been detected in ribosomal proteins L4 and L22 (see chapter 35), both of which assemble with domain I of 23S rRNA (Østergaard et al., 1998). L22 has also been shown to affect a local region of domain II (hairpin 35) where MLS_B drugs also footprint and an erythromycin-resistance mutation has been localized (see chapter 35).

Mutations in a large and highly conserved region of domain IV can produce strong inhibition of peptide bond formation in vitro and severely impair cellular growth (Leviev et al., 1995). The 3′ end of aminoacyl-($2N_3A76$)tRNA cross-linked to one of these sites (U1926) after EF-Tu-catalyzed binding, presumably in a state immediately prior to that where peptide bond formation occurs (Wower et al., unpublished).

At present, it seems likely that other parts of the 50S subunit contribute to the folding and stabilization of the peptidyltransferase cavity, and the region in domain II may be directly involved in facilitating movement of the nascent peptide from the peptidyltransferase cavity into the peptide channel.

LOCATIONS OF THE 3′ TERMINI OF tRNAs IN THE P/P′ SITE

When tRNAs are bound to free ribosomes under equilibrium conditions, they will bind in the most stable state, which will include interactions with the 3′-terminal adenosine. These interactions are weak, contributing only about 0.6 kcal/mol to the total free energy change of 13.4 kcal/mol arising from poly(U)-dependent binding of deacylated tRNA at the P site of *Escherichia coli* ribosomes (Parfenov and Saminsky, 1993). Moreover, kinetic studies of the reaction of the 3′ end of the peptidyl-tRNA with puromycin suggest that this 3′ terminus is partly flexible until three or more amino acids have been added (Panet et al., 1970). In order to investigate the ribosomal groups bordering the 3′-terminal adenosines of azido-derivatized deacylated tRNA and N-Ac-Phe-tRNA bound in the P/P′ site, they were cross-linked to the ribosome by irradiation with UV light at 365 nm, and five "zero-length" cross-links were detected at positions A2062-C2063, m^2G2069, U2506, U2584-U2585, and C2601-A2602 of 23S rRNA, probably at the base residues, as well as with protein

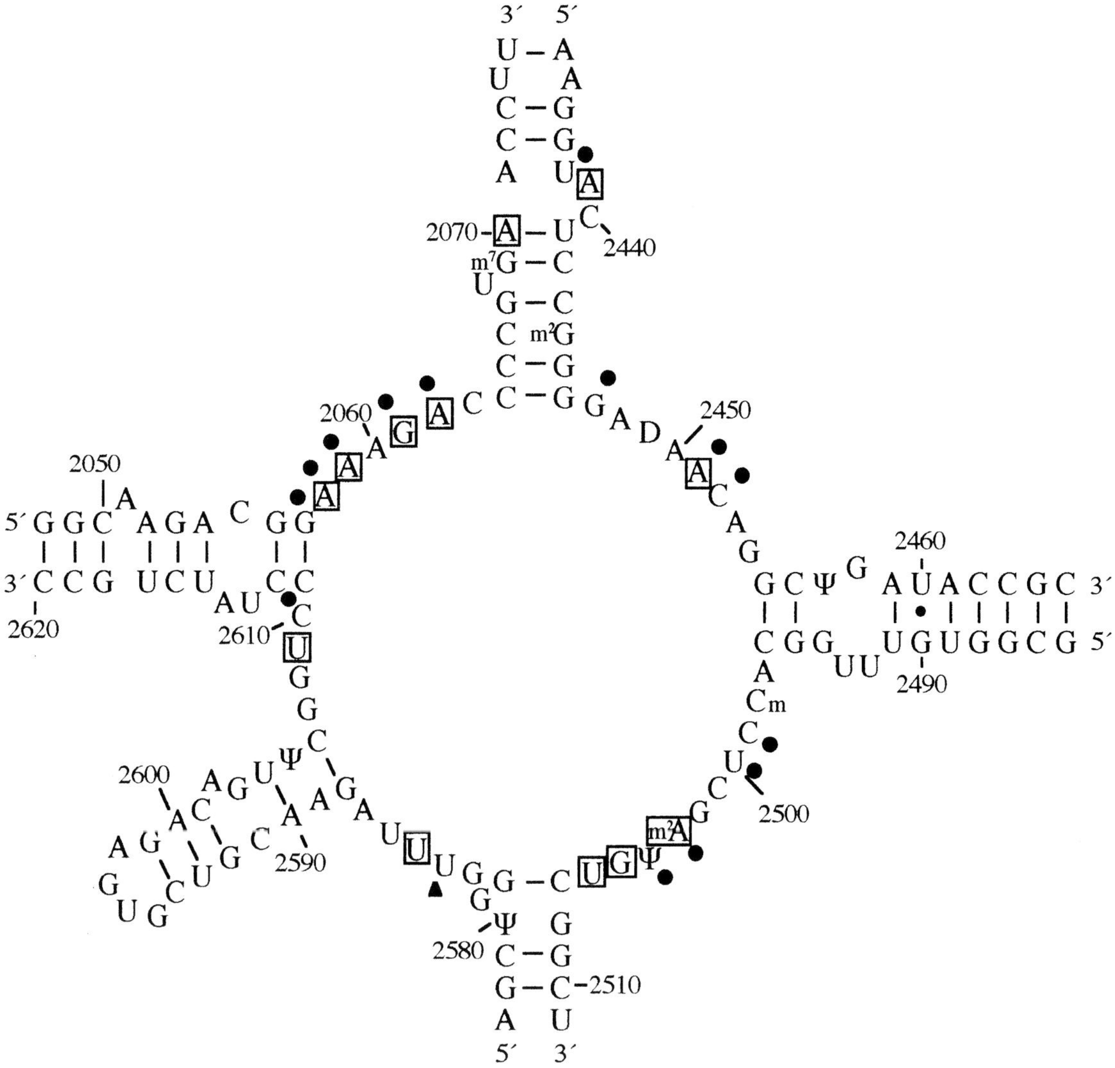

Figure 1. Secondary structure of the peptidyltransferase loop region (*E. coli* sequence) showing the sites of drug resistance (solid circles, base changes; solid triangle, lack of modification) and the nucleotides displaying altered chemical reactivities in the presence of drugs (boxed bases). The data are from Moazed and Noller, 1987; Garrett and Rodriguez-Fonseca, 1995; Rodriguez-Fonseca et al., 1995, and references therein; Lázaro et al., 1996; Tan et al., 1996; Porse and Garrett, 1999b; and Leviev et al., 1994.

L27 (Kirillov et al., 1999; Wower et al., unpublished). As indicated in the introduction to this chapter, some of the cross-linked nucleotides may lie in intermediate states that are occupied while entering or leaving the P′ site of the peptidyltransferase center on (otherwise) free ribosomes.

Removal of the 3′-terminal adenosine from P/P′-site-bound deacylated tRNA leads to enhanced chemical reactivities of U2506 and U2584-U2585 (Moazed and Noller, 1989), and modification-selection experiments with the 3′-terminal fragment of N-Ac-Tyr-tRNA reinforce these results (Bocchetta et al., 1998). Furthermore, the functional importance of U2506 and U2584-U2585 is strongly underlined by the demonstration that all possible mutations at these nucleotides, except U2584C, produce lethal growth phenotypes in *E. coli* and generally produce low levels of peptidyltransferase activity in isolated ribosomes (Porse et al., 1996; Green et al., 1997).

EFFECTS OF ANTIBIOTICS ON THE 3′ END OF P/P′-SITE-BOUND tRNA

Cross-linking of deacylated-($2N_3A76$)tRNA bound in the P/P′ site was tested in the presence of a selection of peptidyltransferase inhibitors, including sparsomycin, chloramphenicol, the streptogramins pristinamycin IA and IIA, gougerotin, lincomycin, and spiramycin, in order to test their capacities to perturb the relative positioning of the 3′ end of P/P′-site-bound tRNA and the *E. coli* ribosome (Kirillov et al., 1999).

Each drug produced major changes, mainly decreases, in cross-linking yields at one or more rRNA sites, but no new cross-links were detected. Nor, with the exception of chloramphenicol, did they alter cross-linking yields to ribosomal protein L27. Each of the drugs affected the U2506 cross-link, where most of them also generate a footprint (Garrett and Rodriguez-Fonseca, 1995); while gougerotin produced a strongly increased yield, all the other drugs produced decreased yields. Moreover, with the exception of pristinamycin IIA, they did not affect the cross-linking yields at U2584-U2585 and C2601-A2602, which also correlates with their not generating a footprint there (Rodriguez-Fonseca et al., 1995). All of the effects were closely similar for both deacylated tRNAs and N-Ac-Phe-tRNAs, indicating that the drugs primarily affect the tRNA. It was concluded that the antibiotics perturb the relative positioning of the 3′-terminal adenosine of the P/P′-site-bound tRNA and the peptidyltransferase loop, possibly by stabilizing a particular ribosomal conformer and preventing further movement of the tRNA (see below).

THE A′ SITE AND PUROMYCIN

When aminoacyl-($2N_3A76$)tRNA is bound to the ribosome and irradiated at 365 nm in the presence of P/P′-site-bound tRNA, the same set of UV-induced cross-links are produced, although they differ in their relative yields. The combined cross-links at U2584-U2585 and C2601-A2602 are much stronger for the bound aminoacyl-($2N_3A76$)tRNA, and the cross-link at U2506 is much stronger for P/P′-site-bound ($2N_3A76$)tRNAs (Kirillov et al., unpublished). Nevertheless, the observation that tRNAs in both positions produced the same sets of rRNA cross-links indicates that their 3′-terminal adenosines are physically close. For the aminoacyl-tRNA, the relative yields of cross-links also depended on whether it was bound enzymatically or nonenzymatically, which implies that it may be positioned more precisely via the ternary complex (Kirillov et al., unpublished).

Given that puromycin has acceptor activity, it has always been assumed to be a structural analogue of the 3′-terminal adenosine and aminoacyl residue of aminoacyl-tRNA and to occupy the site where peptide bond formation occurs. rRNA footprinting of puromycin bound to free ribosomes revealed three protection effects, one at G2553 and others at A2503 (in an archaeal ribosome) and G2505, in addition to an enhanced reactivity at A2439, none of which corresponds to the cross-linking sites for aminoacyl-($2N_3A76$)tRNA that are affected by puromycin (Rodriguez-Fonseca et al., unpublished). G2553 is, however, the site of cross-linking of 4-thio-dT-p-C-p-puromycin (Green et al., 1998) and may correspond to a position where the terminal -C-C sequence attaches, while the other site (A2503) lies in a region where several other antibiotics have been footprinted (see below).

INHIBITORY MECHANISMS

The earlier warning of the limitations of the two-site model for tRNA binding to the ribosome (Woese, 1979) has been reinforced by models showing multiple intermediate states on the 50S subunit (reviewed in Kirillov et al., 1997). Moreover, recent progress in analyzing crystal structures of a functional site on the ribosome (Wimberly et al., 1999) as well as the elongation factors EF-Tu and EF-G in their different functional states (Nyborg and Liljas, 1998) has led to a reassessment of the conformational flexibility of the ribosome that some had foreseen at an early stage (Woese, 1970; Spirin, 1968).

This development has been particularly important for interpreting the inhibitory mechanisms of the ribosomal antibiotics. Those that are partially understood probably all involve the inhibition of functional conformational changes. Thus, in the GTPase-associated site, thiostrepton and micrococcin probably block a conformational transition involving the rRNA and protein L11 (Cundliffe, 1986; Porse and Garrett, 1999a; Porse et al., 1999a). Moreover, in the decoding site, aminoglycosides act on and probably reduce the flexibility of 16S rRNA (see chapter 34), and for the elongation factors, kirromycin and fusidic acid reduce their conformational flexibility (Nyborg and Liljas, 1998). Thus, this mechanism also has to be considered seriously for the peptidyltransferase center. Moreover, there is a strong indication that these putative conformational effects are not confined to one functional site because single-site mutations in the peptidyltransferase center affect antibiotic sensitivity in the GTPase-associated center of the 50S subunit and vice-versa (Mankin et al., 1994).

SPARSOMYCIN—AN INHIBITOR OF A FUNCTIONAL TRANSITION

Recent studies with sparsomycin provide some insight into how the peptidyltransferase drugs may block conformational changes. This antibiotic is a universal and potent inhibitor of peptide bond formation which also selectively acts on several human tumors. It binds strongly to the ribosome only in the presence of an N-blocked donor tRNA substrate (Lázaro et al., 1991), which is then stabilized on the ribosome (Monro and Vázquez, 1967; Moazed and Noller, 1991). On irradiation of complexes of sparsomycin and bacterial, archaeal, or eukaryotic ribosomes complexed with P/P′-site-bound tRNA with low-energy UV light (at 365 nm), the drug cross-links exclusively to nucleotides C2601 and A2602 within the peptidyltransferase loop (Porse et al., 1999c). The same crosslink site is formed with several sparsomycin derivatives modified near the sulfoxy group, which implicates the modified uracil residue of sparsomycin (Fig. 2) in the rRNA cross-link, an inference that is consistent with the wavelength dependence of the cross-linking. The main cross-linking site, A2602 (Fig. 1), is universally conserved, and its chemical reactivity is enhanced in the presence of P/P′-site-bound N-Ac-Phe-tRNA but not deacylated tRNA (Moazed and Noller, 1989). The yield of sparsomycin cross-link was also sixfold higher with N-Ac-Phe-tRNA than with deacylated tRNA, which implies that in the presence of the former, A2602 becomes more accessible so that it can interact more readily with sparsomycin (Porse et al., 1999c) and that A2602 is involved in a functional ribosomal switch. This hypothesis is reinforced by the observation that ribosomal binding of the ternary complex of aminoacyl-tRNA·EF-Tu·GTP (or a nonhydrolyzable analogue) in the presence of deacylated tRNA renders A2602 chemically unreactive (Moazed and Noller, 1989) and that this complex is destabilized by sparsomycin (Hornig et al., 1987). We infer that the conformational switch occurring at A2602 is critically important both for the peptidyl transfer reaction and for sparsomycin inhibition.

A structural rationale for these results is that the modified uracil, which appears to be critical for the inhibitory action of sparsomycin (Fig. 2), interacts with A2602 via base stacking. Since the 3′ terminus of P/P′-site-bound N-Ac-Phe-tRNA is then stabilized on the ribosome, A76 of the tRNA may also stack at this site.

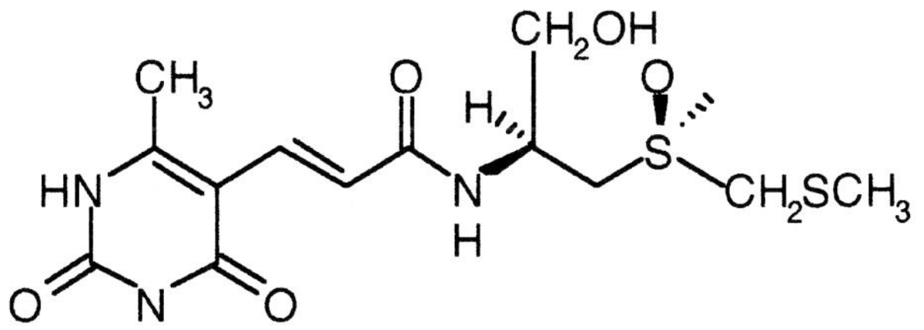

Figure 2. Structure of sparsomycin.

Importantly for the discussion that follows, several peptidyltransferase antibiotics, including puromycin, reduced the yield of the sparsomycin-ribosomal cross-link; the only exceptions were erythromycin and a streptogramin B, both of which belong to the MLS_B antibiotics, which do not primarily act on peptide bond formation (Porse et al., 1999c).

DRUG SITES

Most of the peptidyltransferase antibiotics tested, with sparsomycin as a notable exception, induce changes in the chemical reactivities of nucleotides within the peptidyltransferase loop region at one or more of the nucleotide positions 2058, 2059, 2061, 2062, 2070, 2439, 2451, 2503, 2505, 2506, 2585, and 2609 (Fig. 1), which include the cross-linking positions of the ($2N_3A76$)tRNAs at A2062, U2506, and U2585 (reviewed in Garrett and Rodriguez-Fonseca, 1995, and Kirillov et al., 1999). Moreover, for some of the drugs, single-site mutations within this rRNA region confer increased antibiotic tolerance on the mutant strains (Fig. 1) (Garrett and Rodriguez-Fonseca, 1995). However, despite circumstantial evidence suggesting that many peptidyltransferase drugs interact with rRNA in the ribosome, until recently no rRNA binding had been detected in the absence of ribosomal proteins. This contrasts with the observations for thiostrepton, which binds to free 23S rRNA of the GTPase-associated center (Cundliffe, 1986), and aminoglycosides that interact with a protein-free 16S rRNA fragment from the mRNA decoding center (see chapter 34); it may reflect the fact that proteins are necessary to stabilize a complex rRNA structure. However, recently, the direct interaction of an unmodified streptogramin B drug with small fragments of deproteinized 23S rRNA was detected when UV-induced modifications were produced at U2500 and C2501 in the presence of the drug (Porse et al., 1999b).

It is still difficult to imagine the ribosomal movements that occur in the peptidyltransferase center on peptide bond formation. However, there are likely to be at least two main movements that are coupled, one involving the positioning of the interacting amino and carbonyl groups in the catalytic center and another which facilitates the displacement of the nascent peptide away from the catalytic center. Blocking of either of these conformational transitions (or sets of transitions) could provide a basis for drug inhibitions that occur in the peptidyltransferase center (Porse et al., 1995). Thus, the MLS_B drugs could act primarily on peptide displacement and all of the other drugs could inhibit the ribosomal transition(s) that leads directly to peptide bond formation.

Such a model needs to address two general questions. (i) Why are peptidyltransferase inhibitors so diverse structurally in contrast, for example, to the aminoglycosides that act in the decoding site, and (ii) why do many of the MLS_B drugs inhibit peptide bond formation in in vitro assays? The answer to the first question may be that they are chemically attracted to the peptidyltransferase cavity, where they can bind to different ribosomal conformers at neighboring or overlapping sites on the cavity surface and thereby block critical ribosomal transitions. The answer to the second question may be more complex. Firstly, it is mainly the smaller macrolides that do not inhibit peptide bond formation in vitro, so there may be a steric effect in that only the larger drugs extend into, and affect, the catalytic site (Arévalo et al., 1989). Secondly, the two putative ribosomal transitions are likely to be coupled such that there is feedback from blocking of movement of the nascent peptide to inhibition of the next round of peptide bond formation. This discussion leads to the conclusion that the peptidyltransferase cavity is conformationally labile and contains all the critical nucleotides required for antibiotic binding.

Consideration of the synergistic effects observed with streptogramin A and B antibiotics, which potentially represent the two classes of drugs described above, may yield further insight into their inhibitory mechanisms.

SYNERGISTIC EFFECTS OF STREPTOGRAMINS A AND B

Streptogramin antibiotics contain two active components, A and B (Fig. 3), that inhibit peptide elongation synergistically; individually they are bacteriostatic, whereas together they can be bacteriocidal. The presence of the A component strongly enhances ribosomal binding of the B component (Contreras and Vázquez, 1977), although it is unclear whether the reverse stimulation occurs (Contreras and Vázquez, 1977; Porse et al., 1999b). Although we cannot exclude the possibility that the two drugs interact on the ribosome, the interdependence of their binding may reflect the fact that the A component stabilizes a ribosomal conformer to which the B component binds more strongly (Parfait and Cocito, 1980). This interpretation is reinforced by the partly overlapping rRNA footprints in the peptidyltransferase loop region, both of which are very complex (Porse and Garrett, 1999b). The binding sites of the streptogramin B drug were localized by the UV-induced modifications at positions m^2A2503-Ψ2504 and G2061-A2062 in the peptidyltransferase loop, where the yield of the former increases strongly in the presence of P/P′-site-bound tRNA (Porse et al., 1999b). Moreover, several peptidyltransferase antibiotics strongly affected the cross-linking yields, with chloramphenicol producing a shift in the second modification to A2062-C2063.

A PUTATIVE ANTIBIOTIC BINDING MOTIF

It is likely that most of the nucleotides involved in the antibiotic footprints are functionally related and structurally clustered. The upper (2058, 2059, and 2062) and lower (2503, 2505, 2506, and 2585) sites of the peptidyltransferase loop (Fig. 1) are linked in the streptogramin A and B footprints, and they have also been linked in other antibiotic footprints, in particular those of chloramphenicol and the macrolides (Garrett and Rodriguez-Fonseca, 1995). Building on this, a secondary structural model was proposed for part of the peptidyltransferase loop that is implicated in antibiotic binding, where the 2058-2062 region is juxtaposed with the 2503-2506 region, creating a motif that is stabilized by both coaxial stacking of adjoining helices and base pairing between A2060-G2505 and G2061-Ψ2504 (Fig. 4). This model is supported by several lines of circumstantial evidence from our knowledge of the sequence variability that can be tolerated by the cell in various mutants (Porse and Garrett, 1999b).

One aspect of the model, which may be of functional significance, is that U2506 can form alternative base pairs with A/G2058 or A2059. Evidence for this derives from footprinting results for the A2059G-mutated rRNA of *Halobacterium halobium*. A prediction from the model is that this mutation should favor an equilibrium shift to the G2058-U2506 pair, due to destabilization of the A2059-U2506 interaction, and such a shift is consistent with the chemical inertness of G2058 in this mutant (Porse and Garrett, 1999b). A consequence of the model is that the putative pairings A2060-G2505 and G2061-Ψ2504 will juxtapose A2062 and m^2A2503 (Fig. 4), which correlates with streptogramin B (pristinamycin IA), producing UV-induced modifications both at m^2A2503-Ψ2504 and at G2061-A2062 in *E. coli* (Porse et al., 1999b). Such a structure may also bring together many of the nucleotides that have been implicated in the binding of most of the drugs that act in the peptidyltransferase center, including the MLS_B group (Fig. 1). Thus, if there are two main modes of action for all of these drugs, as suggested above, then

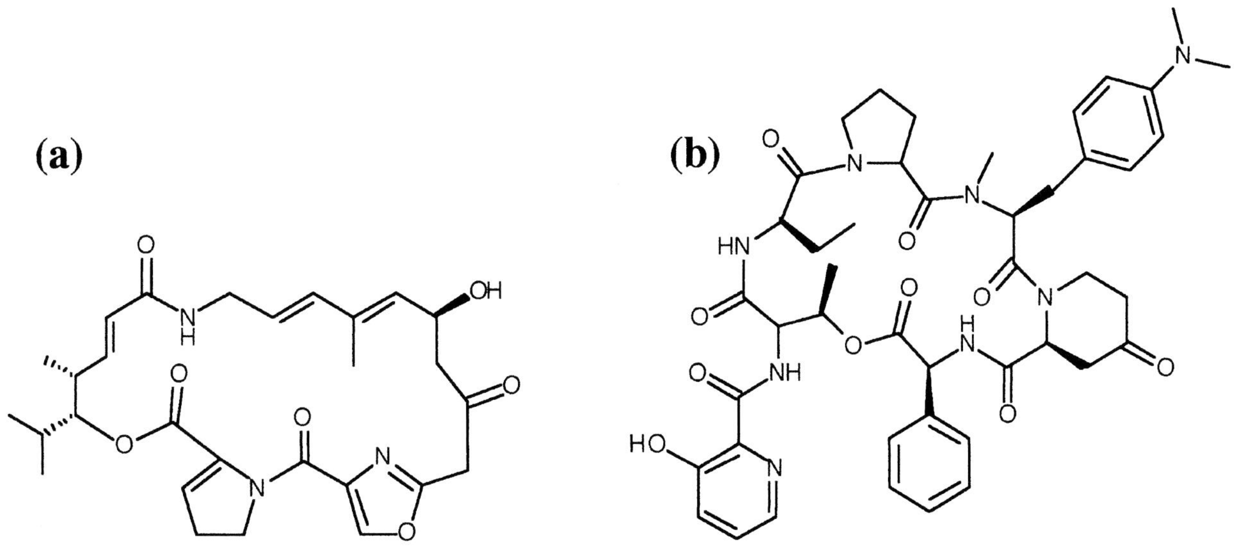

(a)

Pristinamycin IIA

(b)

Pristinamycin IA

Figure 3. Structures of a streptogramin A (pristinamycin IIA) (a) and a streptogramin B (pristinamycin IA) (b).

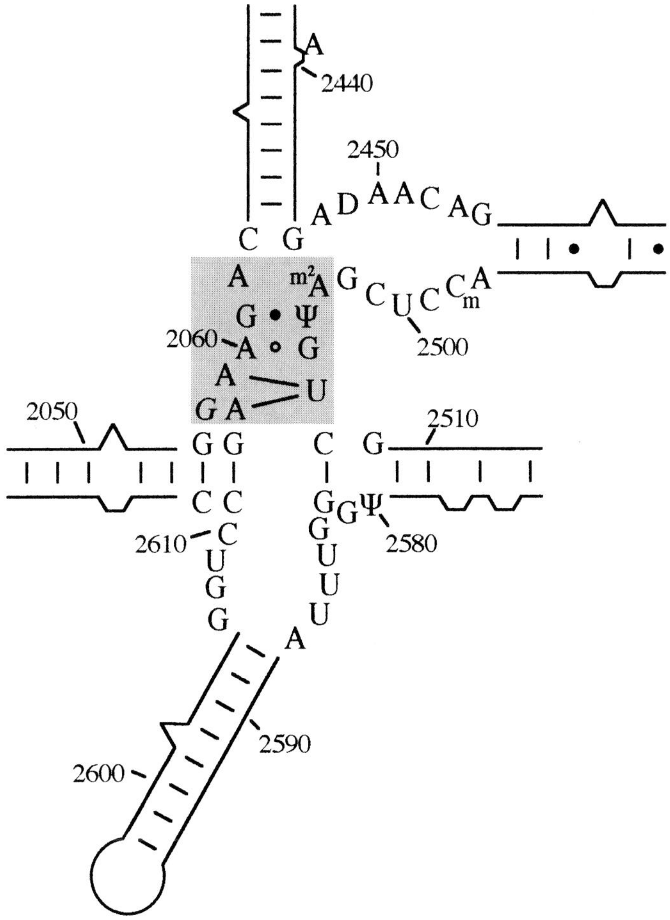

Figure 4. Structure of a putative antibiotic binding motif (*E. coli* sequence) within the peptidyltransferase center. The newly structured region is shaded, and the alternative base pairing at U2506 with A/G2058 or A2059 is considered in the text. The G occurring at position 2058 in archaeal and eukaryotic rRNAs is shown in italics.

their to sites of action may constitute alternating functional conformers of this putative rRNA motif.

CONCLUSION

Antibiotics have provided important probes for blocking and analyzing different functional steps on the ribosome, including initiation, decoding, peptide bond formation, translocation, and termination, as well as the stringent response. Interpretation of their mode of action has been limited by both the lack of structural information on the ribosome and the lack of insight into movements in and on the ribosome. Recently, this has changed for the aminoglycoside binding sites at the decoding site, for the binding sites of kirromycin and fusidic acid in EF-Tu and EF-G, respectively, and for the binding sites of thiostrepton and micrococcin in the GTPase-associated center (Wimberly et al., 1999; Porse et al., 1998; Porse and Garrett 1999a). There has also been a major reevaluation of how antibiotics affect the peptidyltransferase center. Moreover, with the rapid progress in analyzing crystal structures of the 50S subunit at high resolution, we can soon hope to gain more detailed insight into how the drugs affect the ribosomal structure.

REFERENCES

Arévalo, M. A., F. Tejedor, F. Polo, and J. P. G. Ballesta. 1989. Synthesis and biological activity of photoactive derivatives of erythromycin. *J. Med. Chem.* **32**:2200–2204.

Bocchetta, M., L. Xiong, and A. S. Mankin. 1998. 23S rRNA positions essential for tRNA binding in ribosomal functional sites. *Proc. Natl. Acad. Sci. USA* **95**:3525–3530.

Contreras, A., and D. Vázquez. 1977. Synergistic interaction of the streptogramins with the ribosome. *Eur. J. Biochem.* **74**:549–551.

Cundliffe, E. 1986. Involvement of specific portions of rRNA in defined ribosomal functions: a study utilizing antibiotics, p. 586–604. *In* B. Hardesty and G. Kramer (ed.), *Structure, Function, and Genetics of Ribosomes*. Springer-Verlag, New York, N.Y.

Egebjerg, J., N. Larsen, and R. A. Garrett. 1990. Structural map of 23S rRNA, p. 168–179. *In* W. Hill, A. Dahlberg, R. A. Garrett, P. Moore, D. Schlessinger, and J. Warner (ed), *The Ribosome: Structure, Function, and Evolution*. American Society for Microbiology, Washington, D.C.

Garrett, R. A., and C. Rodriguez-Fonseca. 1995. The peptidyltransferase center, p. 327–355. *In* R. A. Zimmermann and A. E. Dahlberg (ed.), *Ribosomal RNA—Structure, Evolution, Processing and Function in Protein Biosynthesis*. CRC Press, Boca Raton, Fla.

Green, R., R. R. Samaha, and H. F. Noller. 1997. Mutations at nucleotides G2251 and U2585 of 23S rRNA perturb the peptidyltransferase center of the ribosome. *J. Mol. Biol.* **266**:40–50.

Green, R., C. Switzer, and H. F. Noller. 1998. Ribosome-catalyzed peptide-bond formation with an A-site substrate covalently linked to 23S rRNA. *Science* **280**:286–289.

Harris, R. J., and R. H. Symons. 1973a. On the molecular mechanism of action of certain **substrates and inhibitors of ribosomal peptidyltransferase.** *Bioorg. Chem.* **2**:266–285.

Harris, R. J., and R. H. Symons. 1973b. A detailed model of the active center of *E. coli* peptidyltransferase. *Bioorg. Chem.* **2**: 286–292.

Hornig, H., P. Woolley, and R. Lührmann. 1987. Decoding at the ribosomal A-site: antibiotics, misreading and energy of aminoacyl-tRNA binding. *Biochimie* **69**:803–813.

Kirillov, S., B. T. Porse, B. Vester, P. Woolley, and R. A. Garrett. 1997. Movement of the 3′-end of tRNA through the peptidyltransferase center and its inhibition by antibiotics. *FEBS Lett.* **406**:223–233.

Kirillov, S. V., B. T. Porse, M. Awayez, and R. A. Garrett. 1999. Peptidyltransferase antibiotics perturb the relative positioning of the 3′-terminal adenosine of P/P′-site-bound tRNA and 23S rRNA in the ribosome. *RNA* **5**:1003–1013.

Kirillov, S. V., B. T. Porse, and R. A. Garrett. Unpublished data.

Lázaro E., L. A. G. M. Van den Broek, A. San Felix, H. C. J. Ottenheijm, and J. P. G. Ballesta. 1991. Biochemical and kinetic characteristics of the interaction of the antitumor antibiotic sparsomycin with prokaryotic and eukaryotic ribosomes. *Biochemistry* **30**:9642–9648.

Lázaro, E., C. Rodriguez-Fonseca, B. Porse, D. Ureña, R. A. Garrett, and J. P. G. Ballesta. 1996. A sparsomycin-resistant mutant of *Halobacterium salinarium* lacks a modification at nucleotide U2603 in the peptidyl transferase center of 23S rRNA. *J. Mol. Biol.* **261**:231–238.

Leviev, I. G., C. Rodriguez-Fonseca, H. Phan, R. A. Garrett, G. Heilek, H. F. Noller, and A. S. Mankin. 1994. A conserved secondary structural motif in 23S rRNA defines the site of interaction of amicetin, a universal inhibitor of peptide bond formation. *EMBO J.* **13**:1682–1686.

Leviev, I., S. Levieva, and R. A. Garrett. 1995. Role for the highly conserved region of domain IV of 23S-like rRNA in subunit-subunit interactions at the peptidyl transferase center. *Nucleic Acids Res.* **11**:1512–1517.

Mankin, A. S., I. Leviev, and R. A. Garrett. 1994. Cross-hypersensitivity effects of mutations in 23S rRNA yield insight into aminoacyl-tRNA binding. *J. Mol. Biol.* **244**:151–157.

Moazed, D., and H. F. Noller. 1987. Chloramphenicol, erythromycin, carbomycin and vernamycin B protect overlapping sites in the peptidyl transferase region of 23S rRNA. *Biochimie* **69**: 879–884.

Moazed, D., and H. F. Noller. 1989. Interaction of tRNA with 23 S rRNA in the ribosomal A, P, and E sites. *Cell* **57**:585–597.

Moazed, D., and H. F. Noller. 1991. Sites of interaction of the -CCA end of peptidyl-tRNA with 23S rRNA. *Proc. Natl. Acad. Sci. USA* **88**:3725–3728.

Monro, R. E., and D. Vázquez. 1967. Ribosome-catalyzed peptidyl transfer: effects of some inhibitors of protein synthesis. *J. Mol. Biol.* **28**:161–165.

Müller, F., I. Sommer, P. Baranov, H. Stark, M. van Heel, M. Rodnina, W. Wintermeyer, and R. Brimacombe. The 3D arrangement of the RNA in the *E. coli* 50S ribosomal subunit. II. Distribution of functionally important sites. *J. Mol. Biol.*, in press.

Nyborg, J., and A. Liljas. 1998. Protein biosynthesis: structural studies of the elongation cycle. *FEBS Lett.* **430**:95–99.

Osswald M., B. Greuer, and R. Brimacombe. 1990. Localization of a series of RNA-protein-crosslink sites in the 23S and 5S rRNA from *E. coli*, induced by treatment of 50S subunits with three different bi-functional reagents. *Nucleic Acids Res.* **18**: 6755–6760.

Østergaard, P., H. Phan, L. B. Johansen, J. Egebjerg, L. Østergaard, B. T. Porse, and R. A. Garrett. 1998. Assembly of proteins and 5S rRNA to transcripts of the major structural domains of 23S rRNA. *J. Mol. Biol.* **284**:227–240.

Panet, A., N. de Groot, and Y. Lapidot. 1970. Substrate specificity of *Escherichia coli* peptidyltransferase. *Eur. J. Biochem.* **15**:222–225.

Parfait, R., and C. Cocito. 1980. Lasting damage to bacterial ribosomes by reversibly bound virginiamycin M. *Proc. Natl. Acad. Sci. USA* **77**:5492–5496.

Parfenov, D. V., and E. M. Saminsky. 1993. Poly(U)-dependent interaction of yeast $tRNA^{Phe}$ and its fragments with *E. coli* ribosomes. II. Location of P-site-binding tRNA centers. *Mol. Biol.* **27**:507–510.

Porse, B. T., and R. A. Garrett. 1999a. Ribosomal mechanics, antibiotics and GTP hydrolysis. *Cell* **97**:423–426.

Porse, B. T., and R. A. Garrett. 1999b. Sites of interaction of streptogramin A and B antibiotics in the peptidyl transferase loop of 23S rRNA and the synergism of their inhibitory mechanisms. *J. Mol. Biol.* **286**:375–387.

Porse, B. T., C. Rodriguez-Fonseca, I. Leviev, and R. A. Garrett. 1995. Antibiotic inhibition of the movement of tRNA substrates through a peptidyltransferase cavity. *Biochem. Cell. Biol.* **73**: 877–885.

Porse, B. T., H. P. Thi-Ngoc, and R. A. Garrett. 1996. The donor substrate site within the peptidyl transferase loop of 23 S rRNA and its putative interactions with the CCA-end of N-blocked aminoacyl-$tRNA^{Phe}$. *J. Mol. Biol.* **264**:472–483.

Porse, B. T., I. Leviev, A. S. Mankin, and R. A. Garrett. 1998. The antibiotic thiostrepton inhibits a functional transition within protein L11 at the ribosomal GTPase center. *J. Mol. Biol.* **276**: 391–404.

Porse, B. T., E. Cundliffe, and R. A. Garrett. 1999a. The antibiotic micrococcin acts on protein L11 in the ribosomal GTPase center. *J. Mol. Biol.* **287**:33–45.

Porse, B. T., S. V. Kirillov, M. J. Awayez, and R. A. Garrett. 1999b. UV-induced modifications in the peptidyl transferase loop of 23S rRNA dependent on the binding of the streptogramin B antibiotic, pristinamycin IA. *RNA* **5**:585–595.

Porse, B. T., S. V. Kirillov, M. Awayez, H. C. J. Ottenheijm, and R. A. Garrett. 1999c. Direct crosslinking of the antitumor antibiotic sparsomycin, and its derivatives, to A2602 in the peptidyl transferase centre of 23S-like rRNA within ribosome-tRNA complexes. *Proc. Natl. Acad. Sci. USA* **96**:9003–9008.

Rodriguez-Fonseca, C., R. Amils, and R. A. Garrett. 1995. Fine structure of the peptidyl transferase centre on 23S-like rRNAs deduced from chemical probing of antibiotic-ribosome complexes. *J. Mol. Biol.* **247**:224–235.

Rodriguez-Fonseca, C., H. Phan, B. T. Porse, S. V. Kirillov, K. Long, R. Amils, and R. A. Garrett. Unpublished data.

Spirin, A. 1968. How does the ribosome work? A hypothesis based on the two subunit construction of the ribosome. *Curr. Mod. Biol.* **2**:115–127.

Tan, G. T., A. DeBlasio, and A. S. Mankin. 1996. Mutations in the peptidyl transferase center of 23S rRNA reveal the site of action of sparsomycin, a universal inhibitor of translation. *J. Mol. Biol.* **261**:222–230.

Wimberly, B. T., R. Guymon, J. P. McCutcheon, S. White, and V. Ramakrishnan. 1999. A detailed view of a ribosomal active site: the structure of the L11-rRNA complex. *Cell* **97**:491–502.

Woese, C. 1970. Molecular mechanics of translation: a reciprocating ratchet mechanism. *Nature* **226**:817–820.

Woese, C. R. 1979. Just so stories and Rube Goldberg machines: speculations on the origin of the protein synthetic machinery, p. 357–373. *In* G. Chambliss, G. R. Craven, J. Davies, K. Davis, L. Kahan, and M. Nomura (ed.), *Ribosomes: Structure, Function and Genetics.* University Park Press, Baltimore, Md.

Wower, J., I. K. Wower, S. V. Kirillov, K. V. Rosen, S. S. Hixon, and R. A. Zimmermann. 1995. Peptidyl transferase and beyond. *Biochem. Cell Biol.* **73**:1041–1047.

Wower, J., I. K. Wower, S. V. Kirillov, K. V. Rosen, S. S. Hixon, and R. A. Zimmermann. Transit of tRNA through the *E. coli* ribosome: crosslinking of the 3′-end of tRNA to specific nucleotides of the 23S rRNA at the A, P and E sites. Unpublished data.

The Ribosome: Structure, Function, Antibiotics, and Cellular Interactions
Edited by R. A. Garrett, S. R. Douthwaite, A. Liljas, A. T. Matheson, P. B. Moore, and H. F. Noller

Chapter 37

Structures and Properties of Ribotoxins

RICHARD KAO and JULIAN DAVIES

INTRODUCTION

Ribosome-inactivating proteins (RIPs) are protein toxins produced by organisms ranging from bacteria to plants which specifically damage eukaryotic and prokaryotic ribosomes, rendering them unable to bind elongation factors, and consequently interfering with the elongation steps in translation. These so-called RIPs are in fact specific degradative enzymes targeting the ribosome. The target site of most RIPs is a universally conserved region (the α-sarcin–ricin loop [SRL]) found in the 3′ region of large-subunit rRNAs (Endo et al., 1987). Two mechanisms have been identified for their toxic action: the α-sarcin-like fungal RIPs from *Aspergillus* catalytically inactivate the large ribosomal subunit (Schindler and Davies, 1977) by acting as specific ribonucleases to cleave a single phosphodiester bond between G4325 and A4326 of the 28S rRNA (Wool, 1997), whereas bacterial and plant RIPs are *N*-glycosidases, which hydrolyze an N-glycosidic linkage between the ribose sugar and the base of A4324 in 28S rRNA (Endo et al., 1987; Obrig et al., 1987). These covalent modifications completely abolish the ability of ribosomes to carry out protein synthesis. RIPs from higher plants can be classified into two categories according to their structures, namely, type I and type II RIPs. Type I RIPs are single-chain peptide toxins of about 30 kDa. Type II RIPs are proteins of approximately 60 kDa composed of two dissimilar subunits linked by disulfide bridges. The A chain of type II RIPs is homologous to that of type I RIPs and acts as the toxic component, whereas the B chain of type II RIPs is a lectin which facilitates the binding and uptake of the toxic A chain into the cell (Fong et al., 1991). Interestingly, the bacterial RIPs are produced by pathogenic gram-negative bacteria, such as *Shigella dysenteriae* and *Escherichia coli* (Shiga toxin-producing *E. coli* in particular), and have caused significant problems for human health in recent years. Although fungal RIPs are not glycosidases as are their plant and bacterial counterparts, they can be classified as type I RIPs owing to their potent toxicities and their actions as single-chain toxins. Table 1 gives a list of representative RIPs.

Fungal Ribotoxins

The name ribotoxin was first proposed to describe a group of fungal RIPs—α-sarcin, restrictocin, and mitogillin—which are among the most potent inhibitors of translation known (Lamy et al., 1992). These toxins, produced by the aspergilli, are basic proteins with molecular masses of ~17 kDa. The ribotoxins were thought to be produced exclusively in aspergilli (*Aspergillus giganteus*, *Aspargillus restrictus*, and *Aspergillus fumigatus*), but recent reports suggest that a family of α-sarcin-like ribotoxins are produced by various fungal species (Martinez-Ruiz et al., 1999; Lin et al., 1997). More recently, other ribotoxins from different strains of *Aspergillus clavatus* (clavin and c-sarcin) and *A. giganteus* (gigantin) have been characterized, including their primary sequences (Parente et al., 1996; Huang et al., 1997; Wirth et al., 1997).

Extensive studies of the regulation, production, and localization of restrictocin in *A. restrictus* have been carried out by Kenealy and coworkers (Brandhorst and Kenealy, 1992; Yang and Kenealy, 1992b). Their findings demonstrated that restrictocin is localized on the surface of conidiophores and the production of restrictocin is correlated with the maturation of the conidia. Brandhorst et al. (1996) demonstrated that the ribotoxin restrictocin deters insects from

Richard Kao and Julian Davies ■ Department of Microbiology and Immunology, The University of British Columbia, 6174 University Blvd., Vancouver, British Columbia V6T 1Z3, Canada.

Table 1. Representative RIPs from various sources

RIP	Mode of action	Producing organism
Plant		
Abrin	Glycosidase	*Abrus precatorius*
Gelonin	Glycosidase	*Gelonium multiflorum*
Ricin	Glycosidase	*Ricinus communis*
Bacterial		
Shiga toxin	Glycosidase	*S. dysenteriae* 1
Shiga-like toxin	Glycosidase	*E. coli* O157:H7
Fungal		
α-Sarcin	Endonuclease	*A. giganteus*
Clavin	Endonuclease	*A. clavatus*
Mitogillin	Endonuclease	*A. restrictus*

feeding on *A. restrictus*, suggesting that the production and localization of restrictocin may have a natural defensive role against insect feeding at times critical to spore formation in *A. restrictus* and that the toxin may have potential as an insect control agent.

Studies of the biosynthesis of α-sarcin in *A. giganteus* showed that α-sarcin is synthesized as an inactive precursor; after the synthesis and the maturation of the toxin it is segregated into membrane compartments (Endo et al., 1993a, 1993b). The translation systems of the producing hosts are sensitive to ribotoxins, with no evidence of the coproduction of any inhibitor (an immunity protein) of the nuclease activity of the ribotoxin (Miller and Bodley, 1988b), as is the case for colicin E3 (Boon, 1972; Bowman et al., 1971). It is thought that the producing host is protected from committing suicide during ribotoxin synthesis by production of the toxin as an inactive proform, which becomes activated on processing through the Golgi system (Endo et al., 1993a, 1993b; Lamy and Davies, 1991). Once the toxin is secreted, fungi are unable to take up the protein and are resistant to high concentrations of exogenous toxin.

Mode of Action of Fungal Ribotoxins

Fungal ribotoxins block protein synthesis by inhibiting both the elongation factor 1 (EF-1)- or EF-Tu-dependent binding of aminoacyl-tRNA and the GTP-dependent binding of EF-2 or EF-G to ribosomes. Eukaryotic ribosomes are extremely sensitive to the fungal ribotoxins; α-sarcin inhibited amino acid incorporation into wheat germ extracts at 0.1 nM (Lamy et al., 1992). The ability of ribotoxins to inhibit protein synthesis always correlates with the production of the α fragment, a 3′-terminal cleavage product from the rRNA of the large ribosomal subunit (Brigotti et al., 1989; Schindler and Davies, 1977). Prokaryotic large ribosomal subunits, such as *E. coli* 50S subunits, are also susceptible to catalytic inactivation by ribotoxins, but they require significantly higher concentrations of the inhibitor (Sanz and Amils, 1984). In fact, all large ribosomal subunits tested in vitro, including ribosomes from ribotoxin-producing species such as *A. giganteus* and *A. restrictus*, are inactivated by coincubation with the ribotoxins (Miller and Bodley, 1988b).

Studies by Hausner and Nierhaus (1988) and Miller and Bodley (1988a) and results from our laboratory (Kao and Davies, 1995) indicate that when isolated *E. coli* 23S rRNA is digested with low concentrations of ribotoxins, a cleavage product identical to the α fragment is obtained, indicating that the tertiary structure of the SRL in large-subunit rRNA is conserved even in the absence of ribosomal proteins and that ribotoxins are specifically targeted to the universally conserved SRL in the presence or absence of other components of the translational apparatus. The unusual catalytic specificity of ribotoxins in recognizing and cleaving only one of more than 7,000 phosphodiester bonds in the ribosome suggests that they evolved as targeted ribonucleases.

Molecular Studies

Recent studies have demonstrated that ribotoxin genes can be cloned and produced in sensitive hosts such as *E. coli* (Better et al., 1992; Endo et al., 1992), *Aspergillis nidulans* (Lamy and Davies, 1991), or *Pichia pastoris* (Martinez-Ruiz et al., 1998) by using regulated expression systems. The mitogillin gene in our studies was chemically synthesized by Better et al. (1992) and cloned into *E. coli* W3110 with the expression vector pING3522, an efficient inducible *E. coli* secretion vector, under the control of the *Salmonella typhimurium araB* promoter. Transcription from this promoter is very tightly regulated. Upon induction by arabinose, the produced mitogillin is released through the cytoplasmic membrane of the host into the culture medium, directed by the *Erwinia carotovora pelB* leader sequence; the secreted mitogillin is biologically active. Such controlled heterogeneous expression of large quantities of ribotoxins has proved to be extremely valuable in molecular studies.

The structural analysis of mitogillin by X-ray crystallography was first initiated by Martinez and Smith (1991), who obtained high-quality crystals; Yang and Moffat (1996) solved and refined the crystal structure of restrictocin to 1.7 Å, and coincidentally, nuclear magnetic resonance spectroscopy was used by Campos-Olivas et al. (1996) to obtain the

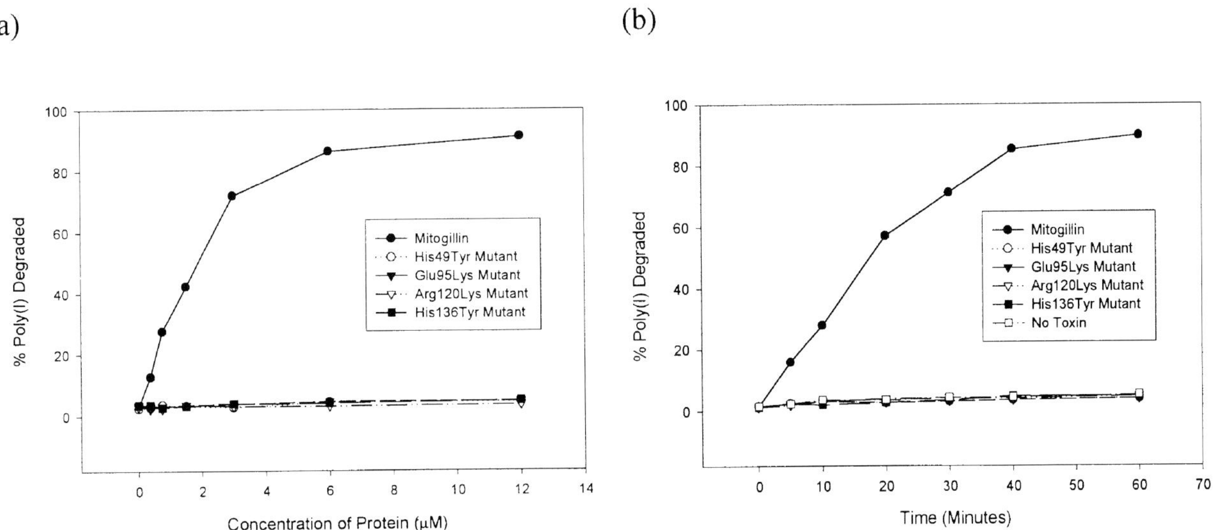

Figure 1. Nonspecific ribonucleolytic activity of mitogillin and its variants on poly(I). (a) Results were plotted as percent of poly(I) degradation versus protein concentration. (b) Results were plotted as percent of poly(I) degradation versus time when 3 μM mitogillin or a variant was used.

structure of α-sarcin, another member of the ribotoxin family.

The extraordinary specificity of ribotoxins is a property of both the ribosome and the ribotoxin. The specific site of cleavage by α-sarcin in 28S rRNA was determined to be on the 3′ side of G4325 (Chan et al., 1983). The structure of a 29-mer synthetic oligoribonucleotide mimicking the SRL of the 28S rRNA has been well characterized (Szewczak et al., 1993; Correll et al., 1998), and a 35-mer oligoribonucleotide, which mimics the SRL of 28S rRNA, has been used to study the substrate specificity of toxins, such as α-sarcin and ricin (Endo et al., 1988). Extensive mutational studies of the 35-mer synthetic oligoribonucleotide have shed light on the rRNA identity elements for the toxins, and the results suggest that although ribotoxins and ricin act on adjacent nucleotides, their rRNA identity elements are different (Wool, 1997).

An understanding of the biology of mitogillin and related fungal ribotoxins at the molecular level has become increasingly important because of their potential application as a component of immunotoxins. This application may be extended by the use of fungal ribotoxin-targeting domains in the development of integral single-chain toxins that possess enzyme activity and can target specific cellular functions.

THE ACTIVE-SITE RESIDUES OF FUNGAL RIBOTOXINS

Mitogillin and related ribotoxins are known to have amino acid sequence similarity to T1-like ribonucleases (Wool, 1984; Lamy et al., 1992), with a unique specificity of interaction with the ribosome causing a single ribonucleolytic cleavage in the large-subunit rRNA. Studies have indicated that the similarities and differences detected in amino acid sequence comparison of ribotoxins and a large family of other guanyl ribonucleases may represent domains or residues essential to ribonucleolytic activity and specificity (Kao and Davies, 1995). The prediction that the ribotoxins are T1-like ribonucleases with additional protein domains extended from the catalytic core of the RNase is further supported by analyses of the crystal structure of restrictocin determined by X-ray analysis (Yang and Moffat, 1996) and the three-dimensional structure of α-sarcin in solution determined by nuclear magnetic resonance (Campos-Olivas et al., 1996). These structural analyses, together with other studies (Mancheno et al., 1995; Kao and Davies, 1995), support the proposal that the ribotoxins mimic the action of T1-like ribonucleases and that residues His49, Glu95, Arg120, and His136 of mitogillin (corresponding to His40, Glu58, Arg77, and His92 in ribonuclease T1) are likely to be involved in the ribonucleolytic activity of the toxin. Mutational studies (Yang and Kenealy, 1992a; Kao and Davies, 1995; Lacadena et al., 1995) have shown that the His136 residue of restrictocin and mitogillin (His137 of α-sarcin) is crucial for the catalytic activities of ribotoxins, but the involvement of other proposed active-site residues was less clear until an extensive mutational study of mitogillin was carried out recently to systematically probe the active site of the fungal ribotoxin mitogillin (Kao et al., 1998).

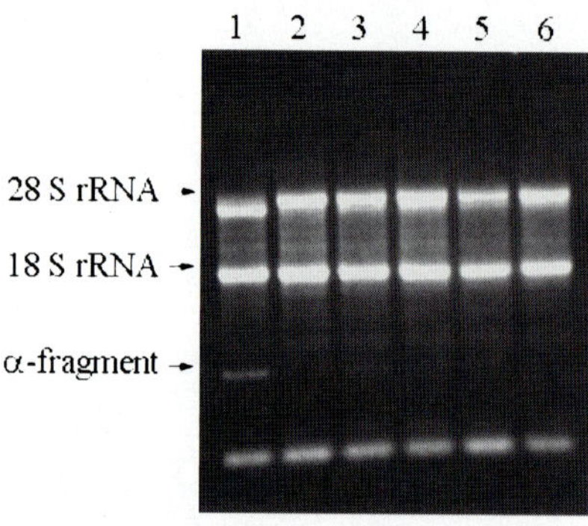

Figure 2. Specific ribonucleolytic activity (in vitro α fragment release) of mitogillin and its variants. Lane 1, mitogillin; lane 2, no toxin; lane 3, His49Tyr mutant; lane 4, Glu95Lys mutant; lane 5, Arg120Lys mutant; lane 6, His136Tyr mutant. The positions of 28S rRNA, 18S rRNA, and the α fragment are indicated.

The mitogillin gene was cloned from *A. fumigatus* cDNA after PCR amplification. Four variant forms of mitogillin (His49Tyr, Glu95Lys, Arg120Lys, and His136Tyr) were isolated by mutagenesis with hydroxylamine followed by an in vivo yeast cytotoxic screen assay. The mitogillin genes, presumably encoding nonfunctional forms as a result of the mutagenesis, were cloned by PCR and fused to the *E. carotovora pelB* leader sequence under the control of the tightly regulated *S. typhimurium araB* promoter and introduced into *E. coli* expression vector pING3522 (Better et al., 1992). When the purified mutant mitogillins were studied in in vitro ribonucleolytic assay systems, it was evident that the His49Tyr, Glu95Lys, Arg120Lys, and His136Tyr mutations severely impair the RNA cleavage capability of mitogillin. These four variants were unable to hydrolyze poly(I) homopolymers efficiently (Fig. 1), and their abilities to cleave rRNA were also adversely affected (Fig. 2). The detection of residual activity in the His49Tyr, Glu95Lys, and Arg120Lys variants and of little activity in the His136Tyr variant suggests that neighboring amino acid residues may partially replace the roles of certain mutated active-site residues. Since these mutants were obtained by an in vivo yeast cytotoxicity screening method, it is likely that the minute residual activities of the variants were insufficient to affect the translational machinery of the host.

Mutational studies of RNase T1 during the last decade were recently reviewed by Steyaert (1997). Most of the available data support the conclusion that Glu58 (corresponding to Glu95 in mitogillin) and His92 (corresponding to His136 in mitogillin) of RNase T1 act as the catalytic base and acid in the transphosphorylation reaction and that His40 (corresponding to His49 in mitogillin) plays the role of the base when Glu58 (Glu95 in mitogillin) is replaced by other residues. It has been suggested that Arg77 (Arg120 in mitogillin) may stabilize the transition state of the catalyzed reaction by neutralizing the neg-

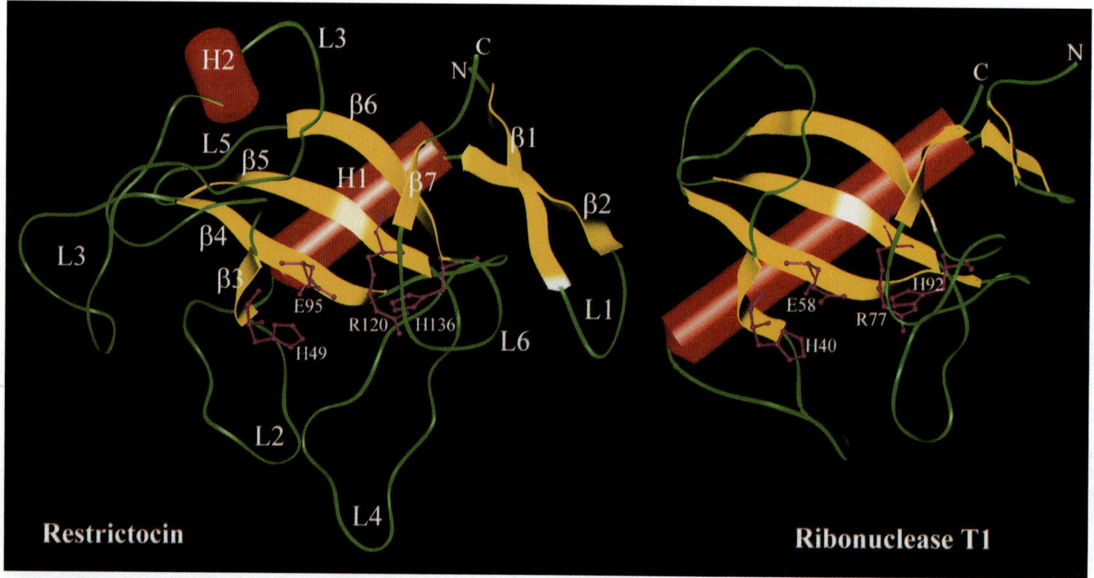

Figure 3. Structural comparison of restrictocin and ribonuclease T1. The secondary structures of the proteins are labeled: $\beta 1$ to $\beta 7$, β-sheets 1 to 7; L1 to L6, loops 1 to 6; H1 and H2, α-helixes 1 and 2. The positions of the catalytically important residues H_{49}, E_{95}, R_{120}, and H_{136} of restrictocin and the corresponding residues H_{40}, E_{58}, R_{77}, and H_{92} of ribonuclease T1 are also indicated. Note the absence of the B1-L1-B2, L3, and L4 domains of mitogillin in ribonuclease T1. The coordinates of restrictocin and ribonuclease T1 are taken from the Protein Data Bank files 1AQZ (Yang and Moffat, 1996) and 1RNT (Arni et al., 1988), respectively.

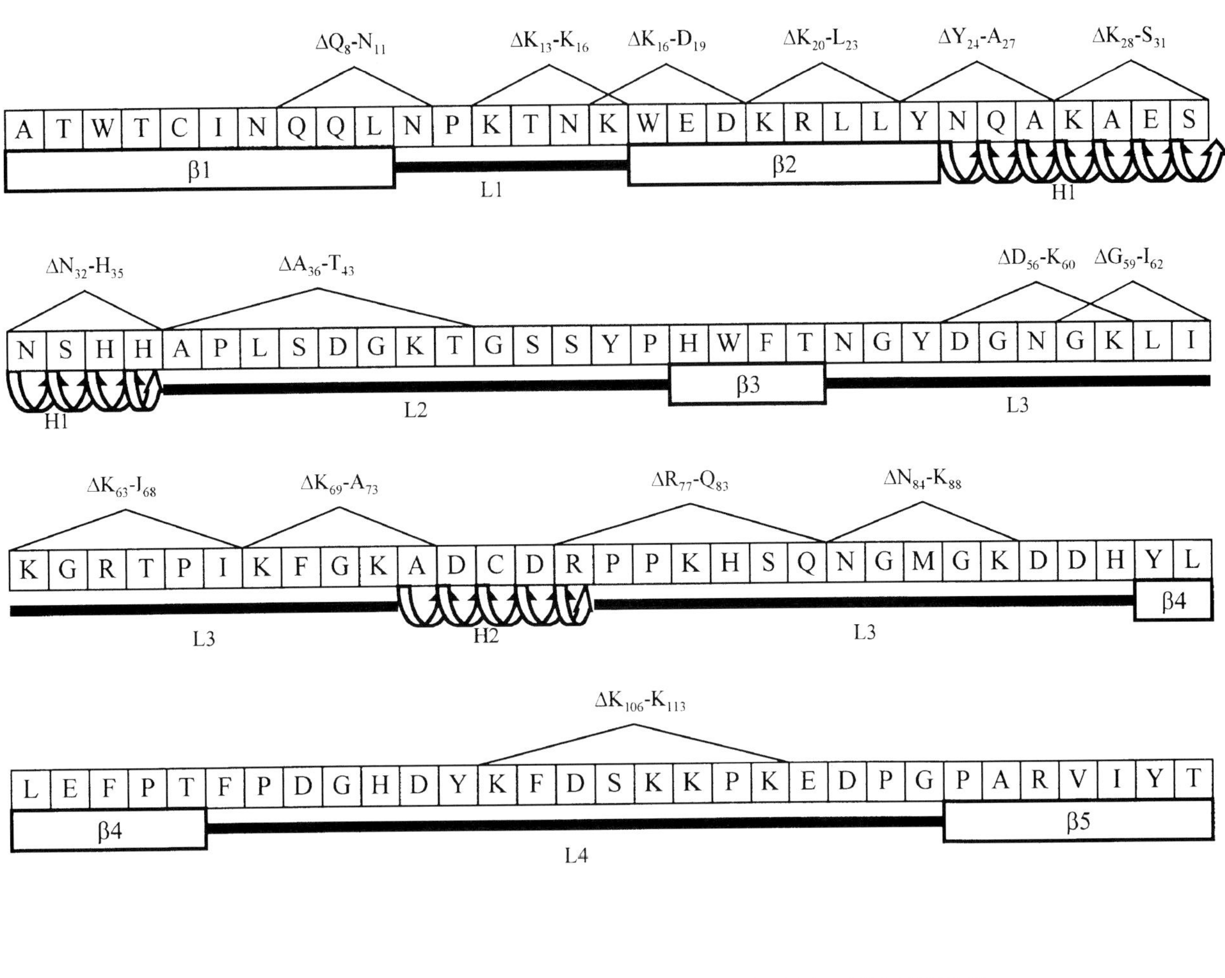

Figure 4. Mutagenesis of mitogillin. The sites of deletions are indicated. Also shown are the secondary structures of mitogillin: β1 to β7, β-sheets 1 to 7; L1 to L6, loops 1 to 6; H1 and H2, alpha helixes 1 and 2.

ative charge developed; however, the precise role of Arg77 is less clear, since a conformationally stable Arg77 variant RNase T1 has not been available for detailed biochemical study (Steyaert, 1997). The strong structural conservation of the core and the putative catalytic residues between fungal ribotoxins and RNase T1 indicates that they share a common catalytic mechanism of RNA cleavage (Yang and Moffat, 1996; Campos-Olivas et al., 1996). Interestingly, Nayak and Batra (1997) described a His49Ala restrictocin variant with elevated ribonucleolytic activity, suggesting that this residue may be involved in specific recognition of the rRNA. However, studies by others (Kao et al., 1998; Martinez del Pozo et al., personal communication) did not support this conclusion. Nevertheless, based on the available structural and biochemical data, it is reasonable to conclude that the His49, Glu95, Arg120, and His136 residues are essential active-site residues for the catalytic activity of mitogillin.

RIBOSOMAL RECOGNITION ELEMENTS IN MITOGILLIN

The recognition elements of the structure of the SRL in 28S rRNA have been studied extensively (Gluck et al., 1992, 1994), and G4319 was proposed to be the identity element for α-sarcin (Gluck and Wool, 1996; Wool, 1997). However, very little is known about the ribosome-targeting elements of the protein toxins. As mentioned above, mitogillin and related ribotoxins have amino acid sequence similarity to T1-like ribonucleases (Lamy et al., 1992), and

Table 2. Production of mutant mitogillin proteins[a]

Mutant mitogillin	Site of deletion	Production of protein
ΔQ_8–N_{11}	Q_8–N_{11}	–
ΔK_{13}–K_{16}	K_{13}–K_{16}	+
ΔK_{16}–D_{19}	K_{16}–D_{19}	+
ΔK_{20}–L_{23}	K_{20}–L_{23}	+
ΔY_{24}–A_{27}	Y_{24}–A_{27}	–
ΔK_{28}–S_{31}	K_{28}–S_{31}	+
ΔN_{32}–H_{35}	N_{32}–H_{35}	–
ΔA_{36}–T_{43}	A_{36}–T_{43}	–
ΔD_{56}–K_{60}	D_{56}–K_{60}	+
ΔG_{59}–I_{62}	G_{59}–I_{62}	+
ΔK_{63}–I_{68}	K_{63}–I_{68}	+
ΔK_{69}–A_{73}	K_{69}–A_{73}	–
ΔR_{77}–Q_{83}	R_{77}–Q_{83}	+
ΔN_{84}–K_{88}	N_{84}–K_{88}	+
ΔK_{106}–K_{113}	K_{106}–K_{113}	+

[a] Production was detected by sodium dodecyl sulfate-polyacrylamide gel electrophoresis of induced culture supernatants followed by Western blotting with rabbit anti-mitogillin antiserum. +, production of protein detected; –, production not detected.

α-sarcin has been shown to behave as a cyclizing ribonuclease, like many other ribonucleases (Lacadena et al., 1998; Perez-Canadillas et al., 1998), but their property of interacting specifically with the ribosome and causing a single ribonucleolytic cleavage in the large-subunit rRNA is unique. Studies have indicated that the similarities and differences detected in the amino acid sequence comparison of ribotoxins and a large family of other guanyl and purine ribonucleases may represent domains or residues key to ribonucleolytic activity and specificity (Kao and Davies, 1995; Kao et al., 1998). The presence of these "extra" protein domains in fungal ribotoxins led to the hypothesis that fungal ribotoxins are a family of naturally engineered toxins with ribosome targeting elements acquired from different ribosome-associated proteins (Kao and Davies, 1999).

Structural comparisons between the three-dimensional structures of restrictocin (99% identity with mitogillin) and T1 ribonuclease reveal strong structural similarities with certain domains in the fungal ribotoxin absent from the T1-like ribonucleases; the most obvious differences are the β-sheet 1 (β1)-loop 1-β2 (B1-L1-B2) region, the loop 3 (L3) region, and the L4 region of restrictocin (Fig. 3). It has been postulated that these domains are inserted elements contributing to the specific, targeted cytotoxicity of this class of proteins.

Deletions (4 to 8 amino acid residues) have been generated by PCR methods in those regions of mitogillia proposed to be inserted elements (Fig. 4). Ten of the 15 deletion mutants produce biologically active proteins with the *E. coli* expression system (Table 2). It is interesting to note that the five nonexpressible mutant mitogillin proteins have deletions located near or extending into elements such as disulfide bridge-forming cysteine residues, proline residues, or helixes, which are likely to be susceptible to host protease activity. The fact that the majority of the deletions obtained retain the catalytic activity of the ribotoxin supports the proposal that these domains have roles in ribotoxin function distinct from the RNase activity.

Deletions in the L4 and B1-L1-B2 regions generate mutants with interesting properties in terms of their ribonucleolytic activities. It has been shown that the octapeptide Lys106 to Lys113 in the L4 region has sequence similarity to various ribosome-associated proteins (Table 3) and that the heptapeptide Thr14 to Lys20 in the B1-L1-B2 region is strongly related to motifs in various elongation factors (Table 4). Although the significance of these relationships has yet to be examined, it is noteworthy that the ΔLys106-Lys113 mutant has lost the ability to recognize and cleave the α-sarcin loop (Kao and Davies, 1995) (Fig. 5) and that the ribonucleolytic activity of the deletion mutants in the B1-L1-B2 region is greatly elevated (Fig. 6 and Table 5). In agreement with the biochemical data, the SRL-restrictocin docking model of Yang and Moffat (1996) indicates that the L4 region of mitogillin is in close proximity

Table 3. Homologous motifs found in ribotoxins and in ribosomal protein S12 from a variety of sources

Protein	Source	Residues	Sequence
Mitogillin	*A. restrictus*	106–113	KFDSKKPK
Restrictocin	*A. restrictus*	106–113	KFDSKKPK
α-Sarcin	*A. giganteus*	107–114	KFDSKKPK
Ribosomal S12	*Thermus aquaticus*	119–126	KYGTKKPK
Ribosomal S12	*Bacillus stearothermophilus*	128–135	KYGAKKPK
Ribosomal S12	*Cryptomonas phi*	115–122	KYGAKKPK
Ribosomal S12	*Zea mays*	116–123	KYGAKKPK
Ribosomal S12	*Streptococcus pneumoniae*	129–136	KYGTKKPK
Mitochondrial S12	*Marchantia polymorphia*	116–123	KYGTKKPK

Table 4. Homologous motifs found in mitogillin and in translation elongation factors

Protein	Source	Residues	Sequence
Mitogillin	*A. restrictus*	14–20	TNKWEDK
Restrictocin	*A. restrictus*	14–20	TNKWEDK
α-Sarcin	*A. giganteus*	15–21	TNKYETK
EF-1α	*Euglena gracilis*	152–158	TNKFDDK
EF-2	Yeast	254–260	TKKWTNK
EF-3	Yeast	140–146	TNKWQEK

to the 28S rRNA identity element (G4319) of the fungal ribotoxin, suggesting that elements in the L4 domain of mitogillin interact with the "bulged" G4319 in the α-sarcin loop to promote specific recognition and cleavage of the substrate.

However, the B1-L1-B2 region of mitogillin is relatively distant from the catalytic center of the nuclease, and the role of this domain in attenuating the catalytic activity of the toxin is intriguing. We propose that this is the result of hydrogen bond formation between amino acid residues in β1 and β2 and residues in β6 in which the catalytic residue His136 is situated (Fig. 7). It is possible that the B1-L1-B2 domain attenuates the catalytic activity of the toxin by keeping the catalytic residue His136 in a configuration that is suboptimal for nucleolytic activity through hydrogen bonding with β6. Upon binding to the ribosome, the interactions between L4 and the α-sarcin loop trigger a conformational change in the protein which disrupts the interactions between the B1-L1-B2 domain and β6, thus positioning the catalytic residue His136 in an optimal environment for cleavage of the phosphodiester bond of the RNA substrate. Evidence in favor of this interpretation comes from the analysis of an Asn7Ala mutant of mitogillin (which presumably eliminates the hydrogen bonding between Asn7 and His136) which shows greatly elevated ribonuclease activity (Table 5). A recent report by Shapiro (1998) shows that the enzymatic potency and specificity of human angiogenin are modulated by hydrogen bonds. This raises the possibility that other substrate-specific ribonucleases may also modulate their specific activities by similar mechanisms.

Deletions in region Asp56 to Lys88 (L3) of mitogillin do not appear to interfere with the ribonucleolytic activity or the specificity of mitogillin. The dramatic decrease in RNase activity in ΔArg77-Gln83 may be due to the deletion of a catalytically important residue(s) or a change in the structure of the toxin, affecting its catalytic activity. The latter interpretation is more likely, since the active site of mitogillin does not indicate the involvement of any residues in region Arg77-Gln83 for catalysis (Kao et al., 1998). It should be noted that residues Arg77 and Gln83 are very close to Cys75 in mitogillin, which forms a disulfide bond with Cys5; deletion of these 7 residues presumably induces a change in the microenvironment of the active site of mitogillin and, as a result, reduces RNA cleavage activity. Based on existing information, the functions of domain L3 of mitogillin cannot be defined further. Yang and Moffat (1996) suggest that this region may influence mitogillin and cell surface receptor recognition or translocation of the toxin through the lipid bilayer of the cell; mutants constructed in this region would be good candidates to elucidate the functions of this "extra" loop of the fungal ribotoxins.

CONCLUSIONS

Characterization of the enzymatic properties of the mutant ribotoxins suggests that their specific cleavage properties are the result of natural genetic engineering in which the ribosome-targeting elements of ribosome-associated proteins were inserted into

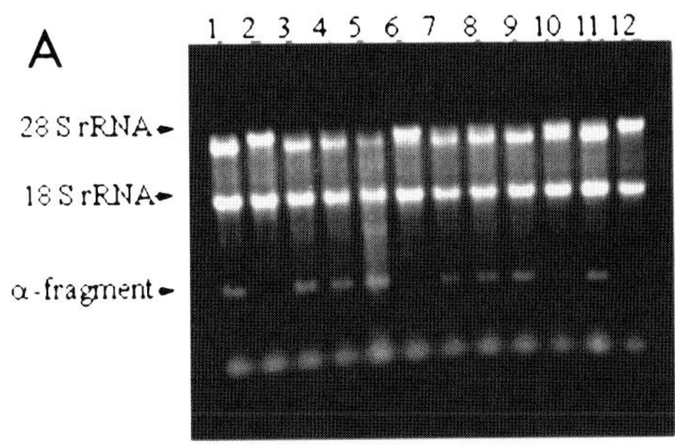

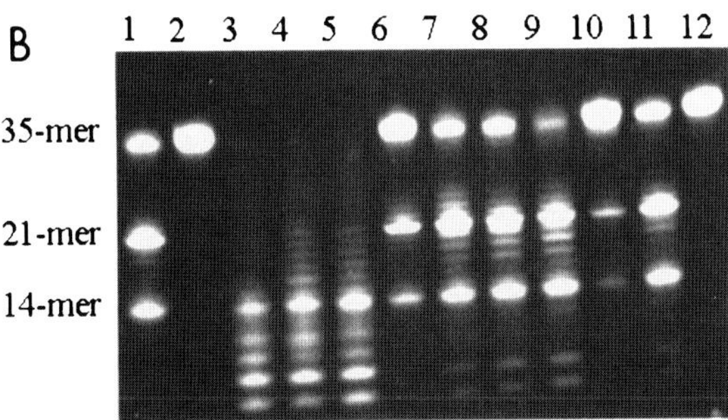

Figure 5. (A) Specific ribonucleolytic activity (in vitro α fragment release) of mitogillin and its variants. The positions of 28S rRNA, 18 S rRNA, and the α fragment are indicated. (B) Synthetic α-sarcin loop cleavage assay. The positions of the 35-mer, 21-mer, and 14-mer are also indicated. Lanes 1, mitogillin; lanes 2, no toxin; lanes 3, ΔK_{13}–K_{16} mutant; lanes 4, ΔK_{16}–D_{19} mutant; lanes 5, ΔK_{20}–L_{23} mutant; lanes 6, ΔK_{28}–S_{31} mutant; lanes 7, ΔD_{56}–K_{60} mutant; lanes 8, ΔG_{59}–I_{62} mutant; lanes 9, ΔK_{63}–I_{68} mutant; lanes 10, ΔR_{77}–Q_{83} mutant; lanes 11, ΔN_{84}–K_{88} mutant; and lanes 12, ΔK_{106}–K_{113} mutant.

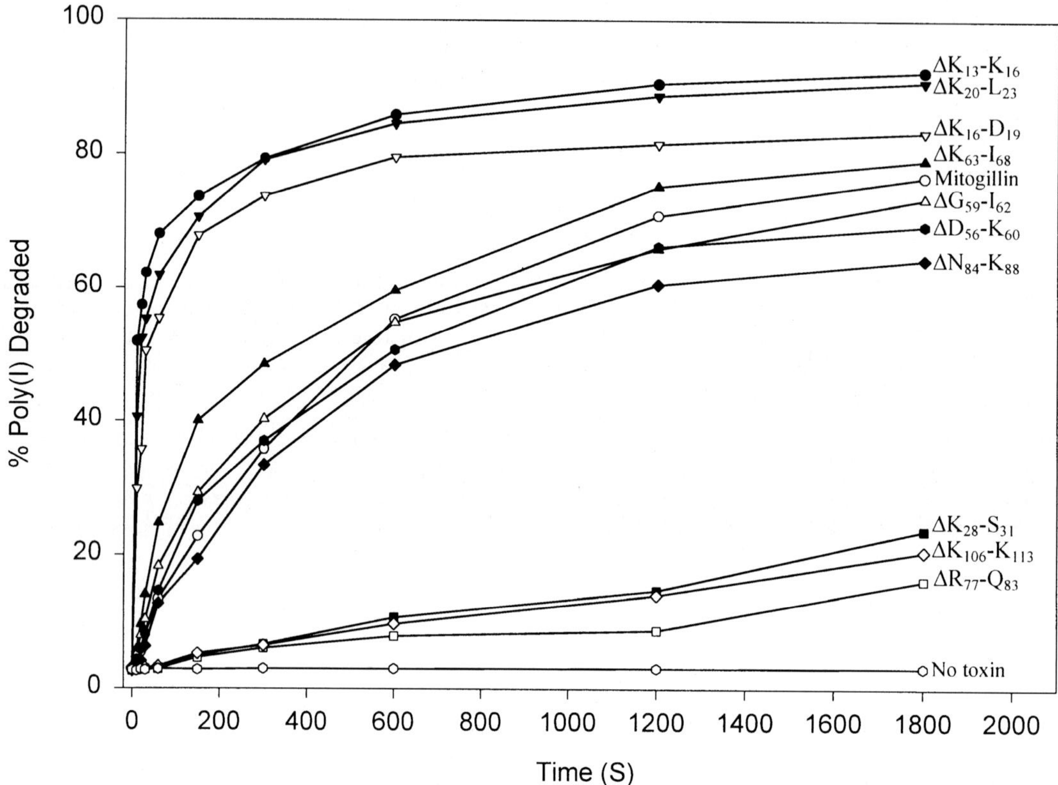

Figure 6. Nonspecific ribonucleolytic activity of mitogillin and its mutants on poly(I) substrate. The results were plotted as percent of poly(I) degradation versus time when 3 μM of mitogillin or a mutant protein was used.

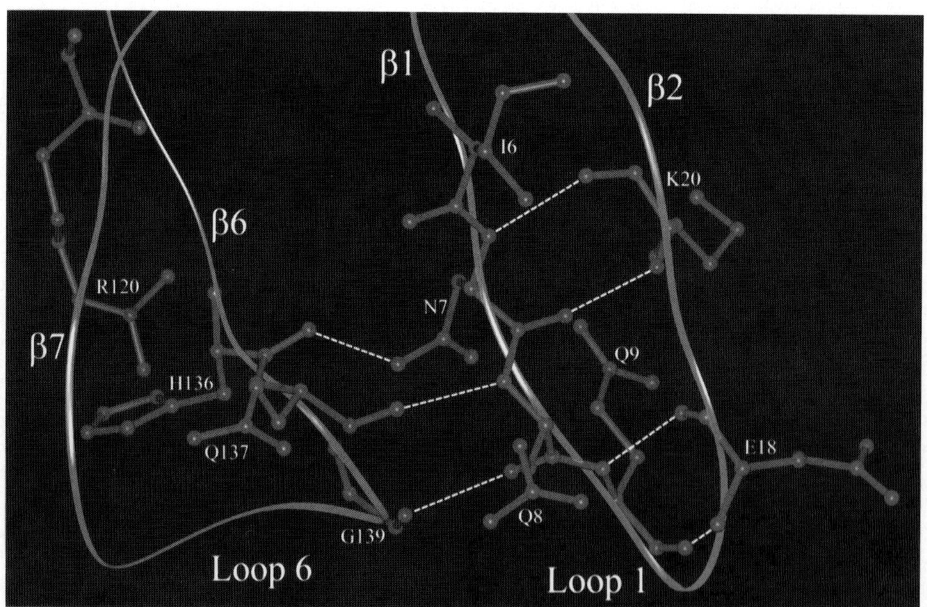

Figure 7. Hydrogen-bonding between B1-L1-B2 and B6-L6-B7 domains of restrictocin. L1 residues 11 to 16 are fully exposed to solvent and are consequently missing in the atomic model. Hydrogen bonds are denoted by dashed lines. Amino acid residues not relevant in forming hydrogen bonds are omitted for clarity.

Table 5. Comparison of the ribonucleolytic activities (initial rate of cleavage) of mitogillin, mutant mitogillins, and ribonuclease T1 on poly(I) homopolymer

Protein	Activity[a]	Relative activity
Mitogillin	0.19 ± 0.02	1.0
ΔK_{13}–K_{16}	5.33 ± 0.28	27.9
ΔK_{16}–D_{19}	4.23 ± 0.23	22.1
ΔK_{20}–L_{23}	5.00 ± 0.24	26.2
ΔK_{28}–S_{31}	0.033 ± 0.002	0.2
ΔD_{56}–K_{60}	0.24 ± 0.02	1.3
ΔG_{59}–I_{62}	0.30 ± 0.02	1.5
ΔK_{63}–I_{68}	0.33 ± 0.01	1.7
ΔR_{77}–Q_{83}	0.027 ± 0.001	0.1
ΔN_{84}–K_{88}	0.26 ± 0.01	1.4
ΔK_{106}–K_{113}	0.039 ± 0.004	0.2
N_7A	10.3 ± 1.1	54.0
RNase T1	16.3 ± 0.84	85.1

[a] The activity is defined as micromolar poly(I) homopolymer degraded per minute (means ± standard deviations for four independent experiments).

nonessential regions of T1-like ribonucleases, such that elements in the L4 region of mitogillin are involved in the specific recognition of the toxin by the ribosome and elements in the B1-L1-B2 region are involved in modulating the catalytic activity of the toxin, and that domain L3 of mitogillin may be involved in functions other than targeting mitogillin to the ribosome. It is evident that these additional elements, which appear to have been engineered naturally into T1-like ribonucleases, are not required for RNase activity (and in fact reduce it) but have functions specific to cytotoxic activity. This suggests the possibility that new types of specific toxins may be engineered by inserting targeting functions (such as rRNA binding domains of the ribotoxins) into nonessential regions of a variety of different degradative enzymes. It is likely that "combined" immunotoxins could be generated in this manner.

We thank A. Martinez-Ruiz and A. Martinez del Pozo for helpful discussions. This work is supported by the Natural Sciences and Engineering Research Council of Canada.

REFERENCES

Arni, R., U. Heinemann, R. Tokuoka, and W. Sarenger. 1988. Three-dimensional structure of the ribonuclease T1 2′-GMP complex at l.9 Å resolution. *J. Biol. Chem.* **263:**15358–15368.

Better, M., S. L. Bernhard, S. P. Lei, D. M. Fishwild, and S. F. Carroll. 1992. Activity of recombinant mitogillin and mitogillin immunoconjugates. *J. Biol. Chem.* **267:**16712–16718.

Boon, T. 1972. Inactivation of ribosomes in vitro by colicin E3 and its mechanism of action. *Proc. Natl. Acad. Sci. USA* **69:**549–552.

Bowman, C. M., J. Sidikaro, and M. Nomura. 1971. Specific inactivation of ribosomes by colicin E3 in vitro and mechanism of immunity in colicingenic cells. *Nat. New Biol.* **234:**133–137.

Brandhorst, T., P. F. Dowd, and W. R. Kenealy. 1996. The ribosome-inactivating protein restrictocin deters insect feeding on Aspergillus restrictus. *Microbiology* **142:**1551–1556.

Brandhorst, T. T., and W. R. Kenealy. 1992. Production and localization of restrictocin in *Aspergillus restrictus*. *J. Gen. Microbiol.* **138:**1429–1435.

Brigotti, M., F. Rambelli, M. Zamboni, L. Montanaro, and S. Sperti. 1989. Effect of alpha-sarcin and ribosome-inactivating proteins on the interaction of elongation factors with ribosomes. *Biochem. J.* **257:**723–727.

Campos-Olivas, R., M. Bruix, J. Santoro, A. Martinez del Pozo, J. Lacadena, J. G. Gavilanes, and M. Rico. 1996. Structural basis for the catalytic mechanism and substrate specificity of the ribonuclease alpha-sarcin. *FEBS Lett.* **399:**163–165.

Chan, Y. L., Y. Endo, and I. G. Wool. 1983. The sequence of the nucleotides at the α-sarcin cleavage site in 28S ribosomal ribonucleic acid. *J. Biol. Chem.* **258:**12768–12770.

Correll, C. C., A. Munishkin, Y. L. Chan, Z. Ren, I. G. Wool, and T. A. Steitz. 1998. Crystal structure of the ribosomal RNA domain essential for binding elongation factors. *Proc. Natl. Acad. Sci. USA* **95:**13436–13441.

Endo, Y., K. Mitsui, M. Motizuki, and K. Tsurugi. 1987. The mechanism of action of ricin and related toxic lectins on eukaryotic ribosomes. The site and the characteristics of the modification in 28S ribosomal RNA caused by the toxins. *J. Biol. Chem.* **262:**5908–5912.

Endo, Y., Y. L. Chan, A. Lin, K. Tsurugi, and I. G. Wool. 1988. The cytotoxins alpha-sarcin and ricin retain their specificity when tested on a synthetic oligoribonucleotide (35-mer) that mimics a region of 28S ribosomal ribonucleic acids. *J. Biol. Chem.* **263:**7917–7920.

Endo, Y., T. Oka, and Y. Aoyama. 1992. An efficient expression system for alpha-sarcin in *Escherichia coli*. *Targeted Diagn. Ther.* **7:**259–269.

Endo, Y., T. Oka, K. Tsurugi, and Y. Natori. 1993a. The biosynthesis of a cytotoxic protein, alpha-sarcin, in a mold *Aspergillus giganteus*. I. Synthesis of prepro- and pro-alpha-sarcin *in vitro*. *Tokushima J. Exp. Med.* **40:**1–6.

Endo, Y., T. Oka, K. Tsurugi, and Y. Natori. 1993b. The biosynthesis of a cytotoxic protein, alpha-sarcin, in a mold *Aspergillus giganteus*. II. Maturation of precursor form of alpha-sarcin *in vivo*. *Tokushima J. Exp. Med.* **40:**7–12.

Fong, W. P., R. N. S. Wong, T. T. M. Go, and H. W. Yeung. 1991. Enzymatic properties of ribosome-inactivating proteins (RIPs) and related toxins. *Life Sci.* **49:**1859–1869.

Gluck, A., and I. G. Wool. 1996. Determination of the 28S ribosomal RNA identity element (G4319) for alpha-sarcin and the relationship of recognition to the selection of the catalytic site. *J. Mol. Biol.* **256:**838–848.

Gluck, A., Y. Endo, and I. G. Wool. 1992. Ribosomal RNA identity elements for ricin A-chain recognition and catalysis. Analysis with tetraloop mutants. *J. Mol. Biol.* **226:**411–424.

Gluck, A., Y. Endo, and I. G. Wool. 1994. The ribosomal RNA identity elements for ricin and for alpha-sarcin: mutations in the putative CG pair that closes a GAGA tetraloop. *Nucleic Acids Res.* **22:**321–324.

Hausner, T. P., and K. H. Nierhaus. 1988. Resistance of both the 5′- and 3′-domain of isolated *Escherichia coli* 23S rRNA against digestion with alpha-sarcin. *Biochem. Int.* **17:**617–627.

Huang, K.-C., Y.-Y. Hwang, L. Hwu, and A. Lin. 1997. Characterization of a new ribotoxin gene (*c-sar*) from *Aspergillus clavatus*. *Toxicon* **35:**383–392.

Kao, R., and J. Davies. 1995. Fungal ribotoxins: a family of naturally engineered targeted toxins? *Biochem. Cell Biol.* **73:**1151–1159.

Kao, R., and J. Davies. 1999. Molecular dissection of mitogillin reveals that the fungal ribotoxins are a family of natural genetically-engineered ribonucleases. *J. Biol. Chem.* **274:**12576–12582.

Kao, R., J. E. Shea, J. Davies, and D. W. Holden. 1998. Probing the active site of mitogillin, a fungal ribotoxin. *Mol. Microbiol.* **29:**1019–1027.

Lacadena, J., J. M. Mancheno, A. Martinez-Ruiz, A. Martinez del Pozo, M. Gasset, M. Onaderra, and J. G. Gavilanes. 1995. Substitution of histidine-137 by glutamine abolishes the catalytic activity of the ribosome-inactivating protein α-sarcin. *Biochem. J.* **309:**581–586.

Lacadena, J., A. Martinez del Pozo, V. Lacadena, A. Martinez-Ruiz, J. M. Mancheno, M. Onaderra, and J. G. Gavilanes. 1998. The cytotoxin alpha-sarcin behaves as a cyclizing ribonuclease. *FEBS Lett.* **424:**46–48.

Lamy, B., and J. Davies. 1991. Isolation and nucleotide sequence of the *Aspergillus restrictus* gene coding for the ribonucleolytic toxin restrictocin and its expression in *Aspergillus nidulans*: the leader sequence protects producing strains from suicide. *Nucleic Acids Res.* **19:**1001–1006.

Lamy, B., J. Davies, and D. Schindler. 1992. The Aspergillus ribonucleolytic toxins (ribotoxins), p. 237–257. *In* A. E. Frankel (ed.), *Genetically Engineered Toxins.* Marcel Dekker, New York, N.Y.

Lin, A., C. Ciou-Jau, and S. S. Tzean. 1997. EMBL/GenBank/DDBJ databases, accession no. AF012812-AF012817.

Mancheno, J. M., M. Gasset, J. Lacadena, A. Martinez Del Pozo, M. Onaderra, and J. G. Gavilanes. 1995. Predictive study of the conformation of the cytotoxic protein alpha-sarcin: a structural model to explain alpha-sarcin-membrane interaction. *J. Theor. Biol.* **172:**259–267.

Martinez, S. E., and J. L. Smith. 1991. Crystallization and preliminary characterization of mitogillin, a ribosomal ribonuclease from *Aspergillus restrictus. J. Mol. Biol.* **218:**489–492.

Martinez del Pozo, A., A. Martinez-Ruiz, and J. G. Gavilanes. Personal communication.

Martinez-Ruiz, A., A. Martinez del Pozo, J. Lacadena, J. M. Mancheno, M. Onaderra, C. Lopez-Otin, and J. G. Gavilanes. 1998. Secretion of recombinant pro- and mature fungal alpha-sarcin ribotoxin by the methylotrophic yeast *Pichia pastoris*: the Lys-Arg motif is required for maturation. *Protein Expr. Purif.* **12:**315–322.

Martinez-Ruiz, A., R. Kao, J. Davies, and A. Martinez del Pozo. 1999. Ribotoxins are a more widespread group of proteins within the filamentous fungi than previously believed. *Toxicon* **37:**1549–1563.

Miller, S. P., and J. W. Bodley. 1988a. Alpha-sarcin cleaves ribosomal RNA at the alpha-sarcin site in the absence of ribosomal proteins. *Biochem. Biophys. Res. Commun.* **154:**404–410.

Miller, S. P., and J. W. Bodley. 1988b. The ribosomes of *Aspergillus giganteus* are sensitive to the cytotoxic action of alpha-sarcin. *FEBS Lett.* **229:**388–390.

Nayak, S. K., and J. K. Batra. 1997. A single amino acid substitution in ribonucleolytic toxin restrictocin abolishes its specific substrate recognition activity. *Biochemistry* **36:**13693–13699.

Obrig, T. G., T. P. Moran, and J. E. Brown. 1987. The mode of action of Shiga toxin on peptide elongation of eukaryotic protein synthesis. *Biochem. J.* **244:**287–294.

Parente, D., G. Raucci, B. Celano, A. Pacilli, L. Zanoni, S. Canevari, E. Adobati, M. I. Colnaghi, F. Dosio, S. Arpicco, L. Cattel, A. Mele, and R. De Santis. 1996. Clavin, a type-1 ribosome-inactivating protein from Aspergillus clavatus IFO 8605. cDNA isolation, heterologous expression, biochemical and biological characterization of the recombinant protein. *Eur. J. Biochem.* **239:**272–280.

Perez-Canadillas, J. M., R. Campos-Olivas, J. Lacadena, A. Martinez del Pozo, J. G. Gavilanes, J. Santoro, J., M. Rico, and M. Bruix. 1998. Characterization of pKa values and titration shifts in the cytotoxic ribonuclease alpha-sarcin by NMR. Relationship between electrostatic interactions, structure, and catalytic function. *Biochemistry* **37:**15865–15876.

Sanz, J. L., and R. Amils. 1984. Sensitivity of thermoacidophilic archaebacteria to α-sarcin. *FEBS Lett.* **171:**63–66.

Schindler, D. G., and J. E. Davies. 1977. Specific cleavage of ribosomal RNA caused by alpha-sarcin. *Nucleic Acids Res.* **4:**1097–1110.

Shapiro, R. 1998. Structural features that determine the enzymatic potency and specificity of human angiogenin: threonine-80 and residues 58-70 and 116-123. *Biochemistry* **37:**6847–6856.

Steyaert, J. 1997. A decade of protein engineering on ribonuclease T1: atomic dissection of the enzyme-substrate interactions. *Eur. J. Biochem.* **247:**1–11.

Szewczak, A. A., P. B. Moore, Y. L. Chang, and I. G. Wool. 1993. The conformation of the sarcin/ricin loop from 28S ribosomal RNA. *Proc. Natl. Acad. Sci. USA* **90:**9581–9585.

Wirth, J., A. Martínez del Pozo, J. M. Mancheño, A. Martínez-Ruiz, J. Lacadena, M. Oñaderra, and J. G. Gavilanes. 1997. Sequence determination and molecular characterization of gigantin, a cytotoxic protein produced by the mould *Aspergillus giganteus* IFO 5818. *Arch. Biochem. Biophys.* **343:**188–193.

Wool, I. G. 1984. The mechanism of action of the cytotoxic nuclease α-sarcin and its use to analyze ribosome structure. *Trends Biochem. Sci.* **9:**14–17.

Wool, I. G. 1997. Structure and mechanism of action of cytotoxic ribonuclease α-sarcin, p. 131–162. *In* G. D. Alessio and J. F. Riordan (ed.), *Ribonucleases: Structures and Functions.* Academic Press, New York, N.Y.

Yang, R., and W. R. Kenealy. 1992a. Effects of amino-terminal extensions and specific mutations on the activity of restrictocin. *J. Biol. Chem.* **267:**16801–16805.

Yang, R., and W. R. Kenealy. 1992b. Regulation of restrictocin production in *Aspergillus restrictus. J. Gen. Microbiol.* **138:**1421–1427.

Yang, X., and K. Moffat. 1996. Insights into specificity of cleavage and mechanism of cell entry from the crystal structure of the highly specific *Aspergillus* ribotoxin, restrictocin. *Structure* **4:**837–852.

The Ribosome: Structure, Function, Antibiotics, and Cellular Interactions
Edited by R. A. Garrett, S. R. Douthwaite, A. Liljas, A. T. Matheson, P. B. Moore, and H. F. Noller

Chapter 38

Structure and Function of the Sarcin-Ricin Domain

IRA G. WOOL, CARL C. CORRELL, and YUEN-LING CHAN

The sarcin-ricin domain (SRD) of *Escherichia coli* 23S rRNA, which includes a near-universal, purine-rich sequence of 12 nucleotides, is the site of action of the eponymous ribotoxins, which come in two flavors: sarcin (Olson et al., 1965; Schindler and Davies, 1977; Endo and Wool, 1982; Wool, 1997) is a ribonuclease that cleaves the phosphodiester bond on the 3′ side of G2661 (Chan et al., 1983); the A chain of ricin (Olsnes and Pihl, 1982), and its analogue pokeweed antiviral protein (PAP), are RNA *N*-glycosylases that depurinate the 5′ adjacent A2660 (Endo and Tsurugi, 1987, 1988; Endo et al., 1987; Marchant and Hartley, 1995). There are more than 4,500 nucleotides in *E. coli* ribosomes, but the toxins catalyze only these single covalent modifications. The modifications inactivate the ribosome and account for the toxicity.

The effect of the toxins on protein synthesis is prima facie evidence that the SRD is necessary for ribosome function (Montanaro et al., 1975; Fernandez-Puentes and Vazquez, 1977), almost certainly because the domain constitutes an important part of the binding site for the elongation factors (Hausner et al., 1987; Moazed et al., 1988). The toxins have no effect on peptide bond formation or on factor-free translocation but disrupt binding to the ribosome of elongation factor G (EF-G) and of the EF-Tu ternary complex (Hausner et al., 1987). The one established function of the SRD is in the binding of the elongation factors; nonetheless, it has been proposed, without direct evidence, that a conformational change in the structure of the domain might drive translocation (Wool et al., 1992). At any rate, the ribotoxins sarcin and ricin have proven valuable, since studies of their mechanisms of action have directed attention to a region of the ribosome where efforts to derive functional correlates of structure have been rewarded.

STRUCTURE IN THE SRD

Synthetic oligoribonucleotides that mimic the SRD first proved useful for biochemical studies of the mechanism of action of the ribotoxins and for analysis of the chemistry of recognition of RNA by the proteins (Endo et al., 1988; Wool et al., 1992; Wool, 1997) and then more recently for a determination of the structure of the RNA. The oligonucleotides maintain a relevant fold, since the specificity of toxin modification of the ribosome and the binding of EF-G are faithfully reproduced.

In the phylogenetically derived secondary structure of *E. coli* 23S rRNA, the SRD (nucleotides 2646 to 2674 [*E. coli* numbering is used throughout]) is a stem-loop; the stem has 7 base pairs, and the loop has 15 nucleotides, of which 12 are near universal. However, a three-dimensional conformation of an oligoribonucleotide (a 29-mer) that mimicked the homologous domain in mammalian 28S rRNA, determined first by nuclear magnetic resonance (NMR) spectroscopy (Szewczak et al., 1991, 1993; Szewczak and Moore, 1995), revealed a compact, more or less continuous, but irregular helix with a bulged guanosine. When this conformation was done in 1993, the only other RNA structures available were tRNAs; moreover, at the time this was the largest RNA whose conformation was established solely by NMR spectroscopy. In the succeeding years there has been a flood of RNA structures, determined either by NMR spectroscopy or by X-ray crystallography, but observations first recognized in the SRD conformation have been confirmed time and again. One is the modular nature of the SRD RNA. Other functionally important RNA domains have the same modules (Pley et al., 1994; Cate et al., 1996; Correll et al., 1997). A second is that non-Watson-Crick base pairs are a

Ira G. Wool, Carl C. Correll, and Yuen-Ling Chan ■ Department of Biochemistry and Molecular Biology, The University of Chicago, Chicago, IL 60637.

commonplace. This makes RNA structure more protean than that of DNA. Noncanonical base pairs can widen the narrow, deep, and therefore relatively inaccessible major groove of A-form RNA so that it is a more hospitable site for the binding of proteins and other ligands. A corollary is that the phylogenetically determined secondary structure of large RNAs can give a false impression of the actual conformation, as it did for the SRD. A reasonable prediction is that the derived secondary structure of rRNA will be correct about regions that are relatively simple, canonical A-form helices but that most of what is depicted as single-stranded sequences will be folded into compact, irregular helices.

The conformation of the mammalian SRD derived from NMR spectroscopy has now been superseded by two three-dimensional structures obtained by X-ray crystallography: the mammalian (rat) SRD at 2.1-Å resolution (Correll et al., 1998) and the *E. coli* domain at 1.11 Å (Correll et al., 1999). Since the *E. coli* structure is at a higher resolution, we shall describe that. But first an aside. The SRD is a garden of earthly delights for the structural biologist. Not only are there three-dimensional NMR conformations (Szewczak et al., 1993; Szewczak and Moore, 1995; Seggerson and Moore, 1998) and X-ray derived structures (Correll et al., 1998, 1999) of the RNA, there are also three-dimensional X-ray structures of four of the domain's most important ligands: an EF-Tu·GTP·aminoacyl-tRNA ternary complex (Nissen et al., 1995), EF-G alone (Ævarsson et al., 1994) and in association with GDP (Czworkowski et al., 1994), the ricin A chain (Katzin et al., 1991; Rutenber et al., 1991), and restrictocin, a close homologue of sarcin (Yang and Moffat, 1996).

General Description

The *E. coli* SRD RNA (Fig. 1a) folds into a distorted hairpin with a GAGA tetraloop, a bulged-G cross-strand A stack, a flexible region, and an A-form helical stem (Correll et al., 1999). Because the resolution is 1.11 Å, the geometry and the hydration of the two common RNA motifs, the GAGA tetraloop and the bulged-G cross-strand A-stack, are seen in greater detail than ever before.

The RNA is stabilized by stacking interactions (Fig. 1b); indeed, an unexpected finding is that there are no stabilizing metal ions in the structure. There are a purine stack (A2660, G2661, and A2662) in the tetraloop, a quadruple G stack with cross-strand interactions (G2659, G2663, G2664, and G2655), and a cross-strand A stack (A2657 and A2665). In cross-strand stacks there is interdigitation of bases in opposite strands, and hence they are likely to confer greater stability than same-strand stacks. Expression in *E. coli* of a 23S rRNA gene with mutations that either reverse the polarity of the pyrimidine and the purine of the C2658·G2663 pair (to G2658·C2663) or replace it with an A2658·U2663 pair is lethal (Chan and Wool, unpublished) (see below). We presume that these mutations are lethal, at least in part, because they disrupt the quadruple G stack and hence destabilize the structure. Oligonucleotides with these mutations have lower melting temperatures (T_ms).

The GAGA Tetraloop

The SRD GAGA tetraloop (Fig. 2) is an archetype GNRA module. The tetraloop is shut off by a C2658·G2663 pair. There is, as in all GNRA motifs, an intraloop G2659·A2662 heteropurinic pair; the pair is unusual because the guanosine has only a single bifurcated hydrogen bond that contacts the base and the backbone of the adenosine. The tetraloop has only two unpaired nucleotides—the PAP A2660 and the sarcin G2661. The Watson-Crick faces of A2660, G2661, and A2662 in the minor groove are exposed and available for recognition of ligands. In addition, there are unique features of the backbone geometry that might facilitate specific recognition.

The Bulged-G Cross-Strand A Stack

The bulged-G cross-strand A-stack motif, which shares the C2658·G2663 pair with the tetraloop, has a sheared A2657·G2664 pair that is different from the G2659·A2662 pair in the tetraloop (Fig. 3a). There is also a reverse-Hoogsteen U2656·A2665 pair and the characteristic stacking of adenosines (A2657 and A2665) (Fig. 3b). The module has the signature feature of the domain, the bulged G2655 that pairs with U2656 and A2665 to form a base triple. G2655 is unstacked on the 5′ side but stacks on the 3′ side with G2664 to create a double cross-strand stack. On the 5′ side of the bulged G2655 the major groove is 2.8 Å wider than in an A-form helix; moreover, between G2655 and A2660, the major groove is shallow and thus provides a potential recognition surface.

The Flexible Region

The designation of this region derives from the sparseness of stabilizing interstrand base contacts; the only direct contact is from the base at A2654 to the backbone at C2666 (Fig. 4). Instead, the A2654·C2666 and U2653·C2667 pairs are knit together by a network of hydrogen bonds mediated by four water molecules. The hydrogen bonds are base-water-backbone and base-water-base; indeed, there

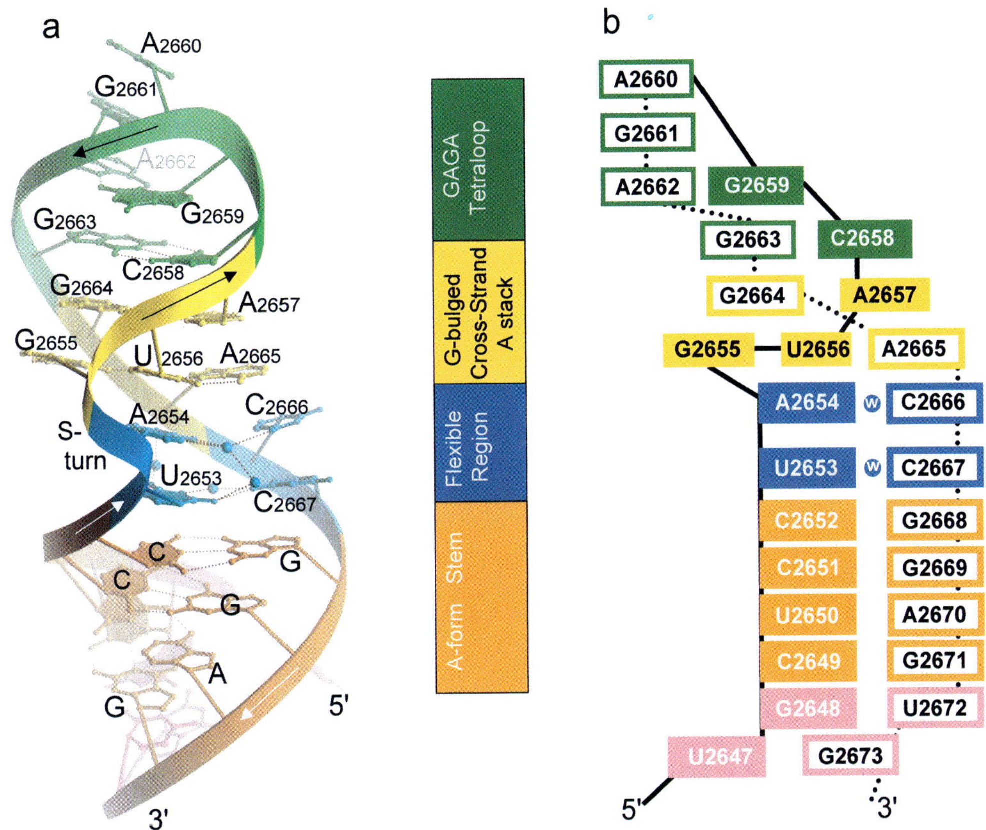

Figure 1. Structure of the *E. coli* SRD. (a) Ribbon diagram with separate modules color coded. (b) Base-stacking diagram; to delineate the cross-strand stacks, the nucleotides on the 5′ side are in solid boxes and those on the 3′ side are in open boxes.

are two of each type. The water-mediated base pairs increase the repertoire beyond the 28 that contain two or three direct hydrogen bonds (Saenger, 1984). Water-mediated base pairs increase hydrogen bonding complexity and enrich surface diversity, which is undoubtedly important for specific RNA-RNA and RNA-protein recognition.

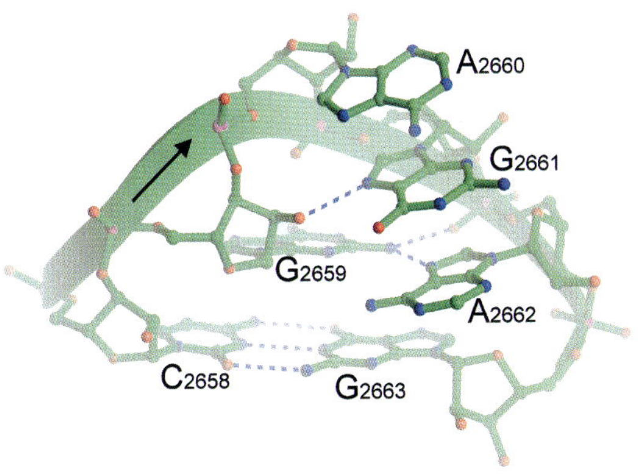

Figure 2. SRD GAGA tetraloop and its closing C·G pair.

There is an S-turn in the backbone of the 5′ strand that winds around the bulged G2655 and that encompasses the flexible-region nucleotides U2653 and A2654 (Fig. 5). The base of A2654 is flipped over. In the bends in the S-turn, primarily at A2654 and at G2655, reversal of chain direction is through distortion of backbone torsion angles and inverted sugar puckers.

Comparison of Rat and *E. coli* SRD RNAs

The conserved regions in the rat and the *E. coli* SRD RNAs adopt similar structures (Fig. 6); the atoms superimpose with a pairwise root mean squared (rms) deviation of 0.6 Å (Correll et al., 1999). The tetraloops and the G-bulged cross-strand A stacks superimpose with a pairwise rms deviation of 0.4 Å. In contrast, the rat and *E. coli* flexible regions differ. This was to be expected, since the former has 3 base pairs whereas the latter has only 2. The major groove on the 5′ side of G2655 is 2 Å narrower in the *E. coli* structure than in the rat structure, albeit still wider than in an A-form helix. Because of the differ-

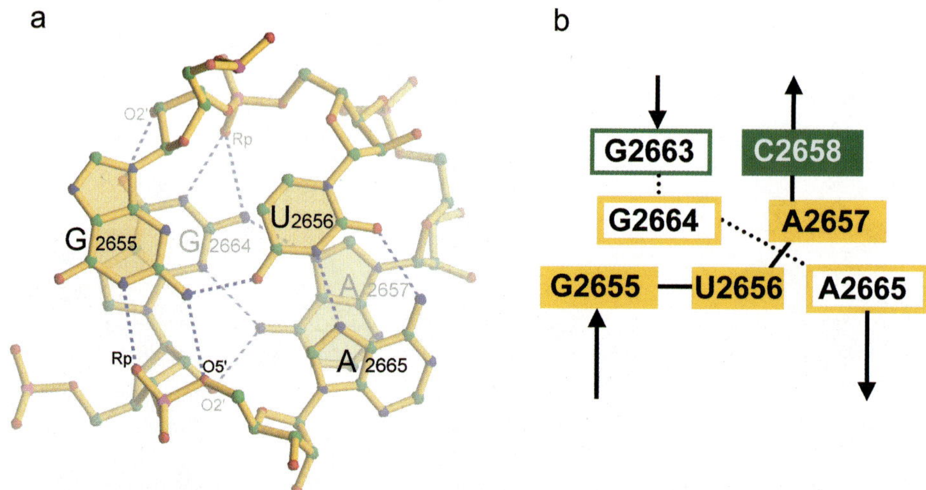

Figure 3. (a) SRD G-bulged cross-strand A stack. (b) Diagram of the A stack; the arrows indicate the chain direction. The significance of the open and closed boxes and of the colors is given in the legend to Fig. 1.

ence in the flexible regions, the position of the entire conserved region in relation to the stem is shifted. When the stems of the rat and the *E. coli* structures are superimposed, the relative positions of N-1 of A2660 and N-7 of G2655 with respect to the homologous nucleotides in the rat structure (A4324 and G4319) change by 4.9 and 7.2 Å, respectively. This is mainly the result of helical translation because the flexible-region base pairs in the rat SRD structure are underwound relative to those in the *E. coli* structure and hence produce less rotation.

FUNCTION IN THE SRD

Function is the dark side of the ribosome (Moore, personal communication). Although there are three-dimensional structures of EF-G alone (Czworkowski et al., 1994) and in a complex with GDP (Ævarsson et al., 1994), the chemistry of the binding of the factor to the ribosome has not been defined. There is, however, evidence that the SRD constitutes an important part of the EF-G binding site (Hausner et al., 1987; Moazed et al., 1988). Proteins frequently protect nucleotides in RNA that form noncanonical base pairs from chemical modification. These nucleotides presumably fold to create a recognition surface by widening the otherwise-inaccessible major groove; noncanonical pairing can also provide unique hydrogen bonding surfaces or distinctively shaped features in the minor groove that proteins might recognize. The SRD RNA has these

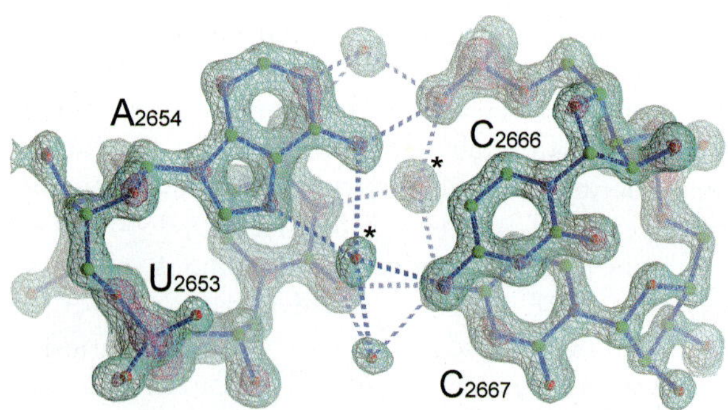

Figure 4. SRD flexible region. The refined model superimposed on the $2F_o$-F_c electron density map showing how the water-mediated base pairs stitch the flexible region nucleotides together. The asterisks designate water molecules that form ideal bridging contacts.

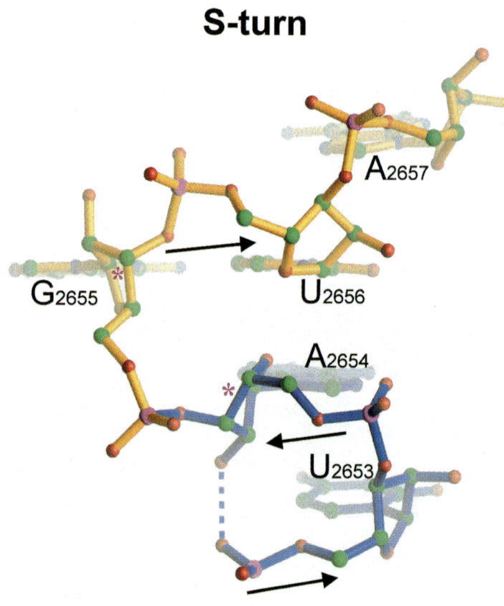

Figure 5. The SRD S-turn. The reversal of chain direction is indicated by arrows; C2′ *endo* sugar puckers are designated by asterisks.

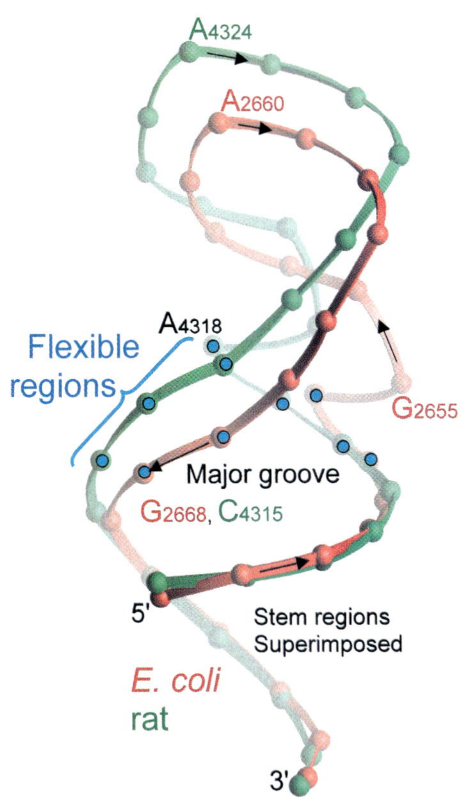

Figure 6. Comparison of the structures of the *E. coli* and rat SRDs. Shown are ribbon diagrams with the helical stems superimposed.

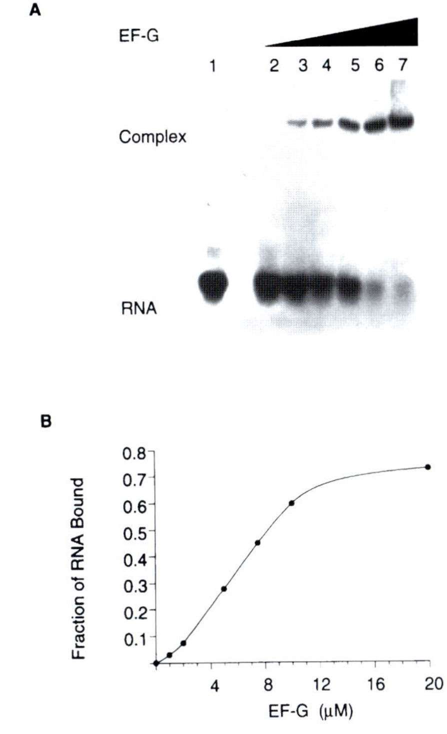

Figure 7. Binding of EF-G to SRD oligoribonucleotides. (A) Gel retardation assay. (B) The results in A plotted to show that binding saturates and that when the reactants are equimolar (10 μM) the complex has 70% of the input RNA.

characteristics. For these reasons, an *E. coli* SRD oligoribonucleotide (a 27-mer) was synthesized and the binding of EF-G to the RNA was assessed in gel retardation assays (Munishkin and Wool, 1997).

EF-G binds to the RNA in a concentration-dependent manner; at the approximate cellular concentration of EF-G (10 μM), binding is near maximal (Fig. 7). Binding saturates when the concentration of EF-G and of the SRD RNA are equimolar (10 μM); at saturation, 70% of the RNA is in complexes with EF-G. The K_d for the binding of EF-G to the RNA is 6.9 μM; the K_d for the binding of EF-G to ribosomes is 0.7 μM; the difference is only an order of magnitude. The interaction is specific by a number of criteria but most convincingly because mutations of a single nucleotide in the oligoribonucleotide abolish EF-G binding (see below).

Binding of EF-G to the SRD RNA neither requires nor is increased by GDPNP, a nonhydrolyzable analogue of GTP. This obviously is not what happens when ribosomes are the substrate. GDP, on the other hand, all but abolishes EF-G binding, a result more in conformity with what occurs physiologically. The results suggest a novel, perhaps a heretical, interpretation of the effect on the structure of EF-G of the putative effector ligands. First, they suggest that GDP induces a conformation of EF-G that is not compatible with stable association with the SRD RNA; this is the orthodox part of the interpretation. The heterodox part is that GTP displaces GDP from EF-G without having a direct effect on EF-G conformation. The assumption is that there are two conformations of EF-G: that the conformation of apo-EF-G and of EF-G·GTP are the same, whereas that of EF-G·GDP is different. In accord with a prediction that derives from this interpretation, GDPNP relieves the GDP inhibition of the binding of EF-G to the SRD RNA.

The light cast by structural studies on the effect, if any, of GDP and GTP on the conformation of EF-G is dim at best. There are no discernible differences in the structures of *Thermus thermophilus* apo-EF-G (Ævarsson et al., 1994) and of EF-G·GDP (Czworkowski et al., 1994) derived from X-ray diffraction of crystals, a result that was surprising, to say the least. There is a reservation about the structure of EF-G·GDP, since the crystals were treated with gluteraldehyde to prevent them from shattering in the X-ray beam; the treatment reduced the resolution to 2.7 Å. Moreover, the crystals were grown without magnesium, which might affect GDP binding. In a later study employing X-ray scattering (Czworkowski and Moore, 1997), EF-G (nucleotide free), EF-G·GDP, EF-G·GTP, and EF-G·GDPNP could not be distinguished; all resembled the structure of crystalline EF-G·GDP. If we accept these results, and there is no apparent reason not to no matter how much they defy expectations, then EF-G, a presumptive member in good standing of the G family, is not a G protein. This is a possibility difficult to accept because of the identity of the G domain in EF-G to that in authentic G-family proteins—EF-Tu (Nissen et al., 1995) and the proto-oncogene RAS (Brünger et al., 1990) are examples. It has been suggested (Czworkowski and Moore, 1997), as a way around the paradox, that a difference in the conformation of EF-G·GDP and EF-G·GTP is only manifest on the ribosome. Another possibility is that there is a conformational transition initiated by GDP and/or GTP but that it cannot be seen at the resolution obtained so far or in the crystal form used for X-ray diffraction.

The three SRD purines (G2655, A2660, and G2661) that are protected from chemical modification by EF-G bound to *E. coli* ribosomes (Moazed et al., 1988) are presumptive EF-G identity elements. Oligoribonucleotides with mutations in the analogues of these purines were prepared, and the binding of EF-G was assessed (Munishkin and Wool, 1997). The nucleotide most critical for the interaction is the G2655 analogue; transversions to cytosine or to uridine abolish the binding of EF-G, and a transition to adenosine increases the K_d by five- to sixfold. Neither transversions nor transitions of the A2660 or the G2661 analogues affect the association. The results suggest that it is the conformation of the SRD rather than nucleotide identity that EF-G recognizes.

There is evidence that the thiostrepton region of *E. coli* 23S rRNA (around A1067) is also involved in EF-G-dependent functions (Bodley et al., 1970; Möller, 1974; Maassen and Möller, 1978; Schmidt et al., 1981; Sköld, 1983; Beauclerk et al., 1984; Moazed et al., 1988). EF-G binds to a synthetic oligoribonucleotide (an 84-mer) that reproduces the sequence of the thiostrepton region (Munishkin and Wool, 1997). Once again, binding is dependent on EF-G concentration but is less efficient than binding to the SRD RNA; at equimolar concentrations (10 μM), only 22% of the RNA is in complexes with EF-G and the K_d is 17.2 μM, about twice that for binding of the factor to SRD RNA.

EF-G binds independently to the SRD and to the thiostrepton region oligoribonucleotides. A 10-fold excess of SRD RNA did not reduce binding of EF-G to the thiostrepton oligonucleotide, and in the reverse experiment, thiostrepton RNA did not interfere with the binding of the factor to the SRD oligonucleotide. Although the SRD and the thiostrepton region are at some distance from each other in the primary sequence of *E. coli* 23S rRNA, we assume that in the folded structure they are close and that together they constitute a good part, if not all, of the EF-G binding domain.

These results mark the achievement of what has been an important goal: the demonstration that a small component of the ribosome can retain a relevant fold and a relevant function. The SRD RNA has only 27 of the 4,566 nucleotides in *E. coli* ribosomes and lacks ribosomal proteins; nonetheless, the binding of EF-G is specific and the K_d is within an order of magnitude of that for binding to intact ribosomes. The results justify, at least as an interim measure, a reductionist approach to the study of the structure and function of the ribosome, an approach that we have called the ribosome in pieces.

MOLECULAR GENETICS IN THE SRD

A number of mutations have been made in the SRD in an *E. coli* 23S rRNA gene (Tapprich and Dahlberg, 1990; Melançon et al., 1992; Bilgin and Ehrenberg, 1994; Marchant and Hartley, 1994, 1995; O'Connor and Dahlberg, 1996; Liu and Liebman, 1996; Macbeth and Wool, 1999a, 1999b; Chan and Wool, unpublished). The mutations have been constructed in a plasmid carrying an *E. coli rrnB* operon. This operon has a transcription unit with the genes for 16S, 23S, and 5S rRNAs. Generally, two plasmids have been used. pSTL102 (Triman et al., 1989) is a high-copy-number plasmid in which transcription of the *rrnB* operon is controlled by a strong endogenous P_1P_2 promoter. Expression from this promoter is constitutive, and 70% of the ribosomes in *E. coli* cells have plasmid-encoded rRNA, despite the *E. coli* chromosome having seven rRNA operons. The second plasmid is pLK45 (Powers and Noller, 1990): in this plasmid the transcription of the *rrnB* operon is regulated by an inducible but weaker λP_L promoter; on induction, only 30% of the ribosomes have plasmid-encoded rRNA. This construct allows the analysis of ribosomes with mutations that are lethal when expressed at high levels; in addition, strains with the mutations can be stored. The drawback is a difficult population problem: 70% of the ribosomes are wild type, and only 30% have the mutation. There are ways, although none entirely satisfactory, to deal with the problem.

GAGA Tetraloop Mutations

Analysis of mutations in synthetic SRD oligoribonucleotides had indicated that the identity element for the ricin A chain was the GAGA tetraloop and that it was necessary and sufficient for recognition of the RNA by the toxin (Glück et al., 1992). The first mutation in the GAGA tetraloop to be constructed (Tapprich and Dahlberg, 1990) was a G2661C transversion at the sarcin cleavage site (this and other mutations can be located in Fig. 1 and 2). The mutation, surprisingly, had no effect on the growth of *E. coli* cells or on the function of ribosomes unless expressed in a strain with a restrictive mutation in ribosomal protein S12. Ribosomes with the two mutations, in G2661 and S12, had a lower rate of elongation, an increase in the accuracy of translation, and an increase in affinity for the EF-Tu·aminoacyl-tRNA·GTP ternary complex. In a separate study (Tapio and Isaksson, 1991), the G2661C mutation slowed translation elongation even in nonrestrictive strains (wild-type S12), although the phenotype was stronger in restrictive strains. The lethal phenotype of a G2661C mutation in a strain with a strong restrictive mutation in S12 was abolished by a mutation in EF-Tu, and the elongation rate was at the same time restored to normal. These observations reinforce the notion of a concerted interplay in support of EF-Tu function between S12 in the 30S and the SRD in the 50S ribosomal subunits. The G2661C mutation also increased the accuracy of translation assessed in vitro and suppressed read-through of nonsense codons and frameshifting in vivo (Melançon et al., 1992). The increase in the fidelity of translation is presumably due to the lowering of the rate of elongation, since accuracy in protein synthesis seems always to be a trade-off with speed. In still another analysis of the effect of the G2661C mutation on the function of the ribosome, it was confirmed that the frequency of translational errors is reduced; however, it was found that this was not the result of an improvement in proofreading but of improved initial selection of aminoacyl-tRNA (Bilgin and Ehrenberg, 1994). Initial selection is defined by the interaction of the ternary complex with the ribosome, whereas proofreading is conditioned by the fate of aminoacyl-tRNA after EF-Tu·GDP is released by the ribosome but before peptide bond formation. The interpretation is that the G2661C mutation improves the accuracy of translation by affecting the interaction of EF-Tu with the SRD. It should be noted that the enhanced fidelity is accompanied by a reduction in k_{cat}/K_m for the binding of the ternary complex to the ribosomal A site. This reinforces, once more, the conclusion that the SRD conditions the dynamics of EF-Tu function.

A cytosine-to-uridine transition mutation (C·G to U·G) in one of the nucleotides in the Watson-Crick pair that closes off the SRD GAGA tetraloop in *Saccharomyces cerevisiae* 25S rRNA decreased fidelity; this was manifest in read-through of stop codons and in shifts in the reading frame (Liu and Liebman, 1996). These effects are also likely to be the consequences of changes in the dynamics of EF-Tu function.

Six mutations (A2660G, G2661A, C2658G, G2663C, C2658G·G2663C, and G2664C) were constructed in the SRD in 23S rRNA, and the assembly of the mutant rRNA into ribosomes was assessed (Marchant and Hartley, 1994). The G2661A transition mutation does not affect the growth of cells or the assembly of 50S subunits, 70S ribosomes, or polysomes. The G2664C mutation does not affect assembly but does reduce the growth rate. The single G2663C and the double C2658G·G2663C mutations decrease the incorporation of 23S rRNA into 50S subunits and substantially reduce formation of 70S couples and of polysomes. No detectable 23S rRNA is found with either the C2658G or A2660G mutations in ribosomal particles. Presumably these mutations affect the transcription of ribosomal DNA or the processing or stability of the transcripts.

Additional mutations in the GAGA tetraloop have now been constructed (Chan and Wool, unpublished): an A2660U transversion of the PAP site nucleotide; a G2663C transversion, which was designed to disrupt the Watson-Crick pair (C2658·G2663) that closes off the tetraloop; and a double mutation, C2658G·G2663C, which reverses the polarity of the pyrimidine and the purine of the closing pair. The three mutations are lethal when expressed at high levels (in pSTL102); when expressed at lower levels (in pLK45), the mutations retard growth, which is most severe in the case of the double mutant. Analysis established that the mutant 23S rRNAs are transcribed and assembled into 50S ribosomal subunits but that the ribosomes do not form polysomes efficiently. More importantly, the ribosomes with mutant 23S rRNA are inactive in poly(U)-directed polyphenylalanine synthesis and in MS2 mRNA-directed synthesis of phage proteins.

The effect of the mutations on the sensitivity of ribosomes to PAP and to sarcin was tested. The test was made possible by silent mutations in a variable region of 23S rRNA that provides an allele-specific priming site. Ribosomes with an A2660U mutation in the SRD in 23S rRNA are resistant to PAP, which ordinarily depurinates the adenosine and inactivates the ribosome; the sensitivity to PAP, is not affected by either the G2663C or the C2658G·G2663C mutation. Similar results were obtained before (Marchant and Hartley, 1995). The findings are consistent with the results of experiments done with oligoribonucleotides (Glück et al., 1992), which had established that a GAGA tetraloop is esssential and sufficient for recognition by the ricin A chain, a toxin with a similar structure and the same mechanism of action as PAP but perhaps not the same RNA identity elements (Marchant and Hartley, 1995). Sensitivity to cleavage by sarcin of the SRD RNA was not affected by the A2660U or G2663C mutations; the C2658G·G2663C double mutation decreased, but did not abolish, sensitivity. The findings are reassuring, since they conform to results obtained in assays in vitro with synthetic oligoribonucleotides.

The most interesting of the three variants is the lethal C2658G·G2663C double mutation. Ordinarily, a reversal of the pyrimidine and the purine in a Watson-Crick pair is considered a neutral change and, hence, unlikely to have a phenotype. Such changes have occurred frequently in rRNA; indeed, they have been important in the phylogenetic determination of secondary structure. However, the reversal of pyrimidine-purine polarity during the evolution of rRNA has been largely, if not exclusively, in A-form helices. Here the reversal is in the closing pair of a GAGA tetraloop, a base pair, moreover, that is shared with a G-bulged cross-strand A stack and that is part of an irregular helix. In addition, this configuration, 5′ purine-3′ pyrimidine, at this location in the SRD has never to anyone's knowledge occurred in nature. A C2658A·G2663U double mutation also has a lethal phenotype (Chan and Wool, unpublished). We suggested earlier that the lethal phenotype of the mutation may derive, at least in part, from the disruption of a quadruple G stack (G2659, G2663, G2664, and G2655). There are other instances where the reversal of the purine and pyrimidine in a Watson-Crick pair affects function. For example, transposition of G2 and C71 in a base pair in the acceptor stem of *E. coli* $tRNA_f^{Met}$ reduces the formylation of Met-$tRNA_f^{Met}$ by the methionyl-tRNA transformylase (Lee et al., 1991; Guillon et al., 1992). No one will be surprised to learn that crystallizing and determining the structure of an SRD oligoribonucleotide with a C2658G·G2663C reversal is high on our wish list.

Bulged-G2655 Mutations

The bulged guanosine at position 2655 is the identity nucleotide for recognition by sarcin (Glück and Wool, 1996) and by EF-G (Munishkin and Wool, 1997) of their binding sites. A set of mutations in G2655 (Fig. 1 and 3) were constructed in 23S rRNA in an *rrnB* operon (Macbeth and Wool, 1999b). When expressed at high levels in pSTL102, G2655 transversion mutations (G2655C or G2655U) are lethal. (The mutations are referred to as G2655Y, since the results are the same for the two guanosine-to-pyrimidine mutations.) To our great surprise, a G2655A transition had no detectable effect on the growth of *E. coli* cells cultured either on agar plates or in liquid medium. This was unexpected, since no organism in the biosphere is known to have an adenosine at position 2655.

To test whether the G2655A mutation might have a subtle effect on growth that could not be detected in conventional assays, we grew the mutant strain in competition with cells transformed with pSTL102 but without a mutation in G2655 (referred to as wild-type cells). Equal numbers of cells harboring the G2655A transition mutation and cells harboring wild-type 23S rRNA in pSTL102 were mixed, grown to stationary phase, and then diluted into fresh medium. This procedure was repeated a number of times. When the culture was 24, 72, or 120 generations removed from the original inoculum, rRNA was extracted from the mixture of cells and the presence of 23S rRNA with a G2655A mutation was determined by primer extension.

The growth of wild-type and mutant strains separately in control experiments established that the mutation is stable for at least 120 generations. However, when the mutant cells were grown in competition with wild-type cells, the fraction in the mixed population with the G2655A mutation was substantially reduced after 72 generations and none could be detected after 120 generations. Obviously cells harboring the G2655A mutation are unable to compete effectively with cells having ribosomes with wild-type 23S rRNA. We presume the mutation is not found in nature because it reduces the fitness of ribosomes and hence of cells.

The G2655Y transversion mutations were expressed in pLK45 to allow a determination of their effect on function. The mutant 23S rRNA is transcribed, processed, and assembled into 50S subunits; there is, however, a decrease in their number relative to 30S subunits. The G2655Y mutations also decreased the total number of polysomes. The decrease is mainly a reduction in the number of smaller polysomes; the increase in the relative proportion of larger polysomes may reflect a slowing of elongation.

The effect of the G2655 mutations (G2655A, G2655U, and G2655C) on sensitivity to sarcin was tested: each confers resistance to the toxin on ribosomes. The findings once again are reassuring, since they too conform to results of assays in vitro with synthetic oligoribonucleotides (Glück and Wool, 1996).

All of the G2655 mutations decrease the capacity of ribosomes to catalyze phage MS2 mRNA-directed protein synthesis. Synthesis by ribosomes with G2655Y transversion mutations is decreased by more than 30%. Thus, there must be a strong presumption that they are inactive, since 70% of the ribosomes have chromosome-encoded wild-type 23S rRNA. Synthesis by ribosomes with the G2655A transition mutation is decreased by 17%; since 30% of the ribosomes in the population have this mutation, the results suggest their activity is decreased by half.

Flexible-Region Mutations

The nucleotides in the flexible region are not stringently conserved; nonetheless, mutations there affect the binding of the elongation factors. For example, point mutations in A2654 and C2666, which form an unusual pair (Fig. 1 and 4), decrease translational fidelity (O'Connor and Dahlberg, 1996). The three possible substitutions of these two nucleotides were constructed, but only the two, C2666U and A2654G, that result in canonical pairs, A2654·U2666 and G2654·C2666, increased stop codon readthrough and frameshifting, presumably by affecting the binding of the EF-Tu ternary complex. An important lesson to be gleaned from these results is that both nucleotide identity and the geometry of a base pair can affect function.

Additional mutations, designed to alter the stability of the flexible region, were constructed (Macbeth and Wool, 1999a). The first was a U2653-C2667 double deletion which removed a water-mediated pyrimidine pair and shortened the SRD. When expressed at a high level in pSTL102, this mutation was lethal. Oligoribonucleotides that have these deletions have higher T_ms; once again, it appears that increased stability is not compatible with flexible-region function. Ribosomes with the U2653-C2667 double-deletion mutation are inactive in the catalysis of MS2 mRNA-directed synthesis of phage proteins. Finally, the K_d for the binding of EF-G to an oligoribonucleotide with the deletions was increased by almost an order of magnitude (from 7.9 to 58.4 μM).

The second flexible-region mutation, a U2653G transversion, creates the potential for a C2653·G2667 canonical pair in place of the water-mediated U2653·C2667 pair (Macbeth and Wool, 1999a). Since this mutation in effect lengthens the A-form helix of the SRD, it is not surprising that the T_m of a mimetic oligoribonucleotide is increased by 9°C. The mutation, whose phenotype is not as severe as that of the U2653-C2667 double deletion, retards growth and impairs the capacity of the ribosomes for protein synthesis. The sensitivity to sarcin of ribosomes with the U2653G mutation is markedly decreased.

BIOLOGY IN THE SRD

In the best of all experimental worlds one combines biochemistry, structure, and genetics—the ménage à trois of every biologist's dream: genetics to understand function, biochemistry to understand mechanism, and structure to tie it all together. There is, once again, biochemical and genetic evidence that the flexible region in the SRD affects the binding of the elongation factors (O'Connor and Dahlberg, 1996; Macbeth and Wool, 1999a, 1999b). However, the elongation factors make additional contacts in the ribosome: EF-G and EF-Tu to the thiostrepton region around nucleotide A1067 in 23S rRNA (Moazed et al., 1988; Saarma et al., 1997; Munishkin and Wool, 1997), EF-G and the EF-Tu ternary complex to proteins and perhaps RNA in the L12 stalk (Hamel et al., 1972; Schrier et al., 1973; Highland and Howard, 1975), and the EF-Tu ternary complex to the 530 loop in 16S rRNA (Powers and Noller, 1993). There are differences in the structures of the flexible regions of the rat and *E. coli* SRDs; the rat domain has an extra helical translation that alters the relation of the tetraloop and of the G-bulged cross-strand A stack to the stem. This may in turn shift the relationship of the elongation factor contacts in the SRD to the other elongation factor binding sites and in that way contribute to the lethal phenotype of the flexible-region mutations and to the species incompatibility of the factors.

Correlation with the structure of the data from experiments on the protection of nucleotides from chemical modification (Moazed et al., 1988; Uchiumi et al., 1999) indicates that the SRD RNA has two distinct surfaces (Fig. 8). One surface is accessible, lies primarily in the major groove, and is likely to bind the elongation factors and toxins. The second is for the most part inaccessible, lies primarily in the minor groove, and is likely to be buried in the ribosome.

The buried surface is defined by the atoms protected from dimethyl sulfate (DMS) modification. N-3 of C2666 and C2667 and N-1 of A2657 and A2662 are protected in the intact 50S ribosomal subunit and when proteins L3 and L6 are bound to 23S rRNA (Uchiumi et al., 1999). L3 and L6 are the only ribosomal proteins that bind to a 135-nucleotide fragment of 23S rRNA that has the SRD. A2662, A2657, C2666, and C2667 form noncanonical pairs and present their Watson-Crick faces to the minor groove. A2662 of the tetraloop and A2657 of the G-bulged cross-strand A stack flank the C2658·G2663 pair, which may also be buried. Atoms in the minor-groove face of this base pair (C2658·G2663) are not modified by DMS; hence, whether the pair is buried was not determined. The other protected atoms, N-

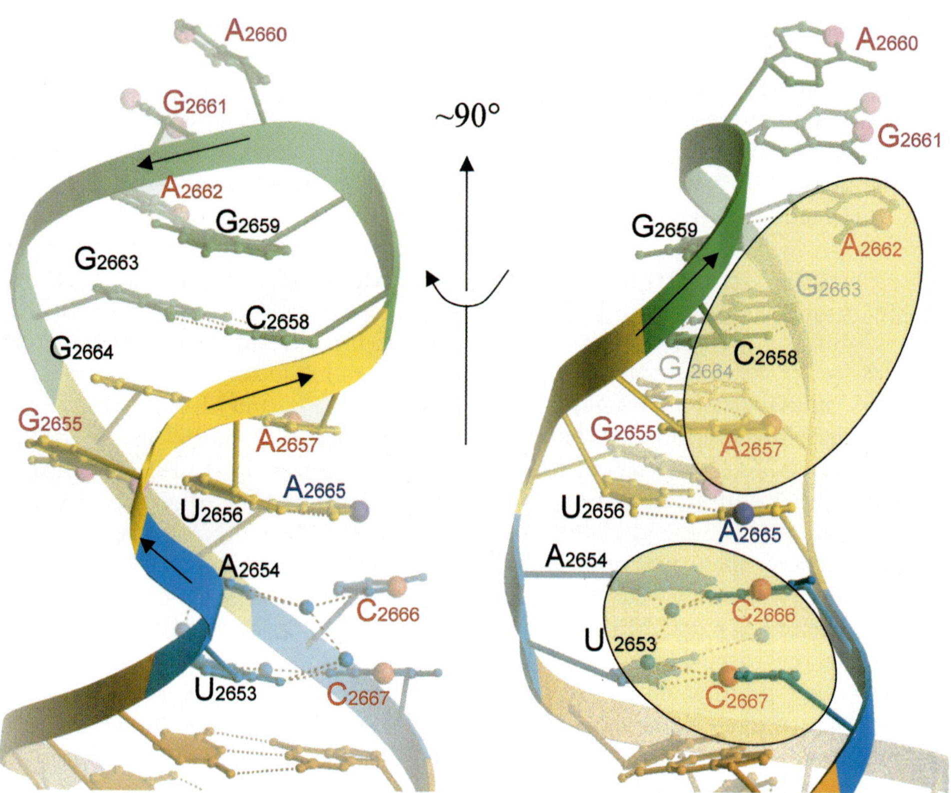

Figure 8. The two faces of the SRD. On the left is an orthogonal view of the solvent-accessible surface of the domain—the surface to which the elongation factors and toxins presumably bind. On the right, the buried surface is designated by yellow ellipses.

3 of C2666 and C2667, are in the flexible region. The protection data imply that the minor groove adjacent to A2662 and A2657 and to C2666 and C2667 is buried. In contrast, the Watson-Crick face of A2665, the intervening nucleotide, is accessible.

The surface opposite the buried face is solvent accessible and may well provide the binding site for the elongation factors and the toxins. G2655, A2660, and G2661 are protected from chemical modification when EF-G is bound to the ribosome (Moazed et al., 1988), suggesting that the elongation factor contacts this face. Binding of EF-G to mutant oligoribonucleotides (Munishkin and Wool, 1997) indicated that the primary determinant in recognition is the conformation of the SRD RNA. Neither transitions nor transversions of A2660 or G2661 affect EF-G binding, and only transversions of G2655 abolished binding. Sarcin recognizes G2655 and then cleaves 11.5 Å away on the 3′ side of G2661 (Glück and Wool, 1996); this suggests that the bulged G2655, the cleavage site (G2661), and the intervening major groove are exposed in the ribosome. The 5′ unstacked surfaces of A2660 and G2655 are available for the binding of elongation factors. One possibility is that EF-G binds to the SRD RNA like a C-clamp (Correll et al., 1998), contacting the 5′ sides of A2660 and G2655. The intervening surface lies primarily in the major groove, with A2660 and G2661 in the minor groove. If the prediction has merit, EF-G would be expected to make contacts to the intervening surface and to protect G2663 and G2664 from chemical modification. However, the method of analysis in one relevant experiment (Uchiumi et al., 1999) would not have detected modification of N-7 of G2663 or G2664 by DMS, since these modifications do not interrupt primer extension by reverse transcriptase (Ehresmann, 1987); hence, whether there is protection of these nucleotides cannot be said to have been determined. N-1 and N-2 of G2663 and G2664 are buried in the structure and would not be modified by kethoxal, either (Moazed et al., 1988).

The EF-Tu·GTP·aminoacyl-tRNA ternary complex bound to the ribosome protects the same 3 bases as does EF-G and, in addition, A2665 (Moazed et al., 1988), which is on the face of the SRD RNA opposite the protected G2655. The protection by the EF-Tu ternary complex of A2665, which lies between two regions buried in the ribosome, may be indirect: the complex may induce a conformational change in the buried surface or in the adjacent region of the ribosome or both. A structural transition might enlarge the tertiary contact surface to include A2665, thereby shielding this minor-groove nucleotide. It is possible that a change in the structure of the buried surface serves as an allosteric switch that transduces a signal generated by the binding of elongation factors to the exposed face. An obvious destination of this putative signal is the peptidyltransferase center.

There are 3 nucleotides in the SRD that are protected from chemical modification by both EF-G and the EF-Tu ternary complex. The implication from this and from many other observations is that the elongation factors bind to the same or an overlapping site. If EF-G and the EF-Tu ternary complex bind to appreciably the same site in the SRD, how then can we account for their cyclic association with the ribosome during protein synthesis? The productive binding of EF-Tu and EF-G to the ribosome is always sequential; the binding of EF-Tu is never followed by the binding of another molecule of EF-Tu, and, of course, the same is true for EF-G. One possibility is that the cyclic binding of the elongation factors is determined by the location on the ribosome of peptidyl-tRNA. The EF-Tu ternary complex may bind to the ribosome only when peptidyl-tRNA is in the P site (posttranslocation), and EF-G may bind only when it is in the A site (pretranslocation). The location of the peptidyl-tRNA is envisioned as favoring or restricting the binding of one of the pair of elongation factors. The restriction might be as simple as steric hindrance by peptidyl-tRNA; alternatively, facilitation of binding could be based in a conformational transition initiated by peptidyl-tRNA or by the elongation factors or by both.

The research was supported by grant GM33702 to I.G.W. from the National Institutes of Health and a Cancer Foundation grant to C.C.C.

We are indebted to our colleagues M. Macbeth, A. Glück, and A. Munishkin, with whom we had the pleasure of working on the research described here.

REFERENCES

Ævarsson, A., E. Brazhnikov, M. Garber, J. Zheltonosova, Y. Chirgadze, S. Al-Karadaghi, L. A. Svensson, and A. Liljas. 1994. Three-dimensional structure of the ribosomal translocase: elongation factor G from *Thermus thermophilus*. *EMBO J.* **13**:3669–3677.

Beauclerk, A. A. D., E. Cundliffe, and J. Dijk. 1984. The binding site for ribosomal protein complex L8 within 23S ribosomal RNA of *Escherichia coli*. *J. Biol. Chem.* **259**:6559–6563.

Bilgin, N., and M. Ehrenberg. 1994. Mutations in 23 S ribosomal RNA perturb transfer RNA selection and can lead to streptomycin dependence. *J. Mol. Biol.* **235**:813–824.

Bodley, J. W., L. Lin, and J. H. Highland. 1970. Studies on translocation. VI. Thiostrepton prevents the formation of a ribosome-G factor-guanine nucleotide complex. *Biochem. Biophys. Res. Commun.* **41**:1406–1411.

Brünger, A. T., M. V. Milburn, L. Tong, A. M. deVos, J. Jancarik, Z. Yamaizumi, S. Nishimura, E. Ohtsuka, and S.-H. Kim. 1990. Crystal structure of an active form of RAS protein, a complex of a GTP analog and the HRAS p21 catalytic domain. *Proc. Natl. Acad. Sci. USA* **87**:4849–4853.

Cate, J. H., A. R. Gooding, E. Podell, K. Zhou, B. L. Golden, C. E. Kundrot, T. R. Cech, and J. A. Doudna. 1996. Crystal structure of a group I ribozyme domain: principles of RNA packing. *Science* **273**:1678–1685.

Chan, Y. L., and I. G. Wool. Unpublished data.

Chan, Y. L., Y. Endo, and I. G. Wool. 1983. The sequence of the nucleotides at the α-sarcin cleavage site in rat 28S ribosomal ribonucleic acid. *J. Biol. Chem.* **258**:12768–12770.

Correll, C. C., B. Freeborn, P. B. Moore, and T. A. Steitz. 1997. Metals, motifs, and recognition in the crystal structure of a 5S rRNA domain. *Cell* **91**:705–712.

Correll, C. C., A. Munishkin, Y. L. Chan, Z. Ren, I. G. Wool, and T. A. Steitz. 1998. Crystal structure of the ribosomal RNA domain essential for binding elongation factors. *Proc. Natl. Acad. Sci. USA* **95**:13436–13441.

Correll, C. C., I. G. Wool, and A. Munishkin. 1999. The two faces of the *Escherichia coli* 23S rRNA sarcin/ricin domain: the structure at 1.11 Å resolution. *J. Mol. Biol.* **292**:275–287.

Czworkowski, J., and P. B. Moore. 1997. The conformational properties of elongation factor G and the mechanism of translocation. *Biochemistry* **36**:10327–10334.

Czworkowski, J., J. Wang, T. A. Steitz, and P. B. Moore. 1994. The crystal structure of elongation factor G complexed with GDP, at 2.7 Å resolution. *EMBO J.* **13**:3661–3668.

Ehresmann, C., F. Baudin, M. Mougel, P. Romby, J. P. Ebel, and B. Ehresmann. 1987. Probing the structure of RNAs in solution. *Nucleic Acids Res.* **15**:9109–9128.

Endo, Y., and K. Tsurugi. 1987. RNA N-glycosidase activity of ricin A-chain. Mechanism of action of the toxic lectin ricin on eukaryotic ribosomes. *J. Biol. Chem.* **262**:8128–8130.

Endo, Y., and K. Tsurugi. 1988. The RNA N-glycosidase activity of ricin A-chain. The characteristics of the enzymatic activity of ricin A-chain with ribosomes and with rRNA. *J. Biol. Chem.* **263**:8735–8739.

Endo, Y., and I. G. Wool. 1982. The site of action of α-sarcin on eukaryotic ribosomes. The sequence at the α-sarcin cleavage site in 28S ribosomal ribonucleic acid. *J. Biol. Chem.* **257**:9054–9060.

Endo, Y., K. Mitsui, M. Motizuki, and K. Tsurugi. 1987. The mechanism of action of ricin and related toxic lectins on eukaryotic ribosomes. *J. Biol. Chem.* **262**:5908–5912.

Endo, Y., Y. L. Chan, A. Lin, K. Tsurugi, and I. G. Wool. 1988. The cytotoxins α-sarcin and ricin retain their specificity when tested on a synthetic oligoribonucleotide (35-mer) that mimics a region of 28S ribosomal ribonucleic acid. *J. Biol. Chem.* **263**:7917–7920.

Fernandez-Puentes, C., and D. Vazquez. 1977. Effects of some proteins that inactivate the eukaryotic ribosome. *FEBS Lett.* **78**:143–146.

Glück, A., and I. G. Wool. 1996. Determination of the 28S ribosomal RNA identity element (G4319) for alpha-sarcin and the relationship of recognition to the selection of the catalytic site. *J. Mol. Biol.* **256**:838–848.

Glück, A., Y. Endo, and I. G. Wool. 1992. Ribosomal RNA identity elements for ricin A-chain recognition and catalysis: analysis with tetraloop mutants. *J. Mol. Biol.* **226**:411–424.

Guillon, J. M., T. Meinnel, Y. Mechulam, C. Lazennec, S. Blanquet, and G. Fayat. 1992. Nucleotides of tRNA governing the specificity of *Escherichia coli* methionyl-$tRNA_f^{Met}$ formyltransferase. *J. Mol. Biol.* **224**:359–367.

Hamel, E., M. Koka, and T. Nakamoto. 1972. Requirement of an *Escherichia coli* 50S ribosomal protein component for effective interaction of the ribosome with T and G factors and with guanosine triphosphate. *J. Biol. Chem.* **247:**805–814.

Hausner, T. P., J. Atmadja, and K. H. Nierhaus. 1987. Evidence that G^{2661} region of 23S rRNA is located at the ribosomal binding sites of both elongation factors. *Biochimie* **69:**911–923.

Highland, J. H., and G. A. Howard. 1975. Assembly of ribosomal proteins L7, L10, L11, and L12, on the 50 S subunit of Escherichia coli. *J. Biol. Chem.* **250:**813–814.

Katzin, B. J., E. J. Collins, and J. D. Robertus. 1991. Structure of ricin A-chain at 2.5 Å. *Proteins* **10:**251–259.

Lee, C. P., B. L. Seong, and U. L. RajBhandary. 1991. Structural and sequence elements important for recognition of *Escherichia coli* formylmethionine tRNA by methionyl-tRNA transformylase are clustered in the acceptor stem. *J. Biol. Chem.* **266:**18012–18017.

Liu, R., and S. W. Liebman. 1996. A translational fidelity mutation in the universally conserved sarcin/ricin domain of 25S yeast ribosomal RNA. *RNA* **2:**254–263.

Maassen, J. A., and W. Möller. 1978. Elongation factor G-dependent binding of a photoreactive GTP analogue to *Escherichia coli* ribosomes results in labeling of protein L11. *J. Biol. Chem.* **253:**2777–2783.

Macbeth, M., and I. G. Wool. 1999a. Characterization of in vitro and in vivo mutations in non-conserved nucleotides in the ribosomal RNA recognition domain for the ribotoxins ricin and sarcin and the translation elongation factors. *J. Mol. Biol.* **285:** 567–580.

Macbeth, M., and I. G. Wool. 1999b. The phenotype of mutations of G2655 in the sarcin/ricin domain of 23 S ribosomal RNA. *J. Mol. Biol.* **285:**965–975.

Marchant, A., and M. R. Hartley. 1994. Mutational studies on the a-sarcin loop of *Escherichia coli* 23S ribosomal RNA. *Eur. J. Biochem.* **226:**141–147.

Marchant, A., and M. R. Hartley. 1995. The action of pokeweed antiviral protein and ricin A-chain on mutants in the α-sarcin loop of *Escherichia coli* 23S ribosomal RNA. *J. Mol. Biol.* **254:** 848–855.

Melançon, P., W. E. Tapprich, and L. Brakier-Gingras. 1992. Single-base mutations at position 2661 of *Escherichia coli* 23S rRNA increase efficiency of translational proofreading. *J. Bacteriol.* **174:**7896–7901.

Moazed, D., J. M. Robertson, and H. F. Noller. 1988. Interaction of elongation factors EF-G and EF-Tu with a conserved loop in 23S rRNA. *Nature* **334:**362–364.

Möller, W. 1974. The ribosomal components involved in EF-G and EF-Tu-dependent GTP hydrolysis, p. 711–731. *In* M. Nomura, A. Tissières, and P. Lengyel (ed.), *Ribosomes*. Cold Spring Harbor Laboratory, Cold Spring Harbor, N.Y.

Montanaro, L., S. Sperti, A. Mattioli, G. Testoni, and F. Stirpe. 1975. Inhibition by ricin of protein synthesis *in vitro*: inhibition of the binding of elongation factor 2 and of adenosine diphosphate-ribosylated elongation factor 2 to ribosomes. *Biochem. J.* **146:**127–131.

Moore, P. B. Personal communication.

Munishkin, A., and I. G. Wool. 1997. The ribosome-in-pieces: binding of elongation factor EF-G to oligoribonucleotides that mimic the sarcin/ricin and thiostrepton domains of 23S ribosomal RNA. *Proc. Natl. Acad. Sci. USA* **94:**12280–12284.

Nissen, P., M. Kjeldgaard, S. Thirup, G. Polekhina, L. Reshetnikova, B. F. C. Clark, and J. Nyborg. 1995. Crystal structure of the ternary complex of Phe-$tRNA^{Phe}$, EF-Tu, and a GTP analog. *Science* **270:**1464–1472.

O'Connor, M., and A. E. Dahlberg. 1996. The influence of base identity and base pairing on the function of the a-sarcin loop of 23 S rRNA. *Nucleic Acids Res.* **24:**2701–2705.

Olsnes, S., and A. Pihl. 1982. Toxic lectins and related proteins, p. 51–105. *In* P. Cohen and S. van Heyningen (ed.), *The Molecular Action of Toxins and Viruses*. Elsevier Biomedical Press, Amsterdam, The Netherlands.

Olson, B. H., J. C. Jennings, V. Roga, A. J. Junek, and D. M. Schuurmans. 1965. Alpha sarcin, a new antitumor agent. II. Fermentation and antitumor spectrum. *Appl. Microbiol.* **13:**322–326.

Pley, H. W., K. M. Flaherty, and D. B. McKay. 1994. Three-dimensional structure of a hammerhead ribozyme. *Nature* **372:** 68–74.

Powers, T., and H. F. Noller. 1990. Dominant lethal mutations in a conserved loop in 16S rRNA. *Proc. Natl. Acad. Sci. USA* **87:** 1042–1046.

Powers, T., and H. F. Noller. 1993. Evidence for functional interaction between elongation factor Tu and 16S ribosomal RNA. *Proc. Natl. Acad. Sci. USA* **90:**1364–1368.

Rutenber, E., B. J. Katzin, S. Ernst, E. J. Collins, D. Mlsna, M. P. Ready, and J. D. Robertus. 1991. Crystallographic refinement of ricin to 2.5 Å. *Proteins* **10:**240–250.

Saarma, U., J. Remme, M. Ehrenberg, and N. Bilgin. 1997. An A to U transversion at position 1067 of 23 S rRNA from Escherichia coli impairs EF-Tu and EF-G function. *J. Mol. Biol.* **272:** 327–335.

Saenger, W. 1984. *Principles of Nucleic Acid Structure*, p. 119–122. Springer-Verlag, New York, N.Y.

Schindler, D. G., and J. E. Davies. 1977. Specific cleavage of ribosomal RNA caused by alpha sarcin. *Nucleic Acids Res.* **4:** 1097–1110.

Schmidt, F. J., J. Thompson, K. Lee, J. Dijk, and E. Cundliffe. 1981. The binding site for ribosomal protein L11 within 23S ribosomal RNA of *Escherichia coli*. *J. Biol. Chem.* **256:**12301–12305.

Schrier, P. I., J. A. Maassen, and W. Moller. 1973. Involvement of 50S ribosomal proteins L6 and L10 in the ribosome dependent GTPase activity of elongation factor G. *Biochem. Biophys. Res. Commun.* **53:**90–98.

Seggerson, K., and P. B. Moore. 1998. Structure and stability of variants of the sarcin-ricin loop of 28 S rRNA: NMR studies of the prokaryotic SRL and a functional mutant. *RNA* **4:**1203–1215.

Sköld, S. E. 1983. Chemical crosslinking of elongation factor G to the 23S RNA in 70S ribosomes from *Escherichia coli*. *Nucleic Acids Res.* **11:**4923–4932.

Szewczak, A. A., and P. B. Moore. 1995. The sarcin/ricin loop, a modular RNA. *J. Mol. Biol.* **247:**81–98.

Szewczak, A. A., Y. L. Chan, P. B. Moore, and I. G. Wool. 1991. On the conformation of the alpha sarcin stem-loop of 28S rRNA. *Biochimie* **73:**871–877.

Szewczak, A. A., P. B. Moore, Y. L. Chan, and I. G. Wool. 1993. The conformation of the sarcin/ricin loop from 28S ribosomal RNA. *Proc. Natl. Acad. Sci. USA* **90:**9581–9585.

Tapio, S., and L. A. Isaksson. 1991. Base 2661 in *Escherichia coli* 23S rRNA influences the binding of elongation factor Tu during protein synthesis *in vivo*. *Eur. J. Biochem.* **202:**981–984.

Tapprich, W. E., and A. E. Dahlberg. 1990. A single base mutation at position 2661 in *E. coli* 23S ribosomal RNA affects the binding of ternary complex to the ribosome. *EMBO J.* **9:**2649–2655.

Triman, K., E. Becker, C. Dammel, J. Katz, H. Mori, S. Douthwaite, C. Yapijakis, S. Yoast, and H. F. Noller. 1989. Isolation of temperature-sensitive mutants of 16S rRNA in *Escherichia coli*. *J. Mol. Biol.* **209:**645–653.

Uchiumi, T., N. Sato, A. Wada, and A. Hachimori. 1999. Interaction of the sarcin/ricin domain of 23 S ribosomal RNA with proteins L3 and L6. *J. Biol. Chem.* **274:**681–686.

Wool, I. G. 1997. Structure and mechanism of action of the cytotoxic ribonuclease α-sarcin, p. 131–162. *In* G. D'Alessio and J. F. Riordan (ed.), *Ribonucleases: Structures and Functions*. Academic Press, San Diego, Calif.

Wool, I. G., A. Glück, and Y. Endo. 1992. Ribotoxin recognition of ribosomal RNA and a proposal for the mechanism of translocation. *Trends Biochem. Sci.* **17:**266–269.

Yang, X., and K. Moffat. 1996. Insights into specificity of cleavage and mechanism of cell entry from the crystal structure of the highly specific *Aspergillus* ribotoxin, restrictocin. *Structure* **4:** 837–852.

X. INITIATION, TERMINATION, AND RECYCLING

X. INITIATION, TERMINATION, AND RECYCLING

The stages of the ribosomal cycle known as initiation, termination, and recycling require several factors. These help to generate the first mRNA-tRNA complex on the 30S subunit to which the 50S subunit is then coupled. After the protein is synthesized, it is released by factors which recognize the stop codon of the message and induce hydrolysis of the protein from the peptidyl-tRNA. The release factor then causes dissociation of the ribosomal subunits such that synthesis of a new protein can commence. These mechansims are all complex, and the steady progress in these areas is summarized in the following chapters.

The Ribosome: Structure, Function, Antibiotics, and Cellular Interactions
Edited by R. A. Garrett, S. R. Douthwaite, A. Liljas, A. T. Matheson, P. B. Moore, and H. F. Noller

Chapter 39

Translation Initiation in Bacteria

CLAUDIO O. GUALERZI, LETIZIA BRANDI, ENRICO CASERTA, ANNA LA TEANA, ROBERTO SPURIO, JERNEJA TOMŠIC, and CYNTHIA L. PON

In this chapter we summarize the most important advances concerning structural and mechanistic aspects of translation initiation in bacteria which have occurred since the appearance of the last reviews on this subject (Gualerzi et al., 1990; Gualerzi and Pon, 1990, 1996; McCarthy and Gualerzi, 1990).

mRNA-RIBOSOME AND fMet-tRNA–RIBOSOME INTERACTIONS AT INITIATION

The small ribosomal subunit (30S) interacts with mRNA and fMet-tRNA in stochastic order to yield a bona fide 30S initiation complex through the rearrangement, kinetically controlled by the three initiation factors, of an unstable kinetic intermediate called the pre-ternary complex (Gualerzi and Pon, 1990). Indeed, aside from the above-mentioned kinetic evidence, the existence of both intermediate binary complexes in the random pathway leading to the 30S initiation complex is well documented: the 30S-mRNA complexes are easily detected (e.g., Ringquist et al., 1993), while evidence for the formation of the more elusive binary complex, 30S–fMet-tRNA, is also striking (e.g., Wu et al., 1996). Furthermore, an intermediate ternary complex having the same properties attributed to the pre-ternary complex, namely, a 30S subunit containing two non-mutually interacting ligands (mRNA and initiator tRNA), has recently been isolated and characterized as the initiation intermediate "frozen" by ribosomal protein (r-protein) S15 during the translational autoregulation of the *rpsO* operon in *Escherichia coli* (Philippe et al., 1993).

Thus, aside from the specificity imposed on the 30S initiation complex by the cognate Watson-Crick (WC) base pairing between initiation codon and the initiator tRNA anticodon, the translation initiation regions (TIRs) of the mRNA and initiator fMet-tRNA are endowed with peculiar properties allowing them to be selected on their own as specific 30S ligands during translation initiation. These properties will be outlined in the following sections.

mRNA Selection and Properties of TIRs

As mentioned above, the first events in translation could be the recognition and binding of the mRNA TIR by the 30S ribosomal subunit. The main elements of a canonical TIR (leaderless mRNAs will be discussed below) include the initiation triplet (most frequently AUG), the purine-rich Shine-Dalgarno (SD) sequence complementary to the 3′-end region of 16S rRNA, and a spacer, of variable length, separating the SD sequence and the initiation triplet (Gualerzi and Pon, 1990; McCarthy and Brimacombe, 1994). Both scholarly and biotechnologically oriented studies have been carried out in recent years with the purpose of identifying the optimal combinations of these TIR elements to maximize mRNA translation (Ringquist et al., 1992; Vellanoweth and Rabinowitz, 1992; Lee et al., 1996; Simmons and Yansura, 1996).

Systematic variations of the TIR elements in the in vivo translation of reporter genes in *E. coli* and *Bacillus subtilis* have produced similar results in two similar analyses (Ringquist et al., 1992; Vellanoweth and Rabinowitz, 1992). The optimal spacing was determined to be 7 to 9 nucleotides, regardless of the identity of the initiation codon, but longer intervals (at least up to 14 nucleotides) were also found to be

Claudio O. Gualerzi, Letizia Brandi, Enrico Caserta, Anna La Teana, Roberto Spurio, Jerneja Tomšic, and Cynthia L. Pon ■ Laboratory of Genetics, Department of Biology MCA, University of Camerino, 62032 Camerino (MC), Italy.

acceptable, while mRNAs with spacings shorter than optimal or rich in secondary structures were translated in *B. subtilis* much less efficiently than in *E. coli*. The preferred (by two- to threefold) initiation codon was found to be AUG in both organisms, and its importance was inversely related to the (presumed) strength of the corresponding SD interaction. Among the rare codons, UUG was slightly better than GUG in *B. subtilis* and slightly worse in *E. coli*, and the nucleotides 3′ to the initiation codon, mainly those belonging to the second codon, were found to influence the rate of initiation, with the magnitude depending on the nature of the initiation codon.

The translational rates measured in these studies clearly indicate that one can attempt to optimize gene expression at the translational level by manipulating the mRNA TIR. This is not an easy task, however, since several variables which can determine the translational rates should be taken into account. Thus, the results of these studies overall were consistent with our kinetic model, in which a variety of rate constants contribute to the process of translation initiation (Gualerzi and Pon, 1990).

Along similar experimental lines, the efficiency of protein secretion in the *E. coli* periplasm has been correlated with the translational efficiency of several TIR variants for the heat-stable enterotoxin II (STII) signal sequence (Simmons and Yansura, 1996). Using mRNA variants covering a 10-fold range of translational efficiency, the authors were able to dramatically improve the secretion of a number of heterologous (but not homologous) proteins and to demonstrate that heterologous-protein secretion in *E. coli* depends on a narrow range of translational levels (Simmons and Yansura, 1996). The differences observed in this study between heterologous and homologous proteins and the sometimes contradictory conclusions reached by similar studies concerning the relevance of specific TIR elements and the consequent criteria for optimization of translation remind us once again that each gene (mRNA) might be endowed with particular, not easily predictable properties.

The combinatorial approach introduced by the SELEX (systematic evolution of ligands by exponential enrichment) technique applied to the identification of the mRNA nucleotide sequences having the highest affinity for 30S subunits yielded, after 11 to 13 rounds, essentially two sets of sequences having 3- to 38-fold-higher affinities for the target (Ringquist et al., 1995). Intact (i.e., +S1) 30S subunits and purified protein S1 generated almost identical consensus sequences, GAUG **ACA** *CGAAAG* CAUC **AAG-GAAC** *CUuucg*, in which the underlined nucleotides form the stems and the shared identical consensus sequences (boldface) form the two loops of a pseudoknot. (The italicized letters represent complementary bases which form base pairs in the pseudoknot. The lowercase letters represent the fixed bases at the 3′ end of the 30-nucleotide randomized region of the starting repertoire used to generate the consensus sequence.) The pseudoknot-containing ligand selected was also found to display significant homology with some naturally occurring TIRs and with a previously described translational enhancer (Loechel et al., 1991). Ligands generated against S1-depleted 30S, on the other hand, contained a consensus SD sequence and exhibited extensive complementarity to the 3′ end of 16S rRNA. Overall, these studies indicated that mRNA-30S interaction was governed by base pairing with 16S rRNA in the absence of S1, while the interaction with intact ribosomes was dominated by the binding properties of S1. The reported presence of an independent high-affinity site for each type of ligand indicates that the two types of binding probably coexist within the same ribosomal particle and is fully consistent with the data indicating the existence (see below) of a dual mRNA binding site on the ribosome (standby and decoding). A recent study, also using a combinatorial in vivo selection approach, identified an mRNA sequence motif which is able to promote efficient expression of a downstream cistron through a termination-reinitiation event in vivo in the complete absence of an SD sequence upstream of the reinitiation AUG triplet (André et al., submitted).

It is a commonplace that initiation of translation in bacteria involves the well-characterized WC SD–anti-SD interaction between mRNA and rRNA; it has been shown nonetheless that translational strength correlates very poorly with the strength of this interaction (e.g., Lee et al., 1996), while mutation or complete deletion of either the SD or the anti-SD sequence abolishes translation neither in vivo nor in vitro and does not affect the selection of the correct reading frame or the overall mechanism by which the initiation complex is made (for reviews, see Gualerzi and Pon, 1990, 1996; Gualerzi et al., 1990; Wu and Janssen, 1996). Furthermore, the in vivo relevance of the SD sequence may be different in different prokaryotes. In fact, besides the more frequent presence of leaderless mRNA in bacteria other than *E. coli* (see below), it has been suggested that, unlike in *E. coli*, where the SD sequence-dependent initiation predominates over an SD sequence-independent mechanism, plant chloroplasts, and very probably their cyanobacterial ancestors, have mainly adopted an SD sequence-independent mechanism to translate their mRNAs (Fargo et al., 1998). This suggestion stems from the observation that the SD-like sequences found in the leaders of many mRNAs from cyanobacteria and chloroplasts are hypervariable in location, size, and base composition and that mutagenic replacement of the putative SD sequences has no effect on the expression of reporter genes in *Chlamydomonas reinhardtii* chloroplasts (Fargo et al., 1998). Thus, the current view of the role and relevance of the SD sequence appears to have become less dogmatic and has fallen more in line with the proposal that the function of this sequence is to anchor the mRNA on the ribosome transiently to allow the kinetic selection of a potential initiation triplet whose local concentration is increased to millimolar concentrations in the proximity of the P decoding site (Gualerzi et al., 1990). More recently it has been suggested that the SD interaction may also help the ribosome to melt unfavorable structures in the mRNA TIR regions. This model, which does not exclude the previous one, stems from an observed coevolution of an inverse relationship between the stability of the TIR secondary structures and the lengths of the corresponding SD sequences (Olsthoorn et al., 1995).

Leaderless mRNAs

The dispensable nature of the SD sequence is further evidenced by the existence (rare in *E. coli* but more frequent in other bacteria) of leaderless mRNAs which begin directly with an AUG initiation triplet (Janssen, 1993). The only essential *cis*-acting recognition element identified with certainty in these mRNAs, which are sometimes very efficiently translated, is the 5′ AUG triplet. Thus, a leaderless chloramphenicol acetyltransferase CAT mRNA from *Streptomyces acrimycini*, whose translation in vivo by *E. coli* was indeed shown to start at the 5′-terminal AUG, conferred chloramphenicol resistance on the cells. Furthermore, while the presence of the AUG start codon proved to be essential for translation, the 5′ addition of an AUGC tetranucleotide to the 5′ AUG caused translation to occur also from the new reading frame defined by the added AUG triplet, allowing for the conclusion that a 5′-terminal start codon is an important recognition feature not only for translation initiation of leaderless mRNAs but also for establishing their reading frames (Wu and Janssen, 1996). To clarify the relationship existing between translation initiation mechanisms on leadered versus leaderless mRNAs and to define better the role of the 5′ initiation triplet as a recognition element for the latter, the translational efficiencies of different types of leadered and leaderless *lacZ* mRNAs were investigated. Upon deletion of the leader from the SD sequence-containing leadered *lacZ* mRNA, expression of the reporter gene in *E. coli* was reduced only twofold; on the other hand, changing the SD sequence (from AGGA to UUUU) within the leader resulted in a more drastic effect (i.e., 15-fold reduction) on the level of translation. Furthermore, replacement of the canonical AUG of the leaderless mRNA by other naturally occurring initiation codons, such as GUG, UUG, and CUG, abolished translation completely, indicating that, unlike leadered mRNAs, which can start with triplets other than AUG, mRNAs lacking the leader are either more dependent on perfect codon-anticodon complementarity or require an AUG initiation codon in a sequence-specific manner to form productive initiation complexes. A mutant initiator tRNA with compensating anticodon mutations improved expression of leadered mRNA but did not restore translation of leaderless mRNA carrying a UAG start codon; it was therefore concluded that a cognate AUG initiation triplet serves as a specific translational signal in these mRNAs while codon-anticodon complementarity per se is not sufficient to ensure the translation of leaderless mRNAs (Van Etten and Janssen, 1998).

The existence of translationally active mRNAs lacking the SD sequence or lacking the 5′ leader altogether has prompted several authors to postulate the existence of *trans*-acting factors and/or additional ribosome recognition elements within the TIRs. Since the latest compilations of these prospective translational enhancers (McCarthy and Gualerzi, 1990; McCarthy and Brimacombe, 1994), two new elements have been added to the list: two contiguous heptanucleotides collectively designated PL have been proposed to base pair with nucleotides 1434 to 1440 and 507 to 513 of 16S rRNA (Kaloyanova et al., 1997) and multimers of the CA dinucleotide, which were reported to stimulate translation of both leadered and leaderless mRNAs (Martin-Farmer and Janssen, 1999).

The search for specific proteins involved in the promotion of mRNA-30S interaction almost invariably ends up identifying r-protein S1, which, at least in *E. coli*, was found to be essential for in vivo translation of all or almost all mRNAs (Sorensen and Fricke, 1998) and which is normally found to be required in vitro, sometimes with the contribution of IF3 (Boni et al., 1991). For instance, an analysis of 30S initiation complex formation with mRNAs (mainly plant virus mRNAs) known to be efficiently and reliably translated in *E. coli* in spite of the absence of an SD sequence demonstrated that, at least for the mRNAs examined, S1 and IF3 are indispensable for translation initiation; S1 in particular seemed to play a key role as a recognition element by binding to sequences within the leaders of (−SD) mRNAs (Tzareva et al., 1994). Another translational *cis*-acting element which is often held responsible for the translation of leaderless mRNAs is the so-called "downstream box" (DB).

The DB: Does It Base Pair with rRNA?

The DB is among the potential *cis*-acting determinants most often credited with initiation site selection. The claims for the existence of such a role for this sequence element are accumulating in the literature at a rate which is perplexingly high in light of the fact that direct proof for its role as a ribosome recognition element is still lacking (Sprengart and Porter, 1997). The DB was originally described as one of many translational enhancers present in several highly expressed *E. coli* and bacteriophage mRNAs (for reviews, see McCarthy and Gualerzi, 1990, and Sprengart and Porter, 1997).

The relevant sequence element is located at various distances downstream from the initiation codon (hence the name) and displays partial complementarity to nucleotides 1469 to 1483 of 16S rRNA (the "anti-DB"). Indeed, it has been proposed that base pairing between DB and anti-DB sequences places the start codon of the mRNA in close contact with the decoding region of 16S rRNA, thereby mediating independent and efficient initiation of translation (Sprengart et al., 1996). The evidence in favor of a participation of the DB sequence in 30S-mRNA interaction is indirect, however, and stems primarily from genetic manipulations which cause an increased or decreased level of translation, depending on whether the DB is removed, added, or shifted in position or its complementarity to 16S rRNA is weakened or strengthened (e.g., Sprengart et al., 1996; Mitta et al., 1997; Etchegaray and Inouye, 1999).

Using three different leaderless transcripts (λ cI, phage P2 gene V, and Tn*1721 tetR* mRNA) and after mutagenic inactivation of the DB element of λ cI mRNA, however, Resch and coworkers (1996) reached the conclusion that the DB element does not influence translation. Furthermore, chemical probing carried out on translation initiation complexes made with λ cI mRNA showed no protection of the bases comprising the putative DB element of this mRNA. In another study, a series of 10 constructs were prepared in which the CAT gene was modified to remove the 5′-terminal nontranslated region and/or its two potential DB elements. The results obtained showed that even after the simultaneous removal of the 5′ leader sequence and of both DB elements of CAT mRNA, expression of CAT under the control of a strong constitutive promoter was sufficient to allow *E. coli* to survive in the presence of up to 20 to 30 μg of chloramphenicol/ml (Odjakova et al., 1998). In agreement with these data, it has been shown that ribosomes carrying an inversion in the anti-DB region of 16S rRNA translate with the same efficiency as wild-type-ribosome mRNAs with and without a DB sequence (O'Connor et al., 1999). These results are not entirely surprising in light of the fact that all existing 16S rRNA models predict that the anti-DB sequence is stably base paired within helix 44 of the rRNA and site-directed mutagenesis has shown that disruption of this base pairing severely affects subunit association (Firpo and Dahlberg, 1998); it seems, therefore, that the anti-DB sequence is not amenable to forming base pairs (at least not WC) with the template. Furthermore, contrary to some claims (Etchegaray and Inouye, 1999), helix 44 is far from the mRNA decoding region in all current models of the 30S subunit (e.g., Mueller and Brimacombe, 1997). Indeed, this helix runs down to the bottom of the interface side of the subunit while the downstream end of the mRNA passes through the hole leading out to the solvent side so that the whole shoulder of the 30S body is lying between the DB element of the mRNA and the anti-DB sequence of helix 44. The topographically difficult task of base pairing DB with anti-DB fully justifies the claim that only 30S subunits lacking protein S2 may have a conformation suitable for this interaction (Shean and Gottesman, 1992).

Although the DB element was originally proposed to act synergistically with the SD sequence to ensure efficient translation of some mRNAs or, when alone, to compensate for the lack of an SD sequence under otherwise normal conditions of translation, more elaborate roles for the DB element have been proposed at both high (Nagai et al., 1991; Morita et al., 1999) and low (Mitta et al., 1997; Etchegaray and Inouye, 1999) extremes of temperature. It has been reported, for instance, that the 14-nucleotide DB element located 12 bases downstream of the initiation codon of the *cspA* mRNA is essential for mRNA translation during cold shock. It was further proposed that in the absence of "cold shock ribosomal factors" it is the presence of the DB sequence which allows cold shock mRNAs to form initiation complexes at low temperature while the non-cold shock mRNAs remain translationally blocked for lack of this sequence (Mitta et al., 1997). More of the same type of evidence purporting a major role of the DB as a translation initiation enhancer in concert with the SD sequence has recently been presented (Etchegaray and Inouye, 1999).

Since cold shock indeed induces alterations of the translational apparatus which favor the translation of cold shock mRNAs (Brandi et al., 1996; Goldenberg et al., 1997) and induces the expression of a fairly large number of RNA binding proteins, including a DEAD box RNA helicase, the possibility that cold-shocked ribosomes may expose the anti-DB sequence, thereby allowing base pairing with the DB element present in cold shock mRNAs, seemed to be worth testing. The same 30S DB mutants used by O'Connor et al. (1999), prepared from both control and cold-shocked cells, were used for these studies. Our results (La Teana et al., unpublished) demonstrated that, under all conditions tested, the anti-DB sequence is not accessible to a complementary oligonucleotide while the anti-SD sequence, which served as a positive control, was found to be fully accessible in all kinds of 30S ribosomal subunits under all conditions. Also, the accessibility to chemical probing of the bases belonging to helix 44 proved to be unaffected by the cold shock. Finally, the 30S DB mutants were found to translate cold shock and non-cold shock mRNAs with the same efficiency as wild-type ribosomes, and cells harboring only the DB-mutated type of 30S subunits (Asai et al., 1999) were found to display almost the same capacity to survive cold shock as the control cells. Taken together, these data support neither the claim that the DB element base pairs with 16S rRNA nor the notion that it is essential for cold adaptation and for the selective translation of cold shock mRNAs.

The Ribosomal mRNA Channel and Initiation Factor-Dependent mRNA Shift

The path of the mRNA through the 30S ribosomal subunit has been defined with remarkable accuracy by chemical probing and "traditional" and site-directed cross-linking (for a review, see Gualerzi and Pon, 1996). The last approach makes use of mRNA analogues carrying 4-thiouridine (4-thioU) residues at selected positions, which, upon UV irradiation, form specific and identifiable cross-links with rRNA and r-proteins within the mRNA-ribosome complexes, thereby allowing the identification of the ribosomal components in close contact with the mRNA, both 5′ and 3′ to the decoding site.

Contrary to an initial (and sometimes still held) belief, initiation factors (IFs) do not grossly affect either the SD–anti-SD interaction or the association constant (K_a) of the 30S subunit-mRNA complex (Gualerzi et al., 1990); they may, however, influence the position of the mRNA on the ribosome. Indeed, it has been suggested that in the presence of IFs the mRNA, which preferentially occupies a standby site corresponding to the region where the SD interaction takes place, is shifted towards another site, presumably closer to the decoding region (Canonaco et al., 1989). Preliminary data in support of this hypothesis included the findings that both diepoxybutane and UV yielded different distributions of mRNA–r-protein cross-links in the presence and absence of IFs and initiator tRNA (Brandt and Gualerzi, 1991). The reported presence of two independent high-affinity sites, one for each type of SELEX-generated ligand, which probably coexist within the same ribosomal particle (Ringquist et al., 1995), is also in good agreement with the presence of a dual mRNA binding site on the ribosome (standby and decoding).

Direct evidence for the occurrence of a partial relocation of the mRNA on the 30S ribosomal subunits under the influence of the factors and a better topographical characterization of the sites occupied by mRNA during different stages of initiation has been obtained by site-directed cross-linking experiments (La Teana et al., 1995). In these experiments the sites of the mRNAs covalently linked to specific positions of 16S rRNA and to individual r-proteins were determined and the extent of their cross-linking was analyzed semiquantitatively, demonstrating not only that the position of the mRNA on the ribosome is readjusted during 30S initiation complex formation under the influence of the IFs (IF3 in particular) but also that this shift, in agreement with that previously suggested (Canonaco et al., 1989), is from an initial standby site to a second site closer to that occupied when the initiation triplet of the mRNA is decoded in the P site. Furthermore, since the overall cross-linking pattern obtained in the presence of IFs is intermediate between those obtained with the 30S-mRNA complex and with the 30S initiation complex, the factors seem to be responsible, to a large extent, for the adjustment of the mRNA which occurs in preparation for the decoding of the initiation triplet in the ribosomal P site by the initiator tRNA. These results are summarized in Fig. 1.

IF2–fMet-tRNA and fMet-tRNA–Ribosome Interactions

The recognition and binding of fMet-tRNA by IF2 play crucial roles in the translation initiation pathway of bacteria. Furthermore, because of its functional importance and kingdom specificity (eukaryotes and archaea use a nonformylated form of initiator Met-tRNA), this interaction represents a potentially ideal target for the development of new antibiotics.

The molecular basis for the functional specificity of initiator $tRNA_f^{Met}$ has been extensively studied, both in vitro and in vivo. The features dictating or allowing interactions and properties characteristic of the initiator tRNA molecule which are not found in elongator tRNAs have been identified mainly by the introduction of mutations altering the specificities of

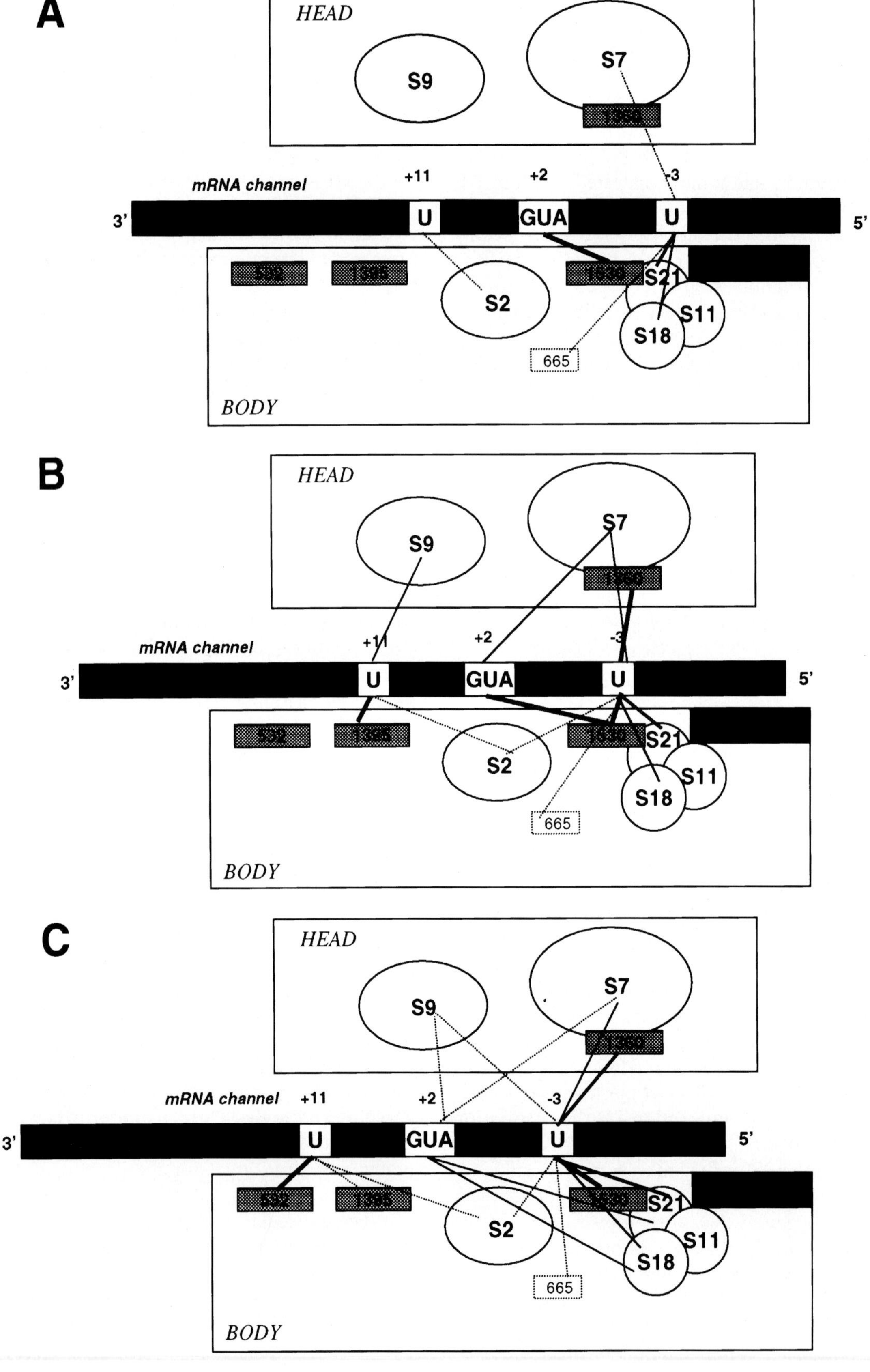
A
HEAD
S9
S7
mRNA channel
+11
+2
-3
3'
U
GUA
U
5'
S2
S21
S11
S18
665
BODY
B
HEAD
S9
S7
mRNA channel
+11
+2
-3
3'
U
GUA
U
5'
S2
S21
S11
S18
665
BODY
C
HEAD
S9
S7
mRNA channel
+11
+2
-3
3'
U
GUA
U
5'
S2
S21
S11
S18
665
BODY

these interactions in both elongator and initiator tRNAs.

The evidence for the importance of formylation in defining the initiator identity of $tRNA_f^{Met}$ is compelling. Indeed, in bacteria, chloroplasts, and mitochondria, the αNH_2 group of Met-$tRNA_f^{Met}$ is specifically modified by the addition of a formyl group through a reaction catalyzed by 10-formyltetrahydrofolate:L-methionyl-tRNA(fMet) *N*-formyltransferase, an enzyme encoded by *fmt*. The depletion of formylation activity which accompanies disruption of this gene in *E. coli* results in a severe growth rate reduction (from 2.3 to 0.28 doublings/h) at 37°C and in complete growth arrest at 42°C (Guillon et al., 1992a).

Neither the formyl group nor the amino acid per se represents an essential recognition element, however. In fact, *N*-acetyl–Phe–tRNA has proven in the past to be an excellent analogue of fMet-tRNA in countless studies, while more recently it has been possible to obtain translation initiation in vivo with fGlu-tRNA, fLys-tRNA, etc., following appropriate genetic manipulations (for a review, see RajBhandary and Chow, 1995). Nonetheless, the chemical nature of the amino acid does not seem to be entirely irrelevant for either the transformylase or IF2. In fact, while the efficiency of formylation decreases in the order Met>Gln>Phe>Val>Lys (Li et al., 1996), IF2 binds fMet-tRNA with higher affinity than fVal-tRNA and fGln-tRNA (Wu and RajBhandary, 1997). These preferences can explain the increased expression in vivo (with respect to the wild-type situation) of a reporter gene (CAT) carrying mutated initiation codons in cells overexpressing IF2 and some (U35A36 and G34) but not other (G34C36) mutated forms of initiator-like tRNAs with matching anticodons (Wu et al., 1996). Since the "native" 30S subunits are saturated with IF2 under normal conditions, it can be envisaged that excess (hyperproduced) IF2 would remain ribosome free and would preferentially subtract from equilibrium the tRNA for which it displays the highest affinity (i.e., fMet-tRNA). The resulting imbalance in the relative levels of initiator and mutated initiator-like tRNAs would in turn favor the formation and subsequent selection (by the 50S subunit) of 30S initiation complexes containing the mutated CAT mRNAs uncompeted by other mRNAs having the same initiation triplet. The wild-type CAT mRNA, on the other hand, besides suffering the competition of all other wild-type mRNAs, would also have a high probability of forming, with the excess mutated tRNAs, noncanonical complexes which would be rejected by IF3 (Gualerzi and Pon, 1990; LaTeana et al., 1993, and references therein).

Since the blockage of the αNH_2-group of Met-$tRNA_f^{Met}$ by the transformylase represents such an important recognition element, the structural features of the initiator tRNA molecule (namely, the 5′ C1, which does not base pair with A72, and base pairs G2-C71, C3-G70, and G4-C69) in the acceptor arm) allowing its specific recognition by the enzyme must be regarded as important determinants of initiation specificity (Guillon et al., 1992b). The crystal structure of an *E. coli* transformylase–fMet-tRNA complex has recently been solved at 2.8-Å resolution. The three-dimensional (3-D) structure shows that the enzyme is away from the anticodon stem and loop and approaches the tRNA from the D-stem side, filling in the inside of the "L" with a loop being wedged into the major groove of the acceptor helix. As a result, the mismatched C1 and A72 bases are separated, with the 3′ arm bent inside the active center (Schmitt et al., 1998). This recognition mechanism is markedly distinct from that of IF2, which binds the acceptor end of fMet-tRNA approaching the molecule from the T-stem side (see below).

Figure 1. Scheme illustrating the cross-linking patterns of 4N/−3 mRNA (an mRNA with a 4-nucleotide spacer and a U residue at position −3) in its binary complex with the 30S ribosomal subunit in the presence of IF1, IF2, and IF3 and in the presence of IFs and fMet-tRNA. Compared to the simplest case of the 30S-mRNA complex (A), in which the mRNA is cross-linked almost exclusively to 1530 of 16S rRNA through its +2 position and to r-proteins S18 and S21 through its −3 position, the cross-linking pattern becomes much more complex in the presence of IFs (B): new, strong cross-links appear between S7 and both +2 and −3, and between S9 and +11, 1395 and +11, and 1360 and −3, while cross-linking of 1530 is partially shifted from +2 to −3. All these changes suggest a "leftward" shift of the mRNA with respect to the 30S subunit. A further rearrangement of the mRNA on the 30S subunit took place in the complete 30S initiation complex (C). Position +11 of the mRNA was no longer cross-linked to S9 and only marginally to 1395 but moved further leftward to cross-link to position 532 of 16S rRNA. Also, position +2 moved away from S7 and 1530, since cross-linking to these two elements was reduced, while +2 and −3 approached S9, with which a weak yet significant cross-linking was established. It should be noted that, in the course of the mRNA shift, the central base of the initiation triplet moves away from 1530 towards 1395, i.e., towards the P site of the subunit. The approximate relative positions of the relevant ribosomal components in the head or body of the 30S ribosomal subunit are shown. In the mRNA (solid bar), the initiation triplet and the positions of the potential cross-linking sites are indicated. Thick lines, major RNA-RNA cross-links; thin lines, major RNA-protein cross-links; dotted lines, quantitatively minor cross-links of either type. The enclosure of 665 within a dotted box indicates the minor nature of this cross-link. (Taken from LaTeana et al., 1995.)

The presence of three consecutive G-C base pairs in the anticodon stems of all initiator tRNAs (including those from eukaryotes and most archaea) is an additional feature distinguishing initiator from elongator tRNAs (Sprinzl et al., 1996). This characteristic G-C stack, which enables initiator tRNA to interact with the ribosomal P site (RajBhandary and Chow, 1995), is believed to impose a rigid and restrained conformation on the 5′ CAU anticodon loop of initiator tRNA, conferring resistance to digestion with the single-strand-specific endonuclease S1. This resistance seems to correlate with initiator function in most (Seong and RajBhandary, 1987) but not all (Mandal et al., 1996) cases. Chemical probing and nuclear magnetic resonance (NMR) spectroscopy data indicate, however, that the anticodon of initiator tRNA is structurally indistinguishable from that of elongator tRNAs (Schweisguth and Moore, 1997) in also having a "U-turn." The U-turns represent an important class of structural motifs in the RNA world, in which a uridine is involved in an abrupt change in the direction of the polynucleotide backbone. In the crystal structure of yeast $tRNA^{Phe}$, the invariant uridine at position 33 (U33), adjacent to the anticodon, stabilizes the exemplar U-turn with three non-WC interactions: hydrogen bonding of the 2′-OH to N-7 of A35 and of the N3-H to A36-phosphate, and stacking between C32 and A35-phosphate.

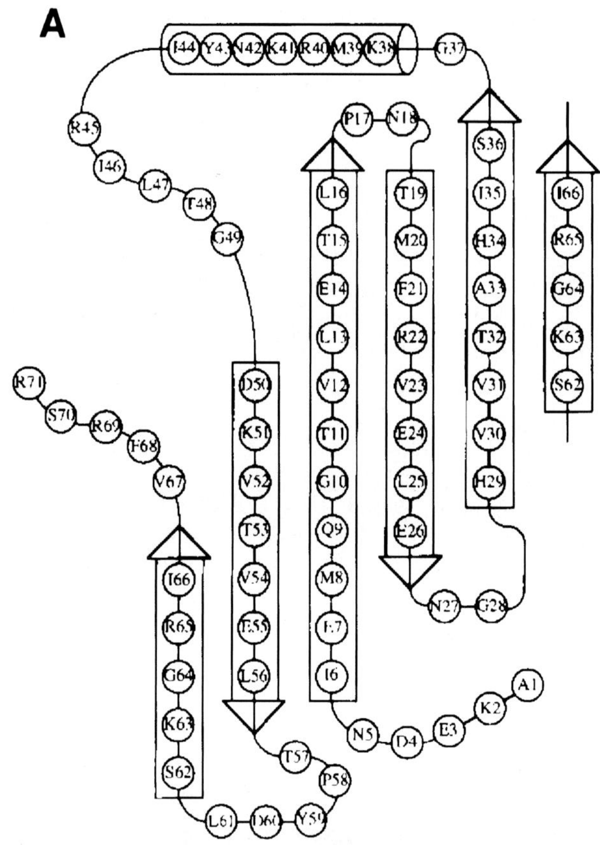

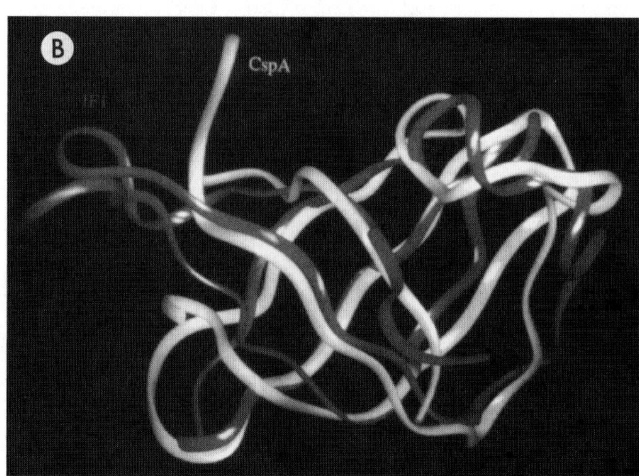

Figure 2. (A) Secondary structure of *E. coli* IF1 (taken from Sette et al., 1997). (B) Comparison of the 3-D structures of *E. coli* IF1 (gray) and cold shock protein CspA (white).

STRUCTURE, FUNCTION, AND RIBOSOMAL LOCALIZATION OF IFs

Structure, Function, and A-Site Location of IF1

IF1 is a small RNA binding protein whose main function is modulating the affinity of IF2 for the ribosome and stimulating the activities of both IF2 and IF3. The finding that IF1 binds to the 30S subunit in a location corresponding to the A site suggests that this factor may also participate in conferring specificity on the formation of the 30S initiation complex by occluding the access of a tRNA to the A site until a 70S initiation complex is formed. Ejection of IF1 from the ribosome at this stage would serve the dual purpose of opening the A site for an incoming aminoacyl-tRNA and weakening the interaction of IF2 with the ribosome in preparation for the ejection of the latter factor.

The 3-D structure of IF1 in solution has been elucidated by heteronuclear NMR spectroscopy (Sette et al., 1997). IF1 was found to contain five β-strands (Fig. 2A) organized in a β-barrel structure (Fig. 2B) with a loop connecting strands 3 and 4 containing a short 3_{10} (three amino acids and 10 hydrogen bonds per turn) helix endowed with high flexibility. The structure of IF1 is very similar to that of the nucleic acid binding proteins of the oligomer binding fold family and, in particular, almost superimposable on that of the *E. coli* CspA and *B. subtilis* CspB cold shock proteins (Fig. 2B). The amino acid residues of IF1 involved in or affected by the binding to the 30S ribosomal subunits have been identified by site-directed mutagenesis and NMR spectroscopy. These residues are highlighted in Fig. 3.

IF2 Properties, Topography, and Effects on Ribosomes

IF2 is an essential protein (Cole et al., 1987) whose main recognized function is the stimulation of fMet-tRNA binding to the ribosomal P site. The affinity of IF2 for the 30S subunit is approximately 2 orders of magnitude greater than for fMet-tRNA and for other aminoacyl (aa)-tRNAs bearing a blocked αNH_2 group (see above). These facts and some experimental evidence previously reviewed (Hershey, 1987; Gualerzi and Pon, 1990) indicate that, unlike EF-Tu, which functions as an aa-tRNA carrier, IF2 is presumably already 30S bound when it interacts with fMet-tRNA and increases the rate of its codon-anticodon interaction at the P site. IF2 also binds GTP and GDP, for which, unlike EF-Tu, it displays similar affinities. The presence and the nature of the guanosine nucleotide ligands do not affect the interaction of IF2 with fMet-tRNA.

Several 16S rRNA fragments and regions cross-linked by *trans*-diamminedichloro platinum(II) to IF2 or having a reduced or increased reactivity by phosphate alkylation with ethylnitrosourea in the presence of the factor have been identified. Most of these fragments are located in the head and the lateral protrusion of the 30S subunit. Overall, these experiments indicate that IF2 does not tightly bind to the 16S rRNA and induces changes in reactivity of the RNA in several positions, most of which are localized within the cleft, the lateral protrusion, and the part of the head facing the protrusion (positions 694, 771, 791, 1225, 1268, 1398, 1401, 1504, and 1527). Other positions (301, 302, 492, and 1428) are spread throughout the particle (Wakao et al., 1991). All the reactivity changes induced by IF2 are still observed in the presence of initiator tRNA and AUG, while the few additional protections induced by the initiator tRNA are mostly centered around the cleft-head-lateral protrusion region, near positions affected by IF2 binding (positions 791, 1222, 1263, 1393, 1395, 1430, 1431, 1504, 1528, and 1529), and enhanced reactivities are observed at positions 708, 709, and

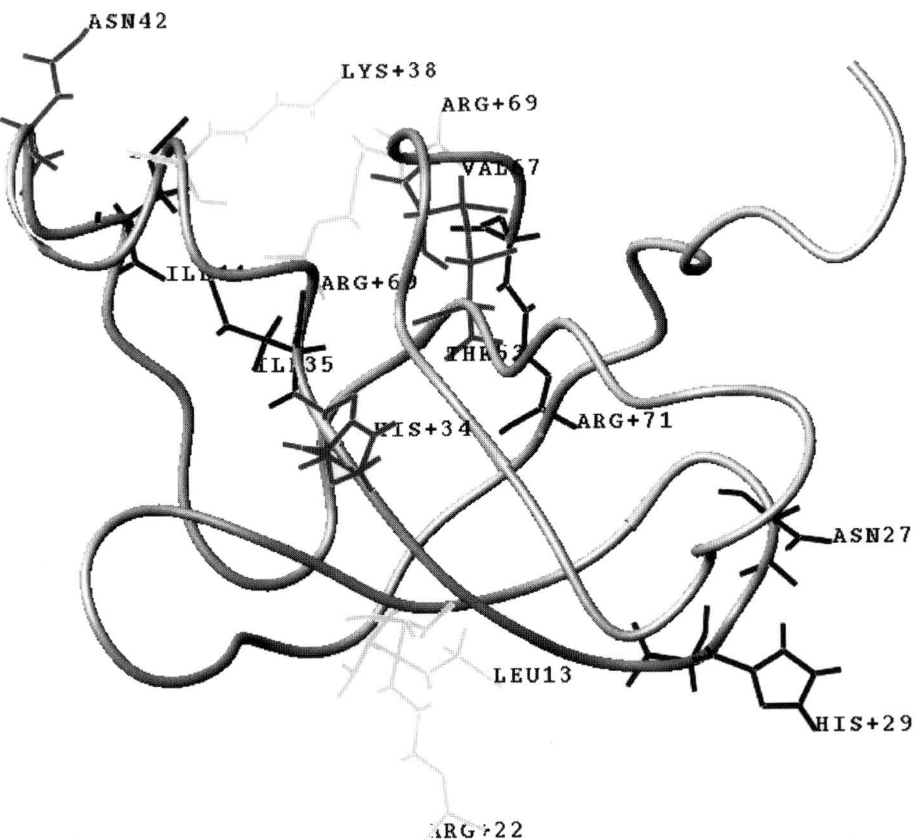

Figure 3. IF1 residues involved in 30S binding. Drawing of IF1 indicating the residues whose signals are broadened (dark gray) or shifted (black) upon binding to the 30S subunit. Residues whose mutagenesis affects ribosome binding are shown in light gray. (From Sette et al., 1997.)

1398 (Romby et al., 1990; Wakao et al., 1990). Also, the interaction between IF2 and initiator fMet-tRNA has been investigated by similar methods. IF2 interacts with the T loop and the minor groove of the T stem of the RNA and induces an increased flexibility in the anticodon arm. In the 30S initiation complex, additional protection is observed in the acceptor stem and in the anticodon arm of the tRNA (Romby et al., 1990).

IF2 and Ribosome-Dependent GTPase

The joining of the 50S ribosomal subunit with the 30S initiation complex gives rise to a 70S initiation complex. At the end of this process, during which the intrinsic (Severini et al., 1991) GTPase of IF2 is activated, fMet-tRNA is found in the P site, amenable to formation of the first peptide bond with the aminoacyl moiety of the aa-tRNA brought to the A site by EF-Tu in response to the second mRNA codon, while IF2 is released from the ribosome (for reviews, see Gualerzi and Pon, 1990, and Spurio et al., 1993). Our understanding of the translation initiation pathway cannot be considered satisfactory without full elucidation of the role played in this process by GTP and by the IF2-dependent GTPase. This role remains very elusive, however.

Unlike the case of the EF-Tu–aa-tRNA interaction, GTP does not influence the interaction of IF2 with fMet-tRNA and the formation of the 30S initiation complex is essentially unaffected by GTP; stimulation by GTP and inhibition by GDP of the latter process have been observed only at extremely low concentrations of ribosomes and IF2 because GTP lowers and GDP increases the K_d of the 30S-IF2 complex (Pon et al., 1985). GTP is also not required for the stability of the 30S initiation complex or for its association with the 50S subunit.

To explain the role of the IF2 GTPase in translation initiation, two main hypotheses have been formulated: (i) release of IF2 from the 70S initiation complex requires or, at least, is accelerated by GTP hydrolysis (Hershey, 1987) and (ii) the energy produced by GTP hydrolysis is required for the adjustment of the initiator tRNA in the P site (La Teana et al., 1996, and references therein). In conflict with these hypotheses, however, the affinity of IF2 for the 70S ribosome is essentially the same in the presence of GTP and GDP or in the absence of any nucleotide (Pon et al., 1985) and a 70S initiation complex artificially depleted of the nucleotide can still form fMet-puromycin (Dubnoff et al., 1972). The latter finding has been confirmed by more recent data, which have indicated that neither GTP nor the mRNA is strictly necessary for the adjustment of the initiator tRNA in the P site; omission of GTP caused only a slight reduction of the rate of fMet-puromycin formation without a significant change in the activation energy, while omission of the template resulted in a requirement for a higher activation energy. Furthermore, not only the puromycin reaction but also the formation of the first peptide bond with an incoming EF-Tu·GTP·aa-tRNA ternary complex was found to occur in the absence of GTP (La Teana et al., 1996). This finding and similar results obtained by fast kinetics analysis (Tomšic et al., submitted) indicate either that GTP is not involved in the release of IF2 from the ribosome or that an inefficient release of IF2 from the ribosome does not impair the binding of the EF-Tu ternary complex to the A site, as would be expected from the premise that IF2, EF-G, and EF-Tu bind to overlapping ribosomal sites (e.g., Luchin et al., 1999, and references therein).

On the other hand, it has been found that in the absence of both GTP and mRNA, essentially no fMet-puromycin was formed, indicating that these components cooperate in the correct alignment of initiator tRNA in the ribosomal P site. These results have been interpreted to mean that several elements (i.e., codon-anticodon and SD interactions, the nature of the initiation triplet, occupancy of the ribosomal mRNA channel, etc.) can contribute to the initial orientation of the fMet-tRNA on the ribosome and that the facility with which the initiator tRNA can be adjusted in the P site depends on how closely this orientation approaches that which is "best fit." The further away the initiator tRNA is from this position, the more indispensable becomes the presence of the other elements. Thus, it can be envisaged that when the mRNA imposes a specific orientation on ribosome- and IF2-bound initiator tRNA through the codon-anticodon interaction, the 3′ acceptor end of fMet-tRNA can be shifted in the P site easily and rapidly via a conformational change of IF2 induced by GTP hydrolysis or through a somewhat slower drifting in the absence of GTP. The omission of the template removes this constraint on the orientation of fMet-tRNA on the ribosome so that IF2 without its GTPase-induced conformational change is no longer sufficient to bring the initiator tRNA into the puromycin-reactive site. Furthermore, since in the absence of GTP an AUG triplet cannot replace a bona fide mRNA, it can be concluded that the occurrence of a codon-anticodon interaction alone is not sufficient to impose the correct orientation on the fMet-tRNA, which is required for its drift into the P site. It is likely that the initiation triplet by itself is too wobbly on the ribosome to be able to constrain the initiator tRNA in an orientation which is suitable for its "spontaneous" adjustment in the P site. Thus, codon-anticodon interaction goes from being optional to necessary but not sufficient, depending on the presence or absence of GTP. Overall, these data indicate that, even if not strictly required for the adjustment of the fMet-tRNA in the P site under normal conditions, the GTPase activity of IF2 plays at least an auxiliary role in this process (La Teana et al., 1996).

A different conclusion concerning the role of the GTPase activity has been reached in a recent study by Luchin et al. (1999). In this work, the authors analyzed the effect of two dominant-negative *E. coli* IF2 G-domain mutants (V400G and H448E). Compared to the wild type, the rate of GTP hydrolysis was found to be 11- and 40-fold lower with the first and the second mutant, respectively, while puromycin reaction was completely inhibited in the H448E mutant and still significant in the case of the V400G mutant. Finally, since both IF2 mutants were found to remain stuck on the ribosome after formation of the 70S initiation complex, the authors concluded that GTP hydrolysis is essential to recycle IF2 off the ribosome and its inactivation accordingly causes a dominant-negative effect.

Working with H301 mutants of *Bacillus stearothermophilus* IF2 (equivalent to *E. coli* H448 of IF2 and to H84 of EF-Tu), we reached different conclusions, however. In fact, in agreement with the data of Luchin et al. (1999), we found that some H301 mutants have a dominant-lethal phenotype. Although some H301 mutants have impaired GTPase activity, others do not, and there is no strict correlation between the lethal effect of the overproduction of some of the mutants and their residual GTPase activity. Furthermore, in agreement with Luchin et al. (1999), we found that all H301 mutants tested were active in fMet-tRNA binding but inactive in fMet-puromycin formation. The amount of IF2 (mutants and wild type) remaining ribosome associated did not correlate with the residual GTPase activities of these proteins. Our conclusion is, therefore, that the conformation of the IF2 molecule is very sensitive to amino acid substitutions around H301 and to ligands binding within the same region. Replacement of H301 with different amino acids, as well as interaction with different GTP analogues, inhibits to various extents GTPase and other IF2 activities for unrelated reasons. All these mutants place the initiator tRNA in a wrong orientation, which prevents fMet-tRNA from reaching the P site, and some of them give rise to dead-end complexes which inhibit translation in vivo as well as in vitro. This would explain the failure of these mutants to support fMet-puromycin formation in spite of the variable degree of their residual GTPase activity and regardless of the fact (see above) that GTP hydrolysis is not required for fMet-puromycin formation. We isolated several intragenic suppressors of the cell lethality caused by overproduction of the H301 mutants. The finding that their secondary mutations had severely impaired the fMet-tRNA binding capacity of IF2 is fully compatible with this interpretation, which, if correct, leaves still unanswered the question concerning the role of IF2 GTPase (Tomsic et al., unpublished).

Structure of the fMet-tRNA Binding Domain of IF2

Our understanding of the molecular basis of this interaction has improved substantially in recent years through the combined efforts of our laboratory, the laboratory of H. Welfle (Berlin, Germany), and the NMR spectroscopy laboratory of R. Boelens (Utrecht, The Netherlands). Previous data had indicated that IF2 interacts with initiator fMet-tRNA via its C-terminal (24.5-kDa) domain. Spectroscopic data and protein-unfolding experiments have subsequently demonstrated that the IF2 C domain (IF2-C) consists of two subdomains having different stabilities and different hydropathicity properties (Misselwitz et al., 1997). Introduction of fluorescent tags by Phe-to-Trp replacements at four specific sites of IF2-C demonstrated that the more stable of the two subdomains corresponds to the N-terminal portion (IF2 C-1) of the C domain, while the less stable one corresponds to the C-terminal part (IF2 C-2), where random and site-directed mutagenesis had identified all IF2 residues implicated in the interaction with fMet-tRNA (Spurio et al., in press). As seen in Fig. 4, the functionally essential residues are C668, K699, R700, Y701, K702, E713, C714, and G715. Genetic manipulation of *infB* allowed us to hyperproduce the corresponding IF2 C-2 subdomain and to demonstrate that this isolated molecule (consisting of 110 amino acids) can bind fMet-tRNA with the same affinity and selectivity as intact IF2 and IF2-C. Additional experiments have also demonstrated that IF2 C-2 can be further trimmed from both the N terminus and the C terminus to yield a 90-amino-acid-long molecule before its fMet-tRNA binding activity is lost (Spurio et al., in press).The 3-D structure in solution of the fMet-tRNA binding domain of IF2 (IF2 C-2) has been elucidated by heteronuclear NMR spectroscopy (Meunier et al., submitted). IF2 C-2 was found to consist of six β-strands forming a closed β-barrel and to be structurally related, in spite of the very low degree of sequence similarity, to domain II of EF-Tu and EF-G (Ævarsson et al., 1994; Nissen et al., 1995). Domain II of EF-Tu constitutes part of the aa-tRNA binding region of this elongation factor, but unlike IF2 C-2, it is unable to autonomously bind aa-

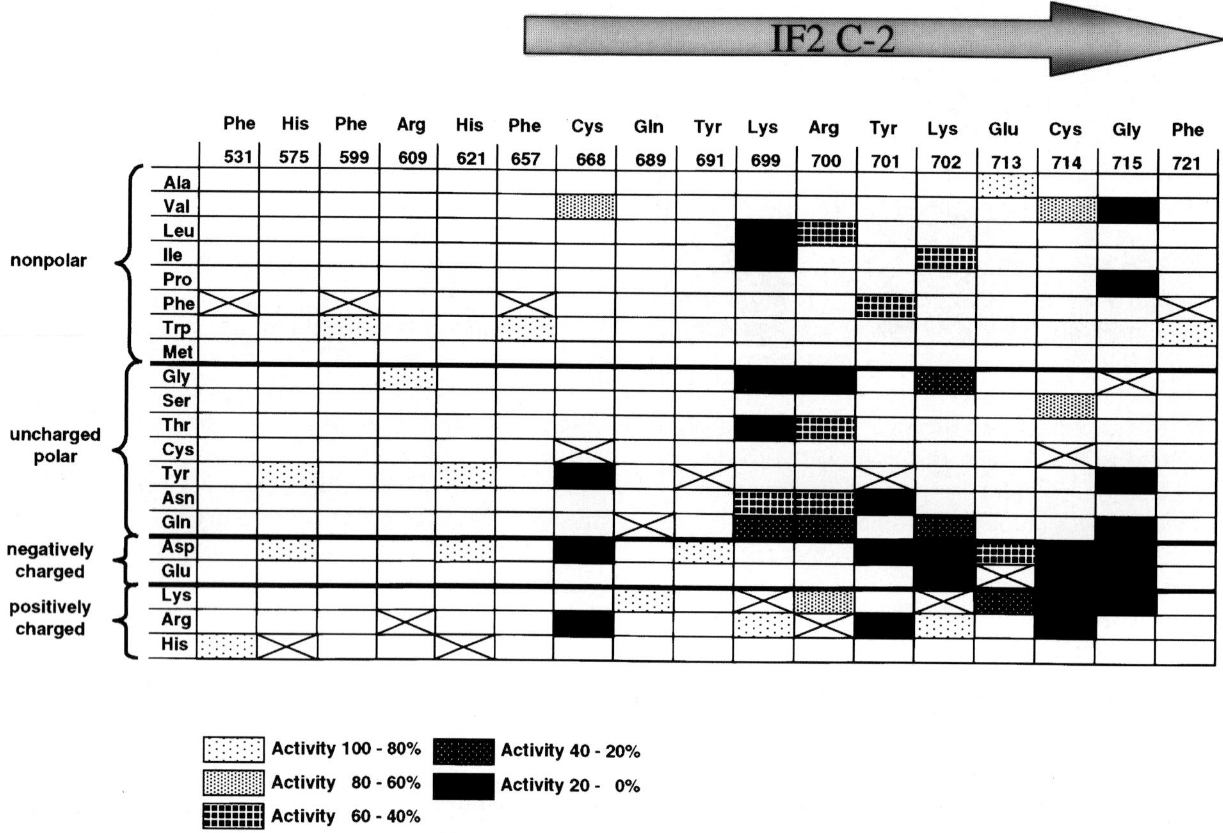

Figure 4. Summary of the results obtained following site-directed mutagenesis of *B. stearothermophilus* IF2. The residual activity in fMet-tRNA binding displayed by each mutant is shown according to the indicated code.

tRNAs, since the participation of all three domains of this factor is required for ternary complex (EF-Tu·GTP·aa-tRNA) formation (Nissen et al., 1995). Domain II of EF-G, on the other hand, has never been reported to interact with aa-tRNAs. Titration of IF2 C-2 with different (potential) ligands was carried out, and the heteronuclear single quantum correlation (HSQC) spectra were recorded to monitor the chemical shift variations of the amides of its various amino acid residues (Guennegues et al., unpublished). Essentially the same amino acids identified by site-directed and random mutagenesis as being involved in fMet-tRNA binding, which are highlighted in the 3-D structure of IF2 C-2 determined by Meunier et al. (submitted) (Fig. 5), were found by NMR spectroscopy to be implicated upon addition of formylmethionine and fMet-$tRNA_f^{Met}$. These amino acids define a continuous patch on the IF2 C-2 surface involved in the interaction with initiator tRNA.

Functional Properties of IF3

In *E. coli*, IF3 consists of 180 amino acids encoded by *infC*, a gene mapped at 37.5 min (Sacerdot et al., 1982) which has been proven to be essential for cell survival (Olsson et al., 1996). Acting as a subunit antiassociation factor, IF3 supplies the pool of free 30S subunits required for translation initiation (Hershey, 1987, and references therein). Furthermore, IF3 ensures efficiency and fidelity in the formation of 30S initiation complexes accelerating this process while favoring the dissociation of the noncanonical complexes (i.e., complexes containing aa-tRNAs other than initiator fMet-tRNA and/or a triplet other than the initiation triplets AUG, GUG, and UUG) (for a review, see Gualerzi and Pon, 1990). This function, which has recently also been demonstrated in vivo following the isolation of IF3 mutants impaired in this activity (Haggerty and Lovett, 1993; Sacerdot et al., 1996; Sussman et al., 1996), can be accounted for by the influence of IF3 on both the on and off rates of codon-anticodon interaction in the P site (La Teana et al., 1993) and, possibly, by the above-mentioned IF3-induced adjustment of the mRNA, which is shifted from the standby site to the P-decoding site on the 30S ribosomal subunit (La Teana et al., 1995).

IF3 is an RNA binding protein, and its binding to the 30S ribosomal subunit was one of the first

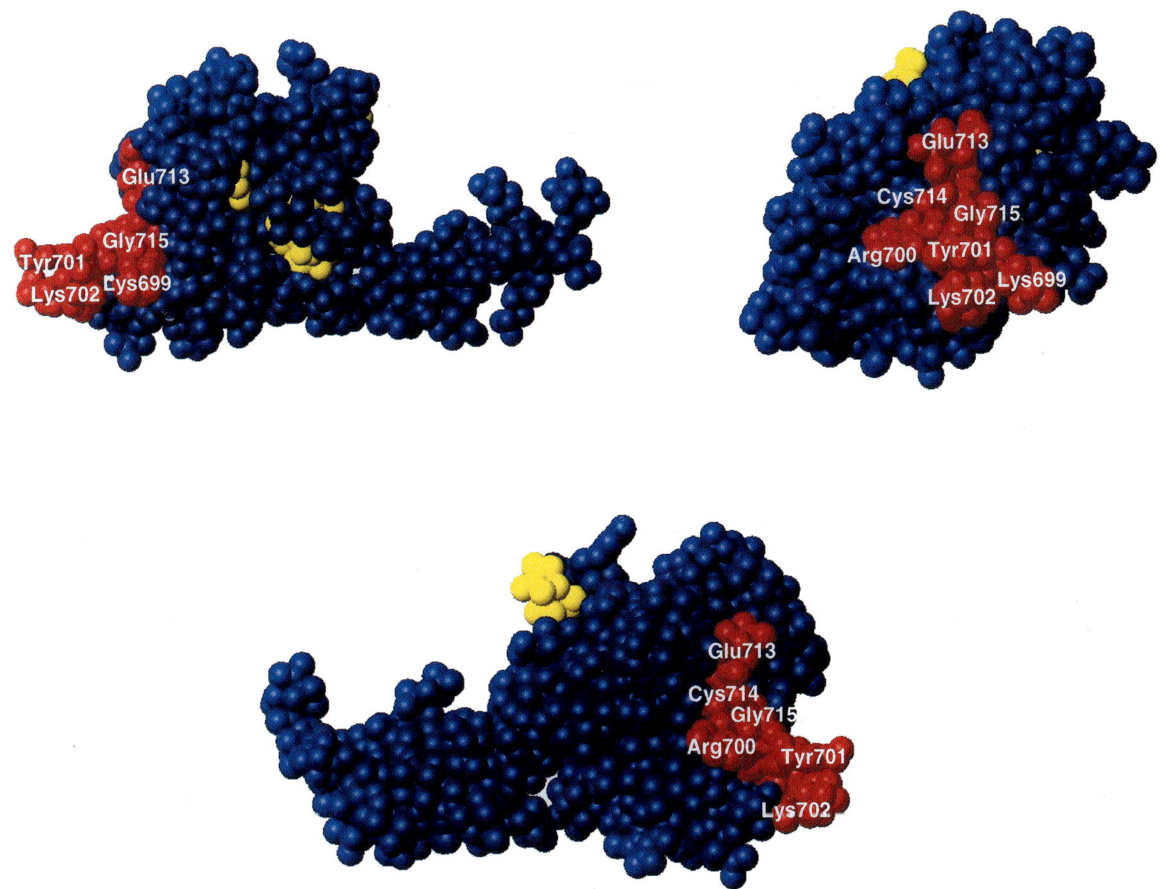

Figure 5. fMet-tRNA binding site of IF2. The locations of the amino acid residues essential (red) and nonessential (yellow) for the interaction of IF2 with fMet-tRNA identified by site-directed mutagenesis are shown within the 3-D structure of IF2 C-2 (blue) determined by multidimensional heteronuclear NMR spectroscopy (Meunier et al., submitted).

functions attributed to the16S rRNA moiety of the ribosome (for a review, see Santer and Dahlberg, 1996). The topographical localization of IF3 on the 30S subunit, determined by immuno-electron microscopy, protein-protein and protein-RNA cross-linking, and the structural and functional properties of the factor, led to the suggestion that IF3 contains two separate binding sites, each interacting with a separate site (on the head and on the platform) of the small ribosomal subunit, thereby bridging the cleft; IF3 was suggested to fulfill its functional role by establishing a "fluctuating" interaction with these two sites, thereby affecting the conformational dynamics of the ribosome (Pon et al., 1982).

IF3 Structure

The two-domain nature of IF3 has recently been confirmed by structural investigations. The 3-D structures of the separate domains of *B. stearothermophilus* IF3 have been elucidated by X-ray crystallography (Biou et al., 1995), and those of *E. coli* IF3 have been elucidated by NMR spectroscopy (Fortier et al., 1994; Garcia et al., 1995a and 1995b). In intact IF3 these two domains are separated by a fairly long (45-Å) and flexible linker (Moreau et al., 1997). Chemical modifications and site-directed mutagenesis demonstrated that the main 30S binding region of IF3 is localized in the C domain (see Sette et al., 1999, and references therein). This conclusion has been confirmed and extended by the results of NMR spectroscopy experiments in which isotopically labeled IF3 was titrated with 30S ribosomal subunits and intensity changes of the cross peaks belonging to the amides of individual amino acids were observed in the 2-D ^{15}N-^{1}H (HSQC) spectra. This approach indicated that the two domains of IF3 are functionally independent, each interacting with different affinity with the ribosomal subunit (Fig. 6). The primary region of IF3 interacting with the 30S subunit was confirmed to be the C domain, which binds primarily through some of its loops and α-helices,

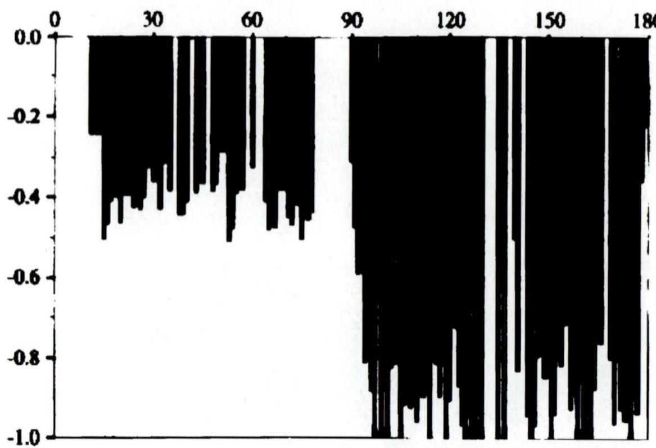

Figure 6. Differential intensity change of the backbone amide resonances of *E. coli* IF3 upon addition of 30S ribosomal subunits. The position of each amino acid in the primary sequence of IF3 is indicated in the abscissa; the ordinate presents the relative intensity change, $(V - V_0)/V_0$, of each assigned cross peak caused by the addition of *E. coli* 30S ribosomal subunits at a stoichiometric ratio, 30S/IF3, of 0.2%. (From Sette et al., 1999.)

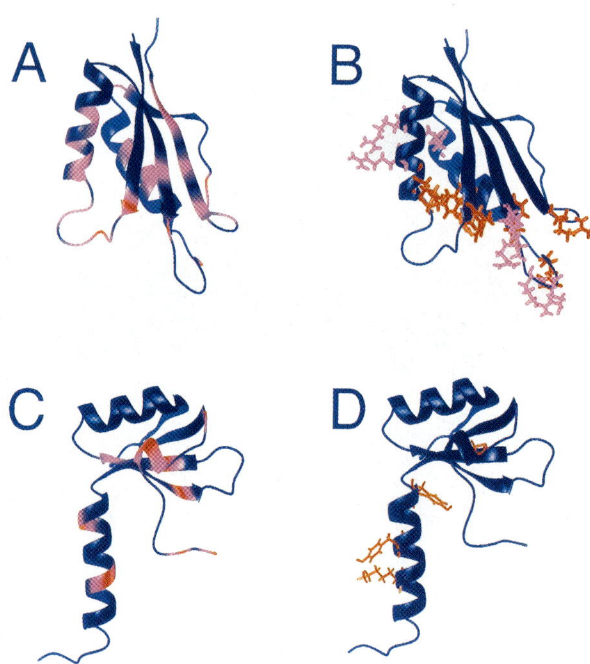

Figure 7. Active sites of *E. coli* IF3. The amino acid residues of the C domain (A and B) and N domain (C and D) of IF3 implicated in the interaction with the 30S ribosomal subunit as identified by NMR (A and C) and by site-directed mutagenesis and chemical modifications (B and D) are indicated with different colors (from blue to red according to relative NMR intensity changes, where red indicates the strongest effects) in the ribbon diagrams of the two domains. The coordinates of the structures were obtained from the Protein Data Bank (PDB). For the C domain, the NMR structure of the *E. coli* protein was used (PDB entry, 1IFE [Garcia et al., 1995a]), and for the N domain, the X-ray structure of *B. stearothermophilus* protein was used (PDB entry, 1TIG [Biou et al., 1995]). (From Sette et al., 1999.)

while the N domain interacts mainly via a small number of residues asymmetrically distributed in the domain. The amino acid residues of both C-terminal (Fig. 7A) and N-terminal (Fig. 7C) domains identified by this NMR approach as being involved in the interaction with the 30S subunit are shown in Fig. 7; the residues of the C domain (Fig. 7B) and N domain (Fig. 7D), which have been implicated by other chemical and genetic methods, are also shown. Comparison of the results indicates the excellent agreement between the two sets of data (see Sette et al., 1999, and references therein).

Docking IF3 to Its Binding Site on the 30S Ribosomal Subunit

It has been suggested (Hartz et al., 1990) that to fulfill its role of initiation fidelity factor, IF3 may "inspect" the 30S initiation complex establishing a direct contact with the anticodon stem and loop of the initiator tRNA, together with part of the initiation codon of the mRNA. This interpretation is in conflict with our previous conclusion, mainly derived from equilibrium perturbation and ligand exchange experiments, that the effect of IF3 is mediated by the 30S ribosomal subunit (Pon and Gualerzi, 1974). The discrepancy does not concern what is ultimately rejected by IF3 but rather how the incorrect complexes are recognized (i.e., through a direct or an indirect interaction of the factor with the ribosome-bound tRNA), and an accurate topographical localization of IF3 on the 30S subunits could turn out to be essential in resolving this problem.

It should be recalled, in summary, that IF3 discriminates against (i) wobbling codon-anticodon interactions between authentic initiator fMet-tRNA and some degenerate initiation codons (e.g., AUU and GUA) but not others (e.g., UUG and GUG), (ii) codon-anticodon interaction between canonical AUG triplet and noncanonical initiator tRNAs, (iii) cognate codon-anticodon interactions at the ribosomal P site involving non-initiator triplets and elongator aa-tRNAs (e.g., UUU and Phe-tRNA) or potential initiation triplets and elongator aa-tRNAs (e.g., GUG and Val-tRNA), and (iv) codon-anticodon interactions involving authentic initiator fMet-tRNA and canonical initiation triplet AUG when the latter is present at the 5′ ends of leaderless mRNAs (Tedin et al., 1999). These properties seem to be difficult to reconcile with a mechanism based on a direct contact and inspection of the initiator tRNA by IF3.

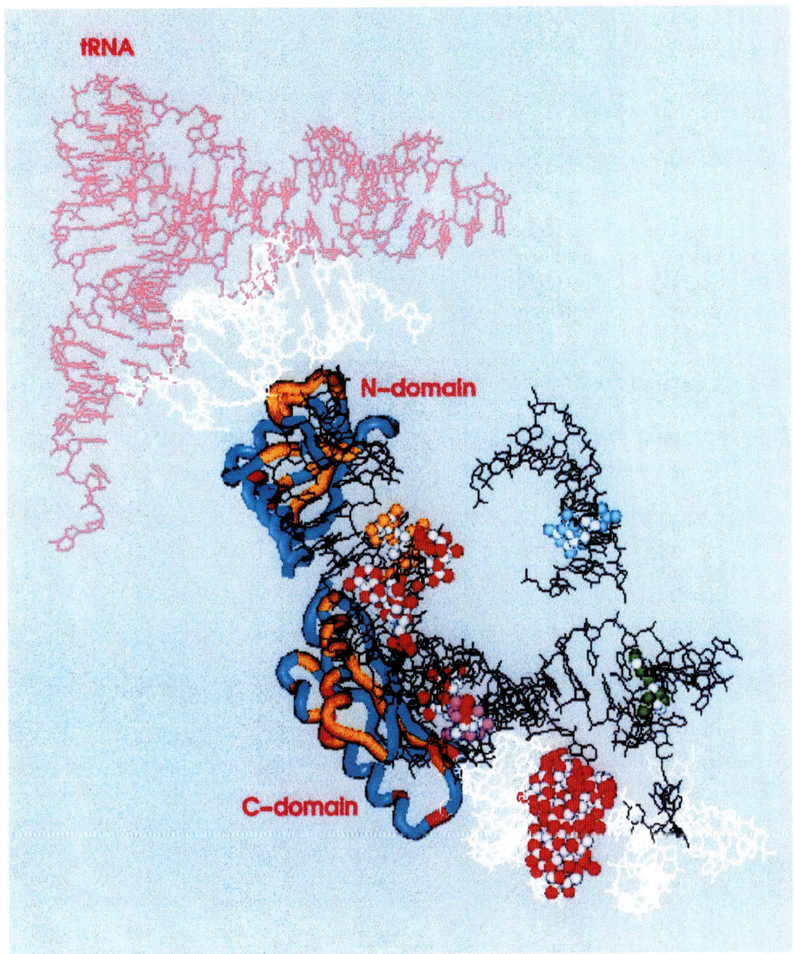

Figure 8. IF3 docking on the 30S ribosomal subunit. The portions of the 16S rRNA helices are shown in black, but those which have been cross-linked to IF3 are shown in white. These are helix 45 in the upper part and helices 25 and 26 in the lower part of the diagram. The other nucleotides indicated are those protected by IF3 from kethoxal (yellow) and CMCT {N-cyclo-hexyl-N′-[2-(N-methylmorpholinio)ethyl]carbodiimide-p-toluene-sulfonate} (orange); G791 (pink) is partially protected from kethoxal and is functionally implicated by mutagenesis in IF3 binding. Also indicated are the nucleotides hyperreactive to dimethyl sulfate (green) and kethoxal (turquoise) and hypersensitive to RNase V1 (red) in the presence of IF3. The figure is taken from Sette et al., 1999, where additional information can be found.

The location of IF3 bound to the 30S subunit of the *Thermus thermophilus* ribosome has been determined by cryo-electron microscopy at 27-Å resolution comparing the 30S·IF3 complex and a control 30S subunit. When the previously solved crystal structure of IF3 was fitted within the prominent lobes of positive density calculated from the difference map of the two electron microscopic reconstructions, IF3 was placed in a position more or less perpendicular to the long axis of the 30S subunit, in a position which was judged to be compatible with the model which implies a direct contact between IF3 and P-site-bound tRNA (McCutcheon et al., 1999).

A preliminary docking of IF3 on the 30S subunit which takes into account all available data concerning the topographical location of IF3, the nature of its active sites, and the spatial distribution of the various 16S rRNA helices suggested by a recent model of the 30S subunit (Mueller and Brimacombe, 1997) would place IF3 in a different orientation on the subunit. While a different orientation on the subunit is by itself irrelevant for the issue being discussed here, the fact that in this model the long IF3 molecule bridges two distant regions of the 16S rRNA and at least one rRNA helix is interposed between the N domain of IF3 and the anticodon stem-loop of the P-site-bound tRNA (Fig. 8) is very important. This would imply that there is no contact on the ribosome between IF3 and the tRNA and would automatically disprove the direct-inspection mechanism. Obviously, additional work is necessary to identify with certainty which part of IF3 interacts with which particular ribosomal component and to refine its docking on the 30S subunit in order to allow a definite choice between the direct and the indirect inspection mechanisms.

The financial support of grants from EC-Biotechnology, MURST-PRIN, CNR (95/95), and CNR (Strategico) to C.O.G and MURST-PRIN and CNR (PF Biotecnologia) to C.L.P. is gratefully acknowledged.

REFERENCES

Ævarsson, A., E. Brazhnikov, M. Garber, J. Zheltonosova, Y. Chirgadze, S. Al-Karadaghi, L. A. Svensson, and A. Liljas. 1994. Three-dimensional structure of the ribosomal translocase: elongation factor G from *Thermus thermophilus*. *EMBO J.* **13:**3669–3677.

André, A., A. Puca, F. Sansone, A. Brandi, G. Antico, and R. A. Calogero. Reinitiation of protein synthesis in *Escherichia coli* can be induced by mRNA cis-elements unrelated to canonical translation initiation signals. Submitted for publication.

Asai, T., D. Zaporojets, C. Squires, and C. L. Squires. 1999. An *Escherichia coli* strain with all chromosomal rRNA operons inactivated: complete exchange of rRNA genes between bacteria. *Proc. Natl. Acad. Sci. USA* **96:**1971–1976.

Biou, V., F. Shu, and V. Ramakrishnan. 1995. X-ray crystallography shows that translational initiation factor IF3 consists of two compact alpha/beta domains linked by an alpha-helix. *EMBO J.***14:**4056–4064.

Boni, I. V., D. M. Isaeva, M. I. Musychenko, and N. V. Tzareva. 1991. Ribosome-messenger recognition: mRNA target sites for ribosomal protein S1. *Nucleic Acids Res.* **19:**155–162.

Brandi, A., P. Pietroni, C. O. Gualerzi, and C. L. Pon. 1996. Post-transcriptional regulation of CspA expression in *Escherichia coli*. *Mol. Microbiol.* **19:**231–240.

Brandt, R., and C. O. Gualerzi. 1991. Ribosome-mRNA contact sites at different stages of translation initiation as revealed by cross-linking of model mRNAs. *Biochimie* **73:**1543–1549.

Canonaco, M. A., C. O. Gualerzi, and C. L. Pon. 1989. Alternative occupancy of a dual ribosomal binding site by mRNA affected by translation initiation factors. *Eur. J. Biochem.* **182:**501–506.

Cole, J. R., C. L. Olsson, J. W. Hershey, M. Grunberg-Manago, and M. Nomura. 1987. Feedback regulation of rRNA synthesis in *Escherichia coli*. Requirement for initiation factor IF2. *J. Mol. Biol.* **198:**383–392.

Dubnoff, J. S., A. H. Lockwood, and U. Maitra. 1972. Studies on the role of guanosine triphosphate in polypeptide chain initiation in *Escherichia coli*. *J. Biol. Chem.* **247:**2884–2894.

Etchegaray, J. P., and M. Inouye. 1999. Translational enhancement by an element downstream of the initiation codon in *Escherichia coli*. *J. Biol. Chem.* **274:**10079–10085.

Fargo, D. C., M. Zhang, N. W. Gillham, and J. E. Boynton. 1998. Shine-Dalgarno-like sequences are not required for translation of chloroplast mRNAs in *Chlamydomonas reinhardtii* chloroplasts or in *Escherichia coli*. *Mol. Gen. Genet.* **257:**271–282.

Firpo, M. A., and A. E. Dahlberg. 1998. The importance of base pairing in the penultimate stem of *Escherichia coli* 16S rRNA for ribosomal subunit association. *Nucleic Acids Res.* **26:**2156–2160.

Fortier, P. L., J. M. Schmitter, C. Garcia, and F. Dardel. 1994. The N-terminal half of initiation factor IF3 is folded as a stable independent domain. *Biochimie* **76:**376–383.

Garcia, C., P. L. Fortier, S. Blanquet, J. Y. Lallemand, and F. Dardel. 1995a. Solution structure of the ribosome-binding domain of *E. coli* translation initiation factor IF3. Homology with the U1A protein of the eukaryotic spliceosome. *J. Mol. Biol.* **254:** 247–259.

Garcia, C., P. L. Fortier, S. Blanquet, J. Y. Lallemand, and F. Dardel. 1995b. ^{1}H and ^{15}N resonance assignments and structure of the N-terminal domain of *Escherichia coli* initiation factor 3. *Eur. J. Biochem.* **228:**395–402.

Goldenberg, D., I. Azar, A. B. Oppenheim, A. Brandi, C. L. Pon, and C. O. Gualerzi. 1997. Role of *Escherichia coli cspA* promoter sequences and adaptation of translational apparatus in the cold shock response. *Mol. Gen. Genet.* **256:**282–290.

Gualerzi, C. O., and C. L. Pon. 1990. Initiation of mRNA translation in prokaryotes. *Biochemistry* **29:**5881–5889.

Gualerzi, C. O., and C. L. Pon. 1996. mRNA-ribosome interaction during initiation of protein synthesis, p. 259–276. *In* R. A. Zimmermann and A. E. Dahlberg (ed.), *Ribosomal RNA. Structure, Evolution, Processing, and Function in Protein Biosynthesis.* CRC Press, Boca Raton, Fla.

Gualerzi, C. O., A. La Teana, R. Spurio, M. A. Canonaco, M. Severini, and C. L. Pon. 1990. Initiation of protein biosynthesis in procaryotes: recognition of mRNA by ribosomes and molecular basis for the function of initiation factors, p. 281–291. *In* W. E. Hill, A. E. Dahlberg, R. A. Garrett, P. B. Moore, D. Schlessinger, and J. R. Warner (ed.), *The Ribosome: Structure, Function, and Evolution.* American Society for Microbiology, Washington, D.C.

Guennegues, M. Unpublished data.

Guillon J. M., Y. Mechulam, J. M. Schmitter, S. Blanquet, and G. Fayat. 1992a. Disruption of the gene for Met-tRNA(fMet) formyltransferase severely impairs growth of *Escherichia coli*. *J. Bacteriol.* **174:**4294–4301.

Guillon, J. M., T. Meinnel, Y. Mechulam, C. Lazennec, S. Blanquet, and G. Fayat. 1992b. Nucleotides of tRNA governing the specificity of *Escherichia coli* methionyl-tRNA(fMet) formyltransferase. *J. Mol. Biol.* **224:**359–367.

Haggerty, T. J., and S. T. Lovett. 1993. Suppression of *recJ* mutations of *Escherichia coli* by mutations in translation initiation factor IF3. *J. Bacteriol.* **175:**6118–6125.

Hartz, D., J. Binkley, T. Hollingsworth, and L. Gold. 1990. Domains of initiator tRNA and initiation codon crucial for initiator tRNA selection by *Escherichia coli* IF3. *Genes Dev.* **4:**1790–1800.

Hershey, J. W. 1987. Protein synthesis, p. 613–647. *In* J. L. Ingraham, K. B. Low, B. Magasanik, M. Schaechter, and H. E. Umbarger (ed.), *Escherichia coli and Salmonella typhimurium: Cellular and Molecular Biology.* American Society for Microbiology, Washington, D.C.

Janssen, G. R. 1993. Eubacterial, archaebacterial, and eucaryotic genes that encode leaderless mRNA, p. 59–67. *In* B. H. Baltz, G. D. Hageman, and P. L. Skatrud (ed.), *Industrial Microorganisms: Basic and Applied Molecular Genetics.* ASM Press, Washington, D.C.

Kaloyanova, D., J. Xu, I. G. Ivanov, and M. G. Abouhaidar. 1997. Gene expression evidence indicates that nucleotides 507–513 and 1434–1440 in 16S rRNA are organized in close proximity on the *Escherichia coli* 30S ribosomal subunit. *Eur. J. Biochem.* **248:**10–14.

La Teana, A., C. L. Pon, and C. O. Gualerzi. 1993. Translation of mRNAs with degenerate initiation triplet AUU displays high initiation factor 2 dependence and is subject to initiation factor 3 repression. *Proc. Natl. Acad. Sci. USA* **90:**4161–4165.

La Teana, A., C. O. Gualerzi, and R. Brimacombe. 1995. From stand-by to decoding site. Adjustment of the mRNA on the 30S ribosomal subunit under the influence of the initiation factors. *RNA* **1:**772–782.

La Teana, A., C. L. Pon, and C. O. Gualerzi. 1996. Late events in translation initiation. Adjustment of fMet-tRNA in the ribosomal P-site. *J. Mol. Biol.* **256:**667–675.

La Teana, A., A. Brandi, M. O'Connor, and C. L. Pon. Translation during cold adaptation does not involve mRNA-rRNA basepairing through the downstream box. Submitted for publication.

Lee, K., C. A. Holland-Staley, and P. R. Cunningham. 1996. Genetic analysis of the Shine-Dalgarno interaction: selection of alternative functional mRNA-rRNA combinations. *RNA* **2:**1270–1285.

Li, S., N. V. Kumar, U. Varshney, and U. L. RajBhandary. 1996. Important role of the amino acid attached to tRNA in formylation and in initiation of protein synthesis in *Escherichia coli. J. Biol. Chem.* **271:**1022–1028.

Loechel, S., J. M. Inamine, and P. C. Hu. 1991. A novel translation initiation region from *Mycoplasma genitalium* that functions in *Escherichia coli. Nucleic Acids Res.* **18:**6905–6911.

Luchin, S., H. Putzer, J. W. Hershey, Y. Cenatiempo, M. Grunberg-Manago, and S. Laalami. 1999. In vitro study of two dominant inhibitory GTPase mutants of *Escherichia coli* translation initiation factor IF2. Direct evidence that GTP hydrolysis is necessary for factor recycling. *J. Biol. Chem.* **274:**6074–6079.

Mandal, N., D. Mangroo, J. J. Dalluge, J. A. McCloskey, and U. L. RajBhandary. 1996. Role of the three consecutive G:C base pairs conserved in the anticodon stem of initiator tRNAs in initiation of protein synthesis in Escherichia coli. *RNA* **2:**473–482.

Martin-Farmer, J., and G. R. Janssen. 1999. A downstream CA repeat sequence increases translation from leadered and unleadered mRNA in *Escherichia coli. Mol. Microbiol.* **31:**1024–1038.

McCarthy, J. E., and R. Brimacombe. 1994. Prokaryotic translation: the interactive pathway leading to initiation. *Trends Genet.* **10:**402–407.

McCarthy, J. E., and C. Gualerzi. 1990. Translational control of prokaryotic gene expression. *Trends Genet.* **6:**78–85.

McCutcheon, J. P., R. K. Agrawal, S. M. Philips, R. A. Grassucci, S. E. Gerchman, W. M. J. Clemons, V. Ramakrishnan, and J. Frank. 1999. Location of translational initiation factor IF3 on the small ribosomal subunit. *Proc. Natl. Acad. Sci. USA* **96:**4301–4306.

Meunier, S., R. Spurio, M. Czisch, R. Wechselberger, M. Guenneugues, C. O. Gualerzi, and R. Boelens. Solution structure of the fMet-tRNA binding domain of *Bacillus stearothermophilus* translation initiation factor IF2. Submitted for publication.

Misselwitz, R., K. Welfe, C. Krafft, C. O. Gualerzi, and H. Welfle. 1997. Translational initiation factor IF2 from *Bacillus stearothermophilus:* a spectroscopic and microcalorimetric study of the C-domain. *Biochemistry* **36:**3170–3178.

Mitta, M., L. Fang, and M. Inouye. 1997. Deletion analysis of *cspA* of *Escherichia coli:* requirement of the AT-rich UP element for cspA transcription and the downstream box in the coding region for its cold shock induction. *Mol. Microbiol.* **26:**321–335.

Moreau, M., E. de Cock, P. L. Fortier, C. Garcia, C. Albaret, S. Blanquet, J. Y. Lallemand, and F. Dardel. 1997. Heteronuclear NMR studies of *E. coli* translation initiation factor IF3. Evidence that the inter-domain region is disordered in solution. *J. Mol. Biol.* **266:**15–22.

Morita, M., M. Kanemori, H. Yanagi, and T. Yura. 1999. Heat-induced synthesis of sigma32 in *Escherichia coli*: structural and functional dissection of rpoH mRNA secondary structure. *J. Bacteriol.* **181:**401–410.

Mueller, F., and R. Brimacombe. 1997. A new model for the three-dimensional folding of *Escherichia coli* 16 S ribosomal RNA. I. Fitting the RNA to a 3D electron microscopic map at 20Å. *J. Mol. Biol.* **271:**524–544.

Nagai, H., H. Yuzawa, and T. Yura. 1991. Interplay of two cis-acting mRNA regions in translational control of sigma 32 synthesis during the heat shock response of *Escherichia coli. Proc. Natl. Acad. Sci. USA* **88:**10515–10519.

Nissen, P., M. Kjeldgaard, S. Thirup, G. Polekhina, L. Reshetnikova, B. F. C. Clark, and J. Nyborg. 1995. Crystal structure of the ternary complex of Phe-tRNA^{Phe}, EF-Tu, and a GTP analog. *Science* **270:**1464–1472.

O'Connor, M., T. Asai, C. L. Squires, and A. E. Dahlberg. 1999. Enhancement of translation by the downstream box does not involve mRNA-rRNA base pairing. *Proc. Natl. Acad. Sci. USA* **96:**8973–8978.

Odjakova, M., A. Golshani, G. Ivanov, M. Abou Haidar, and I. Ivanov. 1998. The low level expression of chloramphenicol acetyltransferase (CAT) mRNA in *Escherichia coli* is not dependent on either Shine-Dalgarno or the downstream boxes in the CAT gene. *Microbiol. Res.* **153:**173–178.

Olsson, C. L., M. Graffe, M. Springer, and J. W. Hershey. 1996. Physiological effects of translation initiation factor IF3 and ribosomal protein L20 limitation in *Escherichia coli. Mol. Gen. Genet.* **250:**705–714.

Olsthoorn, R. C., S. Zoog, and J. van Duin. 1995. Coevolution of RNA helix stability and Shine-Dalgarno complementarity in a translational start region. *Mol. Microbiol.* **15:**333–339.

Philippe, C., F. Eyermann, L. Benard, C. Portier, B. Ehresmann, and C. Ehresmann. 1993. Ribosomal protein S15 from *Escherichia coli* modulates its own translation by trapping the ribosome on the mRNA initiation loading site. *Proc. Natl. Acad. Sci. USA* **90:**4394–4398.

Pon, C. L., and C. Gualerzi. 1974. Effect of initiation factor 3 binding on the 30S ribosomal subunits of *Escherichia coli. Proc. Natl. Acad. Sci. USA* **71:**4950–4954.

Pon, C. L., R. T. Pawlik, and C. Gualerzi. 1982. The topographical localization of IF3 on *Escherichia coli* 30 S ribosomal subunits as a clue to its way of functioning. *FEBS Lett.* **137:**163–167.

Pon, C. L., M. Paci, R. T. Pawlik, and C. O. Gualerzi. 1985. Structure-function relationship in *Escherichia coli* initiation factors. Biochemical and biophysical characterization of the interaction between IF-2 and guanosine nucleotides. *J. Biol. Chem.* **260:**8918–8924.

RajBhandary, U. L., and C. M. Chow. 1995. Initiator tRNAs and initiation of protein synthesis, p. 511–528. *In* D. Söll and U. L. RajBhandary (ed.), *tRNA: Structure, Biosynthesis, and Function.* ASM Press, Washington, D.C.

Resch, A., K. Tedin, A. Grundling, A. Mundlein, and U. Blasi. 1996. Downstream box-anti-downstream box interactions are dispensable for translation initiation of leaderless mRNAs. *EMBO J.* **15:**4740–4748.

Ringquist, S., S. Shinedling, D. Barrick, L. Green, J. Binkley, G. D. Stormo, and L. Gold. 1992. Translation initiation in *Escherichia coli:* sequences within the ribosome-binding site. *Mol. Microbiol.* **6:**1219–1229.

Ringquist, S., M. MacDonald, T. Gibson, and L. Gold. 1993. Nature of the ribosomal mRNA track: analysis of ribosome-binding sites containing different sequences and secondary structures. *Biochemistry* **32:**10254–10262.

Ringquist, S., T. Jones, E. E. Snyder, T. Gibson, I. Boni, and L. Gold. 1995. High-affinity RNA ligands to *Escherichia coli* ribosomes and ribosomal protein S1: comparison of natural and unnatural binding sites. *Biochemistry* **34:**3640–3648.

Romby, P., H. Wakao, E. Westhof, M. Grunberg-Manago, B. Ehresmann, C. Ehresmann, and J. P. Ebel. 1990. The conformation of the initiator tRNA and of the 16S rRNA from *Escherichia coli* during the formation of the 30S initiation complex. *Biochim. Biophys. Acta* **1050:**84–92.

Sacerdot, C., G. Fayat, P. Dessen, M. Springer, J. A. Plumbridge, M. Grunberg-Manago, and S. Blanquet. 1982. Sequence of a 1.26-kb DNA fragment containing the structural gene for *E. coli* initiation factor IF3: presence of an AUU initiator codon. *EMBO J.* **1**:311–315.

Sacerdot, C., C. Chiaruttini, K. Engst, M. Graffe, M. Milet, N. Mathy, J. Dondon, and M. Springer. 1996. The role of the AUU initiation codon in the negative feedback regulation of the gene for translation initiation factor IF3 in *Escherichia coli. Mol. Microbiol.* **21**:331–346.

Santer, M., and A. E. Dahlberg. 1996. Ribosomal RNA: an historical perspective, p. 3–20. *In* R. A. Zimmermann and A. E. Dahlberg (ed.), *Ribosomal RNA. Structure, Evolution, Processing, and Function in Protein Biosynthesis.* CRC Press, Boca Raton, Fla.

Schmitt, E., M. Panvert, S. Blanquet, and Y. Mechulam. 1998. Crystal structure of methionyl-tRNAfMet transformylase complexed with the initiator formyl-methionyl-tRNAfMet. *EMBO J.* **17**:6819–6826.

Schweisguth, D. C., and P. B. Moore. 1997. On the conformation of the anticodon loops of initiator and elongator methionine tRNAs. *J. Mol. Biol.* **267**:505–519.

Seong, B., and U. L. RajBhandary. 1987. *Escherichia coli* formylmethionine tRNA: mutations of GGG:CCC sequence conserved in anticodon stem of initiator tRNAs affect initiation of protein synthesis and conformation anticodon loop. *Proc. Natl. Acad. Sci. USA* **84**:334–338.

Sette, M., P. van Tilborg, R. Spurio, R. Kaptein, M. Paci, C. O. Gualerzi, and R. Boelens. 1997. The structure of the translational initiation factor IF1 from *E. coli* contains an oligomer-binding motif. *EMBO J.* **16**:1436–1443.

Sette, M., R. Spurio, P. van Tilborg, C. O. Gualerzi, and R. Boelens. 1999. Identification of the ribosome binding sites of translation initiation factor IF3 by multidimensional heteronuclear NMR spectroscopy. *RNA* **5**:82–92.

Severini, M., R. Spurio, A. La Teana, C. L. Pon, and C. O. Gualerzi. 1991. Ribosome-independent GTPase activity of translation initiation factor IF2 and of its G-domain. *J. Biol. Chem.* **266**: 22800–22802.

Shean, C. S., and M. E. Gottesman. 1992. Translation of the prophage lambda cI transcript. *Cell* **70**:513–522.

Simmons, L. C., and D. G. Yansura. 1996. Translational level is a critical factor for the secretion of heterologous proteins in *Escherichia coli. Nat. Biotechnol.* **14**:629–634.

Sorensen, M. A., and J. P. S. Fricke. 1998. Ribosomal protein S1 is required for translation of most, if not all, natural mRNAs in *Escherichia coli* in vivo. *J. Mol. Biol.* **280**:561–569.

Sprengart, M. L., and A. G. Porter. 1997. Functional importance of RNA interactions in selection of translation initiation codons. *Mol. Microbiol.* **24**:19–28.

Sprengart, M. L., E. Fuchs, and A. G. Porter. 1996. The downstream box: an efficient and independent translation initiation signal in *Escherichia coli. EMBO J.* **15**:665–674.

Sprinzl, M., C. Steegborn, G. Hubel, and S. Steinberg. 1996. Compilation of tRNA sequences and sequences of tRNA genes. *Nucleic Acids Res.* **24**:68–72.

Spurio, R., M. Severini, A. La Teana, M. A. Canonaco, R. T. Pawlik, C. O. Gualerzi, and C. L. Pon. 1993. Novel structural and functional aspects of translational initiation factor IF2, p. 241–252. *In* K. H. Nierhaus, F. Franceschi, A. R. Subramanian, V. A. Erdmann, and B. Wittmann-Liebold (ed.), *The Translational Apparatus. Structure, Function, Regulation, Evolution.* Plenum Press, New York, N.Y.

Spurio, R., L. Brandi, E. Caserta, C. L. Pon, C. O. Gualerzi, R. Misselwitz, C. Krafft, K. Welfle, and H. Welfle. The C-terminal sub-domain (IF2 C-2) contains the entire fMet-tRNA binding site of initiation factor IF2. *J. Biol. Chem.*, in press.

Sussman, J. K., E. L. Simons, and R. W. Simons. 1996. *Escherichia coli* translation initiation factor 3 discriminates the initiation codon in vivo. *Mol. Microbiol.* **21**:347–360.

Tedin, K., I. Moll, S. Grill, A. Resch, A. Graschopf, C. O. Gualerzi, and U. Blaesi. 1999. Translation initiation factor 3 antagonizes authentic start codon selection on leaderless mRNAs. *Mol. Microbiol.* **31**:67–77.

Tomšic, J., A. Smorlesi, R. Spurio, A. La Teana, C. L. Pon, and C. O. Gualerzi. Unpublished data.

Tomšic, J., L. A. Vitali, T. Daviter, A. Savelsbergh, R. Spurio, P. Striebeck, W. Wintermeyer, M. V. Rodnina, and C. O. Gualerzi. The late events in the translation initiation pathway in eubacteria: a kinetic analysis. Submitted for publication.

Tzareva, N. V., V. I. Makhno, and I. V. Boni. 1994. Ribosome-messenger recognition in the absence of the Shine-Dalgarno interactions. *FEBS Lett.* **337**:189–194.

Van Etten, W. J., and G. R. Janssen. 1998. An AUG initiation codon, not codon-anticodon complementarity, is required for the translation of unleadered mRNA in *Escherichia coli. Mol. Microbiol.* **27**:987–1001.

Vellanoweth, R. L., and J. C. Rabinowitz. 1992. The influence of ribosome-binding-site elements on translational efficiency in *Bacillus subtilis* and *Escherichia coli* in vivo. *Mol. Microbiol.* **6**: 1105–1114.

Vitali, L. A., M. V. Rodnina, R. Spurio, P. Striebeck, W. Wintermeyer, and C. O. Gualerzi. Unpublished data.

Wakao, H., P. Romby, S. Laalami, J. P. Ebel, C. Ehresmann, and B. Ehresmann. 1990. Binding of initiation factor 2 and initiator tRNA to the *Escherichia coli* 30S ribosomal subunit induces allosteric transitions in 16S rRNA. *Biochemistry* **29**: 8144–8151.

Wakao, H., P. Romby, J. P. Ebel, M. Grunberg-Manago, C. Ehresmann, and B. Ehresmann. 1991. Topography of the *Escherichia coli* ribosomal 30S subunit-initiation factor 2 complex. *Biochimie* **73**:991–1000.

Wu, C. J., and G. R. Janssen. 1996. Expression of a streptomycete leaderless mRNA encoding chloramphenicol acetyltransferase in *Escherichia coli. J. Bacteriol.* **179**:6824–6830.

Wu, X. Q., and U. L. RajBhandary. 1997. Effect of the amino acid attached to *Escherichia coli* initiator tRNA on its affinity for the initiation factor IF2 and on the IF2 dependence of its binding to the ribosome. *J. Biol. Chem.* **272**:1891–1895.

Wu, X. Q., P. Iyengar, and U. L. RajBhandary. 1996. Ribosome-initiator tRNA complex as an intermediate in translation initiation in *Escherichia coli* revealed by use of mutant initiator tRNAs and specialized ribosomes. *EMBO J.* **15**:4734–4739.

The Ribosome: Structure, Function, Antibiotics, and Cellular Interactions
Edited by R. A. Garrett, S. R. Douthwaite, A. Liljas, A. T. Matheson, P. B. Moore, and H. F. Noller

Chapter 40

Factor-Mediated Termination of Protein Synthesis: a Welcome Return to the Mainstream of Translation

DANIEL N. WILSON, MARK E. DALPHIN, HERMAN J. PEL, LOUISE L. MAJOR, JOHN B. MANSELL, and WARREN P. TATE

For many years the termination mechanism seemed isolated from the other factor-mediated events of translation—decoding was via proteins rather than RNAs, and the bacterial and eukaryotic mechanisms seemed quite distinct. Since the last ribosome meeting in Victoria, Canada, new information has changed this perception dramatically. We now view termination as an integral part of the cohort of factor-mediated events, and there is clear evidence that the bacterial and eukaryotic mechanisms have much in common. More and more, termination is beginning to resemble a typical translation mechanism with parallels to other phases of protein synthesis. While it is still puzzling that the primary sequences of the bacterial and eukaryotic factors show little homology, mechanistically there are many similarities between the two. Concepts such as structural mimicry between the complexes formed at each stage of protein synthesis emphasize this change in perception. In particular, the termination complexes clearly fit into the same mold as the initiation and elongation complexes; presumably this mold is dictated by the geometry of the active center of the ribosome, where the complexes interact.

In this review we will focus on how the information gathered principally since the last ribosome meeting has enhanced our understanding of the termination phase of protein synthesis.

THE STRUCTURE OF THE RFs

Class I Decoding RFs

Bacterial RF1 and RF2

An expansion in the number of complete genomes analyzed has highlighted the importance of the decoding release factors (RFs). Of particular note are the *Mycoplasma* species, which have a "streamlined" genome and have dispensed with most nonessential genes but have retained one decoding factor, RF1 (Fraser et al., 1995). The codon specific for RF2 has been assigned to Gln. Furthermore, disabling the RF1 or RF2 gene in *Escherichia coli* is lethal, unless a variant factor can undertake both decoding roles (Ito et al., 1998b). Similarly, in mitochondria, one decoding factor has been retained for the termination mechanism: RF1 has been found in *Saccharomyces cerevisiae* (Pel et al., 1992a) and mammalian mitochondria (Lee et al., 1987). Again, the role of RF2 has been made redundant and the UGA codon has been reassigned as Trp.

There are currently 35 full bacterial RF1 and RF2 sequences in the GenBank database. Consistent with the early observations, they show significant similarities at the amino acid level. Comparative sequence analyses have been used to propose domain models for the factors; for example, Pel et al. (1992b)

Daniel N. Wilson, Louise L. Major, John B. Mansell, and Warren P. Tate ■ Department of Biochemistry and Centre for Gene Research, University of Otago, PO Box 56, Dunedin, New Zealand. Mark E. Dalphin ■ Amgen Inc., Amgen Center, Thousand Oaks, CA 91320-1799. Herman J. Pel ■ DSM Food Specialities, DSM Gist 426-0295, P.O. Box 1, 2600 MA Delft, The Netherlands.

separate the RFs into five structural domains, while Ito and colleagues (1996) propose a model of seven domains. On the other hand, Moffat and Tate (1994) have proposed a two-domain model derived from functional studies. Of special interest are a region in domain IV (Pel model) highly conserved among all RFs where there is 41% amino acid identity and the presence of two regions conserved within each RF type that are sequence specific for RF1 or RF2 but not both. Interestingly, a large number of the identified mutations map to one of the RF type-specific regions (Table 1 and Fig. 1). These variants have often been isolated by genetic screens as nonsense suppressors and shown to confer a recessive phenotype. Two mitochondrial RF mutants also map to one of the RF-specific regions (Pel et al., 1992a), and the variant proteins have been shown to be impaired in ribosome binding in vitro (Askarian-Amiri et al., unpublished). While these screens identify "functional" regions in RFs, they reveal little about the domain organization of the bacterial or mitochondrial RFs.

The putative ribosome binding domain in the bacterial RFs correlates with the region proposed to mimic the tRNA anticodon (Nakamura and Ito, 1998). Mutagenesis within this region identified two dominant-lethal sites at positions 207 and 213 in RF2 (Ito et al., 1996) which are suppressed by second-site mutations at positions 150 to 160, suggesting these regions may interact as part of a codon recognition element (Nakamura and Ito, 1998). A cross-link from the first position of the stop codon has been tentatively identified at positions 130 to 140 in the RF2 sequence (Poole and Tate, unpublished). Consistent with this proposal, an RF2 mutation nearby, E167K, exhibits omnipotent termination activity (Ito et al., 1998b) and interchanging these regions between RF1 and RF2 alters their codon specificities (Nakamura and Ito, 1998).

Figure 1. Sequence alignment of the region encompassing the RF type-specific sequences (positions 190 to 217) of selected prokaryotic RFs. Conserved (black boxes) and similar (light gray boxes with black text) residues between all RFs and conserved residues specific for each RF type (dark gray boxes with white text) are presented. Mutations (a to f) clustered within this region are documented in Table 1. The abbreviations are defined in the legend to Fig. 2, except the following: Bfi, *Bacillus firmus*; Hpy, *Helicobacter pylori*; Bbu, *Borrelia burgdorferi*; and Tth, *Thermus thermophilus*.

No variants of RF2 that have changes in the highly conserved region of the RFs have been isolated. Chimeric constructs between RF1 and RF2 identified 10 residues located within or near the highly conserved region of RF2 that, when substituted into RF1, abolished peptidyltransferase activity. A change at one position in this block of 10 residues from threonine, at position 246 in RF2, to serine, the equivalent residue in RF1, restored activity (Wilson, 1999). Cleavage of RF2 by chymotrypsin at a neighboring residue, Arg^{244}, which was proposed to be within an interdomain loop region, resulted in an RF2 molecule which had enhanced ribosomal binding activity but had lost peptidyl-tRNA hydrolysis activity (Moffat and Tate, 1994). Furthermore, unrelated work suggested that Thr^{246} was responsible for the inactivity of recombinant RF2 from *Escherichia coli*, whereas recombinant RF2 from *Salmonella typhimurium* and *Streptomyces coelicolor* having Ala^{246} or Ser^{246} were not compromised in this way (Uno et al., 1996). In fact, of all known bacterial RFs, only *E. coli* RF2 has Thr at position 246, suggesting there may be a regulatory mechanism operating through Thr^{246} in *E. coli*. Detailed analyses of the effects of a number of amino acid substitutions at position 246 showed that ribosomal binding and peptide release activities were affected differentially, depending on the substitution (Wilson and Tate, unpublished). These studies suggest that this region of domain IV in the RFs may act like a conforma-

Table 1. Characterized mutations in bacterial RF1 and RF2

RF	Mutation[a]	Reference
RF1 (*E. coli*)	R137P	Ryden and Isaksson (1984)
RF1 (*S. typhimurium*)	G168D	Olafsson et al. (1996)
	G180S	Elliott and Wang (1991)
	H182Y	
mtRF1	R190K	Pel et al. (1992a, 1992b)
RF1 (*S. cerevisiae*)	P192L	Pel et al. (1993)
RF2 (*E. coli*)	L63F and D79G	Wu et al. (1990)
	E89K	Mikuni et al. (1991)
	D143N	
	L328F	
	F207T	Ito et al. (1996)
	R213I	
	T246A	Uno et al. (1996)
	E167K	Ito et al. (1998b)
RF2 (*S. typhimurium*)	Y144UGA	Kawakami and Nakamura (1990)

[a] All *E. coli* numbering corresponds to the equivalent residue in *E. coli*.

tional switch. Correct recognition of the stop codon by the signal recognition domain may convey, through this switch region, a conformational change in the peptidyl-tRNA hydrolysis domain of the factor necessary for peptide release to occur.

Eukaryotic decoding RF

Although a single eukaryotic RF (eRF) was detected almost 30 years ago and shown to have a single recognition site for all three codons, it was only in 1994 that the correct gene was identified (Frolova et al., 1994). Originally, the gene for eRF was identified as the gene for tryptophanyl-tRNA synthetase (Lee et al., 1990), but it was subsequently shown that the two proteins were distinct (Frolova et al., 1993). The rabbit eRF peptide sequence was shown to be almost identical to peptide from human TB3-1, *Xenopus laevis* Cl1, the *S. cerevisiae sup45* products, and the translated sequence of *Arabidopsis thaliana sup45*-like cDNAs, and so an eRF1 family was proposed (Frolova et al., 1994). Expression of the human and *X. laevis* genes in yeast and in *E. coli* yields protein with termination activities at all three stop codons (Frolova et al., 1994). Although Sup45p has yet to exhibit RF activity in vitro, genetic data show the *X. laevis*, human, or Syrian hamster protein can substitute for Sup45p, supporting the contention that sup45 is a genuine eRF1 (Urbero et al., 1997). A number of other genes have now been identified as belonging to the eRF1 gene family, including genes from four archaebacteria.

Standard methods of sequence comparison fail to detect sequence similarity between the bacterial decoding RFs and the eRF1 family. Conservation of small sequence elements has led to a proposal of a common evolutionary origin, but this contention remains controversial. The fact that omnipotent bacterial RFs have been identified has been taken as further evidence of a common progenitor (Ito et al., 1998b), although the loss of specificity extends to nondiscriminatory decoding of sense codons as well, suggesting a loss of fidelity in codon recognition rather than acquisition of an eRF1-like activity.

Class II Stimulatory RFs

Bacterial RF3

It was over 20 years between the first report of a third bacterial RF and the identification of the gene (*prfC*) encoding RF3 (Grentzmann et al., 1994; Mikuni et al., 1994). RF3 was shown to contain a GTP binding motif similar to EF-Tu and EF-G and thus to belong to the large family of monomeric G proteins. Although RF3 is not essential for cell survival, since *prfC* knockout strains are viable and *Mycoplasma* has dispensed with this gene, RF3 is necessary for optimal translational activity, particularly under environmentally stressful conditions (Mikuni et al., 1994). RF3 has also been shown to enhance the efficiency of decoding of stop signals and in particular the signals that are used by highly expressed genes (Crawford et al., 1999). This suggests that when bacteria require high rates of translation and efficient decoding of stop signals, RF3 makes an important contribution to the translational efficiency.

eRF3

Early experiments had suggested that eRF1 had an associated GTPase activity. However, no GTP binding domain was evident within the eRF1 gene, and indeed, the expressed protein had no GTPase activity (Frolova et al., 1994). An RF3 equivalent was suggested to be present in eukaryotic organisms in addition to eRF1 (Frolova et al., 1994). Sup35p was a prime candidate in yeast because mutations in *sup35* were omnipotent suppressors (Stansfield et al., 1992). Equivalent genes have been identified in *X. laevis* (Zhouravleva et al., 1995) and in humans (Hoshino et al., 1989). Significantly, the presence of a GTP binding motif was detected in the C-terminal region of all eRF3 homologues, which resembles those seen in the translation factors RF3, EF-Tu, EF-G, and EF-1α. The *X. laevis* eRF3 was shown to bind GTP and to have a low-level intrinsic GTPase activity (Zhouravleva et al., 1995). In a quaternary complex with GTP, eRF1, and ribosomes, eRF3-mediated GTPase activity was greatly stimulated (Frolova et al., 1996). It was clear that the missing and predicted eRF3 had been identified. An intriguing question was how these findings could be correlated with the original studies of the eRF, which clearly showed GTPase activity and a release activity influenced by GTP. It seems evident that what was being studied and recognized to be a dimer (Konecki et al., 1977) was in fact a complex of eRF1 and eRF3. This indicated that the two proteins must stay associated throughout purification such that eRF3 was always present during functional characterization of the eRF1 protein.

Furthermore, the ability of *X. laevis* eRF3 to enhance termination activity was shown to be dependent upon an interaction with eRF1 in vitro (Zhouravleva et al., 1995). Similarly Sup35p and Sup45p have been shown to interact in vitro and in vivo, as indicated by the yeast two-hybrid system. Recent work has focused on defining the sites of interaction on the two factors. Nakamura and coworkers identified two distinct regions in *Schizosaccharomyces pombe* eRF1 necessary for interaction with eRF3 from the same organism. The C-terminal 6 to 11 amino acids were identified as being the most important for eRF3 interaction, although a C-terminally truncated eRF1 was still able to complement a temperature-sensitive Sup45p mutant in vivo (Ito et al., 1998a). For human eRF1, an internal and a C-terminal region were identified as important for eRF3 interaction, but surprisingly, the exact positions differed from those identified for the yeast protein. A conserved GILRY sequence (positions 411 to 415) was critical for interaction and GTPase activation of eRF3, whereas the most C-terminal 22 amino acids were not important (Merkulova et al., 1999). Interestingly, the archaebacteria eRF1-like sequences lack these 22 amino acids but have a conserved GILRY motif. Two internal sites in eRF3 are critical for interaction with eRF1. These are distinct from the GTP binding motifs, suggesting the two functions are independent. The ability of eRF1 to affect the nucleotide binding of eRF3 and that of the ribosome to stimulate GTPase activity of eRF3 suggest that these factors fulfil the role of GEF (guanine nucleotide exchange factor) and GAP (GTPase activating protein), respectively (Merkulova et al., 1999).

N-terminally truncated eRF3 can still stimulate eRF1-mediated termination, which has raised the question of whether the N-terminal domain has an alternative function. Mutations in, or overexpression of, Sup35p generates a non-Mendelian [PSI^+] phenotype (Doel et al., 1994; Ter-Avanesyan et al., 1994). Two alternative conformations of Sup35p are believed to exist, the native form and an aberrant form, which is associated with a prion-like phenotype. The interaction between Sup35p molecules that was necessary for the prion-like propagation in [PSI^+] cells was mediated through the N-terminal domain (Patino et al., 1996; Paushkin et al., 1996). The characteristic nonsense suppressor phenotype of [PSI^+] cells results from sequestering of Sup45p into large prion-like aggregates of Sup35p. A selective advantage of [PSI^+] over [psi^-] cells was evident under environmentally stressful conditions (Eaglestone et al., 1999). The ability of [PSI^+] cells to solubilize the large reservoir of Sup35p and Sup45p from large aggregates provides a mechanism with which to control translational termination under conditions of stress (Eaglestone et al., 1999).

THE RF COMPLEX: STRUCTURAL MIMICRY AT THE RIBOSOMAL ACTIVE CENTER

In 1994, as a result of functional studies, a model, designated the tRNA analogue model, was proposed (Moffat and Tate, 1994). In this model the decoding RFs (RF1, RF2, and eRF1) span the two parts of the ribosomal active center, from the decoding site of the small subunit to the peptidyltransferase center of the large subunit. This model proposed a form of structural mimicry between the tRNA and the decoding RF in the ribosomal A site, where the RF was tentatively divided into two domains, one for codon recognition and the other for peptidyl-tRNA hydrolysis. Subsequently, structures of two elongation factors were solved: a ternary complex of EF-Tu·GTP·aminoacyl-tRNA, and an EF-G·GDP complex (reviewed in Nyborg et al., 1996). These structures were remarkable in that they suggested that domains III, IV, and V of EF-G structurally mimicked the tRNA in the ternary complex; in fact, the two structures were almost completely superimposable. This was relevant to the tRNA analogue model because it implied that the decoding RFs also mimicked these domains in EF-G. Indeed, Nyborg et al. (1996) proposed that this molecular mimicry extended to the initiation and termination phases of translation. The extensive structural homology between the C-terminal domains of EF-G and the tRNA prompted a comparison of EF-G and the decoding RFs. The approach was based on the fact that current computer algorithms would not identify single conserved residues at discrete positions in divergent proteins, so the authors identified such residues "by eye" (Ito et al., 1996). As a result, Uno and colleagues (1996) have proposed an anticodon mimicry element consisting of conserved residues located in the RF equivalent to domain IV of EF-G. Certainly a number of mutations that affect RF binding to the ribosome cluster in this region, and directed hydroxyl radicals generated from a tether in this region of RF1 cleaved the rRNA within the decoding site of the small subunit (Nakamura and Ito, 1998), a result comparable to a tether placed in domain IV of EF-G which cleaved at the decoding site residue C1400 (Wilson and Noller, 1998).

Initiation factor 2 (IF2) has also been postulated to mimic the tRNA at the A site while directing the initiator tRNA into the P site (Nyborg et al., 1996). Manual rather than computer-based sequence alignments detected extensive homology between IF2 and EF-G. A region of domain IV of EF-G lacking homology with IF2 was shown to be homologous to IF1 (Brock et al., 1998) and overlapped with the region of domain IV proposed to be the equivalent of the tRNA anticodon mimicry domain in the decoding RF. This concept supports a complex being formed between IF2 and IF1, where IF1, like the decoding RF, interacts with the ribosomal A site.

Independently we had been prompted by the fact that IF1 can protect bases at the decoding site of the 16S rRNA to search for homology between IF1 and the decoding RFs. We were excited to find from a comparison of all known prokaryotic IF1s, RF1s, and RF2s that there was a 37-amino-acid sequence motif with extensive homology between the RFs and the IF1s (Pel and Tate, unpublished). Moreover, the RF1s had identity at some residues and the RF2s had identity at others (Fig. 2). The motif represents >50% of the IF1 sequence and about 10% of the sequences of the RFs. The statistical significance of the alignment was confirmed with the MACAW program. A measure of the significance can be gained by comparing the significance of the region with IF1 alignments and determining whether the significance increases when the RF sequences are added to the alignment. The probability of the region occurring by chance (between *E. coli* and *Bacillus subtilis* IF1s) was 6.1×10^{-12}, and this *P* value decreased (thus, the statistical significance increased) by 100,000 and 2,500 when *E. coli* and *B. subtilis* RF2 sequences, respectively, were included in the alignment. Furthermore, an amino acid profile was identified that will select all RF1s, RF2s, and IF1s from the Swiss-Prot protein database with the exception of a *Mycobacterium* IF1 (which has a Glu in place of a Gly at position 51 [Fig. 2C]). A dot blot illustrates the similarity between positions 40 to 60 in IF1 and 180 to 200 in RF2 (Fig. 3A). No similarity with eRF1 and IF1 was detected with similar analyses.

The structure of IF1 has been determined by nuclear magnetic resonance spectroscopy (Sette et al., 1997) and is classified as a member of an oligomer binding (OB)-fold family of proteins, based on the ability of this structure to bind oligosaccharides and oligonucleotides (Murzin, 1993). The architecture of this motif includes a five-stranded β-sheet coiled to form a closed β-barrel and capped by an α-helix. The region of homology between the RFs and the IFs correlates with β-sheets III (positions 29 to 36), IV (50 to 56), and V (62 to 66) and the α-helix (38 to 44). The predicted secondary structure for the RFs within the region of similarity has been compared with the predicted structure of IF1. Two characteristics of the classical OB-fold motif are absent, the α-helix between β-sheets III and IV which caps the barrel (although this was not predicted in IF1 either) and two of the β-strands. Hence, RFs cannot be regarded as members of the OB-fold family of proteins.

Is it possible that the missing structural elements are contributed from another source? It is notable that when a dot blot of RF3 is compared with one of EF-G there is a region in domain IV of EF-G that has homology with the C-terminal region of RF3 in addition to the major homology of the N-terminal part of RF3 with the G domains and domains I and II (Fig. 3B). A threading analysis comparing RF3 to the EF-G structure gives superimposition of the two structures in the expected domains, but in addition, the RF3 structure correlates significantly with domain III and part of domain IV of EF-G (Fig. 4) (Dalphin and Tate, unpublished). This suggests that together the decoding RF and RF3 might be mimicking the structure in domain IV of EF-G. These studies indicate that RF complexes structurally mimic the elongation phase, but also the initiation phase, of protein synthesis. However, they also indicate that the schematic models of the mimicry complexes previously presented (Moffat and Tate, 1994; Nakamura and Ito, 1998) may be too simplistic with regard to RF3, since the N-terminal region of RF3, together with the decoding RF, may occupy the ribosomal A site.

THE DECODING RF RECOGNIZES THE STOP SIGNAL AT THE A SITE OF THE DECODING CENTER

Few protein-RNA interactions are restricted to only 3 nucleotides in the mRNA. Hence, the tRNA anticodon domain of the class I RFs is likely to be recognizing more than just the stop codon in the A site. Clearly, the stop codons are an important part of the stop signal and serve to orient the signal with respect to the decoding factor in the A site. Many suppression studies indicated that the context had a strong influence on the competition between RF decoding of the stop codon in the A site as stop or sense by near-cognate or cognate suppressor tRNAs. Subsequently, the discovery of recoding provided examples of stop codons regularly failing to terminate protein synthesis, such as at the RF2 frameshift site (Donly et al., 1990). These examples presented clear evidence that the stop signal must be larger than a triplet codon, as displayed in the genetic code (reviewed in Tate and Mannering, 1996).

To investigate this further, we set up a database to analyze the immediate context surrounding naturally occurring termination sites. The database, TransTerm (Brown et al., 1993), allowed us to study the characteristics of the stop signals and compare them with those found at recoding sites. Currently, the database has over 165,000 sequences from 750 organisms and has been converted to a relational database (Dalphin et al., 1998) with a website interface. Analyses of the nucleotide biases at positions adjacent to the stop codon gave characteristic profiles regardless of the organism that was studied—namely, a bias that extended on both sides of the stop codon. This

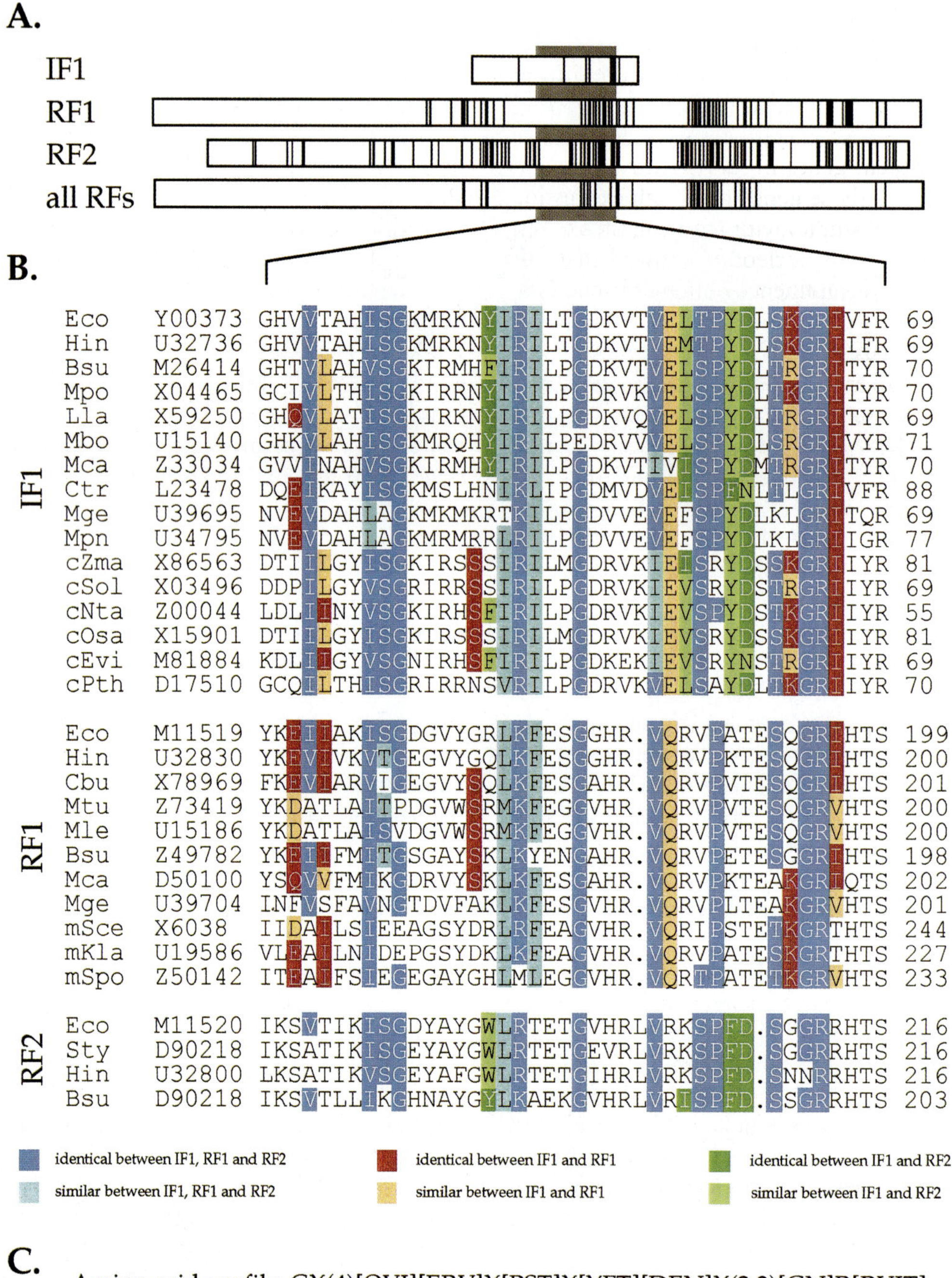

Figure 2. Similarity between prokaryotic and organellar IF1s and class I RFs. The two types of class I RFs, RF1 and RF2, have been considered as separate families, since their stop codon specificities (UAA and UAG for RF1; UAA and UGA for RF2) suggest that significant deviations may exist in their interactions with mRNA and the decoding site. (A) Positions of amino acids identical among all currently available prokaryotic and organellar RFs (all RFs), RF1s (RF1), RF2s (RF2), and IF1s (IF1) are indicated by vertical bars. The alignments were accomplished with the PILEUP program (GCG package, University of Wisconsin Genetics Computer Group). (B) Alignment of distantly related members of the IF1 and RF families. IF1 and RF amino acid residues that are identical or functionally related in at least three IF1s and three RF1s or two RF2s are indicated. The following sets of amino acids are considered to be functionally related: A and G; S, T, and C; L, I, V, M, and F; F, Y, W, and H; K and R; D and E; D and N; and E and Q. Abbreviations: Bsu, *B. subtilis*; Cbu, *Coxiella burnetii*; Ctr, *Chlamydia trachomatis*; Eco, *E. coli*; Hin, *Haemophylus influenzae*; Lla, *Lactococcus lactis*; Mbo, *Mycobacterium bovis*; Mca, *Mycoplasma capricolum*; Mge, *Mycoplasma genitalium*; Mle, *Mycobacterium leprae*; Mpn, *Mycoplasma pneumoniae*; Mpo, *Marchantia polymorpha*; Mtu, *Mycobacterium tumefaciens*. cZma, cSol, cNta, cOsa, cEvi, and cPth are chloroplast IF1s of *Zea mays*, *Spinacia oleracea*, *Nicotiana tabacum*, *Oryza sativa*, *Epifagus virginiana*, and *Pinus thunbergii*, respectively.

would be the expectation if the context of the termination signal involved nucleotides on both sides of the stop codon. Genes that had been characterized as highly expressed in *E. coli* had strong bias at the +4 position (or the base immediately 3′ to the codon), with the large majority having U in this position. In contrast, mammalian genes predominantly had A or G—particularly for those genes with high expression levels. Experimental studies with recoding sites to test the influence of the +4 nucleotide showed that this position has enormous influence on the efficiencies of the stop signals, in vivo and in vitro, in bacteria, yeast, and mammals (Tate and Mannering, 1996). This was consistent with the strong bias seen in this position from the analyses of the natural termination sites and demonstrated that the most efficient signals were the most abundant. Analyses of the positions further downstream indicated that the bias correlated with the termination efficiency of the signal in bacteria (Major et al., 1996) and in mammals (McCaughan et al., unpublished), although the effect was less dramatic than that at the +4 position.

To investigate whether the decoding RF was in close contact with the signal, we used site-directed cross-linking from particular positions, starting with the +1 U common to all stop codons in the A site and extending from the +4 to at least the +10 position. A cross-link from the +1 position indicated close contact between the A-site codon and the factor (Brown and Tate, 1994), and this result was reproduced when the cross-linking moiety was in the +4 (Poole et al., 1997), +5, or +6 position, but not in the +7 to +10 positions (Poole et al., 1998). Furthermore, there was a close correlation between the positions in the mRNA which were able to form cross-links and those which influenced termination efficiency in vivo.

Isaksson and coworkers have performed extensive studies analyzing the effect on nonsense suppression of the sequences upstream of stop codons. The last two sense codons were shown to influence the efficiency of termination. The coding potential of these positions was important rather than the sequences themselves, with particular amino acids enhancing or diminishing the termination efficiency of the stop codon (Mottagui-Tabar et al., 1994; Björnsson et al., 1996). How can these codons affect termination efficiency when they are already complexed

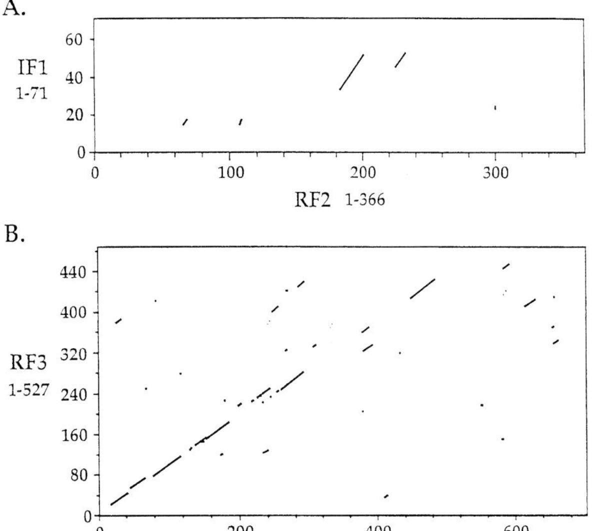

Figure 3. Dot blot of *E. coli* IF1 against RF2 and RF3 against EF-G. The full sequences of IF1 (positions 1 to 71) and RF2 (1 to 366) (A) and RF3 (1 to 527) and EF-G (1 to 703) (B) were compared to determine regions of similarity by using the Genetics Computer Group programs Compare and Dotplot. Dotplot was run at a density of 306.67. The alignments were performed with a stringency value of 15.

with other translation complexes, the tRNAs at the P and E sites? The importance of the nature of the amino acids encoded suggests that the geometry of the termination complex in the active center is being affected. Indeed, the propensity of the amino acid to form an α-helical secondary structure was suggested to enhance the transpeptidation reaction (Lim and Spirin, 1986). The identity of the tRNA bound to the last codon, and the wobble base of this tRNA codon, can also affect the termination efficiency of the subsequent stop codon (Zhang et al., 1996). These observations suggest that the binding site of the RF complex on the ribosome is tightly constrained and that the termination efficiency is compromised by various substrates already in the center, such as the tRNA spanning the two subunits in the P site and the last two amino acids on the nascent polypeptide chain at the peptidyltransferase center.

The influence of sequences upstream and downstream of the stop codon on the stop signal prompted us to determine whether a minimal signal which com-

mSce, mKLa, and mSpo are mitochondrial RFs of *S. cerevisiae*, *Kluyveromyces lactis*, and *S. pombe*, respectively. The GenBank accession numbers are listed after the species names. Blocks of sequence similarity between RFs and IF1s were identified with MACAW. Similar alignments were obtained when the Blosum 62 and PAM250 scoring matrices were used. (C) The amino acid profile specifically selects all RF1s and RF2s and all but one of the IF1s from the current update of the Swiss-Prot protein database.

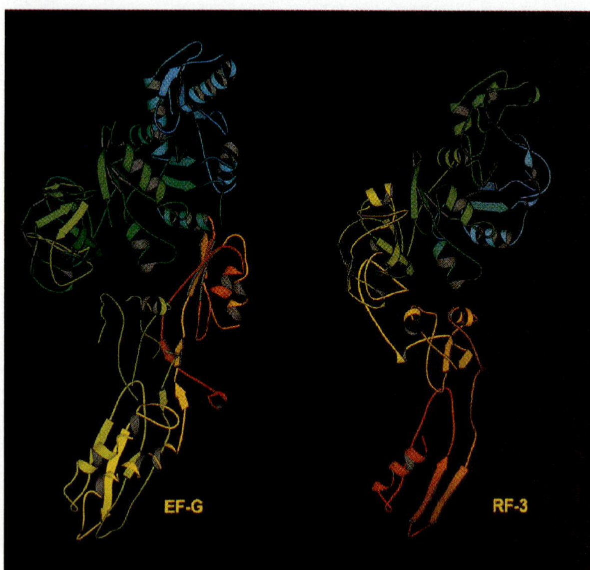

Figure 4. Predicted structure of RF3 based on threading analyses with EF-G. The predicted structure of RF3 is compared with the known three-dimensional structure of EF-G. Threading of RF3 sequences onto EF-G was performed with Swiss-Model (Guex and Peitsch, 1997).

bined both of these elements could be defined. After detailed analyses with TransTerm and combining all of the available experimental data, we predicted a minimal 12-base sequence element as the termination signal for prokaryotes (Major et al., unpublished). The stop signal may be regarded as being three-dimensional in that it reflects not only a direct interaction between the decoding factor and 6 nucleotides of the mRNA (the stop codon and the 3 nucleotides following) but also interactions between the decoding factor and the substrates that define the active center of the ribosome. To examine the combined effects of the upstream and downstream sequences on termination efficiency, signals were designed which included the strongest or the weakest predicted upstream and downstream sequences. In addition, hybrid constructs were designed which contained the strongest predicted upstream sequences coupled with the weakest downstream sequences and vice versa. The predicted signals were tested experimentally by using the 3A′ reporter system developed by Isaksson and coworkers (the data for the UGA stop codon are presented in Fig. 5). This had the advantage that the strengths of the signals could be tested at natural termination sites in vivo in strains lacking suppressor tRNAs but also under greater competition at recoding sites in the presence of specific suppressor tRNAs. Since there are many genes in *E. coli* using UAA- and UGA-containing signals, and since these signals are used for highly expressed genes, the information to design the strongest elements was far more reliable than for UAG-containing signals. There are few UAG-containing signals in the *E. coli* genome, and the only signals containing UAG that are used by highly expressed genes contain tandem stop codons. A separate study of signals containing tandem stop codons has also been undertaken. The efficiencies of the signals containing UAA and UGA codons were exactly as predicted, particularly when in competition with a suppressor tRNA. Even the weakest UAA-containing signals were highly efficient unless they were under competition, and then there was a relatively large difference between the efficiencies of the strongest and weakest signals, with the hybrids falling in between the two extremes. Similarly, UGA-containing signals behaved as predicted from the designs, although in the absence of competition the weak signals performed poorly in termination and the hybrids also gave intermediate levels of readthrough. In contrast, our predictions for strong signals for UAG were not upheld, reflecting the lack of data available to direct the design of the element. However, when we used a tandem stop codon in the signal, reflecting what is found in the two UAG-containing highly expressed genes, they performed according to prediction (Major et al., unpublished).

These data suggest that decoding of the stop signal in the A site of the ribosome occurs through the contact of an anticodon-like region of the RF (predicted by Nakamura and colleagues) with at least 6 nucleotides in the mRNA. The extended termination signal provides interactions that influence the efficiency of the decoding process and thereby the ability of a near-cognate tRNA or a recoding event to compete successfully with the termination event. In contrast, the upstream part of the stop signal (the stop codon) will influence the orientation of the RF complex in the A site.

ORIENTATION OF THE TERMINATION COMPLEX AND DYNAMICS OF THE TERMINATION EVENT

Assays used routinely to study termination have been invaluable for dissection of the individual steps of the termination event and to probe the interactions of the components within the termination complex at the ribosomal A site. However, they have not been adequate to study the dynamics of the termination event. For example, assays developed with mini-mRNAs (with an initiation and termination codon in tandem) displayed variable specific activities for the factor-mediated events (McCaughan et al., 1998). This was deduced to result from an unfavorable ori-

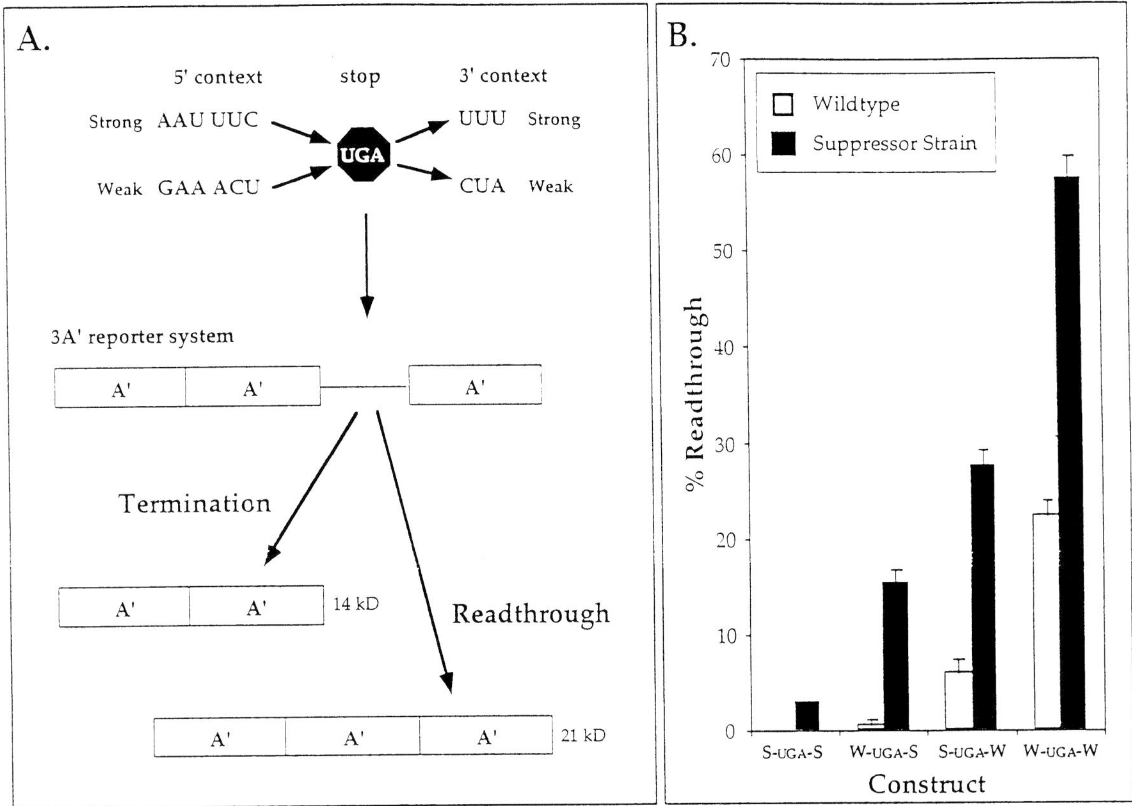

Figure 5. The 3A′ reporter system was used to analyze the upstream and downstream context effects on translations termination. (A) Sequences that were predicted to provide a strong or weak context were placed upstream (5′) and downstream (3′) of the stop codon (in this case, UGA). The termination signals were placed between the second and third A′ domains in the 3A' reporter plasmid. Expression from this plasmid generates two potential products, a termination product (14 kDa) and a readthrough product (21 kDa), which can be separated on the basis of size. (B) Expression of the 3A′ reporter plasmid containing UGA stop signals in wild-type (white) and suppressor (black) tRNA strains is presented as percent readthrough. Combinations of strong (S) and weak (W) upstream and downstream contexts are illustrated (for example, a strong upstream and weak downstream signal is S-UGA-W). The error bars indicate standard errors of the mean.

entation of the RF within the termination complex for release of the model peptide. If the orientation of the stop codon ultimately governs the orientation of the decoding RF, then these start-stop mini-RNAs must incorrectly display the stop codon in the A site. The fact that addition of a triplet stop codon restores release activity to these complexes suggests that the "real" A site is not occupied by the stop codon of the mini-RNA. Splitting the start-stop codons or putting a spacer between them in the mini-RNAs also restores the release activity (McCaughan et al., 1998), supporting the concept of a distorted orientation between the signal and the factor. The ability of various substitutions at position 246 in the "switch region" of RF2 to confer different binding and release phenotypes on the molecule is also indicative of this phenomenon and illustrates the sensitivity of the peptide release mechanism to the positioning of the decoding RF at the A site.

The sophisticated in vitro termination assay developed by Pavlov, Ehrenberg, and coworkers has been a very welcome addition to the battery of techniques available to study the dynamics of termination (Pavlov and Ehrenberg, 1996). This assay is done with mini-mRNAs but requires both initiation and elongation events to proceed before the termination and recycling events occur. Therefore, it mimics all steps of protein synthesis and has enabled a series of elegant studies defining the kinetic influences of the factors and their concentrations and of the template mRNA to be assessed (see chapter 44). The requirement for purified factors from all stages makes this assay inappropriate for routine use; thus, there is a need for an assay of intermediate sophistication that will have some of the power of the Ehrenberg assay system but will be easier for interested laboratories to utilize. Such an assay could incorporate features of the in vitro elongation assay that was established by Nierhaus and coworkers (Nierhaus, 1990; Nierhaus et al., 1995). Providing a single elongation cycle before termination occurs might be sufficient to overcome the inhibitory constraint of the initiation context seen in the assays based on start-stop mini-mRNAs. This would leave a dipeptidyl-tRNA at the P site and might correctly position the stop codon in the ribosomal A site. A stable posttranslocation complex would have the stop codon at the A site and a deacylated tRNA at the E site in addition to the dipeptidyl-tRNA at the P site—a situation closely mimicking the natural events occurring at termination. However, our in vivo studies of the RF2 recoding site show that perturbing the concentration of any one interacting ligand can distort quite markedly the dynamics of the translational event. Clearly it is difficult to simulate the exact balance of concentrations of the competing or influencing ligands for a particular event in vitro.

The complete in vitro translation assay (Pavlov and Ehrenberg, 1996) has been particularly useful in resolving the function of RF3. The discovery that RF3 was a translational G protein was puzzling, considering early work which examined the effect of guanine nucleotides on RF3 activity. Both GTP and GDP had been reported to decrease the binding of the decoding RFs to the ribosome and the rate of dissociation of the termination complex, whereas nucleotide-free RF3 favored termination complex formation and stimulated the release activity of the decoding factor. Was nucleotide-free RF3 acting as an association factor to enhance the binding affinity of the other RFs, analogous to EF-Tu function, whereas the guanine nucleotide form of RF3 was acting as a dissociation factor to recycle the decoding RFs, analogous to EF-G? Structural mimicry could support a role for either of these functions. Certainly, the nucleotide-free but not the nucleotide-associated form of RF3 was shown to bind to the ribosome cooperatively with the decoding RF (Pel et al., 1998). Our threading of the RF3 structure against EF-G, implying that RF3 occupies some of domain IV of EF-G adjacent to the putative anticodon domain of the decoding factor, also implies an interaction between the two classes of factors on the ribosome. No stable interaction between the two classes of factors independent of the ribosome has been demonstrated, unlike the strong association between the eukaryotic equivalents. A critical question is whether RF3 acts before or after hydrolysis of the peptidyl-tRNA. The complete translation system in vitro has shown that RF3 does not enhance the delivery rate of the decoding factor, as predicted by an EF-Tu-like model, nor the rate of peptidyl-tRNA hydrolysis but decreases the ribosome recycling time by stimulating the rate of dissociation of the decoding RF from the ribosome (see chapter 44). This is consistent with RF3 and GTP having no influence on the action of the decoding RF before hydrolysis of the peptidyl-tRNA and with RF3 catalyzing the dissociation of the decoding RF after transpeptidation. Much of the work on RF3 has been carried out with the nucleotide-free form, which represents a transient intermediate state in vivo or may have no physiological relevance. An A-site termination complex containing nucleotide-free RF3 and a decoding RF may indeed produce a more stable interaction between the two factors, allowing a higher probability of a productive release event in the simple triplet in vitro termination assay (Caskey et al., 1968) before the factors dissociate.

This model of RF3 action implies that it would be independent of the decoding function of the class I RFs. Therefore, it was a surprise to us when we found that the presence, absence, or overexpression of RF3 had differential effects on the efficiency of weak and strong stop signals at the RF2 recoding site in vivo. Termination at strong stop signals was enhanced to a greater extent than that at weak signals in the presence of RF3—"the strong got stronger" (Crawford et al., 1999). We interpreted this as indicating that the decoding factor has a higher affinity for strong stop signals, so much so that the dissociation rate compromises optimum recycling of the factor. The role of RF3 in this situation would be to increase the local concentration of the free decoding factor at the stop signals of other translating ribosomes, thus providing a competitive advantage to the termination event over frameshifting. We were able to simulate these effects in an *E. coli* RF3 null strain by modest overproduction (two- to fivefold) of the decoding factor itself (Crawford et al., 1999). This increased the local concentration to levels that might be mediated with RF3 and reinforced the suggestion that there was a differential benefit for strong stop signals over weak signals when under competition with recoding events at the recoding site. The implication of these findings for an organism's expression strategy is that highly expressed genes that have strong stop signals and require high translational efficiency gain benefit from having RF3 in the cell to allow rapid recycling of the pool of decoding factors. The free concentration of the factors is maintained at potential termination sites. Additionally, recoding sites, which often have the weakest stop signals, benefit less from the RF3 recycling function and would be detrimental to the cell only if these recoding events were particularly sensitive to RF3. Organisms which lack RF3 would have less flexibility in their expression patterns than organisms that include this factor in their repertoires.

Another class of recoding sites that have competition between the ligands affecting termination and specialized elongation ligands are the Sec (selenocysteine) incorporation sites. We have studied the formate dehydrogenase H site in detail. The stop signal has a strong upstream context and a weak downstream context, and in most growth conditions termination is overwhelmingly competitive against Sec incorporation. Changing the upstream context to the strongest one possible modestly improves termination, but with a much weaker context the Sec incorporation is significantly improved. In contrast, strengthening the weak downstream context significantly improves the competitiveness of termination. The upstream contexts of these sites reflect the amino acids adjacent to Sec at the active center of the protein; thus, the upstream context in the mRNA may be constrained by the substitutions allowable to maintain enzymatic activity. We have shown that increasing the concentration of either competing ligand, decoding RF2 or Sec-tRNA, at the Sec recoding site influences the outcome by favoring termination or Sec incorporation, respectively. This indicates that despite the secondary structure of the region with a unique function of carrying the SELB (the specialized elongation factor for Sec-tRNA, equivalent to EF-Tu) elongation factor·Sec-tRNA into the ribosome's active center (Heider et al., 1992; Hüttenhofer and Böck, 1998), there is direct competition between the competing ligands (Mansell and Tate, unpublished). Sec incorporation is favored in slow-growth and anaerobic environments. Under these circumstances the amount of decoding RF is low, although concentration may be protected by an altered cell volume (Adamski et al., 1994). However, under such conditions Sec incorporation may be favored over termination if the Sec machinery is not as sensitive to the growth rate as the decoding RF.

THE FATE OF THE POSTTERMINATION COMPLEX

The disassembly of the termination complex in prokaryotes was shown almost 30 years ago to involve two factors, a ribosome-releasing factor, now called ribosome-recycling factor (RRF), and EF-G (Hirashima and Kaji, 1970, 1972). RF3 can substitute for EF-G in RRF-driven reactions (Grentzmann and Kelly, 1997). Clearly, in the absence of RRF, ribosomes scan downstream of the stop signal and reinitiate, and scanning can extend as far as 45 nucleotides (Janosi et al., 1998). Ehrenberg and coworkers have shown that both RF3 and RRF are necessary for fast ribosome recycling, and the effects on the dissociation of the decoding RFs appear to be additive (see chapter 44). However, RRF contributes to a greater extent in decreasing ribosome recycling times than RF3. The cumulative effects of RF3 and RRF suggest they catalyze separate rate-limiting steps. RRF has been shown to influence translational fidelity by decreasing misincorporation of Ile during polyphenylalanine synthesis in vitro (Janosi et al., 1996). The importance of RRF to cellular physiology was seen when the inactivation of the gene generated bacteriocidal or bacteriostatic phenotypes. Furthermore, it has been identified in all prokaryotic genomes, including the *Mycoplasma* species (Fraser et al., 1995), as well as a number of organelle factors. At present, the equivalent in eukaryotes has not revealed itself, suggesting that RRF may be a suitable target for antibiotics (Kaji et al., 1998). On the other hand, this situation is similar to that of the termination field a short time ago, where, although an eRF was known, there seemed little similarity between the proteins and the mechanisms.

NOVEL SITUATIONS REQUIRING TERMINATION FACTORS

A surveillance complex has been identified in yeast which monitors mRNAs containing premature nonsense codons so that they can be targeted for decay. This can prevent the production of truncated proteins that may be toxic by interfering with the function of the normal protein. This process, termed nonsense-mediated decay (NMD), is ubiquitous in all eukaryotes examined (reviewed in Culbertson, 1999). The surveillance complex consists of at least three proteins, Upf1, Upf2, and Upf3 (Weng et al., 1996; He et al., 1997), with homologues in yeast, nematodes, and mammals, including humans. Upf3 seems to bind to all transcripts but may be displaced by full translation by the ribosome. Failure to displace Upf3 leads to formation of a complex with the other Upf components. Of particular interest for termination is the fact that Upf1 has been reported to stimulate termination (Czaplinski et al., 1998). This protein has been shown to interact with eRF1, and the role of Up1 in termination can be separated from its role in NMD (Weng et al., 1996). The current model suggests that Upf1 binding to eRF1 and eRF3 prevents NMD until termination has occurred. In yeast, the fate of the mRNA is dictated by two sequence elements downstream of the termination signal. The downstream element is recognized up to 200 nucleotides 3′ to the stop codon by the surveillance complex, and this targets the mRNA for degradation (Ruiz-Echevarria et al., 1998). A second stabilizer element may act as a counter to this process when it is between the stop signal and the downstream element.

In prokaryotes, potentially damaging protein fragments translated from damaged RNA are handled by another interesting mechanism. Here the 10Sa RNA (tmRNA) acts as both a tRNA and an mRNA for a ribosome stalled at a damaged RNA because its 3′ section has been truncated (Tu et al., 1995). The tmRNA allows insertion of Ala before acting as an mRNA encoding a peptide tag which is added to the protein fragment, signaling its destruction by ATP-dependent proteases. Importantly, this mechanism recovers trapped ribosomes (reviewed in Muto et al., 1998). It is interesting that in all prokaryotes examined, including *Mycoplasma*, the tmRNA gene has been retained (Williams, 1999). A preliminary survey of the stop signal usage by 56 members of this family has indicated that over 80% have the strong +4 base U or G, with a high proportion of tandem termination codons (~30%). One example from *Bacillus megaterium* has a triple stop codon in its signal. Apparently a highly efficient and fail-safe stop signal is important for this system.

The work from our own laboratory described here has been supported by a Howard Hughes Medical Institute International Investigator award to W.P.T., a Human Frontiers of Science Programme grant to W.P.T. (awarded jointly to Y. Nakamura, L. Kisselev, and M. Philippe), and grants from the Marsden Fund and the Health Research Council from New Zealand.

REFERENCES

Adamski, F. M., K. K. McCaughan, F. Jorgensen, C. G. Kurland, and W. P. Tate. 1994. The concentration of polypeptide chain release factors 1 and 2 at different growth rates of *Escherichia coli*. *J. Mol. Biol.* **238:**302–308.

Askarian-Amiri, M. E., H. J. Pel, and W. P. Tate. Unpublished data.

Björnsson, A., S. Mottagui-Tabar, and L. A. Isaksson. 1996. Structure of the C-terminal end of the nascent peptide influences translation termination. *EMBO J.* **15:**1696–1704.

Brock, S., K. Szkaradkiewicz, and M. Sprinzl. 1998. Initiation factors of protein biosynthesis in bacteria and their structural relationship to elongation and termination factors. *Mol. Microbiol.* **29:**409–417.

Brown, C. M., and W. P. Tate. 1994. Direct recognition of mRNA stop signals by *Escherichia coli* polypeptide chain release factor 2. *J. Biol. Chem.* **269:**33164–33170.

Brown, C. M., M. E. Dalphin, P. A. Stockwell, and W. P. Tate. 1993. The translational termination signal database. *Nucleic Acids Res.* **21:**3119–3123.

Caskey, C. T., R. Tompkins, E. Scolnick, T. Caryk, and M. Nirenberg. 1968. Sequential translation of trinucleotide codons for the initiation and termination of protein synthesis. *Science* **162:** 135–138.

Crawford, D.-J. G., K. Ito, Y. Nakamura, and W. P. Tate. 1999. Indirect regulation of translational termination efficiency at highly expressed genes and recoding sites by the factor recycling function of *Escherichia coli* release factor RF3. *EMBO J.* **18:** 727–732.

Culbertson, M. R. 1999. RNA surveillance—unforeseen consequences for gene expression, inherited genetic disorders and cancer. *Trends Genet.* **15:**74–80.

Czaplinski, K., M. J. Ruiz-Echevarria, S. V. Paushkin, X. Han, Y. M. Weng, H. A. Perlick, H. C. Dietz, M. D. Ter-Avanesyan, and S. W. Peltz. 1998. The surveillance complex interacts with the translation release factors to enhance termination and degrade aberrant mRNAs. *Genes Dev.* **12:**1665–1677.

Dalphin, M. E., and W. P. Tate. Unpublished data.

Dalphin, M. E., C. M. Brown, P. A. Stockwell, and W. P. Tate. 1998. The translational signal database, TransTerm, is now a relational database. *Nucleic Acids Res.* **26:**335–337.

Doel, S. M., S. J. McCready, C. R. Nierras, and B. S. Cox. 1994. The dominant *PNM2*− mutation which eliminates the *psi* factor of *Saccharomyces cerevisiae* is the result of a missense mutation in the *SUP35* gene. *Genetics* **137:**659–670.

Donly, B. C., C. D. Edgar, F. M. Adamski, and W. P. Tate. 1990. Frameshift autoregulation in the gene for *Escherichia coli* release factor 2—partly functional mutants result in frameshift enhancement. *Nucleic Acids Res.* **18:**6517–6522.

Eaglestone, S. S., B. S. Cox, and M. F. Tuite. 1999. Translation termination efficiency can be regulated in *Saccharomyces cerevisiae* by environmental stress through a prion-mediated mechanism. *EMBO J.* **18:**1974–1981.

Elliott, T., and X. H. Wang. 1991. *Salmonella typhimurium prfA* mutants defective in release factor 1. *J. Bacteriol.* **173:**4144–4154.

Fraser, C. M., J. D. Gocayna, O. White, M. D. Adams, R. A. Clayton, R. D. Fleischmann, C. J. Bult, A. R. Kerlavage, G. Sutton, J. M. Kelley, J. L. Fritchman, J. F. Weidman, K. V. Small, M. Sandusky, J. Fuhrman, D. Nguyen, T. R. Utterback, D. M. Saudek, C. A. Phillips, J. M. Merrick, J.-F. Tomb, B. A. Dougherty, K. F. Bott, P.-C. Hu, T. S. Lucier, S. N. Peterson, H. O. Smith, C. A. Hutchison III, and J. C. Venter. 1995. The minimal gene complement of *Mycoplasma genitalium*. *Science* **270:**397–403.

Frolova, L. Y., M. E. Dalphin, J. Justesen, R. J. Powell, G. Drugeon, K. K. McCaughan, L. L. Kisselev, W. P. Tate, and A. L. Haenni. 1993. Mammalian polypeptide chain release factor and tryptophanyl-transfer RNA synthetase are distinct proteins. *EMBO J.* **12:**4013–4019.

Frolova, L., X. Le Goff, H. H. Rasmussen, S. Cheperegin, G. Drugeon, M. Kress, I. Arman, A. L. Haenni, J. E. Celis, M. Philippe, J. Justesen, and L. Kisselev. 1994. A highly conserved eukaryotic protein family possessing properties of polypeptide chain release factor. *Nature* **372:**701–703.

Frolova, L., X. Le Goff, G. Zhouravleva, E. Davydova, M. Philippe, and L. Kisselev. 1996. Eukaryotic polypeptide chain release factor eRF3 is an eRF1- and ribosome-dependent guanosine triphosphatase. *RNA* **2:**334–341.

Grentzmann, G., and P. J. Kelly. 1997. Ribosomal binding site of release factors RF1 and RF2—a new translational termination assay *in vitro*. *J. Biol. Chem.* **272:**12300–12304.

Grentzmann, G., D. Brechemierbaey, V. Heurgue-Hamard, L. Mora, and R. H. Buckingham. 1994. Localization and characterization of the gene encoding release factor RF3 in *Escherichia coli*. *Proc. Natl. Acad. Sci. USA* **91:**5848–5852.

Guex, N., and M. C. Peitsch. 1997. SWISS-MODEL and the Swiss-PdbViewer: an environment for comparative protein modeling. *Electrophoresis* **18:**2714–2723.

He, F., A. H. Brown, and A. Jacobson. 1997. Upf1p, Nmd2p, and Upf3p are interacting components of the yeast nonsense-mediated mRNA decay pathway. *Mol. Cell. Biol.* **17:**1580–1594.

Heider, J., C. Baron, and A. Bock. 1992. Coding from a distance: dissection of the mRNA determinants required for the incorporation of selenocysteine into protein. *EMBO J.* **11:**3759–3766.

Hirashima, A., and A. Kaji. 1970. Factor dependent breakdown of polysomes. *Biochem. Biophys. Res. Comm.* **41:**877–883.

Hirashima, A., and A. Kaji. 1972. Factor-dependent release of ribosomes from mRNA. Requirement for two heat stable factors. *J. Mol. Biol.* **65:**43–58.

Hoshino, S., H. Miyazawa, T. Enomoto, F. Hanaoka, Y. Kikuchi, A. Kikuchi, and M. Ui. 1989. A human homologue of the yeast *GST1* gene codes for a GTP-binding protein and is expressed in a proliferation-dependent manner in mammalian cells. *EMBO J.* **8:**3807–3814.

Hüttenhofer, A., and A. Böck. 1998. Selenocysteine inserting RNA elements modulate GTP hydrolysis of elongation factor SelB. *Biochemistry* **37:**885–890.

Ito, K., K. Ebihara, M. Uno, and Y. Nakamura. 1996. Conserved motifs in prokaryotic and eukaryotic polypeptide release factors: tRNA-protein mimicry hypothesis. *Proc. Natl. Acad. Sci. USA* **93:**5443–5448.

Ito, K., K. Ebihara, and Y. Nakamura. 1998a. The stretch of C-terminal acidic amino acids of translational release factor eRF1 is a primary binding site for eRF3 of fission yeast. *RNA* **4:**958–972.

Ito, K., M. Uno, and Y. Nakamura. 1998b. Single amino acid substitution in prokaryote polypeptide release factor 2 permits it to terminate translation at all three stop codons. *Proc. Natl. Acad. Sci. USA* **95:**8165–8169.

Janosi, L., R. Ricker, and A. Kaji. 1996. Dual functions of ribosome recycling factor in protein biosynthesis: disassembling the termination complex and preventing translational errors. *Biochimie* **78:**959–969.

Janosi, L., S. Mottagui-Tabar, L. A. Isaksson, Y. Sekine, E. Ohtsubo, S. Zhang, S. Goon, S. Nelken, M. Shuda, and A. Kaji. 1998. Evidence for *in vivo* ribosome recycling, the fourth step in protein biosynthesis. *EMBO J.* **17:**1141–1151.

Kaji, A., E. Teyssier, and G. Hirokawa. 1998. Disassembly of the post-termination complex and reduction of translational error by ribosome recycling factor (RRF)—a possible new target for antibacterial agents. *Biochem. Biophys. Res. Commun.* **250:**1–4.

Kawakami, K., and Y. Nakamura. 1990. Autogenous suppression of an opal mutation in the gene encoding peptide chain release factor 2. *Proc. Natl. Acad. Sci. USA* **87:**8432–8436.

Konecki, D. S., K. C. Aune, W. P. Tate, and C. T. Caskey. 1977. Characterization of reticulocyte release factor. *J. Biol. Chem.* **252:**4514–4520.

Lee, C. C., K. M. Timms, C. N. A. Trotman, and W. P. Tate. 1987. Isolation of a rat mitochondrial release factor. Accommodation of the changed genetic code for termination. *J. Biol. Chem.* **262:**3548–3552.

Lee, C. C., W. J. Craigen, D. M. Muzny, E. Harlow, and C. Caskey. T. 1990. Cloning and expression of a mammalian peptide chain release factor with sequence similarity to tryptophanyl-tRNA synthetases. *Proc. Natl. Acad. Sci. USA* **87:**3508–3512.

Lim, V. I., and A. S. Spirin. 1986. Stereochemical analysis of ribosomal transpeptidation. Conformation of nascent peptide. *J. Mol. Biol.* **188:**565–574.

Major, L. L., L. A. Isaksson, and W. P. Tate. Unpublished data.

Major, L. L., E. S. Poole, M. E. Dalphin, S. A. Mannering, and W. P. Tate. 1996. Is the in-frame termination signal of the *Escherichia coli* release factor 2 frameshift site weakened by a particularly poor context? *Nucleic Acids Res.* **24:**2673–2678.

Mansell, J. B., and W. P. Tate. Unpublished data.

McCaughan, K. K., M. J. Berry, and W. P. Tate. Unpublished data.

McCaughan, K. K., E. S. Poole, H. J. Pel, J. B. Mansell, S. A. Mannering, and W. P. Tate. 1998. Efficient *in vitro* translational termination in *Escherichia coli* is constrained by the orientations of the release factor, stop signal and peptidyl-tRNA within the termination complex. *Biol. Chem.* **379:**857–866.

Merkulova, T. I., L. Y. Frolova, M. Lazar, J. Camonis, and L. L. Kisselev. 1999. C-terminal domains of human translation termination factors eRF1 and eRF3 mediate their *in vivo* interaction. *FEBS Lett.* **443:**41–47.

Mikuni, O., K. Kawakami, and Y. Nakamura. 1991. Sequence and functional analysis of mutations in the gene encoding peptide-chain-release factor 2 of *Eschericia coli. Biochimie* **73:**1509–1516.

Mikuni, O., K. Ito, J. Moffat, K. Matsumura, K. McCaughan, T. Nobukuni, W. Tate, and Y. Nakamura. 1994. Identification of the *prfC* gene, which encodes peptide-chain release factor 3 of *Escherichia coli. Proc. Natl. Acad. Sci. USA* **91:**5798–5802.

Moffat, J. G., and W. P. Tate. 1994. A single proteolytic cleavage in release factor 2 stabilizes ribosome binding and abolishes peptidyl-tRNA hydrolysis activity. *J. Biol. Chem.* **269:**18899–18903.

Mottagui-Tabar, S., A. Bjornsson, and L. A. Isaksson. 1994. The second to last amino acid in the nascent peptide as a codon context determinant. *EMBO J.* **13:**249–257.

Murzin, A. G. 1993. OB (oligonucleotide/oligosaccharide binding)-fold: common structural and functional solution for non-homologous sequences. *EMBO J.* **12:**861–867.

Muto, A., C. Ushida, and H. Himeno. 1998. A bacterial RNA that functions as both a tRNA and an mRNA. *Trends Biochem. Sci.* **23:**25–29.

Nakamura, Y., and K. Ito. 1998. How protein reads the stop codon and terminates translation. *Genes Cells* **3:**265–278.

Nierhaus, K. H. 1990. The allosteric three-site model for the ribosomal elongation cycle—features and future. *Biochemistry* **29:**4997–5008.

Nierhaus, K. H., D. Beyer, M. Dabrowski, M. A. Schäfer, C. M. T. Spahn, J. Wadzack, K.-U. Bittner, N. Burkhardt, G. Diedrich, R. Jünemann, D. Kamp, H. Voss, and H. B. Stuhrmann. 1995. The elongating ribosome: structural and functional aspects. *Biochem. Cell. Biol.* **73:**1011–1021.

Nyborg, J., P. Nissen, M. Kjeldgaard, S. Thirup, G. Polekhina, B. F. C. Clark, and L. Reshetnikova. 1996. Structure of the ternary complex of EF-Tu: macromolecular mimicry in translation. *Trends Biochem. Sci.* **21:**81–82.

Olafsson, O., J. U. Ericson, R. Vanbogelen, and G. R. Björk. 1996. Mutation in the structural gene for release factor 1 (RF-1) of *Salmonella typhimurium* inhibits cell division. *J. Bacteriol.* **178:**3829–3839.

Patino, M. M., J. J. Liu, J. R. Glover, and S. Lindquist. 1996. Support for the prion hypothesis for inheritance of a phenotypic trait in yeast. *Science* **273:**622–626.

Paushkin, S. V., V. V. Kushnirov, V. N. Smirnov, and M. D. Ter-Avanesyan. 1996. Propagation of the yeast prion-like [*psi*+] determinant is mediated by oligomerization of the SUP35-encoded polypeptide chain release factor. *EMBO J.* **15:**3127–3134.

Pavlov, M. Y., and M. Ehrenberg. 1996. Rate of translation of natural mRNAs in an optimized *in vitro* system. *Arch. Biochem. Biophys.* **328:**9–16.

Pel, H. J., and W. P. Tate. Unpublished data.

Pel, H. J., C. Maat, M. Rep, and L. A. Grivell. 1992a. The yeast nuclear gene *MRF1* encodes a mitochondrial peptide chain release factor and cures several mitochondrial RNA splicing defects. *Nucleic Acids Res.* **20:**6339–6346.

Pel, H. J., M. Rep, and L. A. Grivell. 1992b. Sequence comparison of new prokaryotic and mitochondrial members of the polypeptide chain release factor family predicts a 5-domain model for release factor structure. *Nucleic Acids Res.* **20:**4423–4428.

Pel, H. J., M. Rep, H. J. Dubbink, and L. A. Grivell. 1993. Single point mutations in domain-II of the yeast mitochondrial release factor mRF-1 affect ribosome binding. *Nucleic Acids Res.* **21:**5308–5315.

Pel, H. J., J. G. Moffat, K. Ito, Y. Nakamura, and W. P. Tate. 1998. *Escherichia coli* release factor 3: resolving the paradox of a typical G protein structure and atypical function with guanine nucleotides. *RNA* **4:**47–54.

Poole, E. S., and W. P. Tate. Unpublished data.

Poole, E. S., R. Brimacombe, and W. P. Tate. 1997. Decoding the translational termination signal: the polypeptide chain release factor in *Escherichia coli* crosslinks to the base following the stop codon. *RNA* **3:**974–982.

Poole, E. S., L. L. Major, S. A. Mannering, and W. P. Tate. 1998. Translational termination in *Escherichia coli*: three bases following the stop codon crosslink to release factor 2 and affect the decoding efficiency of UGA-containing signals. *Nucleic Acids Res.* **26:**954–960.

Ruiz-Echevarria, M. J., C. I. Gonzalez, and S. W. Peltz. 1998. Identifying the right stop: determining how the surveillance complex recognizes and degrades an aberrant mRNA. *EMBO J* **17:**575–589.

Ryden, S. M., and L. A. Isaksson. 1984. A temperature sensitive mutant of *Escherichia coli* that shows enhanced misreading of UAG/A and increased efficiency for some tRNA nonsense suppressors. *Mol. Gen. Genet.* **193:**38–45.

Sette, M., P. van Tilborg, R. Spurio, R. Kaptein, M. Paci, C. O. Gualerzi, and R. Boelens. 1997. The structure of the translational initiation factor IF1 from *Escherichia coli* contains an oligomer-binding motif. *EMBO J.* **16:**1436–1443.

Stansfield, I., G. M. Grant, Akhmaloka, and M. F. Tuite. 1992. Ribosomal association of the yeast *SAL4* (*SUP45*) gene product: implications for its role in translation fidelity and termination. *Mol. Microbiol.* **6:**3469–3478.

Tate, W. P., and S. A. Mannering. 1996. Three, four or more: the translational stop signal at length. *Mol. Microbiol.* **21:**213–219.

Ter-Avanesyan, M. D., A. R. Dagkesamanskaya, V. V. Kushnirov, and V. N. Smirnov. 1994. The *SUP35* omnipotent suppressor gene is involved in the maintenance of the non-Mendelian determinant [*psi*+] in the yeast *Saccharomyces cerevisiae*. *Genetics* **137:**671–676.

Tu, G. F., G. E. Reid, J. G. Zhang, R. L. Moritz, and R. J. Simpson. 1995. C-terminal extension of truncated recombinant proteins in *Escherichia coli* with a 10Sa RNA decapeptide. *J. Biol. Chem.* **270:**9322–9326.

Uno, M., K. Ito, and Y. Nakamura. 1996. Functional specificity of amino acid at position 246 in the tRNA mimicry domain of bacterial release factor 2. *Biochimie* **78:**935–943.

Urbero, B., L. Eurwilaichitr, I. Stansfield, J. P. Tassan, X. Le Goff, M. Kress, and M. F. Tuite. 1997. Expression of the release factor eRF1 (Sup45p) gene of higher eukaryotes in yeast and mammalian tissues. *Biochimie* **79:**27–36.

Weng, Y. M., K. Czaplinski, and S. W. Peltz. 1996. Identification and characterization of mutations in the *UPF1* gene that affect nonsense suppression and the formation of the Upf protein complex but not mRNA turnover. *Mol. Cell. Biol.* **16:**5491–5506.

Williams, K. P. 1999. The tmRNA website. *Nucleic Acids Res.* **27:** 165–166.

Wilson, D. N. 1999. A conformation switch region between functional domains regulates activity of RF2. Ph.D. thesis. University of Otago, Dunedin, New Zealand.

Wilson, K. S., and H. F. Noller. 1998. Mapping the position of translational elongation factor EF-G in the ribosome by directed hydroxyl radical probing. *Cell* **92:**131–139.

Wilson, K. S., and W. P. Tate. Unpublished data.

Wu, E. D., H. Inokuchi, and H. Ozeki. 1990. Identification of the mutations in the *prfB* gene of *Escherichia coli K12*, which confer UGA suppressor activity. *Jpn. J. Genet.* **65:**115–119.

Zhang, S. P., M. Ryden-Aulin, and L. A. Isaksson. 1996. Functional interaction between release factor 1 and P site peptidyl-tRNA on the ribosome. *J. Mol. Biol.* **261:**98–107.

Zhouravleva, G., L. Frolova, X. Le Goff, R. Leguellec, S. Inge-Vechtomov, L. Kisselev, and M. Philippe. 1995. Termination of translation in eukaryotes is governed by two interacting polypeptide chain release factors, eRF1 and eRF3. *EMBO J.* **14:** 4065–4072.

The Ribosome: Structure, Function, Antibiotics, and Cellular Interactions
Edited by R. A. Garrett, S. R. Douthwaite, A. Liljas, A. T. Matheson, P. B. Moore, and H. F. Noller

Chapter 41

rRNA Functional Sites and Structures for Peptide Chain Termination

EMANUEL J. MURGOLA, ALEXEY L. ARKOV, NATALYA S. CHERNYAEVA, KLAS O. F. HEDENSTIERNA, and FRANCES T. PAGEL

It is well established that rRNAs play an active role in every aspect of translation. However, until recently, it has been unclear which sites and structures participate specifically in the termination stage and in the mechanisms for recognition of the termination codons, for transmission of the signal to the hydrolytic center, and finally, for hydrolysis of peptidyl-tRNA to free the completed protein. This chapter gives an overview of recent work in our laboratory that has demonstrated the roles of two rRNA sites, one in 16S rRNA (the small subunit) and the other in 23S rRNA (the large subunit), in release factor (RF) binding and in catalysis of peptidyl-tRNA hydrolysis and has implicated five other regions in termination, with one (in 23S rRNA) being a likely part of the hydrolytic center. Our results thus far verify the validity of trying to isolate termination-defective mutants of rRNA by searching for nonsense suppressors that exhibit specificity for one or two of the termination codons, at least under some conditions.

rRNA IN TERMINATION: "IT TAKES TWO TO TANGO"

When the A site of the ribosome encounters a stop codon, UGA, UAA, or UAG, a cytoplasmic protein RF binds to the ribosome and steps are triggered that lead to the hydrolysis of peptidyl-tRNA. In *Escherichia coli*, there are two RFs that are necessary for termination and that function in a codon-dependent manner. RF1 works at UAG and UAA, and RF2 works at UGA and UAA. However, the precise role of rRNA in this process has been unclear (for reviews, see Buckingham et al., 1997; Murgola, 1996; Nakamura et al., 1996; and Tate et al., 1996). Here we present an overview of recent work in our laboratory that has implicated specific rRNA sites and structures in peptide chain termination. The most definitive and significant of our results at the moment is the demonstration that termination requires the participation of RNAs from both ribosomal subunits, that those RNAs utilize highly conserved sites and structures, and that they are intimately involved in the binding of RF2 to the ribosome, the recognition and transmission of the termination signal for hydrolysis, and the catalysis of hydrolysis itself.

SEARCHES FOR TERMINATION-DEFECTIVE MUTANTS

It has been said that, in mutant hunts, "you get what you select for." Thus, it seems reasonable to try to get termination-defective mutants of rRNA by looking for rRNA nonsense suppressors. However, when the ribosome encounters a termination codon in the A site, there is a competition between the termination mechanism (that is, recognition of the termination codon and subsequent, unidentified steps leading to hydrolysis of peptidyl-tRNA) and the potential misreading of the termination codon by a normal tRNA. Therefore, in general, genetic nonsense suppression can be caused either by a defect in peptide chain termination, resulting in decreased competition with a normal tRNA for misreading of the termination codon, or by increased mistranslation (decreased accuracy), resulting in a direct increase in

Emanuel J. Murgola, Alexey L. Arkov, Natalya S. Chernyaeva, Klas O. F. Hedenstierna, and Frances T. Pagel ■ Department of Molecular Genetics (Box 11), The University of Texas M. D. Anderson Cancer Center, 1515 Holcombe Blvd., Houston, TX 77030.

misreading of the termination codon by one or more normal tRNAs. We have reasoned, however, that rRNA mutations that lead to codon-specific nonsense suppression, that is, readthrough of only specific termination codons, and not to suppression of related missense codons are more likely to be primarily defective in chain termination than to represent increased misreading per se by tRNAs (Murgola, 1996; Murgola et al., 1988, 1995). That reasoning was supported recently by in vitro experiments (see below) in which the ribosomes from two UGA-specific rRNA suppressor mutants were shown to be defective in termination, preferentially exhibiting major defects in termination dependent on RF2, which works specifically at UGA.

A TALE OF TWO SUPPRESSORS

RNAs from both ribosomal subunits were implicated in termination in vivo with the isolation and characterization of two UGA-specific rRNA nonsense suppressor mutations, the first, C1054A, in 16S rRNA and the second, G1093A, in 23S rRNA. Further analyses of these mutations both in vivo and in vitro in a realistic termination assay system have now established the roles in termination not only of the particular nucleotides but also of the structures in which they are located, that is, helix 34 of 16S rRNA and the GTPase-associated center of 23S rRNA.

C1054A, a 16S rRNA mutation at nucleotide (nt) 1054, which is a virtually universally conserved nucleotide (Van de Peer et al., 1998) of helix 34 (Fig. 1A), was isolated as a suppressor of UGA nonsense mutations, spontaneously and in the chromosome, and further, was shown not to suppress either UAA or UAG at the same reporter gene positions and under the same growth conditions that revealed UGA suppression (Murgola et al., 1988; Pagel et al., 1997). In addition, A1054 did not suppress missense mutations (in particular, a UGG missense mutation and several other missense codons differing from a stop codon by just 1 nt) or two +1 and one −1 frameshift mutations (Pagel et al., 1997). Since A1054 was codon specific in its nonsense suppression and gave no evidence of affecting misreading events under the conditions examined, the in vivo data were interpreted as indicating a defect in termination (Murgola et al., 1988; Pagel et al., 1997). The 23S rRNA mutation, G1093A, was obtained in a cloned copy of one of the rRNA operons (*rrnB*) replicating in a *mutD* strain (Jemiolo et al., 1995). It was isolated as a UGA-specific suppressor, that is, it did not suppress UAA and UAG nonsense mutations or several missense mutations, including UGG, under the same conditions that revealed UGA suppression. Therefore, as with C1054A in 16S rRNA, it was suggested that A1093 caused a defect in termination (Jemiolo et al., 1995). The virtually universally conserved nucleotide G1093 is located in the GTPase center (Fig. 1B), a highly conserved structure involved in elongation factor G (EF-G)- and EF-Tu-dependent hydrolysis of GTP (Saarma et al., 1997; Thompson, 1996).

Consistent with the in vivo characteristics of the C1054A and G1093A mutations were the recent results (Arkov et al., 1998a, submitted) of experiments in which the ribosomes from those two mutant strains were examined in a realistic in vitro termination assay (Freistroffer et al., 1997; Pavlov et al., 1998). C1054A decreased the effective association rate constant, k_{cat}/K_m, for the ribosome and RF2, and the catalytic rate of peptidyl-tRNA hydrolysis after RF2 was bound to the ribosome. On the other hand, C1054A caused only a very small decrease in k_{cat}/K_m for RF1-dependent termination, which was significantly smaller than that in the presence of RF2. Furthermore, the catalytic rate of peptidyl-tRNA hydrolysis in the presence of RF1 was very close to normal. These in vitro data (Arkov et al., 1998a, submitted) provided an explanation for the UGA-specific suppression caused by C1054A in vivo, namely, that mutant ribosomes read through UGA because they were defective in RF2-dependent termination but that readthrough of UAA and UAG was not seen due to the almost-normal function of RF1 at these codons. Furthermore, other nucleotides of helix 34 have been implicated in termination (Fig. 1A, bottom left). In particular, the mutation C1200U also caused readthrough of nonsense codons (Moine and Dahlberg, 1994; Murgola et al., 1995). However, similar to A1054, it did not suppress several missense mutations, including UGG, AAG, and GAA (Pagel and Murgola, unpublished). In addition, changes of C1192 in the helix caused defects in the binding of both RF1 and RF2 to the ribosome in vitro (Brown et al., 1993). Similar to the effect of C1054A on termination, defects in RF binding caused by the mutations C1192U and C1192G were larger for RF2 than for RF1.

In the same realistic in vitro assays used for the C1054A analyses (Freistroffer et al., 1997; Pavlov et al., 1998), G1093A severely decreased the effective association rate constant, k_{cat}/K_m, for RF2 and the ribosome. At UAA, k_{cat}/K_m for mutant ribosomes was about 20 times lower than for wild-type ribosomes, and at UGA it was 14 times lower. As opposed to the large defect in RF2-dependent termination, G1093A caused only a 46% decrease in k_{cat}/K_m for RF1 at UAA. Furthermore, the mutation decreased the cat-

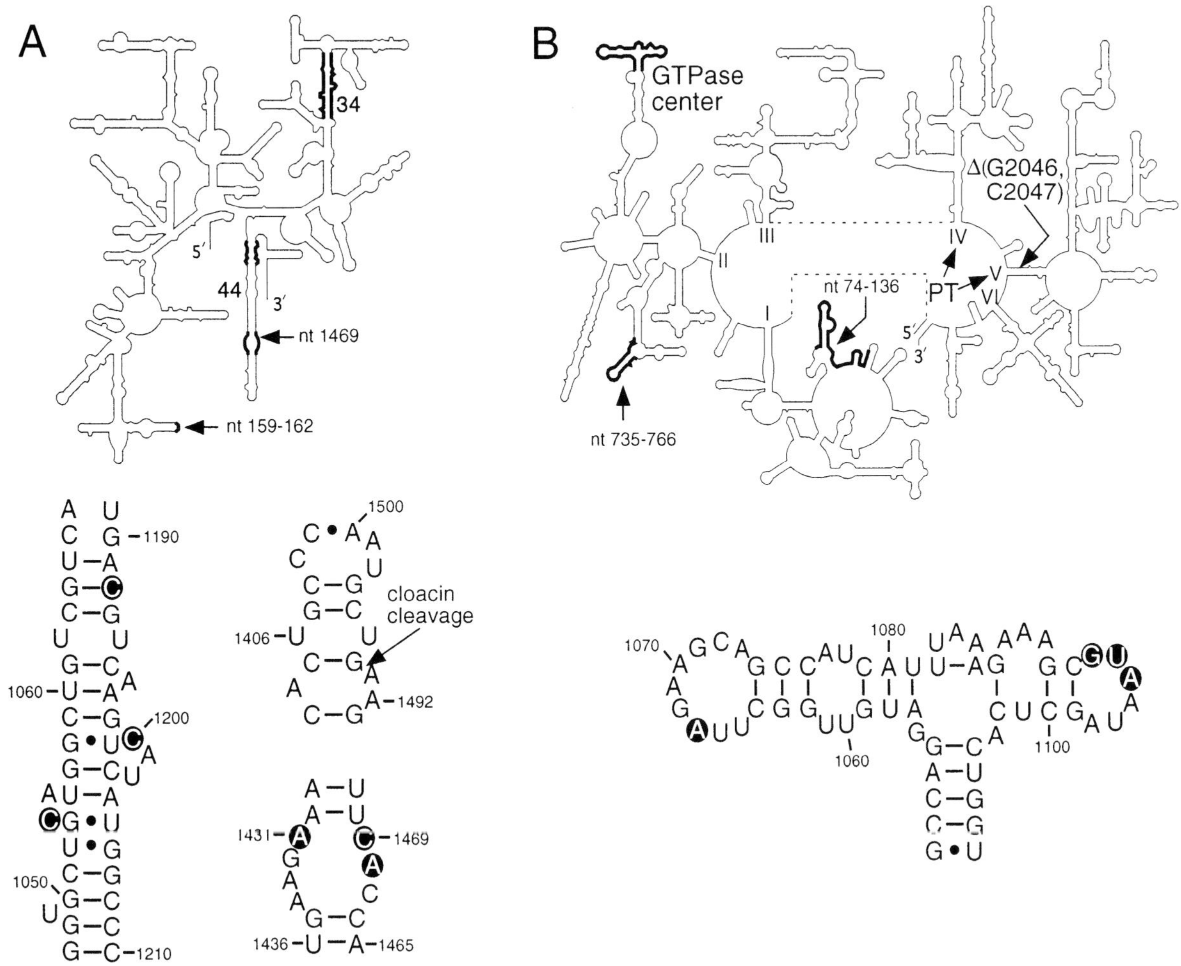

Figure 1. rRNA regions implicated in translation termination. (A) (Top) Secondary structure of 16S rRNA (small subunit). Indicated in boldface are helix 34; the base of helix 44, the major part of which is involved in the A site; a large unpaired region of helix 44 comprising nt 1431 to 1434 and 1467 to 1469; and the GAAA tetraloop at nt 159 to 162. (Bottom left) Detail of helix 34, with nucleotides implicated in termination circled. (Bottom right) Detail of the base, or A-site, portion of helix 44 (upper) and detail of the nonconserved internal loop in helix 44 (lower), with nucleotides implicated in termination circled. (B) (Top) Secondary structure of 23S rRNA (large subunit). Indicated in boldface are the GTPase center and the two UGA suppressor fragment-associated sites, nt 74 to 136 in domain I and nt 735 to 766 in domain II. A suggested general location of the RNA component of peptidyltransferase in domains IV and V (Garrett and Rodriguez-Fonseca, 1996) is indicated (PT). The location of the 2-nt deletion that apparently compensates for termination defects caused by the mutation G1093A in the GTPase center is shown (ΔG2046 C2047). (Bottom) Detail of the GTPase center. The apparent RF2-interactive sites in the GTPase center RNA, based on UGA suppressor mutational analyses (Murgola et al., 1995; Xu and Murgola, 1996, unpublished) and analyses of the mutation G1093A in realistic in vitro termination assays (Arkov et al., 1998a, submitted), are circled.

alytic rate of peptidyl-tRNA hydrolysis after RF2 was bound to the ribosome, but RF1-mediated hydrolysis on the mutant ribosomes was very close to that on the wild-type ribosomes. The major defect in RF2-dependent termination caused by G1093A in vitro (Arkov et al., 1998a, submitted) was consistent with the specific suppression of UGA (RF2-specific stop codon) seen in vivo in the presence of the mutation (Jemiolo et al., 1995). As was mentioned for C1054A, defects in RF1-dependent termination detected in vitro for G1093A might not be sufficiently large to cause UAG and UAA suppression under the same conditions that revealed UGA suppression. The in vitro data indicated that G1093A affected binding of RF2 to the ribosome, thereby suggesting that the G1093 region is a binding site for RF2 (Arkov et al., submitted). Finally, in addition to G1093, other nucleotides in the GTPase center have been implicated in RF2-dependent termination based on UGA-specific suppressor mutations (see below).

There is evidence that, besides helix 34, another region of 16S rRNA, helix 44, is involved in termination. For *E. coli* in particular, it was shown that cloacin DF13 cleavage (between A1493 and G1494) of the part of the helix that contributes to the A site (Fig. 1A) inhibits formation of a complex between ribosomes, UAA, and RF1 and RF2 (Caskey et al., 1977). In addition, the aminoglycoside antibiotics hygromycin and neomycin, which interact with the A-site portion of the helix (Purohit and Stern, 1994), substantially inhibited termination driven by either RF1 or RF2 in vitro (Brown et al., 1993). On the other hand, UAA bound to the A site was cross-linked to helix 44 and RF2 (Tate et al., 1990), demonstrating the close proximity of RF2 to this helix and suggesting a direct functional interaction between RF2 and helix 44 in the A site and/or the stop codon during termination. Since C1054 of helix 34, the A-site portion of helix 44, and the A-site codon are situated very close to each other in the 30S ribosomal subunit (Dontsova et al., 1992; Tate et al., 1990), it is tempting to speculate that the C1054 region of helix 34, the A-site nucleotides of helix 44, and the 3 nt of a stop codon form a structurally unique surface that is specifically recognized by a small portion of RF. The strength of this recognition may be modulated by nucleotides adjacent to the stop codon and P-site-bound tRNA (Arkov et al., 1993; Buckingham et al., 1990; Pavlov et al., 1998; Poole et al., 1998; Zhang et al., 1996).

While the importance of the A-site portion of helix 44 for termination is well substantiated, some recent data indicate involvement of another site in that helix in the termination process, namely, the internal loop comprising nt 1431 to 1434 and 1467 to 1469 (Murgola et al., 1995; Zhao et al., unpublished) (Fig. 1A). This site is in a part of helix 44 that is nonconserved in both sequence and structure. In a screen for nonsense suppressor mutations after PCR random mutagenesis, we found the mutant A1431G, which suppressed UAG and UGA mutations but not UAA. Early evidence for the involvement of this region, though nonconserved, in decoding came from the characterization of the mutation C1469U as a "Ram" type mutation; that is, it reversed the streptomycin dependence of a particular *rpsL* mutation and it increased the frequency of translational errors in a poly(U)-directed in vitro protein synthesis assay (Allen and Noller, 1991). Nucleotide 1469 is on the opposite side of the internal loop from 1431 (Fig. 1A, bottom right). It was interesting, therefore, that a general mistranslation phenotype and a potentially termination-defective phenotype (that is, selective suppression of termination codons) were produced by mutations in the same internal loop. When we tested C1469U for nonsense suppression, it behaved like A1431G, suppressing UAG and UGA mutations but not UAA. By site-directed mutagenesis, we made most of the possible mutations (deletions and base substitutions) at 1431 and 1469, as well as at the highly conserved nucleotide 1468. None of the mutations caused suppression of any of the *trpA* UAA mutations, but they exhibited heterogeneity in UAG and/or UGA suppression, even among changes at a given position (1431 and 1468). Furthermore, the mutant phenotypes of G1431 and U1469 are not caused by formation of a Watson-Crick base pair, and the function of the internal loop in translation depends on both the size of the loop and the identities of the nucleotides on both sides. Cross-linking of nt 1468 to nt 894 would seem to place this site and structure at the decoding center (Wilms et al., 1997).

THE GTPase CENTER

After finding the G1093A mutation in the GTPase center of 23S rRNA (as discussed above), we asked whether we could find other codon-specific suppressors in 23S rRNA, particularly in the vicinity of nt 1093. Hence, we first targeted a particular segment of the rRNA operon *rrnB* cloned into the pBR322-derived plasmid pNO2680 (Murgola et al., 1995). The *Sna*BI-to-I-*Ceu*I segment of *rrnB* was subjected to PCR-directed random mutagenesis. This 1.4-kb segment (nt 524 to 1928) contains all of domains II and III and most of domain IV (of the six domains that constitute 23S rRNA). It included, therefore, nt 1093, the site of the first codon-specific (UGA) nonsense suppressor mutation in 23S rRNA. The further handling of the mutagenesis products has been described previously (Murgola et al., 1995).

This search for nonsense suppressors yielded only UGA suppressors, and only in the GTPase center (Xu and Murgola, unpublished). As with G1093A, all were UGA-specific under ordinary *E. coli* culture conditions, as well as high-expression and high-temperature conditionally lethal. Further mutational analyses of the mutants have led to the following conclusions: (i) the size of the 1093-to-1098 hexaloop, as well as the identity of certain nucleotides there and at nt 1067, is important for normal termination and growth; (ii) the G·A base pair (Huang et al., 1996) closing the 1093-to-1098 hexaloop is not essential for normal growth, peptide chain termination, or assembly into functional ribosomes (Xu and Murgola, 1996); (iii) the hexaloop "U-turn" structure, nt 1094 to 1096, is critical for normal termination; and (iv) nucleotides A1095 and A1067, necessary for the binding to ribosomes of thiostrepton, an antibiotic that inhibits polypeptide RF binding to ribosomes in vitro, are necessary for normal peptide chain termination in vivo. In particular, on the last point, since it was shown by others that base changes at nt 1067 and 1095 result in decreased binding of thiostrepton (in a particular order at each nucleotide), we made those changes and tested them for growth inhibition, temperature-conditional lethality, and termination codon readthrough. For the three base substitutions at each position, the order of increasing non-wild-type phenotypes (that is, UGA suppression and high-temperature lethality) was the order of decreasing thiostrepton binding observed by Rosendahl and Douthwaite (1994). Several deletions and point mutations throughout the region resulted in severe temperature-conditional lethality but no suppression at the permissive temperature. These observations indicate that the suppression observed with other mutations is due to alteration of specific interactions for termination rather than simply disruption of higher-order structures in the region. Consequently, from our mutant analyses in vivo and the results of in vitro termination assays (as discussed in the previous section), which demonstrated not only that G1093A is defective in catalysis of hydrolysis of peptidyl-tRNA but also that it affects the binding of RF2 to the ribosome, we can suggest that the new UGA suppressor mutations implicate 4 nt as RF2-interactive sites, namely, nt 1067, 1093, 1094, and 1095 (Fig. 1B, bottom).

Because of the UGA suppressor specificity of the mutations identified in the GTPase center and the results of in vitro termination assays indicating the involvement of this region not only in catalysis of RF2-dependent peptidyl-tRNA hydrolysis but also in the binding of RF2 to the ribosome (Arkov et al., 1998a, submitted), we asked whether a GTPase center RNA fragment, expressed in vivo from a multicopy plasmid, might "titrate" intracellular RF2 and lead to UGA readthrough, that is, UGA-specific nonsense suppression. To that end, we cloned the 95-nt GTPase center segment from nt 1030 to 1134 (as opposed to the 58-nt segment shown in Fig. 1B, bottom) and expressed it conditionally from the *tac* promoter. What we observed was not UGA suppression but instead UAG-specific suppression (Xu et al., unpublished). We suggested that, RF2 binding aside, the cloned GTPase center titrated the large-subunit protein L11, which binds to this region of 23S rRNA. That interpretation was based on the results of experiments done 17 years ago, in a minimal in vitro termination assay system (Tate et al., 1983), which showed that L11 is essential for UAG termination and somewhat inhibitory of UGA termination. Therefore, in vivo titration of free L11 in the cell should lead to some L11-defective ribosomes and UAG readthrough. Conversely, L11-deficient ribosomes should result in less inhibition of UGA termination and should be observable through an antisuppressor effect (due to better UGA termination) in the presence of a tRNA UGA suppressor.

We tested that hypothesis in several ways (Xu et al., unpublished). First, the UAG suppression effect of expression of the cloned GTPase center was completely reversed by simultaneous introduction of a plasmid-borne wild-type L11 gene but not by introduction of a mutant L11 gene that contained an L11 binding mutation, Gly65Ala, constructed in our laboratory (Chernyaeva and Murgola, unpublished) on the basis of the binding defect demonstrated by Xing et al. (1997). Second, the UAG readthrough effect was also produced independently by introduction of the cloned antisense RNA to either the entire L11 mRNA or just the 5′ part of the L11 mRNA. Third, the UAG readthrough effect was not seen when the cloned GTPase fragment contained a mutation (deletion of nt 1112) in the binding region for the L10·$(L12)_4$ pentamer, which binds to the RNA cooperatively with L11. Finally, the cloned GTPase center RNA also has an antisuppressor effect on the activity of a tRNA UGA suppressor, consistent with better UGA termination. These results verify in vivo the opposite roles of protein L11 in RF1- and RF2-dependent termination, as indicated previously in vitro (Tate et al., 1983).

The results of our functional studies fit very nicely with the recent determinations of the crystal structure of a GTPase center fragment bound either to the RNA binding C-terminal half of ribosomal protein L11 (Conn et al., 1999) or to the entire L11 molecule (Wimberly et al., 1999). For example, those structural analyses showed that both the 1093-to-1098 loop and the 1065-to-1073 loop are situated very close to each other, that each of the termination-interactive nucleotides 1067 and 1095 is the middle nucleotide of a U-turn, and that those nucleotides are entirely uncovered by L11, except for a possible occasional interaction with the N-terminal portion of L11. Considering those observations along with the results of our functional studies in vivo (Xu and Murgola, unpublished) and in vitro (Arkov et al., 1998a, submitted), it seems plausible that both loops interact with a compact region of RF2. Experiments in our laboratory are under way to test the hypothesis of direct interaction between the two loops of the GTPase center and RF2.

The UGA-specific suppression caused by the mutations in the GTPase center indicates that RF1 does not interact with the two loops of the GTPase center. Instead RF1 may interact, as suggested by others also (Tate et al., 1984, 1986), with the amino-terminal half of ribosomal protein L11, which may bind transiently to the 1067 and 1095 loops of the GTPase center (Wimberly et al., 1999). The small decrease in k_{cat}/K_m in the RF1-dependent reactions seen in vitro with G1093A-containing ribosomes (Arkov et al., submitted) may be an indirect effect of this mutation on such a hypothetical L11-RF1 complex.

INTERACTIVE TETRALOOPS

Screenings for nonsense suppressors in 16S rRNA yielded several clones with mutations in the highly conserved GAAA tetraloop at nucleotide positions 159 to 162 (see Fig. 1A, top). To date, all mutational changes found or made in this loop, including several that conform to the GNRA motif that is common among hairpin loops in a variety of structural RNAs, display the same phenotype (readthrough of UGA and UAG, but not UAA, termination codons, and high-temperature lethality). Furthermore, the same GNRA sequences that were tested in the 159-to-162 loop (GAGA, GCAA, GCGA, and GGAA) are found in many other terminal loops in rRNAs. Consequently, if the sole function of this loop were to fulfill some local requirement, such as stabilizing the helix or inducing the correct folding, the mutants that conform to GNRA would be expected to function satisfactorily. As this is not the case, it seems highly likely that this loop is involved in an essential higher-order interaction.

In a current three-dimensional model of 16S rRNA in the ribosome (Mueller and Brimacombe, 1997), the hairpin capped by the 159-to-162 loop forms a projection, called the spur or toe, near the bottom of the small subunit. The position of the spur is such that it does not make contact with the large subunit. It thus seems likely that the function of this loop is to form the binding site for a molecule that is not part of the ribosome. To our knowledge no translation factors or other macromolecules have been implicated in binding to this hairpin. However, directed hydroxyl radical probing with EF-G has shown that the 159-to-162 loop is in proximity to this elongation factor, and electron microscopic visualization of ribosome-bound EF-Tu has shown that the spur is in the vicinity of, although not in direct contact with, this elongation factor (Wilson and Noller, 1998a, and references therein).

Many GNRA loops in other structural RNAs, such as self-splicing introns and RNase P RNA, have been shown to be involved in tertiary interactions with other RNA structures called receptors. The results from the 159-to-162 loop indicated that GNRA loops in rRNAs may be involved in similar interactions. In order to determine if involvement in higher-order interactions is widespread among the GNRA loops in rRNAs, we have compared available rRNA sequences and secondary structures from bacteria to see if terminal helices that are capped by conserved GNRA loops tend to be of conserved length (Hedenstierna et al., submitted). The data were calculated from sequences and secondary structures downloaded from the rRNA World Wide Web server (Van de Peer et al., 1998; De Rijk et al., 1998). The reasoning behind this approach is that if a terminal loop is engaged in an essential interaction, then the length of the corresponding helix is expected to be constrained, because any shortening or lengthening of the helix would either disrupt the interaction or disturb the three-dimensional structure of the molecule. The results of this analysis showed that there is a strong tendency for conserved GNRA loops to be associated with terminal helices that display conserved length. Conversely, if the sequence of a particular loop is not conserved, the length of the corresponding helix tends to be variable. A small number of cases of conserved-length helices capped by conserved UNCG loops was also found, indicating that this type of loop also may be involved in higher-order interactions. It should be noted that although this evidence indicates that many of the GNRA loops are involved in interactions, it does not suggest the identities of the molecules with which these loops interact. The partners in these interactions could therefore be the rRNA of either subunit or any of the ribosomal proteins, or they could be molecules that bind transiently to the ribosome, such as tRNAs or translation factors.

FUNCTIONAL FRAGMENTS

A new approach to the study of rRNA sites and structures involved in termination has led to identification of fragments of *E. coli* 23S rRNA that caused UGA-specific suppression, namely (Fig. 1B, top), a domain I fragment, nt 74 to 136, and the antisense to that domain I fragment (Arkov et al., 1998b), and the antisense to a domain II segment, nt 735 to 766 (Chernyaeva et al., 1999). This approach included screening a library of rRNA fragments (Tenson et al., 1996) to identify those whose expression caused suppression of UGA nonsense mutations in reporter genes. Since the identified fragments caused suppression of UGA but not UAA or UAG, it was suggested that they interfere with RF2-dependent termination. Although precise mechanisms of UGA readthrough by these fragments remain to be elucidated, it seems likely nevertheless that this fragment-screening approach (Arkov et al., 1998b; Chernyaeva et al.,

1999; Tenson et al., 1996), together with targeted expression of cloned segments (such as the GTPase center, as discussed above), will be a valuable addition to the "tool box" for studying ribosome function.

WHERE IS THE HYDROLYTIC SITE?

An important unanswered question is, where is the site that directly catalyzes the hydrolysis of peptidyl-tRNA? Is this site on the RFs and/or on the ribosome? There is evidence that the ribosome contains such a site. An important discovery, which led to this point of view, was made by Caskey et al. (1971), who showed that both bacterial and mammalian ribosomes can carry out hydrolysis of peptidyl-tRNA without RFs in the presence of 30% acetone and tRNA. Mutations in rRNAs that substantially decrease catalysis of peptidyl-tRNA hydrolysis in the presence of just one RF, like C1054A (16S rRNA) and G1093A (23S rRNA), are not likely to affect the hydrolytic site. Mutations affecting such a site would be expected to cause decreases in the catalytic rates of peptidyl-tRNA hydrolysis in the presence of both RF1 and RF2 in vitro and readthrough of all stop codons in vivo. Therefore, characterization of some nonspecific rRNA suppressors of nonsense mutations in an in vitro termination system may lead to identification of the hydrolytic site. Meanwhile, there is indirect evidence that the ribosomal peptidyltransferase (PT) is involved in hydrolysis of peptidyl-tRNA during termination (Caskey and Beaudet, 1972; Caskey et al., 1971; Vogel et al., 1969). Furthermore, 23S rRNA has been implicated in PT activity (Garrett and Rodriguez-Fonseca, 1996; Noller et al., 1992). Therefore, a 23S rRNA component of PT (Fig. 1B, top) may be at least a part of the hydrolytic site. Involvement of the rRNA component of PT in termination has been proposed (Tate and Brown, 1992).

Under physiological conditions, association of either RF1 or RF2 with the ribosome activates the hydrolytic site. The mechanism of this activation is unknown. However, there is genetic evidence for an interaction between the GTPase center in domain II and the entrance to domain V during termination that may be part of the activation process. More specifically, deletion of 2 nt in the entrance to domain V, Δ(G2046 C2047), was found to compensate for the temperature-conditional lethality caused by G1093A in the GTPase center (Fig. 1B, top) (Murgola et al., 1995). Furthermore, this deletion dramatically decreased the very effective UGA suppression caused by G1093A in vivo, suggesting that Δ(G2046 C2047) was able to compensate for termination defects caused by G1093A. In addition, the deletion was found to cause weak UGA suppression on its own, implicating that site in the termination process (Arkov and Murgola, unpublished). Collectively, these data are consistent with the possibility of an interaction between the GTPase center and the entrance to domain V that may be important for proper termination. Since domain V has been implicated in peptidyltransferase activity (Garrett and Rodriguez-Fonseca, 1996), one may speculate that an interaction between the GTPase center (domain II) and the entrance to domain V is important for the transformation of peptidyltransferase into the ribosomal peptidyl-tRNA hydrolase. Consistent with this hypothesis of a direct interaction between domain II and domain V is the proximity of these domains to each other (Bogdanov et al., 1995; Stiege et al., 1983). However, an indirect interaction between the GTPase center and the entrance to domain V, perhaps through RF2, is also possible.

SUMMARY

Progress has been made in identifying some of the sites and structures of rRNA involved in translation termination. Available data make it possible to propose a model for the involvement of rRNAs in RF2-dependent termination. Figure 2 shows possible interactions between RF2 and different rRNA regions as well as an rRNA-rRNA interaction that may be involved in transmission of a stop signal (UGA) from the small-subunit portion of the A site at helices 34 and 44 (16S rRNA) to the hydrolytic center of the large subunit (23S rRNA). It is possible that when UGA enters the A site of the ribosome, RF2 directly binds to the UGA, as suggested by the results of cross-linking experiments (Tate et al., 1990), and to helices 34 and 44 (Arkov et al., 1998a, submitted; Brown et al., 1993; Murgola, 1996; Murgola et al., 1988; Pagel et al., 1997). As discussed above, UGA, helix 34, and helix 44 together may form a unique structure specifically recognized by RF2. RF2 may also interact with the GTPase center, as suggested by results obtained in vivo (Jemiolo et al., 1995; Murgola, 1996; Murgola et al., 1995; Xu and Murgola, 1996, unpublished) and in vitro (Arkov et al., 1998a, submitted). After RF2 binds to the rRNA regions, the GTPase center may directly interact with domain V of 23S rRNA (Murgola et al., 1995; Arkov and Murgola, unpublished), which is a likely place for the hydrolytic center, as discussed above. This RNA-RNA interaction may activate the hydrolytic center, resulting in hydrolysis of peptidyl-tRNA. Alternatively,

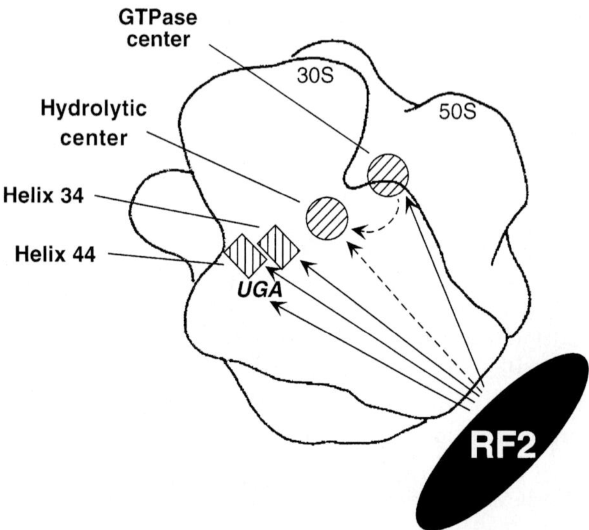

Figure 2. Possible functional interactions during transmission of a stop signal in the presence of RF2. The approximate ribosome locations of rRNA regions implicated in RF2-dependent termination are shown as either hatched rectangles (helix 34 and helix 44 of 16S rRNA) or hatched circles (23S rRNA regions, namely, the GTPase center and the hydrolytic center). A complete ribosome assembled from a 30S and a 50S subunit is viewed from the solvent side of the 30S subunit. Outlines of the 30S and 50S subunits are according to Frank et al. (1991). Locations of helix 34, helix 44, and the A-site UGA codon were deduced from their proximities to the anticodon of the A-site tRNA, whose ribosome location has been presented (Wilson and Noller, 1998b, and references therein). The GTPase center is shown according to Wilson and Noller (1998a). The hydrolytic center is assumed to be situated near the acceptor end of the P-site tRNA, whose location has been presented (Wilson and Noller, 1998b, and references therein). Functional interactions between RF2 and RNA elements strongly suggested by the studies referred to in the text are shown as continuous arrows originating from RF2. Other interactions, for which there are few data available, are shown as dashed arrows.

RF2 may interact with and activate the hydrolytic center (Fig. 2).

OUTLOOK

While some progress in studying the involvement of rRNAs in translation termination has been made, we are yet very far from understanding the mechanism of the termination process. It is very unclear what exactly RF recognizes, what the hydrolytic center is, and how interaction of the RF with the ribosome triggers the hydrolytic center to release a polypeptide. However, satisfactory answers to these exciting problems should not be long in coming since genetic approaches to finding termination-defective mutants have proven to be successful, as indicated by the use of realistic in vitro termination assays.

The model presented in Fig. 2 summarizes existing data and suggests future experiments both in vivo and in vitro. In vitro experiments must include examining interactions of rRNA segments and domains with relevant purified proteins, as well as assaying representative mutants, beginning with several others in the GTPase center and helix 34, in realistic in vitro termination systems. In vivo functional analyses will require the isolation of new mutants whose phenotypes should be indicative of specific defects in the termination process.

We have suggested that an effective operational phenotype for potential termination-defective mutants is codon-specific nonsense suppression, at least under some conditions. It should be emphasized, however, that such a phenotype can be explained in other ways, less likely but indicative of other classes of very interesting mutants. Nevertheless, we have suggested that most of such mutants will be defective in termination—either only in termination or in both termination and accuracy of decoding. The first two such codon-specific nonsense suppressors to be analyzed in detail in vitro as well as in vivo were shown to be defective in termination and, in particular, to be altered in the binding of RF2 to the ribosome. Another consideration is that, as many have thought over the years and as we have supported above, it is likely that some configuration of the PT center accomplishes the hydrolysis of peptidyl-tRNA. Thus, it should be possible to incorporate predictable characteristics of an altered PT center into termination phenotypes to be sought. We have made a start on getting appropriate phenotypes, but now a little more creative imagination may be required.

We appreciated very much the collaboration with Janet L. Siefert (Rice University) and George E. Fox (University of Houston) on the comparative sequence analysis of GNRA loops in rRNAs, the initial contribution of Song Q. Zhao to the experimental work on the GAAA loop at nt 159 to 162 of 16S rRNA, and the unpublished dissertation results of Song Q. Zhao and Wenbing Xu on the bulge region of helix 44 in 16S rRNA and the GTPase center in 23S rRNA, respectively. We thank Måns Ehrenberg and members of his laboratory, in particular, David Freistroffer, Reza Karimi, and Michael Pavlov, for the very fruitful collaboration on the in vitro analyses of rRNA mutants. We are grateful also to Alexander Mankin for our collaborations on identification of translationally functional rRNA fragments. Finally, in carrying out the "fragment approach," A.L.A. and E.J.M. appreciated the inspiration of Patsy Cline!

This work was supported by grant GM21499 from the U.S. National Institute of General Medical Sciences to E.J.M.

REFERENCES

Allen, P., and H. F. Noller. 1991. A single base substitution in 16S ribosomal RNA suppresses streptomycin dependence and increases the frequency of translation errors. *Cell* **66:**141–148.

Arkov, A. L., and E. J. Murgola. Unpublished data.

Arkov, A. L., S. V. Korolev, and L. L. Kisselev. 1993. Termination of translation in bacteria may be modulated via specific interaction between peptide chain release factor 2 and the last peptidyl-tRNA$^{Ser/Phe}$. *Nucleic Acids Res.* **21:**2891–2897.

Arkov, A. L., D. V. Freistroffer, M. Ehrenberg, and E. J. Murgola. 1998a. Mutations in RNAs of both ribosomal subunits cause defects in translation termination. *EMBO J.* **17:**1507–1514.

Arkov, A. L., A. Mankin, and E. J. Murgola. 1998b. An rRNA fragment and its antisense can alter decoding of genetic information. *J. Bacteriol.* **180:**2744–2748.

Arkov, A. L., D. V. Freistroffer, M. Y. Pavlov, M. Ehrenberg, and E. J. Murgola. Mutations in conserved regions of ribosomal RNAs decrease the productive association of peptide-chain release factors with the ribosome during translation termination. Submitted for publication.

Bogdanov, A. A., O. A. Dontsova, S. S. Dokudovskaya, and I. N. Lavrik. 1995. Structure and function of 5S rRNA in the ribosome. *Biochem. Cell Biol.* **73:**869–876.

Brown, C. M., K. K. McCaughan, and W. P. Tate. 1993. Two regions of the *Escherichia coli* 16S ribosomal RNA are important for decoding stop signals in polypeptide chain termination. *Nucleic Acids Res.* **21:**2109–2115.

Buckingham, R. H., P. Sörensen, F. T. Pagel, K. A. Hijazi, B. H. Mims, D. Brechemier-Baey, and E. J. Murgola. 1990. Third position base changes in codons 5′ and 3′ adjacent UGA codons affect UGA suppression in vivo. *Biophys. Biochim. Acta* **1050:** 259–262.

Buckingham, R. H., G. Grentzmann, and L. Kisselev. 1997. Polypeptide chain release factors. *Mol. Microbiol.* **24:**449–456.

Caskey, C. T., and A. L. Beaudet. 1972. Antibiotic inhibitors of peptide chain termination, p. 326–336. *In* E. Muñoz, F. García-Ferrandiz, and D. Vazquez (ed.), *Molecular Mechanisms of Antibiotic Action on Protein Biosynthesis and Membranes.* Elsevier Scientific Publishing Company, Amsterdam, The Netherlands.

Caskey, C. T., A. L. Beaudet, E. M. Scolnick, and M. Rosman. 1971. Hydrolysis of fMet-tRNA by peptidyl transferase. *Proc. Natl Acad. Sci. USA* **68:**3163–3167.

Caskey, C. T., L. Bosch, and D. S. Konecki. 1977. Release factor binding to ribosome requires an intact 16S rRNA 3′ terminus. *J. Biol. Chem.* **252:**4435–4437.

Chernyaeva, N. S., and E. J. Murgola. Unpublished data.

Chernyaeva, N. S., E. J. Murgola, and A. S. Mankin. 1999. Suppression of nonsense mutations induced by expression of an RNA complementary to a conserved segment of 23S rRNA. *J. Bacteriol.* **181:**5257–5262.

Conn, G. L., D. E. Draper, E. E. Lattman, and P. G. Gittis. 1999. Crystal structure of a conserved ribosomal protein-RNA complex. *Science* **284:**1171–1174.

De Rijk, P., A. Caers, Y. Van de Peer, and R. De Wachter. 1998. Database on the structure of large ribosomal subunit RNA. *Nucleic Acids Res.* **26:**183–186.

Dontsova, O., S. Dokudovskaya, A. Kopylov, A. Bogdanov, J. Rinke-Appel, N. Jünke, and R. Brimacombe. 1992. Three widely separated positions in the 16S RNA lie in or close to the ribosomal decoding region; a site-directed cross-linking study with mRNA analogues. *EMBO J.* **11:**3105–3115.

Frank, J., P. Penczek, R. Grassucci, and S. Srivastava. 1991. Three-dimensional reconstruction of the 70S *Escherichia coli* ribosome in ice: the distribution of ribosomal RNA. *J. Cell Biol.* **115:**597–605.

Freistroffer, D. V., M. Y. Pavlov, J. MacDougall, R. H. Buckingham, and M. Ehrenberg. 1997. Release factor RF3 in *E. coli* accelerates the dissociation of release factors RF1 and RF2 from the ribosome in a GTP-dependent manner. *EMBO J.* **16:**4126–4133.

Garrett, R. A., and C. Rodriguez-Fonseca. 1996. The peptidyl transferase center, p. 327–355. *In* R. A. Zimmermann and A. E. Dahlberg (ed.), *Ribosomal RNA: Structure, Evolution, Processing, and Function in Protein Biosynthesis.* CRC Press, Boca Raton, Fla.

Hedenstierna, K. O. F., J. L. Siefert, G. E. Fox, and E. J. Murgola. Submitted for publication.

Huang, S., Y.-X. Wang, and D. E. Draper. 1996. Structure of a hexanucleotide RNA hairpin loop conserved in ribosomal RNAs. *J. Mol. Biol.* **258:**308–321.

Jemiolo, D. K., F. T. Pagel, and E. J. Murgola. 1995. UGA suppression by a mutant RNA of the large ribosomal subunit. *Proc. Natl. Acad. Sci. USA* **92:**12309–12313.

Moine, H., and A. E. Dahlberg. 1994. Mutations in helix 34 of *Escherichia coli* 16S ribosomal RNA have multiple effects on ribosome function and synthesis. *J. Mol. Biol.* **243:**402–412.

Mueller, F., and R. Brimacombe. 1997. A new model for the three-dimensional folding of *Escherichia coli* 16 S ribosomal RNA. I. Fitting the RNA to a 3D electron microscopic map at 20 Å. *J. Mol. Biol.* **271:**524–544.

Murgola, E. J. 1996. Ribosomal RNA in peptide chain termination, p. 357–369. *In* R. A. Zimmermann and A. E. Dahlberg (ed.), *Ribosomal RNA: Structure, Evolution, Processing, and Function in Protein Biosynthesis.* CRC Press, Boca Raton, Fla.

Murgola, E. J., K. A. Hijazi, H. U. Göringer, and A. E. Dahlberg. 1988. Mutant 16S ribosomal RNA: a codon-specific translational suppressor. *Proc. Natl Acad. Sci. USA* **85:**4162–4165.

Murgola, E. J., F. T. Pagel, K. A. Hijazi, A. L. Arkov, W. Xu, and S. Q. Zhao. 1995. Variety of nonsense suppressor phenotypes associated with mutational changes at conserved sites in *Escherichia coli* ribosomal RNA. *Biochem. Cell Biol.* **73:**925–931.

Nakamura,Y., K. Ito, and L. A. Isaksson. 1996. Emerging understanding of translation termination. *Cell* **87:**147–150.

Noller, H. F., V. Hoffarth, and L. Zimniak. 1992. Unusual resistance of peptidyl transferase to protein extraction procedures. *Science* **256:**1416–1419.

Pagel, F. T., and E. J. Murgola. Unpublished results.

Pagel, F. T., S. Q. Zhao, K. A. Hijazi, and E. J. Murgola. 1997. Phenotypic heterogeneity of mutational changes at a conserved nucleotide in 16S ribosomal RNA. *J. Mol. Biol.* **267:**1113–1123.

Pavlov, M. Y., D. V. Freistroffer, V. Dinçbas, J. MacDougall, R. H. Buckingham, and M. Ehrenberg. 1998. A direct estimation of the context effect on the efficiency of termination. *J. Mol. Biol.* **284:**579–590.

Poole, E. S., L. L. Major, S. A. Mannering, and W. P. Tate. 1998. Translational termination in *Escherichia coli:* three bases following the stop codon crosslink to release factor 2 and affect the decoding efficiency of UGA-containing signals. *Nucleic Acids Res.* **26:**954–960.

Purohit, P., and S. Stern. 1994. Interaction of a small RNA with antibiotic and RNA ligands of the 30S subunit. *Nature* **370:**659–662.

Rosendahl, G., and S. Douthwaite. 1994. The antibiotics micrococcin and thiostrepton interact directly with 23S rRNA nucleotides 1067A and 1095A. *Nucleic Acids Res.* **22:**357–363.

Saarma, U., J. Remme, M. Ehrenberg, and N. Bilgin. 1997. An A to U transversion at position 1067 of 23S rRNA from Escherichia coli impairs EF-Tu and EF-G function. *J. Mol. Biol.* **272:** 327–335.

Stiege, W., C. Glotz, and R. Brimacombe. 1983. Localisation of a series of intra-RNA cross-links in the secondary and tertiary structure of 23S RNA, induced by ultraviolet irradiation of *Escherichia coli* 50S ribosomal subunits. *Nucleic Acids Res.* **11:** 1687–1706.

Tate, W., B. Greuer, and R. Brimacombe. 1990. Codon recognition in polypeptide chain termination: site directed crosslinking of termination codon to *Escherichia coli* release factor 2. *Nucleic Acids Res.* 18:6537–6544.

Tate, W. P., and C. M. Brown. 1992. Translational termination—"stop" for protein synthesis or "pause" for regulation of gene expression. *Biochemistry* 31:2443–2450.

Tate, W. P., H. Schulze, and K. H. Nierhaus. 1983. The *Escherichia coli* ribosomal protein L11 suppresses release factor 2 but promotes the release factor 1 activities in peptide chain termination. *J. Biol. Chem.* 259:12816–12820.

Tate, W. P., M. J. Dognin, M. Noah, M. Stöffler-Meilicke, and G. Stöffler. 1984. The NH_2-terminal domain of *Escherichia coli* ribosomal protein L11. *J. Biol. Chem.* 259:7317–7324.

Tate, W. P., K. K. McCaughan, C. D. Ward, V. G. Sumpter, C. N. A. Trotman, M. Stöffler-Meilicke, P. Maly, and R. Brimacombe. 1986. The ribosomal binding domain of the Escherichia coli release factors. *J. Biol. Chem.* 261:2289–2293.

Tate, W. P., E. S. Poole, and S. A. Mannering. 1996. Hidden infidelities of the translational stop signal. *Prog. Nucleic Acids Res. Mol. Biol.* 52:293–335.

Tenson, T., A. DeBlasio, and A. Mankin. 1996. A functional peptide encoded in the *Escherichia coli* 23S rRNA. *Proc. Natl. Acad. Sci. USA* 93:5641–5646.

Thompson, J. 1996. Ribosomal RNA, translocation, and elongation factor-associated GTP hydrolysis, p. 311–325. *In* R. A. Zimmermann and A. E. Dahlberg (ed.), *Ribosomal RNA: Structure, Evolution, Processing, and Function in Protein Biosynthesis.* CRC Press, Inc., Boca Raton, Fla.

Van de Peer, Y., A. Caers, P. De Rijk, and R. De Wachter. 1998. Database on the structure of small ribosomal subunit RNA. *Nucleic Acids Res.* 26:179–182.

Vogel, Z., A. Zamir, and D. Elson. 1969. The possible involvement of peptidyl transferase in the termination step of protein biosynthesis. *Biochemistry* 8:5161–5168.

Wilms, C., J. W. Noah, D. Zhong, and P. Wollenzien. 1997. Exact determination of UV-induced crosslinks in 16S ribosomal RNA in 30S ribosomal subunits. *RNA* 3:602–612.

Wilson, K. S., and H. F. Noller. 1998a. Mapping the position of translational elongation factor EF-G in the ribosome by directed hydroxyl radical probing. *Cell* 92:131–139.

Wilson, K. S., and H. F. Noller. 1998b. Molecular movement inside the translational engine. *Cell* 92:337–349.

Wimberly, B. T., R. Guymon, J. P McCutcheon, S. W. White, and V. Ramakrishnan. 1999. A detailed view of a ribosomal active site: the structure of the L11-RNA complex. *Cell* 97:491–502.

Xing, Y., D. GuhaThakurta, and D. E. Draper. 1997. The RNA binding domain of ribosomal protein L11 is structurally similar to homeodomains. *Nat. Struct. Biol.* 4:24–27.

Xu, W., and E. J. Murgola. 1996. Functional effects of mutating the closing G·A base-pair of a conserved hairpin loop in 23S ribosomal RNA. *J. Mol. Biol.* 264:407–411.

Xu, W., and E. J. Murgola. Unpublished data.

Xu, W., N. S. Chernyaeva, and E. J. Murgola. Unpublished data.

Zhang, S., M. Rydén-Aulin, and L. A. Isaksson. 1996. Functional interaction between release factor one and P-site peptidyl-tRNA on the ribosome. *J. Mol. Biol.* 261:98–107.

Zhao, S. Q., F. T. Pagel, and E. J. Murgola. Unpublished data.

The Ribosome: Structure, Function, Antibiotics, and Cellular Interactions
Edited by R. A. Garrett, S. R. Douthwaite, A. Liljas, A. T. Matheson, P. B. Moore, and H. F. Noller

Chapter 42

Genetic Probes to Bacterial Release Factors: tRNA Mimicry Hypothesis and Beyond

YOSHIKAZU NAKAMURA, YOICHI KAWAZU, MAKIKO UNO, KUNIYASU YOSHIMURA, and KOICHI ITO

The termination of protein synthesis takes place on ribosomes as a response to a stop rather than a sense codon in the decoding site (A site). Extraribosomal proteins designated polypeptide release factors (RFs) play an essential role in this process. Although the termination process and RF activity were discovered in vitro in the late 1960s, much of the mechanism and the protein features have remained obscure. The pioneer works by Caskey, Capecchi, Nirenberg, and their colleagues have uncovered the involvement of codon-specific RFs, RF1 and RF2, of *Escherichia coli* in vitro (Scolnick et al., 1968; Capecchi and Klein, 1969; Caskey et al., 1969; Goldstein and Caskey, 1970). Bacterial RF1 and RF2 participate in recognition of UAG or UAA and UGA or UAA, respectively, and are now known as class I RFs (reviewed in Nakamura and Ito, 1998). The fact that two RFs from prokaryotes exhibit codon specificity suggested that they should interact directly with the codon. Although this has long been thought, evidence has been lacking for such a direct contact until recently except for a cross-linking experiment with a thiouracil residue in a stop codon within a designed RNA (Tate et al., 1990). Indeed, an alternative view has been proposed that there need not be direct contact between the factor and the stop codon but rather that the codon may be decoded by rRNA at a site on the ribosome distant from the usual decoding site for sense codons (Murgola et al., 1988) or a double-stranded complex of codon and rRNA may be recognized by the RF at the decoding site (reviewed in Tate et al., 1996). These opinions serve to illustrate that the mechanism of stop signal recognition by RFs, whether direct or indirect, was poorly understood until recently.

After 3 decades of investigation, most of the genes for RFs in prokaryotes and eukaryotes have been identified, and we are now aware of the essential requirement for two classes of RFs for translation termination: a class I factor, codon-specific RFs (RF1 and RF2 in prokaryotes and eRF1 in eukaryotes), and a class II factor, nonspecific RFs (RF3 in prokaryotes and eRF3 in eukaryotes) that bind guanine nucleotides and stimulate class I RF activity. In bacteria, the genes for RF1, RF2, and RF3 are designated *prfA*, *prfB*, and *prfC*, respectively. Unlike bacterial RF1 and RF2, the known eukaryotic eRF1s recognize all three stop codons (Konecki et al., 1977; Frolova et al., 1994). The recent increases in genetic knowledge and tools have enabled us to study RF functions and structures in systematic and comprehensive ways by using reverse genetics and phylogenetic comparisons. Several relevant reviews of termination of protein synthesis or polypeptide RFs have been written in recent years (Tate and Brown, 1992; Nakamura et al., 1995, 1996; Tate and Mannering, 1996; Tate et al., 1996; Buckingham et al., 1997; Nakamura and Ito, 1998), that readers can refer to for details not described here. This chapter will not repeat the well-documented facts or summaries but will focus on the genetic background leading to the "RF-tRNA mimicry hypothesis" (see below) (Ito et al., 1996) and the progress made after its proposal by using genetic probes of bacterial RFs. It should be a surprising fact that it took more than 30 years after the discovery of the mechanism of sense-codon decoding by tRNA before we began to understand the mechanism of stop codon decoding by protein RFs.

Yoshikazu Nakamura, Yoichi Kawazu, Makiko Uno, Kuniyasu Yoshimura, and Koichi Ito ■ Department of Tumor Biology, The Institute of Medical Science, The University of Tokyo, Minato-ku, Tokyo 108-8639, Japan.

RF-tRNA MIMICRY

A major breakthrough in this long-standing coding problem was achieved in our laboratory by the discovery of universally conserved structures in prokaryotic and eukaryotic RFs that led to a novel hypothesis of "molecular mimicry" between RFs and tRNA. The idea began to emerge when we cloned and sequenced the *prfC* gene coding for RF3 (see below). Its primary protein sequence resembled the N-terminal part of the elongation factor EF-G rather than that of EF-Tu, which prompted us to speculate that, among proteins that interact with RF3, there should be a factor equivalent to the C-terminal part of EF-G at the primary sequence level (Mikuni et al., 1994; Kawazu et al., 1995; Ito et al., 1996; Nakamura et al., 1996). RF1 and RF2 were strong candidates. Although the structural data greatly facilitated the mimicry hypothesis (see below), these initial ideas stimulated us to test the primary sequence comparisons of RFs from different organisms and other translation factors. Thus, the RF-tRNA mimicry hypothesis was born after the discovery of universally conserved motifs in class I RFs, which led us to propose an RF seven-domain model and to designate domains A through G. (Note that these are sequence-based designations and do not represent structurally defined domains.) Two of them, domains D and E, also have primary (and secondary) sequence homology with the C-terminal portion, domain IV, of EF-G (Ito et al., 1996) (Fig. 1). The three-dimensional structural study of *Thermus thermophilus* EF-G revealed that domains III to V appear to mimic the shapes of the acceptor stem, the anticodon helix, and the T stem of tRNA, respectively (Ævarsson et al., 1994; Czworkowski et al., 1994; Nissen et al., 1995). Therefore, it appears that domains D and E of the RF, assuming their homology with domain IV of EF-G, constitute a tRNA mimicry domain necessary for RF binding to the ribosomal A site (Ito et al., 1996; reviewed in Nakamura and Ito, 1998). Furthermore, the three-dimensional structures of the ternary complex of Phe-tRNA, *Thermus aquaticus* elongation factor EF-Tu, and the nonhydrolyzable GTP analogue, guanosine-5′-(β·γ-imido)triphosphate (GDPNP), are almost completely superimposable with that of the EF-G–GDP complex, suggesting common requirements for their structures and functions on the ribosome (Nissen et al., 1995).

The RF-tRNA mimicry model explains not only the common steps during elongation and termination processes, including the requirement for RFs for binding to the A site of the ribosome, but also, by assuming an anticodon mimicry element in the protein, the coding problem of how RFs have the ability to recognize the stop codon. Two alleles in domain D of *E. coli* RF2, amino acid positions 207 and 213, have been suggested to be engaged in peptide anticodon activity because of their differentially conserved features in RF1 and RF2, mutational toxicity (by substitutions, F207T and R213I), and the predicted topology at the tip of the anticodon stem mimicry, according to the known mimicry of domain IV of EF-G (Ito et al., 1996). We assumed that, unless these two alleles are not the direct constituents of the anticodon mimicry element, they could be located at or near the putative anticodon element, which we have previously referred to as site 1 (reviewed in Nakamura and Ito, 1998). This putative anticodon element, which we refer to as the peptide anticodon, should not be a structural mimic of the anticodon loop of tRNA despite being functionally similar, since the mechanism of mRNA recognition by the RF protein is completely distinct from that by tRNA. Given that the same polypeptide loci of RF1 and RF2 are used for recognition of UAA or UAG and UAA or UGA, respectively, it is reasonable to speculate that the shared as well as differentially conserved amino acids in these two RFs could be responsive to the common basal recognition of the terminator triplets as well as to the discrimination of purine bases at the second and third positions. Genetic evidence for the latter prediction has been obtained by peptide-swapping experiments between RF1 and RF2 (Ito et al., in press).

OMNIPOTENT-RELEASE VARIANT OF RF2

An intriguing RF2 variant that acquired omnipotent decoding activity has been isolated as a plasmid-borne *prfB* mutant that restored the growth of a temperature-sensitive *prfA1* strain of *E. coli* (Ito et al., 1998b). This omnipotent activity is caused by a single Glu-to-Lys mutation at position 167 of RF2, and we refer to this variant as RF2* (Fig. 1). In both in vivo and in vitro polypeptide termination assays, RF2* catalyzed UAG/UAA termination, as does RF1, as well as UGA termination, showing that the mutation did not switch stop codon selectivity from RF2 to RF1 but conferred on RF2 omnipotent recognition activity. We assumed that Glu 167 may not be part of the peptide anticodon of RF2 but rather the E167K mutation abolished the putative third-base discriminator function of RF2. To our knowledge, RF2* is the first protein that was altered in stop codon selectivity, providing direct genetic evidence for the existence of the peptide anticodon in RF2. This omnipotent alteration is mapped within domain C at or near the region designated site 2 (Nakamura and

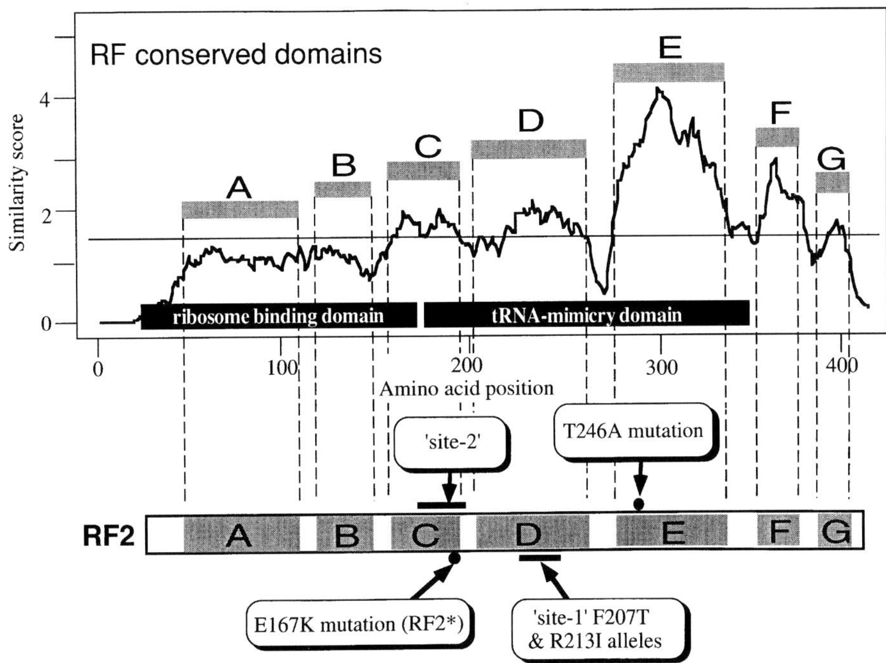

Figure 1. Average similarity plot of prokaryotic RF sequences, including *S. cerevisiae* mRF1 (top), and locations of phenotypically important mutations affected in RF2 (bottom). The proposed seven domains A through G are assigned (Ito et al., 1996), and the proposed domain functions are indicated. The average similarity score along the entire sequence is shown by the horizontal line. The amino acid position refers to the coordinate of the similarity alignment described (Ito et al., 1996), which is distinct from the actual amino acid positions of RF1 and RF2.

Ito, 1998) (Fig. 1), whose alterations are known to influence the mutant phenotypes at site 1. Therefore, it is plausible that domain C (site 2) functions as an effector domain required for correct interaction between the codon and the peptide anticodon on the ribosome and that E167K and other mutations at site 2 might block this process.

To our understanding, prokaryotes have two class I RFs while eukaryotes have only one. *Mycoplasma genitalium* is a bacterium with a small genome, and it has only one factor, RF1 (Fraser et al., 1995). We assume that an equivalent to RF2 disappeared during evolution, since *Mycoplasma* reassigned the UGA codon to the glutamine codon, making RF2 harmful. With regard to the "one- or two-RF scenario," it is surprising that RF2 can revert (easily) to an omnipotent factor with a single-amino-acid substitution (E167K) and that this omnipotent variant RF2* restored the growth of a chromosomal RF1-RF2 double-knockout strain (Ito et al., 1998b). This indicated that one omnipotent RF is sufficient for bacterial growth, similar to the single omnipotent eukaryotic factor, representing the first genetic conversion of a dual-RF system to a unitary system in prokaryotes. Given the strictly conserved two RFs in prokaryotes, there should have been some evolutionary bias rather than a simple "frozen accident" to force prokaryotes to maintain two RFs.

OMNIPOTENT NONSENSE SUPPRESSOR VARIANTS OF RF1 AND RF2

Of the proposed seven domains of prokaryotic RF1 and RF2, the C-terminal tRNA mimicry portion, including domains D and E, is highly conservative, whereas the N-terminal portion, including domains A and B, is much less conservative (Fig. 1). It is obvious that RF1 and RF2 are more than simple tRNA mimicry proteins, since they can activate hydrolysis of peptidyl-tRNA and can bind to the ribosome in the absence of an EF-Tu-like vehicle protein. In other words, a truncated tRNA mimicry polypeptide cannot bind to the ribosome or activate peptidyl-tRNA hydrolysis by itself. These activities must be encoded in domains other than the tRNA mimicry portion. To gain insights into the domain(s) responsible for ribosome binding, we have employed genetic selection of novel RF2 mutants that exert amber (UAG) suppressor activity, contrary to the widely accepted view that mutations in RF1 and RF2 cause nonsense suppression of UAG (amber) and UGA (opal) codons, respectively, and that they do not exert a reciprocal (cross)-suppression phenotype. We referred to this allele as *csu* for cross suppression (Yoshimura et al., 1999). The rationale for the selection of *csu* mutants was based on the assumption that, given that there is a codon-independent step common to RF1 and RF2

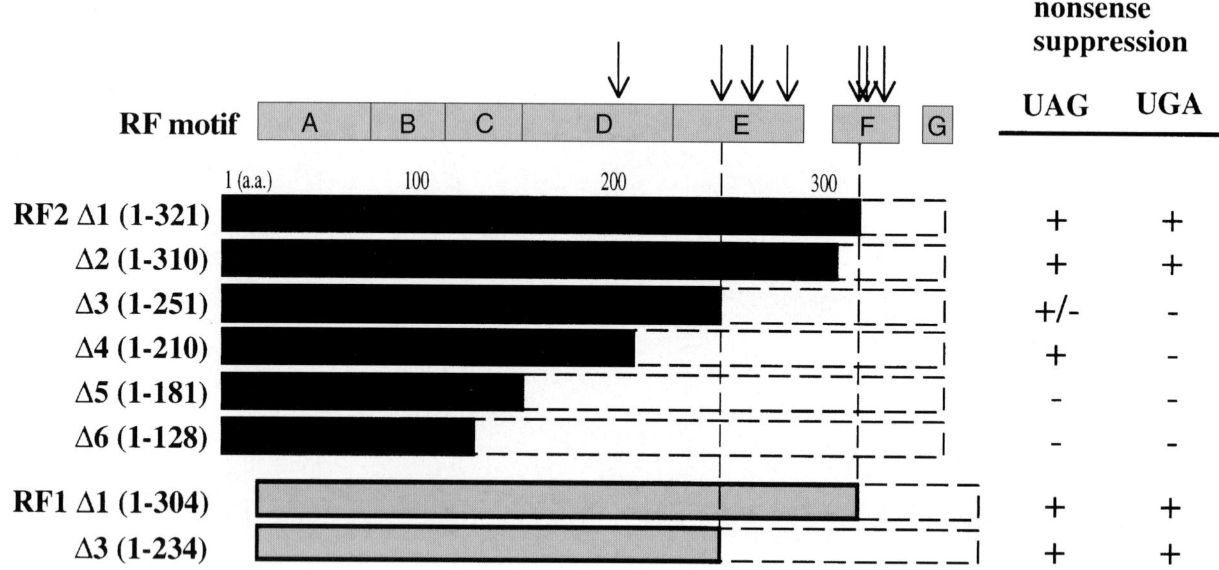

Figure 2. Truncated RF1 and RF2 polypeptides possessing the cross-suppression (Csu) activity monitored by the readthrough assay of UAG and UGA alleles in *lacZ* (Yoshimura et al., 1999). The positions of *csu* alleles are indicated by arrows. a.a., amino acid(s); +, present; −, absent; +/−, weakly present.

during the sequential interaction of RF with the ribosome, a mutation interfering with each step could give rise to a Csu phenotype.

Nine *csu* mutants were selected on the plasmid bearing the RF2 gene as amber suppressors by using a *lacZ* (UAG) reporter: five were missense and four were nonsense mutations. Moreover, RF2 C-terminal deletions equivalent to the nonsense alleles exerted not only amber suppression activity but also opal suppression activity; the equivalent RF1 segments also showed both suppression phenotypes (Fig. 2). All the *csu* mutations were mapped to the C-terminal half of RF2 and coincided strikingly with the highly conserved amino acids. These findings suggest that N-terminal domains of RF1 and RF2 facilitate codon-independent binding to the ribosome and that *csu* mutations affect this conserved function. One explanation is to assume that the first step in the sequence of interactions between the ribosome and RF is a codon-independent formation of an initial complex (the RF-ribosome initial binding model [Fig. 3]), comparable to the initial binding step of the EF-Tu·GTP·aminoacyl-tRNA ternary complex to the ribosome preceding codon recognition (Rodnina et al., 1996; Pape et al., 1998) and that altered *csu* variants may block the formation of proper initial complexes preceding the stop codon recognition.

RF1/2 + ribosome $\underset{1'}{\overset{1}{\rightleftharpoons}}$ initial binding complex $\underset{2'}{\overset{2}{\rightleftharpoons}}$ codon recognition $\overset{3}{\rightarrow}$ peptide release

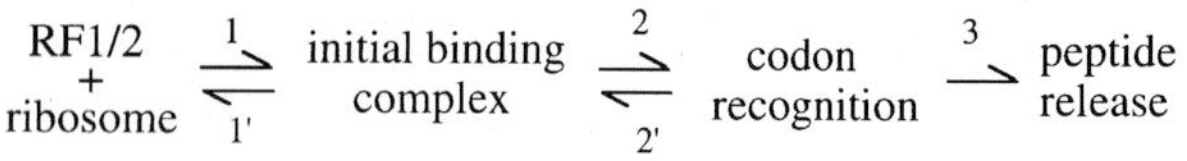

Figure 3. RF-ribosome initial binding hypothesis. The predicted sequence of interactions between the ribosome and RF is shown. We propose initial binding to the ribosome of RF preceding stop codon recognition, and *csu* mutations may block or interfere with either step 1′ or 2.

EXCEPTIONAL Thr 246 IN *E. COLI* RF2

Amino acids in domain E are the most highly conserved in prokaryotic RFs. Position 246 of RF2 is one of these conservative alleles in prokaryotic RF1 and RF2 proteins, including *Saccharomyces cerevisiae* mRF1, and is Ala or Ser (Fig. 1). Intriguingly, *E. coli* RF2 is an exception with Thr at this position. Due to this abnormal residue, *E. coli* RF2 terminates translation very weakly at UAA or UGA, and an excess of *E. coli* RF2 was toxic to cells (Uno et al., 1996). Upon substitution of Ala for Thr 246 of *E. coli* RF2, the protein acquired increased release activity for UAA and UGA and lost the toxicity, indicating that *E. coli* RF2 activity is potentially weak due to the distortion and that the amino acid at position 246 plays a crucial role, not for codon discrimination but for polypeptide release directly or indirectly (Uno et al., 1996).

Amino acid acceptability or requirements for position 246 were examined by screening active RF2 clones from *E. coli* RF2 pools randomized at position 246 as well as by saturation genetics (Uno et al., unpublished). Active RF2 clones contained Thr, Ser, Ala, and Gly (Table 1). These RF2 derivatives showed polypeptide release activity at UGA and UAA in the

Table 1. Mutagenesis of Thr 246 of *E. coli* RF2

Amino acid[a]	Activity[b]		Growth interference[c]
	In vivo	In vitro	
Thr	+	+	No
Ala	+	+	No
Ser	+	+	No
Gly	+	+	No
Leu	−	−	No
Ile	−	−	No
Val	−	−	No
Glu	−	−	Yes
Asp	−	−	Yes

[a] Amino acids at position 246 of *E. coli* RF2 were manipulated by site-directed mutagenesis as well as by genetic selection from randomized trinucleotide collections (Uno et al., unpublished).
[b] In vivo activity of RF2 was tested by complementation of the temperature-sensitive (*prfB286*) or null (*prfB*::Cm[r]) RF2 allele. In vitro activity was tested by in vitro fMet release assay with the purified derivatives. +, present; −, absent.
[c] Growth interference was monitored by growth of transformants upon expression of the altered RF2 from plasmid.

in vitro termination reaction with a 9-mer mini-mRNA. Several other amino acids, including Leu, Ile, and Val, were not active despite being similar in mass. Therefore, position 246 requires restricted small and relatively hydrophilic amino acids.

We have previously raised the possibility of post translational modification of Thr 246 to account for the abnormal presence of Thr at this position (Uno et al., 1996). However, substitutions of Glu and Asp, which mimic a negative charge, for Thr 246 not only inactivated the release activity but conferred the growth interference phenotype on RF2 (Table 1). Therefore, the introduction of a negative charge to position 246 by phosphorylation may not account for the presence of Thr 246 (unpublished). Although the biological meaning of this exception is not immediately obvious, we speculate that inactive or interfering amino acids induce disordered topology and interfere with the normal peptidyl-tRNA hydrolysis. Moffat and Tate (1994) have found that chymotrypsin cleavage of RF2 at position 244 (in domain E) stabilizes the codon-specific RF-ribosome interaction and abolishes peptidyl-tRNA hydrolysis activity. It remains to be investigated whether this locus, including positions 244 and 246, interacts with the peptidyl-transferase center.

RF3 PROTEIN ACTION

Unlike RF1 and RF2, a third factor, RF3, has received little attention since its initial characterization in the 1970s; its biological significance in protein synthesis has been a long-standing puzzle. After 2 decades of silence, the gene for RF3 was discovered simultaneously by two groups (Mikuni et al., 1994; Grentzmann et al., 1994). Very similarly, eukaryotic counterparts of bacterial RF3, eRF3, were identified only recently in *Xenopus laevis*, humans, and yeast (Zhouravleva et al., 1995; reviewed in Buckingham et al., 1997), in spite of the initial indication of a requirement for GTP for polypeptide release in the early pioneer work with rabbit reticulocytes (Konecki et al., 1977). The primary protein sequence of RF3 resembled the N-terminal part of EF-G, including G domains as well as the extra G′ domain that was missing in most other G proteins (Mikuni et al., 1994; Kawazu et al., 1995), suggesting its functional and structural similarity to EF-G.

Assuming that RF1 and RF2 are tRNA analogues, one can speculate that RF3 is an "EF-G-like" translocase protein or an "EF-Tu-like" vehicle protein (Ito et al., 1996). Therefore, we have predicted that the RF1- or RF2-RF3 (or eRF1-eRF3) complex, if formed appropriately, will resemble EF-G or the EF-Tu·GTP·aminoacyl-tRNA ternary complex (Ito et al., 1996). However, there are now several lines of evidence that bacterial RF3 may not be a functional and/or structural mimic of EF-Tu and may also be functionally distinct from eukaryotic eRF3 (reviewed in Nakamura and Ito, 1998). For instance, eRF3 and eRF1 bind in vivo and in vitro and exist as a heterodimer in *S. cerevisiae* cell lysates (Stansfield et al., 1995; Zhouravleva et al., 1995; Ito et al., 1998a), while bacterial RF3 does not bind stably to RF1 or RF2 (Ito et al., 1996); eRF3 shows considerable C-terminal homology to EF-1α (reviewed in Stansfield and Tuite, 1994), while RF3 has a long C-terminal polypeptide compared with EF-Tu, which shows significant homology with domain III and part of domain IV of EF-G (Kawazu et al., 1995). Therefore, RF3 is closer to EF-G than to EF-Tu in prokaryotes while eRF3 is functionally closer to EF-Tu than to EF-G in eukaryotes (reviewed in Nakamura and Ito, 1998).

Ehrenberg and colleagues have reconstituted a complete in vitro translation system to measure repeated rounds of translation of short synthetic mRNAs and have found that bacterial RF3 functions to recycle RF1 and RF2 after the release of nascent polypeptides in a GTP-dependent manner and that fast recycling of ribosomes requires both RF3 and RRF (a ribosome-recycling factor [reviewed in Janosi et al., 1996]) (Freistroffer et al., 1997; Pavlov et al., 1997). This is the first clear discussion of the role of RF3 in vitro. This is not inconsistent with our previous prediction of the functional similarity between RF3 and EF-G (Ito et al., 1996; Nakamura et al., 1996), given that recycling of RF1 and -2 and translocation of peptidyl-tRNAs on the ribosome may share the same conformational transition of the ribosome induced by the actions of RF3 and EF-G, respectively. Nevertheless, it remains to be examined whether this RF1- and -2-recycling activity of RF3 is sufficient to account for the in vivo phenotypes of altered RF3 variants or overproduced RF3 (see below). Also, it remains to be investigated whether eRF3 catalyzes recycling of eRF1 and/or ribosomal recycling in vitro or has another novel function(s) in vivo with a prion-like N-terminal extension (reviewed in Stansfield and Tuite, 1994, and Lindquist, 1997).

UP-FUNCTION DERIVATIVES OF RF3

Although RF3 does not form a stable heterodimer complex, there are several lines of genetic and biochemical evidence for its interaction with RF1 and RF2 on the decoding site of the ribosome. Defects in RF3 cause misreading of all three stop codons, and an excess of RF3 stimulates the polypeptide termination activities of RF1 and RF2 (Mikuni et al., 1994; Grentzmann et al., 1994; Kawazu et al., 1995; Matsumura et al., 1996; Yanofsky et al., 1996). Mutations designated *sra* are suppressor alleles in the RF3 gene that are able to restore temperature-sensitive lethal defects in RF1 and RF2 (Matsumura et al., 1996). These *sra* variants of RF3 enhance the activities of not only the defective RF1 or RF2 products but also the wild-type products in vivo (Matsumura et al., 1996). In vitro termination analysis with purified *sra* derivatives clearly demonstrated enhanced *N*-formylmethionine (fMet) release at the UAA codon compared with that of the wild-type RF3 (Fig. 4). Nevertheless, the rate of stimulation varied depending on the messenger oligonucleotides employed in the reaction. All four *sra* RF3 variants enhanced fMet release more clearly in the multiround in vitro termination system composed of a 9-mer mini-messenger (5′-UUC AUG UAA-3′), f[^{3}H]Met-tRNA^f, RF1, and the ribosome isolated from *E. coli* (Fig. 4A) rather than in the classical single-round termination system containing separate AUG and UAA triplets instead of the single minimessenger (Fig. 4B). Two of these alleles, *sra-1* and *sra-4*, were also shown to enhance termination in the frameshift termination assay much more than the wild-type RF3 at the strong termination signals (including the base following the triplet), such as UAAU and UGAU, but less at the weak termination signals, such as UGAC (Crawford et al., 1999). These findings are interpreted as indicating that a higher rate of association of RF1 and RF2, and hence a lower rate of dissociation, with the

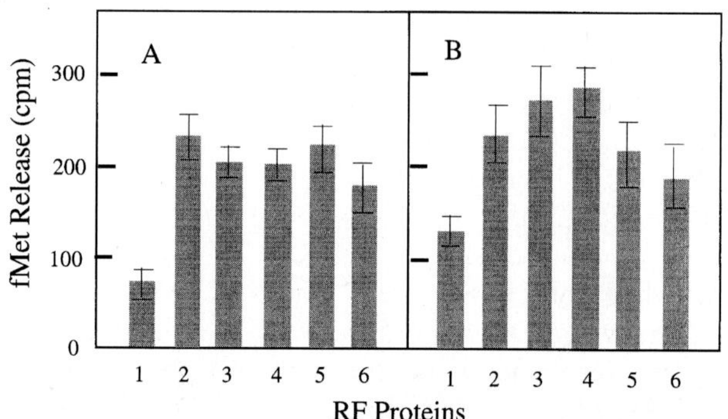

Figure 4. Stimulation of the polypeptide release activity of RF1 by *sra* variants of RF3 in vitro. f[^{3}H]Met release from the f[^{3}H]Met-tRNA^f·oligonucleotide·ribosome complex upon addition of RF1 and RF3 variants was determined essentially as described previously (Yoshimura et al., 1999) except for the messenger oligonucleotide(s). (A) 9-mer minimessenger, 5′-UUC AUG UAA-3′. (B) AUG initiator and UAA terminator triplets. Proteins: 1, RF1 alone; 2, RF1 plus *sra1* RF3; 3, RF1 plus *sra2* RF3; 4, RF1 plus *sra3* RF3; 5, RF1 plus *sra4* RF3; 6, RF1 plus wild-type RF3. Experiments were performed independently at least seven times, and the values are expressed with standard deviations (Kawazu et al., unpublished).

ribosome is programmed at strong stop signals (Poole et al., 1995) and that *sra* RF3 variants can efficiently accelerate the recycling of RF1 and RF2 at the strong signals (Crawford et al., 1999). Although the precise mechanism of stimulation by *sra* alleles remains to be investigated, these findings fit the recycling function of RF3 predicted by Ehrenberg and colleagues. However, we assume that the simple recycling model of RF3, even if true, may not represent the whole story of RF3 because it is not sufficient to account for several phenomena revealed by the genetic probes, for examples, why *sra* variants of RF3 are able to stimulate RF1 (or RF2) activity in the classical single-round fMet release reaction (Fig. 4B), why up-function alterations are relatively easily created by single *sra* substitutions in broad portions of RF3 (Matsumura et al., 1996), and how cooperative binding of RF3 (a guanine nucleotide free form) and RF2 to the ribosome is achieved (Pel et al., 1998). Further studies of RF3 and the role of GTP hydrolysis are needed.

PERSPECTIVES

The molecular mimicry between EF-G, the EF-Tu·GTP·aminoacyl-tRNA ternary complex, and perhaps the (e)RF1·(e)RF3·GTP complex makes it likely that these complexes fit into a similar pocket in the ribosome during decoding. This resemblance between part of the translation factor and tRNA represents a novel concept of molecular mimicry between nucleic acid and protein (Nissen et al., 1995). Consistent with this discovery, functional mimicry of a major autoantigenic epitope of the human insulin receptor by RNA has been described (Doudna et al., 1995) and protein mimicry of DNA has been shown in the crystal structure of the uracil-DNA glycosylase-uracil glycosylase inhibitor protein complex (Mol et al., 1995; Savva and Pear, 1995), as well as in the nuclear magnetic resonance structure of transcription factor TBP-$TAF_{II}230$ complex (Liu et al., 1998). Nature must have evolved this "art" of molecular mimicry between protein and tRNA or DNA to satisfy common requirements for structure and function on the ribosome or other regulatory systems.

The RF-tRNA mimicry hypothesis has proven useful to solve the problem of how RF recognizes the stop codon. The key breakthrough was to predict a putative anticodon mimic, i.e., the peptide anticodon, in RF, which has been successfully clarified by the subsequent genetic studies in our laboratory. Although the emerging peptide anticodon of RF functionally mimics the anticodon of tRNA (Ito et al., in press), their structures may not be exactly the same because of the magnitude of complexity required to generate a tRNA-anticodon equivalent with amino acids. The answer to whether and how much the shape of RF mimics tRNA must await the structural solution by X-ray crystallography or nuclear magnetic resonance analysis. Irrespective of the shape of RF, the accumulating genetic and other relevant data support the existence of the peptide anticodon in RFs and its close proximity to the tip of the anticodon stem mimic, as we have predicted in the mimicry hypothesis. Clearly, further study beyond RF-tRNA mimicry will be necessary to elucidate the fundamental mechanism by which protein reads the genetic code and triggers peptidyl-tRNA hydrolysis on the ribosome. The whole picture should be much more complex and dynamic. We trust that the genetic probes will be of great importance and again prove powerful in solving these remaining problems.

We are grateful to the past and present members of this laboratory as well as to other colleagues who contributed the studies described here.

This work was supported in part by grants from The Ministry of Education, Science, Sports and Culture, Japan; the Human Frontier Science Program (awarded in 1993 and 1997); and the Basic Research for Innovation Biosciences Program of Bio-oriented Technology Research Advancement Institution (BRAIN).

REFERENCES

Ævarsson, A., E. Brazhnikov, M. Garber, J. Zheltonosova, Y. Chirgadze, S. Al-Karadaghi, L. A. Svensson, and A. Liljas. 1994. Three-dimensional structure of the ribosomal translocase: elongation factor G from *Thermus thermophilus*. *EMBO J.* **13:**3669–3677.

Buckingham, R. H., G. Grentzmann, and L. Kisselev. 1997. Polypeptide chain release factors. *Mol. Microbiol.* **24:**449–456.

Capecchi, M. R., and H. A. Klein. 1969. Characterization of three proteins involved in polypeptide chain termination. *Cold Spring Harbor Symp. Quant. Biol.* **34:**469–477.

Caskey, T., E. Scolnick, R. Tompkins, J. Goldstein, and G. Milman. 1969. Peptide chain termination, codon, protein factor, and ribosomal requirements. *Cold Spring Harbor Symp. Quant. Biol.* **34:**479–488.

Crawford, D.-J. G., K. Ito, Y. Nakamura, and W. P. Tate. 1999. Indirect regulation of termination efficiency at highly expressed genes and recoding sites by the factor recycling function of *Escherichia coli* release factor RF3. *EMBO J.* **18:**727–732.

Czworkowski, J., J. Wang, T. A. Steitz, and P. B. Moore. 1994. The crystal structure of elongation factor G complexed with GDP, at 2.7 Å resolution. *EMBO J.* **13:**3661–3668.

Doudna, J. A., T. R. Cech, and B. A. Sullenger. 1995. Selection of an RNA molecule that mimics a major autoantigenic epitope of human insulin receptor. *Proc. Natl. Acad. Sci. USA* **92:**2355–2359.

Fraser, C. M., J. D. Gocayne, O. White, M. D. Adams, R. A. Clayton, R. D. Fleischmann, C. J. Bult, A. R. Kerlavage, G. Sutton, J. M. Kelley, J. L. Fritchman, J. F. Weidman, K. V. Small, M. Sandusky, J. Fuhrmann, D. Nguyen, T. R. Utterback, D. M. Saudek, C. A. Phillips, J. M. Merrick, J.-F. Tomb, B. A. Dougherty, K. F. Bott, P.-C. Hu, T. S. Lucier, S. N. Peterson, H. O. Smith, C. A. Hutchison III, and J. C. Venter. 1995. The minimal gene complement of *Mycoplasma genitalium*. *Science* **270:**397–403.

Freistroffer, D. V., M. Y. Pavlov, J. MacDougall, R. H. Buckingham, and M. Ehrenberg. 1997. Release factor RF3 in *E. coli* accelerates the dissociation of release factors RF1 and RF2 from the ribosome in a GTP-dependent manner. *EMBO J.* **16:**4126–4133.

Frolova, L., X. Le Goff, H. H. Rasmussen, S. Cheperegin, G. Drugeon, M. Kress, I. Arman, A.-L. Haenni, J. E. Celis, M. Philippe, J. Justesen, and L. Kisselev. 1994. A highly conserved eukaryotic protein family possessing properties of polypeptide chain release factor. *Nature* **372:**701–703.

Goldstein, J. L, and C. T. Caskey. 1970. Peptide chain termination: effect of protein S on ribosomal binding of release factors. *Proc. Natl. Acad. Sci. USA* **67:**537–543.

Grentzmann, G., D. Brechemier-Baey, V. Heurgue, L. Mora, and R. H. Buckingham. 1994. Localization and characterization of the gene encoding release factor RF3 in *Escherichia coli*. *Proc. Natl. Acad. Sci. USA* **91:**5848–5852.

Ito, K,, K. Ebihara, M. Uno, and Y. Nakamura. 1996. Conserved motifs of prokaryotic and eukaryotic polypeptide release factors: tRNA-protein mimicry hypothesis. *Proc. Natl. Acad. Sci. USA* **93:**5443–5448.

Ito, K., K. Ebihara, and Y. Nakamura. 1998a. The stretch of C-terminal acidic amino acids of translational release factor eRF1 is a primary binding site for eRF3 of fission yeast. *RNA* **4:**958–972.

Ito, K., M. Uno, and Y. Nakamura. 1998b. Single amino acid substitution in prokaryote polypeptide release factor 2 permits it to terminate translation at all three stop codons. *Proc. Natl. Acad. Sci. USA* **95:**8165–8169.

Ito, K., M. Uno, and Y. Nakamura. Stop codons in messenger RNA deciphered by a tripeptide anticodon. *Nature*, in press.

Janosi, L., H. Hara, S. Zhang, and A. Kaji. 1996. Ribosome recycling by ribosome recycling factor (RRF)—an important but overlooked step of protein biosynthesis. *Adv. Biophys.* **32:**121–201.

Kawazu, Y., K. Ito, K. Matsumura, and Y. Nakamura. 1995. Comparative characterization of release factor RF-3 genes of *Escherichia coli, Salmonella typhimurium*, and *Dichelobacter nodosus. J. Bacteriol.* 177:5547–5553.

Kawazu, Y., K. Ito, and Y. Nakamura. Unpublished data.

Konecki, D. S., K. C. Aune, W. Tate, and C. T. Caskey. 1977. Characterization of reticulocyte release factor. *J. Biol. Chem.* 252:4514–4520.

Lindquist, S. 1997. Mad cows meet psi-chotic yeast: the expansion of the prion hypothesis. *Cell* 89:495–498.

Liu, D., R. Ishima, K. I. Tong, S. Bagby, T. Kokubo, D. R. Muhandiram, L. E. Kay, Y. Nakatani, and M. Ikura. 1998. Solution structure of a TBP-$ATF_{II}230$ complex: protein mimicry of the minor groove surface of the TATA box unwound by TBP. *Cell* 94:573–583.

Matsumura, K., K. Ito, Y. Kawazu, O. Mikuni, and Y. Nakamura. 1996. Suppression of temperature-sensitive defects of polypeptide release factors RF-1 and RF-2 by mutations or by an excess of RF-3 in *Escherichia coli. J. Mol. Biol.* 258:588–599.

Mikuni, O., K. Ito, J. Moffat, K. Matsumura, K. McCaughan, T. Nobukuni, W. Tate, and Y. Nakamura. 1994. Identification of the *prfC* gene, which encodes peptide-chain-release factor 3 of *Escherichia coli. Proc. Natl. Acad. Sci. USA* 91:5798–5802.

Moffat, J. G., and W. P. Tate. 1994. A single proteolytic cleavage in release factor 2 stabilizes ribosome binding and abolishes peptidyl-tRNA hydrolysis activity. *J. Biol. Chem.* 269:18899–18903.

Mol, C. D., A. S. Arvai, R. J. Sanderson, G. Slupphaug, B. Kavli, H. E. Krokan, D. W. Mosbaigh, and J. A. Tainer. 1995. Crystal structure of human uracil-DNA glycosylase in complex with a protein inhibitor: protein mimicry of DNA. *Cell* 82:701–708.

Murgola, E. J., K. A. Hijiazi, H. U. Goringer, and A. E. Dahlberg. 1988. Mutant 16S rRNA: a codon specific translational suppressor. *Proc. Natl. Acad. Sci. USA* 85:4162–4165.

Nakamura, Y., and K. Ito. 1998. How protein reads the stop codon and terminates translation. *Genes Cells* 3:263–278.

Nakamura, Y., K. Ito, K. Matsumura, Y. Kawazu, and K. Ebihara. 1995. Regulation of translation termination: conserved structural motifs in bacterial and eukaryotic polypeptide release factors. *Biochem. Cell Biol.* 73:1113–1122.

Nakamura, Y., K. Ito, and L. A. Isaksson. 1996. Emerging understanding of translation termination. *Cell* 87:147–150.

Nissen, P., M. Kjeldgaard, S. Thirup, G. Polekhina, L. Reshetnikova, B. F. C. Clark, and J. Nyborg. 1995. Crystal structure of the ternary complex of Phe-$tRNA^{Phe}$, EF-Tu, and a GTP analog. *Science* 270:1464–1472.

Pape, T., W. Wintermeyer, and M. V. Rodnina. 1998. Complete kinetic mechanism of elongation factor Tu-dependent binding of aminoacyl-tRNA to the A site of the *E. coli* ribosome. *EMBO J.* 17:7490–7497.

Pavlov, M. Y., D. V. Freistroffer, J. MacDougall, R. H. Buckingham, and M. Ehrenberg. 1997. Fast recycling of *Escherichia coli* ribosomes requires both ribosome recycling factor (RRF) and release factor RF3. *EMBO J.* 16:4134–4141.

Pel, H. J., J. G. Moffat, K. Ito, Y. Nakamura, and W. P. Tate. 1998. *Escherichia coli* release factor-3: resolving the paradox of a typical G protein structure and atypical function with guanine nucleotides. *RNA* 4:47–54.

Poole, E. S., C. M. Brown, and W. P. Tate. 1995. The identity of the base following the stop codon determines the efficiency of *in vivo* translational termination in *Escherichia coli. EMBO J.* 14:151–158.

Rodnina, M. V., T. Pape, R. Fricke, L. Kuhn, and W. Wintermeyer. 1996. Initial binding of the elongation factor Tu·GTP·aminoacyl-tRNA complex preceding codon recognition on the ribosome. *J. Biol. Chem.* 271:646–652.

Savva, R., and L. H. Pear. 1995. Nucleotide mimicry in the crystal structure of the uracil-DNA glycosylase-uracil glycosylase inhibitor protein complex. *Nat. Struct. Biol.* 2:752–757.

Scolnick, E., R. Tompkins, T. Caskey, and M. Nirenberg 1968. Release factors differing in specificity for terminator codons. *Proc. Natl. Acad. Sci. USA* 61:768–774.

Stansfield, I., and M. Tuite. 1994. Polypeptide chain termination in *Saccharomyces cerevisiae. Curr. Genet.* 25:385–395.

Stansfield, I., K. M. Jones, V. V. Kushnirov, A. R. Dagkesamanskay, A. I. Poznyakovski, S. V. Paushkin, C. R. Nierras, B. S. Cox, M. D. Ter-Avanesyan, and M. F. Tuite. 1995. The products of the *SUP45* (eRF1) and *SUP35* genes interact to mediate translation termination in *Saccharomyces cerevisiae. EMBO J.* 14:4365–4373.

Tate, W. P., and C. M. Brown. 1992. Translational termination: "stop" for protein synthesis or "pause" for regulation of gene expression. *Biochemistry* 31:2443–2450.

Tate, W. P., and S. A. Mannering. 1996. Three, four or more: the translational stop signal at length. *Mol. Microbiol.* 21:213–219.

Tate, W., B. Greuer, and R. Brimacombe 1990. Codon recognition in polypeptide chain termination: site directed crosslinking of termination codon to *Escherichia coli* release factor 2. *Nucleic Acids Res.* 18:6537–6544.

Tate, W. P., E. S. Poole, and S. A. Mannering. 1996. Hidden infidelities of the translational stop signal. *Prog. Nucleic Acids Res.* 52:293–335.

Uno, M., K. Ito, and Y. Nakamura. 1996. Functional specificity of amino acid at position 246 in the tRNA mimicry domain of bacterial release factor 2. *Biochimie* 78:935–943.

Uno, M., K. Ito, and Y. Nakamura. Unpublished data.

Yanofsky, C., V. Horn, and Y. Nakamura. 1996. Loss or overproduction of polypeptide release factor 3 influences expression of the tryptophanase operon of *Escherichia coli. J. Bacteriol.* 178:3755–3762.

Yoshimura, K., K. Ito, and Y. Nakamura. 1999. Amber (UAG) suppressors affected in UGA/UAA-specific polypeptide release factor 2 of bacteria: genetic prediction of initial binding to ribosome preceding stop codon recognition. *Genes Cells* 4:253–266.

Zhouravleva, G., L. Frolova, X. Le Goff, R. Le Guellec, S. Inge-Vechtomov, L. Kisselev, and M. Philippe. 1995. Termination of translation in eukaryotes is governed by two interacting polypeptide chain release factors, eRF-1 and eRF-3. *EMBO J.* 14: 4065–4072.

The Ribosome: Structure, Function, Antibiotics, and Cellular Interactions
Edited by R. A. Garrett, S. R. Douthwaite, A. Liljas, A. T. Matheson, P. B. Moore, and H. F. Noller

Chapter 43

Ribosome-Recycling Factor: an Essential Factor for Protein Synthesis

AKIRA KAJI and GO HIROKAWA

The protein synthesis termination step catalyzed by the termination codon-specific peptide release factors (RFs) (Tate et al., 1993; Buckingham et al., 1997) leaves the posttermination complex as shown in Fig. 1B. Disassembly of the posttermination complex is catalyzed by two factors: elongation factor G (EF-G) (Hirashima and Kaji, 1972b, 1973) or RF3 (Grentzmann et al., 1998) and ribosome-recycling factor (RRF; originally called ribosome-releasing factor [Janosi et al., 1994]). In addition, RRF maintains translational fidelity during chain elongation (Janosi et al., 1996b). Discovered in 1970 (Hirashima and Kaji, 1970), RRF is an essential factor (Janosi et al., 1994). In this chapter, we review key experimental findings about RRF up to May 1999 (see brief reviews in Kaji et al., 1998, and Janosi et al., 1996b, and the historical background in Janosi et al., 1996a).

BASIC PROPERTIES OF RRF

Escherichia coli RRF is a basic protein with a molecular mass of approximately 20 kDa consisting of 185 amino acids (Ichikawa and Kaji, 1989). Purification consists of five steps (Hirashima and Kaji, 1972b). It is coded for by the gene *frr* situated in the 4-min region, 1.1 kb downstream from the gene coding for EF-Ts (Ichikawa et al., 1989). The gene *frr* has a very strong promoter (Shimizu and Kaji, 1991) with minimal expression under laboratory conditions (Ichikawa et al., 1989) but elevated in some pathogens infecting animals (Vizcaino et al., 1996; Lowe et al., 1998). The molar amount of RRF in a cell is approximately one-half the total molar amount of ribosome, and approximately 30% of total cellular RRF is bound to ribosomes (Ryoji, 1981).

RRF ASSAY AND RELEASE OF tRNA DURING THE RRF REACTION

As an in vitro system closest to the natural disassembly of the posttermination complex, we chose the system programmed by R17 amB2 phage RNA (Gussin, 1966; Robinson et al., 1969). The amB2 phage has UAG at the seventh codon of the coat cistron. The 4th amino acid in this cistron is Asn, and the 8th, 9th, 10th, and 11th amino acids are Phe, Val, Leu, and Val, respectively. In the experiment described in Table 1 (Ogawa and Kaji, 1975), amB2 phage RNA was labeled with ^{3}H and ribosomes were bound to the initiation sites of this RNA together with fMet-tRNA. To this complex, Ala, Ser, Asn, Phe, Thr, RF1, and other components necessary to bring ribosomes to the sixth codon (UAG) were added. The release of ribosomes from the labeled R17 phage RNA by RRF was monitored by sucrose gradient centrifugation. As shown in Table 1, when RRF was omitted, release of R17 RNA was reduced to 1/20 that of the complete system, indicating that RRF indeed releases ribosomes from natural mRNA at the natural termination codon. The ribosome has to be at the termination codon for the release because removal of Asn from the reaction mixture significantly reduced the RRF reaction.

A similar termination complex can be made with f[^{35}S]Met-tRNA and 9 nucleotides [^{32}P]UUC AUG UAA labeled with ^{32}P at the 5′ end (Grentzmann et al., 1998). RF2 converted this to the posttermination complex. Physical release of mRNA from this complex dependent on RRF, EF-G, and RF2 or RF1 could be observed by separating the mRNA from the posttermination complex by Microcon 10 filters (Amicon) (Grentzmann et al., 1998).

Akira Kaji and Go Hirokawa ■ Department of Microbiology, School of Medicine, University of Pennsylvania, 319B Johnson Pavilion, 3610 Hamilton Walk, Philadelphia, PA 19104.

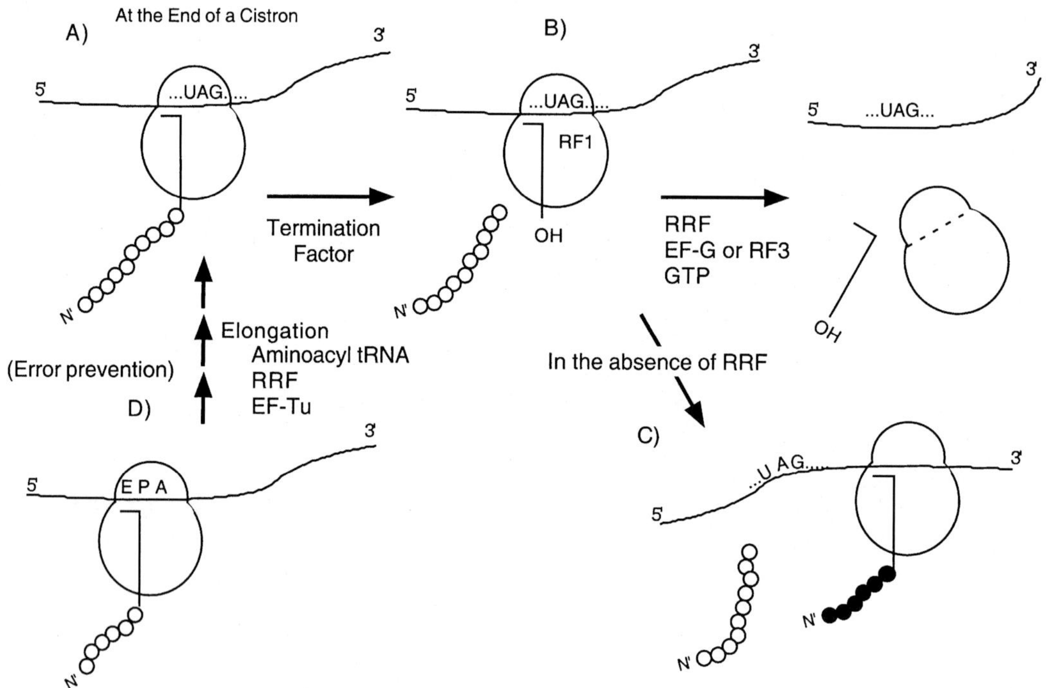

Figure 1. Translational steps involving RRF. (A) Termination complex. (B) Posttermination complex. (C) Unscheduled reinitiation of translation in the absence of RRF. (D) Elongation steps where RRF reduces translational error. Open circles, normal peptide chain; solid circles, peptide chain produced by the unscheduled translation downstream from the termination codon due to the absence of RRF. (Reprinted from Kaji et al., 1998, with permission.)

As a routine assay procedure, polyribosomes treated with puromycin can be used as a model substrate for RRF. Since termination factors (RFs) are proposed to bind at the A site (Tate et al., 1983, 1990), the resulting posttermination complex (as shown in Fig. 1B) would have deacylated tRNA on the P or E site. It is likely that the A site is empty with RF1 because RF3 removes it from the posttermination complex, while the status of the A site remains obscure with RF2 because the experimental data reported are contradictory as to recycling of RF2 by RF3 (Freistroffer et al., 1997; Grentzmann et al., 1998) in the posttermination complex. We therefore assume that the posttermination complex has an empty A site. In the puromycin-treated polyribosome, each ribosome presumably has an empty A site and bound deacylated tRNA at the P site, a structure similar to that of the posttermination complex. To ensure that the A site is empty, tetracycline (Kaji and Ryoji, 1979; Schnappinger and Hillen, 1996) was added immediately before polysome isolation. It is clear from Fig. 2 that 70S ribosomes were released by RRF and EF-G from this substrate (Hirashima and Kaji, 1970, 1972a). It should be noted that most of the polyribosomes observed in Fig. 2a were recovered as 70S ribosomes (Fig. 2c). The absorbances at the tops of the tubes in Fig. 2a and b are due to added nucleotides and released tRNA. Therefore, under these experimental conditions, little ribosomal subunits were observed. Because this assay was readily available in the 1970s, discovery, isolation, purification, characterization, and cloning of RRF have been carried out by this method (Hirashima and Kaji, 1972b, 1973; Ichikawa and Kaji, 1989; Ichikawa et al., 1989).

During the RRF reaction, all of the bound tRNA was released (Table 2). The maximum release of

Table 1. Effect of removal of asparagine on release of ribosome-bound amB2 R17 RNA and peptide from ribosomes[a]

Components[b]	Ribosome-bound mRNA (pmol)	mRNA released (pmol)
Complete system	0.72	1.46
+ Asparaginase, −Asn, −RRF	2.18	0
+ Asparaginase, −Asn	1.64	0.54
−RRF	2.11	0.07

[a] Adapted from Ogawa and Kaji, 1975, with permission.
[b] The complete system (0.5 ml) contained the initiation complex of ribosomes and labeled amB2 R17 RNA, EF-G (13 μg), EF-T (mixture of EF-Tu and EF-Ts; 52 μg), a mixture of tRNA from *E. coli* (50 μg), crude peptide release factor RF1 (144 μg), ATP (0.4 mM), phosphoenolpyruvate (4 mM), GTP (0.2 mM), pyruvate kinase (15 μg), amino acid mixture (alanine, serine, phenylalanine, threonine, and asparagine), and RRF (0.6 μg). The mixtures were incubated at 33°C for 15 min and analyzed on sucrose gradients. Where indicated, asparaginase (0.4 μg) was added.

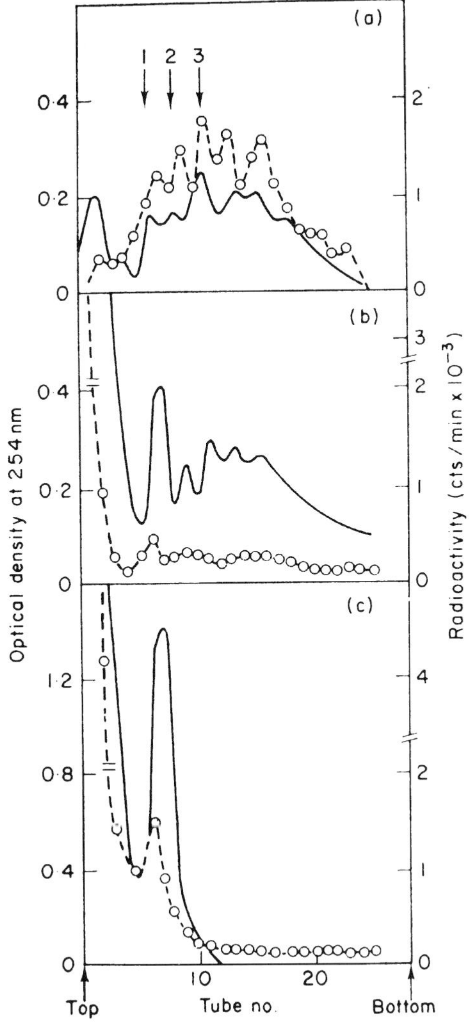

Figure 2. Conversion of polysomes to monosomes by RRF. (a) Sedimentation profile of polysomes with nascent peptide (labeled with ^{14}C amino acids [open circles]). Solid line represents optical density at 260 nm of fractions measured using ISCO. 1, 2, and 3, are mono-, di-, and trisomes, respectively. (b) Polysomes in panel a treated with puromycin. (c) Polysomes in panel b treated with RRF, EF-G, and GTP. Sedimentation was from left to right in the sucrose gradient centrifugation. (Reprinted from Hirashima and Kaji, 1972a, with permission.)

tRNA was observed in the complete system, and omission of one or more of the components decreased the release of tRNA (Hirashima and Kaji, 1973).

ROLE OF EF-G AND THE RRF BINDING SITE OF THE RIBOSOME

The RRF reaction is dependent on GTP and either EF-G (Hirashima and Kaji, 1973) or RF3 (Grentzmann et al., 1998) in a 1:1 stoichiometric relationship (Hirashima and Kaji, 1972b). Since EF-G and RF3 are ribosome-dependent GTPases (Conway and Lipmann, 1964; Grentzmann et al., 1998), they must act similarly with RRF. When the substrate was incubated for 10 min with EF-G followed by the addition of fusidic acid (Tanaka, 1995), the RRF reaction did not take place. Similarly, preincubation of the substrate with either RRF or EF-G followed by removal of the factors used for the preincubation did not prepare the substrate for the other factor (Hirashima and Kaji, 1973). In support of this notion that two factors have to be present simultaneously, a recent experiment of ours showed that the binding of RRF to ribosomes depends on EF-G (unpublished).

Table 2. Release of tRNA from ribosomes during release of ribosomes from mRNA[a]

Reaction mixture	tRNA released (cpm)
Polysomes alone	222
Polysomes, PM	423
Polysomes, PM, RRF	239
Polysomes, PM, EFG	371
Polysomes, PM, EFG, GTP	642
Polysomes, PM, RRF, EFG, GTP	1,048
Total tRNA bound to polysomes	953

[a] The amounts of tRNA (expressed as [^{14}C]aminoacyl-tRNA) released from 2.08 A_{260} units of polysomes are expressed as their capacities to accept ^{14}C-amino acids. The amount of [^{14}C]aminoacyl-tRNA is shown as the radioactivity insoluble in cold (0°C) trichloroacetic acid. PM, puromycin. (Adapted from Hirashima and Kaji, 1973, with permission.)

We postulate that RRF must interact with ribosomes near the A site. As described below, RRF helps maintain translational fidelity during the chain elongation step (Janosi et al., 1996b), suggesting that it interacts with ribosomes at or near the A site, the decoding site. Furthermore, inhibition by an excess of RF1 (which presumably binds to the A site) during an in vitro oligopeptide synthesis was reversed by addition of an excess of RRF (Pavlov et al., 1997b). Lastly, EF-G, which is required for the ribosomal binding of RRF, binds near or at the A site of the ribosome (Agrawal et al., 1998; Wilson and Noller, 1998b; Richman and Bodley, 1972), supporting the notion that RRF must bind near the A site.

The exact molecular mechanism of the RRF reaction remains to be elucidated. This is the only reaction in protein synthesis where two factors are simultaneously required. Since EF-G and GTP can change ribosomal conformation and cause relative movement between the subunits (Oefverstedt et al., 1994; Rodnina and Wintermeyer, 1998; Spahn and Nierhaus, 1998; Wilson and Noller, 1998a; Czworkowski and Moore, 1996; Moazed and Noller, 1989), the configurational change of the ribosome must be an important element of the RRF reaction. This notion is supported by the early observation that the ribosomes released by RRF and EF-G are much more prone to dissociate than 70S ribosomes isolated in a conventional manner (tightly coupled ribosomes [Spedding, 1990]) (see Behavior of Ribosomes of the Posttermination Complex: Possible Role of IF3 below).

Since RRF must participate in the chain elongation step as an error reducer (see The Role of RRF during Peptide Chain Elongation below), it may even participate in the translocation with EF-G. The important difference between the posttermination complex and the elongation complex is that the former has deacylated tRNA instead of peptidyl-tRNA. This difference leads to disassembly on the one hand and to translocation on the other. In fact, even in the presence of peptidyl-tRNA, RRF may accidentally disassemble the elongating complex (Heurgué-Hamard et al., 1998).

BEHAVIOR OF RIBOSOMES OF THE POSTTERMINATION COMPLEX: POSSIBLE ROLE OF IF3

The fate of ribosomes at the posttermination complex is a historical question which was hotly debated many years ago, with the parties involved not realizing that they were debating the reaction catalyzed by RRF (Mangiarotti and Schlessinger, 1966; Schlessinger et al., 1967; Kohler et al., 1968; Phillips et al., 1969; Algranati et al., 1969; Kelley and Schaechter, 1969; Kaempfer, 1970, 1971; Davis, 1971; Subramanian and Davis, 1971). One school of thought held that the ribosomes are released as subunits and the association of subunits into 70S ribosomes is prevented by IF3 (Kaempfer, 1970). The other view was that ribosomes are released as 70S ribosomes, as we see under our assay conditions. IF3 dissociates the 70S ribosomes into their subunits for the initiation (Davis, 1971). In this connection, it should be noted that a decrease of the protein synthesis rate results in the accumulation of 70S ribosomes while the amount of the ribosomal subunit remains unchanged (Joklik and Becker, 1965; Hogan and Korner, 1968; Kohler et al., 1968). This suggests that the RRF reaction may be a control point for in vivo protein synthesis, especially in view of the long stay of ribosomes at the termination codon (Björnsson and Isaksson, 1996).

The early unsettled dispute regarding the fate of ribosomes at the posttermination complex suggests that the outcome of the RRF reaction could vary depending on the environment surrounding the posttermination complex. The environment includes Mg^{2+} and polysome concentrations, the amount of IF3, the nucleotide sequence surrounding the termination codon, and the nature of ribosomes. For example, in the translational coupling of the lysis and the coat gene of group II RNA coliphage GA1 (Inokuchi et al., 1986), ribosomes are not released by RRF from the posttermination complex (Inokuchi et al., 1998). In the in vitro synthesis of an oligopeptide, fMet-Phe-Thr-Ile, coded for by a short synthetic mRNA (Pavlov et al., 1997a), ribosomes may not leave mRNA. With a strong Shine-Dalgarno sequence nearby, the ribosomes (or 30S subunits) in the posttermination complex are in contact with the ribosomal binding sequence because ribosomes can span mRNA at least 25 nucleotides upstream from the P site (Gold, 1988; Steitz, 1969).

As another example of various behaviors of ribosomes at the posttermination complex, van Duin's group proposed that 30S subunits, which were not released at the termination codon of the coat cistron, will start scanning up and down along the phage RNA until they catch the initiation site of the lysis protein, which is 35 (phage fr) or 47 (MS2) nucleotides upstream of the termination codon (Adhin and van Duin, 1990; Schmidt et al., 1987). Martin and Webster observed the transient appearance of the 30S subunit-f2 RNA complex, but the 30S subunits were eventually released. It should be emphasized that in the same paper Martin and Webster also provided an example of ribosomes being released from mRNA as 70S ribosomes (see Fig. 6 in Martin and Webster, 1975).

As can be seen from the above discussion, it is possible that, depending on the conditions, the RRF reaction may leave the 30S subunit on mRNA with subsequent release of 30S subunits from mRNA by other factors under certain circumstances. However, this may not be a general phenomenon because early work on bacterial polysomes isolated from growing bacteria, as quoted at the beginning of this section, never showed evidence for the 30S subunits remaining on mRNA.

To take care of the 30S subunits left on mRNA by RRF as discussed above, IF3 (Subramanian and Davis, 1970; McCarthy and Gualerzi, 1990) may complete the disassembly process. Indeed, IF3 removes aminoacyl-tRNA from the complex of the 30S subunit, tRNA, and the mRNA presumably by disassembling the complex as shown in Table 3 (Gualerzi et al., 1971). In this experiment, various complexes of 30S subunits, synthetic polynucleotides, and the corresponding aminoacyl-tRNA were prepared and the disassembly of the complex was measured upon addition of purified IF3 by following the release of aminoacyl-tRNA. All complexes, except for the initiation complex, were disassembled by IF3. This action of IF3 was later confirmed by Martin and

Table 3. IF_3 disassembles the complex of 30S subunits with mRNA except for the initiation complex[a]

Conditions in complex formation			aa-tRNA (pmol) remaining on 30S		% of bound aa-tRNA released
aa-tRNA[b]	Codon	Additions	$-IF_3$	$+IF_3$	
Phe-tRNA	Poly(U)	GTP, EF-T	4.20	1.55	63.0
NA Phe-tRNA	Poly(U)		3.93	1.82	53.0
NA Phe-tRNA	Poly(U)	GTP, IF1, IF2	2.28	0.87	61.9
Val-tRNA	Poly(AUG)		1.45	1.05	27.4
Lys-tRNA	Poly(A)		3.68	1.66	55.0
fMet-tRNA	Poly(AUG)	GTP, IF1, IF2	4.85	4.70	3.2

[a] Adapted from Gualerzi et al., 1971, with perrmission.
[b] aa, aminoacyl; NA, *N*-acetyl.

Webster, who also observed the release of 30S subunits from mRNA by initiation factors (see Table 2 in Martin and Webster, 1975).

DEPENDENCE OF PROTEIN SYNTHESIS ON RRF

The stimulatory effect of RRF on protein synthesis (up to seven- to eightfold) was observed when the ribosome concentration was limiting, as shown in Table 4 (Ryoji et al., 1981). The stimulation by RRF was greater with the limited amount of ribosomes because RRF recycles spent ribosomes. Earlier, Kung et al. (1977) observed a net increase (fourfold) of β-galactosidase synthesis by RRF. These studies have recently been confirmed in a complete in vitro translation system with pure factors and aminoacyl-tRNA synthetases. The recycling time of translation was shortened significantly by the addition of RRF and RF3 (Pavlov et al., 1997a).

FATE OF RIBOSOMES OF THE POSTTERMINATION COMPLEX IN THE ABSENCE OF RRF: ISOLATION AND CHARACTERIZATION OF MUTANT *E. COLI* WITH TEMPERATURE-SENSITIVE RRF

With the in vitro R17 amB2 phage coat protein synthesis system described above, we found that, in the presence of RRF, most of the product was the NH_2-terminal hexapeptide. In the absence of RRF, however, the hexapeptide synthesis was decreased but a coat-like protein was synthesized. The NH_2-terminal amino acid sequence of this coat-like protein was Phe·Val·Leu·Val, corresponding exactly to the 8th through 11th residues of the R17 coat protein (Weber, 1967). We concluded that ribosomes remaining on the mRNA reinitiated unscheduled translation from the triplet next to the termination codon, as shown in Fig. 1C.

To confirm the above observation in vivo, we used LJ14, one of 12 temperature-sensitive RRF mutant *E. coli* strains we had isolated. The growth characteristics of LJ14 are shown in Fig. 3 (Janosi et al., 1998). Those cultures exposed to 43°C at zero time showed a decrease in viable counts, suggesting that the inactivation of RRF in the lag phase is bactericidal. To study the reinitiation in vivo, we constructed plasmids which had the *lac* promoter, the *lacZ* ribosome binding site, and a short open reading frame (ORF), followed by the *lacZ* coding sequence (from the 10th authentic *lacZ*), either in zero frame (pPEN2363) or in −2 frame (pPEN2369), without any initiation signals.

As shown in Fig. 4, the plasmids with the reporter gene were placed in LJ14. At 39°C, β-galactosidase (Fig. 1C) was induced in LJ14 with an in-frame *lacZ'*-*'lacZ* reporter plasmid. The induction was minimal at 31°C, where most of the temperature-sensitive RRF should be active. The MC1061 strain with wild-type RRF (Fig. 4, middle) and LJ14 harboring pPEN907 carrying wild-type *frr* (Fig. 4, right) showed only negligible β-galactosidase activity. Sim-

Table 4. Stimulation of T4 lysozyme synthesis in vitro by RRF[a]

Amt of ribosome (A_{260} units)	Lysozyme activity[b] (A_{450} units/min)		Effect of RRF[c]
	−RRF	+RRF	
0.065	1.1×10^{-3}	7.8×10^{-3}	×7.7
0.108	7.5×10^{-3}	20.5×10^{-3}	×2.7
0.432	13.2×10^{-3}	24.5×10^{-3}	×1.9

[a] Adapted from Ryoji et al., 1981, with permission.
[b] T4 lysozyme was synthesized in vitro under the direction of T4 late mRNA in the presence of various amounts of ribosomes.
[c] Amount of increase (fold).

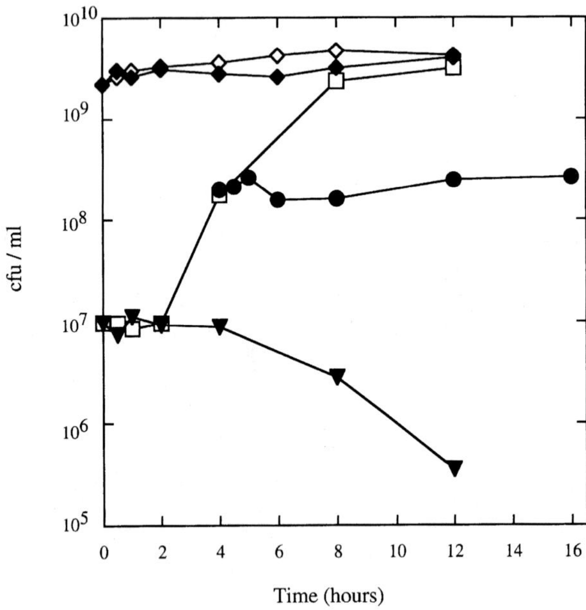

Figure 3. Growth curve of *E. coli* LJ14 carrying temperature-sensitive RRF. Triangles, overnight culture (grown at 32°C) diluted and exposed to 43°C at time zero; open squares, same as triangles, but the culture was grown at 32°C; solid circles, same as open squares but the culture was exposed to 43°C at 4 h; solid diamonds, overnight culture not diluted and cultured at 32°C; open diamonds, same as solid diamonds but the culture was exposed to 43°C at time zero. Viable counts were plotted against the time of culture. (Reprinted from Janosi et al., 1998, with permission.)

ilar results were obtained with pPEN2369, the −2 frame reporter plasmid, and another construct with a −1 frame reporter gene (data not shown). The optimum reinitiation for the in-frame construct was at 35°C, while it was at 39°C with the out-of-frame construct (Janosi et al., 1998). This suggests that ribosomes in the posttermination complex may undergo thermal agitation to change the reading frame in vivo. The amino-terminal analysis of the induced β-galactosidase showed that ribosomes (perhaps as 70S ribosomes [Petersen et al., 1978]) can slide downstream as many as 10 to 45 nucleotides before they randomly start translation of the mRNA. With a separate plasmid construct that enables us to measure both upstream ORF translation and reinitiated unscheduled translation, we found that almost 100% of the ribosomes reading the upstream ORF reinitiated unscheduled protein synthesis downstream from the ORF after inactivation of RRF (Janosi et al., 1998).

Those ribosomes which reinitiated the unscheduled translation were eventually released from the end of the mRNA as monosomes (Hirokawa et al., 1998). This conclusion was based on the following experimental results: upon in vivo inactivation of RRF, (i) polysomes are rapidly converted to monosomes and (ii) these monosomes are inactive for in vitro protein synthesis programmed by natural mRNA (unpublished).

THE ROLE OF RRF DURING PEPTIDE CHAIN ELONGATION

As shown in Table 5 (Janosi et al., 1996b), the misincorporation of Ile into polyphenylalanine synthesized under the direction of polyuridylic acid (Davies et al., 1964; Kaji and Kaji, 1965) was significantly stimulated if RRF was removed from the in vitro protein synthesis system. A pronounced effect (close to a ninefold increase in error in the absence of RRF) was observed when the noncognate amino acid was leucine or a mixture of amino acids. RRF did not influence the large error (10-fold increase) induced by streptomycin (Janosi et al., 1996b). Recent collaborative preliminary work with M. Ehrenberg of Uppsala suggested that removal of RRF in the in vitro system programmed by a synthetic mRNA (AUG-UUC-UUC) may increase the error measured by the relative formation of fMet-Phe-Leu-tRNA to that of fMet-Phe-Phe-tRNA (unpublished). In support of this concept, in collaboration with Y. Sekine of Tokyo University, we found that β-galactosidase synthesized in vivo at the semipermissive temperature of LJ14 is much more labile at 50°C than normal β-galactosidase. This is probably due to abnormal amino acid sequences in the enzyme resulting from the error-prone translational process (unpublished). It is possible that RRF reduces translational error by interfering with the binding of noncognate aminoacyl-tRNA (Takeda et al., 1968) to the ribosome, though the recent kinetic experiments on the binding of aminoacyl-tRNA make this possibility appear remote (Pape et al., 1998).

AMINO ACID SEQUENCES OF RRFs AND CHARACTERIZATION OF OTHER PROKARYOTIC RRFs

As of May 1999, approximately 30 RRF homologues had been sequenced, and the results are summarized in Fig. 5 (the alignment [Corpet, 1988] and hydrophilicities [Hopp and Woods, 1981] were calculated). All living organisms (including the smallest, *Mycoplasma genitalium* [Fraser et al., 1995]) whose genomes have been sequenced, except archaea, have a homologue of RRF. Archaea, in between eukaryotes and prokaryotes in their phylogenetic development, have a protein synthesis system similar to that

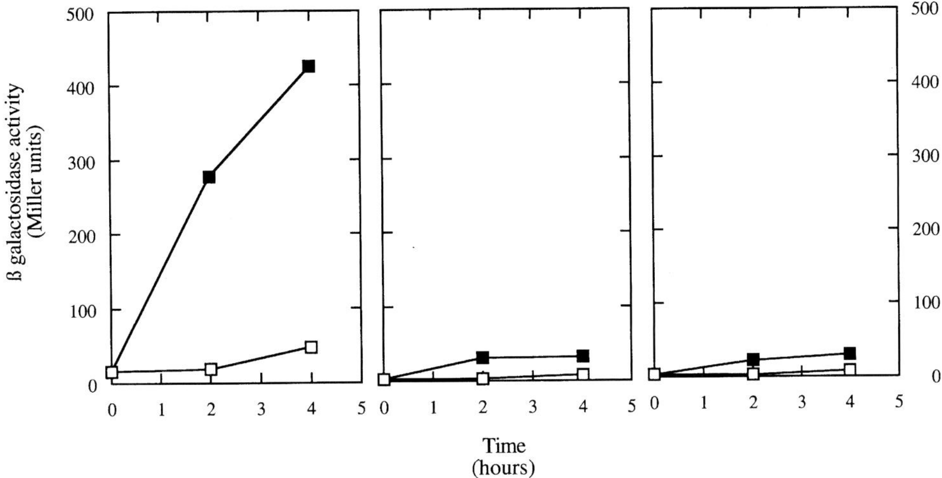

Figure 4. In vivo inactivation of temperature-sensitive RRF induces the translation of the reporter gene (β-galactosidase) situated in frame downstream from the termination codon of the upstream ORF. The reporter gene, without the initiation signals, is carried by a plasmid, pPEN2363. pPEN2363 is in LJ14 carrying temperature-sensitive RRF (left panel), in wild-type *E. coli* (middle panel), and in LJ14 harboring a plasmid carrying the wild-type *frr* (right panel). The culture was grown at 31°C, and the temperature was shifted to 39°C at time zero (solid squares). Open squares control at 31°C. β-galactosidase activity is plotted against the time after the temperature shift up. (Reprinted from Janosi et al., 1998, with permission.)

of eukaryotic cytoplasm. The absence of RRF in archaea suggests that RRF homologues are not involved in cytoplasmic protein synthesis in eukaryotes. The similarity (Devereux et al., 1984) of these RRF homologue sequences to that of *E. coli* RRF ranges from 35.4% (*Saccharomyces cerevisiae*) to 79.5% (*Haemophilus influenzae*). It can be noted from Fig. 5 that highly conserved (close to 90%) residues indeed exist among the RRFs of all organisms in the region between arginine 110 and arginine 132 as well as between proline 103 and isoleucine 82. Since the former includes the most hydrophilic region, it is possible that this region plays an important functional role.

We found that, similarly to other *Pseudomonas aeruginosa* genes (Hungerer et al., 1995; Kokjohn and Miller, 1987), the *P. aeruginosa* RRF gene functions in *E. coli*. Thus, we cloned *P. aeruginosa frr* from the *P. aeruginosa* genomic cosmid library by complementation of LJ2221 (RecA minus), carrying temperature-sensitive RRF, by using the conjugal-

Table 5. RRF promotes fidelity of translation

Experimental conditions[a]	Translational error
Leucine as noncognate	
No added RRF	0.043
2 μg of RRF	0.011
Amino acid mixture as noncognate	
No added RRF	0.0085
2 μg of RRF	0.00051

[a] Poly-uridylic acid was used as mRNA (Janosi et al., 1996b).

mating method. *P. aeruginosa frr* encoded a protein of 185 amino acid residues that had extended homology (59.6% identity) with *E. coli* RRF (Ohnishi et al., 1999). As shown in Fig. 6, a plasmid, pMO2925, carrying *P. aeruginosa frr* could complement the temperature-sensitive growth of LJ2221 when 1 mM IPTG (isopropyl-β-D-thiogalactopyranoside, the inducer of *P. aeruginosa* RRF synthesis) was added. A cluster of five genes, *rpsB-tsf-pyrH-frr-cdsA*, which are widely preserved among prokaryotes, were located within 30 to 32 min on the *P. aeruginosa* chromosome (Holloway et al., 1994). The purified *P. aeruginosa* RRF expressed in large amounts in *E. coli* cross-reacted with polyclonal antibody against *E. coli* RRF.

For structural studies of RRF, we recently cloned *frr* and purified RRF from a thermophilic bacterium, *Thermotoga maritima* (Huber et al., 1986). *T. maritima frr* encoded a protein of 185 amino acid residues that had 38.9% identity with *E. coli* RRF. In contrast to *P. aeruginosa* RRF, it was bactericidal in *E. coli*. The bactericidal effect of *T. maritima* RRF was reversed by the simultaneous expression of the cloned *E. coli* RRF within the same cell. This suggests that *T. maritima* RRF inhibits the *E. coli* RRF reaction competitively, resulting in bactericidal action (unpublished).

Since the expression of *Staphylococcus aureus* RRF is elevated during infection (Lowe et al., 1998), we collaborated with R. Deresiewicz of Harvard Medical School to create temperature-sensitive *S. aureus* mutants carrying temperature-sensitive RRFs.

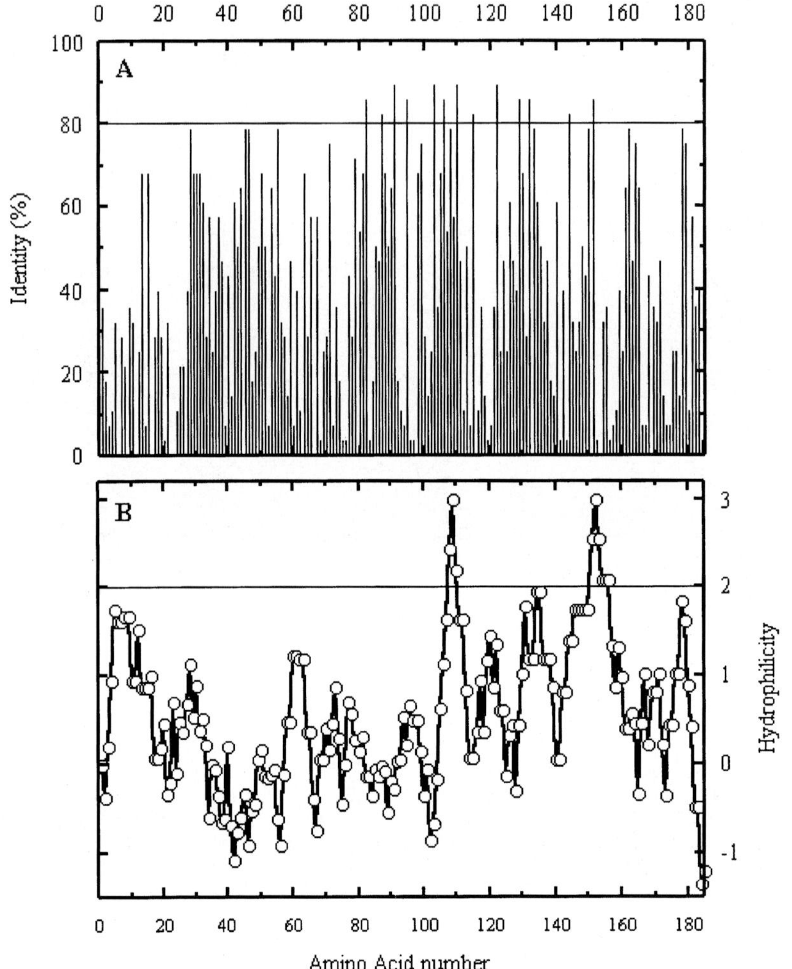

Figure 5. Sequence analysis of 29 RRFs of various species. (A) Average percent identity of each residue expressed as a bar. A horizontal line is drawn at 80% identity. (B) Hydrophilicity of each residue of *E. coli* RRF.

Because of the extensive homology between *S. aureus* RRF and *E. coli* RRF (46.2% identity and 59.3% similarity), we found that the amino acids responsible for temperature sensitivity are common between them while they are not functionally exchangeable (unpublished).

EUKARYOTIC RRF

On the basis of the data obtained in collaboration with Jacques Joyard of Grenoble, France, we propose that eukaryotic RRF is an organelle protein which is probably involved in protein synthesis within the organelles in much the same way as bacterial RRF functions (Rolland et al., 1999). This conclusion is further supported by the fact that translation in chloroplasts is similar to that in prokaryotes (Harris et al., 1994; Danon, 1997) and that yeast (Kanai et al., 1998) and human (Zhang and Spremulli, 1998) RRFs may be mitochondrial proteins. The 1,109-bp nucleotide sequence isolated from the cDNA library of spinach codes for a putative RRF (called RRFHCP) of 271 amino acid residues (30,431 Da). The C-terminal sequence (residues 87 to 271) has 46% identity (66% homology) with the sequence of *E. coli* RRF. The N terminus of the deduced amino acid sequence of RRFHCP is compatible with mitochondria and chloroplast targeting sequences (Nakai and Kanehisa, 1992). Spinach has a single gene for RRFHCP, and expression of this gene appears to be almost restricted to photosynthetic tissues but is light independent.

From the amino-terminal sequence of the processed RRFHCP (mature RRFCH), we conclude that the mature RRFHCP is composed of 193 amino acid residues (21,838 Da). As shown in Fig. 7, the mature RRFHCP is not functional in *E. coli* and it is toxic to LJ2221 carrying temperature-sensitive RRF even at the permissive temperature but not to wild-type *E.*

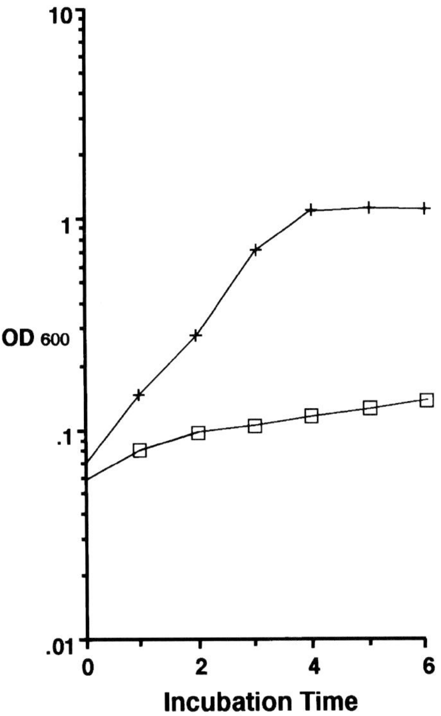

Figure 6. *P. aeruginosa* RRF can function in *E. coli*. Complementation of LJ2221 carrying temperature-sensitive RRF at 42°C by *P. aeruginosa* RRF. LJ2221 harboring a plasmid (open squares, empty vector pMW118; plus signs, pMO2925 carrying *P. aeruginosa frr* under the control of the *lac* promoter) was grown at 32°C to the log phase. At time zero, the culture temperature was shifted to 42°C and the culture was incubated with (plus signs) and without (open squares) the addition of IPTG. The optical density of the culture is plotted against the time after the temperature shift up. (Adapted from Ohnishi et al., 1999.)

coli. Upon induction of RRFHCP by IPTG, the CFU of the strain carrying temperature-sensitive RRF but not wild-type *E. coli* decreased. In addition, the inhibitory effect of RRFHCP was much more pronounced than that on wild-type RRF in the in vitro RRF reaction (Rolland et al., 1999). We postulate that temperature-sensitive RRF, even at the permissive temperature, has a weaker affinity for the substrate than wild-type RRF, resulting in the selective toxicity of mature RRFHCP on temperature-sensitive RRF but not on wild-type RRF.

Genes encoding yeast mitochondrial homologues of prokaryotic RRF and EF-G have been identified (Vambutas et al., 1991; Ouzounis et al., 1995). In collaboration with R. Wek of Indiana University School of Medicine, we showed that disruption of the chromosomal copy of the yeast RRF homologue is not lethal and induces a petite phenotype on glucose but not on glycerol (Teyssier et al., 1998). The yeast RRF homologue is exclusively localized in mitochondria. The *frr*-deficient yeast grown on glucose contained very little mitochondrial structure, and the mitochondrial DNA was damaged. This phenotype could not be reversed by adding the plasmid carrying the yeast *frr*. On the other hand, we were able to construct normal haploid yeast deficient in the *frr* homologue but harboring a plasmid carrying the wild-type yeast *frr* homologue. Starting from this strain, we recently succeeded in constructing temperature-sensitive yeast with temperature-sensitive RRF. The growth of this yeast strain is temperature sensitive on glycerol but not on glucose, suggesting the need for RRF for maintenance of mitochondria whose protein synthesis is dependent on factors that are coded for by nuclear genes (Pel and Grivell, 1994, and unpublished data).

STRUCTURAL BIOLOGY OF RRF

We recently solved the crystal structure of *T. maritima* RRF in collaboration with A. Liljas of Lund University, Sweden (unpublished). With this three-dimensional structure, the sequence alignment, the hydrophilicity of the surface (Fig. 5), the charge distribution on the surface of RRF, and various mutants isolated in our laboratory described below, we hope to identify functional regions of RRF.

The 18 temperature-sensitive isolates were represented by 12 independent genotypes in the 558 nucleotides encoding RRF (185 amino acids) (Janosi et al., 1998). In the N-terminal portion, there are temperature-sensitive mutations at the 1st, 9th, and 21st amino acids. Allele *frr-8* lost 12 N-terminal amino acids, yet there is only moderate thermolability. This suggests that the 12 N-terminal amino acids are not important for RRF thermostability. In a region further downstream, between codons Lys^{21} and Val^{117}, temperature-sensitive mutations did not occur, suggesting that slight alterations in this region of the protein may be lethal or have no effect on thermostability. The two alleles *frr-13* and *frr-16* code for C-terminal truncations which are 5 and 4 amino acids long, respectively. This proves that the intact C-terminal region is not essential for RRF activity.

The nucleotide changes responsible for the intragenic suppressor have been studied. Without exception, the mutation which was responsible for the temperature-sensitive *frr* phenotype reverted to the wild-type sequence in all the intragenic revertants. An additional two mutations in the revertants did not influence the RRF activity and are therefore functionally silent mutations.

To isolate a number of lethal mutations of RRF, the screening used for isolation of temperature-

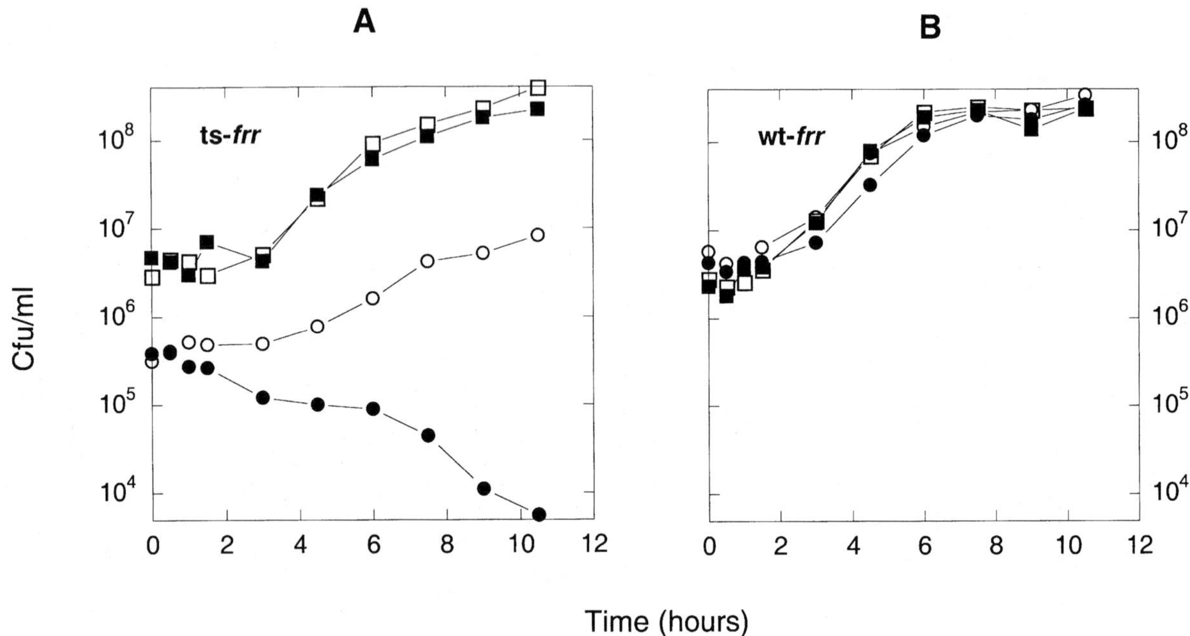

Figure 7. Plant RRF kills *E. coli* carrying temperature-sensitive (ts) RRF but not wild-type *E. coli*. (A) *E. coli* LJ14 (with tsRRF) harboring a plasmid carrying plant *frr* was grown to the stationary phase and diluted at time zero. One culture (solid circles) but not the other (open circles) received IPTG to induce the plant gene at time zero and was incubated at 32°C. An identical experiment was performed with LJ14 carrying the same vector plasmid which was empty (no plant gene) with (closed squares) and without (open squares) IPTG. (B) Identical to panel A except that *E. coli* was wild type. Viable counts were plotted against the time after the addition of IPTG. (Reprinted from Rolland et al., 1999, with permission.)

sensitive RRF mutants (Janosi et al., 1998) was extended to identify 149 lethally mutated *frr* genes that carry either a small deletion(s) or an amino acid alteration(s). In addition, we have 15 independently isolated suppressor mutations of the temperature-sensitive mutants giving clues about other biological components which interact with RRF (unpublished).

RRF, AN IDEAL NEW TARGET FOR ANTIMICROBIAL AGENTS

The bactericidal and bacteriostatic effects of inhibition of RRF suggest that RRF may be an ideal target for antibacterial agents. Irreversible inhibition of RRF is bactericidal, as shown in Fig. 7 (Rolland et al., 1999). Even the reversible inactivation of temperature-sensitive RRF caused bacterial death in the lag phase (Fig. 3) (Janosi et al., 1998). As described in Eukaryotic RRF above, the *frr* homologues in eukaryotes are organelle proteins. As pointed out in Amino Acid Sequences of RRFs and Characterization of Other Prokaryotic RRFs above, archaea, whose protein synthesis system is of the eukaryotic type (Bult et al., 1996; Klenk et al., 1997), do not have an *frr* homologue. This strongly suggests that inhibition of eukaryotic RRF should not influence the major cytoplasmic protein synthesis. Thus, we predict that antibacterial agents which target bacterial RRF will have only slight, tolerable side effects like those observed with the currently used antibiotics, such as tetracycline and erythromycin, which inhibit mitochondrial protein synthesis (Pious and Hawley, 1972; Doersen and Stanbridge, 1982). Below, we describe additional advantages of targeting RRF.

First, as stated above, the three-dimensional structure has been elucidated. Computer-based rational drug design is now possible. Second, all pathogenic bacteria have RRF. Third, RRFs of all pathogens are similar but not identical, making it possible to design an inhibitor either specific to each pathogen or with a wide spectrum. Fourth, the expression of RRFs of pathogens appears to be heightened during infection (Lowe et al., 1998). This suggests strongly the importance of RRF for pathogenesis. Fifth, it has been reported that the major antibacterial antibody detected upon infection of animals by *Brucella melitensis* was anti-RRF (Vizcaino et al., 1996). This opens the possibility that the infecting organism may be easily determined by examining the nature of the anti-RRF antibody present in the patient so that a specific anti-RRF drug may be administered.

NOMENCLATURE OF RRF

An ad hoc nomenclature subcommittee of the International Union of Biochemistry gave the name RF4 to RRF (Clark et al., 1996). However, this name, implying release factor 4, is not appropriate because it confuses RRF with RF1 through RF3. Because of this, the committee agreed not to use this term, and the name RRF will be retained.

REFERENCES

Adhin, M. R., and J. van Duin. 1990. Scanning model for translational reinitiation in eubacteria. *J. Mol. Biol.* **213:**811–818.

Agrawal, R. K., P. Penczek, R. A. Grassucci, and J. Frank. 1998. Visualization of elongation factor G on the *Escherichia coli* 70S ribosome: the mechanism of translocation. *Proc. Natl. Acad. Sci. USA* **95:**6134–6138.

Algranati, I. D., N. S. Gonzalez, and E. G. Bade. 1969. Physiological role of 70S ribosomes in bacteria. *Biochemistry* **62:**574–580.

Björnsson, A., and L. A. Isaksson. 1996. Accumulation of a mRNA decay intermediate by ribosomal pausing at a stop codon. *Nucleic Acids Res.* **24:**1753–1757.

Buckingham, R. H., G. Grentzmann, and L. Kisselev. 1997. Polypeptide chain release factors. *Mol. Microbiol.* **24:**449–456.

Bult, C. J., O. White, G. J. Olsen, L. Zhou, R. D. Fleischmann., G. G. Sutton, J. A. Blake, L. M. FitzGerald, R. A. Clayton, J. D. Gocayne, A. R. Kerlavage, B. A. Dougherty, J. F. Tomb, M. D. Adams, C. I. Reich, R. Overbeek, E. F. Kirkness, K. G. Weinstock, J. M. Merrick, A. Glodek, J. L. Scott, N. S. M. Geoghagen, and J. C. Venter. 1996. Complete genome sequence of the methanogenic archaeon, *Methanococcus jannaschii. Science* **273:**1058–1073.

Clark, B. F. C., M. Grunberg-Manago, N. K. Gupta, J. W. B. Hershey, A. G. Hinnebusch, R. J. Jackson, U. Maitra, M. B. Mathews, W. C. Merrick, R. E. Rhoads, N. Sonenberg, L. L. Spremulli, H. Trachsel, and H. O. Voorma. 1996. Prokaryotic and eukaryotic translation factors. *Biochimie* **78:**1119–1122.

Conway, T. W., and F. Lipmann. 1964. Characterization of a ribosome-linked guanosine triphosphatase in *Escherichia coli* extracts. *Proc. Natl. Acad. Sci. USA* **52:**1462–1469.

Corpet, F. 1988. Multiple sequence alignment with hierarchical clustering. *Nucleic Acids Res.* **16:**10881–10890.

Czworkowski, J., and P. B. Moore. 1996. The elongation phase of protein synthesis. *Prog. Nucleic Acids Res. Mol. Biol.* **54:**293–332.

Danon, A. 1997. Translation regulation in the chloroplast. *Plant Physiol.* **115:**1293–1298.

Davies, J., W. Gilbert, and L. Gorini. 1964. Streptomycin, suppression, and the code. *Proc. Natl. Acad. Sci. USA* **51:**883–890.

Davis, B. D. 1971. Role of subunits in the ribosome cycle. *Nature* **321:**153–157.

Devereux, J., P. Haeberli, and O. Smithies. 1984. A comprehensive set of sequence analysis programs for the VAX. *Nucleic Acids Res.* **12:**387–395.

Doersen, C. J., and E. J. Stanbridge. 1982. Erythromycin inhibition of cell proliferation and in vitro mitochondrial protein synthesis in human HeLa cells is pH dependent. *Biochim. Biophys. Acta* **698:**62–69.

Fraser, C. M., J. D. Gocayne, O. White, M. D. Adams, R. A. Clayton., R. D. Fleischmann, C. J. Bult, A. R. Kerlavage, G. Sutton, J. M. Kelley, J. L. Fritchman, J. F. Weidman, K. V. Small, M. Sandusky, J. Furhman, D. Nguyen, T. R. Utterback, D. M. Saudek, C. A. Phillips, J. M. Merrick, J.-F. Tomb, B. A. Dougherty, K. F. Bott, P.-C. Hu, T. S. Lucier, S. N. Peterson, H. O. Smith, C. A. Hutchison III, and J. C. Venter. 1995. The minimal gene complement of *Mycoplasma genitalium. Science* **270:**397–403.

Freistroffer, D. V., M. Y. Pavlov, J. MacDougall, R. H. Buckingham, and M. Ehrenberg. 1997. Release factor RF3 in *E. coli* accelerates the dissociation of release factors RF1 and RF2 from the ribosome in a GTP-dependent manner. *EMBO J.* **16:**4126–4133.

Gold, L. 1988. Posttranscriptional regulatory mechanisms in *Escherichia coli. Annu. Rev. Biochem.* **57:**199–233.

Grentzmann, G., P. J. Kelly, S. Laalami, M. Shuda, M. A. Firpo, Y. Cenatiempo, and A. Kaji. 1998. Release factor RF-3 GTPase activity acts in disassembly of the ribosome termination complex. *RNA* **4:**973–983.

Gualerzi, C., C. L. Pon, and A. Kaji. 1971. Initiation factor dependent release of aminoacyl-tRNAs from complexes of 30S ribosomal subunits, synthetic polynucleotide and aminoacyl tRNA. *Biochem. Biophys. Res. Commun.* **45:**1312–1319.

Gussin, G. N. 1966. Three complementation groups in Bacteriophage R17. *J. Mol. Biol.* **21:**435–453.

Harris, E. H., J. E. Boynton, and N. W. Gillham. 1994. Chloroplast ribosomes and protein synthesis. *Microbiol. Rev.* **58:**700–754.

Heurgué-Hamard, V., R. Karimi, L. Mora, J. MacDougall, C. Leboeuf, G. Grentzmann, M. Ehrenberg, and R. H. Buckingham. 1998. Ribosome release factor RF4 and termination factor RF3 are involved in dissociation of peptidyl-tRNA from the ribosome. *EMBO J.* **17:**808–816.

Hirashima, A., and A. Kaji. 1970. Factor dependent breakdown of polysomes. *Biochem. Biophys. Res. Commun.* **41:**877–883.

Hirashima, A., and A. Kaji. 1972a. Factor-dependent release of ribosomes from messenger RNA—requirement for two heat-stable factors. *J. Mol. Biol.* **65:**43–58.

Hirashima, A., and A. Kaji. 1972b. Purification and properties of ribosome-releasing factor. *Biochemistry* **11:**4037–4044.

Hirashima, A., and A. Kaji. 1973. Role of elongation factor G and a protein factor on the release of ribosomes from messenger ribonucleic acid. *J. Biol. Chem.* **248:**7580–7587.

Hirokawa, G., L. Janosi, Y. Sekine, R. Ricker, and A. Kaji. 1998. Possible reasons for lethal effect of *in vivo* inactivation of Ribosome Recycling Factor (RRF), abstr. p. 142. *In Cold Spring Harbor Translational Control Meeting.* Cold Spring Harbor Laboratory, Cold Spring Harbor, N.Y.

Hogan, B. L. M., and A. Korner. 1968. Ribosomal subunits of landschutz ascites cells during changes in polysome distribution. *Biochim. Biophys. Acta* **169:**129–138.

Holloway, B. W., U. Römling, and B. Tümmler, 1994. Genomic mapping of *Pseudomonas aeruginosa* PAO. *Microbiology* **140:** 2907–2929.

Hopp, T. P., and K. R. Woods. 1981. Prediction of protein antigenic determinants from amino acid sequences. *Proc. Natl. Acad. Sci. USA* **78:**3824–3828.

Huber, R., T. A. Langworthy, H. Konig, M. Thomm, C. R. Woese, U. Sleytr, and K. Stetter. 1986. *Thermotoga maritima* sp. nov. represents a new genus of unique extremely thermophilic eubacteria growing up to 90° C. *Arch. Microbiol.* **144:**324–333.

Hungerer, C., B. Troup, U. Römling, and D. Jahn, 1995. Regulation of the *hemA* gene during 5-aminolevulinic acid formation in *Pseudomonas aeruginosa. J. Bacteriol.* **177:**1435–1443.

Ichikawa, S., and A. Kaji. 1989. Molecular cloning and expression of ribosome releasing factor. *J. Biol. Chem.* **264:20054–20059.**

Ichikawa, S., M. Ryoji, Z. Siegfried, and A. Kaji 1989. Localization of the ribosome-releasing factor gene in the *Escherichia coli* chromosome. *J. Bacteriol.* **171:**3689–3695.

Inokuchi, Y., R. Takahashi, T. Hirose, S. Inayama, and A. B. Jacobson. 1986. The complete nucleotide sequence of the group II RNA coliphage GA. *J. Biochem.* **99:**1169–1180.

Inokuchi, Y., A. Hirashima, L. Janosi, and A. Kaji. 1998. Nonparticipation of ribosome recycling factor (RRF) in the lysis gene expression of RNA coliphage GA, abstr. p. 442. *In 21st Japanese Molecular Biology Meeting.* Japanese Society for Molecular Biology.

Janosi, L., I. Shimizu, and A. Kaji. 1994. Ribosome recycling factor (ribosome releasing factor) is essential for bacterial growth. *Proc. Natl. Acad. Sci. USA* **91:**4249–4253.

Janosi, L., H. Hara, S. Zhang, and A. Kaji. 1996a. Ribosome recycling by ribosome recycling factor (RRF)—an important but overlooked step of protein biosynthesis. *Adv. Biophys.* **32:**121–201.

Janosi, L., R. Ricker, and A. Kaji. 1996b. Dual functions of ribosome recycling factor in protein biosynthesis: disassembling the termination complex and preventing translational errors. *Biochimie* **78:**959–969.

Janosi, L., S. Mottagui-Tabar, L. A. Isaksson, Y. Sekine, E. Ohtsubo, S. Zhang, S. Goon, S. Nelken, M. Shuda, and A. Kaji. 1998. Evidence for in vivo ribosome recycling, the fourth step in protein biosynthesis. *EMBO J.* **17:**1141–1151.

Joklik, W. K., and Y. Becker. 1965. Studies on the genesis of polyribosomes. *J. Mol. Biol.* **13:**496–510.

Kaempfer, R. 1970. Dissociation of ribosomes on polypeptide chain termination and origin of single ribosomes. *Nature* **228:** 534–537.

Kaempfer, R. 1971. Control of single ribosome formation by an initiation factor for protein synthesis. *Proc. Natl. Acad. Sci. USA* **68:**2458–2462.

Kaji, A., and M. Ryoji. 1979. Mechanism of action of antibacterial agents—tetracycline. *Antibiotics* **1:**304–328.

Kaji, A., E. Teyssier, and G. Hirokawa. 1998. Disassembly of the post-termination complex and reduction of translational error by ribosome recycling factor (RRF)—a possible new target for antibacterial agents. *Biochem. Biophys. Res. Commun.* **250:**1–4.

Kaji, H., and A. Kaji. 1965. Specific binding of sRNA to ribosomes: effect of streptomycin. *Proc. Natl. Acad. Sci. USA* **54:** 213–219.

Kanai, T., S. Takeshita, H. Atomi, K. Umemura, M. Ueda, and A. Tanaka. 1998. A regulatory factor, Fil1p, involved in derepression of the isocitrate lyase gene in Saccharomyces cerevisiae. A possible mitochondrial protein necessary for protein synthesis in mitochondria. *Eur. J. Biochem.* **256:**212–220.

Kelley, W. S., and M. Schaechter. 1969. Magnesium ion-dependent dissociation of polysomes and free 70 s ribosomes in *Bacillus megaterium*. *J. Mol. Biol.* **42:**599–602.

Klenk, H.-P., R. A. Clayton, J.-F. Tomb, O. White, K. E. Nelson, K. A. Ketchum, R. J. Dodson, M. Gwinn, E. K. Hickey, J. D. Peterson, D. L. Richardson, A. R. Kerlavage, D. E. Graham, N. C. Kyrpides, R. D. Fleischmann, J. Quackenbush, N. H. Lee, G. G. Sutton, S. Gill, E. F. Kirkness, B. A. Dougherty, K. McKenney, M. D. Adams, B. Loftus, S. Peterson, C. L. Reich, L. K. McNeil, J. H. Badger, A. Glodek, L. Zhou, R. Overbeek, J. D. Gocayne, J. F. Weidman, L. McDonald, T. Utterback, M. D. Cotton, T. Spriggs, P. Artiach, B. P. Kaine, S. M. Sykes, P. W. Sadow, K. P. D'Andrea, C. Bowman, C. Fujii, S. A. Garland, T. M. Mason, G. J. Olsen, C. M. Fraser, H. O. Smith, C. R. Woese, and J. C. Venter. 1997. The complete genome sequence of the hyperthermophilic, sulphate-reducing archaeon *Archaeoglobus fulgidus*. *Nature* **390:**364–370.

Kohler, R. E., E. Z. Ron, and B. D. Davis. 1968. Significance of the free 70s ribosomes in *Escherichia coli* extracts. *J. Mol. Biol.* **36:**71–82.

Kokjohn, T. A., and R. V. Miller. 1987. Characterization of the *Pseudomonas aeruginosa recA* analog and its protein product: *rec-102* is a mutant allele of the *P. aeruginosa* PAO *recA* gene. *J. Bacteriol.* **169:**1419–1508.

Kung, H.-F., B. V. Treadwell, C. Spears, P.-C. Tai, and H. Weissbach. 1977. DNA-directed synthesis *in vitro* of β-galactosidase: requirement for a ribosome release factor. *Proc. Natl. Acad. Sci. USA* **74:**3217–3221.

Lowe, A. M., D. T. Beattie, and R. L. Deresiewicz. 1998. Identification of novel staphylococcal virulence genes by *in vivo* expression technology. *Mol. Microbiol.* **27:**967–976.

Mangiarotti, G., and D. Schlessinger. 1966. Polyribosome metabolism in *Escherichia coli*. I. Extraction of polyribosomes and ribosomal subunits from fragile, growing *Escherichia coli*. *J. Mol. Biol.* **20:**123–143.

Martin, J., and R. E. Webster. 1975. The *in vitro* translation of a terminating signal by a single *Escherichia coli* ribosome. *J. Biol. Chem.* **250:**8132–8139.

McCarthy, J. E. G., and C. Gualerzi. 1990. Translational control of prokaryotic gene expression. *Trends Genet.* **6:**78–85.

Moazed, D., and H. F. Noller. 1989. Intermediate states in the movement of transfer RNA in the ribosome. *Nature* **342:**142–148.

Nakai, K., and M. Kanehisa. 1992. A knowledge base for predicting protein localization sites in eukaryotic cells. *Genomics* **14:**897–911.

Oefverstedt, L.-G., K. Zhang, S. Tapio, U. Skoglund, and L. A. Isaksson. 1994. Starvation in vivo for aminoacyl-tRNA increases the spatial separation between the two ribosomal subunits. *Cell* **79:**629–638.

Ogawa, K., and A. Kaji. 1975. Requirement for ribosome-releasing factor for the release of ribosomes at the termination codon. *Eur. J. Biochem.* **58:**411–419.

Ohnishi, M., L. Janosi, M. Shuda, H. Matsumoto, T. Hayashi, Y. Terawaki, and A. Kaji. 1999. Molecular cloning, sequencing, purification, and characterization of *Pseudomonas aeruginosa* ribosome recycling factor, RRF. *J. Bacteriol.* **181:**1281–1291.

Ouzounis, C., P. Bork, G. Casari, and C. Sander. 1995. New protein functions in yeast chromosome VIII. *Protein Sci.* **4:**2424–2428.

Pape, T., W. Wintermeyer, and M. V. Rodnina. 1998. Complete kinetic mechanism of elongation factor Tu-dependent binding of aminoacyl-tRNA to the A site of the *E. coli* ribosome. *EMBO J.* **17:**7490–7497.

Pavlov, M. Y., D. Freistroffer, J. MacDougall, R. H. Buckingham, and M. Ehrenberg. 1997a. Fast recycling of *Escherichia coli* ribosomes requires both ribosome recycling factor (RRF) and release factor RF3. *EMBO J.* **16:**4134–4141.

Pavlov, M. Y., D. V. Freistroffer, V. Heurgué-Hamard, R. H. Buckingham, and M. Ehrenberg. 1997b Release factor RF3 abolishes competition between release factor RF1 and ribosome recycling factor (RRF) for a ribosome binding site. *J. Mol. Biol.* **273:**389–401.

Pel, H. J., and L. A. Grivell. 1994. Protein synthesis in mitochondria. *Mol. Biol. Rep.* **19:**183–194.

Petersen, H. U., E. Joseph, A. Ullman, and A. Danchin. 1978. Formylation of initiator tRNA methionine in procaryotic protein synthesis: in vivo polarity in lactose operon expression. *J. Bacteriol.* **135:**453–459.

Phillips, L. A., B. Hotham-Iglewski, and R. M. Franklin. 1969. Polyribosomes of *Escherichia coli*. *J. Mol. Biol.* **40:**279–288.

Pious, D. A., and P. Hawley. 1972. Effect of antibiotics on respiration in human cells. *Pediatr. Res.* **6:**687–692.

Richman, N., and J. W. Bodley. 1972. Ribosomes cannot interact simultaneously with elongation factors EF-Tu and EF-G. *Proc. Natl. Acad. Sci. USA* **69:**686–689.

Robinson, W. E., R. H. Frist, and P. Kaesberg. 1969. Genetic coding: oligonucleotide coding for first six amino acid residues of the coat protein of R17 bacteriophage. *Science* **166:**1291–1293.

Rodnina, M. V., and W. Wintermeyer. 1998. Form follows function: structure of an elongation factor G-ribosome complex. *Proc. Natl. Acad. Sci. USA* **95:**7237–7239.

Rolland, N., L. Janosi, M. A. Block, A. Shuda, E. Teyssier, C. Miege, C. Cheniclet, J. Carde, A. Kaji, and J. Joyard. 1999. Plant ribosome recycling factor homologue is a chloroplastic protein and is bactericidal in Escherichia coli carrying temperature-sensitive ribosome recycling factor. *Proc. Natl. Acad. Sci. USA* **96:**5464–5469.

Ryoji, M. 1981. Studies on the roles of ribosome releasing factor of Escherichia coli. Ph.D. thesis. University of Pennsylvania School of Medicine, Philadelphia, Pa.

Ryoji, M., J. W. Karpen, and A. Kaji. 1981. Further characterization of ribosome releasing factor and evidence that it prevents ribosomes from reading through a termination codon. *J. Biol. Chem.* **256:**5798–5801.

Schlessinger, D., G. Mangiarotti, and D. Apirion. 1967. The formation and stabilization of 30S and 50S ribosome couples in Escherichia coli. *Proc. Natl. Acad. Sci. USA* **58:**1782–1789.

Schmidt, B. F., B. Berkhout, G. P. Overbeek, A. van Strien, and J. van Duin. 1987. Determination of the RNA secondary structure that regulates lysis gene expression in bacteriophage MS2. *J. Mol. Biol.* **195:**505–516.

Schnappinger, D., and W. Hillen. 1996. Tetracyclines: antibiotic action, uptake, and resistance mechanisms. *Arch. Microbiol.* **165:** 359–369.

Shimizu, I., and A. Kaji. 1991. Identification of the promoter region of the ribosome-releasing factor cistron (*frr*). *J. Bacteriol.* **173:**5181–5187.

Spahn, C. M. T., and K. H. Nierhaus. 1998. Models of the elongation cycle: an evaluation. *Biol. Chem.* **379:**753–772.

Spedding, G. 1990. *Ribosomes and Protein Synthesis: a Practical Approach.* IRL, Oxford, United Kingdom.

Steitz, J. A. 1969. Polypeptide chain initiation: nucleotide sequences of the three ribomosal binding sites in bacteriophage R17 RNA. *Nature* **224:**957–964.

Subramanian, A. R., and B. D. Davis. 1970. Activity of initiation factor F_3 in dissociating *Escherichia coli* ribosomes. *Nature* **228:** 1273–1275.

Subramanian, A. R., and B. D. Davis. 1971. Rapid exchange of subunits between free ribosomes in extracts of *Escherichia coli. Proc. Natl. Acad. Sci. USA* **68:**2453–2457.

Takeda, Y., I. Suzuka, and A. Kaji. 1968. Comparative studies on specific and nonspecific binding of transfer ribonucleic acid to ribosomes. *J. Biol. Chem.* **243:**1075–1081.

Tanaka, N. 1995. Fusidic acid. *Antibiotics* **3:**436–447.

Tate, W. P., H. Hornig, and R. Luhrmann. 1983. Recognition of termination codon by release factor in the presence of a tRNA-occupied A site. *J. Biol. Chem.* **258:**10360–10365.

Tate, W. P., B. Kastner, C. D. Edgar, K. K. McCaughan, and R. Traut. 1990. The ribosomal domain of the bacterial release factors. *Eur. J. Biochem.* **187:**543–548.

Tate, W. P., F. M. Adamski, C. M. Brown, M. E. Dalphin, J. P. Gray, J. A. Horsfield, K. K. McCaughan, J. G. Moffat, R. J. Powell, K. M. Timms, and C. N. A. Trotman. 1993. Translational stop signals: evolution, decoding for protein synthesis and recoding for alternative events, p. 253–262. *In* K. H. Nierhaus, F. Franceschi, A. R. Subramanian, V. A. Erdmann, and B. Wittmann-Liebold (ed.), *The Translational Apparatus: Structure, Function, Regulation, Evolution.* Plenum Press, New York, N.Y.

Teyssier, E., N. Rolland, L. Janosi, M. A. Block, M. Shuda, C. Miege, J. Joyard, and A. Kaji. 1998. Ribosome recycling factor homologues in eukaryotes, abstr. p. 268. *In Cold Spring Harbor Laboratory Translational Control Meeting.*

Vambutas, A., S. H. Ackerman, and A. Tzagoloff. 1991. Mitochondrial translational-initiation and elongation factors in *Saccharomyces cerevisiae. Eur. J. Biochem.* **201:**643–652.

Vizcaino, N., A. Cloeckaert, G. Dubray, and M. Zygmunt. 1996. Cloning, nucleotide sequence, and expression of the gene coding for a ribosome releasing factor-homologous protein of *Brucella melitensis. Infect. Immun.* **64:**4834–4837.

Weber, K. 1967. Amino acid sequence studies on the tryptic peptides of the coat protein of the bacteriophage R17. *Biochemistry* **6:**3144–3154.

Wilson, K. S., and H. F. Noller. 1998a. Molecular movement inside the translational engine. *Cell* **92:**337–349.

Wilson, K. S., and H. F. Noller. 1998b. Mapping the position of translational elongation factor EF-G in the ribosome by directed hydroxyl radical probing. *Cell* **92:**131–139.

Zhang, Y. L., and L. L. Spremulli. 1998. Identification and cloning of human mitochondrial translational release factor 1 and the ribosome recycling factor. *Biochim. Biophys. Acta* **1443:**245–250.

The Ribosome: Structure, Function, Antibiotics, and Cellular Interactions
Edited by R. A. Garrett, S. R. Douthwaite, A. Liljas, A. T. Matheson, P. B. Moore, and H. F. Noller

Chapter 44

Mechanism of Bacterial Translation Termination and Ribosome Recycling

MÅNS EHRENBERG, VILDAN DINCBAS, DAVID FREISTROFFER, VALERIE HEURGUÉ-HAMARD, REZA KARIMI, MICHAEL PAVLOV, and RICHARD H. BUCKINGHAM

FACTORS REQUIRED FOR PEPTIDE RELEASE—RF1 AND RF2

The identification of protein factors required by ribosomes for the last stages of protein synthesis began in the late 1960s (Capecchi, 1967a; Caskey et al., 1968). In bacteria, it was found that two protein release factors (RFs) were needed for specific recognition of the three stop codons, one (RF1) responding to UAG and UAA and a second (RF2) responding to UAA and UGA (Scolnick et al., 1968). The action of one or another of these proteins leads to the hydrolysis of the ester linkage between tRNA and polypeptide on the ribosome and release of the complete polypeptide. Two in vitro assays were employed in the first searches for RFs, the hexapeptide release assay, based on the translation of an amber mutant of R17 RNA (Capecchi, 1967a, 1967b), and the formylmethionine release assay (Caskey et al., 1968), named after the molecule released.

The gene encoding RF1 (*prfA*), located at 27 min on the *Escherichia coli* chromosome, was cloned by screening for antisuppressor activity in the presence of an amber suppressor tRNA (Weiss et al., 1984), whereas *prfB*, which encodes RF2 and maps to 62 min, was identified by screening a protein overexpression library with antibodies raised against the factor (Caskey et al., 1984). The proteins show extensive similarity in amino acid sequence to each other and to other bacterial RFs but very little similarity to the eukaryotic RF eRF1, a single, larger protein able to recognize all three stop codons (Frolova et al., 1994; Konecki et al., 1977). A recent careful search for sequences conserved among RFs from the different living kingdoms has identified one region of conserved sequence common to all three kingdoms which contains an absolutely conserved GGQ motif (Frolova et al., 1999) but which is not compatible with an earlier suggested alignment between the eukaryotic and bacterial factors (Ito et al., 1996). Why two factors appear to be needed in bacteria to decode three stop codons whereas one suffices in eukaryotes is not clear, all the more so as a single mutation in *E. coli* RF2 can widen the specificity of stop codon recognition to include UAG (Ito et al., 1998b). The cost of this change in terms of the misrecognition of sense codons as stop codons may be a clue to the evolutionary advantage of maintaining two factors, but this needs further investigation.

The precise mechanism leading to hydrolysis of the ester linkage that allows release of the polypeptide chain is unknown, but it is generally considered to involve the peptidyltransferase activity of the 50S ribosomal subunit, modified by the presence of the RFs on the ribosome, rather than being a hydrolytic activity of the factor itself. Thus, it has long been known that ribosomes can hydrolyze peptidyl-tRNA under certain conditions in the absence of RFs (Caskey et al., 1971). For previous reviews of the peptide release mechanism, see Craigen et al., 1990; Tate et al., 1990; Buckingham et al., 1997; and Nakamura and Ito, 1998.

RF3, THE GTPase OF TERMINATION

A third protein was discovered that stimulated formylmethionine release by RF1 or RF2 but itself lacked codon specificity (Goldstein et al., 1970; Mil-

Måns Ehrenberg, Vildan Dincbas, David Freistroffer, Valerie Heurgué-Hamard, Reza Karimi, and Michael Pavlov ■ Department of Cell and Molecular Biology, BMC, Box 596, S-751 24 Uppsala, Sweden. **Richard H. Buckingham** ■ UPR 9073 du CNRS, Institut de Biologie Physico-Chimique, 13 Rue Pierre et Marie Curie, Paris 75005, France.

man et al., 1969). For many years it remained uncertain whether the real role of this factor, RF3, was concerned with the process of translation termination at all. Thus, the effect of adding RF3 (then called S) to the fMet release assay was to reduce the K_m for the nonsense trinucleotide (Caskey et al., 1969), an action of doubtful relevance to normal termination. The addition of RF3 (S) reduced the K_m for the nonsense trinucleotide without affecting the catalytic constant (k_{cat}) of the reaction (Caskey et al., 1969). The identification and cloning of the gene (*prfC*) for RF3 (Grentzmann et al., 1994; Mikuni et al., 1994) opened the way both to in vivo studies of the role of the factor and to the characterization of its activity in purified systems for protein synthesis and polypeptide chain termination. In vivo, *prfC* can be inactivated without lethal effect, though strains lacking the protein show a clear defect in growth (Grentzmann et al., 1994; Mikuni et al., 1994), particularly at low temperatures. The sequence of RF3 revealed extensive amino acid similarity to the translation factors EF-Tu and EF-G, including the GTP binding motifs characteristic of these and other GTPases (Bourne et al., 1991; Grentzmann et al., 1994; Mikuni et al., 1994). This was consistent with observations that guanine nucleotides modified the action of RF3 in the formylmethionine release assay, although curiously, the nucleotides were found to abolish stimulation of release by RF3 rather than the reverse (Milman et al., 1969).

The elucidation of the role of RF3 in translation termination and the demonstration that GTP hydrolysis was needed for RF3 action became possible with the development of improved purified in vitro systems for protein synthesis (Freistroffer et al., 1997). In the absence of RF, translation of synthetic mRNAs containing a Shine-Dalgarno sequence, an appropriately spaced AUG codon, and a short coding sequence followed by a stop codon leads to ribosomal termination complexes paused with the stop codon in the ribosomal A site and a peptidyl-tRNA bound in the P site. These complexes can be isolated by gel filtration and used as substrates for purified RF, thereby allowing the classical methods of enzyme kinetics to be applied to the termination reaction (Freistroffer et al., 1997).

When RF1 or RF2 was added to release complexes in stoichiometric (or larger) amounts, peptide release was found to be rapid and unaffected by the presence of RF3, with or without guanine nucleotides (Fig. 1). In contrast to this, peptide release catalyzed by limiting amounts of RF1 or RF2 was strongly stimulated by RF3, provided that GTP was also added, suggesting that RF3 accelerated the recycling of RF1 and RF2 between molecules of termination complex, a possibility first proposed by Goldstein and Caskey (1970). This conclusion was supported by measurements of release over very short periods, during which the action of RF1 (or RF2) is restricted to a single cycle of peptidyl-tRNA hydrolysis. Under these conditions, the addition of RF3 and GTP had little or no effect on the amount of peptide released, again indicating that RF3 did not affect the k_{cat} of the release reaction. Further experiments showed that RF3 in the presence of GTP did not affect the k_{cat}/K_m for peptide release, confirming that the action of RF3 was restricted to stimulating the dissociation of RF1 or RF2 following peptide release (Freistroffer et al., 1997).

The *prfC* gene was originally identified and cloned because transposon insertions inactivating the gene acted as nonsense suppressors, i.e., they allowed phenotypic suppression of opal mutations (Mikuni et al., 1994) or amplified suppression due to the presence of weak tRNA nonsense suppressors (Grentzmann et al., 1994). These cloning strategies were based on the idea that RF3 should directly stimulate RF1 and RF2 action. It now seems likely that the effects of inactivating *prfC* in vivo are due to reductions in the concentration of free RF1 and RF2 in the cell, resulting from the slow dissociation of the factors from the ribosome when RF3 is absent. The properties of the protein that first led to its isolation seem to be related to a form of RF3 with only transient existence in the cell, the form with no bound guanine nucleotide. In vitro experiments by Pel et al. (1998) suggest that nucleotide-free RF3, unlike RF3·GTP, can stimulate RF2 binding to ribosomes, and they offer an explanation of the paradoxical effects of guanine nucleotides on the behavior of RF3 in the fMet release assay. Nucleotide-free RF3 may mimic a transition state of RF3 in which the protein is able to interact with the decoding RF on the ribosome (Pel et al., 1998) (see below).

The eukaryotic family of GTPases involved in termination (eRF3) may play the same role as bacterial RF3, but this has not been established. Unlike RF3 in *E. coli*, eRF3 is essential for viability, at least in *Saccharomyces cerevisiae*, though this is not necessarily indicative of a functional distinction between the factors. Two significant differences have nevertheless been demonstrated. The GTPase activity associated with RF3 requires only the presence of ribosomes, whereas that of eRF3 needs eRF1 as well (Freistroffer et al., 1997; Frolova et al., 1996). This interaction between the codon-specific RF and the GTPase component RF3 constitutes the second difference between the bacterial and eukaryotic systems: the factors can form a stable complex in the absence of ribosomes in eukaryotes but not in bacteria,

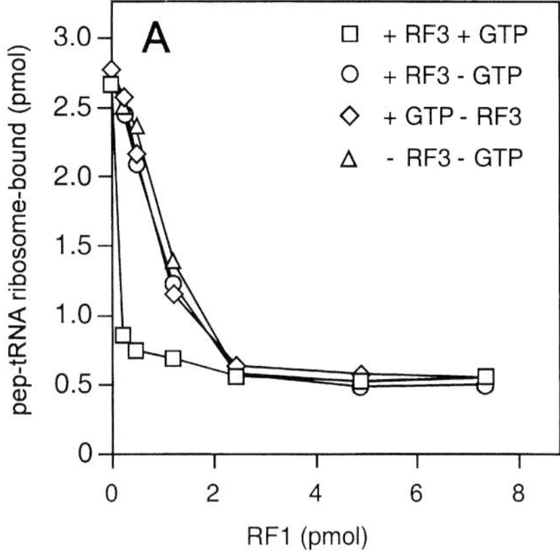

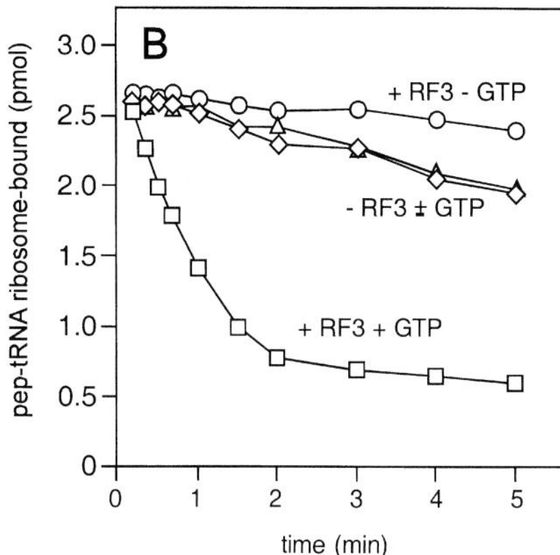

Figure 1. Stimulation of RF1 recycling by RF3. Termination complex containing fMet-Phe-Thr-Ile-tRNA^{Ile} bound in the P site with a UAA codon in the A site was prepared as described by Freistroffer et al. (1997). **(A)** The amount of peptidyl-tRNA remaining bound to ribosomes after 2 min of incubation with RF1, with and without 0.9 μM RF3 and with or without 0.2 mM GTP, is shown as a function of the amount of RF1 present. **(B)** The amount of peptidyl-tRNA remaining bound to ribosomes after incubation with 0.25 pmol of RF1 is shown as in panel A as a function of time of incubation. Stimulation of RF1 recycling is seen to require the presence of GTP. The data are from Freistroffer et al., 1997.

though the nature and the biological significance of this interaction are still controversial (Ebihara and Nakamura, 1999; Frolova et al., 1998; Ito et al., 1998a; Merkulova et al., 1999; Paushkin et al., 1997).

RIBOSOME-RECYCLING TIME AND THE ROLES OF RF3 AND RRF

While the experiments of Freistroffer et al. (1997) show clearly that RF3 stimulates RF1 and RF2 recycling between different termination complexes, they emphasize only one aspect of the role of the factor, that of maintaining adequate concentrations of free RF1 and RF2 in the cell. A second aspect is related to the length of time that RF1 and RF2 remain bound to the ribosome. A different type of in vitro experiment measures the time taken by the ribosome to complete an entire cycle, from initiation of translation on the mRNA through translation of the coding sequence, release of the peptide chain, and return to the point of the next translation initiation. When performed in vitro with purified components, such experiments allow the effect on the ribosome-recycling time of omitting one or more components of the system to be determined quantitatively (Pavlov et al., 1997a). In particular, such experiments have allowed the contributions of two proteins, RF3 and RRF (ribosome-recycling factor), to be studied in detail. RRF was first isolated and characterized by Hirashima and Kaji (1970, 1972) as a factor that catalyzed the breakdown of isolated polysomes when added together with elongation factor EF-G and GTP. The gene (*frr*) encoding RRF was subsequently localized and cloned (Ichikawa and Kaji, 1989; Ichikawa et al., 1989). Although the factor is conserved in all bacteria so far sequenced (Janosi et al., 1996a, 1998) and is known to be essential for viability in *E. coli* (Janosi et al., 1994), no eukaryotic cytoplasmic protein with significant sequence similarity to bacterial RRF has so far been found, and it is unknown whether a functional eukaryotic homologue exists.

When all components needed for the translation of a short synthetic mRNA (encoding fMet-Phe-Leu) are present, ribosomes recycle about every 6 s (Pavlov et al., 1997a). In the absence of RRF, the recycling time becomes 28 s; in the absence of RF3 (but in the presence of 1 μM RRF) the recycling time is 14 s; and in the absence of both proteins the recycling time increases to 39 s. The increase associated with the absence of RF3 is interpreted as being due to the slow dissociation of RF1 from the ribosome (Fig. 2). Thus, an important function of RF3 may be to accelerate the overall termination reaction and avoid interference between ribosomes, such as queuing on mRNA upstream of stop signals. The action of RRF is seen to depend on the concentration of EF-G (Fig. 2) (Pavlov et al., 1997a). This is consistent with the conclusions of Hirashima and Kaji (1972) that EF-G is required for RRF function.

A more detailed study of the ribosome-recycling time as a function of the concentrations of RF1, RF3, and RRF (Pavlov et al., 1997b) shows that RF1, having dissociated from the posttermination complex, may rebind and inhibit a subsequent step in the ribosome cycle. Here again, RF3 reduces the effect of such rebinding by inducing the rapid dissociation of

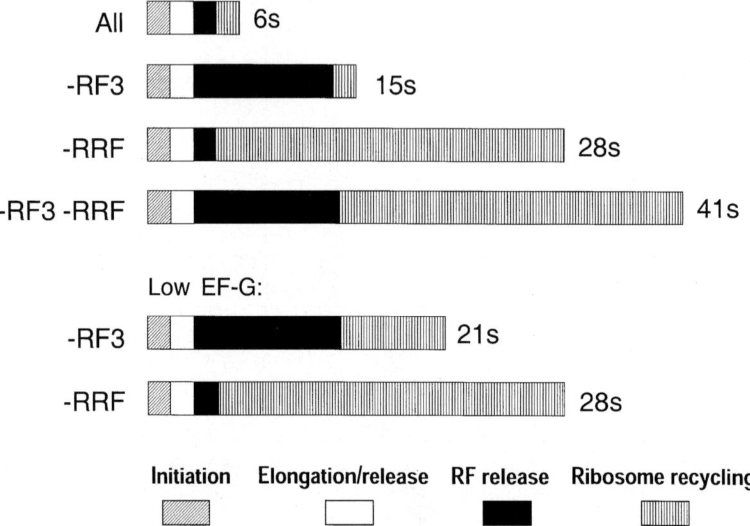

Figure 2. Rapid recycling of ribosomes requires RF3, RRF, and EF-G. Ribosome-recycling times for the translation of a short mRNA encoding fMet-Phe-Leu were measured in vitro as described by Pavlov et al. (1997a) and are shown at the right of each horizontal bar in seconds. For purposes of interpretation, the overall recycling time in the presence of all components is divided arbitrarily into equal periods for initiation, elongation and release, RF release, and the remaining events in ribosome recycling. The horizontal bars interpret the increased recycling time when RRF, RF3, or both factors are omitted in terms of increased times required for RF release and/or ribosome recycling. The two lowest bars show the effect of reducing the concentration of EF-G from 0.8 to 0.3 μM, which increases the overall recycling time in the presence but not in the absence of RRF. The data are from Pavlov et al., 1997a.

RF1. The effect of RF3 on the ribosomal recycling of different RFs is strongly correlated with their affinity for the ribosome. Thus, the acceleration of recycling by RF3 is most pronounced with RF1, which binds tightly to the ribosome, whereas in the case of a form of RF2 synthesized by an overproducing strain, which binds weakly to the ribosome, the acceleration of recycling by RF3 is rather modest (Pavlov et al., 1998). In the absence of RF3, the inhibitory action of RF1 on ribosome recycling can be reduced by increasing the RRF concentration (Fig. 3), suggesting that whatever the action of RRF may be, it is incompatible with the presence of RF1 on the ribosome. The extensive pause in the ribosome cycle seen in these experiments when RRF is absent may explain why RRF, unlike RF3, is essential to viability in *E. coli* (Janosi et al., 1994, 1996b).

The interpretation of these in vitro experiments is supported by in vivo studies of the effects of overproducing RF1 in mutant strains affected in RF3 or RRF (Pavlov et al., 1997b). The *frr* mutant employed was isolated by the use of a selective screen for *E. coli* cells able to compensate for a reduced activity of the enzyme peptidyl-tRNA hydrolase, which hydrolyzes peptidyl-tRNA dissociating from the ribosome before normal termination (Heurgué-Hamard et al., 1998). This mutant strain has about 10% of the normal level of RRF (MacDougall, unpublished), a reduction which has a small though significant negative effect on growth. Overproducing RF1 in a wild-type background has no clear effect on growth; however, in an RF3-deficient strain, overproduction partly restores growth. This might be expected in view of the role of RF3 in recycling RF1: in RF3-deficient cells, the level of free RF1 should be reduced due to slow release from the ribosome. In contrast, RF1 overproduction inhibits growth in a partially RRF-deficient strain (Pavlov et al., 1997b). This is also consistent with the predictions of the in vitro experiments, where RF1 rebinding to ribosomes is seen to lead to an increased delay in ribosome recycling when the level of RRF is low.

RELEASE OF DEACYLATED tRNA FROM THE POSTTERMINATION RIBOSOMAL COMPLEX

After removal of the polypeptide from peptidyl-tRNA by RFs, the ribosome is in a state with deacylated tRNA in the P site and an RF in the A site. In order for the ribosome to recycle back to a new initiation event, the deacylated tRNA must be removed from the P site and the two ribosomal subunits must come apart so that the 30S subunit can properly position a new mRNA and an initiator tRNA. We will first summarize some experiments that study the dissociation of deacylated tRNA without regard for

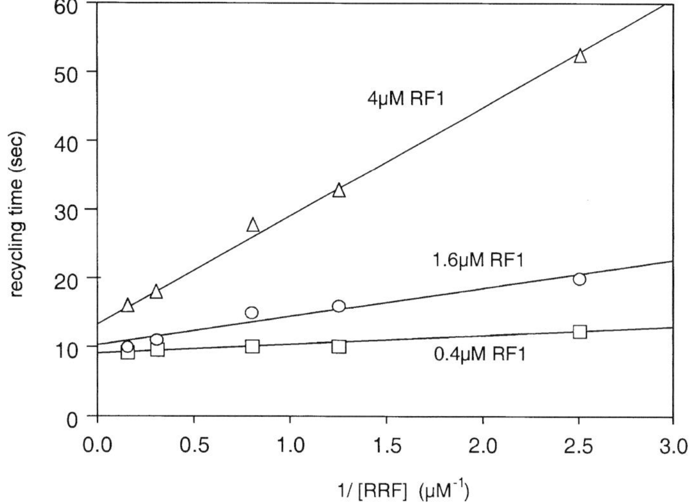

Figure 3. RRF restores rapid ribosome recycling inhibited by high concentration of RF1. Ribosome-recycling times for the synthesis of fMet-Phe-Leu were measured in vitro as described by Pavlov et al. (1997b) at three concentrations of RF1 in the absence of RF3. The recycling times are shown as functions of the reciprocal of the RRF concentration present in the translation system. The data, drawn from Pavlov et al., 1997b, show that the increase in recycling time due to RF1 rebinding to the posttermination complex can be countered by increasing the concentration of RRF in the translation system.

whether this takes place from 70S ribosomes or 30S ribosomal subunits, then consider the question of when subunit dissociation occurs, and finally return to the release of deacylated tRNA in light of the answer.

Fast dissociation of deacylated tRNA from the P site after termination with RF1 or RF2 requires the four translation factors RF3, RRF, EF-G, and IF3 (Karimi et al., 1999). When protein synthesis is terminated with the antibiotic puromycin rather than with RFs, the maximal rate of dissociation of deacylated tRNA does not require RF3 but only RRF, EF-G, and IF3 (Fig. 4). This suggests that tRNA dissociation is accomplished by the action of the last three factors, provided that RF1 or RF2 has been removed from the ribosome by RF3 or that termination was accomplished in an RF-independent way, i.e., with puromycin. This is in line with the conclusion (Pavlov et al., 1997b) that RRF and either one of the RFs RF1 and RF2 compete for a common binding site on the ribosome. The result shown in Fig. 4 further suggests that RF3 plays no other role in ribosomal recycling than to catalyze removal of RF1 or RF2 from the ribosome, in contrast to a recent suggestion by Grentzmann et al. (1998).

Dissociation of tRNA from the ribosome after termination by RRF, EF-G, and IF3 requires GTP hydrolysis as well as the presence of GTP. This conclusion is supported by the observation that the three factors do not accelerate tRNA dissociation in the absence of guanine nucleotide or in the presence of the noncleavable GTP analogue GDPNP and only weakly in the presence of GDP (Karimi et al., 1999). This result also suggests that the mechanism by which deacylated tRNA is removed from the ribosome cannot be translocation from the P site to the E site by the joint action of EF-G and RRF, as suggested by a number of authors (Janosi et al., 1996a; Nakamura et al., 1996; Pavlov et al., 1997b). It was observed by Rod-

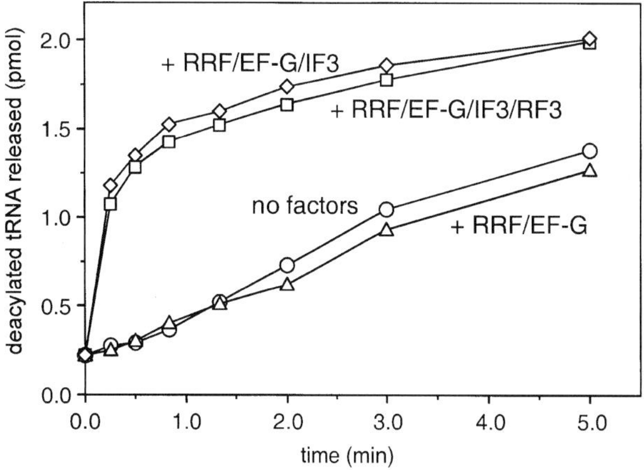

Figure 4. RRF, EF-G, and IF3 release deacylated tRNA from the posttermination complex. Termination complexes containing fMet-Phe-Thr-Ile-$tRNA^{Ile}$ bound in the P site with a UAA codon in the A site were prepared as described by Karimi et al. (1999), and the tetrapeptide was released by reaction with puromycin. The dissociation of deacylated $tRNA^{Ile}$ was monitored by incubation of the posttermination complexes in the presence of an excess of Ile-tRNA synthetase and other components needed for the rapid aminoacylation of $tRNA^{Ile}$ with [^{14}C]Ile, with or without the addition of factors RRF, EF-G, IF3, and RF3 as shown. The data are from Karimi et al., 1999.

nina et al. (1997) as well as by Karimi (unpublished) that EF-G and GDPNP quite efficiently catalyze translocation of peptidyl-tRNA from the A site to the P site and that this leads to rapid removal of the deacylated tRNA, originally in the P site, from the ribosome via the E site. Since dissociation of deacylated tRNA from the P site by RRF, EF-G, and IF3 after termination is strictly dependent on GTP, it is very unlikely that the mechanism can be translocation.

RRF AND EF-G CATALYZE RIBOSOME RECYCLING BY DISSOCIATING 50S SUBUNITS FROM THE 70S-mRNA-tRNA POSTTERMINATION COMPLEX

The question of how ribosomes leave the mRNA following translation termination has been the subject of much debate since the mid-1960s. Subunit exchange between ribosomes was demonstrated experimentally by Kaempfer (1968), but whether ribosomes left the mRNA as 70S particles, as suggested by Davis (1971) and others, or whether the subunits dissociated separately, as proposed by Martin and Webster (1975), and which protein factors are associated with which step has remained unclear. The involvement of RRF and EF-G in the overall process was clearly established (Hirashima and Kaji, 1970, 1972; Ogawa and Kaji, 1975; Subramanian and Davis, 1973), and ideas concerning the mechanism were recently summarized in terms of six possible alternative pathways (Janosi et al., 1996a). A dissociation factor believed to be responsible for splitting the ribosome into subunits was discovered by Subramanian et al. (1968, 1969). It was later identified with initiation factor IF3 (Sabol et al., 1970; Subramanian and Davis, 1970) and shown not to interact with 70S ribosomes but to bind to 30S particles and act as an anti-association factor (Kaempfer, 1972).

Our experiments in vitro with the purified translation system show that ribosome recycling from initiation via protein elongation and termination of protein synthesis depends strictly on the presence of IF3 (Pavlov, unpublished). However, under conditions where there is an excess of preinitiated 30S subunits containing mRNA and initiator tRNA as well as IF1 and IF2, the maximal rate of recycling of 50S subunits between the 30S subunits in the making of short oligopeptides requires RRF and EF-G but not IF3 (Fig. 5) (Karimi et al., 1999). This implies that RRF and EF-G together split the ribosome into its subunits after termination and that this step is the overture to ribosome recycling back to initiation of translation from the posttermination state. How this dramatic event may take place was suggested by experiments showing that EF-G and the noncleavable GTP analogue GDPNP form a very stable complex with the posttermination ribosome as well as with the 50S subunit in the presence but not in the absence of RRF (Karimi et al., 1999). This indicates that the primary function of RRF is to lock EF-G firmly on the ribosome. Subsequently, when GTP is hydrolyzed and EF-G changes to a conformation with very weak binding to the ribosome (Brot et al., 1971; Karimi, unpublished), RRF effectively prevents the otherwise-fast dissociation of EF-G and instead channels the dramatic increase in standard free energy of the ribosome·EF-G complex to rapid subunit separation.

IF3 CATALYZES DISSOCIATION OF DEACYLATED tRNA FROM THE 30S-mRNA-tRNA POSTTERMINATION COMPLEX

The finding that dissociation of deacylated tRNA from the posttermination ribosome requires not only ribosome splitting by RRF and EF-G but also the action of IF3 implies that tRNA remains firmly bound to the 30S subunit after ribosome splitting and that IF3 is instrumental in its removal. This was verified in a model experiment with 30S particles programmed with mRNA coding for fMet-Phe in complex with deacylated $tRNA^{Phe}$. In the absence of IF3, but in the presence of EF-G and RRF, the tRNA remains bound to 30S for long periods. As shown by Karimi et al. (1999), the addition of IF3 rapidly removes $tRNA^{Phe}$ and makes it available for charging with Phe-tRNA synthetase (Fig. 6). This suggests that IF3 enters the ribosome cycle already in its posttermination phase and that the factor may remain on the 30S subunit, facilitating initiator tRNA binding, until it finally leaves the ribosome at a time that may coincide with 70S formation (Pon and Gualerzi, 1986). The ability of IF3 to remove deacylated tRNA from 30S ribosomal subunits did not come as a surprise, as the factor selectively removes other forms of tRNA from non-initiation complexes involving 30S subunits, such as charged tRNA (Gualerzi et al., 1971), charged and N-blocked tRNA (Risuleo et al., 1976), and even tRNA fragments that lack domains, including the amino acid acceptor stem (Hartz et al., 1990). This activity seems generally to have been considered necessary to avoid the accumulation of "noncanonical initiation complexes." However, our experiments suggest that the activity is also required by the cell to release deacylated tRNA from the posttermination 30S complex that arises at every termination event in protein synthesis.

Experiments demonstrating that mRNAs with short open reading frames (ORFs) and very strong

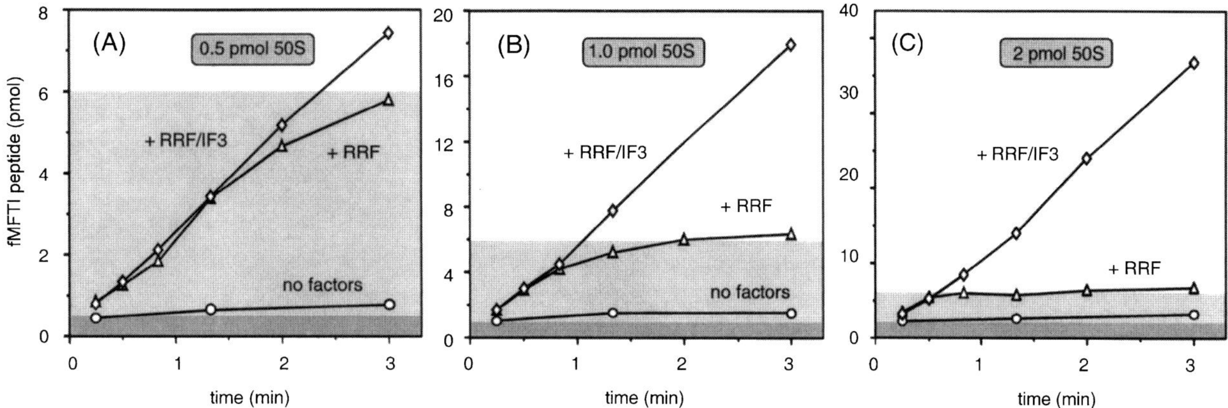

Figure 5. RRF and EF-G are required for recycling of the 50S ribosomal subunit, and IF3 is required for recycling of the 30S ribosomal subunit. The synthesis of fMet-Phe-Thr-Ile was measured as described by Karimi et al. (1999). The translation system contained a constant amount (6 pmol) of active preinitiated 30S subunits and different amounts of active 50S subunits as indicated and as shown by the upper limits of the light-gray (30S) and medium-gray (50S) shaded areas. The experiments were carried out in the absence of RRF and IF3 (○), with 1 μM RRF (△), and with 1 μM (each) of RRF and IF3 (◇). In the absence of RRF and IF3, the amount of tetrapeptide synthesized is limited by the amount of 50S subunit added. In the presence of RRF (and EF-G) but the absence of IF3, the amount of tetrapeptide synthesized is limited by the amount of preinitiated 30S subunits, showing that under these conditions the 50S subunit can recycle. In the presence of both RRF (and EF-G) and IF3, the amount of tetrapeptide synthesized becomes greater than the amount of preinitiated 30S subunits, indicating that both subunits can recycle. The data are from Karimi et al., 1999.

ribosome binding sequences have very high probabilities of participating in new rounds of protein synthesis without ever leaving the 30S subunit (Pavlov et al., 1997a, 1997b) suggest that mRNA dissociation from the ribosome cannot be the primary target for EF-G and RRF as previously suggested (Janosi et al., 1996a). It seems more likely that when deacylated tRNA has left the 30S subunit, the mRNA may either dissociate or reinitiate on the same 30S particle, possibly without the aid of additional factors. The overall model for the events following peptide release which allow the ribosome to reinitiate translation is shown in Fig. 7.

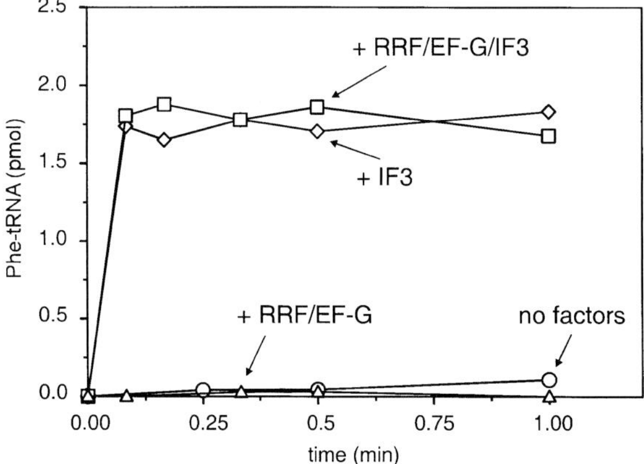

Figure 6. IF3 removes deacylated tRNA from the 30S subunit. 30S ribosomal subunits were programmed with a short mRNA encoding fMet-Phe-Leu and with deacylated $tRNA^{Phe}$ in the P site, as described by Karimi et al. (1999). The complexes were then incubated in the absence of RRF, EF-G, and IF3 (○), with 1 μM RRF and 1 μM EF-G (△), with 1 μM IF3 (◇), and with 1 μM RRF, 1 μM EF-G, and 1 μM IF3 (□).The dissociation of deacylated $tRNA^{Phe}$ was monitored by charging the tRNA in the presence of an excess of Phe-tRNA synthetase, [^{14}C]Phe, and other components needed for rapid aminoacylation. The data are from Karimi et al., 1999.

The fate of the mRNA after termination and ribosome recycling is highly relevant for translation of multicistronic mRNAs. A number of in vivo experiments suggest that ribosomes can diffuse long distances along mRNAs after termination and in the presence of RRF and EF-G (Adhin and van Duin, 1990; Draper, 1996; Sarabhai and Brenner, 1967). It is therefore not unlikely that repeated initiations on 30S subunits without the mRNA leaving may originate in rapid diffusion of the mRNA from its termination back to its initiation position. If such a diffusion mechanism is responsible, then one would predict that the probability of an mRNA returning to its previous initiation site following termination would become progressively lower as the length of the ORF increases. This is indeed the case. The probability of reinitiating without leaving the 30S ribosomal subunit for mRNAs coding for di-, tetra-, hepta-, 20-, and 30-meric oligopeptides was seen to vary from 99 to 75% (Dincbas, unpublished). Although other mechanisms are still possible for reinitiation without mRNA dissociation, the diffusion model presently appears to be the most likely explanation for these data.

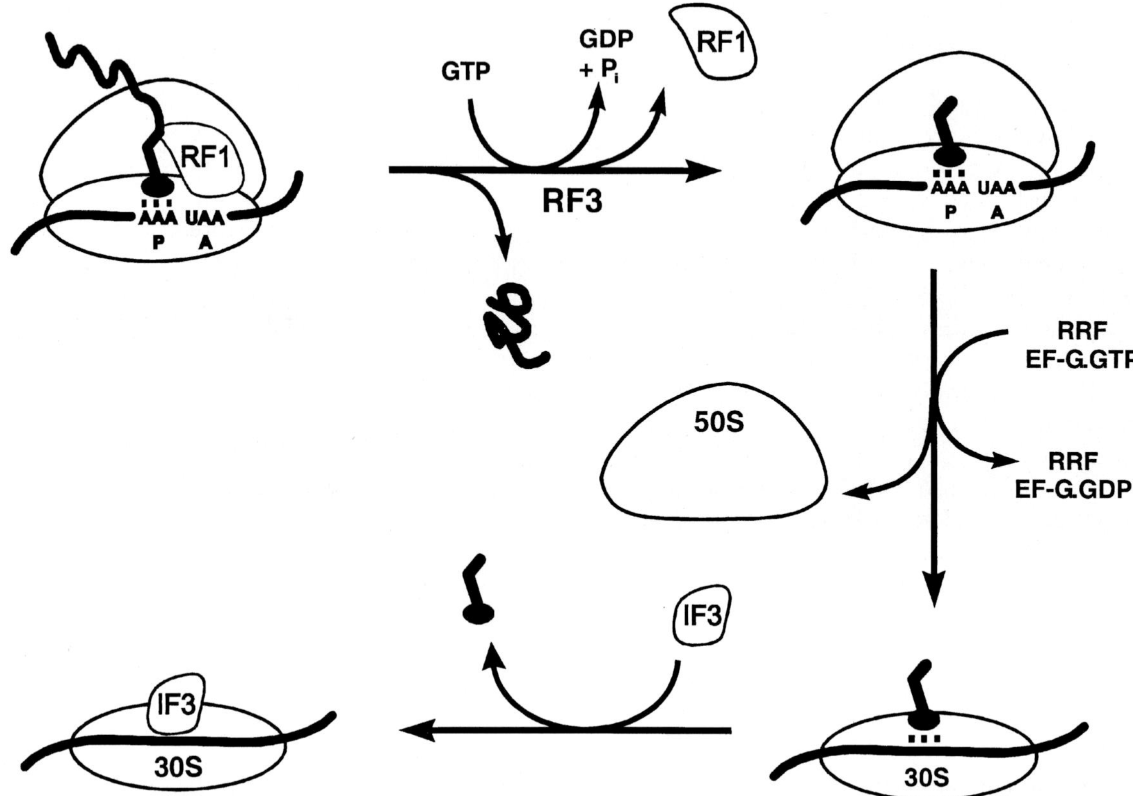

Figure 7. Model for the events following peptide release. RF3 catalyzes the dissociation of RF1 (or RF2) from the A site after peptide release (Freistroffer et al., 1997). RRF and EF-G then bind to the 70S posttermination complex and, in a GTP-requiring reaction, provoke the dissociation of the 50S subunit. Finally, IF3 displaces the deacylated tRNA from the 30S posttermination complex (Karimi et al., 1999).

RIBOSOME RECYCLING WITH ACTIVE AND DEFICIENT RRF

The in vitro data described above suggest that when termination of protein synthesis is followed by the action of RRF, EF-G, and IF3 the 30S particle remains attached to the mRNA, allowing the latter to diffuse long distances along the subunit. This suggests that in vivo experiments indicating the occurrence of diffusion of mRNA on ribosomes after termination (Adhin and van Duin, 1990; Draper, 1996; Sarabhai and Brenner, 1967) in fact report the relative movement of mRNA and the 30S subunit rather than of mRNA and the whole ribosome. This would mean that reinitiation on the same mRNA requires a Shine-Dalgarno sequence followed by a start codon, just as for de novo initiation following the binding of an external mRNA to the 30S subunit. However, if the function of RRF is impaired so that the ribosome remains as a 70S particle on the mRNA after termination of translation, a radically different scenario may be expected.

A series of temperature-sensitive mutants of RRF were selected by Janosi et al. (1998), and their phenotypes were inspected. In one type of experiment, the *lac* promoter was followed by a small ORF ending in a stop codon with the *lacZ* coding sequence lacking a ribosome binding site and initiation codon further downstream. It was shown that RRF deficiency enhanced β-Gal synthesis when the reporting *lacZ* gene was in frame with the ORF as well as when it was in frames -1 and -2. In another type of experiment a modified version of the 3A′ reporter system (Bjornsson and Isaksson, 1996) was used to further analyze where reinitiation takes place after termination at a stop codon under conditions of RRF deficiency. Reinitiations in all reading frames were seen at downstream distances from the stop signal ranging from 7 to 45 codons. No reinitiation was seen either downstream of an authentic initiation signal or upstream of the stop codon. In light of the biochemical findings described above, these in vivo results may tentatively be explained as follows.

Under conditions of RRF deficiency, ribosomes that reach a stop codon terminate normally but then pause for relatively long times, since their splitting is delayed. The pausing allows piling up of ribosomes upstream of the leading ribosome anchored at the stop codon. When deacylated tRNA leaves the pausing ribosome, it can move along mRNA as a 70S unit, initially quickly and unidirectionally, pushed by its upstream neighbor, and later by diffusion, until its A site is filled by a ternary complex that fits the codon that happens to be exposed. Aminoacyl-tRNA is subsequently translocated into the P site, and a reinitiated round of protein synthesis takes place. The pushing mode of ribosome movement, probably lasting until the next ribosome in line reaches the stop codon and pauses, is likely to prevent reinitiation immediately downstream of the stop signal, in accordance with the observations of Janosi et al. (1998). The suggested queuing will also effectively stop reinitiation upstream of the stop codon. When the diffusing ribosome reaches an authentic start signal, it will pause again until initiation takes place, which explains why no initiations occurred downstream of the initiation codon.

SOME UNRESOLVED PROBLEMS AND CONCLUDING REMARKS

In the presence of GTP, RF3 accelerates recycling of RF1 and RF2 by enhancing their rates of dissociation after termination of translation (Freistroffer et al., 1997; Pavlov et al., 1997a, 1998). However, the mechanism of RF3 action and the role of GTP hydrolysis remain obscure. In our original model of RF3 action, an RF3·GDP complex tightly bound to the ribosome was postulated, since it was found that RF3 without guanine nucleotide or with GDP strongly inhibits recycling of RF1 and RF2 (Freistroffer et al., 1997). More recent experiments have confirmed all results reported by Freistroffer et al. (1997) with one exception. They show, in particular, that while RF3 in the absence of guanine nucleotide does indeed inhibit recycling of RF1 or RF2, addition of GDP relieves this inhibition (Fig. 8). These results are consistent with the cooperative binding of RF3 and RF1 or RF2 to the ribosome in the absence of guanine nucleotides as observed by Pel et al. (1998) and contradict the model suggested by Freistroffer et al. (1997). Accordingly, the question of how GTP hydrolysis is used to remove RF1 or RF2 from the ribosome after termination remains unanswered. One interesting possibility is that the tight complex between the posttermination ribosome, RF1 or RF2, and RF3 is a functional intermediate in the RF3-driven release of RF1 or RF2.

The recent demonstration that RRF and EF-G together split the ribosome after termination with and dissociation of RF1 or RF2 in a reaction that

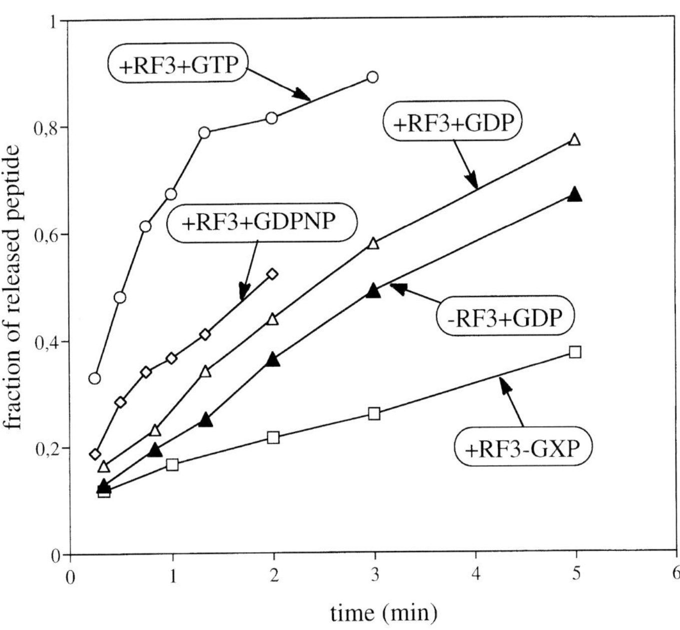

Figure 8. Stimulation of RF2 recycling by RF3 in the presence of different guanine nucleotides. A catalytic amount of RF2 (10 mM) was added to termination complexes (90 nM) containing fMet-Phe-Thr-Ile-tRNA^{Ile} bound in the P sites of ribosomes with a UGA(U) stop signal in the A site, and the fraction of the released fMet-Phe-Thr-Ile tetrapeptide was measured as a function of time. RF3 (0.9 μM) and guanine nucleotides (0.2 mM) were added where indicated. The figure shows that the rate of RF2 recycling in the presence of GTP (○) is much faster than with GDPNP (◇) and that RF3 added with GDP (△) is slightly stimulatory. In contrast, RF3 added without guanine nucleotide (□) inhibits RF2 recycling compared to the control in the absence of RF3 (▲).

requires GTP hydrolysis (Karimi et al., 1999) brings to the foreground the question of how sets of translation factors like IF1 and IF2 (Karimi et al., 1998) or EF-G, RRF, and RF3 (Heurgué-Hamard et al., 1998) stimulate drop-off of peptidyl-tRNAs. Do these two different sets of translation factors operate in principally similar ways, where ribosome splitting is a fundamental step in the drop-off reaction, or do the peptidyl-tRNAs dissociate from intact 70S particles?

Future attempts to explain these and other obscure aspects of the translational mechanism stand to benefit greatly from the combined application of functional biochemistry and high-resolution structural analysis. The potential synergistic gains from this combination of approaches exist now as never before in the history of ribosome research.

This work was supported by the Swedish Research Council for Engineering Sciences, the Swedish Natural Science Research Council, the Centre National pour la Recherche Scientifique (UPR9073), l'Association pour la Recherche sur le Cancer, the Fondation pour la Recherche Medicale, INTAS, and the Human Capital and Mobility Programme of the European Community.

REFERENCES

Adhin, M. R., and J. van Duin. 1990. Scanning model for translational reinitiation in eubacteria. *J. Mol. Biol.* **213:**811–818.

Bjornsson, A., and L. A. Isaksson. 1996. Accumulation of a mRNA decay intermediate by ribosomal pausing at a stop codon. *Nucleic Acids Res.* **24:**1753–1757.

Bourne, H. R., D. A. Sanders, and F. McCormick. 1991. The GTPase superfamily: conserved structure and molecular mechanism. *Nature* **349:**117–127.

Brot, N., C. Spears, and H. Weissbach. 1971. The interaction of transfer factor G, ribosome and guanosine nucleotide in the presence of fusidic acid. *Arch. Biochem. Biophys.* **143:**286–296.

Buckingham, R. H., G. Grentzmann, and L. Kisselev. 1997. Polypeptide chain release factors. *Mol. Microbiol.* **24:**449–456.

Capecchi, M. R. 1967a. Polypeptide chain termination in vitro: isolation of a release factor. *Proc. Natl. Acad. Sci. USA* **58:**1144–1151.

Capecchi, M. R. 1967b. A rapid assay for polypeptide chain termination. *Biochem. Biophys. Res. Commun.* **28:**773–778.

Caskey, C. T., R. Tompkins, E. Scolnick, T. Caryk, and M. Nirenberg. 1968. Sequential translation of trinucleotide codons for the initiation and termination of protein synthesis. *Science* **162:** 135–138.

Caskey, C. T., E. Scolnick, R. Tompkins, J. Goldstein, and G. Milman. 1969. Peptide chain termination: codon, protein factor, and ribosomal requirements. *Cold Spring Harbor Symp. Quant. Biol.* **34:**479.

Caskey, C. T., A. L. Beaudet, E. M. Scolnick, and M. Rosman. 1971. Peptidyl transferase hydrolysis of fmet-tRNA. *Proc. Natl. Acad. Sci. USA* **68:**3163–3167.

Caskey, C. T., W. C. Forrester, W. P. Tate, and C. D. Ward. 1984. Cloning of the *Escherichia coli* release factor 2 gene. *J. Bacteriol.* **158:**365–368.

Craigen, W. J., C. C. Lee, and C. T. Caskey. 1990. Recent advances in peptide chain termination. *Mol. Microbiol.* **4:**861–865.

Davis, B. D. 1971. Role of subunits in the ribosome cycle. *Nature* **231:**153–157.

Dincbas, V. Unpublished data.

Draper, D. E. 1996. Translational initiation, p. 902–908. *In* F. C. Neidhardt, R. Curtiss III, J. L. Ingraham, E. C. C. Lin, K. B. Low, B. Magasanik, W. S. Reznikoff, M. Riley, M. Schaechter, and H. E. Umbarger (ed.), *Escherichia coli and Salmonella: Cellular and Molecular Biology*. ASM Press, Washington, D.C.

Ebihara, K., and Y. Nakamura. 1999. C-terminal interaction of translational release factors eRF1 and eRF3 of fission yeast: G-domain uncoupled binding and the role of conserved amino acids. *RNA* **5:**739–750.

Freistroffer, D. V., M. Y. Pavlov, J. MacDougall, R. H. Buckingham, and M. Ehrenberg. 1997. Release factor RF3 in *E. coli* accelerates the dissociation of release factors RF1 and RF2 from the ribosome in a GTP dependent manner. *EMBO J.* **16:**4126–4133.

Frolova, L., X. Le Goff, H. H. Rasmussen, S. Cheperegin, G. Drugeon, M. Kress, I. Arman, A. L. Haenni, J. E. Celis, M. Philippe, J. Justesen, and L. L. Kisselev. 1994. A highly conserved eukaryotic protein family possessing properties of polypeptide chain release factor. *Nature* **372:**701–703.

Frolova, L., X. Le Goff, G. Zhouravleva, E. Davydova, M. Philippe, and L. L. Kisselev. 1996. Eukaryotic polypeptide chain release factor eRF3 is a guanosine triphosphatase. *RNA* **2:**334–341.

Frolova, L. Y., J. L. Simonsen, T. I. Merkulova, D. Y. Litvinov, P. M. Martensen, V. O. Rechinsky, J. H. Camonis, L. L. Kisselev, and J. Justesen. 1998. Functional expression of eukaryotic polypeptide chain release factors 1 and 3 by means of baculovirus/insect cells and complex formation between the factors. *Eur. J. Biochem.* **256:**36–44.

Frolova, L. Y., R. Y. Tsivkovskii, G. F. Sivolobova, N. Y. Oparina, O. I. Serpinski, V. M. Blinov, S. I. Tatkov, and L. L. Kisselev. 1999. Mutations in the highly conserved GGQ motif of class 1 polypeptide release factors abolish ability of human eRF1 to trigger peptidyl-tRNA hydrolysis. *RNA* **5:**1014–1020.

Goldstein, J., G. Milman, E. Scolnick, and T. Caskey. 1970. Peptide chain termination. VI: Purification and site of action of S. *Proc. Natl. Acad. Sci. USA* **65:**430–437.

Goldstein, J. L., and C. T. Caskey. 1970. Peptide chain termination: effect of protein S on ribosomal binding of release factors. *Proc. Natl. Acad. Sci. USA* **67:**537–543.

Grentzmann, G., D. Brechemier-Baey, V. Heurgué, L. Mora, and R. H. Buckingham. 1994. Localization and characterization of the gene encoding release factor RF3 in *Escherichia coli*. *Proc. Natl. Acad. Sci. USA* **91:**5848–5852.

Grentzmann, G., P. J. Kelly, S. Laalami, M. Shuda, M. A. Firpo, Y. Cenatiempo, and A. Kaji. 1998. Release factor RF-3 GTPase activity acts in disassembly of the ribosome termination complex. *RNA* **4:**973–983.

Gualerzi, C., C. L. Pon, and A. Kaji. 1971. Initiation factor dependent release of aminoacyl-tRNAs from complexes of 30S ribosomal subunits, synthetic polynucleotide and aminoacyl tRNA. *Biochem. Biophys. Res. Commun.* **45:**1312–1319.

Hartz, D., J. Binkley, T. Hollingsworth, and L. Gold. 1990. Domains of initiator tRNA and initiation codon crucial for initiator tRNA selection by *Escherichia coli* IF3. *Genes Dev.* **4:**1790–1800.

Heurgué-Hamard, V., R. Karimi, L. Mora, J. MacDougall, C. Leboeuf, G. Grentzmann, M. Ehrenberg, and R. H. Buckingham. 1998. Ribosome release factor RF4 and termination factor RF3 are involved in dissociation of peptidyl-tRNA from the ribosome. *EMBO J.* **17:**808–816.

Hirashima, A., and A. Kaji. 1970. Factor dependent breakdown of polysomes. *Biochem. Biophys. Res. Commun.* **41:**877–883.

Hirashima, A., and A. Kaji. 1972. Factor-dependent release of ribosomes from messenger RNA: requirement for two heat-stable factors. *J. Mol. Biol.* **65:**43–58.

Ichikawa, S., and A. Kaji. 1989. Molecular cloning and expression of ribosomal releasing factor. *J. Biol. Chem.* **264:**20054–20059.

Ichikawa, S., M. Ryoji, Z. Siegfried, and A. Kaji. 1989. Localization of the ribosome-releasing factor gene in the *Escherichia coli* chromosome. *J. Bacteriol.* **171:**3689–3695.

Ito, K., K. Ebihara, M. Uno, and Y. Nakamura. 1996. Conserved motifs in prokaryotic and eukaryotic polypeptide release factors: tRNA-protein mimicry hypothesis. *Proc. Natl. Acad. Sci. USA* **93:**5443–5448.

Ito, K., K. Ebihara, and Y. Nakamura. 1998a. The stretch of C-terminal acidic amino acids of translational release factor eRF1 is a primary binding site for eRF3 of fission yeast. *RNA* **4:**958–972.

Ito, K., M. Uno, and Y. Nakamura. 1998b. Single amino acid substitution in prokaryote polypeptide release factor 2 permits it to terminate translation at all three stop codons. *Proc. Natl. Acad. Sci. USA* **95:**8165–8169.

Janosi, L., I. Shimizu, and A. Kaji. 1994. Ribosome recycling factor (ribosome releasing factor) is essential for bacterial growth. *Proc. Natl. Acad. Sci. USA* **91:**4249–4253.

Janosi, L., H. Hara, S. Zhang, and A. Kaji. 1996a. Ribosome recycling by ribosome recycling factor (RRF): an important but overlooked step of protein synthesis. *Adv. Biophys.* **32:**121–201.

Janosi, L., R. Ricker, and A. Kaji. 1996b. Dual function of ribosome recycling factor in protein biosynthesis: disassembling the termination complex and preventing translational errors. *Biochimie* **78:**959–969.

Janosi, L., S. Mottagui-Tabar, L. A. Isaksson, Y. Sekine, E. Ohtsubo, S. Zhang, S. Goon, S. Nelken, M. Shuda, and A. Kaji. 1998. Evidence for in vivo ribosome recycling, the fourth step in protein biosynthesis. *EMBO J.* **17:**1141–1151.

Kaempfer, R. 1968. Ribosomal subunit exchange during protein synthesis. *Proc. Natl. Acad. Sci. USA* **61:**106–113.

Kaempfer, R. 1972. Initiation factor IF-3: a specific inhibitor of ribosomal subunit association. *J. Mol. Biol.* **71:**583–598.

Karimi, R. Unpublished data.

Karimi, R., M. Y. Pavlov, V. Heurgue-Hamard, R. H. Buckingham, and M. Ehrenberg. 1998. Initiation factors IF1 and IF2 synergistically remove peptidyl-tRNAs with short polypeptides from the P-site of translating Escherichia coli ribosomes. *J. Mol. Biol.* **281:**241–252.

Karimi, R., M. Pavlov, R. H. Buckingham, and M. Ehrenberg. 1999. Novel roles for classical factors at the interface between translation termination and initiation. *Molec. Cell* **3:**601–609.

Konecki, D. S., K. C. Aune, W. Tate, and C. T. Caskey. 1977. Characterization of reticulocyte release factor. *J. Biol. Chem.* **252:**4514–4520.

MacDougall, J. Unpublished results.

Martin, J., and R. E. Webster. 1975. The *in vitro* translation of a terminating signal by a single *Escherichia coli* ribosome. *J. Biol. Chem.* **250:**8132–8139.

Merkulova, T. I., L. Y. Frolova, M. Lazar, J. Camonis, and L. L. Kisselev. 1999. C-terminal domains of human translation termination factors eRF1 and eRF3 mediate their in vivo interaction. *FEBS Lett.* **443:**41–47.

Mikuni, O., K. Ito, J. Moffat, K. Matsumura, K. McCaughan, T. Nobukuni, W. Tate, and Y. Nakamura. 1994. Identification of the *prfC* gene, which encodes peptide chain release factor 3 of *Escherichia coli. Proc. Natl. Acad. Sci. USA* **91:**5798–5802.

Milman, G., J. Goldstein, E. Scolnick, and T. Caskey. 1969. Peptide chain termination. III. Stimulation of *in vitro* termination. *Proc. Natl. Acad. Sci. USA* **63:**183–190.

Nakamura, Y., and K. Ito. 1998. How protein reads the stop codon and terminates translation. *Genes Cells* **3:**265–278.

Nakamura, Y., K. Ito, and L. A. Isaksson. 1996. Emerging understanding of translation termination. *Cell* **87:**147–150.

Ogawa, K., and A. Kaji. 1975. Requirement for ribosome releasing factor for the release of ribosomes at the termination codon. *J. Biochem.* (Tokyo) **58:**411–419.

Paushkin, S. V., V. V. Kushnirov, V. N. Smirnov, and M. D. Ter-Avanesyan. 1997. Interaction between yeast Sup45p (eRF1) and Sup35p (eRF3) polypeptide chain release factors: implications for prion-dependent regulation. *Mol. Cell. Biol.* **17:**2798–2805.

Pavlov, M. Unpublished observations.

Pavlov, M. Y., D. Freistroffer, J. MacDougall, R. H. Buckingham, and M. Ehrenberg. 1997a. Fast recycling of *E. coli* ribosomes requires both ribosome recycling factor (RRF) and release factor RF3. *EMBO J.* **16:**4134–4141.

Pavlov, M. Y., D. V. Freistroffer, V. Heurgué-Hamard, R. H. Buckingham, and M. Ehrenberg. 1997b. Release factor RF3 abolishes competition between release factor RF1 and ribosome recycling factor (RRF) for a ribosome binding site. *J. Mol. Biol.* **273:**389–401.

Pavlov, M. Y., D. V. Freistroffer, V. Dincbas, J. MacDougall, R. H. Buckingham, and M. Ehrenberg. 1998. A direct estimation of the context effect on the efficiency of termination. *J. Mol. Biol.* **284:**579–590.

Pel, H. J., J. G. Moffat, K. Ito, Y. Nakamura, and W. P. Tate. 1998. Escherichia coli release factor-3—resolving the paradox of a typical G-protein structure and atypical function with guanine-nucleotides. *RNA* **4:**47–54.

Pon, C. L., and C. O. Gualerzi. 1986. Mechanism of translational initiation in prokaryotes. IF3 is released from ribosomes during and not before 70 S initiation complex formation. *FEBS Lett.* **195:**215–219.

Risuleo, G., C. Gualerzi, and C. Pon. 1976. Specificity and properties of the destabilization, induced by initiation factor IF-3, of ternary complexes of the 30-S ribosomal subunit, aminoacyl-tRNA and polynucleotides. *Eur. J. Biochem.* **67:**603–613.

Rodnina, M. V., A. Savelsbergh, V. I. Katunin, and W. Wintermeyer. 1997. Hydrolysis of GTP by elongation factor G drives tRNA movement on the ribosome. *Nature* **385:**37–41.

Sabol, S., M. A. Sillero, K. Iwasaki, and S. Ochoa. 1970. Purification and properties of initiation factor F3. *Nature* **228:**1269–1273.

Sarabhai, A., and S. Brenner. 1967. A mutant which reinitiates the polypeptide chain after chain termination. *J. Mol. Biol.* **27:**145–162.

Scolnick, E. M., R. Tompkins, C. T. Caskey, and M. Nirenberg. 1968. Release factors differing in specificity for terminator codons. *Proc. Natl. Acad. Sci. USA* **61:**768–774.

Subramanian, A. R., and B. D. Davis. 1970. Activity of initiation factor F3 in dissociating *Escherichia coli* ribosomes. *Nature* **228:**1273–1275.

Subramanian, A. R., and B. D. Davis. 1973. Release of 70 S ribosomes from polysomes in *Escherichia coli. J. Mol. Biol.* **74:**45–56.

Subramanian, A. R., E. Z. Ron, and B. D. Davis. 1968. A factor required for ribosome dissociation in *Escherichia coli. Proc. Natl. Acad. Sci. USA* **61:**761–767.

Subramanian, A. R., B. D. Davis, and R. J. Beller. 1969. The ribosome dissociation factor and the ribosome-polysome cycle. *Cold Spring Harbor Symp. Quant. Biol.* **34:**223–230.

Tate, W. P., C. M. Brown, and B. Kastner. 1990. Codon recognition by the polypeptide release factor, p. 393–401. *In* W. E. Hill, A. Dahlberg, R. A. Garrett, P. B. Moore, D. Schlessinger, and J. R. Warner (ed.), *The Ribosome: Structure, Function, and Evolution.* American Society for Microbiology, Washington, D.C.

Weiss, R. B., J. P. Murphy, and J. A. Gallant. 1984. Genetic screen for cloned release factor genes. *J. Bacteriol.* **158:**362–364.

XI. IN CONCLUSION

The Ribosome: Structure, Function, Antibiotics, and Cellular Interactions
Edited by R. A. Garrett, S. R. Douthwaite, A. Liljas, A. T. Matheson, P. B. Moore, and H. F. Noller

Chapter 45

Concluding Remarks for the Helsingør Ribosome Conference, 13 to 17 June 1999

PETER B. MOORE

By tradition, ribosome meetings terminate with a summary talk delivered by a senior member of the field. This is not the first time I have been asked to perform this duty, and it concerns me that the community is trying to send me a message: namely, that I am history. This is a strange ritual, when you think about it. You were all here. You all heard what was said, and nothing I say is likely to change your minds about what it all means. Nevertheless, obedient to the wishes of the organizers and deferring to the dictates of tradition, I will tell you what I heard and what it means to me.

It is clear that 1999 marks a watershed in the history of this field. On 1 January 1999, it was not possible to place even a single residue in the ribosome with atomic resolution. As of 17 June 1999, the structure of an entire domain of the 30S subunit has been worked out at that resolution, as well as a significant part of the 50S subunit. Who knows where we will be by 31 December 1999? It would not surprise me if a reasonably accurate structural model of the entire ribosome were to become available in the next 12 months, and the impact is certain to be enormous.

There will be a tendency for this remarkable development to be credited entirely to the handful of people currently active in the area of whole-ribosome crystallography, but in my view, it would be a serious error to take so limited a view. This achievement is the culmination of a collective quest of 35 years' standing to which many of this conference's participants have made major contributions; it is a collective achievement.

In the late 1960s and early 1970s, at the first meetings of this group, I can remember arguing with Charles Kurland about the chemical composition of the ribosome. A tear comes to the eye thinking about those halcyon, bygone days. The work done then made the 1970s possible—a decade, by the way, in which many thought this field was dying, but informed by hindsight, we can see that this was by no means the case.

The 1970s was the decade of sequences, and I feel it necessary to remind the young that in that era, if you wanted a sequence, you did not send 100 μg of protein to the local Sequences-R-Us outlet and get the result back by e-mail the following day. Instead, you did it yourself, with your own hands, using milligrams of protein you had prepared from nonoverproducing cells. It took courage to sequence ribosomal proteins in the ribosome in those days, but fortunately there were people around equal to the challenge. I am reminded of that era by Brigitte Wittmann-Liebold's presence at this meeting. By the end of the decade, most of the ribosomal proteins from *Escherichia coli* had been sequenced, and Harry Noller and his colleagues had produced the first 16S and 23S rRNA sequences. In addition, by using electron microscopy, neutron scattering, and a multitude of less physical approaches, the overall shape of the ribosome was determined, as well as the placement of many of its components and its active sites. Finally, in the closing years of the decade, Ada Yonath began her remarkable effort to crystallize the ribosome, the success of which has led us to the happy circumstances in which we find ourselves today. What more can you expect from a single decade?

The 15 years from 1980 to 1995 saw developments that in my view were more evolutionary than revolutionary. An enormous amount of sequence information accumulated. The low-resolution structure-mapping experiments begun in the 1970s were

Peter B. Moore ■ Department of Chemistry, Yale University, New Haven, CT 06520.

largely concluded. Cross-linking and footprinting methods were refined, and a rich body of data was generated about the spatial relationships between components that has residue resolution in many instances. Electron microscopists interested in ribosomes continued developing techniques for reconstructing three-dimensional images of single biological particles, and at the same time, nuclear magnetic resonance spectroscopists and X-ray crystallographers began producing high-resolution structures of individual ribosomal components in increasing numbers. About 1995, these currents started converging. The electron microscopists began generating images of unprecedented resolution, and groups outside Berlin began responding to the challenge posed by Yonath's crystals. In 12 months or so, it seems likely that this epic voyage will be completed and the data obtained by so many over so many years will be consolidated into a single, unified picture of how the ribosome is organized.

Over the course of this meeting a number of people have expressed sentiments to me that I have interpreted, perhaps uncharitably, as having the following subtext: "What is to become of me now that it is all over?" To even imagine that it is "all over" is to be guilty of grievous error. To paraphrase Winston Churchill, this is not the end. This is not even the beginning of the end. But it is certainly the end of the beginning. All we are going to have in 12 months is a single frame from the long, complicated movie we will have to see if we are to fully understand the ribosomal steps of protein synthesis. Crystallographers are going to be trying to obtain structures for ribosomes in different functional states for years to come, but I will be surprised if they solve more than a handful of them. Electron microscopists must carry on. Their technique is far better adapted to capturing transient states than X-ray crystallography. Molecular biologists and biochemists must also persist. The techniques they have perfected will be more powerful than ever when both experimental designs and data interpretation are informed by an accurately understood structure. Only by a blend of all these approaches will the missing frames be obtained and the movie completed.

So the answer I give to those concerned with their place in the new order is simple. Yes, the field is going to change; it is going to get better. The next 10 years in this field should be amazing. I am already having the time of my scientific life, and I look forward to the future with happy anticipation. If you do not feel that way, you should search for a way your course can be reshaped so that you do too. It should not be hard to do.

The only cloud I see on the horizon results from the fact that not enough people are working on ribosome enzymology. The field needs more people like Wolfgang Wintermeyer, Marina Rodnina, and Måns Ehrenberg, who are willing to work full time on the biochemistry of protein synthesis. It is not that I find their work unreliable but rather that the system they are exploring is so complicated and so hard to work with. As we all know, the enzymatic activity of ribosome preparations varies from preparation to preparation, decays with preparation age, and is diabolically dependent on environmental conditions. The potential for conflict and misunderstanding about even the simplest experimental facts is very great, and unless the facts can be agreed upon, it is going to be hard to reach a consensus about how this system works. Without more hands, it is going to be hard to reach the level of understanding we require in a timely manner. The reason timing is suddenly very important is that unless we understand what happens during protein synthesis, the structural biologists, who dominate the field today, won't know what it is their structures must explain.

In closing, I turn to the happiest part of my task, which is thanking the organizers of this conference for the efforts they have made on behalf of all of its participants. This has been an outstanding meeting, put on with panache in a marvelous location. Since I am a member of the Organizing Committee, these comments will seem self-serving to some, but I assure you that they are not. The contributions made by the members of the Organizing Committee to the hard work that made this meeting possible were inversely proportional to the distance between Copenhagen and their places of residence. Thus, the sole function of the overseas members was to encourage the Danish members, Steve Douthwaite and, above all, Roger Garrett, to do what they were going to do anyway. Credit for the success of this enterprise belongs entirely to them.

INDEX